RESIDENTIAL CONSTRUCTION ACADEMY

# Carpentry

Third Edition

FLOYD VOGT

Australia • Brazil • Canada • Mexico • Singapore • Spain • United Kingdom • United States

**Residential Construction Academy: Carpentry, Third Edition**
**Floyd Vogt**

Vice President, Career and Professional Editorial: Dave Garza

Director of Learning Solutions: Sandy Clark

Senior Acquisitions Editor: Jim DeVoe

Managing Editor: Larry Main

Product Manager: Jennifer Jacobson, Ohlinger Publishing Services

Editorial Assistant: Cris Savino

Vice President, Career and Professional Marketing: Jennifer Baker

Marketing Director: Deborah Yarnell

Marketing Manager: Katie Hall

Marketing Coordinator: Mark Pierro

Production Director: Wendy Troeger

Production Manager: Mark Bernard

Content Project Manager: David E. Plagenza

Art Director: Casey Kirchmayer

Technology Project Manager: Joe Pliss

Cover Image: small image: ©istockphoto.com and large image: ©GettyImages.com/ Mark A. Paulda



Library of Congress Control Number: 2011928792

ISBN-13: 978-1-111-30826-1

ISBN-10: 1-111-30826-8

**Delmar**
5 Maxwell Drive
Clifton Park, NY 12065-2919
USA

Cengage Learning is a leading provider of customized learning solutions with office locations around the globe, including Singapore, the United Kingdom, Australia, Mexico, Brazil, and Japan. Locate your local office at: **international.cengage.com/region**

Cengage Learning products are represented in Canada by Nelson Education, Ltd.

To learn more about Delmar, visit **www.cengage.com/delmar**

Purchase any of our products at your local college store or at our preferred online store **www.cengagebrain.com**

**Notice to the Reader**

Publisher does not warrant or guarantee any of the products described herein or perform any independent analysis in connection with any of the product information contained herein. Publisher does not assume, and expressly disclaims, any obligation to obtain and include information other than that provided to it by the manufacturer. The reader is expressly warned to consider and adopt all safety precautions that might be indicated by the activities described herein and to avoid all potential hazards. By following the instructions contained herein, the reader willingly assumes all risks in connection with such instructions. The publisher makes no representations or warranties of any kind, including but not limited to, the warranties of fitness for particular purpose or merchantability, nor are any such representations implied with respect to the material set forth herein, and the publisher takes no responsibility with respect to such material. The publisher shall not be liable for any special, consequential, or exemplary damages resulting, in whole or part, from the readers' use of, or reliance upon, this material.

Printed in the United States of America
1 2 3 4 5 6 7 15 14 13 12 11

# Table of Contents

SECTION 1

## Tools and Materials ........ 1

CHAPTER 1

### Hand Tools ........ 3



CHAPTER 2

### Hand Power Tools ........ 26

SECTION 2

# Rough Carpentry ....111

CHAPTER 6

## Building Plans and Codes ......... 113



CHAPTER 7

## Building Layout ......................... 138

CHAPTER 8

## Concrete Form Construction .... 162



CHAPTER 9

## Floor Framing ........ 203

CHAPTER 10

## Wall and Ceiling Framing.........249



CHAPTER 11

## Scaffolds, Ladders, and Sawhorses...........302

CHAPTER 12

## Roof Framing

SECTION 3

## Exterior Finish



CHAPTER 13

## Windows and Doors

CHAPTER 14

## Roofing .......... 440



CHAPTER 15

## Siding and Decks .......... 469

CHAPTER 18

## Interior Finish .......................... 590



CHAPTER 19

## Stair Framing and Finish .......... 639

# Preface

## About the Residential Construction Academy Series

One of the most pressing problems confronting the building industry today is the shortage of skilled labor. The construction industry must recruit an estimated 200,000 to 250,000 new craft workers each year to meet future needs. This shortage is expected to continue well into the next decade because of projected job growth and a decline in the number of available workers. At the same time, the training of available labor is becoming an increasing concern throughout the country. This lack of training opportunities has resulted in a shortage of 65,000 to 80,000 skilled workers per year. This challenge is affecting all construction trades and is threatening the ability of builders to construct quality homes.

These challenges led to the creation of the innovative *Residential Construction Academy Series*. The *Residential Construction Academy Series* is the perfect way to introduce people of all ages to the building trades while guiding them in the development of essential workplace skills, including carpentry, electrical wiring, HVAC, plumbing, masonry, and facilities maintenance. The products and services offered through the *Residential Construction Academy* are the result of cooperative planning and rigorous joint efforts between industry and education. The program was originally conceived by the National Association of Home Builders (NAHB)—the premier association of more than 200,000 member groups in the residential construction industry—and its workforce development arm, the Home Builders Institute (HBI).

For the first time, construction professionals and educators created national skills standards for the construction trades. In the summer of 2001, NAHB, through the HBI, began the process of developing residential craft standards in six trades: carpentry, electrical wiring, HVAC, plumbing, masonry, and facilities maintenance. Groups of employers from across the country met with an independent research and measurement organization to begin the development of new craft training standards. Care was taken to assure representation of builders and remodelers, residential and light commercial, custom single family and high production or volume builders. The guidelines from the National Skills Standards Board were followed in developing the new standards. In addition, the process met or exceeded American Psychological Association standards for occupational credentialing.

Next, through a partnership between HBI and Delmar / Cengage Learning, learning materials—textbooks, videos, and instructor's curriculum and teaching tools—were created to teach these standards effectively. A foundational tenet of this series is that students *learn by doing.* Integrated into this colorful, highly illustrated text are Procedure sections designed to help students apply information through hands-on, active application. A constant focus of the *Residential Construction Academy* is teaching the skills needed to be successful in the construction industry and constantly applying the learning to real-world applications.

The newest programming component to the Residential Construction Academy Series is industry Program Credentialing and Certification for both instructors and students by the Home Builders Institute. National Instructor Certification ensures consistency in instructor teaching/training methodologies and knowledge competency when teaching to the industry's national skills standards. Student Certification is offered for each trade area of the Residential Construction Academy Series in the form of rigorous testing. Student Certification is tied to a national database that will provide an opportunity for easy access for potential employers to verify skills and competencies. Instructor and Student Certification serve the basis for Program Credentialing offered by HBI. For more information on HBI Program Credentialing and Instructor and Student Certification, please go to **http://www.hbi.org/certification.**

# About this Book

A home is an essential part of human life. It provides protection, security, and privacy to its occupants. It is viewed as the single-most important possession a family can own. This book is written for students who want to learn how to build a home.

This book is organized in four sections: Tools and Materials, Rough Carpentry, Exterior Finish, and Interior Finish. These sections and the chapters within them are presented in the order in which a home is constructed.

We begin with an understanding of the tools and building materials used in residential construction. Hand and power tools provide the means to shape the material into the desired form. The choices of material are vast and change as technology provides better products to meet the needs of the industry. Fasteners hold it all together.

Rough framing creates the outline of the building. The learner will understand how a building grows and seems to come alive where empty space once existed. Carpenters begin where the masons leave off by installing the floor and walls. Workers are required to work above the ground using scaffolds and ladders; safety is a constant focus for the learner throughout the book. Once skill at working aboveground is achieved, the roof is erected. After the outline of the building is completed, it is ready for the finishes that make the building weathertight and comfortable.

The learner then moves on to the exterior finish that covers the frame, protecting it from the effects of weather and seasonal changes. Each locale has its particular climatic influence. Some regions are cold and homes there must be well insulated; others are warm but subject to high winds and hurricanes. Other areas are somewhere in between. The exterior finish defends the home and makes an architectural statement of style.

Interior finish provides flat surfaces ready for decor. It completes the boundary between the outside and the warmth or coolness of the inside. The learner is introduced to many different types of materials that may be used for finishing the interior and exterior building surfaces.

This book is designed to present information in a step-by-step fashion. Learners are expected to understand the skills and techniques of earlier chapters before going on to new material. The learners' knowledge of construction grows with the home. Yet, just as it is OK to cut lumber using your right hand or your left, variations in construction techniques do exist. It should be understood at the outset that there is more than one way to do most tasks. The techniques adopted for this book are time tested and chosen for their simplicity and straightforward approach, making the presentation of this information as easy as possible.

Life as a construction worker is a noble profession. Workers have the opportunity to work with their hands in a creative manner. Taking materials delivered to the site and generating a structure brings someone's dream to fruition. Anyone with the desire to do so can learn the skills to accomplish the tasks of home construction. It is to those students who choose to follow our forefathers into the field of construction that this book is dedicated.

## New to This Edition

The third edition of *RCA: Carpentry* features a new introductory chapter that covers industry-related topics such as business practices, ethical and legal issues, career goals, customer service, and construction trends, as well as trade-specific career goals and career profiles of fellow professionals working in the field.

In addition, the new edition features many content and organizational changes. Coverage of power tools is now split into two chapters, Chapter 2 "Hand Power Tools" and Chapter 3 "Stationary Power Tools." Likewise, coverage of blueprints, codes, and building layout is now split into two chapters—Chapter 6, "Blueprint Reading and Building Codes," and Chapter 7 "Building Layout." Furthermore, the chapter on insulation and wall finish (formerly Chapter 14) has been reorganized into two chapters—Chapter 16 "Insulation and Ventilation" and Chapter 17 "Wall Finish." Additional information on heat loss and insulation can be found in Chapter 16. Many chapters have a new section on estimating materials. Formulas are included that introduce the logic of determining the quantities of material for a particular job.

New **Green Tip** boxes and **Green Checklists** highlight tips and topics on new "green" construction methods.

A new end-of-chapter feature, **Know Your Codes** is an application-based feature that prompts students to research their local and regional codes, providing excellent practice for a critical on-the-job skiil.

A new **What's Wrong with This Picture?** feature highlights basic tool use and common mistakes.

An answer key for the chapter-ending Review Questions is now included.

# Features

This innovative series was designed with input from educators and industry professionals and informed by the curriculum and training objectives established by the standards committee. The following features aid learning:

**LEARNING FEATURES** such as the **Introduction, Objectives,** and **Glossary** set the stage for the coming body of knowledge and help learners identify key concepts and information. These learning features serve as a road map for continuing through the chapter. Learners also may use them as an on-the-job reference.

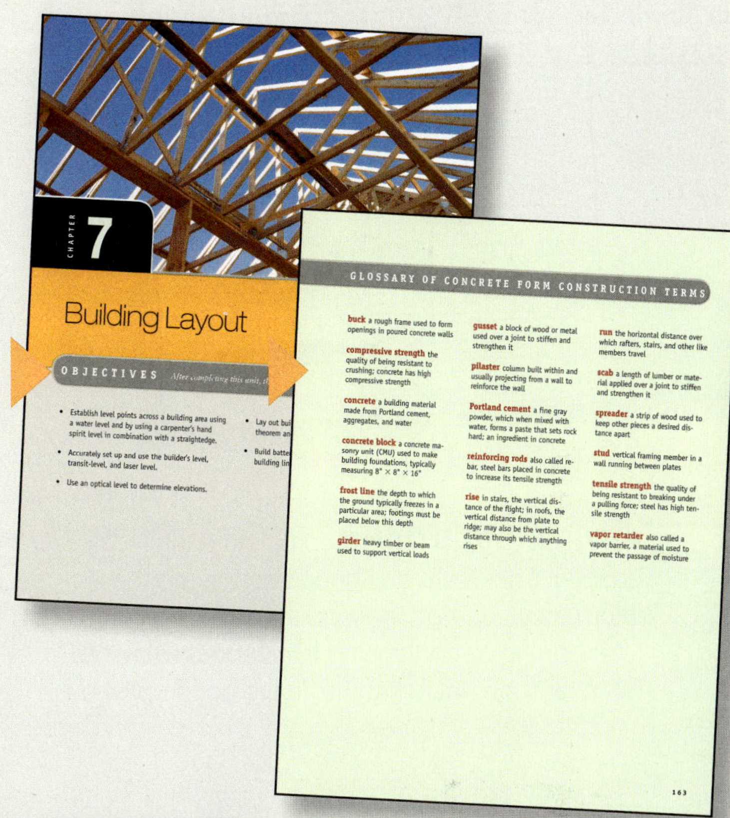
CHAPTER 7

Building Layout

OBJECTIVES

- Establish level points across a building area using a water level and by using a carpenter's hand spirit level in combination with a straightedge.
- Accurately set up and use the builder's level, transit-level, and laser level.
- Use an optical level to determine elevations.

GLOSSARY OF CONCRETE FORM CONSTRUCTION TERMS

**buck** a rough frame used to form openings in poured concrete walls

**compressive strength** the quality of being resistant to crushing; concrete has high compressive strength

**concrete** a building material made from Portland cement, aggregates, and water

**concrete block** a concrete masonry unit (CMU) used to make building foundations, typically measuring 8" × 8" × 16"

**frost line** the depth to which the ground typically freezes in a particular area; footings must be placed below this depth

**girder** heavy timber or beam used to support vertical loads

**gusset** a block of wood or metal used over a joint to stiffen and strengthen it

**pilaster** column built within and usually projecting from a wall to reinforce the wall

**Portland cement** a fine gray powder, which when mixed with water, forms a paste that sets rock hard; an ingredient in concrete

**reinforcing rods** also called rebar, steel bars placed in concrete to increase its tensile strength

**rise** in stairs, the vertical distance of the flight; in roofs, the vertical distance from plate to ridge; may also be the vertical distance through which anything rises

**run** the horizontal distance over which rafters, stairs, and other like members travel

**scab** a length of lumber or material applied over a joint to stiffen and strengthen it

**spreader** a strip of wood used to keep other pieces a desired distance apart

**stud** vertical framing member in a wall running between plates

**tensile strength** the quality of being resistant to breaking under a pulling force; steel has high tensile strength

**vapor retarder** also called a vapor barrier, a material used to prevent the passage of moisture

163

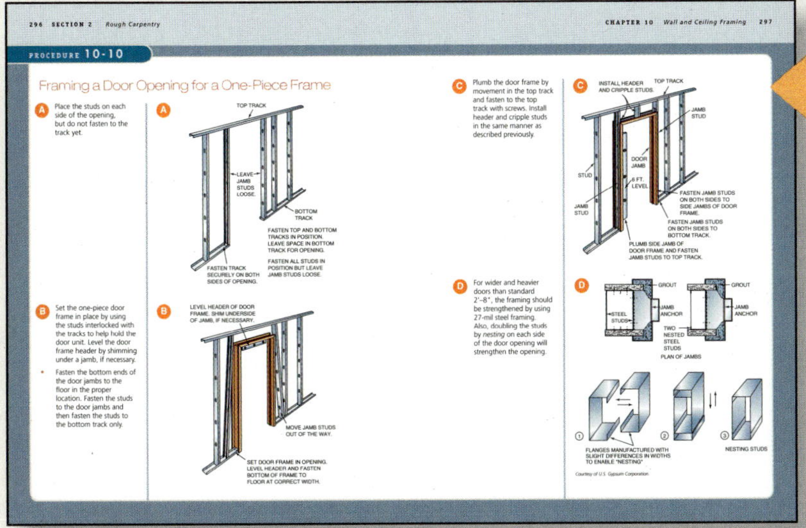
296 SECTION 2 Rough Carpentry — CHAPTER 10 Wall and Ceiling Framing 297

PROCEDURE 10-10

Framing a Door Opening for a One-Piece Frame

A Place the studs on each side of the opening, but do not fasten to the track yet.

B Set the one-piece door frame in place by using the studs interlocked with the tracks to help hold the door unit. Level the door frame header by shimming under a jamb, if necessary.

- Fasten the bottom ends of the door jambs to the floor in the proper location. Fasten the studs to the door jambs and then fasten the studs to the bottom track only.

C Plumb the door frame by movement in the top track and fasten to the top track with screws. Install header and cripple studs in the same manner as described previously.

D For wider and heavier doors than standard 2'–8", the framing should be strengthened by using 27-mil steel framing. Also, doubling the studs by *nesting* on each side of the door opening will strengthen the opening.



Courtesy of U.S. Gypsum Corporation

**ACTIVE LEARNING** is a core concept of the *Residential Construction Academy Series.* Information is heavily illustrated to provide a visual of new tools and tasks encountered by learners. Chapters also contain a **Procedures** section that takes the information and applies it so that learning is accomplished through doing. In the **Procedures,** various tasks in home construction are grouped in a step-by-step approach. The overall effect is a clear view of the task, making learning easier.

**SAFETY** is emphasized throughout the text to instill safety as an attitude among learners. Safe job site practices by all workers are essential; if one person acts in an unsafe manner, all workers on the job are at risk for injury. Learners will come to appreciate that safety is a blend of ability, skill, and knowledge that should be continuously applied to all they do in the construction industry.

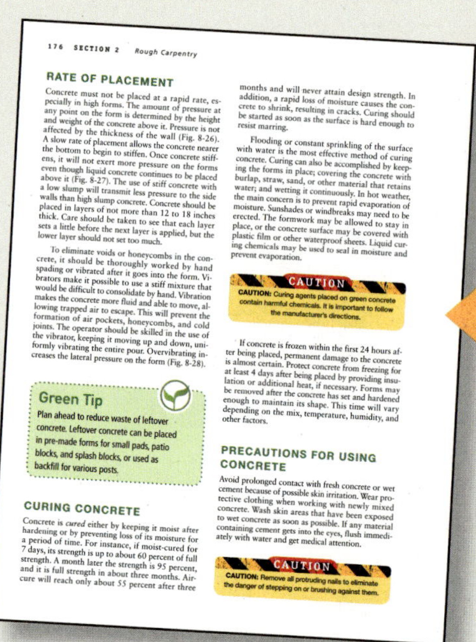

176 SECTION 2 Rough Carpentry

RATE OF PLACEMENT

Concrete must not be placed at a rapid rate, especially in high forms. The amount of pressure at any point on the form is determined by the height and weight of the concrete above it. Pressure is not affected by the thickness of the wall (Fig. 8-26). A slow rate of placement allows the concrete nearer the bottom to begin to stiffen. Once concrete stiffens, it will not exert more pressure on the forms even though liquid concrete continues to be placed above it (Fig. 8-27). The use of stiff concrete with a low slump will transmit less pressure to the side walls than high slump concrete. Concrete should be placed in layers of not more than 12 to 18 inches thick. Care should be taken to see that each layer sets a little before the next layer is applied, but the lower layer should not set too much.

To eliminate voids or honeycombs in the concrete, it should be thoroughly worked by hand spading or vibrated after it goes into the form. Vibrators make it possible to use a stiff mixture that would be difficult to consolidate by hand. Vibration makes the concrete more fluid and able to move, allowing trapped air to escape. This will prevent the formation of air pockets, honeycombs, and cold joints. The operator should be skilled in the use of the vibrator, keeping it moving up and down, uniformly vibrating the entire pour. Overvibrating increases the lateral pressure on the form (Fig. 8-28).

Green Tip

Plan ahead to reduce waste of leftover concrete. Leftover concrete can be placed in pre-made forms for small pads, patio blocks, and splash blocks, or used as backfill for various posts.

CURING CONCRETE

Concrete is *cured* either by keeping it moist after hardening or by preventing loss of its moisture for a period of time. For instance, if moist-cured for 7 days, its strength is up to about 60 percent of full strength. A month later the strength is 95 percent, and it is full strength in about three months. Air-cure will reach only about 55 percent after three months and will never attain design strength. In addition, a rapid loss of moisture causes the concrete to shrink, resulting in cracks. Curing should be started as soon as the surface is hard enough to resist marring.

Flooding or constant sprinkling of the surface with water is the most effective method of curing concrete. Curing can also be accomplished by keeping the forms in place; covering the concrete with burlap, straw, sand, or other material that retains water; and wetting it continuously. In hot weather, the main concern is to prevent rapid evaporation of moisture. Sunshades or windbreaks may need to be erected. The formwork may be allowed to stay in place, or the concrete surface may be covered with plastic film or other waterproof sheets. Liquid curing chemicals may be used to seal in moisture and prevent evaporation.

CAUTION

CAUTION: Curing agents placed on green concrete contain harmful chemicals. It is important to follow the manufacturer's directions.

If concrete is frozen within the first 24 hours after being placed, permanent damage to the concrete is almost certain. Protect concrete from freezing for at least 4 days after being placed by providing insulation or additional heat, if necessary. Forms may be removed after the concrete has set and hardened enough to maintain its shape. This time will vary depending on the mix, temperature, humidity, and other factors.

PRECAUTIONS FOR USING CONCRETE

Avoid prolonged contact with fresh concrete or wet cement because of possible skin irritation. Wear protective clothing when working with newly mixed concrete. Wash skin areas that have been exposed to wet concrete as soon as possible. If any material containing cement gets into the eyes, flush immediately with water and get medical attention.

CAUTION

CAUTION: Remove all protruding nails to eliminate the danger of stepping on or brushing against them.

**CAUTION** features highlight safety issues and urgent safety reminders for the trade.

**REVIEW QUESTIONS** complete each chapter. These are designed to reinforce the information learned in the chapter and to give learners the opportunity to think about what has been learned and what they have accomplished.

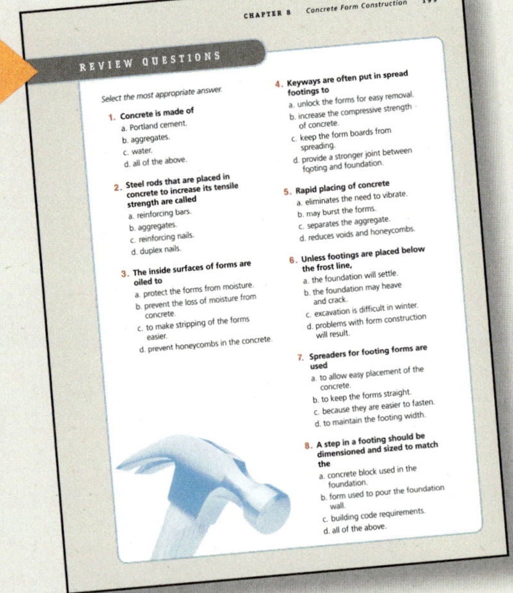

CHAPTER 8 Concrete Form Construction 199

REVIEW QUESTIONS

*Select the most appropriate answer.*

1. **Concrete is made of**
   a. Portland cement.
   b. aggregates.
   c. water.
   d. all of the above.
2. **Steel rods that are placed in concrete to increase its tensile strength are called**
   a. reinforcing bars.
   b. aggregates.
   c. reinforcing nails.
   d. duplex nails.
3. **The inside surfaces of forms are oiled to**
   a. protect the forms from moisture.
   b. prevent the loss of moisture from concrete.
   c. to make stripping of the forms easier.
   d. prevent honeycombs in the concrete.
4. **Keyways are often put in spread footings to**
   a. unlock the forms for easy removal.
   b. increase the compressive strength of concrete.
   c. keep the form boards from spreading.
   d. provide a stronger joint between footing and foundation.
5. **Rapid placing of concrete**
   a. eliminates the need to vibrate.
   b. may burst the forms.
   c. separates the aggregate.
   d. reduces voids and honeycombs.
6. **Unless footings are placed below the frost line,**
   a. the foundation will settle.
   b. the foundation may heave and crack.
   c. excavation is difficult in winter.
   d. problems with form construction will result.
7. **Spreaders for footing forms are used**
   a. to allow easy placement of the concrete.
   b. to keep the forms straight.
   c. because they are easier to fasten.
   d. to maintain the footing width.
8. **A step in a footing should be dimensioned and sized to match the**
   a. concrete block used in the foundation.
   b. form used to pour the foundation wall.
   c. building code requirements.
   d. all of the above.

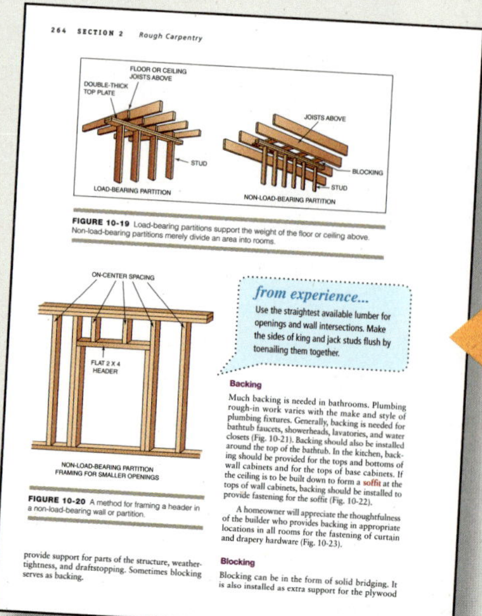

264 SECTION 2 Rough Carpentry



FIGURE 10-19 Load-bearing partitions support the weight of the floor or ceiling above. Non-load-bearing partitions merely divide an area into rooms.



FIGURE 10-20 A method for framing a header in a non-load-bearing wall or partition.

provide support for parts of the structure, weathertightness, and draftstopping. Sometimes blocking serves as backing.

*from experience...*

Use the straightest available lumber for openings and wall intersections. Make the sides of king and jack studs flush by toenailing them together.

Backing

Much backing is needed in bathrooms. Plumbing rough-in work varies with the make and style of plumbing fixtures. Generally, backing is needed for bathtub faucets, showerheads, lavatories, and water closets (Fig. 10-21). Backing should also be installed around the top of the bathtub. In the kitchen, backing should be provided for the tops and bottoms of wall cabinets and for the tops of base cabinets. If the ceiling is to be built down to form a soffit at the tops of wall cabinets, backing should be installed to provide fastening for the soffit (Fig. 10-22).

A homeowner will appreciate the thoughtfulness of the builder who provides backing in appropriate locations in all rooms for the fastening of curtain and drapery hardware (Fig. 10-23).

Blocking

Blocking can be in the form of solid bridging. It is also installed as extra support for the plywood

**FROM EXPERIENCE** offers tricks of the trade and mentoring wisdom that make a particular task a little easier for the novice to accomplish.

**GREEN TIPS** and **GREEN CHECKLISTS** highlight tips and topics on new "green" construction methods.

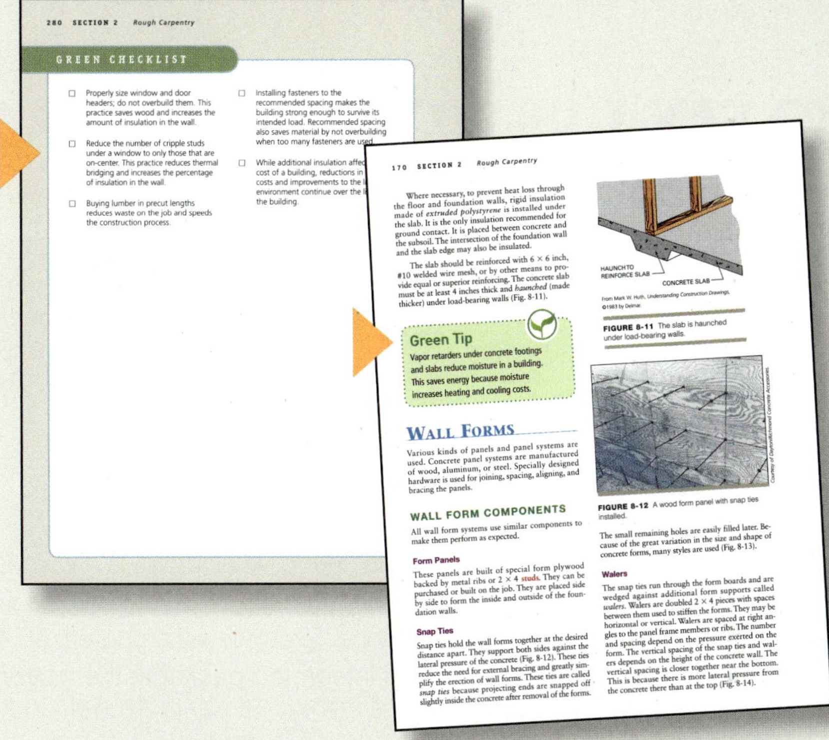
280 SECTION 2 *Rough Carpentry*

GREEN CHECKLIST

- ☐ Properly size window and door headers; do not overbuild them. This practice saves wood and increases the amount of insulation in the wall.
- ☐ Reduce the number of cripple studs under a window to only those that are on-center. This practice reduces thermal bridging and increases the percentage of insulation in the wall.
- ☐ Buying lumber in precut lengths reduces waste on the job and speeds the construction process.
- ☐ Installing fasteners to the recommended spacing makes the building strong enough to survive its intended load. Recommended spacing also saves material by not overbuilding when too many fasteners are used.
- ☐ While additional insulation affec cost of a building, reductions in costs and improvements to the l environment continue over the l the building.

170 SECTION 2 *Rough Carpentry*

Where necessary, to prevent heat loss through the floor and foundation walls, rigid insulation made of ***extruded polystyrene*** is installed under the slab. It is the only insulation recommended for ground contact. It is placed between concrete and the subsoil. The intersection of the foundation wall and the slab edge may also be insulated.

The slab should be reinforced with 6 × 6 inch, #10 welded wire mesh, or by other means to provide equal or superior reinforcing. The concrete slab must be at least 4 inches thick and *haunched* (made thicker) under load-bearing walls (Fig. 8-11).

HAUNCHTO REINFORCE SLAB

CONCRETE SLAB

From Mark W. Huth, *Understanding Construction Drawings*. ©1983 by Delmar.

**FIGURE 8-11** The slab is haunched under load-bearing walls.

**Green Tip**

Vapor retarders under concrete footings and slabs reduce moisture in a building. This saves energy because moisture increases heating and cooling costs.

Courtesy of Dayton/Richmond Concrete Accessories.

**FIGURE 8-12** A wood form panel with snap ties installed.

WALL FORMS

Various kinds of panels and panel systems are used. Concrete panel systems are manufactured of wood, aluminum, or steel. Specially designed hardware is used for joining, spacing, aligning, and bracing the panels.

WALL FORM COMPONENTS

All wall form systems use similar components to make them perform as expected.

**Form Panels**

These panels are built of special form plywood backed by metal ribs or 2 × 4 studs. They can be purchased or built on the job. They are placed side by side to form the inside and outside of the foundation walls.

**Snap Ties**

Snap ties hold the wall forms together at the desired distance apart. They support both sides against the lateral pressure of the concrete (Fig. 8-12). These ties reduce the need for external bracing and greatly simplify the erection of wall forms. These ties are called *snap ties* because projecting ends are snapped off slightly inside the concrete after removal of the forms.

The small remaining holes are easily filled later. Because of the great variation in the size and shape of concrete forms, many styles are used (Fig. 8-13).

**Walers**

The snap ties run through the form boards and are wedged against additional form supports called *walers*. Walers are doubled 2 × 4 pieces with spaces between them used to stiffen the forms. They may be horizontal or vertical. Walers are spaced at right angles to the panel frame members or ribs. The number and spacing depend on the pressure exerted on the form. The vertical spacing of the snap ties and walers depends on the height of the concrete wall. The vertical spacing is closer together near the bottom. This is because there is more lateral pressure from the concrete there than at the top (Fig. 8-14).

**KNOW YOUR CODES** is an application-based feature that prompts students to research their local and regional codes, providing excellent practice for a critical on-the-job skill.

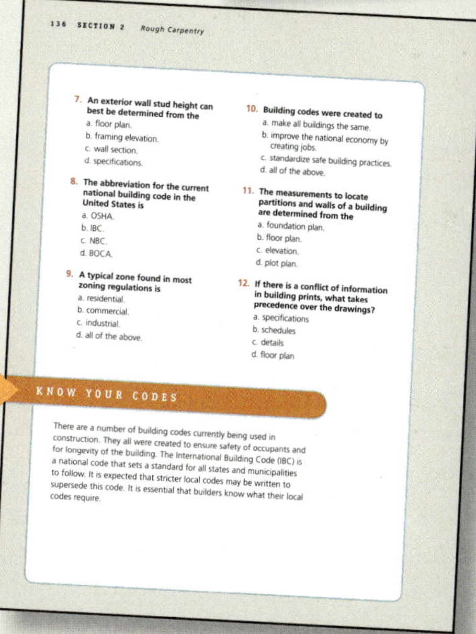
136 SECTION 2 *Rough Carpentry*

7. **An exterior wall stud height can best be determined from the**
   a. floor plan.
   b. framing elevation.
   c. wall section.
   d. specifications.
8. **The abbreviation for the current national building code in the United States is**
   a. OSHA.
   b. IBC.
   c. NBC.
   d. BOCA.
9. **A typical zone found in most zoning regulations is**
   a. residential.
   b. commercial.
   c. industrial.
   d. all of the above.
10. **Building codes were created to**
    a. make all buildings the same.
    b. improve the national economy by creating jobs.
    c. standardize safe building practices.
    d. all of the above.
11. **The measurements to locate partitions and walls of a building are determined from the**
    a. foundation plan.
    b. floor plan.
    c. elevation.
    d. plot plan.
12. **If there is a conflict of information in building prints, what takes precedence over the drawings?**
    a. specifications
    b. schedules
    c. details
    d. floor plan

KNOW YOUR CODES

There are a number of building codes currently being used in construction. They all were created to ensure safety of occupants and for longevity of the building. The International Building Code (IBC) is a national code that sets a standard for all states and municipalities to follow. It is expected that stricter local codes may be written to supersede this code. It is essential that builders know what their local codes require.

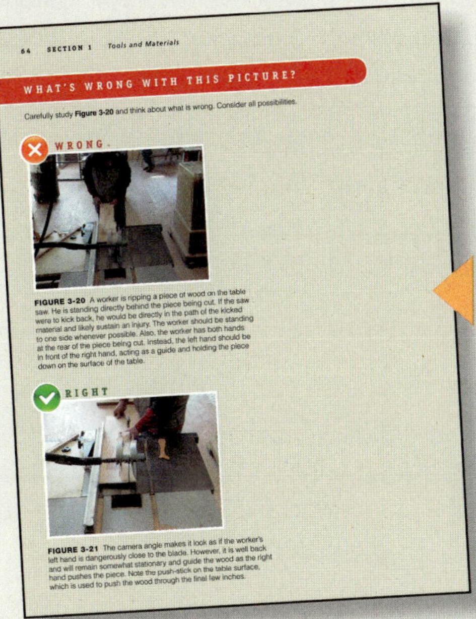
64 SECTION 1 *Tools and Materials*

WHAT'S WRONG WITH THIS PICTURE?

Carefully study **Figure 3-20** and think about what is wrong. Consider all possibilities.

WRONG

**FIGURE 3-20** A worker is ripping a piece of wood on the table saw. He is standing directly behind the piece being cut. If the saw were to kick back, he would be directly in the path of the kicked material and likely sustain an injury. The worker should be standing to one side whenever possible. Also, the worker has both hands at the rear of the piece being cut. Instead, the left hand should be in front of the right hand, acting as a guide and holding the piece down on the surface of the table.

RIGHT

**FIGURE 3-21** The camera angle makes it look as if the worker's left hand is dangerously close to the blade. However, it is well back and will remain somewhat stationary and guide the wood as the right hand pushes the piece. Note the push-stick on the table surface, which is used to push the wood through the final few inches.

**WHAT'S WRONG WITH THIS PICTURE?** feature highlights basic tool use and common mistakes.

# Turnkey Curriculum and Teaching Material Package

We understand that a text is only one part of a complete, turnkey educational system. We also understand that instructors want to spend their time on teaching, not preparing to teach. The *Residential Construction Academy Series* is committed to providing thorough curriculum and prepatory materials to aid instructors and alleviate some of their heavy preparation commitments. An integrated teaching solution is ensured with the text, Instructor Resources™, print *Instructor's Resource Guide,* and student videos.

## Instructor Resources

Delmar/Cengage Learning's Instructor Resources is a complete guide to classroom management. The Instructor Resources CD-ROM contains the Instructor's Resources Guide, PowerPoint presentations, ExamView Computerized Test Bank, and Image Library. Designed as a complete and integrated package, the Instructor Resources also provides suggestions for when and how to use the package components. An **Instructor's Guide** is also available.

## PowerPoint

The author has created a series of PowerPoint presentations that give thorough, step-by-step overviews of the crucial topics of the course. These presentations can be used to introduce or review the topics.

## DVDs

The *Carpentry DVD Series* is an integrated part of the *Residential Construction Academy: Carpentry* package. This DVD series steps viewers through the process of constructing a home. The series contains a set of eight, 20-minute videos that cover everything from the basics of obtaining the building permit to the detail work of interior trim. Special geographic considerations are addressed to more accurately reflect building practices throughout the United States. In addition, the videos offer such features as Carpenter's Tips and Safety Tips full of practical advice from the experts.

The complete set includes: Video #1–Building Layout; Video #2–Form & Concrete Placement; Video #3–Sub-Flooring & Wood-Bearing Walls; Video #4–Truss Installation; Video #5–Window & Door Installation; Video #6–Interior Trim; Video #7–Interior Partition Framing; Video #8–Stair Construction.

## Workbook

Designed to accompany *Residential Construction Academy: Carpentry, Third Edition,* this Workbook is an extension of the core text, and provides additional review questions and problems designed to challenge and reinforce the student's comprehension of the content presented in the core text.

## Performance Based Testing

Performance based testing is designed to measure actual tasks performed in a real world environment. These tasks, common in the residential construction industry, for carpentry, house wiring, plumbing, and HVAC are covered in this new CD-ROM tool. Included in the practical exams are process and product tasks and behavior-based assessment, which will examine the steps in applying skills, while product assessment will evaluate an actual work product.

## Online Companion

The Online Companion is an excellent supplement for students. It features many useful resources to support the *Carpentry* book, videos, and CDs. Linked from the Student Materials section of http://www.residentialacademy.com, the Online Companion includes chapter quizzes, an online glossary, product updates, related links, and more. Visit: http://www.delmarlearning.com/companions/index.asp?isbn=1401813437.

# About the Author

The author of this text, Floyd Vogt, is a sixth-generation carpenter/contractor. He was raised in a family with a small business devoted to all phases of home construction and began working in the family business at age 15.

After completing a B.A. in chemistry from the State University of New York College at Oneonta, Mr. Vogt returned to the field as a self-employed remodeler. In 1985, he began teaching at State University of New York at Delhi, New York, **www.delhi.edu.** He has taught many courses, including Light Framing, Advanced Framing, Math, Energy Efficient Construction, Finish Carpentry, Finish Masonry, and Estimating. He is currently a Professor in the Construction Management, Design and Build Bachelors degree program at Delhi. Course responsibilities include Residential Construction, Auto CAD, Construction Seminar and Physical Science Applications. He has served as a carpentry regional coordinator for Skills-USA and post-secondary Skills-USA student advisor. He is currently a member of a local town Planning Board.

## Compliance with Apprenticeship, Training, Employer, and Labor Services (ATELS)

These materials are in full compliance with the Apprenticeship, Training, Employer, and Labor Services (ATELS) requirements for classroom training.

# Acknowledgments


## Carpentry National Skill Standards

The NAHB and HBI would like to thank the many individuals, members, and companies that participated in the creation of the Carpentry National Skill Standards. Special thanks are extended to the following individuals and companies:

Karen Butts
Vinyl Siding Institute

Kevin Eddy
Les Eddy & Sons General Contractors, Inc.

Tim Faller
Field Training Services

Fred Humphreys
Home Builders Institute

Bob Jenkins
Maryland Correctional Training Center

Eric Listou
Top Quality Remodeling

Mark Martin
Penobscot Job Corp Center

Jack Sanders
Home Builders Institute

David Sitton
Beazer Homes

Ed Snider
Beazer Homes

David VanCise
Indian River Community College

Floyd Vogt
Delhi College of Technology

Ray Wasdyke
Wasdyke Associates

In addition to the standards committee, many other people contributed their time and expertise to the project. They have spent hours attending focus groups, reviewing and contributing to the work. Delmar Learning and the author extend our sincere gratitude to:

Greg Fletcher
Kennebec Valley Technical College

Tim Lockley
George Jr. Republic Vo-Tech

Mark Martin
Penobscot Job Corp Center

David McCosby
New Castle School of Trades

Douglas D. Park-Sample
Umpqua Community College

James A. Smith
Dauphin County Technical School

Ted Stodder
College of the Redwoods

Lester Stackpole
Eastern Maine Technical College

Earl T. Torgerson
Bismarck State College

David VanCise
Indian River Community College

The author would like to thank every person, too numerous to name here, who had a role in making this book come into being, from publishers to editors of all sorts and trades people willing to share their experiences. Special thanks to the folks at Ohlinger Publishing Services for a super job well done; thanks, Jennifer. Finally, thanks to my wife Pamela, whose continuous support makes any chore easier to complete.

# Introduction

## ORGANIZATION OF THE INDUSTRY

The residential construction industry is one of the biggest sectors of the American economy. According to the U.S. Department of Labor, construction is one of the Nation's largest industries, with 7.2 million wage and salary jobs and 1.8 million self-employed and unpaid family workers in 2008. About 64 percent of wage and salary jobs in construction were in the specialty trade contractors sector, primarily plumbing, heating and air-conditioning, electrical, and masonry. The National Association of Home Builders (NAHB) reports that home building traditionally accounts for 50–55 percent of the construction industry. Opportunities are available for people to work at all levels in the construction industry, from those who handle the tools and materials on the job site to the senior engineers and architects who spend most of their time in offices. Few people spend their entire lives in a single occupation, and even fewer spend their lives working for one employer. You should be aware of all the opportunities in the construction industry so that you can make career decisions in the future, even if you are sure of what you want to do at this time.

### Construction Personnel

The occupations in the construction industry can be divided into four categories:

- Unskilled or semiskilled labor
- Skilled trades or crafts
- Technicians
- Design and management

## Unskilled or Semiskilled Labor

Construction is labor intensive. That means it requires a lot of labor to produce the same dollar value of end products by comparison with other industries, where labor may be a smaller part of the picture. Some construction laborers are construction workers who have not reached a high level of skill in a particular trade and are not registered in an apprenticeship program. Other laborers may also be skilled construction workers opting to work in a role with reduced responsibility and increased simplicity. Laborers are sometimes assigned the tasks of moving materials, running errands, and working under the close supervision of a skilled worker. Their work keeps them in excellent physical condition.

Some laborers specialize in working with a particular trade, such as mason's tenders or carpenter's helpers (Fig. I-1). The mason's tender knows

**FIGURE I-1** Some laborers specialize in working as a mason's tender.

how to mix mortar for particular conditions, can erect scaffolding, and is familiar with the bricklayer's tools. Many laborers go on to acquire skills and become skilled workers. Laborers who specialize in a particular trade are often paid slightly more than completely unskilled laborers.

## Skilled Trades

A *craft* or *skilled trade* is an occupation working with tools and materials and building structures. The building trades are the crafts that deal most directly with building construction (see Fig. I-2).

The building trades are among the highest paying of all skilled occupations. However, work in the building trades can involve working in cold conditions in winter or hot sun in the summer. Also, job opportunities will be best in an area where a lot of construction is being done. The construction industry is growing at a high rate nationwide. Generally, plenty of work is available to provide a comfortable living for a good worker.

### *Apprenticeship*

The skill needed to be employed in the building trades is often learned in an apprentice program. Apprenticeships are usually offered by trade unions,

- Carpenter
  - Framing carpenter
  - Finish carpenter
  - Cabinetmaker
- Plumber
  - New construction
  - Maintenance and repair
- Roofer
- Electrician
  - Construction electrician
  - Maintenance electrician
- Mason
  - Bricklayer (also lays concrete blocks)
  - Cement finisher
- HVAC technician
- Plasterer
  - Finish plaster
  - Stucco plaster
- Tile setter
- Equipment operator
- Drywall installer
  - Installer
  - Taper
- Painter

**FIGURE I-2** Building trades.

trade associations, technical colleges, and large employers. *Apprentices* attend class a few hours a week to learn the necessary theory. The rest of the week they work on a job site under the supervision of a *journeyman* (a skilled worker who has completed the apprenticeship and has experience on the job). The term "journeyman" is a gender neutral term that has been used for decades. It is worth noting that many highly skilled building trades' workers are women. Apprentices receive a much lower salary than journeymen, often about 50 percent of what a journeyman receives. The apprentice wage usually increases as stages of the apprenticeship are successfully completed. By the time the apprenticeship is completed, the apprentice can be earning as much as 95 percent of what a journeyman earns. Many apprentices receive college credit for their training. Some journeymen receive their training through school or community college and on-the-job training. In one way or another, some classroom training and some on-the-job supervised experience are usually necessary to reach journeyman status. Not all apprentice programs are the same, but a typical apprenticeship lasts four or five years and requires between 100 and 200 hours per year of classroom training along with 1200 to 1500 hours per year of supervised work experience.

## Technicians

*Technicians* provide a link between the skilled trades and the professions. Technicians often work in offices, but their work also takes them to construction sites. Technicians use mathematics, computer skills, specialized equipment, and knowledge of construction to perform a variety of jobs. Figure I-3 lists several technical occupations.

Most technicians have some type of college education, often combined with on-the-job experience, to prepare them for their technical jobs. Community colleges often have programs aimed at preparing people to work at the technician level in construction. Some community college programs are intended especially for preparing workers for the building trades, while others have more of a construction management focus. Construction management courses give the graduate a good overview of the business of construction. The starting salary for a construction technician is about the same as for a skilled trade, but the technician can be more certain of regular work and will have better opportunities for advancement.

| Technical Career | Some Common Jobs |
|---|---|
| Surveyor | Measures land, draws maps, lays out building lines, and lays out roadways |
| Estimator | Calculates time and materials necessary for project |
| Drafter | Draws plans and construction details in conjunction with architects and engineers |
| Expeditor | Ensures that labor and materials are scheduled properly |
| Superintendent | Supervises all activities at one or more job sites |
| Inspector | Inspects project for compliance with local building codes at various stages of completion |
| Planner | Plans for best land and community development |

**FIGURE I-3** Technicians.

## Design and Management

Architecture, engineering, and contracting are the design and management professions. The *professions* are those occupations that require more than four years of college and a license to practice. Many contractors have less than four years of college, but they often operate at a very high level of business, influencing millions of dollars, and so they are included with the professions here. These construction professionals spend most of their time in offices and are not frequently seen on the job site.

Architects usually have a strong background in art, so they are well prepared to design attractive, functional buildings. A typical architect's education includes a four-year degree in fine art, followed by a master's degree in architecture. Most of their construction education comes during the final years of work on the architecture degree.

Engineers generally have more background in math and science, so they are prepared to analyze conditions and calculate structural characteristics. There are many specialties within engineering, but civil engineers are the ones most commonly found in construction. Some civil engineers work mostly in road layout and building. Other civil engineers work mostly with structures in buildings. They are sometimes referred to as structural engineers.

*Contractors* are the owners of the businesses that do most of the building. In larger construction firms, the principal (the owner) may be more concerned with running the business than with supervising construction. Some contractors are referred to as general contractors and others as *subcontractors* (Fig. I–4). The general contractor is the principal construction company hired by the owner to construct the building. A general contractor might have only a skeleton crew, relying on subcontractors for most of the actual construction. The general contractor's superintendent coordinates the work of all the subcontractors.

It is quite common for a successful journeyman to start his or her own business as a contractor, specializing in the field in which he or she was a journeyman. These are the subcontractors that sign on to do a specific part of the construction, such as framing or plumbing. As the contractor's company grows and the company works on several projects at one time, the skilled workers with the best ability to lead others may become foremen. A foreman is a working supervisor of a small crew of workers in a specific trade. All contractors have to be concerned with business management. For this reason, many successful contractors attend college and get a degree in construction management. Most states require contractors to have a license to do contracting in their state. Requirements vary from state to state, but a contractor's license usually requires several years of experience in the trade and a test on both trade information and the contracting business.

## *An Overall View of Design and Construction*

To understand the relationships between some of the design and construction occupations, we shall look at a scenario for a typical housing development. The first people to be involved are the community planners and the real estate *developer*. The real estate developer has identified a 300-acre tract on which he would like to build nearly 1000 homes, which he will later sell at a good profit. The developer must work with the city planners to ensure that the use he has planned is acceptable to the city. The city planner is responsible for ensuring that all building in the city fits the city's development plan and zoning ordinances. On a project this big, the developer might even bring in a planner of his own to help decide where parks and community buildings should be located and how much parking space they will need.

As the plans for development begin to take shape, it becomes necessary to plan streets and to start designing houses to be built throughout the

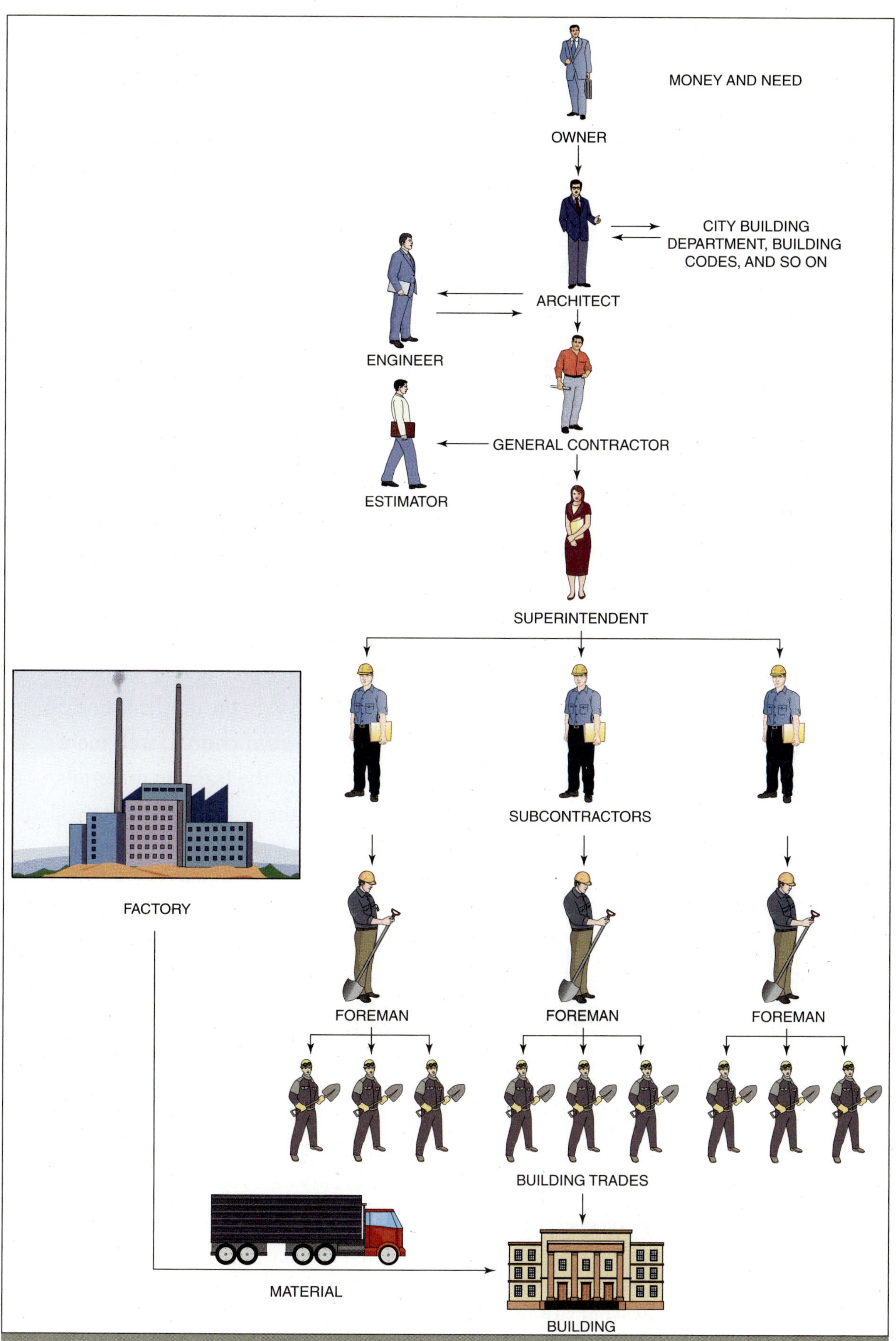


**FIGURE I-4** Organization of the construction industry.

development. A civil engineer is hired to plan and design the streets. The civil engineer will first work with the developer and planners to lay out the locations of the streets, their widths, and drainage provisions to get rid of storm water. The civil engineer also considers soil conditions and expected traffic to design the foundation for the roadway.

An architectural firm, or perhaps a single architect, will design the houses. Typically, several stock plans are used throughout a development, but many homeowners wish to pay extra to have a custom home designed and built. In a custom home, everything is designed for that particular house. Usually the homeowner, who will eventually live in the house, works with the architect to specify the sizes, shapes, and locations of rooms, interior and exterior trim, type of roof, built-in cabinets and appliances, use of outdoor spaces, and other special features. Architects specialize in use of space, aesthetics (attractive appearance), and livability features. Most architectural features do not involve special structural considerations, but when they do, a structural engineer is employed to analyze the structural requirements and help ensure that the structure will adequately support the architectural features.

One part of construction that almost always involves an engineer is the design of roof trusses. Roof trusses are the assemblies that make up the frame of the roof. Trusses are made up of the top chords, bottom chords, web members, and gussets. The engineer considers the weight of the framing materials, the weight of the roof covering, the anticipated weight of any snow that will fall on the roof in winter, and the span (the distance between supports) of the truss to design trusses for a particular purpose. The architect usually hires the engineer for this work, and so the end product is one set of construction drawings that includes all the architectural and engineering specifications for the building. Even though the drawings are sometimes referred to as architectural drawings, they include work done by architects, engineers, and their technicians. Building codes require an architect's seal on the drawings before work can begin. The architect will require an engineer to certify certain aspects of the drawings before putting the architect's seal on them.

## FORMS OF OWNERSHIP

Construction companies vary in size from small, one-person companies to very large international organizations that do many kinds of construction. However, the size of the company does not necessarily indicate the form of

| Forms of Ownership | What it Means | Advantages | Disadvantages |
|---|---|---|---|
| **Sole Proprietorship** | A sole proprietorship is a business whose owner and operator are the same person. | The owner has complete control over the business and there is a minimum of government regulation. If the company is successful, the owner receives high profits. | If the business goes into debt the owner is responsible for that debt. The owner can be sued for the company, and the owner suffers all the losses of the company. |
| **Partnership**<br><br>**(*General* and *Limited Liability Partnership* *(LLP)*)** | A partnership is similar to a sole proprietorship, but there are two or more owners.<br><br>*General*: In a general partnership, each partner shares the profits and losses of the company in proportion to the partner's share of investment in the company.<br><br>*LLP*: A limited liability partner is one who invests in the business, receives a proportional share of the profit or loss, but has limited liability. | *General Partnership*: The advantage is that the partners share the expense of starting the business and partnerships are not controlled by extensive government regulations.<br><br>*LLP*: A limited liability partner can only lose his or her investment. | *General Partnership*: Each partner can be held responsible for all the debts of the company.<br><br>*LLP*: Every LLP must have one or more general partners who run the business. The general partners in an LLP have unlimited liability and they can be personally sued for any debts of the company. |
| **Corporation** | In a corporation, a group of people own the company. Another, usually smaller, group of people manage the business. The owners buy shares of stock. A share of stock is a share or a part of the business. The value of each share increases or decreases according to the success of the company. | In a corporation, no person has unlimited liability. The owners can only lose the amount of money they invested in stock. The owners of a corporation are not responsible for the debts of the corporation. The corporation itself is the legal body and is responsible for its own debts. | The government has stricter regulations for corporations than for the other forms of ownership. Also, corporations are more expensive to form and to operate than are proprietorships and partnerships. |

**FIGURE I-5** Three types of ownership.

ownership. Three types of ownership and the advantages and disadvantages of each are shown in Figure I-5.

## UNIONS AND CONTRACTORS' ASSOCIATIONS

The construction industry contains thousands of organizations of people with common interests and goals. Whole directories of these organizations are available in libraries and on the Internet. Two categories of construction organizations are of particular importance to construction students: craft unions and contractors' associations.

## Unions

A *craft union*, usually just called a "union," is an organization of workers in a particular building trade. Workers' unions were first formed in the 1800s when factory workers were being forced to work extreme hours under unsafe conditions—and for very low wages. Although working conditions in both factories and construction have improved dramatically, unions continue to serve a valuable role in the construction industry. Figure I-6 lists several national construction craft unions.

Union members pay dues to be members of the union. Dues money pays for the benefits the union provides for its members. Most unions have an

International Association of Bridge, Structural, Ornamental and Reinforcing Iron Workers (http://www.ironworkers.org/)

International Association of Heat and Frost Insulators and Asbestos Workers (http://www.insulators.org/)

International Brotherhood of Boilermakers, Iron Ship Builders, Blacksmiths, Forgers and Helpers (http://www.boilermakers.org/)

International Brotherhood of Electrical Workers (http://www.ibew.org/)

International Brotherhood of Teamsters (http://www.teamster.org/)

International Union of Bricklayers and Allied Craftworkers (http://www.bacweb.org/)

International Union of Elevator Constructors (http://www.iuec.org/)

International Union of Operating Engineers (http://www.iuoe.org/)

International Union of Painters and Allied Trades (http://www.iupat.org/)

Laborers' International Union of North America (http://www.liuna.org/)

Operative Plasterers' and Cement Masons' International Association of the United States and Canada (http://www.opcmia.org/)

Sheet Metal Workers' International Association (http://www.smwia.org/)

United Association of Journeymen and Apprentices of the Plumbing and Pipefitting Industry of the United States and Canada (http://www.ua.org/)

United Brotherhood of Carpenters and Joiners of America (http://www.carpenters.org/)

United Union of Roofers, Waterproofers and Allied Workers (http://www.unionroofers.com/)

Utility Workers Union of America (http://www.uwua.org/)

**FIGURE I-6** Construction craft unions.

apprenticeship program that includes both classroom instruction and on-the-job supervised work experience. Some of the members' dues pay for instructors, classroom space, and training supplies. Unions usually provide a pension for members who have worked in the trade. Because they represent a large block of members, unions can be a powerful force in influencing government to do such things as pass worker safety laws, encourage more construction, and support technology that is good for construction. Unions negotiate with employers (contractors) to establish both a pay rate and working conditions for their members. It is quite typical to find that union members enjoy a higher hourly pay rate than nonunion workers in the same trade.

## Contractors' Associations

Associations of contractors include just about every imaginable type of construction contractor. Figure I-7 lists only a small number of the largest associations that have apprenticeship programs. Some contractors' associations are formed to represent only nonunion contractors; a few represent only union contractors; and others represent both. Many associations of nonunion contractors were originally formed because the contractor members felt a need to work together to provide some of the benefits that union contractors receive—such as apprentice training and a lobbying voice in Washington, D.C.

| |
|---|
| Air Conditioning Contractors of America (http://www.acca.org) |
| Air Conditioning Heating and Refrigeration Institute (http://www.ahrinet.org/) |
| Associated Builders and Contractors (http://www.abc.org) |
| National Association of Home Builders (http://www.nahb.org) |
| Home Builder's Institute (http://www.hbi.org) |
| Independent Electrical Contractors Association (http://www.ieci.org) |
| National Electrical Contractors Association (http://www.necanet.org) |
| National Utility Contractors Association (http://www.nuca.com) |
| Plumbing-Heating-Cooling Contractors Association (http://www.phccweb.org) |
| The Associated General Contractors (AGC) of America (http://www.agc.org) |

**FIGURE I-7** These are only a few of the largest construction associations.

## BUILDING CODES

Most towns, cities, and counties have building codes. A *building code* is a set of regulations (usually in the form of a book) that ensure that all buildings in that jurisdiction (area covered by a certain government agency) are of safe construction. Building codes specify such things as minimum size and spacing of lumber for wall framing, steepness of stairs, and fire rating of critical components. The local building department enforces the local building codes. States usually have their own building codes, and state codes often require local building codes to be at least as strict as the state code. Most small cities and counties adopt the state code as their own, meaning that the state building code is the one enforced by the local building department.

Until recently, three major model codes were published by independent organizations. (A model code is a suggested building code that is intended to be adopted as is or with revisions to become a government's official code.) Each model code was widely used in a different region of the United States. By themselves model codes have no authority. They are simply a model that a government agency can choose to adopt as their own or modify as they see fit. In 2009, the International Code Council published a new model code called the *International Building Code*. They also published the *International Residential Code* to cover home construction (Fig. I-8). Since publication of the first *International Building Code,* states have increasingly adopted it as their building code.

Other than the building code, many codes govern the safe construction of buildings: plumbing codes, fire protection codes, and electrical codes. Most workers on the job site do not need to refer to the codes much during construction. It is the architects and engineers who design the buildings that usually see that the code requirements are covered by their designs. Plumbers and electricians do, however, need to refer to their respective codes frequently. Especially in residential construction, it is common for the plans to indicate where fixtures and outlets are to be located, but the plumbers and electricians must calculate loads and plan their work so it meets the requirements of their codes. The electrical and plumbing codes are updated frequently, so the workers in those trades spend a certain amount of their time learning what is new in their codes.

**FIGURE I-8** The *International Building Code* and the *International Residential Code.*

## WORKING IN THE INDUSTRY

Often success in a career depends more on how people act or how they present themselves to the world than it does on how skilled they are at their job. Most employers would prefer to have a person with modest skills but a great work ethic than a person with great skills but a weak ethic.

### Ethics

*Ethics* are principles of conduct that determine which behaviors are right and wrong. The two aspects of ethics are values and actions. *Values* have to do with what we believe to be right or wrong. We can have a very strong sense of values, knowing the difference between right and wrong, but not act on those values. If we know what is right but we act otherwise, we lack ethics. To be ethical, we must have good values and act accordingly.

We often hear that someone has a great work ethic. That simply means that the person has good ethics in matters pertaining to work. Work ethic is the quality of putting your full effort into your job and striving to do the

| A person with a strong work ethic: |
|---|
| • shows up to work a few minutes early instead of a few minutes late. |
| • looks for a job to do as soon as the previous one is done. (This person is sometimes described as a self-starter.) |
| • does every job as well as possible. |
| • stays with a task until it is completely finished. |
| • looks for oppurtunities to learn more about the job. |
| • cooperates with others on the job. |
| • is honest with the employer's materials, time, and resources. |

**FIGURE I–9** Characteristics of a good work ethic.

best job you can. A person with a strong work ethic has the qualities listed in Figure I-9. Good work ethics become habits, and the easiest way to develop good work ethics is to consciously practice them.

## Common Rationalizations

We judge ourselves by our best intentions and our best actions. Others judge us by our last worst act. Conscientious people who want to do their jobs well often fail to consider their behavior at work. They tend to compartmentalize ethics into two parts: private and occupational. As a result, sometimes good people think it is okay to do things at work that they know would be wrong outside of work. They forget that everyone's first job is to be a good person. People can easily fall prey to rationalizations when they are trying to support a good cause. "It is all for a good cause" is an attractive rationale that changes how we see deception, concealment, conflicts of interest, favoritism, and violations of established rules and procedures. In making tough decisions, do not be distracted by rationalizations.

Good work ethics yield great benefits. As little children, most of us learned the difference between right and wrong. As adults, when we do what we know is right, we feel good about ourselves and what we are doing. On the other hand, doing what we know is wrong is depressing. We lose respect for ourselves, knowing that what we have done is not something we would want others to do to us. Employers recognize people with a good work ethic. They are the people who are always doing something productive, their work turns out better, and they seem cheerful most of the time. Which person do you think an employer will give the most opportunities to: a person who is always busy and whose work is usually well done or a person who seems glum and must always be told what to do next?

## Working on a Team

Constructing a building is not a job for one person acting alone (Fig. I-10). The work at the site requires cooperative effort by carpenters, masons, plumbers, painters, electricians, and others. Usually several workers from each of these trades collaborate. A construction project without teamwork would have lots of problems. For example, one carpenter's work might not match up with another carpenter's work. There could be too much of some materials and not enough of others. Walls may be enclosed before the electrician runs the wiring in them.

Teamwork is very important on a construction site, but what does being a team player on a construction team mean? Effective team members have the best interests of the whole team at heart. Each team member has to carry his or her own load, but it goes beyond that. Sometimes a team member might have to carry more than his or her own load, just because that is what is best for the team. If you are installing electrical boxes and the plumber says one of your boxes is in the way of a pipe, it might be in the best interests of the project to move the electrical box. That would mean you would have to undo work you had just completed and then redo it. It is, after all, a lot easier to relocate an outlet box than to reroute a sink drain.

The following are six traits of an effective team:

- *Listening.* Team members listen to one another's ideas. They build on teammates' ideas.
- *Questioning.* Team members ask one another sincere questions.
- *Respect.* Team members respect one another's opinions. They encourage and support the ideas of others.
- *Helping.* Team members help one another.
- *Sharing.* Team members offer ideas to one another and tell one another what they have learned.
- *Participation.* Team members contribute ideas, discuss them, and play an active role together in projects.

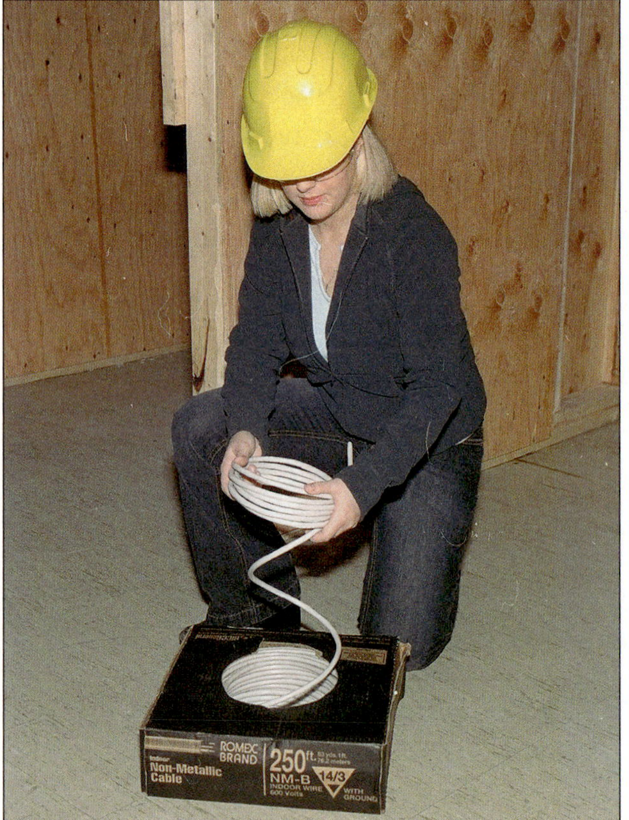


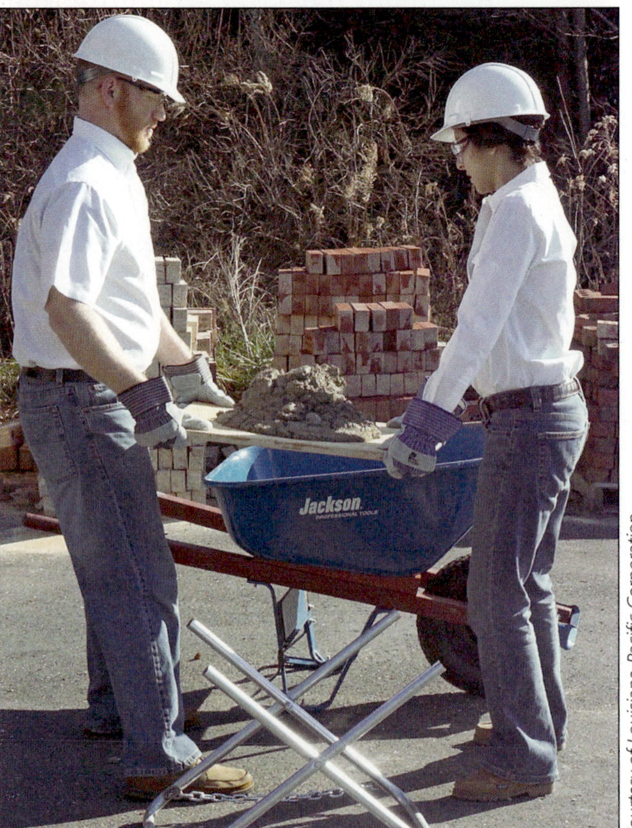


Courtesy of Louisiana-Pacific Corporation.

**FIGURE I-10** Work on the job requires cooperative efforts by different individuals from different trade areas.

## Communication

How could members function as a team without communication? Good communication is one of the most important skills for success in any career. Employers want workers who can communicate effectively; but more importantly, you must be able to communicate with others to do your job well and to be a good team member. How many of the six traits of an effective team require communication?

Many forms of communication exist, but the most basic ones are speaking, listening, writing, reading, and body language. If you master those five forms of communication, you will probably succeed in your career.

### *Speaking*

To communicate well through speech, you need a reasonably good vocabulary. It is not necessary, or even desirable, to fill your speech with a lot of flowery words that do not say much or that you do not really understand. What is necessary is to know the words that convey what you want the listener to hear, and it is equally necessary to use good enough grammar so those words can be communicated properly. Using the wrong word or using it improperly can cause two serious problems: For one thing, if you use the wrong word, you will not be saying what you intended to say. This is also often true if you use a great word wrong since you still might not be saying what you thought you were saying. For another thing (the second serious problem), using a poor choice of words or using bad grammar gives the listener the impression that you are poorly educated or that maybe you just do not care about good communication skills. As a businessperson, you will find that communicating is critical to earning respect as a professional as well as to gaining people's business. Three important steps of effective communication are:

- Looking your listeners in the eye.
- Asking yourself if you think they understand what you are saying. If it is important, ask them if they understand.
- Trying a different approach if they do not understand.

### *Listening*

Good listening is an important skill. Have you ever had people say something to you, and after they were finished and gone, you wondered what they said or you missed some of the details? Perhaps they were giving you directions or telling you about a school assignment. If only you could listen to them again! If possible, try paraphrasing. Paraphrasing means to repeat what they said but in different words. If someone gives you directions, wait until the person is finished. Then, repeat the directions to the person, so he or she can tell you if you are correct. Look at the speaker and form a mental picture of what the speaker is saying. Make what the speaker is saying important to you. Good listening can mean hearing and acting on a detail of a job that will result in giving a competitive edge in bidding.

### *Writing*

Writing is a lot like speaking, except you do not have the advantage of seeing if the person seems to understand or of asking if the person understands. That means you really have to consider your reader. If you are giving instructions, keep them as simple as possible. If you are reporting something to a supervisor, make your report complete, but do not take up his or her time with unrelated trivia. Penmanship, spelling, and grammar count. Always use good grammar to ensure that you are saying what you intend and that your reader will take you seriously. Use standard penmanship, and make it as neat as possible. Do not invent new ways of forming letters, and do not try to make your penmanship ornate. You will only make it harder to read. If you are unsure of how to spell a word, look it up in a dictionary. Next time, you will know the word and will not have to look it up. After you write something, read it, thinking about how your intended reader will take it. Make changes if necessary. Your writing is important! Sole proprietors have to demonstrate good writing skills in proposals and contracts. If either of these is poorly written, it can cost the business a lot of money.

### *Reading*

You will have to read at work. That is a fact no matter what your occupation. You will have to read building specifications, instructions for use of materials and tools, safety notices, and notes from the boss (Fig. I-11). To develop reading skills, find something you are interested in and spend at least 10 or 15 minutes every day reading it. You might read the sports section of the newspaper, books about your hobby, hunting and fishing magazines, or anything else that is interesting to you. What is important is that you read. Practicing reading will make you a better reader. It will also make you a better writer and a better speaker. When you come across a word you do not know how to pronounce or you do not know the meaning of, look it

**FIGURE I-11** Copies of Material Data Safety Sheets are one example of relevant materials.

up or ask someone for help. You will find that you learn pronunciation and meaning very quickly, and your communication skills will improve faster than you expect. In practically no time, you will not need help very often.

## *Body Language*

*Body language* is an important form of communication. How you position your body and what you do with your hands, face, and eyes all convey a lot of information to the person you are communicating with. Whole books are written about how body language is used to communicate and how to read body language. We will only discuss a couple of key points here.

When you look happy and confident, the message you convey is that you are honest (you have nothing to hide or to worry about) and you probably know what you are talking about. If you look unhappy, unsure of yourself, or uninterested, your body language tells the other person to be wary of what you are saying—something is wrong. The following are a few rules for body language that will help you convey a favorable message:

- Look the other person in the eye. Looking toward the floor makes you look untrustworthy. Looking off in space makes you seem uninterested in the other person.
- Keep your hands out of your pockets, and do not wring your hands. Just let your hands rest at your sides or in your lap if you are sitting. An occasional hand gesture is okay, but do not overdo it.
- Dress neatly. Even if you are wearing work clothes, you can be neat. Faddish clothes, extra baggy or extra tight fitting clothes, and T-shirts with offensive messages on them all distract from the real you.

- Speak up. How loudly you speak might not seem like body language, but it has a lot to do with how people react to you. If they have to strain to hear what you are saying, they will think that either you are not confident in what you are saying or you are angry and not to be trusted. If you see your listeners straining to hear you or if they frequently ask you to repeat what you are saying, speak a little louder.

## Customer Service

In any industry, you will only be as successful as you are good at building your reputation for doing quality work and to the degree that your customers are happy with you and your job. On the job site, your customer might be a crew chief, a foreman, a subcontractor, or a contractor. If you are the contractor or subcontractor, the customer will be whoever hired you. It doesn't actually matter who hired you, though—your role will always be to do the very best job you can for whomever it is that you are working.

Good customer service also includes providing a good value for your fees, being honest, communicating clearly, being cooperative, and looking to provide the best possible experience your customer can have in working with you. Just as when you practice good ethics, when you provide great customer service you will enjoy your job much more. You will be proud of your work, others will want to hire you more often, and your career will be much easier to build. Think about how you like being treated when you are a customer—and always try to treat your customers at least as well.

## Lifelong Learning

Lifelong learning refers to the idea that we all need to continue to learn throughout our entire lives. We have greater opportunities to learn and greater opportunities to move up a career ladder today. Our lives are filled with technology, innovative new materials, and new opportunities. People change not only jobs, but entire careers several times during their working life. Those workers who do not understand the new technology in the workplace, along with those who do not keep up with the changes in how their company is managed, are destined to fall behind economically. There is little room in a fast-paced company of this century for a person whose knowledge and skills are not growing as fast as the company. To keep up with new information and to develop new skills for the changing workplace, everyone must continue to learn throughout life.

## CONSTRUCTION TRENDS

Every industry has innovations, and construction is no exception. As a construction professional, it is important to be aware of new technologies, new methods, and new ways of thinking about your work. This is as important for a worker's future employment as being aware of safety and ethical business practices. Some of the key technological trends include disaster mitigation, maintenance, building modeling, and green building.

### Disaster Mitigation

Both new and existing buildings need to be strengthened and improved to deal with earthquakes, floods, hurricanes, and tornados. Actions like improving wall bracing or preparing moisture management reduces damage and improves safety when these events occur. These actions are increasingly required by building regulations (especially in disaster-prone areas) and requested by property owners and insurers.

### Maintenance

Unlike single natural disasters, preventing long-term wear and tear is also an important industry trend. Property owners are more concerned about the costs, effort, and time required to repair and to maintain their homes and buildings. As such, there has been significant research into materials that are more durable, construction assemblies that manage moisture, air, and elements better, and overall higher quality construction work.

### Building Modeling

One of the biggest new trends in construction technology doesn't include construction materials at all: it includes being able to design, simulate, and manage buildings with the use of computer and information technology. Some of these tools, like Computer-Aided Drafting (CAD) and Computer-Aided Manufacturing (CAM), have been around for decades. Others, like energy modeling and simulation software or project management tools, are being used more and more. Still others, like Building Information Modeling (BIM), are gathering many of these previous tools into single computing platforms. In all cases, the ability to use computers and professional software is becoming mandatory among workers.

## Green Building

Probably the biggest trend in the construction industry over the last decade has been *green building*—that is, planning, design, construction, and maintenance practices that try to minimize a building's impact on the environment throughout its use. Although a set definition of green building is still evolving, everyone agrees on a few key concepts that are important and that in themselves are also major construction trends.

### *Occupant Health and Safety*

The quality of indoor air is influenced by the kinds of surface paints and sealants that are used as well as the management of moisture in plumbing lines, HVAC equipment, and fixtures. Long-term maintenance and care by homeowners and remodelers also can shape the prevalence of pests, damage, and mold. Builders and remodelers are becoming more aware of the products and assemblies they use that could have an effect on indoor environments.

### *Water Conservation and Efficiency*

Many builders and property owners are attempting to collect, efficiently use, and reuse water in ways that save the overall amount being used. From using collectors of rainwater to irrigate lawns, to installing low-flow toilets and water-conserving appliances, to feeding used "greywater" from sinks and showers into secondary non-occupant water needs, water efficiency is a trend in all green building but especially where water shortages or droughts are prevalent.

### *Low-Impact Development*

Builders concerned with the effect of the construction site on the land, soil, and water underneath are incorporating storm water techniques, foundation and pavement treatments, and landscaping preservation methods to minimize disturbances to the land and surrounding natural environments.

### *Material Efficiency*

Builders are becoming more aware of the amount of waste coming from construction sites, and inefficiency in the amount of materials (like structural members) that they install in buildings. Many of the materials that are used in construction also do not come from naturally renewable sources or from recycled content materials. Using materials from preferred sources, using them wisely, and then appropriately recycling what is left is a big industry trend (Fig. I-12).

Courtesy of Carl Seville.

**FIGURE I-12** Construction site waste recycling.

## *Energy Efficiency and Renewable Energy Sources*

The most widely known of all green building trends involves the kind and amount of energy that buildings use. Oftentimes, builders can incorporate the use of renewable energy sources (like solar photovoltaics) or passive solar orientation into their designs. Then, the combination of good building envelope construction and efficient equipment and appliances can all reduce utility costs for property owners, much like the maintenance trend reduces repair costs (Fig. I-13).

**FIGURE I-13** Duct insulation increases system energy efficiency.

There are many ways to keep track of the latest trends in the construction industry. Trade or company's journals, on-line resources and blogs, and the latest research coming out of government and university laboratories are several ways to keep informed and up-to-date on the latest industry trends.

## JOB OPPORTUNITIES

Carpentry is the trade in which a craftsperson works and shapes wood for structural and aesthetic purposes. It is an old and noble profession. The ancient Greeks and Romans employed carpenters to shape wood for everyday use and also for great cathedrals. Carpenters have been steadily employed throughout the ages. As in the past, today they work with all species of wood, enjoying the look and feel of it and appreciating how the aroma of wood enhances the experience. Nowadays, a variety of other materials is available. Plastics, metal, masonry, and wood composite materials have been created to look and act like wood. The carpenter is expected to adapt to these new materials.

Carpenters work under all sorts of environmental conditions. Some carpenters, like cabinetmakers, work inside. Others like framers, work primarily outdoors. Work is often physically demanding and sometimes strenuous. Carpenters stand and support materials, walk and carry things, bend and pick up, climb, and stretch. They use their bodies all day long.

The carpentry trade is a dangerous one. Health risks may be present when using materials, tools, or simply by being on the job. Materials, such as plywood, may be abrasive or sharp to skin. Other materials, such as framing lumber, may be heavy to lift. Sanding, cutting, and grinding some materials produce noxious dust and odors. Tools used in the everyday life of a carpenter, for example saws and chisels, will cut skin more easily than other construction material. Proper tool use is critical. The actual work environment often creates situations where workers are at risk. They may slip, fall, or have something fall on them. Carpentry and construction jobs are dangerous jobs, but they can always be safe when workers take sensible precautions.

Carpentry jobs are found everywhere people need a home or building. Carpentry is considered a grouping of many different jobs. In residential construction, all work done in building a home that is not masonry, electrical, plumbing, or heating is typically done by a carpenter. These tasks include concrete form work, framing, siding, roofing, windows and doors, insulation, drywall and finishing, trim work, and staircases. Carpentry is a broad field of many different tasks.

Carpentry jobs may be with small firms in which carpentry crews perform many tasks, such as building an entire house from start to finish. Other jobs with larger firms often involve duties where carpenters perform only a few tasks, such as constructing concrete forms or steel studs and drywall construction. Carpenters may even be self-employed. These aspects result in employment opportunities for carpenters to work anywhere that construction is taking place.

Teamwork is vitally important in carpentry. Many tasks are performed only by a small group like raising a framed wall or lifting a heavy beam into position. Other tasks may be performed by just one person working alone. But when two people join in a task working together, work proceeds more quickly and more easily than it would if two people worked alone on the same project.

## LOOKING FORWARD

As the economy fluctuates, the carpenter must be flexible. During times where homes are in high demand, carpentry work for new construction predominates. When the market turns, new home demand drops, and existing homes are often remodeled. Carpenters are needed to perform tasks in new construction and in remodeling projects of all kinds. Carpenters who are able to switch from residential to remodeling work have the best opportunity to stay employed.

Salaries for various carpentry jobs vary from region to region. Rural areas typically pay less than urban areas.

| **CONSTRUCTION JOBS** | |
|---|---|
| **Title** | **Annual salary** |
| Laborer | $23,500–$38,500 |
| Framer | $27,500–$43,000 |
| Roofer | $28,500–$46,000 |
| Floorer | $27,000–$48,500 |
| Cabinetmaker | $29,500–$44,000 |
| Dry Wall finisher | $31,000–$50,500 |
| Construction foreman | $39,500–$62,000 |

CARPENTRY CAREER PROFILE

# Michael E.C. Surguy

## TITLE

Owner, Michael E. C. Surguy Carpenter Contractors Inc., New Providence, NJ

## EDUCATION

Michael graduated from high school in 1987 and admits that he did not apply himself. He attended vocational school studying carpentry, started framing, and worked in the field for several years.

When he was 26 years old, with several years of labor experience to build on, Michael decided to take two years off and study. He entered State University of New York at Delhi as the oldest student in the program that year. "I went there because I wanted to learn," he explained. "Going to class was one thing, but I brought blueprints from old employers so I could study them and my instructors helped me out with the math." Michael obtained his Associate degree in Occupational Studies from Delhi having developed his specialized carpentry and woodworking skills.

## HISTORY

Today, Michael is the sole owner of his company, which does basic framing to complex remodeling and additions working on $3–4 million homes. Recognized by the Community Builders Association of New Jersey, in 2004, as a recipient of their Sub-Contractor of the Year Award, Surguy is one of New Jersey's specialized carpentry contractors.

## ON THE JOB

In a typical day, Michael is on the job. "In my business and doing the types of jobs I do, I need to be on the job. I don't just set up my guys and leave. I have a tool belt on every day. People pay me to be there and work. If I'm not there, the work doesn't get done."

## CHALLENGES

Michael knows that his job isn't for everyone. "When I started out, I lived at my parents' house. I was lucky. I made nothing for the first three years, but my guys always got paid. I got my tools and paid my insurance. If I hadn't lived at home, there's no way I could have done it. Nobody tells you this," he explains. Michael described building his reputation from doing jobs that he thought he could do and being honest with his bids, breaking each job down, figuring out the cost of materials and employees. At first, Michael took on small jobs which turned into bigger jobs. He started out as sole-proprietor and then became a corporation with employees. "Now it's a different ballgame. You have to have liability insurance. It's a business."

## IMPORTANCE OF EDUCATION

Michael persevered in school, committed to learning what he knew he needed to learn. "When I finished Delhi, I did a lot of studying on my own with trigonometry. I studied a lot at night on my own and took night classes. I bought carpentry books and studied out of those." When he was starting his business, he prepared himself to be a one-man operation. "I knew that when I started out, I wouldn't have a crew. I can probably pull a 5-guy job with just 2 guys. I kept a smaller crew but produced just as much work."

## FUTURE OPPORTUNITIES

Michael expressed a great deal of concern over what this economy has done and will continue to do to the home building business. He hopes to make it through these tough times by diversifying.

## WORDS OF ADVICE

Michael recalls the advice of his favorite instructor at Delhi: "You can either learn the right way. Or you can learn the wrong way. But if you learn the wrong way, you will always work the wrong way." And to this he adds advice from his own experience: "You have to do work to get work. If you don't do work, people won't know you."

CARPENTRY CAREER PROFILE

# Korie A. Bishop

## TITLE

Construction Project Manager, Stop & Shop Supermarket Company, Quincy, MA

## EDUCATION

Korie Bishop wanted to apply her strength in math and science to a college major. She considered architecture and engineering and decided that civil engineering would offer her the most opportunities. She later earned an Associate's Degree in civil engineering technology and later a Bachelor's Degree in civil engineering. In college, Korie learned she enjoyed structural aspects of civil engineering more than the environmental aspects. "I liked the steel and concrete aspects," she said. "I wanted a job in this field, but in the mid-90s there weren't structural jobs in my region."

## HISTORY

Korie was offered a job with The Whiting-Turner Contracting Company, a Baltimore based company that had a project in Burlington, VT. "It was a good opportunity to get into the building fields." As Project Engineer, Korie managed multi-million dollar industrial construction projects—manufacturing facilities called "clean-rooms" where microchips are manufactured. "Where the people in the white suits work!" she explained.

"From day one I was given a lot of responsibility." Korie explained how building a clean-room requires building all the systems. "I learned a lot on the job. It was a very hands-on position. As a design engineer, I would have been in the office, but I was in the field. Every day was different. I coordinated a lot of people—architects, engineers, customer—and I met a lot of people."

In 2005, after ten years with Whiting-Turner, Korie began looking for a new opportunity in the construction field that required less travel. "I was lucky and found a position with Stop & Shop as a Construction Project Manager," which she explained was one step up from her previous position. In this role, as the project "owner," she is the one who hires the general contractor in the construction or remodel of stores. Her territory is Connecticut, upstate New York, and Western Massachusetts.

## ON THE JOB

Korie manages about three or four projects at a time. Korie describes her typical day: "I head out early to one my projects and meet with the general contractor and subcontractors; the architect will also attend. I'm a go-between. There's a weekly status meeting for each job. I'll answer questions and go over paperwork." After the meetings, Korie will work from job site, return to her office, or work from home.

Korie might also attend a pre-planning meeting for a project that is starting soon. She explained that remodeling a grocery store, as opposed to building a new store is a whole different type of construction and requires more operations meetings. "If it's a remodel, I'll review the plans with the store operations team. I'll work with the store manager to discuss how to work with limited impact to the customers." Korie explained the importance of doing the work at night and making sure that in the morning the store looks like a store and not a construction zone.

## CHALLENGES

Korie described the challenges of working with a lot of different people and that being educated as engineer meant she did not necessarily learn people skills. "This is something that doesn't come naturally to a lot of engineers." Korie explained that learning how to read people and how to communicate is a "side-skill" that she had to develop.

Korie also described the organizational challenges of her job. "I have to keep good files and take good notes. I have to keep track of all the things I have going on. I always have lots of irons in the fire. Things move quickly, and I'm usually working out of my car or a job trailer."

## IMPORTANCE OF EDUCATION

Korie described how education benefits those in this business. "Education provides a foundation for the thought processes involved in any business. In school, you have to solve problems and be organized. These are skills you'll use going forward. If you're able to get through school successfully, you can do any job successfully."

(CONTINUED)

## CARPENTRY CAREER PROFILE (CONTINUED)

### FUTURE OPPORTUNITIES

Korie said that opportunities exist at Stop & Shop Supermarket Company to move up within the construction management department, and while she aspires to move up, she is happy where she is. "Not one morning do I feel like I don't want to go to work," she said.

### WORDS OF ADVICE

"Don't be afraid to ask questions about things you don't know. I have learned so much from people in the field. These tradesmen, a plumber or electrician, have been doing this for decades, and they love to be in the role of an educator. People will respect you for asking questions, and as a boss, when people ask me questions it shows me they are interested and want to do a good job."

# John Slaughter

## TITLE

President and CEO, Retroactive Home Construction, Inc., Sacramento, CA

## EDUCATION

Although John intended to pursue a degree in architecture upon entering Texas A&M, he earned a degree in Building Construction in 1988. "I changed my major right away. The construction group was more my style. I learned the basics of how to do everything."

## HISTORY

After graduation, he got a job as a laborer for a massive home builder in the Sacramento, California area. "My job was entry level—keeping houses clean and hauling lumber. I moved into reading plans and quality control. I did that for about two years." John was promoted several times and, in 2003, he changed companies.

He left the public corporation career to join a large local builder as Vice President of Construction and Operations, which allowed him to work directly for the owner. The company built communities from Yuba City to Bakersfield,

an area extending about 350 miles. "It was a large company, a wonderful job, and I loved every minute of it." Unfortunately, the downturn in the economy halted the growth in home construction in California seen in recent decades and led to John's company going bankrupt from having too much debt in landholdings.

In 2008, John started his own home-building company. He started with one home that had been foreclosed upon before completion. John is currently working with banks, presently bidding on homes that have been left incomplete. "As much as this economy is sad and depressing, there is opportunity."

## ON THE JOB

Being self-employed allows him to set his own schedule. In a typical day, he meets with subcontractors and walks homes. He also spends time learning

(CONTINUED)

## CARPENTRY CAREER PROFILE (CONTINUED)

about accounting processes, invoicing, lien releases, and liability insurance—all aspects of the homebuilding industry that personnel in other departments usually handle.

### CHALLENGES

John enjoys the freedom that comes with having his own company, but admits the downsides. "There's no steady paycheck or benefits, and I'm always wondering about the next job."

### IMPORTANCE OF EDUCATION

John described how fortunate he was to find his path through a technical college and earn his degree, an achievement that has come in handy over the years. "Like a lot of people at that age, I was feeling my way through life . . . The degree has really helped me in my career. When it came time for opportunities and promotions, the company I was at always noticed that I had my degree, and it made a difference."

John is required to have a contractor's license to do what he is doing now. "I have had the license for 17 years, but this is the first time I have been using it." John also explained how an education can get your foot in the door and the value of working within an association like the NAHB (National Association of Home Builders) or the BIA (Building Industry Association).

### FUTURE OPPORTUNITIES

In the short term, John is working to get more houses and grow his company. "By getting more business, I can hire help and get even more business. I can keep growing the company." John is also pursuing an MBA degree with University of Phoenix. "More education is never a bad thing. Having my MBA will make me a little more credible in the future."

### WORDS OF ADVICE

John shared pieces of advice he received in his life that he has never forgotten. "First, watch what everyone else around you is doing, and do something better. Ask yourself what is missing, what you can do better to be different." John also stressed that ethics and honesty, such as honest bids that include a fair profit, and treating people with respect—while they sound clichés—are crucial to success. "You have got to really treat people right."

# Jeffrey S. Rainforth

## TITLE

Vice President, Phelps Construction Group, Boonton, New Jersey

## EDUCATION

As a boy, Jeff Rainforth knew exactly what he wanted to do, to build and be a builder. He did carpentry work in the summers, framing houses, siding and roofing for home builders. Admitting that he didn't push himself in high school, Jeff and his parents thought State University of New York College Of Technology at Delhi was a good option for him. His major was construction technology. "Coming out of high school I figured I would take a couple years of classes and then go back to building houses." His plans changed, however. His Delhi instructors pushed him and showed him different perspectives on construction and advanced framing principles. Jeff buckled down with his studies and started taking more advanced math and science courses, which he discovered came easy to him.

Jeff graduated from Delhi in 1995 and transferred to The College of Environmental Science & Forestry at Syracuse University graduating in 1997 with a degree in construction management and civil engineering. He went to work as a project manager for a large general contractor where he had interned—a firm which builds $1 to $50 million projects. In this position, Jeff's responsibilities included planning site logistics, developing and maintaining project schedules, and coordinating subcontractors. He also became an OSHA trainer.

While in this role, Jeff was project manager on jobs that included the New Jersey Devils marketing center, the New York football Giants executive office expansion and locker room renovations, the restoration of buildings on Ellis Island and, likely his proudest achievement, the high-profile renovations of the Statue of Liberty after she was closed

(CONTINUED)

CARPENTRY CAREER PROFILE (CONTINUED)

for security purposes following 9/11. "It was a very intense, demanding project, but certainly the most rewarding project I will probably ever be involved with."

After 11 years, Jeff and several colleagues started their own company that now employs 16 people and builds projects ranging from $1–$15 million. After a few smaller size projects, they were awarded their first high profile project—building the NJ Devils' executive offices. "We made the impossible happen and completed their 18,000 square foot offices in ten weeks. This project put our new company on the map," he said proudly.

As Vice President, Jeff manages projects, estimates, schedules the work, and lines up more business. His company will tackle projects including building, renovating, and restoring schools, malls, hotels, churches, post offices, libraries, police headquarters, office buildings and warehouses.

## ON THE JOB

Jeff described how his company builds five to seven projects at a time while estimating several others, and what he does in a typical day, which includes speaking to subcontractors, estimating work, managing building projects, and building new relationships and networking.

## IMPORTANCE OF EDUCATION

Jeff admitted that he undervalued education at the start but came to appreciate it through the experience of going to school and meeting new people like his instructors. "I also learned that education doesn't end at your graduation ceremony," he said.

## FUTURE OPPORTUNITIES

Very satisfied about where he is in his career, Jeff said, "I was part of building my company from the ground up. I'm very proud of the team that we have put together. I look forward to the challenges and opportunities that lay ahead."

## CHALLENGES

Jeff described the greatest challenge as keeping up with change whether there are changes on the job, new technology and materials, or new management tools and software to plan and schedule construction projects.

## WORDS OF ADVICE

"Don't ever sell yourself short. Continue to challenge yourself to go further and achieve your goals. Don't be afraid to ask questions and learn from your peers' experiences."

SECTION 1

# Tools and Materials

CHAPTER 1

# Hand Tools

**OBJECTIVES** *After completing this unit, the student should be able to:*

- Identify and describe the hand tools the carpenter commonly uses.
- Use hand tools in a safe and appropriate manner.
- Maintain hand tools in suitable working condition.

## GLOSSARY OF HAND TOOL TERMS

**crosscut** a cut made across the grain of lumber

**dado** a wide cut, partway through and across the grain of lumber

**groove** a wide cut, partway through and running with the grain of lumber

**heel** the back end of objects, such as a handsaw or hand plane

**kerf** the width of a cut made with a saw

**level** horizontal; perpendicular to the force of gravity

**plumb** vertical; aligned with the force of gravity

**plumb bob** a pointed weight attached to a line for testing plumb

**rabbet** an L-shaped cutout along the edge or end of lumber

**square** a tool used to mark a layout and mark angles, particularly 90-degree angles; a term used to describe when two lines or sides meet at a 90-degree angle; the amount of roof covering that will cover 100 square feet of roof area

**toe** the forward end of tools, such as a hand saw and hand plane

**whet** the honing of a tool by rubbing the tool on a flat sharpening stone

One of the many benefits of working in the field of construction is the diversity of tools available. Tools are the means by which construction happens. Their use requires the operator to be knowledgeable in how to manipulate the tool safely. This applies to hand tools as well as power tools.

Safety is an attitude—an attitude of acceptance of a tool and all its operational requirements. Safety is a blend of ability, skill, and knowledge that should always be applied when working with tools.

Knowing how to choose the proper tool and how to keep it in good working condition are essential. A tradesperson should never underestimate the importance of tools or neglect their proper use and care. Tools should be kept clean and in good condition. If steel parts get wet on the job, dry them as soon as possible and coat them with light oil to prevent rusting.

Carpenters are expected to have their own hand tools and keep them in good working condition. Tools vary in quality, which is related to cost. Generally, expensive tools are of better quality than inexpensive tools. For example, inferior tools cannot be brought to a sharp, keen edge and will dull rapidly. They will bend or break under normal use.

High-quality tools are worth the expense. The condition of a tool reveals the attitude of the owner toward his or her profession.

# Layout Tools

Many layout tools must be used by the carpenter. They are used, among other things, to measure distances, lay out lines and angles, test for depths of cuts, and set vertical and level pieces.

## MEASURING TOOLS

The ability to take measurements quickly and accurately must be mastered early in the carpenter's training. Practice reading the rule or tape to gain skill in fast and precise measuring. The United States is one of only a few industrialized countries that do not use the metric system of measure. The metric system is actually easier to use, once it is understood. Linear metric measure centers on the meter, which is slightly longer than a yard. Meters are broken into smaller pieces and grouped into large units. A decimeter is 1/10 of a meter. A centimeter (1/100 of a meter) and millimeter (1/1,000 of a meter) are used instead of inches and fractions. Simply by moving the decimal, a measurement that is 1.5 meters can also be stated as 15 decimeters, 150 centimeters, and 1,500 millimeters. The prefix *kilo-* represents 1,000 times larger, and *kilometer* is used instead of *mile*. In metric measure, a 2 × 4 is 39 mm (millimeters) × 89 mm.

### Rules and Tapes

*Rules* used in construction in the United States are divided into feet, inches, and usually 16ths of an inch (Fig. 1-1). Most rules and tapes used by the carpenter have increments of 16 inches, clearly marked in red, and 19.2 inches, indicated by small black diamonds. These highlights are used to help in laying out spaced framing members.

Most carpenters use *pocket tapes* (Fig. 1-2), which are available in 6- to 35-foot lengths.

*Steel tapes* of 50- and 100-foot lengths are commonly used to lay out longer measurements. The end of the tape has a steel ring with a folding hook attached. The hook may be unfolded to go over the edge of an object. It may also be left in the folded position and the ring placed over a nail when extending the tape. Remember to place the nail so that the *outside* of the ring, which is the actual end of the tape, is to the mark (Fig. 1-3). Rewind the tape when not using it. Keep it dry.

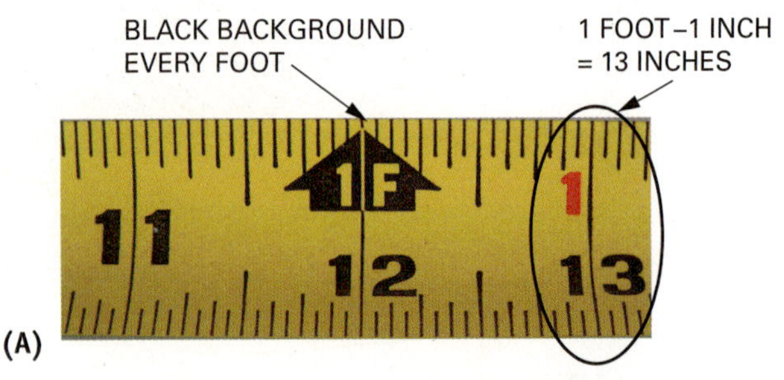


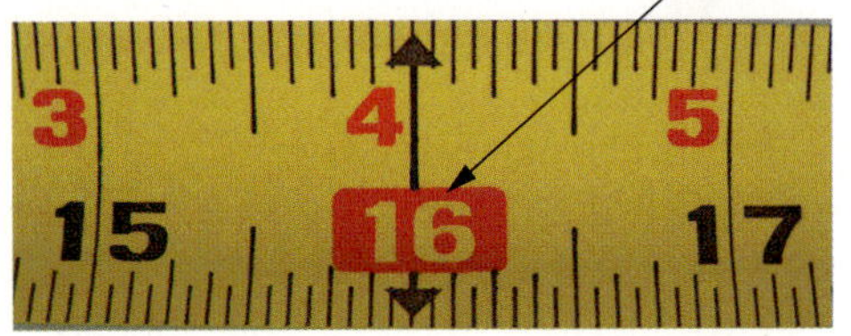

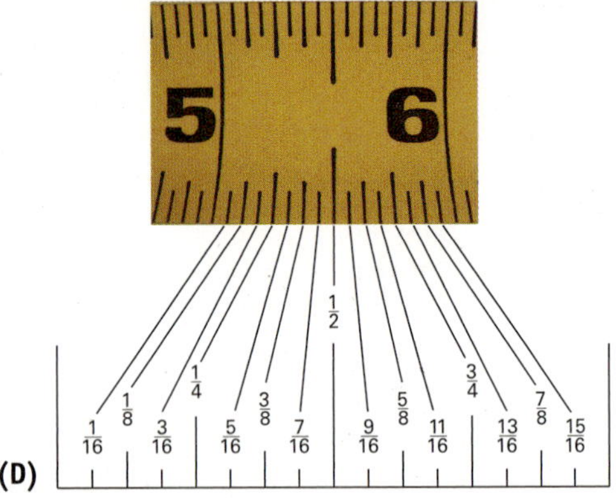


**FIGURE 1-1** A standard English scale showing fractions of an inch.

**FIGURE 1-2** Pocket tape.

**FIGURE 1-3** Steel tape.

**CAUTION:** Steel tapes can cut skin as they are retracted. They also are easily broken if the tape is folded and creased.

## SQUARES

The carpenter has the use of a number of different kinds of squares to lay out for square and other angle cuts.

### Combination Squares

The *combination square* (Fig. 1-4) consists of a movable blade, 1 inch wide and 12 inches long, that slides along the body of the square. It is used

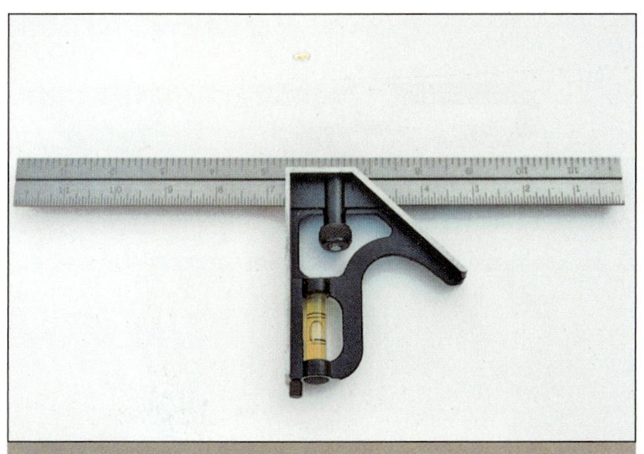

**FIGURE 1-4** The body and blade of a combination square are adjustable.

to lay out or test 90- and 45-degree angles. Hold the body of the square against the edge of the stock and mark along the blade (Fig. 1-5). It can function as a depth gauge to lay out or test the depth of **rabbets**, **grooves**, and **dadoes**. It can also be used with a pencil as a marking gauge to draw lines parallel to the edge of a board. Lines may also be gauged by holding the pencil and riding the finger along the edge of the board. Finger gauging takes practice but, once mastered, saves a lot of time. Be sure to check the edge of the wood for slivers first.

## Speed Squares

Some carpenters prefer to use a triangular-shaped square known by the brand name *Speed Square* (Fig. 1-6). Speed squares are made of one piece of plastic and aluminum alloy and are available in

(A)

(B)

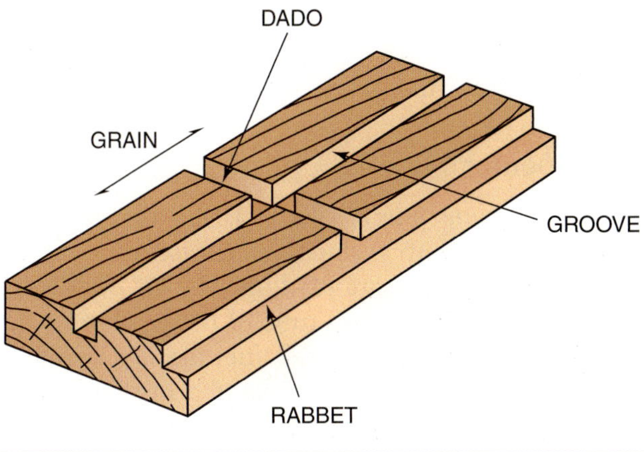

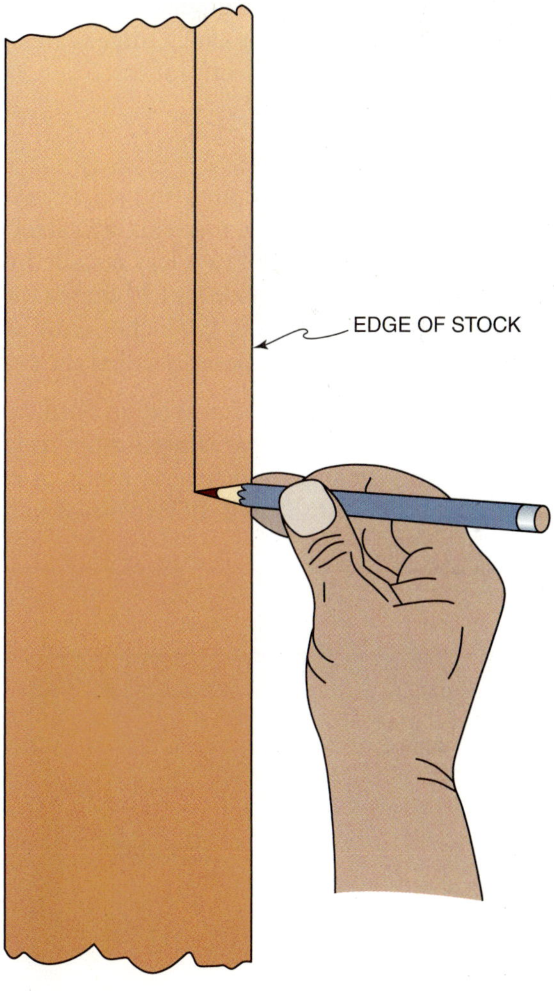

**FIGURE 1-5** The combination square is useful for squaring and as a marking gauge. A pencil held in one hand is a quick way to draw a parallel line. Check the wood first to reduce the potential for splinters.

**FIGURE 1-6** Speed Squares are used for layout of rafters and other angles.

different sizes. They can be used to lay out angles, particularly the 90- and 45-degree angles. A degree scale allows angles to be laid out. Other scales on the square may be used to lay out rafters.

### Framing Squares

The *framing square,* often called the *steel square* (Fig. 1-7), is an L-shaped tool made of thin steel or aluminum. The longer of the two legs is called the *blade* or *body;* it is 2 inches wide and 24 inches long. The shorter leg, the *tongue,* is 1½ inches wide and 16 inches long. The outside corner is called the *heel.*

Many different tables are stamped on both sides of the square. The *rafter table* is used most today; it is used to find the length of several kinds of rafters. (How to use this table will be explained in the chapters on roof framing.) The framing square is useful for laying out bridging, stair framing, and squaring longer lines.

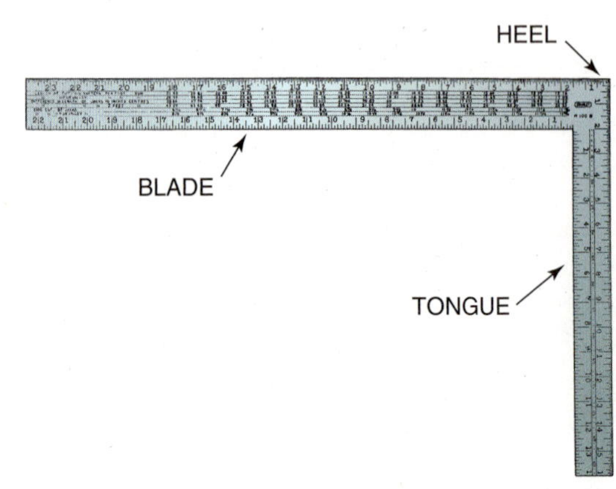


*Courtesy of Stanley Tools.*

**FIGURE 1-7** Framing square.

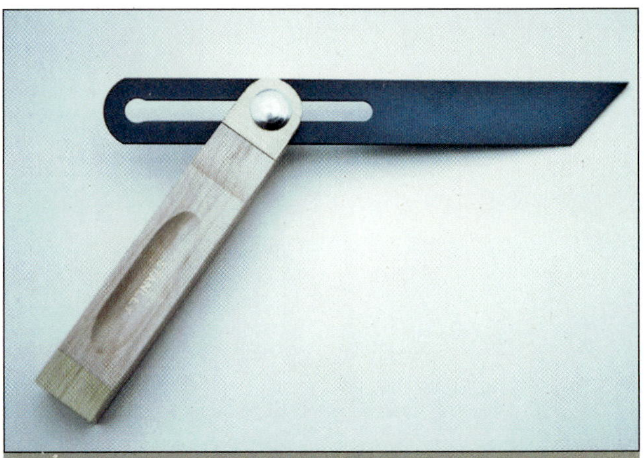

**FIGURE 1-8** Sliding T-bevel.

### Sliding T-Bevels

The *sliding T-bevel,* sometimes called a *bevel square* (Fig. 1-8), consists of a body and a sliding blade that can be turned to any angle and locked in position. It is used to lay out or test angles other than those laid out with squares.

## LEVELS

In construction, the term **level** is used to indicate that which is horizontal, and the term **plumb** is used to mean the same as vertical. An important point to remember is that level and plumb lines, or objects, must be straight throughout their entire length or height. Parts of a structure may have their end points level or plumb with each other. If they are not straight in between, however, they are not level or plumb for their entire length (Fig. 1-9).

### Carpenter's Levels

The *carpenter's level* (Fig. 1-10) is used to test both level and plumb surfaces. Accurate use of the level depends on accurate reading. The air bubble in the slightly crowned glass tube of the level must be exactly centered between the lines marked on the tube. The tubes of a level are oriented in two directions for testing level and plumb. **Note:** Care must be taken not to drop the level because this could break the glass or disturb the accuracy of the level. To check a level for accuracy, place it on a nearly level or plumb object that is firm. Note the exact position of the level on the object. Read the level carefully

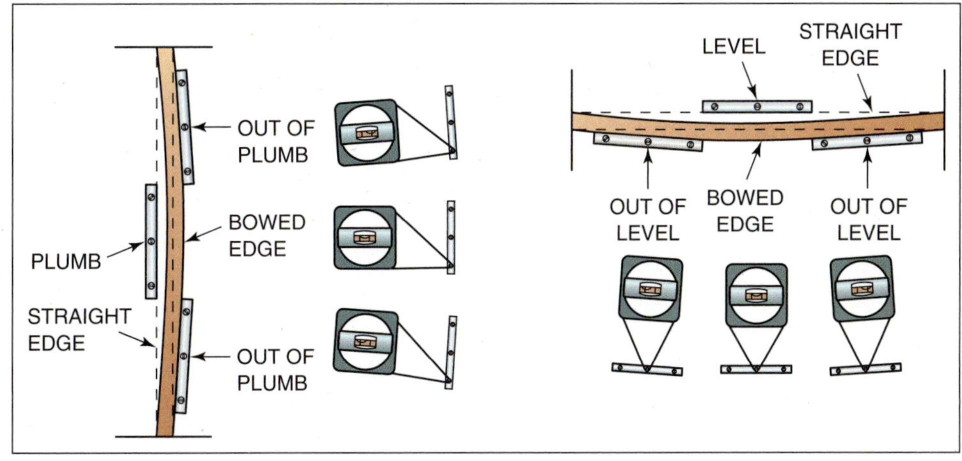


**FIGURE 1-9** To be level or plumb for their entire length, pieces must be straight from end to end.

and remember where the bubble is located within the lines on the bubble tube. Rotate the level along its vertical axis and reposition it in the same place on the object (Fig. 1-11). If the bubble reads the same as the previous measurement, then the level is accurate.

AIR BUBBLE

TUBE HAS A SLIGHT CURVATURE—
AIR BUBBLE RISES TO TOP OF CURVE.

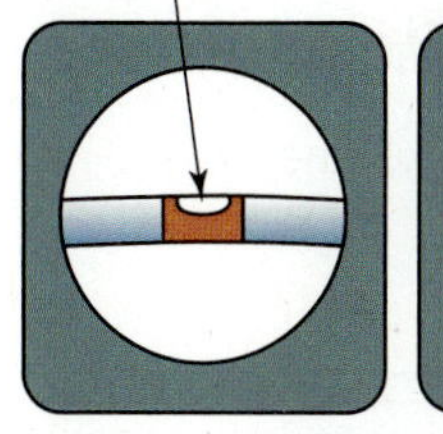

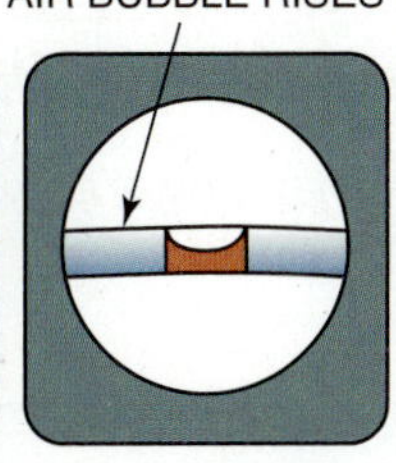

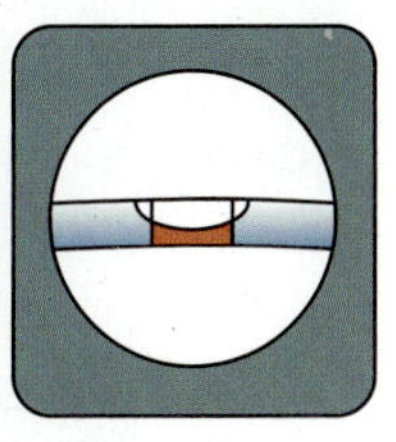

| IN HOT WEATHER OR IF LAYING IN THE SUN, THE LIQUID IN THE TUBE EXPANDS MAKING THE AIR BUBBLE SMALLER. | AT MODERATE TEMPERATURE, THE AIR BUBBLE SHOULD FIT EXACTLY BETWEEN THE TWO LINES ON THE TUBE. | AT COLDER TEMPERATURES, THE LIQUID IN THE TUBE CONTRACTS MAKING THE AIR BUBBLE LARGER. |
|---|---|---|

REGARDLESS OF CONDITIONS, THE AIR BUBBLE MUST BE CENTERED BETWEEN THE TWO LINES ON THE TUBE.

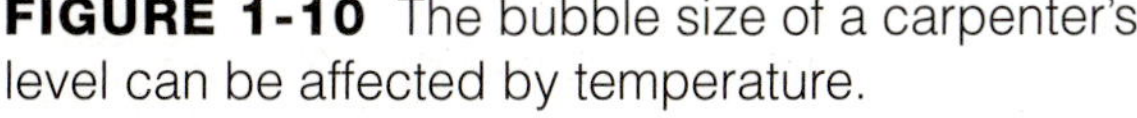

**FIGURE 1-10** The bubble size of a carpenter's level can be affected by temperature.

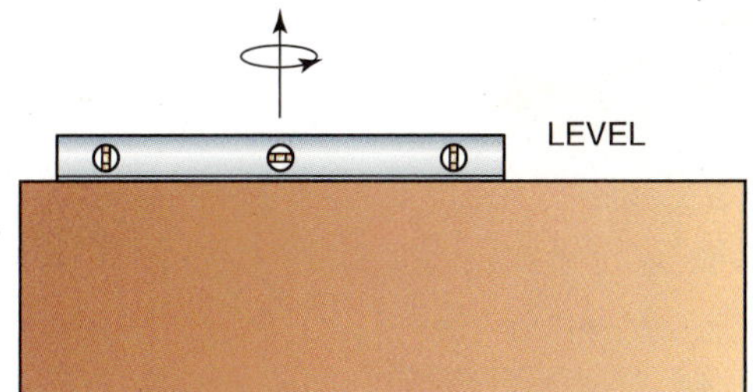

**FIGURE 1-11** If the bubbles read the same, before and after rotating, the level is accurate.

### Plumb Bobs

The **plumb bob** (Fig. 1-12) is very accurate and is used frequently for testing and establishing plumb lines. Suspended from a line, the plumb bob hangs absolutely vertical when it stops swinging. However, it is difficult to use outside when the wind is blowing. The plumb bob is useful for quick and accurate plumbing of posts, studs, door frames, and other vertical members of a structure (Fig. 1-13). It can be suspended from a great height to establish a point that is plumb over another.

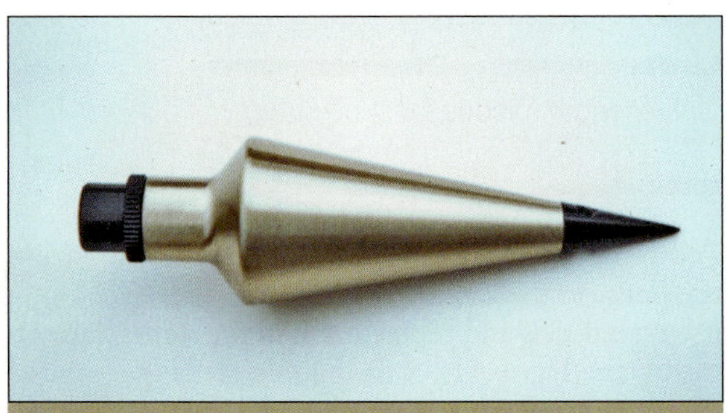

**FIGURE 1-12** Plumb bob.

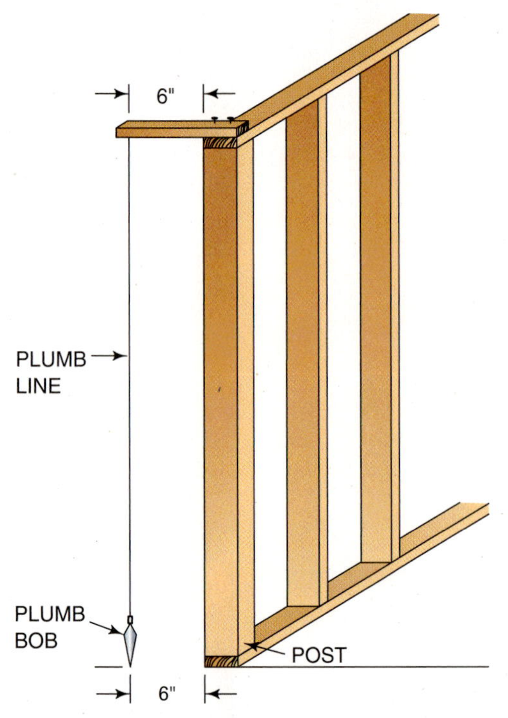


**FIGURE 1-13** The post is plumb when the distance between it and the plumb line is the same.

## CHALK LINES

Long straight lines are laid out by using a *chalk line*. A line coated with chalk dust is stretched tightly between two points and snapped against the surface (Fig. 1-14). The chalk dust is dislodged from the line and remains on the surface. The chalk line and reel together are called a *chalk box* (Fig. 1-15). The box is filled with chalk dust that comes in a number of colors, red and blue being the most common. Care should be taken when using red as it tends to be permanent and will bleed through many paints.

### Chalk Line Techniques

When unwinding and chalking the line, keep it off the surface until the line is ready to be snapped. Otherwise, many other lines will be made on the surface simply by laying the line down on the surface. Make sure lines are stretched tightly before

**FIGURE 1-14** Snapping a chalk line.

**FIGURE 1-15** Chalk line reel.

snapping in order to produce a neat and straight line. If the line sags, take the sag out by supporting the line near the center. It often takes two people to stretch and snap a line. One holds the string taut, and the other presses the center of the line to the deck. This same person then can snap the line on both sides of the center. Keep the line from getting wet. A wet chalk line is practically useless.

## SCRIBERS

*Wing dividers* can be used as a compass to lay out circles and arcs and as dividers to space off equal distances. *Scribers* are similar but have a pencil attached to one end (Fig. 1-16). Scribing is the technique of laying out stock to fit against an irregular surface. For easier and more accurate scribing, heat and bend the end of the solid metal leg outward (Fig. 1-17).

**FIGURE 1-16** Scribing is laying out a piece to fit against an irregular surface.

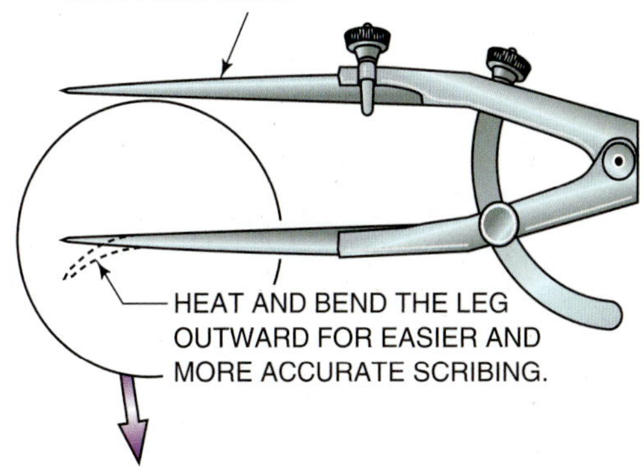


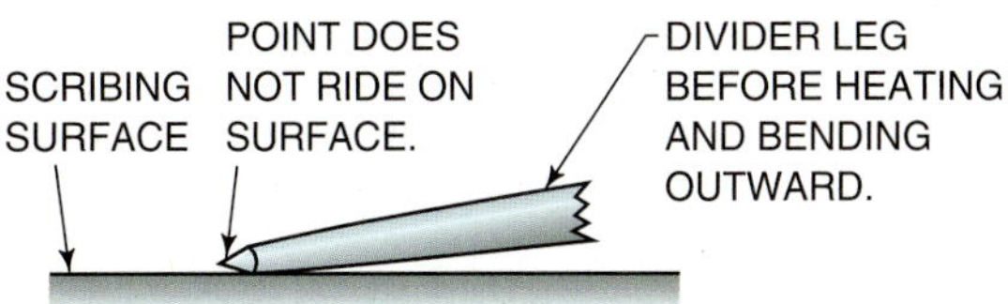


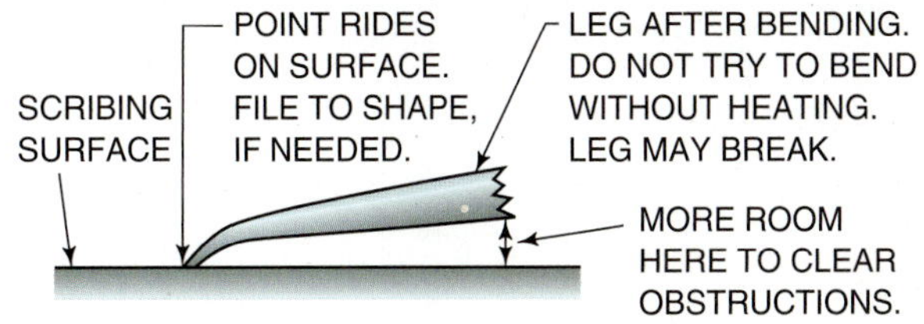


**FIGURE 1-17** Adjusting one of the metal legs of a scriber makes it a more accurate tool.

# EDGE-CUTTING TOOLS

Edge cutting tools include chisels, planes, snips and knives. Each cuts with a single sharp edge.

## WOOD CHISEL

The *wood chisel* (Fig. 1-18) is used to cut recesses in wood for things such as door hinges and locksets. Chisels are sized according to the width of the blade and are available in widths of ⅛ inch to 2 inches. Most carpenters can do their work with a set consisting of chisels that are ¼, ½, ¾, 1, and 1½ inches

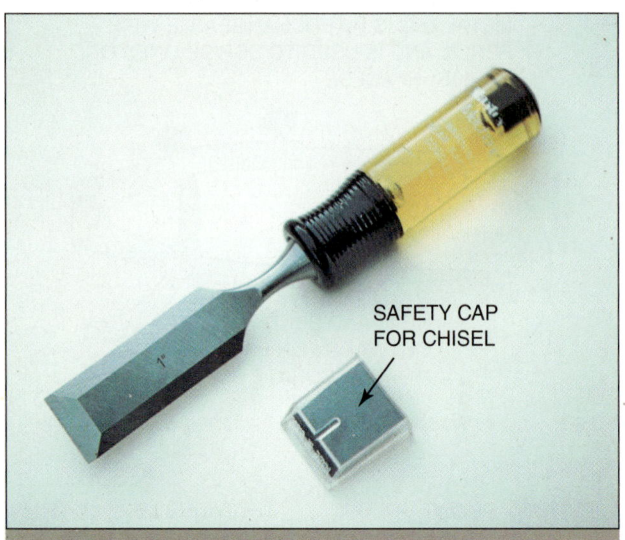


**FIGURE 1-18** Wood chisel.

**FIGURE 1-19** Keep both hands behind the chisel's cutting edge.

in size. Be sure to keep both hands behind the cutting edge (Fig. 1-19).

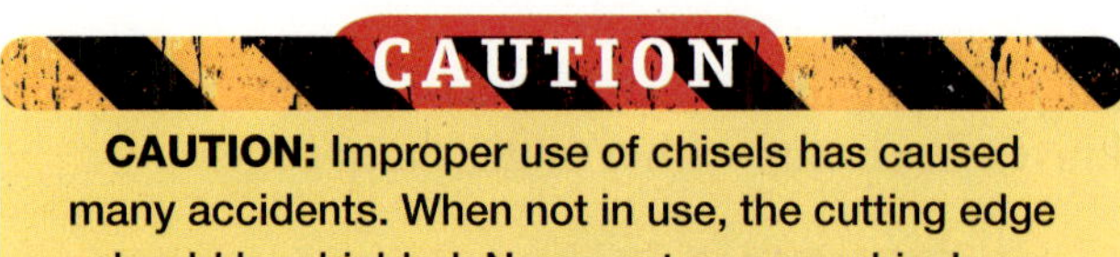


**CAUTION:** Improper use of chisels has caused many accidents. When not in use, the cutting edge should be shielded. Never put or carry chisels or other sharp or pointed tools in pockets.

## BENCH PLANES

*Bench planes* (Fig. 1-20) come in several sizes. They are used for smoothing rough surfaces and shaping work down to the desired size. Bench planes are given names according to their length. The longest is called the *jointer.* In declining order are the *fore, jack,* and *smooth* planes. The jack plane is 14 inches long and of all the bench planes is considered the best for all-around work.

**FIGURE 1-20** A general purpose bench plane.

## BLOCK PLANES

*Block planes* are small planes designed to be held in one hand. They are often used to smooth the edges of short pieces and for trimming end grain to make fine joints (Fig. 1-21). Block planes have a smaller blade angle than bench planes, and the cutting edge bevel is on the top side. On bench planes, the cutting edge is on the bottom (Fig. 1-22). Most carpenters prefer the low angle block plane because it seems to have a smoother cutting action and because it fits into the hand more comfortably.

### Using Planes

When planing, have the stock securely held. Always plane with the grain. When starting, push forward while applying pressure downward on the **toe** (front). When the **heel** (back) clears the end, apply pressure downward on both ends while pushing forward. When the opposite end is approached, relax pressure on the toe and continue pressure on the heel until the cut is complete (Fig. 1-23). This

(A)
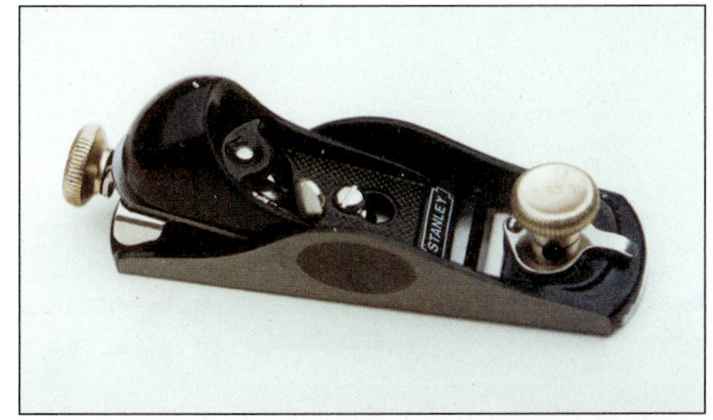


(B)
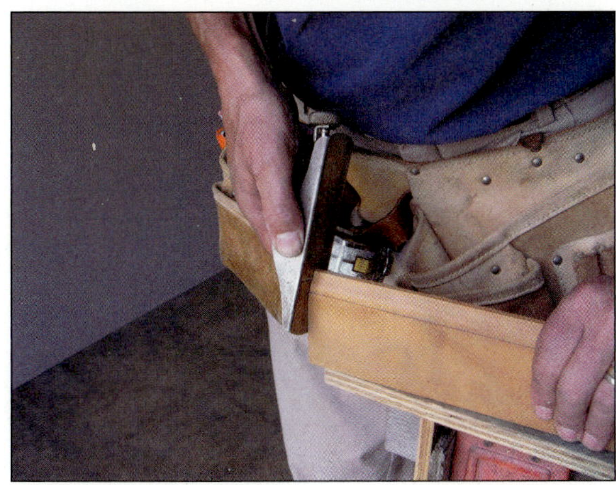

(a) Courtesy of Stanley Tools.

**FIGURE 1-21** The block plane is small and often has a low blade angle.

method prevents tilting the plane over the ends of the stock and helps ensure a straight, smooth edge.

### Honing Chisels and Plane Irons

To produce a keen edge, chisels and plane irons must be **whetted** (sharpened) by using an oilstone or waterstone. Hold the tool on a well-oiled stone so that the bevel rests flat. Move the tool back and forth across the stone for a few strokes. Then, make a few strokes with the flat side of the chisel or plane iron held absolutely flat on the stone. Continue whetting in this manner until as keen an edge as possible is obtained. To obtain a keener edge, repeat the procedure on a finer stone or on a piece of leather. The edge is sharp when, after the bevel has been whetted but before it is turned over, no wire edge can be felt on the flat side (Fig. 1-24).

## TIN SNIPS

*Tin snips* are used to cut thin metal, such as roof flashing and metal roof edging. Three styles of *aviation snips* are available for straight metal cutting and for left and right curved cuts (Fig. 1-25). The color of the handles denotes the differences in the design of the snips. Yellow handles are for straight cuts, green are for cutting curves to the right, and red are for cutting to the left.

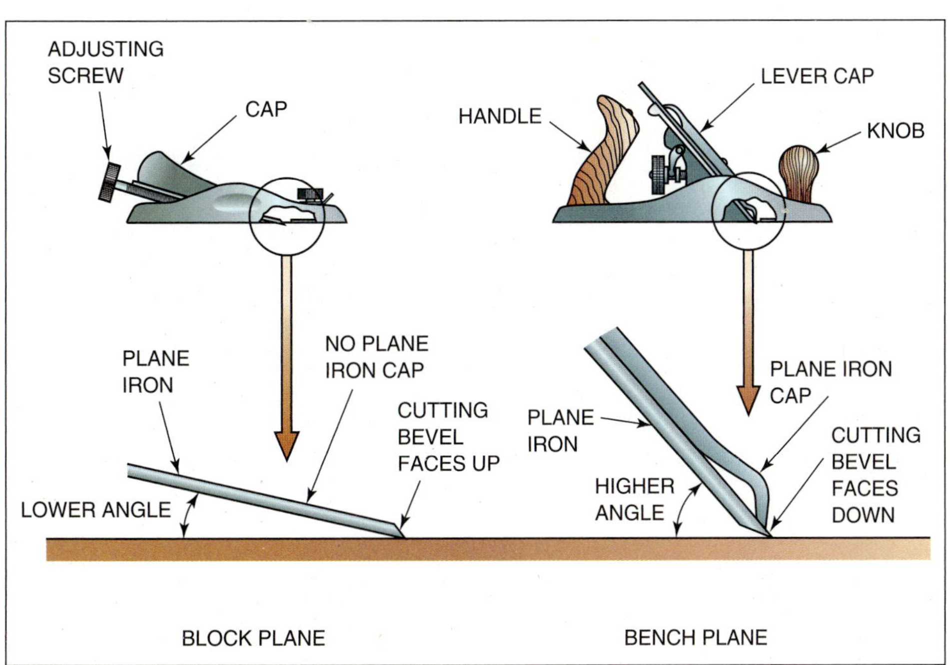


**FIGURE 1-22** Difference between block and bench planes.

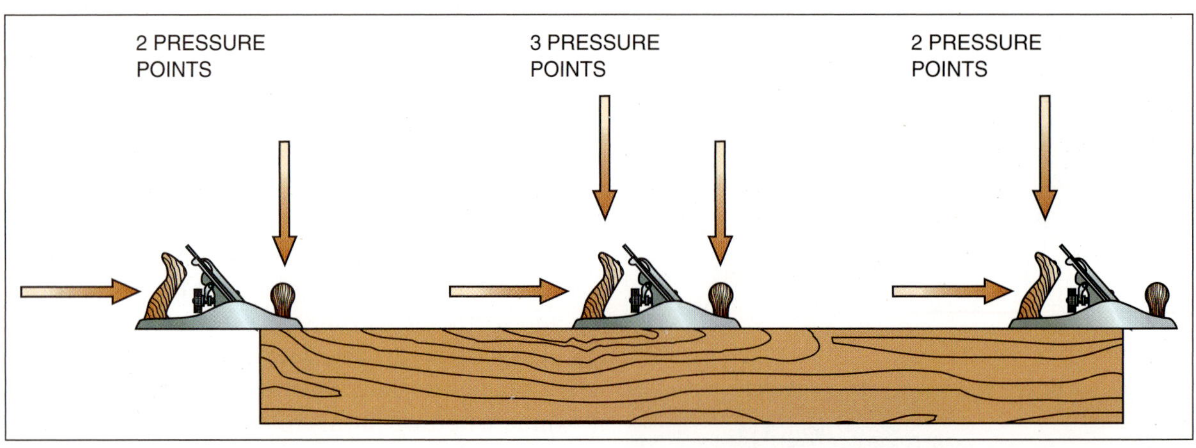


**FIGURE 1-23** Correct method of planing edges.

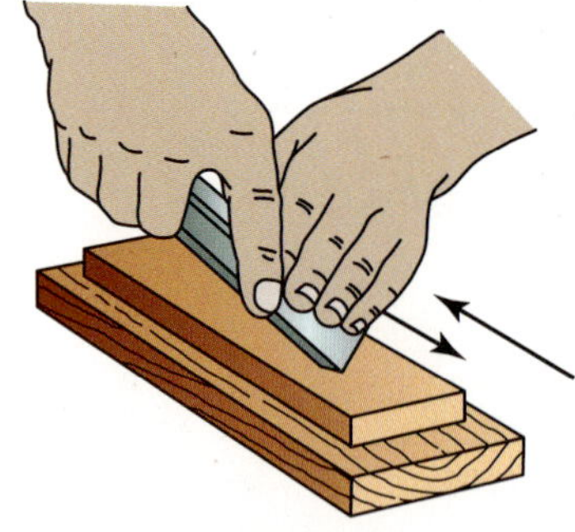

**FIGURE 1-24** Whetting a plane iron.

**FIGURE 1-25** Metal shears and right- and left-cutting aviation snips.

## UTILITY KNIFE

A *utility knife* (Fig. 1-26) is a universal cutting tool. These tools are frequently used for cutting gypsum board, softboards, and a variety of finish materials. Blades may be sharpened on a whetstone or replaced entirely.

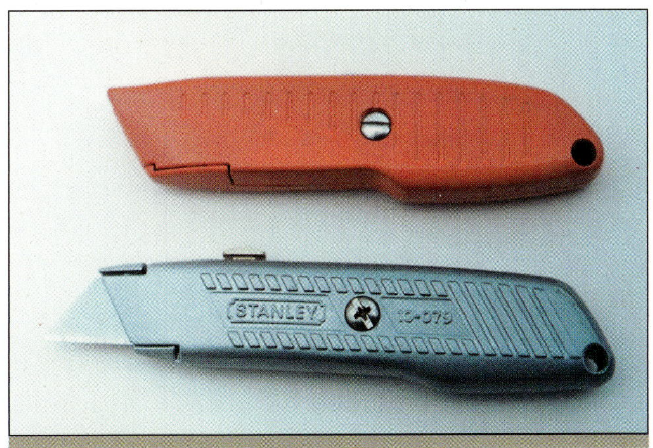


**FIGURE 1-26** Utility knife.

# Tooth-Cutting Tools

The carpenter uses several kinds of saws to cut wood, metal, and other materials. Each one is designed for a particular purpose.

## HANDSAWS

Handsaws (Fig. 1-27) used to cut across the grain of lumber are called **crosscut** *saws. Ripsaws* are designed to be used to cut with the grain. The difference in the cutting action is in the shape of the teeth. The crosscut saw has teeth shaped like knives. These teeth cut across the wood fibers to give a smoother action and surface. The ripsaw has teeth shaped like rows of tiny chisels that cut the wood ahead of them (Fig. 1-28). Another design to handsaw teeth, called a *shark tooth saw,* makes the teeth longer and able to cut in both directions of blade travel. To keep the saw from binding, the teeth are *set*—that is, alternately bent—to make the saw cut or **kerf** wide enough to give clearance for the blade.

Courtesy of Stanley Tools.

**FIGURE 1-27** Handsaws are still useful on the job site. Some handsaws are made with deeper teeth, which are designed to cut in both directions.

### Using Handsaws

Stock is hand sawed with the face side up because the back side is splintered along the cut by the action of the saw going through the stock. This is not important on rough work. However, on finish work, it is essential to identify the face side of a piece and to make all layout lines and saw cuts with the face side up.

To make any cut, the saw kerf is made on the waste side of the layout line by cutting away part of the line and leaving the rest. Start the cut by using a thumb to guide the blade of the saw. Make sure the thumb is above the teeth (Fig. 1-29). Gently pull the saw in an upstroke to create more surface area into which the teeth can bite. Then begin the cut by pushing the saw into a downstroke. Back your thumb away.

CAUTION

**CAUTION:** When using a handsaw, take care that the saw does not jump out and cut your hand.

## COMPASS AND KEYHOLE SAWS

The *compass saw* is used to make curved cuts. The *keyhole saw* is similar to the compass saw, except its blade is narrower for making curved cuts of a smaller diameter (Fig. 1-30). When cuts are started inside a piece, a hole is needed to start the cut. Sometimes, in soft material, the point of the saw blade can be simply pushed through the material to start.

## COPING SAW

The *coping saw* (Fig. 1-31) is used primarily to cut molding into a coped joint. A *coped joint* is made by cutting and fitting the end of a molding piece against the face of another. (Coping is explained in detail in Chapter 18.) The coping saw is also used to make any small, irregular curved cuts in wood or other soft material.

The blades may be installed with the teeth either pointing away from or toward the handle. The saw cuts in the direction the teeth are pointed.

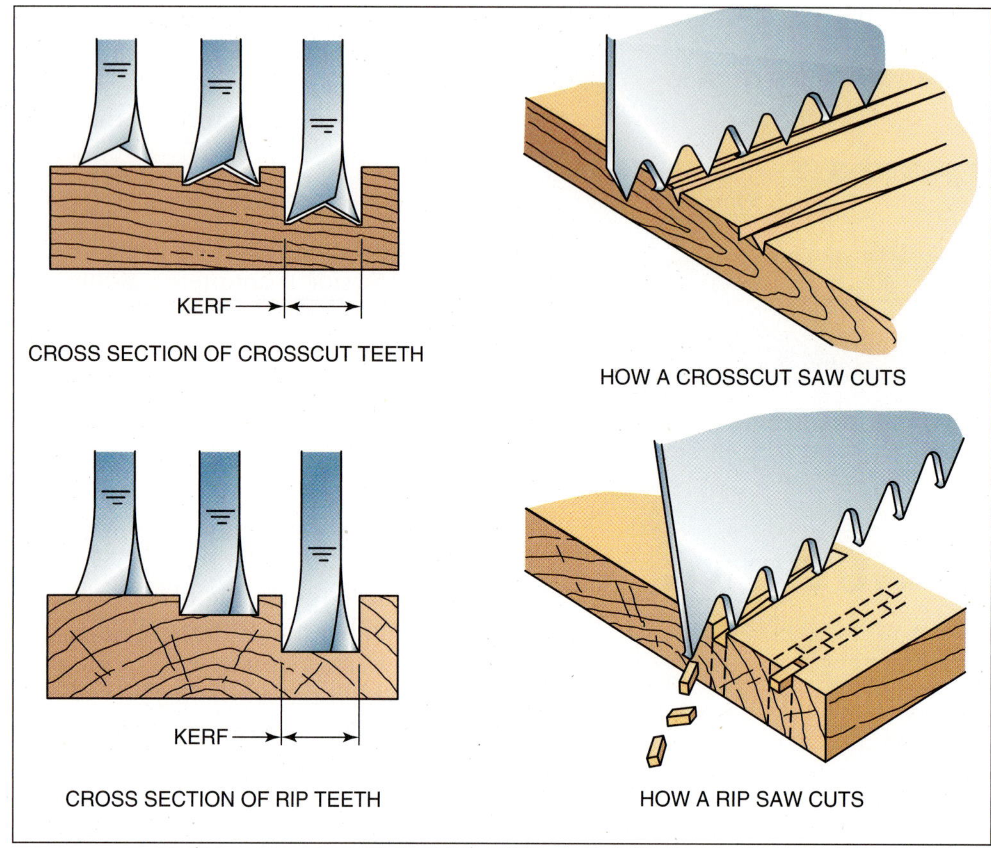


*Courtesy of Disston.*

**FIGURE 1-28** Cutting action of rip- and crosscut saws.

**FIGURE 1-29** Starting a cut with a handsaw.

## HACKSAWS

*Hacksaws* (Fig. 1-32) are generally used to cut thin material such as metal. Coarse-toothed blades are used for fast cutting in thicker material and fine-toothed blades are used for smooth cutting and for cutting thinner material.

**FIGURE 1-30** Compass saw.

## WALLBOARD SAWS

The *wallboard saw* (Fig. 1-33) is similar to the compass saw but is designed especially for gypsum board. The point is designed to be pushed into the material to start the cut.

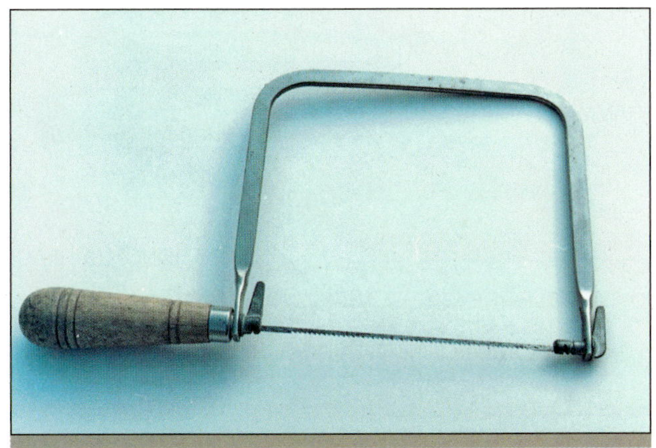

**FIGURE 1-31** Coping saw.

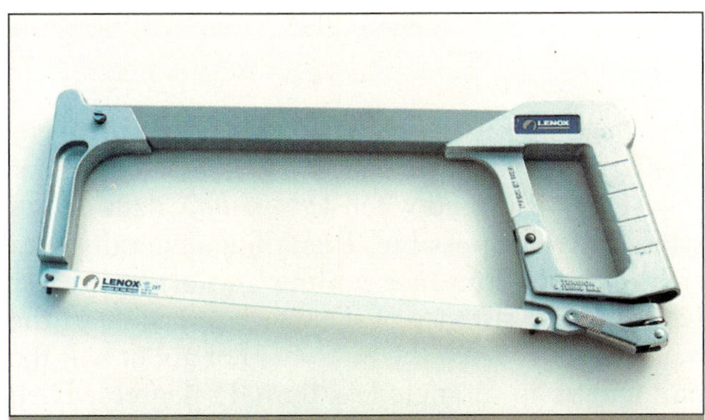

**FIGURE 1-32** Hacksaw.

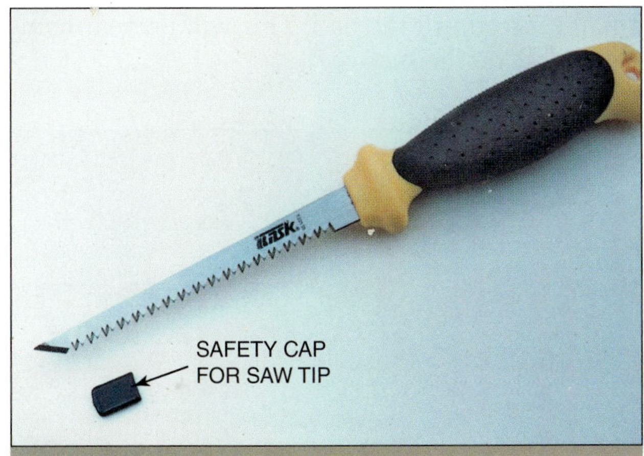


**FIGURE 1-33** Wallboard saw.

# FASTENING TOOLS

Fastening tools drive nails or screws to hold material in position.

## HAMMERS

The carpenter's *claw hammer* is available in a number of styles and weights. The claws may be straight or curved. Head weights range from 7 to 32 ounces. Most popular for general work is the 16-ounce, curved claw hammer (Fig. 1-34). For rough work, a long-handled 20- to 32-ounce *framing hammer* (Fig. 1-35) is often used.

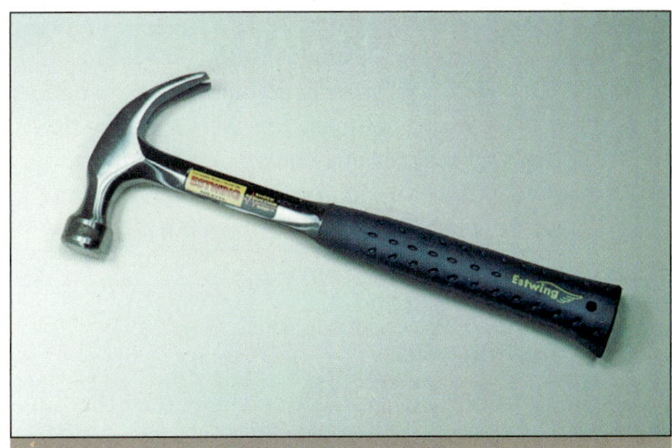

**FIGURE 1-34** Curved claw hammer.

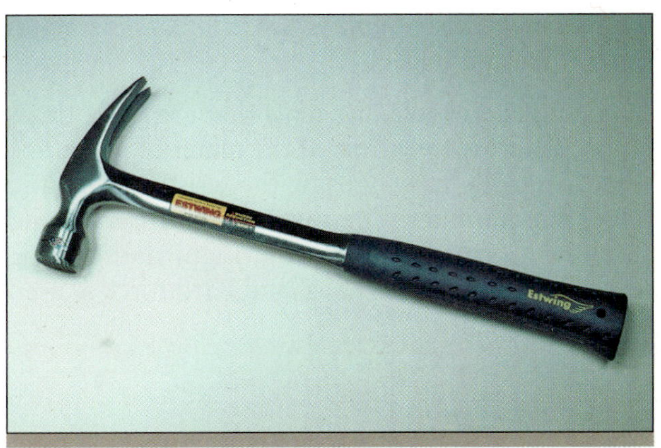

**FIGURE 1-35** Framing hammer.

## NAIL SETS

*Nail sets* (Fig. 1-36) are used to set nail heads below the surface for finishing. The most common sizes are 1⁄32, 2⁄32, and 3⁄32 inch. The size refers to the diameter of the tip. The surface of the tip is concave to prevent it from slipping off the nail head. If the tip becomes flattened, the nail set is less useful.

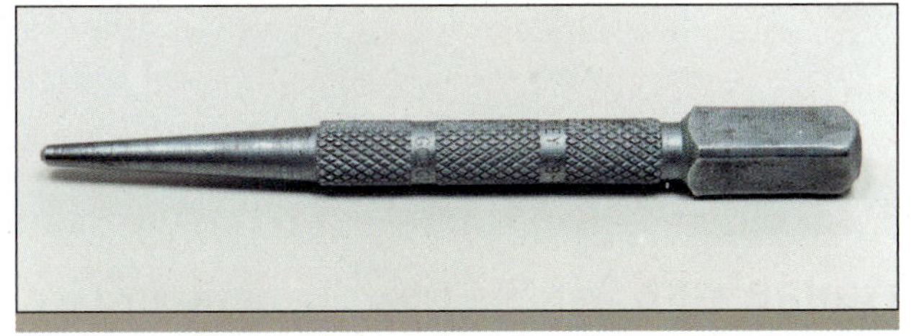

**FIGURE 1-36** Nail set.

**CAUTION**

**CAUTION:** Do not set hardened nails or hit the tip of the nail set with a hammer. This will cause the tip to flatten and may produce chips of flying metal. Always wear eye protection when driving or setting nails. Do not hit the side of the nail set.

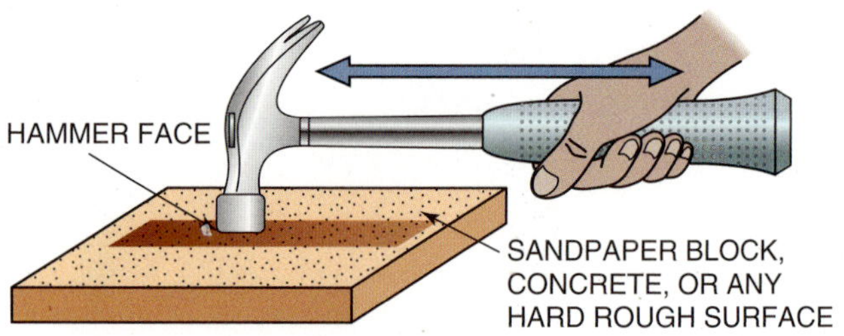


**FIGURE 1-38** Roughing up the hammerhead face helps keep the hammerhead from glancing off the nail.

## Nailing Techniques

Hold the hammer firmly, close to the end of the handle, and hit the nail squarely. Wear eye protection when using a hammer (Fig. 1-37). If the hammer frequently glances off the nail head, try cleaning the hammer face (Fig. 1-38). Be careful not to glance the hammer off the nail, which will damage material surface and fingers.

As a general rule, use nails that are three times longer than the thickness of the material being fastened. Start the nail by using the wrist to raise and lower the hammer. Then use the entire arm and shoulder to finish the driving. Light nail driving force is used on finish nails, and hard force is used on framing nails.

**FIGURE 1-37** Wear eye protection when driving nails.

*Toenailing* is the technique of driving nails at an angle to fasten the end of one piece to another (Fig. 1-39). It is used when *face nailing,* nails driven straight, is not possible. Toenailing generally uses smaller nails than face nailing and offers greater withdrawal resistance of the pieces joined. Start the nail a distance from the board end of about half the nail length at an angle less than 45 degrees. Then bend the nail to more than 45 degrees, closer to 60 degrees, and drive the nail (Fig. 1-43). If appearance is important, drive nails almost home and then use a nail set to finish the nail. This will prevent hammer marks on the surface.

**FIGURE 1-39** Toenailing is the technique of driving nails at an angle.

## *from experience...*

Blunting or cutting off the point of the nail also helps prevent splitting the wood (Fig. 1-40). When nailing along the length of a piece, stagger the nails from edge to edge, rather than in a straight line. This helps prevent splitting the board and provides greater strength (Fig. 1-41). Drive nails at an angle for greater holding power. Additionally, this will help prevent the points of long nails from protruding the backside of thinner material (Fig. 1-42).

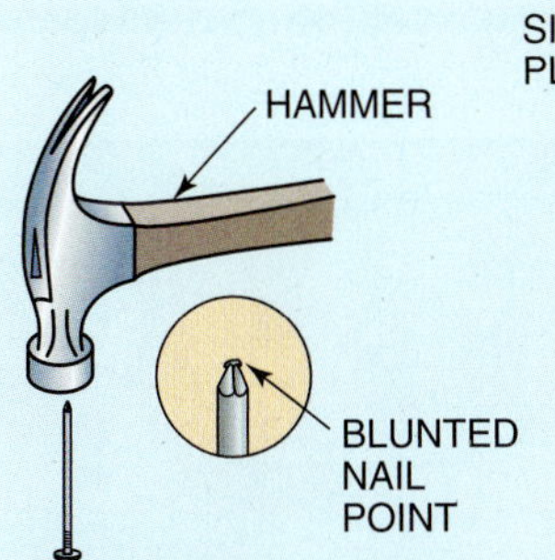


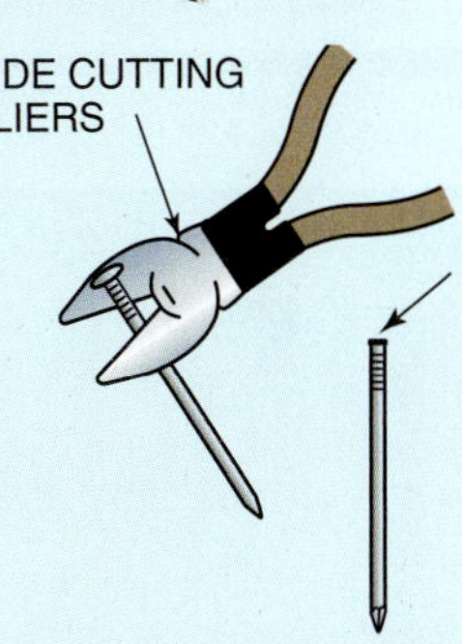


SOMETIMES BLUNTING THE POINT OF A NAIL WILL PREVENT SPLITTING THE STOCK. TAP THE POINT LIGHTLY SO AS NOT TO BEND THE NAIL.

IF A TWIST DRILL IS NOT HANDY TO PREDRILL NAIL HOLES, A NAIL OF THE SAME GAUGE WITH ITS HEAD CUT OFF MAY BE USED.

**FIGURE 1-40** Methods to avoid splitting wood.

**FIGURE 1-41** Stagger nails for greater strength and to avoid splitting the stock.

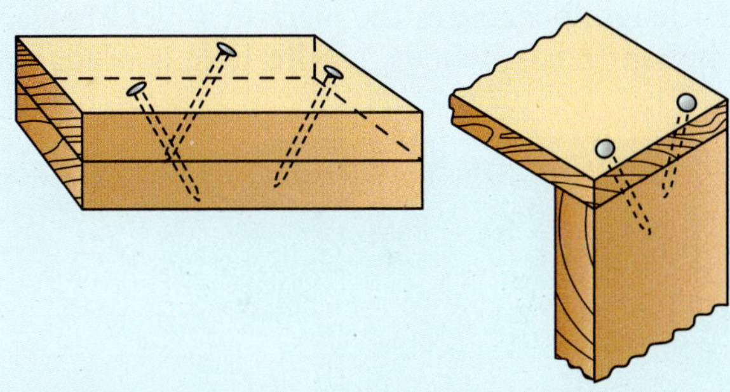

**FIGURE 1-42** Driving nails at an angle increases holding power.

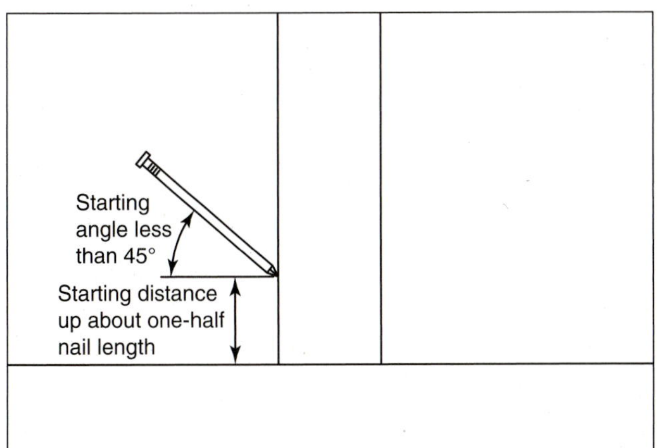


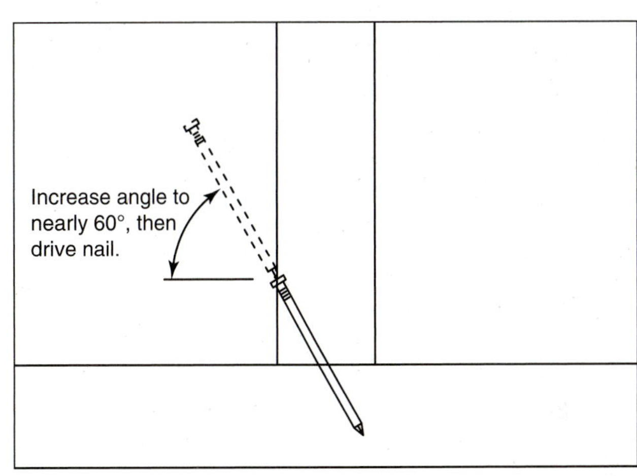


**FIGURE 1-43** Technique for proper toenailing.

## SCREWDRIVERS

*Screwdrivers* are manufactured to fit all types of screw heads. The *slotted* screwdriver has a straight tip to drive common screws, and the *Phillips* screwdriver has a cross-shaped tip (Fig. 1-44). Other screwdrivers include the Robertson screwdriver, which has a squared tip. Slotted screwdrivers are sized by the length of the blade and by the head type. Lengths generally run from 3 to 12 inches. Screwdrivers should fit snugly, without play, into the slot of the screw being driven (Fig. 1-45). The correct size screwdriver helps ensure that the screw will be driven without slipping out of the slot.

*Screwdriver bits* (Fig. 1-46) are available in many shapes and sizes to accommodate a variety of screws. They are designed to drive a screw using a drill or screw gun.

### Screwdriving Techniques

If possible, select screws so that two-thirds of their length penetrates the piece in which they are gripping. In hardwoods, a pilot hole is needed to prevent the wood from splitting. Be sure the hole is large enough so the screw will not break and small enough so the screw will grip.

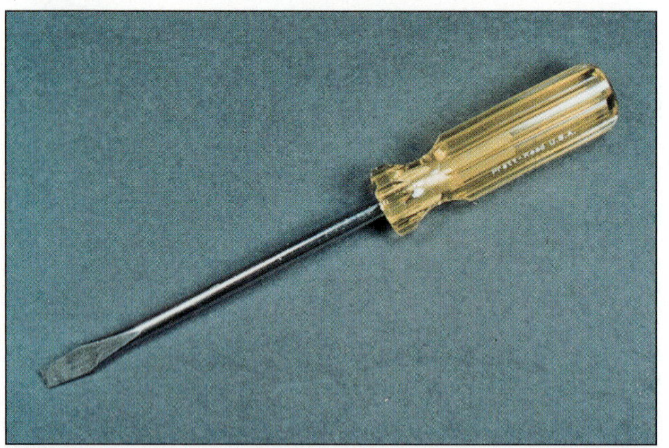

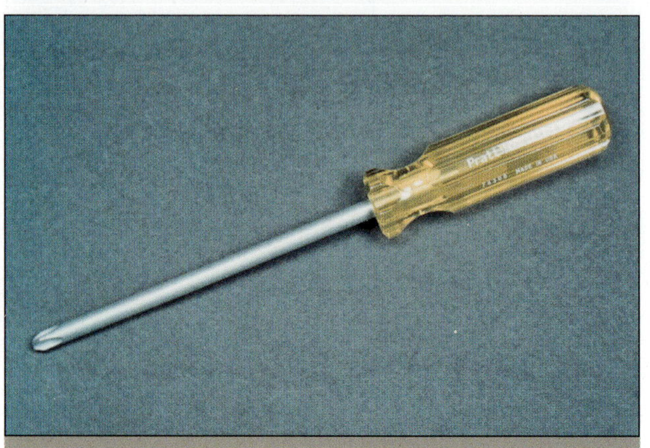

**FIGURE 1-44** Slotted and Phillips screwdrivers.

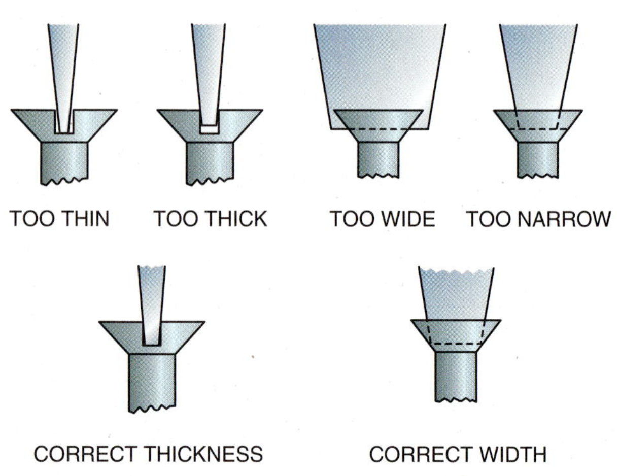


**FIGURE 1-45** Select the correct size screwdriver for the screw being driven.

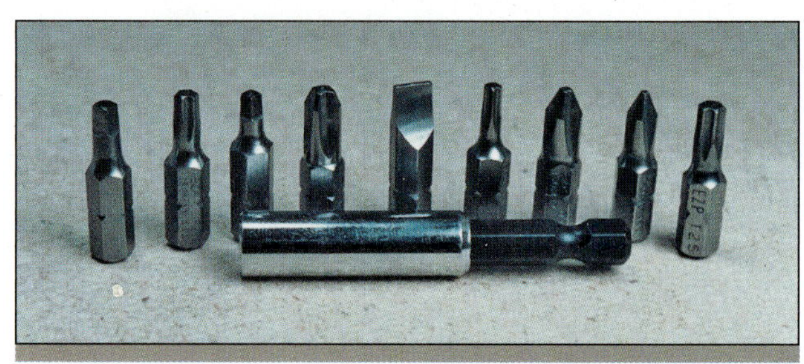

**FIGURE 1-46** Screwgun drive bits for various screw head styles.

# DISMANTLING TOOLS

Dismantling tools are used to take down staging and scaffolding, concrete forms, and other temporary structures. In addition, they are used for tearing out sections of a building before remodeling. Care should be used so as not to damage the material any more than necessary.

## HAMMERS

In addition to fastening, hammers are often used for pulling nails to dismantle parts. To increase leverage and make nail pulling easier, place a small block of wood under the hammerhead (Fig. 1-47).

**FIGURE 1-47** Pull a nail more easily by placing a block of wood under the hammer.

## BARS AND PULLERS

The *wrecking bar* (Fig. 1-48) is used to withdraw large nails and to pry when dismantling parts of a structure (Fig. 1-49). They are available in lengths from 12 to 36 inches, with the 30-inch size being most common.

Carpenters need a small *flat bar,* similar to those shown in Figure 1-50, to pry small work and pull small nails. To extract nails that have been driven home (all the way in), a *nail claw,* commonly called a *cat's paw,* is used (Fig. 1-51).

**CAUTION**

**CAUTION: Hammers are often used to drive bars and pullers into place. This is acceptable because bars and pullers are made of a softer steel that tend not to be brittle. But, over time, the steel is deformed, creating sharp edges that can easily cut skin.**

**FIGURE 1-48** Wrecking bar.

**FIGURE 1-49** Using a wrecking bar to pry stock loose.

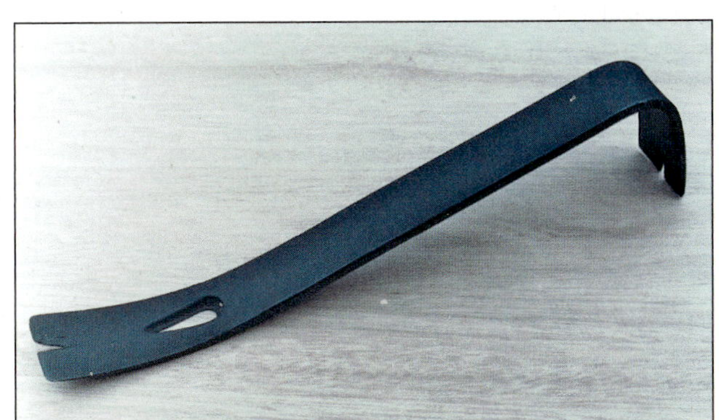

**FIGURE 1-50** Flat bars.

## HOLDING TOOLS

An *adjustable wrench* is often used to turn nuts, lag screws, bolts, and other objects (Fig. 1-52). The wrench is sized by its overall length. The 10-inch adjustable wrench is most widely used.

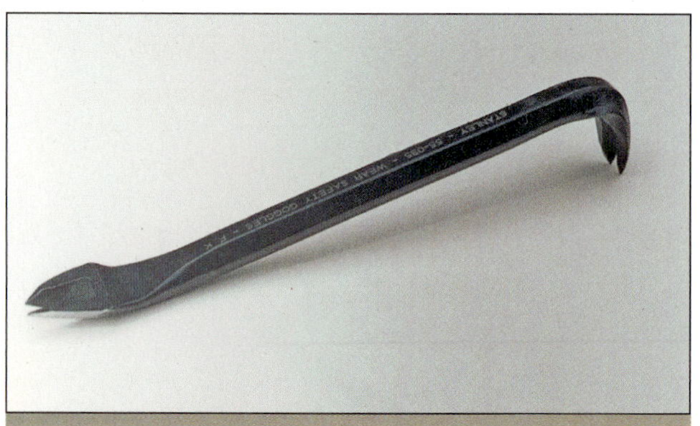

**FIGURE 1-51** Nail claw.

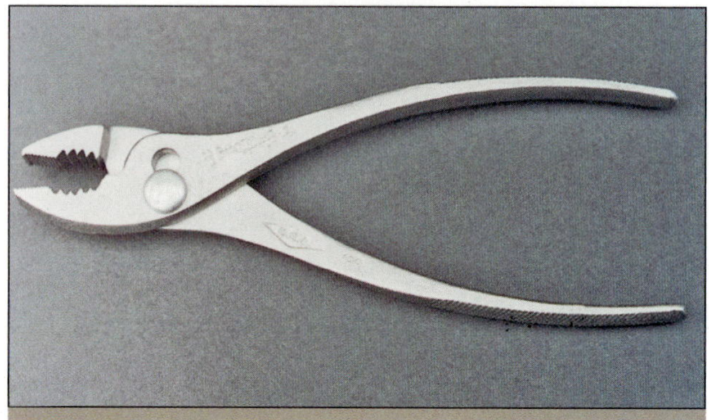

**FIGURE 1-53** Combination pliers.

**FIGURE 1-52** Adjustable wrench.

**FIGURE 1-54** (A) Spring clamp, (B) C clamp, (C) quick clamp, (D) wood screw.

Many kinds and styles are manufactured, and the *combination pliers* (Fig. 1-53) is designed for general use. A pair of pliers is often used for extracting, turning, and holding objects.

*C clamps* (Fig. 1-54) are useful for holding objects together while they are being fastened, holding temporary guides, applying pressure to glued joints, and many other purposes. The size is designated by the throat opening.

*Spring clamps* are a quick way to hold material together. They are opened like scissors, and a spring causes them to close. Plastic tips on the ends reduce damage to the material.

*Wood screws* are hardwood blocks with threaded rods through them. They supply ample holding power over a large surface area, which prevents marring.

*Quick clamps* are designed for speed and ease of operation.

## SUMMARY

- Layout tools are used either to mark the material or to assist in marking materials in preparation for cutting.
- Edge-cutting tools have large cutting surfaces that separate the waste material away in chunks or strips.
- Tooth-cutting tools cut by removing small bits of material along a cut line.
- Fastening tools are used to join materials together by driving fasteners.
- Dismantling tools are used to separate materials and remove fasteners.

# REVIEW QUESTIONS

*Select the most appropriate answer.*

1. **A safe worker attitude that promotes a safe job site comes from**
   a. ability.
   b. skill.
   c. knowledge.
   d. all of the above.

2. **The term *square* refers to**
   a. a steel tool used for layout and marking of right angles.
   b. two things that are perpendicular.
   c. a four-sided figure with equal sides and interior angles.
   d. all of the above.

3. **The term *deci-* in the metric system refers to**
   a. decimals.
   b. 1/10.
   c. fractions.
   d. 1/100.

4. **When stretching a steel tape to lay out a measurement, place the ring so the**
   a. 1-inch mark is on the starting line.
   b. end of the steel tape is on the starting line.
   c. inside of the ring is on the starting line.
   d. outside of the ring is on the starting line.

5. **In construction, the term *plumb* means perfectly**
   a. horizontal.
   b. level.
   c. straight.
   d. vertical.

6. **A large, L-shaped squaring tool that has tables stamped on it is called a**
   a. framing square.
   b. Speed Square.
   c. bevel square.
   d. combination square.

7. **The layout tool that may be adjusted to serve as a marking gauge is a**
   a. framing square.
   b. Speed Square.
   c. bevel square.
   d. combination square.

8. **To adjust a carpenter's level into a level position at which the bubble is touching the right line, the**
   a. right side should be raised.
   b. left side should be raised.
   c. left side should be lowered.
   d. entire level should be raised.

**9. When snapping a long chalk line, care should be taken to**

a. dampen the string.
b. keep the string from sagging.
c. hold the string loosely.
d. let the string touch the surface as it unwinds.

**10. The tool used to mark material to conform to an irregular surface is called a**

a. pen.
b. chisel.
c. scriber.
d. chalk line.

**11. To prevent a nail set from slipping off the nail head of a finish nail, its tip is**

a. flat and smooth.
b. convex.
c. concave.
d. checkered.

**12. The name of the one-handed plane with a low blade angle is the**

a. block plane.
b. bench plane.
c. chisel.
d. plane iron.

**13. The color of aviation snips is to help the tradesperson know**

a. which hand to use.
b. which direction curves may easily be cut.
c. the manufacturer.
d. what material may be easily cut.

**14. When using a handsaw, the cut is often started by first**

a. using a downstroke.
b. adjusting the saw so as to cut off the marked line.
c. using an upstroke.
d. keeping the other hand 6 inches away from the saw.

## WHAT'S WRONG WITH THIS PICTURE?

Carefully study **Figure 1-55** and think about what is wrong. Consider all possibilities.

### WRONG

**FIGURE 1-55** The carpenter is chiseling toward his hand. This is a potentially harmful situation. Furthermore, the workpiece is apparently not properly secured; otherwise, it would not be necessary to hold on to it. In the event the piece being worked on cannot be secured, the hand should be behind the chisel, not in the path of the cut.

### RIGHT

**FIGURE 1-56** This is a much better situation, chiseling away from oneself and securing the workpiece in a vise.

CHAPTER 2

# Hand Power Tools

**OBJECTIVES** *After completing this unit, the student should be able to:*

- State general safety rules for operating power tools.
- Describe and safely use the following: circular saws, saber saws, reciprocating saws, drills, hammer-drills, screwdrivers, planes, routers, sanders, staplers, nailers, and power-actuated drivers.

## GLOSSARY OF HAND POWER TOOL TERMS

**bevel** the sloping edge or side of a piece at any angle other than a right angle

**chamfer** an edge or end bevel that does not go all the way across the edge or end

**compound miter** a bevel cut across the width and also through the thickness of a piece

**fence** a guide for ripping lumber on a table saw

**miter** the cutting of the end of a piece at any angle other than a right angle

The sound of construction has changed over the years. The rhythmic whoosh of a handsaw has virtually been replaced with the whir and ring of a circular saw. Power tools have been created to increase the productivity of most job site tasks.

The number and style of power tools available today for the carpenter is vast, and the list continues to grow. Power tools enable the carpenter to do more work in less time with less effort.

However, with increased speed and production comes an increase in personal risk. This danger can come from a spectrum of human shortcomings ranging from a lack of knowledge and skill to overconfidence and carelessness. Safe operation of power tools requires knowledge and discipline.

Learn the safe operation techniques from the manufacturer's recommended instructions before operating any tool. Once you understand these procedures, do them right every time the tool is used.

Don't take chances. Life is too short as it is.

Being aware of the dangers of operating power tools is the first step in avoiding accidents. This begins with eye and ear protection.

CAUTION

**CAUTION:** General Safety Rules

- Have a complete understanding of a tool before attempting to operate it.
- To avoid potentially fatal electrical shock, make sure the tool is properly grounded and connected to a *ground fault circuit interrupter (GFCI).* This device monitors the flow of electricity and will trip if as little as 5 milliamperes leak out of the tool. It shuts off the current before any chance of electrical shock can occur.
- Always wear eye and ear protection while operating power tools. Skin grows back, but eyes and ears do not.
- Never use a power tool with a frayed or worn-out cord.
- Use the proper wire size extension cord. Do not use excessively long cords, or a voltage drop will occur at the tool, causing poor tool performance, overheating, and possible damage to the tool.
- Place extension cords out of the way of job site traffic. This will prevent the cords from being damaged and causing tripping accidents.
- Always unplug the tool when making adjustments to the cutters or blades.
- Watch for loose clothing that might become caught in the tool.
- Make sure you have complete control of the tool. Secure the material being worked to improve tool control.
- To ensure safety, use sharp cutters and blades. They cut faster, cleaner, and with less stress on the tool and the operator. Use the safety guards as designed by the tool manufacturer.
- Do not allow your attention to be distracted when operating tools. Stay alert and develop an attitude of care and concern for yourself, others, the tool, and the material being worked.

## CIRCULAR SAWS

Sometimes called the *Skilsaw,* the portable electric circular saw (Fig. 2-1) is used often by the carpenter. A retractable safety guard is provided over the blade, extending under the base. The saw is adjustable for depth of cut, and the base may be tilted for making **bevel** and **miter** cuts. **Compound miters** are angle cuts across the width and also through the thickness (Fig. 2-2). Saws are manufactured in many styles and sizes. The size is determined by the diameter of the blade, which ranges from 4½ to 16 inches. The blade may be driven directly by the motor or through a worm gear (Fig. 2-3). To remove the blade, first disconnect the saw by unplugging it.

**FIGURE 2-1** Cutting dimension lumber with a portable circular saw.

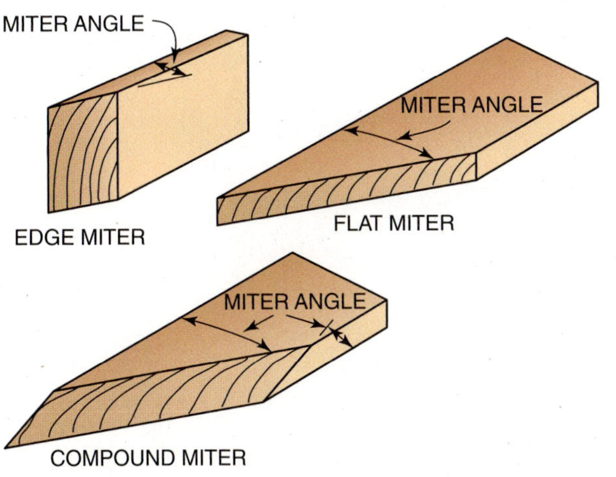


**FIGURE 2-2** Edge, flat, and compound miters.

(A)

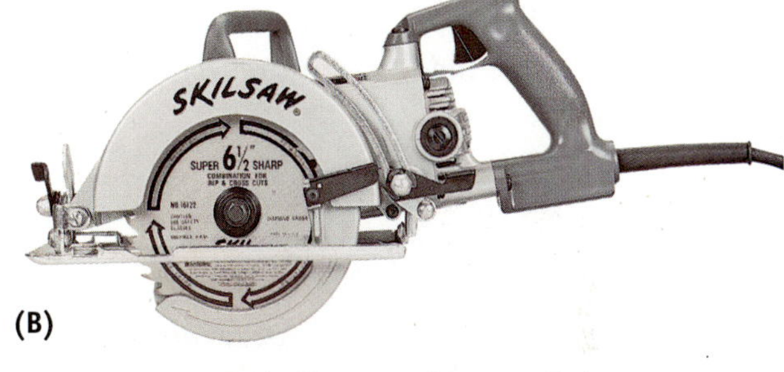


(B)

*(a) Courtesy of Porter Cable, (b) Courtesy of S-B Power Tools.*

**FIGURE 2-3** Direct drive and worm gear drive portable electric circular saws.

Loosen the bolt that holds the blade in place by rotating the bolt in the same direction as the rotation of the blade. To tighten, turn the bolt in a direction opposite to the rotation of the blade. Most saws have a lock for the motor while loosening or tightening the bolt. Circular saw blades are available in a number of styles. The shape and number of teeth per inch determine their cutting action. Carbide-tipped blades are used more than high-speed steel blades. They stay sharper longer when cutting material that dulls ordinary blades quickly.

**CAUTION**

**CAUTION: Make sure the saw blade is installed with the teeth pointing in the correct direction. The teeth cut up toward the base as it rotates.**

## USING THE PORTABLE CIRCULAR SAW

Follow a safe and established procedure:

- Mark the stock to be cut. Wear eye and ear protection.
- Make sure the work is securely held and that the waste will fall away and not bind the saw blade (Fig. 2-4). Never cut a board that is propped between supports.
- Adjust the depth of cut so that the blade just cuts through the work, not more than ⅛ inch. Never expose the blade any more than is necessary (Fig. 2-5).

**CAUTION**

**CAUTION:** Make sure the guard operates properly. Be aware that the guard may possibly stick in the open position. Never wedge the guard back in the open position.

- Hold the saw with two hands, resting the forward end of the base on the work. With the blade back from the material, pull the trigger to start the saw. When it has reached full speed, advance the saw into the work.

**CAUTION**

**CAUTION:** Any deviation from the line may cause the saw to bind or kick back. Do not force the saw into a turn or a curve. Back out and start the cut again. In case the saw does bind, stop the motor and bring the saw back to where it will run free. Recut and continue following the line closely.

- Watch the line and the saw to follow a straight path. Cut as close to the line as possible with the saw cutting on the waste portion of the board.

**CAUTION**

**CAUTION:** Keep the saw clear of your body until the saw blade has completely stopped. Always remember, it can still cut other things as long as the blade is spinning. Let the waste piece drop clear and release the switch.

**FIGURE 2-4** Saw cuts are made over the end of supports so the waste will fall clear and not bind the blade.

**FIGURE 2-5** The blade of the saw is adjusted for depth only enough to cut through the work. (*Guard has been held open for clarity.*)

**FIGURE 2-6** Retracting the guard of the portable circular saw by hand.

- Sometimes, when cutting an angle, it may be necessary to retract the guard beforehand, holding with a thumb (Fig. 2-6). Release the handle after the cut has been started and continue as previously directed.

## MAKING PLUNGE CUTS

Many times, it is necessary to make internal cuts in the material, such as for openings in floors, walls, and countertops. To make these cuts with a circular saw, the saw must be plunged into the material.

- Lay out the cut to be made. Wear eye and ear protection.
- Adjust the saw for depth of cut.
- Hold the guard open and tilt the saw up with the front edge of the base resting on the work.
- Move the saw blade over, and in line with, the cut to be made.
- Making sure the teeth of the blade are not touching the work, start the saw.
- Lower the blade slowly into the work by rotating the saw with the front edge of the base as a pivot.

**CAUTION**

**CAUTION:** Do not move the saw backwards while cutting as it may cause severe damage to the operator and material when it runs backwards up out of the cut.

**FIGURE 2-7** Making a plunge cut with a portable circular saw. First retract the guard, place the front edge of the saw base on the material, and then pivot the running saw slowly into the material.

- Follow the line carefully, until the entire base rests squarely on the material (Fig. 2-7).
- Advance the saw into the corner. Release the switch and wait until the saw stops before removing it from the cut.

# SABER SAWS

The *saber saw* (Fig. 2-8) is sometimes called a *jigsaw*. It is widely used to make curved cuts. There are many styles and varieties of saber saws. Some saws can be switched from straight up-and-down strokes to several orbital (circular) motions to provide the most effective cutting action for various materials. The base of the saw may be tilted to make bevel cuts. Many blades are available for fine

**FIGURE 2-8** The jigsaw is used to make cuts along curved lines.

or coarse cutting in wood or fine cutting in metal. Wood-cutting blades have teeth that are from 6 to 12 points to the inch. Blades with coarse teeth (fewer points to the inch) cut faster but rougher. Blades with more teeth to the inch may cut slower but produce a smoother cut surface.

## USING THE SABER SAW

Follow a safe and established procedure:

- Outline the cut to be made. Wear eye and ear protection.
- Secure the work by hand, tacking, clamping, or some other method.
- Hold the base of the saw firmly on the work. With the blade clear, pull the trigger. Push the saw into the work, following the line closely.
- Make the saw cut into the waste side of the line, and cut as close to the line as possible without completely removing it. Maintain firm downward pressure on the saw to reduce vibration and improve cutting speed.
- Keep the saw moving forward, holding the base down firmly on the work.
- Turn the saw as necessary, but not too much, in order to follow the line to be cut. Feed the saw into the work only as fast as it will cut. Do not force it.

## MAKING PLUNGE CUTS

Plunge cuts may also be made with the saber saw in a manner similar to that used with the circular saw:

- Lay out the material to be cut. Wear eye and ear protection.
- Tilt the saw up on the forward end of its base with the blade in line and clear of the work (Fig. 2-9).
- Start the motor, holding the base steady. Gradually and slowly lower the saw until the blade penetrates the work and the base rests firmly on it.
- Back up about an inch, turn the corner by cutting a small arc, and cut along the other side and into the corner. Continue in this manner until all the sides of the opening are cut. Turn the saw around and cut in the opposite direction to cut out the corners.

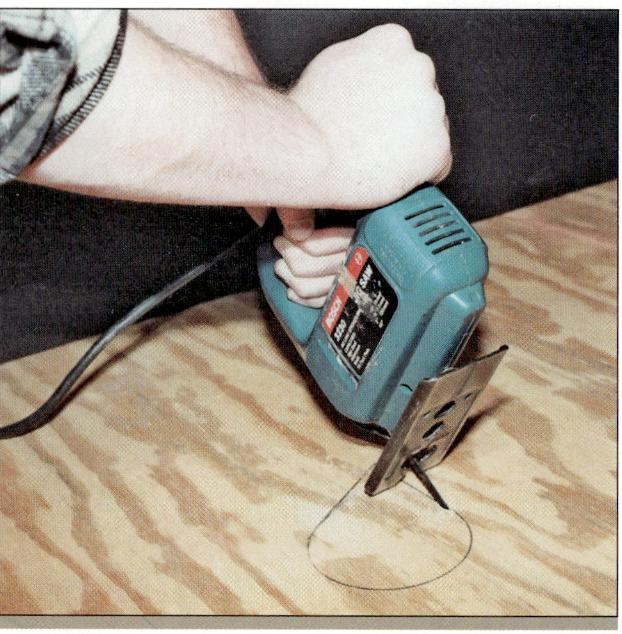

**FIGURE 2-9** Making an internal cut by plunging the saber saw.

CAUTION

**CAUTION:** Make sure the tool comes to a complete stop before withdrawing it from the material being cut.

CAUTION

**CAUTION:** Hold the saw firmly to prevent it from jumping when the blade makes contact with the material and to make a successful plunge cut. Cut along the line into the corner.

# RECIPROCATING SAWS

The *reciprocating saw* (Fig. 2-10), sometimes called a *Sawzall,* is used primarily for roughing in work. This work consists of cutting holes and openings for such things as pipes, heating and cooling ducts, and roof vents. Most models have a variable speed from 0 to 2,400 strokes per minute. Like saber saws, some models may be switched to several orbital cutting strokes from a straight back-and-forth (reciprocal) cutting action.

Common blade lengths run from 4 to 12 inches. They are available for cutting practically any type

**FIGURE 2-10** Using the reciprocating saw to cut an opening in the subfloor.

of material, such as wood, metal, plaster, fiberglass, and ceramics.

## USING THE RECIPROCATING SAW

The reciprocating saw is used in a manner similar to the saber saw. The difference is that the reciprocating saw is heavier and more powerful. It can be used more efficiently to cut through rough, thick material, such as walls, when remodeling. With a long blade, it can be used to cut flush with a floor or along the side of a stud. Follow a safe and established procedure:

- Outline the cut to be made. Wear eye and ear protection.
- Hold the base or shoe of the saw firmly against the work.
- Pull trigger and ease the saw into the cut.

Holes may be predrilled for the blade to make cutouts; insert the blade and start the motor.

# DRILLS AND DRIVERS

Portable power drills, manufactured in a great number of styles and sizes, are widely used to drill holes and drive fasteners in all kinds of construction materials. Battery-powered cordless models are used because of their convenience or when electrical power is not available.

## DRILLS

The drills used in the construction industry are classified as light-duty or heavy-duty. Light-duty drills usually have a *pistol-grip* handle. Heavy-duty drills may have a *spade-shaped* or *D-shaped* handle (Fig. 2-11).

The size of a drill is determined by the capacity of the *chuck*, its maximum opening. The chuck is that part of the drill that holds the cutting tool. The most popular sizes for light-duty models are ¼ and ⅜ inch. Heavy-duty drills have a ½-inch chuck or larger. Most drills have variable speed and reversible controls. Speed of rotation can be controlled from 0 to maximum rpm (revolutions per minute) by varying the pressure on the trigger switch. Slow speeds are desirable for drilling larger holes or metal.

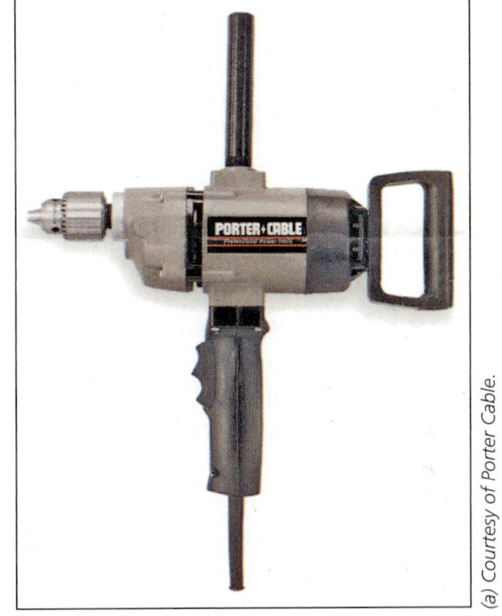


(A)

(a) Courtesy of Porter Cable.

(B)

**FIGURE 2-11** Portable power drills are available in a number of styles.

Faster speeds are used for drilling smaller holes and driving fasteners. A reversing switch changes direction of the rotation for removing screws or withdrawing bits and drills from holes.

## BITS AND TWIST DRILLS

*Twist drills* range in size from 1/16 to 1 inch in increments of 1/64 inch. These drills are particularly useful for drilling holes for screws and are also used in power drills (Fig. 2-12). High-speed steel or coated twist bits may also be used on soft steel. Use a slower drill speed with firm downward pressure when drilling metal to reduce heat buildup.

*Auger bits* (Fig. 2-13) are available with coarse or fine *feed screws.* Bits with coarse feed screws are used for fast boring in rough work. Fine feed bits are used for slower boring in finish work. A full set of auger bits ranges in size from 1/4 to 1 inch, graduated in 1/16-inch increments. The bit size is designated by the number of 1/16-inch increments in its diameter.

For instance, a #12 bit has 12 sixteenths. Therefore, it will bore a 3/4-inch diameter hole. To bore holes over 1 inch in diameter, an *expansive bit* (Fig. 2-14) may be used. With two interchangeable and adjustable cutters, holes up to 3 inches in diameter may be bored. This tool is handy for boring 2 1/8-inch holes in doors for locksets. *Speed bits,* sometimes called *spade bits,* are flat bits designed to make fast holes in wood (Fig. 2-15). They come in sizes from 1/4 to 2 inches. For a hole with a cleaner edge in finish work, the *power bore* bit may be used.

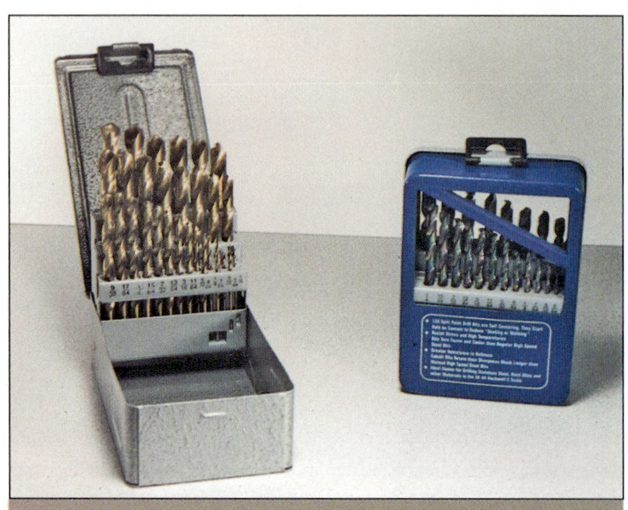

**FIGURE 2-12** A set of twist drills from 1/16 to 1/4 inch in increments of sixteenths.

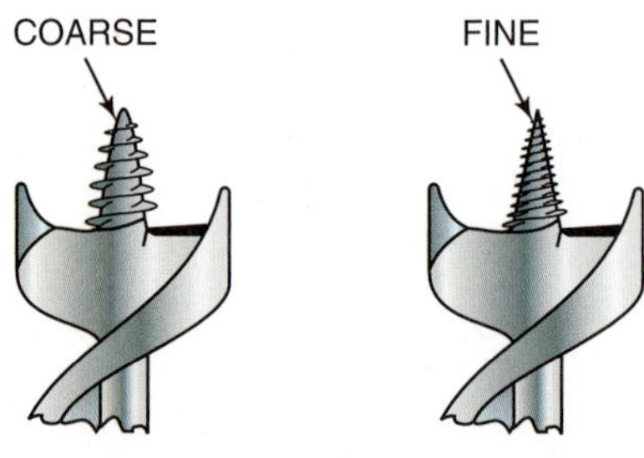


**FIGURE 2-13** Coarse and fine auger bit feed screws.

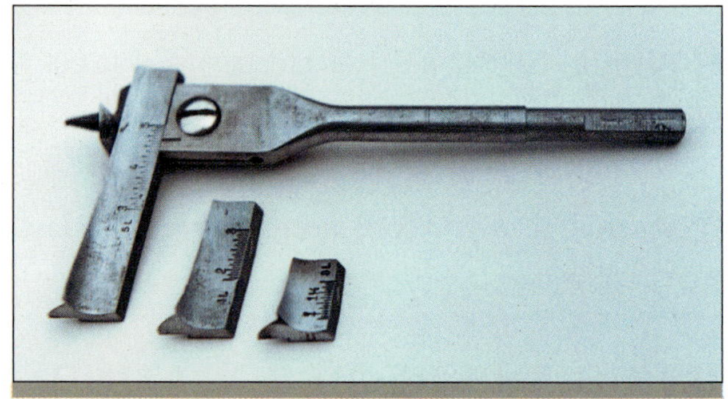

**FIGURE 2-14** Expansive bit.

Notice that none of these types has a center point that is threaded.

**CAUTION**

**CAUTION:** Never use bits with threaded center points in electric drills when boring deep holes. A threaded shank will draw the bit into the work. This makes it difficult to withdraw the bit and may cause the operator to lose control of the drill.

## BORING TECHNIQUES

To avoid splintering the back side of a piece when boring all the way through, reduce the travel speed and stop just when the feed screw point comes through. Finish by boring from the back side. Care must be taken not to strike any nails or other objects that might cause blunting and shortening of the spurs. If the spurs become too short, the auger bit is ruined.

Power drill speeds should be fairly high when drilling wood with moderate pressure on the material. As the bit nears the bottom of the hole, care should be taken to reduce pressure so the bit finishes the hole completely before breaking out the bottom.

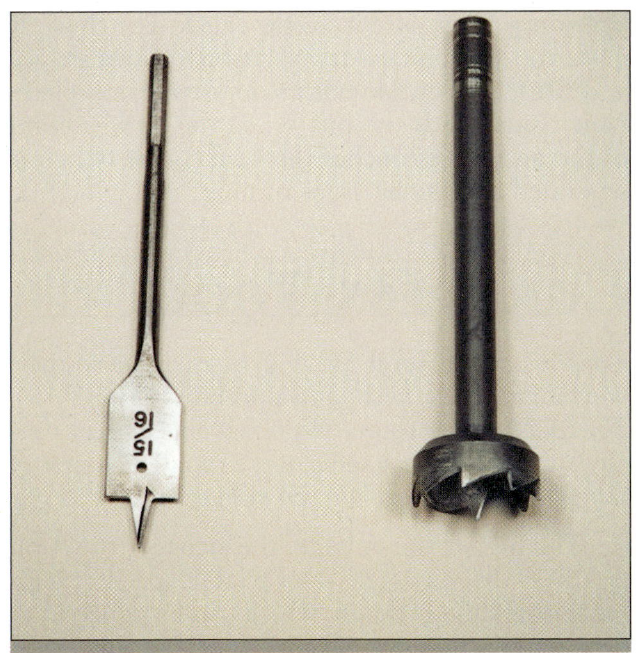

**FIGURE 2-15** Speed and power bore bits are used to drill larger holes in wood and similar material.

### Other Drill Accessories

Occasionally, carpenters may use *hole saws*. These cut holes through thin material from 5/8 inch to 6 inches in diameter (Fig. 2-16). *Masonry drill bits* have carbide tips for drilling holes in concrete, brick, tile, and other masonry. They are frequently used in portable power drills. They are more efficiently used in *hammer-drills*.

**CAUTION**

**CAUTION: Hold small pieces securely by clamping or other means. When drilling through metal, especially, the drill has a tendency to hang up when it penetrates the underside. If the piece is not held securely, the hang-up will cause the piece to rotate with the drill. It could then hit anything in its path and possibly cause serious injury to a person before power to the drill can be shut off.**

## USING PORTABLE ELECTRIC DRILLS

Follow a safe and established procedure:

- Mark the hole to be made. Wear eye and ear protection.
- Select the proper size bit or drill and insert it into the drill. Tighten the chuck with the chuck key or by holding the chuck of a keyless chuck.
- For accuracy, holes may be center-punched, which keeps the drill from wandering off center. Place the bit on the center of the hole to be drilled.
- Start the motor and apply pressure as required, but do not force the bit.
- Drill into the stock, being careful not to wobble the drill. Failure to hold the drill steady may result in breakage of small twist drills.

**FIGURE 2-16** Hole saws actually saw holes in material, leaving a circular center plug.

**CAUTION**

**CAUTION: Remove the bit from the hole frequently while drilling to clear the chips. Failure to do this may result in the drill binding and twisting from the operator's hands. Be ready to release the trigger switch instantly if the drill does bind.**

## HAMMER-DRILLS

*Hammer-drills* (Fig. 2-17) are similar to other drills. However, they can deliver as much as 50,000 hammer blows per minute on the drill point. Most popular are the 3/8- and 1/2-inch sizes. Most models have a variable speed of from 0 up to 2,600 rpm.

## SCREWGUNS

*Screwguns* (Fig. 2-18) are used extensively for fastening gypsum board to walls and ceilings with screws. They are similar in appearance to the

**FIGURE 2-17** The hammer-drill is used to make holes in concrete.

Courtesy of Porter Cable.

**FIGURE 2-18** The screw gun or drywall driver is used to fasten wallboard with screws.

light-duty drills, except for the chuck. The chuck is made to receive screwdriver bits of various shapes and sizes. A screwgun has an adjustable nosepiece, which surrounds the bit. When the forward end of the nosepiece touches the surface, the clutch is separated and the bit stops turning.

# Cordless Tools

*Cordless* power tools are widely used due to their convenience, strength, and durability (Fig. 2-19). The tools' power source is a removable battery usually attached to the handle of the tool. The batteries range in voltage from 4 to 24 volts.

The higher the voltage, the stronger the tool. Cordless drills come with variable-speed reversing motors and a clutch, which can be adjusted to drive screws to a desired torque and depth. Cordless circular saws can cut dimension lumber and are very handy as trim saws.

Downtime is practically eliminated with spare batteries and improvements in the chargers, which can quick charge in as little as fifteen minutes. Improved chargers can charge batteries of various voltages. They stop charging when the correct voltage level is reached, before the battery begins to overheat. This prolongs the life of the battery.

Cordless tools tend to make the job site safer by eliminating extension cords. Yet the extension cord sometimes serves as a safety line for tools that fall. Take care to set the tool down in a safe place after each use, particularly when working on scaffolds.

Courtesy of Porter Cable.

**FIGURE 2-19** Cordless tools kit with battery charger.

FIGURE 2-20 Cordless drills.

Cordless drills come with variable-speed reversing motors and a positive clutch, which can be adjusted to drive screws to a desired torque and depth (Fig. 2-20). Some models allow for a hammer-drilling mode when drilling masonry. Cordless circular saws are powerful and able to easily cut dimension lumber as well as trim material (Fig. 2-21). Sizes include 4½- and 6½-inch blades. Cordless jigsaws and reciprocating saws offer the same features as corded models. They also cut and handle in much the same manner with the same performance (Fig. 2-22). Impact drivers simplify driving screws and lags (Fig. 2-23). They act like pneumatic impact tools driving with a hammering action. Drive bits tend to stay in the screw slots of the screw better, reducing spin out. Spin out is when the driving bit spins in the drive slot, rounding the corners and making the screw more difficult to drive.

FIGURE 2-21 Cordless circular saws.

FIGURE 2-22 Reciprocating saws are commonly called (A) jigsaw and (B) sawzall.

**FIGURE 2-23** Cordless impact driver and accessories.

## Green Tip

Batteries, particularly from portable power equipment, should be recycled. This keeps the toxic material inside batteries away from the environment and helps reduce other raw material manufacturing costs.

**CAUTION**

**CAUTION:** It is easy to think that because these tools are battery powered, they are safer to use than higher voltage tools with cords. While they are safe to use, the operator should never forget the proper techniques and requirements for using the tools. Always wear personal protection devices.

# Portable Power Planes

Portable power planes make some planing jobs much easier for the carpenter. The plane is used primarily to smooth and straighten long edges, such as fitting doors in openings. It is manufactured in lengths of up to 18 inches. The electric motor powers a cutter head that may measure up to 3¾ inches wide.

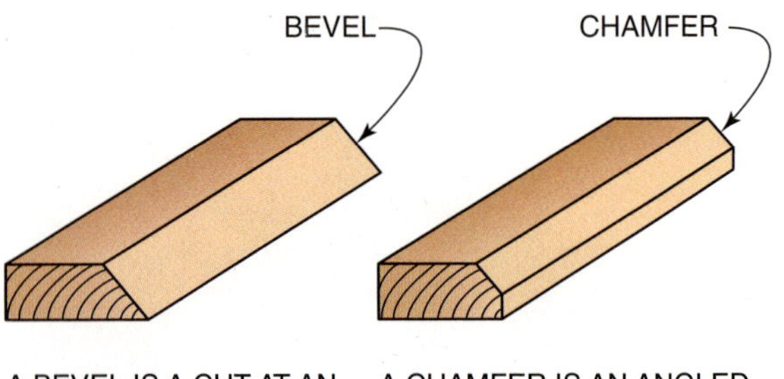


**FIGURE 2-24** A bevel and chamfer.

The planing depth, or the amount that can be taken off with one pass, can be set for 0 up to ⅛ inch. An adjustable **fence** allows planing square, beveled edges to 45 degrees called **chamfers** (Fig. 2-24).

**CAUTION**

**CAUTION:** Extreme care must be taken when operating power planes. There is no retractable guard, and the high-speed cutterhead is exposed on the bottom of the plane. Keep the tool clear of your body until it has completely stopped. Keep extension cords clear of the tool.

## OPERATING POWER PLANES

Follow a safe and established procedure:

- Outline the cut to be made. Wear eye and ear protection.
- Secure the piece being planed.
- Set the side guide to the desired angle, and adjust the depth of cut.
- Hold the toe (front) firmly on the work, with the plane cutterhead clear of the work.
- Start the motor. With steady, even pressure, make the cut through the work for the entire length. Guide the angle of the cut by holding the guide against the side of the stock.
- Apply pressure to the toe of the plane at the beginning of the cut. Apply even pressure after the heel is on the piece. Move pressure to the heel (back) at the end of the cut to prevent tipping the plane over the ends of the work.

# ROUTERS

*Routers* are available in many models, ranging from 1/4 hp (horsepower) to more than 3 hp with speeds of 18,000 to 30,000 rpm (Fig. 2-25). These tools have high-speed motors that enable the operator to make clean, smooth-cut edges with a variety of shapes and sizes of bits (Fig. 2-26). An adjustable base is provided to control the depth of cut.

**FIGURE 2-25** Using a portable electric router.

When operating the router, it is important to be mindful of the bit at all times. Watch what you are cutting and keep the router moving. Stalling the movement of the router will cause the bit to burn or melt the material. A light-duty specialized type of router is called a *laminate trimmer.* It is used almost exclusively for trimming the edges of plastic laminates (Fig. 2-27), a thin, hard material used primarily as a decorative covering for kitchen and bathroom cabinets and countertops.

## USING THE ROUTER

Before adjusting or touching the cutters of the router, make sure power is disconnected. Follow a safe and established procedure:

- Wear eye and ear protection.
- Select the correct bit for the type of cut to be made.

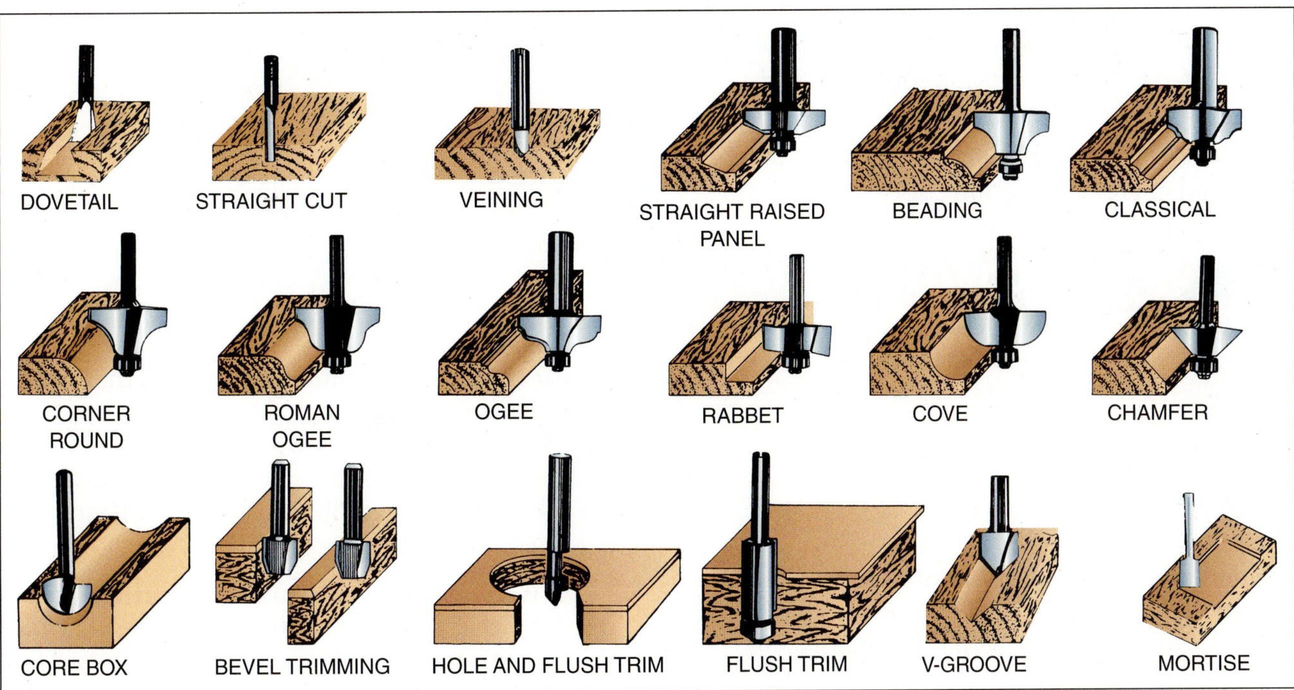


*Courtesy of Stanley Tools.*

**FIGURE 2-26** Router bit selection guide.

**FIGURE 2-27** The laminate trimmer is used to trim the edges of plastic laminates.

**FIGURE 2-28** Using a portable electric belt sander.

- Insert the bit into the chuck. Make sure the chuck grabs at least twice the shaft diameter of the bit. Adjust the depth of cut.
- Clamp the work securely in position. Plug in the cord.
- Lay the base of the router on the work with the router bit clear of the work. Start the motor.
- Advance the bit into the cut, pulling the router in a direction that is against the rotation of the bit. To rout the outside edges and ends, the router is moved counterclockwise around the piece. When making internal cuts, the router is moved in a clockwise direction.

CAUTION

**CAUTION: Finish the cut, keeping the router clear of your body until it has stopped. Always be aware that the router bit is unguarded.**

# SANDERS

Sanders are designed to smooth wood surfaces in preparation for a finish. They come in many styles and can use sand paper with varying grits.

## BELT SANDERS

The *belt sander* is used frequently for sanding cabinetwork and interior finish (Fig. 2-28). The size of the belt determines the size of the sander. Belt widths range from 2½ to 4 inches. Belt lengths vary from 16 to 24 inches. The 3-inch by 21-inch belt sander is a popular, lightweight model. Some sanders have a bag to collect sanding dust.

**FIGURE 2-29** Sanding belts should be installed in the proper direction. Use the arrow on the back of the belt as a guide.

Sanding belts are usually installed by retracting the forward roller of the belt sander (Fig. 2-29). Install the belt over the rollers. Then release the forward roller to its operating position. The forward roller can be tilted slightly to keep the sanding belt centered as it is rotating. Stand the sander on its back end. Hold it securely and start it. Turn the adjusting screw one way or

**FIGURE 2-30** The belt should be centered on its rollers by using the tracking screw.

the other to track the belt and center it on the roller (Fig. 2-30).

## USING THE BELT SANDER

It is wise to practice on scrap stock until enough experience is gained to ensure an acceptable sanded surface. Belt sanders can remove a lot of material in a short time. Care must be taken to sand squarely on the sander's base pad.

CAUTION

**CAUTION:** Make sure the switch of the belt sander is off before plugging the cord into a power outlet. Some trigger switches can be locked into the "on" position. If the tool is plugged in when the switch is locked in this position, the sander will travel at high speed across the surface. This could damage the work and tool and/or injure anyone in its path.

Follow a safe and established procedure:

- Wear eye and ear protection.
- Secure the work to be sanded.
- Make sure the belt is centered on the rollers and is tracking properly.
- Holding the tool with both hands, start the machine.
- Place the pad of the sander flat on the work. Pull the sander back and lift it just clear of the work at the end of the stroke.
- Bring the sander forward. Continue sanding using a skimming motion that lifts the sander just clear of the work at the end of every stroke. Sanding in this manner prevents overheating the sander, the belt, and the material being sanded. It allows debris to be cleared from the work. The operator can also see what has been done.
- Do not sand in one spot too long. Be careful to keep the sander flat, not tilting the sander in any direction. Always sand with the pad flat on the work. Do not exert excessive pressure. The weight of the sander is enough. Always sand with the grain to produce a smooth finish.
- Make sure the sander has stopped before setting it down. It is a good idea to lay it on its side to prevent accidental traveling.

CAUTION

**CAUTION:** Be careful to keep the electrical cord clear of the tool. Because of the constant movement of the sander, the cord may easily get tangled in the sander if the operator is not alert.

## FINISHING SANDERS

The finishing sander (Fig. 2-31) is used for the final sanding of interior work. These tools are manufactured in many styles and sizes. They are available in cordless models. Finishing sanders either have an orbital motion, an oscillating (straight back-and-forth) motion, or a combination of motions

Courtesy of Robert Bosch Tool Corporation.

**FIGURE 2-31** A portable electric orbital finishing sander.

controlled by a switch. The *random orbital sander* has a design that randomly moves the center of the rotating paper at high speed. This allows the paper to sand in all directions at once. The straight line motion is slower but produces no cross-grain scratches on the surface.

### Using the Finish Sander

Follow a safe and established procedure:

- Wear eye and ear protection.
- Select the desired grit sandpaper. Attach it to the pad, making sure it is tight. A loose sheet will tear easily.
- Start the motor and sand the surface evenly, *slowly* pushing and pulling the sander with the grain. Let the action of the sander do the work. Do not use excessive pressure as this may overload the machine and burn out the motor. Always hold the sander flat on its pad.

# Pneumatic Staplers and Nailers

*Pneumatic staplers* and *nailers* are commonly called *guns* (Fig. 2-32). They are used widely for quick fastening of framing, subfloors, wall and roof sheathing, roof shingles, exterior finish, and interior trim. A number of manufacturers make a variety of models in several sizes for special fastening jobs. Remember to wear eye and ear protection.

Courtesy of Paslode.

**FIGURE 2-32** Pneumatic nailers and staplers are widely used to fasten building parts.

**CAUTION**

**CAUTION: Pneumatic tools have devices that make the tool safer to operate. In spite of these features, personal injury and death have occurred with these tools. Take extreme care to learn and follow the recommended operating procedures.**

## NAILING GUNS

The heavy-duty *framing gun* (Fig. 2-33) drives smooth-shank, headed nails up to 3¼ inches, ring-shank nails up to 2⅜ inches, and screw shank nails up to 3 inches. A light-duty version (Fig. 2-34) drives smooth-shank nails up to 2⅜ inches and ring-shank nails up to 1¾ inches to fasten light framing, subfloor, sheathing, and similar components of a building. Nails come glued in strips for easy insertion into the magazine of the gun (Fig. 2-35). The *finish nailer* (Fig. 2-36) drives finish nails from 1 to 2½ inches long. It can be used for the application of practically all kinds of exterior and interior finish work. It sets or flush drives nails as desired. The *brad nailer* (Fig. 2-37) drives brads ranging in length from ½ inch to 1⅝ inches. It is used to fasten small moldings and trim, cabinet door frames and panels, and other miscellaneous finish carpentry. The *roofing nailer* (Fig. 2-38) is designed for fastening asphalt and fiberglass roof shingles. It drives five different sizes of wide, round-headed roofing nails from ⅞ to 1¾ inches. The nails come in coils of 120 (Fig. 2-39), which are easily loaded in a nail canister.

## STAPLERS

Like nailing guns, *staplers* are manufactured in a number of models and sizes. A popular tool is the *roofing stapler* (Fig. 2-40), which may be used to fasten roofing shingles. It comes in several models and drives 1-inch wide-crown staples in lengths from ¾ inch to 1½ inches. The staples, like nails, come glued together in strips (Fig. 2-41) for quick and easy reloading. Most stapling guns can hold up to 150 staples.

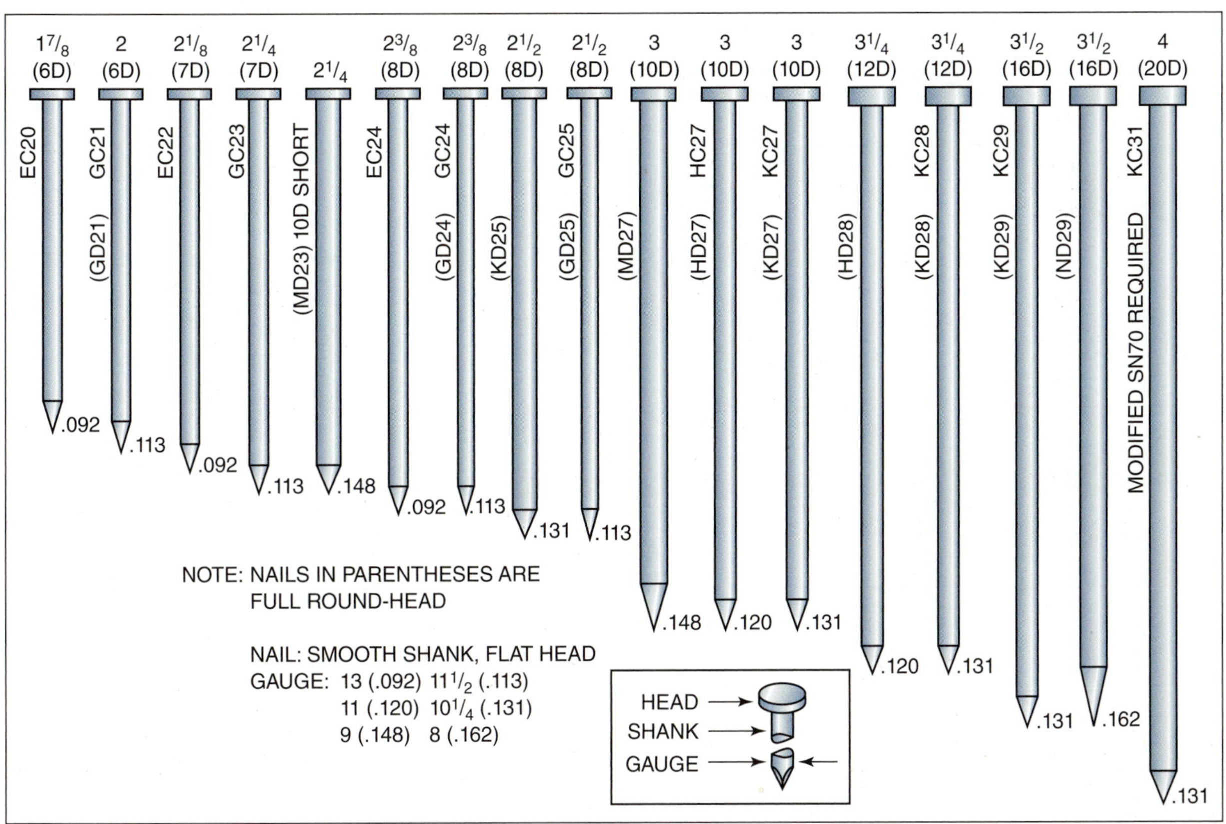


*Courtesy of Senco Products, Inc.*

**FIGURE 2-33** Heavy-duty framing nailers are used for floor, wall, and roof framing.

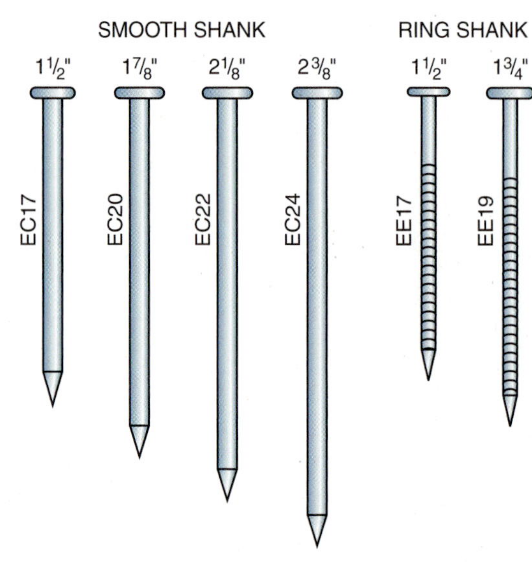


**FIGURE 2-34** A light-duty nailer is used to fasten light framing, subfloors, and sheathing.

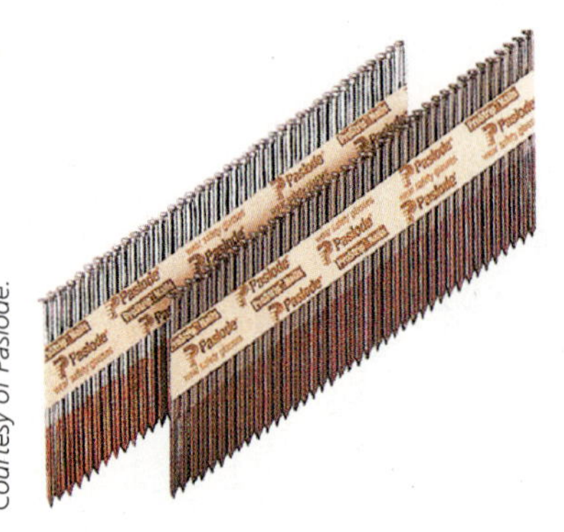

**FIGURE 2-35** Both headed and finish nails used in nailing guns come glued together in strips.

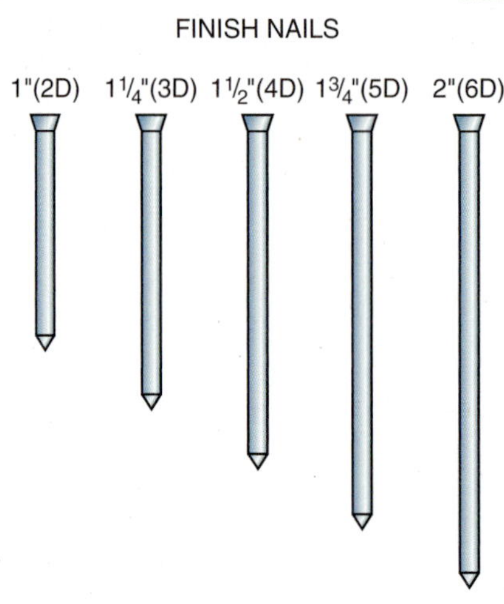


**FIGURE 2-36** The finish nailer is used to fasten all kinds of interior trim.

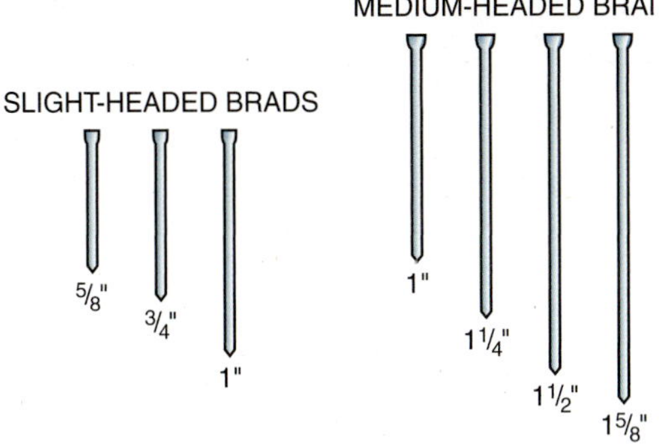


**FIGURE 2-37** A light-duty brad nailer is used to fasten thin molding and trim.

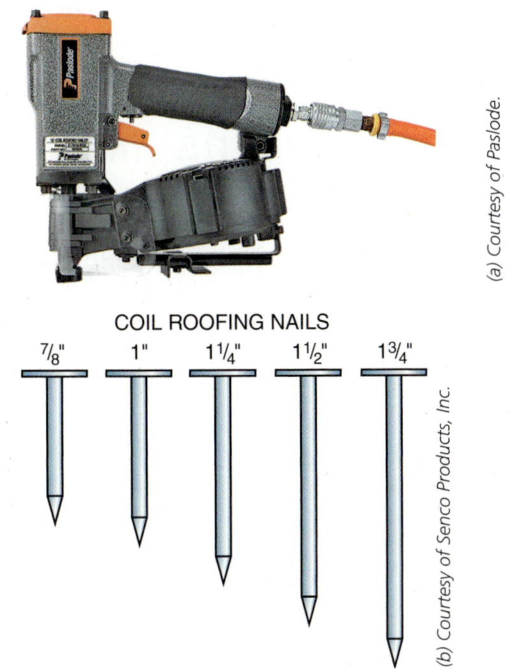


**FIGURE 2-38** A coil roofing nailer is used to fasten asphalt roof shingles.

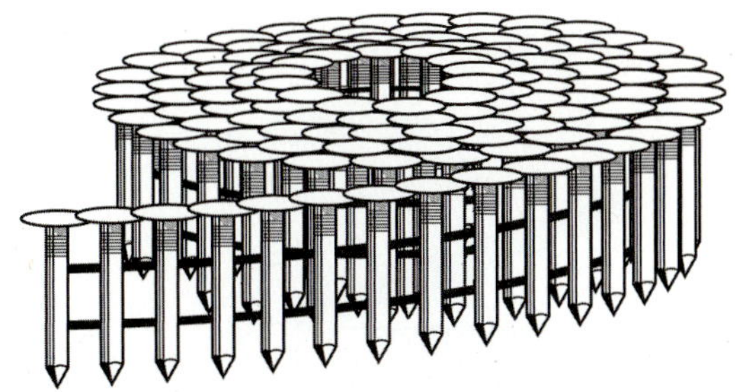

*Courtesy of Paslode.*

**FIGURE 2-39** Roofing nails come in coils for use in the roofing nailer.

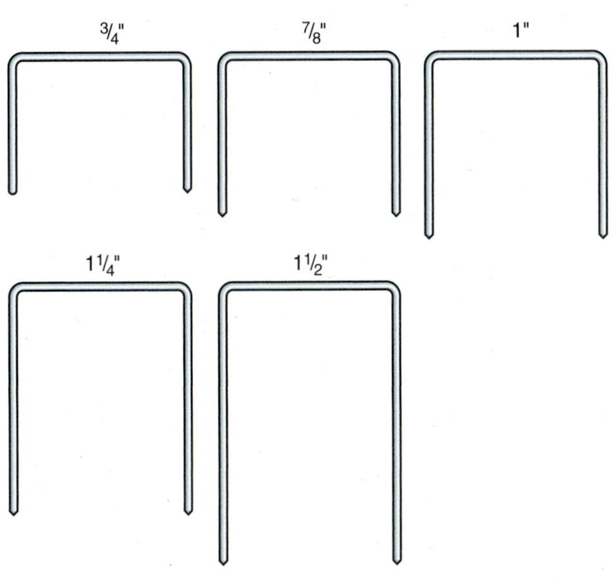


**FIGURE 2-40** The wide-crown stapler, being used to fasten roof shingles, may be used to fasten a variety of materials.

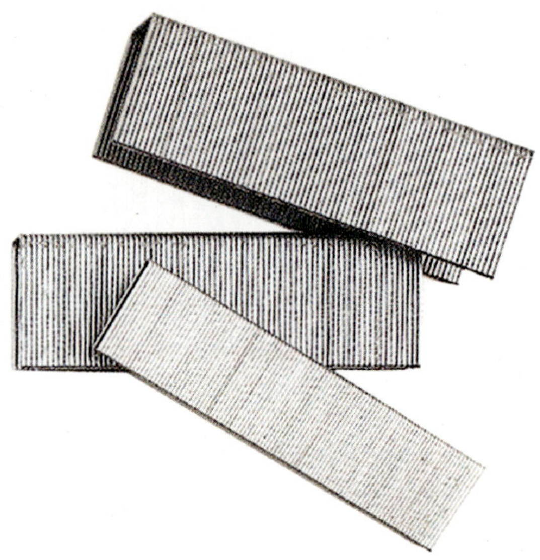

Courtesy of Paslode.

**FIGURE 2-41** Staples, like nails, come glued together in strips for use in stapling guns.

## CORDLESS GUNS

Conventional pneumatic staplers and nailers are powered by compressed air. The air is supplied by an *air compressor* (Fig. 2-42) through long lengths of air hose stretched over the construction site. The development of *cordless nailing* and stapling guns (Fig. 2-43) eliminates the need for air compressors and hoses. The cordless gun uses a disposable fuel cell. A battery and spark plug power an internal combustion engine that forces a piston down to drive the fastener.

Courtesy of Porter Cable.

**FIGURE 2-42** The development of cordless nailers and staplers eliminates the need for air compressors.

Courtesy of Paslode.

**FIGURE 2-43** Each cordless gun comes in its own case with battery, battery charger, safety glasses, instructions, and storage for fuel cells.

## USING STAPLERS AND NAILERS

Because of the many designs and sizes of staplers and nailers, you should study the manufacturer's directions and follow them carefully. Use the right nailer or stapler for the job at hand. Make sure all safety devices are working properly, and always wear eye protection. A spring-loaded safety device allows the tool to operate only when this device is firmly depressed on a work surface and the trigger is pulled. Follow a safe and established procedure:

- Wear eye and ear protection.
- Load the magazine with the desired size staples or nails.

- Connect the air supply to the tool. For those guns that require it, make sure there is an oiler in the air supply line, adequate oil to keep the gun lubricated during operation, and an air filter to keep moisture from damaging the gun. Use the recommended air pressure.

CAUTION

**CAUTION:** Exceeding the recommended air pressure may cause damage to the gun or burst air hoses, possibly causing injury to workers.

- Press the trigger and tap the nose of the gun to the work. When the trigger is depressed, a fastener is driven each time the nose of the gun is tapped to the work. The fastener may also be safely driven by first pressing the nose of the gun to the surface and then pulling the trigger.
- Upon completion of fastening, disconnect the air supply. Be sure to remove your finger from the trigger each time you complete a nailing run.

CAUTION

**CAUTION:** Never leave an unattended gun with the air supply connected. Always keep the gun pointed toward the work. Never point it at other workers or fire a staple except into the work. A serious injury can result from horseplay with the tool.

# Powder-Actuated Drivers

*Powder-actuated drivers* (Fig. 2-44) are used to drive specially designed pins into masonry or steel. They are used in a manner similar to firing a gun. Powder charges of various strengths drive the pin when detonated.

Drivepins are available in a variety of sizes. Powder charges are color-coded according to strength. The strength of the charge must be selected with great care. Because of the danger in operating these guns, many states require certification of the operator. Certificates may be obtained from the manufacturer's representative after a brief training course.

**FIGURE 2-44** Powder-actuated drivers are used for fastening into masonry or steel.

## USING POWDER-ACTUATED DRIVERS

Follow a safe and established procedure:

- Study the manufacturer's directions for safe and proper use of the gun.
- Wear eye and ear protection.
- Make sure the drivepin will not penetrate completely through the material into which it is driven. This has been the cause of fatal accidents.
- To prevent ricochet hazard, make sure the recommended shield is in place on the nose of the gun. Many different shields are available for special fastening jobs.
- Select the proper fastener for the job. Consult the manufacturer's drivepin selection chart to determine the correct fastener size and style.

- Select a powder charge of necessary strength. Always start with the weakest charge that will do the job. Load the driver with the pin first and the cartridge second. Keep the tool pointed at the work.
- Press the tool hard against the work surface, and pull the trigger. The resulting explosion drives the pin.
- Eject the spent cartridge and clean the tool as needed.

CAUTION

**CAUTION:** If the gun does not fire, hold it against the work surface for at least 30 seconds. Then remove the cartridge according to the manufacturer's directions. If the cartridge is stuck, do not attempt to pry it out with a knife or screwdriver; most cartridges are rim-fired and could explode.

## SUMMARY

- Stationary power tools are heavy duty tools designed to repetitively shape and cut materials.
- Table saws crosscut and rip by pushing material through the saw.
- Miter saws cut in a chopping action at many angles for joining many finish materials.
- Band saws cut in a continuous cutting downward action allowing curved cuts.
- Jointers smooth material edge surfaces by passing material over the cutter.
- Drill press is designed for making precise holes.
- Shaper is a heavy-duty router mounted in a table.
- Bench sanders allow repetitive smoothing of edges and surfaces.

## GREEN CHECKLIST

- ☐ Batteries, particularly from portable power equipment, should be recycled. This keeps the toxic material inside batteries away from the environment and helps reduce other raw material manufacturing costs.

## REVIEW QUESTIONS

*Select the most appropriate answer.*

**1. To use a power tool properly, the operator should always**

a. wear eye protection.
b. wear ear protection.
c. understand the manufacturer's recommended instructions.
d. all of the above.

**2. The guard of the portable electric saw should never be**

a. lubricated.
b. adjusted.
c. retracted by hand.
d. wedged open.

**3. When selecting an extension cord for a power tool,**

a. use a longer cord to keep the cord from heating up.
b. keep the cord evenly spread out around the work area.
c. use one with a GFCI.
d. all of the above.

**4. When using a power tool for cutting, the operator should wear**

a. safety contact lenses and steel-toed work boots.
b. ear and eye protection.
c. stereo headphones.
d. all of the above.

**5. Sharp tools**

a. put less stress on the operator than duller tools.
b. cut slower than duller tools.
c. are more dangerous to use than duller tools.
d. all of the above.

**6. The saber or jig saw is used primarily for making**

a. curved cuts.
b. compound miters.
c. cuts in drywall.
d. long straight cuts.

7. **The saw primarily used for rough-in work is the**
   a. saber saw.
   b. jig saw.
   c. router.
   d. reciprocating saw.

8. **The size of an electric drill is determined by its**
   a. ampere rating.
   b. horsepower.
   c. weight.
   d. chuck capacity.

9. **The tool that is best suited for drilling metal as well as wood is the**
   a. auger bit.
   b. high-speed twist drills.
   c. expansive bit.
   d. speed bit.

10. **To produce a neat and clean hole in wood,**
    a. use a fast-spinning, sharp bit.
    b. use a slower travel speed.
    c. finish the hole by drilling from the back side.
    d. all of the above.

11. **When using the router to shape four outside edges and ends of a piece of stock, the router is guided in a**
    a. direction with the grain.
    b. clockwise direction.
    c. counterclockwise direction.
    d. all of the above.

12. **The tool where some codes require the operator to be certified because of the potential danger is the**
    a. powder-actuated driver.
    b. hammer drill.
    c. cordless pneumatic nailer.
    d. screwgun.

13. **The saw arbor nuts that hold circular saw blades in position are loosened by rotating the nut**
    a. clockwise.
    b. with the rotation of the blade.
    c. counterclockwise.
    d. against the rotation of the blade.

14. **The tool best suited to make the final sanding of interior work is the**
    a. belt sander.
    b. orbital sander.
    c. pneumatic stapler.
    d. all of the above.

15. **The pneumatic tool best suited for fastening thin trim pieces is the**
    a. heavy-duty nailer.
    b. light-duty nailer.
    c. brad nailer.
    d. roofing nailer.

## KNOW YOUR CODES

Some local codes require user certification for operation of powder-actuated drivers. The tool uses a gunpowder charge to drive the fastener. Make certain you know if your local code has this requirement.

## WHAT'S WRONG WITH THIS PICTURE?

Carefully study **Figure 2-45** and think about what is wrong. Consider all possibilities.

### WRONG

**FIGURE 2-45** This person is preparing to cut a piece of lumber that is supported between two sawhorses. When the saw cut is approximately three-fourths of the way through the board, the center of the board will sag, and the saw will bind. The high-speed rotation of the saw blade will cause the saw to kick back toward the person holding the saw. This may result in a serious injury.

### RIGHT

**FIGURE 2-46** In order to avoid the problems described in Figure 2-45, the piece that is being cut off should be allowed to freely fall away from the blade. The proper technique is shown here.

CHAPTER 3

# Stationary Power Tools

**OBJECTIVES** *After completing this unit, the student should be able to:*

- State general safety rules for operating power tools.
- Describe and adjust a table saw, miter saw, band saw, jointer, drill press, shaper, and bench sander.
- Safely operate a table saw, miter saw, band saw, jointer, drill press, shaper, and bench sander.

## GLOSSARY OF STATIONARY POWER TOOL TERMS

**crosscut** a cut made across the grain of lumber

**fence** a guide for ripping lumber on a table saw

**miter gauge** a guide used on the table saw for making miters and square ends

**rip** sawing lumber in the direction of the grain

Carpentry is the art of shaping and assembling woodwork. Stationary power tools are designed to assist carpenters in that endeavor. They are heavy duty tools designed to repetitively cut and shape wood with consistent results. They usually have fences that allow each piece cut to be the same as the piece before it.

Stationary power tools are similar to handheld models. The table and miter saws have circular saws attached to tables or frames. The band saw efficiently cuts curves in a vertical motion similar to the jig saw. The jointer is a hand plane. The drill press is a hand drill. The shaper is a router attached to a table. The bench sander is a belt sander. All these tools are designed to optimize the cutting operation for fast, professional results.

However, with increased speed and production comes an increase in personal risk. Safe operation of power tools requires knowledge and discipline. Learn the safe operation techniques from the manufacturer's recommended instructions before operating any tool. Be aware of the dangers of operating power tools. This begins with eye and ear protection.

CAUTION

**CAUTION:** General Safety Rules

- Be trained and competent in the use of stationary power tools before attempting to operate them without supervision.
- Make sure power is disconnected when touching the cutters.
- Make sure saw blades are sharp and suitable for the operation. Ensure that safety guards are in place and that all guides are in proper alignment and secured.
- Wear eye protection and appropriate, properly fitted clothing at all times.
- Keep the work area clear of scrap that might present a tripping hazard.
- Keep stock clear of saw blades before starting a machine.
- Do not allow your attention to be distracted while operating power tools. Watch the cutting operation closely.
- Turn off the power and make sure a machine has stopped before leaving the area.

Additional safety precautions that apply to specific operations are given when those operations are described.

## Table Saws

The size of the *table saw* (Fig. 3-1) is determined by the diameter of the saw blade. It may measure up to 16 inches. A commonly used table saw on the construction site is the 10-inch model. The blade is

**FIGURE 3-1** The 10-inch contractor's saw is frequently used on the job site.

adjusted for depth of cut and tilted up to 45 degrees by means of hand wheels. Blades are secured to the machine with an arbor nut. To remove the blade, first unplug the saw. Then while holding the arbor with a wrench, rotate the nut in the same direction the blade rotates when cutting.

A *rip fence* is used to guide material while ripping it to size. A miter gauge is used to make crosscuts. These features should not be used in the same operation. This is unsafe because the smaller cut off piece could bind in saw and fly out toward the operator with great force.

A guard should always be placed on the blade to protect the operator. Exceptions to this include some table saw operations, such as dadoes, rabbets, and cuts, in which the blade does not penetrate the entire stock thickness. A general rule is that if the guard can be used for any operation on the table saw, it should be used. Therefore, good habits are important to make and keep. Never reach over the blade.

**CAUTION**

**CAUTION:** Stand to either side of the blade. Never stand directly in back of the saw blade. Make sure no one else is in line with the saw blade in case of kickback. Be prepared to turn the saw blade off at any moment should the blade bind. A splitter and anti-kickback device should be used when the cutting operation allows it.

## RIPPING OPERATIONS

To rip stock to width, follow a safe and established procedure:

- Wear eye and ear protection.
- Measure from the rip fence to the point of a saw tooth set closest to the fence. Lock the fence in place. Check and adjust the rip fence measuring scale, if necessary.
- Adjust the height of the blade to about ¼ inch above the stock to be cut. Some manufacturers recommend setting the blade at full height inside the blade guard to allow the blade to run cooler and cut more easily.
- With the stock clear of the blade, turn on the power.
- Hold the stock against the fence with the left hand. Push the stock forward with the right hand, holding the end of the stock (Fig. 3-2).

**FIGURE 3-2** Using the table saw to rip lumber (guard has been removed for clarity).

Push the stock firmly, listening to the saw to determine appropriate feed speed. The blade speed should always be allowed to run at full.

- As the end approaches the saw blade, let the stock slip through the left hand, removing it from the work. If the stock is of sufficient width (at least 5 inches wide), finish the cut with the right hand pushing the end all the way through the saw. Otherwise use a *push stick* (Fig. 3-3) to finish the cut.

(A)

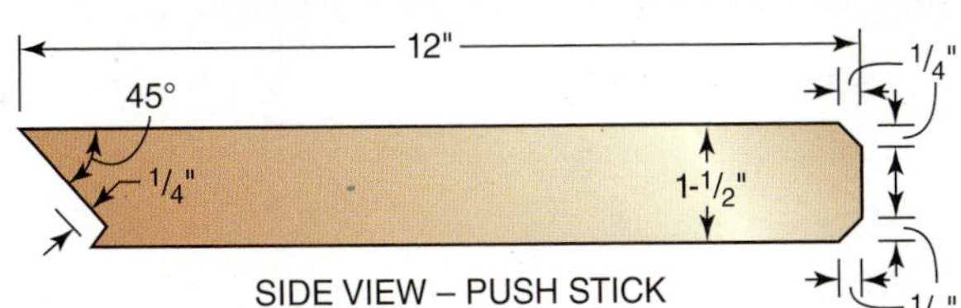


NOTE: PUSH STICK MUST BE THIN ENOUGH TO PASS BETWEEN THE RIP FENCE AND SAW BLADE WITH CLEARANCE.

(B)

**FIGURE 3-3** (A) Use a push stick to rip narrow pieces (guard has been removed for clarity). (B) Push stick design.

**CAUTION**

**CAUTION:** Make sure the stock is pushed all the way through the saw blade. Leaving the cut stock between the fence and a running saw blade may cause a kickback, injuring anyone in its path. Use a push stick when ripping narrow pieces. Do not pick small waste pieces from the saw table when the saw is running. Remove them with a stick or wait until the saw has stopped. Always use the rip fence for ripping operations. Never make freehand cuts. Never reach over a running saw blade.

## CROSSCUTTING OPERATIONS

For most **crosscutting** operations, the **miter gauge** is used. It slides in grooves on the table surface. It may be turned and locked in any position up to 45 degrees (Fig. 3-4).

**CAUTION**

**CAUTION:** The miter gauge should not be used at the same time with the rip fence. This could allow the small piece to bind between the blade and fence, which will come flying out back at the operator.

*Dadoing* is done in a similar manner as crosscutting, except with the use of a dado set (Fig. 3-5). The dado set is used to cut only partway through the stock thickness.

**FIGURE 3-4** Using the miter gauge as a guide to crosscut (guard has been removed for clarity).

**FIGURE 3-5** Cutting dadoes using a dado head.

To crosscut stock to length, follow a safe and established procedure:

- Lay out the desired length of stock.
- Wear eye and ear protection.
- Set desired blade and miter gauge angles.
- Hold stock firmly against miter gauge back from the blade.
- Turn on motor and, using two hands, ease the stock into the blade.
- Stand to the side of the line of the blade.
- Finish cut and remove the stock before returning the miter gauge to the start position.
- Keep the blade clear of debris.

# MITER SAWS

The *miter saw,* also called a *power miter box,* comes in sizes ranging from 8½ to 12 inches. The circular saw blade and motor, mounted above the base, are pushed down using a chopping action to make the cut (Fig. 3-6). The saw blade may be pivoted 45 degrees to the right or left. The *compound miter saw* will also allow the blade to tilt left and right up to 45 degrees to make compound miter cuts. The *sliding compound miter saw* allows the blade to slide forward. This saw, although used for finish work, can also easily cut a 2 × 12 at 90 degrees and a 2 × 10 at 45 degrees and is often used to cut framing material. Most models make it possible to cut crown molding on the flat instead of having to hold the piece at an angle (Fig. 3-7). The saw is set to

**FIGURE 3-6** The blade of the compound miter saw may be tilted.

**FIGURE 3-7** The power miter saw may be used to cut compound miters on ceiling molding held flat to the saw base.

cut a compound miter by setting the saw to two angles. These angle settings are easy to find with small indents in the angle slides.

### CUTTING OPERATIONS

To crosscut stock to length with a power miter saw, follow a safe and established procedure:

- Unlock the storage clamps and allow the saw to rise up into ready position.
- Lay out the desired length. Wear eye and ear protection.
- Place and hold firmly the stock against the fence with one hand, which should be out of the line of cut by as much as possible.
- If using a sliding model, with the other hand pull the saw forward and down close to the material.
- Start the motor and push the saw gently down into the material cutting the waste side of the layout line.
- If using a sliding model, push the saw gently back to finish the cut.
- Release the trigger and allow the saw to come to the rest position again. More cutting and operating techniques are described in Chapter 18, Interior Finish.

## BAND SAWS

The band saw is a tool designed to cut irregular shapes in various thicknesses of material (Fig. 3-8). It is usually operated freehand, moving the stock to

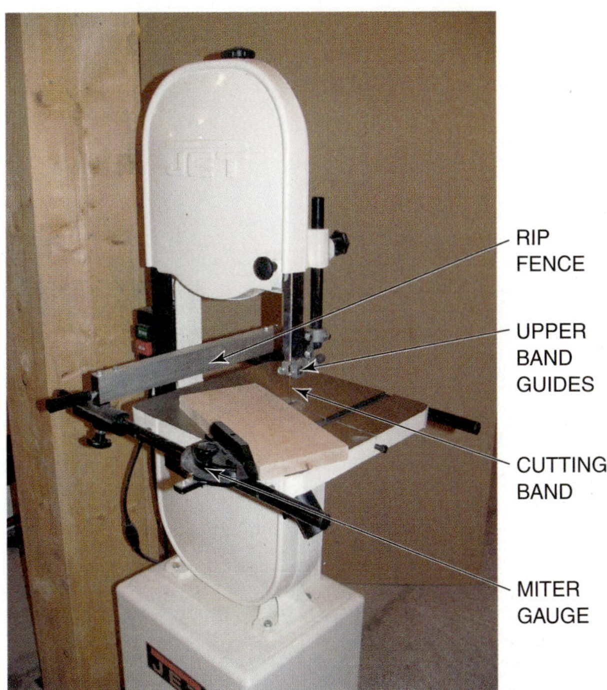


**FIGURE 3-8** The band saw can saw curves, irregular shapes, as well as straight cuts.

**FIGURE 3-9** The band saw may be used to resaw boards into thinner pieces.

follow curved lines. It can also be set up to use a miter gauge for straight cuts. In addition, the band saw is well suited to resaw thick boards into thinner ones with a small saw kerf (Fig. 3-9).

The saw is named for its cutting blade, which is a continuous thin steel band with teeth. Band saws have two large diameter wheels, one above the table and one below (Fig. 3-10). These wheels rotate

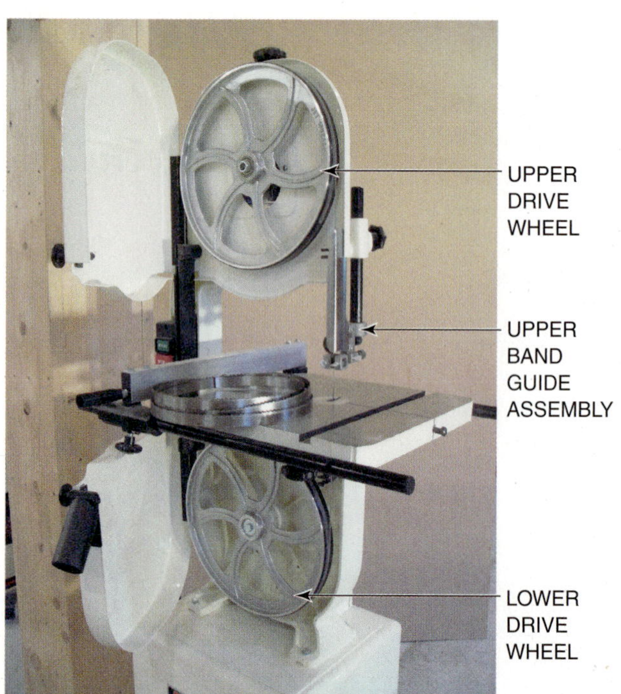


**FIGURE 3-10** The band saw has a continuous cutting band spun on two large wheels.

**FIGURE 3-11** The guides help keep the band in place during cutting operations.

the band into a cutting action that is always down, toward the table. The band is held straight for cutting by guide blocks and wheels. A set of guides is located above and below the table. The upper guide may be adjusted in height and should be positioned about ¼ inch above the stock being cut (Fig. 3-11). This will keep the blade straight during the cut.

Band saw sizes are determined by the diameter of the drive wheels. They typically range from 10 inches to 18 inches in diameter. Band saw blades have varying numbers of teeth per inch and band widths ranging from ⅛ inch to ¾ inch. This allows the bands to cut differently for the work needed. Fast rough cutting is done with wide course teeth, while fine work requires a narrow, fine-toothed blade. Metal may also be cut with a band saw when set up with a metal cutting blade spinning at a slower rpm.

**CAUTION**

**CAUTION:** Take care to avoid backing out of long cuts. Backing out could cause the band to come off the drive wheels.

The width of the band also affects the turning radius on the cut (Fig. 3-12). A narrow band of ⅛ inch will allow a ¼-inch radius cut, where the ¾-inch width will cut only along a 2½-inch radius.

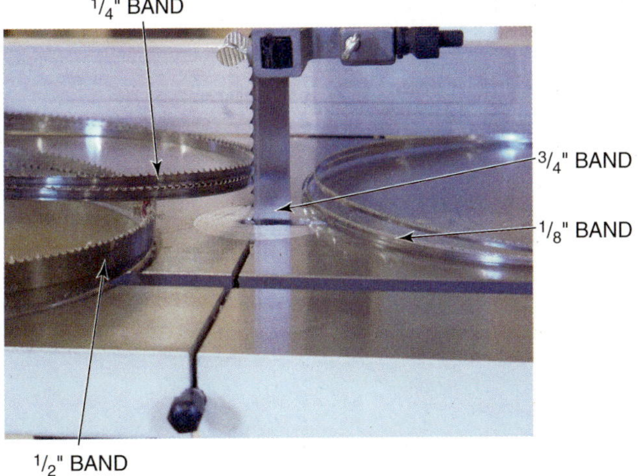


**FIGURE 3-12** Bands come in various widths.

# Jointers

A *jointer* is used to smooth stock surfaces to remove the marks made by sawing (Fig. 3-13). Its size is determined by the length of the cutting knives or roughly the width of the table, typically 4, 6, or 8 inches. A spring loaded safety shield swings in place to cover the portion of the cutter head not being used.

The jointer is constructed with two tables: infeed and outfeed. The two tables are always kept parallel to each other. The offset height of the tables determines the depth of the cut (Fig. 3-14). The infeed table is adjustable in height by turning a wheel under the table. The outfeed table is stationary and set at a height equal to the extreme cutting height of the cutting head. This allows the jointed stock to have full support after it has passed over the cutting head.

**FIGURE 3-13** Jointers are used to smooth board surfaces.

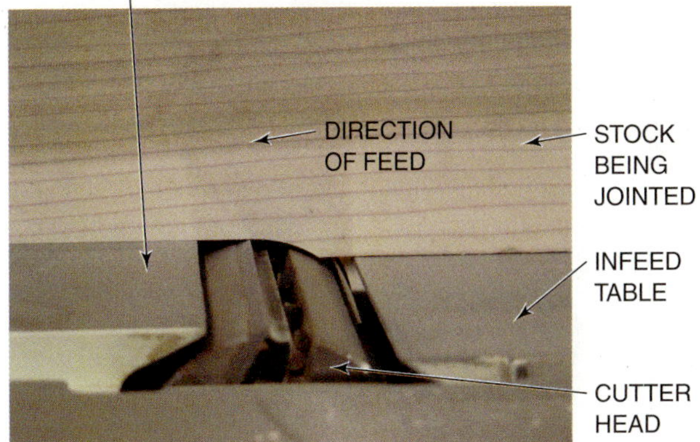


**FIGURE 3-14** The outfeed table is set to the height of the cutters, and the infeed table is adjustable for depth of cut. (Safety shield removed for clarity.)

CAUTION

**CAUTION:** Check the manufacturer's recommendations for the maximum allowable depth of cut.

Safe operation requires that the operator secure the stock with two hands. Begin with the stock resting on the infeed table and fence. Concentrate on keeping the stock against the table and fence during the entire pass, preventing it from wobbling. Then slowly ease the stock toward the safety shield and into the cutter head (Fig. 3-15). At some point halfway through the cutting pass, the operator's hands must shift to secure the stock to the outfeed table and fence. Take care to keep the stock moving and firmly held against the fence.

CAUTION

**CAUTION:** Avoid using your fingers to hold the stock directly over the cutters. Lift your hands over the area of the cutter head from infeed side to outfeed side. Accidents have happened in which cutters have kicked the stock out of the operator's hands. Always think ahead.

**FIGURE 3-15** Safe operation requires full attention to the machine and slow, steady feeding of the stock.

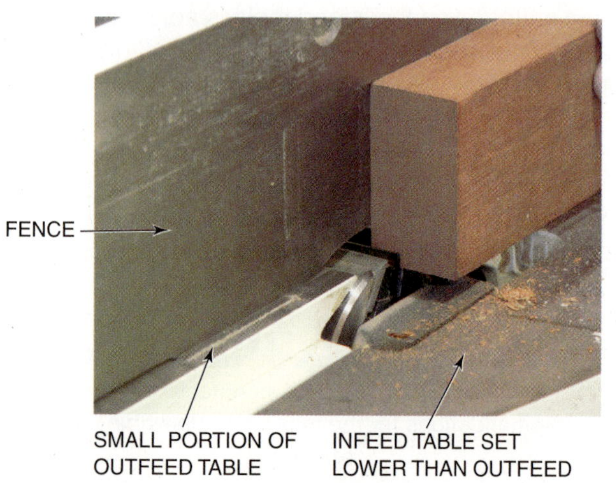


**FIGURE 3-16** Jointers can cut chamfers and dados. Safety shield is often removed for this operation.

Other operations of the jointer include chamfers and rabbets (Fig. 3-16). The fence may be tilted to joint bevels of various angles. Rabbets are performed by adjusting the fence over the cutter head such that only a small portion of it is available for cutting. Be sure to replace the safety shield after performing rabbet cuts.

# Drill Press

A *drill press* is a precision drilling tool consisting of a base, column, table, and head (Fig. 3-17). The base and column support the table and head. Power and control of the drill press are done from the head while the table supports the material being worked.

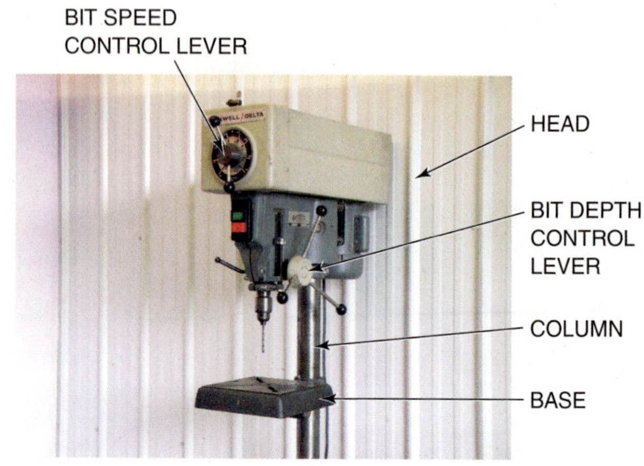


**FIGURE 3-17** A drill press is used to make repetitive and precise holes.

Bit speed is adjustable by typically changing the belt location on the cone-shaped pulleys. The smaller pulley on the motor will cause the bit to rotate slower. Wood is best bored by using a high speed with light pressure on the bit. Metal is best drilled by using a very slow speed with higher pressure on the bit. Special cutting oil is recommended when drilling metal to keep the bit cool and speed up the removal of material in the hole.

CAUTION

**CAUTION:** Take care to secure the work to the table before drilling. If the bit binds in the hole, the entire piece may spin out of control.

# Shaper

A *shaper* is used to make molding and decorative edges in irregular surfaces (Fig. 3-18). It is a heavy-duty router attached upside down to a base. It has a removable rip fence and vertical shaft projecting up through the center of the table that adjusts to the desired height. The spindle shaft is typically ½ inch in diameter that spins about 9,000–18,000 RPM. This high speed allows for a smooth finished

**FIGURE 3-18** Shapers are used to make decorative edges in wood.

product. There are literally hundreds of shapes and styles of bits available.

The stock may be tooled by using the fence as a guide or by using a collar on the bit. The fence provides a quick set-up for straight edges such as raised panel doors or trim. A collar may be located on the bit itself, which rides along any irregular shape. This allows a decorative edge by cutting only the amount designed in the bit.

> **CAUTION**
>
> **CAUTION:** Great care must be taken when setting up the shaper. Check the cutters carefully to be sure they are sharp and in proper working condition. If a tip should break off while the cutter is spinning, the result could cause serious injury.

## Bench Sander

*Bench sanders* in the form of a drum, disk, and belt are designed to smooth almost any shaped piece of work (Fig. 3-19). The material is moved over the sander to smooth and remove material fast.

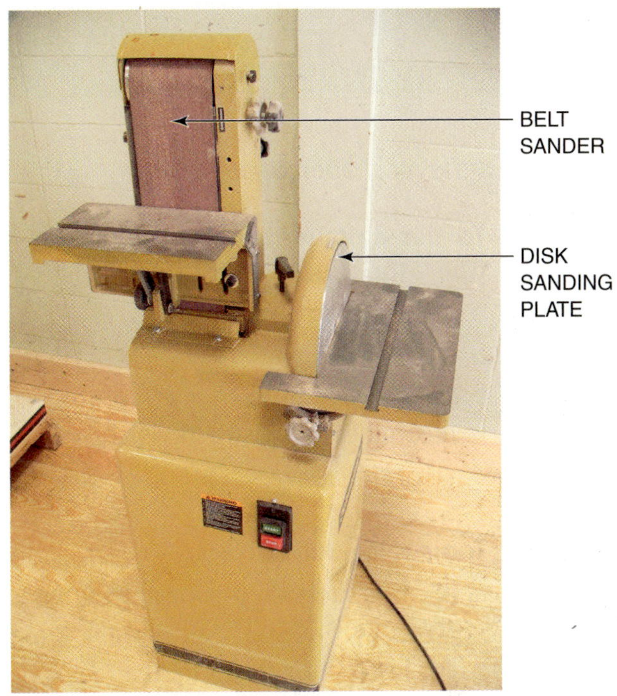


**FIGURE 3-19** Bench sanders are used to sand material quickly and easily.

## Summary

- Stationary power tools are heavy duty tools designed to repetitively shape and cut materials.
- Table saws crosscut and rip by pushing material through the saw.
- Miter saws cut in a chopping action at many angles for joining many finish materials.
- Band saws cut in a continuous downward action allowing curved cuts.
- Jointers smooth material edge surfaces by passing material over the cutter.
- Drill press is designed for making precise holes.
- Shaper is a heavy-duty router mounted in a table.
- Bench sanders allow repetitive smoothing of edges and surfaces.

## REVIEW QUESTIONS

*Select the most appropriate answer.*

**1. To use a power tool properly, the operator should always**

a. wear eye protection.
b. wear ear protection.
c. understand the manufacturer's recommended instructions.
d. all of the above.

**2. Tools with dull cutters**

a. put less stress on the operator than sharp tools.
b. cut faster than sharp tools.
c. are more dangerous to use than sharper tools.
d. all of the above.

**3. The table saw arbor nut that holds a circular saw blade in position is loosened by rotating the nut**

a. clockwise.
b. with the rotation of the blade.
c. counterclockwise.
d. against the rotation of the blade.

**4. The features of a table saw that should not be used at the same time are**

a. rip fence and push stick.
b. miter gauge and rip fence.
c. push stick and miter gauge.
d. miter gauge and a dado head.

**5. The table saw feature used to make large grooves in material is**

a. dado head.
b. miter gauge.
c. rip fence.
d. rabbet head.

**6. The table saw guide used for cutting with the grain is called a**

a. rip fence.
b. miter gauge.
c. tilting arbor.
d. ripping jig.

**7. The table saw tool that should not be used at the same time as a rip fence is a**

a. blade guard.
b. push stick.
c. dado head.
d. miter gauge.

**8. The tool suited to rip a board into thinner pieces with a small kerf is the**

a. miter saw.
b. table saw.
c. Jointer.
d. Band saw.

9. **When using the table saw for ripping operations, a push stick should be used if the ripped width is narrower than**
   a. 3 inches.
   b. 4 inches.
   c. 5 inches.
   d. 6 inches.

10. **The saw most often used for cutting interior trim is a**
   a. power miter box.
   b. saber saw.
   c. table saw.
   d. portable circular saw.

11. **Most models of compound miter saws allow for**
   a. angle left and right.
   b. angle side to side.
   c. cutting ceiling molding on the flat.
   d. all of the above.

12. **Band saw sizes are determined by**
   a. drive wheel diameters.
   b. depth of cut.
   c. width of blade.
   d. length of the band cutting blade.

13. **Band saws are designed to cut in a[n]**
   a. downward direction only.
   b. upward direction only.
   c. either downward or upward direction.
   d. neither downward or upward direction.

14. **The portion of a jointer that is always adjusted level with the top edge of the cutter is the**
   a. infeed table.
   b. outfeed table.
   c. fence.
   d. safety shield.

15. **The stationary power tool best suited to make a panel for a raised panel door is the**
   a. jointer.
   b. bench sander.
   c. drill press.
   d. shaper.

## WHAT'S WRONG WITH THIS PICTURE?

Carefully study **Figure 3-20** and think about what is wrong. Consider all possibilities.

### WRONG

**FIGURE 3-20** A worker is ripping a piece of wood on the table saw. He is standing directly behind the piece being cut. If the saw were to kick back, he would be directly in the path of the kicked material and likely sustain an injury. The worker should be standing to one side whenever possible. Also, the worker has both hands at the rear of the piece being cut. Instead, the left hand should be in front of the right hand, acting as a guide and holding the piece down on the surface of the table.

### RIGHT

**FIGURE 3-21** The camera angle makes it look as if the worker's left hand is dangerously close to the blade. However, it is well back and will remain somewhat stationary and guide the wood as the right hand pushes the piece. Note the push-stick on the table surface, which is used to push the wood through the final few inches.

# Wood and Wood Products

**OBJECTIVES** *After completing this unit, the student should be able to:*

- Define hardwood and softwood and give examples of some common kinds.
- State the grades and sizes of lumber.
- Calculate linear foot and compute square foot and board foot measure.
- Describe the composition, kinds, sizes, and several uses of: plywood, oriented strand board, particleboard, hardboard, medium-density fiberboard, and softboard.
- Describe the uses and sizes of: laminated veneer lumber, parallel strand lumber, laminated strand lumber, wood I-beams, and glue-laminated beams.

## GLOSSARY OF WOOD AND WOOD PRODUCT TERMS

**air dried** technique of removing water from lumber using natural wind currents

**annular rings** the rings seen when viewing a cross-section of a tree trunk; each ring constitutes one year of tree growth

**board** lumber usually less than 2 inches thick

**board foot** a measure of lumber volume that equals 1 foot square and 1 inch thick or any equivalent lumber volume. The letter M is used to represent 1000 board feet.

**cambium layer** a layer just inside the bark of a tree where new cells are formed

**coniferous** trees that are cone-bearing; also known as *evergreen* trees

**crown** a warp in a board along its length forming an arc.

**deciduous** trees that shed leaves each year

**dimension** a term used to define a measurement of an item; also used to refer to all 2x lumber used in framing

**dry kiln** large ovens used to remove water from lumber

**finger joint** a process in which shorter lengths are glued together using deep, thin V grooves, resulting in longer lengths

**header** framing members placed at right angles to joists, studs, and rafters to form and support openings

**heartwood** the wood in the inner part of a tree, usually darker and containing inactive cells

**lumber** general term for wood that is cut from a log to form boards, planks, and timbers

**medullary rays** bands of cells radiating from the cambium layer to the pith of a tree to transport nourishment toward the center

**millwork** any wood products that have been manufactured, such as moldings, doors, windows, and stairs, for use in building construction; sometimes called *joinery*

**on center (OC)** the distance from the center of one structural member to the center of the next one

**panel** a large sheet of building material that usually measures 4 × 8 feet

**pith** the small, soft core at the center of a tree

**plain-sawed** a method of sawing lumber that produces flat-grain where annular rings tend to be parallel to the width of the board

**quarter-sawed** a method of sawing lumber that produces a close grain pattern where the annular rings tend to be perpendicular to the width of the board

**sapwood** the outer part of a tree just beneath the bark containing active cells

**sawyer** a person whose job is to cut logs into lumber

**tempered** treated in a special way to make a material harder and stronger

**timbers** large pieces of lumber over 5 inches in thickness and width

The construction material most often associated with a carpenter is wood. Its flexibility and durability make it the first choice in many applications in home construction. Wood is easy to tool and work with, pleasing to look at and smell, and strong enough to last a long time.

Lumber is manufactured from the renewable resources of the forest. Trees are harvested and sawn into lumber in many shapes and sizes with a variety of characteristics. It is necessary to understand the nature of wood to get the best results from its use. With this knowledge, the carpenter can protect lumber from decay, select it for appropriate use, work it with proper tools, and join and fasten it to the best advantage.

The industry of engineered lumber has emerged to reduce manufacturing waste and improve the strength of wood products. These improvements to manufacturing efficiency have led to better stewardship of the natural resource. Products have been developed from reconstituted wood in the form of large panels, commonly called plywood, and other shapes for use as structural members of a building. They are collectively referred to as engineered lumber.

The types and styles of wood and wood products are abundant. To be able to use these materials to the best efficiency, the tradesperson must understand them. This understanding begins with an ability to measure and calculate quantities accurately.

# Wood

The carpenter works with wood more than any other material and must understand its characteristics in order to use it intelligently. Wood is a remarkable substance. It can be cut, shaped, or bent into just about any form. It is a fairly efficient insulating material. There are many kinds of wood that vary in strength, workability, elasticity, color, grain, and texture. It is important to keep these qualities in mind when selecting wood. For instance, baseball bats, diving boards, and tool handles are made from hickory and ash because of their greater ability to bend without breaking (elasticity). Oak and maple are used for floors because of their hardness and durability. Redwood, cedar, cypress, and teak are used in exterior situations because of their resistance to decay. Cherry, mahogany, and walnut are typically chosen for their beauty. With proper care, wood will last indefinitely. This material of beauty and warmth has thousands of uses. Wood is one of our greatest natural renewable resources.

Wood is made up of many hollow cells held together by a natural substance called *lignin*. The size, shape, and arrangement of these cells determine the strength, weight, and other properties of wood.

Tree growth takes place in the **cambium layer**, which is just inside the protective shield of the tree called the *bark*. The tree's roots absorb water that passes upward through the **sapwood** to the leaves, where it is combined with carbon dioxide from the air. Sunlight causes these materials to change into food, which is then carried down and distributed toward the center of the trunk through the **medullary rays**.

As the tree grows outward from the **pith** (center), the inner cells become inactive and turn into **heartwood**. Heartwood is the central part of the tree and usually is darker in color and more durable than sapwood. The heartwood of cedar, cypress, and redwood, for instance, is extremely resistant to decay and is used extensively for outdoor furniture, patios, and exterior siding. Used for the same purposes, sapwood decays more quickly.

Each growing season, the tree adds new layers to its trunk. Wood grows rapidly in the spring; it is rather porous and light in color. In summer, tree growth is slower; the wood is denser and darker, forming distinct rings. Because these rings are formed each year, they are called **annular rings** (Fig. 4-1). By counting the dark rings, the age of a tree can be determined. By studying the width of the rings, periods of abundant rainfall and sunshine or periods of slow growth can be discerned. Some trees, like the Douglas fir, grow rapidly to great heights and have very wide and pronounced annular rings. Mahogany, which grows in a tropical climate where the weather is more constant, has annular rings that are not so contrasting and sometimes are hardly visible.

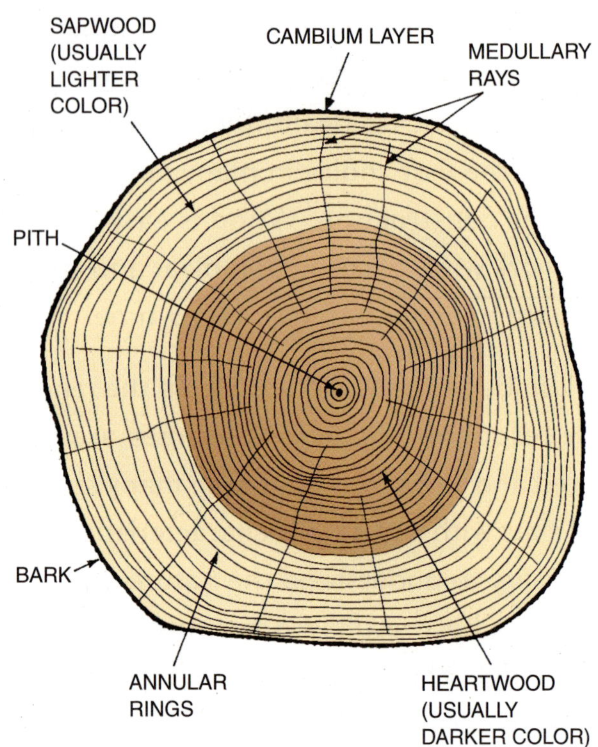


*Courtesy of Western Wood Products Association.*

**FIGURE 4-1** A cross section of a tree showing its structure.

## HARDWOODS AND SOFTWOODS

Woods are classified as either hardwood or softwood. There are different methods of classifying these woods. The most common method of classifying wood is by its source. Hardwood comes from **deciduous** trees that shed their leaves each year. Softwood is cut from **coniferous**, or cone-bearing, trees, commonly known as *evergreens* (Fig. 4-2). In this method of classifying wood, some of the softwoods may actually be harder than the hardwoods. For instance, fir, a softwood, is harder and stronger

**FIGURE 4-2** Hardwood is from broad-leaf trees, softwood from cone-bearing trees.

than basswood, a hardwood. Some common hardwoods are ash, birch, cherry, hickory, maple, mahogany, oak, and walnut. Some common softwoods are pine, fir, hemlock, spruce, cedar, cypress, and redwood.

The best way to learn the different kinds of wood is by working with them. Each time you handle a piece of wood, examine it. Look at the color and the grain; feel if it is heavy or light, if it is soft or hard; and smell it for a characteristic odor. After studying the characteristics of the wood, ask or otherwise find out the kind of wood you are holding, and remember it. In this manner, after a period of time, those kinds of wood that are used regularly on the job can be identified easily.

Courtesy of American Wood Dryers.

**FIGURE 4-4** Drying lumber in a kiln.

## MANUFACTURE OF LUMBER

When logs arrive at the sawmill, the bark is removed first. Then a huge saw slices the log into large planks, which are passed through another series of saws. These saws slice, edge, and trim the planks into various dimensions, and the pieces become **lumber**.

Once trimmed of all uneven edges, the lumber is stacked according to size and grade and taken outdoors where *sticking* takes place. Sticking is the process of restacking the lumber on small cross-sticks that allow air to circulate between the pieces (Fig. 4-3).

This air-seasoning process may take six months to two years due to the large amount of water found in lumber. For this reason, lumber is often placed in **dry kilns** to speed up the drying processing (Fig. 4-4). Once dry, the rough lumber is surfaced to standard sizes and shipped.

**FIGURE 4-3** Lumber is stickered and stacked to allow air drying.

The long, narrow surface of a piece of lumber is called its *edge,* the long, wide surface is termed its *side,* and its extremities are called *ends.* The distance across the edge is called its *thickness,* across its side is called its *width,* and from end to end is its *length* (Fig. 4-5).

### Plain-Sawed Lumber

A common way of cutting lumber is called the **plain-sawed** method. This method produces a distinctive grain pattern on the wide surface (Fig. 4-6). This method of sawing is the least expensive and produces greater widths. However, plain-sawed lumber shrinks more during drying and warps easily. Plain-sawed lumber is sometimes called *slash-sawed* lumber.

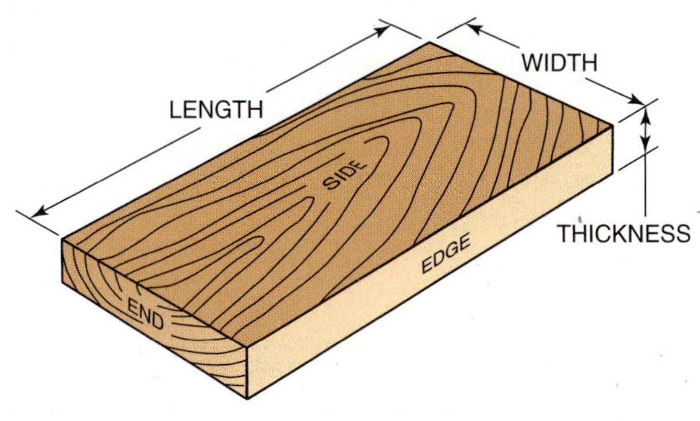


**FIGURE 4-5** Lumber surfaces are distinguished by specific names.

(A)

(B)

**FIGURE 4-6** (A) Typical sawing pattern for plain-sawed lumber. (B) Surfaces of plain-sawed lumber.

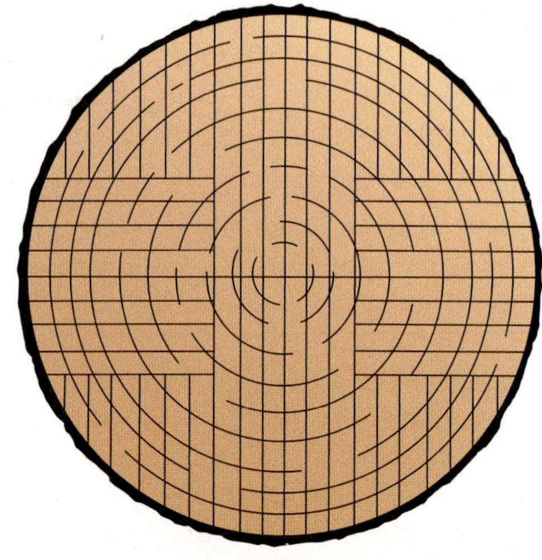

(A)

(B)

**FIGURE 4-7** (A) Typical sawing pattern for quarter-sawed lumber. (B) Surfaces of quarter-sawed lumber.

### Quarter-Sawed Lumber

Another less often used method of cutting the log, called quarter-sawing, produces pieces in which the annular rings are at or almost at right angles to the wide surface. Quarter-sawed lumber has less of a tendency to warp and shrinks less and more evenly when it dries. A distinctive and desirable grain pattern is produced in some wood (Fig. 4-7), and it tends to be more durable.

### Combination Sawing

Most logs are cut into a combination of plain-sawed and quarter-sawed lumber. With computers and laser-guided equipment, the sawyer determines how to cut the log with as little waste as possible in the shortest amount of time to get the desired amount and kinds of lumber (Fig. 4-8).

## Green Tip

Manufacturing wood products from logs uses less energy than manufacturing metal or masonry products.

### Moisture Content and Shrinkage

When a tree is first cut down, it contains a great amount of water. Lumber, when first cut from the log, is called *green lumber* and is very heavy because most of its weight is water. A piece 2 inches thick, 6 inches wide, and 10 feet long may contain as much as 4¼ gallons of water weighing about 35 pounds (Fig. 4-9).

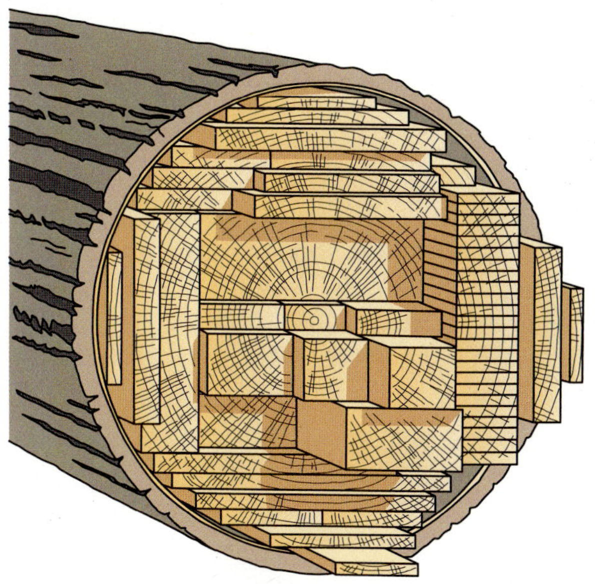

Courtesy of Western Wood Products Association.

**FIGURE 4-8** Combination-sawed lumber.

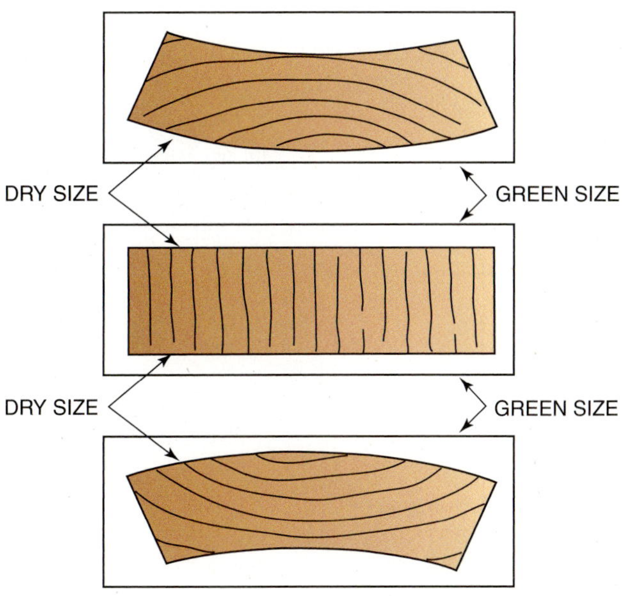


**FIGURE 4-10** Lumber shrinks in the direction of the annular rings.

Green lumber should not be used in construction. As green lumber dries, it shrinks considerably and unequally as the large amount of water leaves it. When it shrinks, it usually warps, depending on the way it was cut from the log (Fig. 4-10). The use of green lumber in construction results in cracked ceilings and walls, squeaking floors, sticking doors, and many other problems caused by shrinking and warping of the lumber as it dries. Therefore, lumber must be dried to the equilibrium moisture content (MC) before it can be surfaced and used.

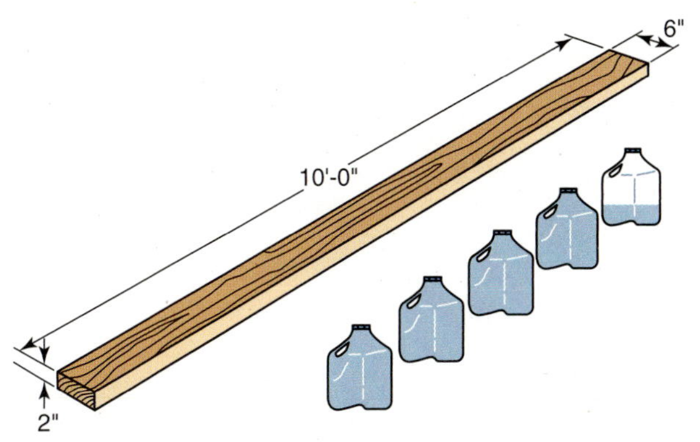


**FIGURE 4-9** Green lumber contains a large amount of water.

At this point, lumber shrinks and swells only a little with changes in the moisture content of the air. Realizing that lumber undergoes certain changes when moisture is absorbed or lost, the experienced carpenter uses techniques to deal with this characteristic of wood (Fig. 4-11).

## Lumber Storage

Lumber should be delivered to the job site so materials are accessible in the proper sequence; that is, those that are to be used first are on the top and those to be used last are on the bottom.

Lumber stored at the job site should be adequately protected from moisture and other hazards. A common practice that must be avoided is placing unprotected lumber directly on the ground. Use short lengths of lumber running at right angles to the length of the pile and spaced close enough to keep the pile from sagging and coming into contact with the ground. The base on which the lumber is to be placed should be fairly level to keep the pile from falling over.

Protect the lumber with a tarp or other type of cover. Leave enough room at the bottom and top of the pile for circulation of air. Keep the piles in good order. Lumber spread out in a disorderly fashion can cause accidents and subject the lumber to stresses that may cause warping.

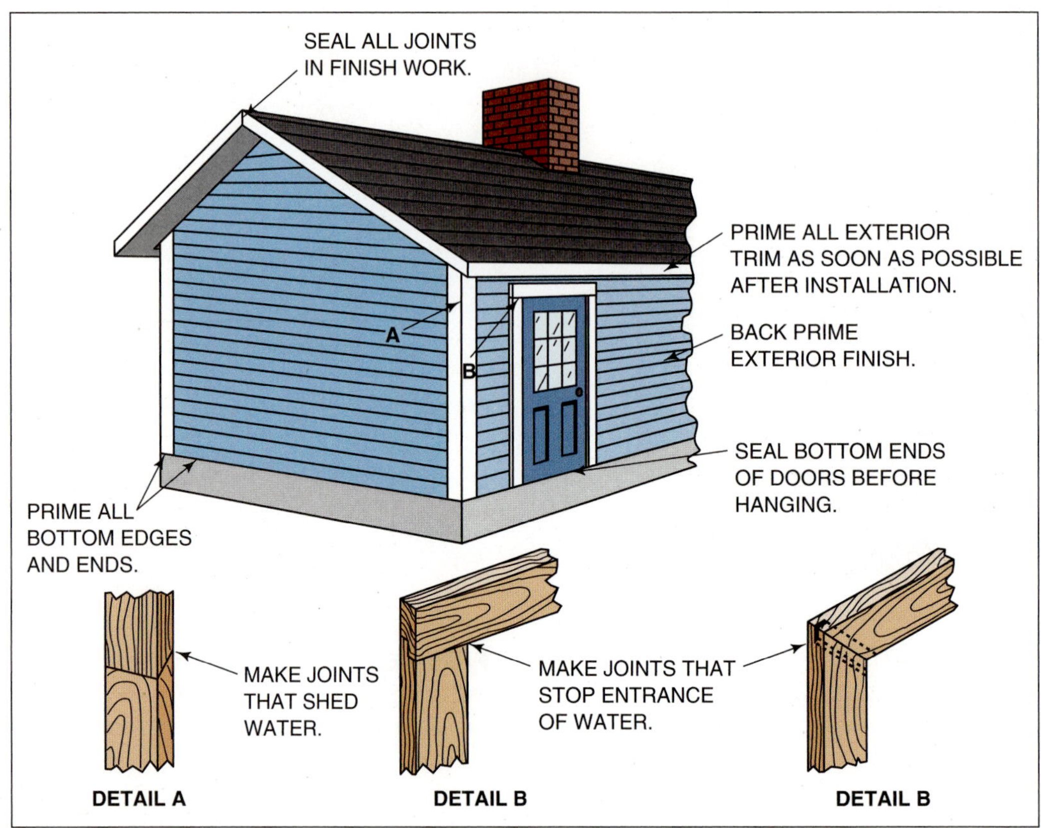


**FIGURE 4-11** Techniques to prevent water from getting in behind the wood surface.

> **CAUTION**
>
> **CAUTION:** Lumber piles can be the source of serious accidents. Lumber improperly stacked is not only unsightly, but it can topple over without warning. A properly stacked pile of lumber is pleasing to look at and safe. Also, the quantity of material in the pile is easier to determine.

## Lumber Defects

A defect in lumber is any fault that detracts from its appearance, function, or strength. One type of defect is called a *warp*. Warps are caused by, among other things, drying lumber too fast, careless handling and storage, or surfacing the lumber before it is thoroughly dry. Warps are classified as *crooks, bows, cups,* and *twists* (Fig. 4-12).

A mild type of warp is called a **crown**. A crown is an arc along the length of a board. Lumber used in framing may have slight to moderate crowns. This usually has minimal effect because framing members are assembled with crowns projecting in

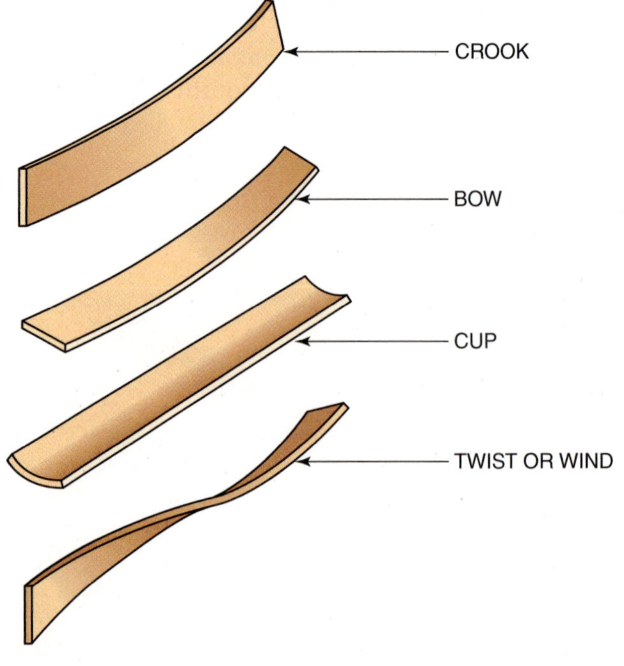


**FIGURE 4-12** Kinds of warps.

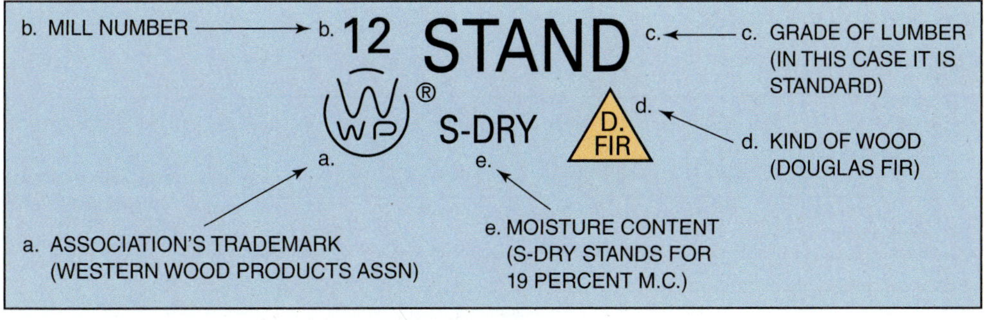


*Courtesy of Western Wood Products Association.*

**FIGURE 4-13** Typical softwood lumber grade stamp.

the same direction. Lumber used for finish work should be straight.

*Knots* are cross sections of branches in a trunk of a tree. Knot defects affect the strength of a board in varying degrees. If they are small and tight the board remains strong, whereas large loose knots weaken a board significantly. *Pitch pockets* are small cavities that hold pitch, which sometimes oozes out. A *wane* is bark on the edge of lumber or the surface from which the bark has fallen.

### Lumber Grades and Sizes

Wood products associations establish sizes and grades of lumber. A grade stamp of the association placed on the lumber is an assurance that lumber grade standards have been met. Member mills use the association grade stamp to indicate strict quality control. A typical grade stamp is shown in Figure 4-13.

The Western Wood Products Association (WWPA) and the Northeast Lumber Manufacturers Association (NELMA) grade lumber in three categories: **boards** (under 2 inches thick), **dimension** (2 to 4 inches thick), and **timbers** (5 inches and thicker). The board group is divided into boards, sheathing, and form lumber. The dimension group is divided into light framing, studs, structural light framing, and structural joists and planks. Timbers are divided into beams and stringers (Fig. 4-14). The National Hardwood Lumber Association establishes hardwood grades. Select and firsts and seconds *(FAS)* are the best grades of hardwood. The length and width of the board determines whether it is select or FAS. The less restrictive grades in decreasing order are #1 common, #2 common, #3A, and #3B.

### Lumber Sizes

Rough lumber that comes directly from the sawmill is close in size to what it is called, *nominal size*. There are slight variations to nominal size because of the heavy machinery used to cut the log into lumber. When rough lumber is planed, it is reduced in thickness and width to standard and uniform sizes. Its nominal size does not change even though the actual size does. Therefore, when *dressed* (surfaced), although a piece may be called a 2 × 4, its actual size is 1½ inches (38 mm) by 3½ inches (89 mm). The same applies to all surfaced lumber; the nominal size (what it is called) and the actual size are not the same (Fig. 4-15). Hardwood lumber is usually purchased in the rough and straightened, smoothed, and sized as needed by the carpenter or wood worker.

## BOARD MEASURE

Softwood lumber is usually purchased by specifying the number of pieces—thickness (in inches) × width (inches) × length (feet), i.e., 35 2" × 6" × 16'—in addition to the grade. Often, when no particular lengths are required, the thickness, width, and total number of linear feet (length in feet) are ordered. The length of the pieces then may vary and are called *random lengths*. Another method of purchasing softwood lumber is by specifying the thickness, width, and total number of **board feet**. Lumber purchased in this manner may also contain random lengths.

Softwood, in large quantities, and hardwood lumber are priced and sold by the board foot. A board foot is a measure of the volume of lumber. It is equivalent to a piece 1 inch thick, 12 inches wide, and 1 foot long. A piece of lumber 1 inch thick and 6 inches wide must be 2 feet long to equal 1 board foot. A board 2 inches thick has twice as many board feet as a board 1 inch thick of the same width and length (Fig. 4-16). Use the nominal dimensions in the calculations, not the actual dimensions.

## Grade Selector Charts

### Boards

| | | |
|---|---|---|
| APPEARANCE GRADES | SELECTS | B & BETTER (IWP—SUPREME)*<br>C SELECT (IWP—CHOICE)<br>D SELECT (IWP—QUALITY) |
| | FINISH | SUPERIOR<br>PRIME<br>E |
| | PANELING | CLEAR (ANY SELECT OR FINISH GRADE)<br>NO. 2 COMMON SELECTED FOR KNOTTY PANELING<br>NO. 3 COMMON SELECTED FOR KNOTTY PANELING |
| | SIDING (BEVEL BUNGALOW) | SUPERIOR<br>PRIME |
| BOARDS SHEATHING & FORM LUMBER | | NO. 1 COMMON (IWP—COLONIAL)<br>NO. 2 COMMON (IWP—STERLING)<br>NO. 3 COMMON (IWP—STANDARD)<br>NO. 4 COMMON (IWP—UTILITY)<br>NO. 5 COMMON (IWP—INDUSTRIAL) |
| | | **ALTERNATE BOARD GRADES**<br>SELECT MERCHANTABLE<br>CONSTRUCTION<br>STANDARD<br>UTILITY<br>ECONOMY |

SPECIFICATION CHECK LIST

- ☐ Grades listed in order of quality.
- ☐ Include all species suited to project.
- ☐ Specify lowest grade that will satisfy job requirement.
- ☐ Specify surface texture desired.
- ☐ Specify moisture content suited to project.
- ☐ Specify (WWP) grade stamp. For finish and exposed pieces, specify stamp on back or ends.

| Western Red Cedar | |
|---|---|
| FINISH<br>PANELING<br>AND CEILING | CLEAR HEART<br>A<br>B |
| BEVEL<br>SIDING | CLEAR HEART—V.G. HEART<br>A—BEVEL SIDING<br>B—BEVEL SIDING<br>C—BEVEL SIDING |

*Idaho White Pine carries its own comparable grade designations.

### Dimension/All Species 2" to 4" thick (also applies to finger-jointed stock)

| | | |
|---|---|---|
| **LIGHT FRAMING**<br>2" to 4" Thick<br>2" to 4" Wide | CONSTRUCTION<br>STANDARD<br>UTILITY | This category for use where high strength values are NOT required; such as studs, plates, sills, cripples, blocking, etc. |
| **STUDS**<br>2" to 4" Thick<br>2" and Wider | STUD | An optional all-purpose grade. Characteristics affecting strength and stiffness values are limited so that the "Stud" grade is suitable for vertical framing members, including load bearing walls. |
| **STRUCTURAL LIGHT FRAMING**<br>2" to 4" Thick<br>2" to 4" Wide | SELECT STRUCTURAL<br>#$1_T$Btr.*<br>NO. 1<br>NO. 2<br>NO. 3 | These grades are designed to fit those engineering applications where higher bending strength ratios are needed in light framing sizes. Typical uses would be for trusses, concrete pier wall forms, etc. |
| **STRUCTURAL JOISTS & PLANKS**<br>2" to 4" Thick<br>5" and Wider | SELECT STRUCTURAL<br>#$1_T$Btr.*<br>NO. 1<br>NO. 2<br>NO. 3 | These grades are designed especially to fit in engineering applications for lumber five inches and wider, such as joists, rafters and general framing uses. |

*Douglas fir/Larch$_T$ Hem-Fir only (DF–L)

### Timbers 5" and thicker

| | | | |
|---|---|---|---|
| **BEAMS & STRINGERS**<br>5" and thicker<br>Width more than 2" greater than thickness | SELECT STRUCTURAL<br>NO. 1<br>NO. 2**<br>NO. 3** | **POSTS & TIMBERS**<br>5" x 5" and larger<br>Width not more than 2" greater than thickness | SELECT STRUCTURAL<br>NO. 1<br>NO. 2**<br>NO. 3** |

**Design values are not assigned.

*Courtesy of Western Wood Products Association.*

**FIGURE 4-14** Softwood lumber grades.

| BOARD AND DIMENSION LUMBER SIZES (IN INCHES) | | | | | |
|---|---|---|---|---|---|
| NOMINAL | SURFACED SIZE | NOMINAL | SURFACED SIZE | NOMINAL | SURFACED SIZE |
| 1X2 | 3/4 X 1 1/2 | 2X2 | 1 1/2 X 1 1/2 | | |
| 1X3 | 3/4 X 2 1/2 | 2X3 | 1 1/2 X 2 1/2 | | |
| 1X4 | 3/4 X 3 1/2 | 2X4 | 1 1/2 X 3 1/2 | 4X4 | 3 1/2 X 3 1/2 |
| 1X6 | 3/4 X 5 1/2 | 2X6 | 1 1/2 X 5 1/2 | 4X6 | 3 1/2 X 5 1/2 |
| 1X8 | 3/4 X 7 1/4 | 2X8 | 1 1/2 X 7 1/4 | 4X8 | 3 1/2 X 7 1/4 |
| 1X10 | 3/4 X 9 1/4 | 2X10 | 1 1/2 X 9 1/4 | 4X10 | 3 1/2 X 9 1/4 |
| 1X12 | 3/4 X 11 1/4 | 2X12 | 1 1/2 X 11 1/4 | 4X12 | 3 1/2 X 11 1/4 |

**FIGURE 4-15** Softwood lumber sizes.

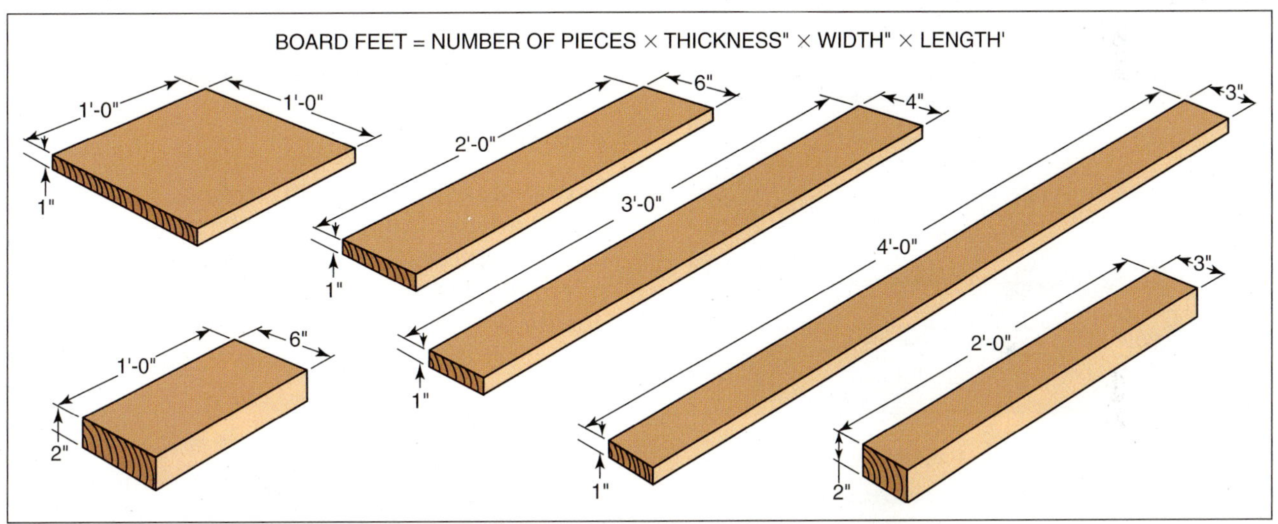


**FIGURE 4-16** Each piece contains one board foot.

To calculate the number of board feet, use the formula: number of pieces × thickness in inches × width in inches × length in feet ÷ 12 = number of board feet. For example: The board feet in sixteen 2 × 4s, each of which is 8 feet long, is 16 pieces × 2 inches × 4 inches × 8 feet ÷ 12 = 85⅓ board feet.

# Engineered Panels

The term *engineered panels* refers to man-made products in the form of large reconstituted wood sheets, sometimes called **panels** or boards. The panels are widely used in the construction industry. It is important to know the kinds and uses of various engineered panels in order to use them to the best advantage.

## PLYWOOD—APA-RATED PANELS

Many sawmills belong to associations that inspect, test, and allow mills to stamp the product to certify that it conforms to government and industrial standards. The grade stamp assures the consumer that the product has met the rigid quality and performance requirements of the association (Fig. 4-17).

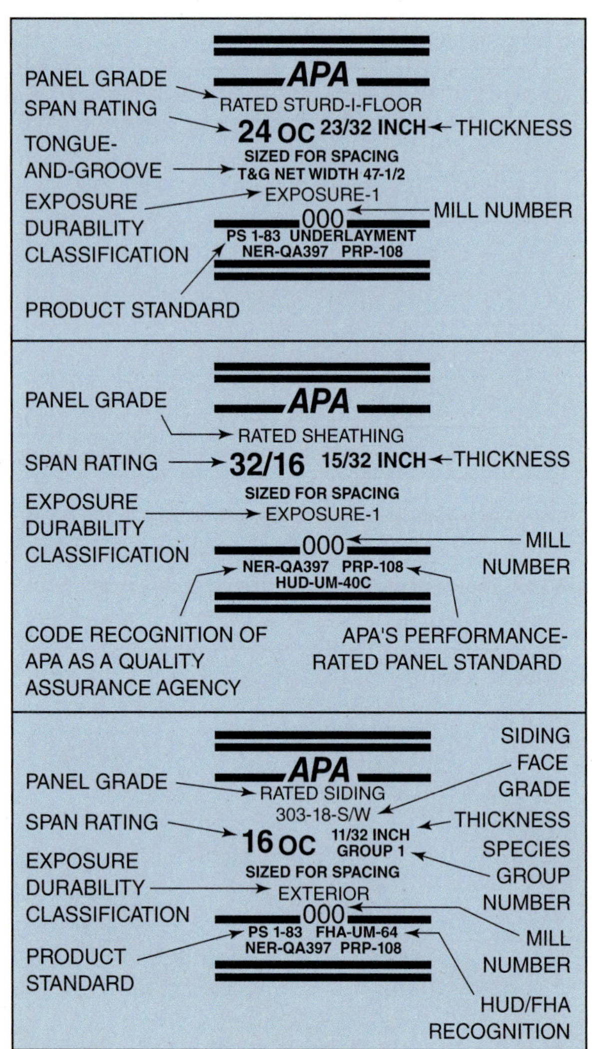


Courtesy of APA—The Engineered Wood Association.

**FIGURE 4-17** The grade stamp is assurance of a high-quality, performance-rated panel.

One of the most extensively used engineered panels is plywood (Fig. 4-18). Plywood is a sandwich of wood. Most plywood panels are made up of sheets of veneer (thin pieces) called *plies*. These plies, arranged in layers, are bonded under pressure with glue to form a very strong panel. Because of its construction, plywood remains stable with changes of humidity and resists shrinking and swelling.

## Green Tip

Plywood manufacturing process uses the best quality veneer logs for outer surfaces while lesser quality logs are used for the inner cores. This efficiency results in the best appearance while maintaining superior strength.

Courtesy of APA—The Engineered Wood Association.

**FIGURE 4-18** APA performance-rated panels.

### Veneer Grades

In declining order, the letters *A, B, C plugged, C,* and *D* are used to indicate the appearance quality of panel veneers. Two letters are found in the grade stamp of veneered panels. One letter indicates the quality of one face, while the other letter indicates the quality of the opposite face.

### Strength Grades

Softwood veneers are made of many different kinds of wood. These woods are classified in groups according to their strength. Group 1 is the strongest. Douglas fir and southern pine are in Group 1 and are used to make most of the softwood plywood. The group number is also shown in the grade stamp.

### Oriented Strand Board

*Oriented strand board (OSB)* is a nonveneered performance-rated structural panel composed of small oriented (lined up) strand-like wood pieces arranged in three to five layers with each layer at right angles to the other (Fig. 4-19). The cross-lamination of the layers achieves the same advantages of strength and stability as in plywood.

### Performance Ratings

A performance-rated panel meets the requirements of the panel's end use. The four end uses for which panels are rated are single-layer flooring; exterior

Courtesy of Louisiana Pacific Corporation.

**FIGURE 4-19** Oriented strand board being used for wall sheathing.

siding; sheathing used for roofs, floors, and walls; and special high-strength applications. Names given to designate end uses are *APA-Rated Sturd-I-Floor, APA-Rated Siding, APA-Rated Sheathing,* and *Structural I* (Fig. 4-20). Panels are tested to meet standards in areas of resistance to moisture, strength, and stability.

### Exposure Durability

APA performance-rated panels are also manufactured in three exposure durability classifications: *Exterior, Exposure 1,* and *Exposure 2.* Panels marked Exterior are designed for permanent exposure to the weather or moisture. Exposure 1 panels are intended for use when long delays in construction may cause the panels to be exposed to the weather before being protected.

Panels marked Exposure 2 are designed for use when only moderate delays in providing protection from the weather are expected. The exposure durability of a panel may be found in the grade stamp.

### Span Ratings

The span rating in the grade stamp on APA-rated sheathing appears as two numbers separated by a slash, such as 32/16 or 48/24. The left number denotes the maximum recommended spacing of supports when the panel is used for roof or wall sheathing.

The right number indicates the maximum recommended spacing of supports when the panel is used for subflooring. A panel marked 32/16, for example, may be used for roof sheathing over rafters not more than 32 inches **on center** or for subflooring over joists not more than 16 inches on center. The span ratings on APA-Rated Sturd-I-Floor and APA-Rated Siding appear as a single number.

# Nonstructural Panels

All the rated-panel products discussed previously may be used for nonstructural applications. In addition, other plywood products, grade stamped by the American Plywood Association, are available for nonstructural use. They include sanded and touch-sanded plywood panels and specialty hardwood plywood panels.

## HARDWOOD PLYWOOD

Hardwood plywood is available with hardwood face veneers, of which the most popular are birch, oak, and lauan. Beautifully grained hardwoods are sometimes matched in a number of ways to produce interesting face designs. Hardwood plywood is used in the interior of buildings for such things as wall paneling, built-in cabinets, and fixtures.

## PARTICLEBOARD

Particleboard is a reconstituted wood panel made of wood flakes, chips, sawdust, and planer shavings (Fig. 4-21). These wood particles are mixed with an adhesive, formed into a mat, and pressed into sheet form. The kind, size, and arrangement of the wood particles determine the quality of the board.

The quality of particleboard is indicated by its density (hardness), which ranges from 28 to 55 pounds per cubic foot. Nonstructural particleboard is used in the construction industry for the construction of kitchen cabinets and countertops, and for the core of veneer doors and similar panels.

## FIBERBOARDS

Fiberboards are manufactured as *high-density, medium-density,* and *low-density* boards.

APA RATED SHEATHING
Typical Trademark

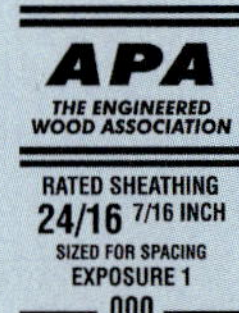


Specially designed for subflooring and wall and roof sheathing. Also good for a broad range of other construction and industrial applications. Can be manufactured as OSB, plywood, or a composite panel. BOND CLASSIFICATIONS: Exterior, Exposure 1, Exposure 2. COMMON THICKNESSES: 5/16, 3/8, 7/16, 15/32, 1/2, 19/32, 5/8, 23/32, 3/4.

APA STRUCTURAL I RATED SHEATHING(c)
Typical Trademark

APA
THE ENGINEERED WOOD ASSOCIATION
RATED SHEATHING
32/16 15/32 INCH
SIZED FOR SPACING
EXPOSURE 1
000
STRUCTURAL I RATED
DIAPHRAGMS-SHEAR WALLS
PANELIZED ROOFS
PRP-108 HUD-UM-40

Unsanded grade for use where shear and cross-panel strength properties are of maximum importance, such as panelized roofs and diaphragms. Can be manufactured as OSB, plywood, or a composite panel. BOND CLASSIFICATIONS: Exterior, Exposure 1. COMMON THICKNESSES: 5/16, 3/8, 7/16, 15/32, 1/2, 19/32, 5/8, 23/32, 3/4.

APA RATED STURD-I-FLOOR
Typical Trademark

Specially designed as combination subfloor-underlayment. Provides smooth surface for application of carpet and pad and possesses high concentrated and impact load resistance. Can be manufactured as OSB, plywood, or a composite panel. Available square edge or tongue-and-groove. BOND CLASSIFICATIONS: Exterior, Exposure 1, Exposure 2. COMMON THICKNESSES: 19/32, 5/8, 23/32, 3/4, 1, 1-1/8.

APA RATED SIDING
Typical Trademark

For exterior siding, fencing, etc. Can be manufactured as plywood, as a composite panel or as an overlaid OSB. Both panel and lap siding available. Special surface treatment such as V-groove, channel groove, deep groove (such as APA Texture 1-11), brushed, rough sawn and overlaid (MDO) with smooth- or texture-embossed face. Span Rating (stud spacing for siding qualified for APA Sturd-I-Wall applications) and face grade classification (for veneer-faced siding) indicated in trademark. BOND CLASSIFICATION: Exterior. COMMON THICKNESSES: 11/32, 3/8, 7/16, 15/32, 1/2, 19/32, 5/8.

(a) Specific grades, thicknesses and bond classifications may be in limited supply in some areas. Check with your supplier before specifying.

(b) Specify Performance Rated Panels by thickness and Span Rating. Span Ratings are based on panel strength and stiffness. Since these properties are a function of panel composition and configuration as well as thickness, the same Span Rating may appear on panels of different thickness. Conversely, panels of the same thickness may be marked with different Span Ratings.

(c) All plies in Structural I plywood panels are special improved grades and panels marked PS 1 are limited to Group 1 species. Other panels marked Structural I Rated qualify through special performance testing.

*Courtesy of APA—The Engineered Wood Association.*

**FIGURE 4-20** Guide to APA performance-rated panels.

*Courtesy of Weyerhaeuser Company.*

**FIGURE 4-21** Particleboard is made from wood flakes, shavings, resins, and waxes.

*Hardboards* are high-density fiberboards. They are sometimes known by the trademark *Masonite*. Some panels are **tempered**, or coated with oil and baked to increase hardness, strength, and water resistance. The most popular thicknesses of hardboard range from 1/8 to 3/8 inch. The most popular sheet size is 4 feet by 8 feet, although sheets may be ordered in practically any size.

Hardboard may be used inside or outside. It is widely used for exterior siding and interior wall paneling. It is also used extensively for cabinet backs and drawer bottoms. It can be used wherever a dense, hard panel is required.

*Medium-density fiberboard (MDF)* is manufactured in a manner similar to that used to make hardboard except that the fibers are not pressed as tightly together. It is available in thicknesses ranging from 3/16 to 1½ inches and comes in widths of 4 feet and 5 feet. Lengths run from 6 to 18 feet. MDF may be used for case goods, drawer parts, kitchen cabinets, cabinet doors, signs, and some interior wall finish.

*Low-density fiberboard* is called *softboard*. Softboard is light and contains many tiny air spaces because the particles are not compressed tightly. The most common thicknesses range from ½ to 1 inch. The most common sheet size is 4 feet by 8 feet, although many sizes are available. Because of their lightness, softboard panels are used primarily for insulating or sound control purposes. They are used extensively as decorative panels in suspended ceilings and as ceiling tiles (Fig. 4-22).

Courtesy of Armstrong World Industries.

**FIGURE 4-22** Softboards are used extensively for decorative ceiling panels.

### OTHER

Many more products are used in the construction industry, in addition to those already mentioned. *Sweet's Architectural File* is an excellent resource to become better acquainted with the thousands of building material products on the market. This reference is well known by architects, contractors, and builders, and is revised and published annually. *Sweet's* may be found online at http://www.sweets.com.

## ENGINEERED LUMBER

Engineered lumber products are reconstituted wood products and assemblies designed to replace traditional structural lumber (Fig. 4-23). Engineered lumber products consume less wood and can be made from smaller trees. Traditional lumber processes typically convert 40 percent of a log to structural solid lumber. Engineered lumber processes convert up to 75 percent of a log into structural lumber. In addition, the manufacturing processes of engineered lumber consume far less energy than those of solid lumber. Engineered lumber products have greater strength and consequently can span greater distances.

**CAUTION**

**CAUTION:** Engineered lumber can be slippery compared to standard lumber. Be careful when working on or around these products.

Courtesy of Weyerhaeuser Company.

**FIGURE 4-23** There are several types of engineered lumber.

**FIGURE 4-24** Close-up view of LVL.

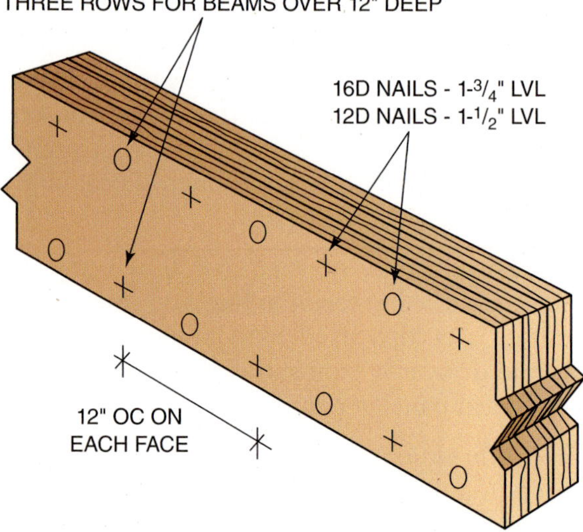


*Courtesy of Louisiana Pacific Corporation.*

**FIGURE 4-25** Recommended nailing pattern for fastening LVL beams together.

## LAMINATED VENEER LUMBER

*Laminated veneer lumber,* commonly called LVL, is one of several types of engineered lumber products (Fig. 4-24). Laminated veneer lumber is manufactured up to 3½ inches thick, 18 inches wide, and 80 feet long. The usual thicknesses are 1½ inches and 1¾ inches. LVL widths are usually 9¼, 11¼, 11⅞, 14, 16, and 18 inches. LVL beams may be fastened together to make a thicker and stronger beam (Fig. 4-25).

Laminated veneer lumber is intended for use as high-strength, load-carrying beams to support the weight of construction over window and door openings, and in floor and roof systems of residential and light commercial wood frame construction (Fig. 4-26). It can be cut with regular tools and requires no special fasteners.

## PARALLEL STRAND LUMBER

*Parallel strand lumber (PSL),* commonly known by its brand name Parallam (Fig. 4-27), is designed to replace large dimension lumber (beams, planks, and posts). PSL comes in many thicknesses and widths and is manufactured up to 66 feet long. PSL is available in square and rectangular shapes for use as posts and beams. Solid 3½ inch thicknesses are

*Courtesy of Weyerhaeuser Company.*

**FIGURE 4-26** LVL is designed to be used for load-carrying beams.

**FIGURE 4-27** Close-up view of PSL.

compatible with 2 × 4 wall framing. Parallel strand lumber can be used wherever there is a need for a large beam or post.

**Green Tip**

Engineered lumber maximizes the efficiency of wood resources by reducing manufacturing waste. Sawn lumber has about 50% waste while OSB production, for example, has about 8% waste.

The differences between PSL and solid lumber are many. Solid lumber beams may have defects, like knots, checks, and shakes, which weaken them, but PSL is consistent in strength throughout its length. PSL is readily available in longer lengths and its surfaces are sanded smooth, eliminating the need to cover them by boxing in the beams.

## LAMINATED STRAND LUMBER

*Laminated strand lumber (LSL)* is commonly known by its brand name TimberStrand (Fig. 4-28). At present, LSL is being manufactured from surplus, overmature aspen trees that usually are not large, strong, or straight enough to produce ordinary wood products. It is used for a wide range of **millwork**, such as doors, windows, and virtually any product that requires high-grade lumber. It is also used for truck decks, manufactured housing, and some structural lumber, such as window and door **headers**.

## WOOD I-BEAMS

Wood I-beams are engineered wood assemblies that use an efficient "I" shape, common in steel beams, which gives them tremendous strength in relation to their size and weight (Fig. 4-29). Consequently, they are able to carry heavy loads over long distances while using considerably less wood than solid lumber of a size necessary to carry the same load over the same span.

(A)

(B)

**FIGURE 4-28** (A) Close-up view of LSL. (B) Laminated strand lumber is used for non-load-bearing situations such as rim joists.

**FIGURE 4-29** Wood I-beams are available in many sizes.

The flanges of the beam may be made of laminated veneer lumber or specially selected **finger-jointed** solid wood lumber (Fig. 4-30). The web of the beam may be made of plywood, laminated veneer lumber, or oriented strand board.

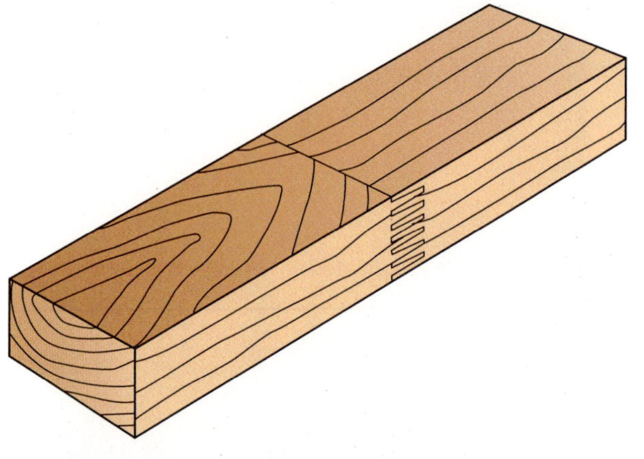

**FIGURE 4-30** Finger-joints are used to join the ends of short pieces of lumber to make a longer piece.

Wood I-beams may have webs of various thicknesses, and flanges may vary in thickness and width, depending on intended end use and the manufacturer. Beam depths are available from 9¼ to 30 inches (Fig. 4-31). Wood I-beams are available up to 80 feet long.

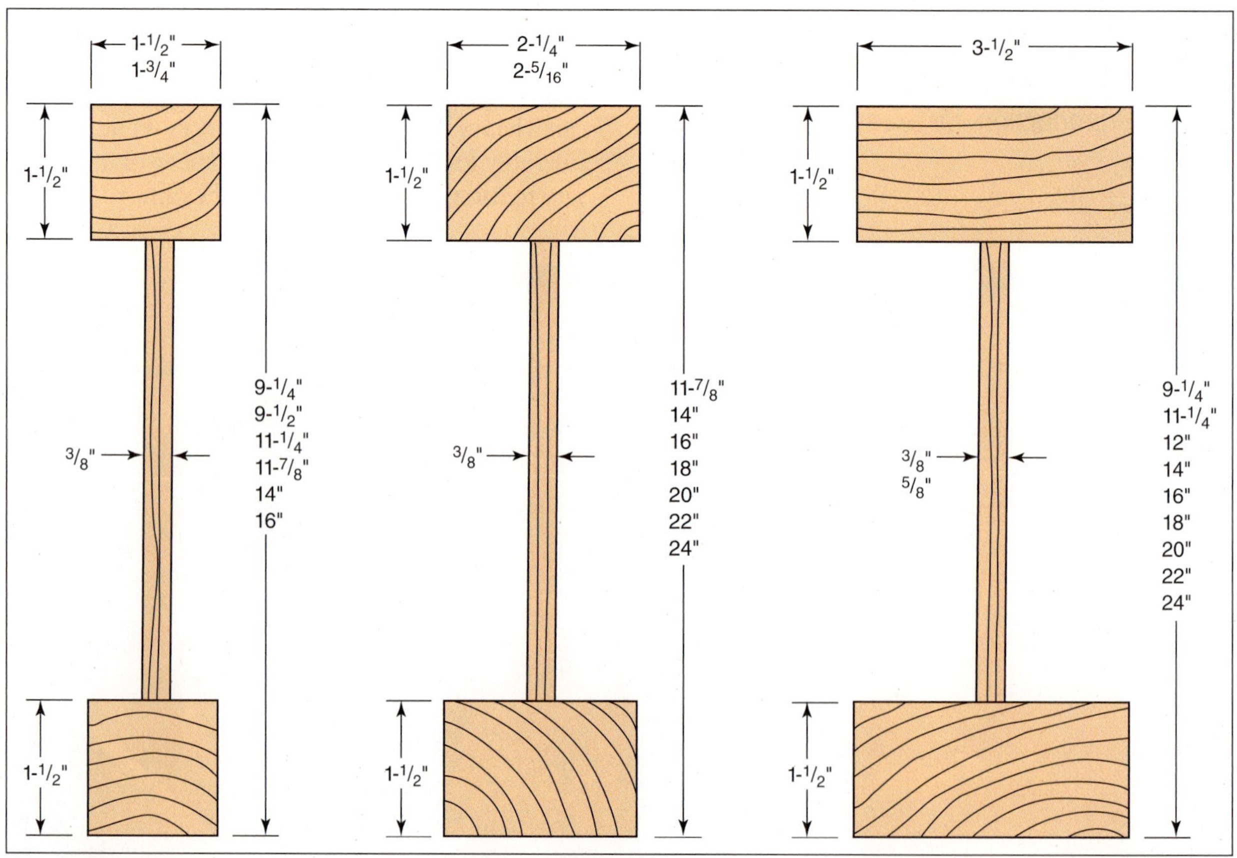


**FIGURE 4-31** Wood I-beam sizes. Sizes vary with the manufacturer.

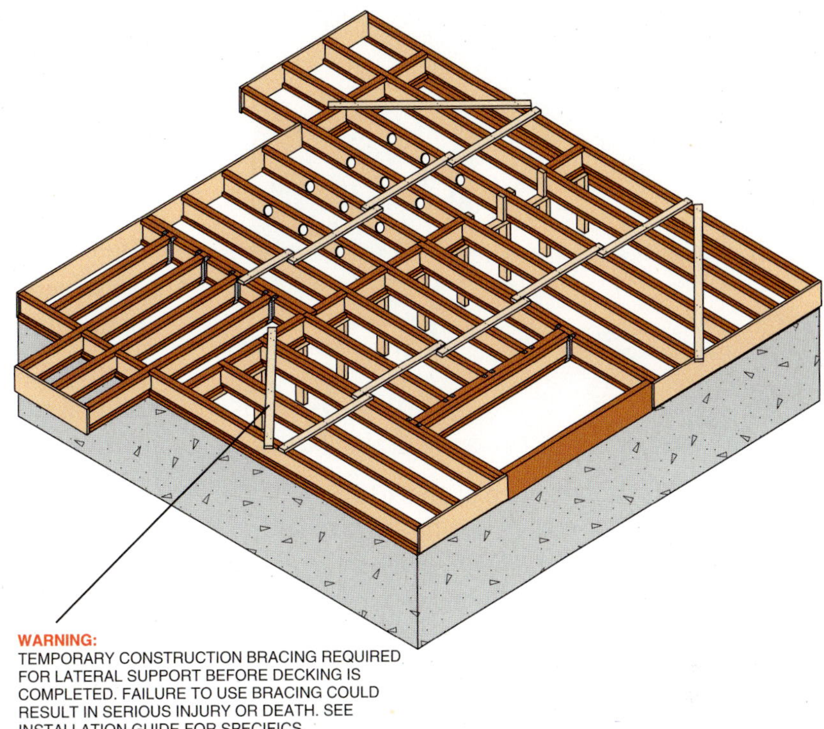


*Courtesy of Louisiana Pacific Corporation.*

**FIGURE 4-32** Wood I-beams are used as roof rafters, as well as for floor joists and window and door headers.

Wood I-beams are intended for use in residential and commercial construction as floor joists, roof rafters, and headers for window, entrance door, and garage door openings (Fig. 4-32). Window and door headers are beams that support the load above wall openings.

## GLUE-LAMINATED LUMBER

Glue-laminated lumber, commonly called *glulam,* is constructed of solid lumber glued together, side against side, to make beams and joists of large dimensions that are stronger than natural wood of the same size (Fig. 4-33). They are used for structural purposes yet are decorative as well, and in most cases their surfaces are left exposed to show the natural wood grain (Fig. 4-34).

*Courtesy of APA—The Engineered Wood Association.*

**FIGURE 4-33** Glue-laminated lumber is commonly called *glulam.*

### Green Tip

Engineered lumber is more dimensionally stable than sawn lumber from the effects of shrinking and swelling due to moisture. This reduces callbacks.

Courtesy of Weyerhaeuser Company.

**FIGURE 4-34** The appearance of an exposed beam is important. Some glulam beams are manufactured for appearance, as well as strength.

# Summary

- Wood is a universal building material used in construction.
- Engineered panels are used extensively to create strong, rigid frames and components.
- Nonstructural panels are used for interior applications sheltered from weather.
- Engineered lumber may replace wood as a building material for strong and stiff structures.

## GREEN CHECKLIST

- ☐ Manufacturing wood products from logs uses less energy than manufacturing metal or masonry products.
- ☐ Plywood manufacturing process uses the best quality veneer logs for outer surfaces while lesser quality logs are used for the inner cores. This efficiency results in the best appearance while maintaining superior strength.
- ☐ Engineered lumber maximizes the efficiency of wood resources by reducing manufacturing waste. Sawn lumber has about 50% waste while OSB production, for example, has about 8% waste.
- ☐ Engineered lumber is more dimensionally stable than sawn lumber from the effects of shrinking and swelling due to moisture. This reduces callbacks.

## REVIEW QUESTIONS

*Select the most appropriate answer.*

**1. Softwood is so named because it is made from trees that have**

a. soft bark.
b. soft wood.
c. cones.
d. leaves.

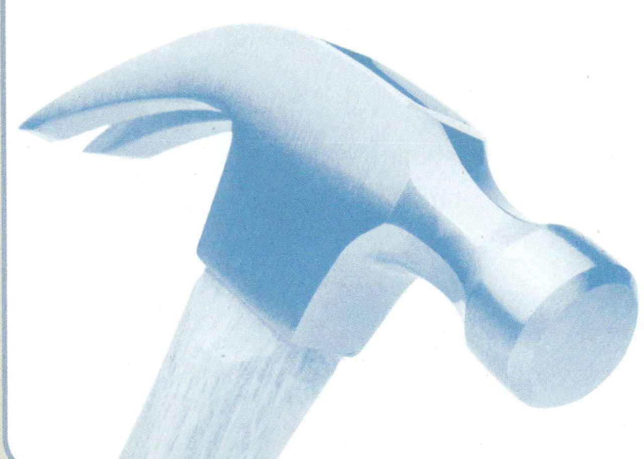

**2. When the grain of wood is close together, like quarter-sawed lumber, it tends to _____ than plain-sawed lumber.**

a. warp more
b. shrink more
c. be more durable
d. be more often produced

**3. Lumber is called "green" when**

a. it is stained by fungi.
b. the tree is still standing.
c. it is first cut from the log.
d. it has dried and is ready for paint.

**4. A commonly used and abundant softwood is**

a. ash.
b. fir.
c. basswood.
d. birch.

**5. The number of board feet in 100–2 × 6–12′ is**

a. 100.

b. 144.

c. 1200.

d. 14,400.

**6. The part of a log that contains the darker, more durable wood is**

a. cambium layer.

b. sapwood.

c. heartwood.

d. medullary rays.

**7. The best-appearing face veneer of a softwood plywood panel is indicated by the letter**

a. A.

b. B.

c. E.

d. Z.

**8. Which is the better plywood for exterior wall sheathing?**

a. APA Structural Rated Sheathing, Exposure 1

b. APA A-C, Exterior

c. APA-Rated Sturd-I-Floor, Exposure 2

d. CD, Plugged, Exterior

**9. Particleboard not rated as structural may be used for**

a. countertops.

b. subflooring.

c. wall sheathing.

d. roof sheathing.

**10. Much of the softboard used in the construction industry is for**

a. underlayment for wall-to-wall rugs.

b. roof covering.

c. decorative ceiling panels.

d. interior wall finish.

**11. The recommended grade of plywood in which only the appearance of one side is important for interior applications, such as built-ins and cabinets, is**

a. AA.

b. BC.

c. CD.

d. AD.

**12. Laminated veneer lumber is used as**

a. headers.

b. beams.

c. girders.

d. all of the above.

**13. Parallel strand lumber is manufactured to be used as or replace a**

a. stud.

b. rafter.

c. post or beam.

d. joist.

**14. The web of wood I-beams may be made of**

a. hardboard.

b. particleboard.

c. solid lumber.

d. strand board.

15. **The flanges of wood I-beams are generally made from**
    a. glue-laminated lumber.
    b. laminated veneer lumber.
    c. parallel strand lumber.
    d. laminated strand lumber.

## KNOW YOUR CODES

Lumber types, sizes and grades are determined by engineers who work from the architect's drawings. These material choices should always be followed to ensure the building performs as expected.

## WHAT'S WRONG WITH THIS PICTURE?

Carefully study **Figure 4-35** and think about what is wrong. Consider all possibilities.

### WRONG

**FIGURE 4-35** This view is looking down on a subfloor that is being assembled on top of engineered I-beam-type joists. The floor beneath is the crawl space floor. The subfloor material being used is particleboard, which is not structurally adequate to use in this situation. Particleboard, although very dense, will eventually sag in between each of the joists.

### RIGHT

**FIGURE 4-36** Plywood and oriented strand board are widely accepted choices for use in this application.

# Fasteners

**OBJECTIVES** *After completing this unit, the student should be able to name and describe the following commonly used fasteners and select them for appropriate use:*

- Nails
- Screws
- Lag screws
- Bolts
- Solid wall anchors
- Hollow wall anchors
- Adhesives

# GLOSSARY OF FASTENER TERMS

**anchor** a device used to fasten structural members in place

**box nail** a thin nail with a head, usually coated with a material to increase its holding power

**brad** a thin, short, finishing nail

**duplex nail** a double-headed nail used for temporary fastening, such as in the construction of wood scaffolds

**electrolysis** the decomposition of one of two unlike metals in contact with each other in the presence of water

**face nail** method of driving a nail straight through a surface material into a supporting member

**finish nail** a thin nail with a small head designed for setting below the surface of finish material

**galvanized** protected from rusting by a coating of zinc

**mastic** a thick adhesive

**penny (d)** a term used in designating nail sizes

**toenail** method of driving a nail diagonally through a surface material into supporting member

The simplicity of fasteners can be misleading to students of construction. It is easy to believe only that nails are driven, screws are turned, and sticky stuff is used to glue. While this tends to be true, joining material together so it will last a long time is more challenging. Many times, a fastener is used for just one type of material. Some fasteners should never be used with certain materials. The fastener selected often separates a high-quality job from a shoddy one.

Fasteners have been evolving for centuries. Today, they come in many styles, shapes, and sizes, requiring different fastening techniques. It is important that the carpenter know what fasteners are available, which securing technique should be employed, and how to wisely select the most appropriate fastener for various materials under different conditions.

# NAILS

Hundreds of kinds of nails are manufactured for just about any kind of fastening job. They differ according to purpose, shape, material, coating, and other characteristics. They may be driven as **toenails** or as **face nails**. Nails are made of aluminum, brass, copper, steel, and other metals. Some nails are hardened so that they can be driven into masonry without bending. Only the most commonly used nails are described in this chapter (Fig. 5-1).

Uncoated steel nails are called *bright* nails. Various coatings may be applied to reduce corrosion, increase holding power, and enhance appearance. To prevent rusting, steel nails are coated with zinc. These nails are called **galvanized** nails.

When fastening metal that is going to be exposed to the weather, use nails of the same material, if possible. For example, when fastening aluminum, copper, or galvanized iron, use nails made of the same metal. Otherwise, a reaction with moisture and the two different metals, called **electrolysis**, will cause one of the metals to disintegrate faster over time.

CAUTION

**CAUTION:** Always wear eye protection when driving nails. Some nails are hard and brittle at the same time. Pieces of them can break off and become shrapnel. All nails can fly out when they are being started into the material.

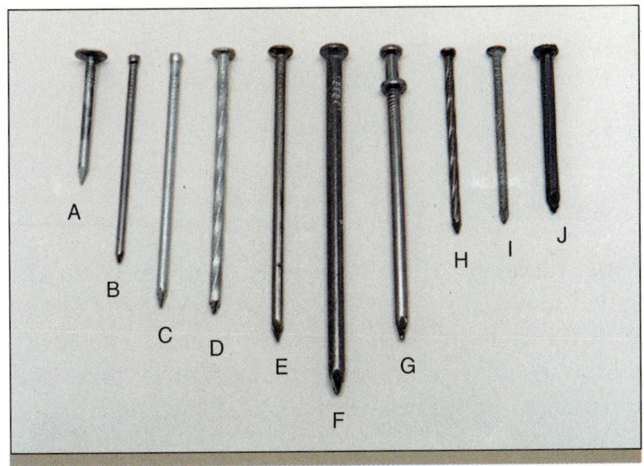


**FIGURE 5-1** Kinds of commonly used nails (A) roofing, (B) finish, (C) galvanized finish, (D) galvanized spiral shank, (E) box, (F) common, (G) duplex, (H) spiral shank, (I) coated box, and (J) masonry.

When fastening some woods, such as cedar, redwood, and oak, exposed to the weather, use stainless steel nails. Otherwise, a reaction between the acid in the wood and bright nails causes dark, ugly stains to appear around the fasteners.

## NAIL SIZES

The sizes of some nails are designated by the **penny** system. The origin of this system of nail measurement is not clear. Although many people think it should be discarded, it is still used in the United States. Some believe it originated years ago when one hundred nails of a specific length cost a certain number of pennies. Of course, the larger nails cost more per hundred than smaller ones, so nails that cost 8 pennies were larger than those that cost 4 pennies. The symbol for penny is *d;* perhaps it is the abbreviation for *denarius,* an ancient Roman coin.

In the penny system, the shortest nail is 2d and 1 inch long. The longest nail is 60d and 6 inches long (Fig. 5-2). A sixpenny nail is written as 6d and is 2 inches long. Eventually, a carpenter can determine the penny size of nails just by looking at them. As a general rule, select a nail that is three times longer than the material being fastened.

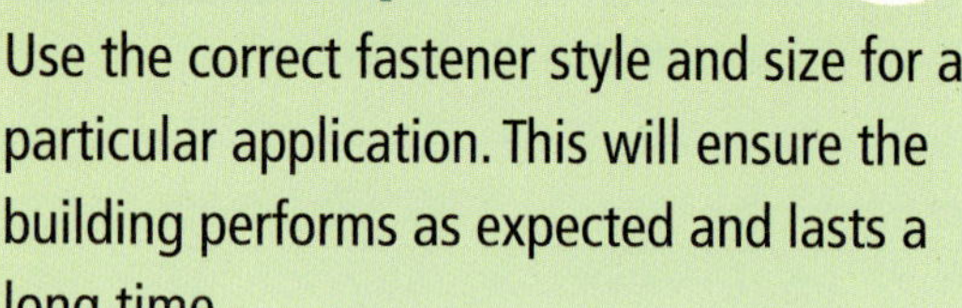

### Green Tip

Use the correct fastener style and size for a particular application. This will ensure the building performs as expected and lasts a long time.

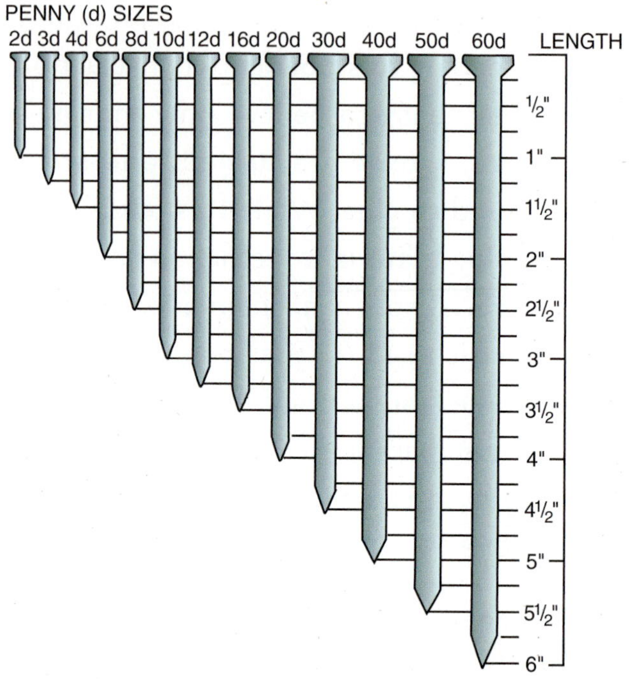


**FIGURE 5-2** Some nails are sized according to the penny system.

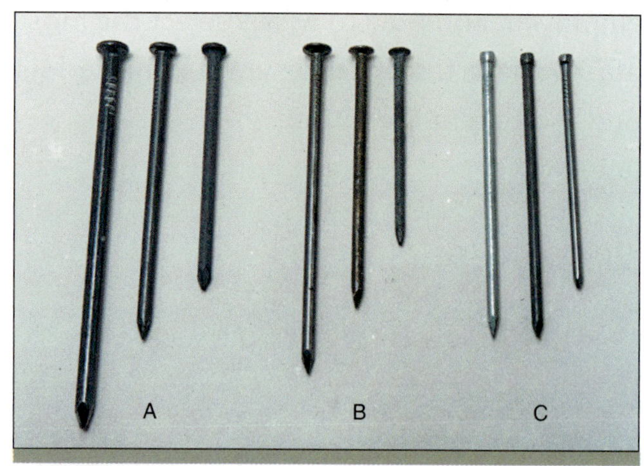


**FIGURE 5-3** Most widely used nails are the (A) common, (B) box, and (C) finish nails.

# KINDS OF NAILS

Most nails, cut from long rolls of metal wire, are called *wire nails. Cut nails* are wedge-shaped pieces stamp-cut from thin sheets of metal. The most widely used wire nails are the common, box, and finish nails (Fig. 5-3).

## COMMON NAILS

*Common nails* are made of heavy gauge wire and have a medium-sized head. They have a pointed end and a smooth shank. A barbed section just under the head increases the holding power of common nails.

## BOX NAILS

**Box nails** are similar to common nails, except they are thinner. Because of their small gauge, they can be used close to edges and ends with less danger of splitting the wood. Many box nails are coated with resin cement to increase their holding power.

## FINISH NAILS

**Finish nails** are of light gauge with a very small head. They are used mostly to fasten interior trim. The small head is sunk into the wood with a nail set and covered with a filler. The small head of the finish nail does not detract from the appearance of a job as much as would a nail with a larger head.

## CASING NAILS

*Casing nails* are similar to finish nails. The head is cone-shaped and slightly larger than that of the finish nail, but smaller than that of the common nail. Many carpenters prefer them to fasten exterior finish. The shank is the same gauge as that of the common nail.

## DUPLEX NAILS

On temporary structures, such as wood scaffolding and concrete forms, the **duplex nail** is often used. The lower head ensures that the piece is fastened tightly. The projecting upper head makes it easy to pry the nail out when the structure is dismantled (Fig. 5-4).

**FIGURE 5-4** Duplex nails are used on temporary structures.

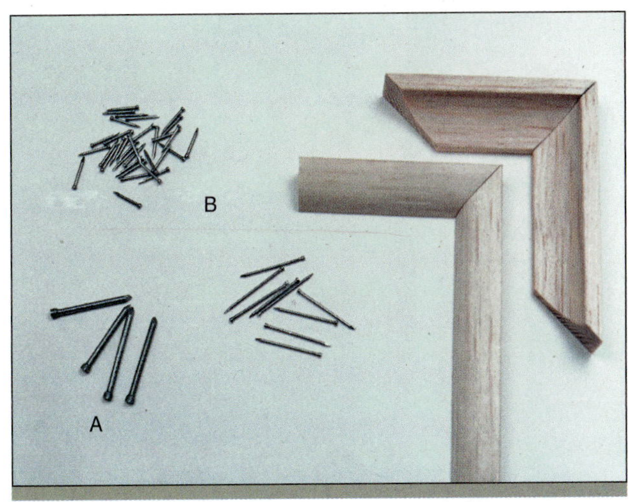


**FIGURE 5-5** (A) Finish nails and (B) brad nails.

## BRADS

**Brads** are small finishing nails (Fig. 5-5). They are sized according to length in inches and gauge. Usual lengths are from ½ inch to 1½ inches, and gauges run from #14 to #20. The higher the gauge number, the thinner the brad. Brads are used for fastening thin material, such as small molding.

## ROOFING NAILS

*Roofing nails* are short nails of fairly heavy gauge with wide, round heads. They are used for purposes such as fastening roofing material and softboard wall sheathing. The large head holds thin or soft material more securely and is coated to prevent rusting. The shank is usually barbed to increase holding power. Usual sizes run from ¾ inch to 2 inches.

## MASONRY NAILS

*Masonry nails* may be cut nails or wire nails (Fig. 5-6). These nails are made from hardened steel to prevent them from bending when being driven into concrete or other masonry. Great care should be exercised when driving them because masonry nails are also very brittle and may break and shatter. Always wear safety glasses.

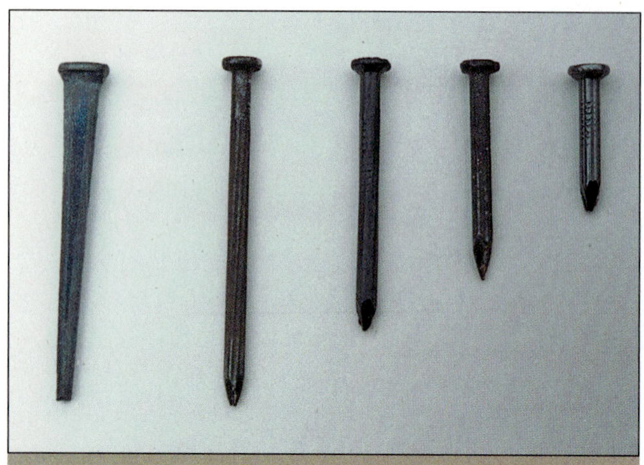

**FIGURE 5-6** Masonry nails are made of hardened steel.

WOOD SCREWS

| LENGTH | GAUGE NUMBERS | | | | | | | | | | | | | | | | |
|---|---|---|---|---|---|---|---|---|---|---|---|---|---|---|---|---|---|
| 1/4 INCH | 0 | 1 | 2 | 3 | | | | | | | | | | | | | |
| 3/8 INCH | | | 2 | 3 | 4 | 5 | 6 | 7 | | | | | | | | | |
| 1/2 INCH | | | 2 | 3 | 4 | 5 | 6 | 7 | 8 | | | | | | | | |
| 5/8 INCH | | | | 3 | 4 | 5 | 6 | 7 | 8 | 9 | 10 | | | | | | |
| 3/4 INCH | | | | | 4 | 5 | 6 | 7 | 8 | 9 | 10 | 11 | | | | | |
| 7/8 INCH | | | | | | | 6 | 7 | 8 | 9 | 10 | 11 | 12 | | | | |
| 1 INCH | | | | | | | 6 | 7 | 8 | 9 | 10 | 11 | 12 | 14 | | | |
| 1 1/4 INCH | | | | | | | | 7 | 8 | 9 | 10 | 11 | 12 | 14 | 16 | | |
| 1 1/2 INCH | | | | | | | 6 | 7 | 8 | 9 | 10 | 11 | 12 | 14 | 16 | 18 | |
| 1 3/4 INCH | | | | | | | | | 8 | 9 | 10 | 11 | 12 | 14 | 16 | 18 | 20 | |
| 2 INCH | | | | | | | | | 8 | 9 | 10 | 11 | 12 | 14 | 16 | 18 | 20 | |
| 2 1/4 INCH | | | | | | | | | | 9 | 10 | 11 | 12 | 14 | 16 | 18 | 20 | |
| 2 1/2 INCH | | | | | | | | | | | | | 12 | 14 | 16 | 18 | 20 | |
| 2 3/4 INCH | | | | | | | | | | | | | | 14 | 16 | 18 | 20 | |
| 3 INCH | | | | | | | | | | | | | | | 16 | 18 | 20 | |
| 3 1/2 INCH | | | | | | | | | | | | | | | | 18 | 20 | 24 |
| 4 INCH | | | | | | | | | | | | | | | | 18 | 20 | 24 |

WHEN YOU BUY SCREWS, SPECIFY (1) LENGTH, (2) GAUGE NUMBER, (3) TYPE OF HEAD – FLAT, ROUND OR OVAL, (4) MATERIAL – STEEL, BRASS, BRONZE, ETC., AND (5) FINISH – BRIGHT, STEEL, CADMIUM, NICKEL OR CHROMIUM PLATED.

**FIGURE 5-7** Wood screw sizes.

# SCREWS

Wood screws are used when greater holding power is needed or when the work being fastened must at times be removed. For example, door hinges must be applied with screws because nails would pull loose after awhile, and the hinges may, at times, need to be removed. When ordering screws, specify the length, gauge, type of head, coating, kind of metal, and screwdriver slot.

## SCREW SIZES

Screws are made in many different sizes. Usual lengths range from ¼ inch to 4 inches. Gauges run from 0 to 24 (Fig. 5-7). Unlike brads and some nails, the higher the gauge number, the greater the diameter of the screw. The lower gauge numbers are for shorter, thinner screws. Higher gauge numbers are for longer screws. All screw lengths are not available in every gauge.

## KINDS OF SCREWS

A wood screw is identified by the shape of the screwhead and screwdriver slot. For example, a screw may be called a *flat head Phillips* or a *round head common screw*. Three of the most common shapes of screwheads are the *flat head, round head,* and *oval head*. Screw lengths are measured from the point to that part of the head that sets flush with the wood when fastened (Fig. 5-8). Many other screws are available that are designed for special purposes. Like nails, screws come in a variety of metals and coatings.

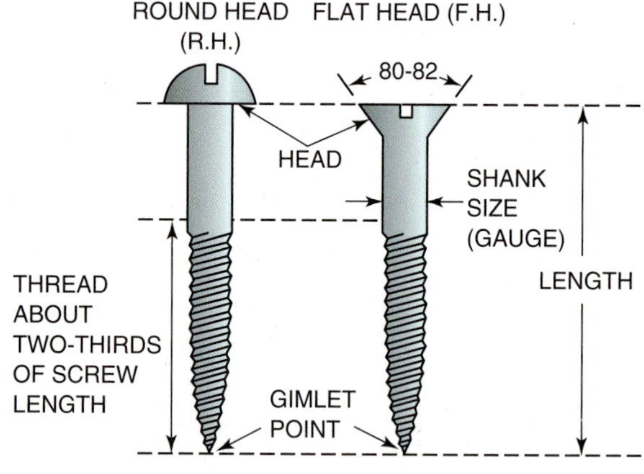


**FIGURE 5-8** Common kinds of screws and screw terms.

A screwhead that is made with a straight, single slot is called a *common screw*. A Phillips head screw has a crossed slot. There are many other types of screwdriver slots, each with a different name (Fig. 5-9).

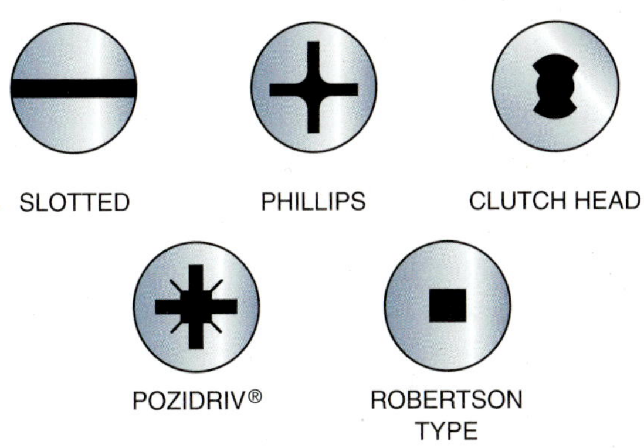


**FIGURE 5-9** Common kinds of screw slots.

## SHEET METAL SCREWS

The threads of sheet metal screws extend for the full length of the screw and are much deeper. Sheet metal screws are used for fastening thin metal. They are also recommended for fastening hardboard and particleboard because their deeper thread grabs better in softer and fiberless material.

Another type of screw, used with power screwdrivers, is the *self-tapping screw*, which is used extensively to fasten metal framing. This screw has a cutting edge on its point to eliminate predrilling a hole (Fig. 5-10). Drill points are available in various lengths and must be equal to the thickness of the metal being fastened.

**CAUTION**

**CAUTION:** Care should be taken when screws are driven with drills or screwguns. Metal slivers can be inflicted by spinning screws when they are held between the finger and thumb.

## LAG SCREWS

*Lag screws* (Fig. 5-11) are similar to wood screws, except that they are larger and have a square or hex head designed to be turned with a wrench instead. This fastener is used when great holding power is needed to join heavy parts and where a bolt cannot be used.

Lag screws are sized by diameter and length. Diameters range from ¼ inch to 1 inch, with lengths from 1 inch to 12 inches and up. Shank and pilot holes to receive lag screws are drilled in the same manner as for wood screws. Place a flat washer under the head to prevent the head from digging into the wood as the lag screw is tightened down. Predrill the hole or apply a little wax to the threads to allow the screw to turn more easily and to reduce the risk of the head twisting off.

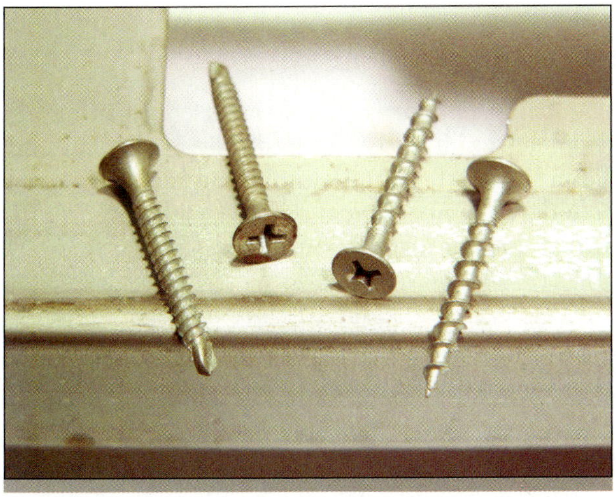

**FIGURE 5-10** Self-tapping screws have either piercing or drill-point tips.

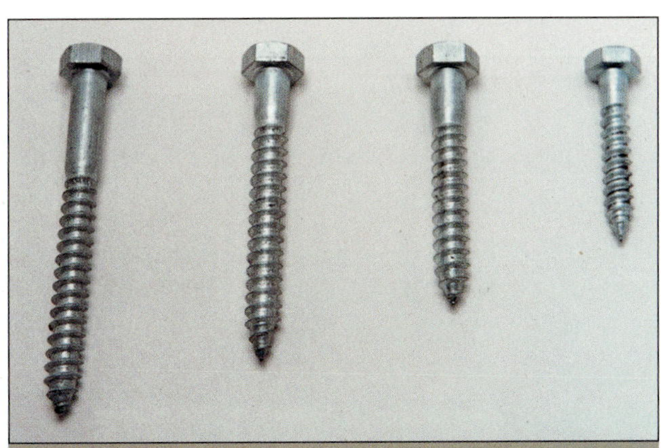

**FIGURE 5-11** Lag screws are large screws with a square or hex head.

# Bolts

Bolts have a large head similar to screws and a threaded end that will accept a nut. Many kinds are available for special purposes, but only a few are generally used.

## KINDS OF BOLTS

Commonly used bolts are the carriage, machine, and stove bolts (Fig. 5-12).

*Carriage Bolts.* The carriage bolt has a square section under its oval head. The square section is embedded in wood, which helps prevent the bolt from turning as the nut is tightened.

*Machine Bolts.* The machine bolt has a square or hex head. This is held with a wrench to keep the bolt from turning as the nut is tightened.

*Stove Bolts.* Stove bolts have either round or flat heads with a screwdriver slot. They are usually threaded all the way up to the head.

## BOLT SIZES

Bolt sizes are specified by diameter and length. Carriage and machine bolts range from ¾ inch to 20 inches in length and from 3⁄16 to ¾ inch in diameter. Stove bolts are small by comparison to other bolts. They commonly come in lengths from ⅜ inch to 6 inches and from ⅛ to ⅜ inch in diameter. Drill holes for bolts the same diameter as the bolt. Use flat washers under the head (except for carriage bolts) and under the nut to prevent the nut from cutting into the wood and to distribute the pressure over a wider area. Use wrenches of the correct size to tighten the bolt. Be careful not to overtighten carriage bolts, which will damage the material being fastened. The head need only be drawn snug, not pulled below the surface.

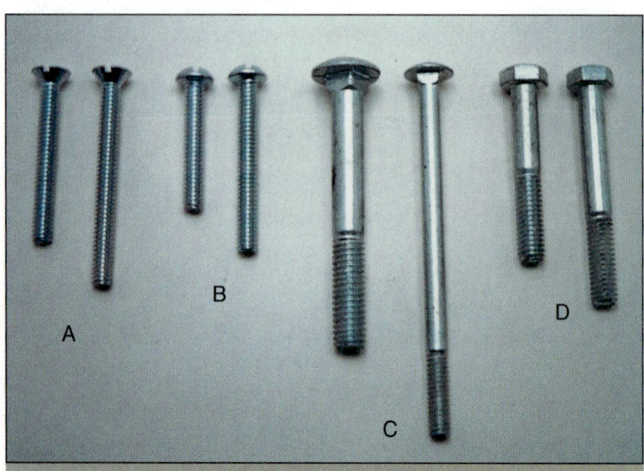


**FIGURE 5-12** Commonly used bolts include (A) flat-head stove, (B) round-head stove, (C) carriage, and (D) machine.

# Anchors

Special kinds of fasteners used to attach parts to solid masonry and hollow walls and ceilings are called **anchors**. There are hundreds of types available. Those most commonly used are described in this chapter.

## SOLID WALL ANCHORS

*Solid wall anchors* may be classified as heavy-, medium-, or light-duty. Heavy-duty anchors are used to install things, such as machinery, hand rails, dock bumpers, and storage racks. Medium-duty anchors may be used for hanging pipe and ductwork, securing window and door frames, and installing cabinets. Light-duty anchors are used for fastening such things as junction boxes, bathroom fixtures, closet organizers, small appliances, smoke detectors, and other lightweight objects.

### Heavy-Duty Anchors

The *wedge anchor* (Fig. 5-13) is used when high resistance to pullout is required. The anchor and hole diameter are the same, simplifying installation. The hole depth is not critical as long as the minimum is drilled. Proper installation requires cleaning out the hole (Fig. 5-14).

The *sleeve anchor* (Fig. 5-15) and its hole size are also the same, but the hole depth need not be exact. After the anchor is inserted in the hole, it is expanded by tightening the nut. The *drop-in anchor* (Fig. 5-16) consists of an expansion shield and a cone-shaped, internal expander plug. The hole must be drilled at least equal to the length of the anchor. A setting tool, supplied with the anchors, must be used to drive and expand the anchor. This anchor takes a machine screw or bolt. Insert the anchor into a chuck adapter and drill hole in concrete. Remove

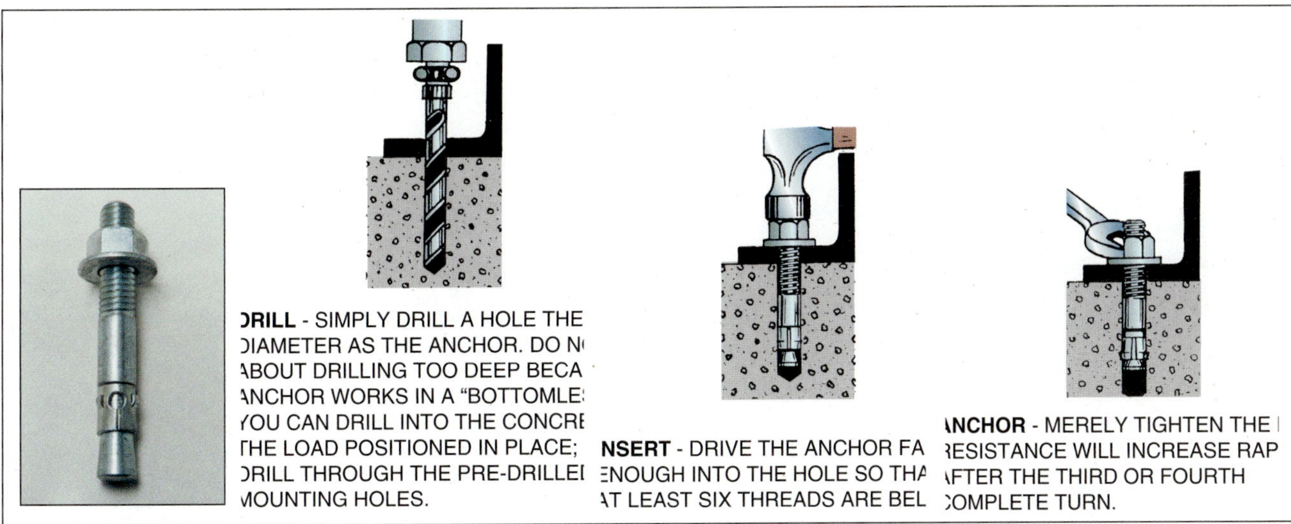


Line art courtesy of U.S. Anchor, Pompano Beach, FL.

**FIGURE 5-13** The wedge anchor has high resistance to pullout.

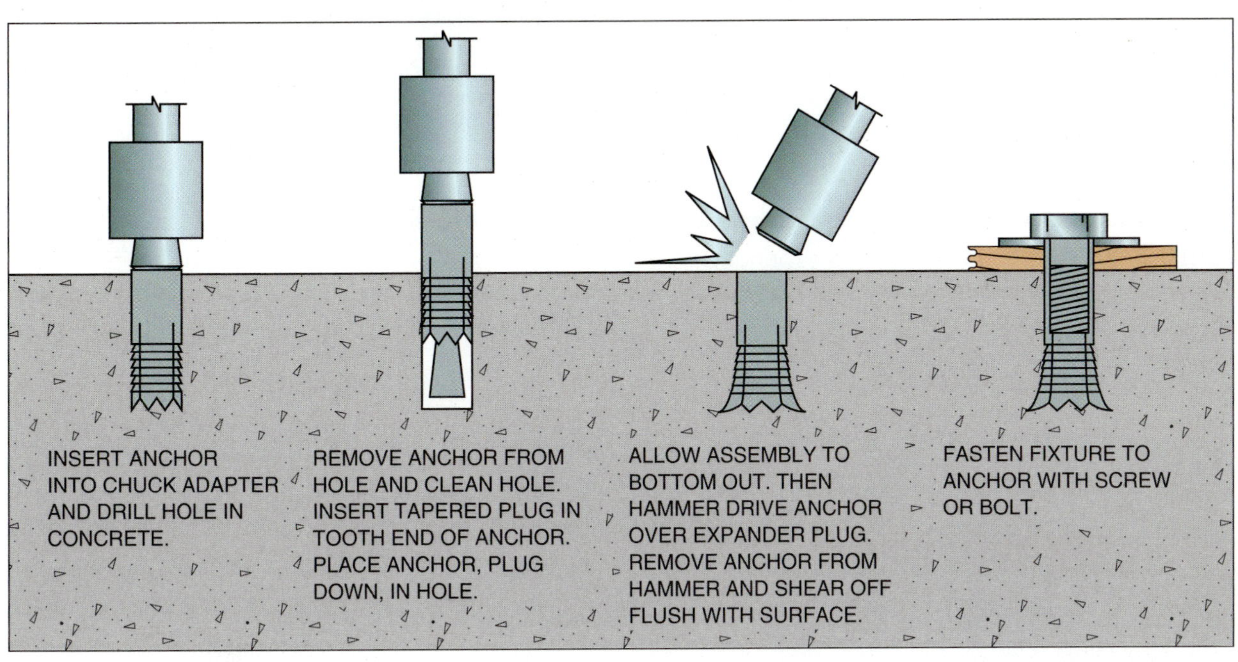


Courtesy of U.S. Anchor, Pompano Beach, FL.

**FIGURE 5-14** The self-drilling anchor requires no predrilled hole.

the anchor from the hole and clean the hole. Insert the tapered plug into the tooth end of the anchor. Place the anchor and plug down in hole. Allow the assembly to bottom out. Then use a hammer to drive the anchor over the expander plug. Remove the anchor from the hammer and shear off flush with the surface. Fasten the fixture to the anchor with a screw or a bolt.

### Medium-Duty Anchors

*Split fast anchors* (Fig. 5-17) are one-piece steel with two sheared expanded halves at the base. When driven, these halves are compressed and exert immense outward force on the inner walls of the hole as they try to regain their original shape. They come in both flat and round head styles.

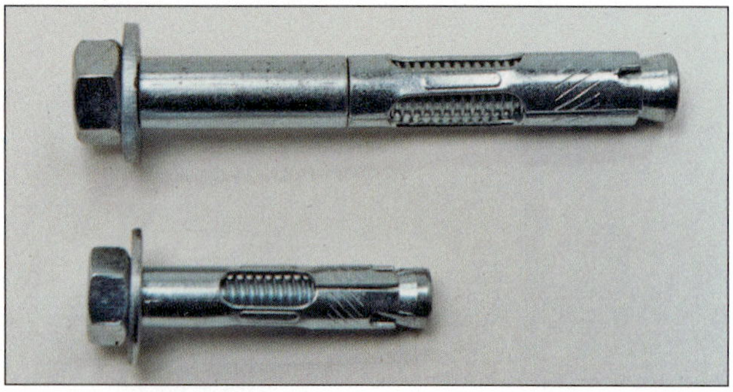

**FIGURE 5-15** Sleeve anchors eliminate the problem of exact hole depth requirements.

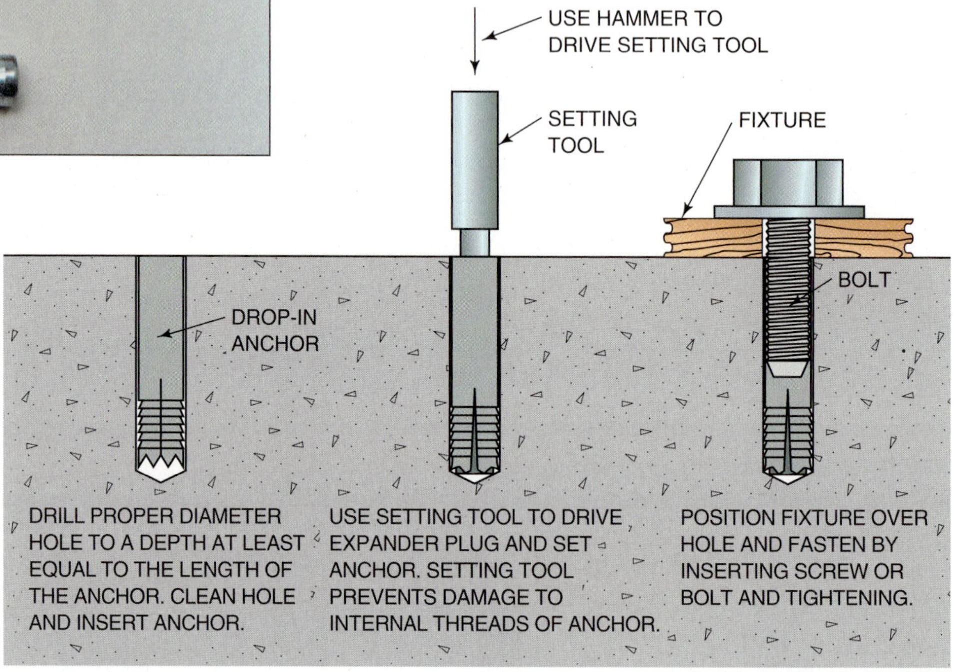

*Line art courtesy of U.S. Anchor, Pompano Beach, FL.*

**FIGURE 5-16** The drop-in anchor is expanded with a setting tool.

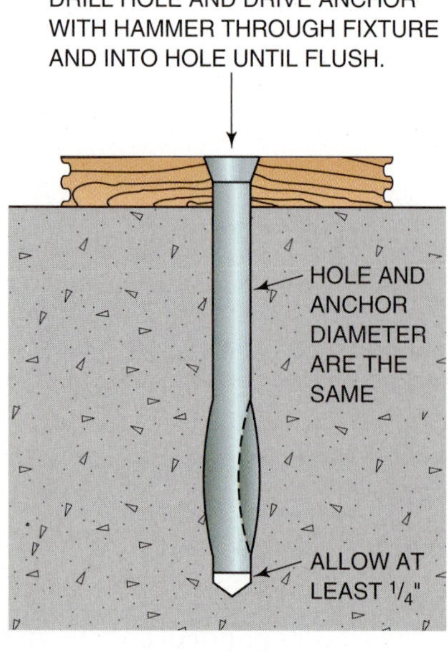


**FIGURE 5-17** The split fast is a one-piece, all-steel anchor for hard masonry.

*Single* and *double expansion anchors* (Fig. 5-18) are used with machine screws or bolts. Drill a hole of the recommended diameter to a depth equal to the length of the anchor. Place the anchor into the hole, flush with or slightly below the surface. Position the object to be fastened and bolt it into place. Once fastened, the object may be unbolted, removed, and refastened, if desired.

The *lag shield* (Fig. 5-19) is used with a lag screw. It is inserted into a hole of the recommended diameter and a depth equal to the length of the shield plus ½ inch or more. The lag screw length is determined by adding the length of the shield, the thickness of the material to be fastened, plus ¼ inch. The tip of the lag screw must protrude from the bottom of the anchor to ensure proper expansion. As the fastener is threaded in, the shield expands tightly and securely in the drilled hole.

The *concrete screw* (Fig. 5-20) uses specially fashioned high and low threads that cut into a properly sized hole in concrete. Screws come in

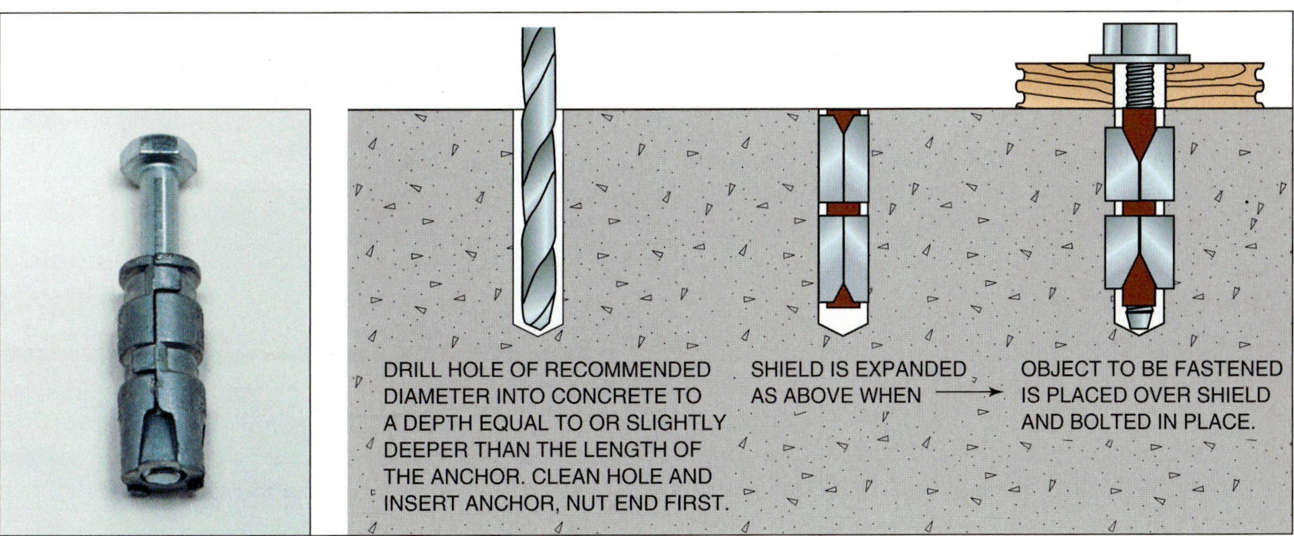


*Line art courtesy of U.S. Anchor, Pompano Beach, FL.*

**FIGURE 5-18** Two opposing wedges of the double expansion anchor pull toward each other, expanding the full length of the anchor body.

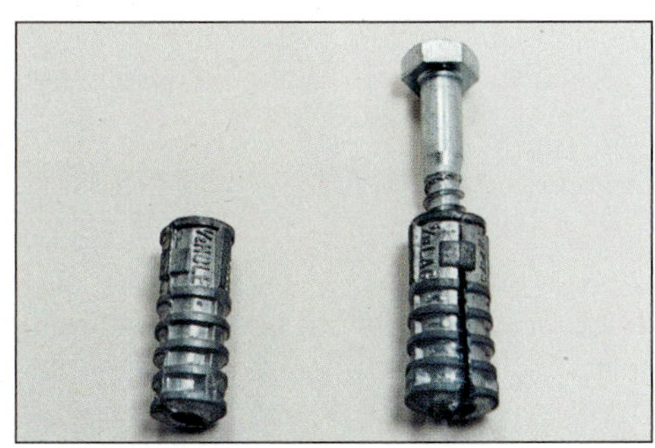

**FIGURE 5-19** Lag shields are designed for light- to medium-duty fastening in masonry.

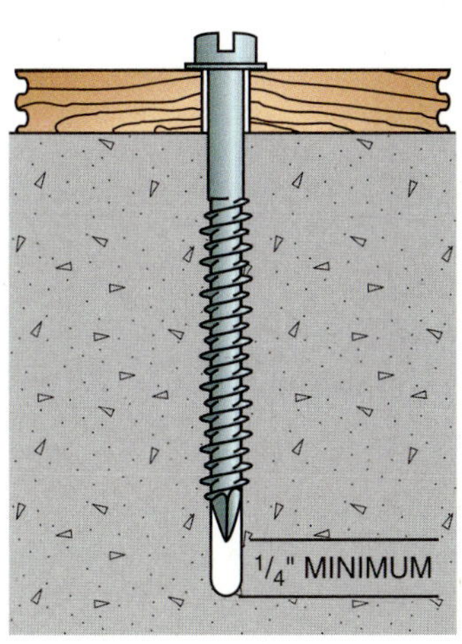


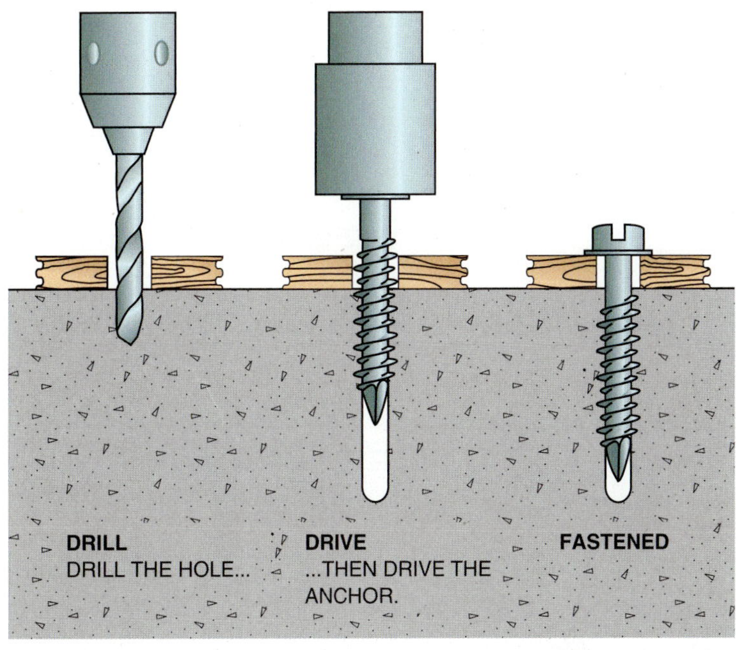


*Courtesy of U.S. Anchor, Pompano Beach, FL.*

**FIGURE 5-20** The concrete screw system eliminates the need for an anchor when fastening into concrete.

3/16- and 1/4-inch diameters and up to 6 inches in length. The hole diameter is important to the performance of the screw. It is recommended that a minimum of 1-inch and a maximum of 1¾-inch embedment be used to determine the fastener length. The concrete screw system eliminates the need for plastic or lead anchors.

### Light-Duty Anchors

Three kinds of drive anchors are commonly used for quick and easy fastening in solid masonry. The *hammer drive anchor* (Fig. 5-21) has a body of zinc alloy containing a steel expander pin. In the *aluminum drive anchor,* both the body and the pin are aluminum to avoid the corroding action of electrolysis. The *nylon nail anchor* uses a nylon body and a threaded steel expander pin. All are installed in a similar manner.

*Lead* and *plastic anchors,* also called *inserts* (Fig. 5-22), are commonly used for fastening lightweight fixtures to masonry walls. These anchors have an unthreaded hole into which a wood or sheet metal screw is driven. The anchor is placed into a hole of the recommended diameter and 1/4 inch or more deeper than the length of the anchor. As the screw is turned, the threads of the screw cut into the soft material of the insert. This causes the insert to expand and tighten in the drilled hole. Ribs on the sides of the anchors prevent them from turning as the screw is driven.

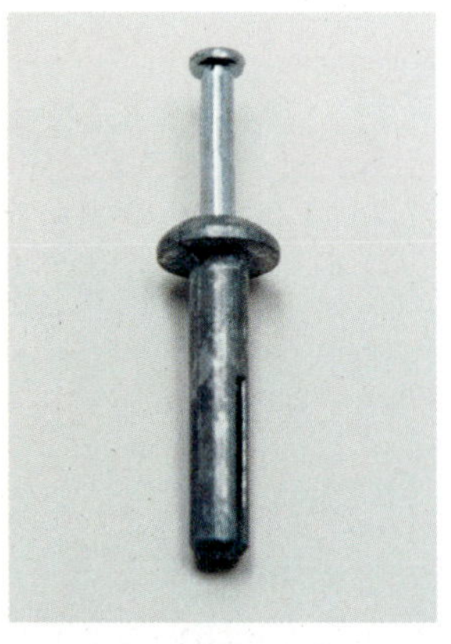

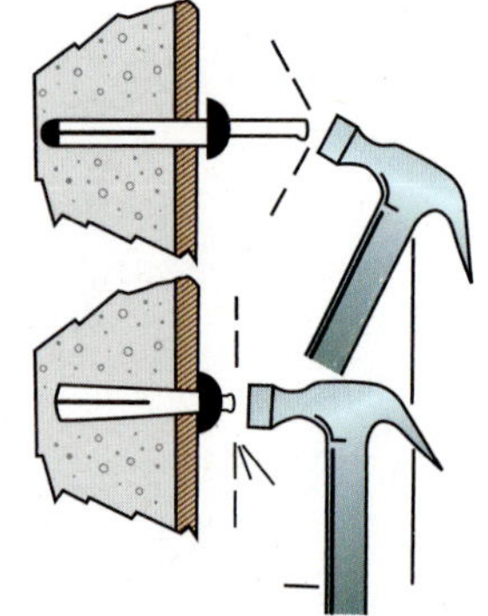

*Line art courtesy of U.S. Anchor, Pompano Beach, FL.*

**FIGURE 5-21** Hammer drive anchors come assembled for quick and easy fastening.

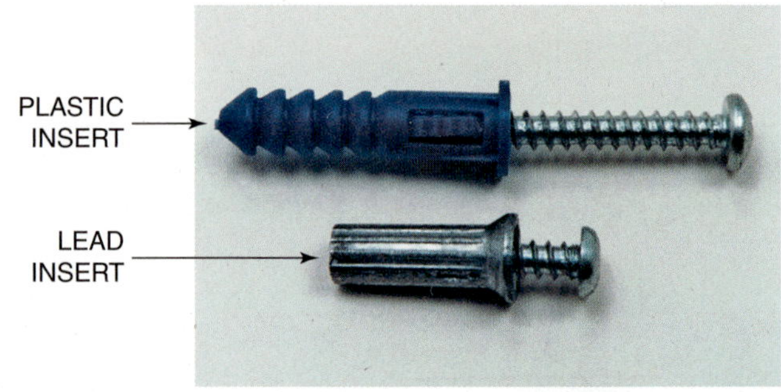


**FIGURE 5-22** Lead and plastic anchors or inserts are used for light-duty fastening.

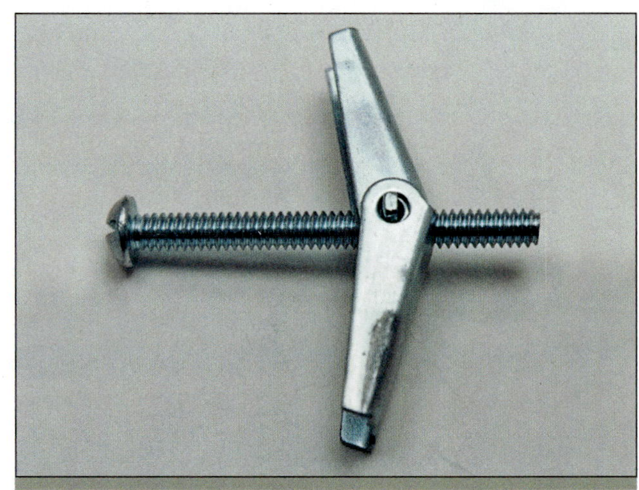

**FIGURE 5-23** Toggle bolts are used for fastening in hollow walls.

## HOLLOW WALL FASTENERS

### Toggle Bolts

*Toggle bolts* (Fig. 5-23) may have a wing or a tumble toggle. The wing toggle is fitted with springs, which cause it to open. The hole must be drilled large enough for the toggle of the bolt to slip through. A disadvantage of using toggle bolts is that, if removed, the toggle falls off inside the wall.

### Plastic Toggles

The *plastic toggle* (Fig. 5-24) consists of four legs attached to a body that has a hole through the center and fins on its side to prevent turning during installation. The legs collapse to allow insertion into the hole. As sheet metal screws are turned through the

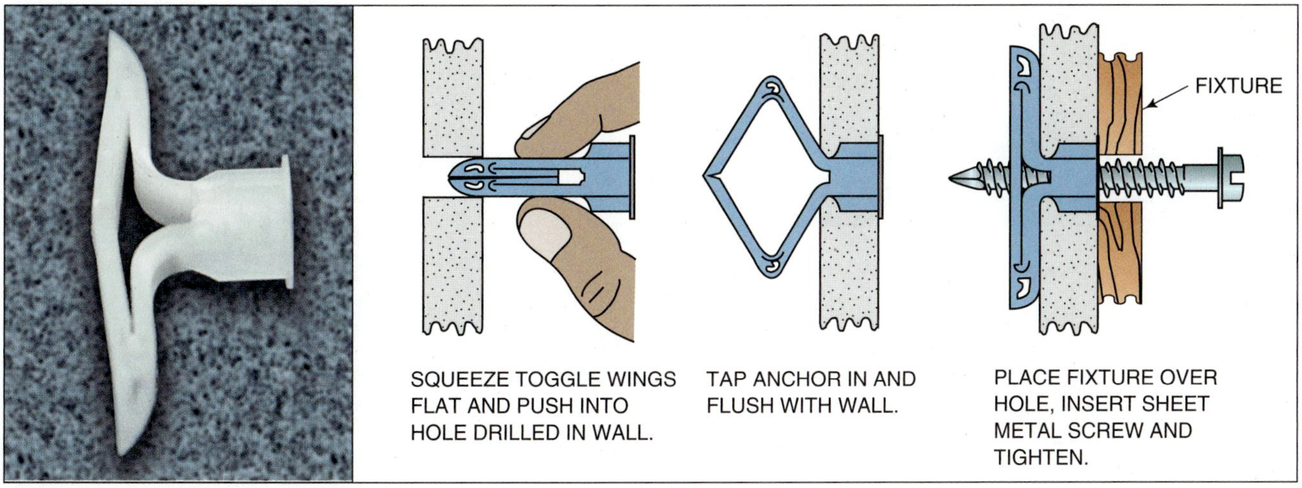

SQUEEZE TOGGLE WINGS FLAT AND PUSH INTO HOLE DRILLED IN WALL.

TAP ANCHOR IN AND FLUSH WITH WALL.

PLACE FIXTURE OVER HOLE, INSERT SHEET METAL SCREW AND TIGHTEN.

FIXTURE

*Line art courtesy of U.S. Anchor, Pompano Beach, FL.*

**FIGURE 5-24** The plastic toggle is a unique removable and reusable hollow wall anchor.

body, they draw in and expand the legs against the inner surface of the wall.

### Expansion Anchors

Hollow wall *expansion anchors* are commonly called *molly screws* (Fig. 5-25). The anchor consists of an expandable sleeve, a machine screw, and a fiber washer. The collar on the outer end of the sleeve has two sharp prongs that grip into the surface of the wall material. This prevents the sleeve from turning when the screw is tightened to expand the anchor. After expanding the sleeve, the screw is removed, inserted through the part to be attached, and then replaced into the anchor. Some types require that a hole be drilled, while other types have pointed ends that may be driven through the wall material.

Installed fixtures may be removed and refastened or replaced by removing the anchor screw without disturbing the anchor. Anchors are manufactured for various wall board thicknesses. Make sure to use the right size anchor for the wall thickness in which the anchor is being installed.

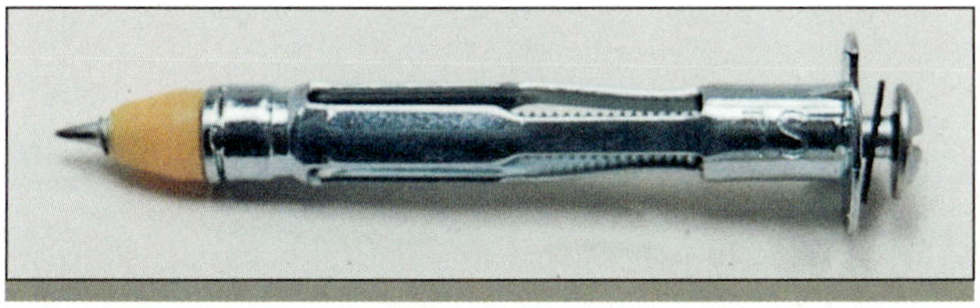

**FIGURE 5-25** Hollow wall expansion anchors are commonly called *molly screws*.

### Conical Screws

The deep threads of the *conical screw* anchor (Fig. 5-26) resist stripping out when screwed into gypsum board, strand board, and similar material. After the plug is seated flush with the wall, the fixture is placed over the hole and fastened by driving a screw through the center of the plug.

### Universal Plugs

The *universal plug* (Fig. 5-27) is made of nylon and is used for a number of hollow wall and some solid wall applications. A hole of proper diameter is drilled. The plug is inserted, and the screw is driven to draw or expand the plug.

## CONNECTORS

Widely used in the construction industry are devices called *connectors*. Connectors are metal pieces formed into various shapes to join wood to wood, or wood to concrete or other masonry. They are called specific names depending on their function.

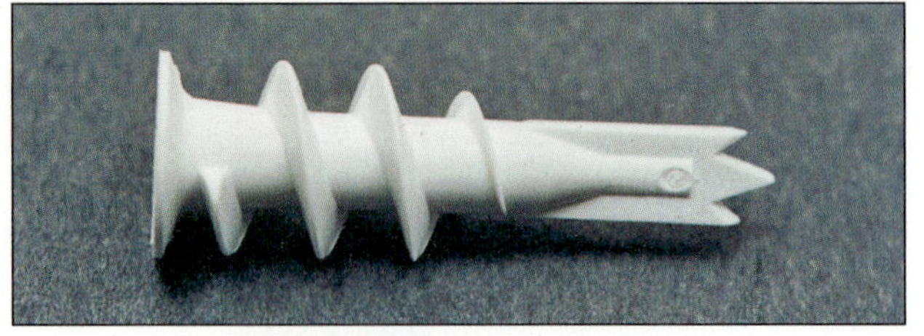

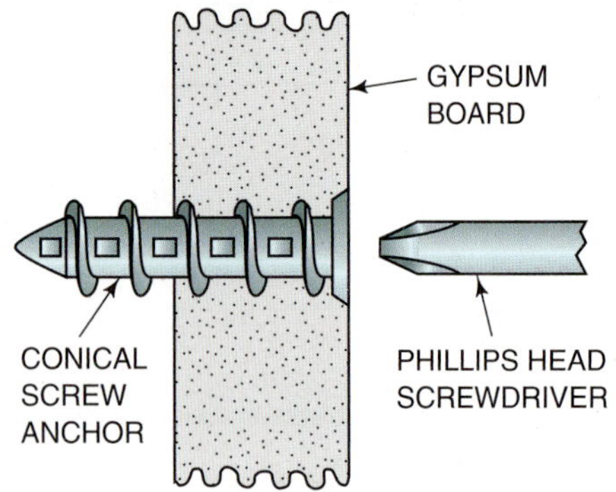


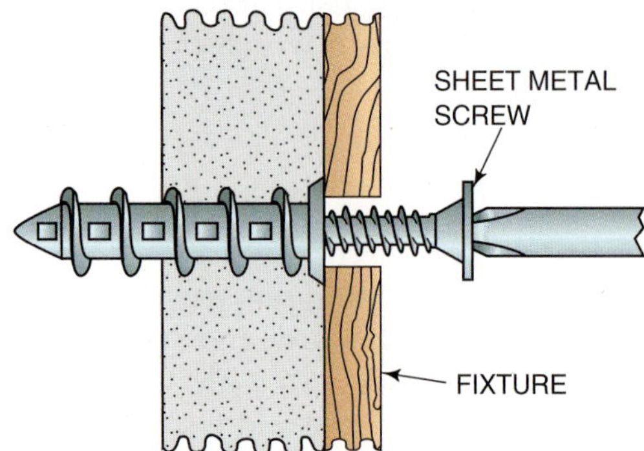


*Line art courtesy of U.S. Anchor, Pompano Beach, FL.*

**FIGURE 5-26** The conical screw anchor is a self-drilling, hollow wall anchor for lightweight fastenings.

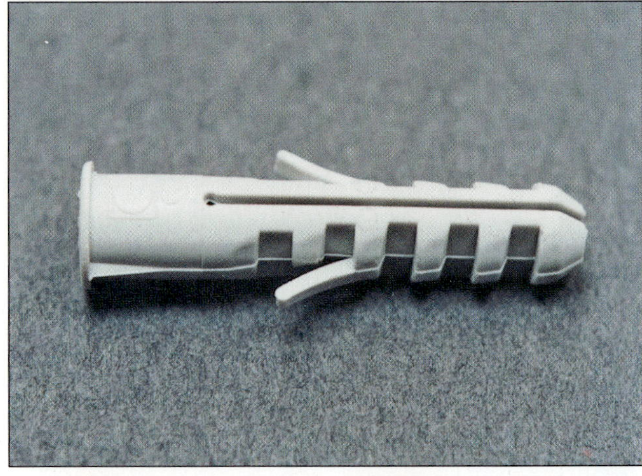

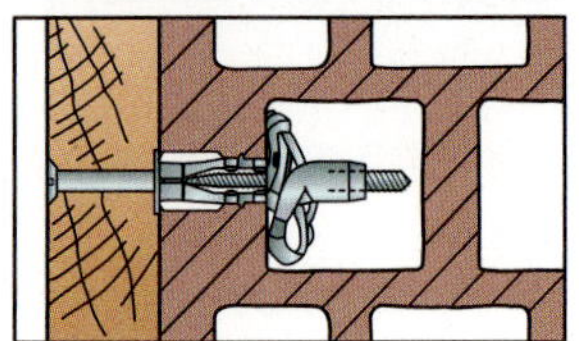


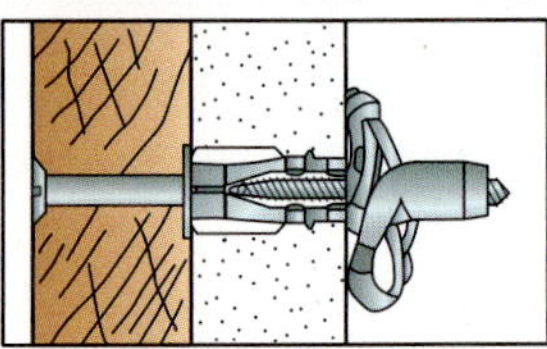


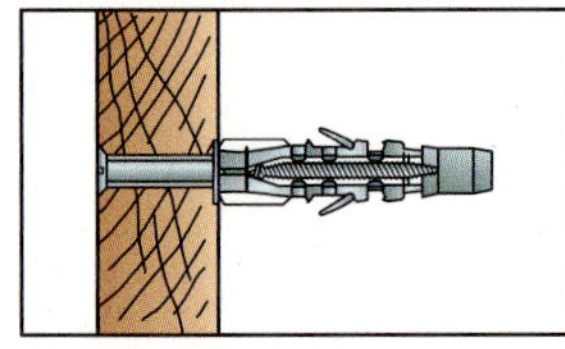


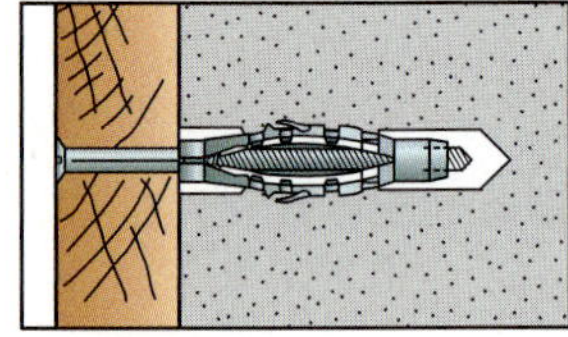


*Line art courtesy of U.S. Anchor, Pompano Beach, FL.*

**FIGURE 5-27** The universal plug is used for many types of hollow wall fastening.

### WOOD-TO-WOOD

*Framing anchors* and *seismic* and *hurricane ties* (Fig. 5-28) are used to join parts of a wood frame. *Post* and *column caps* and *bases* are used at the top and bottom of those members (Fig. 5-29). *Joist hangers* and *beam hangers* are available in many sizes and styles (Fig. 5-30). It is important to use anchor nails of the proper style, size, and quantity in each hanger.

### WOOD-TO-CONCRETE

Some wood-to-concrete connectors are *sill anchors, anchor bolts*, and *hold-downs* (Fig. 5-31). A *girder hanger* and a *beam seat* (Fig. 5-32) make beam-to-foundation wall connections. *Post bases* come in various styles. They are used to anchor posts to concrete floors or footings. Many other specialized connectors are used in frame construction. Some are described in the framing sections of this book.

## ADHESIVES AND GLUES

Glue is mostly used for interior finish work. A number of **mastics** (heavy, paste-like adhesives) are used throughout the construction process. Adhesives generally come in tubes that are applied with a caulking gun.

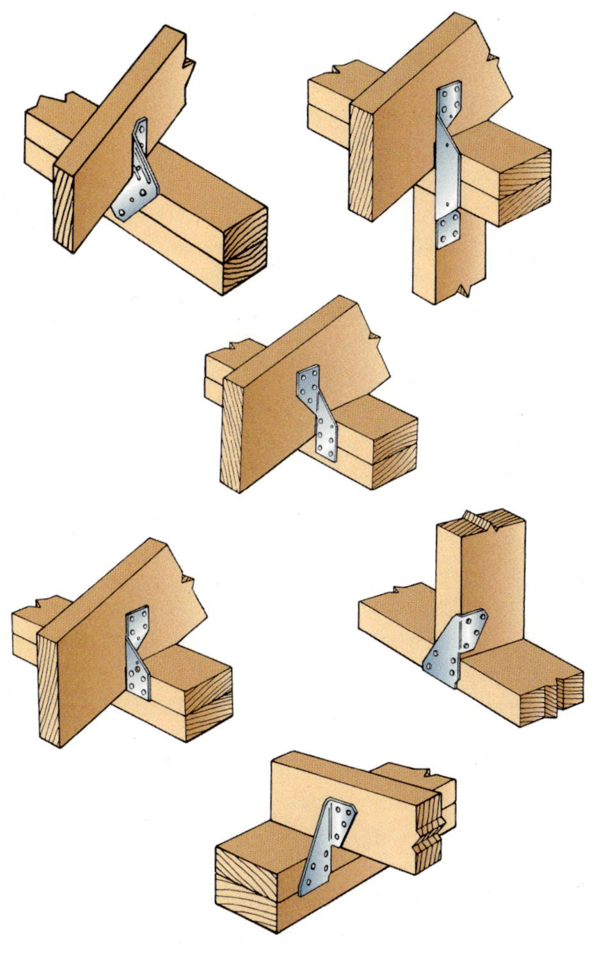

Courtesy of Simpson Strong-Tie Company.

**FIGURE 5-28** Framing ties and anchors are manufactured in many unique shapes.

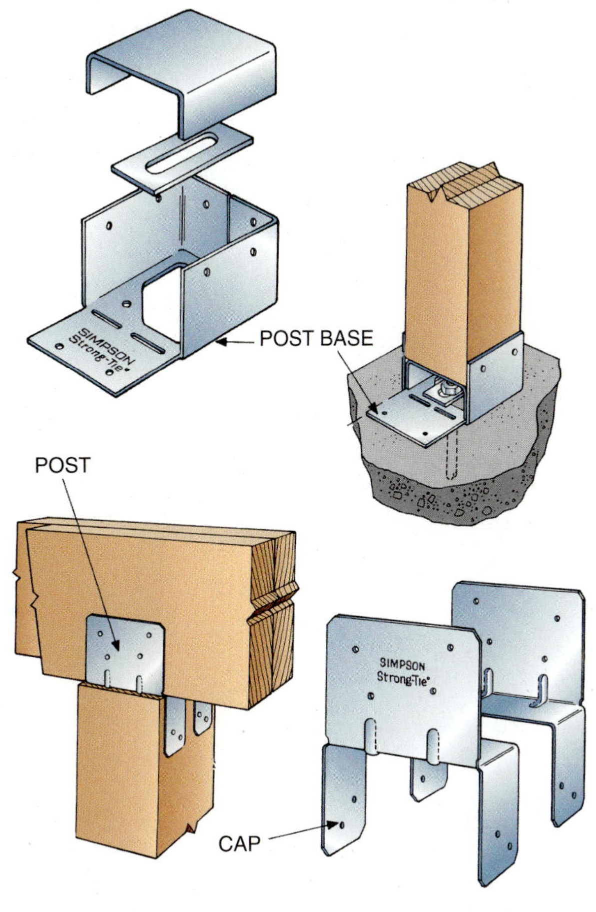


Courtesy of Simpson Strong-Tie Company.

**FIGURE 5-29** Caps and bases help fasten tops and bottoms of posts and columns.

## WHITE AND YELLOW GLUE

Most of the glue used by the carpenter is the so-called white glue or yellow glue. Yellow glue is faster setting, so joints should be made quickly after applying the glue. They are available under a number of trade names and are excellent for joining wood parts not subjected to moisture. Some varieties of glue are moisture resistant but should not be used for exterior applications.

## CONTACT CEMENT

*Contact cement* is so named because pieces coated with it bond on contact and need not be clamped under pressure. It is extremely important that pieces are positioned accurately before contact is made. Contact cement is widely used to apply

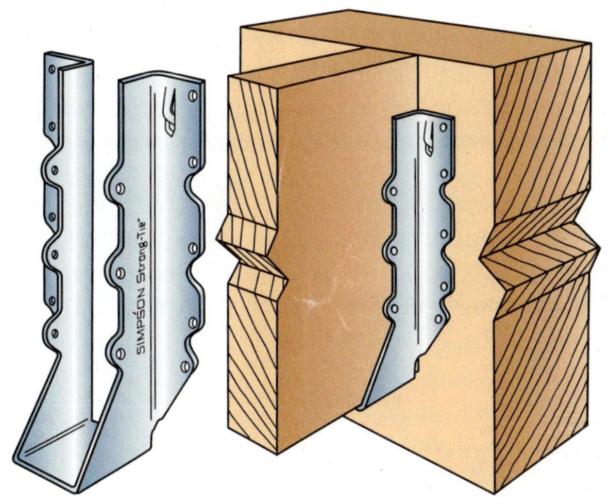


Courtesy of Simpson Strong-Tie Company.

**FIGURE 5-30** Hangers are used to support joists and beams.

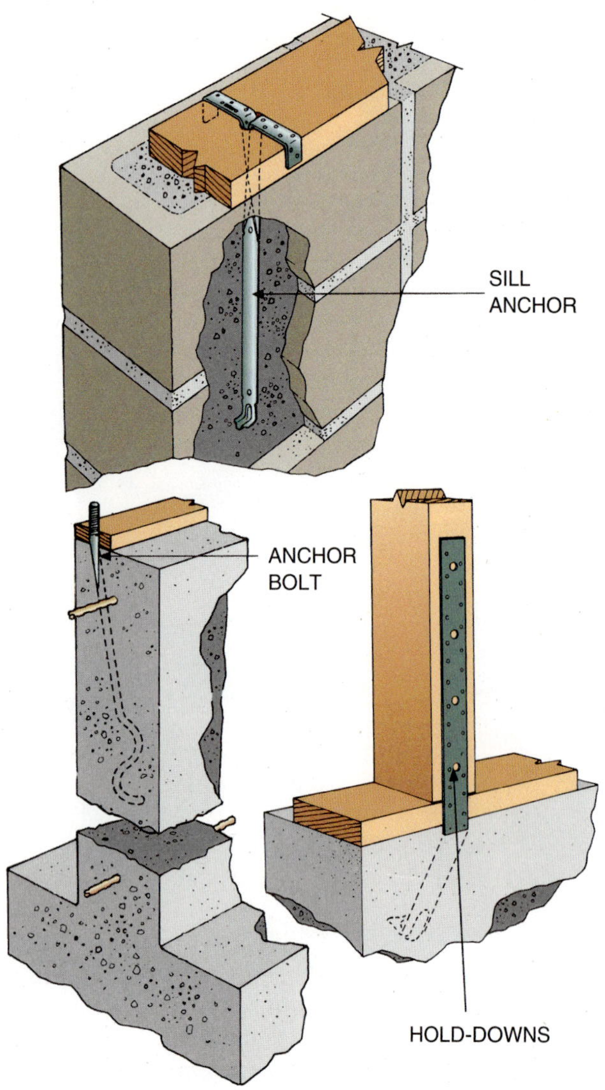


Courtesy of Simpson Strong-Tie Company.

**FIGURE 5-31** Sill anchors, anchor bolts, and hold-downs connect frame members to concrete.

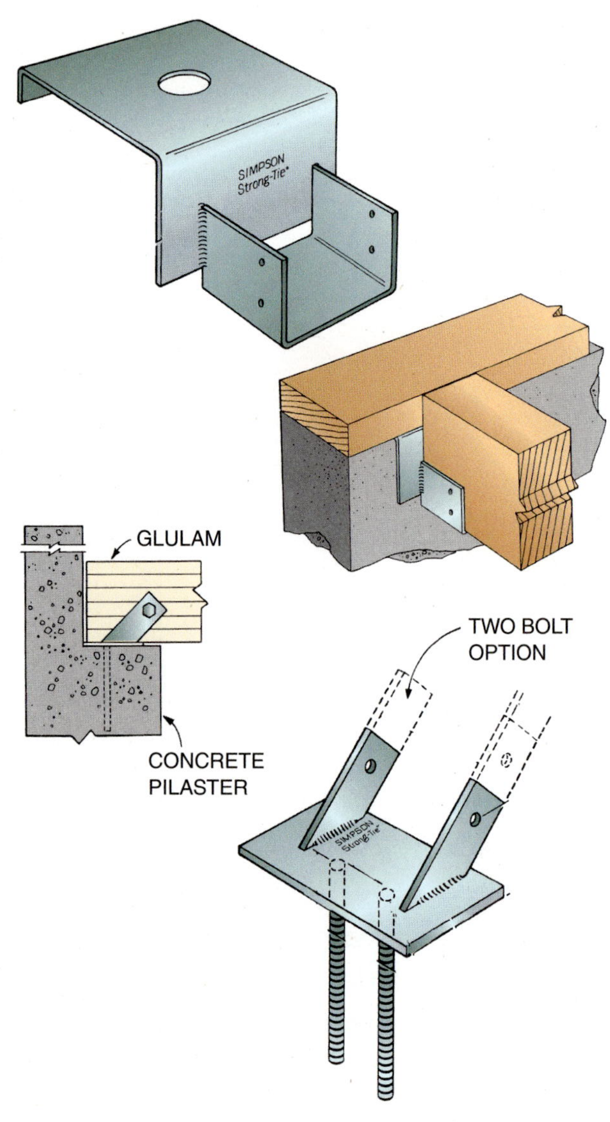


Courtesy of Simpson Strong-Tie Company.

**FIGURE 5-32** Girder and beam seats provide support from concrete walls.

plastic laminates for kitchen countertops. It is also used to bond other thin or flexible material that otherwise might require elaborate clamping devices.

## MASTICS

Several types of mastics are used throughout the construction trades. They come in cans or cartridges used in hand or air guns. These adhesives prove a stronger bond, and fewer fasteners are needed.

### Construction Adhesive

One type of mastic is called *construction adhesive.* It is used in a glued floor system, described in Chapter 9 on floor framing (Fig. 5-33). It can be used in cold weather, even on wet or frozen wood. It is also used on stairs to increase stiffness and eliminate squeaks.

### Panel Adhesive

*Panel adhesive* is used to apply wall paneling, foam insulation, gypsum board, and hardboard to wood, metal, and masonry. It is usually dispensed with a caulking gun.

**FIGURE 5-33** Applying panel adhesive to joists with a caulking gun for plywood subfloor.

### Troweled Mastics

Other types of mastics may be applied by hand for such purposes as installing vinyl base, vinyl floor tile, or ceramic wall tile. A notched trowel is usually used to spread the adhesive. The depth and spacing of the notches along the edges of the trowel determine the amount of adhesive left on the surface.

It is important to use a trowel with the correct notch depth and spacing. Failure to follow recommendations will result in serious consequences. Too much adhesive causes the excess to squeeze out onto the finished surface. This leaves no alternative but to remove the applied pieces, clean up, and start over. Too little adhesive may result in loose pieces.

## SUMMARY

- Nails are the first line of fasteners used to join building material.
- Nails come in a variety of sizes and materials for use in all situations.
- Screws provide excellent strength of fastening while also being easily removed.
- Screws come in a variety of sizes and materials for use in all situations.
- Bolts join materials with fine threads and nuts.
- Anchors come in many degrees of strength for a variety of applications.
- Connectors provide strength when joining large wood, steel, and masonry members together.
- Adhesives are designed to glue materials together with many strengths and flexibilities.

## GREEN CHECKLIST

- ☐ Use the correct fastener style and size for a particular application. This will ensure the building performs as expected and lasts a long time.

## REVIEW QUESTIONS

*Select the most appropriate answer.*

**1. The term that describes driving nails straight into the material is called**

a. face nailing.
b. back nailing.
c. toenailing.
d. all of the above.

**2. The length of an eight penny nail is**

a. 1½ inches.
b. 2 inches.
c. 2½ inches.
d. 3 inches.

**3. Fasteners coated with zinc to retard rusting are said to be**

a. coated.
b. dipped.
c. electroplated.
d. galvanized.

**4. Brads are**

a. types of screws.
b. small box nails.
c. small finishing nails.
d. kinds of stove bolts.

**5. Casing nails are used to fasten**

a. interior finish.
b. exterior finish.
c. door jambs.
d. roof shingles.

**6. Care should be taken when driving masonry nails because masonry nails are**

a. made of high-strength masonry.
b. thinner than box nails.
c. brittle.
d. sharp.

7. **On temporary structures, such as concrete forms, nails used for easy removal are**
   a. common nails.
   b. duplex nails.
   c. galvanized nails.
   d. brads.

8. **As a general rule, how should the length of a nail compare to the thickness of the material being fastened?**
   a. The same
   b. Twice as long
   c. 2½ times as long
   d. 3 times as long

9. **The large screw used in wood and sometimes with lead shields is called a**
   a. toggle screw.
   b. lag screw.
   c. conical screw.
   d. flat head screw.

10. **The name of a heavy-duty anchor used in masonry is the**
    a. wedge anchor.
    b. drop-in anchor.
    c. sleeve anchor.
    d. all of the above.

11. **Wood-to-wood framing anchors should be fastened in place with**
    a. roofing nails because of their larger head.
    b. common nails because of their extra length.
    c. anchor nails because of their special size.
    d. casing nails because of their proper sized head.

12. **When wood is attached to masonry, the fastener of choice depends on the**
    a. strength of the masonry.
    b. species of wood used.
    c. intended load to be placed on the anchor.
    d. frost level for the geographic area.

13. **A mastic used to attach floor systems is called**
    a. yellow glue.
    b. contact cement.
    c. construction adhesive.
    d. all of the above.

14. **Screws with self-tapping tips are**
    a. sheet metal.
    b. lags.
    c. common.
    d. Phillips.

15. **An anchor not used in hollow walls such as drywall is**
    a. conical screw.
    b. toggle bolt.
    c. split fast anchor.
    d. universal plug.

## KNOW YOUR CODES

Correct sized, proper type, and quantity of fasteners are important. When the load on a building or building section is analyzed, engineers determine the fastener specifications. It is very important the carpenter follow these requirements. Safety of the occupants is at stake.

## WHAT'S WRONG WITH THIS PICTURE?

Carefully study **Figure 5-34** and think about what is wrong. Consider all possibilities.

### WRONG

**FIGURE 5-34** The nail type is a poor choice. The nail shown in the photograph is a finish nail and should not be used to apply sheathing. If placed under stress, the small head would pull through the plywood instead of holding it to the framing member. Other problems shown include sheathing that is not aligned with the edge and the use of a drywall screw, seen here in the upper right corner.

### RIGHT

**FIGURE 5-35** The proper size and type of nail to use on ½" plywood would be a 6d common as shown here.

SECTION 2

# Rough Carpentry

# Building Plans and Codes

**OBJECTIVES** *After completing this unit, the student should be able to:*

- Describe and explain the function of the various kinds of drawings contained in a set of blueprints.
- Demonstrate how specifications are used.
- Identify various types of lines and read dimensions.
- Identify and explain the meaning of symbols and abbreviations used on a set of prints.
- Read and interpret plot, foundation, floor, and framing plans.
- Define and explain the purpose of building codes and zoning laws.
- Explain the requirements for obtaining a building permit and the duties of a building inspector.

# GLOSSARY OF BUILDING PLANS AND CODES TERMS

**detail** close-up view of a plan or section

**elevation** a drawing in which the height of the structure or object is shown; also, the height of a specific point in relation to another reference point

**isometric** a drawing in which three surfaces of an object are seen in one view, with the base of each surface drawn at a 30-degree angle

**orthographic** a multiview drawing

**plan** in an architectural drawing, an object drawn as viewed from above

**pythagorean theorem** a mathematical expression that states the sum of the square of the two sides of a right triangle equals the square of the diagonal side

**section** drawing showing a vertical cut-view through an object or part of an object

Before construction can begin, workers must be able to picture how the building will look. The owners and architects have their ideas, dreams, and desires for the building. It is up to the tradesperson to make those visions come alive.

The method adopted for this communication is called the *blueprint*, whose name comes from an early processing of paper that turned the entire sheet blue except for where lines and words were drawn. These prints embody the ideas of the designer, and in order for these dreams to come true, the tradesperson must be able to interpret and understand what the prints represent.

Also, methods of construction vary from one locale to another. To ensure that buildings are built according to safety and design standards for a particular locale, building codes and ordinances were developed. The set of prints represent the decisions for future building made from these variations. The tradesperson must be able to read and interpret a set of prints.

# BUILDING PLANS

An *architect* designs buildings and creates drawings to reflect those designs. These drawings use various kinds of lines, dimensions, notes, symbols, and abbreviations to describe a structure to be built. The method of drawing, lettering, and dimensioning may vary slightly with the drafter's style. Different kinds of drawings are required for the construction of a building. These are called plans, elevations, sections, and details. When put together, they constitute a *set of prints.*

Today most architectural plans, also called architectural drawings or construction drawings, are done using a CADD (computer-aided drafting and design) program. CADD drawings are produced faster and more easily than drawings done by hand. Changes to the plan can be made faster, allowing them to be easily customized to the desires of the customer. Once the plans are finished, they are sent to a plotter or copier, which prints as many copies as needed.

Prints are drawn using symbols and a standardized language. To adequately read and interpret these prints, the builder must be able to understand the language of architectural plans. While this is not necessarily difficult, it can be confusing.

# TYPES OF VIEWS

The architect can choose from any of several types of views to describe a building with clarity, and often multiple views are prepared. Because many people and many different trades are needed to build a house, different views are required that contain different information.

## PICTORIAL VIEW

Pictorial drawings are usually three-dimensional (3D) *perspective* or **isometric** views (Fig. 6-1). The lines in a perspective view diminish in size as they converge toward vanishing points located on a line called the *horizon.* In an isometric drawing, the horizontal lines are drawn at 30-degree angles. All lines are drawn to actual scale. They do not diminish or converge as in perspective drawings.

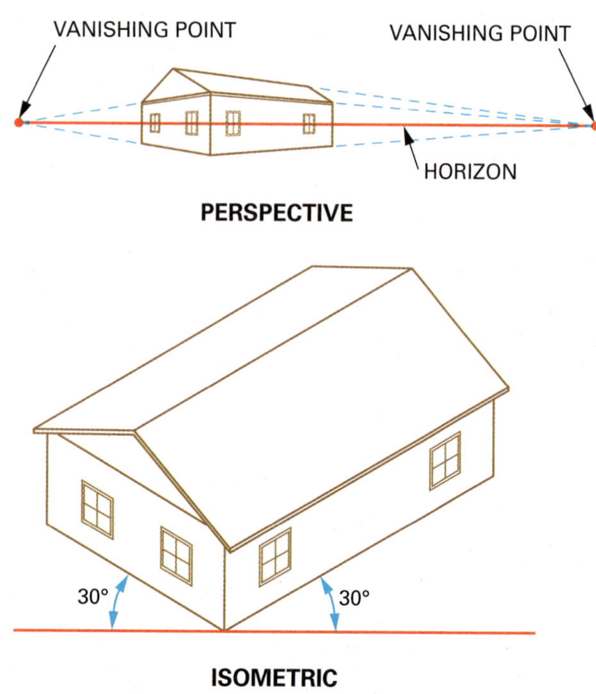


**FIGURE 6-1** Pictorial views used in architectural drawings.

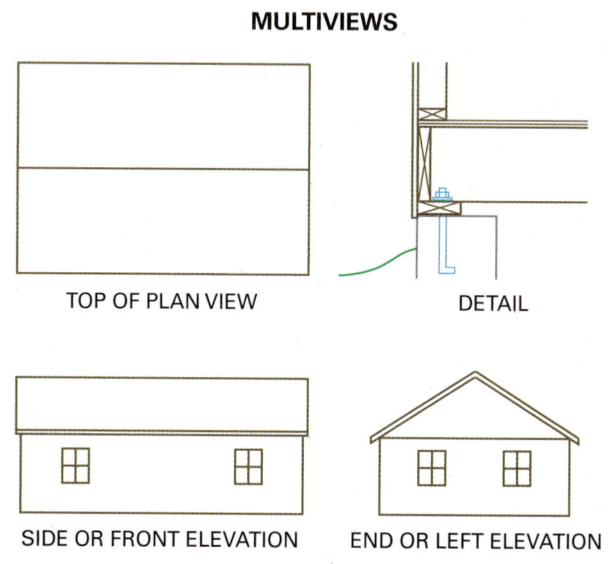


**FIGURE 6-3** The two-dimensional views used in architectural drawings.

**FIGURE 6-2** A presentation drawing is usually a perspective view.

A *presentation drawing* is usually a perspective view. It shows the building from a desirable vantage point to display its most interesting features (Fig. 6-2). Walks, streets, shrubs, trees, vehicles, and even people may be drawn. Presentation drawings are often colored for greater eye appeal. Presentation drawings provide little information for construction purposes. They are usually used as a marketing tool to show the client how the completed building will look.

## MULTIVIEW DRAWINGS

Different kinds of *multiview* drawings, also called **orthographics**, are required for the construction of a building. Multiview drawings are two-dimensional (2D) and offer separate views of the building (Fig. 6-3). The bottom view is never used in architectural drawings.

These different views are called **plan view**, **elevation**, **section view**, and **details**. When put together, they constitute a set of architectural plans. *Plan views* show the building from above, looking down. There are many kinds of plan views for different stages of construction (Fig. 6-4). *Elevations* show the building as seen from the street. They can show front, back, right side, and left side. *Section views* show a cross section as if the building were sliced open to reveal its skeleton (Fig. 6-5). *Details* are blown up, with zoomed-in views of various items to show a closer view.

## PLAN VIEWS

***Plot Plan.*** The *plot plan* shows information about the lot, such as property lines and directions and measurements for the location of the building, walks, and driveways (Fig. 6-6). It shows elevation heights and direction of the sloping ground. The drawing simulates a view looking down from a considerable height. It is made at a small scale because of the relatively large area it represents. This view helps the builder locate the building on the site and helps local officials to estimate the impact of the project on the lot.

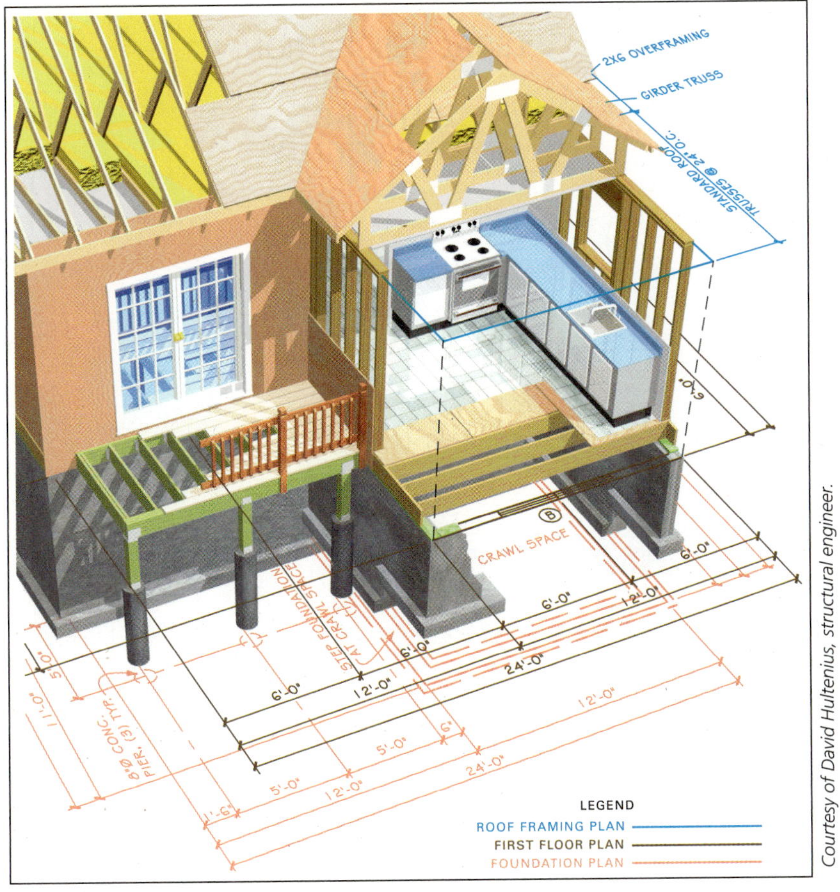


Courtesy of David Hultenius, structural engineer.

**FIGURE 6-4** Plan views are horizontal cut views through a building.

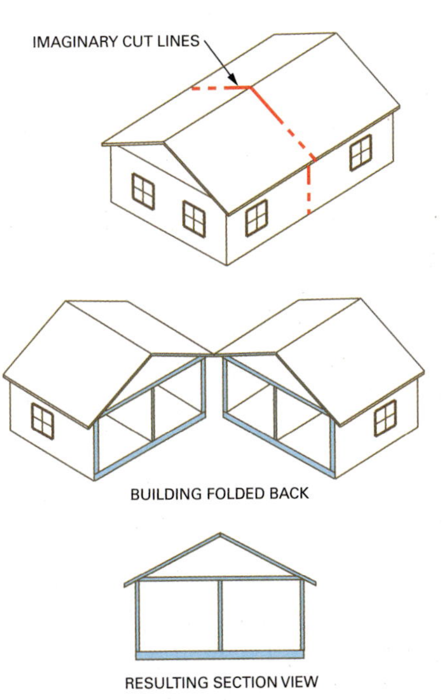


**FIGURE 6-5** Section views are cutaway views.

## Green Tip

Orienting a building to address the sun will affect a lifetime of heating and cooling dollars. Warm climates should include natural shading and northerly exposures. Cooler areas should orient longer building sides and larger glass toward southerly exposures.

***Foundation Plan.*** The *foundation plan* (Fig. 6-7) shows a horizontal cut through the foundation walls. It shows the shape and dimensions of the foundation walls and footings. Windows, doors, and stairs located in this level are included. First-floor framing material size, spacing, and direction are sometimes included.

***Floor Plan.*** The **floor plan** is a view of the horizontal cut made about 4 to 5 feet above the floor. It shows

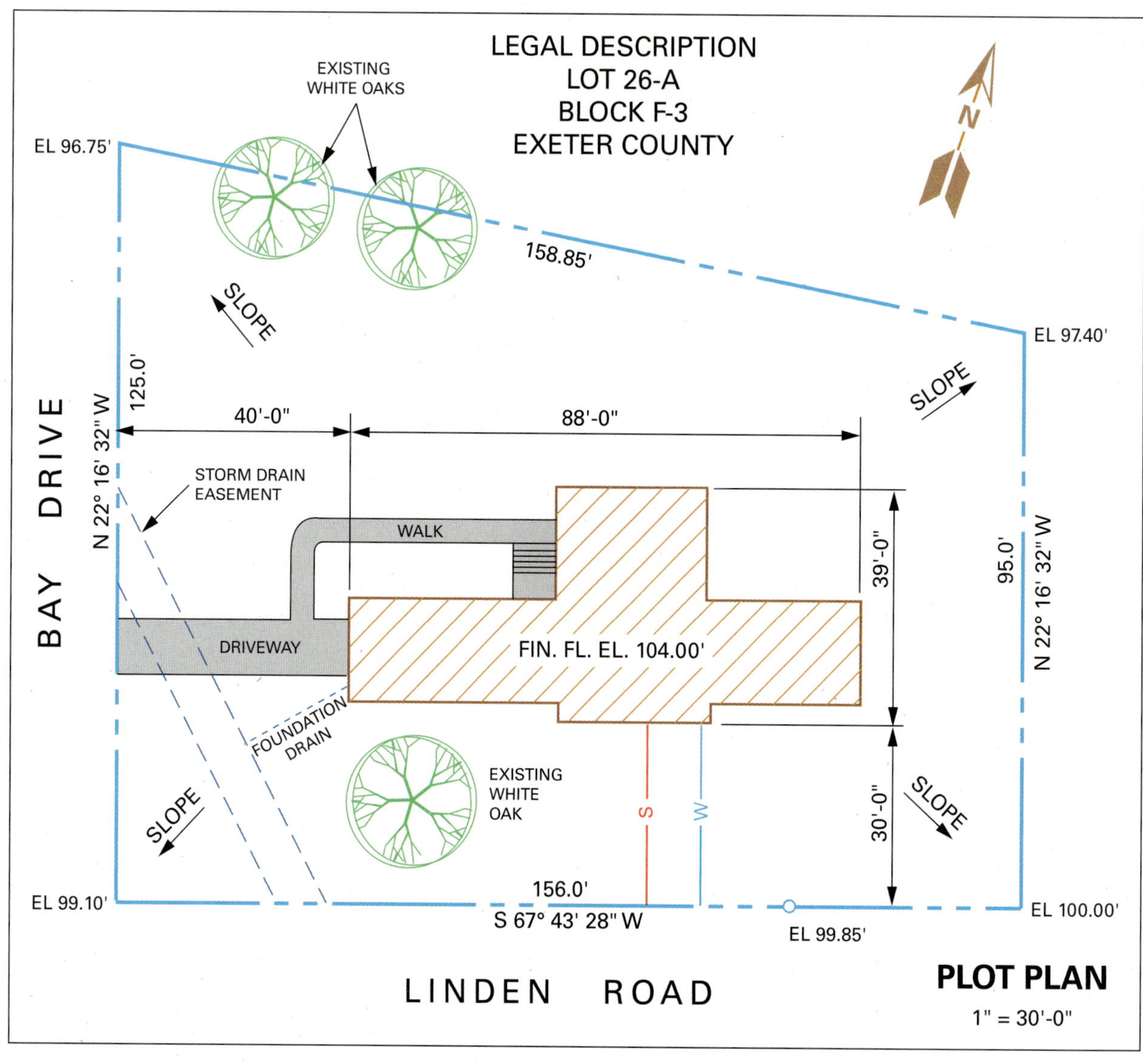


**FIGURE 6-6** A plot plan is a view of construction from about 500 feet above.

the locations of walls and partitions, windows, and doors and fixtures appropriate for each room (Fig. 6-8). Dimensions are included.

*Framing Plan.* *Framing plans* are not always found in a set of architectural drawings. When used, they may be of the floor or roof framing. They show the support beams and girders as well as the size, direction, and spacing of the framing members (Fig. 6-9).

## ELEVATIONS

*Elevations* are a group of drawings (Fig. 6-10) that show the shape and finishes of all sides of the exterior of a building. *Interior elevations* are drawings of certain interior walls. The most common are kitchen and bathroom wall elevations. They show the design and size of cabinets built on the wall (Fig. 6-11). Other walls that have special features, such as a fireplace, may require an elevation drawing. Occasionally found in some sets of plans are *framing elevations*. Similar to framing plans, they show the spacing, location, and sizes of wall framing members. No further description of framing drawings is required to be able to interpret them.

## SECTION VIEWS

A set of architectural plans may have many *section* views. Each is designed to reveal the structure or skeleton view of a particular part of the building (Fig. 6-12). A section reference line is found on the plans or elevations to identify the section being viewed.

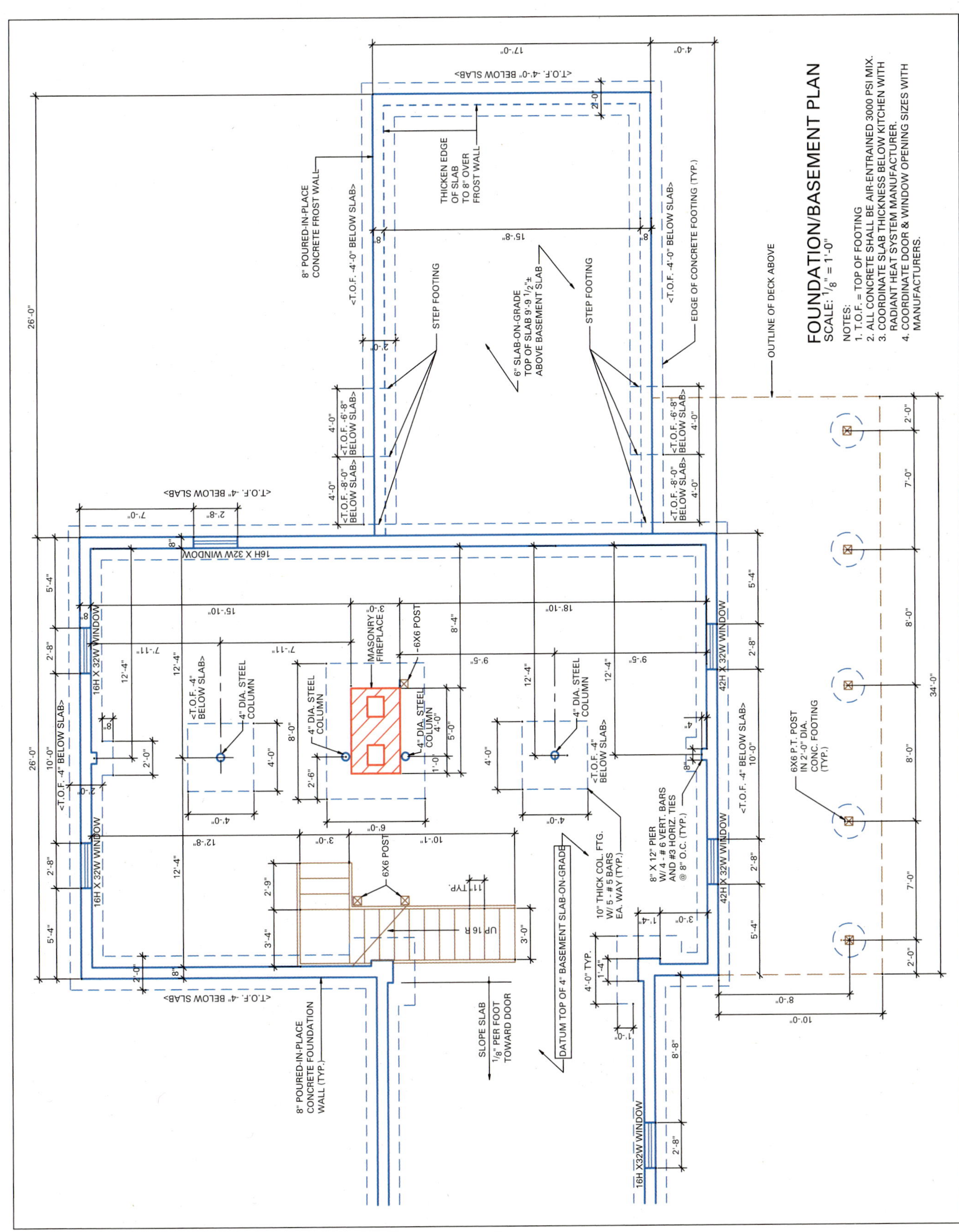


Courtesy of David Hultenius, structural engineer.

**FIGURE 6-7** A foundation plan.

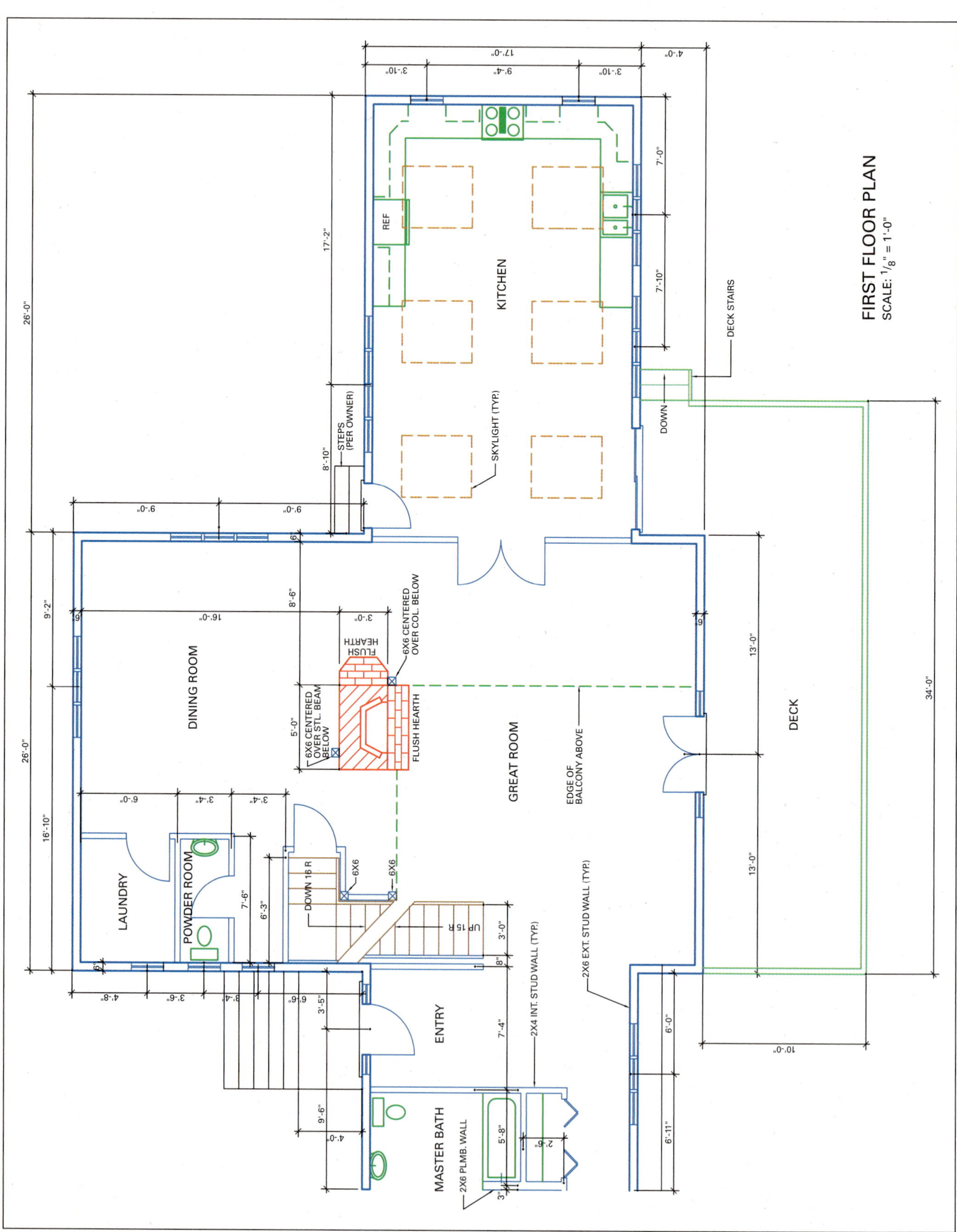


Courtesy of David Hultenius, structural engineer.

**FIGURE 6-8** A floor plan.

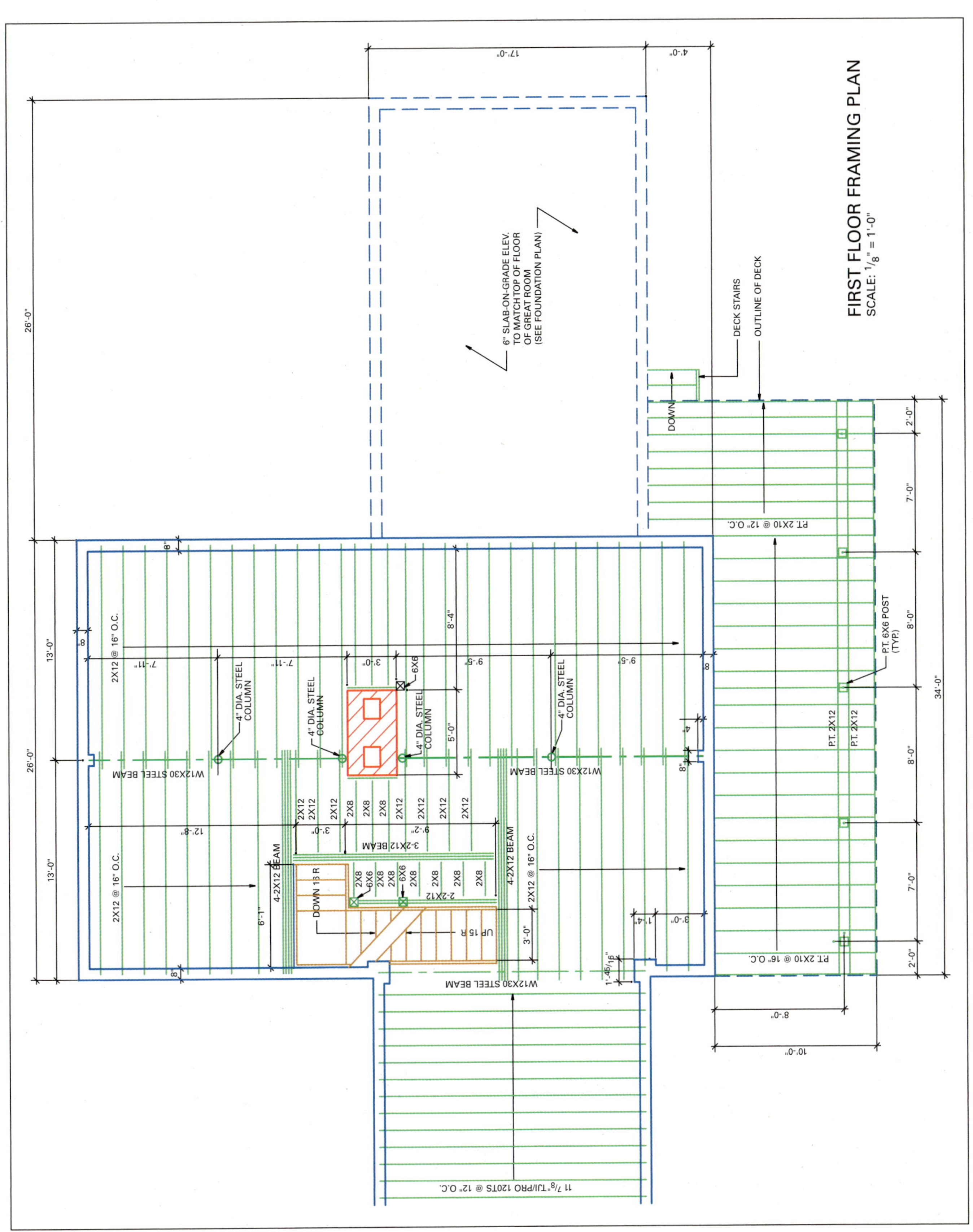

Courtesy of David Hultenius, structural engineer.

**FIGURE 6-9** A first-floor framing plan.

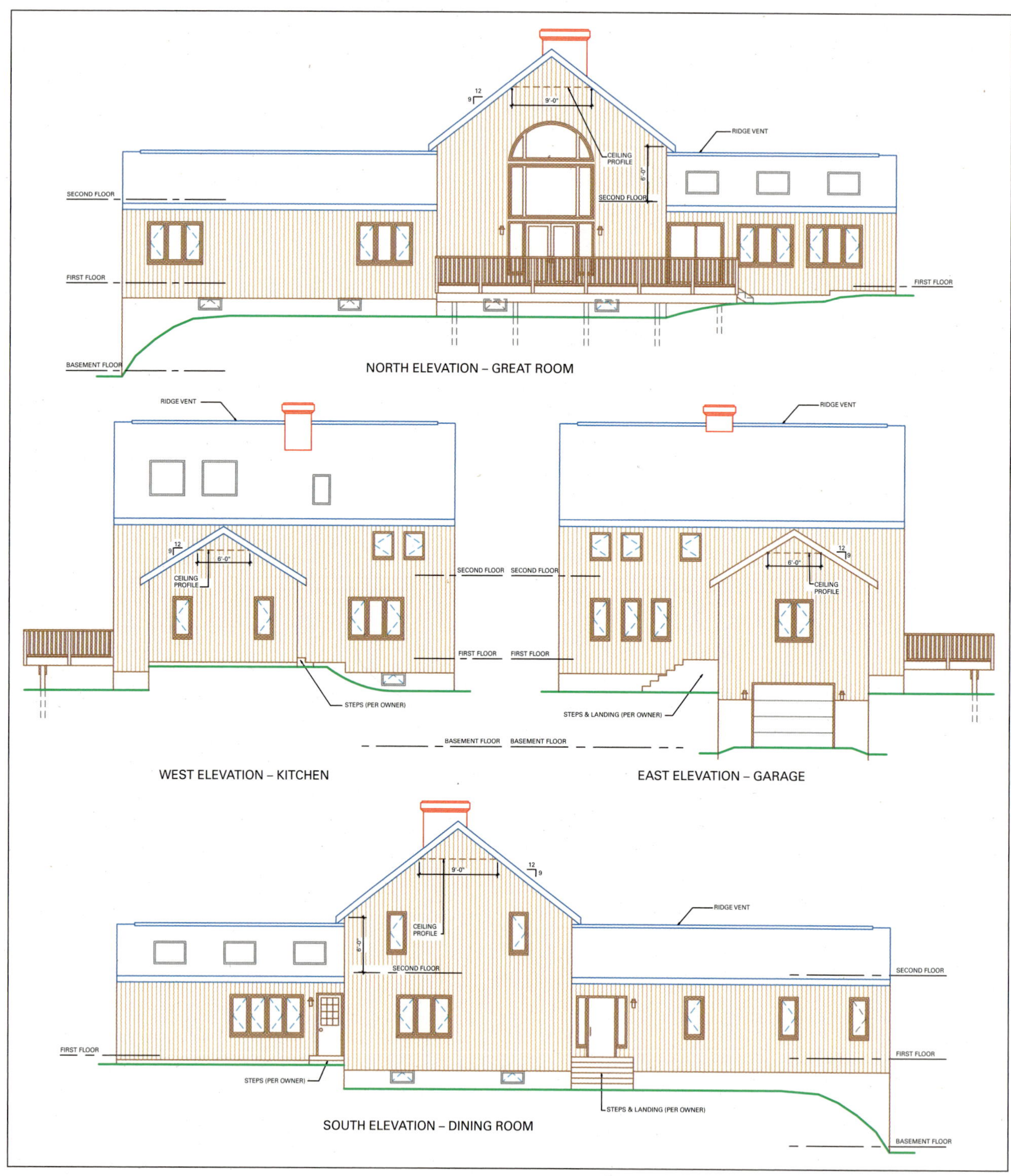


Courtesy of David Hultenius, structural engineer.

**FIGURE 6-10** Elevations.

## DETAILS

To make parts of the construction more clear, it is usually necessary to draw *details*. Details are small parts drawn at a very large scale, even full size if necessary. Their existence is revealed on the plan and elevation views using a symbol (Fig. 6-13). The location of the symbol shows the location of the vertical cut through the building. This symbol may have different shapes, yet all have numbers or letters that refer to the page where the detail is shown.

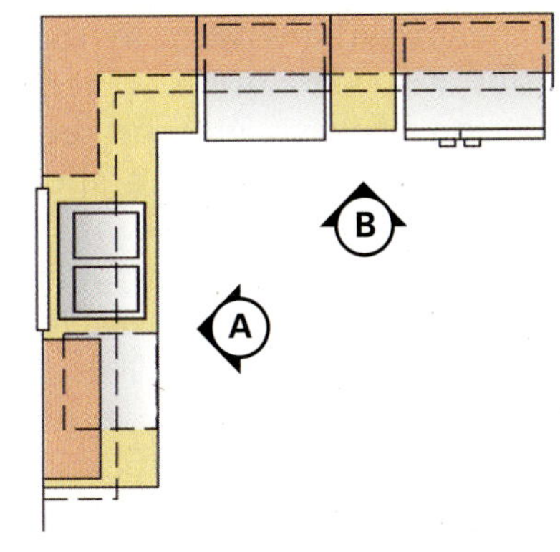


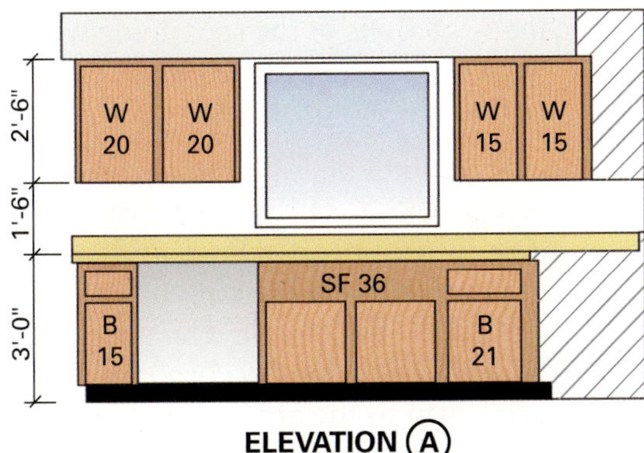


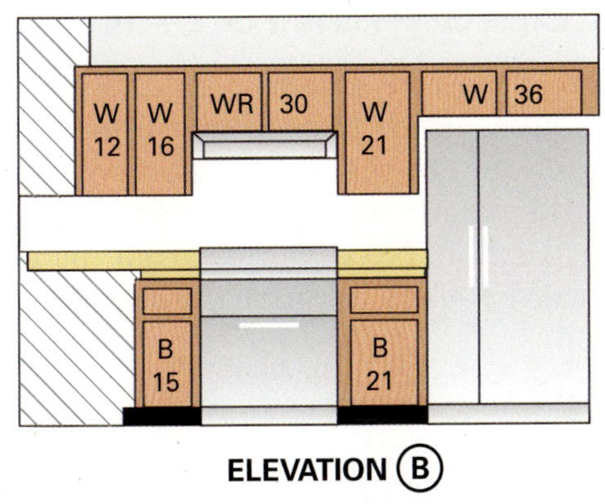


**FIGURE 6-11** Interior wall elevations.

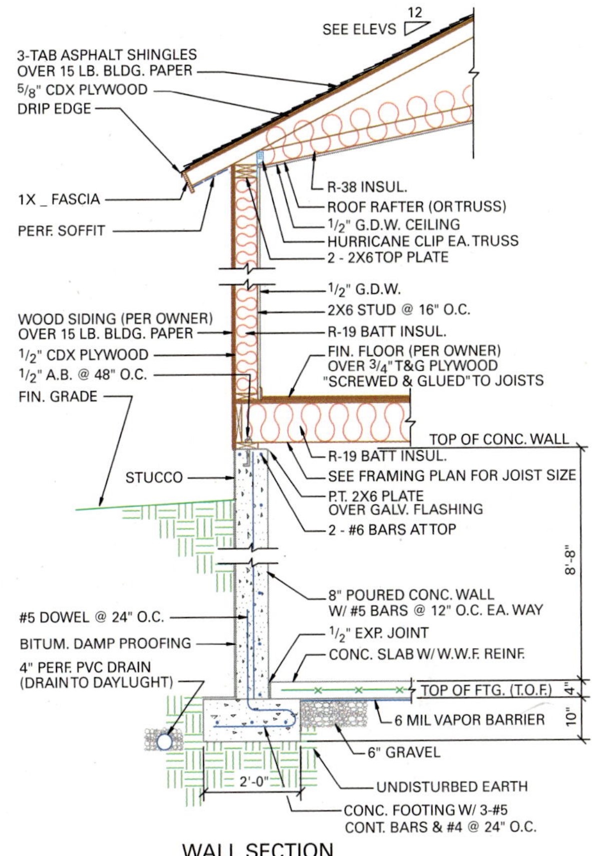


*Courtesy of David Hultenius, structural engineer.*

**FIGURE 6-12** A section is a view of a vertical cut through part of the construction.

## OTHER DRAWINGS

Drawings relating to electrical work, plumbing, heating, and ventilating may be on separate sheets in a set of prints. For smaller projects, separate plans are not always needed. All necessary information can usually be found on the floor plan.

The carpenter is responsible for building to accommodate wiring, pipes, and ducts. He or she should be able to read these plans with some degree of proficiency to understand the work involved.

## SCHEDULES

In addition to drawings, printed instructions are included in a set of drawings. *Window schedules* and *door schedules* (Fig. 6-14) give information about the location, size, and kind of windows and doors to be installed in the building. Each unit is given a number or letter. A corresponding number or letter is found on the floor plan to show the location of the unit. Windows may be identified by letters and doors by numbers. A *finish schedule* (Fig. 6-15) gives information on the kind of finish material to be used on the floors, walls, and ceilings of the individual rooms.

## SPECIFICATIONS

**Specifications**, commonly called *specs*, are written to give information that cannot be completely provided in the drawings or schedules. They supplement

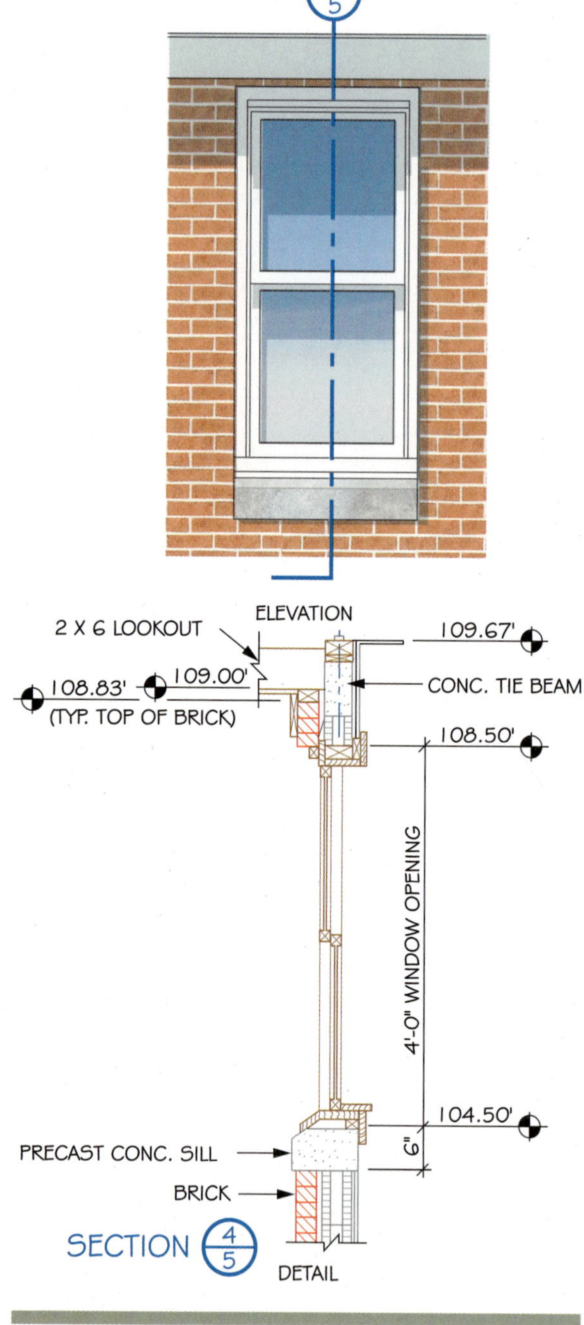


**FIGURE 6-13** Detail of a window.

the working drawings with more complete descriptions of the methods, materials, and quality of construction. If there is a conflict, the specifications take precedence over the drawings. Any conflict should be pointed out to the architect so corrections can be made.

The amount of detail contained in the specs will vary, depending on the size of the project. On small jobs, they may be written by the architect. On larger jobs, a **specifications** writer, trained in the construction process, may be required. For complex commercial projects, a specifications guide, used by spec writers, has been developed by the Construction Specifications Institute (CSI). The guide has sixteen major divisions, each containing a number of subdivisions (Fig. 6-16). An example of the content and the manner in which specifications are written, using the specification guide, is shown in Figure 6-17.

For residential and some light commercial construction, many sections of the spec guide would not apply. A shortened version is then used, eliminating divisions 12,000, 13,000, and 14,000.

Notations or notes are often made on the sheets of the drawings (Fig. 6-18). They assist the reader by providing information where and when it is needed. This saves time in looking up information in the specifications document. The ability to read notations and specifications accurately is essential in order to conform to the architect's design.

## BUILDING PRINTS LANGUAGE

Carpenters must be able to read and understand the combination of lines, dimensions, symbols, and notations on the drawings. Only then can they build exactly as the architect has designed the construction. No deviation from the building plans may be made without the approval of the architect.

### Scales

Full-sized drawings of a building are made to *scale.* This means that each line in the drawing is reduced proportionally to a size that clearly shows the information and can be handled conveniently. The scale of a drawing is clearly stated.

The most commonly used scale found on prints is ¼ inch equals 1 foot. This is indicated as ¼" = 1'-0". This means that every ¼ inch on the drawing will equal 1 foot in the building. Floor plans and exterior elevations for most residential buildings are drawn at this scale. To show the location of a building on a plot plan, the scale is reduced to fit the drawing on the paper. The architect may use a scale of 1⁄16" = 1'-0". To show section views and details more clearly, larger scales of ½" = 1'-0" or up to 3" = 1'-0" are used (Fig. 6-19).

| WINDOW SCHEDULE (SIZE OF OPENING FOR FRAME) | | | | |
|---|---|---|---|---|
| TYPE | HEIGHT | WIDTH | STYLE | MATERIAL |
| 1 | $4^0$ | $2^0$ | DOUBLE HUNG | VINYL |
| 2 | $6^0$ | $4^0$ | DOUBLE HUNG | VINYL |
| 3 | $6^0$ | $3^0$ | DOUBLE HUNG | VINYL |
| 4 | $3^0$ | $4^0$ | CASEMENT | VINYL |
| 5 | $5^0$ | $3^0$ | AWNING | VINYL |
| 6 | $6^0$ | $8^0$ | AWNING | VINYL |
| 7 | $4^0$ | $3^0$ | DOUBLE HUNG | VINYL |

| DOOR SCHEDULE | | | |
|---|---|---|---|
| TYPE | HEIGHT | STYLE | MATERIAL |
| A | $2^8 \times 6^8$ | ENTRY SOLID CORE | WOOD |
| B | $2^0 \times 6^8$ | LOUVER HINGED | WOOD |
| C | $2^0 \times 6^8$ | HOLLOW CORE | WOOD |
| D | $5^0 \times 6^8$ | FRENCH DOORS | WOOD & GLASS |
| E | $3^0 \times 6^8$ | SOLID CORE FULL LIGHT | VINYLCLAD |
| F | $3^0 \times 6^8$ | SOLID CORE | WOOD |
| G | $5^0 \times 6^8$ | SLIDING GLASS | GLASS & ALUM. |
| H | $3^0 \times 6^8$ | SOLID CORE HALF LIGHT | VINYL & GLASS |

*Courtesy of PTEC-Clearwater-Architectural Drafting Department.*

**FIGURE 6-14** A typical window and door schedule.

| FINISH SCHEDULE | | | | | | | |
|---|---|---|---|---|---|---|---|
| ROOM | WALLS | PAINT COLORS | BASE | FLOOR | CEILING | CORNICE | REMARKS |
| LIV. RM. | DRY WALL | BONE | WOOD | OAK | PLASTER | WOOD | BOOKCASE |
| DIN. RM. | " | " | " | " | " | PICT. MLDG | CLIPBD. |
| KITCHEN | " | EGG SHELL | TILE | VINYL | " | —— | —— |
| HALL | " | " | WOOD | OAK | " | WOOD | SEE DTL. |
| ENTRY | " | " | " | " | " | —— | —— |

**FIGURE 6-15** A typical finish schedule.

## Types of Lines

Some lines in an architectural drawing look darker than others. They are broader so they stand out clearly from other lines. This variation in width is called *line contrast*. This technique, like all architectural drafting standards, is used to make the drawing easier to read and understand (Fig. 6-20).

Lines that outline the object being viewed are broad, solid lines called *object lines*. To indicate an object not visible in the view, a *hidden line*

DIVISION 1—GENERAL REQUIREMENTS
- 01010 Summary of Work
- 01100 Alternatives
- 01150 Measurement & Payment
- 01200 Project Meetings
- 01300 Submittals
- 01400 Quality Control
- 01500 Temporary Facilities & Controls
- 01600 Material & Equipment
- 01700 Project Closeout

DIVISION 2—SITE WORK
- 02010 Subsurface Exploration
- 02100 Clearing
- 02110 Demolition
- 02200 Earthwork
- 02250 Soil Treatment
- 02300 Pile Foundations
- 02350 Caissons
- 02400 Shoring
- 02500 Site Drainage
- 02550 Site Utilities
- 02600 Paving & Surfacing
- 02700 Site Improvements
- 02800 Landscaping
- 02850 Railroad Work
- 02900 Marine Work
- 02950 Tunneling

DIVISION 3—CONCRETE
- 03100 Concrete Formwork
- 03150 Forms
- 03200 Concrete Reinforcement
- 03250 Concrete Accessories
- 03300 Cast-in-Place Concrete
- 03350 Specially Finished (Architectural) Concrete
- 03360 Specially Placed Concrete
- 03400 Precast Concrete
- 03500 Cementitious Decks
- 03600 Grout

DIVISION 4—MASONRY
- 04100 Mortar
- 04150 Masonry Accessories
- 04200 Unit Masonry
- 04400 Stone
- 04500 Masonry Restoration & Cleaning
- 04550 Refractories

DIVISION 5—METALS
- 05100 Structural Metal Framing
- 05200 Metal Joists
- 05300 Metal Decking
- 05400 Lightgage Metal Framing
- 05500 Metal Fabrications
- 05700 Ornamental Metal
- 05800 Expansion Control

DIVISION 6—WOOD & PLASTICS
- 06100 Rough Carpentry
- 06130 Heavy Timber Construction
- 06150 Trestles
- 06170 Prefabricated Structural Wood
- 06200 Finish Carpentry
- 06300 Wood Treatment
- 06400 Architectural Woodwork
- 06500 Prefabricated Structural Plastics
- 06600 Plastic Fabrications

DIVISION 7—THERMAL & MOISTURE PROTECTION
- 07100 Waterproofing
- 07150 Dampproofing
- 07200 Insulation
- 07300 Shingles & Roofing Tiles
- 07400 Preformed Roofing & Siding
- 07500 Membrane Roofing
- 07570 Traffic Topping
- 07600 Flashing & Sheet Metal
- 07800 Roof Accessories
- 07900 Sealants

DIVISION 8—DOOR & WINDOWS
- 08100 Metal Doors & Frames
- 08200 Wood & Plastic Doors
- 08300 Special Doors
- 08400 Entrances & Storefronts
- 08500 Metal Windows
- 08600 Wood & Plastic Windows
- 08650 Special Windows
- 08700 Hardware & Specialties
- 08800 Glazing
- 08900 Window Walls/Curtain Walls

DIVISION 9—FINISHES
- 09100 Lath & Plaster
- 09250 Gypsum Wallboard
- 09300 Tile
- 09400 Terrazzo
- 09500 Acoustical Treatment
- 09540 Ceiling Suspension Systems
- 09550 Wood Flooring
- 09650 Resilient Flooring
- 09680 Carpeting
- 09700 Special Flooring
- 09760 Floor Treatment
- 09800 Special Coatings
- 09900 Painting
- 09950 Wall Covering

DIVISION 10—SPECIALTIES
- 10100 Chalkboards & Tackboards
- 10150 Compartments & Cubicles
- 10200 Louvers & Vents
- 10240 Grilles & Screens
- 10260 Wall & Corner Guards
- 10270 Access Flooring
- 10280 Specialty Modules
- 10290 Pest Control
- 10300 Fireplaces
- 10350 Flagpoles
- 10400 Identifying Devices
- 10450 Pedestrian Control Devices
- 10500 Lockers
- 10530 Protective Covers
- 10550 Postal Specialties
- 10600 Partitions
- 10650 Scales
- 10670 Storage Shelving
- 10700 Sun Control Devices (Exterior)
- 10750 Telephone Enclosures
- 10800 Toilet & Bath Accessories
- 10900 Wardrobe Specialties

DIVISION 11—EQUIPMENT
- 11050 Built-in Maintenance Equipment
- 11100 Bank & Vault Equipment
- 11150 Commercial Equipment
- 11170 Checkroom Equipment
- 11180 Darkroom Equipment
- 11200 Ecclesiastical Equipment
- 11300 Educational Equipment
- 11400 Food Service Equipment
- 11480 Vending Equipment
- 11500 Athletic Equipment
- 11500 Industrial Equipment
- 11600 Laboratory Equipment
- 11630 Laundry Equipment
- 11650 Library Equipment
- 11700 Medical Equipment
- 11800 Mortuary Equipment
- 11830 Musical Equipment
- 11850 Parking Equipment
- 11860 Waste Handling Equipment
- 11870 Loading Dock Equipment
- 11880 Detention Equipment
- 11900 Residential Equipment
- 11970 Theater & Stage Equipment
- 11990 Registration Equipment

DIVISION 12—FURNISHINGS
- 12100 Artwork
- 12300 Cabinets & Storage
- 12500 Window Treatment
- 12550 Fabrics
- 12600 Furniture
- 12670 Rugs & Mats
- 12700 Seating
- 12800 Furnishing Accessories

DIVISION 13—SPECIAL CONSTRUCTION
- 13010 Air Supported Structures
- 13050 Integrated Assemblies
- 13100 Audiometric Room
- 13250 Clean Room
- 13350 Hyperbaric Room
- 13400 Incinerators
- 13440 Instrumentation
- 13450 Insulated Room
- 13500 Integrated Ceiling
- 13540 Nuclear Reactors
- 13550 Observatory
- 13600 Prefabricated Structures
- 13700 Special Purpose Rooms & Buildings
- 13750 Radiation Protection
- 13770 Sound & Vibration Control
- 13800 Vaults
- 13850 Swimming Pools

DIVISION 14—CONVEYING SYSTEMS
- 14100 Dumbwaiters
- 14200 Elevators
- 14300 Hoists & Cranes
- 14400 Lifts
- 14500 Material Handling Systems
- 14570 Turntables
- 14600 Moving Stairs & Walks
- 14700 Tube Systems
- 14800 Powered Scaffolding

DIVISION 15—MECHANICAL
- 15010 General Provisions
- 15050 Basic Materials & Methods
- 15180 Insulation
- 15200 Water Supply & Treatment
- 15300 Waste Water Disposal & Treatment
- 15400 Plumbing
- 15500 Fire Protection
- 15600 Power or Heat Generation
- 15650 Refrigeration
- 15700 Liquid Heat Transfer
- 15800 Air Distribution
- 15900 Controls & Instrumentation

DIVISION 16—ELECTRICAL
- 16010 General Provisions
- 16100 Basic Materials & Methods
- 16200 Power Generation
- 16300 Power Transmission
- 16400 Service & Distribution
- 16500 Lighting
- 16600 Special Systems
- 16700 Communications
- 16850 Heating & Cooling
- 16900 Controls & Instrumentation

**FIGURE 6-16** Construction Specifications Institute, CSI format for specifications.

**Division 6-Wood & Plastics**

Section 06200—Finish Carpentry

*General:* This section covers all finish woodwork and related items not covered elsewhere in these specifications. The contractor shall furnish all materials, labor, and equipment necessary to complete the work, including rough hardware, finish hardware, and specialty items.

*Protection of Materials:* All millwork (finish woodwork*) and trim is to be delivered in a clean and dry condition and shall be stored to insure proper ventilation and protection from dampness. Do not install finish woodwork until concrete, masonry, plaster, and related work is dry.

*Materials:* All materials are to be the best of their respective kind. Lumber shall bear the mark and grade of the association under whose rules it is produced. All millwork shall be kiln-dried to a maximum moisture content of 12%.

1. Exterior trim shall be select grade white pine, S4S.
2. Interior trim and millwork shall be select grade white pine, thoroughly sanded at the time of installation.

*Installation:* All millwork and trim shall be installed with tight fitting joints and formed to conceal future shrinkage due to drying. Interior woodwork shall be mitered or coped at corners (cut in a special way to form neat joints*). All nails are to be set below the surface of the wood and concealed with an approved putty or filler.

*(explanations in parentheses have been added to aid the student.)

**FIGURE 6-17** Sample specifications following the CSI format.

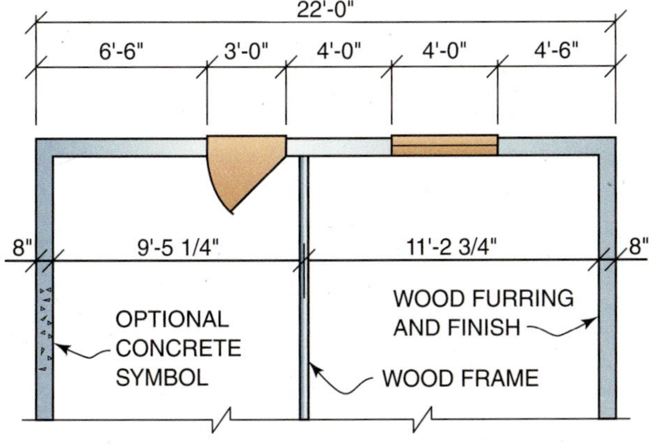


**FIGURE 6-18** Some typical general notes.

consisting of short, fine, uniform dashes is used. Hidden lines are used only when necessary. Otherwise, the drawing becomes confusing to read.

*Centerlines* are indicated by a fine, long dash, then a short dash, then a long dash, and so on. They show the centers of doors, windows, partitions, and

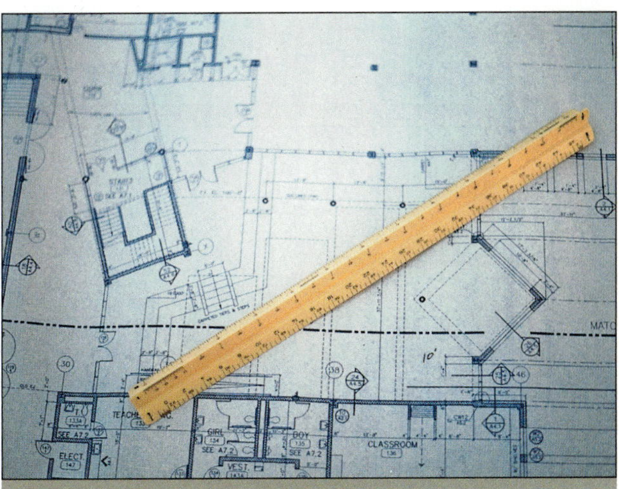

**FIGURE 6-19** The architect's scale.

similar parts of the construction. A *section reference* or *cutting-plane line* is sometimes a broad line consisting of a long dash followed by two short dashes. At its ends are arrows. The arrows show the direction in which the cross-section is viewed. Letters identify the cross-sectional view of that specific part of the building. More elaborate methods of labeling section reference lines are used in larger, more complicated sets of plans (Fig. 6-21). Section drawings may be on the same page as the reference line or on other pages.

A *break line* is used in a drawing to terminate part of an object that, in actuality, continues. It can be used only when there is no change in the drawing at the break. Its purpose is to shorten the drawing to better use space.

A *dimension line* is a fine, solid line used to indicate the location, length, width, or thickness of an object. It is terminated with arrowheads, dots, or slashes (Fig. 6-22). *Extension lines* are fine, solid lines projecting from an object to show the extent of a dimension. A *leader line* is a fine solid line. It terminates with an arrowhead and points to an object from a notation.

## Dimensions

Dimension lines on a set of print are generally drawn as continuous lines. The dimension appears above and near the center of the line. All dimensions on vertical lines should appear above the line when the print is rotated 1/4 turn clockwise. Extension lines are drawn from the object, but not touching the object, so that the end point of the dimension is

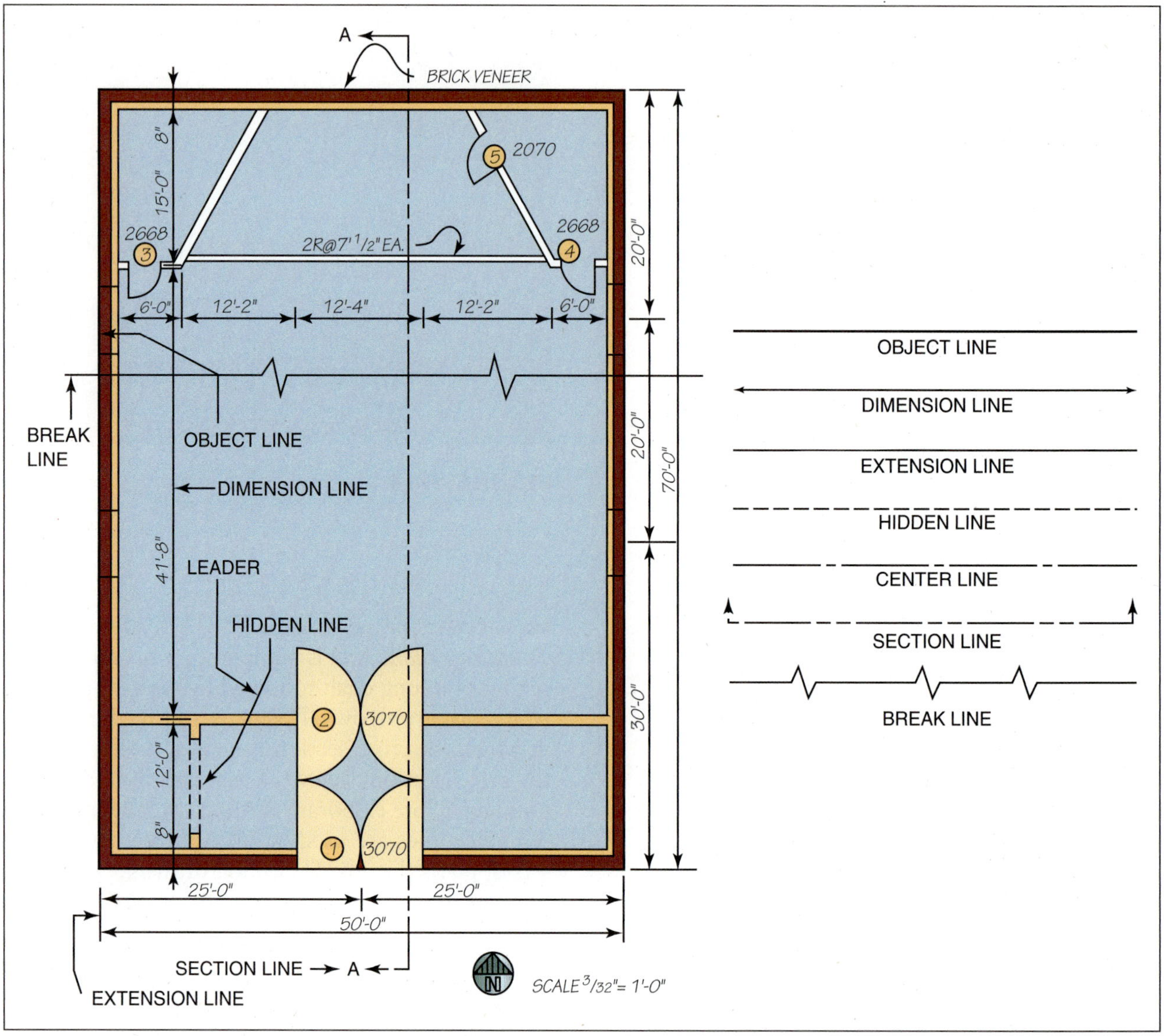

**FIGURE 6-20** Types of lines on architectural drawings.

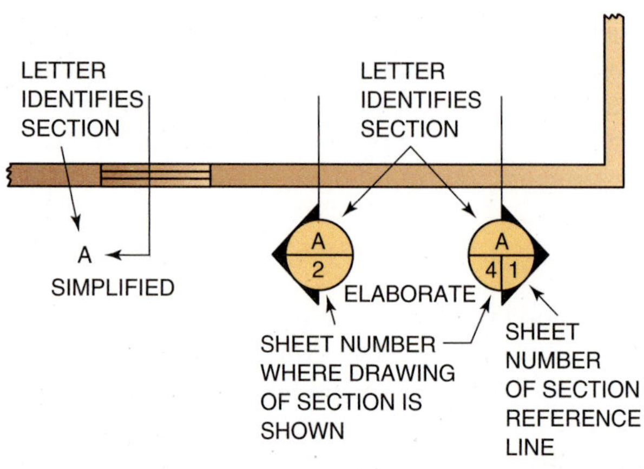


**FIGURE 6-21** Several ways of labeling section reference lines.

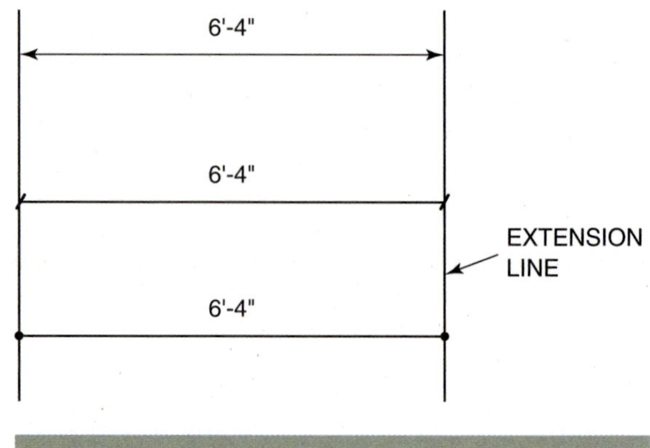


**FIGURE 6-22** Several methods of terminating dimension lines.

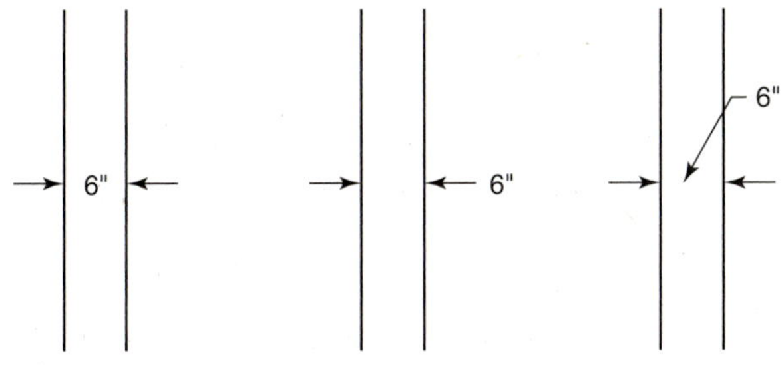


**FIGURE 6-23** Dimensioning small spaces.

clearly defined. When the space is too small to permit dimensions to be shown clearly, they may be drawn as shown in Figure 6-23. Dimensions on architectural blueprints are given in feet and inches, such as 3'-6", 4'-8", and 13'-7". Dimensions of under 1 foot are given in inches, as 10", 8", and so on.

## Symbols

Symbols are used on drawings to represent objects in the building, such as doors, windows, cabinets, plumbing, and electrical fixtures. Others are used in regard to the construction, such as for walls, stairs, fireplaces, and electrical circuits. They may be used for identification purposes, such as those used for section reference lines. The symbols for various construction materials, such as lumber, concrete, sand, and earth (Fig. 6-24), are used when they make the drawing easier to read.

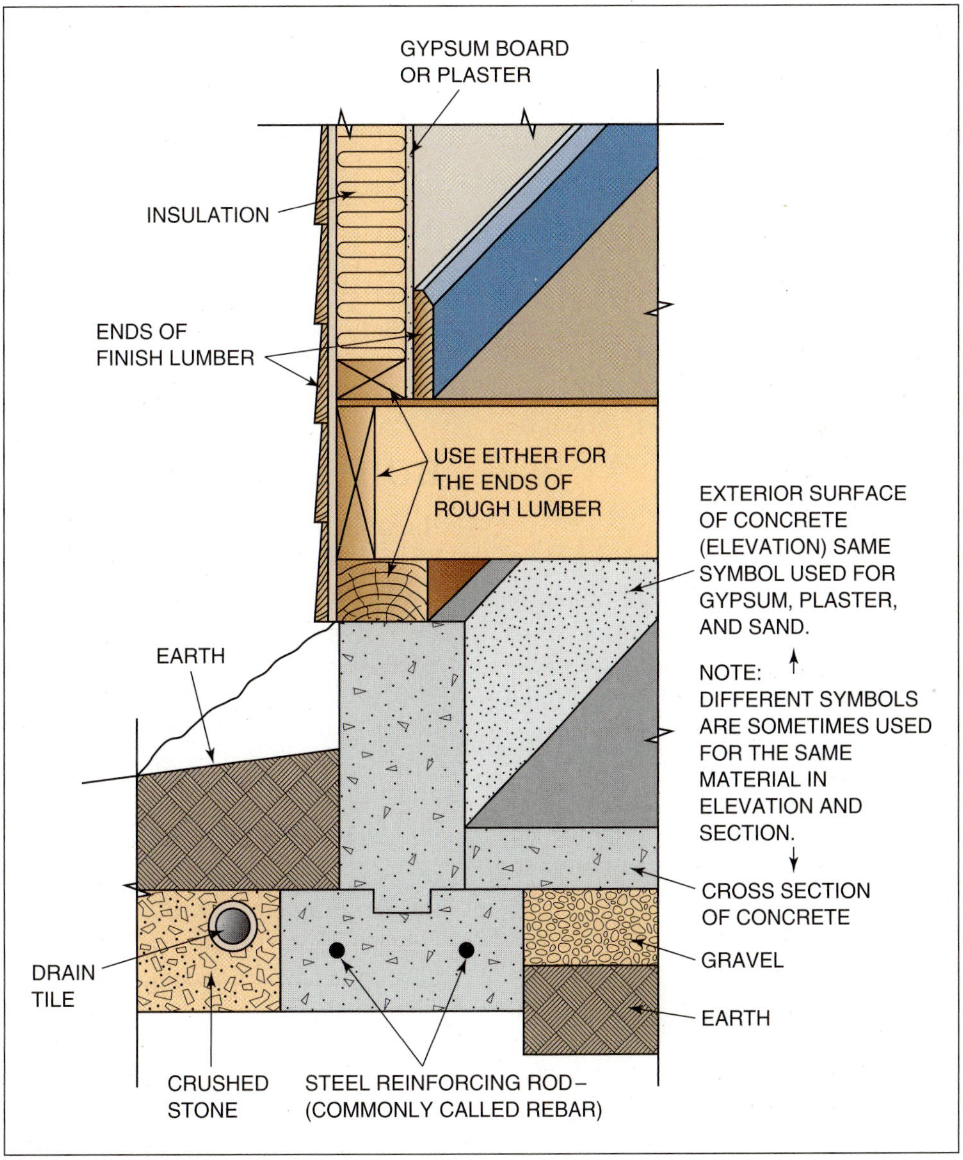

**FIGURE 6-24** Symbols for commonly used construction materials.

### Abbreviations

Architects find it necessary to use abbreviations on drawings to conserve space. Only capital letters, such as DR for door, are used. Several words may use the same abbreviation, such as W for west, width, or water. The location of these abbreviations is the key to their meaning. A list of commonly used abbreviations is shown in Figure 6-25.

# Codes

Cities and towns often have local laws governing many aspects of new construction and remodeling, in addition to state and national building codes. These laws and codes may seem restrictive to some but are designed to protect the consumer and the community. Codes and regulations provide for safe, properly designed buildings in a planned environment.

## ZONING REGULATIONS

*Zoning regulations* deal, generally speaking, with keeping buildings of similar size and purpose in areas for which they have been planned. They also regulate the space in each of the areas. The community is divided into areas called *zones,* shown on zoning maps.

### Zones

The names given to different zones vary from community to community. A large city may have 30 or more zoning districts. There may be several *single family residential zones.* Other areas may be zoned as *multifamily residential.* Other residential zones may be set aside for mobile home parks and those that allow a combination of residences, retail stores, and offices. Other zones may be designated for the central business district, various kinds of commercial districts, and different industrial zones.

### Lots

Zoning laws regulate buildings and building sites. Most cities specify a *minimum lot size* for each zone and a *maximum ground coverage* by the structure. The *maximum height* of the building for each zoning district is stipulated. A *minimum lot width* is usually specified. These distances are called *setbacks.* They are usually different for front, rear, and side. Some communities require a certain amount of landscaped area, called *green space,* to enhance the site. In some residential zones, as much as half the lot must be reserved for green space.

### Nonconforming Buildings

Because some cities were in existence before the advent of zoning laws, many buildings and businesses may not be in their proper zone. They are called *nonconforming.* It would be unfair to require that buildings be torn down or to stop businesses to meet the requirements of zoning regulations. Nonconforming businesses or buildings are allowed to remain, but restrictions are placed on rebuilding. If partially destroyed, they may be allowed to rebuild, depending on the amount of destruction. If 75 percent or more is destroyed, they are not usually allowed to rebuild in the same manner or for the same purpose in the same zone.

Any hardships imposed by zoning regulations may be relieved by a *variance.* Variances are granted by a zoning board of appeals within each community. A public hearing is held after a certain period of time. The general public and, in particular, those abutting the property, are notified. The petitioner must prove certain types of hardship specified in the zoning laws before the zoning variance can be granted.

## BUILDING CODES

*Building codes* regulate the design and construction of buildings by establishing minimum safety standards. They prevent things such as roofs being ripped off by high winds, floors collapsing from inadequate support, buildings settling because of a poor foundation, and tragic deaths from fire due to lack of sufficient exits from a building. In addition to building codes, other codes govern the mechanical, electrical, and plumbing trades. Some communities have no building codes, while others write their own.

It is important to have a general knowledge of the building code used by a particular community. Construction superintendents and contractors must have extensive knowledge of the codes.

### National Building Codes

Recently, a national building code, called the International Building Code (IBC), was adopted in an attempt to standardize the minimum building requirements. It was created with the assistance of BOCA (Building Officials and Codes Administrators

|  |  |
|---|---|
| Access Panel | AP |
| Acoustic | ACST |
| Acoustical Tile | AT |
| Aggregate | AGGR |
| Air Conditioning | AIR COND |
| Aluminum | AL |
| Anchor Bolt | AB |
| Angle | ANG |
| Apartment | APT |
| Approximate | APPROX |
| Architectural | ARCH |
| Area | A |
| Area Drain | AD |
| Asbestos | ASB |
| Asbestos Board | AB |
| Asphalt | ASPH |
| Asphalt Tile | AT |
| Basement | BSMT |
| Bathroom | B |
| Bathtub | BT |
| Beam | BM |
| Bearing Plate | BRG PL |
| Bedroom | BR |
| Blocking | BLKG |
| Blueprint | BP |
| Boiler | BLR |
| Book Shelves | BK SH |
| Brass | BRS |
| Brick | BRK |
| Bronze | BRZ |
| Broom Closet | BC |
| Building | BLDG |
| Building Line | BL |
| Cabinet | CAB |
| Calking | CLKG |
| Casing | CSG |
| Cast Iron | CI |
| Cast Stone | CS |
| Catch Basin | CB |
| Cellar | CEL |
| Cement | CEM |
| Cement Asbestos Board | CEM AB |
| Cement Floor | CEM FL |
| Cement Mortar | CEM MORT |
| Center | CTR |
| Center to Center | C to C |
| Center Line | or CL |
| Center Matched | CM |
| Ceramic | CER |
| Channel | CHAN |
| Cinder Block | CIN BL |
| Circuit Breaker | CIR BKR |
| Cleanout | CO |
| Cleanout Door | COD |
| Clear Glass | CL GL |
| Closet | C, CL, or CLO |
| Cold Air | CA |
| Cold Water | CW |
| Collar Beam | COL B |
| Concrete | CONC |
| Concrete Block | CONC B |
| Concrete Floor | CONC FL |
| Conduit | CND |
| Construction | CONST |
| Contract | CONT |
| Copper | COP |
| Counter | CTR |
| Cubic Feet | CU FT |
| Cut Out | CO |
| Detail | DET |
| Diagram | DIAG |
| Dimension | DIM |
| Dining Room | DR |
| Dishwasher | DW |
| Ditto | DO |
| Double-Acting | DA |
| Double Strength Glass | DSG |
| Down | DN |
| Downspout | DS |
| Drain | D or DR |
| Drawing | DWG |
| Dressed and Matched | D & M |
| Dryer | D |
| Electric Panel | EP |
| End to End | E to E |
| Excavate | EXC |
| Expansion Joint | EXP JT |
| Exterior | EXT |
| Finish | FIN |
| Finished Floor | FIN FL |
| Firebrick | FBRK |
| Fireplace | FP |
| Fireproof | FPRF |
| Fixture | FIX |
| Flashing | FL |
| Floor | FL |
| Floor Drain | FD |
| Flooring | FLG |
| Fluorescent | FLUOR |
| Flush | FL |
| Footing | FTG |
| Foundation | FND |
| Frame | FR |
| Full Size | FS |
| Furring | FUR |
| Galvanized Iron | GI |
| Garage | GAR |
| Gas | G |
| Glass | GL |
| Glass Block | GL BL |
| Grille | G |
| Gypsum | GYP |
| Hardware | HDW |
| Hollow Metal Door | HMD |
| Hose Bib | HB |
| Hot Air | HA |
| Hot Water | HW |
| Hot Water Heater | HWH |
| I Beam | I |
| Inside Diameter | ID |
| Insulation | INS |
| Interior | INT |
| Iron | I |
| Jamb | JB |
| Kitchen | K |
| Landing | LDG |
| Lath | LTH |
| Laundry | LAU |
| Laundry Tray | LT |
| Lavatory | LAV |
| Leader | L |
| Length | L, LG, or LNG |
| Library | LIB |
| Light | LT |
| Limestone | LS |
| Linen Closet | L CL |
| Lining | LN |
| Living Room | LR |
| Louver | LV |
| Main | MN |
| Marble | MR |
| Masonry Opening | MO |
| Material | MATL |
| Maximum | MAX |
| Medicine Cabinet | MC |
| Minimum | MIN |
| Miscellaneous | MISC |
| Mixture | MIX |
| Modular | MOD |
| Mortar | MOR |
| Moulding | MLDG |
| Nosing | NOS |
| Obscure Glass | OBSC sL |
| On Center | OC |
| Opening | OPNG |
| Outlet | OUT |
| Overall | OA |
| Overhead | OVHD |
| Pantry | PAN |
| Partition | PTN |
| Plaster | PL or PLAS |
| Plastered Opening | PO |
| Plate | PL |
| Plate Glass | PL GL |
| Platform | PLAT |
| Plumbing | PLBG |
| Plywood | PLY |
| Porch | P |
| Precast | PRCST |
| Prefabricated | PREFAB |
| Pull Switch | PS |
| Quarry Tile Floor | QTF |
| Radiator | RAD |
| Random | RDM |
| Range | R |
| Recessed | REC |
| Refrigerator | REF |
| Register | REG |
| Reinforce or Reinforcing | REINF |
| Revision | REV |
| Riser | R |
| Roof | RF |
| Roof Drain | RD |
| Room | RM or R |
| Rough | RGH |
| Rough Opening | RO |
| Rubber Tile | R TILE |
| Scale | SC |
| Schedule | SCH |
| Screen | SCR |
| Scuttle | S |
| Section | SECT |
| Select | SEL |
| Service | SERV |
| Sewer | SEW |
| Sheathing | SHTHG |
| Sheet | SH |
| Shelf and Rod | SH & RD |
| Shelving | SHELV |
| Shower | SH |
| Sill Cock | SC |
| Single Strength Glass | SSG |
| Sink | SK or S |
| Soil Pipe | SP |
| Specification | SPEC |
| Square Feet | SQ FT |
| Stained | STN |
| Stairs | ST |
| Stairway | STWY |
| Standard | STD |
| Steel | ST or STL |
| Steel Sash | SS |
| Storage | STG |
| Switch | SW or S |
| Telephone | TEL |
| Terra Cotta | TC |
| Terrazzo | TER |
| Thermostat | THERMO |
| Threshold | TH |
| Toilet | T |
| Tongue and Groove | T & G |
| Tread | TR or T |
| Typical | TYP |
| Unfinished | UNF |
| Unexcavated | UNEXC |
| Utility Room | URM |
| Vent | V |
| Vent Stack | VS |
| Vinyl Tile | V TILE |
| Warm Air | WA |
| Washing Machine | WM |
| Water | W |
| Water Closet | WC |
| Water Heater | WH |
| Waterproof | WP |
| Weather Stripping | WS |
| Weephole | WH |
| White Pine | WP |
| Wide Flange | WF |
| Wood | WD |
| Wood Frame | WF |
| Yellow Pine | YP |

**FIGURE 6-25** Commonly used abbreviations.

International, Inc.), ICBO (International Conference of Building Officials), and the SBCCI (Southern Building Code Congress International, Inc.). States are being encouraged to use it with the understanding that stricter state and local codes will supersede the IBC. The goal of these codes is to ensure safe, affordable housing for the nation.

In Canada, the *National Building Code* sets the minimum standard. Some provinces augment this code with more stringent requirements and publish the combination as a *Provincial Building Code.* A few cities have charters, which allow them to publish their own building codes.

### Use of Residential Codes

In addition to structural requirements, major areas of residential codes include:

- Exit facilities, such as doors, halls, stairs, and windows as emergency exits, and smoke detectors.
- Room dimensions, such as ceiling height and minimum area.
- Light, ventilation, and sanitation, such as window size and placement, maximum limits of glass area, fans vented to the outside, requirements for baths, kitchens, and hot and cold water.

## BUILDING PERMITS

A *building permit* is needed before construction can begin. Application is made to the office of the local building official. The building permit application form (Fig. 6-26) requires a general description of the construction, legal description and location of the property, estimated cost of construction, and information about the applicant. Drawings of the proposed construction are submitted with the application. The type and kind of drawings required depend on the complexity of the building. For commercial work, usually five sets of plot plans and two sets of other drawings are required. The drawings are reviewed by the building inspection department. If all is in order, a permit (Fig. 6-27) is granted upon payment of a fee. The fee is usually based on the estimated cost of the construction. Electrical, mechanical, plumbing, water, and sewer permits are usually obtained by subcontractors. The permit card must be displayed on the site in a conspicuous place until the construction is completed.

## INSPECTIONS

*Building inspectors* visit the job site to perform code inspections at various intervals. These inspections may include:

1. *Setback inspection* to verify the building location meets the local codes.
2. *Foundation inspection* takes place after the trenches have been excavated and forms erected and ready for the placement of concrete. No reinforcing steel or structural framework of any part of any building may be covered without an inspection and a release.
3. *Floor frame inspection* before the subfloor is applied.
4. *Frame inspection* takes place after the roof, framing, fire blocking, and bracing are in place, and all concealed wiring, pipes, chimneys, ducts, and vents are complete.
5. *Utility inspection* for items such as electric, gas, sewer or septic systems, water or well, etc.
6. *Final inspection* occurs when the building is finished. A certificate of occupancy or completion is then granted.

It is the contractor's responsibility to notify the building official when the construction is ready for a scheduled inspection. If all is in order, the inspector signs the permit card in the appropriate space and construction continues. If the inspector finds a code violation, it is brought to the attention of the contractor or architect for compliance.

These inspections ensure that construction is proceeding according to approved plans. They also make sure construction is meeting code requirements. This protects the future occupants of the building and the general public. In most cases, a good rapport exists between inspectors and builders, enabling construction to proceed smoothly and on schedule.

# CITY OF ANYWHERE, USA

## APPLICATION FOR BUILDING PERMIT

RADON GAS FEE ____________

**FOR OFFICE USE ONLY**

Permit Type ____________ Permit # ____________

Permit Class of Work ____________ Log # ____________

Permit Use Code ____________ Issue Date ____________

Lot ________ Block ________ Sub ____________ Permit Cost ____________

Fire Zone: IN _____ OUT _____ Zone: ________ T.I.F. Due (Y/N or NA) ____________

Utility Notification 1. FL Power ________ B of A (Y/N) ____________ Case No.
2. Peoples Gas ________ E.D.C. (Y/N) ____________ Case No.
**NOTE:** Items with* 3. Water Dept. ________ C.R.A. (Y/N) ____________ Case No.
must be entered in computer. H.P.C. (Y/N) ____________ Case No.

*Plat Page ________ *Sec ________ *Township ________ *Range ________ Zone ________

*Dept of Commerce Code ____________ *Const. Type ________ Protected _____ Unprotected _____

*Additional Permits Required:
Building ________ Plumbing ________ No. of W.C. ________ No. of Meters ________

Electrical ________ Mechanical ________ Gas ________ Fire Sprk. ________ Landscape ________

Park/Paving ________ Total Spaces ________ Handicap ________

*Flood Zone ____________ *Setbacks: Front ________ Left Side ________ Right Side ________
Rear ________ Other Requirements ____________
Threshold Building YES _____ NO _____ ____________

**Special Notes/Comments to Inspector:** ____________

**APPLICANT PLEASE FILL OUT THIS SECTION**

JOB ADDRESS ____________ Suite or Apt. No. ________

CONTRACTOR ____________ Cert./Reg. No. ________ Telephone ________

PROPERTY OWNER'S Name ____________ Address ____________

City ____________ State _____ Zip ________ Telephone ________

Building Description: Total Sq. Ft. ____________ Estimated Job Value ____________
LF-SF or Dimensions ____________ Building Use ____________
Valuation of Work ____________ Former Use ____________
No. of Units ________ No. of Suites ________ No. of Stories ________

**Special Notes or Comments:** ____________

**PHONE 555-1234 FOR ALL INSPECTIONS** HCS-12 Rev. 6-1-88

*(OVER)*

**FIGURE 6-26** A typical form used to apply for a building permit.

ANYWHERE, USA
DEPARTMENT OF HOUSING & CONSTRUCTION SERVICES

# BUILDING PERMIT

THIS PERMIT BECOMES INVALID IF NO INSPECTIONS HAVE BEEN MADE DURING ANY 3 MONTH PERIOD.

**Flood Elevation - ________ Lowest Floor Minimum Required**

- ☐ New Construction
- ☐ Grounds Improvements
- ☐ Utility Building
- ☐ Reroofing
- ☐ Moving
- ☐ Fences
- ☐ Pool
- ☐ Other ________________
- ☐ Siding
- ☐ Walls

**Permit No.** ______________ (ZONE) ______________

**Job Address** ______________________________

**Lot** ________ **Blk.** ________ **Sub.** ________________

**Date** ______________________________

This permit covers building construction only. Additional permits are required for electric, plumbing, gas and/or mechanical installations.

| BUILDING | | | ELECTRICAL | | | PLUMBING | | | MECHANICAL-GAS | | |
|---|---|---|---|---|---|---|---|---|---|---|---|
| Type of Inspection | Date | Inspector | Type of Inspection | Date | Inspector | Type of Inspection | Date | Inspector | Type of Inspection | Date | Inspector |
| | | | | | | | | | | | |
| | | | | | | | | | | | |
| | | | | | | | | | | | |
| | | | | | | | | | | | |
| | | | | | | | | | | | |
| | | | | | | | | | | | |
| | | | | | | | | | | | |
| | | | | | | | | | | | |
| | | | | | | | | | | | |

NOTE: Building, Electrical, Plumbing and Mech/Gas Inspections shall be dated and initialed by inspectors before walls and ceilings are covered.

| BUILDING OK TO COVER | | ELECTRICAL OK TO COVER | | PLUMBING OK TO COVER | | MECH/GAS OK TO COVER | |
|---|---|---|---|---|---|---|---|
| Date | Inspector | Date | Inspector | Date | Inspector | Date | Inspector |
| | | | | | | | |

NOTE: This card shall remain posted at the job site until all final inspections have been dated and initialed by inspectors.

| BUILDING FINAL OK | | ELECTRICAL FINAL OK | | PLUMBING FINAL OK | | MECH/GAS FINAL OK | |
|---|---|---|---|---|---|---|---|
| Date | Inspector | Date | Inspector | Date | Inspector | Date | Inspector |
| | | | | | | | |

THIS CARD MUST BE POSTED IN AN EASILY SEEN LOCATION.

For INSPECTIONS, call 555-1234. ■ For other information, call 555-6789.

**FIGURE 6-27** A building permit. Communities use different kinds of forms.

## SUMMARY

- Blueprints communicate to the builder the architect's intentions for buildings.
- Codes guide designers and builders in the methods and practices of safe construction.

## GREEN CHECKLIST

☐ Orienting a building to address the sun will affect a lifetime of heating and cooling dollars. Warm climates should include natural shading and northerly exposures. Cooler areas should orient longer building sides and larger glass toward southerly exposures.

## REVIEW QUESTIONS

*Select the most appropriate answer.*

**1. A view of a set of prints that looks from the top downward is called**

a. an elevation.
b. a perspective.
c. a plan.
d. a section.

**2. A view of a set of prints showing a vertical cut through the construction is called**

a. an elevation.
b. a detail.
c. a plan.
d. a section.

**3. The most commonly used scale for floor plans is**

a. ¼″ = 1′-0″.
b. ¾″ = 1′-0″.
c. 1½″ = 1′-0″.
d. 3″ = 1′-0″.

**4. Centerlines drawn on a set of prints are indicated by**

a. a series of short, uniform dashes.
b. a series of long and short dashes.
c. a long dash followed by two short dashes.
d. a solid, broad, dark line.

**5. On what drawing would the setback of the building from the property lines be found?**

a. Floor plan
b. Plot plan
c. Elevation
d. Foundation plan

**6. The abbreviation GYP found on a set of prints means**

a. Gypsum board or drywall.
b. Grounded for Your Protection.
c. Galvanized Yellow Pine.
d. Gross Yearly Production.

**7. An exterior wall stud height can best be determined from the**

a. floor plan.

b. framing elevation.

c. wall section.

d. specifications.

**8. The abbreviation for the current national building code in the United States is**

a. OSHA.

b. IBC.

c. NBC.

d. BOCA.

**9. A typical zone found in most zoning regulations is**

a. residential.

b. commercial.

c. industrial.

d. all of the above.

**10. Building codes were created to**

a. make all buildings the same.

b. improve the national economy by creating jobs.

c. standardize safe building practices.

d. all of the above.

**11. The measurements to locate partitions and walls of a building are determined from the**

a. foundation plan.

b. floor plan.

c. elevation.

d. plot plan.

**12. If there is a conflict of information in building prints, what takes precedence over the drawings?**

a. specifications

b. schedules

c. details

d. floor plan

## KNOW YOUR CODES

There are a number of building codes currently being used in construction. They all were created to ensure safety of occupants and for longevity of the building. The International Building Code (IBC) is a national code that sets a standard for all states and municipalities to follow. It is expected that stricter local codes may be written to supersede this code. It is essential that builders know what their local codes require.

## WHAT'S WRONG WITH THIS PICTURE?

Carefully study **Figure 6-28** and think about what is wrong. Consider all possibilities.

### WRONG

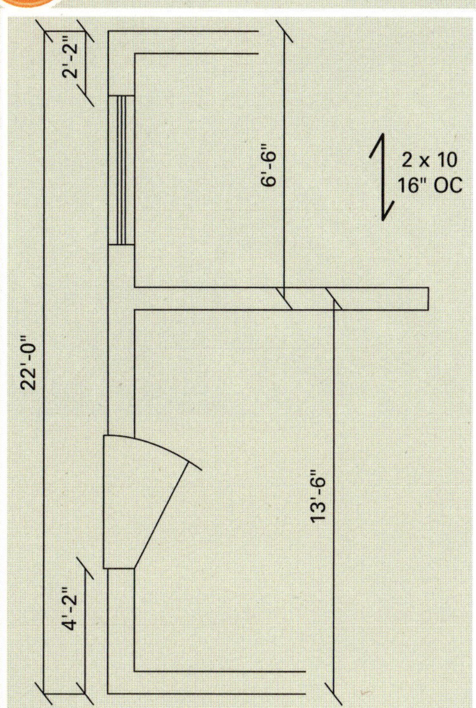


**FIGURE 6-28** This view shows a partial floor plan of a structure. This plan is either not drawn to scale, or the partition wall, window, and door are not drawn in correct relationship to each other. The numbers do not add up. The 22' width of the building does not add up to the sum of 13'-6" and 6'-6". In addition, the window is shown as 2'-2" from the outside wall. However, in the plan it looks centered in the 6'-6" area (which would place it approximately 3'-3" from the outside wall). The door also looks centered in the room while its dimensions indicate otherwise, and the door's scale does not look correct.

### RIGHT

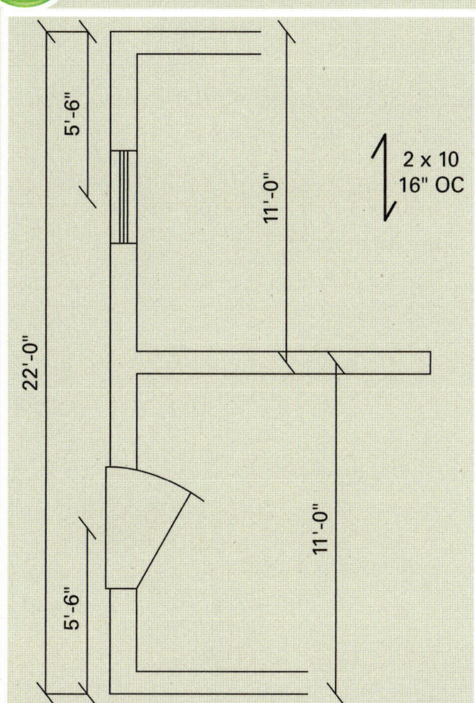


**FIGURE 6-29** These dimensions appear to be much more realistic, to scale, and the numbers add up correctly. Plans in today's world are often drawn using computer-assisted drafting and design (CADD). CADD software often will eliminate dimensioning errors. However, computers cannot account for every situation, and all plans need to be double-checked.

CHAPTER 7

# Building Layout

**OBJECTIVES** *After completing this unit, the student should be able to:*

- Establish level points across a building area using a water level and by using a carpenter's hand spirit level in combination with a straightedge.
- Accurately set up and use the builder's level, transit-level, and laser level.
- Use an optical level to determine elevations.
- Lay out building lines by using the Pythagorean theorem and check the layout for accuracy.
- Build batter boards and accurately establish building lines with string.

## GLOSSARY OF BUILDING LAYOUT TERMS

**foundation** that part of a wall on which the major portion of the structure is erected

**laser** a concentrated, narrow beam of light; optical leveling and plumbing instrument used in building construction

**ledger** a temporary or permanent supporting member for joists or other members running at right angles; horizontal member of a set of batter boards

The first step in construction is to locate where the building will be placed on the lot. Its elevation relative to the surrounding land must be determined as well. Some areas of the country require a licensed surveyor do the building layout. These companies may use high-tech equipment such as GPS (global positioning system) guided instruments. Codes in other areas allow the builder/carpenter to perform layout of the building location.

The layout process starts with locating the corners and lines of the property. They are denoted in a variety ways such as a stake, concrete marker, or tree. Setback distances of the building from the property line are shown on the plot plan.

Stakes then are placed where the building corners will be positioned. Strings are suspended on wood frames to show the location and elevation of the building foundation. Accuracy in layout allows for smooth transitions from one phase of construction to another, saving time, effort, and money.

# BUILDING LAYOUT

Building layout requires accurate placement and leveling of building lines over the length and width of the structure. The carpenter must be able to set up, adjust, and use a variety of leveling and layout tools.

## LEVELING TOOLS

Several tools, ranging from simple to state-of-the-art, are used to level the layout. More sophisticated leveling and layout tools should be handled with care to maintain their accuracy. Keep them in the carrying case when not being used or when being transported in a vehicle over long distances. If the instrument gets wet, dry it before returning it to its case.

### Levels and Straightedges

If no other tools are available, a *carpenter's hand level* and a long *straightedge* may be used together to level across a building area. Use a long level and select a long length of lumber. Make sure it is straight and wide enough that it will not sag when placed on edge and supported only on its ends. Place one end on the first stake or surface to be leveled and the other end on the top of another stake driven partially in the ground. Drive the second stake until it is level. Be precise. Move the straightedge and repeat the process. Continue moving the straightedge from stake to stake until the desired distance is leveled (Fig. 7-1).

This can be an accurate, although time-consuming, method of leveling over a long distance. If you want to level to the corners of a rectangle, start by driving a stake near the center to the desired height. Then level from the center stake to each corner in the manner described (Fig. 7-2).

### Water Levels

A *water level* is an accurate tool, dating back centuries. It is used for leveling from one point to another.

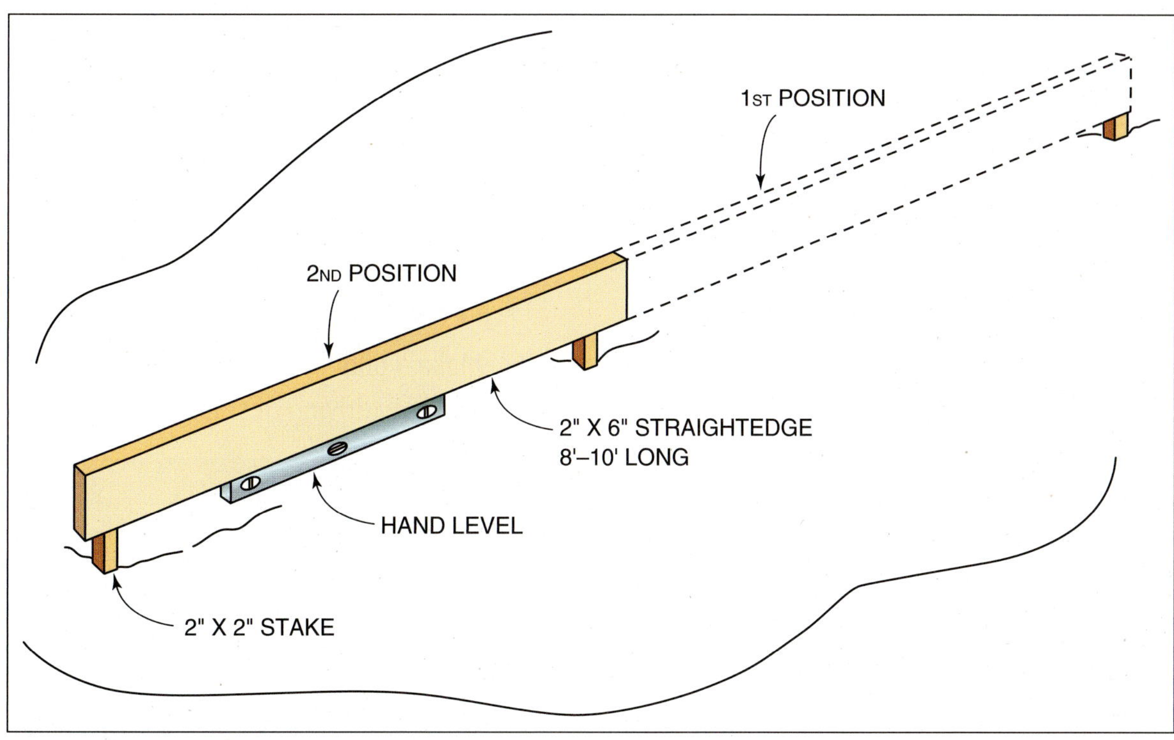


**FIGURE 7-1** Leveling with a straightedge from stake to stake.

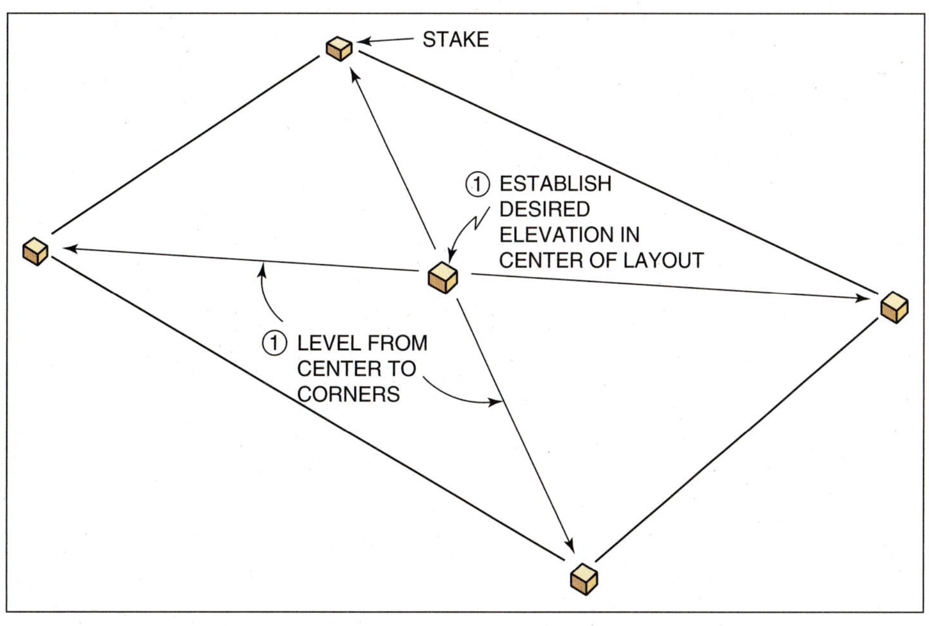


**FIGURE 7-2** Leveling from the center of a building layout to the corners.

Its accuracy, within a pencil line, is based on the fact that water seeks its own level (Fig. 7-3). One commercial model consists of 50 feet of small-diameter, clear vinyl tubing and a small tube storage container. A built-in reservoir holds the colored water that fills the tube. The reservoir is held to the starting

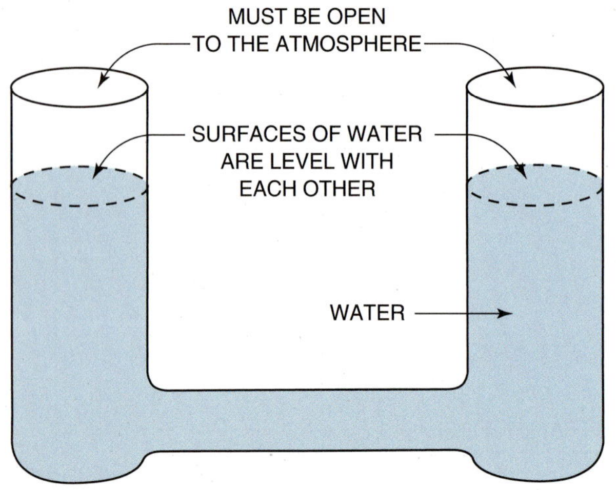


**FIGURE 7-3** Water seeks its own level. Both ends of the water level must be open to the atmosphere.

elevation. The other end is moved down until the water level is seen, and then the surface is marked at the level of the water in the tube (Fig. 7-4).

## Builder's Levels

The *builder's level* (Fig. 7-5) consists of a telescope to which a *spirit level* is mounted. The telescope is fixed in a horizontal position. It can rotate 360 degrees for measuring horizontal angles but cannot be tilted up or down.

## Transit-levels

The *transit level* (Fig. 7-6) is similar to the builder's level; however, its telescope can be tilted up and down 45 degrees in each direction.

### *from experience...*

Make a pilot hole in the earth for straight and easy driving of stakes.

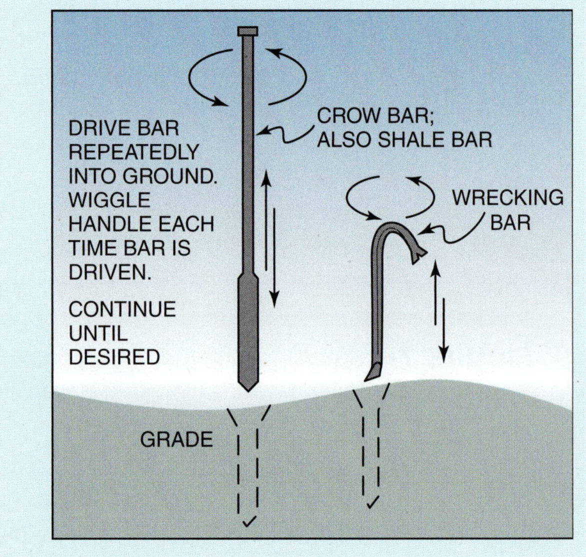


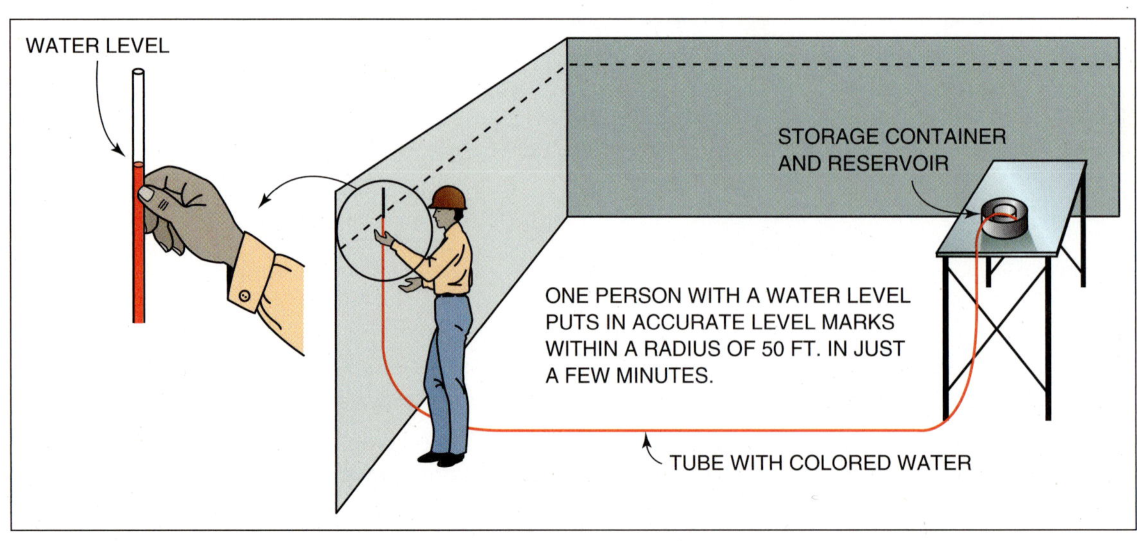


**FIGURE 7-4** The water level is a simple and accurate tool to use.

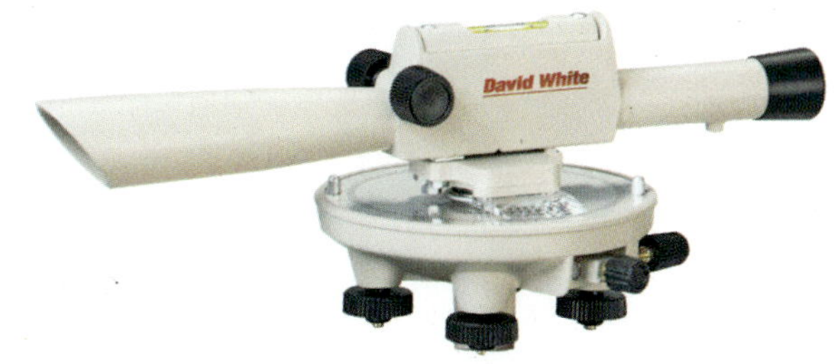


Courtesy of David White.

**FIGURE 7-5** The builder's level.

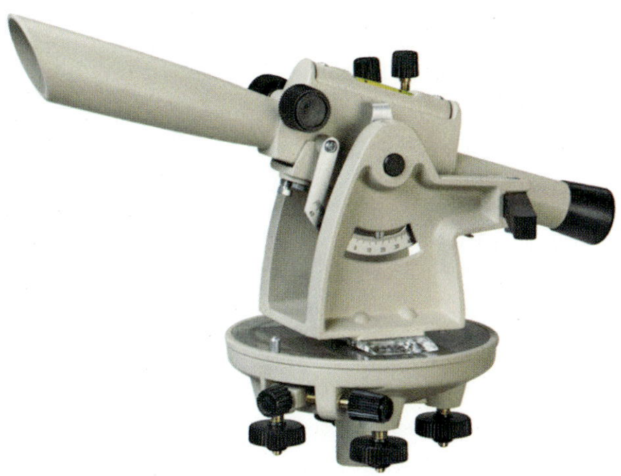

Courtesy of David White.

**FIGURE 7-6** The telescope of the transit level may be moved up and down 45 degrees each way.

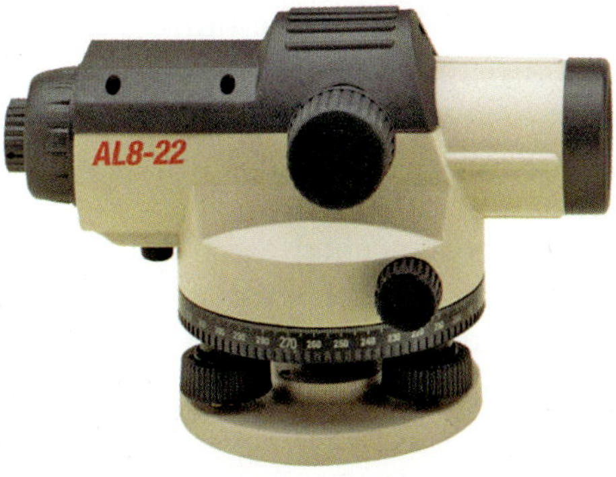


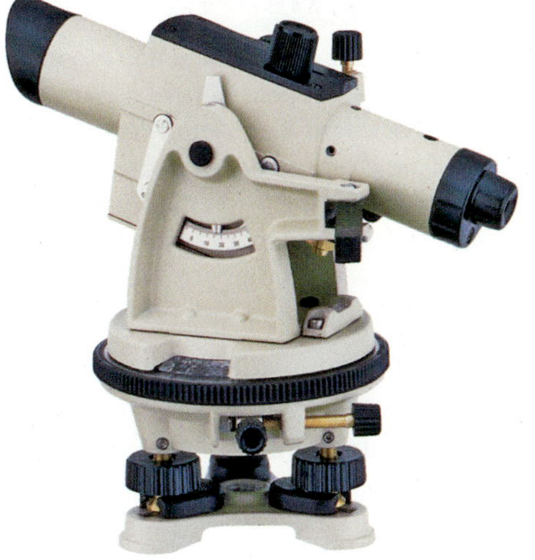

Courtesy of David White.

**FIGURE 7-7** Automatic builder's level and transit level.

### Automatic Levels

*Automatic levels* and *automatic transit-levels* (Fig. 7-7) are similar to those previously described except that they have an internal *compensator.* This compensator uses gravity to maintain a true level line of sight. Even if the instrument is jarred, the line of sight stays true because gravity does not change.

### Using Optical Levels

Before the level can be used, it must be placed on a *tripod* or some other solid support and leveled.

## SETTING UP AND ADJUSTING THE OPTICAL LEVEL

The telescope is adjusted to a level position by means of four leveling screws that rest on a base leveling plate. Open and adjust the legs of the tripod to a convenient height. Spread the legs of the tripod well apart, and firmly place its feet into the ground with the tripod top nearly level. Lift the instrument from its case by the frame. Note how it is stored so it can be replaced in the case in the same position. Make sure the horizontal clamp screw is loose so the telescope revolves freely. Secure the instrument to the tripod.

**CAUTION**

**CAUTION:** Care must be taken not to damage the instrument. Never use force on any parts of the in strument. All moving parts turn freely and easily by hand. Excessive pressure on the leveling screws may damage the threads of the base plate. Unequal tension on the screws will cause the instrument to wobble on the base plate, resulting in leveling errors.

Accurate leveling of the builder's level is important. Line up the telescope directly over two opposite leveling screws. Turn the screws in opposite directions with forefingers and thumbs. Move the thumbs toward or away from each other, as the case may be, to center the bubble in the spirit level (Fig. 7-8). The bubble will always move in the same direction as your left thumb is moving when it rotates a leveling screw.

CAUTION

**CAUTION:** On a smooth surface, it is essential that the points on the feet hold without slipping. Make small holes or depressions into which the tripod points will fit. Or, insert screw eyes at the lower inside of the tripod legs and attach wire or a light chain to the three screw eyes (Fig. 7-9).

Rotate the telescope 90 degrees over the other two opposite leveling screws and repeat the procedure. Make sure each of the screws has the same, but not too much, tension. Return to the original position, check, and make minor adjustments. Continue adjustments until the bubble remains exactly centered when the instrument is revolved in a complete circle.

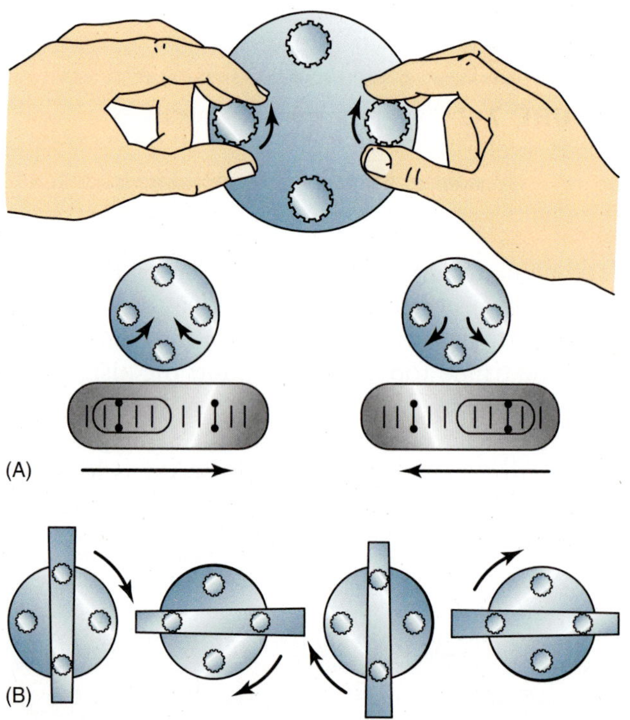


**FIGURE 7-8** (A) Level the instrument by moving thumbs toward or away from each other. (B) The instrument is level when the bubble remains centered as the telescope is revolved in a complete circle.

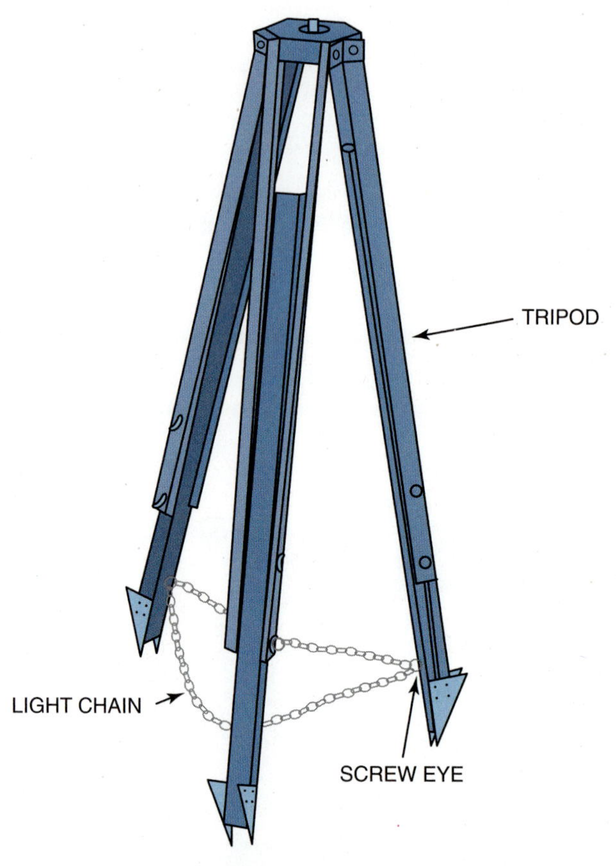


**FIGURE 7-9** Make sure the feet of the tripod do not slip on smooth or hard surfaces.

CAUTION

**CAUTION:** Do not leave a set-up instrument unattended near moving equipment.

## Sighting the Level

To sight an object, rotate the telescope and sight over its top, aiming it at the object. Look through the telescope. Focus it by turning the focusing knob one way or the other, until the object becomes clear. Keep both eyes open. This eliminates squinting, does not tire the eyes, and gives the best view through the telescope. When looking into the telescope, vertical and horizontal cross hairs are seen. They enable the target to be centered properly (Fig. 7-10). Center the cross hairs on the object by moving the telescope left or right.

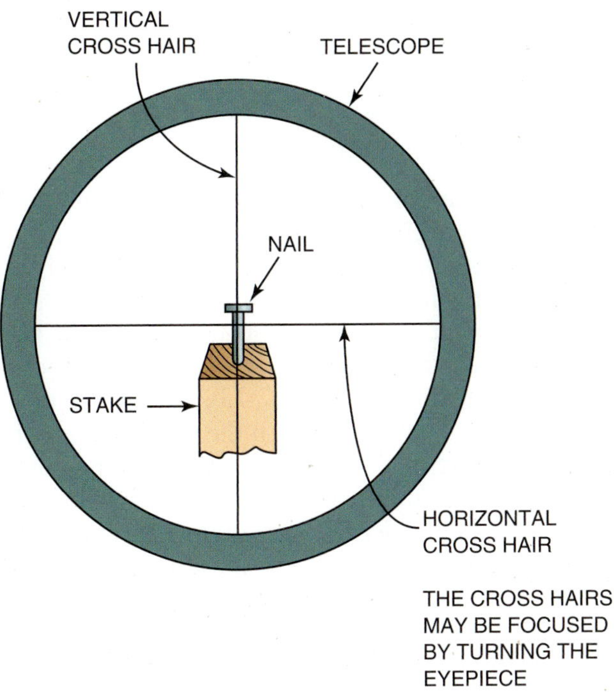


**FIGURE 7-10** When looking in the telescope, vertical and horizontal cross hairs are seen.

> **CAUTION**
>
> **CAUTION:** If the lenses need cleaning, dust them with a soft brush or rag. Do *not* rub the dirt off. Rubbing may scratch the lens coating.

## USING THE OPTICAL LEVEL

When the instrument is leveled, a given point on the line of sight is exactly level with any other point. Any line whose points are the same distance below or above this line of sight is also level (Fig. 7-11).

To level one point with another, a helper must hold a *target* on the point to be leveled. A reading is taken. The target is then moved to selected points that are brought to the same elevation by moving those points up or down to get the same reading.

Tape is often used as a target. The end of the tape is placed on the point to be leveled. The tape is then moved up or down until the same mark is read on the tape as was read at the starting point.

The simplest target is a plain 1 × 2 strip of wood. The end of the stick is held on the starting point of the

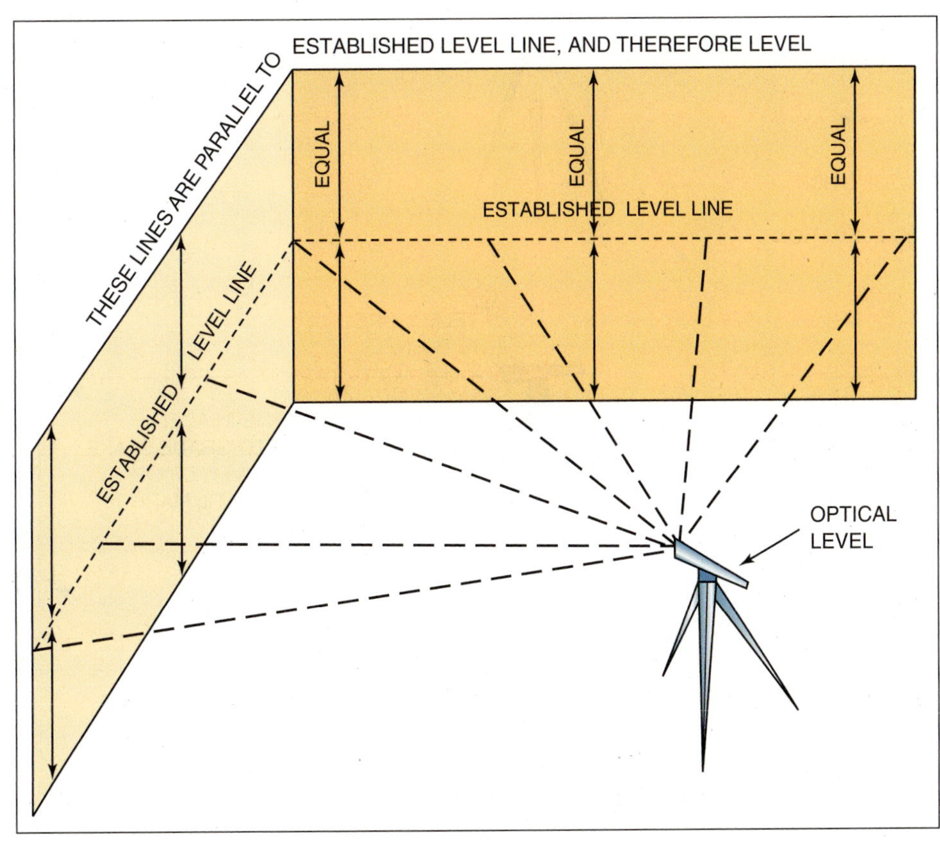


**FIGURE 7-11** Any line parallel to the established level line is also level.

desired elevation. The line of sight is marked on the stick. The end of the stick is then placed on top of various points. They are moved up or down to bring the mark to the same height as the line of sight (Fig. 7-12). A stick of practically any length can be used.

For longer sightings, the *leveling rod* is used because of its clearer graduations. Various rods are manufactured of wood or fiberglass for several leveling purposes. They are made with two or more sections that extend easily and lock into place. Rods vary in length from two-section rods extending 9'-0" up to seven-section rods extending 25'-0".

## Establishing Elevations

Many points on the job site, such as the depth of excavations, the height of foundation footing and walls, and the elevation of finish floors, are required to be set at specified elevations or grades. These elevations are established by starting from the *benchmark*. The benchmark is a point of designated elevation that is accessible at all times during the construction.

## Laying out a Horizontal Angle

Using a plumb bob, position the transit or builder's level directly over a stake that will become the vertex of the angle. Vertex is the 90 degree corner. Next, level the instrument and loosen the horizontal clamping screw. Rotate the instrument until the vertical cross hair is nearly in line with a distant point on one side of the angle (Fig. 7-13). Tighten the horizontal clamp screw. Then turn the tangent screw to line up the vertical cross hair exactly with the point. By hand, turn the horizontal circle scale to zero. Loosen the clamp screw. Swing the telescope to the desired angle and tighten the horizontal clamping screw. Have a helper drive a stake that lines up with the vertical cross hair. Then drive a nail in the stake exactly where it aligns with the cross hair.

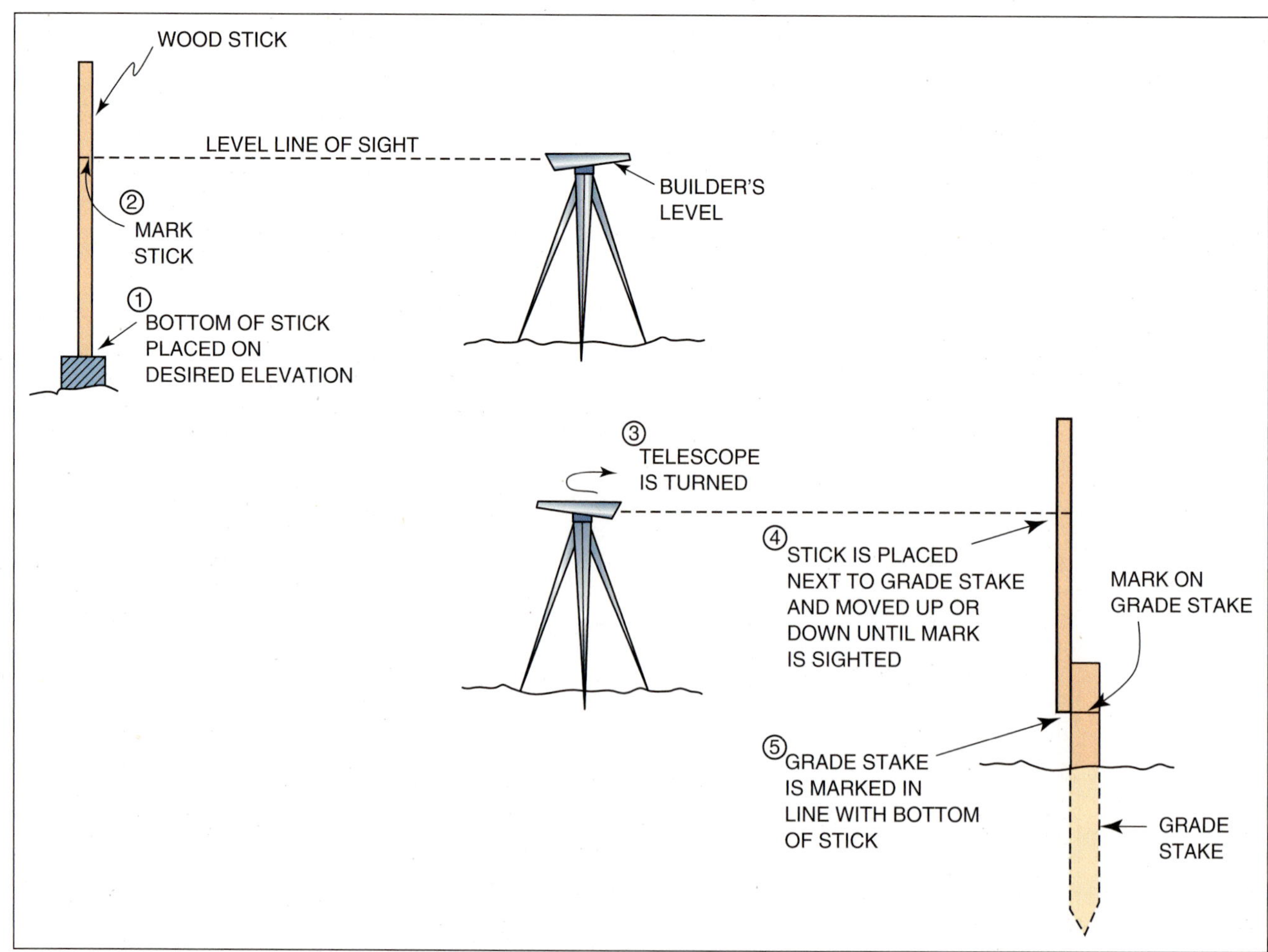


**FIGURE 7-12** Using a stick for a target.

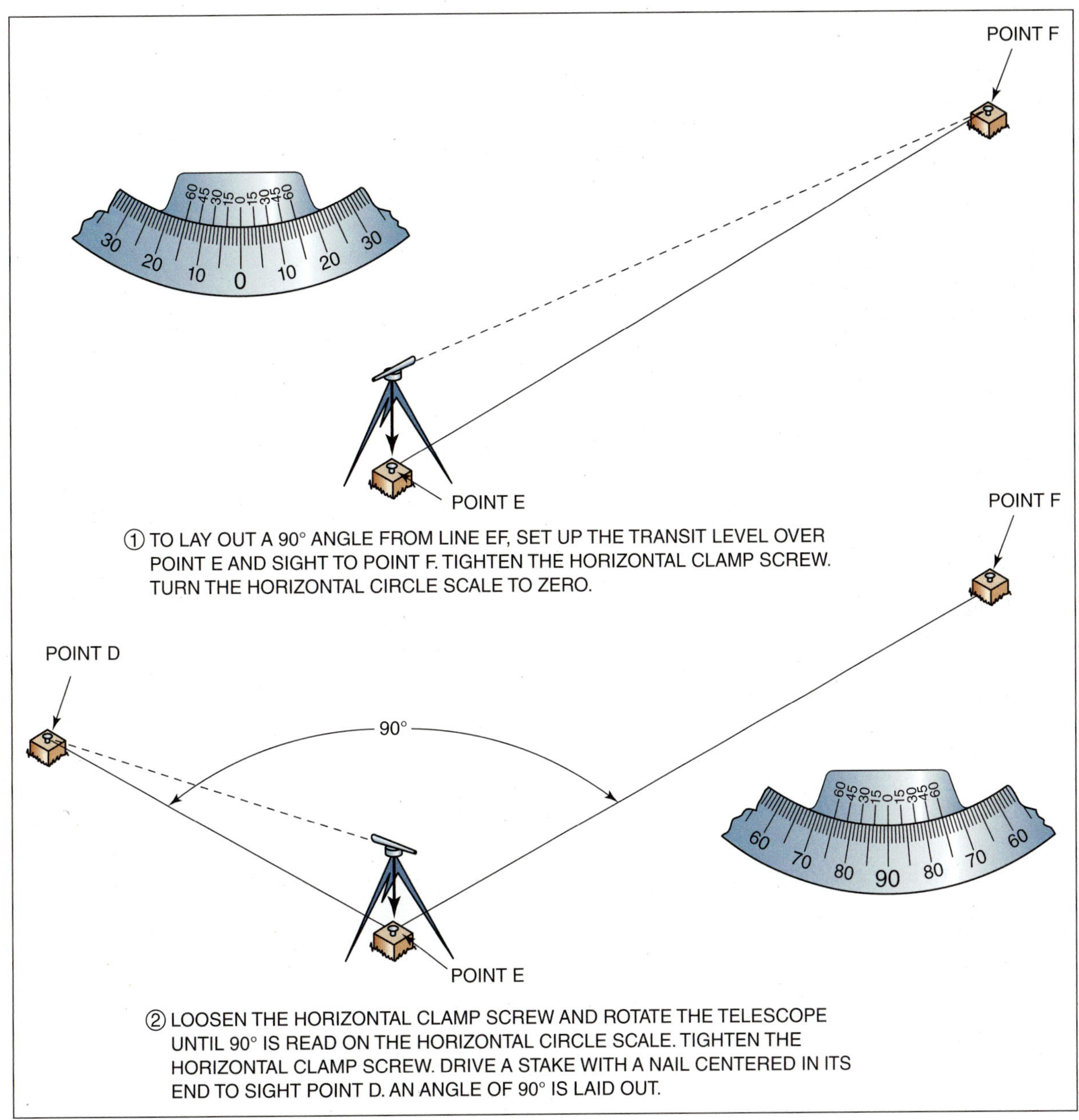


**FIGURE 7-13** Laying out a right angle.

> *from experience...*
>
> Errors and confusion can occur if the level lines are near the center of the stick. Clearly mark the top or bottom of the stick to reduce the risk of turning the stick over.

## LASER LEVELS

A **laser** is a device that releases a narrow beam of light. The light beam is perfectly straight. Unless interrupted by an obstruction or otherwise disturbed, the light beam can be seen over long distances. The *laser level* has been developed for the construction industry to provide more accurate and efficient layout work (Fig. 7-14).

Courtesy of Trimble.

**FIGURE 7-14** Laser levels have been developed for use in the construction industry.

## Kinds and Uses of Laser Levels

Several manufacturers make laser levels in a number of different models. Low-priced units are leveled and adjusted manually. More expensive models automatically adjust and maintain level. Power sources include batteries and AC/DC converters for 110 or 220 volts.

## Establishing and Determining Elevations

Once the unit is leveled, the laser is turned on. It will emit a light beam, about ⅜ inch in diameter. The beam can be rotated through a full 360 degrees, creating a level *plane* of sight. As it rotates, it establishes equal points of elevation over the entire job site, similar to a line of sight being rotated by the telescope of an optical instrument (Fig. 7-15). It may also be set to emit a straight line.

Lasers are difficult to see outdoors in bright sunlight. To detect the beam, a battery-powered electronic sensor target, also called a *receiver* or *detector,* is attached to the leveling rod or stick. Most sensors have a visual display with selectable audio to indicate when it is close to or on the beam (Fig. 7-16). In addition to electronic sensor targets, specially designed targets are used for interior work, such as installing ceiling grids and leveling computer floors.

The procedures for establishing and determining elevations with laser levels are similar to those with optical instruments.

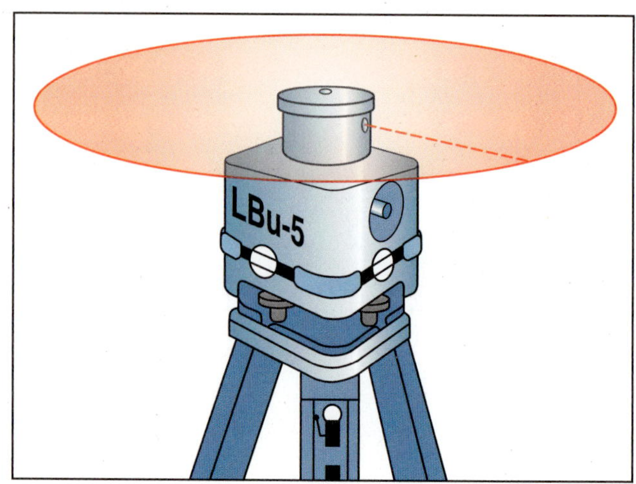


Courtesy of Laser Alignment, Inc.

**FIGURE 7-15** The laser beam rotates 360 degrees, creating a level plane of light.

Courtesy of Laser Alignment, Inc.

**FIGURE 7-16** An electronic target senses the laser beam. An audio provides tones to match the visual display.

APERTURE LABEL

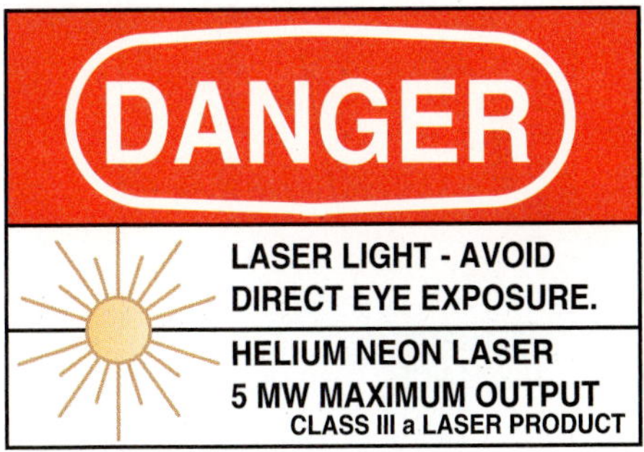


WARNING LABEL

*Courtesy of Laser Alignment, Inc.*

**FIGURE 7-17** Warning labels must be attached to every laser instrument.

### Laser Safety

With a little common sense, the laser can be used safely. All laser instruments are required to have warning labels attached (Fig. 7-17). Only trained persons should set up and operate laser instruments. The following are safety precautions for laser use:

- Never stare directly into the laser beam or view it with optical instruments.
- When possible, set the laser up so it is above or below eye level.
- Turn the laser off when not in use.
- Do not point the laser at others.

# LOCATING THE BUILDING

Layout begins with the dimensions of the building and their location on the site. This information is determined from the plot plan. Surveyors are sometimes asked to determine the building location. It is often the carpenter's responsibility to lay out building lines.

## STAKING THE BUILDING

Find the survey markers that locate the corners of the property. Do not guess where the property lines are. It may be necessary to stretch and secure lines between each corner. From the plot plan, determine the distances the building is to be from the property lines (Fig. 7-18). Measure in on each side from the front property line the specified front setback. Drive stakes on each end. Stretch a line between the stakes to mark the front line of the building (Fig. 7-19). Along the front building line, measure in from the side property line the specified side setback distance of the building. Drive the first building corner stake. Put a nail in the top of the stake to mark the exact side setback (Fig. 7-20). From this nail, measure the dimension of the building length along the front building line. Drive the second stake and set a nail in the top of the stake, marking the exact length of the building (Fig. 7-21).

The third stake on the rear building line may be located by using one of at least three methods. The first is to use an optical instrument such as a transit-level, as shown in Figure 7-13. The instrument is set up directly over the first stake and sighting the second stake with the crosshairs. Then the telescope is rotated 90 degrees. Using a tape measure and the cross hairs, the third stake is located similar to point D in Figure 7-13. The second method is to use the 3-4-5 method. This process uses multiples of 3-4-5 to create larger right triangles. Multiplying each by the same number creates a larger triangle that also has a right angle (Fig. 7-22).

The third method is faster and more accurate. It uses two tapes measuring the building width and the diagonal of the building at the same time. To determine the diagonal of the building, use the **Pythagorean theorem**: $C^2 = A^2 + B^2$.

For example, if a building width is 31'-9" and the length is 42'-6" is plugged into the Pythagorean Theorem, C is the diagonal measurement. First the feet-inches number must be converted to a decimal. Then these numbers are entered into the formula. Lastly, the resulting square root is converted back to feet-inches to the nearest fraction. Typically, the sixteenth is the fraction used in carpentry.

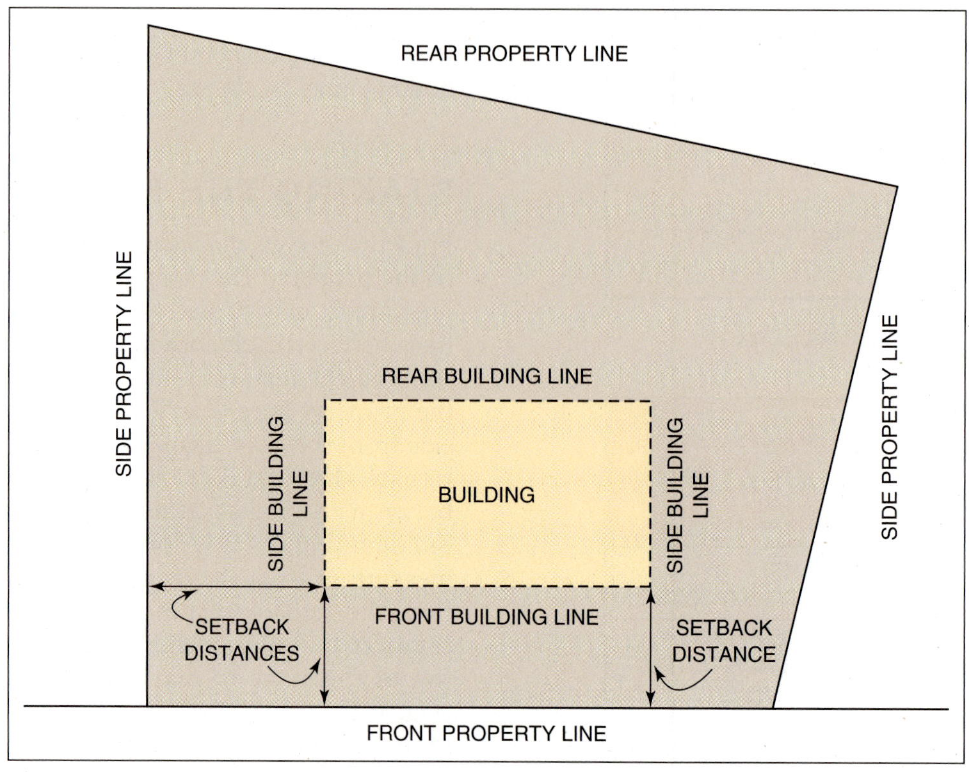


**FIGURE 7-18** Locating the building on a lot from the dimensions found on the plot plan.

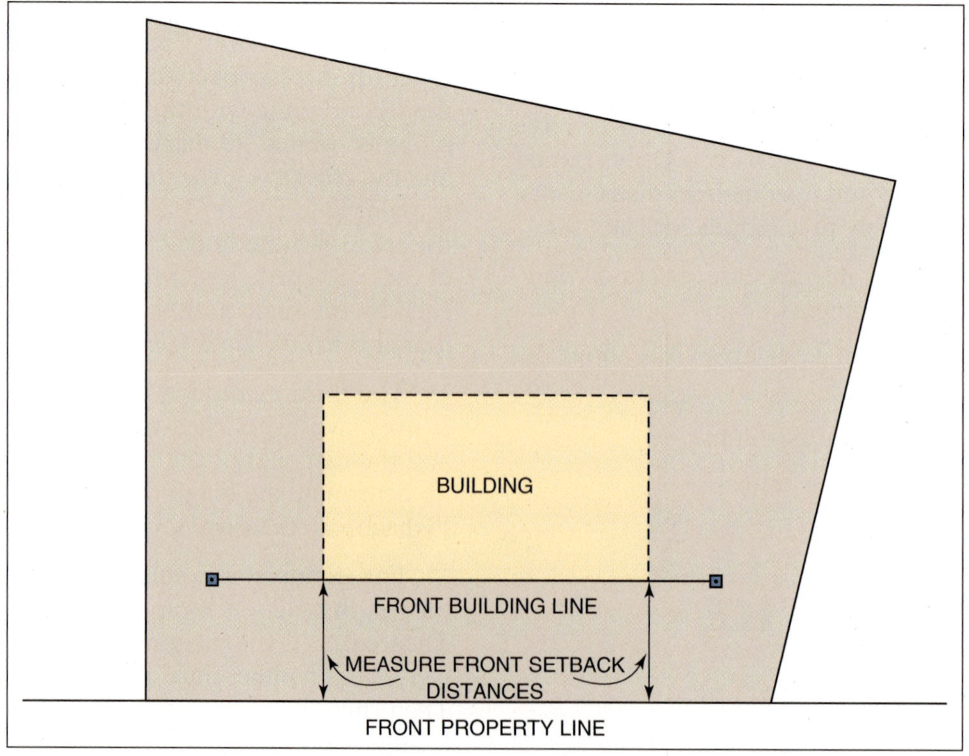


**FIGURE 7-19** Locating the front building line.

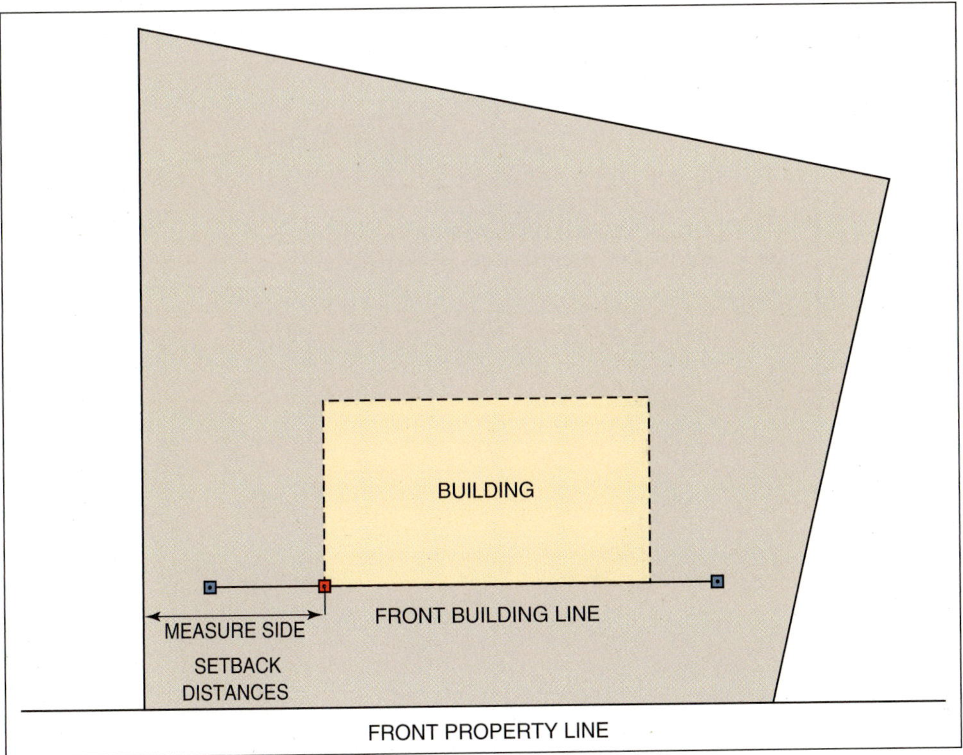


**FIGURE 7-20** Measuring from the side property line to locate the first building corner stake.

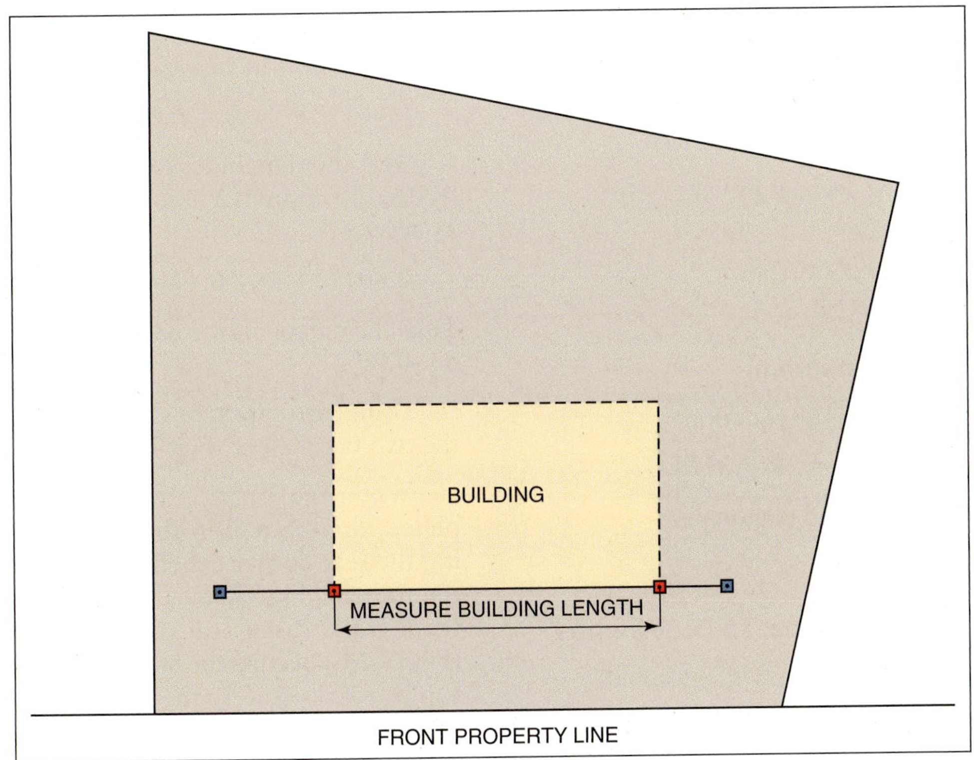


**FIGURE 7-21** Measuring along the building line to locate the second corner stake.

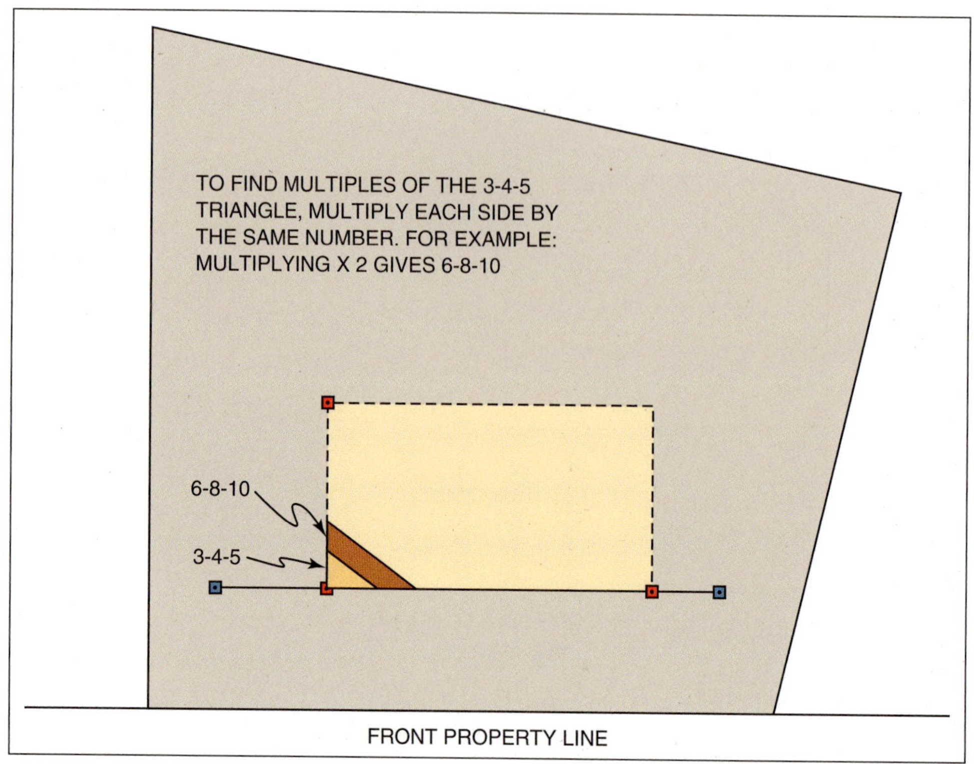


**FIGURE 7-22** Using the 3-4-5 right triangle method to lay out a 90-degree angle.

### Converting Feet-Inches to a Decimal Number

Divide the inches portion by 12 (12 inches in a foot) and add this to the whole number of feet.

31'-9" converts to decimal by:
$9 \div 12 + 31 = 31.75'$

42'-6" converts to decimal by:
$6 \div 12 + 42 = 42.5'$

### Using Pythagorean Theorem

Enter these numbers into the equation.

$C^2 = 31.75^2 + 42.5^2 = 2814.3125$

Take the square root to find diagonal

$C = 53.05009'$

### Converting Decimal Number to Feet-Inches

Subtract whole feet to leave only the decimal.

$53.05009' - 53 = 0.05009'$

Multiply resulting decimal portion by 12 to convert to inches.

$0.05009' \times 12$ inches per foot $= 0.601131"$

Subtract whole inches leaving a remainder.

$0.601131" - 0 = 0.601131"$

Convert the remainder to a fraction by multiplying by 16, the desired denominator. Round the fraction as necessary.

$0.601131" \times 16 = 9.6/16 = 10/16 = 5/8"$.

Thus the diagonal of a 31'-9" by 42'-6" rectangle is 53'-0 5/8".

Using two tapes, position the third stake where the two tapes cross (Fig. 7-23).

When the location of the third stake is completed, drive a nail in the top of the stake, marking the rear corner exactly. Using two tapes, locate the fourth stake by measuring the building length from stake three and the width from stake two (Fig. 7-24). Secure the stake and drive a nail in its top to mark exactly the other rear corner. Check the accuracy of the work by measuring the lengths, widths, and the diagonals corner to corner. Make adjustments if the measurements differ while maintaining accuracy of the width and length measurements (Fig. 7-25).

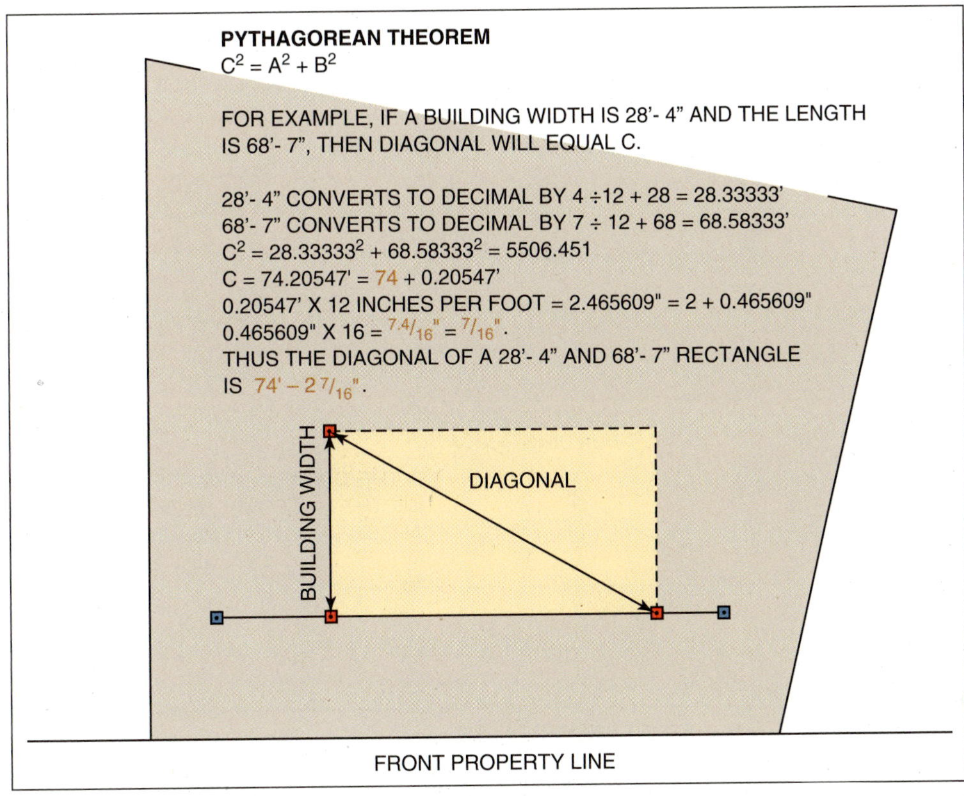


**FIGURE 7-23** Using the Pythagorean Theorem to position the third corner stake. Two tapes are used to measure the width of the building and the calculated diagonal.

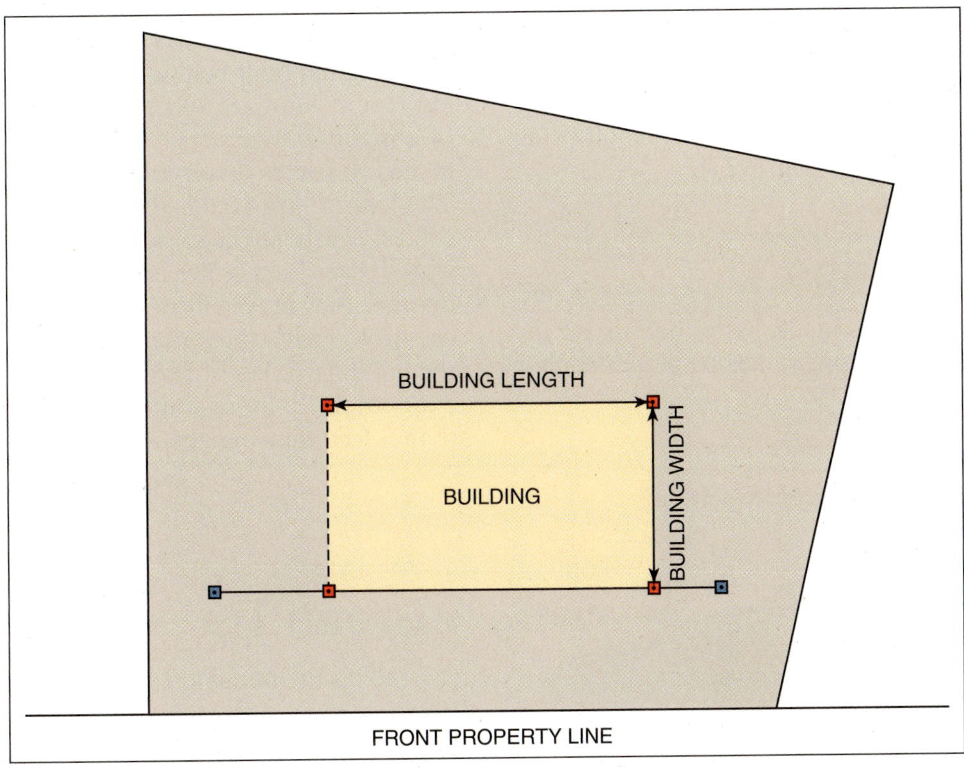


**FIGURE 7-24** The last corner stake is positioned by measuring, with two tapes, the width and length of the building.

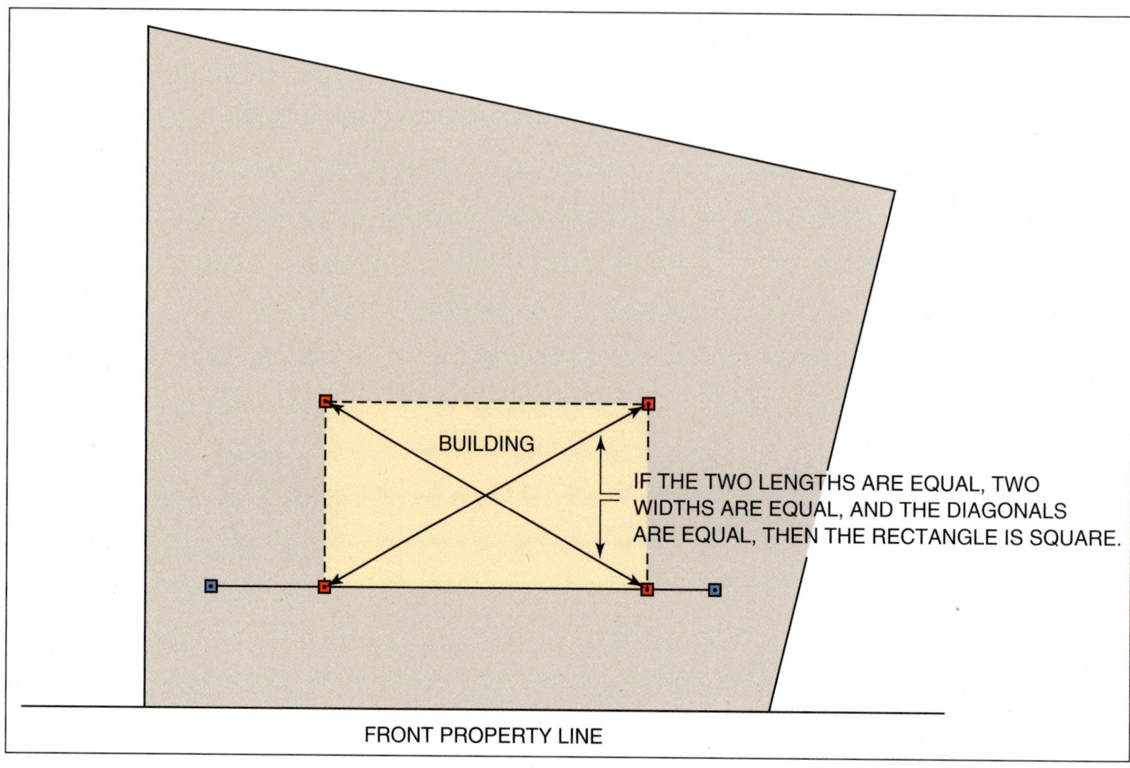


**FIGURE 7-25** If the length and width measurements are accurate and the diagonal measurements are equal, then the corners are square.

All measurements must be made on the level. If the land slopes, the tape is held level with a plumb bob suspended from it (Fig. 7-26). Irregular-shaped buildings are laid out using the fundamental principles outlined previously (Fig. 7-27).

## BATTER BOARDS

Before the excavation is made, batter boards are installed to allow the layout stakes to be reinstalled quickly after the excavation is made. They are wood frames built behind the stakes to which building layout lines are secured. Batter boards consist of horizontal members called **ledgers**. These are attached to stakes driven into the ground (Fig. 7-28). The ledgers are fastened in a level position to the stakes, nearly at the same height as the **foundation** wall. Batter boards are erected in such a place and manner that they will not be disturbed during excavation. Drive the batter board stakes back from the building stakes at least equal to the depth of the excavation. In loose soil or when stakes are higher than 3 feet, they may be braced (Fig. 7-29).

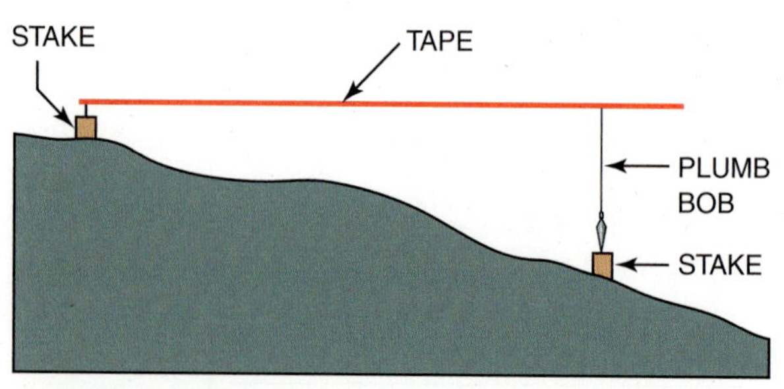


**FIGURE 7-26** For building layouts on sloping land, measurements must be taken on the level.

### Green Tip

Grading the building site to drain water naturally away from the building saves energy. Dry buildings are easier to heat and cool than damp ones.

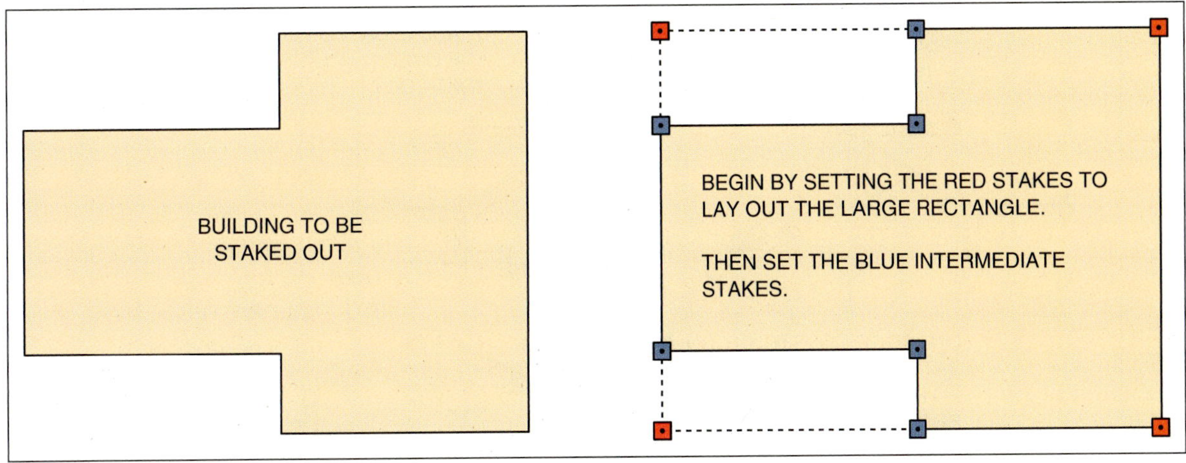


**FIGURE 7-27** Locate and drive corner stakes of the large rectangle first, then intermediate ones, when laying out an irregular building.

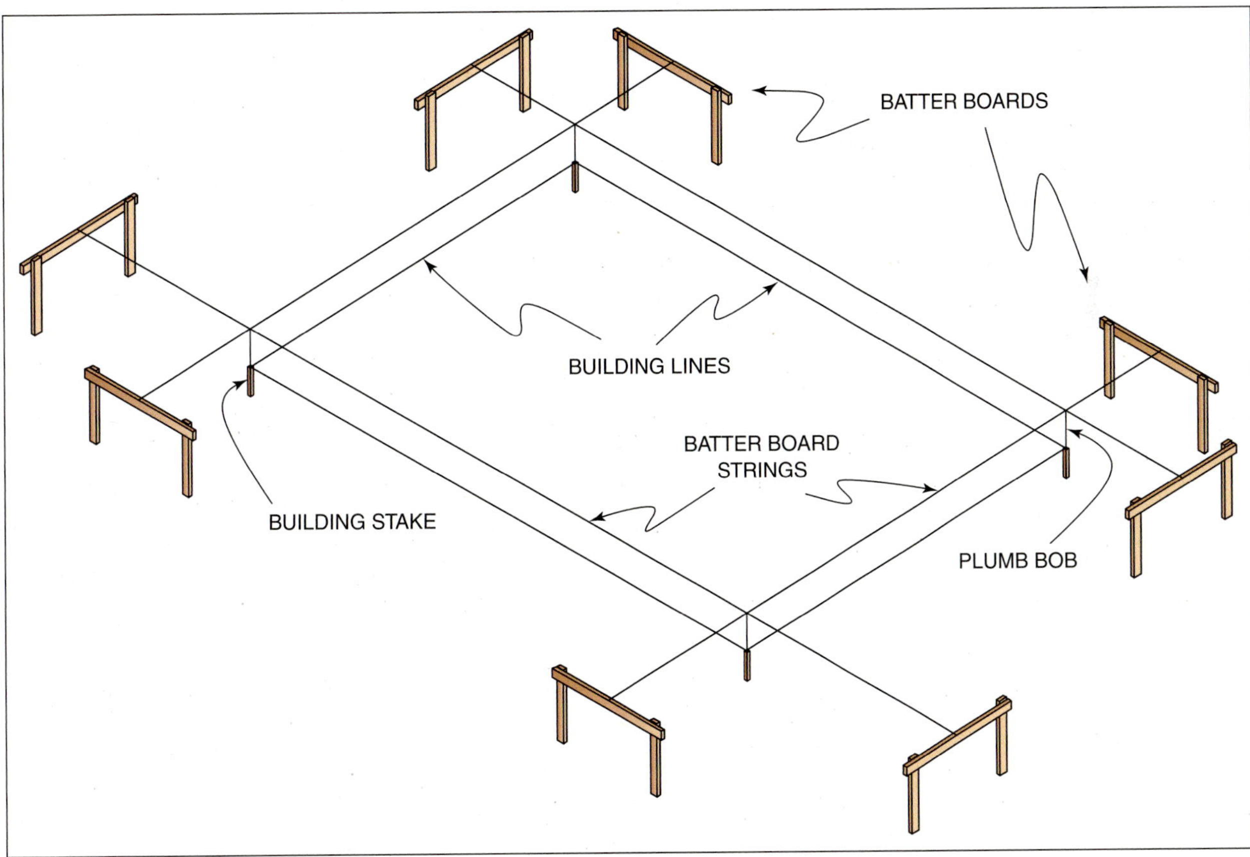


**FIGURE 7-28** Batter boards are installed behind the building stakes and nearly level with the top of the foundation.

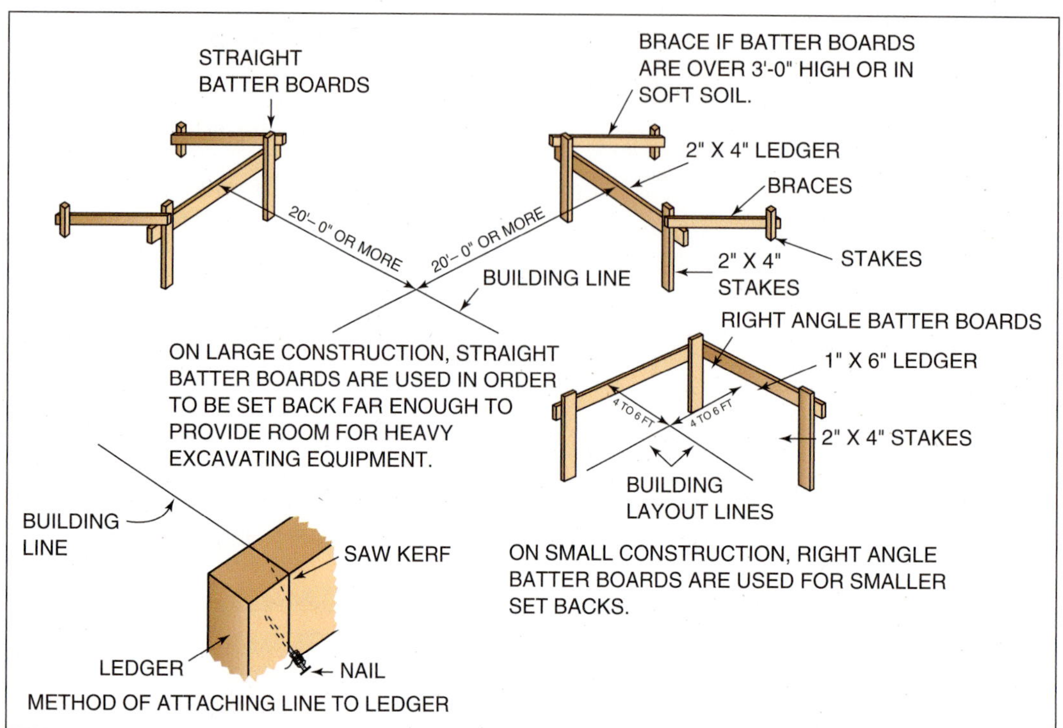


**FIGURE 7-29** Batter boards are placed back far enough so they will not be disturbed during excavation operations.

Set up the builder's level about center on the building location. Sight to the benchmark and determine the height of the ledgers. Sight and mark each batter board stake at the specified elevation. Attach ledgers to the stakes so that the top edge of each ledger is on the mark.

Stretch lines between batter boards directly plumb over the nailheads of the original corner stakes. The position of the lines is located by suspending a plumb bob directly over the nailheads of the building corner stakes. Check the accuracy of the layout by again measuring the diagonals to see if they are equal. If not, make the necessary adjustment until they are equal (Fig. 7-30).

When the lines are accurately located, make a saw cut on the outside corner of the top edge of the ledger. This prevents the layout lines from moving when stretched and secured. Be careful not to make the saw cut below the top edge (Fig. 7-31).

During the excavation, the batter board strings and corner stakes are removed while leaving the batter boards intact. Later, the corner stakes are repositioned in the excavation by first reinstalling the batter board strings. Then use a plumb bob to relocate the corner stakes down inside the excavation (Fig. 7-32).

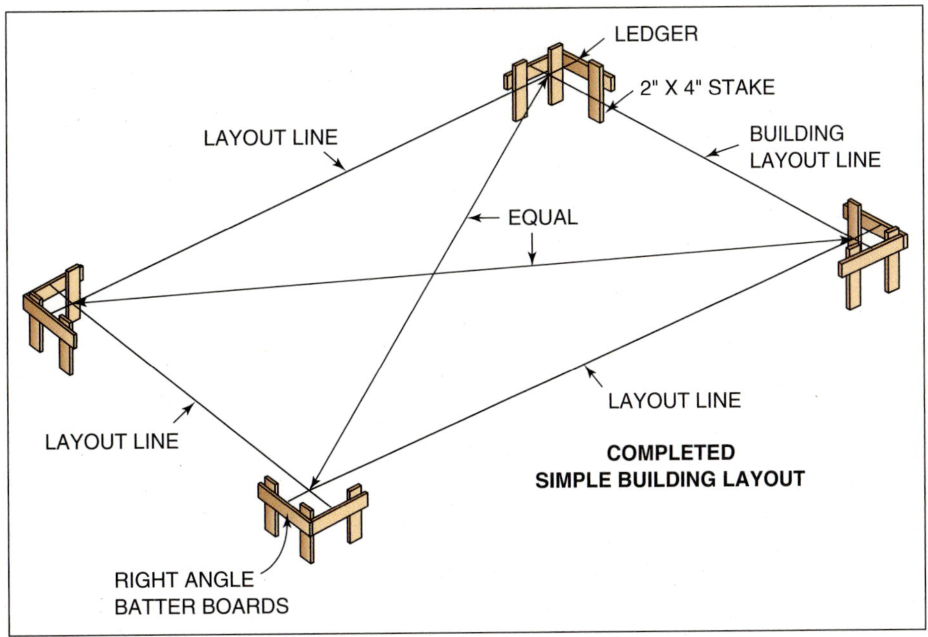


**FIGURE 7-30** Check the length and width and then the diagonals for accuracy of the building layout.

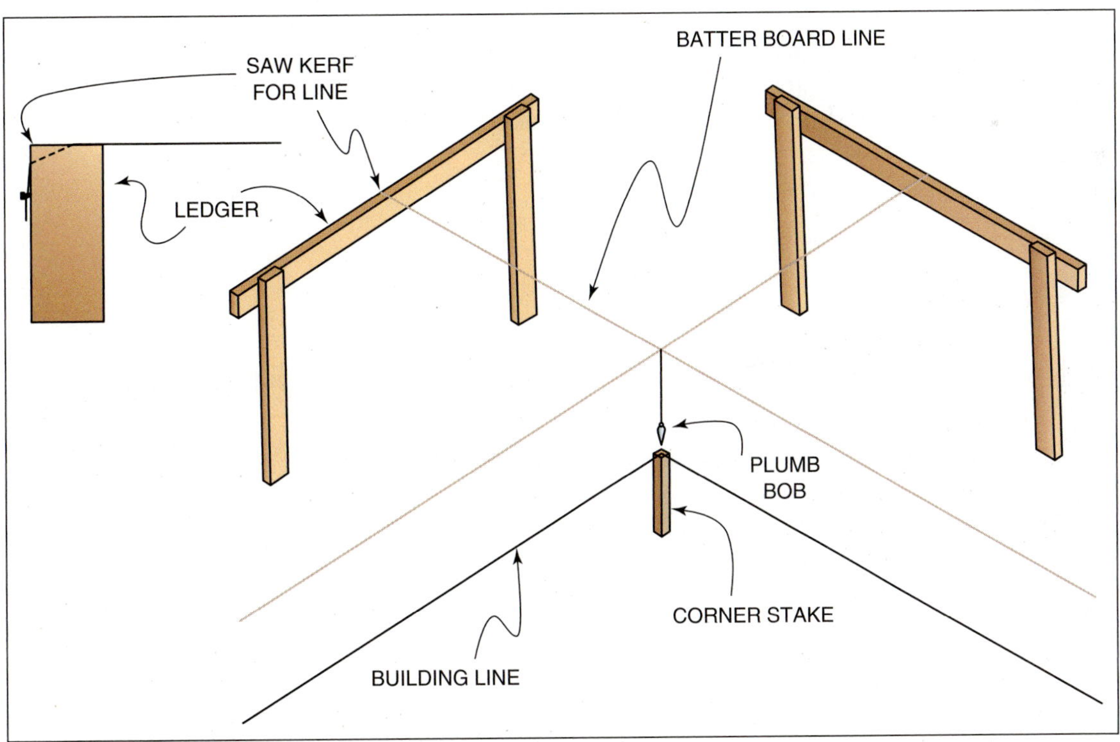


**FIGURE 7-31** Typical batter board construction and method of locating layout lines on batter boards by suspending a plumb bob directly over corner stakes.

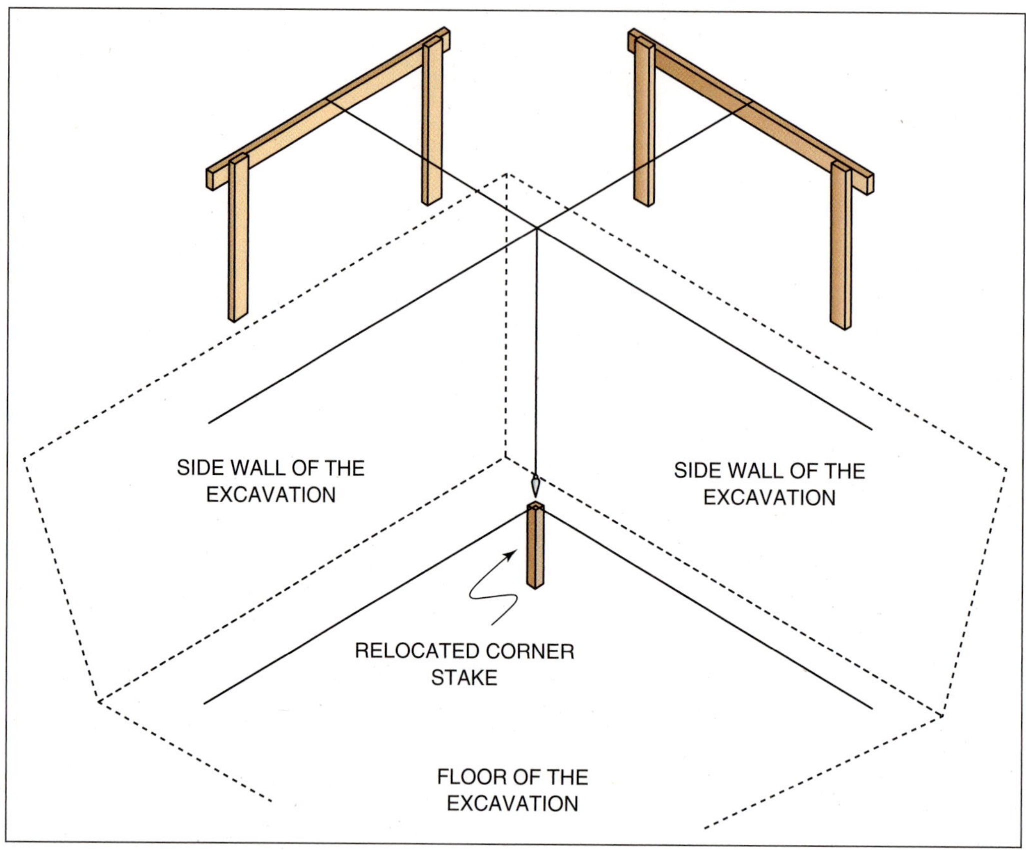


**FIGURE 7-32** Relocating the corner stakes deep in the excavation using the batter board strings and a plumb bob.

# SUMMARY

- Building layout employs special tools and equipment to maintain straight, level, and square.
- Locating the building is the practice of placing a building in the proper position on a building site.

## GREEN CHECKLIST

- ☐ Grading the building site to drain water naturally away from the building saves energy. Dry buildings are easier to heat and cool than damp ones.

## REVIEW QUESTIONS

*Select the most appropriate answer.*

1. **A useful tool to set stakes of a small building is**
   a. straight edge and a level.
   b. water level.
   c. automatic builder's level.
   d. all of the above.

2. **The principle that makes a water level work well is that water**
   a. is available in all areas of the world.
   b. evaporates when it is spilled.
   c. tends to seek level by itself.
   d. all of the above.

3. **The optical builder's level and transit level differ in that the builder's level cannot read**
   a. horizontal angles.
   b. vertical angles.
   c. elevation differences.
   d. all of the above.

4. **When leveling a builder's level with four leveling screws, rotate the screws**
   a. in the same direction as your thumbs.
   b. in the same direction.
   c. clockwise.
   d. counter clockwise.

5. **The horizontal cross hair of a transit level is used for**
   a. laying out horizontal angles.
   b. laying out straight horizontal lines.
   c. plumbing walls and posts.
   d. reading elevations.

6. **The reference for establishing elevations on a construction site is called the**
   a. starting point.
   b. reference point.
   c. benchmark.
   d. sight mark.

**7. The first thing to do when laying out a 90-degree horizontal angle with a builder's level is to**

a. level the instrument.

b. adjust instrument and tripod over a point using a plumb bob.

c. turn the horizontal circle scale to zero.

d. secure the horizontal tangent screw.

**8. When using a builder's level and a target, Stake A is read to be 5'-3" and stake B is 7'-8". Which statement is true?**

a. Stake A is 2-5" higher than Stake B.

b. Stake A is 2-5" lower than Stake B.

c. Stake B is 7'-8" above Stake A.

d. Stake B is 5'-3" below Stake A.

**9. When laying out a rectangular building, the third stake is best set using**

a. a tape measure from the second stake.

b. Pythagorean Theorem.

c. water level.

d. all of the above.

**10. Convert 54'-10" to decimal feet.**

a. 54.10'

b. 54.625'

c. 54.833

d. 54.987

**11. The diagonal of a rectangle whose dimensions are 30.1 feet × 40.1 feet may be estimated to be slightly more than**

a. 45 feet.

b. 50 feet.

c. 60 feet.

d. 70 feet.

**12. The diagonal of a rectangle whose dimensions are 32'-6" × 48'-9" is**

a. 58.590'

b. 58.646'

c. 58.715'

d. 58.770'

**13. What is the measurement in feet-inches to the nearest sixteenth of 15.59021'**

a. 15'-6¼"

b. 15'-7 1/16"

c. 15'-9 7/16"

d. 15'-9 5/16"

**14. Batter boards should be constructed to be**

a. less than 10 feet away from the excavation.

b. sturdy and permanent.

c. in a place that will not be disturbed during excavation.

d. all of the above.

**15. The strings of a batter board system are used to**

a. represent the top outside edge of the foundation wall.

b. allow quick replacement of building corner stakes after excavation.

c. make the placement of the building easier.

d. all of the above.

## WHAT'S WRONG WITH THIS PICTURE?

Carefully study **Figure 7-33** and think about what is wrong. Consider all possibilities.

### WRONG

**FIGURE 7-33** Two workers are taking a lunch break while sitting on the batter boards at an excavation site. The snow is not the problem but only an unfortunate reality in some northern climates. While these batter boards are correctly located several feet back from the area to be excavated, the problem is the workers sitting on them. Batter boards, once located, should never be unnecessarily touched. They are set in precise locations, and leaning on them could throw off their placement, which could, in turn, result in improper positioning of the footer or foundation wall.

### RIGHT

**FIGURE 7-34** Note the undisturbed batter boards, as should always be the case.

CHAPTER 8

# Concrete Form Construction

**OBJECTIVES** *After completing this unit, the student should be able to:*

- Construct forms for footings, slabs, walks, and driveways.
- Construct concrete forms for foundation walls.
- Lay out and build concrete forms for stairs.
- Explain techniques used for the proper placement and curing of concrete.
- Describe the composition of concrete and factors affecting its strength, durability, and workability.
- Explain the reasons for making a slump test.
- Explain the reasons for reinforcing concrete and describe the materials used.
- Estimate quantities of concrete.

## GLOSSARY OF CONCRETE FORM CONSTRUCTION TERMS

**buck** a rough frame used to form openings in poured concrete walls

**compressive strength** the quality of being resistant to crushing; concrete has high compressive strength

**concrete** a building material made from Portland cement, aggregates, and water

**concrete block** a concrete masonry unit (CMU) used to make building foundations, typically measuring 8" × 8" × 16"

**frost line** the depth to which the ground typically freezes in a particular area; footings must be placed below this depth

**girder** heavy timber or beam used to support vertical loads

**gusset** a block of wood or metal used over a joint to stiffen and strengthen it

**pilaster** column built within and usually projecting from a wall to reinforce the wall

**Portland cement** a fine gray powder, which when mixed with water, forms a paste that sets rock hard; an ingredient in concrete

**reinforcing rods** also called rebar, steel bars placed in concrete to increase its tensile strength

**rise** in stairs, the vertical distance of the flight; in roofs, the vertical distance from plate to ridge; may also be the vertical distance through which anything rises

**run** the horizontal distance over which rafters, stairs, and other like members travel

**scab** a length of lumber or material applied over a joint to stiffen and strengthen it

**spreader** a strip of wood used to keep other pieces a desired distance apart

**stud** vertical framing member in a wall running between plates

**tensile strength** the quality of being resistant to breaking under a pulling force; steel has high tensile strength

**vapor retarder** also called a vapor barrier, a material used to prevent the passage of moisture

Concrete as a building material is unique. The installation time is determined by how fast the concrete cures. Once it cures, it cannot be easily undone, changed, or moved. It must be placed right the first time. If a form fails to hold, costly labor, time, and materials are lost. Construction of concrete forms is the responsibility of the carpenter.

These forms must meet specified dimensions and be strong enough to withstand tremendous pressure. After the concrete is placed, time must be spent in dismantling the forms and cleanup. Therefore, form construction should be strong yet easy to remove.

Understanding the characteristics of concrete is essential for the construction of reliable concrete forms. Only then can the correct handling of freshly mixed material and the fine quality of hardened concrete be achieved.

# THE NATURE OF CONCRETE

Concrete can be formed into practically any shape for construction of buildings, bridges, dams, and roads. Improvements over the years have created a product that is strong, durable, and versatile (Fig. 8-1). **Concrete** also has many decorative and aesthetic uses. It is a mixture of **Portland cement**, fine and coarse aggregates, water, and various admixtures. Aggregates are fillers, usually sand, gravel, or stone. Admixtures are materials added to the mix to achieve certain desired qualities. Any water used to make concrete should be clean and even drinkable. Other water may be used but should be tested first to make sure it is acceptable. The amount of water used largely determines the quality of the concrete. Most concrete used in construction is ready-mixed concrete delivered to the job in a large truck (Fig. 8-2). It is sold by the cubic yard. The truck contains a large revolving drum, capable of holding up to 10 cubic yards.

Concrete should be placed on stable subsoil. All topsoil in the area in which the slab is to be placed must be removed. A base for the slab consisting of 4 to 6 inches of gravel, crushed stone, or other approved material must be well compacted in place. The soil under the slab may be treated with chemicals for control of termites, but caution is advised. Such treatment should be done only by those thoroughly trained in the use of these chemicals.

Courtesy of Portland Cement Association.

**FIGURE 8-1** Concrete is widely used for foundations and structural support.

**FIGURE 8-2** Ready-mixed concrete is delivered in trucks.

## Green Tip

Some areas of the United States recycle concrete. Concrete is broken into the sizes of large aggregates and used in new concrete, replacing crushed stone.

# FOOTINGS

Footings are the first part of a building to be installed. They are built in several styles depending on the building's requirements.

## FOOTING TYPES

The *footing* for a foundation provides a base on which to spread the load of a structure over a wider area of the soil. For foundation walls, the most typical type is a continuous or spread footing (Fig. 8-3). To provide support for columns and posts, pier footings of square, rectangular, circular, or tapered shape are sometimes used.

In most cases, the spread and pier footings are formed separately from the foundation wall. The footing width is usually twice the wall thickness, and the footing depth is often equal to the wall thickness (Fig. 8-4). Usually, these footings are strengthened by reinforcing rods of specified size and spacing.

### Frost Line

Footings must be located below the frost line. The frost line is the point below the surface to which the ground usually freezes in winter. Because water expands when frozen, foundations whose footings are above the frost line will heave and may crack when the ground freezes. In extreme northern climates, footings must be placed as much as 6 feet below the surface (Fig. 8-5). In tropical climates, footings need to reach only solid subsoil, with no consideration given to frost.

***For step-by-step instructions on CONSTRUCTING FOOTING FORMS, see Procedure 8-1 on pages 186–188.***

### Keyways

A keyway is usually formed in the footing by pressing 2 × 4 lumber into the fresh concrete (Fig. 8-6). The keyway form is beveled on both edges for easy removal after the concrete has set. The purpose of a keyway is to provide a lock between the footing and the foundation wall. This joint helps the foundation wall resist the lateral or side-to-side pressure of the backfilled earth against it.

### Stepped Wall Footings

When the foundation is to be built on sloped land, it is sometimes necessary to *step* the footing. The footing is formed at different levels to save material. When stepped footing forms are built, the thickness of the footing must be maintained. The vertical and horizontal footing distances are adjusted so that a whole number of concrete blocks or concrete forms can easily be placed into that section of the footing without cutting. The vertical part of each step should not exceed the footing thickness. The horizontal part of the step must be at least twice the vertical part (Fig. 8-7). End blocks are placed between the forms to retain the concrete at each step.

### Column Footings

Concrete for footings, supporting columns, posts, fireplaces, chimneys, and similar objects is usually placed at the same time as the wall footings. The size and shape of the column footing vary according to what it has to support. The dimensions are determined from the foundation plan.

The forms for this type of footing are usually built by nailing form pieces together in square, rectangular, or tapered shapes to the specified size (Fig. 8-8). Stakes are driven on all sides. Forms are usually fastened in a position so that the top edges are level with the wall footing forms.

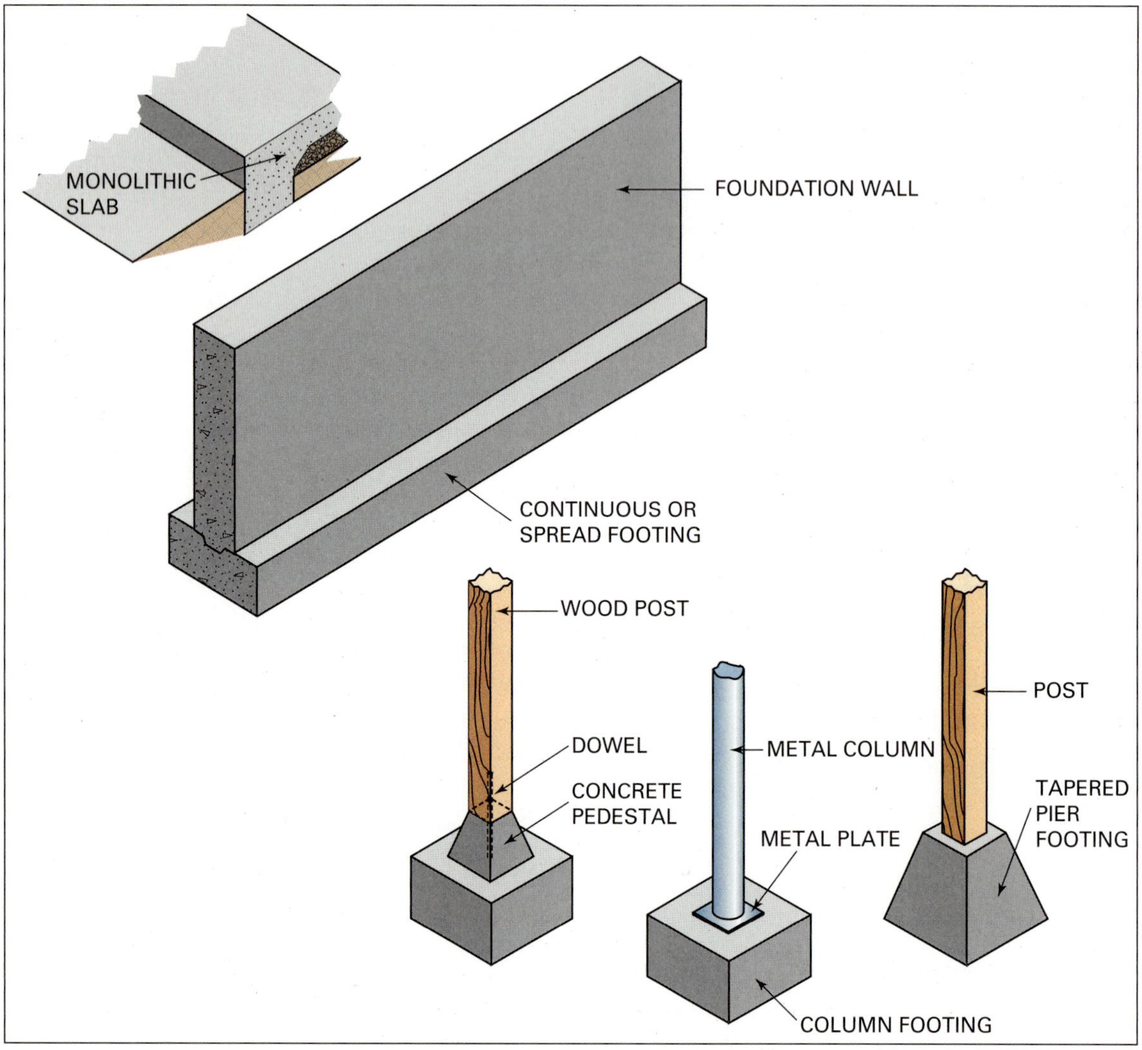

MONOLITHIC SLAB
FOUNDATION WALL
CONTINUOUS OR SPREAD FOOTING
WOOD POST
DOWEL
CONCRETE PEDESTAL
METAL COLUMN
METAL PLATE
COLUMN FOOTING
POST
TAPERED PIER FOOTING

**FIGURE 8-3** Several types of footings are constructed to support foundations.

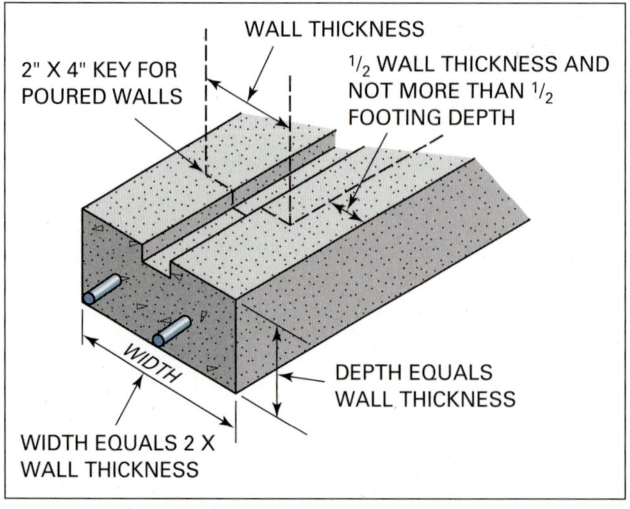


**FIGURE 8-4** Typical footing for residential construction.

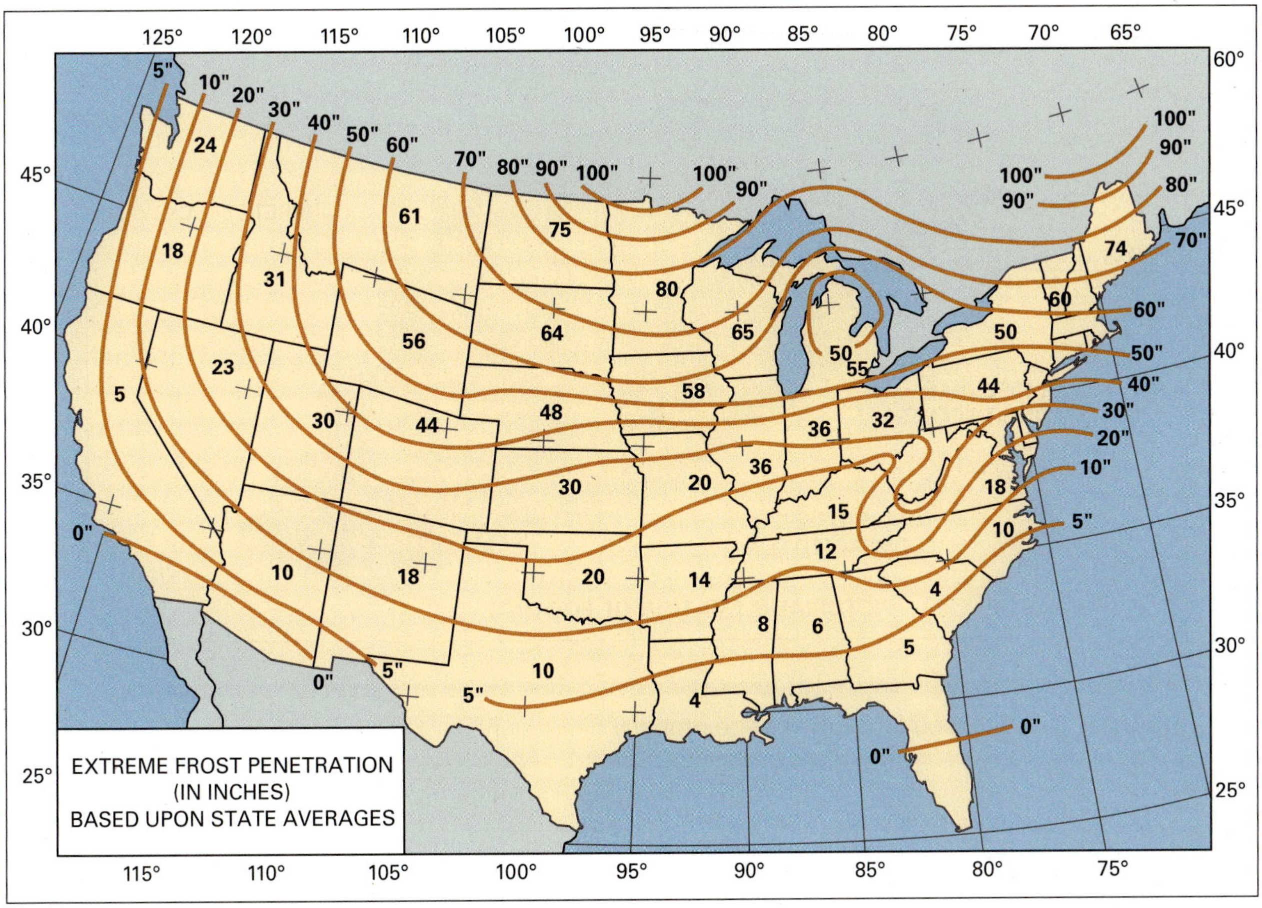

*Courtesy of U.S. Army Corps of Engineers.*

**FIGURE 8-5** Frost line penetration in the United States.

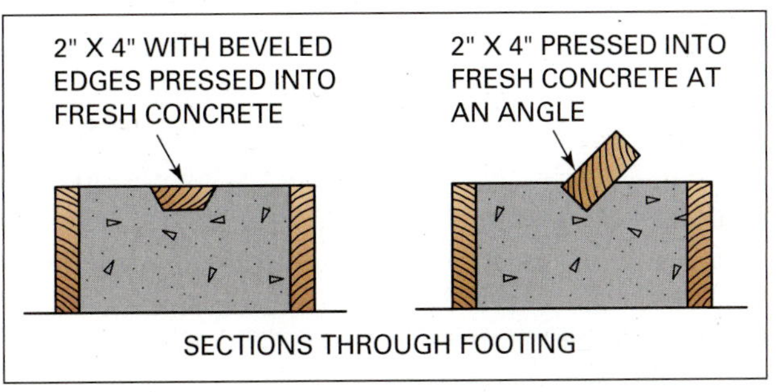


**FIGURE 8-6** Methods of forming keyways in the footing.

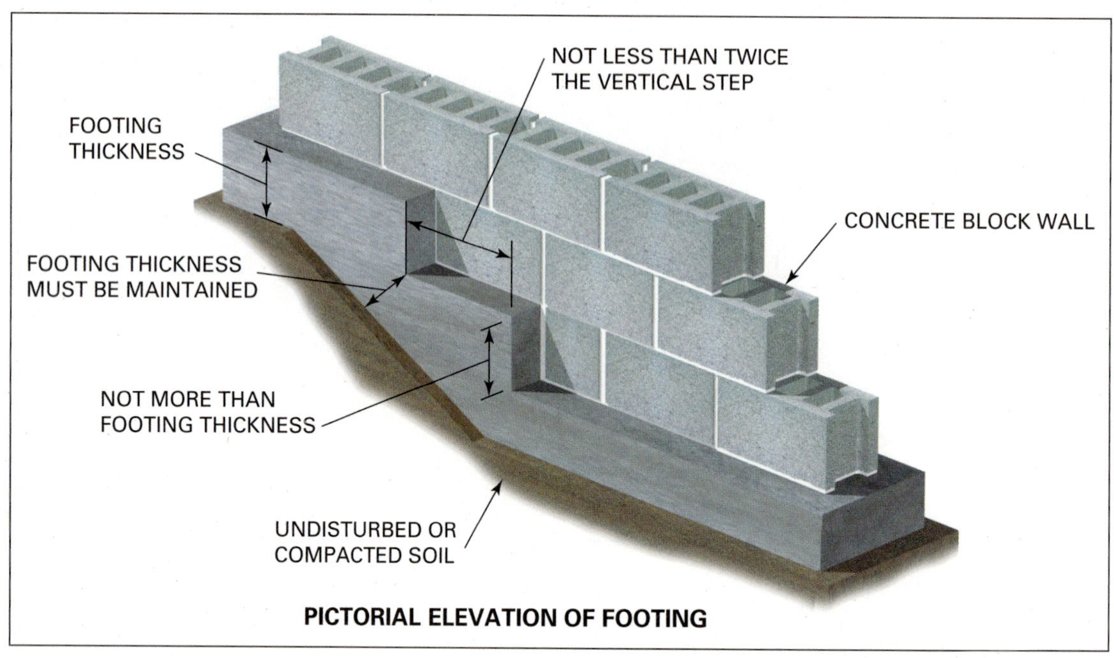


**FIGURE 8-7** Stepped footings must be properly dimensioned.

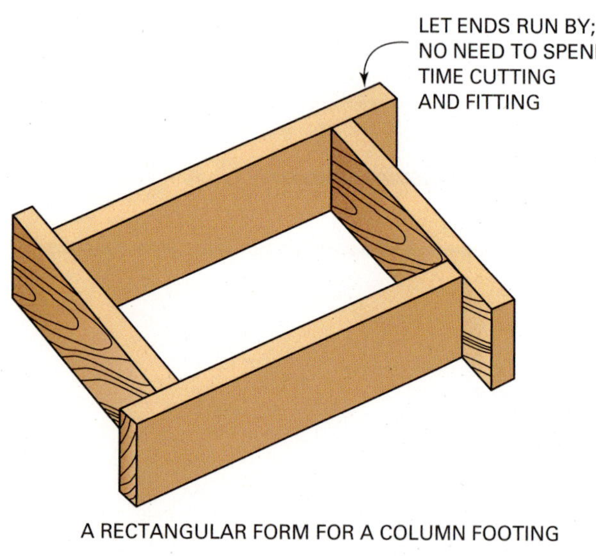


**FIGURE 8-8** Construction of column footing form.

# Slab Forms

Building forms for slabs, walks, and driveways is similar to building continuous footing forms, except the inside form boards are omitted. Generally, forms for floor slabs are built level, whereas walks and driveways are sloped to shed water. Usually, 2 × 4 or 2 × 6 lumber is used for the sides of the form.

Slabs should be isolated from other building members to protect the slab from cracking with expansion joints. Expansion joints are typically made of a rigid foam- or asphalt-impregnated sheathing material. They are positioned between the slab and other objects, such as foundation walls, before concrete is placed.

## SLAB-ON-GRADE

In warm climates, where frost penetration into the ground is not very deep, little excavation is necessary. The first floor may be a concrete slab placed directly on the ground. This is commonly called *slab-on-grade* construction (Fig. 8-9). With improvements in the materials and methods of construction, insulated slabs-on-grade are being used more often in colder climates.

## MONOLITHIC SLABS

A combined slab and foundation/footing is called a *monolithic slab* (Fig. 8-10). This type of slab is also referred to as a *thickened edge slab*. It consists of a slab with a footing around the perimeter. The slab and footing concrete are placed at the same time. Forms for monolithic slabs are constructed using the same procedure as for slabs.

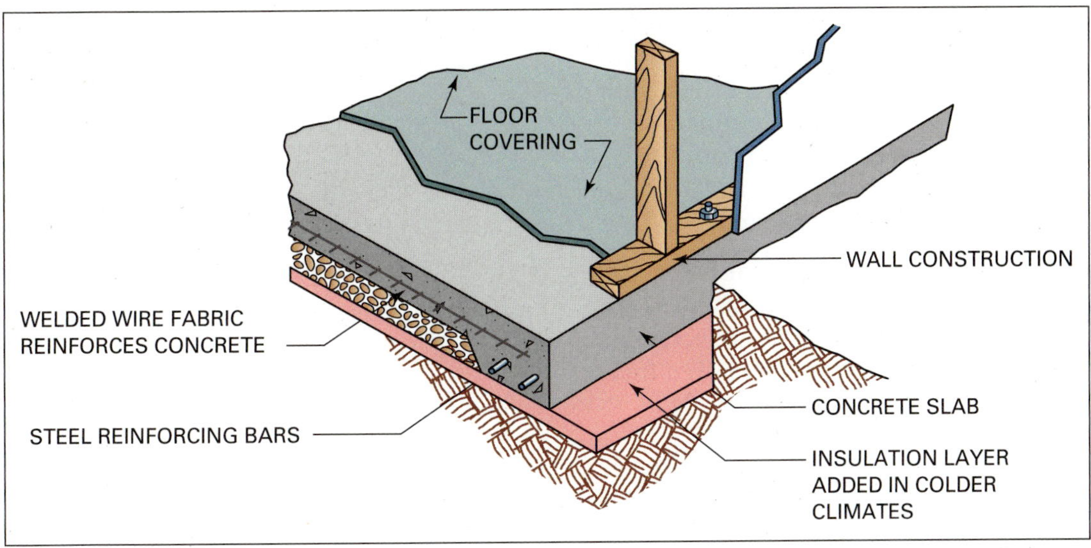


From Mark W. Huth, *Understanding Construction Drawings,* ©1983 by Delmar.

**FIGURE 8-9** Slab-on-grade foundation.

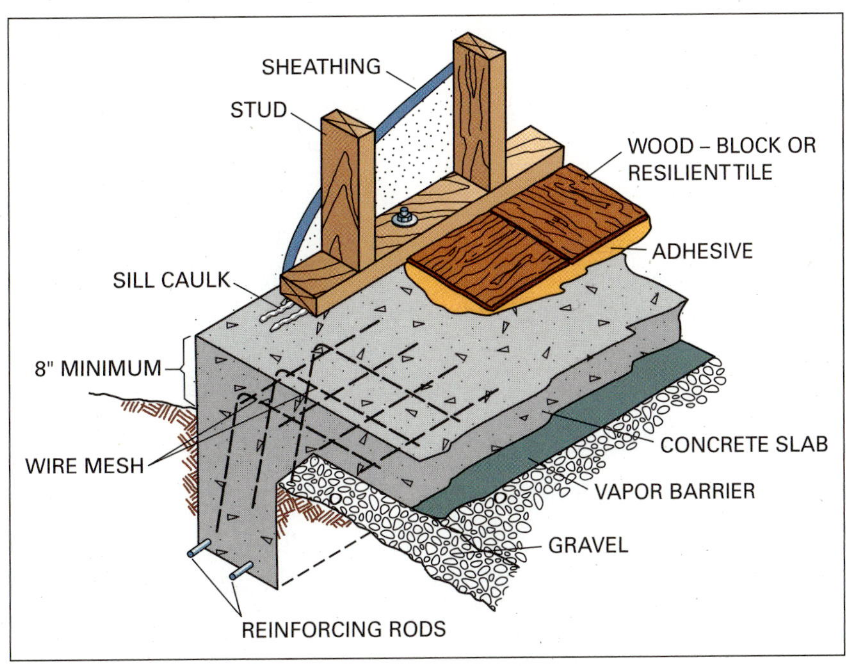


**FIGURE 8-10** Monolithic slab-on-grade.

## SLAB PROTECTION

Concrete exposed to living space of the building should be protected from moisture and cold. A **vapor retarder**, also called a *vapor barrier,* is placed under the concrete slab to prevent soil moisture from rising through the slab. It also provides a barrier to soil gases, such as radon and methane, entering the living space. Concrete is hard but porous and often develops cracks. This barrier should be a heavy plastic film, such as 6-mil polyethylene or other equal material. It should be strong enough to resist puncturing during the placing of the concrete. Joints in the vapor retarder must be lapped at least 4 inches and sealed. A layer of sand may be applied to protect the membrane during concrete placement.

Where necessary, to prevent heat loss through the floor and foundation walls, rigid insulation made of *extruded polystyrene* is installed under the slab. It is the only insulation recommended for ground contact. It is placed between concrete and the subsoil. The intersection of the foundation wall and the slab edge may also be insulated.

The slab should be reinforced with 6 × 6 inch, #10 welded wire mesh, or by other means to provide equal or superior reinforcing. The concrete slab must be at least 4 inches thick and *haunched* (made thicker) under load-bearing walls (Fig. 8-11).

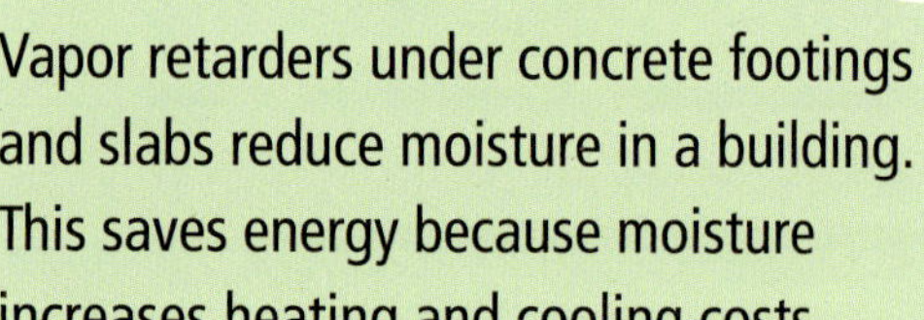

**Green Tip**

Vapor retarders under concrete footings and slabs reduce moisture in a building. This saves energy because moisture increases heating and cooling costs.

From Mark W. Huth, *Understanding Construction Drawings*, ©1983 by Delmar.

**FIGURE 8-11** The slab is haunched under load-bearing walls.

# Wall Forms

Various kinds of panels and panel systems are used. Concrete panel systems are manufactured of wood, aluminum, or steel. Specially designed hardware is used for joining, spacing, aligning, and bracing the panels.

## WALL FORM COMPONENTS

All wall form systems use similar components to make them perform as expected.

### Form Panels

These panels are built of special form plywood backed by metal ribs or 2 × 4 **studs**. They can be purchased or built on the job. They are placed side by side to form the inside and outside of the foundation walls.

### Snap Ties

Snap ties hold the wall forms together at the desired distance apart. They support both sides against the lateral pressure of the concrete (Fig. 8-12). These ties reduce the need for external bracing and greatly simplify the erection of wall forms. These ties are called *snap ties* because projecting ends are snapped off slightly inside the concrete after removal of the forms.

*Courtesy of Dayton/Richmond Concrete Accessories.*

**FIGURE 8-12** A wood form panel with snap ties installed.

The small remaining holes are easily filled later. Because of the great variation in the size and shape of concrete forms, many styles are used (Fig. 8-13).

### Walers

The snap ties run through the form boards and are wedged against additional form supports called *walers*. Walers are doubled 2 × 4 pieces with spaces between them used to stiffen the forms. They may be horizontal or vertical. Walers are spaced at right angles to the panel frame members or ribs. The number and spacing depend on the pressure exerted on the form. The vertical spacing of the snap ties and walers depends on the height of the concrete wall. The vertical spacing is closer together near the bottom. This is because there is more lateral pressure from the concrete there than at the top (Fig. 8-14).

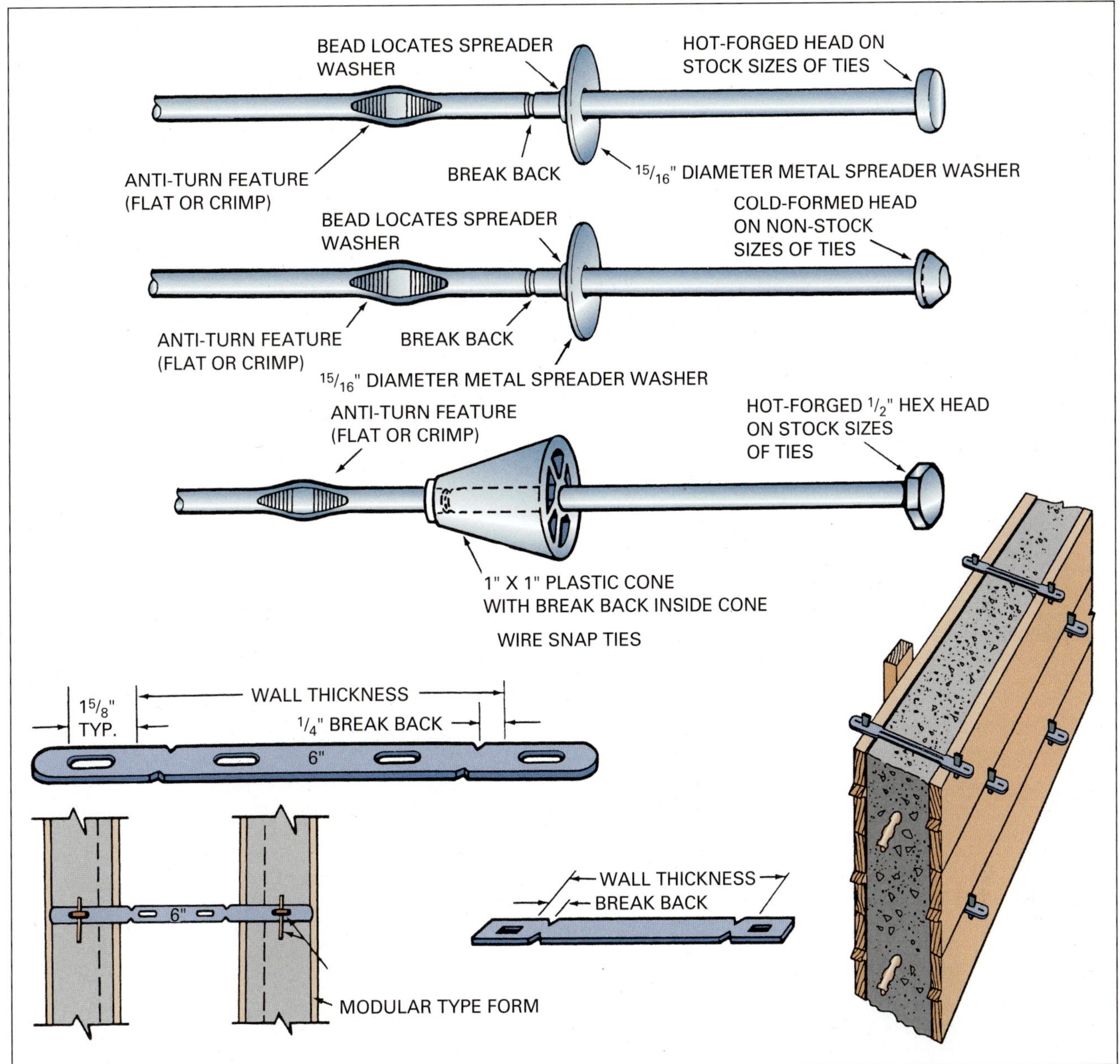


Courtesy of Dayton/Richmond Concrete Accessories.

**FIGURE 8-13** A large variety of snap ties are manufactured.

***For step-by-step instructions on ERECTING WALL FORMS, see Procedure 8-2 on pages 189–196.***

## CONCRETE FORMING SYSTEMS

A concrete forming system consists of manufactured panels and components for concrete form construction. The panels may be made of steel or a combination of steel and wood. The panels are tied together with metal wedges (Fig. 8-15). Forming system hardware is specially designed for use with these systems.

Starting at the outside corner, the panels are wedged together by attaching to angle irons (Fig. 8-16). Walers are easily installed on wall forms when the forming system hardware is used (Fig. 8-17). Adjustable braces are available. Only a few are required to hold the wall straight (Fig. 8-18).

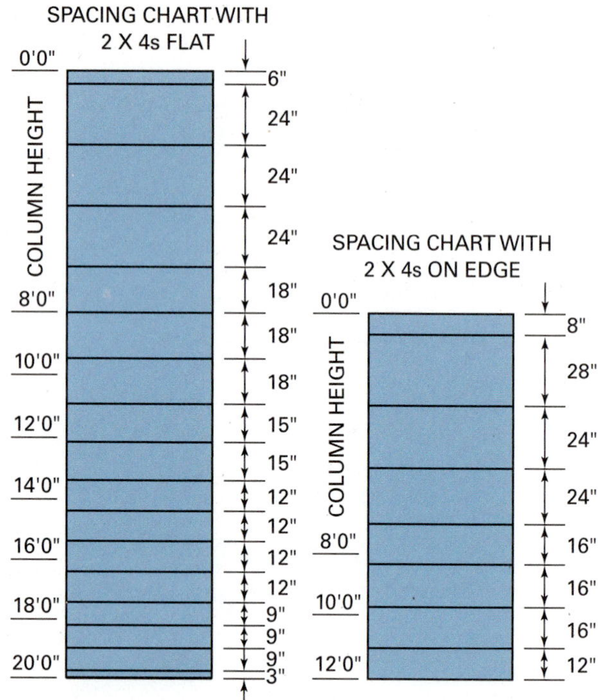


**FIGURE 8-14** Horizontal stiffeners are placed closer together near the bottom than they are at the top. Dimensions are for purposes of illustration only.

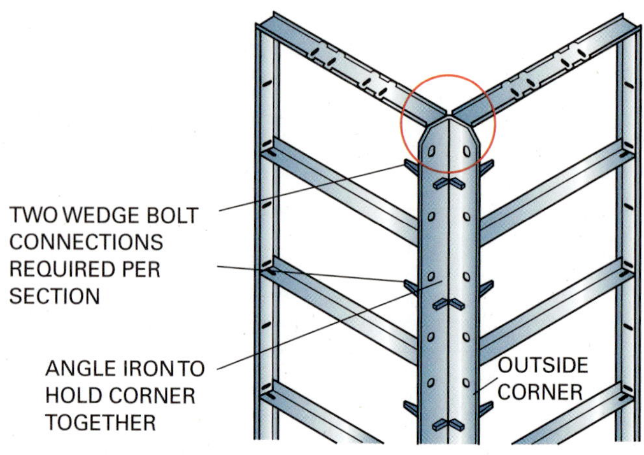


**FIGURE 8-16** Outside corners are created by attaching panels to angle irons with wedges.

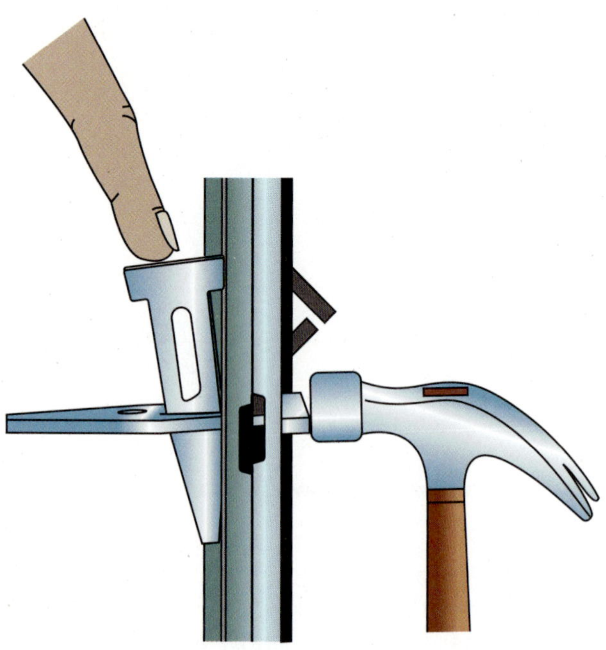

Courtesy of Dayton Superior Corporation.

**FIGURE 8-15** Only a hammer is needed to tie panels or corners together when metal wedges are used. Tapping to the side as the wedge is pushed down adequately tightens the form, leaving it also loose enough to be easily removed later.

Courtesy of Dayton Superior Corporation.

**FIGURE 8-17** Walers are easily installed on forming systems when special hardware is used.

**FIGURE 8-18** Adjustable braces are available with forming systems.

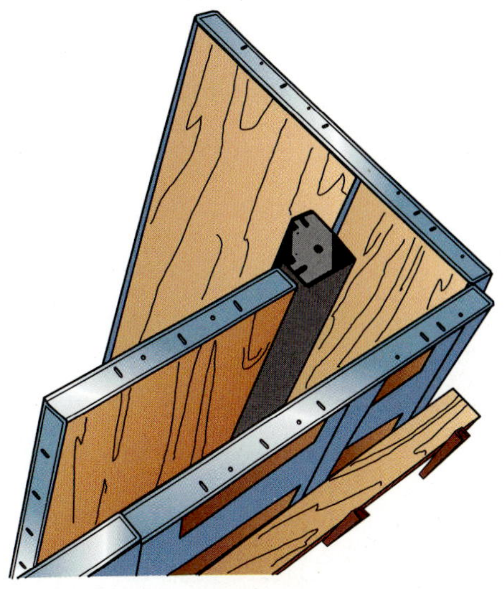

**FIGURE 8-19** Steel inside corners of a forming system make fast erection for formwork possible.

Steel inside corners are used with wall panels for fast and easy forming of the inside corner (Fig. 8-19). Forming systems require a considerable initial outlay of funds. However, the investment is saved many times over because of the reduced labor costs and reusability of the forming system's components. These systems are also available for rent.

## Stair Forms

It may be necessary to refer to Chapter 19 on stair construction for the definition of stair terms, types of stairs, stair layout, and methods of stair construction. Concrete stairs may be suspended or supported by earth (Fig. 8-20). Each type may be constructed between walls or have open ends. Stairs are either formed between two existing walls or between side wall forms. The side wall forms are often larger than needed. They must be plumb and firmly braced in position. During concrete placement, the bottom side of the stair must be supported by earth or a form. The bottom support for the stairs is graded or formed to provide proper thickness to the stairs. Concrete should not be overly thick because that would waste a relatively expensive material.

**Rise** is the vertical distance of a step, the height of each step. It is also called the *riser.* **Run** is the horizontal distance of a step, roughly the width of each step. It is called the *tread.* Layout may be done with a level and a chalk line (Fig. 8-21).

Riser form boards are beveled on the bottom edge to permit the mason to trowel the entire surface of the tread. Otherwise, the bottom edge of the riser form will leave its impression in the concrete tread (Fig. 8-22).

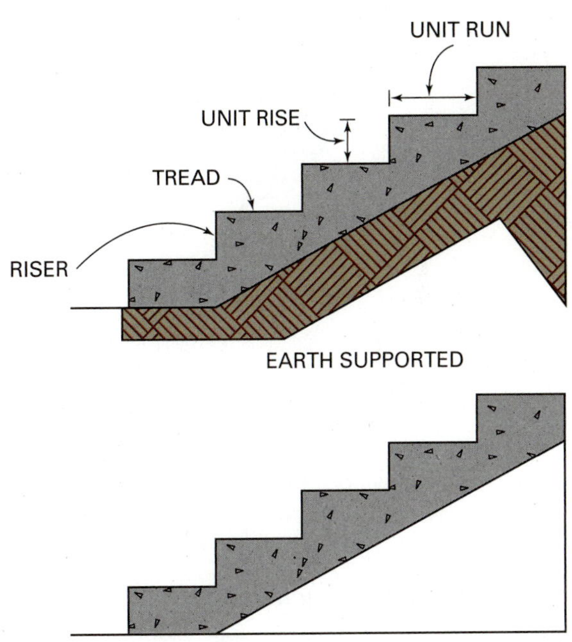


**FIGURE 8-20** Types of concrete stairs. Stairs may be formed with both ends closed or open, or with one end closed and the other end open.

Use only as many fasteners as needed to speed up form removal later. Screws or duplex nails should be used to make stripping the form easier.

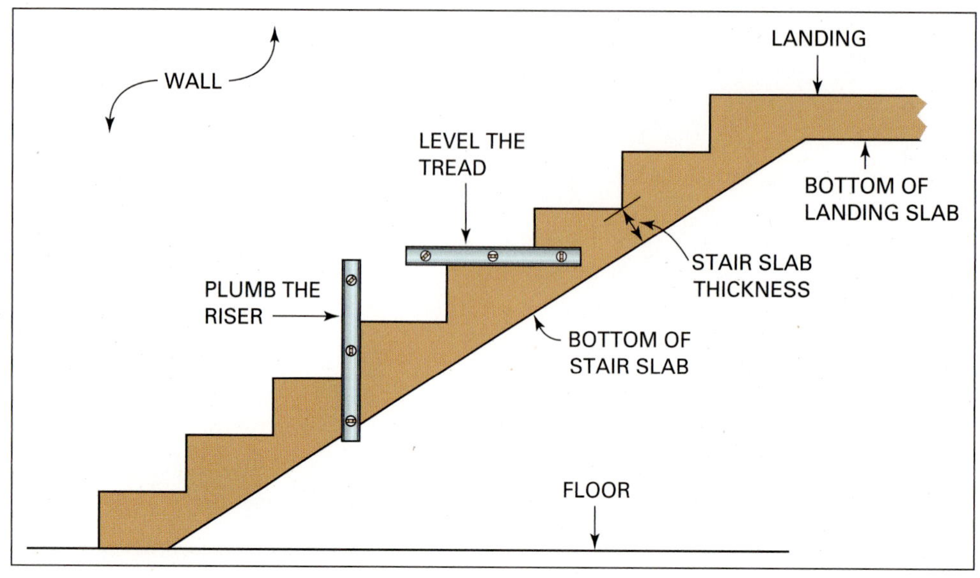


**FIGURE 8-21** Make the layout for suspended stairs on the inside of the wall forms.

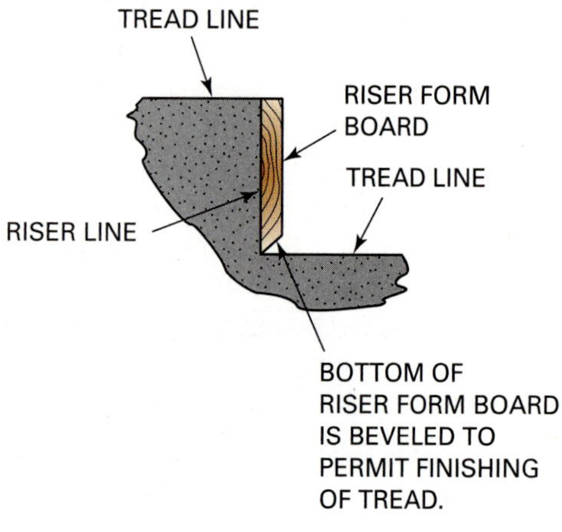


**FIGURE 8-22** Cross section of a riser form showing a beveled bottom.

***For step-by-step instructions on CONSTRUCTING CONCRETE STAIR FORMS, see Procedure 8-3 on pages 197–198.***

# Economy and Conservation in Form Building

Economical concrete construction depends on the reuse of forms. Forms should be designed and built to facilitate stripping and reuse. Use panels to build forms whenever possible. Use only as many nails as necessary for strength to make stripping forms easier.

Care must be taken when stripping forms to prevent damage to the panels so they can be reused. Stripped forms should be cleaned of all adhering concrete and stacked neatly. Long lengths of lumber can often be used in forming without cutting. Random-length walers can extend beyond the forms. There is no need to spend a lot of time cutting lumber to exact length (see Fig. 6-8). The important thing is to form the concrete to specified dimensions without spending too much time in unnecessary fitting.

# Concrete Reinforcement

Concrete has high ***compressive strength***. This means that it resists well to being crushed. It has, however, low ***tensile strength***, or is not as resistant to bending or pulling apart. Steel reinforcing bars, called *rebars,* are used in concrete to increase its tensile strength. Such concrete is then called *reinforced concrete.*

## REBARS

Rebars used in construction have a surface with ridges that increase the bond between the concrete and the steel. They come in standard sizes, identified by numbers that indicate the diameter in eighths. For instance, a #6 rebar has a diameter of 6/8 or 3/4 inch (Fig. 8-23). Metric rebars are measured in millimeters.

Slab reinforcement may be done with #3 bars placed in an 18" grid or #4 bars placed in a 24" grid. The size, location, and spacing of rebars are determined by concrete engineers and shown on the plans. Rebars are placed inside the form before the concrete is placed.

### WIRE MESH

Welded wire mesh is used to reinforce concrete floor slabs resting on the ground, driveways, and walks. It is identified by the gauge and spacing of the wire. Common gauges are #6, #8, and #10. The wire is usually spaced to make a 6-inch grid (Fig. 8-24). Welded wire mesh is laid in the slab above the vapor retarder before the concrete is placed.

| BAR # | BAR Ø INCHES | METRIC SIZES (MM) | BAR WEIGHT LBS PER 100 LIN FT. |
|---|---|---|---|
| 2 | 1/4 | 6 | 17 |
| 3 | 3/8 | 10 | 38 |
| 4 | 1/2 | 13 | 67 |
| 5 | 5/8 | 16 | 104 |
| 6 | 3/4 | 19 | 150 |
| 7 | 7/8 | 22 | 204 |
| 8 | 1 | 26 | 267 |

**FIGURE 8-23** Numbers and sizes of commonly used reinforcing steel bars.

| MESH SIZE INCHES | MESH GAUGE | MESH WEIGHT LBS PER 100 SQ. FT. |
|---|---|---|
| 6X6 | #6 | 42 |
| 6X6 | #8 | 30 |
| 6X6 | #10 | 21 |

**FIGURE 8-24** Size, gauge, and weight of commonly used welded wire mesh.

## PLACING CONCRETE

The inside surfaces of the form should be clean and brushed or sprayed with form oil to make form removal easier. The steel reinforcement should be free of oil. Also, before concrete is placed, the forms and subgrade are moistened with water. This is done to prevent rapid loss of water from the concrete, thereby improving the cure.

Concrete is *placed,* not poured. Water should never be added so that concrete flows into forms without working it. Adding extra water alters the water–cement ratio, reducing the strength of the concrete.

*Slump tests* are performed by inspectors on the job to determine the consistency of the concrete. The concrete sample for a test should be taken 2 to 3 minutes before the concrete is placed and completed. A *slump cone* is filled with concrete according to a proper procedure. It is then carefully removed by tipping the cone over and lifting off the cone. The cone is gently placed beside the concrete, and the amount of slump measured (Fig. 8-25). The concrete truck should get as close as possible. Concrete should not be pushed or dragged any more than necessary. It should not be dropped more than 4 to 6 feet. Drop chutes should be used in high forms to prevent the buildup of dry concrete on the side of the form or reinforcing bars above the level of the placement. Drop chutes also prevent separation caused by concrete striking and bouncing off the side of the form.

**FIGURE 8-25** A slump test shows the wetness or dryness of a concrete mix.

## RATE OF PLACEMENT

Concrete must not be placed at a rapid rate, especially in high forms. The amount of pressure at any point on the form is determined by the height and weight of the concrete above it. Pressure is not affected by the thickness of the wall (Fig. 8-26). A slow rate of placement allows the concrete nearer the bottom to begin to stiffen. Once concrete stiffens, it will not exert more pressure on the forms even though liquid concrete continues to be placed above it (Fig. 8-27). The use of stiff concrete with a low slump will transmit less pressure to the side walls than high slump concrete. Concrete should be placed in layers of not more than 12 to 18 inches thick. Care should be taken to see that each layer sets a little before the next layer is applied, but the lower layer should not set too much.

To eliminate voids or honeycombs in the concrete, it should be thoroughly worked by hand spading or vibrated after it goes into the form. Vibrators make it possible to use a stiff mixture that would be difficult to consolidate by hand. Vibration makes the concrete more fluid and able to move, allowing trapped air to escape. This will prevent the formation of air pockets, honeycombs, and cold joints. The operator should be skilled in the use of the vibrator, keeping it moving up and down, uniformly vibrating the entire pour. Overvibrating increases the lateral pressure on the form (Fig. 8-28).

**Green Tip**

Plan ahead to reduce waste of leftover concrete. Leftover concrete can be placed in pre-made forms for small pads, patio blocks, and splash blocks, or used as backfill for various posts.

## CURING CONCRETE

Concrete is *cured* either by keeping it moist after hardening or by preventing loss of its moisture for a period of time. For instance, if moist-cured for 7 days, its strength is up to about 60 percent of full strength. A month later the strength is 95 percent, and it is full strength in about three months. Air-cure will reach only about 55 percent after three months and will never attain design strength. In addition, a rapid loss of moisture causes the concrete to shrink, resulting in cracks. Curing should be started as soon as the surface is hard enough to resist marring.

Flooding or constant sprinkling of the surface with water is the most effective method of curing concrete. Curing can also be accomplished by keeping the forms in place; covering the concrete with burlap, straw, sand, or other material that retains water; and wetting it continuously. In hot weather, the main concern is to prevent rapid evaporation of moisture. Sunshades or windbreaks may need to be erected. The formwork may be allowed to stay in place, or the concrete surface may be covered with plastic film or other waterproof sheets. Liquid curing chemicals may be used to seal in moisture and prevent evaporation.

**CAUTION**

**CAUTION:** Curing agents placed on green concrete contain harmful chemicals. It is important to follow the manufacturer's directions.

If concrete is frozen within the first 24 hours after being placed, permanent damage to the concrete is almost certain. Protect concrete from freezing for at least 4 days after being placed by providing insulation or additional heat, if necessary. Forms may be removed after the concrete has set and hardened enough to maintain its shape. This time will vary depending on the mix, temperature, humidity, and other factors.

## PRECAUTIONS FOR USING CONCRETE

Avoid prolonged contact with fresh concrete or wet cement because of possible skin irritation. Wear protective clothing when working with newly mixed concrete. Wash skin areas that have been exposed to wet concrete as soon as possible. If any material containing cement gets into the eyes, flush immediately with water and get medical attention.

**CAUTION**

**CAUTION:** Remove all protruding nails to eliminate the danger of stepping on or brushing against them.

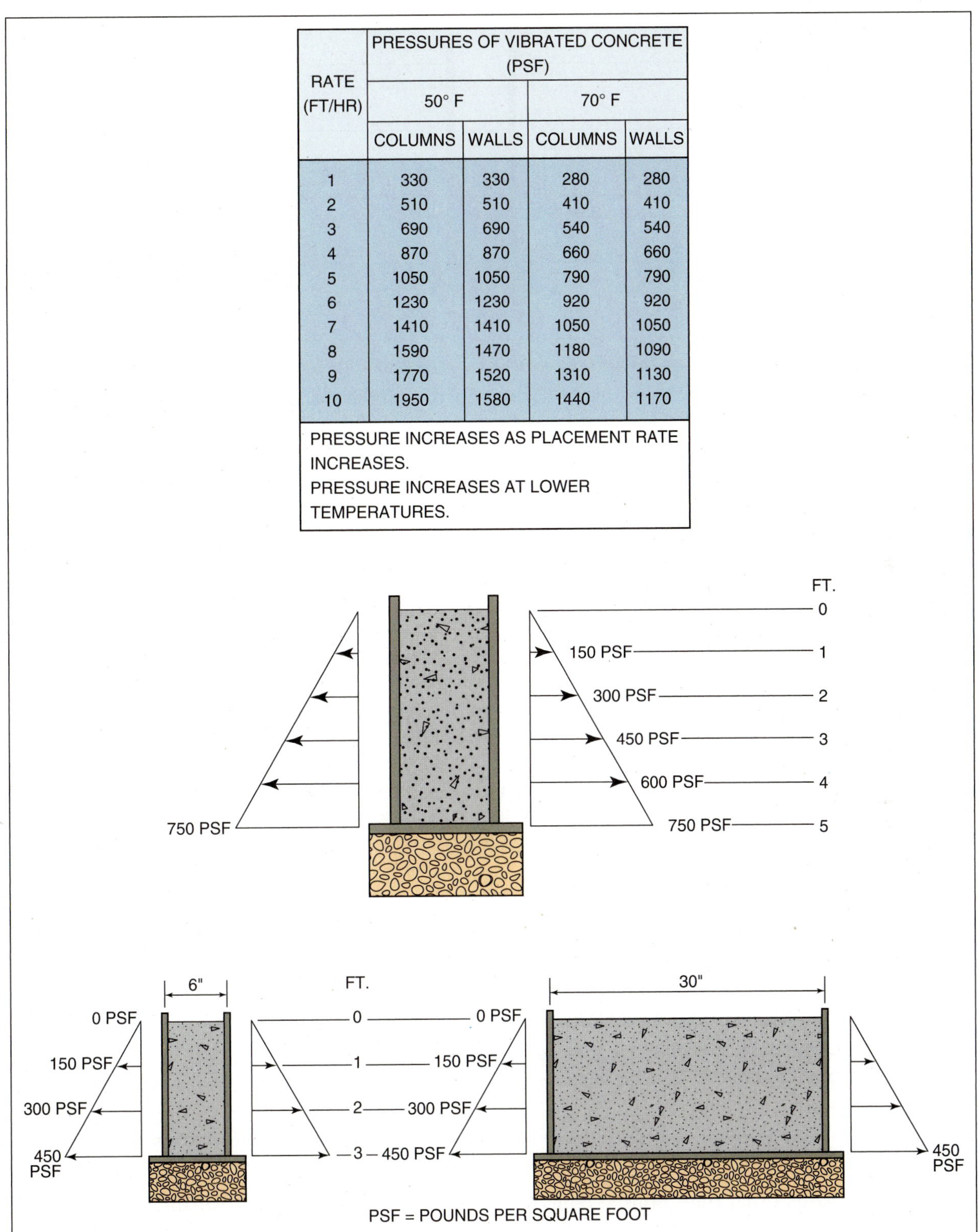

| RATE (FT/HR) | PRESSURES OF VIBRATED CONCRETE (PSF) | | | |
|---|---|---|---|---|
| | 50° F | | 70° F | |
| | COLUMNS | WALLS | COLUMNS | WALLS |
| 1 | 330 | 330 | 280 | 280 |
| 2 | 510 | 510 | 410 | 410 |
| 3 | 690 | 690 | 540 | 540 |
| 4 | 870 | 870 | 660 | 660 |
| 5 | 1050 | 1050 | 790 | 790 |
| 6 | 1230 | 1230 | 920 | 920 |
| 7 | 1410 | 1410 | 1050 | 1050 |
| 8 | 1590 | 1470 | 1180 | 1090 |
| 9 | 1770 | 1520 | 1310 | 1130 |
| 10 | 1950 | 1580 | 1440 | 1170 |

PRESSURE INCREASES AS PLACEMENT RATE INCREASES.
PRESSURE INCREASES AT LOWER TEMPERATURES.



**FIGURE 8-26** The height of concrete being poured affects the amount of pressure against the forms. Pressure is not affected by the thickness of the wall.

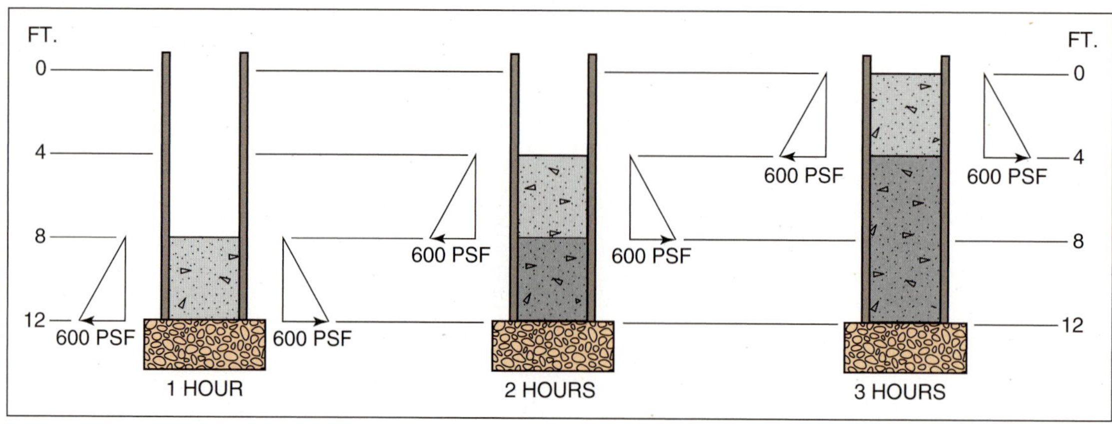


**FIGURE 8-27** Once concrete stiffens, it does not exert more pressure on the form even though liquid concrete continues to be placed above it.

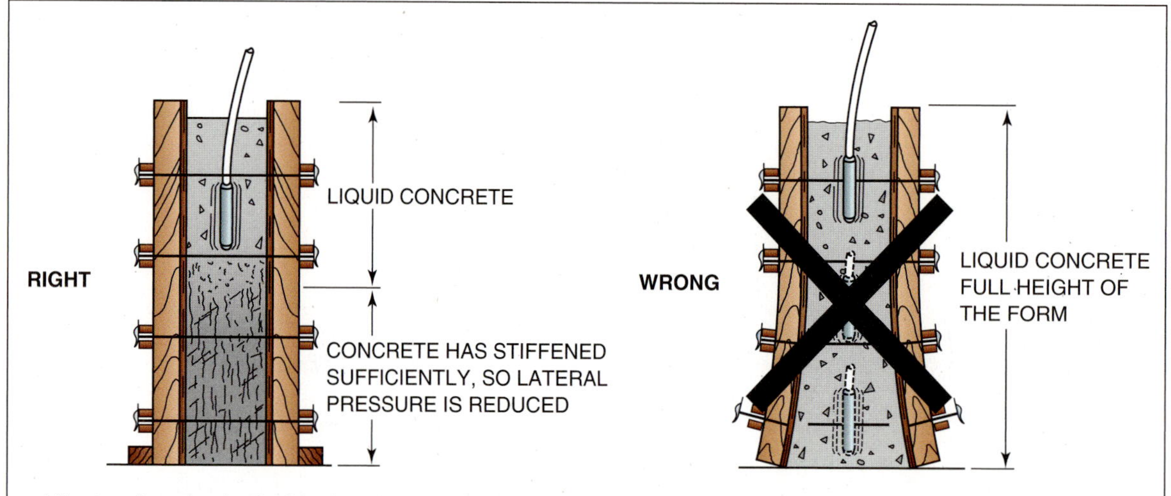


Vibrate only to the depth of the freshly placed concrete. Inserting the vibrator too far will cause the concrete at the bottom of the form to remain in a liquid state longer than expected. This will result in higher than expected lateral form pressure and may cause the form to fail. The depth of vibration should just penetrate the previous layer of concrete by a few inches.

*Courtesy of Dayton/Richmond Concrete Accessories.*

**FIGURE 8-28** Avoid excessive vibration of concrete.

# Foundation Types

Types of foundations other than poured concrete exist to meet various industry demands. They each provide the structural support necessary for the building while having different characteristics.

## CONCRETE BLOCK FOUNDATION

The concrete block foundation is made by stacking together concrete masonry units (CMUs) in an array. CMUs are typically placed in a staggered bond. This provides maximum wall strength (Fig. 8-29).

A CMU is sometimes called concrete block or cinder block. They come in regular and light-weight versions with varying widths. They all measure 8 inches high and typically 16 inches long; some are 18 inches long. Each measurement is smaller to allow for a ⅜-inch mortar joint. CMU widths range from 4 to 12 inches in 2-inch increments (Fig. 8-30).

Most CMUs are called *stretchers*, which have the familiar shape of a CMU. Some are constructed with flush ends for building square openings and

**FIGURE 8-29** Concrete block walls are made with concrete masonry units (CMU).

**FIGURE 8-30** Concrete masonry units are made in varying widths.

corners. These are called sash or corner blocks (Fig. 8-31). They also have a slot midway for easy splitting to make a half block.

## ALL-WEATHER WOOD FOUNDATION

All-weather wood foundations are made of pressure-treated lumber and plywood (Fig. 8-32). This lumber is specially treated to resist decay when buried below grade. It can be used to support light-frame buildings, such as houses, schools, apartments, and office buildings. Framing techniques are the same as for walls and floors (see Chapters 7 and 8).

**FIGURE 8-31** Concrete masonry units are made in two styles.

## PRECAST CONCRETE FOUNDATIONS

Precast concrete foundations provide a fast alternative to traditional foundation types. They are manufactured off-site and delivered on truck. They are then craned into place by a specially trained crew (Fig. 8-33). Setup time is typically one day.

Sections are insulated with 1 inch of extruded polystyrene foam and have a ribbed construction that allows for additional insulation. This makes the foundation more energy efficient than most foundation types. Each section is bolted together, and each seam is caulked to produce a guaranteed watertight seal (Fig. 8-34). A slab is later placed as a floor to support the bottom of the foundation from moving toward the inside during the backfilling of the foundation.

### Green Tip

Masonry can store solar energy when installed in areas of direct sunlight. This helps balance the daily high and low room temperatures.

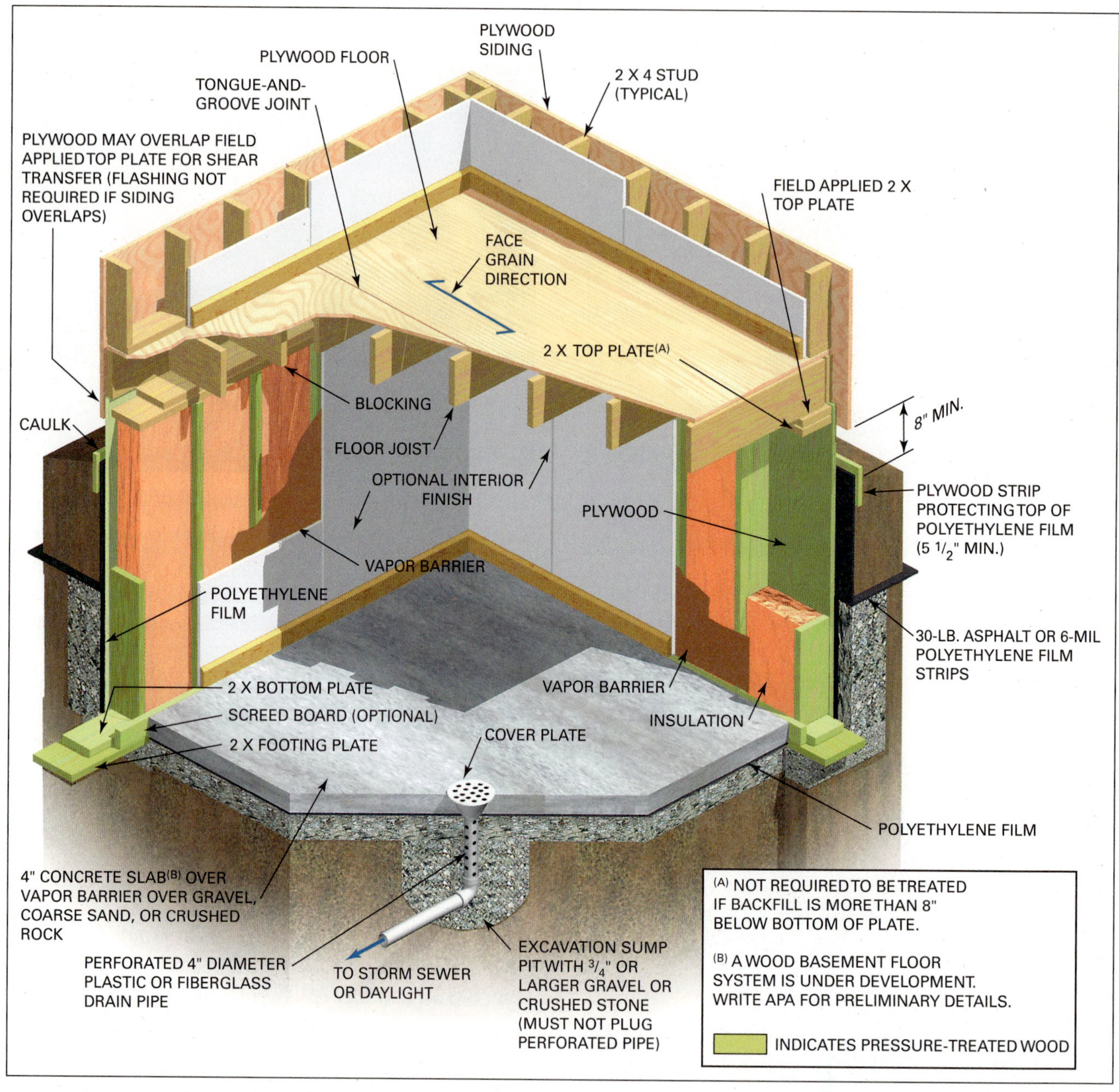


**FIGURE 8-32** All-weather wood foundations are made with pressure-treated lumber.

## INSULATED CONCRETE FORMS

A rigid foam forming system called insulated concrete forms (ICFs) uses 2-inch thick expanded polystyrene (Fig. 8-35). Units are manufactured into blocks with a metal or plastic tie system. The block sizes vary depending on manufacturer, ranging from whatever wall thickness needed × 16–18" high × 3–4 feet long (Fig. 8-36). The blocks have fastening strips included on their surfaces to make it easy to screw on a wall finish later.

Assembly begins one course at a time with corner and straight blocks. The next course is set so the vertical seams are broken (Fig. 8-37). This helps interlock the blocks for strength. Rebar is installed as needed after each block layer is installed. Special reinforced tape is sometimes used to help hold the course of block together.

**FIGURE 8-33** Precast concrete foundations are delivered on trucks and craned into place.

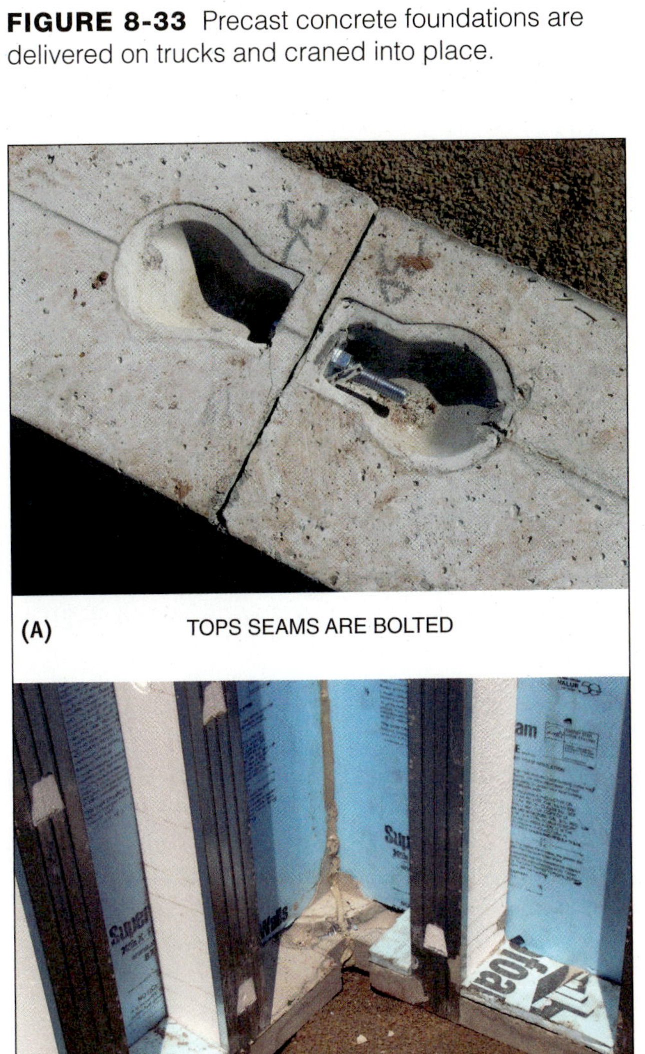

(A) TOPS SEAMS ARE BOLTED

(B) ALL JOINTS ARE CAULKED

**FIGURE 8-34** Precast concrete panels are fastened together to make them weather and watertight.

**FIGURE 8-35** Insulated concrete forms are used for foundations and walls.

**FIGURE 8-36** Insulated concrete forms are assembled from smaller units.

**FIGURE 8-37** Insulated concrete form units are stagger bonded for strength.

The success of this system is in the method of placing the concrete. Placement of the concrete is done in levels or lifts. Only one course is filled at a time, giving the concrete time to hydrate or harden slightly. Typically, by the time placement of the first lift is done for the entire perimeter, the concrete is ready to support another lift.

# Estimating Concrete Construction Quantities

Estimating the material in concrete construction involves form boards and panels, rebar, and volume of concrete. Remember that if the number of pieces does not come out even, round them up to the next whole number of pieces.

## FORM BOARD

Form boards are sold by the piece of a certain size, which is determined from the architectural plans. Slab-on-grade form boards are estimated by dividing the slab perimeter by the board length. Footing form boards are estimated by multiplying the perimeter by 2, then dividing that by the board length. For example if the perimeter is 144 feet and the boards are 12 feet long, then the number of boards is 144 × 2 ÷ 12 = 24 boards. A minimal amount of waste is added for overlap and cuts. This extra is the **waste factor.**

## WASTE FACTOR

The waste factor changes depending on the material, but the process of including it in the estimate is the same. Multiply the quantity needed times (one + the waste percentage). This will increase the quantity by the additional material needed. For example, if the quantity is 100 pieces and the waste factor is 15 percent, then multiply 100 by (1 + 0.15 or 1.15) = 115 pieces.

## FOUNDATION WALL FORMS

To find the foundation wall forms, multiply the perimeter by 2, then divide by the width of the panel. This will give the number of panels. For example, if the perimeter is 144 feet, then the number of 2-foot panels is 144 × 2 ÷ 2 = 144 panels.

## REBAR

Rebar is placed in concrete to provide tensile strength. The quantity depends on where it is installed. Two rows of rebar are typically placed in footings. Poured foundation wall rebar is installed in a grid, typically 2' × 2' OC (on center). Rebar is purchased by pieces 20 feet long or by a specific length.

*Footing Rebar.* The number of pieces of rebar may be found by doubling the perimeter or length of footings. Extra is needed since rebar must be overlapped and tied together at their ends. This extra should be about 10 percent more. For example, if the perimeter is 144 feet and the rebar piece length is 20 feet, the estimated footing rebar is 144 × 2 ÷ 20 × 1.10 = 15.8 = 16 pieces.

Poured walls require rebar installed in a 2 × 2 grid pattern comprised of vertical and horizontal rows of bar.

*Wall Horizontal Rebar.* Horizontal rebar quantity is the perimeter × number of rows of bar ÷ 20 feet per piece. The number of rows is the wall height ÷ grid spacing; then subtract 1 because a row is not needed at the top of the wall. Add 10 percent for overlap. For example, if the perimeter is 144 feet, grid spacing is 2 feet OC, and the wall height is 8 feet, then the horizontal rebar is 8 ÷ 2 −1 = 3 rows × 144 ÷ 20 × 1.10 = 23.6 = 24 bars.

*Wall Vertical Rebar.* Vertical rebar quantity is the number of vertical bars plus one per corner. The number of vertical bars is the perimeter ÷ grid spacing. For example, if the perimeter is 144 feet and the grid is 2 feet OC on a rectangular building, then the number of vertical rebars is 144 ÷ 2 = 72 + 4 = 76 pieces.

## CONCRETE

Ready-mix concrete is sold by the cubic yard (CY) or cubic meter. These amounts reflect the volume of concrete needed. To determine the volume of concrete needed for a job, multiply the width × length × thickness. Since measuring is typically done in feet, the result of this multiplication comes out as cubic feet. Cubic feet are converted to cubic

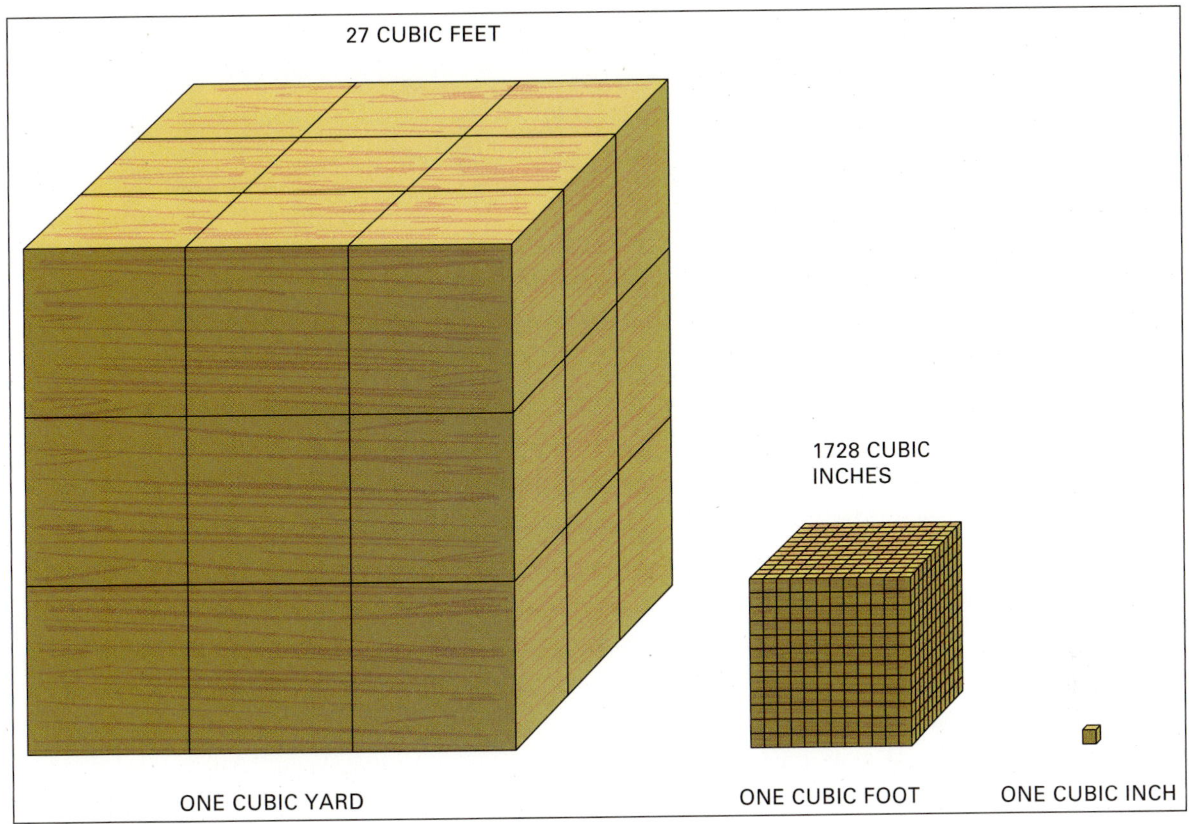


**FIGURE 8-38** Relative sizes of cubic measurements.

yards by dividing cubic feet by 27 (the number of cubic feet in 1 cubic yard) (Fig. 8-38).

For example, how many cubic yards will be needed for a wall that is 8 inches thick, 8 feet high, and 36 feet long. First, convert the 8 inches of thickness to feet: 8 ÷ 12 = 0.6666667 feet. Note here that the 6 is repeated forever, but the calculator rounds the last displayed digit to 7. Then calculate the volume by multiplying thickness × width × length, or 0.6666667 × 8 × 36 = 192 cubic feet. Dividing by 27 yields 7.1 CY.

The actual amount of concrete needed is often not the same as the amount estimated. Slight variations in the forms may cause large errors in the calculated volume. The quantities from the ready-mix company are often close, but not perfect. Also, some spillage will occur. For these reasons, the amount of concrete ordered should be a little more than estimated.

The amount of waste factor depends on the type of forms used. Wall forms have smooth, uniform sides, and the calculated concrete quantities can be very close to actual quantities. Slabs, on the other hand, have a bottom surface that is irregular, making the actual thickness measurement difficult to determine. Therefore, take several thickness measurements and average them. Also, add a higher waste factor than you would for wall forms. In general, the waste factor amount is 5 to 10 percent. (Fig. 8-39) shows a table of the shorthand versions of these estimating techniques with an example.

| Estimate the materials for a foundation of a rectangular 30′ × 56′ building. Walls 8′ tall and 10″ thick, the footing is 2′ wide by 12″ thick, and slab is 5″ thick. | | | |
|---|---|---|---|
| **Item** | **Formula** | **Waste factor** | **Example** |
| **Footing form boards** | footing PERM × 2 sides ÷ 12' = NUM of 12' boards | | 172' × 2 ÷ 12' = 28.6 ⇒ 29 boards |
| **Slab form boards** | slab PERM ÷ board LEN = NUM of boards | | 172' ÷ 12' = 14.3 ⇒ 15 boards |
| **Wall forms** | wall PERM × 2 sides ÷ 2' WID of form = NUM of form panels | | 172 × 2 ÷ 2' = 172 panels |
| **Rebar footing** | PERM × 2 ÷ 20' rebar LEN × waste = NUM of bars. | Add 10% for bar overlap | 172' × 2 ÷ 20' × 1.10 =18.9 ⇒ 19 rebars |
| **Rebar walls 2' × 2' grid horizontal** | PERM × [wall HGT ÷ 2 FT grid − 1] ÷ 20' rebar LEN × waste = NUM of 20' HOR bars | Add 10% for bar overlap | [8' ÷ 2 − 1] × 172' ÷ 20 × 1.10 = 28.3 ⇒ 29 rebars |
| **Rebar walls 2' × 2' grid vertical** | PERM ÷ 2 ft grid + 1 per corner = NUM of VERT bars | | 172 ÷ 2 + 4 = 90 vertical rebars |
| **Concrete footing (*Be sure all measurements are in terms of feet*)** | footing width × footing depth × PERM ÷ 27 = CY | Add 5% | 2' × 1' × 172' ÷ 27 × 1.05 5 5.6 1 5 3/4 CY |
| **Concrete slab** | slab WID × slab LEN × slab thickness ÷ 27 = CY concrete | Add 10% | 30' × 56' × 5"/12 ÷ 27 × 1.10 = 28.5 ⇒ 28 3/4 CY |
| **Concrete wall** | PERM × wall HGT × wall thickness ÷ 27 = CY | Add 5% | 172' × 8 × 10"/12 ÷ 27 × 1.05 = 44.5 ⇒ 44 3/4 CY |

**FIGURE 8-39** Example of estimating form and concrete materials with formulas.

## SUMMARY

- Concrete is comprised of materials that enter a chemical reaction to form a hard and durable building material.
- Concrete forms are constructed to hold loose concrete in the intended shape long enough for it to set.
- Slab forms hold the perimeter of flat surface concrete.
- Wall forms shape concrete into vertical surfaces.
- Concrete forming systems allow the form materials to be reused many times.
- Concrete reinforcement is necessary to give concrete tensile strength.
- Concrete placement must not overload the form.
- Foundation types vary in style to accommodate different environmental stresses.
- Estimating concrete and form materials is necessary for efficient purchases.

## GREEN CHECKLIST

- ☐ Some areas of the United States recycle concrete. Concrete is broken into sizes of large aggregate and used in new concrete, replacing crushed stone.
- ☐ Vapor retarders under concrete footings and slabs reduce moisture in a building. This saves energy because moisture increases heating and cooling costs.
- ☐ Plan ahead to reduce waste of leftover concrete. Leftover concrete can be placed in pre-made forms for small pads, patio blocks, and splash blocks, or used as backfill for various posts.
- ☐ Masonry can store solar energy when installed in areas of direct sunlight. This helps balance the daily high and low room temperatures.

PROCEDURE 8-1

# Constructing Footing Forms

**A** Stretch lines back on the batter boards in line with the outside edge of the footing. This is done by measuring from the saw kerf on the batter board ledger toward the outside of the building, one-half the wall thickness. Suspend a plumb bob from the batter board lines at each corner.

**A**

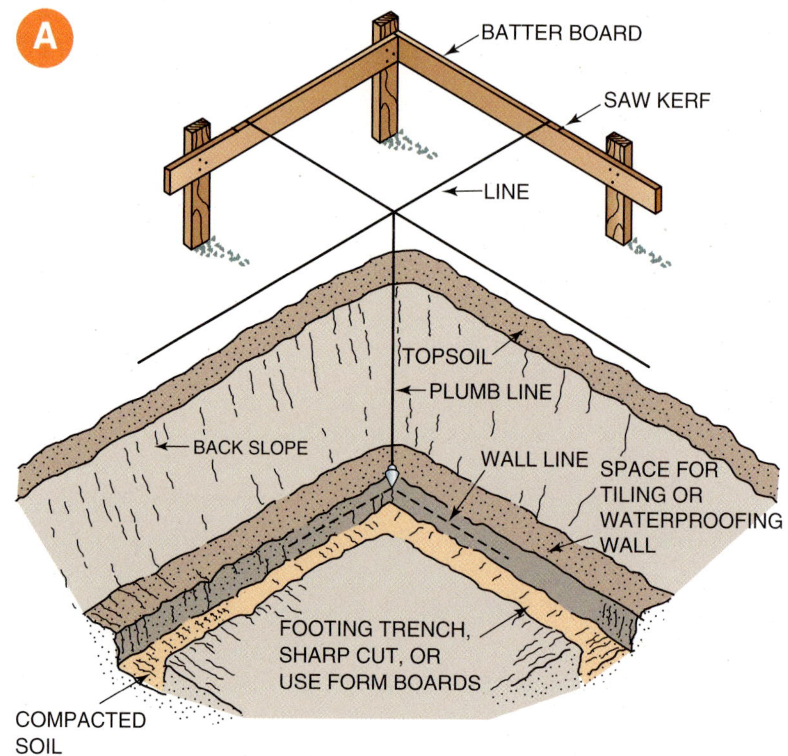

**B** Drive corner stakes to the correct elevation and stretch lines between the stakes, if desired.

- Begin forming with the outside form board by using stakes to hold the sides in position. Space the stakes 4 to 6 feet apart or as necessary to hold the form straight. Fasten the sides by driving nails through the stakes. Using duplex nails will speed up form removal. Keep the top inside corner of this form as close as possible to the line without touching it. Be sure the form does not touch the line. If the form touches the line at any point, the line is moved and is no longer straight. Continue forming the outside of the footing in this manner all around.
- Before erecting the inside forms, cut a number of **spreaders.** These are nailed to the top edges of the form. They tie the two sides together and keep them the correct distance apart. Their length is the width of the footing plus twice the form board thickness.

**B**

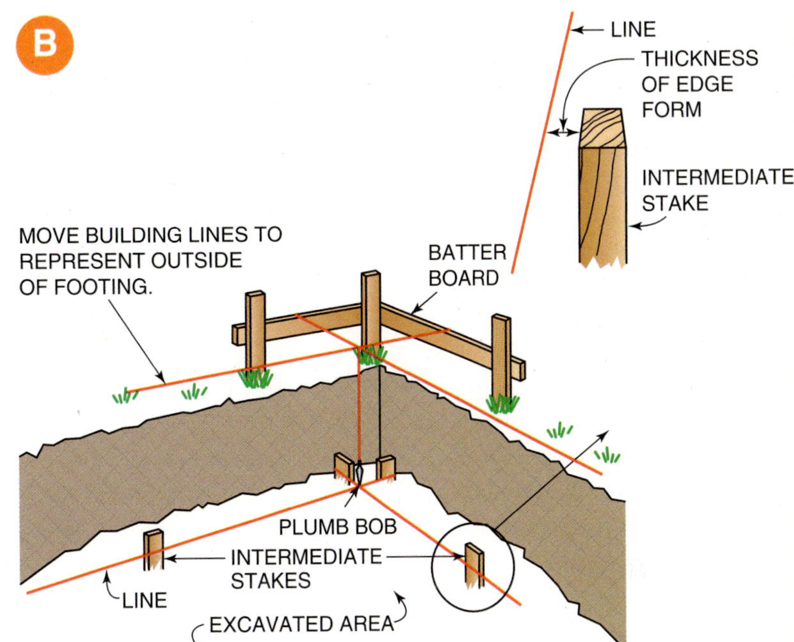

PROCEDURE 8-1

# Constructing Footing Forms (continued)

**C** Erect the inside forms in a manner similar to the outside forms. Place stakes for the inside forms opposite those holding the outside form. Level across from the outside form to determine the height of the inside form.

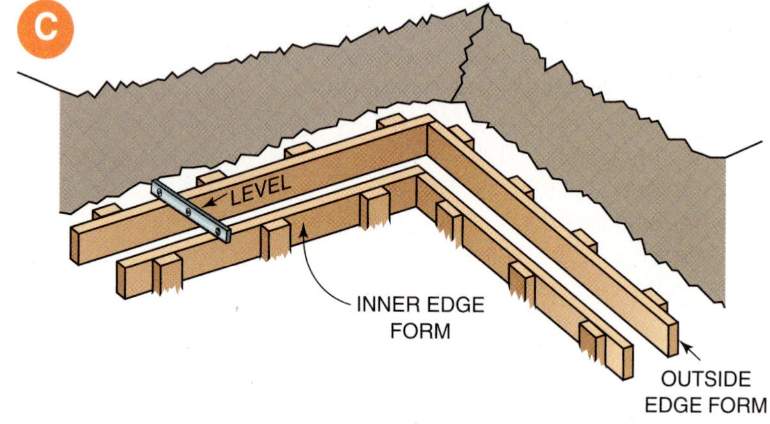


**D** Fasten the spreaders across the form at intervals necessary to hold the form the correct distance apart. Brace the stakes where necessary to hold the forms straight. In many cases, no bracing is necessary. Footing forms are sometimes braced by shoveling earth or placing large stones against the outside of the forms.

- After the footing concrete has hardened sufficiently, remove and clean the footing forms. The salvaged forms can be reused. Never leave any form board behind to be buried because it will draw termites.

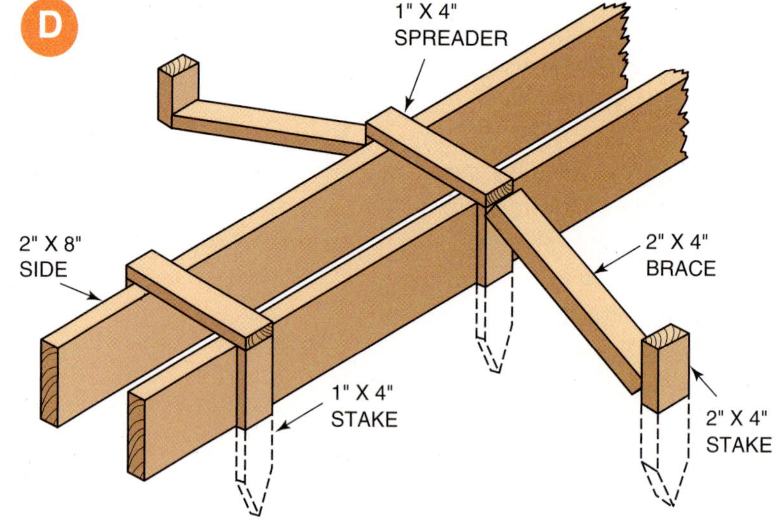

PROCEDURE 8-2

# Erecting Wall Forms

**A** Stretch lines again on the batter boards aligned with the outside of the foundation wall. These strings should be located in the original saw kerfs on the batter board ledger.

- Suspend a plumb bob from the layout lines to the footing. Mark footing at each corner that is plumb with the batter board layout lines.
- Snap a chalk line on the top of the footing between the corner marks outlining the outside of the foundation wall.

**A**

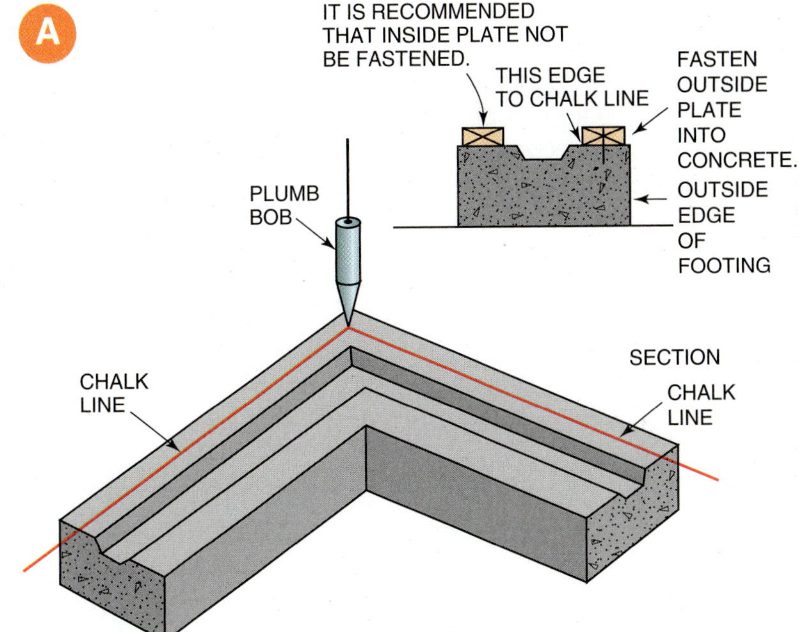

PROCEDURE **8-2**

# Erecting Wall Forms (continued)

**B** Set panels directly on the concrete footing to the chalk line or on 2 × 4 or 2 × 6 lumber plates. Plates are recommended because they provide a positive online wall pattern. They also tend to level out rough areas on the footing. Plates function to locate the position and size of pilasters, changes in wall thickness, and corners. Secure plates to the footing using masonry nails.

- Stack the number of panels necessary to form the inside of the wall in the center of the excavation. Lay the panels needed for the outside of the wall around the walls of the excavation. The face of all panels should be oiled or treated with a chemical-releasing agent. This provides a smooth face to the hardened concrete and makes stripping of the forms easy.

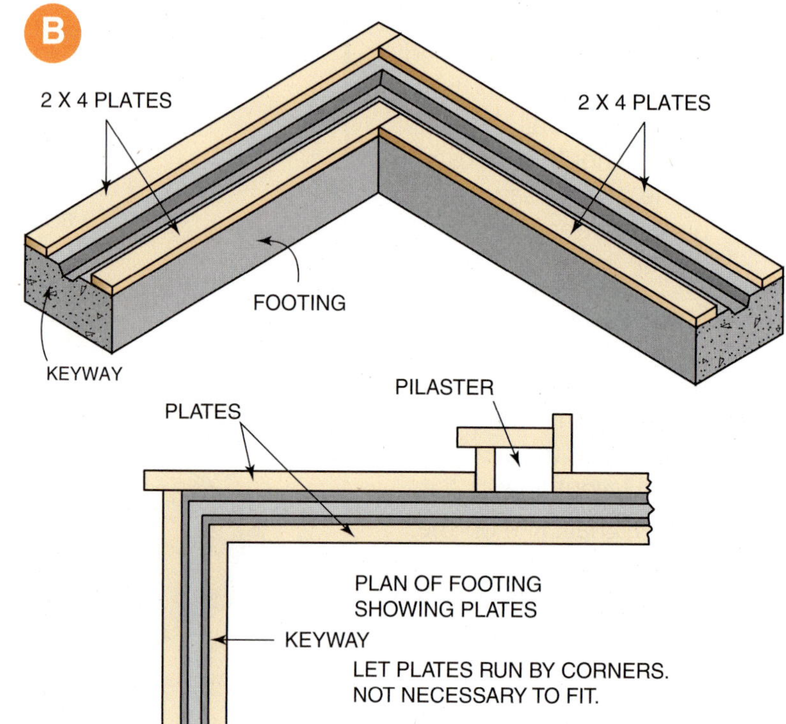

**CAUTION**

**CAUTION:** Form panels are heavy, and care should be taken while lifting them.

**C** Erect the outside wall forms first. Set all corner panels in place by nailing into the plate with duplex nails. Make sure the corners are plumb by testing with a hand level.

- Fill in between the corners with panels, keeping the same width panels opposite each other. Place snap ties in the slots between panels as work progresses. Use filler panels as necessary to complete each wall section. Brace the wall temporarily as needed. Place snap ties in the intermediate holes. Be careful not to leave out any snap ties.
- If the concrete is to be reinforced, tie the rebars in place, as detailed on the prints.

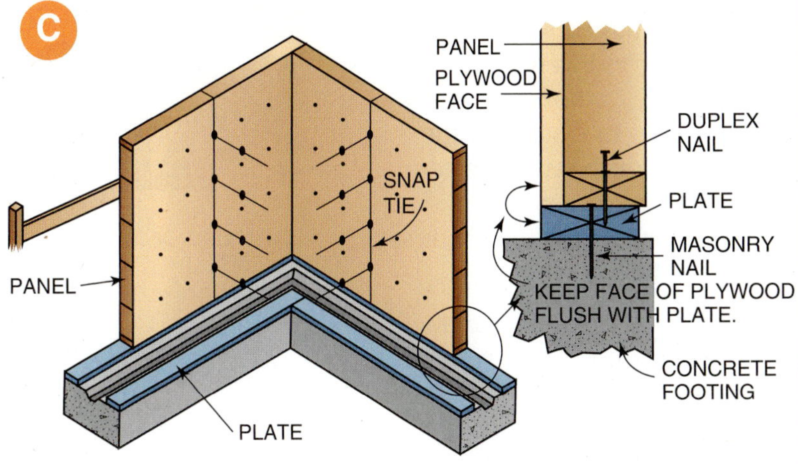


**D** Erect the panels for the inside of the wall. Keep joints between panels opposite to those for the outside of the wall. Insert the other end of the snap ties between panels and in intermediate holes as panels are erected.

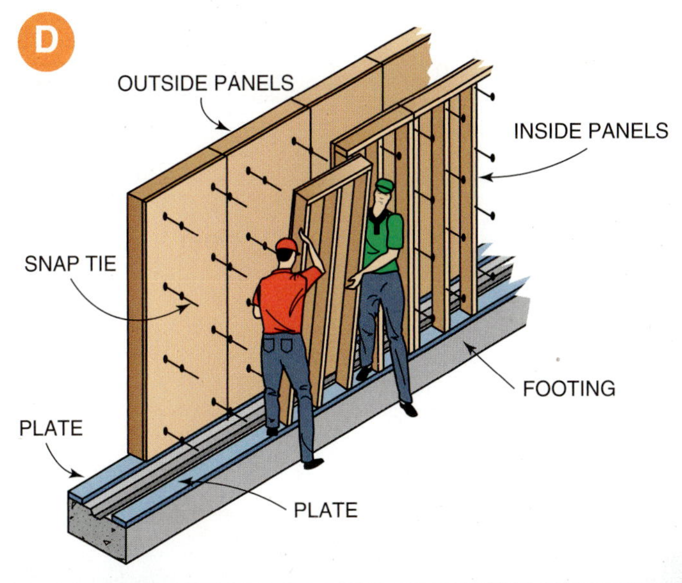

PROCEDURE 8-2

# Erecting Wall Forms (continued)

**E** Install the walers. Let the snap ties come through them and wedge into place. Do not cut the walers to length; let the ends of the walers extend by the corners of the formwork. This will reduce waste.

- Do not attempt to draw up warped wales with a wedge or overtighten the wedge in any manner. Overtightening will cause the metal spreader washers to bend out of shape or will shatter the plastic cones, resulting in a decreased and incorrect wall thickness.
- The optimum wedge position is when the snap tie head is at the midpoint of the wedge. You may place the snap tie head higher on the wedge, as long as it is not overtightened. However, the snap tie head must not be positioned lower than the midpoint, as this will place it on a section of the wedge that has not been designed to carry the rated load.
- As the concrete is consolidated, internal vibrations may cause the steel wedges that have not been nailed into position to loosen, bounce around, and eventually fall off, resulting in premature form failure.

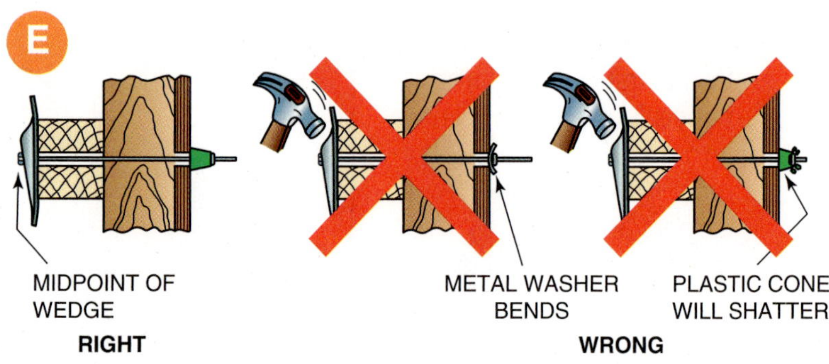


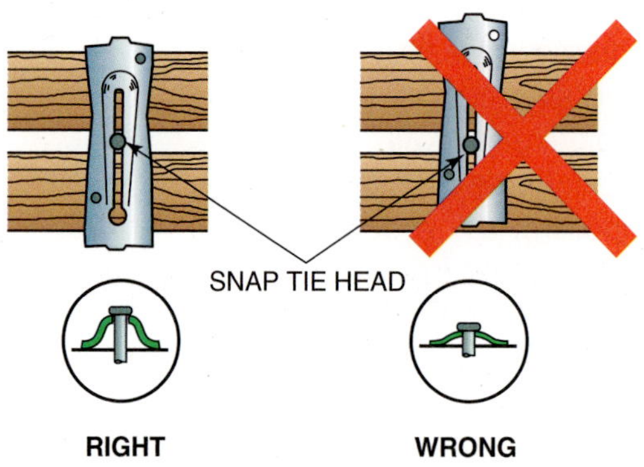


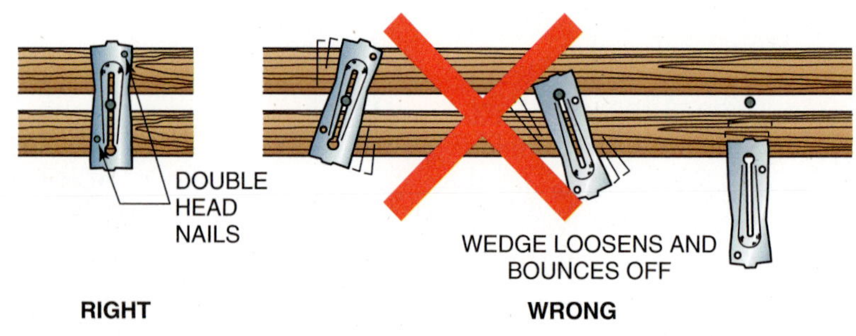

F Reinforce the corners with vertical 2 × 4s. This is called *yoking* the corners. Care must be taken when installing and driving snap tie wedges.

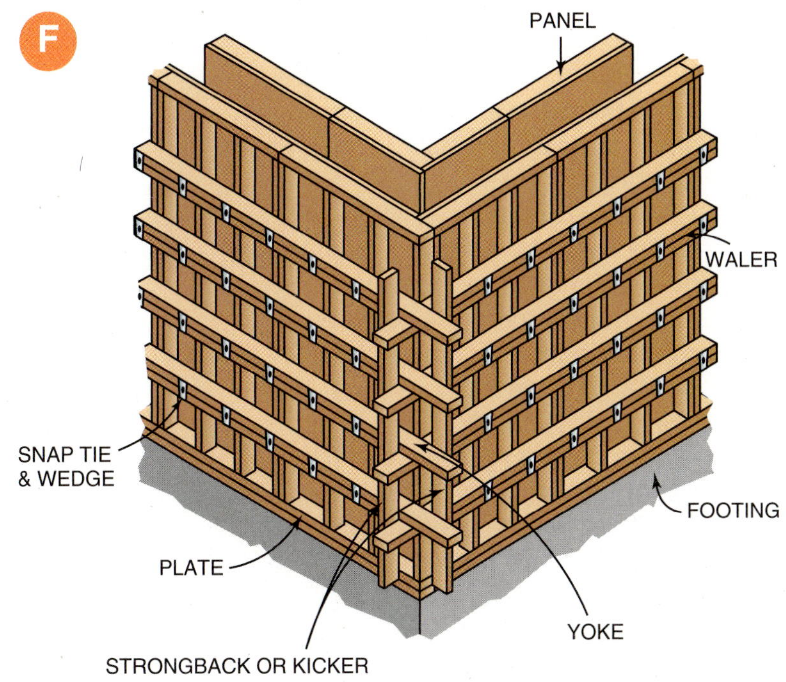


G If necessary, form the wall at intervals for the construction of **pilasters**, thickened portions of the wall. They strengthen the wall or provide support for beams. They may be constructed on the inside or outside of the wall. In the pilaster area, longer snap ties are necessary.

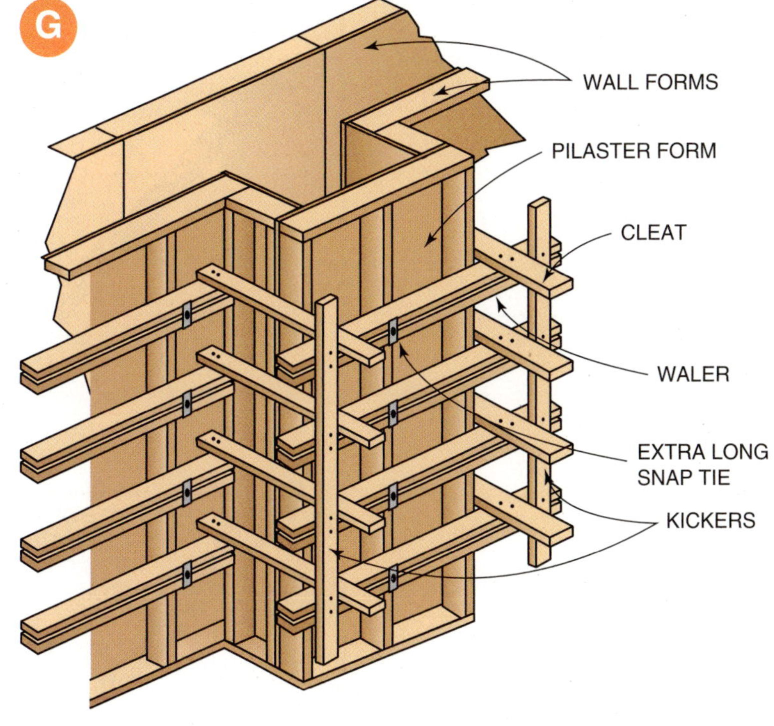

PROCEDURE 8-2

# Erecting Wall Forms (continued)

**H** Brace the walls inside and outside as necessary to straighten them. Wall forms are easily straightened by sighting by eye along the top edge from corner to corner. Another method of straightening is by stretching a line from corner to corner at the top of the form over two blocks of the same thickness. Move the forms until a test block of equal thickness passes just under the line.

**I** A special adjustable form brace and aligner is sometimes used for positioning and holding wall forms.

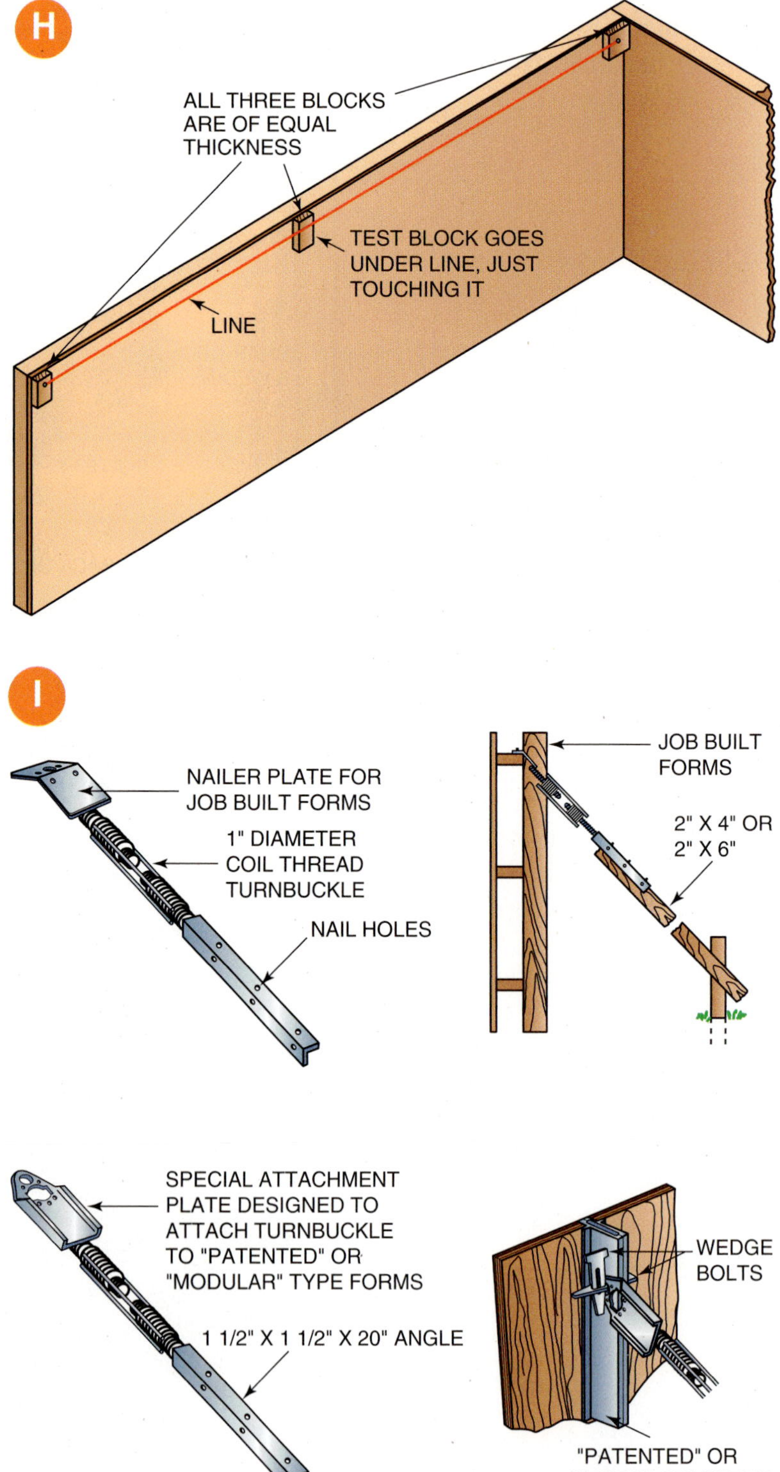


*Courtesy of Dayton/Richmond Concrete Accessories.*

J The top surface of the concrete must be level and smooth to make attaching the wood frame easier. This is done by using a screed. In this case the screed may be small block of wood worked in a sawing motion over the top of the concrete freshly placed concrete. Sometimes the height of the form is higher than the concrete. Chalk lines are snapped on the inside of the forms at the desired height. Concrete is placed to the line. Grade nails may be driven part way in at intervals along the chalk line to serve as guides to leveling the concrete. Another method is to fasten strips of wood on the inside forms along the chalk line and use a screed notched to go over the strips.

J

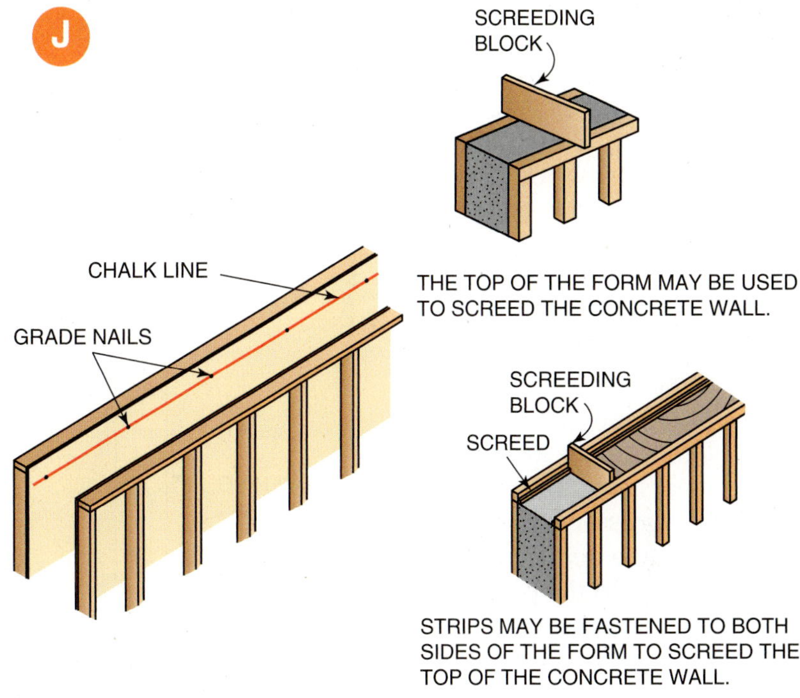


**CAUTION**

**CAUTION: Do not overvibrate the concrete. This will, at the least, cause the aggregate to separate and flow to the bottom or, in the worst case, cause the form to rupture.**

K Set the anchor bolts into the fresh concrete. Various styles and sizes are manufactured. Care must be taken to set the anchor bolts at the correct height and at specified locations. Bolts should be spaced 6 to 8 feet apart and between 6 and 12 inches from each end of all sill plates.

K

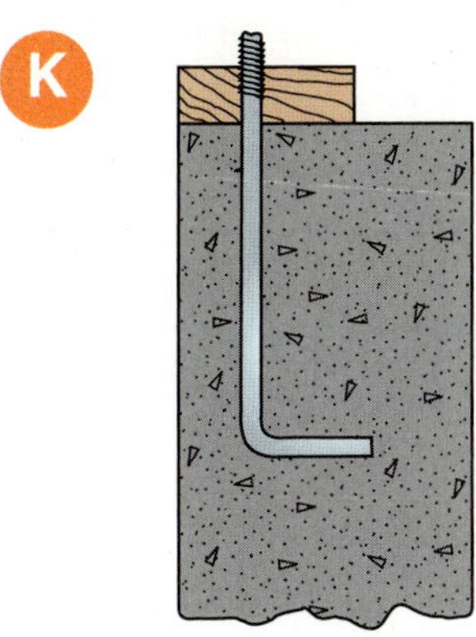

ANCHOR TYPE A
L-BOLT

*Courtesy OF SIMPSON STRONG-TIE COMPANY.*

PROCEDURE 8-2

# Erecting Wall Forms (continued)

**L** Install *blockouts* for larger openings. Blockouts are also called **bucks.** They are installed between the inside faces of the form panels. The blockout is usually made of 2-inch stock. Its width is the same as the thickness of the foundation wall. Intermediate support pieces and braces may be necessary in bucks for large openings to withstand the pressure of the concrete against them.

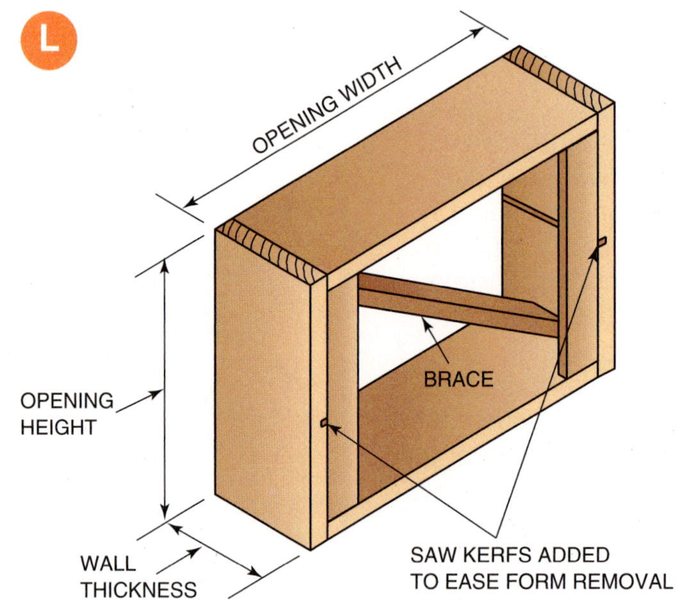


**M** Install forms for *girder pocket* blockouts. These will create recesses in the top of the foundation wall to receive the ends of **girders** or beams. The pocket should be at least 1 inch wider than the girder and ½ inch deeper than the minimum bearing of the girder, which is usually 4 inches.

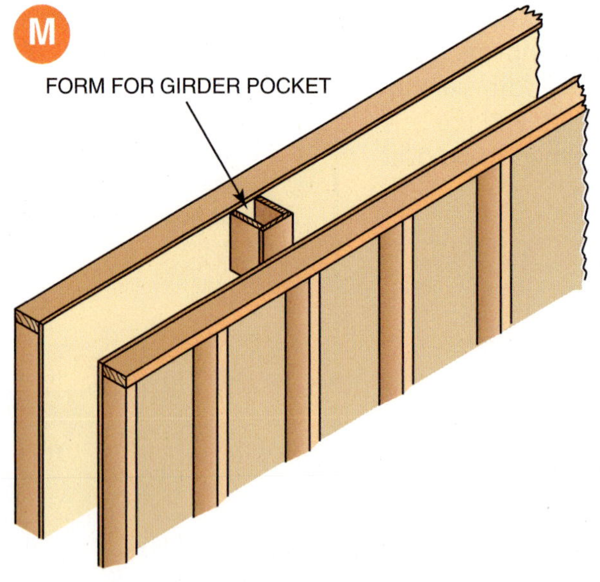

PROCEDURE 8-3

## Constructing Concrete Stair Forms

**A** Determine the *rise* and *run* of each step and lay them out on the inside of the existing walls or form. See Chapter 16 on stair construction for stair layout techniques.

- Secure cleats (short strips of wood) to the side wall form at each riser location. These will support riser ends. Allowances must be made for the thickness of the riser form board. Cleats may be omitted if forming between existing walls because step forms can be wedged in place.
- For earth-supported stairs, place fill that is free of organic matter and compacted in place to reduce settling.

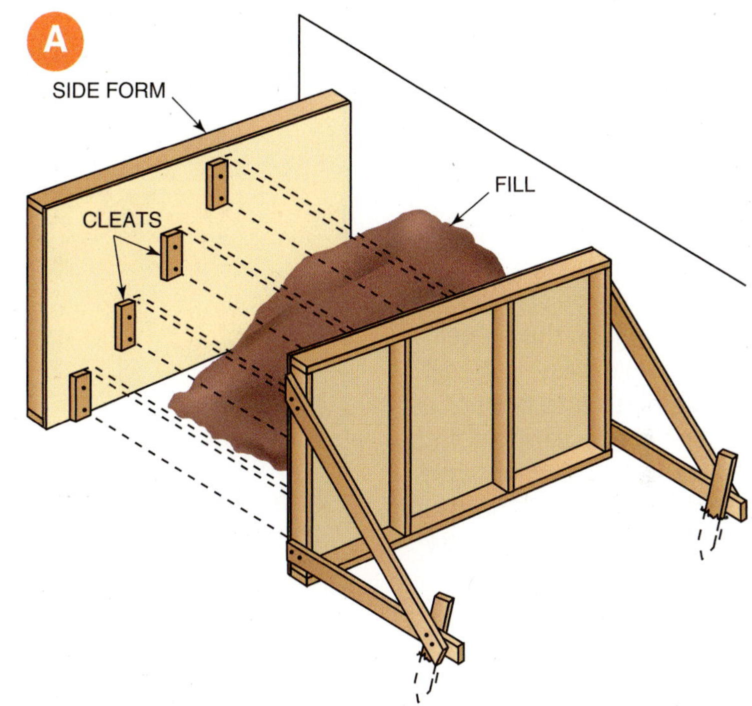

PROCEDURE 8-3

# Constructing Concrete Stair Forms (continued)

**B** For suspended stairs, construct a support made of plywood, joists, shores, and stair horses. Be sure to allow for the thickness of the plywood deck, the width of the supporting joists, and the depth of the *stair horses* to maintain the proper stair thickness. Brace all members into position, and secure all shore and horse joints with a **scab** or **gusset** plate. Scabs are short lengths of narrow boards fastened across a joint to strengthen it. Wedge the shoring as necessary to bring the surface of the plywood to the layout line. Fasten all wedges in place so they will not move.

- Oil all of the wood and plywood to facilitate stripping.
- Install the rebars or reinforcing as specified by the prints.
- Install riser boards that are ripped to width to correspond to the height of each riser. Align boards to the layout lines on the side wall or form. Position the riser top and bottom edges to the tread layout lines.

**C** Brace the risers from top to bottom at mid-span. This keeps them from bowing outward due to the pressure of the concrete.

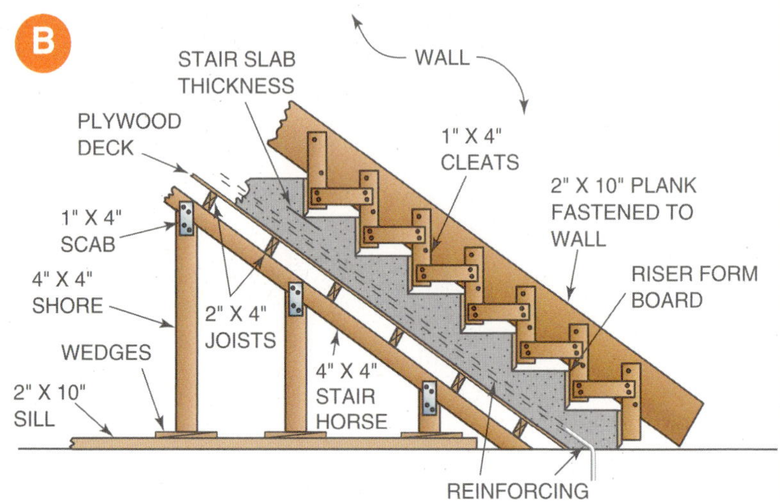


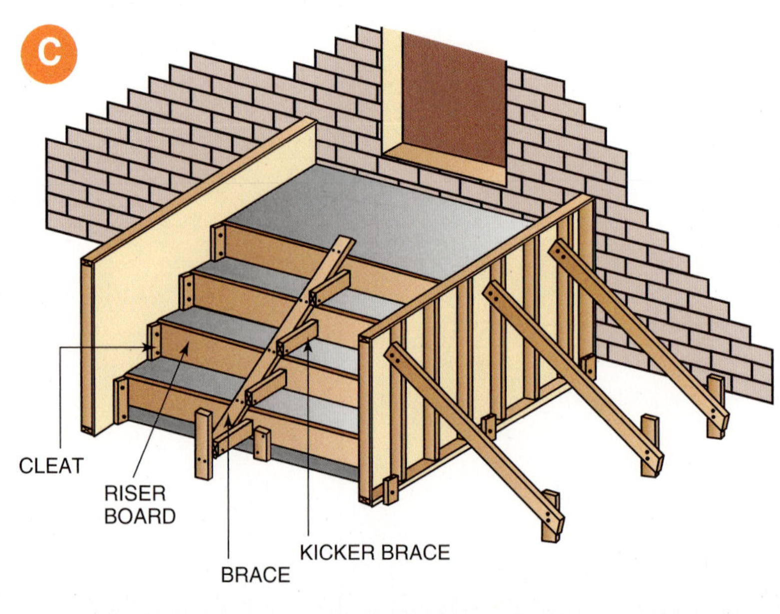

## REVIEW QUESTIONS

*Select the most appropriate answer.*

1. **Concrete is made of**
   a. Portland cement.
   b. aggregates.
   c. water.
   d. all of the above.

2. **Steel rods that are placed in concrete to increase its tensile strength are called**
   a. reinforcing bars.
   b. aggregates.
   c. reinforcing nails.
   d. duplex nails.

3. **The inside surfaces of forms are oiled to**
   a. protect the forms from moisture.
   b. prevent the loss of moisture from concrete.
   c. to make stripping of the forms easier.
   d. prevent honeycombs in the concrete.

4. **Keyways are often put in spread footings to**
   a. unlock the forms for easy removal.
   b. increase the compressive strength of concrete.
   c. keep the form boards from spreading.
   d. provide a stronger joint between footing and foundation.

5. **Rapid placing of concrete**
   a. eliminates the need to vibrate.
   b. may burst the forms.
   c. separates the aggregate.
   d. reduces voids and honeycombs.

6. **Unless footings are placed below the frost line,**
   a. the foundation will settle.
   b. the foundation may heave and crack.
   c. excavation is difficult in winter.
   d. problems with form construction will result.

7. **Spreaders for footing forms are used**
   a. to allow easy placement of the concrete.
   b. to keep the forms straight.
   c. because they are easier to fasten.
   d. to maintain the footing width.

8. **A step in a footing should be dimensioned and sized to match the**
   a. concrete block used in the foundation.
   b. form used to pour the foundation wall.
   c. building code requirements.
   d. all of the above.

**9. The horizontal surface of a stepped footing must be at least**
- a. 4 feet.
- b. twice the vertical step distance.
- c. the vertical step distance.
- d. the thickness of the footing.

**10. The typical order of installation of manufactured forming system is**
- a. inside forms, outside forms, reinforcing bars, then snap ties.
- b. inside forms, reinforcing bars, snap ties, then outside forms.
- c. outside forms, snap ties, inside forms, then reinforcing bars.
- d. outside forms, snap ties, reinforcing bars, then inside forms.

**11. When erecting concrete formwork made of wood,**
- a. drive extra nails for added strength.
- b. drive all nails home.
- c. use 8d nails.
- d. use duplex nails.

**12. Walers are used on forms to**
- a. stiffen.
- b. plumb.
- c. level.
- d. brace.

**13. A slab should be protected from**
- a. freezing before it cures.
- b. curing too fast.
- c. moisture after it cures by a vapor retarder.
- d. all of the above.

**14. Before concrete cures, it should be protected from**
- a. overheating.
- b. freezing.
- c. excessive vibrations.
- d. all of the above.

**15. The volume of concrete for a 6″ slab that measures 24′ × 36′ is**
- a. 16 cubic yards.
- b. 36 cubic yards.
- c. 192 cubic yards.
- d. 432 cubic yards.

**16. The procedure for erecting footing forms that comes after locating the corner stakes using a plumb bob and batter boards is to install the**
- a. spreaders.
- b. outside form boards.
- c. inside form boards.
- d. reinforcing bars.

**17. Overvibrating concrete while placing it in wall forms causes**
- a. voids and honeycombs.
- b. the aggregate to rise to the top.
- c. extra side pressure on the forms that could cause form failure.
- d. all of the above.

**18. Accessories placed in a foundation wall form to create spaces are called**
- a. blockouts.
- b. bucks.
- c. girder pocket forms.
- d. all of the above.

**19. The type of foundation made with CMUs is**

a. precast concrete.
b. all-weather wood.
c. insulated concrete.
d. concrete block.

**20. The type of foundation delivered to the site with a crane is**

a. precast concrete.
b. all-weather wood.
c. insulated concrete.
d. concrete block.

## KNOW YOUR CODES

Concrete design for strength depends on the intended purpose of the concrete. For example, sidewalk slabs exposed to severe cold temperature and heavy traffic need to be harder than residential footing concrete. Always follow the specifications in the building print documents.

## WHAT'S WRONG WITH THIS PICTURE?

Carefully study **Figure 8-40** and think about what is wrong. Consider all possibilities.

### WRONG

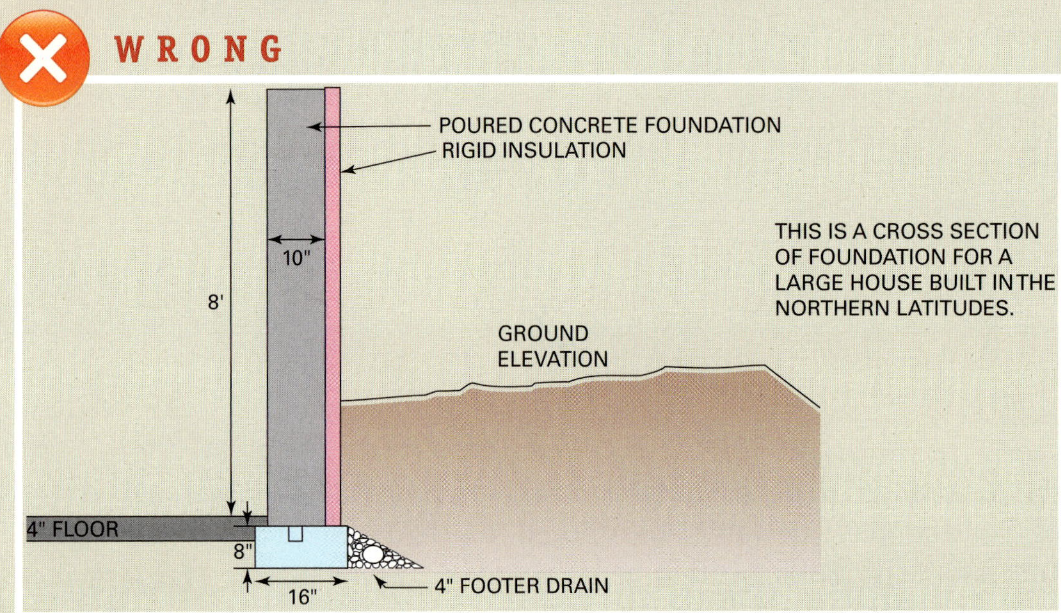


**FIGURE 8-40** In order to avoid frost in northern latitudes, the bottom of the footing should be located approximately 4 feet beneath the ground. Although measurements are not given, the diagram shows that much less than half of the 8-foot wall is below grade. Also, the ground is sloping toward the house, but it should slope away to channel moisture away from the foundation. Finally, the width of the footing should be at least double the wall width, and the thickness (height) of the footing should be at least the thickness of the wall.

### RIGHT

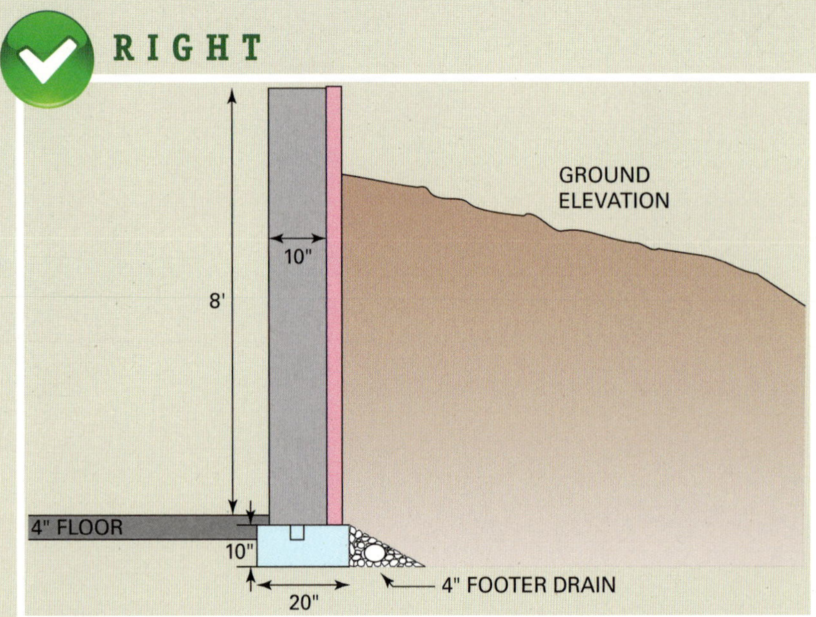


**FIGURE 8-41** Ground level in this diagram is much higher against the foundation wall. The soil slopes away from the house to promote drainage, and the footing dimensions are sized properly to support the wall.

CHAPTER 9

# Floor Framing

**OBJECTIVES** *After completing this unit, the student should be able to:*

- Describe platform, balloon, and post-and-beam framing, and identify framing members of each.
- Describe several energy and material conservation framing methods.
- Build and install girders, erect columns, and lay out sills.
- Lay out and install floor joists.
- Frame openings in floors.
- Lay out, cut, and install bridging.
- Apply subflooring.
- Describe methods to prevent destruction by wood pests.

## GLOSSARY OF FLOOR FRAMING TERMS

**anchor bolt** long metal fasteners with a threaded end used to secure materials to concrete

**balloon frame** a type of frame in which studs are continuous from foundation sill plate to roof

**band joist** the member used to stiffen the ends of floor joists where they rest on the sill

**bridging** diagonal braces or solid wood blocks between floor joists used to distribute the load imposed on the floor

**column** a large vertical member used to support a beam or girder

**dimension lumber** a term used to describe wood that is sold for framing and general construction

**draftstops** also called firestops; material used to reduce the size of framing cavities in order to slow the spread of fire; in floor framing, a draftstop consists of full-width dimension lumber blocking between studs

**flush** a term used to describe when surfaces or edges are aligned with each other

**girders** heavy beams that support the inner ends of floor joists

**header** members placed at right angles to joists, studs, and rafters to form and support openings in a wood frame

**joist** horizontal framing members used in a spaced pattern that provide support for the floor or ceiling system

**linear feet** a measurement of length

**masonry** any construction of stone, brick, tile, concrete, plaster, and similar materials

**plate** top or bottom horizontal member of a wall frame

**platform frame** method of wood frame construction in which walls are erected on a previously constructed floor deck or platform

**post** a vertical member used to support a beam or girder

**pressure-treated** treatment given to lumber that applies a wood preservative under pressure

**ribbon** a narrow board cut into studs of a balloon frame to support floor joists

**sheathing** boards or sheet material that are fastened to joists, rafters, and studs and on which the finish material is applied

**shims** a thin, wedge-shaped piece of material used behind pieces for the purpose of straightening them or for bringing their surfaces flush

**sill** first horizontal wood member resting on the foundation supporting the framework of a building; also, the lowest horizontal member in a window or door frame

**sill sealer** material placed between the foundation and the sill to prevent air leakage

**subfloor** material used as the first floor layer on top of joists

**tail joists** shortened on center joists running from a header to a sill or girder

**termite shields** metal flashing plate over the foundation to protect wood members from termites

**trimmer joist** a joist or stud placed at the sides of an opening running parallel to the main framing members

Floor framing begins, for some, the most exciting part of construction. It is the time when a pile of material turns into something special. For carpenters, this stage is loosely called framing.

Wood-frame construction is widely used for residential and light commercial construction. Wood-framed homes are durable and, if properly maintained, will last indefinitely. Many existing wood-frame structures are hundreds of years old. Because of the ease with which wood can be cut, shaped, fitted, and fastened, many different architectural styles are possible. In addition to single-family homes, wood-frame construction is used for all kinds of low-rise buildings, such as apartments, condominiums, offices, motels, warehouses, and manufacturing plants.

The floor frame is the first section installed after the foundation is completed. The floor system has many components, which include sills, girders, joists, bridging, and subfloor. Other framing members are used to create openings and add strength to the building. In order to assemble a floor frame, a carpenter must be able to identify all the components, accurately locate their positions, and cut each member to fit.

## Types of Frame Construction

There are several methods of framing a building. New designs using engineered lumber are increasing the height and width to which wood-frame structures can be built.

### PLATFORM FRAME CONSTRUCTION

The **platform frame**, sometimes called the *western frame,* is most commonly used in residential construction (Fig. 9-1). In this type of construction, the floor is built and then walls are erected on top of it. When more than one story is built, the second-floor platform is erected on top of the walls of the first story. A platform frame is easier to erect. At each floor level, a flat surface is provided on which to work. A common practice is to assemble wall frame units on the floor and then tilt the units up into place (Fig. 9-2).

#### Effects of Shrinkage

Lumber shrinks mostly across width and thickness and not so much along its length. A disadvantage of the platform frame is the potential for settling caused by the shrinkage of the large number of horizontal load-bearing frame members. However, because of the equal amount of horizontal lumber, the shrinkage is more or less equal throughout the building. To minimize shrinkage, only framing lumber with the proper moisture content should be used.

### BALLOON FRAME CONSTRUCTION

In the **balloon frame**, the wall studs and first-floor joists rest on the sill. The second-floor joists rest on a 1 × 4 **ribbon** that is cut in **flush** with the inside edges of the studs (Fig. 9-3). This type of construction is used less often, but a substantial number of structures built with this type of frame are still in use.

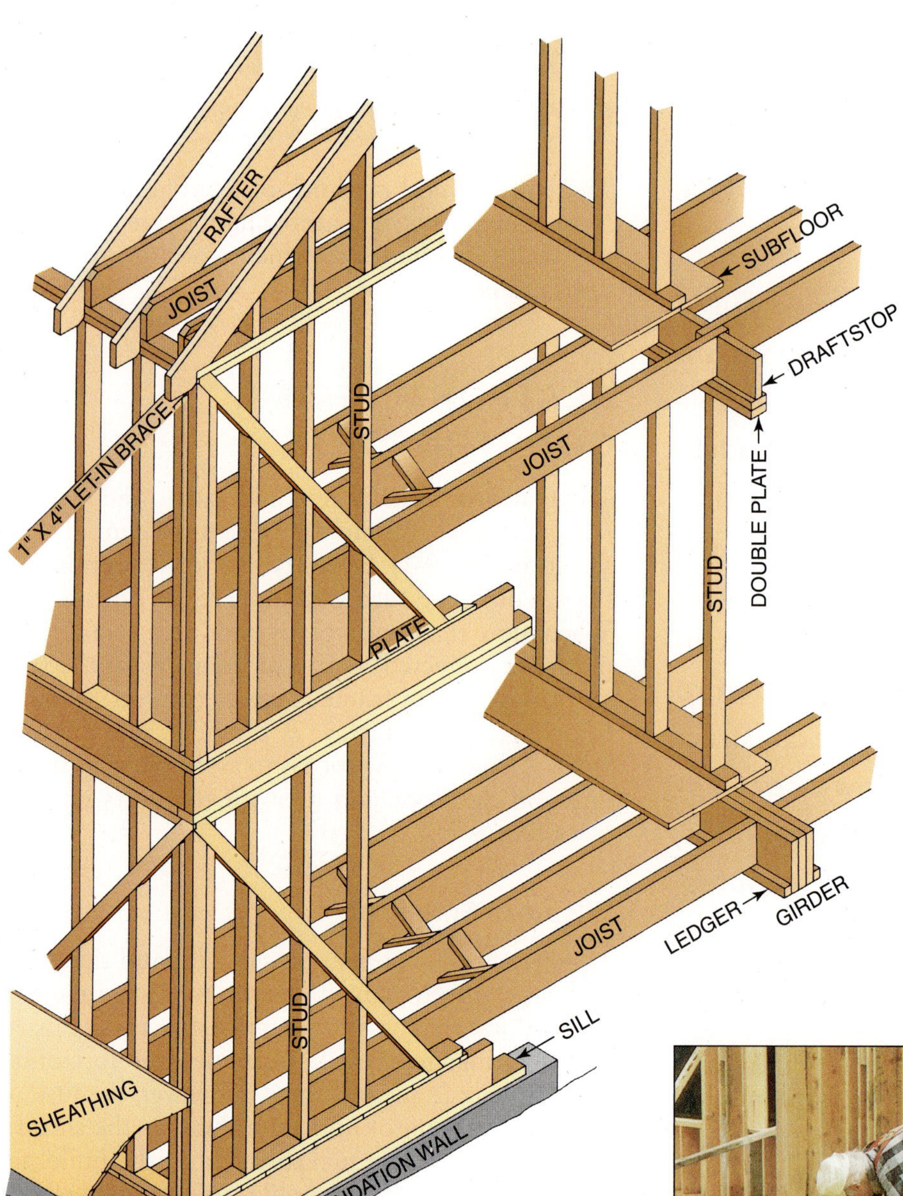

**FIGURE 9-1** Platform frame construction showing the location and names of its members.

**FIGURE 9-2** Some carpenters specialize in framing.

## Draftstops

In a balloon frame, the studs run from sill to plate. Draftstops, sometimes called *firestops,* must be installed in the walls in several locations. A draftstop is an approved material used in the space between frame members to prevent the spread of fire for a certain period of time. In a wood frame, a draftstop in a wall might consist of dimension lumber blocking between studs. In the platform frame, the wall plates act as draftstops (Fig. 9-4).

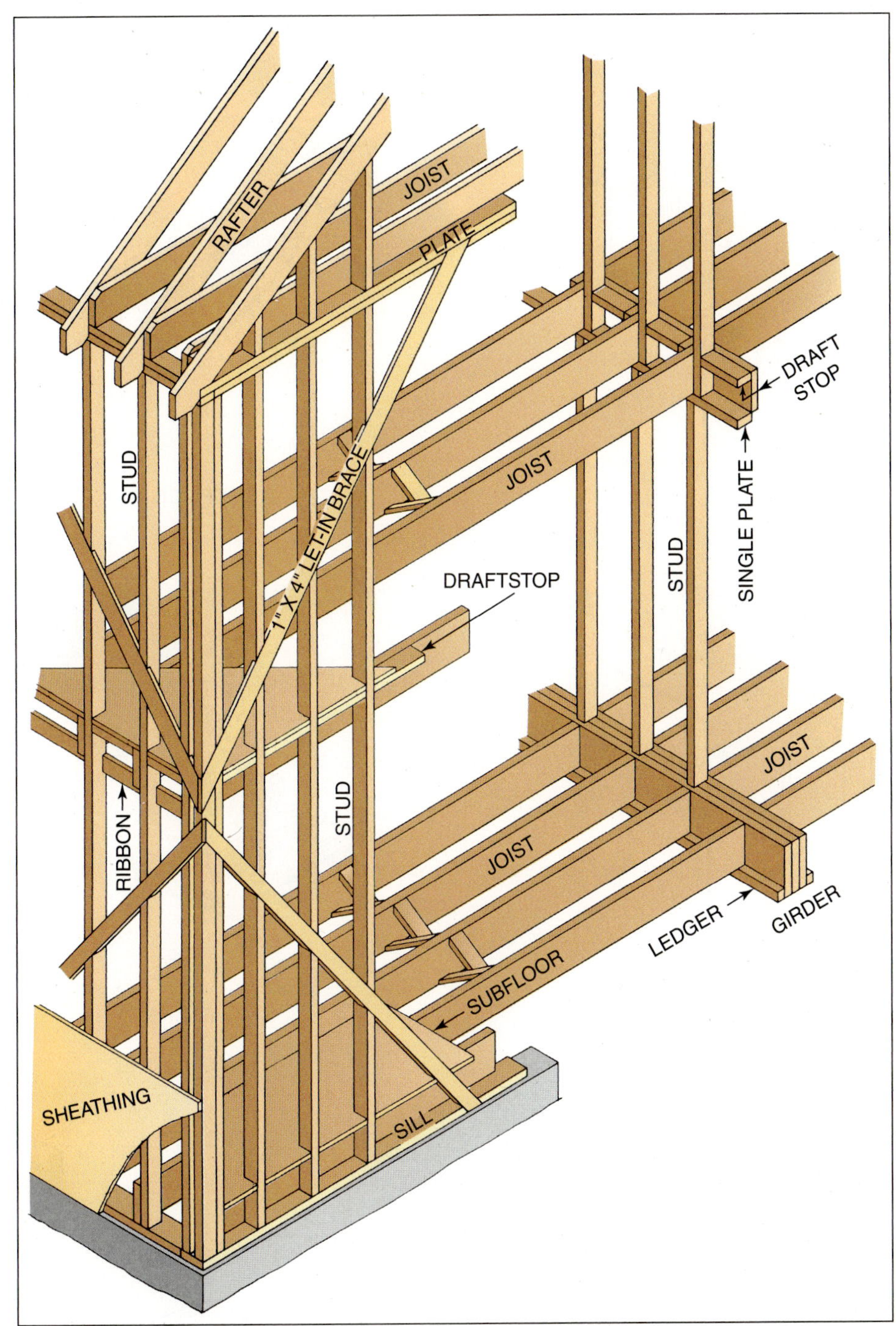


**FIGURE 9-3** The location of the members of a balloon frame.

## POST-AND-BEAM FRAME CONSTRUCTION

The *post-and-beam* frame uses fewer but larger pieces that are widely spaced (Fig. 9-5).

### Floors

APA-Rated Sturd-I-Floor 48 on center (OC), which is 1³⁄₃₂ inches thick, may be used on floor joists that are spaced 4 feet OC instead of matched boards (Fig. 9-6). In addition to being nailed, the plywood

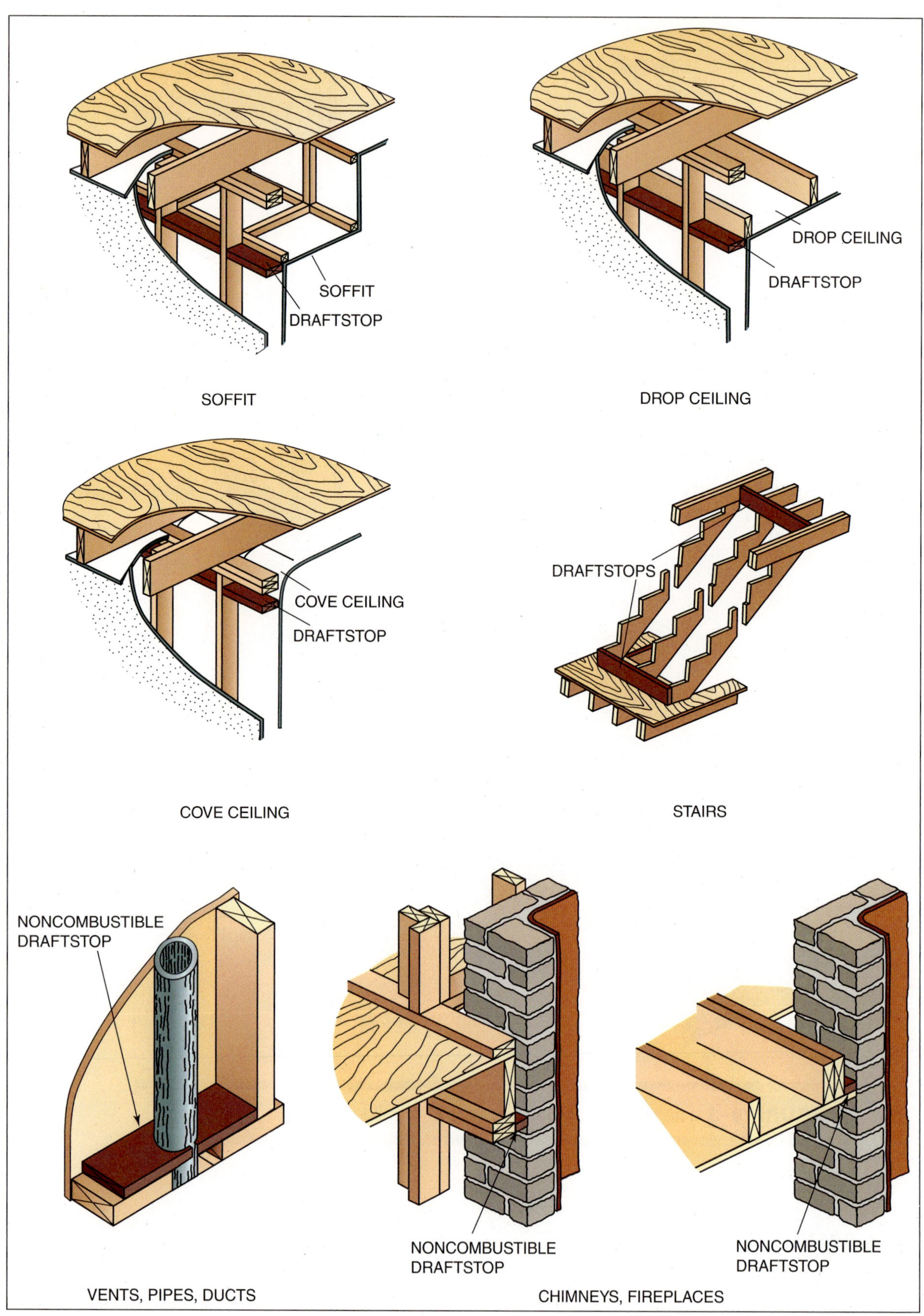


*Courtesy of Western Wood Products Association.*

**FIGURE 9-4** Draftstops slow the spread of fire.

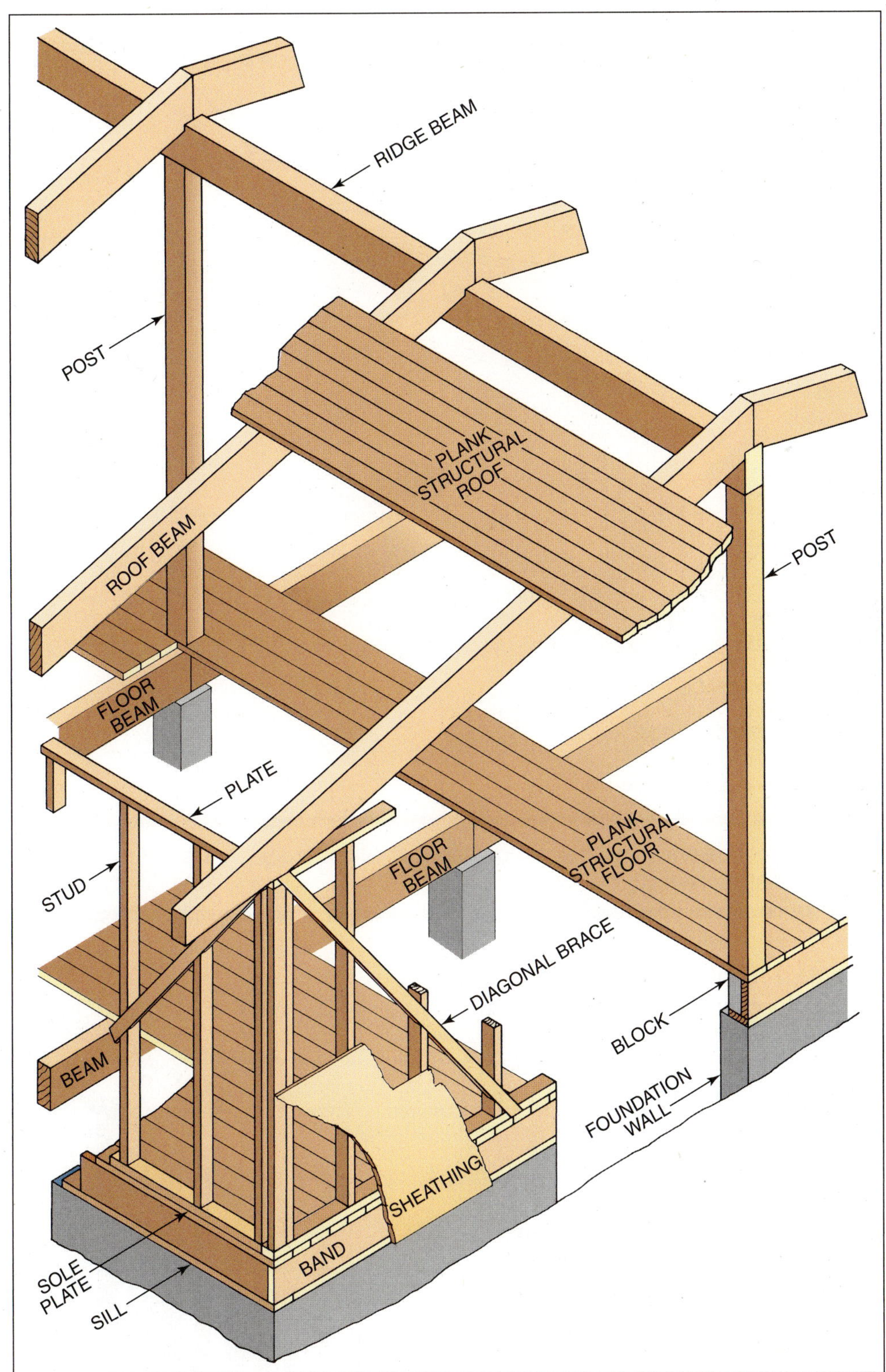


**FIGURE 9-5** The post-and-beam frame.

panels are glued to the floor beams with construction adhesive applied with caulking guns. The use of matched planks allows the floor beams to be more widely spaced.

## Walls

Exterior walls of a post-and-beam frame may be constructed with widely spaced **posts**. This allows wide expanses of glass to be used from floor to

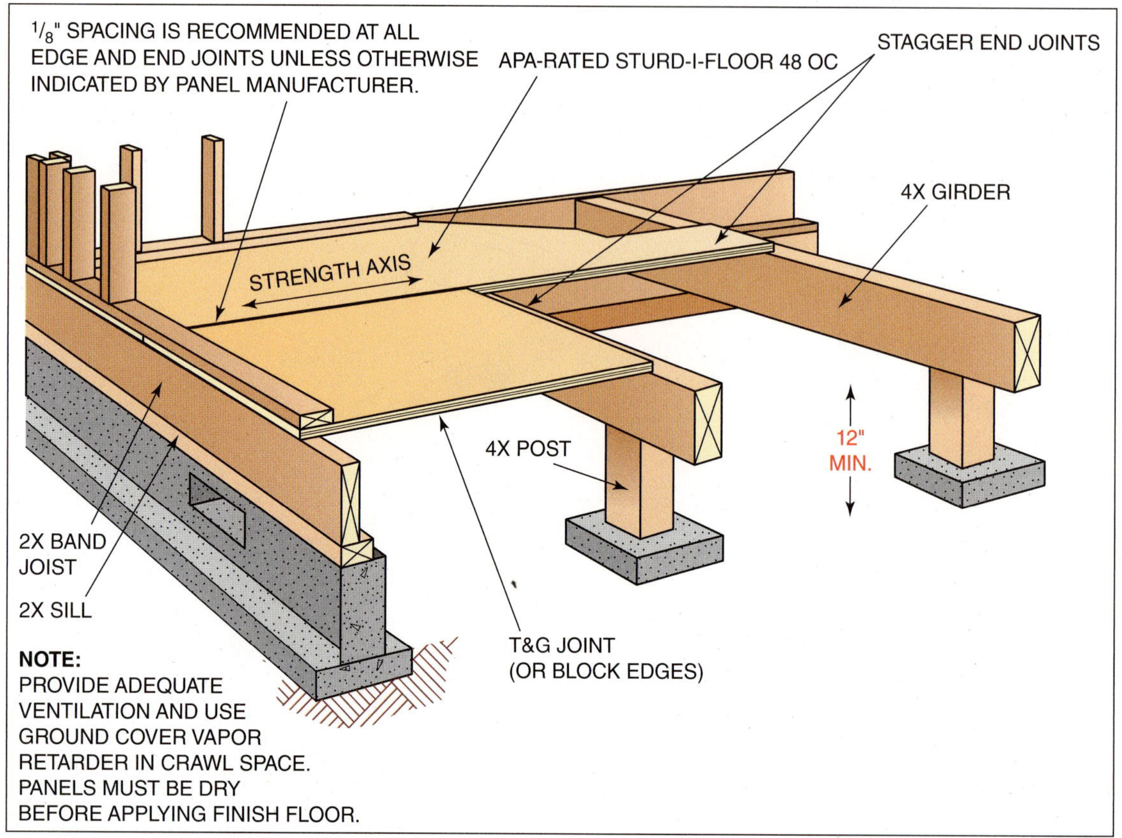


*Courtesy of APA—The Engineered Wood Association.*

**FIGURE 9-6** Floor beams may be spaced 4 feet OC. Member spacing is reduced to 32 inches OC for stiffer floors.

ceiling. Usually some sections between posts in the wall are studded at close intervals, as in platform framing. This provides for door openings, fastening for finish, and wall **sheathing**. In addition, close spacing of the studs permits the wall to be adequately braced (Fig. 9-7).

## Roofs

The post-and-beam frame roof is widely used. The exposed roof beams and sheathing on the underside are attractive. Usually, the bottom surface of the roof planks is left exposed to serve as the finished ceiling. Roof planks come in 2-, 3-, and 4-inch nominal thicknesses. The roof is insulated on top of the deck in order not to spoil the appearance of the exposed beams and deck on the underside (Fig. 9-8). Because of the fewer number of pieces used, a well-planned post-and-beam frame saves material and labor costs. Metal connectors are used to join members of the frame (Fig. 9-9).

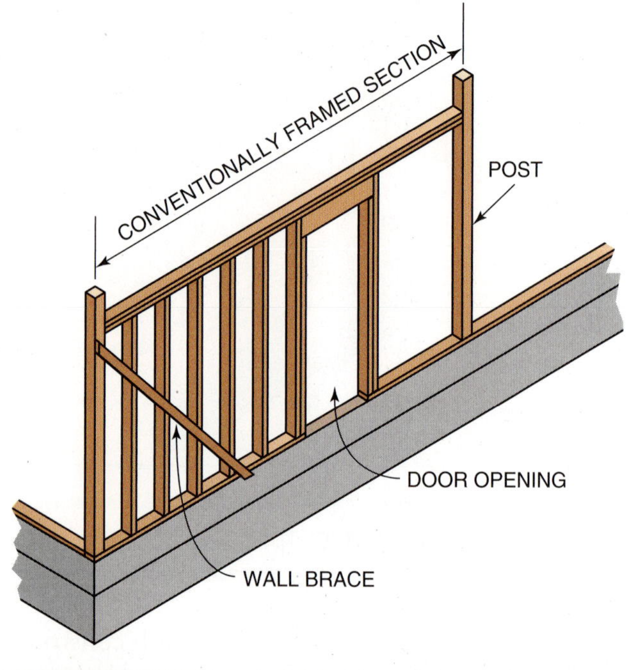


**FIGURE 9-7** Sections of the exterior walls of a post-and-beam wall may need to be conventionally framed.

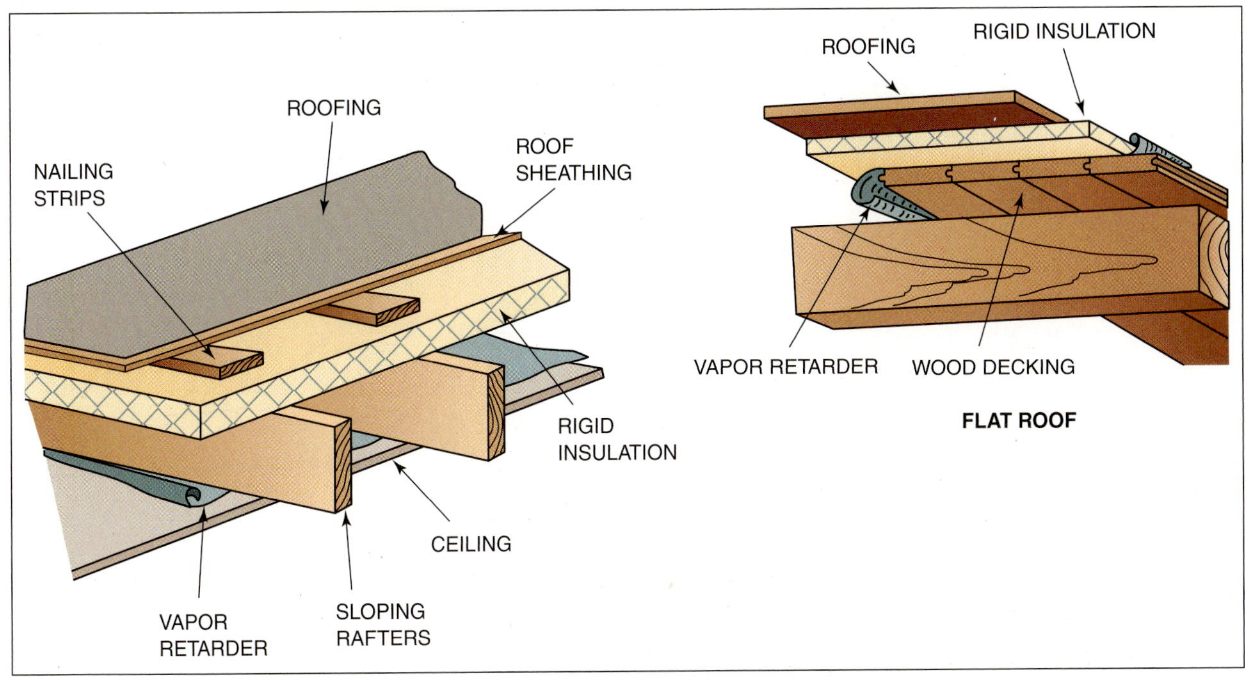


**FIGURE 9-8** Method of installing rigid insulation on roofs.

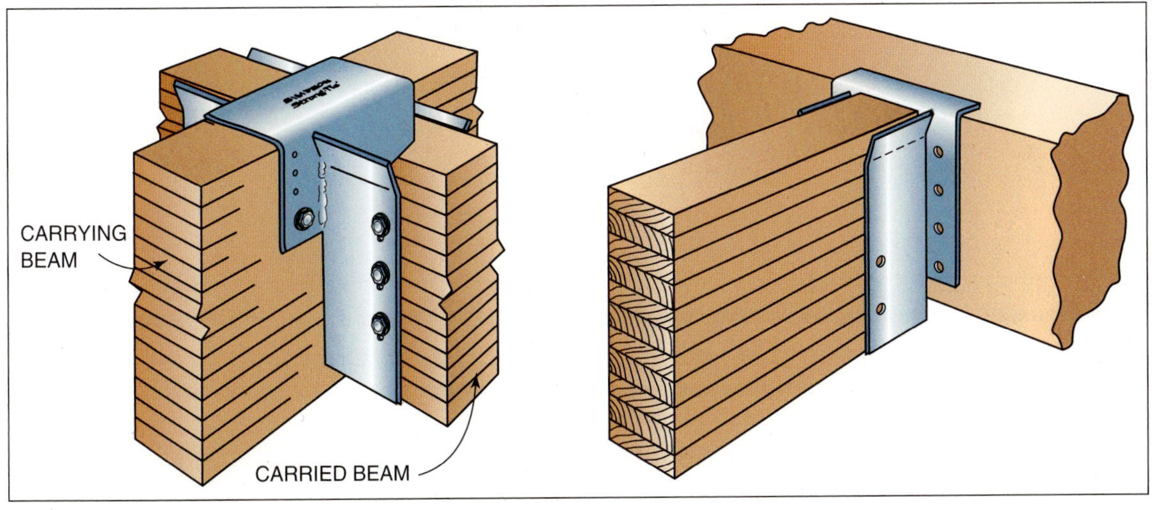


*Courtesy of Simpson Strong-Tie Company.*

**FIGURE 9-9** Metal connectors are specially made to join glulam beams.

## ENERGY AND MATERIAL CONSERVATION FRAMING METHODS

There has been much concern and thought about conserving energy and materials in building construction. Several systems have been devised that differ slightly from conventional framing methods. They conserve energy and use less material and labor. Check state and local building codes for limitations.

### The 24-Inch Module Method

One method uses plywood over lumber framing spaced on a 24-inch module. All framing, floors, walls, and roofs are spaced 24 inches OC. Joists, studs, and rafters line up with each other (Fig. 9-10).

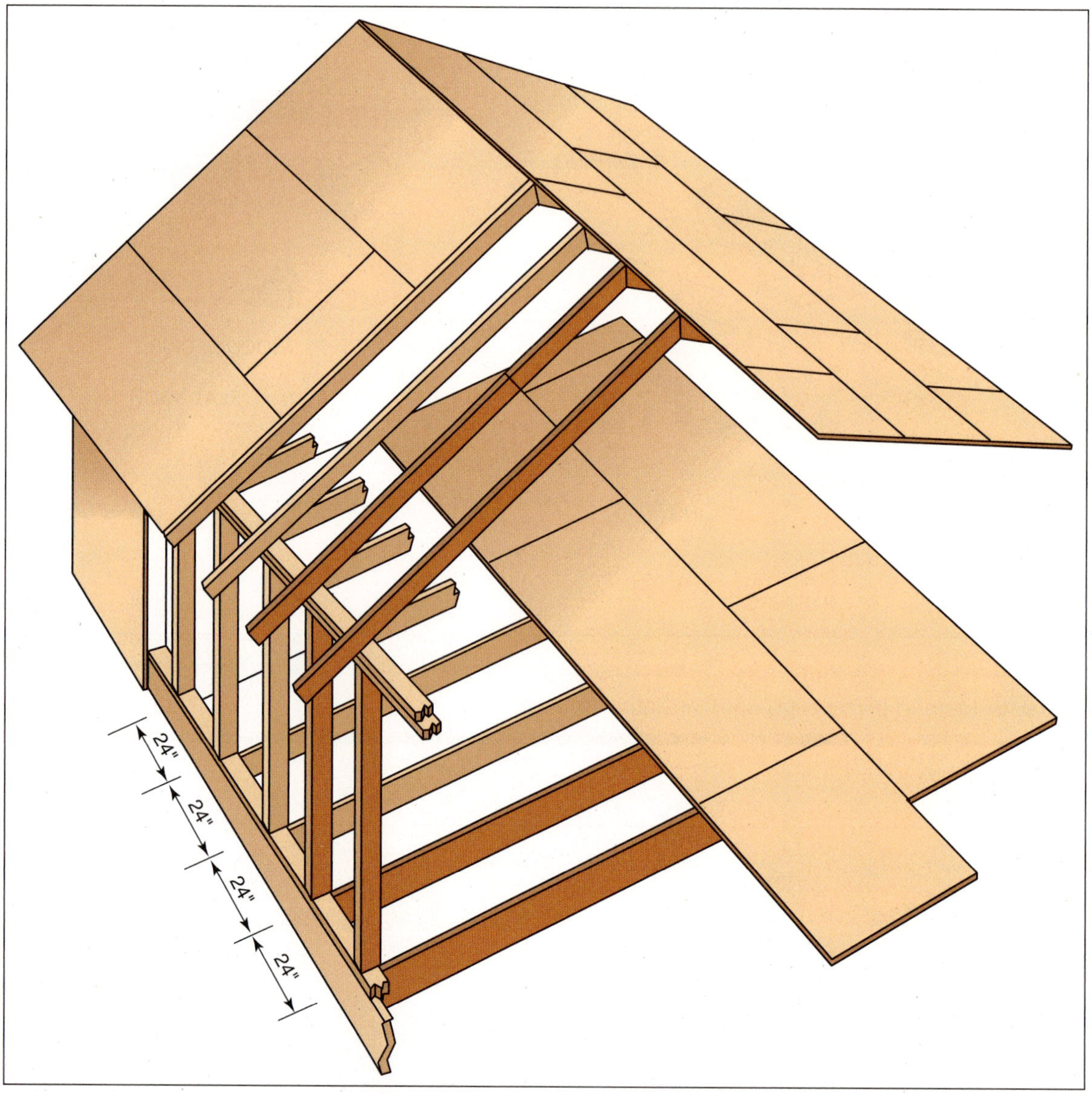

**FIGURE 9-10** The 24-inch module system of framing uses less material and labor.

*Floors.* For maximum savings, a single layer of ¾-inch tongue-and-grooved plywood is used over joists spaced 24 inches OC. In-line floor joists are used to make installation of the plywood floor easier (Fig. 9-11). The use of adhesive when fastening the plywood floor is recommended. Gluing increases stiffness and prevents squeaky floors (Fig. 9-12).

**Green Tip**

Building dimensions in four-foot modules reduces material waste since standard building panels are 4' × 8'.

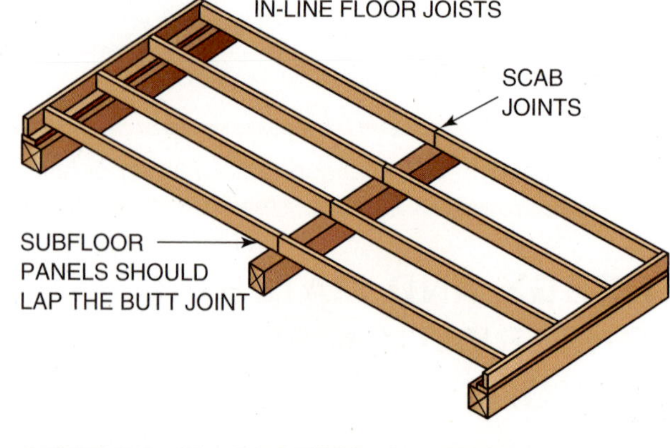


**FIGURE 9-11** In-line floor joists simplify installation of plywood subflooring.

Courtesy of APA—The Engineered Wood Association.

**FIGURE 9-12** Using adhesive when fastening subflooring makes the floor frame stiffer and stronger.

*Walls.* Studs up to 10 feet long with 24-inch spacing can be used in single-story buildings. Stud height for two-story buildings should be limited to 8 feet. A single layer of plywood acts as both sheathing and exterior siding. Wall openings are laid out so that at least one side of the opening falls on an on-center stud (Fig. 9-13).

*Roofs.* Rafters or trusses are spaced 24 inches OC in line with the studs under plywood roof sheathing.

### The Arkansas System

An energy-saving construction system developed by the Arkansas Power and Light Company uses 2 × 6 wall studs spaced 24 inches OC. This

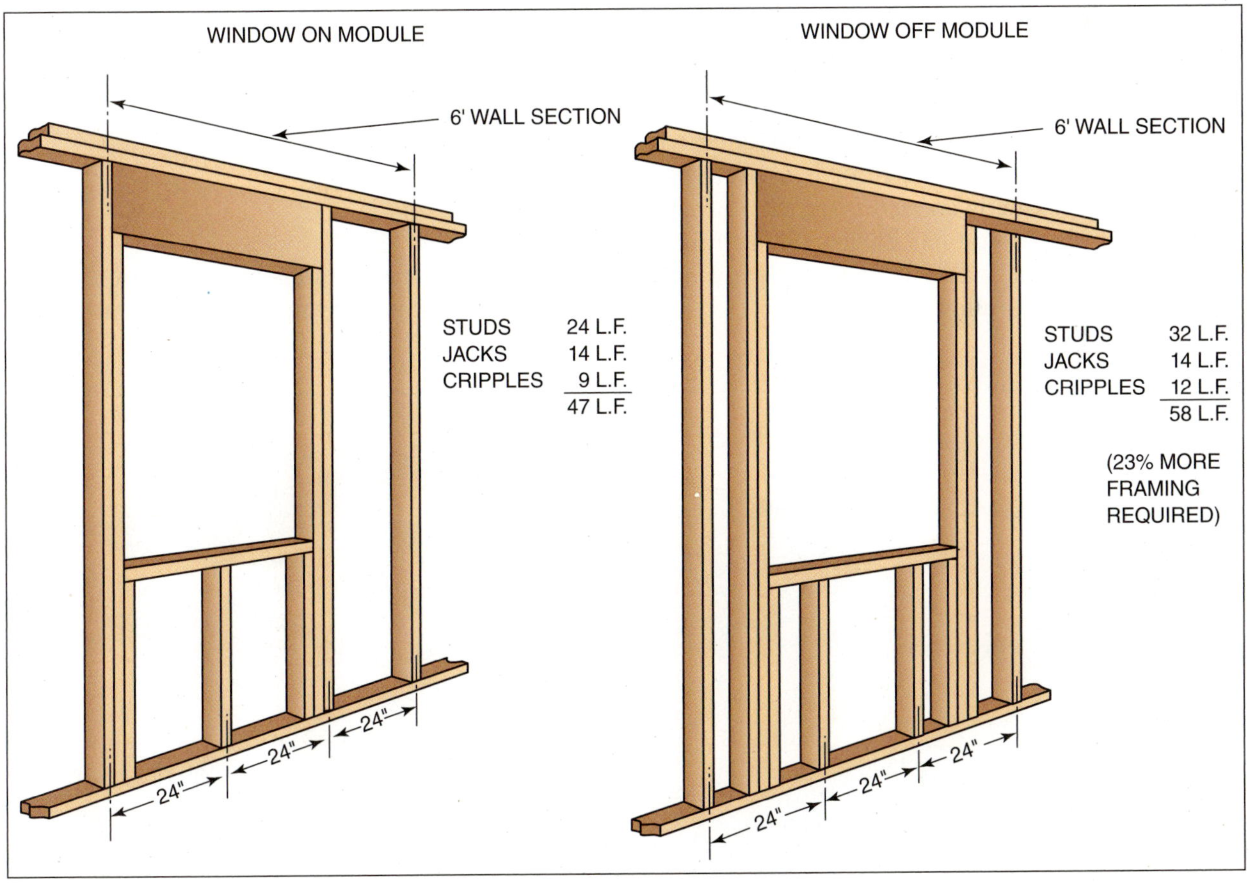

**FIGURE 9-13** To conserve materials, locate wall openings so they fall on the spacing module.

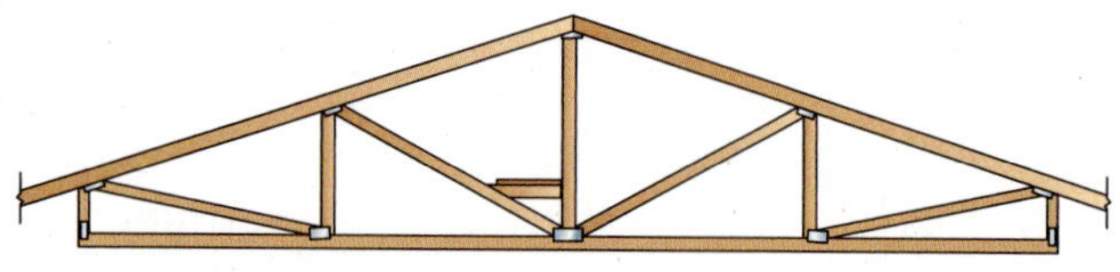

**FIGURE 9-14** Modified truss design accommodates 12-inch ceiling insulation without compressing at eaves.

permits using 6-inch insulation in the exterior walls. A modified truss accommodates 12 inches of insulation in the ceiling without compressing it at the eaves (Fig. 9-14).

# Layout and Construction of the Floor Frame

A floor frame consists of members fastened together to support the loads a floor is expected to bear. The floor frame is started after the foundation has been placed and has hardened. A straight and level floor frame makes it easier to frame and finish the rest of the building.

Because platform framing is used more than any other type, this section describes how to lay out and construct a platform frame. The knowledge gained in this section can be used to lay out and construct any type of floor frame.

## DESCRIPTION AND INSTALLATION OF FLOOR FRAME MEMBERS

In the usual order of installation, the floor frame consists of girders, posts or **columns**, sill plates, joists, bridging, and subfloor (Fig. 9-15).

### Description of Girders

**Girders** are heavy beams that support the inner ends of the floor joists. Several types are commonly used.

*Kinds of Girders.* Girders may be made of solid wood or built up of two or more pieces of dimension lumber. Laminated veneer lumber or glulam beams may also be used as girders (Fig. 9-16). Sometimes, wide flange, I-shaped steel beams are used.

*Built-Up Girders.* If built-up girders of dimension lumber or engineered lumber are used, a minimum

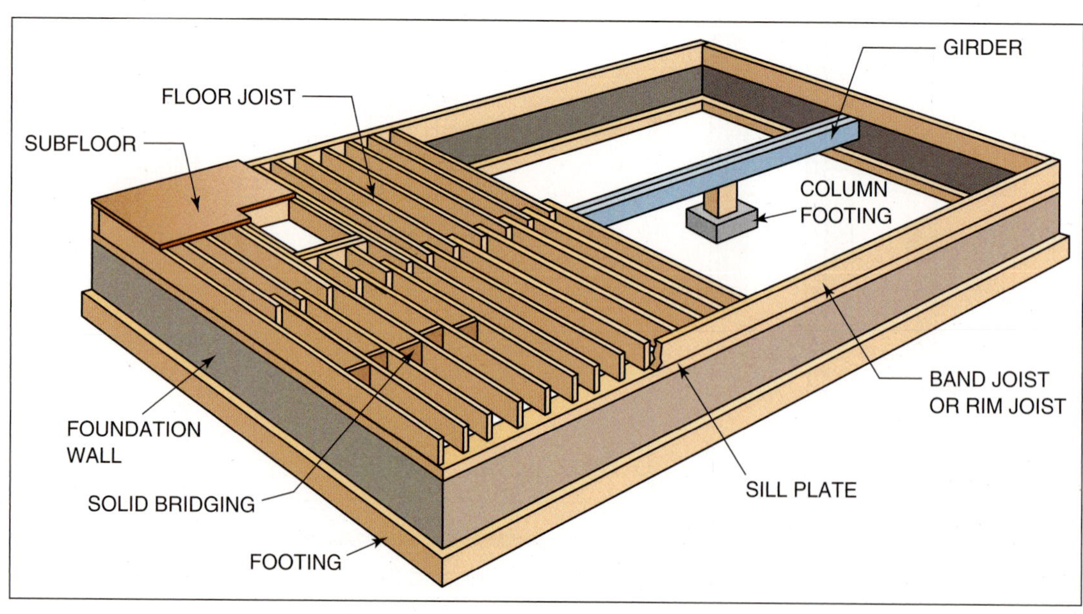


From Architectural Drafting & Design, 2003, Delmar.

**FIGURE 9-15** A floor frame of platform construction.

**FIGURE 9-16** A large glulam beam is used for a girder.

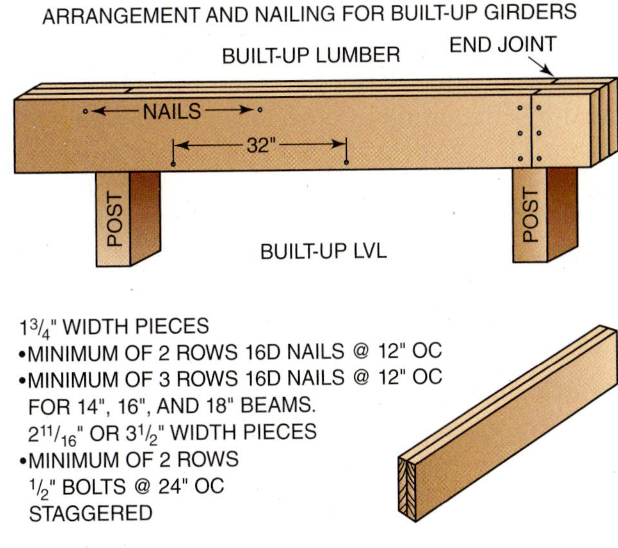


**FIGURE 9-17** Spacing of fasteners and joints of built-up girders.

of three members are fastened together with three 3½-inch or 16d nails at each end. The other nails are staggered not farther than 32 inches apart from end to end (Fig. 9-17). Sometimes ½-inch bolts are required. Applying glue between the pieces makes the bond stronger.

*Girder Location.* A pocket formed in the foundation wall usually supports the ends of the girder (Fig. 9-18). The pocket should provide at least a

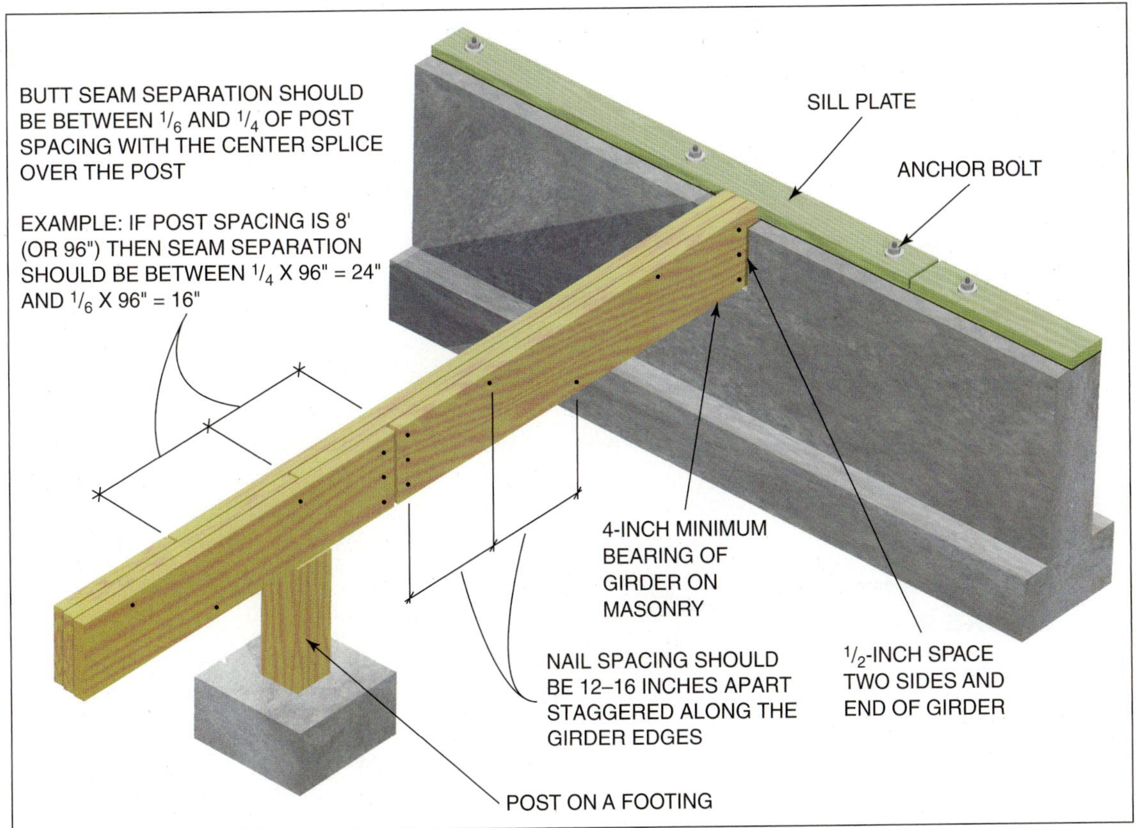


**FIGURE 9-18** Spacing of fasteners and seams of a built-up girder made with dimension lumber.

4-inch bearing for the girder. It should be wide enough to provide ½-inch clearance on both sides and the end. This allows any moisture to be evaporated by circulation of air. Thus, no moisture will get into the girder, which could cause decay of the timber. The pocket is formed deep enough to provide for shimming the girder to its designated height. Use steel **shims** as needed to adjust the height; wood shims are not suitable for use under girders.

***For step-by-step instructions on how to INSTALL GIRDERS, see Procedure 9-1 on page 230.***

## SILLS

**Sills**, also called *mudsills* or *sill plates*, are horizontal members of a floor frame. They lie directly on the foundation wall and provide a bearing for *floor joists*. It is often required that the sill be made with a decay-resistant material, such as pressure-treated lumber, redwood, black locust, or cedar. Sills may consist of a single 2 × 6, or doubled 2 × 6s (Fig. 9-19).

The sill is attached to the foundation wall with **anchor bolts**. Their size, type, and spacing are specified on the blueprints. To seal up irregularities between the foundation wall and the sill, a **sill sealer** is used. It comes 6 inches wide and in rolls of 50 feet. It compresses when the weight of the structure is upon it.

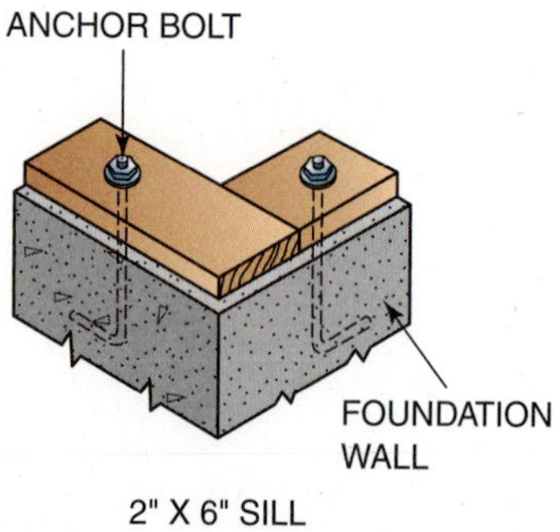


**FIGURE 9-19** Sill details at the corner. A bolt should be located 6 to 12 inches from the ends of each sill.

***For step-by-step instructions on INSTALLING SILLS, see Procedure 9-2, pages 231–232.***

## FLOOR JOISTS

Floor joists are horizontal members of a frame. They rest on and transfer the load to sills and girders. In residential construction, nominal 2-inch thick lumber placed on edge has been traditionally used. Wood I-beams, with lengths up to 80 feet, are being specified more often today (Fig. 9-20). In commercial work, steel or a combination of steel and wood trusses is frequently used.

Joists are generally spaced 16 inches OC in conventional framing. They may be spaced 12, 19.2, or 24 inches OC, depending on the type of construction and the intended load.

### Joist Framing at the Sill

Joists should rest on at least 1½ inches of bearing on wood and 3 inches on **masonry**. In platform construction, the ends of floor joists are capped with a **band joist**, also called a *rim joist, box header,* or *joist header*. In a balloon frame, joists are cut flush with the outside edge of the sill (Fig. 9-21). The use of wood I-beams requires sill construction as recommended by the manufacturer for satisfactory performance of the frame (Fig. 9-22).

### Joist Framing at the Girder

If joists are lapped over the girder, the minimum amount of lap is 3 inches, and the maximum overhang is 12 inches. There is no need to lap wood I-beams. They come in lengths long enough to span the building. However, they may need to be supported by girders depending on the span and size of the I-beam. Web stiffeners should be applied to the beam ends if the hanger does not reach the top flange of the beam.

Sometimes, to gain more headroom, joists may be framed into the side of the girder. There are a number of ways to do this (Fig. 9-23). Joist hangers must be used to support wood I-beams.

**FIGURE 9-20** Engineered lumber makes a strong floor system.

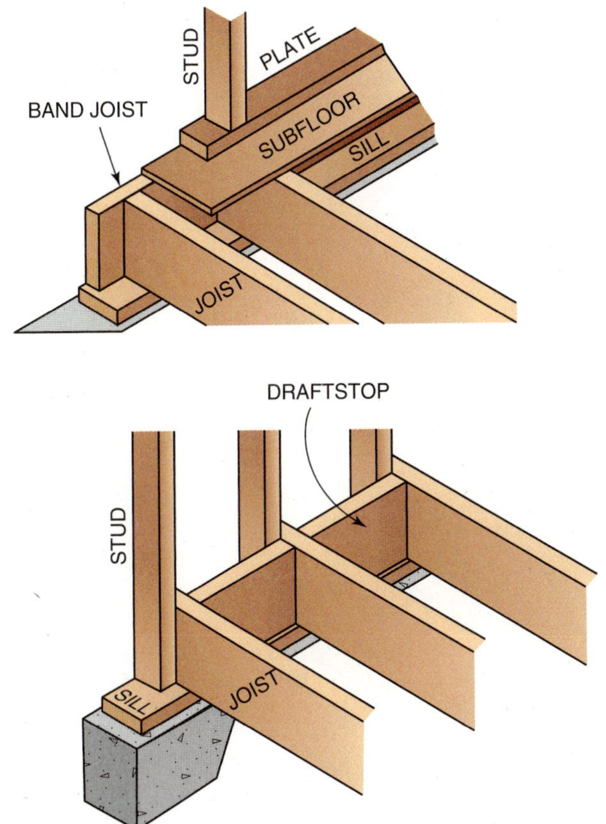


**FIGURE 9-21** Typical framing at the sill with balloon framing.

### Notching and Boring of Joists

Notches in the bottom or top of sawn lumber floor joists should not exceed one-sixth of the joist depth. Notches should not be located in the middle one-third of the joist span. Notches on the ends should not exceed one-fourth of the joist depth.

Holes bored in joists for piping or wiring should not be larger than one-third of the joist depth. They should not be closer than 2 inches to the top or bottom of the joist (Fig. 9-24).

Some wood I-beams are manufactured with 1½-inch perforated knockouts in the web at approximately 12 inches OC along its length. This allows easy installation of wiring and pipes. To cut other size holes in the web, consult the manufacturer's specifications guide. Do not cut or notch the flanges of wood I-beams.

***For step-by-step instructions on LAYING OUT FLOOR JOISTS AND FLOOR OPENINGS, see Procedure 9-3, pages 233–235.***

## DOUBLING FLOOR JOISTS

For added strength, floor joists are doubled and must be securely fastened together. Floor openings

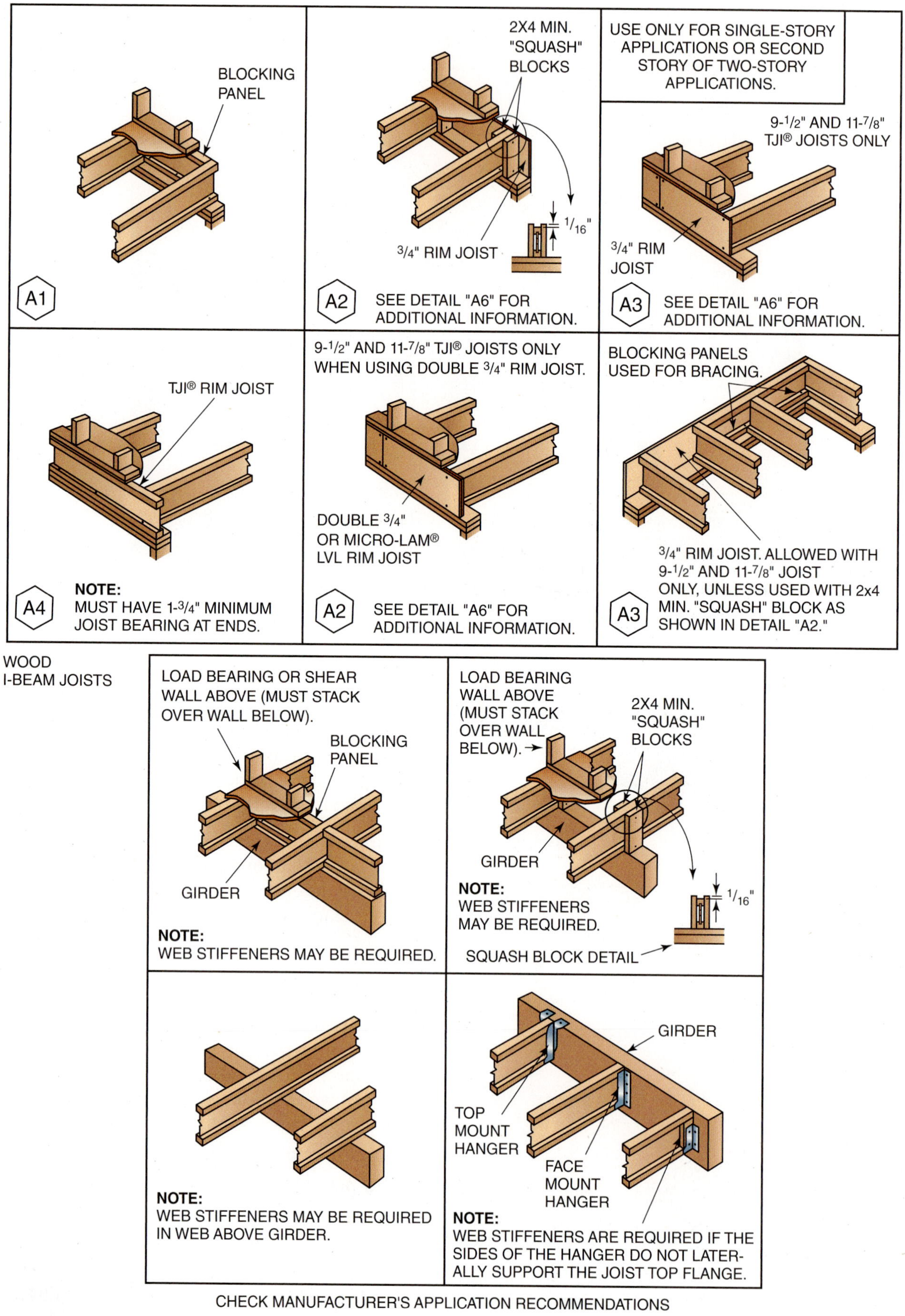

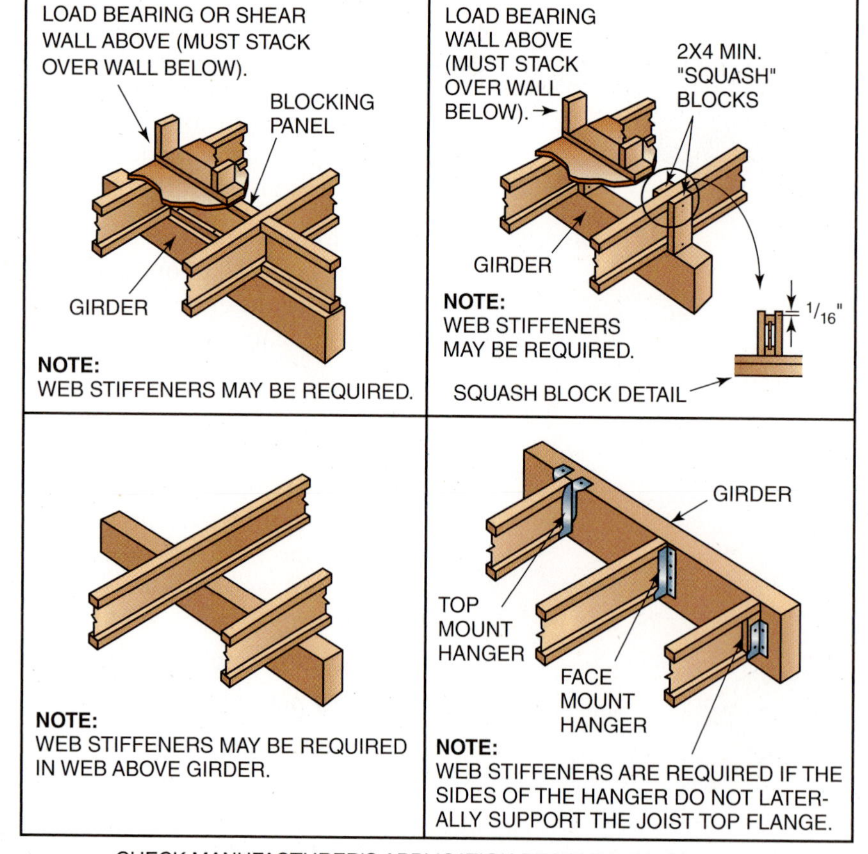


**FIGURE 9-22** Wood I-beam framing details at the sill.

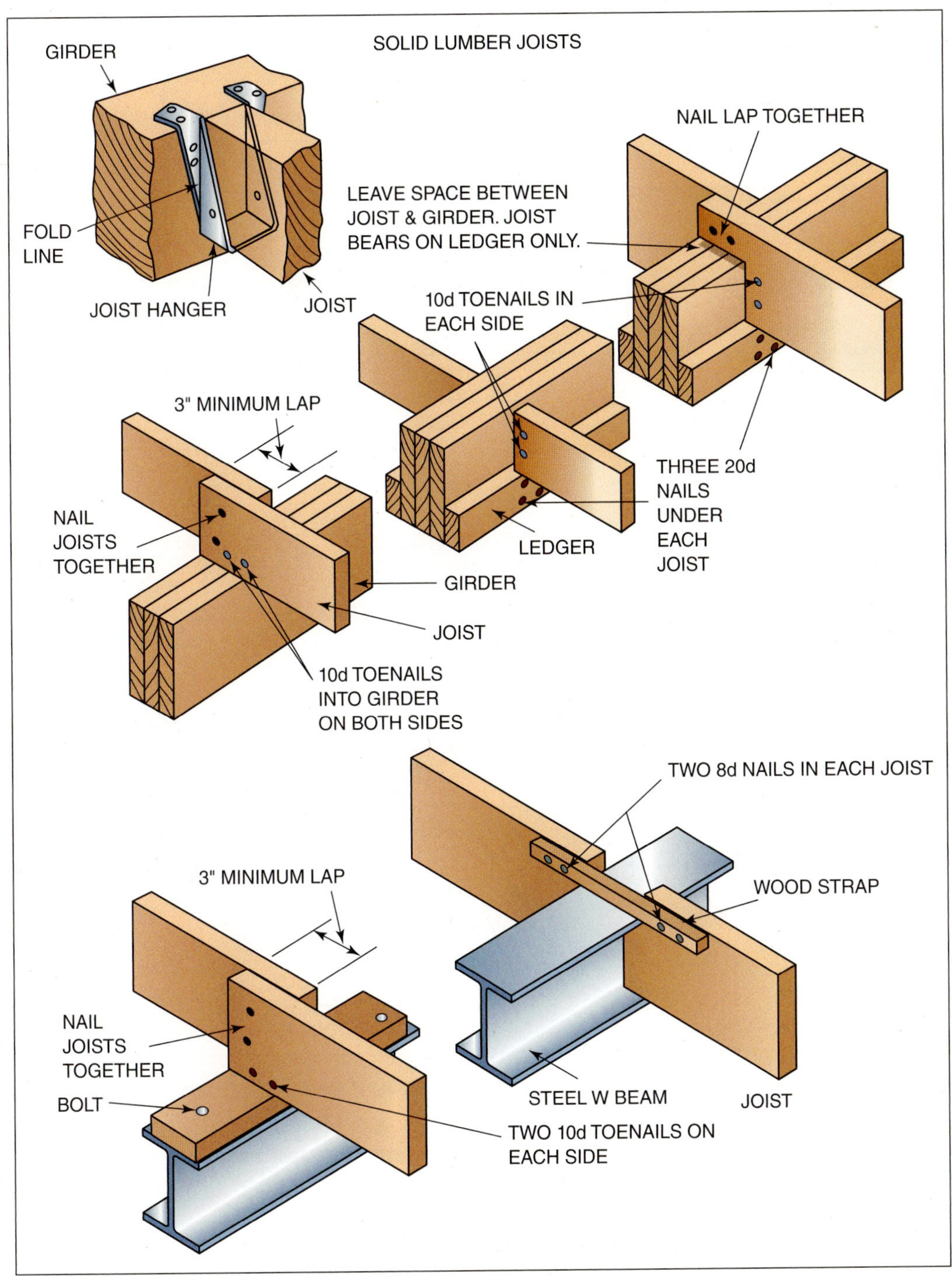


**FIGURE 9-23** Joist framing details at the girder.

cut more than one joist, therefore all sides of the opening must be doubled. Top edges of doubled joists should be flush or even. In many cases, the top edges do not align flush with each other. They must be brought flush before they can be nailed together.

To bring them flush, toenail down through the top edge of the higher one, at about the joist's mid span. At the same time, squeeze both together tightly by hand. Use as many toenails as necessary, spaced where needed, to bring the top edges flush (Fig. 9-25). Usually no more than two or three nails

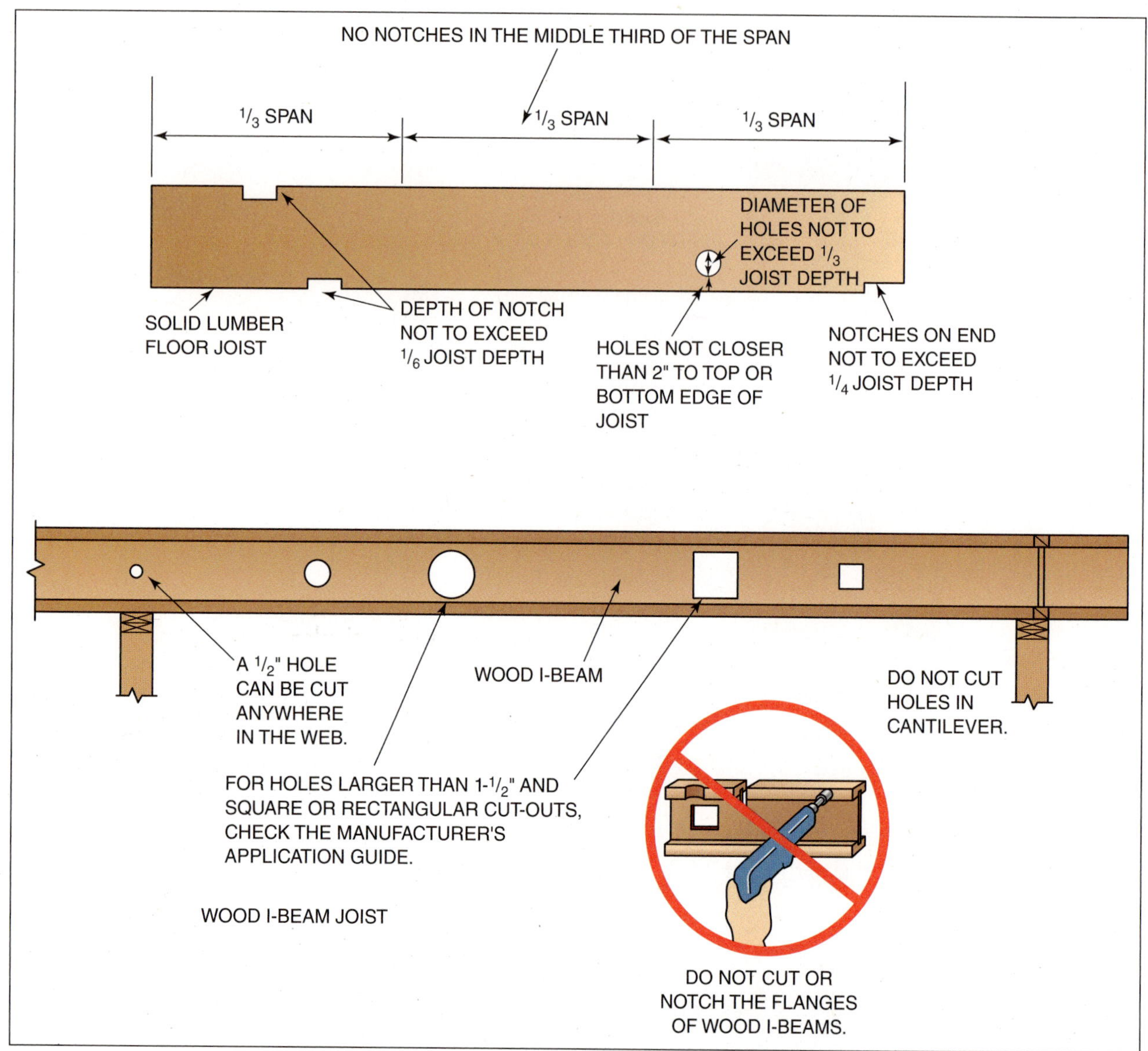


**FIGURE 9-24** Allowable notches, holes, and cutouts in floor joists.

are needed. Then, fasten the two pieces securely together. Drive nails from both sides, staggered from top to bottom, about 2 feet apart. Angle nails slightly so they do not protrude.

## CANTILEVERED JOISTS

When the floor system extends beyond the foundation, the floor joists become cantilevered joists. Cantilevered joists are supported at the inside end by a doubled joist or girder (Fig. 9-26). The outside support is not at the end of the joist. Overhangs are generally not more than 24", and joist length must be at least three times longer than the overhang. The joist hangers should be turned upside down from normal installation.

## WOOD CONNECTORS

Many types and styles of connectors are used to secure structural components of a building (Fig. 9-27). The size, quantity, and location of nails are crucial. Follow the manufacturer's recommendations.

***For step-by-step instructions on FRAMING FLOOR OPENINGS, see Procedure 9-4, pages 236–237.***

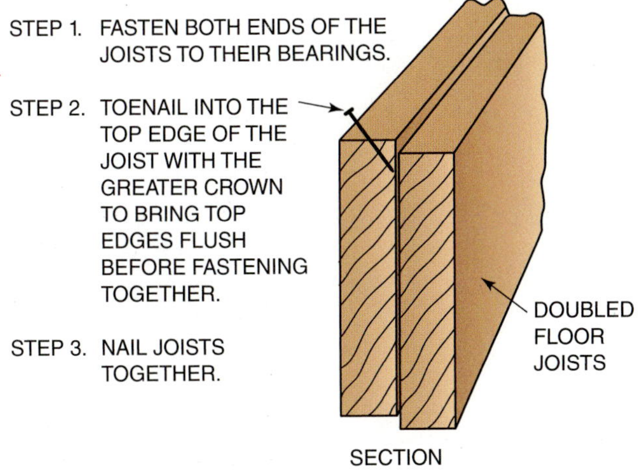


**FIGURE 9-25** Technique for aligning the top edges of dimension lumber.

*For step-by-step instructions on INSTALLING FLOOR JOISTS, see Procedure 9-5, pages 238–239.*

*For step-by-step instructions on INSTALLING THE BAND JOIST, see Procedure 9-6 on page 240.*

## BRIDGING

**Bridging** is installed in rows between floor joists at mid-span between supports such as girders and sills. One row of bridging is needed for joists 8 to 16 feet long. Two rows are needed for joists 16 to 24 feet long. Its purpose is to distribute a concentrated load on the floor over a wider area.

Bridging may be solid wood, wood cross-bridging, or metal cross-bridging (Fig. 9-28). Usually, solid wood bridging is the same size as the floor joists. It is installed in an offset fashion to permit end nailing. Solid blocking is required between joists over supporting girders and walls.

Wood cross-bridging should be at least nominal 1 × 3 lumber with two 6d nails at each end. It is placed in double rows that cross each other in the joist space.

Metal cross-bridging is available in different lengths for the particular joist size and spacing.

*For step-by-step instructions on LAYING OUT AND CUTTING WOOD CROSS-BRIDGING, see Procedure 9-7 on page 241.*

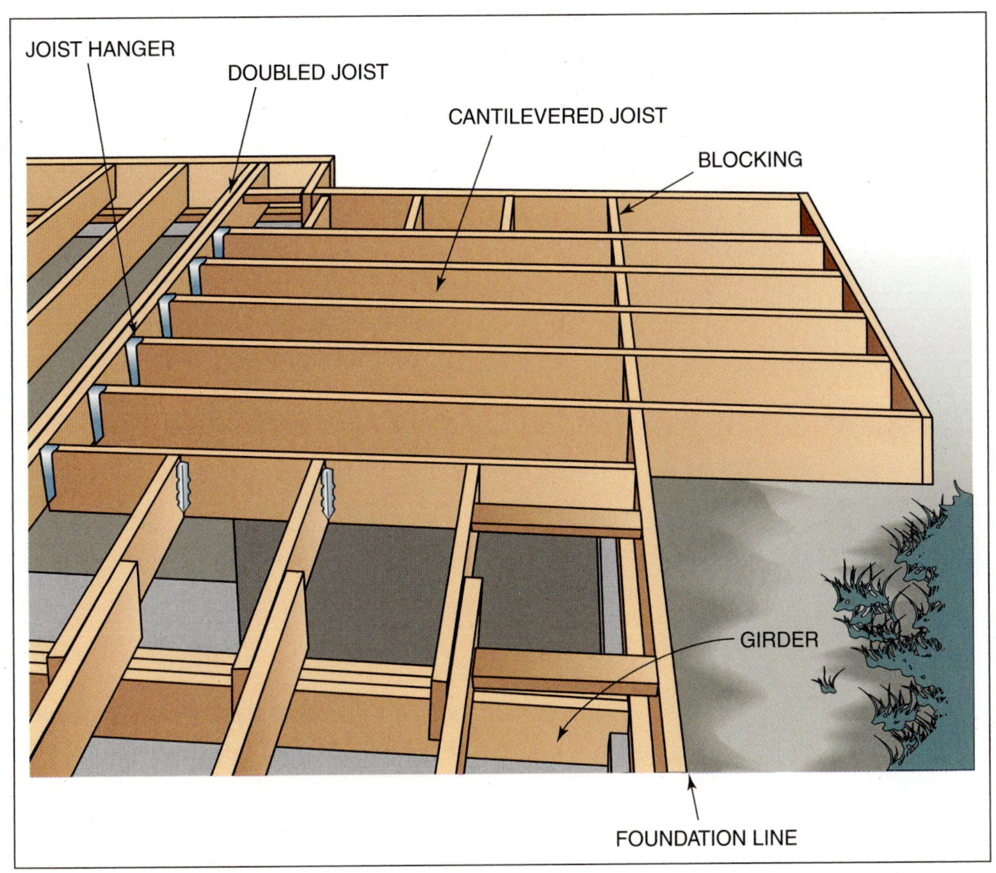


**FIGURE 9-26** Cantilevered joists overhang the foundation.

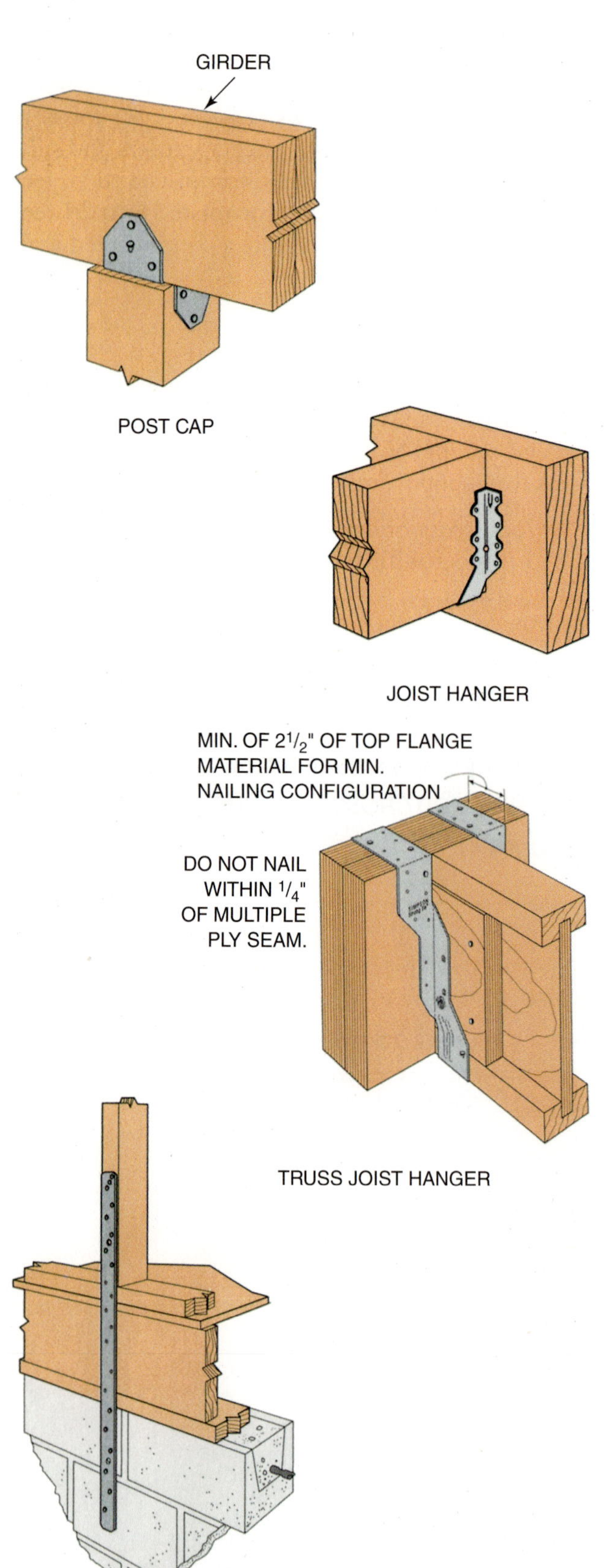


*Courtesy of Simpson Strong-Tie Company.*

**FIGURE 9-27** Typical connectors used in floor framing.

**FIGURE 9-28** Types of bridging.

***For step-by-step instructions on INSTALLING BRIDGING AND BLOCKING, see Procedure 9-8, pages 242–243.***

## COLUMNS

Girders may be supported by framed walls, wood posts, or steel columns (Fig. 9-29). Metal plates are used at the top and bottom of the columns to distribute the load over a wider area. After the floor joists are installed and before any more weight is placed on the floor, the temporary posts supporting the girder are replaced with permanent posts or columns.

***For step-by-step instructions on INSTALLING COLUMNS, see Procedure 9-9 on page 244.***

## SUBFLOORING

Subflooring is used over joists to form a working platform. This is also a base for finish flooring, such as hardwood flooring, or underlayment for carpet or resilient tiles. Sturd-I Floor panels are used when a single-layer subfloor/underlayment system is desired. Blocking is required under the joints of these panels unless tongue-and-groove edges are used.

***For step-by-step instructions on APPLYING PLYWOOD SUBFLOORING, see Procedure 9-10 on page 245.***

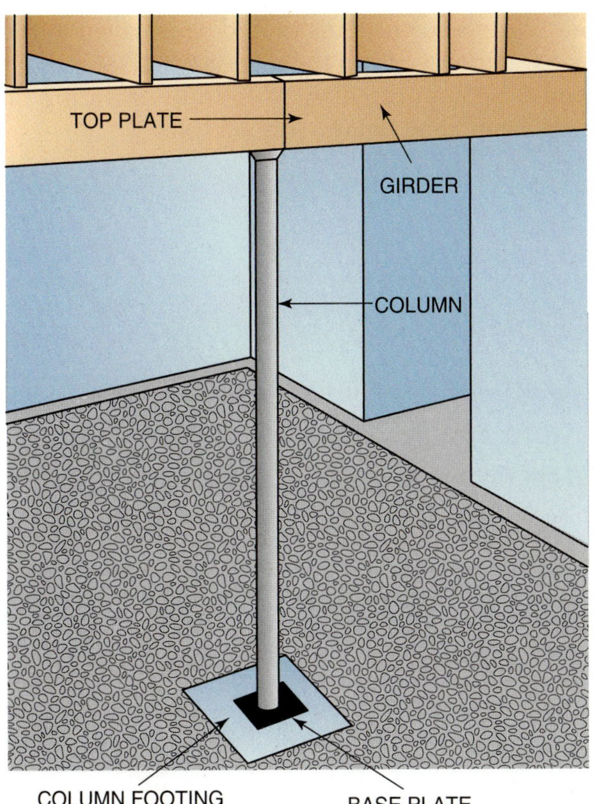


**FIGURE 9-29** Steel columns are often used to support girders and beams.

# TERMITES

Of all the destructive wood pests, termites are the most common. Termites play a beneficial role in their natural habitat, but they cause tremendous economic loss annually. They break down dead or dying plant material to enrich the soil. However, when termites feed on wood structures, they become pests. They attack wood throughout most of the country, but they are more prevalent in warmer sections (Fig. 9-30). Buildings should be designed and constructed to minimize termite attack.

## TECHNIQUES TO PREVENT TERMITES

All the techniques used for the prevention of termite attack are based on keeping the wood in the structure dry (equilibrium moisture content) and making

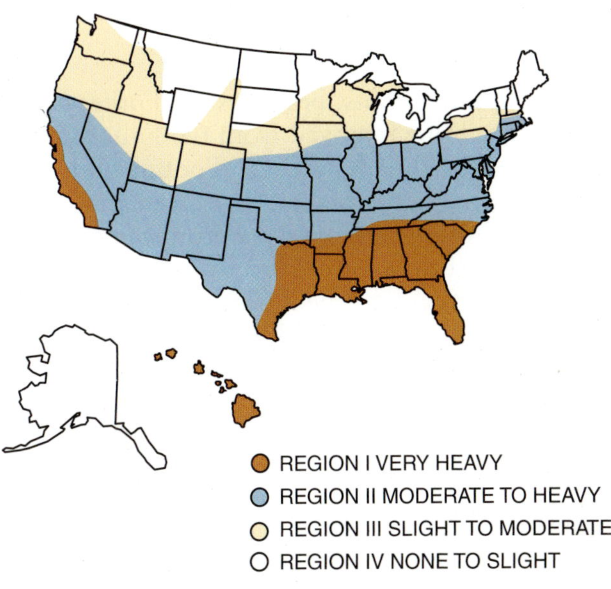


**FIGURE 9-30** Degree of subterranean termite hazard in the United States.

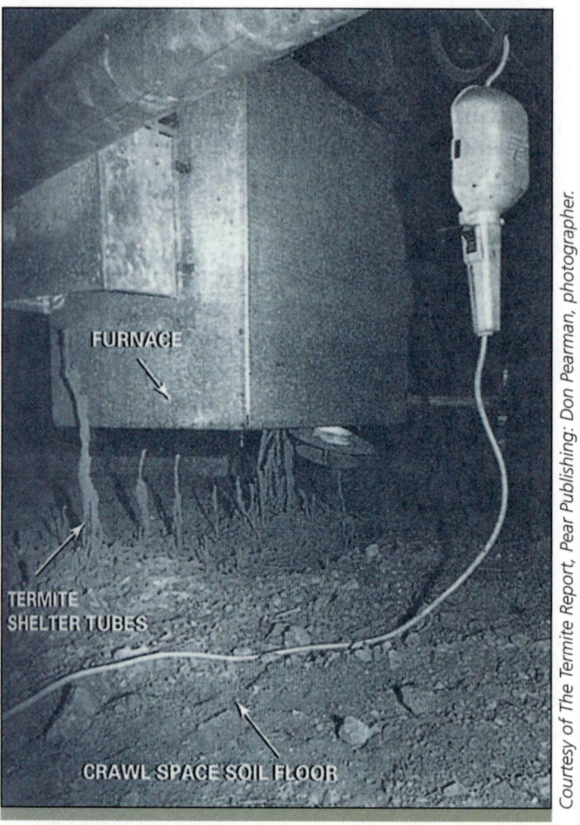


**FIGURE 9-31** Termites can build unsupported shelter tubes as high as 12 inches. Heat attracts them in their attempt to reach wood.

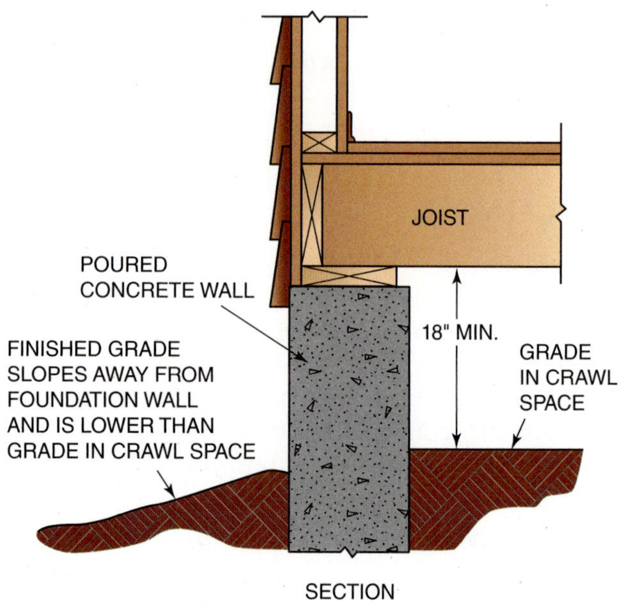


**FIGURE 9-32** The finished grade in the crawl space and around the outside of the foundation.

it as difficult as possible for termites to get to the wood (Fig. 9-31).

### The Site

All tree stumps, roots, branches, and other wood debris should be removed from the building site. Do not bury debris on the site. Footing and wall-form planks, boards, stakes, spreaders, and scraps of lumber should be removed from the area before backfilling around the foundation. Lumber scraps should not be buried anywhere on the building site. None should be left on the ground beneath or around the building after construction is completed.

The site should be graded to slope away from the building on all sides. The outside finished grade should always be equal to or below the level of the soil in crawl spaces. This ensures that water is not trapped underneath the building (Fig. 9-32).

Perforated drain pipe should be placed around the foundation, alongside the footing. This will drain water away from the foundation (Fig. 9-33).

Gutters and downspouts should be installed to lead roof water away from the foundation. Downspouts should be connected to a separate drain pipe to facilitate moving the water quickly.

## Green Tip

Insects will not consume or destroy dry wood. Buildings last longer if they are kept dry.

### Exterior Woodwork

Wall siding usually extends no more than 2 inches below the top of foundation walls. It should be at least 6 inches above the finished grade. Porch supports should be placed not closer than 2 inches from the building to prevent hidden access by termites. Wood steps should rest on a concrete base that extends at least 6 inches above the ground. Door jambs, posts, and similar wood parts should never extend into or through concrete floors.

**FIGURE 9-33** Perforated drain pipe is placed alongside the foundation footing to drain water away from the building.

## Termite Shields

If **termite shields** are properly designed, constructed, installed, and maintained, they will force termites into the open. This will reveal any tubes constructed around the edge and over the upper surface of the shield (Fig. 9-34). However, research has shown that termite shields have not been effective in preventing termite infestations. Because of improper installation and infrequent inspection, they are not presently recommended by government agencies for detection and prevention of termite attack. However, check local building codes that may mandate their use.

## Use of Pressure-Treated Lumber

Lumber in which preservatives are forced into the wood's cells under pressure is commonly called **pressure-treated lumber** (Fig. 9-35). Termites

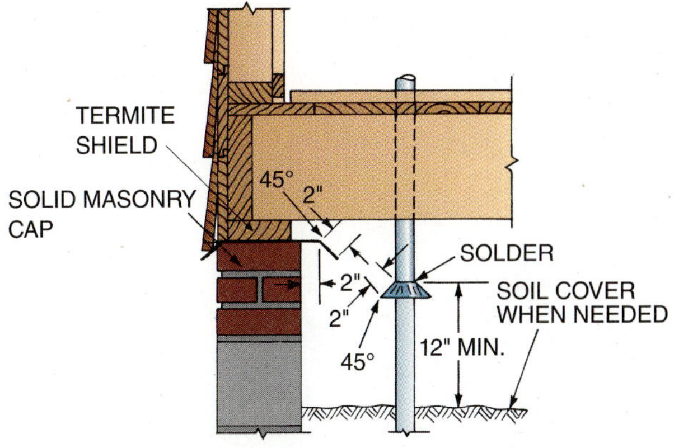


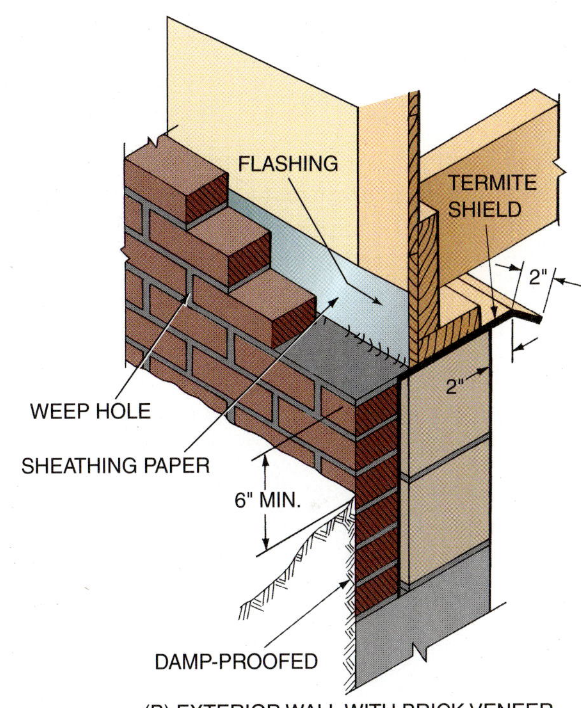


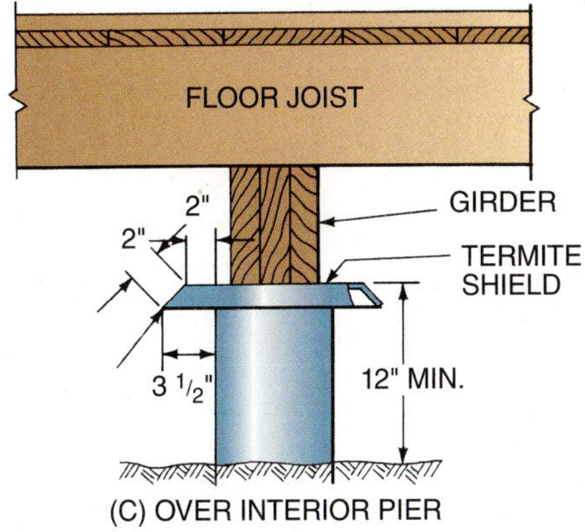


**FIGURE 9-34** Typical installation of termite shields.

Courtesy of Weyerhaeuser Company.

**FIGURE 9-35** Preservatives are forced into lumber under pressure in large cylindrical tanks.

generally will not eat treated lumber. They will tunnel over it to reach untreated wood. Their shelter tubes then may be exposed to view and their presence easily detected upon inspection. Generally, building codes require the use of pressure-treated lumber for the following structural members:

- Wood joists or the bottom of structural floors without joists that are located closer than 18 inches to exposed soil.
- Wood girders that are closer than 12 inches to exposed soil in crawl spaces or unexcavated areas.
- Sleeper, sills, and foundation plates on a concrete or masonry slab that is in direct contact with the soil.

# Estimating Material Quantities for Floor Framing

Estimating the material in floor framing involves girder material, sills, anchor bolts, OC and rim joists, bridging, and subfloor. Remember that if the number of pieces does not come out even, round them up to the next whole number of pieces.

## GIRDER

The amount of girder material required is determined by the girder length. If the girder is steel, then the quantity required is merely the actual length of the girder. If, however, the girder is built up, then a calculation must be performed to arrive at the amount of girder material required.

For a built-up girder, multiply the length of the girder by the number of plies in the girder. Then divide by the length of each piece of girder material to get the number of pieces to purchase. For example, if a 45-foot-long girder is made of four plies and made from material that is 12 feet long, then 45 × 4 = 180 **linear feet**. Then divide 180 by 12 to arrive at 15 pieces. Add one more for waste to make sure butt seams are properly aligned.

## SILLS

The number of sill pieces depends on the building perimeter and the length of each sill piece. Take the building perimeter and divide by the sill piece length. For example, if a building is 28' × 48' and sill pieces are 12 feet long, then the number of pieces is calculated as follows:

2 × (28 + 48) = 152 feet (perimeter) ÷ 12 = 12.667 or 13 pieces. Remember to add sill material if steel girder is used.

## ANCHOR BOLTS

The number of anchor bolts depend on the number of sill pieces needed for the perimeter. There should be at least three bolts per sill, plus one per corner to compensate for the fact that sills are usually cut at each corner. Each piece must have at least one bolt at their ends. For example, if the sill pieces are calculated to be 13, and there are four corners, then the bolts are 13 × 3 bolts per piece + 4 = 43 bolts.

## FLOOR JOISTS

To determine the number of floor joists to order, first determine if they are full length across the building or lapped at the girder. The quantity of lapped floor joists is twice that of full-length joists. To find the number of full-length joists, divide the length of the building by the spacing in terms of feet. For example, if the spacing is 16 inches, then it is 16 ÷ 12 = 1.333 feet. So a 40-foot-long building divided by 1.333 = 30 pieces. Add one joist to start. Double this if joists span only to the girder. Also add joists for every parallel partition above that will need extra support.

## BAND JOISTS

The band joist material is the twice the building length divided by the length of material used. For example, if the length is 60 feet and the material used is 12 feet long, then the band joist material is 60 × 2 sides ÷ 12 = 10 pieces. Add extra for headers as needed.

## BRIDGING

Bridging quantity depends on the style of bridging. The linear feet of solid bridging is simply the length of the building times the number of rows of bridging. Linear feet of cross-bridging is determined by taking the number of joists times 3 feet for 16-inch OC joists. This number is arrived at because 3 feet of bridging is needed for each joist cavity. Four feet is needed for 19.2-inch OC, and 5 feet is needed for 24-inch OC. Then multiply times the number of rows of bridging needed. For example, if two rows of bridging are needed for 30 full-length 16-inch OC joists, then 30 × 3 feet per joist space × 2 rows = 180 linear feet of bridging must be purchased.

## SUBFLOOR

Subfloor sheets are determined by multiplying the number of sheets in a row of subfloor by the number of rows across the building. For example, if the building is 30 × 50 feet and standard 4' × 8' panels are used, the number of sheets in a row is

**Estimate the materials for a floor system of a rectangular 30′ × 56′ building. A two-ply LVL full length girder, 16″ OC joists lapped at the girder, 12′ sill and band joists, 5 parallel partitions and one row of bridging each side of girder.**

| Item | Formula | Waste factor | Example |
|---|---|---|---|
| **Girder Built-up** | girder LEN × NUM of plys ÷ LEN of each ply = NUM of boards | | 30′ × 2 ÷ 30′ = 2 LVL boards |
| **Sill** | PERM ÷ LEN of each piece = NUM of boards | Add steel girder sill | 172′ ÷ 12 = 14.3 = 15 boards |
| **Anchor Bolts** | NUM Sills × 3 + one per corner = NUM of bolts | | 15 × 3 + 4 = 49 bolts |
| **Joists-Band joist** | building LEN × 2 sides ÷ length of each board = NUM of boards | | 56′ × 2 ÷ 12 = 9.3 = 10 boards |
| **Joists-OC (On Center)** | (building LEN ÷ OC spacing in ft + 1) × 2 if lapped + parallel partition supports = NUM joists | | (56′ ÷ 16⁄12 + 1) × 2 + 5 = 91 joists |
| **Bridging-Solid** | building LEN × NUM of rows ÷ LEN joists = NUM of boards | | 56′ × 2 ÷ 16 = 7 boards |
| **Bridging-Wood Cross** | building LEN ÷ OC spacing in ft × ft per OC spacing) × NUM rows = lineal ft bridging | | 56′ ÷ 16⁄12 × 3 × 2 = 252 lineal ft bridging |
| **Bridging-Metal** | building LEN ÷ OC spacing in ft × 2 × NUM rows = NUM of pieces | | 56′ ÷ 16⁄12 × 2 × 2 = 168 NUM of pieces |
| **Subfloor-sheets** | building LEN ÷ 8 (ft per sheet) = round up to nearest ½ sheet = NUM pieces per row<br>Building WID ÷ 4 (ft per sheet) = round up to nearest ½ sheet = NUM of rows<br>NUM of pieces per row × NUM of rows = NUM of pieces | | 56′ ÷ 8 = 7 pieces per row<br>30′ ÷ 4 = 7.5 rows<br>7 × 7.5 = 52.5 = 53 pieces |

**FIGURE 9-36** Example of estimating floor framing materials with formulas.

the building length divided by 8 (feet per sheet) = 50 ÷ 8 = 6.25 sheets per row. Rounding this up to the nearest one-half sheet makes it 6.5 sheets. The number of rows across the building is building width divided by 4 (feet per sheet) = 30 ÷ 4 = 7.5 rows. Leave this at the nearest one-half sheet. The number of pieces then is 6.5 × 7.5 = 48.75 or 49 sheets. See Figure 9-36 for a listing of the shorthand versions of these estimating techniques with another example.

# SUMMARY

- Types of framing vary to suit the design requirements.
- Layout and construction of the floor frame must follow a procedure to ensure proper strength and durability.
- Termites must be eliminated from a wood frame to maintain a long life.

## GREEN CHECKLIST

- ☐ Building dimensions in four-foot modules reduces material waste since standard building panels are 4 × 8'.
- ☐ Insects will not consume or destroy dry wood. Buildings last longer if they are kept dry.

PROCEDURE 9-1

# Installing Girders

*Steel girders often come in one piece and may be set in place with heavy equipment. Wood girders are usually built up and erected in sections. Start by building one section.*

**A** Set one end of the girder in the pocket in the foundation wall.

- Place and fasten the other end on a braced temporary support. Continue building and erecting sections until the girder is completed to the opposite pocket.
- A solid wood girder is installed in the same manner as a built-up girder. Half-lap joints are made directly over posts or columns.
- Sight the girder by eye from one end to the other and place wedges under the temporary supports to straighten the girder.
- Permanent posts or columns are usually installed after the girder has some weight imposed on it by the floor joists.

**B** Temporary posts should be strong enough to support the weight imposed on them until permanent ones are installed.

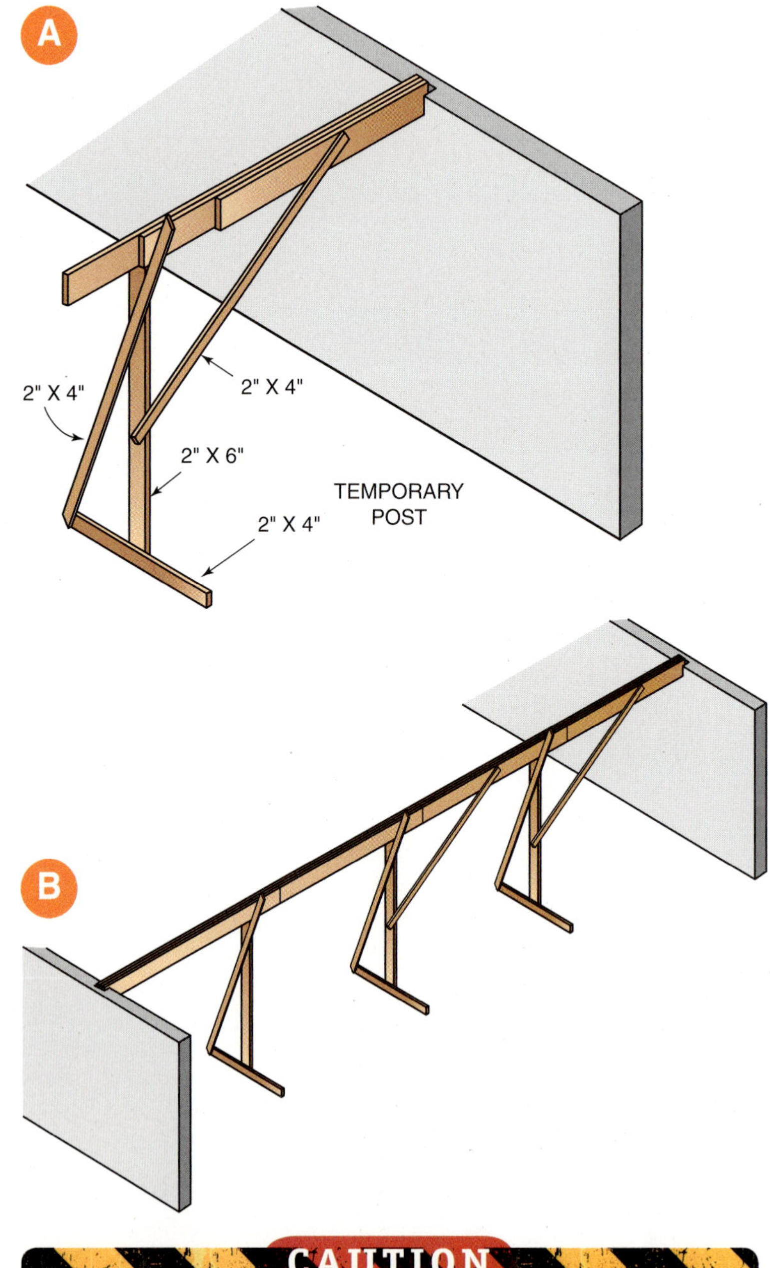


CAUTION

**CAUTION:** Girders are heavy. Care must be taken while lifting them. Also, make sure that any temporary posts and bracing used to support the girders are secure enough to keep them from falling.

PROCEDURE 9-2

# Installing Sills

*Sills must be installed so they are straight, level, and to the specified dimension of the building. The level of all other framing members depends on the care taken with the installation of the sill. The outside edge of the sill may be flush with, set back from, or overhanging the outside of the foundation wall.*

CAUTION

**CAUTION:** Follow these safety rules when handling pressure-treated lumber.

- Wear eye protection and a dust mask when sawing or machining treated wood.
- When the work is completed, wash areas of skin contact thoroughly before eating or drinking.
- Clothing that accumulates sawdust should be laundered separately from other clothing and before reuse.
- Dispose of treated wood by ordinary trash collection or burial.
- Do not burn treated wood. The chemicals retained in the ash could pose a health hazard.

*Courtesy of the Southern Pine Marketing Council.*

PROCEDURE 9-2

# Installing Sills (continued)

- First remove washers and nuts from the anchor bolts.
- Snap a chalk line on the top of the foundation wall in line with the inside edge of the sill.
- Cut the sill sections to length.

**A** Hold the sill in place against the anchor bolts to avoid covering the chalk line. Square lines across the sill on each side of the bolts. Measure the distance from the center of each bolt to the chalk line. Transfer this distance at each bolt location to the sill by measuring from the inside edge.

- Bore holes in the sill for each anchor bolt. Bore the holes at least 1⁄8-inch oversize to allow for adjustments.
- Place the sill sections in position over the anchor bolts after installing the sill sealer. The inside edges of the sill sections should be on the chalk line.
- Replace the nuts and washers and level the sill by shimming where necessary. Tighten the nuts snugly, being careful not to overtighten the nuts, especially if the concrete wall is still green (not thoroughly dry and hard). This may crack the wall. The general rule is finger tight, with one-half turn more.

**B** If the inside edge of the sill plate comes inside the girder pocket, notch the sill plate around the end of the girder. Raise the ends of the wood girder so it is flush with the top of the sill plate. Lower a steel girder for extra plate.

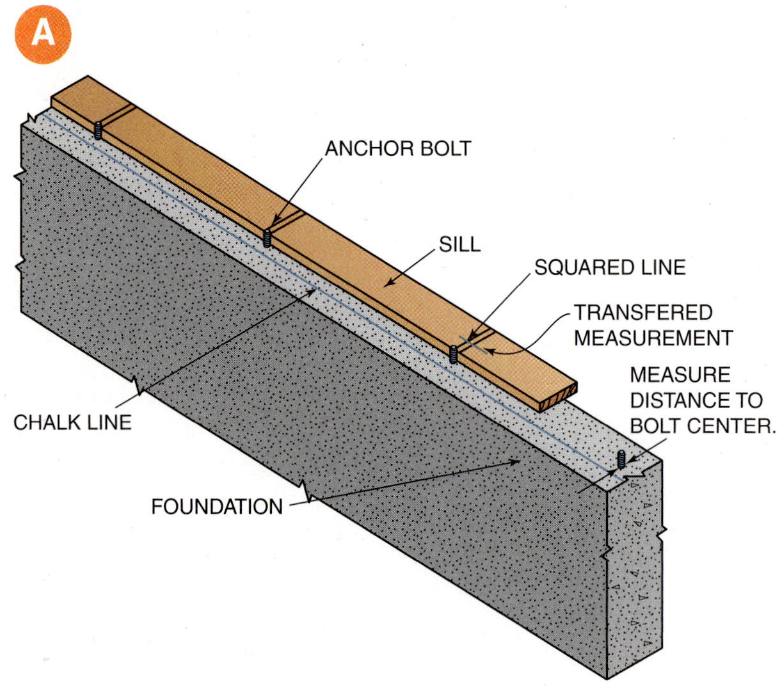


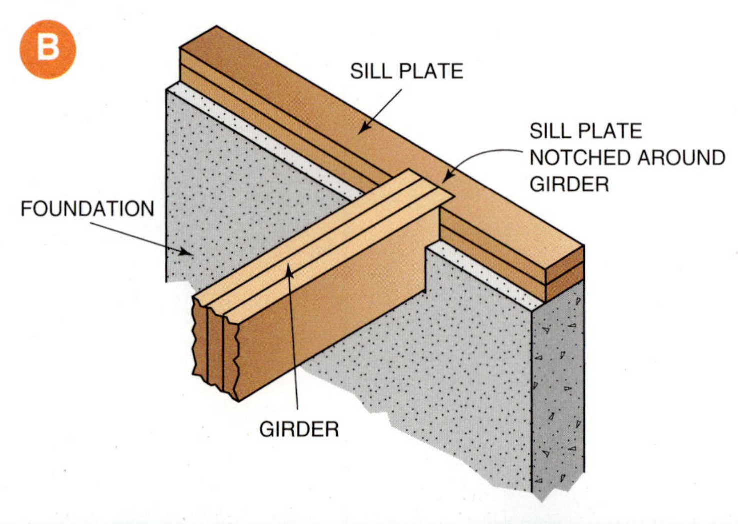

PROCEDURE 9-3

# Laying Out Floor Joists and Floor Openings

*The locations of floor joists are often marked on the sill plate. After studying the plans, locate the openings and special framing before laying out the OC joists. This reduces confusion by minimizing extra marks.*

**A** A squared line marks the side of the joist. An *X* to one side of the line indicates on which side of the line the joist is to be placed.

- Mark the sill plate, where joists are to be doubled, on each side of large floor openings.
- Lay out for partition supports, or wherever doubled floor joists are required.

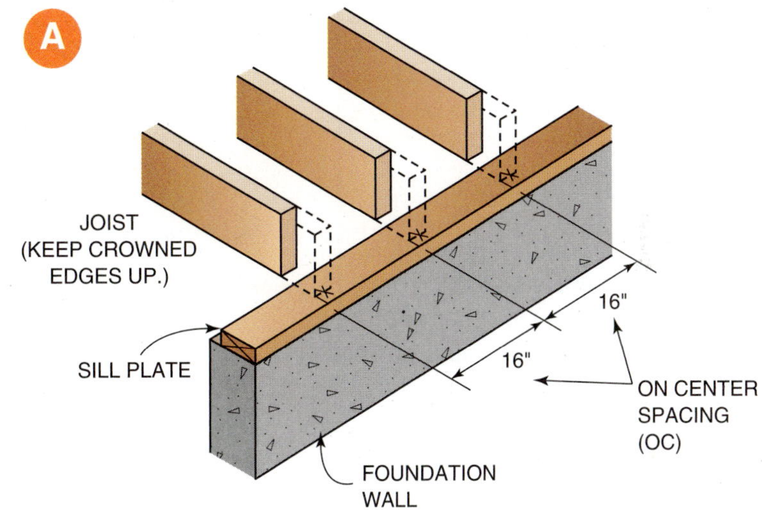


**B** Floor joists must be laid out so that the ends of *plywood subfloor* sheets fall directly on the center of floor joists. Start the joist layout by measuring the joist spacing from the end of the sill.

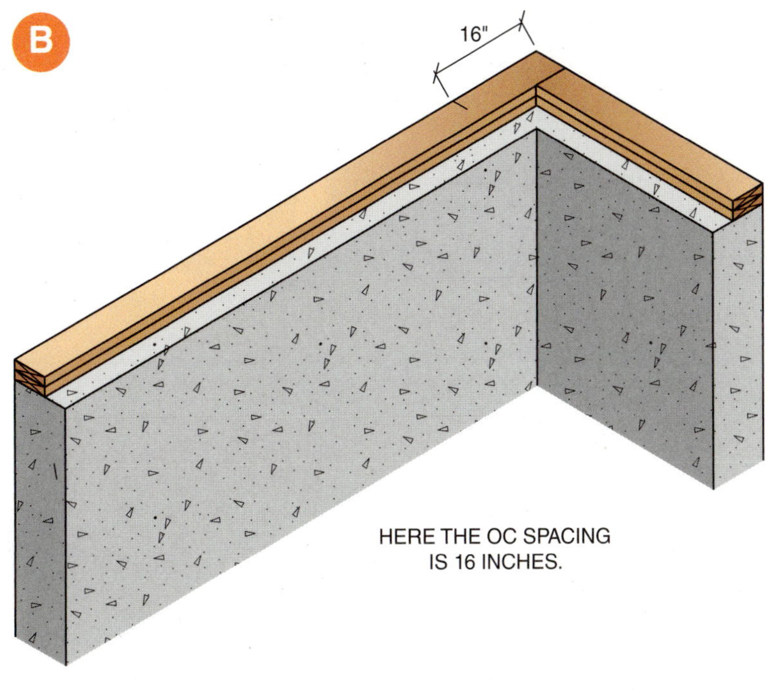

PROCEDURE 9-3

# Laying Out Floor Joists and Floor Openings (continued)

**C** Measure back one-half the thickness of the joist. Square a line across the sill. This line indicates the side of the joist closest to the corner. Place an *X* on the side of the line on which the joist is to be placed.

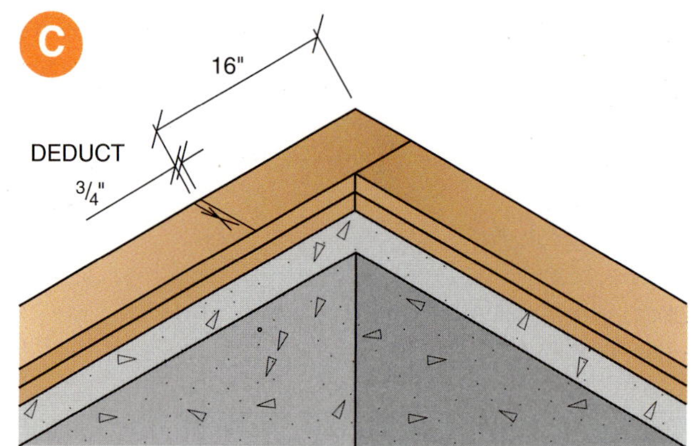


JOIST THICKNESS IS USUALLY 1 1/2 INCHES

**D** From the squared line, measure and mark the spacing of the joists along the length of the building. Place an *X* on the same side of each line as for the first joist location.

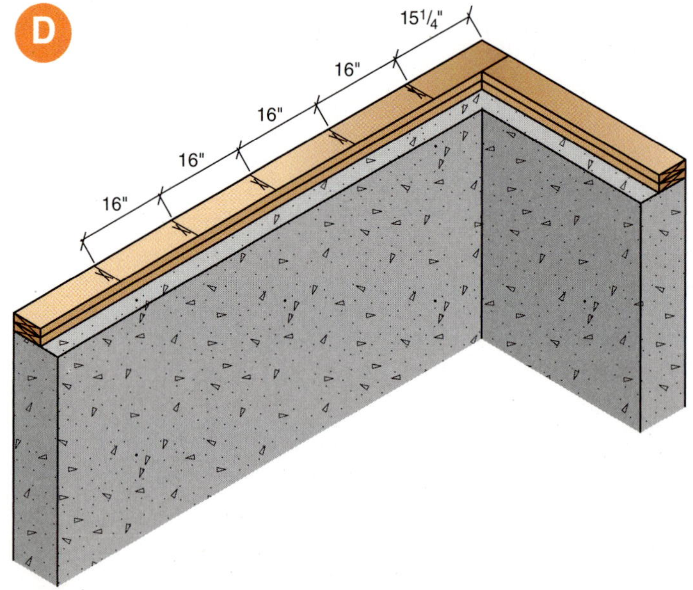


**E** When measuring for the spacing of the joists, use a tape stretched along the length of the building.

- Most tapes have prominent markings for 16- and 19.2-inch spacing. Using a tape in this manner is more accurate. Measuring and marking each space by stepping off generally causes a gain in the spacing. If the spacing is not laid out accurately, the plywood subfloor may not fall in the center of some floor joists. Time will then be lost either cutting the plywood back or adding nailing strips of lumber to the floor joists.

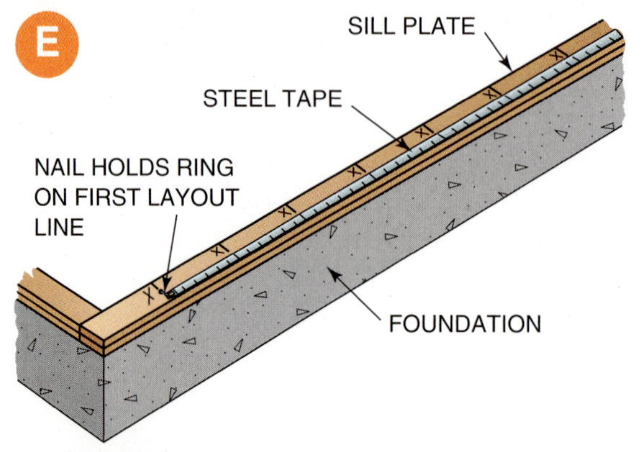

**F** Identify the layout marks that are not for full-length floor joists. Shortened floor joists at the ends of floor openings are called **tail joists.** They are usually indicated by changing the *X* to a *T*.

- Check the mechanical drawings to make adjustments in the framing to allow for the installation of mechanical equipment.

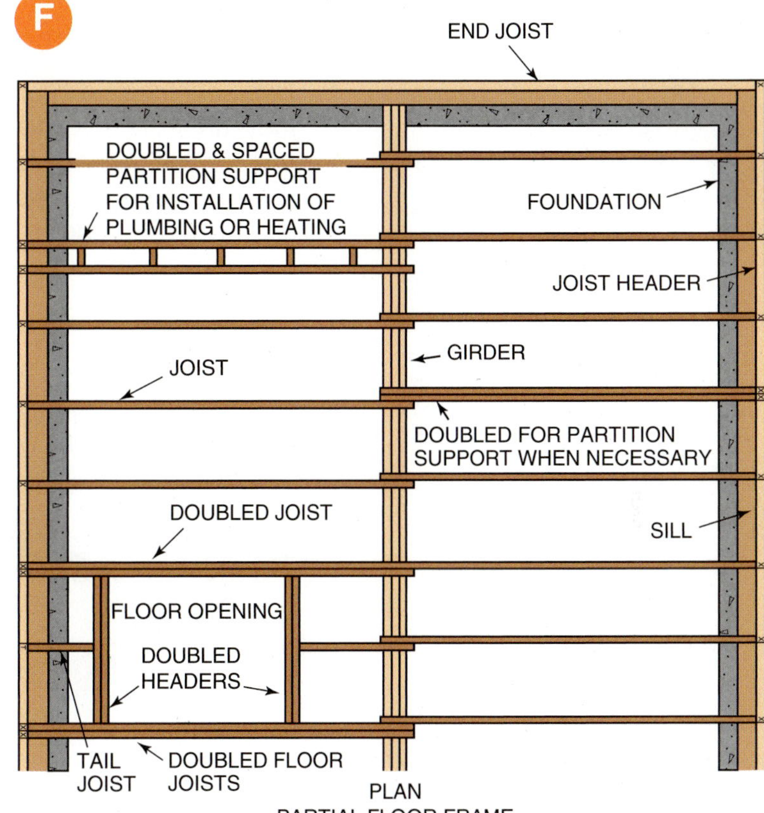


**G** Lay out the floor joists on the girder and on the opposite wall in a similar manner. If the joists are in-line, *X*s are made on the same side of the layout marks on both the girder and the sill plate on the opposite wall. If the joists are lapped, an *X* is placed on both sides of the layout marks at the girder and on the opposite side of the layout marks on the other wall.

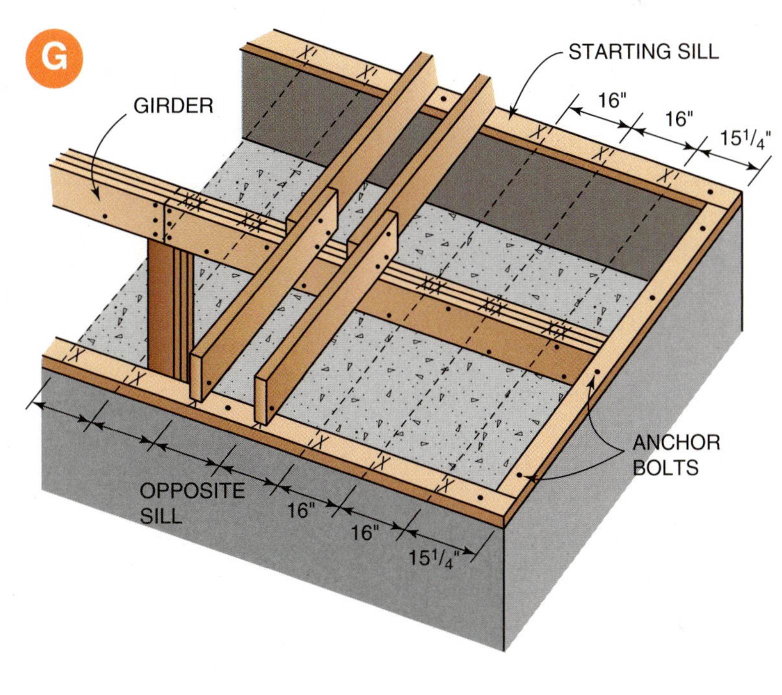

PROCEDURE 9-4

# Framing Floor Openings

*Large openings in floors should be framed before floor joists are installed. This is because room is needed for end nailing. The order of assembly is important to ensure that all members are face nailed for maximum speed of installation and fastener strength.*

**A** Fasten the inside **trimmer joists** in place. Trimmer joists are full-length joists that create the inside of the opening.

- Mark the location of the **header** on the trimmers. Headers are members of the opening that run at right angles to the floor joists.
- Cut four headers to length by taking the measurement at the sill between the trimmers. Take this measurement at the sill or girder where the trimmers are fastened rather than at the opening. Because the lumber is not often straight, the measurement between trimmers taken at the opening may not be accurate.

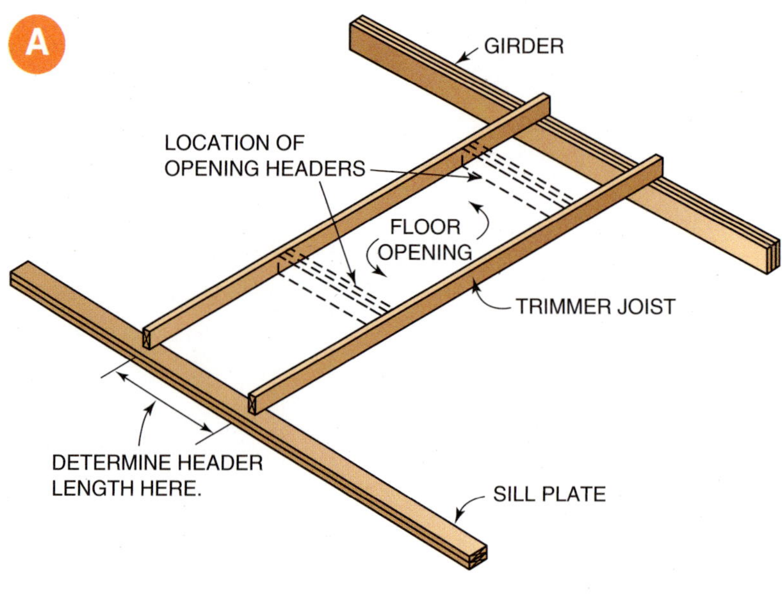


**B** Place two headers, one for each end of the opening, on the sill between the trimmers. Transfer the layout of the tail joists on the sill to the headers. Fasten the first header on each end of the opening in position by driving nails through the side of the trimmer into the ends of the headers. Be sure this first header is the header that is farthest from the center of the floor opening.

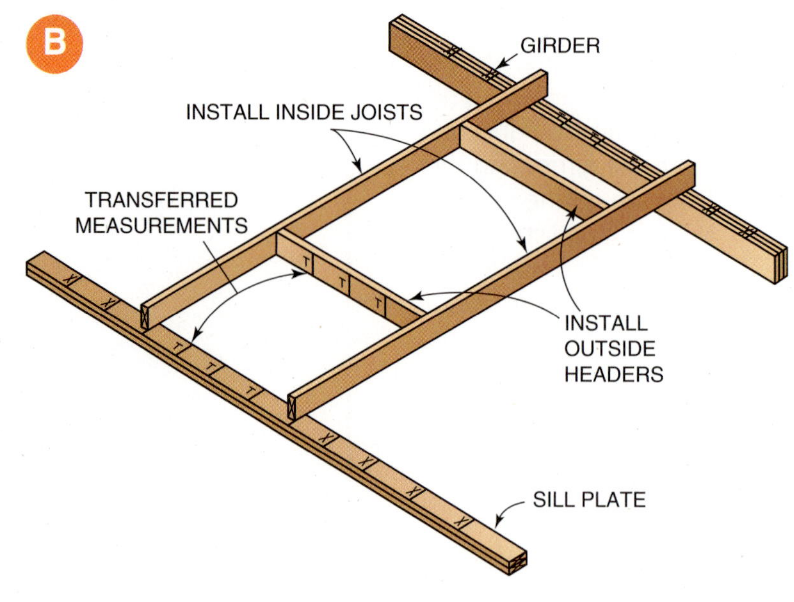

**C** Fasten the tail joists in position by face nailing through the header and toenailing into girder or sill. Remember, the tail joists are shortened OC joists.

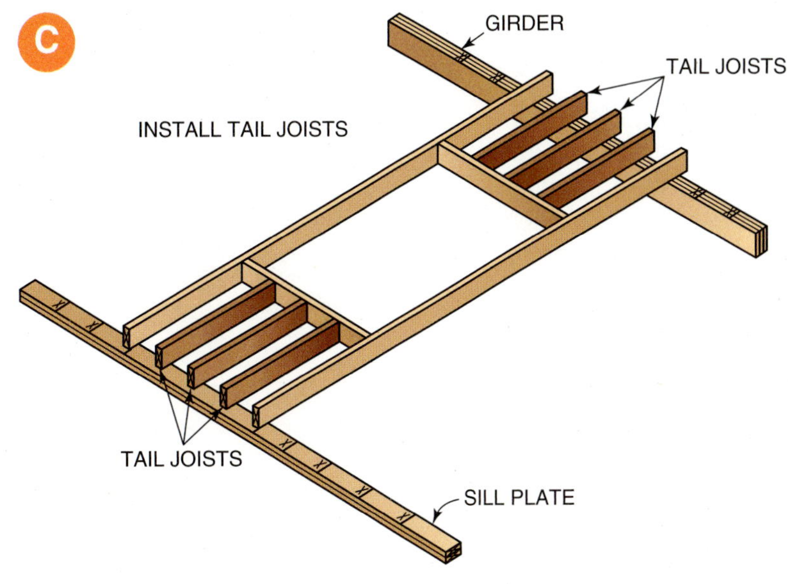


**D** Install inside header, face nailing into outside header and inside trimmer.

- Finally, double up the trimmer joists, and then add full-length joists to complete the frame.

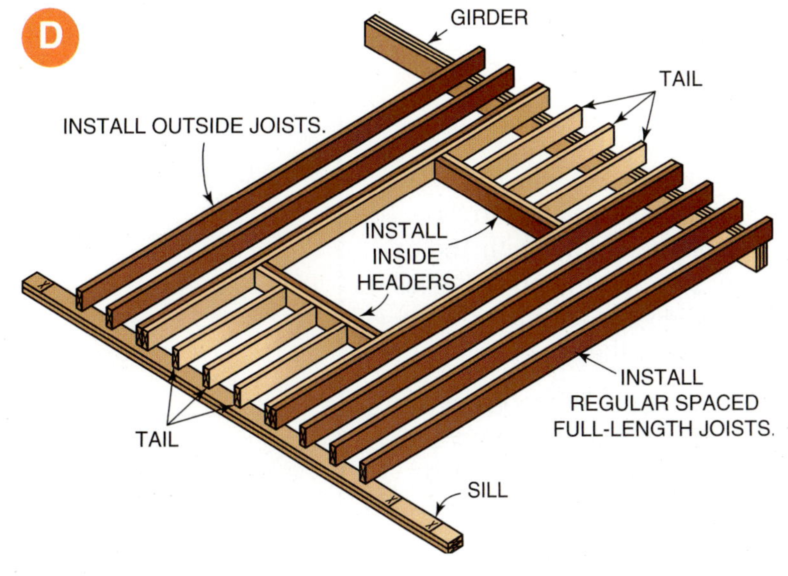

PROCEDURE 9-5

# Installing Floor Joists

**A** Stack the necessary number of full-length floor joists at intervals along both walls. Each joist is sighted along its length by eye. Any joist with a severe crook or other warp should not be used. Joists are installed with the crowned edge up.

**A**

**B** Keep the end of the floor joist back from the outside edge of the sill plate by the thickness of the band joist.

**B**

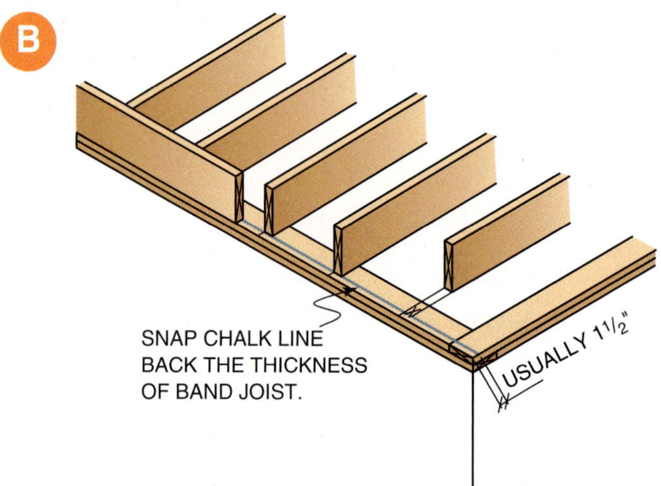


**C** Toenail the joists to the sill and girder with 10d or 3-inch common nails. Nail the joists together if they lap at the girder.

- When all floor joists are in position, they are sighted by eye from end to end and straightened. They may be held straight by strips of 1 × 3s tacked to the top of the joists at approximately in the middle of the joist span.

**C**

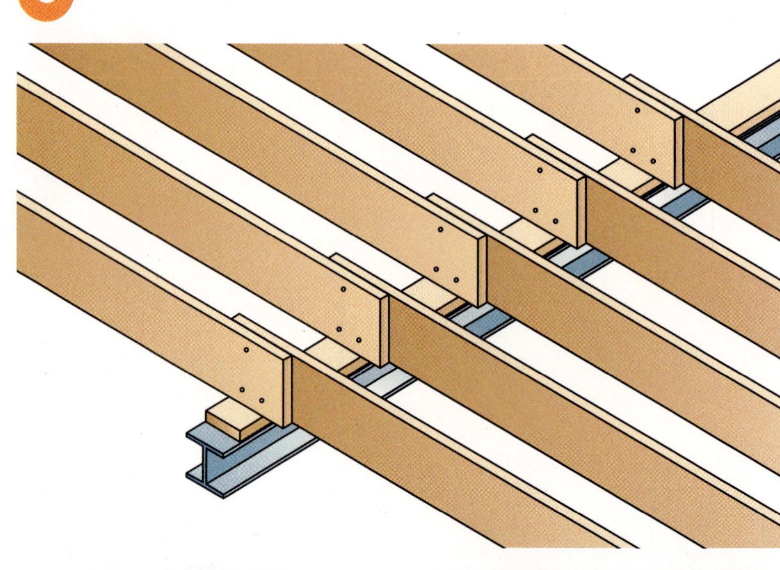

D Engineered wood I-beams are installed using standard tools. They can be easily cut to any required length at the job site. A minimum bearing of 1½ inches is required at joist ends and 3½ inches over the girder. The wide, straight wood flanges on the joist make nailing easier, especially with pneumatic framing nailers.

- Nail joists at each bearing with a minimum of three 8d common nails. Keep nails at least 1½ inches from the joist ends to avoid splitting.

*Courtesy of Boise Cascade.*

**CAUTION**

**CAUTION:** Engineered wood I-beams are more flexible than standard lumber until they are permanently installed and braced. Install strapping on top to stiffen the joists before installing subfloor. Walk only on the strapping or bridging.

PROCEDURE 9-6

# Installing the Band Joist

*After all the openings have been framed and all floor joists are fastened, install the band joist. This closes in the ends of the floor joists. Band joists are usually made of the same size lumber as the floor joists.*

**A** Fasten the band joist into the end of each floor joist with three 16d common nails. If wood I-beams are used as floor joists, drive one nail into the top and bottom flanges. Use 8d common nails to toenail the band joist to the sill plate at 6-inch intervals.

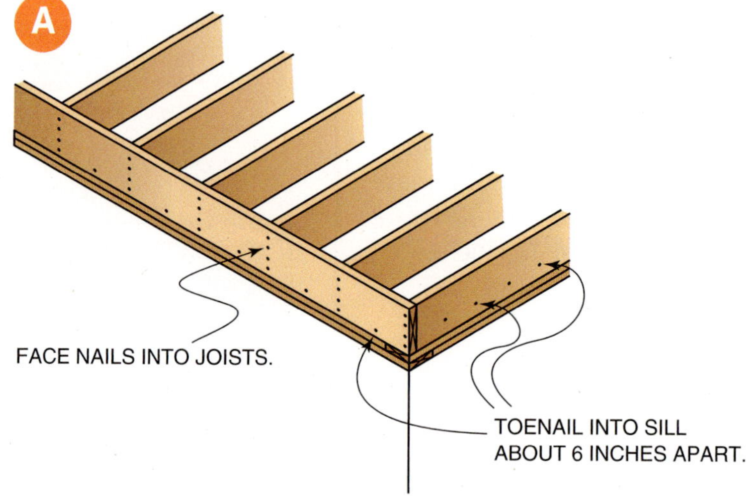

B *Wood Cross-Bridging.* To install wood cross-bridging, start two 6d nails in one end of the bridging before putting it into place.

- Then place it flush with the top of the joist on one side of the line and drive the nails home. Nail only the top end. The bottom ends are not nailed until the subfloor is fastened down.
- Within the same joist cavity or space, fasten another piece of bridging to the other joist. Leave a small space between the bridging pieces where they form the *X* to minimize floor squeaks.
- Continue installing bridging in the other spaces, but alternate so that the top ends of the bridging pieces are opposite each other where they are fastened to the same joist.
- *Metal Cross-Bridging.* Metal cross-bridging is fastened in a manner similar to that used for wood cross-bridging. The method of fastening may differ according to the style of the bridging. The bridging is usually fastened to the top of the joists through predrilled holes in the bridging. Because the metal is thin, nailing to the top of the joists does not interfere with the subfloor.
- Some types of metal cross-bridging have steel prongs that are driven into the side of the floor joists.

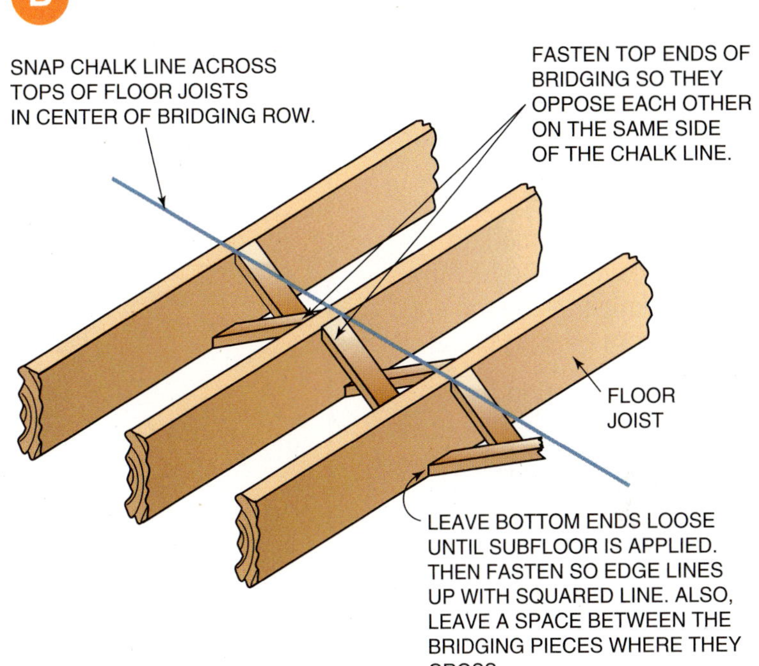

PROCEDURE 9-9

# Installing Columns

**A** Check the girder for level, and adjust as needed. Measure accurately from the column footing to the bottom of the girder. Transfer this mark to the column. Deduct the thickness of the top and bottom bearing plates.

- To mark around the column so it has a square end, wrap a sheet of paper around it. Keeping the edges even, mark along the edge of the paper.
- Install the column's plumb breakline under the girder and centered on the footing. Fasten the top bearing plate to the girder with lag screws.
- If the girder is steel, the plates are then bolted or welded to the girder. The bottoms of the columns are held in place when the finish concrete basement floor is placed around them. If there is to be no floor, then the bottom plate must be anchored to the footing.

**B** Wood posts are installed in a similar manner, except their bottoms are placed on a pedestal footing.

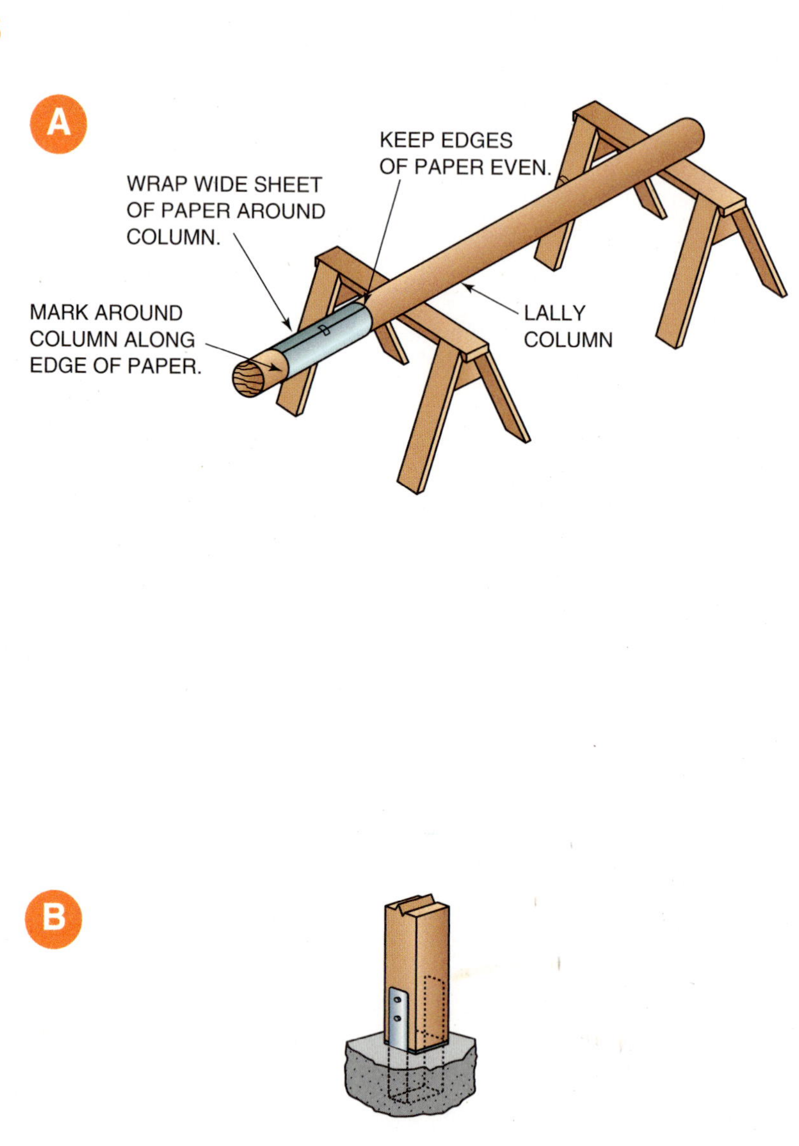


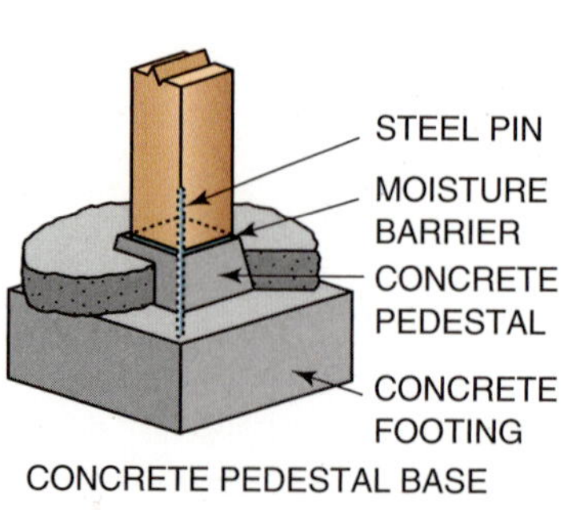


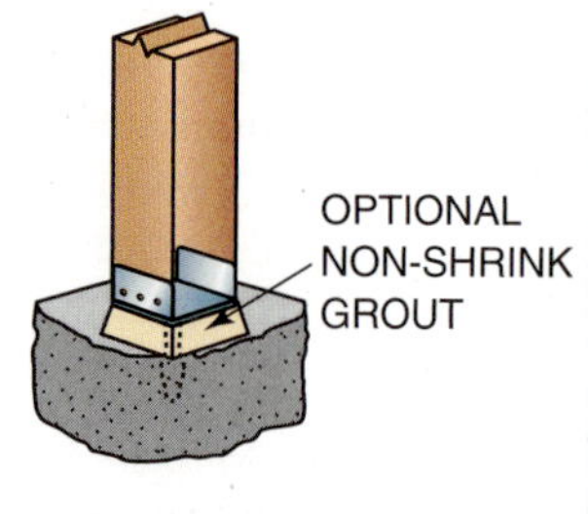


WOOD POST BASE DETAILS

PROCEDURE 9-10

# Applying Plywood Subflooring

**A** Starting at the corner from which the floor joists were laid out, measure in 4 feet from the rim joist. Tongue-and-groove plywood subfloor is only 47½ inches wide.

- Snap a line across the tops of the floor joists from one end to the other. Start with a full panel to the chalk line. Position and nail the corners of the panel.
- Align the OC joists to their correct spacing before nailing. This will make sure the joists are straight.

**B** Start the second row with a halfsheet to stagger the end joints. Continue with full panels to finish the row. Leave a ⅛-inch space between panel edges. Leave a 1⁄16-inch space at all panel end joints to allow for expansion. All end joints are made over joists.

- Minimum nail spacing is 6 inches apart along supported edges and 12 inches apart at intermediate supports.
- Continue laying and fastening plywood sheets in this manner until the entire floor is covered.

**A**

*Courtesy of Louisiana Pacific Corporation.*

**B**

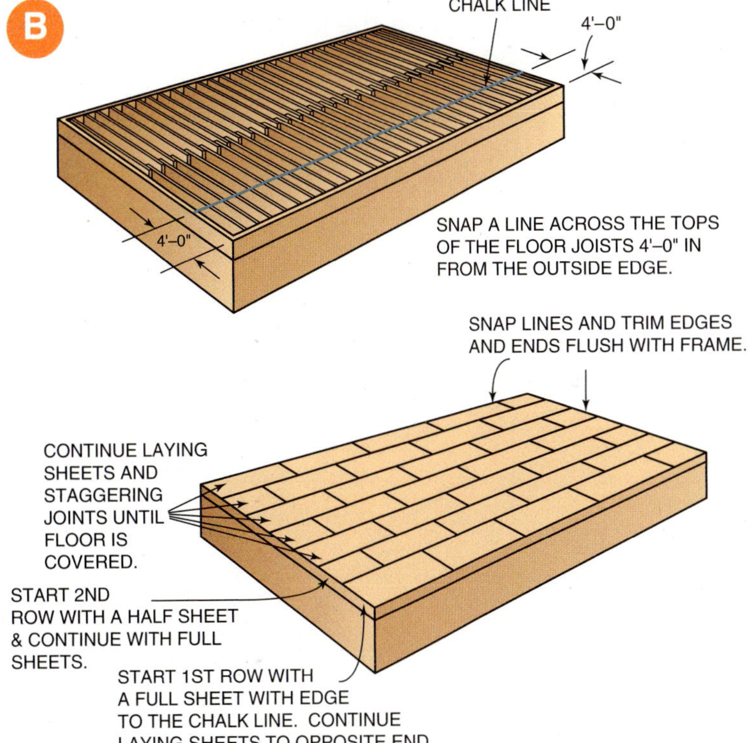

# REVIEW QUESTIONS

*Select the most appropriate answer.*

**1. A platform frame is easy to erect because**

a. only one-story buildings are constructed with this type of frame.
b. each platform may be constructed on the ground.
c. at each level a flat surface is provided on which to work.
d. fewer framing members are required.

**2. A heavy member that supports the inner ends of floor joists is called a**

a. pier.
b. girder.
c. stud.
d. sill.

**3. The member of a floor frame that is fastened directly on the foundation wall is called a**

a. pier.
b. girder.
c. stud.
d. sill.

**4. A quick and accurate method of marking a square end on a round column is to**

a. use a square.
b. measure down from the existing end several times.
c. use a pair of dividers.
d. wrap a piece of paper around it.

**5. Butt seams of members of a built-up wood girder should be between**

a. $\frac{1}{6}$ and $\frac{1}{5}$ of post spacing.
b. $\frac{1}{6}$ and $\frac{1}{4}$ of post spacing.
c. $\frac{1}{5}$ and $\frac{1}{3}$ of post spacing.
d. $\frac{1}{4}$ and $\frac{1}{3}$ of post spacing.

**6. To protect against termites, keep wood girders in crawl spaces and other concealed areas above the ground at least**

a. 8 inches.
b. 12 inches.
c. 18 inches.
d. 24 inches.

**7. When the end of floor joists rest on a supporting member, they should have a bearing of at least**

a. 4 inches.
b. $3\frac{1}{2}$ inches.
c. $2\frac{1}{2}$ inches.
d. $1\frac{1}{2}$ inches.

**8. If floor joists lap over a girder, they should have a minimum lap of**

a. 2 inches.
b. 3 inches.
c. 6 inches.
d. 8 inches.

**9. It is important when installing floor joists to**

a. toenail them to the sill with at least two 8d nails.

b. have the crowned edges up.

c. face nail them to a band joist with at least three 8d nails.

d. all of the above.

**10. For the best order of installation of the members of a floor opening, the next member installed after the inside trimmer is the**

a. outside trimmer.

b. tail joists.

c. inside header.

d. outside header.

**11. Holes bored in a floor joist should be**

a. no closer than 2 inches to the edge of the joist.

b. no larger than one-third of the joist width.

c. no larger than necessary.

d. all of the above.

**12. The bearing points of a girder should be at least**

a. equal to girder depth (cross-sectional height).

b. 4 inches long.

c. 5 inches long.

d. 6 inches long.

**13. The minimum nail spacing for engineered panel subfloor on 16 OC floor joists is**

a. 6 inches on the edge and 6 inches on intermediate supports.

b. 6 inches on the edge and 8 inches on intermediate supports.

c. 6 inches on the edge and 12 inches on intermediate supports.

d. 8 inches on the edge and 8 inches on intermediate supports.

**14. To lay out the OC joists, the first one is set back a distance equal to**

a. one-half the joist thickness.

b. the width of a joist.

c. the thickness of a joist.

d. always ¾ inch.

**15. Pressure treatment is done on lumber to improve its**

a. decay resistance.

b. pressure resistance.

c. nail-holding strength.

d. all of the above.

## KNOW YOUR CODES

The size and type of joists and girders are crucial to the long-term life of the building. Joists and girders are carefully chosen to be both strong and stiff. Always follow the specifications given in the building prints.

## WHAT'S WRONG WITH THIS PICTURE?

Carefully study **Figure 9-37** and think about what is wrong. Consider all possibilities.

### WRONG

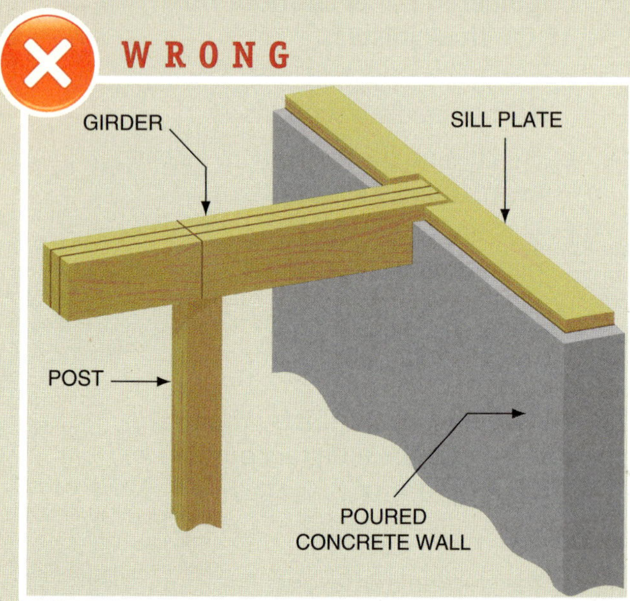


**FIGURE 9-37** The diagram shows a built-up girder. Look closely at the girder, and notice that all of the splices are located over the post. While this may intuitively make sense, in actuality the best construction practices indicate that the splices should be staggered between ⅙ and ¼ of the support post spacing (or post-to-wall spacing), with the center girder piece having a splice directly over the center of the post.

### RIGHT

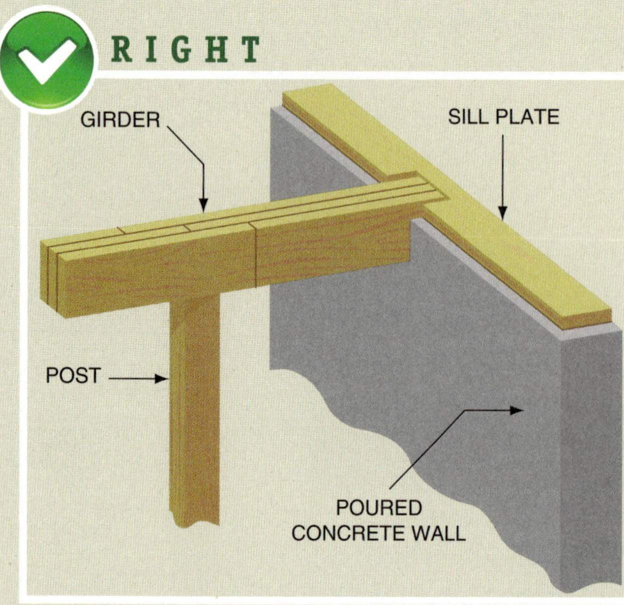


**FIGURE 9-38** This diagram correctly shows the girder as having staggered splices.

CHAPTER 10

# Wall and Ceiling Framing

**OBJECTIVES** *After completing this unit, the student should be able to:*

- Identify and describe the function of each part of the wall frame.
- Determine the length of exterior wall studs.
- Describe four different types of walls used in residential framing.
- Determine the rough opening width and height for windows and doors.
- Lay out the wall plates for partition intersections, openings, and OC studs.
- Describe several methods of framing corner and partition intersections.
- Assemble and construct a wall section.
- Erect and temporarily brace a wall section plumb and straight.
- Describe the function of and install blocking and backing.
- Apply wall sheathing.
- Lay out, cut, and install ceiling joists.
- Identify and describe the components of nonstructural steel wall framing.
- Install a steel door buck.
- Estimate the materials needed for walls and ceiling framing.

# GLOSSARY OF WALL AND CEILING FRAMING TERMS

**backing** strips or blocks of wood installed in walls or ceilings for the purpose of fastening or supporting trim or fixtures

**blocking** pieces of dimension lumber installed between joist and studs for the purposes of providing a nailing surface for intersecting framing members

**gable end** the triangular-shaped section on the end of a building formed by the common rafters and the top plate line

**gypsum board** a sheet product made by encasing gypsum in a heavy paper wrapping used to create the wall surface; also called drywall.

**joist hanger** metal stirrups used to support the ends of joists that do not rest on top of a support member

**load-bearing** term used to describe a structural member that carries weight from another part of the building

**soffit** the underside trim member of a cornice or any such overhanging assembly

With the floor frame complete, construction is moved up onto the deck. During the wall-framing phase of construction, the building begins to take shape. Walls and partitions are laid out to locate positions and openings that occur in them. Exterior walls are constructed to the correct height, braced plumb, and straightened. Window and door rough openings are framed to specified sizes. Interior rough framing is performed with the installation of partitions, backing, blocking, and ceiling joists.

The term rough work generally refers to framing but does not imply that the work is crude. This work will eventually be covered by other material. Careful construction of the rough frame makes application of the finish work less problematic.

This text has adopted a particular method and style of framing. The student should always be aware that variations in construction procedures can and do occur. There are varying ways to frame. The decision on which way to use depends largely on past practice and the techniques used in a particular area, but the finished product is the same. Assuming that all techniques are safe and follow the plans, no one method is thought to be more correct than another. They are just different.

# Parts of a Wall Frame

The wall frame consists of a number of different parts. Carpenters must know the name, function, location, and usual size of each member. Sometimes the names given to certain parts of a structure may differ according to the geographical area. For that reason, some members may be identified with more than one term. An exterior wall frame consists of plates, studs and cripple studs, headers and sills, trimmer or jack studs, corner posts, partition intersections, ribbons, and corner braces (Fig. 10-1).

## PLATES

The top and bottom horizontal members of a wall frame are called *plates*. The bottom member is called a *sole plate*. It is also referred to as the *bottom plate* or *shoe*. The top members are called *top plates*. In a balloon frame, the sole plate is not used. Instead, the studs rest directly on the sill plate.

## STUDS

*Studs* are vertical members of the wall frame. They run full length between plates. Jack or *trimmer* studs are shortened studs that line the sides of an opening. They extend from the bottom plate up to the top of the opening to support the header. On-center studs that must be cut to allow for an opening are called *cripple* studs. They are located above and below an opening and extend from the plates to the opening. Studs are usually 2 × 4s, but 2 × 6s are sometimes used in exterior walls to allow for thicker insulation. Studs are usually spaced 16 and sometimes 24 inches OC.

## HEADERS

*Headers* run at right angles to studs. They form the top of wall openings. Headers must be strong enough to support the load above the opening. The depth of the header depends on the width of the opening. As the width of the opening increases, so must the strength of the header. Check drawings, specifications, codes, or manufacturer's literature for header sizes. Header depth is often made to completely fill in the space above an opening (Fig. 10-2).

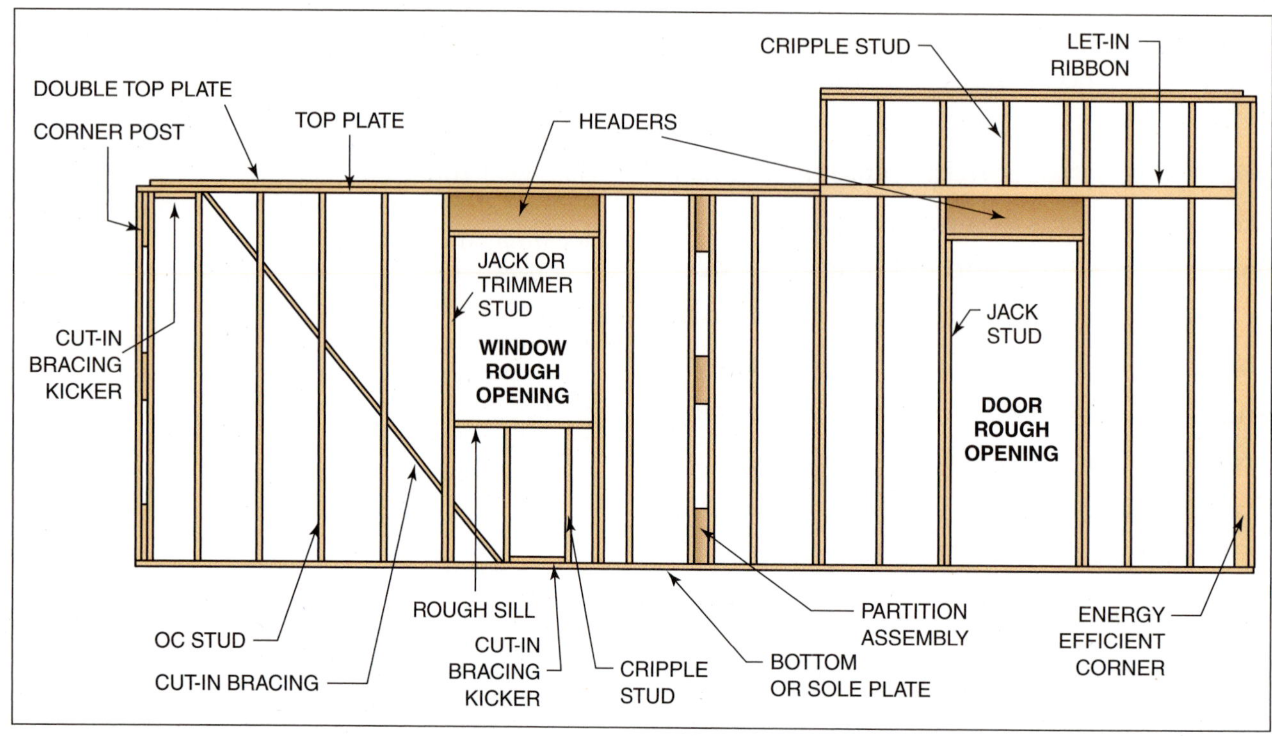


**FIGURE 10-1** Parts of an exterior wall frame.

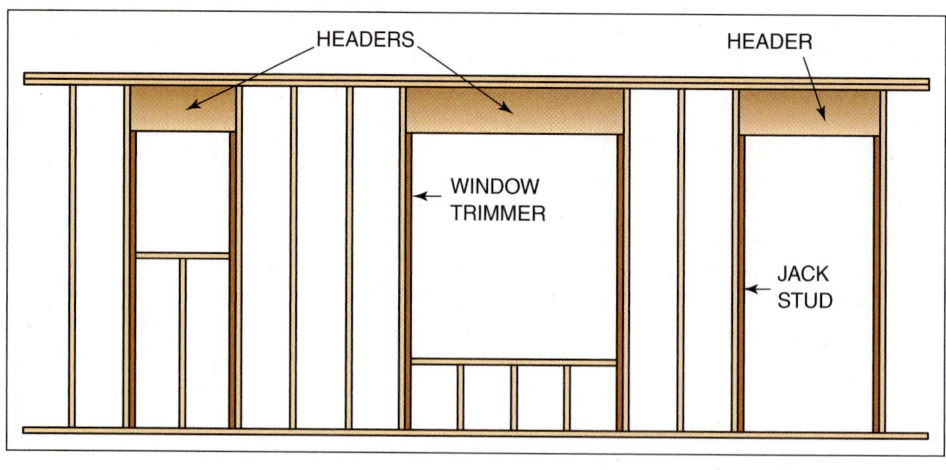


**FIGURE 10-2** It is common practice to use the same header depth for all the openings. Single and double ply headers are used as needed.

## Kinds of Headers

Headers are made in many different ways, depending on desired strength and available materials (Fig. 10-3). Figures 4-26 and 4-33 show the use of laminated veneer lumber, parallel strand lumber, and glulam beams as opening headers. The use of engineered lumber permits the spanning of wide openings, such as double garage door openings.

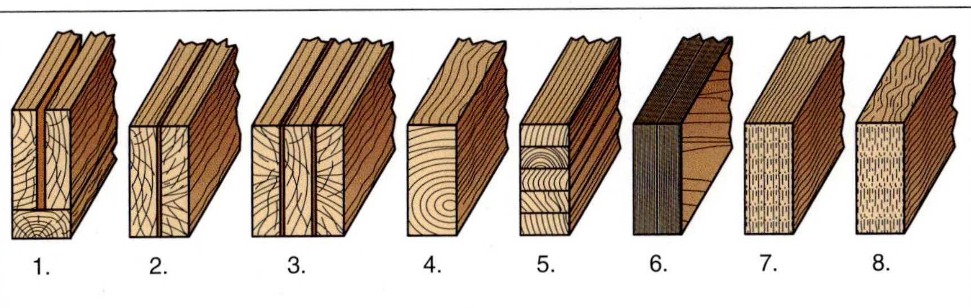


1. A BUILT-UP HEADER WITH A 2 X 4 OR 2 X 6 LAID FLAT ON THE BOTTOM.
2. A BUILT-UP HEADER WITH A 1/2" SPACER SANDWICHED IN BETWEEN.
3. A BUILT-UP HEADER FOR A 6" WALL.
4. A HEADER OF SOLID SAWN LUMBER.
5. GLULAM BEAMS ARE OFTEN USED FOR HEADERS.
6. A BUILT-UP HEADER OF LAMINATED VENEER LUMBER.
7. PARALLEL STRAND LUMBER MAKES EXCELLENT HEADERS.
8. LAMINATED STRAND LUMBER IS USED FOR LIGHT DUTY HEADERS.

**FIGURE 10-3** Types of solid and built-up headers.

## ROUGH SILLS

Forming the bottom of a window opening at right angles to the studs are members called *rough sills.* They secure the top ends of cripple studs and carry little load.

## TRIMMERS (JACKS)

*Trimmers* or *jacks* are shortened studs that support the headers. They are fastened to the full-length studs, often called *king studs,* on either side of the opening. In window openings, they may fit snugly between the header and rough sill. Some codes, however, require that the trimmers run full length (Fig. 10-4). In door openings, the trimmers, sometimes called *liners,* fit between the header and the sole plate (Fig. 10-5).

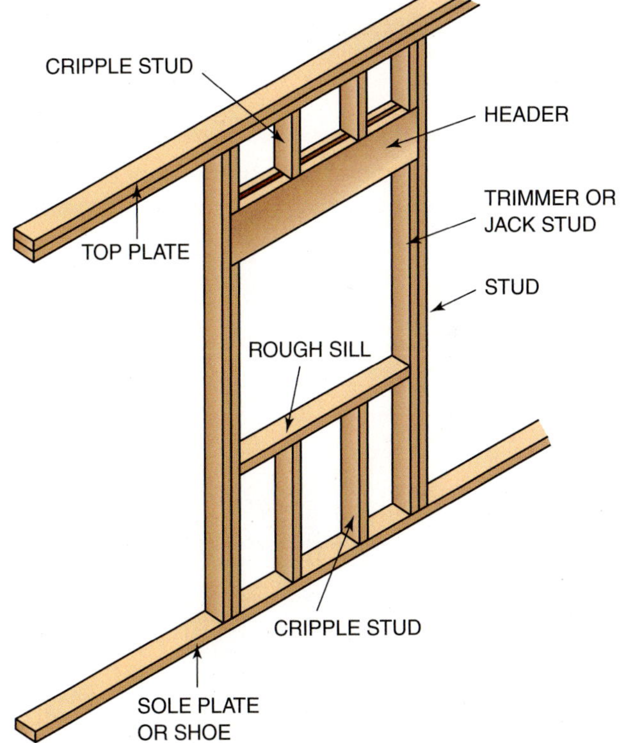


**FIGURE 10-4** Typical framing for a window opening.

### Green Tip

Properly size window and door headers; do not overbuild them. This practice saves wood and increases the amount of insulation in the wall.

## CORNER POSTS

Corner posts are the same length as studs. Corner posts are built in a number of ways. Historically, the corners were made solid. Today, corners are constructed to allow for more insulation to penetrate the corners (Fig. 10-6).

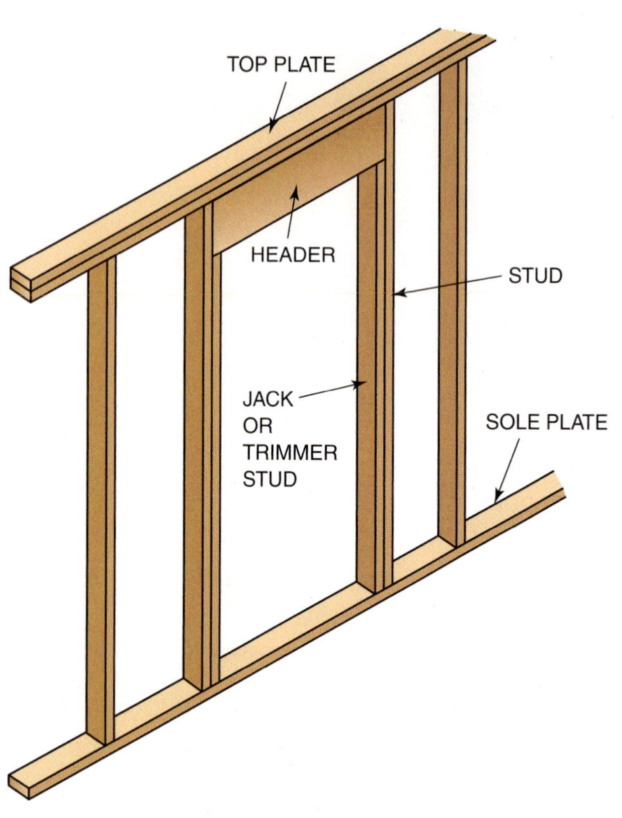


**FIGURE 10-5** Typical framing for a door opening.

## Green Tip

Reduce the number of cripple studs under a window to only those that are on-center. This practice reduces thermal bridging and increases the percentage of insulation in the wall.

## PARTITION INTERSECTIONS

Wherever interior partitions meet an exterior wall, extra framing is needed. These are sometimes called *partition studs* or *assemblies*. This provides for fastening of interior wall covering in the corners formed by the intersecting walls. In most cases, the partition stud is made of two studs nailed to the edge of 2 × 4 blocks, 12 to 18 inches long. One block is placed at the bottom, one at the top, and one about center on the studs.

Because exterior walls are usually insulated and the space behind the partition stud is difficult to insulate, other methods are used. One of these methods is to nail a continuous 2 × 6 or 8 backer to a full-length stud of the intersecting wall. The edges

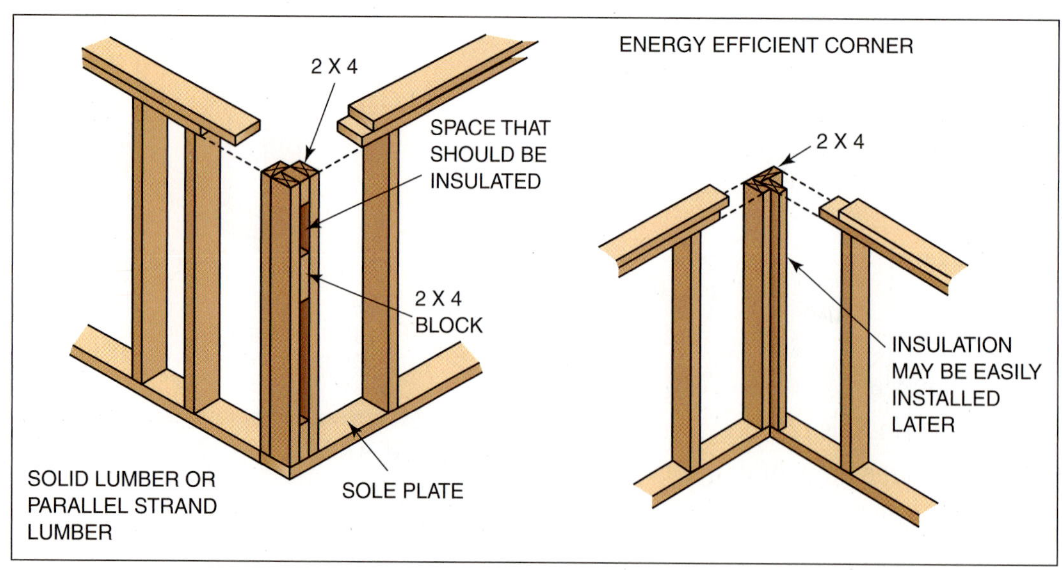


**FIGURE 10-6** Methods of making corner posts.

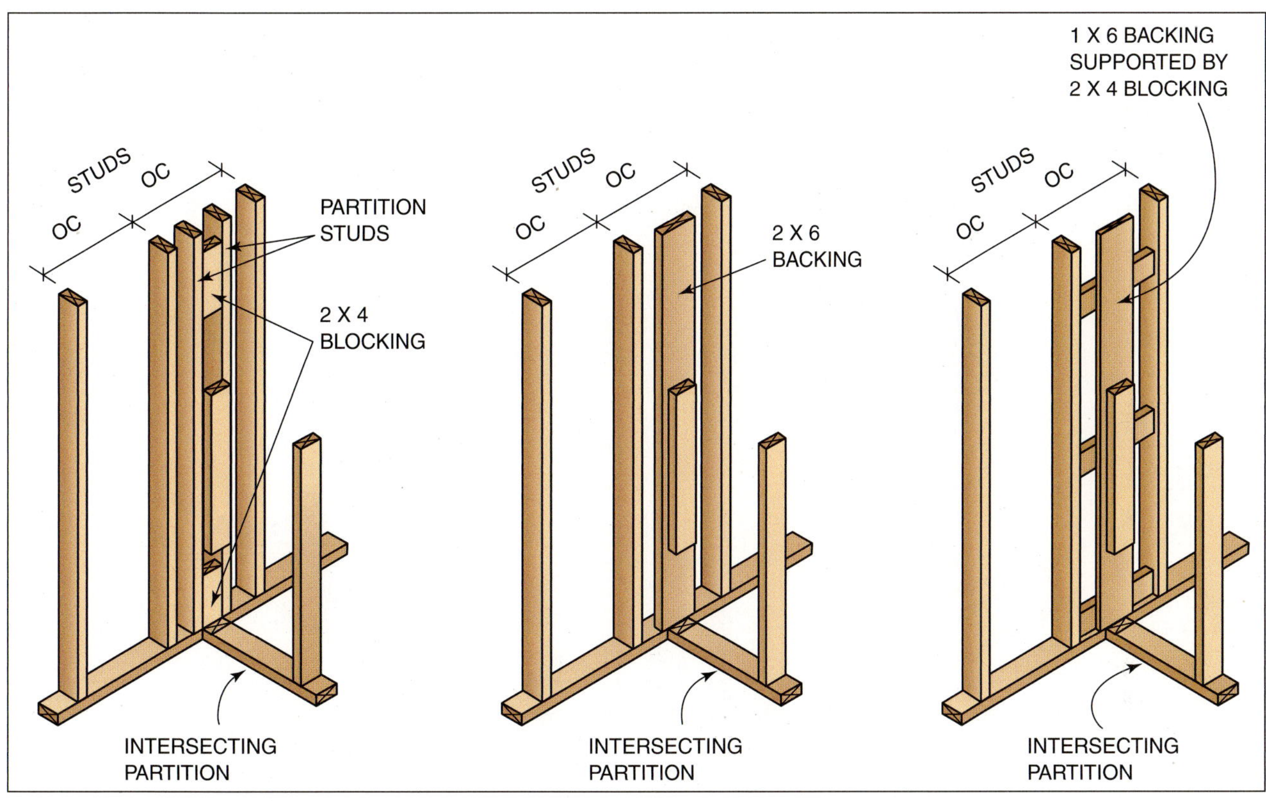


**FIGURE 10-7** Partition intersections are constructed in several ways.

of the backer project an equal distance beyond the edges of the stud. Another method sets blocking back from the inside edge of the stud the thickness of a 1 × 6 board. A 1 × 6 board is then fastened vertically on the inside of the wall so that it is centered on the partition (Fig. 10-7).

## RIBBONS

Ribbons are horizontal members of the exterior wall frame in balloon construction. They are used to support the second-floor joists. The inside edge of the wall studs is notched so that the ribbon lays flush with the edge (Fig. 10-8). Ribbons are usually made of 1 × 4 stock. Notches in the stud should be made carefully so the ribbon fits snugly in the notch. This prevents the floor joists from settling. If the notch is cut too deep, the stud will be unnecessarily weakened.

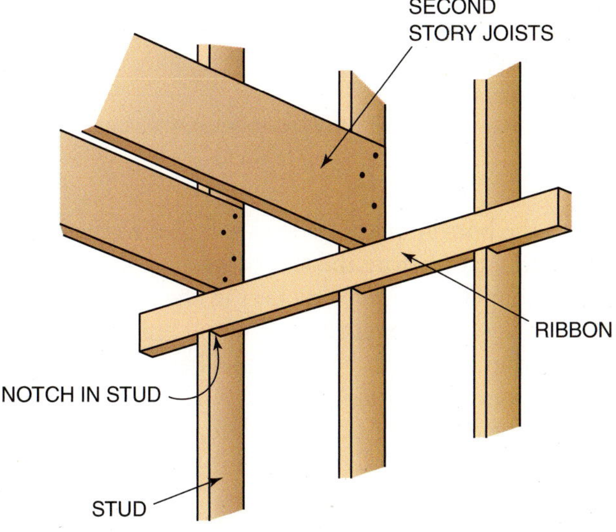


**FIGURE 10-8** Ribbons are used to support floor joists in a balloon frame.

## CORNER BRACES

Generally, no wall bracing is required if rated panel wall sheathing is used. In other cases, such as when insulating board sheathing is used, walls are braced with metal or wood wall bracing. Metal bracing comes in gauges of 22 to 16 in flat, T, or L shapes. They are about 1½ inches wide and run diagonally from the

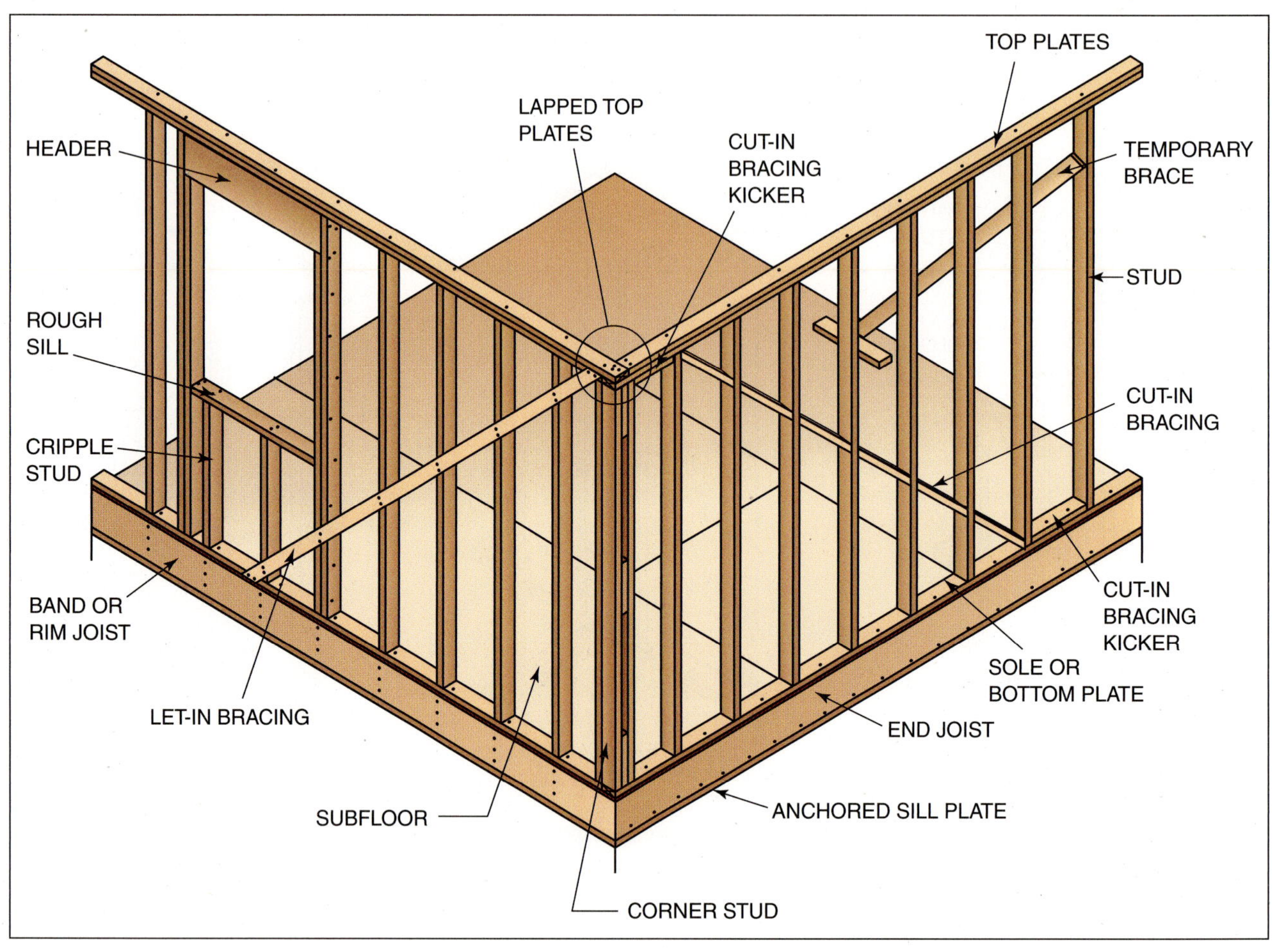


**FIGURE 10-9** Wood wall bracing may be cut in or let in. Note, structural sheathing is required in seismic and high wind areas.

top to the bottom plates. They are nailed to the plates and stud edges before the sheathing is applied. The T and L shapes require a saw kerf in the stud to allow them to lay flat when installed. Wood bracing may be 2 × 4s that are cut in between the studs. Let-in bracing is a continuous diagonal 1 × 4. These are notched into the face of the studs, top plate, and sole plate at each corner of the building (Fig. 10-9).

# Exterior Wall Framing

Plans or blueprints usually indicate the height from finished floor to finished ceiling. This dimension is found in a wall section. It is needed to determine the length of wall studs.

***For step-by-step instructions on DETERMINING THE LENGTH OF STUDS, see Procedure 10-1, pages 281–282.***

## DETERMINING THE SIZE OF ROUGH OPENINGS

A *rough opening (RO)* is an opening framed in the wall in which to install doors and windows. The width and height of rough openings are not usually indicated in the plans. It is the carpenter's responsibility to determine the rough opening size for the particular unit from the information given in the door and window schedule. The window schedule contains the kind, style, manufacturer's model number, size of each unit, and rough opening dimensions.

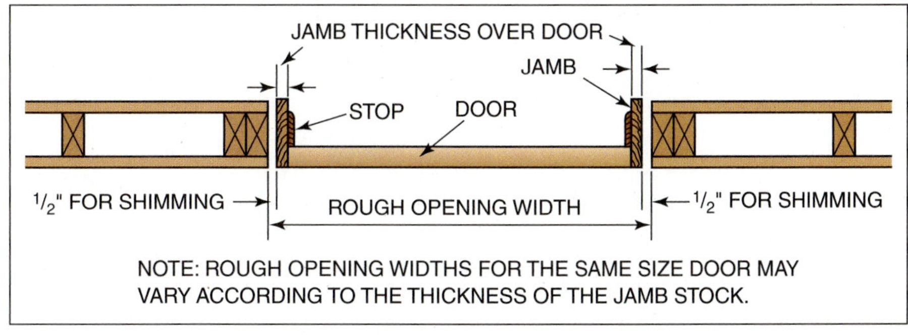


**FIGURE 10-10** Determining the rough opening width of a door opening.

## Rough Opening Sizes for Exterior Doors

The rough opening for an exterior door must be large enough to accommodate the door, door frame, and space for shimming the frame to a level and plumb position. Usually, ½ inch is allowed for shimming at the top and both sides between the door frame and the rough opening (Fig. 10-10). Typically, the door rough opening width is between 2 and 2½ inches larger than the actual door width. Rough door heights are 2 to 3 inches taller than the actual door height.

To calculate the RO height, the type of the door threshold and finish thickness must be known. Thresholds may be hardwood, metal, or a combination of wood and metal (Fig. 10-11). Typically, this dimension establishes the header height for all windows and doors in the house. Doors may be purchased as prehung units, in which the entire unit is set into the opening. Sometimes a door frame without a door is set into the opening. The door may be hung in the frame later. In either case, the rough opening is calculated in the same manner.

***For step-by-step instructions on DETERMINING THE SIZE OF ROUGH OPENINGS, see Procedure 10-2 on page 283.***

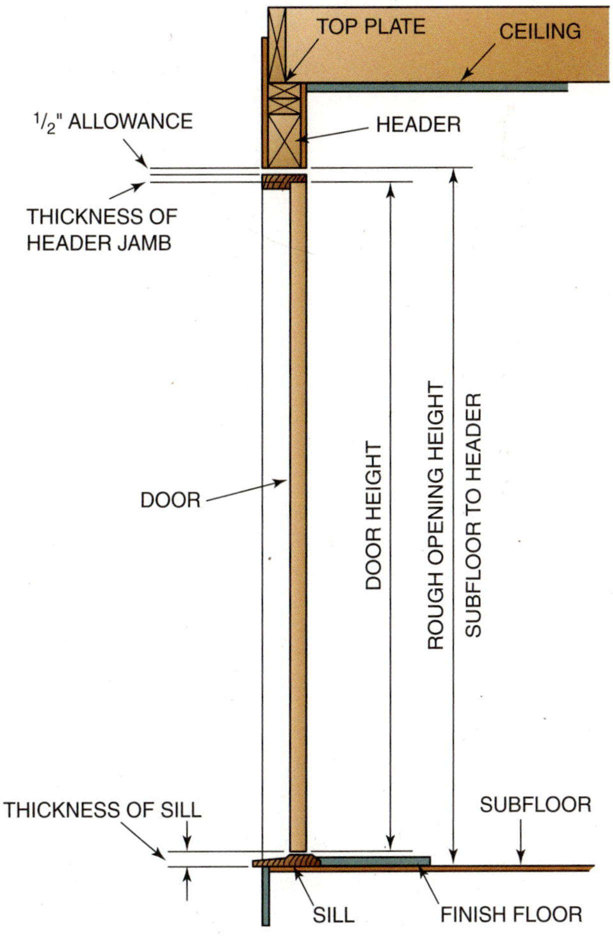


**FIGURE 10-11** Determining the rough opening height of an exterior door opening.

## Rough Opening Sizes for Windows

Many kinds of windows are manufactured by a number of firms. Because of the number of styles, sizes, and variety of construction methods, it is best to consult the manufacturer's catalog to obtain the rough opening sizes. These catalogs show the style and size of the window unit. They also give the ROs for each unit (Fig. 10-12). Catalogs are available from the lumber company that sells the windows. The header height is typically the same as for the doors.

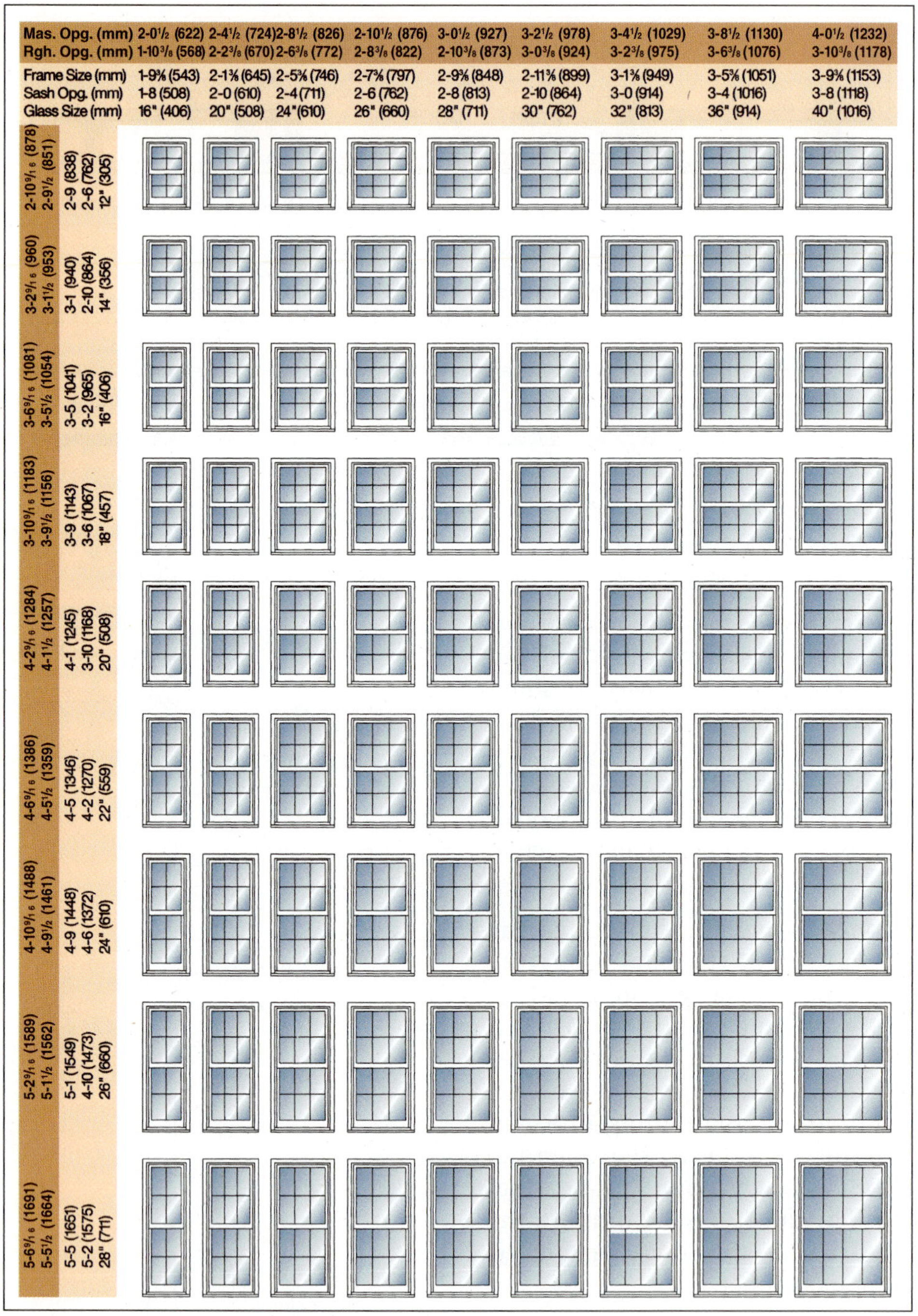

| Mas. Opg. (mm)<br>Rgh. Opg. (mm)<br>Frame Size (mm)<br>Sash Opg. (mm)<br>Glass Size (mm) | 2-0½ (622)<br>1-10 3/8 (568)<br>1-9 3/8 (543)<br>1-8 (508)<br>16" (406) | 2-4½ (724)<br>2-2 3/8 (670)<br>2-1 3/8 (645)<br>2-0 (610)<br>20" (508) | 2-8½ (826)<br>2-6 3/8 (772)<br>2-5 3/8 (746)<br>2-4 (711)<br>24" (610) | 2-10½ (876)<br>2-8 3/8 (822)<br>2-7 3/8 (797)<br>2-6 (762)<br>26" (660) | 3-0½ (927)<br>2-10 3/8 (873)<br>2-9 3/8 (848)<br>2-8 (813)<br>28" (711) | 3-2½ (978)<br>3-0 3/8 (924)<br>2-11 3/8 (899)<br>2-10 (864)<br>30" (762) | 3-4½ (1029)<br>3-2 3/8 (975)<br>3-1 3/8 (949)<br>3-0 (914)<br>32" (813) | 3-8½ (1130)<br>3-6 3/8 (1076)<br>3-5 3/8 (1051)<br>3-4 (1016)<br>36" (914) | 4-0½ (1232)<br>3-10 3/8 (1178)<br>3-9 3/8 (1153)<br>3-8 (1118)<br>40" (1016) |
|---|---|---|---|---|---|---|---|---|---|
| 2-10 9/16 (878)<br>2-9½ (851)<br>2-9 (838)<br>2-6 (762)<br>12" (305) | | | | | | | | | |
| 3-2 9/16 (960)<br>3-1½ (953)<br>3-1 (940)<br>2-10 (864)<br>14" (356) | | | | | | | | | |
| 3-6 9/16 (1081)<br>3-5½ (1054)<br>3-5 (1041)<br>3-2 (965)<br>16" (406) | | | | | | | | | |
| 3-10 9/16 (1183)<br>3-9½ (1156)<br>3-9 (1143)<br>3-6 (1067)<br>18" (457) | | | | | | | | | |
| 4-2 9/16 (1284)<br>4-1½ (1257)<br>4-1 (1245)<br>3-10 (1168)<br>20" (508) | | | | | | | | | |
| 4-6 9/16 (1386)<br>4-5½ (1359)<br>4-5 (1346)<br>4-2 (1270)<br>22" (559) | | | | | | | | | |
| 4-10 9/16 (1488)<br>4-9½ (1461)<br>4-9 (1448)<br>4-6 (1372)<br>24" (610) | | | | | | | | | |
| 5-2 9/16 (1589)<br>5-1½ (1562)<br>5-1 (1549)<br>4-10 (1473)<br>26" (660) | | | | | | | | | |
| 5-6 9/16 (1691)<br>5-5½ (1664)<br>5-5 (1651)<br>5-2 (1575)<br>28" (711) | | | | | | | | | |

*Courtesy of Marvin Windows and Doors.*

**FIGURE 10-12** Sample of a manufacturer's catalog showing rough opening sizes for window units.

## Green Tip

Buying lumber in precut lengths reduces waste on the job and speeds the construction process.

## DETERMINING WALL TYPE

Before layout can begin, the carpenter must first determine if the wall to be laid out is load-bearing or non-load-bearing. It should be noted here that exterior walls are referred to as *walls,* and interior walls are referred to as *partitions*. The **load-bearing** walls (LBWs) usually are built first. They support the ceiling and roof. Non-load-bearing walls (NLBWs) are end walls that usually run parallel with the joists. Interior walls are divided further into load-bearing partitions (LBPs) and non-load-bearing partitions (NLBPs) (Fig. 10-13).

Each type of wall has a slightly different layout characteristic.

It is important to remember that all centerline dimensions for openings are measured from the building line, the outside edge of the exterior framing. Layout must take this fact into account (Fig. 10-14). Figure 10-15 notes the similarities and differences of laying out walls and partitions.

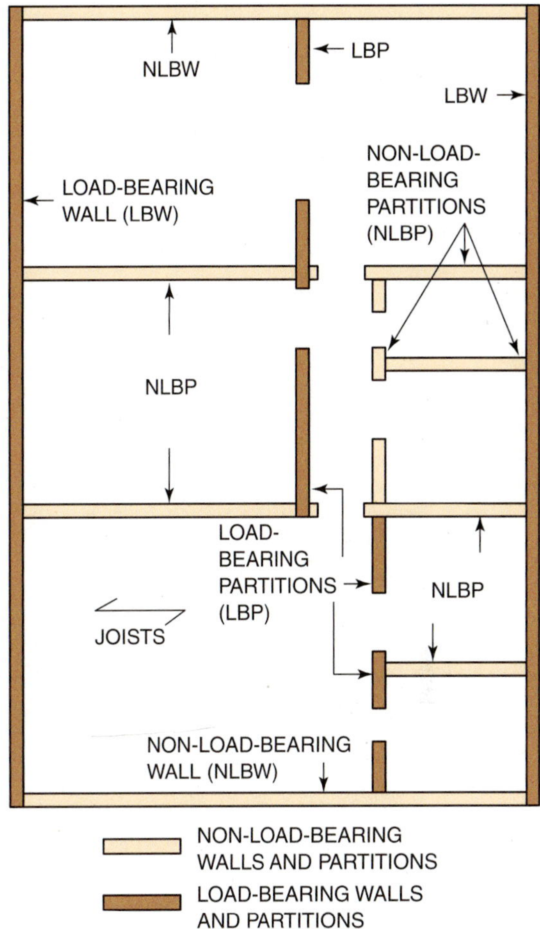


**FIGURE 10-13** Load-bearing walls and partitions run perpendicular to the joists. Non-load-bearing walls and partitions run parallel with the joists. Some minor non-load-bearing partitions may also run perpendicular.

## LAYING OUT THE PLATES

Layouts for the floor joists, exterior walls, ceiling joists, and rafters should all start from the same corner of the building. For example, work from the north end toward the south end when laying out joists, wall studs, and the roof system. This will help keep the studs of both sides of the building aligned over the joists. On building prints, openings, partitions, and wall intersections are usually dimensioned to their centerlines.

***For step-by-step instructions on WALL PLATE LAYOUT, see Procedure 10-3, pages 284–286.***

## ASSEMBLING AND ERECTING WALL SECTIONS

The typical method of framing a wall is to precut the wall frame members, assemble the wall frame on the subfloor, and then stand the frame up. When the frame is erected, the corners are plumbed and temporarily braced. The walls are then straightened and braced between corners. To prevent problems with the installation of the finish work later, it is important to keep the edges of the frame members flush wherever they join each other.

### Precutting Wall Frame Members

Studs may be purchased that are precut to length, giving a rough ceiling height of 8'–1". This is a standard for most homes. For other wall heights, lumber may be cut to length using a power miter saw. Set a stop at the desired distance from the saw blade to cut duplicate lengths. Reject any studs that are severely warped. These rejects may be cut into shorter lengths for blocking, for example, when making corner posts and partition intersections or between studs.

Make up the necessary number of corner posts and partition intersection components (Fig. 10-16).

Cut all headers and rough sills. Their length can be determined from the layout on the plates. Remember that headers are always longer than the RO widths because they sit on top of the jack studs. Cut them accordingly.

It may be necessary to place identifying marks on headers, rough sills, jacks, and trimmers if rough openings are different sizes. Doing so will assist in locating the window or door unit to be placed in each rough opening.

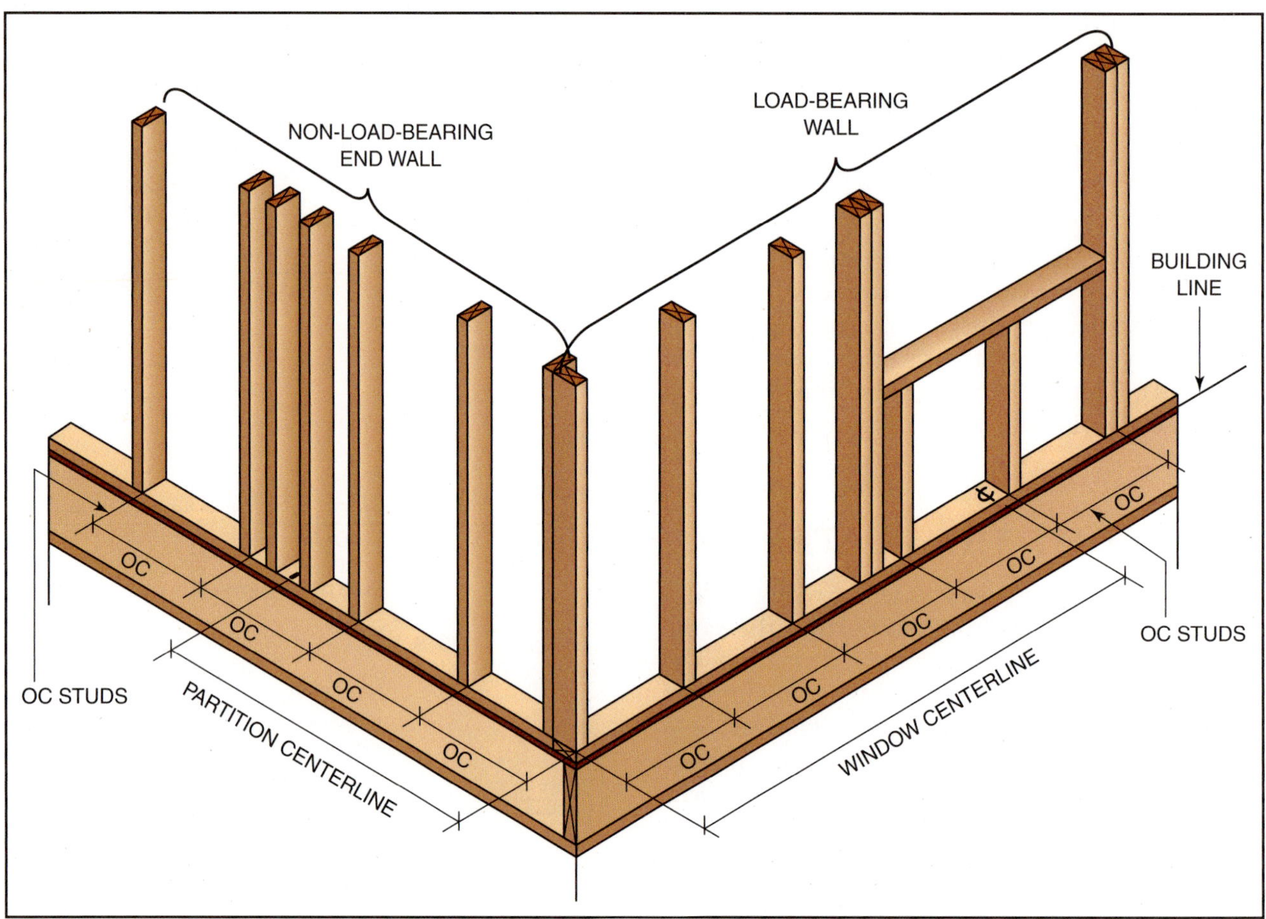


**FIGURE 10-14** Layout for on-center studs and centerlines for openings are measured from the building line.

## Assembling Wall Sections

Few studs are perfectly straight from end to end. A stud that is installed with its crowned edge out next to one with its crowned edge in will certainly present problems later. Crowning each stud up will make assembly easier. Fasten each stud, corner post, and partition intersection in the proper position.

Some builders erect the walls plumb, temporally brace, and apply sheathing later. Other builders install permanent bracing and sheathing while the wall is still laying on the deck. In this case, the wall section should be square. If plywood sheathing is to be used, then no extra bracing is needed because the plywood provides ample rigidity to the wall frame.

***For step-by-step instructions on ASSEMBLING WALL SECTIONS, see Procedure 10-4, pages 287–290.***

## Erecting and Temporarily Bracing Wall Sections

Lifting the wall section is usually done by the framing crew. Lifting jacks may be used as needed, but each wall section should be constructed only as large as the crew can safely and easily lift. Temporary braces usually remain in position until the exterior sheathing is applied. They remain until it is absolutely necessary to remove them. For this reason, the ends of the braces should not extend past the stud.

***For step-by-step instructions on ERECTING AND BRACING WALL SECTIONS, see Procedure 10-5 on page 291.***

## Alternative Methods to Assemble and Erect Exterior Walls

There are several variations of constructing a wall. One such method begins by fastening the sole plate

**Layout Variations for Walls and Partitions**

| | Measure to OC Studs | Measure to Centerlines of Openings |
|---|---|---|
| Load-bearing wall (LBW) | from end of plate | from end of plate |
| Non-load-bearing wall (NLBW) | include width of abutting wall and sheathing thickness | include width of abutting wall |
| Load-bearing partition (LBP) | include width of abutting wall | include width of abutting wall |
| Non-load-bearing partition (NLBP) | from end of plate | include width of abutting wall |

WALLS

1. LOAD-BEARING WALL (LBW)

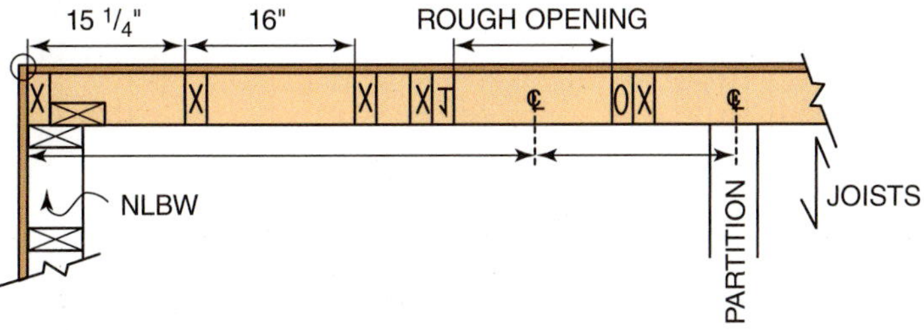


2. NON-LOAD-BEARING WALL (NLBW)

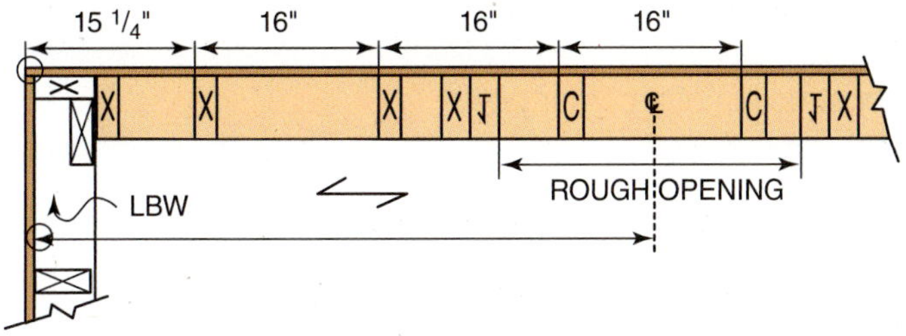


PARTITIONS

3. LOAD-BEARING PARTITION (LBP)

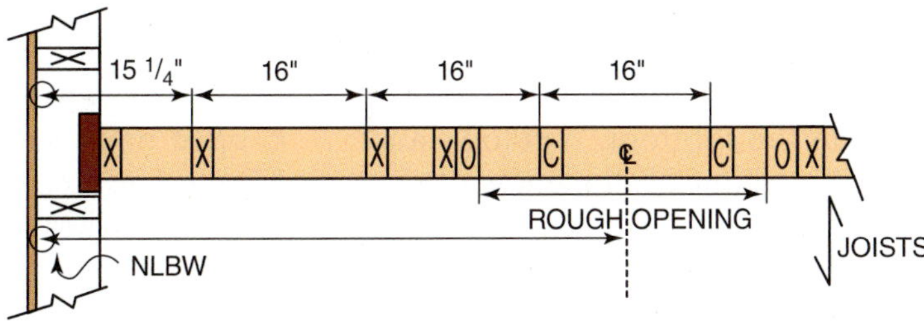


4. NON-LOAD-BEARING PARTITION (NLBP)

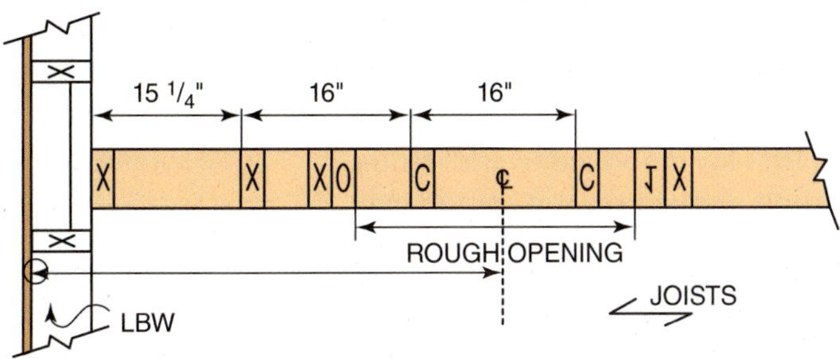


**FIGURE 10-15** Similarities and differences in 16 inches OC wall layout.

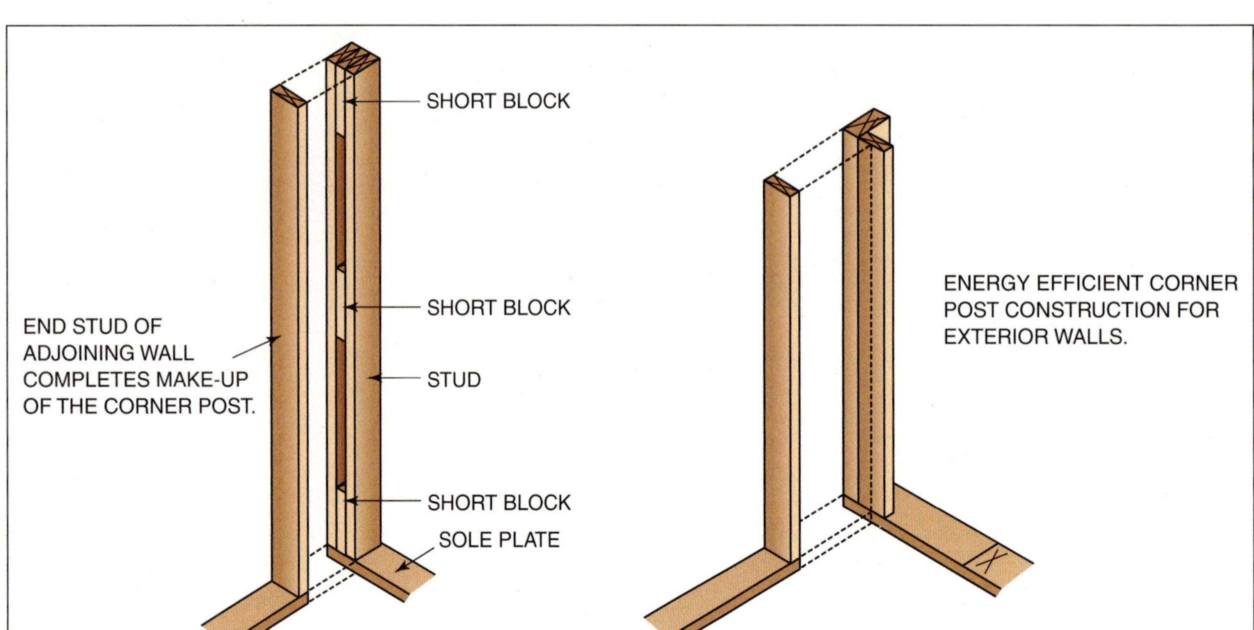


**FIGURE 10-16** Construction of a corner post.

securely to the subfloor. To the top plate, all full-length studs are nailed and then raised into position. The bottoms of the studs are toenailed to the sole plate. In this method, the rough openings are framed later as is the wall sheathing. This method is advantageous when the framing crew is small.

Another method consists of framing and sheathing the entire wall on the subfloor. Windows and doors are installed. Sometimes complete wall sections with doors, windows, and siding installed are prefabricated in the shop, transported to the job site, and erected in position. A disadvantage to this method is that the wall is heavy to raise into position and may require special equipment.

## WALL SHEATHING

*Wall sheathing* is the first layer applied to wall studs. It most often has siding applied on top, but it may also serve as siding. It usually consists of 4' × 8' sheets of material, rated plywood panels, fiberboard, rigid insulation board, or other material such as **gypsum board**. Before plywood wood was available, boards were used.

The sheathing panel may be applied horizontally or vertically. APA-rated plywood panels that serve as wall bracing must be nailed with a maximum nail spacing of 6 inches apart on the edges and 12 inches apart on intermediate studs (Fig. 10-17).

Rated sheathing panels are often used in combination with nonstructural panels. When panels are applied vertically on both sides of each corner, no other wall bracing is necessary. Nonstructural sheathing panels require roofing nails or plastic-capped nails that penetrate into the stud at least 1 inch. The larger head increases the holding power on the soft sheathing material.

### Green Tip

Installing fasteners to the recommended spacing makes the building strong enough to survive its intended load. Recommended spacing also saves material by not overbuilding when too many fasteners are used.

### Application of Sheathing Panels

Sheathing panels are applied in a manner similar to applying subfloor, but they may also be applied vertically. For horizontal applications, use a chalk

APA Panel Wall Sheathing (a)
(APA-RATED SHEATHING panels continuous over two or more spans.)

| Panel Span Rating | Maximum Stud Spacing (in.) | Nail Size (b) (c) | Maximum Nail Spacing (in.) | |
|---|---|---|---|---|
| | | | Supported Panel Edges | Intermediate Supports |
| 12/0, 16/0, 20/0 or wall- 16oc | 16 | 6d for panels $^{1}/_{2}$" thick or less; 8d for thicker panels | 6 | 12 |
| 24/0, 24/16, 32/16 or wall- 24oc | 24 | | | |

(a) See requirements for nailable panel sheathing when exterior covering is to be nailed to sheathing.
(b) Common, smooth, annular, spiral-thread, or galvanized box.
(c) Other code-approved fasteners may be used.

*Courtesy of APA—The Engineered Wood Association.*

**FIGURE 10-17** Selection and fastening guide for APA-rated panel wall sheathing.

line and fasten the first row of panels so their edges are to the line. Start the next row with a half-panel to stagger the joints. Panels should be installed with tight-fitting seams and as few seams as possible. This will make the house more airtight. Apply air barrier material to walls after they are erected, or tape seams of sheathing with sheathing tape.

### Shearwalls

In some locales, wind and seismic forces require the use of a *shearwall.* A shearwall is a heavily reinforced wall section designed to improve the lateral (side-to-side) strength. Shearwalls are sometimes used around garage door openings where the wall is narrow (Fig. 10-18). They are anchored to the foundation and the header or framing above. Care must be taken to install them according to the manufacturer's recommendations.

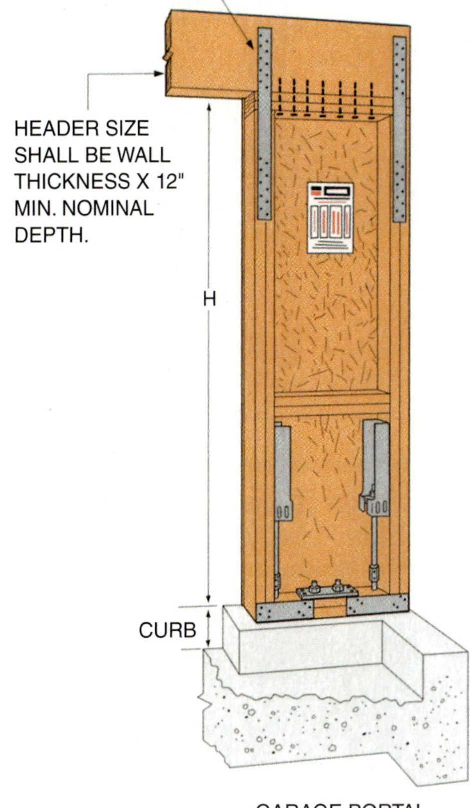


*Courtesy of Simpson Strong-Tie Company.*

**FIGURE 10-18** Shearwalls are installed to stiffen a wall section.

# Interior Framing

*Load-bearing partitions* support the inner ends of ceiling joists or second floor joists above. They are usually placed directly over the girder or the bearing partition in the lower level. Partitions are laid out and erected in a manner similar to exterior walls. The framing of *Non-load bearing partitions* may be left until later so the roof can be made tight as soon as possible (Fig. 10-19). Note that if roof trusses are used, bearing partitions are not needed. The load that is imposed by the trusses is transmitted to the exterior walls only. (Details on roof trusses are given in Chapter 12.) The rough opening sizes are determined in the same fashion as for exterior doors.

## NON-LOAD BEARING PARTITIONS

Non-load bearing partitions carry no load. They divide the floor area into rooms. Openings may be framed with single member headers. Because the wall carries no load, headers are usually 2 × 4s (Fig. 10-20). Non-load bearing partitions may be built after the ceiling joists are installed.

## BACKING AND BLOCKING

**Backing** is used to provide support for later fastening of trim or fixtures. **Blocking** is installed to

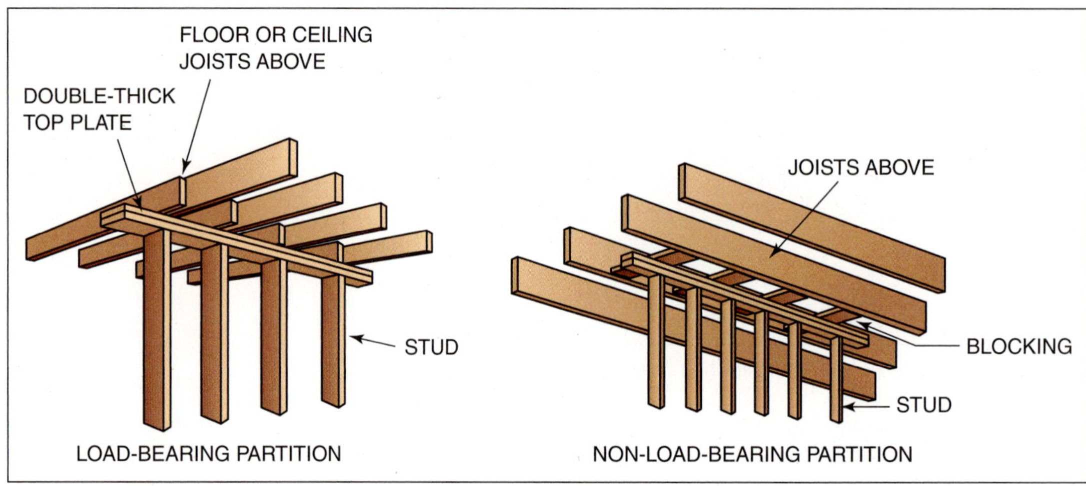


**FIGURE 10-19** Load-bearing partitions support the weight of the floor or ceiling above. Non-load-bearing partitions merely divide an area into rooms.

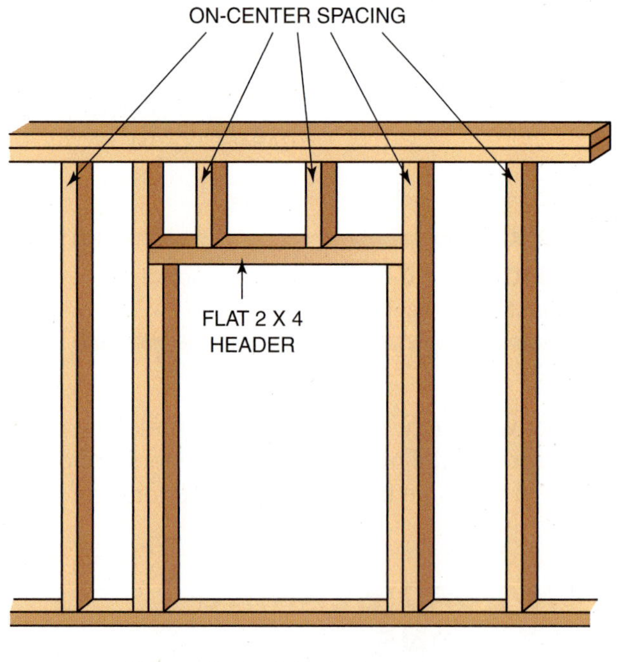


**FIGURE 10-20** A method for framing a header in a non-load-bearing wall or partition.

provide support for parts of the structure, weather-tightness, and draftstopping. Sometimes blocking serves as backing.

*from experience...*

Use the straightest available lumber for openings and wall intersections. Make the sides of king and jack studs flush by toenailing them together.

### Backing

Much backing is needed in bathrooms. Plumbing rough-in work varies with the make and style of plumbing fixtures. Generally, backing is needed for bathtub faucets, showerheads, lavatories, and water closets (Fig. 10-21). Backing should also be installed around the top of the bathtub. In the kitchen, backing should be provided for the tops and bottoms of wall cabinets and for the tops of base cabinets. If the ceiling is to be built down to form a soffit at the tops of wall cabinets, backing should be installed to provide fastening for the soffit (Fig. 10-22).

A homeowner will appreciate the thoughtfulness of the builder who provides backing in appropriate locations in all rooms for the fastening of curtain and drapery hardware (Fig. 10-23).

### Blocking

Blocking can be in the form of solid bridging. It is also installed as extra support for the plywood

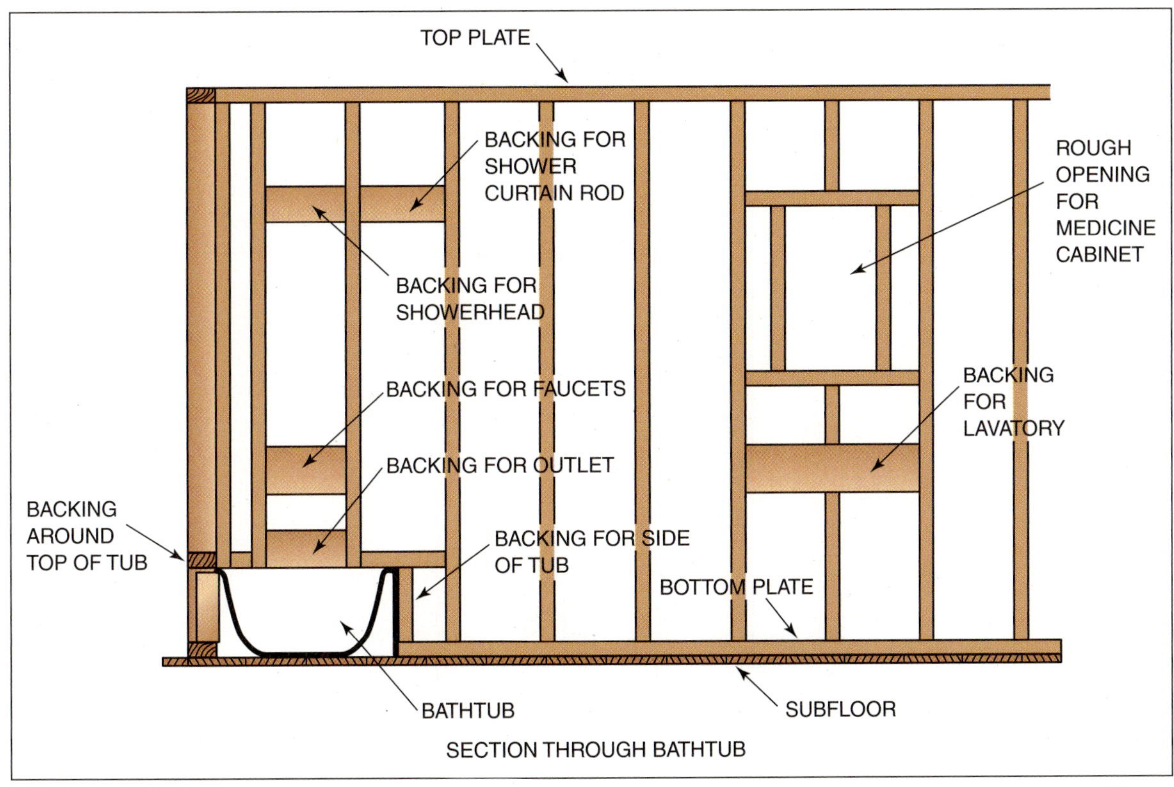


**FIGURE 10-21** Considerable backing is needed in bathrooms.

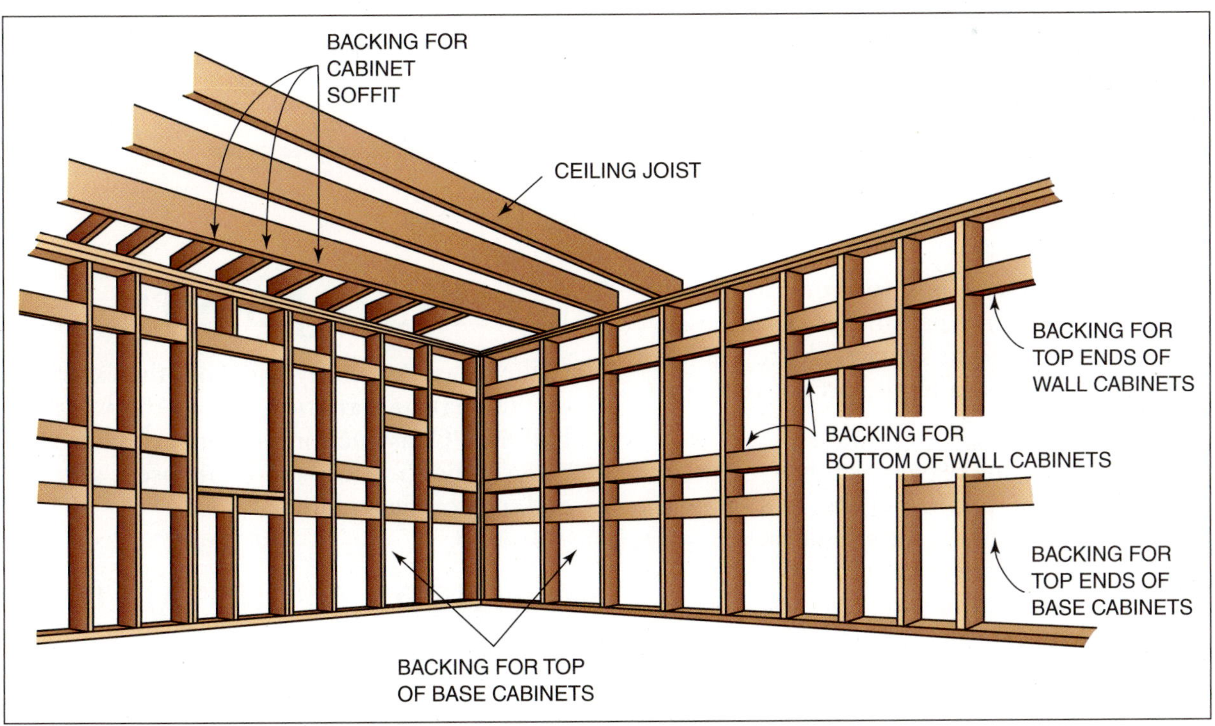


**FIGURE 10-22** Location and purpose of backing in kitchens.

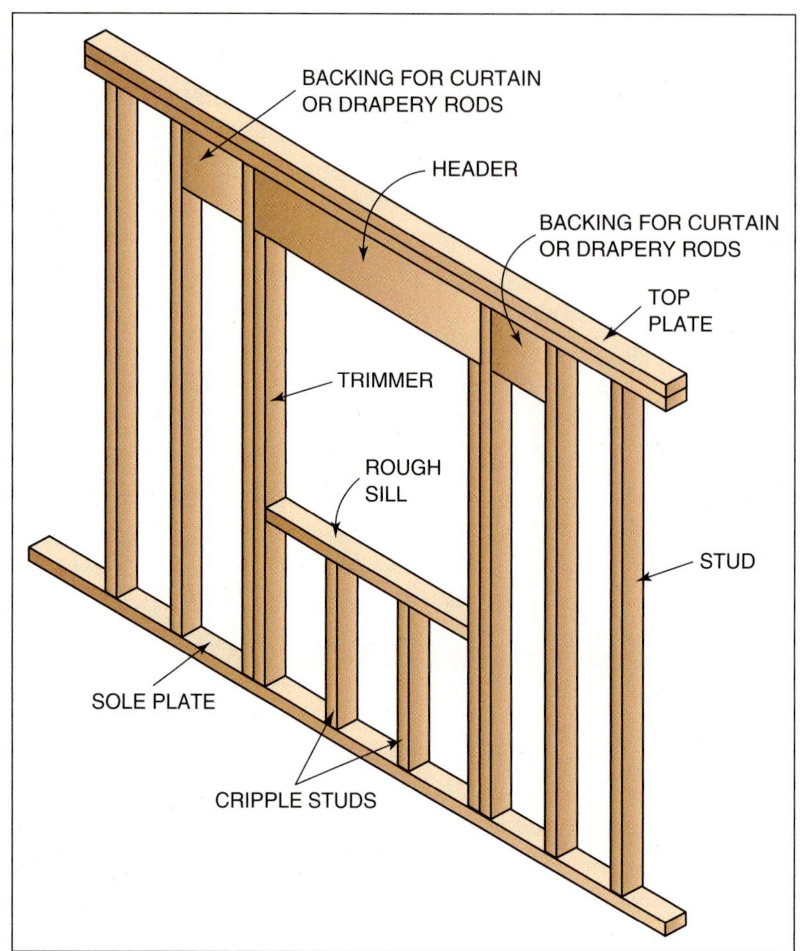


**FIGURE 10-23** An experienced builder will install backing in all rooms for curtain and drapery hardware.

panel edges that are not tongue-and-groove (Fig. 10-24). Ladder-type blocking is necessary between ceiling joists to support the top ends of partitions that run parallel to and between joists (Fig. 10-25).

Draftstop blocking between studs is required in walls over 8'–1" high. The purpose is to slow the spread of fire by reducing the size of the stud cavity. It also serves to stiffen the studs and strengthen the overall structure. Blocking for these purposes may be installed in staggered fashion (Fig. 10-26).

***For step-by-step instructions on INSTALLING BACKING AND BLOCKING, see Procedure 10-6 on page 292.***

# Ceiling Joists

Ceiling joists generally run from the exterior walls to the bearing partition across the width of the building. Construction design varies according to geographic location, materials used, and the size and style of the building.

## METHODS OF INSTALLING CEILING JOISTS

In a conventionally framed roof, the rafters and the ceiling joists form a triangle. Framing a triangle is a common method of creating a strong and rigid building.

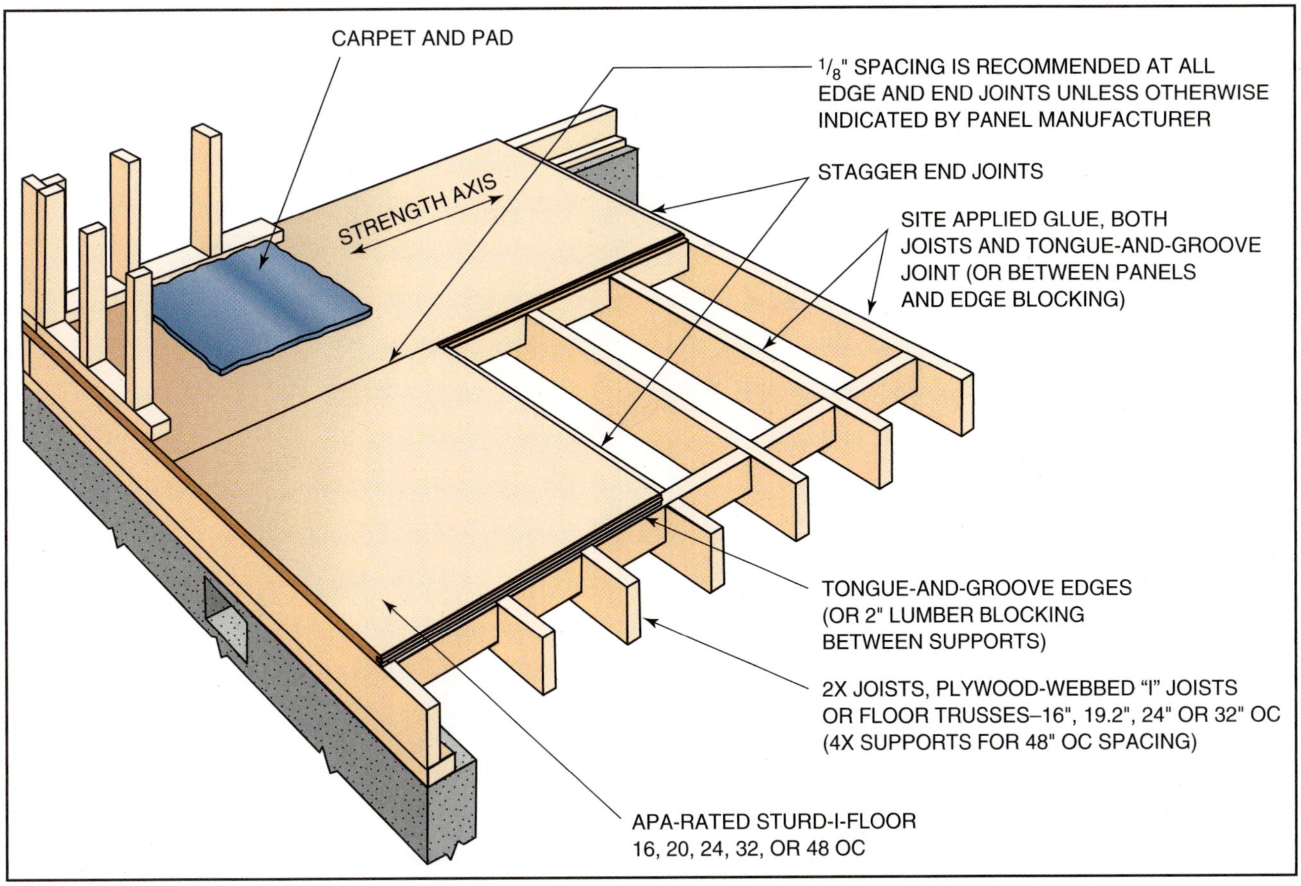

**FIGURE 10-24** The edges of rated panels used as a combination of subfloor and underlayment, or as sheathing, must be tongue-and-groove or supported by blocking.

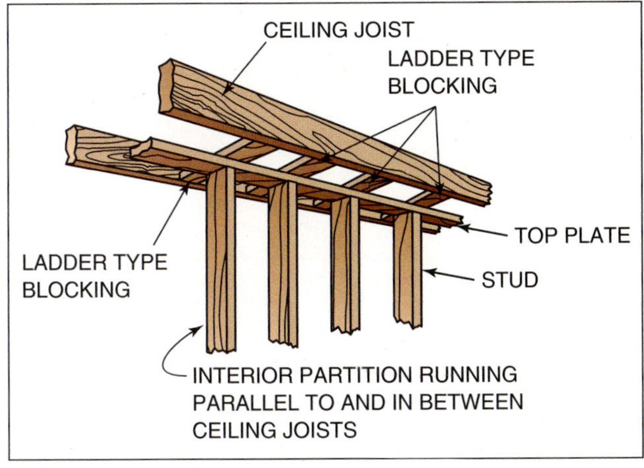


**FIGURE 10-25** Ladder-type blocking must be provided for some interior partitions not supported by furring strips.

The load of the roof and weather are transferred from the roof to the exterior walls (Fig. 10-27). The rafters are located over the studs, and the ceiling joists are fastened to one side of the rafters (Fig. 10-28). This binds the rafters and ceiling joists together into a rigid triangle and keeps the walls from spreading outward due to the weight of the roof.

Ceiling joists may be made from engineered lumber and purchased in long lengths so that the rafter-ceiling joist triangle is easily formed. The location for the ceiling joists on each exterior wall is on the same side of the rafters. Typically, the ceiling joist lengths are half of the building width and therefore must be lapped or spliced over a beam or bearing partition.

Sometimes the ceiling joists are installed in-line. Their ends butt each other at the centerline of the bearing partition. The joint must be scabbed to

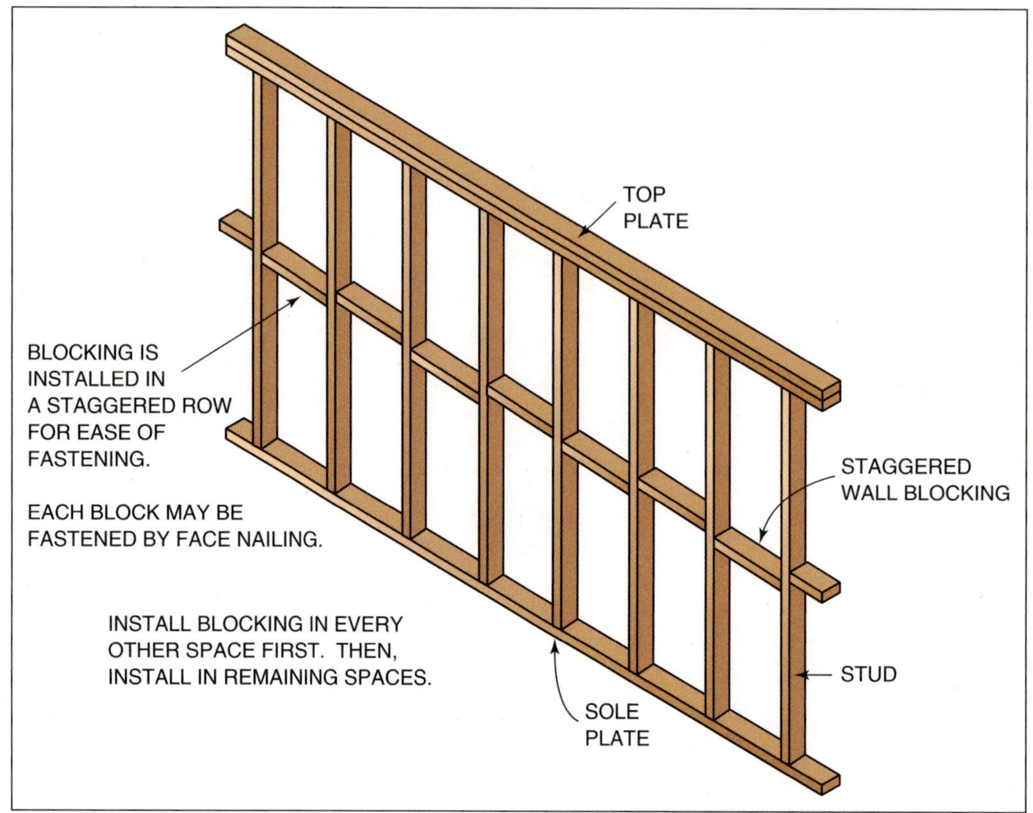


**FIGURE 10-26** Blocking used for stiffening walls and draftstopping may be installed in a staggered fashion.

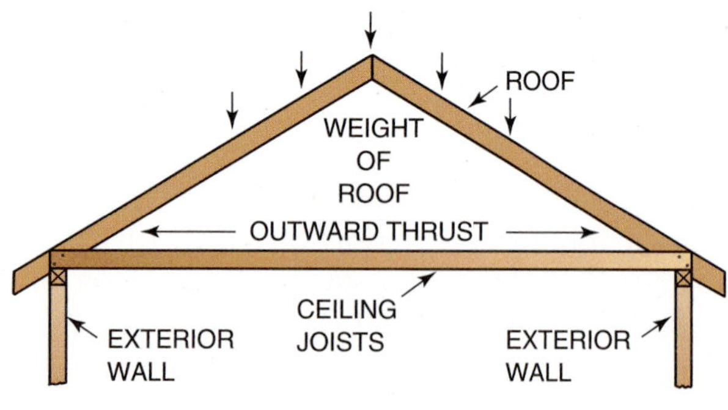


**FIGURE 10-27** The weight of the roof exerts pressures that tend to thrust the walls outward. Ceiling joists tie the frame together into a triangle, which resists the outward thrust.

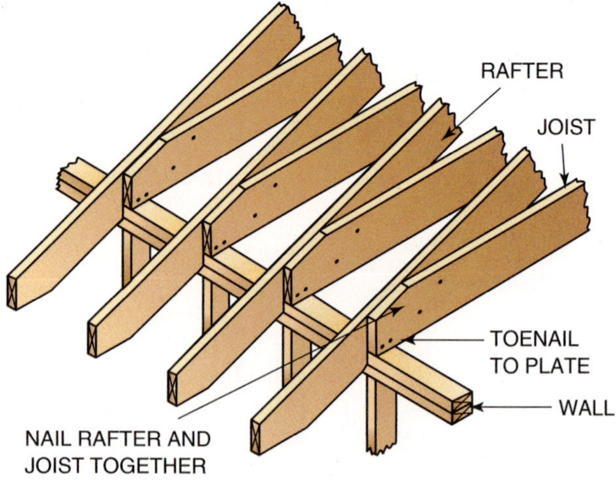


**FIGURE 10-28** Ceiling joists are located so they can be fastened to the side of rafters.

tie the joints together (Fig. 10-29). *Scabs* are short boards fastened to the side of the joist and centered on the joint. They should be a minimum of 24 inches long.

Most of the time, the ceiling joists lap over a bearing partition in the same manner as for floor joists. This puts a stagger in the line of the ceiling joist and consequently in the rafters as well. This stagger is most visible at the ridgeboard (Fig. 10-30).

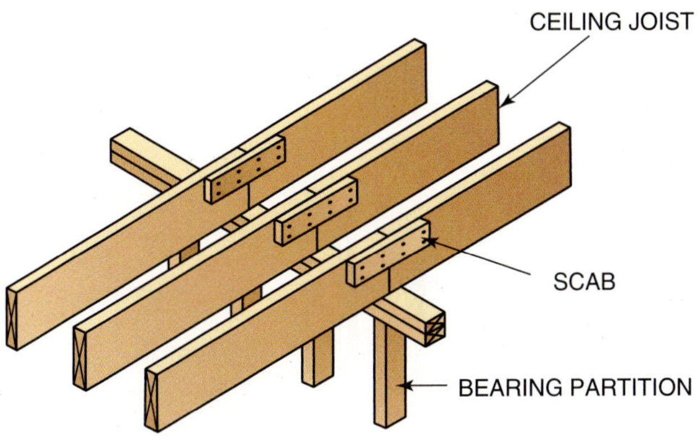


**FIGURE 10-29** The joint of in-line ceiling joists must be scabbed at the bearing partition.

The layout lines for rafters and ceiling joists are measured from the outside end wall onto the top plate. This measurement is the same for both exterior walls and the load-bearing partition, with the only difference being the side of the line on which the ceiling joists and rafters are placed (Fig. 10-31).

## Cutting the Ends of Ceiling Joists

The ends of ceiling joists on the exterior walls usually project above the top of the rafter. This is especially true when the roof has a lower slope. The ceiling joist ends must be cut to the slope of the roof, flush with or slightly below the top edge of the rafter.

Lay out the cut by using a framing square. Cut one joist for a pattern. Use the pattern to mark the rest. Make sure when laying out the joists that you sight each for a crown. Make the cut on the crowned edge so that edge is up when the joists are installed.

Cut the taper on the ends of all ceiling joists before installation. Make sure the length of the taper cut does not exceed three times the depth of the joist. Also, make sure that the end of the joist remaining after cutting is at least one half the joist's width (Fig. 10-32).

## Framing Ceiling Joists to a Beam

In some cases, the bearing partition does not run the length of the building because of large room areas.

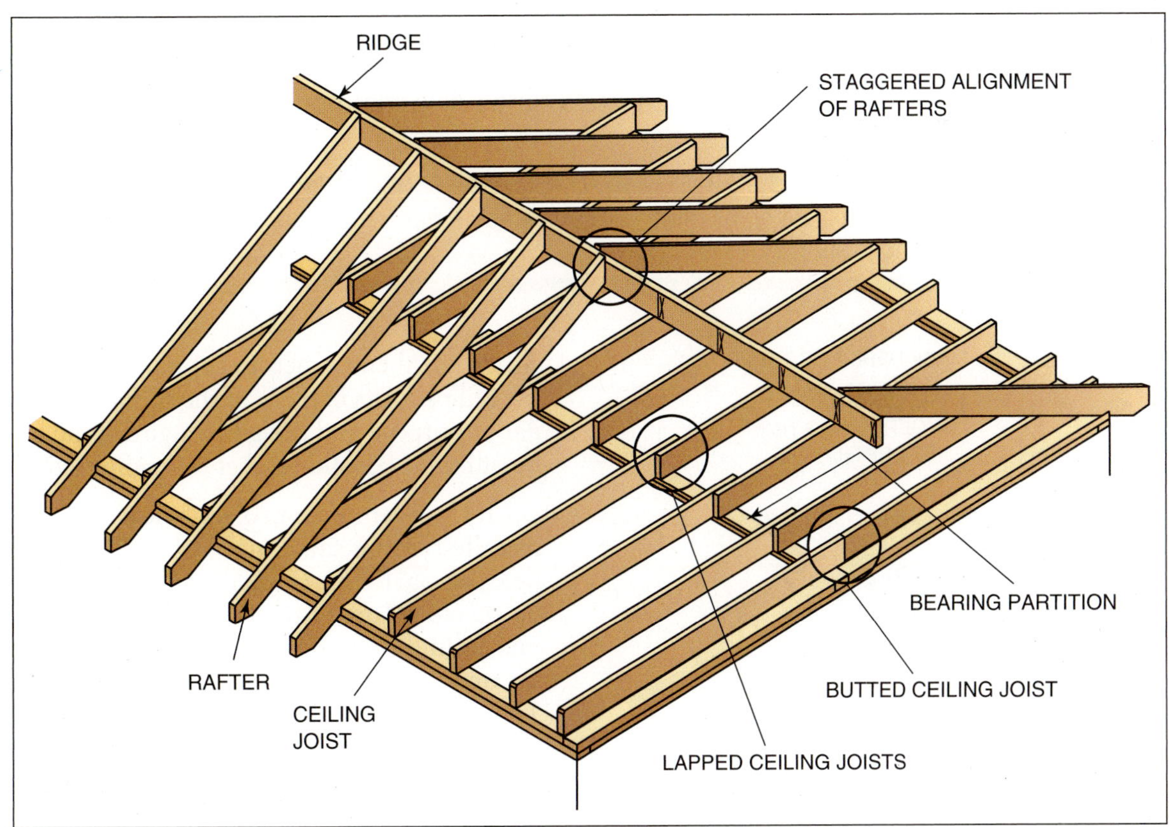


**FIGURE 10-30** When ceiling joists are lapped, it causes a stagger in the rafters, which is visible at the ridge.

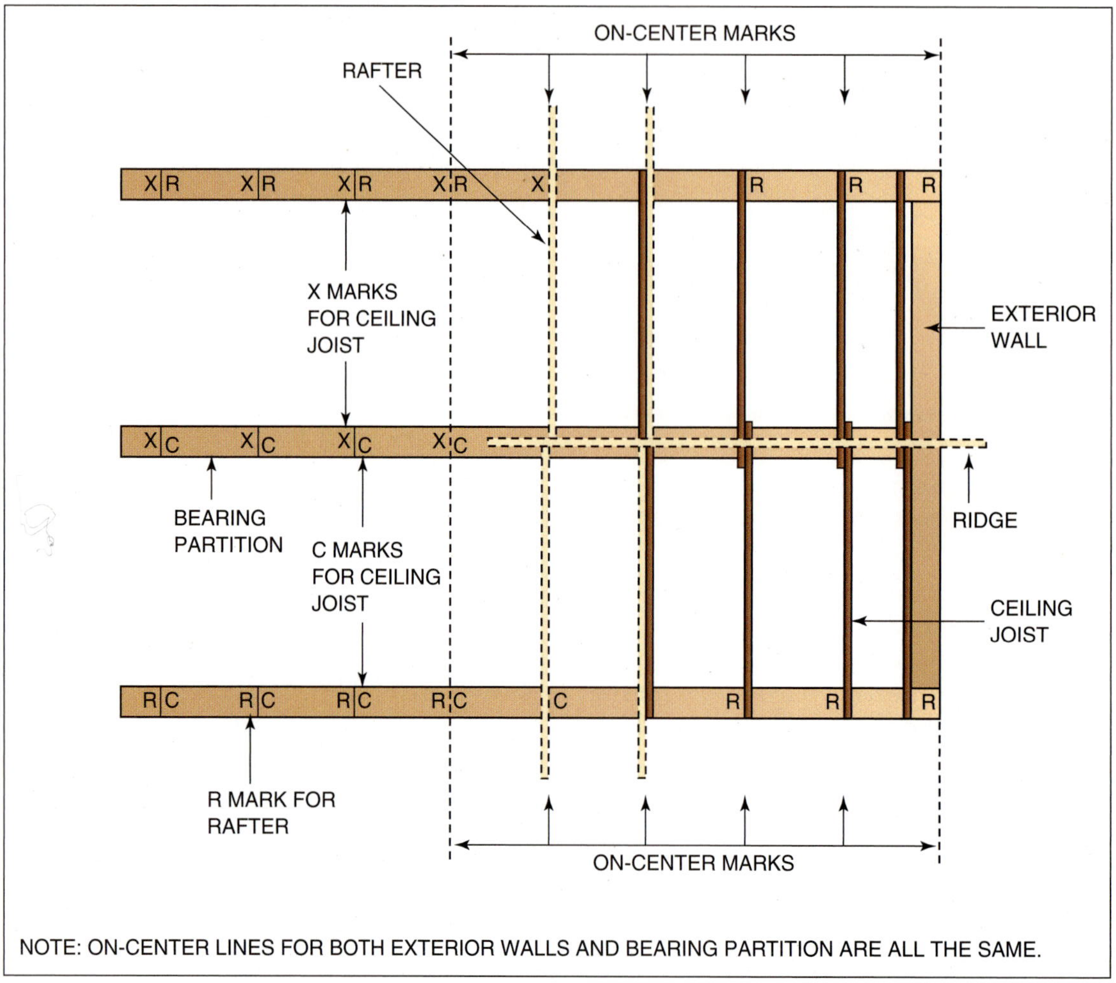

**FIGURE 10-31** Layout lines on all plates are of the same measurements; only the position of the rafters and ceiling joists vary.

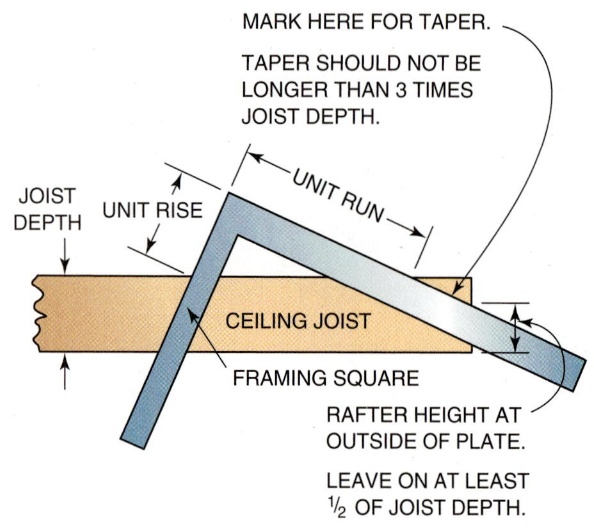


**FIGURE 10-32** Using a framing square to mark the slope of a tapered cut on a ceiling joist.

A beam is then needed to support the inner ends of the ceiling joists in place of the supporting wall. Similar in purpose and design to a girder, the beam may be of built-up, solid lumber or engineered lumber.

If the beam is to project below the ceiling, it is installed in the same manner as a girder or load-bearing partition (Fig. 10-33). If the beam is to be raised in order to make a flush and continuous ceiling below, then the ends of the beam are set on top of the bearing partition and end wall. The joists are butted to the beam and may be supported by a ledger strip or by metal **joist hangers** (Fig. 10-34).

## Openings

Openings in ceiling joists may be provided for installation of chimneys, attic access (scuttle), or disappearing stairs. Large openings are framed in the same manner

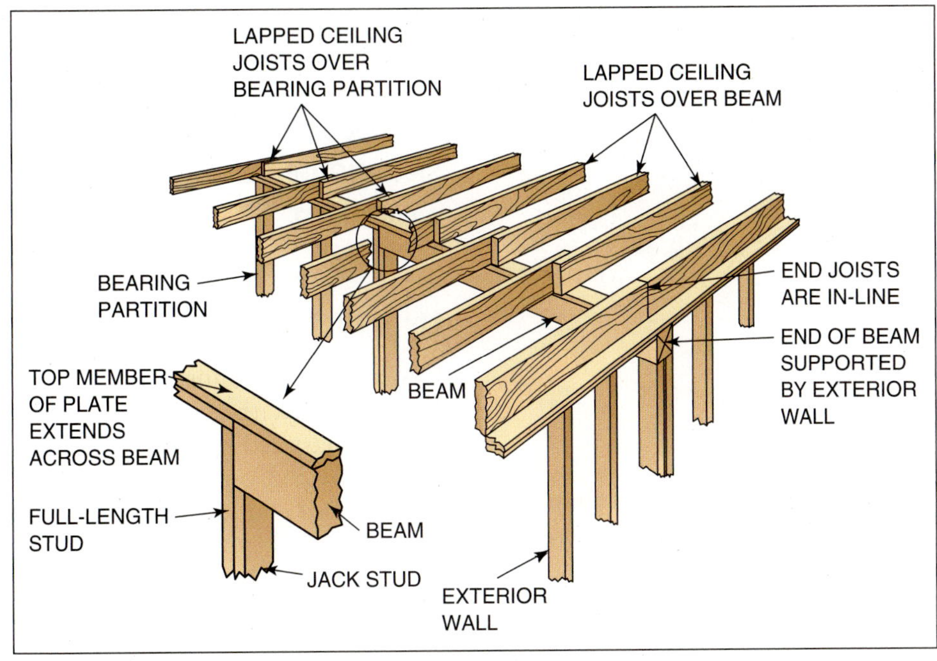


**FIGURE 10-33** Framing ceiling joists over a beam.

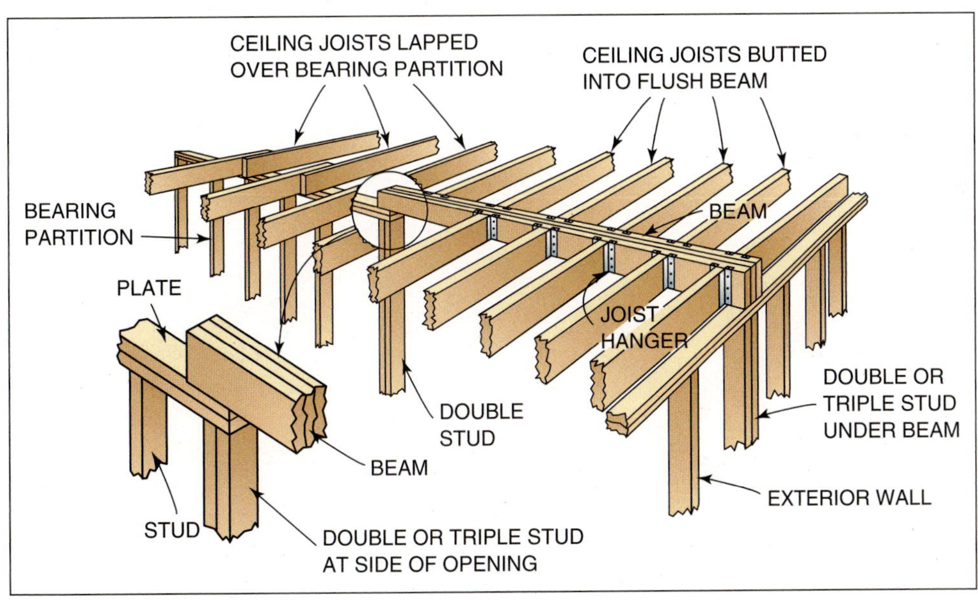


**FIGURE 10-34** Framing ceiling joists into a beam for a flush ceiling.

as for floor joists. For small openings, there is no need to double the joists or headers (Fig. 10-35).

## Ribbands and Strongbacks

Ribbands are 2 × 4s installed flat on top of ceiling joists. They are placed at mid-span to stiffen the joists and to keep the spacing uniform. They should be fastened with 16d nails and long enough to be attached to the end walls. With the addition of a 2 × 6 installed on edge, the ribband becomes a *strongback*. It is used when extra support and stiffness is required on the ceiling joists (Fig. 10-36).

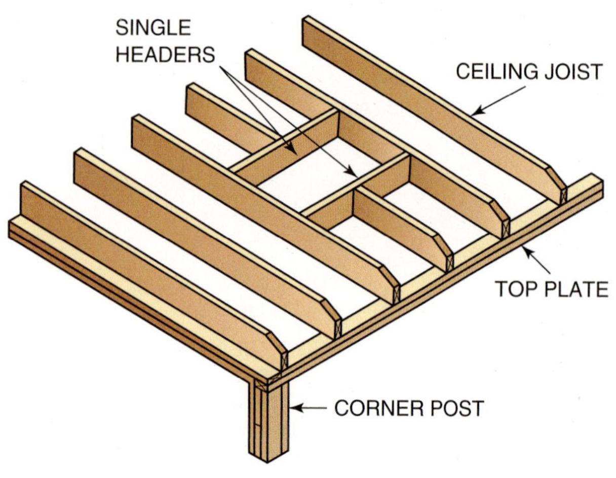


**FIGURE 10-35** Joists and headers need not be doubled for small ceiling openings.

## LAYOUT AND SPACING OF CEILING JOISTS

Ceiling joists are installed alongside the rafters and securely fastened to them. This intersection should rest over top of the OC studs of the exterior walls and load-bearing partition. Start the layout of the ceiling joists from the same corner of the building as were the floor joists and wall studs.

***For step-by-step instructions on the LAYOUT AND SPACING OF CEILING JOISTS, see Procedure 10-7 on page 293.***

## INSTALLING CEILING JOISTS

Ceiling joists are installed in a manner similar to floor joists, with several possible differences. The first and last ceiling joists on each end of the building are placed on the inside of the wall. The outside face of the joist is flush with the inside face of the wall. They are installed in-line with their inner ends butting each other regardless of how the other joists are laid out. This will allow for easy installation of the **gable end** studs and provide bearing surface for the ceiling finish (Fig. 10-37).

Joists are toenailed into position with at least two 10d (3-inch) nails. Reject any badly warped joists, and remember to install all crowns up. Lapped joists at the partition should be toenailed with three 10d nails into the plate and face-nailed together with at least two 10d (3-inch) nails.

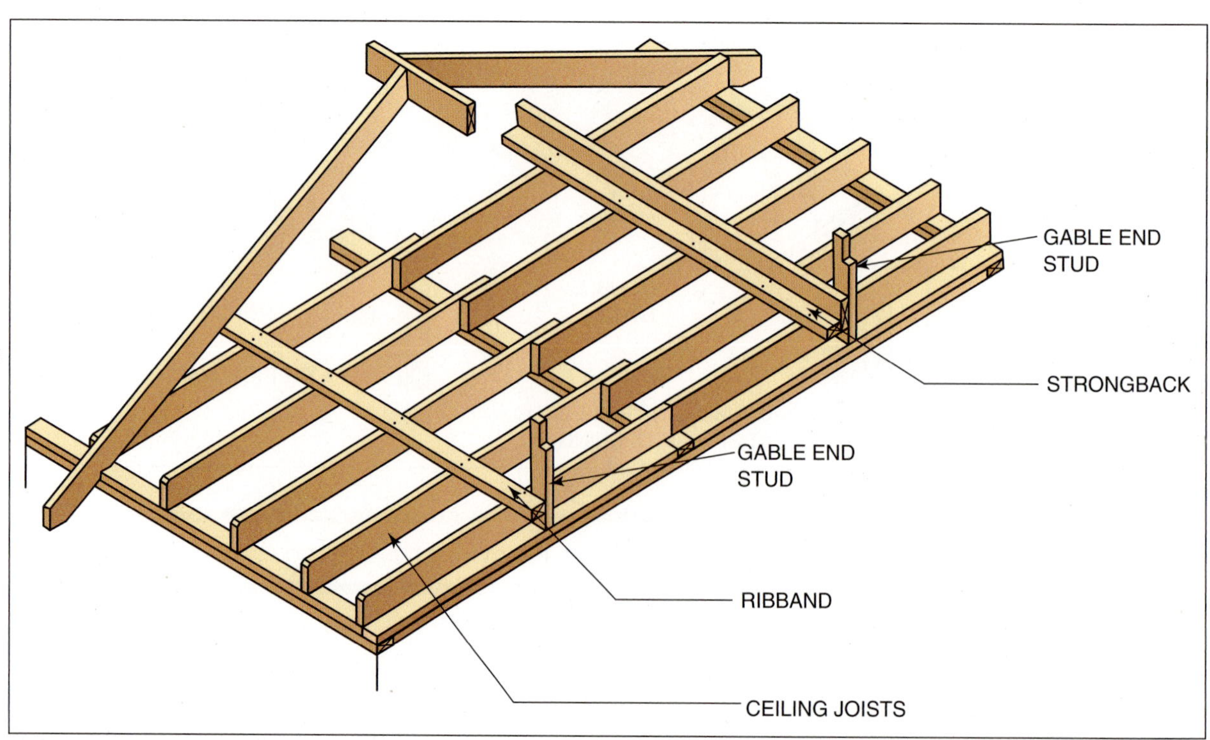


**FIGURE 10-36** Ribbands and strongbacks are sometimes used to stiffen ceiling joists.

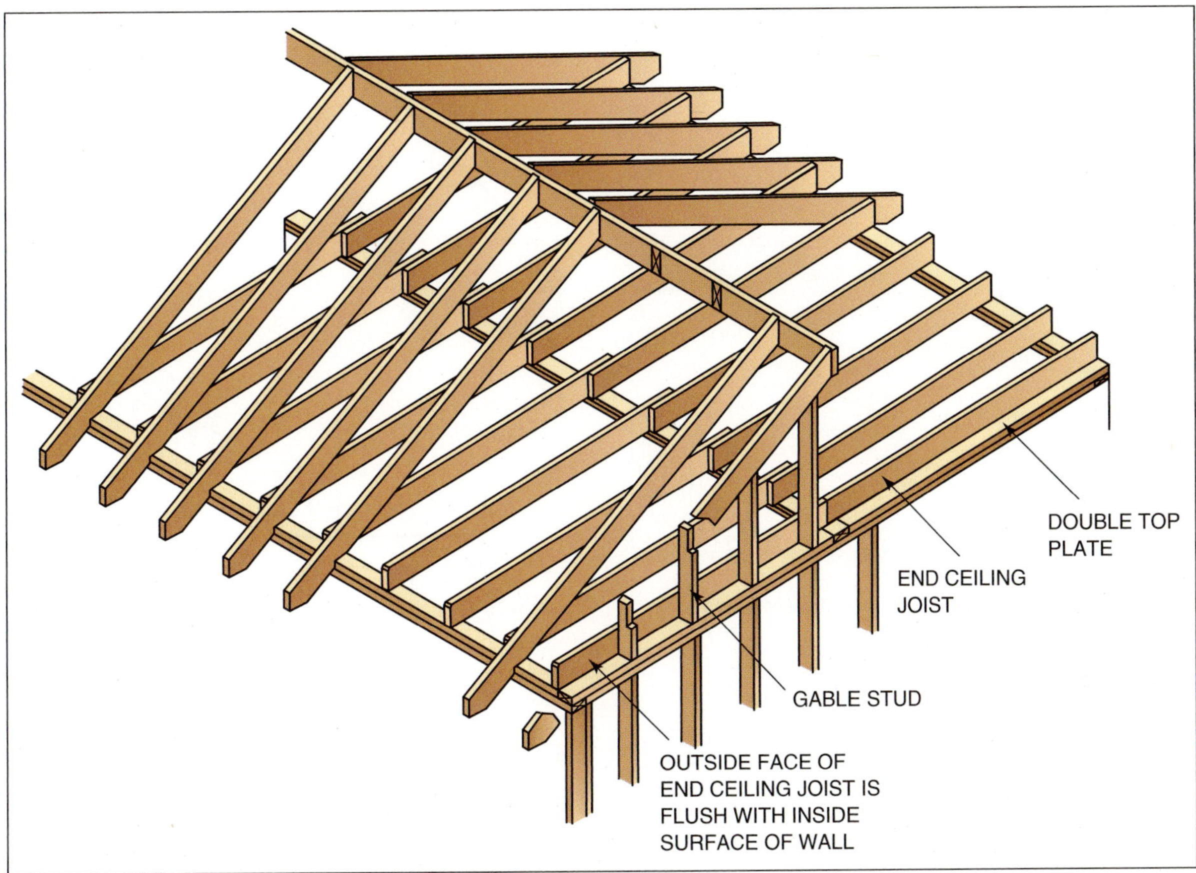


**FIGURE 10-37** The end ceiling joist is located with its outside face flush with the inside wall.

# STEEL FRAMING

Steel framing is used for structural framing and interior non-load-bearing partitions. Steel may be used for floor joists, walls studs, ceiling joists and may also fabricated into roof trusses (Fig. 10-38). Specialists in light steel are required when framing is structural and extensive. However, carpenters often frame interior non-load-bearing partitions and apply *furring channels* of steel. Discussion here is limited to their installation. The size and spacing of steel framing members should be determined from the drawings. The framing of steel interior partitions is similar to the framing of wood partitions. Different kinds of fasteners are used, and some special tools may be helpful.

**FIGURE 10-38** Steel may be used for load and non-load-bearing building framing.

## STEEL FRAMING COMPONENTS

All steel framing components are coated with zinc to resist corrosion. The main parts of an interior steel framing system are studs, track, channels, and accessories.

### Studs

For interior non-load-bearing applications, studs are manufactured from 18-, 27-, and 33-mil steel (or 25-, 22-, and 20-gauge respectively). The stud

**FIGURE 10-39** Steel studs are installed between top and bottom track.

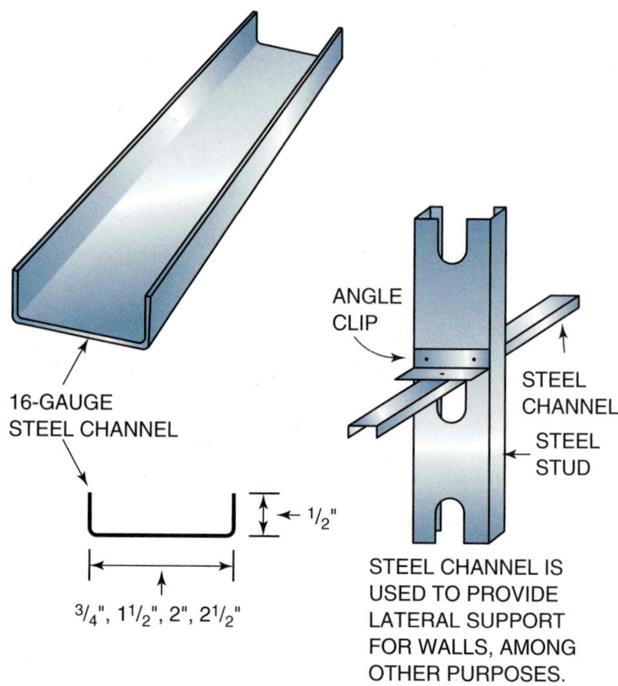


**FIGURE 10-40** Steel channel is used to stiffen the framing members of walls and ceilings.

*web* has punchouts at intervals through which to run pipes and conduits. Studs come in widths of 3½, 5½, 8, 10, and 12 inches, with 1¼ and 1⅝ leg thickness. Studs are available in stock lengths of 8, 9, 10, 12, and 16 feet (Fig. 10-39). Custom lengths up to 28 feet are also available.

### Track

The top and bottom plates of a steel-framed wall are called *runners* or *track*. They are installed on floors and ceilings to receive the studs. They are manufactured in gauges, widths, and leg thicknesses to match studs. Track is available in standard lengths of 10 feet.

### Channels

Steel *cold-rolled channels (CRCs)* are formed from 54-mil steel. They are available in several widths. They come in lengths of 10, 16, and 20 feet. Channels are used in suspended ceilings and walls. When used for lateral bracing of walls, the channel is inserted through the stud punchouts. It is fastened with welds or clip angles to the studs (Fig. 10-40).

### Furring Channels

*Furring channels* or hat track are hat-shaped pieces made of 18- and 33-mil steel. Their overall cross section size is ⅞ inch by 2⁹⁄₁₆ inches. They are available in lengths of 12 feet (Fig. 10-41). Furring channels are applied to walls and ceilings for the screw attachment of gypsum panels.

## FRAMING STEEL PARTITIONS

Lay out steel-framed partitions as you would wood-framed partitions. To cut metal framing to length, tin snips may be used on 18-mil steel. A *chop saw* or *chop box,* a specially designed power miter box with a metal-cutting saw blade, is the preferred tool. With this tool, an entire bundle of studs may be cut at one time. Use ⅜-inch self-drilling pan head screws, or crimp the track and stud together using a crimping tool designed for steel studs.

Tracks are usually fastened into concrete with powder-driven fasteners. Small concrete nails or masonry screws may also be used. Fasten into wood with 1¼-inch, oval-head screws. Attach the track with two fasteners about 2 inches from each end and a maximum of 24 inches on center (OC) in between.

Place all full-length studs in position within the track with the open side facing in the same direction. The web punchouts should be aligned vertically. This provides space for lateral bracing as well as plumbing and wiring to be installed (Fig. 10-42). Fasten all studs to top and bottom track. A magnetic level that holds itself to the stud is very helpful.

***For step-by-step instructions on INSTALLING METAL FRAMING, see Procedure 10-8 on page 294.***

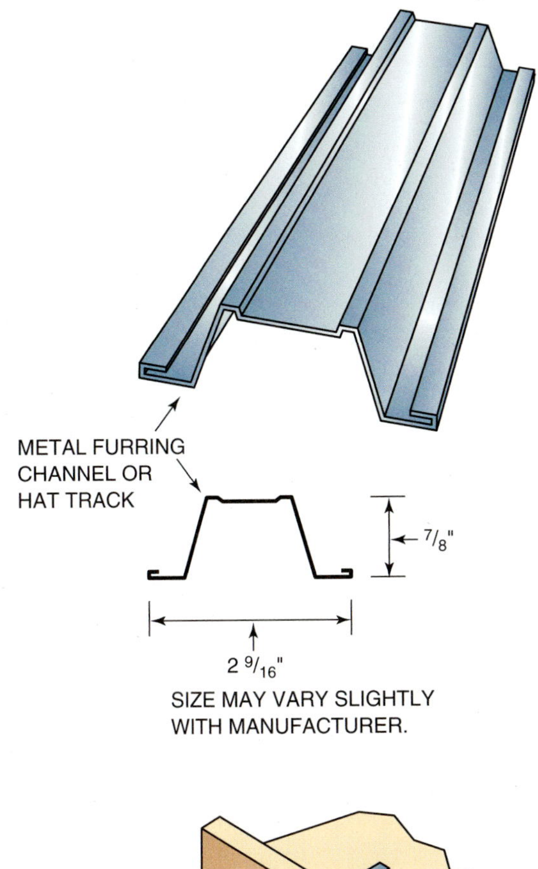


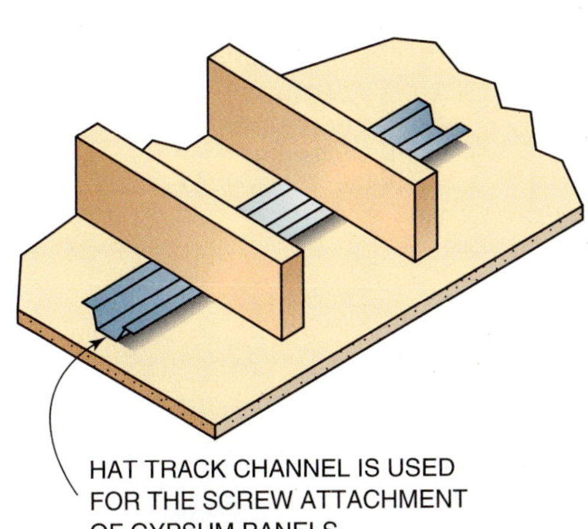


**FIGURE 10-41** Furring channels or hat track are used in both ceiling and wall installations.

**FIGURE 10-42** Steel studs and joists come in several widths, lengths, and gauges.

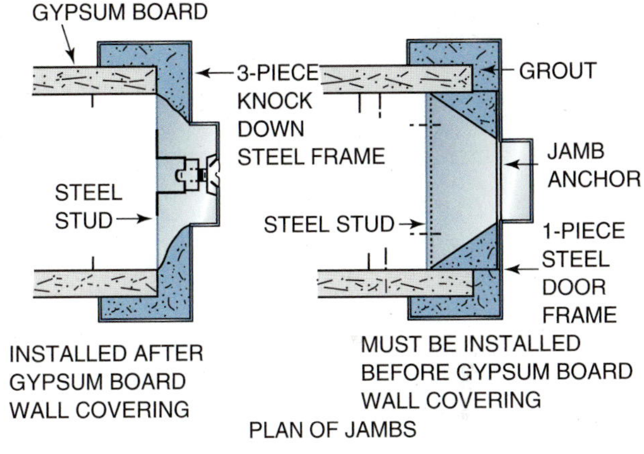


*Courtesy of U.S. Gypsum Corporation.*

**FIGURE 10-43** In a steel-studded wall, either one-piece or three-piece, knocked-down metal door frames are used.

## Framing Wall Openings

The method of framing door openings depends on the type of door frame used. A one-piece metal door frame must be installed before the gypsum board is applied. A three-piece, knocked-down frame is set in place after the wall covering is applied (Fig. 10-43). Window openings are framed in the same manner. However, the opening is typically lined with wood. Wood jack studs, sills and head liners are used to make attachment of window trim easier (Fig. 10-44). Cripples are placed both above and below.

***For step-by-step instructions on FRAMING A DOOR OPENING FOR A THREE-PIECE FRAME AND A ONE-PIECE FRAME, see Procedures 10-9 and 10-10 on pages 295–297.***

# METAL FURRING

Metal furring may be used on ceilings and walls by attaching them at right angles to joists or studs.

**FIGURE 10-44** Window and door openings may be lined with wood for easier fastening of finish materials.

They may be attached vertically or horizontally to masonry walls. Spacing of metal furring channels should not exceed 24 inches OC.

## Ceiling Furring

Metal furring channels may be attached directly to structural ceiling members or may be suspended from them. For direct attachment, saddle tie with double-strand 18-gauge wire to each member. Leave a 1-inch clearance between ends of furring and the adjoining walls. Metal furring channels may be spliced by overlapping the ends at least 8 inches. Tie each end with wire (Fig. 10-45).

## Wall Furring

Wall furring may be installed vertically or horizontally. Secure the channels by staggering the fasteners from one side to the other not more than 24 inches OC (Fig. 10-46). For horizontal application on walls, attach furring channels not more than 4 inches from the floor and ceiling. Fasten in the same manner as vertical furring.

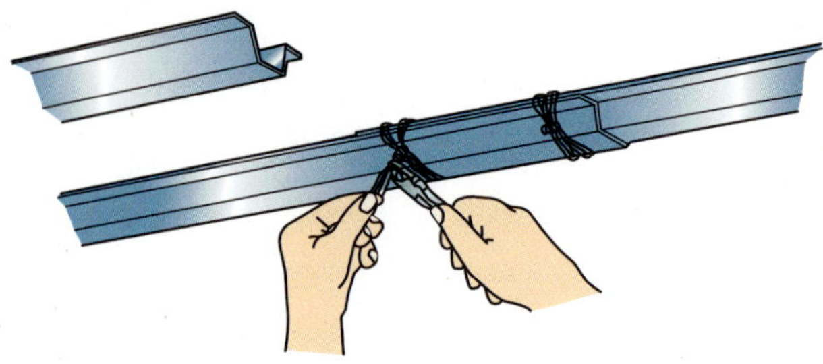

*Courtesy of U.S. Gypsum Corporation.*

**FIGURE 10-45** Method of splicing furring channels.

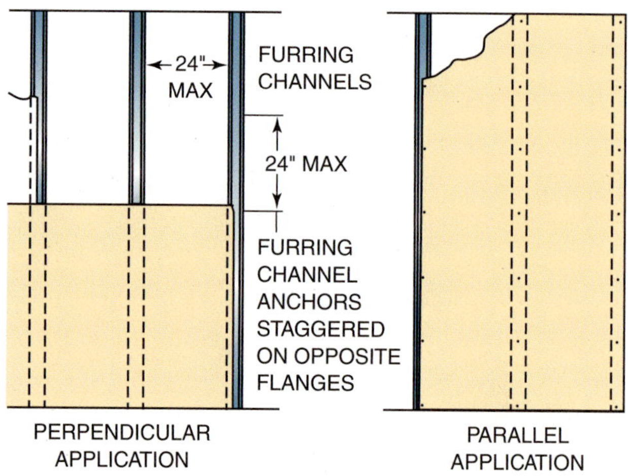

*Courtesy of U.S. Gypsum Corporation.*

**FIGURE 10-46** Furring channels can be attached directly to masonry walls.

## Structural Insulated Panels

Another method of constructing a wall or roof section is to use *structural insulated panels (SIPs)*. They consist of two outer skins and an inner core of an insulating material (Fig. 10-47). This sandwich of materials forms a rigid unit. Most panels use *oriented strand board (OSB)* for their facings. OSB makes panels available in large sizes up to 12 × 36-foot sheets. The cores of SIPs are made of foam insulation often molded from expanded polystyrene, but sometimes from extruded polystyrene or urethane foam.

The insulating core and the two skins of an SIP are not structurally strong by themselves. But when molded together under pressure, these materials create a structural unit that is stiff and strong. The result is that no frame is necessary to support the panels. Panel manufacturers supply splines, connectors, adhesives, and fasteners to erect their systems (Fig. 10-48).

Courtesy of Insulspan.

**FIGURE 10-47** Structural insulated panels are made from thin panel skins and a foam core.

SIPs are produced in thicknesses from 4½ inches to 12¼ inches. Panel sizes typically range from 4' × 8' up to 9' × 24'. Their R-values depend on the foam core and the thickness of the panel. R-value is a rating for a materials resistance to conduction heat. For example, a 4½-inch-thick panel has an R-value of 16, and a 12¼-inch-thick panel has an R-value of 53.

### Green Tip

While additional insulation affects the cost of a building, reductions in energy costs and improvements to the living environment continue over the life of the building.

Blank panels are available, but SIPs are typically prefabricated at the manufacturing plant with all windows and doors precut. They are delivered to the jobsite ready to install with a shop drawing of how to assemble the panels by numbers written on the panels. This feature allows for faster assembly than wood-framed walls and roofs. SIPs have electrical chases precut in the core for insertion of electrical wiring.

Large panels require equipment to set them into position from the truck. Equipment typically used is a forklift, boom truck, or crane depending on the structure being built and site conditions (Fig. 10-49).

Fire performance of SIPs is similar to other wood frame structures. Residential structures are typically required to meet a 15-minute fire performance

Courtesy of Structural Insulated Panel Association.

**FIGURE 10-48** Structural insulated panels are assembled without the traditional wood framework.

Courtesy of Structural Insulated Panel Association.

**FIGURE 10-49** Structural insulated panels are typically assembled using heavy equipment.

*Courtesy of Structural Insulated Panel Association.*

**FIGURE 10-50** Structural insulated panels are sealed at each seam.

standard by applying ½-inch common drywall. Thicker and multiple layers of drywall provide additional fire rating for light commercial and multifamily requirements.

Building with SIPs offer several advantages and benefits over stick framing. It is easy to achieve an airtight home with SIPs (Fig. 10-50). This increases the energy performance of the building. Overall, SIPs use less wood than stick framing. The OSB panels are made from small, fast-growing trees and so provide wood more efficiently than sawn lumber. They also reduce the amount of waste produced during home construction.

# Estimating Materials For Exterior Walls

Estimating the amount of material needed for exterior walls involves studs, wall plates, headers, rough sills, wall sheathing, and temporary bracing. Remember that if the number of pieces does not come out even, round them up to the next whole number of pieces.

## WALL STUDS

Wall studs are estimated from the total linear feet of exterior wall. If they are spaced 16 inches OC, then figure one stud for every linear foot of wall. This allows for the extras needed for corner posts, partition intersections, trimmers, door jacks, and blocking.

## WALL PLATES

For wall plates, multiply the total linear feet of wall by three (one sole plate and two top plates). Then divide by the desired length of the plate material, typically 12 feet. Add 5 percent for waste. For example, if the perimeter is 144 feet and the plate material is to be 12 feet long, the plates needed will be $144 \times 3 \div 12' \times 1.05 = 37.8 = 39$ plates.

## HEADERS

For headers, add up the width of each opening. Add ½ foot per header to allow for extra length of header to sit on jack studs. Divide the total by the length of material being ordered. Add 10 percent for waste.

## ROUGH SILLS

Lineal feet of rough sill material would be the same number as the header material needed for windows.

## WALL SHEATHING

For wall sheathing, first find the total area to be covered. This is found by multiplying the total linear feet of wall by the wall height. Make sure the wall height allows for the area covering the box headers. Deduct the area of any large openings, like garage doors, but retain the area of normal-size openings of windows and doors.

To find the number of sheathing panels, divide the total wall area to be covered by the number of square feet in each sheet. A 4' × 8' panel sheet has 32 square feet. Add about 5 percent for waste. For example, if the perimeter of a rectangular building is 144 feet and the box header is 1 foot wide, the wall sheathing is:

$$144 \times (8 + 1) = 1{,}296 \div 32 \times 1.05 = 42.5 = 43 \text{ sheets.}$$

## TEMPORARY WALL BRACING

Temporary bracing is needed to hold wall sections plumb and secure. Material used is often a 2 × 4 − 12′. The number of pieces may be estimated as one piece per 20 feet of wall plus one per corner. For example, if the perimeter of a rectangular building is 144 feet, the quantity of temporary bracing is:

$$144 \div 20 + 4 = 11.2 = 12 \text{ pieces.}$$

| Estimate the materials for an exterior wall of a rectangular building that measures 30′ × 62′. The box header is one foot wide, six 3246 windows and three 3′-0″ × 6′-8″ doors. | | | |
|---|---|---|---|
| **Item** | **Formula** | **Waste factor** | **Example** |
| **Wall OC Studs** | wall LEN in ft = NUM of studs | | 2 × (30 + 62) = 184′ = 184 PC |
| **Wall plates** | wall LEN × 3 ÷ 12′ = NUM 12′ plates | 5% | 184′ × 3 ÷ 12′ × 1.05 = 48.3 = 49 PC |
| **Headers *win* & *door*** | total of window & door WID plus 6″ per unit | 5% | [6 (3′-2″ + 6″) + 3 (3′-0″ +6″)] × 1.05 = 32.5 = 33 LF |
| **Rough sills** | total of window WID plus 6″ per unit | 5% | 6 (3′-2″ + 6″) × 1.05 = 23.1 = 24 LF |
| **Wall sheathing** | wall area ÷ 32 | 5% | 184′ × (8′ + 1′) ÷ 32 × 1.05 = 54.3 = 55 PC |
| **Temporary braces** | one per 20 feet of wall plus one per corner | | 184 ÷ 20 + 4 = 13.2 = 14 PC |

**FIGURE 10-51** Example of estimating wall framing materials with formulas.

See Figure 10-51 for a listing of the shorthand versions of these estimating techniques with another example.

## SUMMARY

- Walls are comprised of many parts.
- Exterior wall framing must be constructed strong to withstand weather load and to allow for openings.
- Interior wall framing divides the building shell into rooms.
- Ceiling joists support the ceiling finish and insulation layer.
- Steel framing has similar parts as a wood frame but with different names and uses fastening methods.
- Structural insulating panels are prefabricated panels with an insulation layer sandwiched between two surface layers.

## GREEN CHECKLIST

- ☐ Properly size window and door headers; do not overbuild them. This practice saves wood and increases the amount of insulation in the wall.
- ☐ Reduce the number of cripple studs under a window to only those that are on-center. This practice reduces thermal bridging and increases the percentage of insulation in the wall.
- ☐ Buying lumber in precut lengths reduces waste on the job and speeds the construction process.
- ☐ Installing fasteners to the recommended spacing makes the building strong enough to survive its intended load. Recommended spacing also saves material by not overbuilding when too many fasteners are used.
- ☐ While additional insulation affects the cost of a building, reductions in energy costs and improvements to the living environment continue over the life of the building.

PROCEDURE 10-1

# Determining the Length of Studs

*Precut studs are purchased at 92 ½″ long for 8′ ceilings and 104 ½″ for 9′ ceilings. If other ceiling heights are desired, the stud length must be calculated.*

**A** To determine the stud length, the thickness of the finished floor and the ceiling thickness below the joist must be known. To find the height from the sill to the top of the top plate, add the:

- Specified finished floor-to-ceiling height
- Thickness of the ceiling below the ceiling joist. Include the thickness of furring strips in the ceiling, if used.
- Thickness of the finish floor
- Deduct the total thickness of the top plates and the sole plate to find the length of the stud.

For example, if in the plan, the finish floor-to-ceiling height is found to be 7'–9", the finish floor is ¾-inch hardwood, and the ceiling finish is ½-inch drywall, then.

- Add ½ inch for the ceiling thickness and ¾ inch for the finished floor to the finished ceiling height: 7'–9" + ¾" + ½" = 7'–10 ¼"
- Deducting the combined thickness of the top plates and sole plate (usually 4½ inches) gives the stud length: 7'–10¼" – 4½" = 7'–5¾"

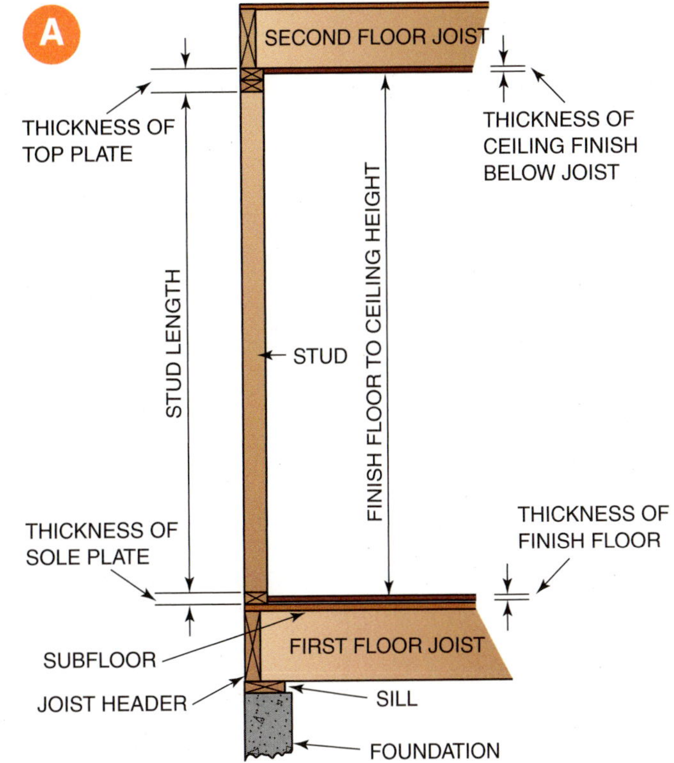


CAUTION

**CAUTION:** It is easy to make mistakes while working with fractions. Being accurate takes practice and discipline. Check your work to be sure it is correct. Catching a mistake during calculations saves costly changes later.

PROCEDURE 10-1

# Determining the Length of Studs (continued)

**B** Studs of balloon frame construction extend from the sill plate to the top plate of the uppermost story. To find the height from the sill to the top of the top plate, add the:

- Finish floor to ceiling heights of all the stories
- Thickness of both finish ceilings
- Thickness of both finish floors and subfloors
- Width of all floor joists

Then deduct the total thickness of the top plates.

For example, in the plans of a two-story house, the finish floor-to-ceiling heights are each found to be 8'–0", the finish floors are each ¾" hardwood, and the finish ceilings are each ½-inch drywall. Floor joists are 2 × 10s, and the subfloor is ⅝-inch. The calculations proceed as follows:

- Finish floor to ceiling heights of all the stories 8'–0" + 8'–0"
- Add the thickness of both finish ceilings + ½" + ½"
- Add the thickness of both subfloors and finish floors + ⅝" + ⅝" + ¾" + ¾"
- Add the width of all floor joists + 9¼" + 9¼" = 17'–10¼"
- Deduct the total thickness of the top plates gives the stud length 17'–10¼" – 3" = 17'–7¼"

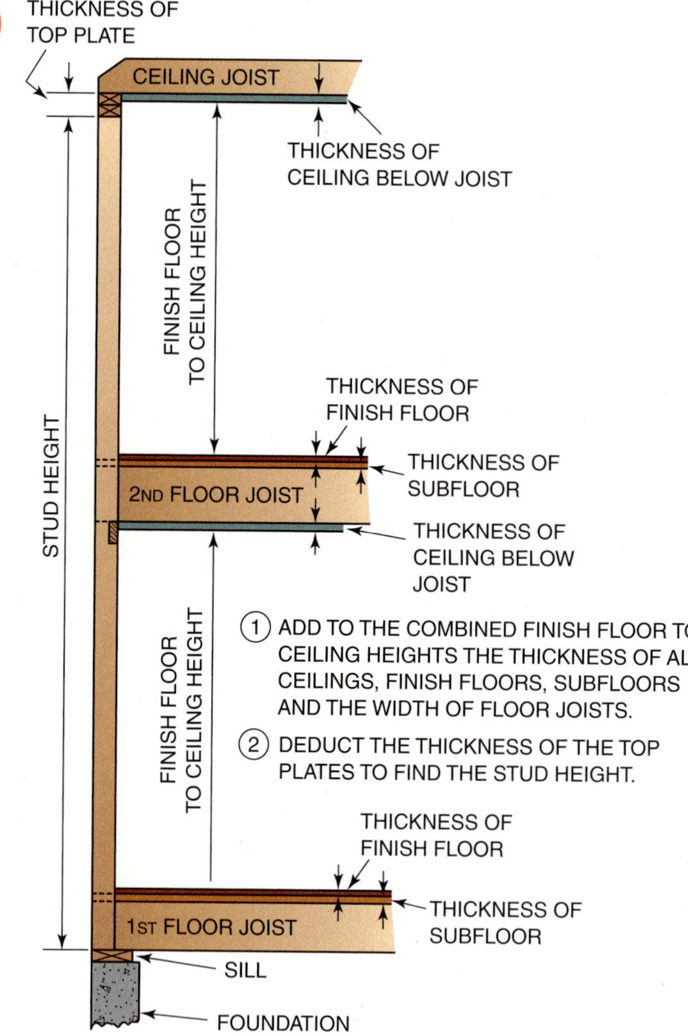

PROCEDURE 10-2

# Determining the Size of Rough Openings

**A** To determine the RO height, add:

- Sill thickness (or finished floor thickness plus a clearance space [½ to 1 inch] under the door)
- Door height
- Head jamb thickness
- Shim space between jamb and header (usually ½ inch)

For example, what is the RO height for a 2'–4" × 6'–8" door with ¾-inch jamb and a 1-inch clearance above a ¾-inch hardwood floor?

- Sill thickness (or finished floor thickness plus a clearance space under the door) + ¾" + 1"
- Add door height + 6'–8"
- Add head jamb thickness + ¾"
- Add shim space (usually ½ inch) + ½" = RO height is ¾" + 1" + 6'–8" + ¾" + ½" = 6'–11"

To determine the jack stud length, deduct the sole plate thickness (usually 1 ½") from RO height.

6'–11" minus 1 ½" = 6'– 9 ½"

**A**

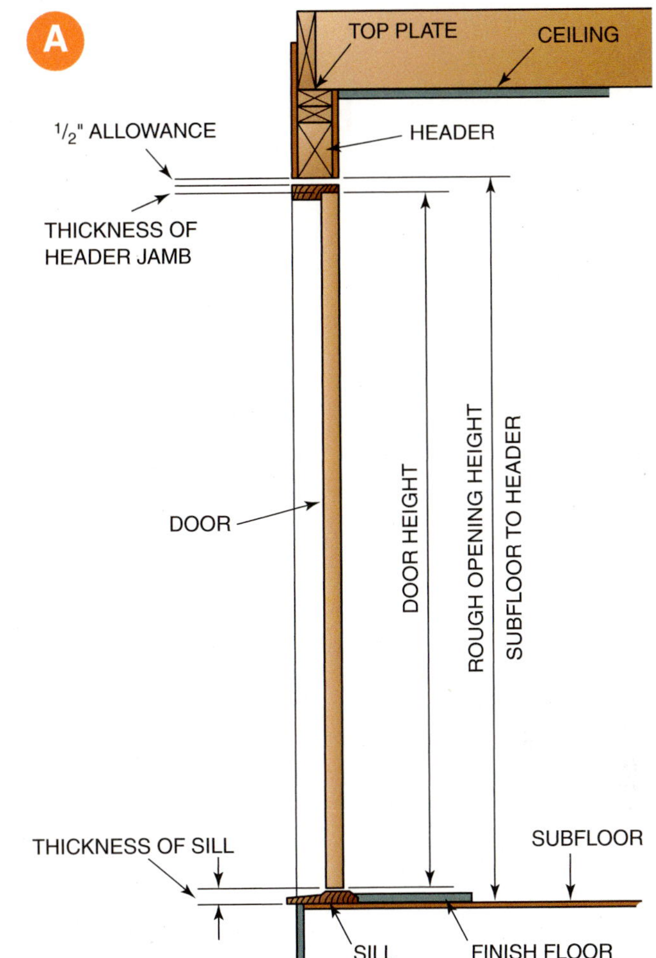

PROCEDURE 10-3

# Laying Out Wall Plate

**A** To begin the wall plate layout, first measure in at the corners, on the subfloor, the thickness of the exterior wall. Snap lines on the subfloor between the marks.

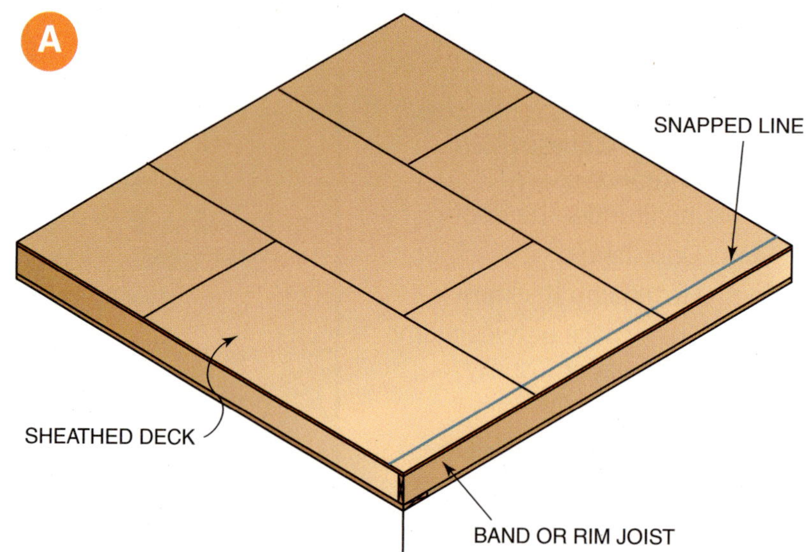


**B** Using straight lengths of lumber for the plates, cut, place, and tack two plates on the deck aligned with the chalk line, one for the sole plate and one for the top plate.

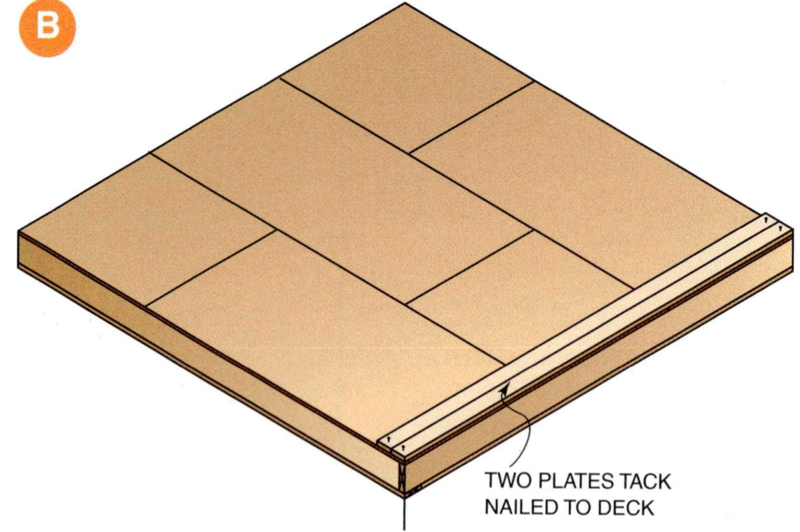

**C** Make sure the butt seams of the plates are centered on a stud. Cut enough plates for the length of the wall section.

**D** From the building prints, determine the centerline dimension of all the openings in the wall. Mark them on the sole plate by using a short line with a *C* over it.

**E** Measure and mark the rough opening width. Recheck the rough opening measurement to be sure it is correct and centered on the centerline. Mark a *T*, *J*, or *O* for the jack stud on the side of each RO line away from the opening center. To distinguish the different openings, use a *T* for window trimmers and a *J* for door jacks. It makes little difference what marks are used as long as the wall assembler understands what they represent.

**F** Measure back from the squared lines, away from the center, the stud thickness, which is usually 1½ inches for the jack stud. Square these lines across the plates. Mark *X*s on the side of the line away from the center. These are for the full-length studs on each side of the openings.

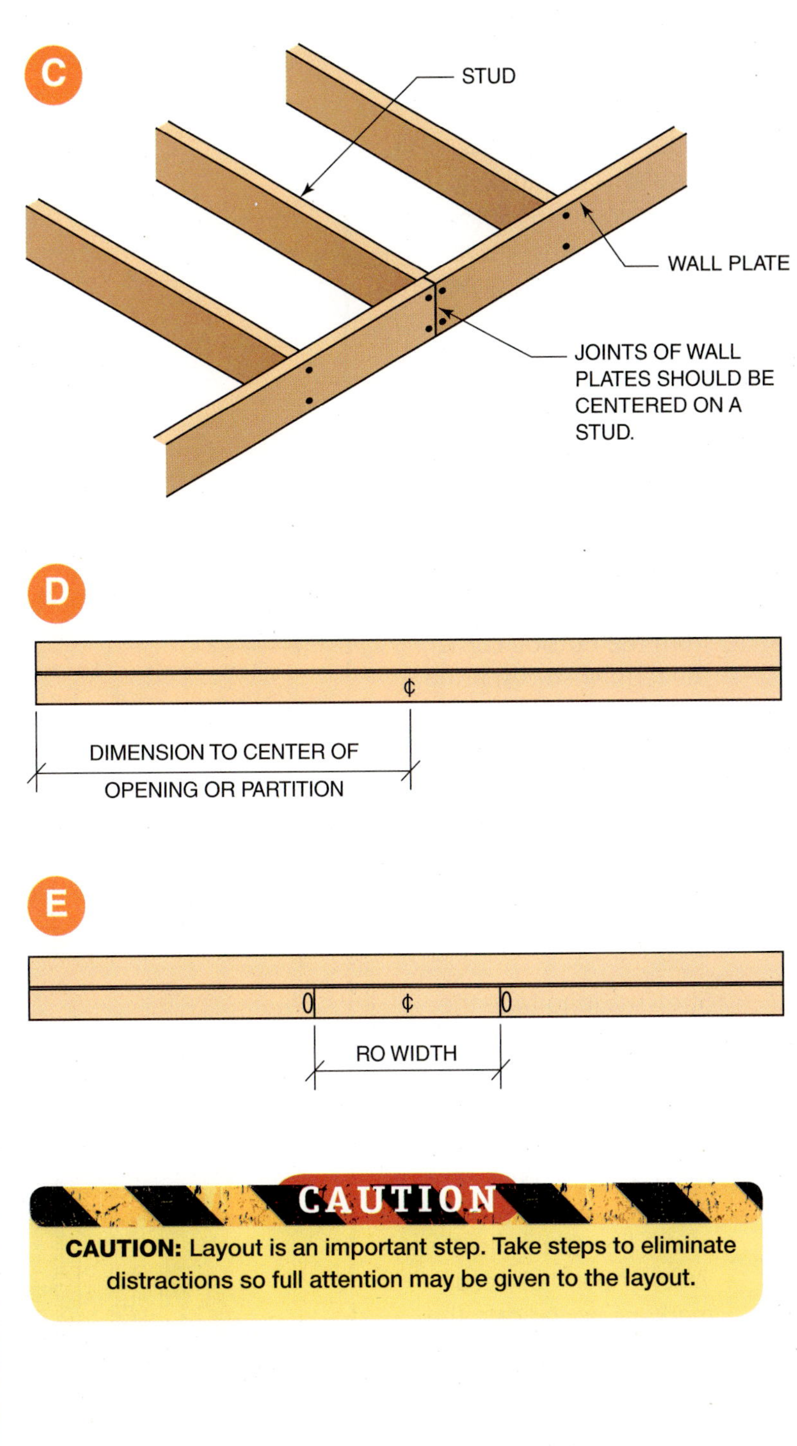

PROCEDURE 10-3

# Laying Out Wall Plate (continued)

**G** Mark the centerlines of all partitions and intersecting the walls on the sole plate. From the centerlines, measure in each direction one-half the partition stud width. Square lines across the sole plate. Mark *X*s on the side of the lines away from center for the location of partition studs.

**H** Mark the on-center stud location by measuring in, from the outside corner, the regular stud spacing. From this mark, measure back one-half the stud thickness. Square this line across the sole plate. Place an *X* on the side of the line where the stud center will be located.

**I** Stretch a long tape the entire length of the wall section from the layout line of the first stud. Mark and square lines across the two plates for each stud. Place *X*s on the same side of the line as the first line. Where openings occur, mark a *C* to indicate the location of cripple studs.

**J** The wall section is then built from the wall plate layout.

G

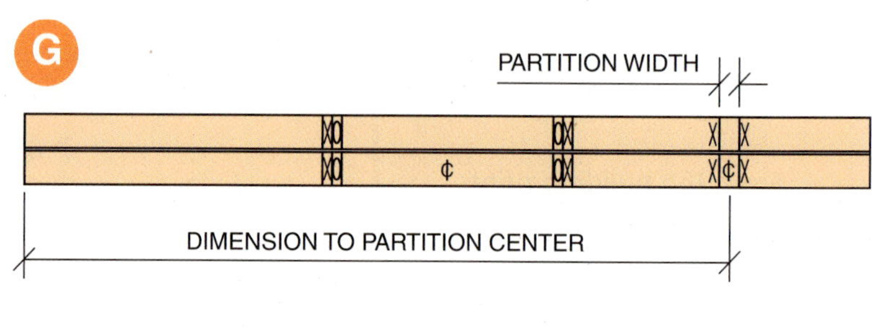


H

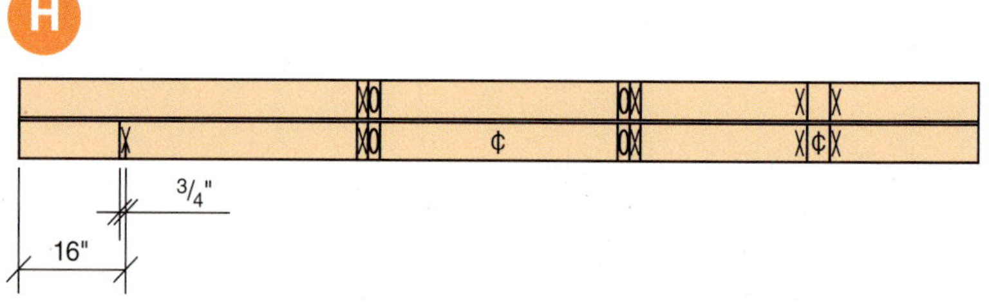


I

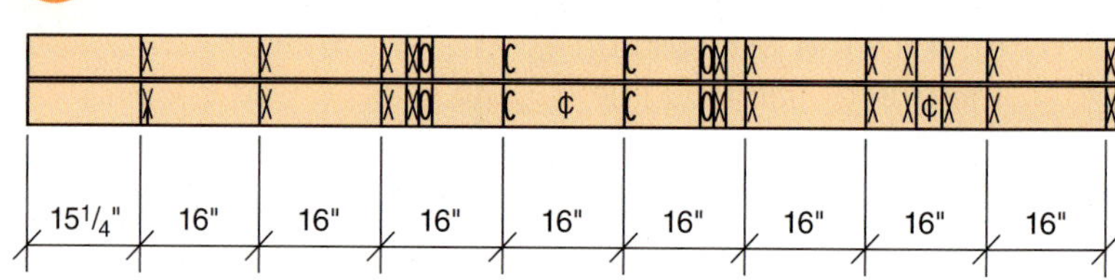


J

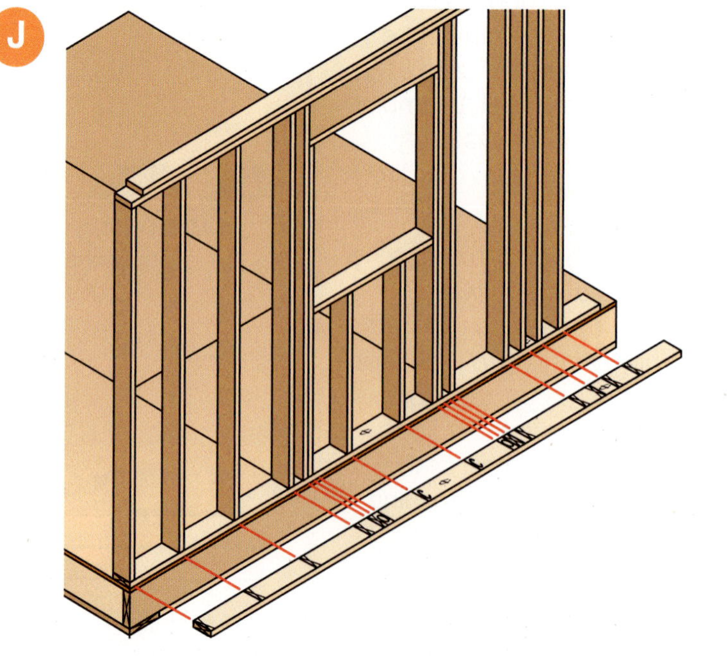

PROCEDURE 10-4

# Assembling Wall Sections

*Drive two 12d or 16d (3¼- or 3½-inch) nails through the plates into the ends of all framing members. A pneumatic framing nailer makes the work faster and easier.*

**A** After the layout is completed, the wall section is assembled.

- Pull the tack nails from the plates, turn up on edge, and separate them a distance equal to the stud length. Take care not to rotate either of the plates lengthwise, making certain the layout lines on top and bottom plates line up.

**A**

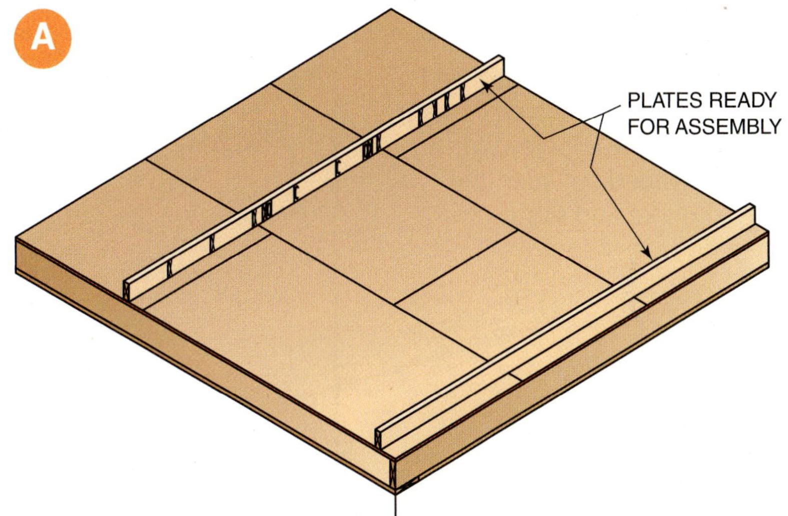


**CAUTION**

**CAUTION:** Careful framing, where the members are all installed flush and even, makes the installation of the finish materials smoother and neater.

**B** Frame the openings first to make assembly easier. Place the jack studs, headers, and rough sills in position. If headers butt the top plate, nail through the plate into the header. Fasten the rough sills in position to the jack studs.

**B**

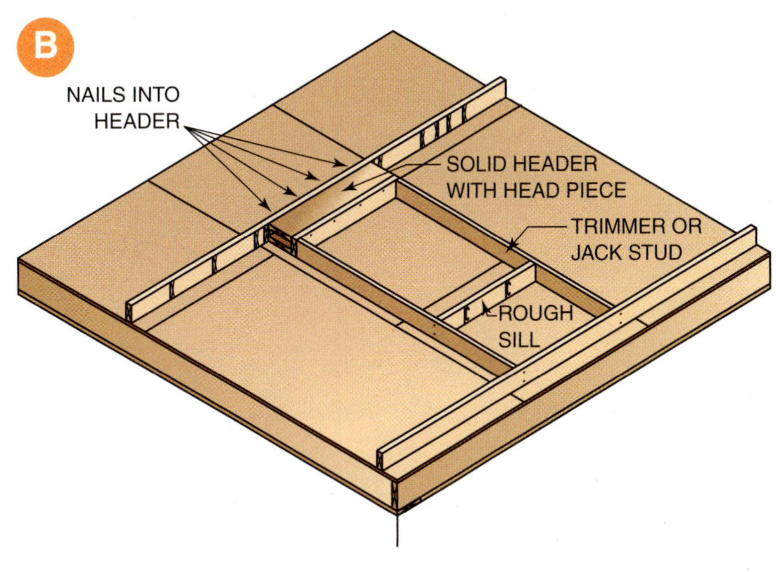

PROCEDURE 10-4

# Assembling Wall Sections (continued)

**C** Fasten a king or full-length stud to each jack stud by driving nails in a staggered fashion about 12 inches apart. Drive the nails at a slight angle so they do not protrude. Also, fasten through the plates. Next, nail the cripple studs as needed between the plates and sills.

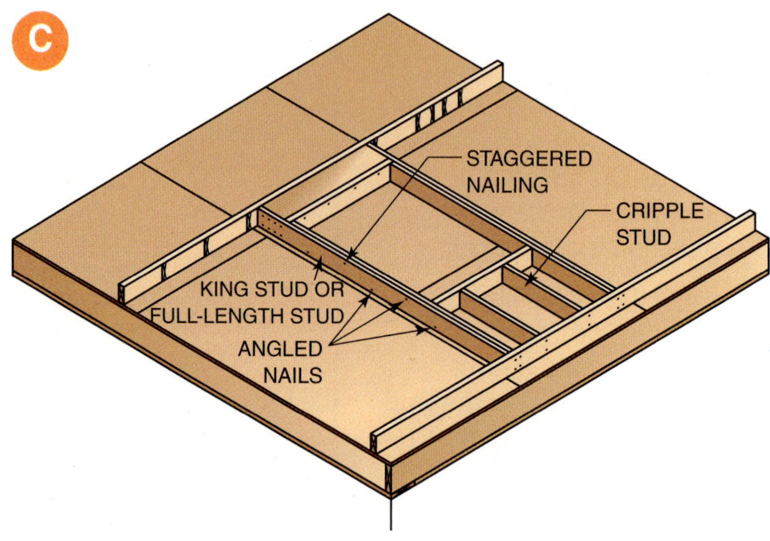


**D** Install the partition studs and then the remaining full studs. Install a doubled top plate. Recess or extend the doubled top plate to make a lap joint. This should be done at all corners and intersections. Take care to nail the doubled top plate into the top plate such that all nails are located above the studs. This will ensure that any holes drilled for wiring or plumbing into the top plates will not hit a nail.

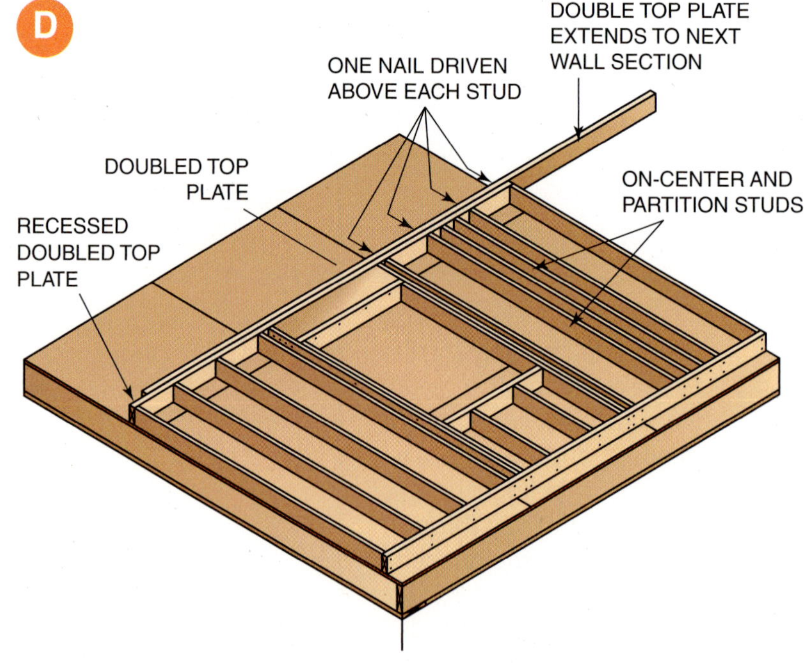

**E** Square the wall section by first aligning the bottom edge of the sole plate to the chalk line on the subfloor. Adjust also the ends of the plate lengthwise into their proper position. Toenail the sole plate to the subfloor with 10d (3-inch) nails spaced about every 6 to 8 feet through what will be the top side of the sole plate when the wall is in its final position. These nails will also help hold the sole plate in position when the wall is erected and may be removed later.

**F** Square may be checked by measuring both diagonals from corner to corner. When they are equal, the section is square. Temporarily brace wall section in its square form. Permanent wind bracing may be achieved with rated wall sheathing, let-in or cut-in bracing. Check local codes for suitability.

- Rated panels are installed in a similar manner as with subfloor and fastened with 6d or 8d fasteners. They may be installed horizontally or vertically.

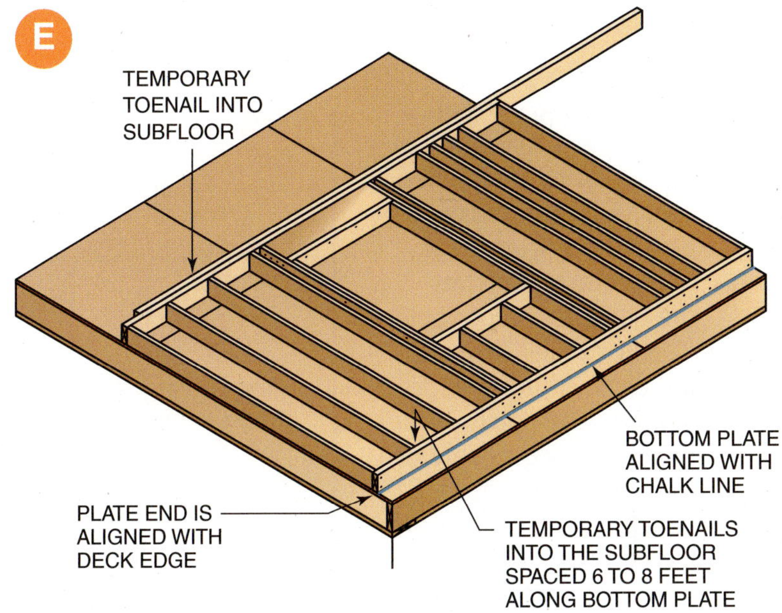


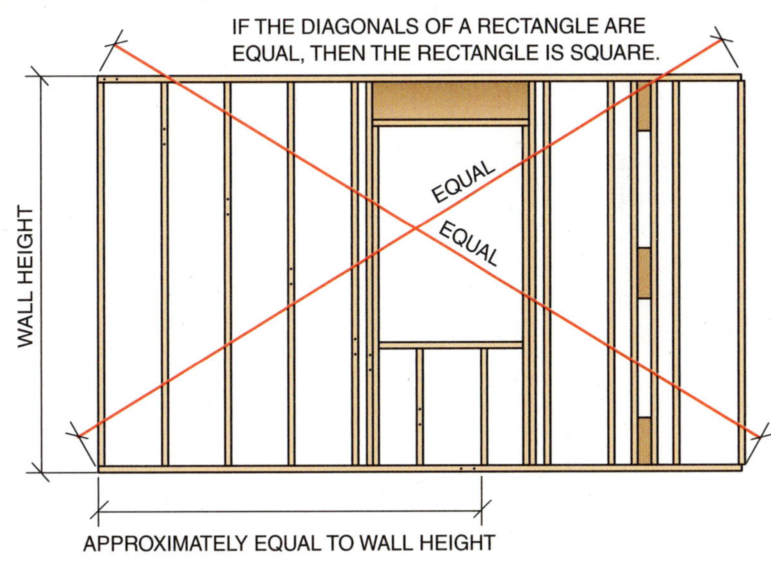

PROCEDURE 10-4

# Assembling Wall Sections (continued)

**G** Let-in bracing is installed by placing a full length of 1 × 4 along the snapped line. Mark the studs and plates along each edge of the brace. Remove the brace. Use a portable electric circular saw with the blade set for the depth of the notch. Make multiple saw cuts between the layout lines. Use a straight claw hammer or a wood chisel to trim the remaining waste from the notch. Fasten the brace in the notches by using two 8d common nails in each framing member.

- Cut-in bracing is installed in a similar fashion as blocking, but the pieces are cut at a 45-degree angle. The kickers are important and should not be omitted (see Fig. 10-1).

**G**

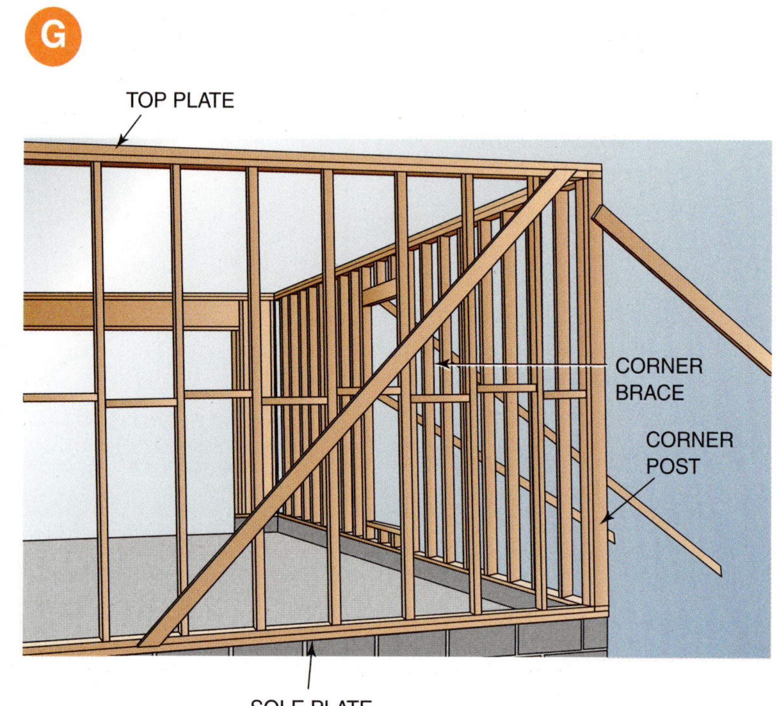

PROCEDURE 10-5

# Erecting and Bracing Wall Sections

**A** Remove the toenails from the top plate while leaving the toenails in the bottom plate. Lift the wall section into place, plumb, and temporarily brace. Use as many braces as required, remembering that the winds may gust and safety always comes first. Install braces at both corners and at intermediate points as needed. Braces may be fastened with one end nailed to the side of a stud and the lower end nailed to a 2 × 4 block that has been nailed to the subfloor.

- Check to be sure that the sole plate is still on the chalk line. Then nail the sole plate to the band or floor joists below every 16 inches along the length. Fasten end studs in the corners together to complete the construction of the corner post.

**B** Straighten the wall by using a string with three blocks of equal thickness. Fasten a block to the side of each end of the top plate. Stretch a line tightly between the blocks so that these blocks hold the string off the wall. Use the third block as a gauge, adjusting for plumb. Adjust the wall in or out with each temporary brace until the gauge block just clears the line when held against the top plate. It must come as close as possible to the line *without* moving the line.

**A**

**CAUTION**

**CAUTION:** A gust of wind can cause the wall to fall off the building. Be careful to maintain good control of the wall section at all times. Also, wall frames can be heavy. Be sure to lift using mostly the leg muscles, keeping your back as straight as possible.

**B**

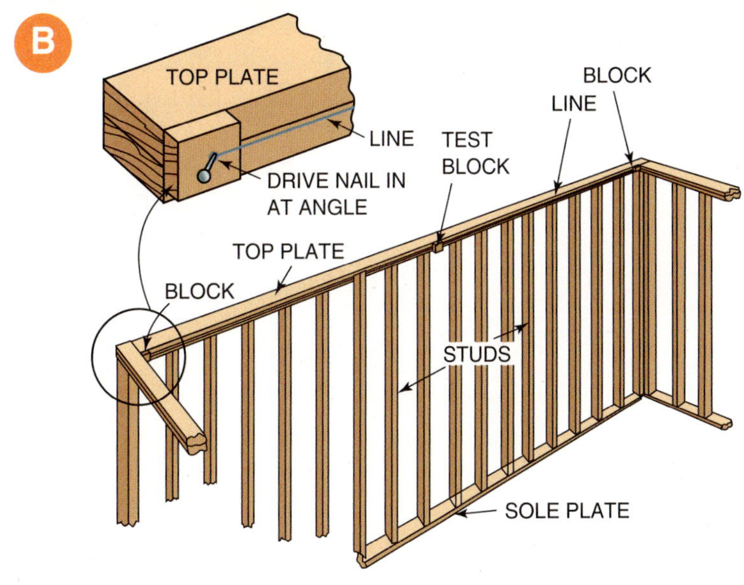

PROCEDURE 10-6

# Installing Backing and Blocking

**A** Marking squared lines on the face of the studs aligned with the chalk line makes installation easier.

- Blocking may be installed in a staggered row by face nailing through the studs into each end of each block. Straight line blocking will require angled face nailing or toenailing on one end.
- Fasten pieces in every other space first. Then go back and fill in. This prevents gaining on the stud space and bowing the studs.

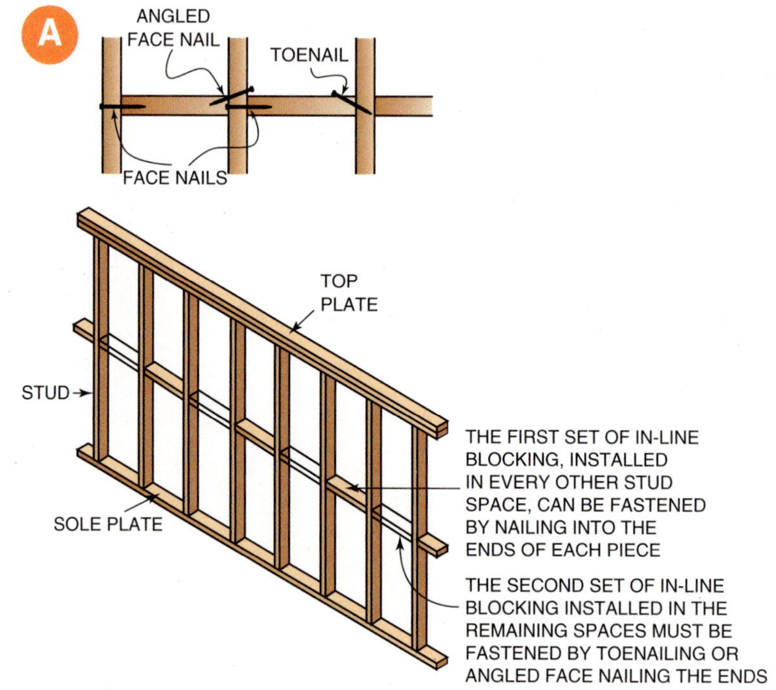


**B** Backing may be installed in a continuous length by notching the studs and fastening into its edges. It may also be installed as pieces between studs.

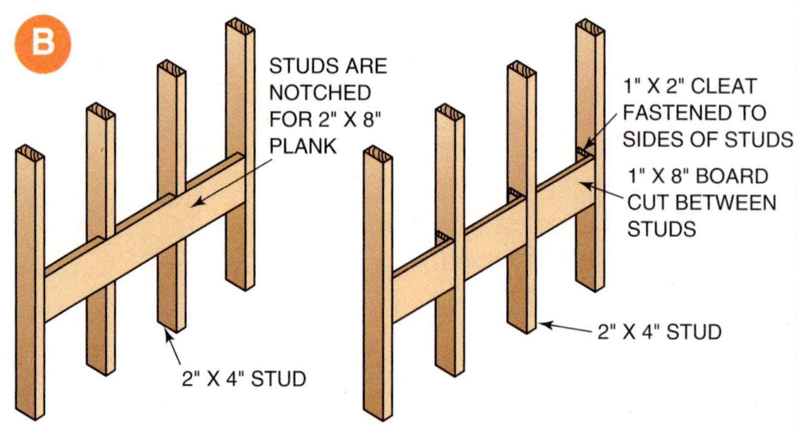

PROCEDURE 10-7

# Layout and Spacing of Ceiling Joists

**A** Lay out spacing lines for the ceiling joists on the doubled top plate. If the studs are accurately located, layout lines can be squared up from the stud onto the top of the doubled top plates.

- For the first wall, mark an *R* for the rafter on one side of the layout line (which side of the line is not critical) and an *X* or a *C* on the other side.
- The layout letters on the opposite wall are determined by the type of ceiling joist construction. When ceiling joists are lapped, place the *R* of the second wall on the opposite side of the line as from the first wall. This will allow for the stagger and lap of rafters and ceiling joists. Mark the *X* and *C* accordingly. When ceiling joists are in-line, place the *R* on the same side of the line as the first wall and mark the *X* and *C*.
- The layout letters on the load-bearing partition must reflect the decisions made on the outside walls. Joists bearing on the partition must stay on the same side of the line as they are on the wall. Note the *X* is made to position one ceiling joist, and the *C* is made for the other.

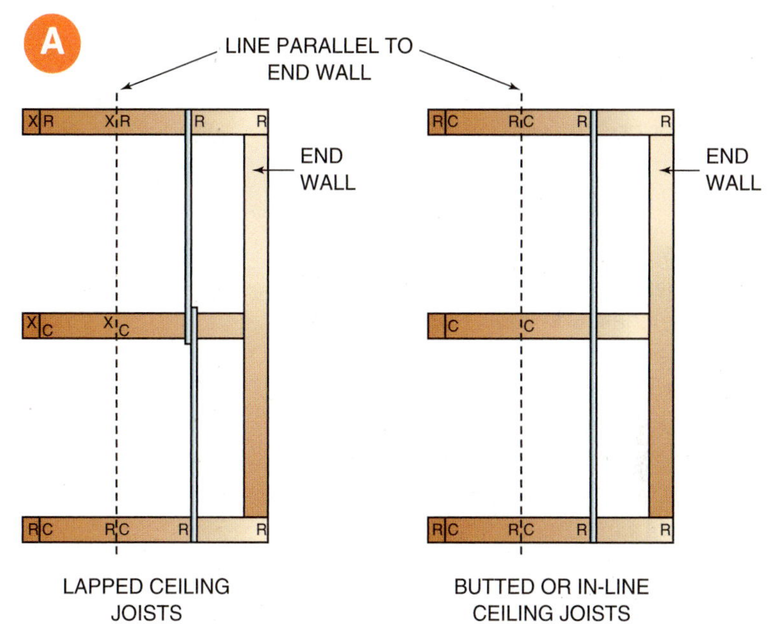

PROCEDURE 10-8

# Installing Metal Framing

A Snap chalk lines on the floor. Plumb up to the ceiling from the partition ends. Snap lines on the ceiling. Check to make sure that partitions will be plumb.

- Fasten track to floor and ceiling so one edge is to the chalk line. Make sure both floor and ceiling track are on the same side of the line. Leave openings in floor track for door frames. Allow for the width of the door plus the thickness of the door frame. At corners, extend one track to the end, then butt or overlap the other track. It is not desirable or necessary to make mitered joints.

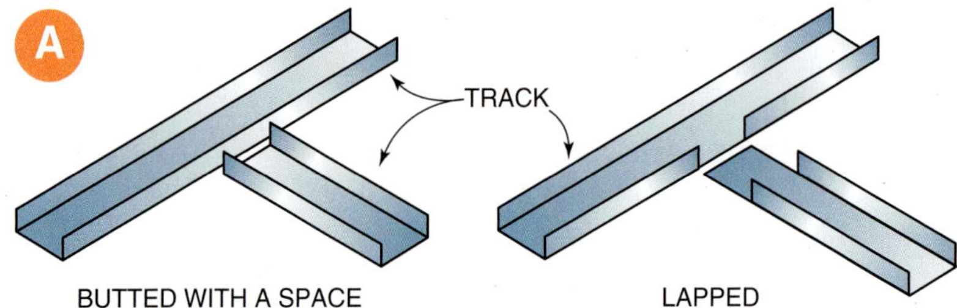


B Lay out the stud spacing and the wall openings on the bottom track. Install the first stud plumb. The top track can be laid out from this first stud, or each stud may be plumbed from the bottom plate as it is installed.

- Cut the necessary number of full-length studs. For ease of installation, cut them about ¼ inch short of the exact length.
- Install studs at partition intersections and corners, fastening to bottom and top track. If moisture may be present where a stud butts an exterior wall, place a strip of asphalt felt between the stud and the wall.

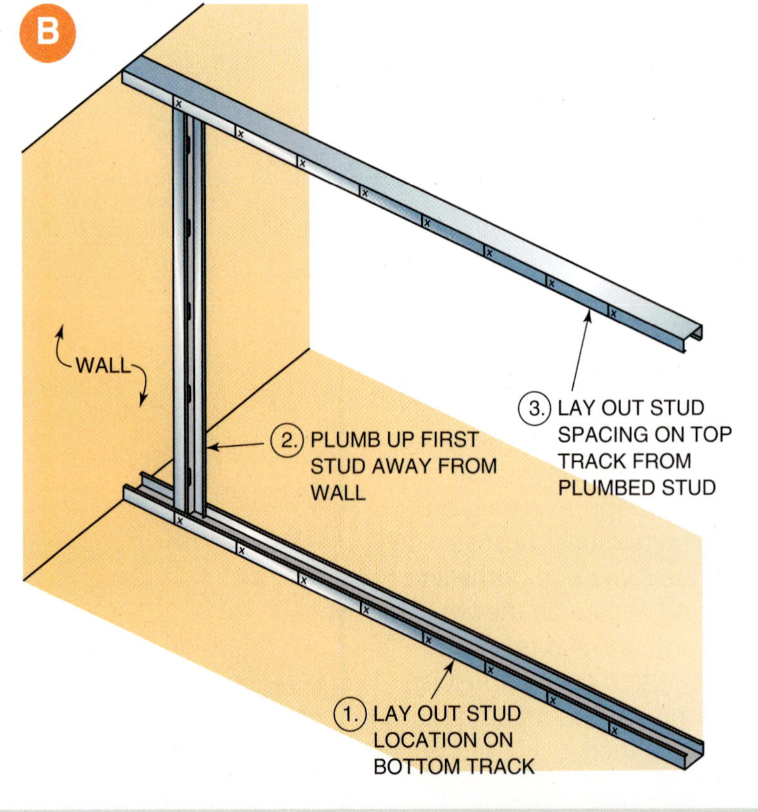

PROCEDURE 10-9

# Framing a Door Opening for a Three-Piece Frame

The three-piece door unit may be installed later after the wall is framed.

**A** First, place full-length studs on each side of the opening in a plumb position. Fasten securely to the bottom and top plates.

- Cut a piece of track for use as a header. Cut it 2 inches longer than the width of the opening to allow for fastening to the studs. Fasten the fabricated header to the studs at the proper height.
- Install the door frame as per the manufacturer's instructions.

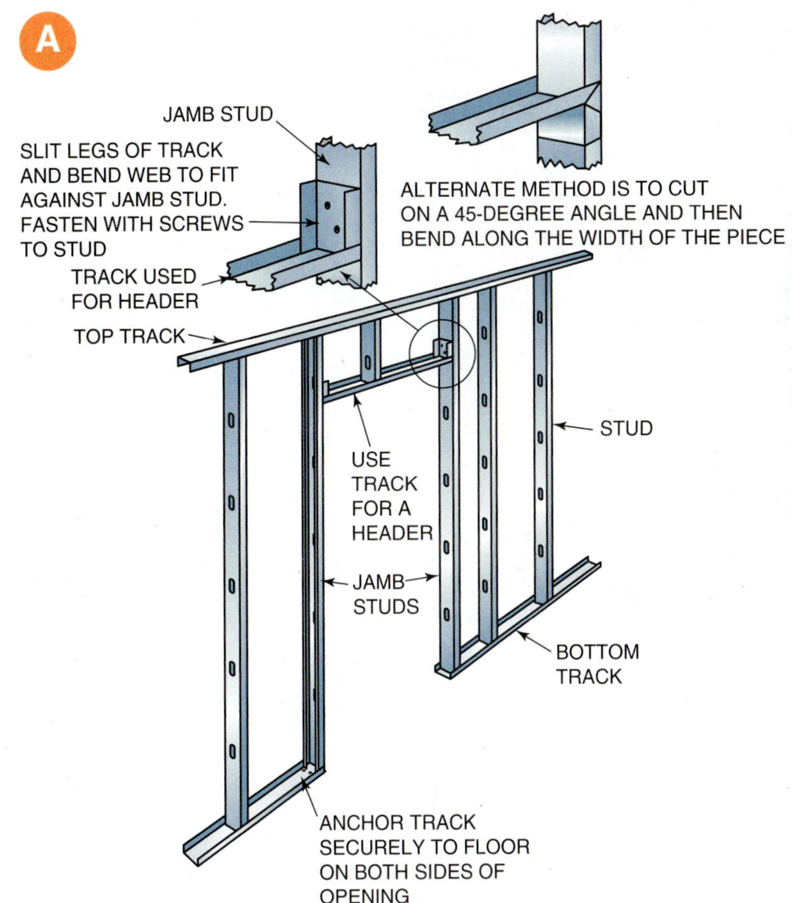

PROCEDURE 10-10

# Framing a Door Opening for a One-Piece Frame

**A** Place the studs on each side of the opening, but do not fasten to the track yet.

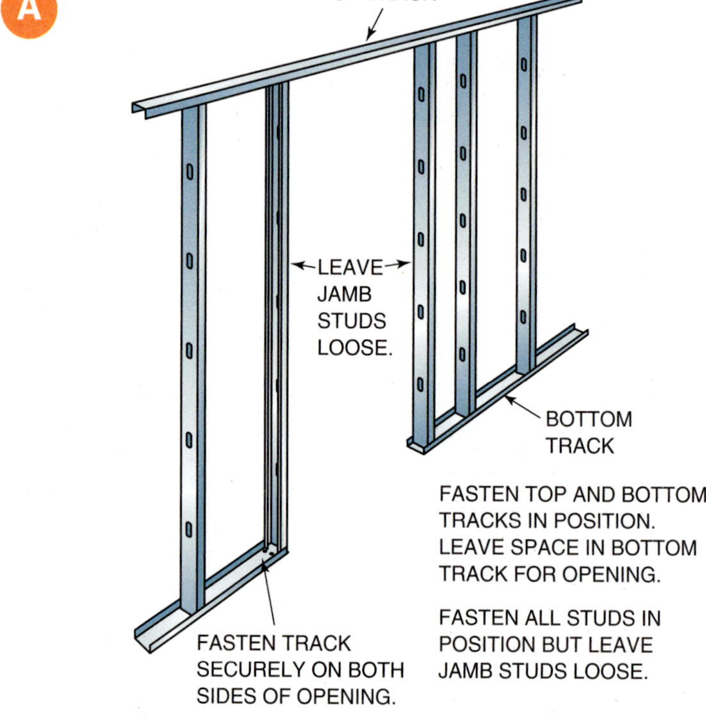


**B** Set the one-piece door frame in place by using the studs interlocked with the tracks to help hold the door unit. Level the door frame header by shimming under a jamb, if necessary.

- Fasten the bottom ends of the door jambs to the floor in the proper location. Fasten the studs to the door jambs and then fasten the studs to the bottom track only.

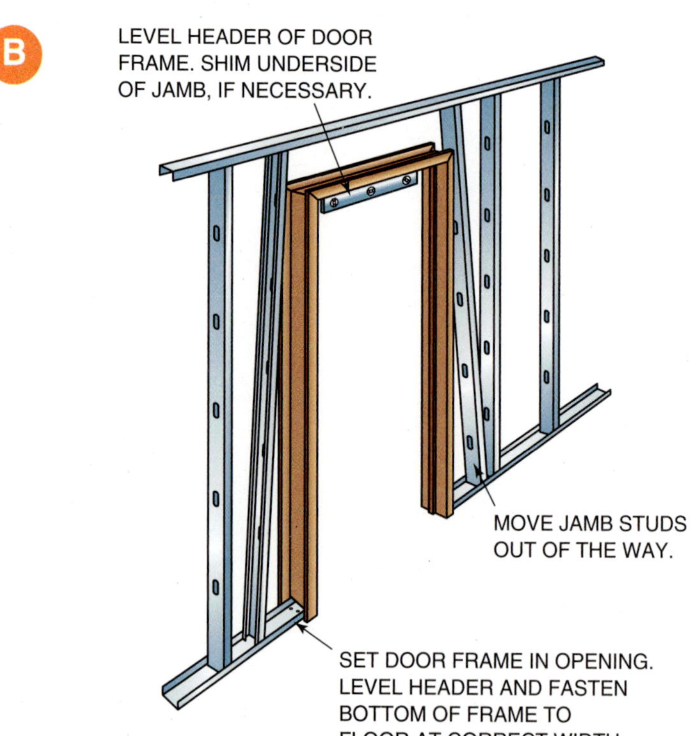

**C** Plumb the door frame by movement in the top track and fasten to the top track with screws. Install header and cripple studs in the same manner as described previously.

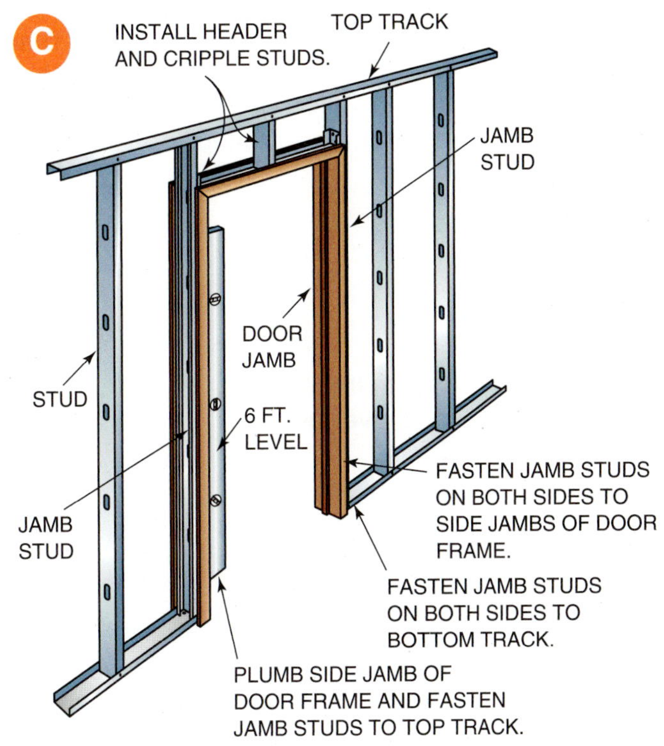


**D** For wider and heavier doors than standard 2'–8", the framing should be strengthened by using 27-mil steel framing. Also, doubling the studs by *nesting* on each side of the door opening will strengthen the opening.

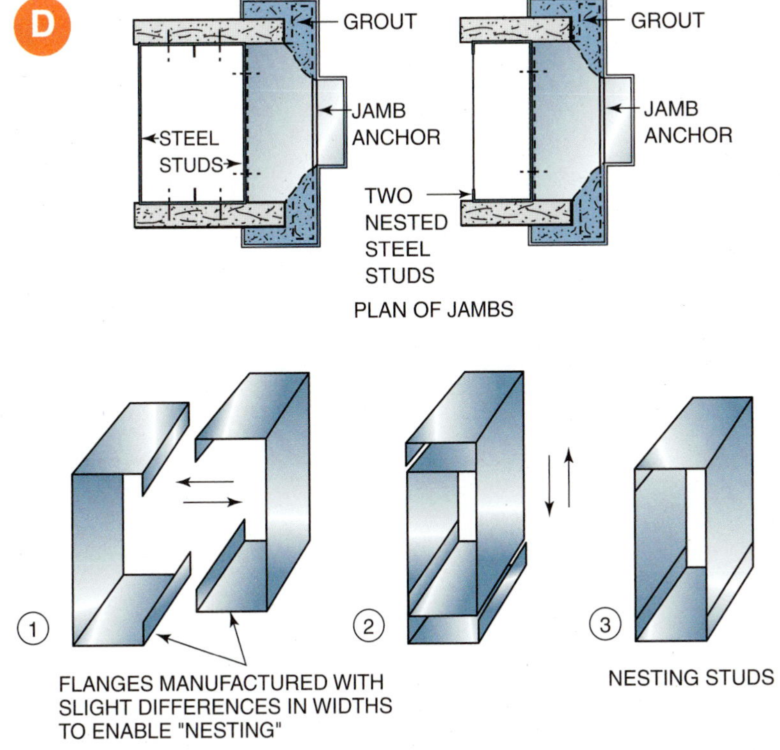


*Courtesy of U.S. Gypsum Corporation.*

# REVIEW QUESTIONS

*Select the most appropriate answer.*

1. **The top and bottom horizontal members of a wall frame are called**
   a. headers.
   b. plates.
   c. trimmers.
   d. sills.

2. **The horizontal wall member supporting the load over an opening is called a**
   a. header.
   b. rough sill.
   c. plate.
   d. truss.

3. **Shortened studs above and below openings are called**
   a. shorts.
   b. lams.
   c. cripples.
   d. stubs.

4. **Diagonal cut-in bracing requires the installation of**
   a. kickers.
   b. backing.
   c. blocking.
   d. 1 × 4s.

5. **The finish floor-to-ceiling height in a platform frame is specified to be 7'–10". The finish floor is ¾-inch thick, and the ceiling material is ½-inch thick. A single sole plate and a double top plate are used, each of which has an actual thickness of 1½ inches. What is the stud length?**
   a. 7'–5¾"
   b. 7'–6¾"
   c. 7'–8¼"
   d. 7'–10½"

6. **A jamb is ¾-inch thick. Allowing ½-inch on each side for shimming the frame, what is the rough opening width for a door that is 2'–8" wide?**
   a. 2'–9½"
   b. 2'–10½"
   c. 2'–11½"
   d. 3'–10½"

7. **When plates for walls and partitions are laid out, measurements for centerlines of openings start from the**
   a. end of the plate.
   b. outside edge of the abutting wall.
   c. building line.
   d. nearest intersecting wall.

8. **The first OC stud is set back**
   a. a distance that is usually ¾ inch.
   b. ½ stud thickness.
   c. to allow the first sheathing piece to be installed flush with the first stud.
   d. all of the above.

9. **A corner stud that allows for ample room for insulation in the corner uses**
   a. three small blocks.
   b. a stud that is rotated from the others in the wall.
   c. three full studs nailed as a post.
   d. all of the above.

10. **Exterior walls are usually straightened before ceiling joists are installed by**
    a. using a two-foot carpenter's level.
    b. using a line stretched between two blocks and testing with a gauge block.
    c. using a plumb bob dropped to the sole plate at intervals along the wall.
    d. by sighting along the length of the wall using a builder's level.

11. **Blocking and backing are installed**
    a. using up scrap lumber first.
    b. as a nail base for cabinets.
    c. as secure parallel partitions to ceiling joists.
    d. all of the above.

12. **Bearing partitions**
    a. have a single top plate.
    b. carry no load.
    c. are constructed like exterior walls.
    d. are erected after the roof sheathing is installed.

13. **The doubled top plate of the bearing partition**
    a. laps the plate of the exterior wall.
    b. is a single member wider than the bottom plate.
    c. butts the top plate of the exterior wall.
    d. is applied after the ceiling joists are installed.

14. **What is the rough opening height of a door opening for a 6'–8" door if the finish floor is ¾-inch thick, ½-inch clearance is allowed between the door and the finish floor, and the jamb thickness is ¾ inch?**
    a. 6'–9"
    b. 6'–9½"
    c. 6'–10"
    d. 6'–10½"

15. **The type of plywood typically used for wall sheathing is**
    a. CDX.
    b. AC.
    c. BC.
    d. hardwood.

16. **Ceiling joists are typically installed**
    a. with their end joints lapped on the bearing partition.
    b. full length along the building length.
    c. after rafters are installed.
    d. with blocks placed between them at the load-bearing partition.

**17. The ends of ceiling joists are cut to the slope of the roof**

a. for easy application of the wall sheathing.

b. so they will not project above the rafters.

c. so their crowned edges will be down.

d. after they are fastened in position.

**18. When you work with steel framing, it can be noted that**

a. special fasteners are needed.

b. top plates are usually doubled.

c. studs are also called track.

d. stud location is not that important.

**19. Estimate the number of 16-inch OC exterior wall studs needed for a rectangular house that measures 28 × 48 × 8 feet high.**

a. 76

b. 152

c. 1344

d. 10,752

**20. Estimate the number of pieces of wall sheathing needed for a rectangular house that measures 28 × 48 × 8 feet high. Figure an extra foot of material to cover the box header. Neglect the openings and gable end. Add 5 percent for cutting waste.**

a. 42

b. 43

c. 44

d. 45

## KNOW YOUR CODES

Regions of the country with seismic activity and high wind have more forces on the building than other regions. Walls in high stress areas are constructed and fastened to withstand these forces. Walls are also built to lower heating and cooling demands. Energy codes for various regions of the country will adjust the wall thickness and composition to improve thermal efficiency. Colder regions use 2 × 6 wall studs to allow space for more insulation. Other regions will use 2 × 4's. Make sure to check the building prints and local codes for your area.

## WHAT'S WRONG WITH THIS PICTURE?

Carefully study **Figure 10-52** and think about what is wrong. Consider all possibilities.

### WRONG

**FIGURE 10-52** The two leftmost vertical framing members are studs, and the rightmost member is a cripple stud. The two horizontal members are the double top plates. The (lower) top plate does not break on top (middle) of a stud. This is poor building technique and should be avoided.

### RIGHT

**FIGURE 10-53** This picture shows a stud with the first plate correctly breaking on the center of it and the second top plate spanning the splice. The other visible member above the plate is a roof truss.

CHAPTER 11

# Scaffolds, Ladders, and Sawhorses

**OBJECTIVES** *After completing this unit, the student should be able to:*

- Identify and describe the safety concerns when using scaffolds.
- Erect and dismantle metal scaffolding in accordance with recommended, safe procedures.
- Follow a recommended procedure to inspect a scaffold for safety.
- Describe the recommended capacities of various parts of a scaffold.
- Construct a scaffold work platform.
- Identify and describe the components of a fall protection system.
- Describe the safety concerns when using mobile metal tubular scaffolds.
- Build safe staging areas using roof brackets.
- Safely set up, use, and dismantle pump jack scaffolding.
- Describe the safe use of ladders, ladder jacks, stepladders, and sawhorses.

# GLOSSARY OF SCAFFOLD, LADDER, AND SAWHORSE TERMS

**cleat** a small strip of wood applied to support a shelf or similar piece

**competent person** designated person on a job site who is capable of identifying hazardous or dangerous situations and has the authority to take prompt corrective measures to eliminate them

**crib** heavy wood blocks and framing used as a foundation for scaffolding

**erectors** workers whose responsibilities include safe assembly of scaffolding

**users** people who work on scaffolding

Scaffolds, ladders, and sawhorses aid the carpenter in working against gravity. Scaffolds, sometimes called *staging*, raise the work area to a desired level, thereby allowing work to continue at great heights. Ladders are the means by which these raised work areas are accessed. Sawhorses can create a raised surface from which to work. Much of the work done on a construction site would be impossible without these aids.

These construction aids pose built-in problems. They are involved in most job site accidents. Because scaffolds and ladders are temporary by nature, they can easily be overlooked as a serious threat to personal safety. For this and other reasons, contractors are carefully watched by safety organizations to ensure their job sites are a safe place to work. But safety begins with each worker.

Safety requires that every worker have the proper attitude, taking steps to work smart and safe at all times. Workers must be responsible for their own safety as well as those around them. Safety is a group effort, and all must comply for safety to exist on the job.

## Scaffolds

Scaffolds are an essential component of construction, as they allow work to be performed at various elevations. However, they also can create one of the most dangerous working environments. All workers on the scaffold must wear hard hats.

The U.S. Occupational Safety and Health Administration (OSHA) reports that in construction, falls are the number one killer, and 40 percent of those injured in falls had been on the job less than one year. A recent survey of scaffold accidents summarizes the problem (Fig. 11-1). A scaffold fatality and catastrophe investigation conducted by OSHA revealed that the largest percentage, 47 percent, was due to equipment failure. In most instances, OSHA found the equipment did not just break; it was broken due to improper use and erection. Failures at the anchor points, allowing either the scaffold parts or its anchor points to break away, were often involved in these types of accidents. Other factors were improper, inadequate, and improvised construction and inadequate fall protection. The point of this investigation is that accidents do not just happen; they are caused.

OSHA regulations on the fabrication of frame scaffolds are found in the Code of Federal Regulations 1926.450, 451, and 452. Workers must understand these regulations thoroughly before any scaffold is erected and used. Furthermore, safety codes that are more restrictive than OSHA, such as those in Canada, California, Michigan, and Washington, should be consulted. Scaffolds must be strong enough to support workers, tools, and materials. They must also provide a safety margin. The standard safety margin requirement is that all scaffolds must be capable of supporting at least four times the maximum intended load.

Those who erect scaffolding must be familiar with the different types and construction methods of scaffolding to provide a safe working platform for all workers. The type of scaffolding depends

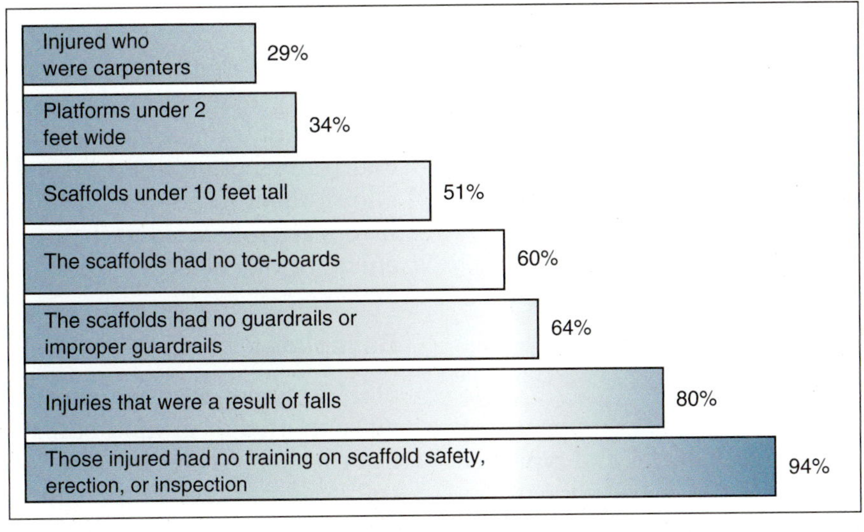


**FIGURE 11-1** Recent accident statistics involving scaffolding.

on its location, the kind of work being performed, the distance above the ground, and the load it is required to support. All workers deserve to be able to return to their families after work without injury.

The regulations on scaffolding enforced by OSHA make it clear that before erecting or using a scaffold, the worker must be trained about the hazards surrounding the use of such equipment. OSHA has not determined the length of training that should be required. Certainly that would depend on the expertise of the student in training. Proof of training must be in writing.

Employers are responsible for ensuring that workers are trained to erect and use scaffolding. One level of training is required for workers, such as painters, to work from the scaffold. A higher level of training is required for workers involved in erecting, disassembling, moving, operating, repairing, maintaining, or inspecting scaffolds.

The employer is required to have a **competent person** to supervise and direct the scaffold erection. This individual must be able to identify existing and predictable hazards in the surroundings or working conditions that are unsanitary, hazardous, or dangerous to employees. This person also has authorization to take prompt corrective measures or eliminate such hazards. A competent person has the authority to take corrective measures and stop work, if necessary, to ensure that scaffolding is safe to use.

## METAL TUBULAR FRAME SCAFFOLD

Metal tubular frame scaffolding consists of manufactured end frames with folding cross braces, adjustable screw legs, baseplates, platforms, and guardrail hardware (Fig. 11-2). Frame scaffolds are

**FIGURE 11-2** A typical metal tubular frame scaffold.

easy to assemble, which can lead to carelessness. Because untrained erectors may think scaffolds are just stacked up, serious injury and death can result from a lack of training.

## METAL SCAFFOLD COMPONENTS

There are many parts and pieces that make up a metal tubular scaffold.

### End Frames

End frames are erected in sections that consist of two end frames and two cross braces and typically come in 5 × 7 modules. They can be wide or narrow. Some are designed for rolling tower scaffolds, while other frames have an access ladder built into the end frame (Fig. 11-3).

### Cross Braces

Cross braces rigidly connect one scaffold member to another member. Cross braces connect the bottoms and tops of frames. This diagonal bracing keeps the end frames plumb and provides the rigidity that allows them to attain their designed strength. The braces are connected to the end frames using a variety of locking devices (Fig. 11-4).

### Baseplates

OSHA regulations require the use of baseplates on all supported or ground-based scaffolds (Fig. 11-5) in order to transfer the load of scaffolding, material, and workers to the supporting surface. It is extremely important to distribute this load over an area large enough to reduce the pounds–per-square-inch load on the ground.

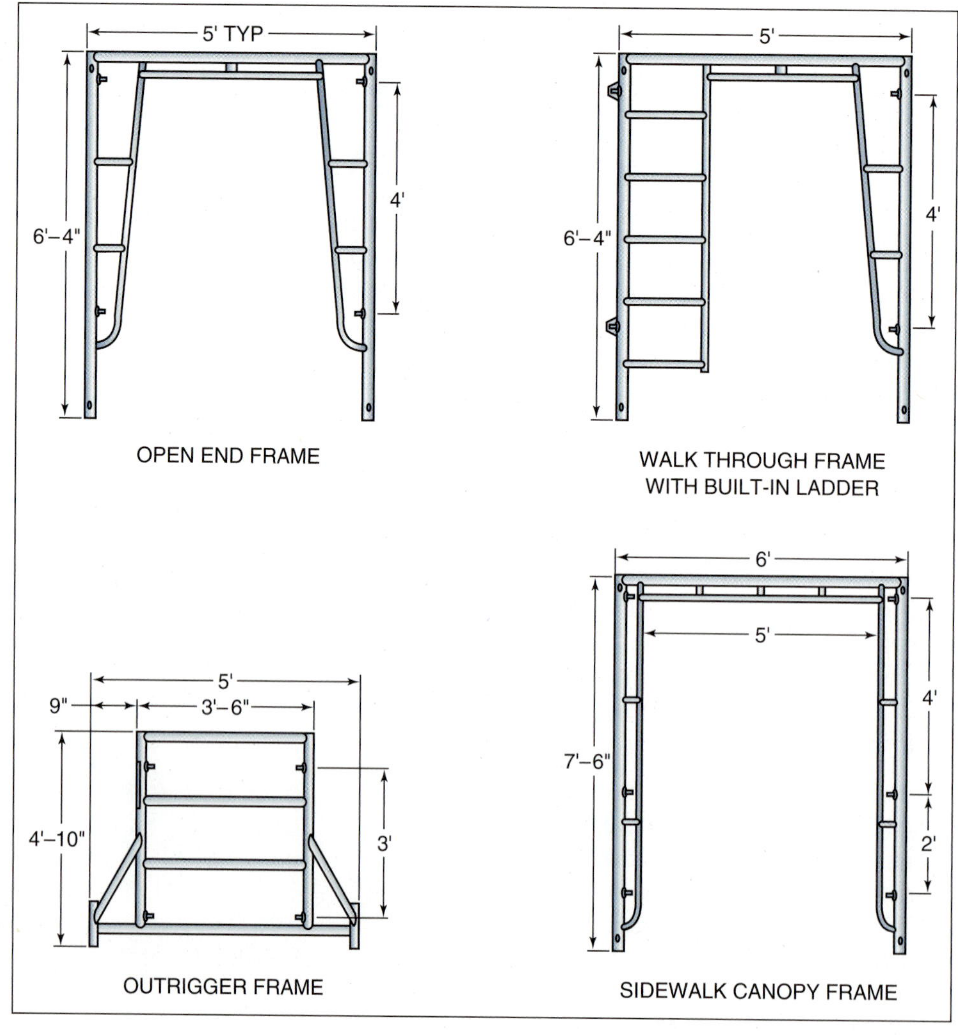


**FIGURE 11-3** Four examples of typical metal tubular end frames.

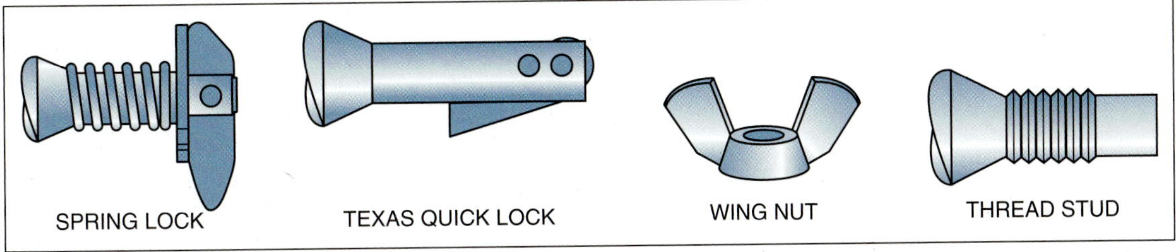


**FIGURE 11-4** Locking devices used to connect cross braces to end frames.

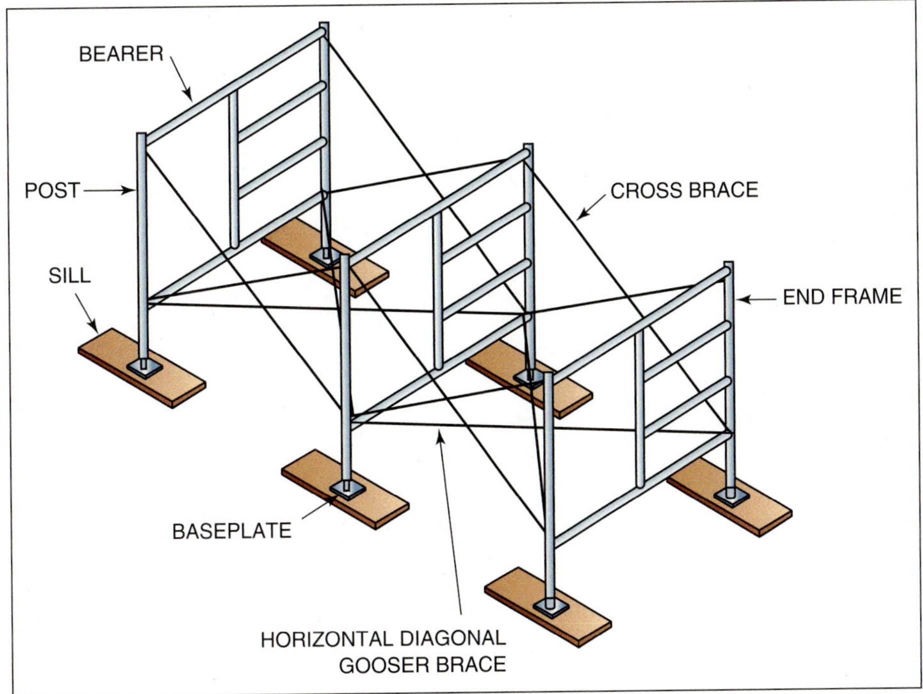


**FIGURE 11-5** Typical baseplate setup for a metal tubular frame scaffold.

## Mud Sill

If the scaffold sinks into the ground when it is being used, accidents will occur. Therefore, baseplates should sit on and be nailed to a mud sill (Fig. 11-6). A *mud sill* is typically a 2 × 10 board approximately 18 to 24 inches long. On soft soil, it may need to be longer and thicker.

## Screwjack Legs

To level an end frame while erecting a frame scaffold, screwjacks may be used. At least one-third of the screwjack must be inserted in the scaffold leg. Lumber may be used to crib up the legs of the scaffold (Fig. 11-7). Cribbing height is restricted to equal the length of the mud sill. Therefore, using a 19-inch long, 2 × 10 mud sill, the crib height is limited to 19 inches. OSHA also prohibits the use of concrete blocks to level scaffolding.

## Plank

Staging planks rest on the bearers. They are laid with the edges close together so the platform is tight. There should be no spaces through which tools or materials can fall. All planking should be scaffold grade or its equivalent. Planking may have the ends banded with steel to prevent excessive checking. Overlapped planks should extend at least 6 inches beyond the bearer. End planks should not overhang the bearer by more than 12 inches (Fig. 11-8).

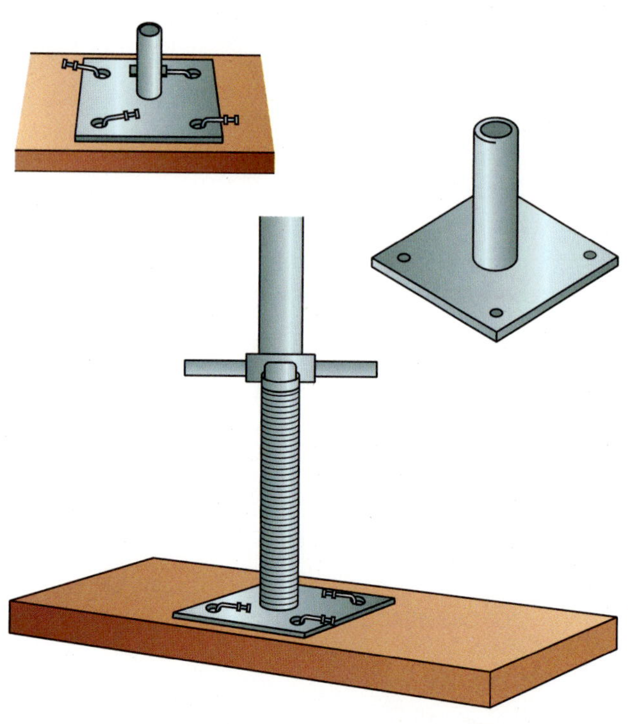

**FIGURE 11-6** Baseplates should be nailed to the mud sill.

## Guardrails

Guardrails are installed on all open sides and ends of scaffolds that are more than 10 feet in height. A guardrail system is a vertical fall-protection barrier consisting of, but not limited to, toprails, midrails, toeboards, and posts (Fig. 11-9). This system prevents employees from falling off a scaffold platform or walkway.

*Top Rail.* Guardrail systems must have a top rail capacity of 200 pounds applied downward or horizontally. The toprail must be between 38 and 45 inches above the work deck.

*Midrail.* The midrail is installed midway between the upper guardrail and the platform surface. The midrail must have a capacity of 150 pounds applied downward or horizontally.

*Toeboard.* If workers are on different levels of the scaffold, toeboards must be installed as an overhead protection for lower-level workers. Toeboards are typically 1 × 4 boards installed under the midrail at the platform. If materials or tools are stacked up higher than the toeboards, screening must be installed.

## Coupling Pins

Coupling pins are used to stack the end frames on top of each other (Fig. 11-10). They have holes in them that match the holes in the end frame legs; these holes allow locking devices to be installed. Workers must ensure the coupling pins are designed for the scaffold frames in use.

## Uplift Protection

The scaffold end frames and platforms must have uplift protection installed when a potential for uplift exists. Installing locking devices through the legs of the scaffold and the coupling pins provides this protection (Fig. 11-11). If the platforms are not equipped with uplift protection devices,

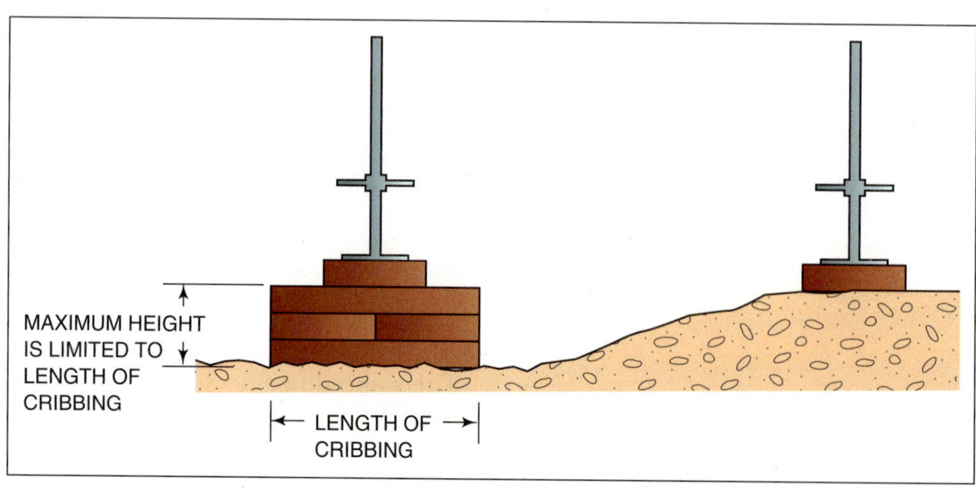


**FIGURE 11-7** Cribbing may be used to level the ground under the scaffold.

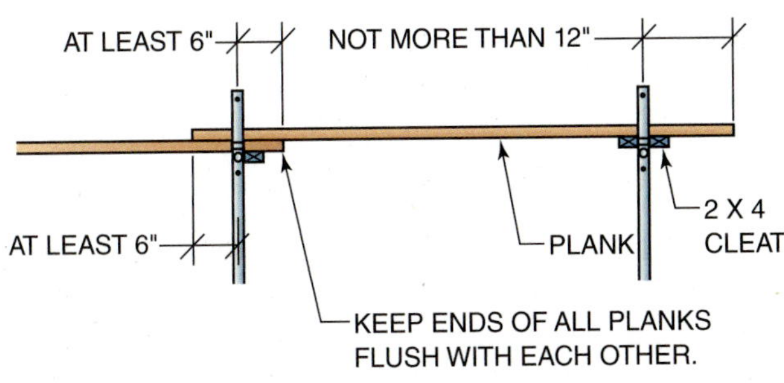


**FIGURE 11-8** Recommended placement for scaffold plank.

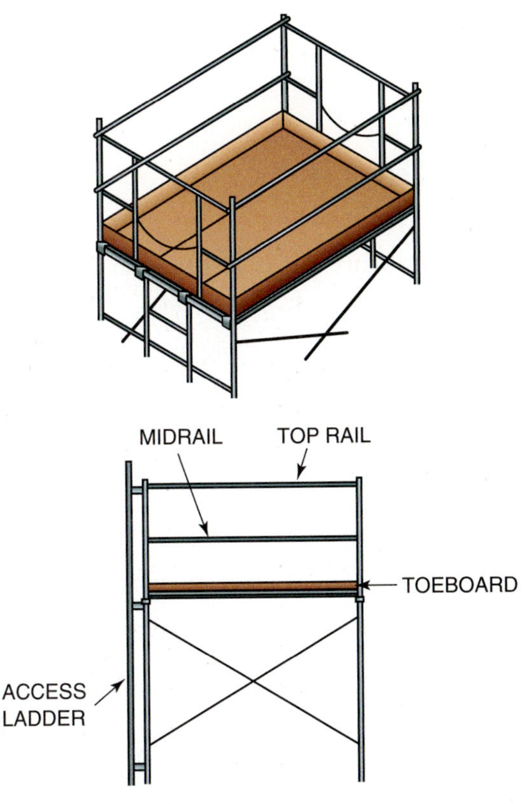


**FIGURE 11-9** Typical guardrail system for a metal tubular frame scaffold.

they can be tied down to the frames with #9 steel-tie wire.

## Scaffold Ladders

OSHA requires safe access onto the scaffold for both erectors and users of the scaffolds. Workers can climb end frames only if they meet OSHA regulations.

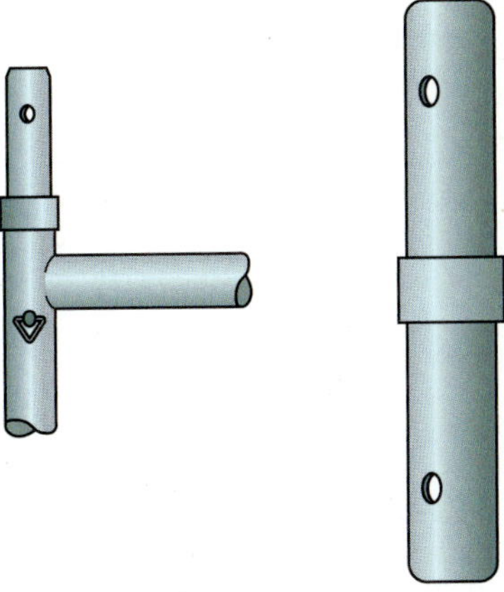

**FIGURE 11-10** Coupling pins to join end frames.

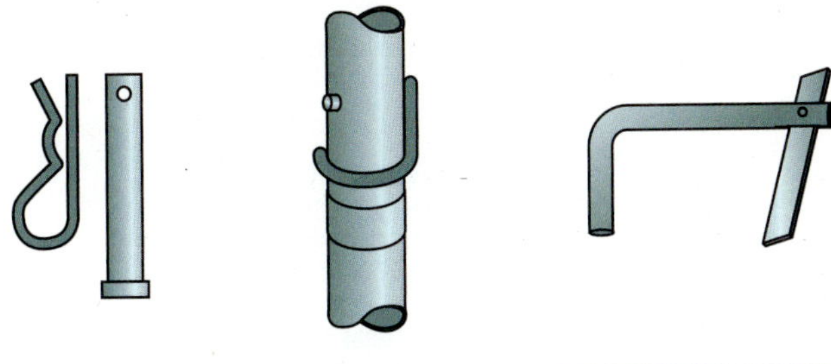

**FIGURE 11-11** Coupling locking devices to prevent scaffold uplift.

Frames may be used as a ladder only if they are designed as such. Frames meeting such design guidelines must have level horizontal members that are parallel and are not more than 16¾ inches apart vertically (Fig. 11-12). Scaffold erectors may climb end frame rungs that are spaced up to 22 inches. Platform planks should not extend over the end frames where end frame rungs are used as a ladder access point. The cross braces should never be used as a means of access or egress.

Attached ladders and stair units may be used (Fig. 11-13). A rest platform is required for every 35 feet of ladder.

## Side Brackets and Hoist Arms

Side brackets are light-duty (35 pounds-per-square-foot maximum) extension pieces used to increase the working platform (Fig. 11-14). They are designed to hold personnel only and are not to be used for material storage. When side brackets are used, the scaffold must have tie-ins, braces, or outriggers to

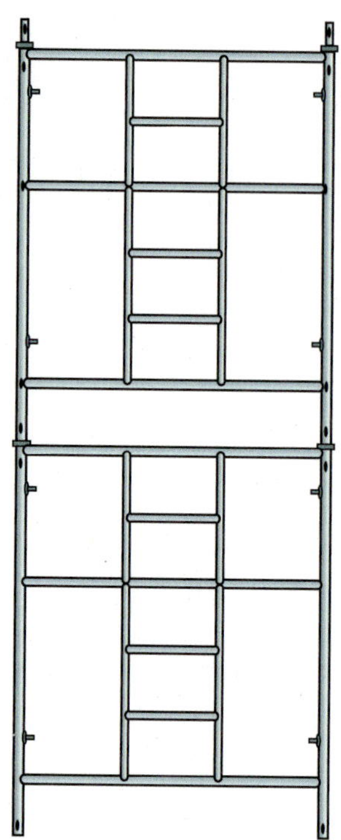


**FIGURE 11-12** The rungs of an end frame designed for a scaffold user access ladder must be spaced no more than 16¾ inches apart.

prevent the scaffold from tipping. Hoist arms and wheel wells are sometimes attached to the top of the scaffold to hoist scaffold parts to the erector or material to the user of the scaffold (Fig. 11-15). The load rating of these hoist arms and wheel wells is typically no more than 100 pounds. The scaffold must be secured from overturning at the level of the hoist arm, and workers should never stand directly under the hoist arm when hoisting a load. They should stand a slight distance away, but not too far to the side, as this will increase the lateral or side loading force on the scaffold.

## SCAFFOLD INSPECTION

Almost half of all scaffold accidents, according to the U.S. Bureau of Labor Statistics, involve defective scaffolds or defective scaffold parts. This statistic means ongoing visual inspection of scaffold parts must play a major role in safe scaffold erection and use. OSHA requires that the competent person inspect all scaffolds at the beginning of every work shift. Visual inspection of scaffold parts should take place at least five times: before erection, during erection, during scaffold use, during dismantling, and before scaffold parts are put back in storage. All damaged parts should be red-tagged and removed from service and then repaired or destroyed as required. Things to look for during the inspection process include the following:

- Broken and excessively rusted welds
- Split, bent, or crushed tubes
- Cracks in the tube circumference
- Distorted members
- Excessive rust
- Damaged brace locks
- Lack of straightness
- Excessively worn rivets or bolts on braces
- Split ends on cross braces
- Bent or broken clamp parts
- Damaged threads on screwjacks
- Damaged caster brakes
- Damaged swivels on casters
- Corrosion of parts
- Metal fatigue caused by temperature extremes
- Leg-ends filled with dirt or concrete

## SCAFFOLD CAPACITY

Stress on a scaffold frame comes from a variety of sources. All parts of a scaffold must be able to withstand these stresses to adequate safety standards.

### Safety Margin

All scaffolds and their components must be capable of supporting, without failure, their own weight and at least four times the maximum intended load applied or transmitted to them. Erectors and users of scaffolding must never exceed this safety factor.

Erectors and users of the scaffold must know the maximum intended load and the load-carrying capacities of the scaffold they are using. The erector must also know the design criteria, maximum

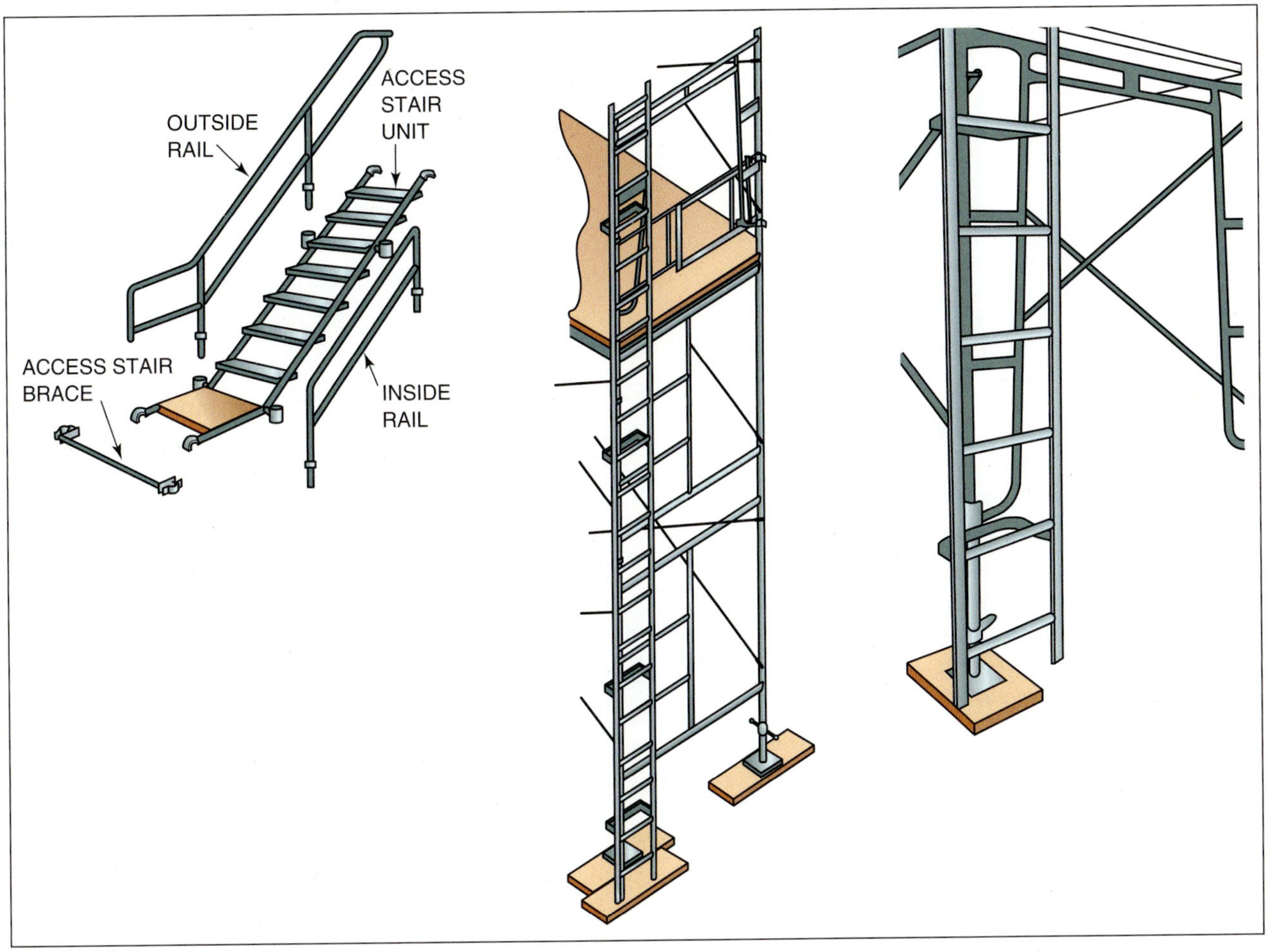

ACCESS
STAIR
UNIT
OUTSIDE
RAIL
ACCESS STAIR
BRACE
INSIDE
RAIL

**FIGURE 11-13** Typical access ladder and stairway.

2 X 4 CLEATS

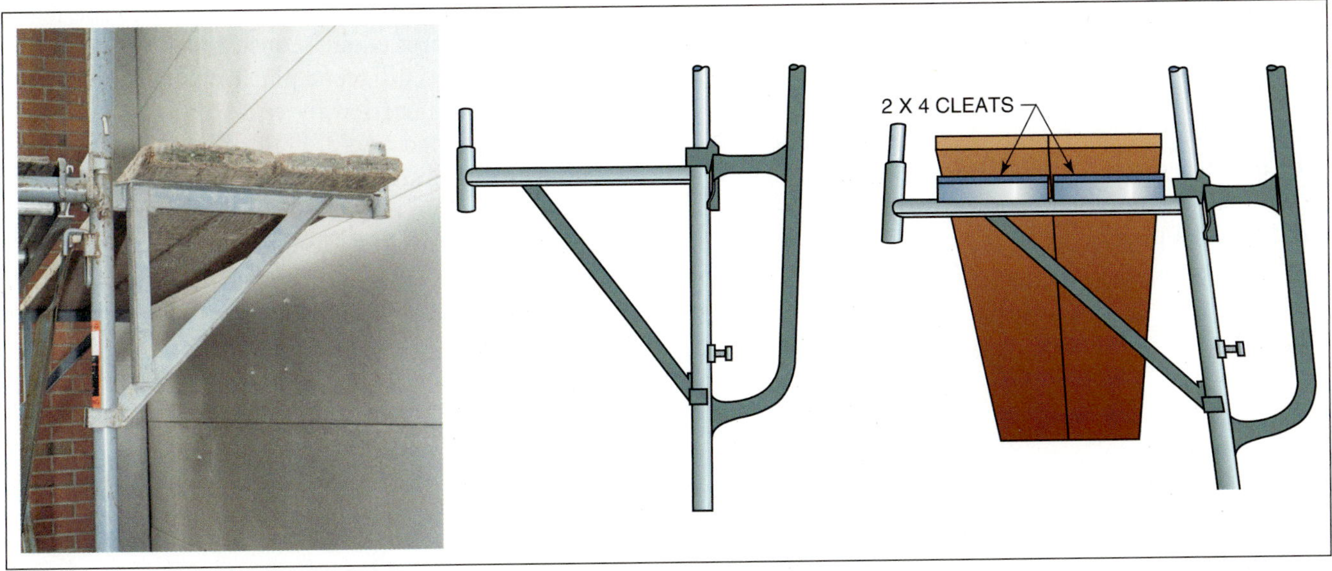

**FIGURE 11-14** Side brackets used to extend a scaffold work platform. These brackets should only be used for workers and never for material storage.

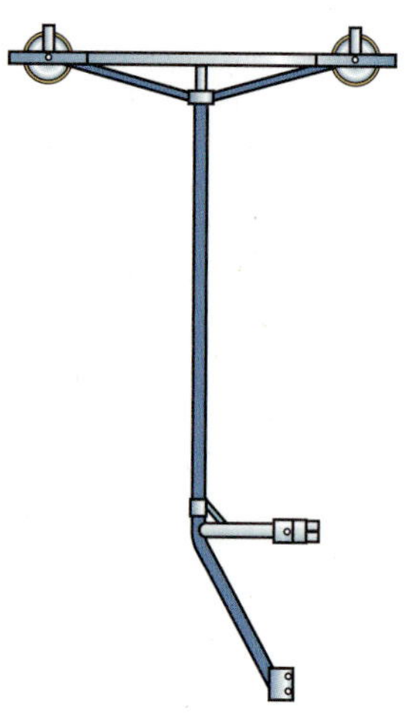

**FIGURE 11-15** Hoist that attaches to the top of a scaffold used to raise material and equipment.

intended load-carrying capacity, and intended use of the scaffold.

When erecting a frame scaffold, the erector should know the load-carrying capacities of its components. The rated leg capacity of a frame may never be exceeded on any leg of the scaffold. Also, the capacity of the top horizontal member of the end frame, called the *bearer,* may never be exceeded.

### Scaffold Coverings

If the scaffold is covered with weatherproofing plastic or tarps, the lateral pressure applied to the scaffold will dramatically increase. Consequently, the number of tie-ins attached to prevent overturning must be increased. Additionally, any guy wires added for support will increase the downward pressure and weight of the scaffold.

### Tie-ins

OSHA regulations state that supported scaffolds that have a ratio greater than four-to-one (4:1) of the height to narrow base width must be restrained from tipping by guying, tying, bracing, or equivalent means. Guys, ties, and braces must be installed at locations where horizontal members support both inner and outer legs. Guy, ties, and braces must be installed according to the scaffold manufacturer's recommendations or at the closest horizontal member to the 4:1 height. For scaffolds greater than 3 feet wide, the vertical locations of horizontal members are repeated every 26 feet. The top guy, tie, or brace of completed scaffolds must be placed no further than the 4:1 height from the top. Such guys, ties, and braces must be installed at each end of the scaffold and at horizontal intervals not to exceed 30 feet.

The tie or standoff should be able to take pushing and pulling forces so the scaffold does not fall into or away from the structure.

### Component Loads

It is possible to overload the bottom legs of the scaffold without overloading the bearer or top horizontal member of any frame. It is also possible to overload the bearer or top horizontal member of the frame scaffold and not overload the leg of that same scaffold. Erectors must pay careful attention to the load capacities of all scaffold components.

*Baseplate.* The supported scaffold poles, legs, posts, frames, and uprights should bear on baseplates, mud sills, or other adequate, firm foundation. Because the mud sills have more surface area than baseplates, sills distribute loads over a larger area of the foundation. Sills are typically wood and come in many sizes. Erectors should choose a size according to the load and the foundation strength required.

*Mud Sills.* Mud sills made of 2 × 10-inch full thickness or nominal lumber should be 18 to 24 inches long and centered under each leg (Fig. 11-16).

*Unequal Loading.* The loads exerted onto the legs of a scaffold are not equal. Consider a scaffold with two loads on two adjacent platforms (Fig. 11-17). Half of load A is carried by end frame #1 and the other half is carried by #2. Half of load B is carried by end frames #2 and #3. End frame #2 carries two half loads, which equals one full load. This is twice the load of end frames #1 and #3. At no time should the manufacturer's load rating for their scaffolding be exceeded.

## SCAFFOLD PLATFORMS

The scaffolding's work area must be fully planked between the front uprights and the guardrail supports in order for the user to work from the scaffold. The plank should not have more than a 1-inch gap between them unless it is necessary to fit around uprights, such as a scaffold leg. If the platform is planked as fully as possible, the remaining gap between the last plank and the uprights of the guardrail system must not exceed 9½ inches. Scaffold

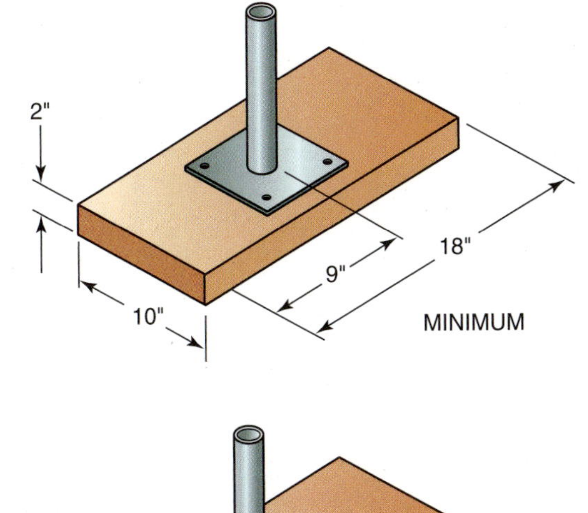


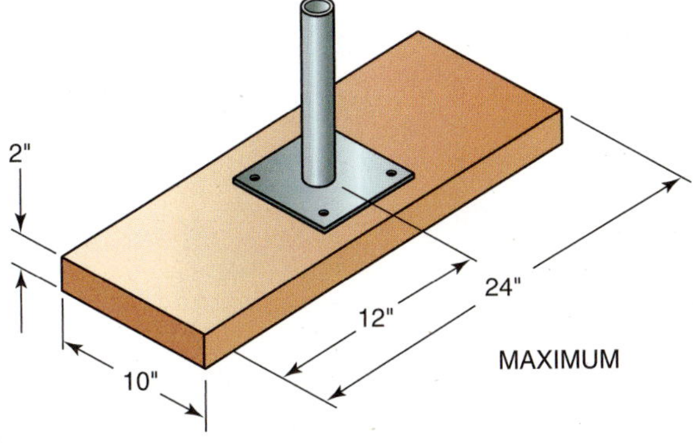


**FIGURE 11-16** Baseplates should be centered on the mud sills.

platforms must be at least 18 inches wide with a guardrail system in place. In areas where they cannot be 18 inches wide, they will be as wide as is feasible. The platform is allowed to be as much as 14 inches away from the face of the work. Planking for the platforms, unless cleated or otherwise restrained by hooks or equivalent means, should extend over the centerline of their support at least 6 inches and no more than 12 inches. If the platform is overlapped to create a long platform, the overlap shall occur only over supports and should not be less than 12 inches unless the platforms are nailed together or otherwise restrained to prevent movement. When fully loaded with personnel, tools, and material, the wood plank used to make the platform must never deflect more than 1/60th of its span. In other words, a 2 × 10-inch plank that is 12 feet long and is sitting on two end frames spaced 10 feet apart should not deflect more than 1/60th of the span or 2 inches, whichever is less.

Solid sawn wood planks should be scaffold-grade lumber, as set out by the grading rules for the species of lumber being used. A recognized lumber grading association, such as the Western Wood Products Association (WWPA) or the National Lumber Grades Authority (NLGA), establishes these grading rules. A grade should be stamped on the scaffold-grade plank, indicating that it meets OSHA and industry requirements for scaffold planks. Two of the most common wood species used for scaffold planks are southern yellow pine and Douglas fir.

OSHA does not require wood scaffold planks to bear grade stamps. The erector may use "equivalent" planks, which are determined equivalent by visually inspecting or test loading the wood plank in accordance with grading rules.

Scaffold platforms are usually rated for the intended load. Light-duty scaffolds are designed at 25 pounds per square foot, medium-duty scaffolds are rated at 50 pounds per square foot, and heavy duty at 75 pounds per square foot. The maximum span of a plank is tabulated in Figure 11-18. Using this chart, the maximum load that could be put on a nominal 2" plank (1½ inch actual) with a span of 7 feet is 25 pounds per square foot. Note that a load of 50 pounds per square foot would require a span of no more than 6 feet. Fabricated planks and platforms are often used in lieu of solid sawn wood planks. These planks and platforms include fabricated wood planks that use a pin to secure the lumber sideways, oriented strand board planks, fiberglass composite planks, aluminum-wood decked planks, and high-strength galvanized steel planks. The loading of fabricated planks or platforms should be obtained from the manufacturer and never exceeded. Scaffold platforms must be inspected for damage before each use.

## SCAFFOLD ACCESS

A means of access must be provided to any scaffold platform that is 2 feet above or below a point of access. Such means include a hook-on or attachable ladder, a ramp, or a stair tower and are determined by the competent person on the job.

If a ladder is used, it should extend 3 feet above the platform and be secured both at the top and bottom. Hook-on and attachable ladders should be specifically designed for use with the type of scaffold used, have a minimum rung length of 11½ inches, and have uniformly spaced rungs with a maximum spacing between rung length of 16¾ inches. Sometimes a stair tower can be used for access to the work platform, usually on larger jobs (Fig. 11-19). A ramp can also be used as access to the scaffold or the work platform. When using a ramp, remember that a guardrail system or fall protection is required at 6 feet above a lower level.

**FIGURE 11-17** The inner end frames, such as #2, often carry twice the load of the end frames located at the end of the scaffold.

| | Maximum permissible plank span | |
|---|---|---|
| Maximum intended load | Full thickness, undressed lumber | Nominal thickness lumber |
| Lbs/sq ft | Feet | Feet |
| 25 | 10 | 8 |
| 50 | 8 | 6 |
| 75 | 6 | ----- |

**FIGURE 11-18** Maximum spacing of planks based on the load rating of the scaffold.

The worker using the scaffold can sometimes access the work platform by using the end frames of the scaffold itself. According to regulations, the end frame must be specifically designed and constructed for use as ladder rungs. The rungs can run up the center or to one side of the end frame; some have the rungs all the way across the end frame.

Scaffold users should never climb any end frame unless the manufacturer of that frame designated it to be used for access.

## SCAFFOLD USE

Scaffolds must not be loaded in excess of their maximum intended load or rated capacities,

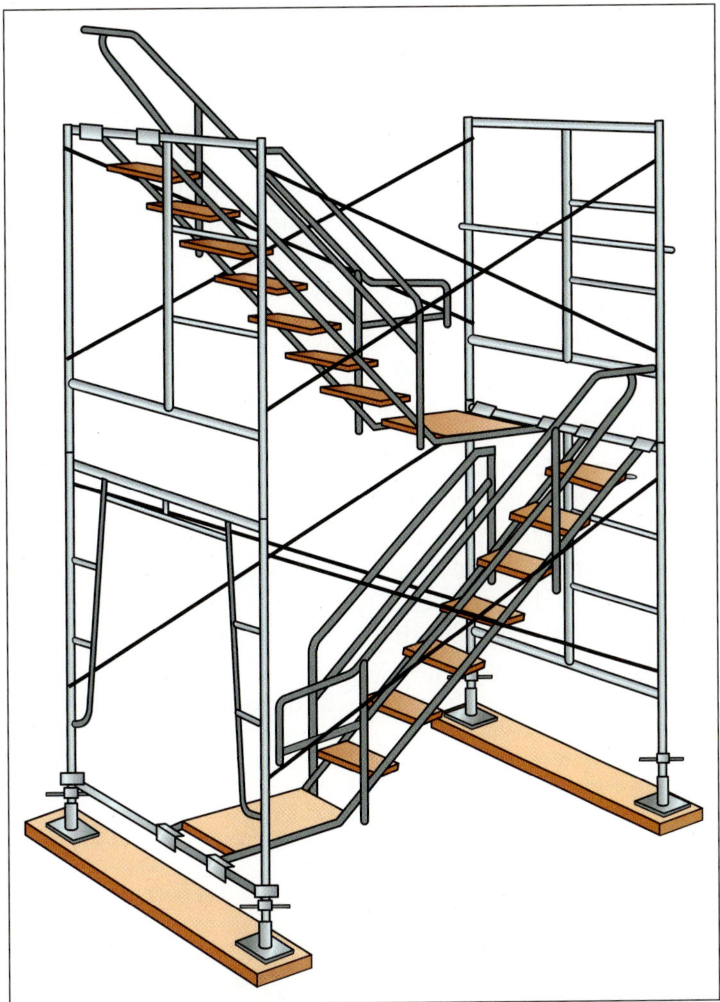

**FIGURE 11-19** Scaffold access may be provided by a stair tower.

whichever is less. Workers must know the capacity of scaffolds they are erecting and using. Before the beginning of each work shift, or after any occurrence that could affect a scaffold's structural integrity, the competent person must inspect all scaffolds on the job. Employees must not work on scaffolds covered with snow or ice except to remove the snow or ice. Generally, work on or from scaffolds is prohibited during storms or high winds. Debris must not be allowed to accumulate on the platforms.

Makeshift scaffold devices, such as boxes or barrels, must not be used on the scaffold to increase workers' working height. Step ladders should not be used on the scaffold platform unless they are secured according to OSHA regulations.

## FALL PROTECTION

Current OSHA standards on scaffolding require fall protection when workers are working at heights above 10 feet. This regulation applies to both the user of the scaffold and the erector or dismantler of the scaffold. These regulations allow the employer the option of a guardrail system or a personal fall protection system. The fall protection system most often used is a complete guardrail system. A guardrail system has a top rail 38 to 45 inches above the work deck, with a midrail installed midway between the top rail and the platform. The work deck should also be equipped with a toeboard. These requirements are for all open sides of the scaffold, except for those sides of the scaffold that are within 14 inches of the face of the building. A typical personal fall protection system consists of five related parts: the harness, lanyard, lifeline, rope grab, and anchor (Fig. 11-20). The failure of any one part means failure of the system. Therefore, constant monitoring of a lifeline system is a critical responsibility. It is easy for a system to lose its integrity almost immediately, even on first use.

OSHA recognizes that sometimes fall protection may not be possible for erectors. As the scaffold increases in length, the personal fall-arrest system may not be feasible because of its fixed anchorage and the need for employees to traverse the entire length of the scaffold. Additionally, fall protection may not be feasible due to the potential for lifelines to become entangled or to create a tripping hazard for erectors or dismantlers as they traverse the scaffold. Do not use the scaffold components as an anchor point of the fall-protection harness. OSHA puts the responsibility of when to use fall protection, both for the user of the scaffold and for the erector, on the competent person.

## FALLING OBJECT PROTECTION

According to industry standards and OSHA requirements, workers must wear hard hats during the process of erecting a scaffold. In addition to hard hats, protection from potential falling objects may be required. When material on the scaffold could fall on workers below, some type of barricade must be installed to prevent that material from falling. OSHA lists toeboards as part of the falling object protection for the workers below the scaffold. The toeboard can serve two functions: it keeps material on the scaffold and keeps the workers on the scaffold platform if they happen to slip.

## DISMANTLING SCAFFOLDS

Many guidelines and rules for erection also apply to scaffold dismantling. However, dismantling requires additional precautions to ensure the scaffold will

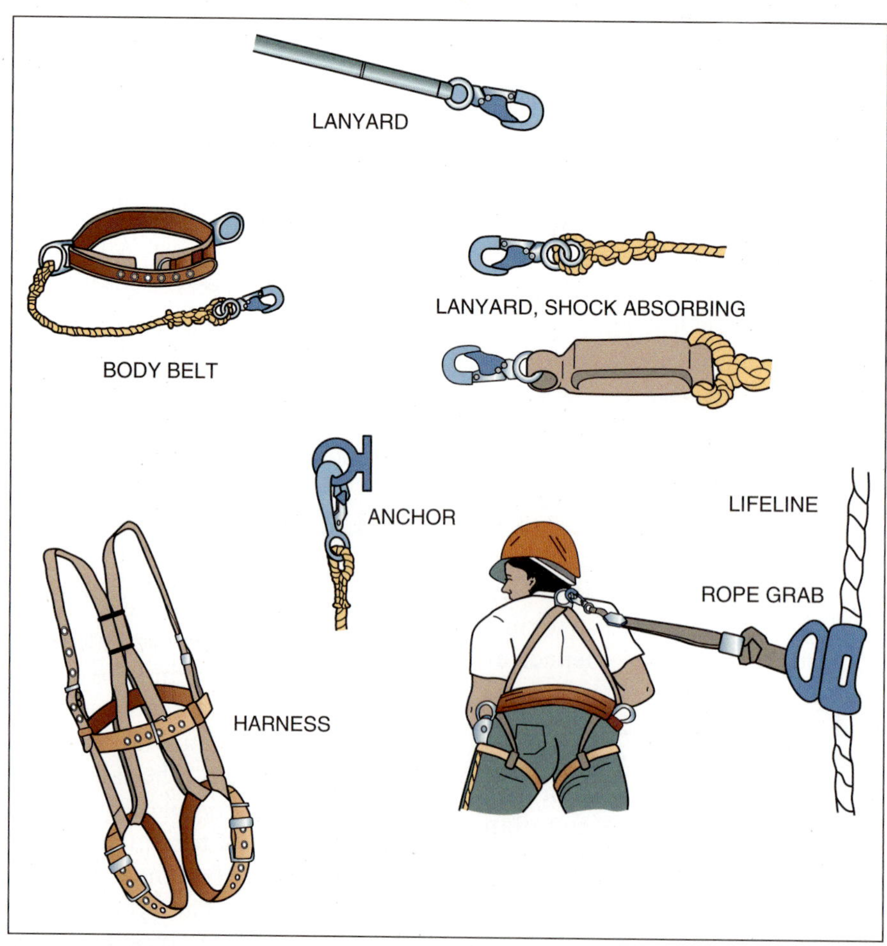


**FIGURE 11-20** Components of a personal protection system.

come down in a controlled, safe, and logical manner. Important factors to consider include the following:

1. Check every scaffold before dismantling. Any loose or missing ties or bracing must be corrected.
2. If a hoist is to be used to lower the material, the scaffold must be tied to the structure at the level of the hoist arm to dispel any overturning effect of the wheel and rope.
3. The erector should be tied off for fall protection, as required by the regulations, unless it is infeasible or a greater hazard to do so.
4. Start at the top and work in reverse order, following the step-by-step procedures for erection. Leave the work platforms in place as long as possible.
5. Do not throw planks or material from the scaffold. This practice will damage the material and presents overhead hazards for workers below.
6. Building tie-ins and bracing can only be removed when the dismantling process has reached that level or location on the scaffold. An improperly removed tie can cause the entire scaffold to overturn.
7. Remove the ladders or the stairs only as the dismantling process reaches that level. Never climb or access the scaffold by using the cross braces.
8. As the scaffold parts come off the scaffold, they should be inspected for any wear or damage. If a defective part is found, it should be tagged for repair and not used again until inspected by the competent person.
9. Dismantled parts and materials should be organized, stacked, and placed in bins or racks out of the weather.
10. Secure the disassembled scaffold equipment to ensure that no unauthorized, untrained employees use it. All erectors must be trained, experienced, and under the supervision and direction of a competent person.
11. Always treat the scaffold components as if a life depends on them. The next time the scaffold is erected, someone's life will depend on its soundness.

***For step-by-step instructions on ERECTING A SCAFFOLD, see Procedure 11-1 on pages 325–329.***

## MOBILE SCAFFOLDS

The rolling tower, or mobile scaffold, is widely used for small jobs, generally not more than 20 feet in height (Fig. 11-21). The components of the mobile scaffold are the same as those for the stationary frame scaffold, with the addition of casters

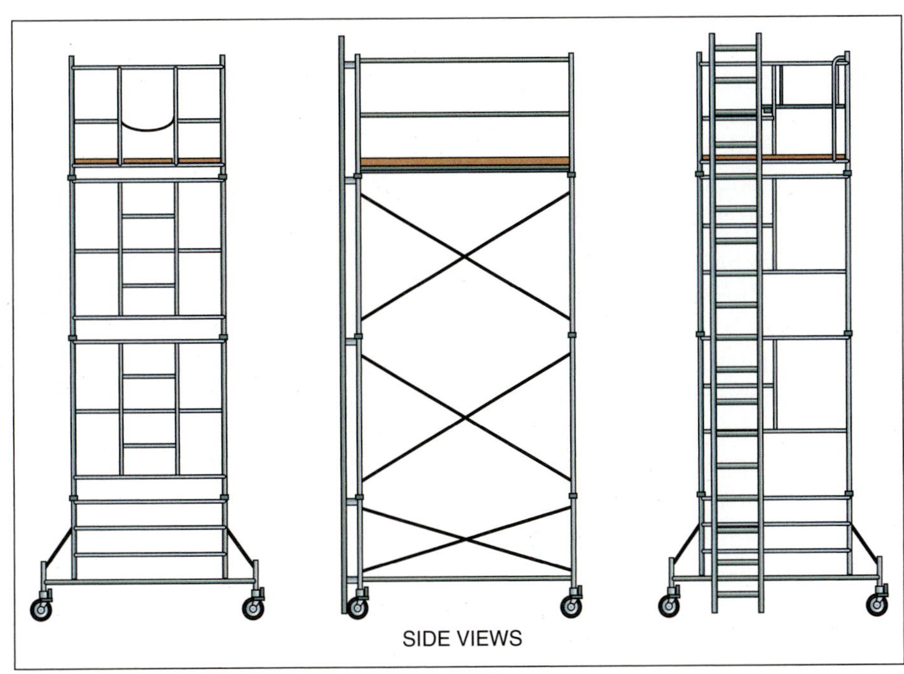


**FIGURE 11-21** Typical setup for a mobile scaffold.

**FIGURE 11-22** Casters replace baseplates to transform a metal tubular frame scaffold into a mobile scaffold.

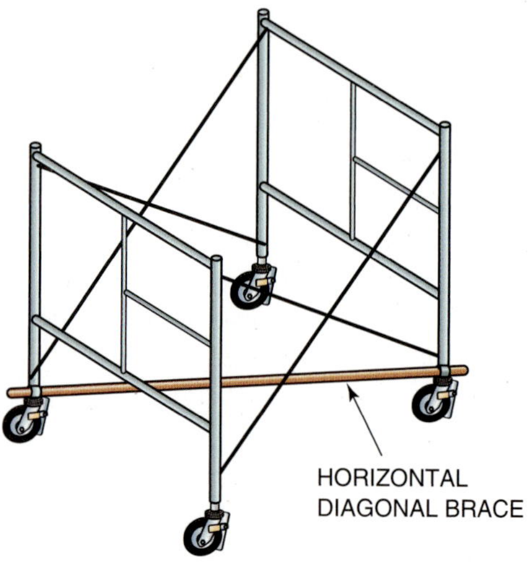


**FIGURE 11-23** The horizontal diagonal brace (or gooser) is used to keep the tower square when it is rolled.

(Fig. 11-22) and horizontal diagonal bracing. There are additional restrictions on rolling towers as well.

The height of a rolling tower must never exceed four times the minimum base dimension. For example, if the frame sections are 5' × 7', the rolling tower can be only 20 feet high. If the tower exceeds this height-to-base ratio, it must be secured to prevent overturning. When outriggers are used on a mobile tower, they must be used on both sides. Casters on mobile towers must be locked with positive wheel swivel locks or the equivalent to prevent movement of the scaffold while it is stationary. Casters typically have a load capacity of 600 pounds each, and the legs of a frame scaffold can hold 2,000 to 3,000 pounds each. Care must be taken not to overload the casters. Never put a cantilevered work platform, side bracket, or hoist arm on the side or end of a mobile tower. Mobile towers can tip over if used incorrectly.

Mobile towers must have horizontal, diagonal, or gooser braces at the base to prevent racking of the tower during movement (Fig. 11-23). Metal hook planks also help prevent racking if they are secured to the frames. The force to move the scaffold should be applied as close to the base as practical, but not more than 5 feet above the supporting surface. The casters must be locked after each movement before beginning work again. Employees are not allowed to ride on rolling tower scaffolds during movement unless the height-to-base width ratio is two-to-one or less. Before the scaffold is moved, each employee on the scaffold must be made aware of the move. Caster and wheel stems shall be pinned or otherwise secured in scaffold legs or adjustment screws. The surface on which the mobile tower rolls must be free of holes, pits, and obstructions and must be within 3 degrees of level. Use a mobile scaffold on firm floors only.

## Pump Jack Scaffolds

Pump jack scaffolds consist of 4" × 4" poles, a pump jack mechanism, and metal braces for each pole. The braces are attached to the pole at intervals and near the top. The arms of the bracket extend from both sides of the pole at 45-degree angles. The arms are attached to the sidewall or roof to hold the pole steady (Fig. 11-24). The scaffold is raised by pressing on the foot pedal of the pump jack. The mechanism has brackets on which to place the scaffold plank.

Other brackets hold a guardrail or platform. Spinning a lever allows the staging to be moved downward (Fig. 11-25).

Pump jack scaffolds are used widely for siding, where staging must be kept away from the walls, and when a steady working height is desired. However, pump jack scaffolds have their limitations. They should not be used when the working load exceeds 500 pounds. No more than two persons are permitted at one time between any two supports. Wood poles must not exceed 30 feet in height. Braces must be installed at a maximum vertical spacing of not more than 10 feet. In order to pump the scaffold past a brace location, temporary braces are used. The temporary bracing is installed about 4 feet above the original bracing. Once the scaffold

**FIGURE 11-24** Components of a pump jack system.

**FIGURE 11-25** Pump jacks area lowered by simultaneously pressing a foot lever and turning a hand crank.

is past the location of the original brace, it can be reinstalled. The temporary brace is then removed. Wood pump jack poles are constructed of two 2 × 4s nailed together. The nails should be 3-inch or 10d, and no less than 12 inches apart, staggered uniformly from opposite outside edges.

## Roofing Brackets

Roofing brackets are used when the pitch of the roof is too steep for carpenters to work without slipping (Fig. 11-26). Usually any roof with more than 4 on 12 slope requires roofing brackets. Roofing brackets are made of metal. Some are adjustable for roofs of different pitches. A metal plate at the top of the bracket has three slots in which to drive nails to fasten the bracket to the roof. The bottom of the slot is round and large enough to slip over the nail head. This enables removal of the bracket from the fasteners without pulling the nails. The bracket is simply tapped upward from the bottom and then lifted over the nailheads. The nails that remain are then driven home.

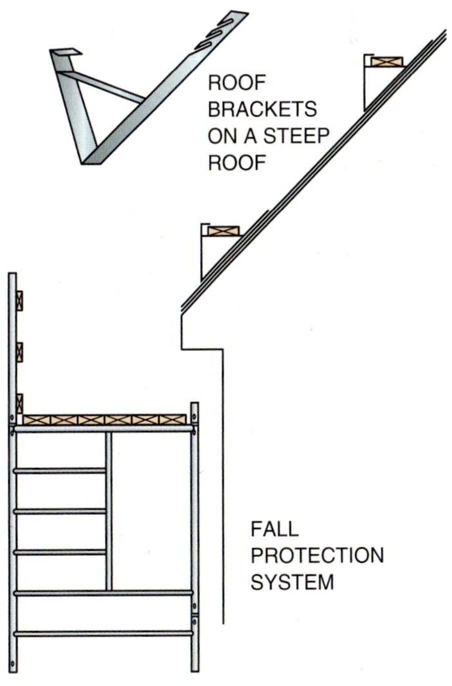


**FIGURE 11-26** Roofing brackets are used when the roof pitch is too steep for carpenters to work without slipping.

## APPLYING ROOF BRACKETS

Roof brackets are used when the roof is being shingled, typically on steep-pitched roofs. They keep the worker from slipping and also hold the roofing materials. Apply roof brackets in rows. Space them out so that they can be reached without climbing off the roof bracket staging below. On asphalt-shingled roofs, place the brackets at about 6- to 8-foot horizontal intervals. The top end of the bracket should be just below the next course of shingles. Nail the bracket over a joint or cutout in the tab of the shingle course below. No joint or cutout in the course above should fall in line with the nails holding the bracket. Otherwise, the roof will leak. Use three 3¼-inch or 12d common nails driven home with at least one nail in a rafter. Open the brackets so the top member is approximately level or slightly leaning toward the roof. Place a staging plank on the top of the brackets.

Overlap them as in wall scaffolds. Keep the inner edges against the roof for greater support. A toeboard made of 1 × 6 or 1 × 8 lumber is usually placed flat on the roof with its bottom edge on top of the brackets. This protects the new roofing from the workers' toes when the roofing has progressed that far (Fig. 11-27). After the shingles are applied, the bracket is tapped on the bottom upward along the slope of the roof to release it from the nails. Raise the shingle and drive the nails home so they do not stick up and damage the shingles.

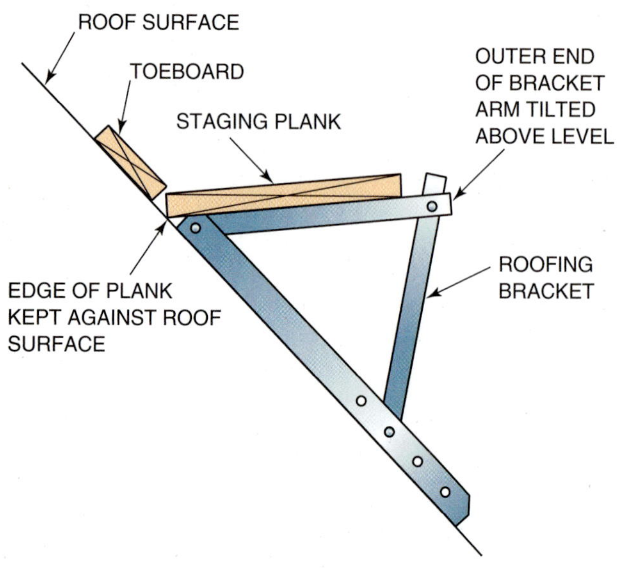


**FIGURE 11-27** The placement of a toeboard and plank used on roof brackets.

# LADDERS

Carpenters must often use ladders to work from or to reach working platforms above the ground. Most commonly used ladders are the stepladder and the extension ladder. They are usually made of wood, aluminum, or fiberglass. Make sure all ladders are in good condition before using them.

CAUTION

**CAUTION:** Before raising an extension ladder, always check for overhead obstructions, particularly power lines. Never use an aluminum ladder near overhead electrical wires.

## EXTENSION LADDERS

To raise an extension ladder, place its feet against a solid object. Pick up the other end. Walk forward under the ladder, pushing upward on each rung until the ladder is upright (Fig. 11-28). With the ladder

**FIGURE 11-28** Raising an extension ladder.

vertical and close to a wall, extend the ladder by pulling on the rope with one hand while holding the ladder upright with the other. Raise the ladder to the desired height. Make sure the spring-loaded hooks are over the rungs on both sides. Lean the top of the ladder against the wall. Move the base out until the distance from the wall is about one-fourth the vertical height. This will give the proper angle to the ladder. The proper angle for climbing the ladder can also be determined, as shown in Figure 11-29. If the ladder is used to reach a roof or working platform, it must extend above the top support by at least 3 feet. When the ladder is in position, shim one leg, if necessary, to prevent wobbling, and secure the top of the ladder to the building. Face the ladder when climbing. Grasp the rungs with both hands, making three-point contact at all times (Fig. 11-30).

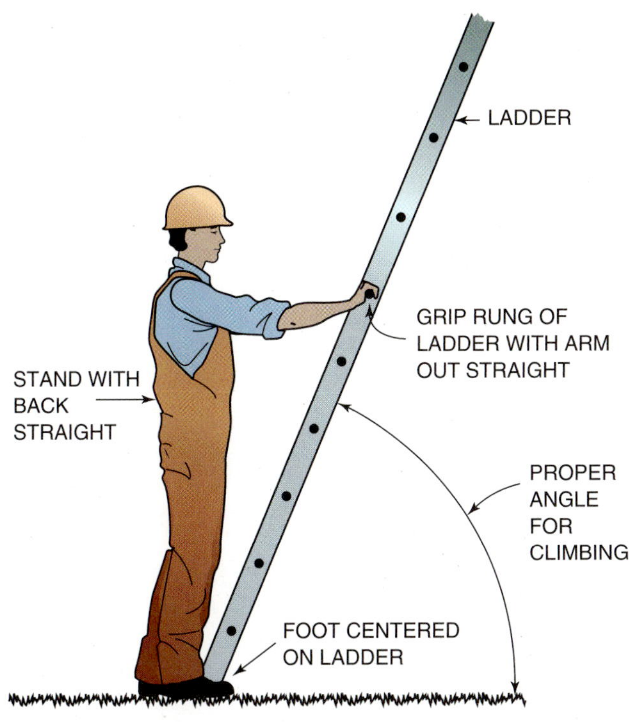


**FIGURE 11-29** Technique for finding the proper ladder angle before climbing.

**FIGURE 11-30** Face the ladder when climbing. Hold on with both hands.

## STEPLADDERS

When using a stepladder, open the legs fully so the brackets are straight and locked. Make sure the ladder does not wobble. If necessary, place a shim under the leg to steady the ladder. Never work above the second step from the top. Do not use the ledge in back of the ladder as a step. The ledge is used to hold tools and materials only. Do not use a folded stepladder that leans against a wall or object. Move the ladder as necessary to avoid overreaching. Make sure all materials and tools are removed from the ladder before moving it.

**FIGURE 11-31** Ladder jacks are used to support scaffold plank for short-term, light repair work.

### LADDER JACKS

*Ladder jacks* are metal brackets installed on ladders to hold scaffold plank. At least two ladders and two jacks are necessary for a section. Ladders should be heavy-duty, free from defects, and placed no more than 8 feet apart. They should have devices to keep them from slipping. The ladder jack should bear on the side rails in addition to the ladder rungs. If bearing on the rungs only, the bearing area should be at least 10 inches on each rung. No more than two persons should occupy any 8 feet of ladder jack scaffold at any one time. The platform width must not be less than 18 inches. Planks must overlap the bearing surface by at least 10 inches (Fig. 11-31).

## CONSTRUCTION AIDS

Sawhorses, work stools, ladders, and other construction aids are sometimes custom-built by the carpenter on the job or in the shop.

### SAWHORSES

*Sawhorses* are used on practically every construction job. They support material that is being laid out or cut to size. Unless they are being used as supports for a trestle scaffold, sawhorses are usually made with a 2 × 4 or 2 × 6 top, 1 × 6 legs, and $\frac{3}{8}$- or $\frac{1}{2}$-inch plywood leg braces. Sawhorses are constructed in a number of ways according to the preference of the individual. However, they should be of sufficient width, a comfortable working height, and light enough to be moved easily from one place to another. A typical sawhorse is 36 inches wide with 24-inch legs (Fig. 11-32). A tall person may wish to make the leg 26 inches long.

### JOB-MADE LADDERS

At times, it is necessary to build a ladder on the job. These are usually short, straight ladders no more than 24 feet in length. The side rails are made of clear, straight-grained 2 × 4 stock spaced 15 to 20 inches apart. Cleats or rungs are cut from 2 × 4 stock and inset into the edges of the side rails not more than $\frac{3}{4}$ inch. Filler blocks are used on the rails between the cleats. Cleats must be uniformly spaced at 12 inches top to top (Fig. 11-33).

## SCAFFOLD SAFETY

The safety of those working at a height depends on properly constructed scaffolds. Those who have the responsibility of constructing scaffolds must be thoroughly familiar with the sizes, spacing, and fastening of scaffold members and other scaffold construction techniques. Safety is an attitude and must become a way of life on the job.

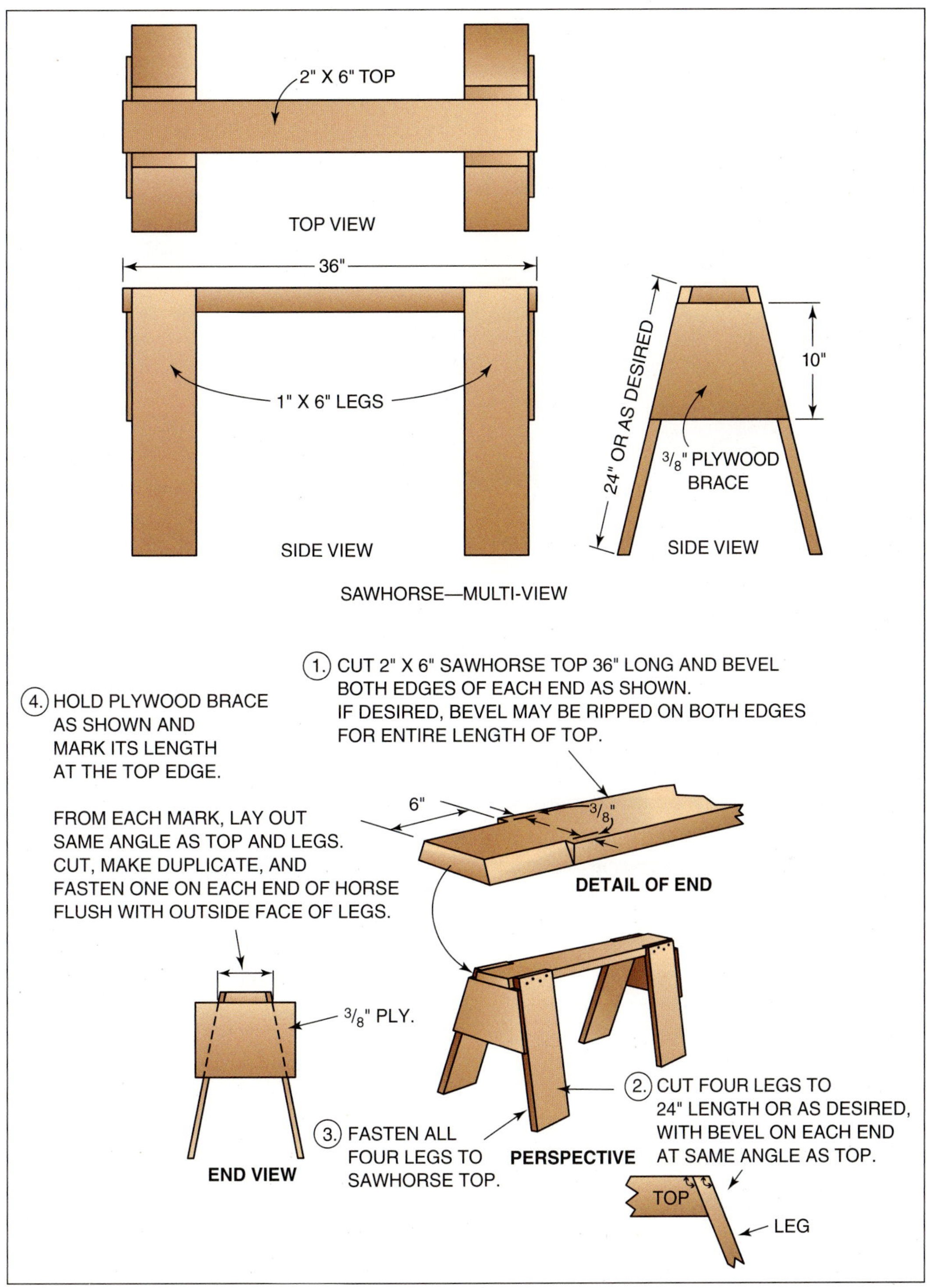


**FIGURE 11-32** Construction of a typical sawhorse.

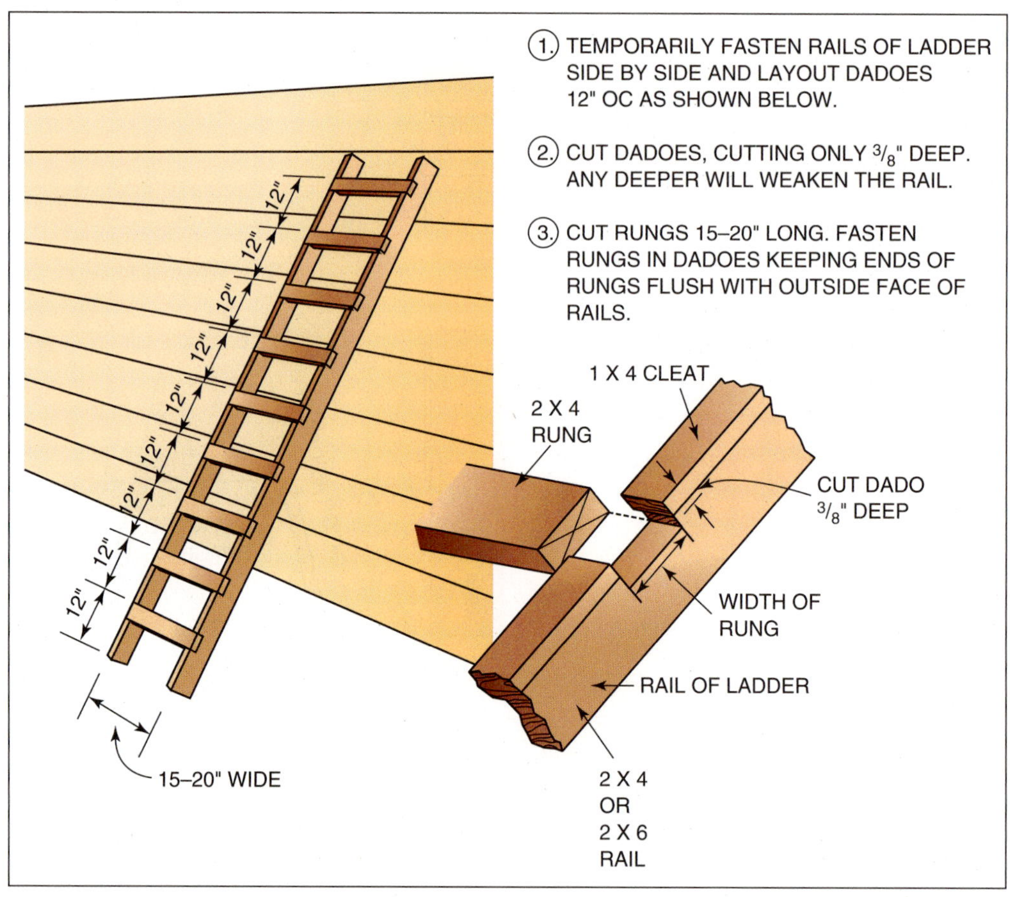


**FIGURE 11-33** Constructing a job-built ladder.

# SUMMARY

- Scaffolds are necessary for construction yet are dangerous if used improperly.
- Metal tubular frame scaffolds are made of strong reusable components.
- Pump jack scaffolds provide a quick and easily height-adjustable work surface.
- Roofing brackets provide a way of working on steep roofs.
- Ladders are essential tools that allow workers access to many different elevations.
- Construction aids can be made on the job to assist the carpenter in a variety of tasks.
- Scaffold safety depends on proper erection and use of scaffold equipment.

PROCEDURE 11-1

# Erecting a Scaffold

## Preassembly Inspection

All workers on the scaffold must wear hard hats.

- Inspect all scaffold components delivered to the job site. Defective parts must not be used.
- Ensure that the foundation of the scaffold is stable and sound, able to support the scaffold, and four times the maximum intended load without settling or displacement.
- Always start erecting the scaffold at the highest elevation, which will allow the scaffold to be leveled without any excavating. Install cribbing, screwjacks, or shorter frames under the regular frames to level the section.
- Verify that the scaffold is level and plumb regularly during erection.

## Assembly

- **A** Lay out the location of baseplates and screwjacks on mud sills. The end frames must be properly spaced for the guardrails and cross braces to be properly installed.
- Stand one of the end frames up and attach the cross braces to each side, making sure the correct length cross braces have been selected for the job. Connect the other end of the braces to the second end frame.

A

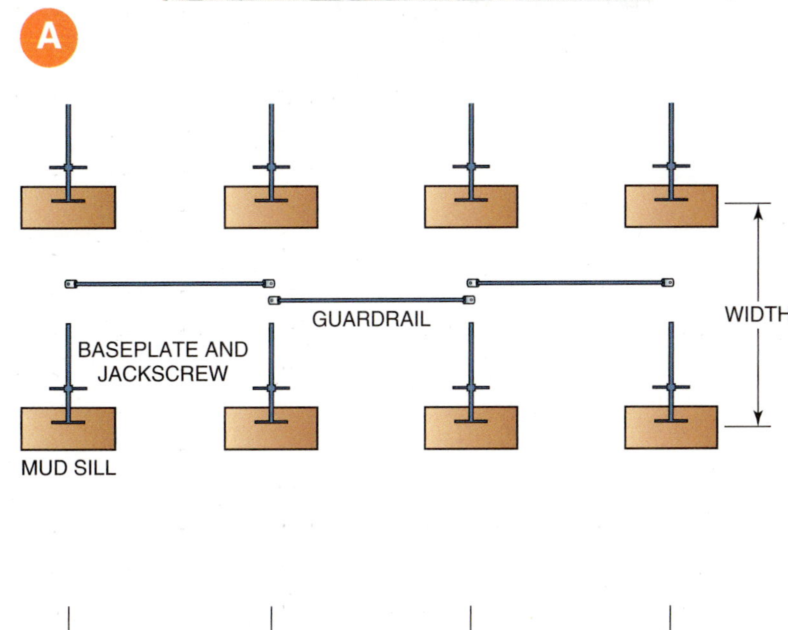

PROCEDURE 11-1

## Erecting a Scaffold (continued)

**B** All scaffold legs must be braced to at least one other leg. Make sure that all brace connections are secure. If any of these mechanisms are not in good working order, replace the frame with one that has properly functioning locks.

**C** Use a level to plumb and level each frame. Remember that OSHA requires that all tubular welded frame scaffolds be plumb and level. Adjust screwjacks or cribbing to level the scaffold.

- As each frame is added, keep the scaffold bays square with each other. Repeat this procedure until the first horizontal scaffold run is erected. Remember, if the first level of scaffolding is plumb and level, the remaining levels will be more easily assembled.

**C**

**D** Place planks on top of the end frames. All planking must meet OSHA requirements and be in good condition. If planks that do not have hooks are used, they must extend over their end supports by at least 6 inches and not more than 12 inches. A **cleat** should be nailed to both ends of wood planks to prevent plank movement. Platform laps must be at least 12 inches, and all platforms must be secured from movement. Hooks on planks also have uplift protection installed on the ends. It is a good practice to plank each layer fully as the scaffold is erected. If the deck is to be used only for erecting, then a minimum of two planks can be used. However, full decking is preferred, as it is a safer method for the erector.

**D**

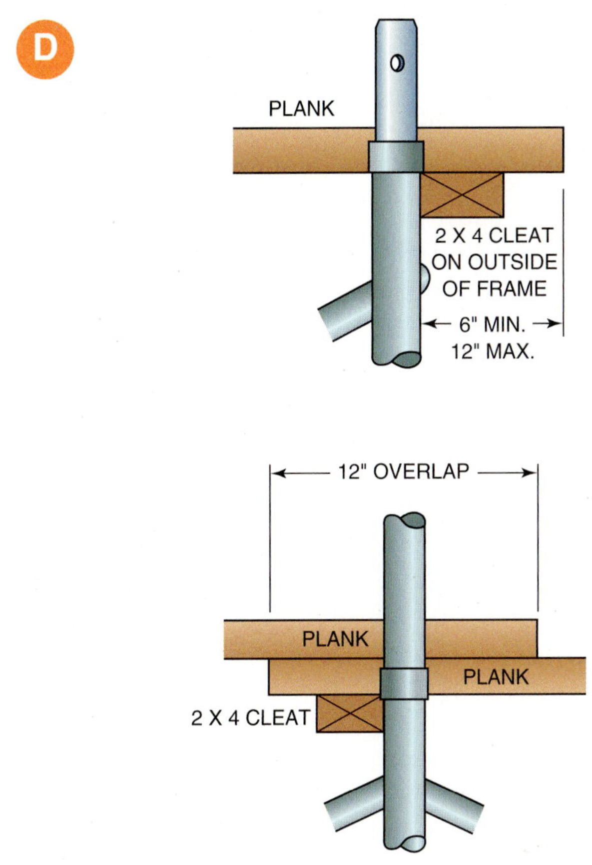

PROCEDURE 11-1

# Erecting a Scaffold (continued)

**E** Access may be on the end frame, if it is so designed, or an attached ladder. If the ladder is bolted to a horizontal member, the bolt must face downward.

**F** The second level of frames may be hung temporarily over the ends of the first frames and then installed onto the coupling pins of the first-level frames. Special care must be taken to ensure proper footing and balance when lifting and placing frames. OSHA requires erector fall protection—a full body harness attached to a proper anchor point on the structure—when it is feasible and not a greater hazard to do so.

- Install uplift protection pins through the legs and coupling pins. Wind, side brackets, and hoist arms can cause uplift, so it is a good practice to pin all scaffold legs together.
- The remaining scaffolding is erected in the same manner as the first.

**E**

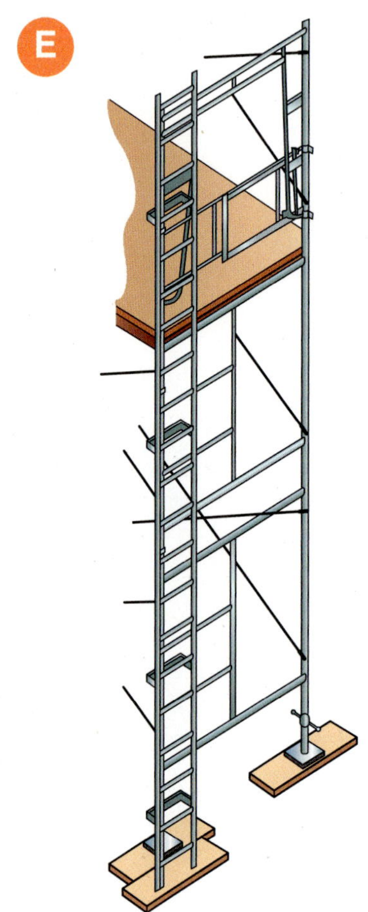

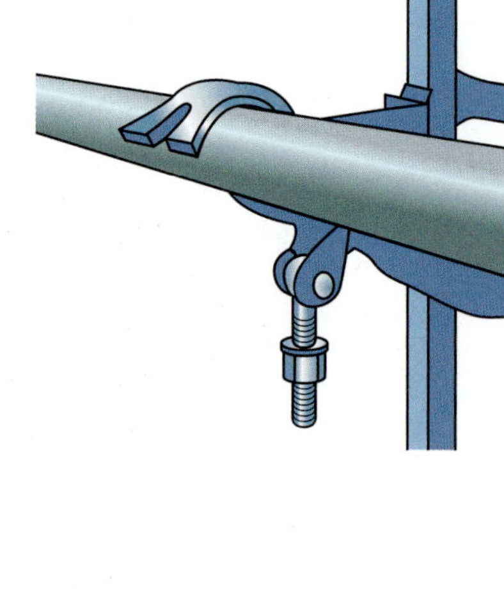

**F**

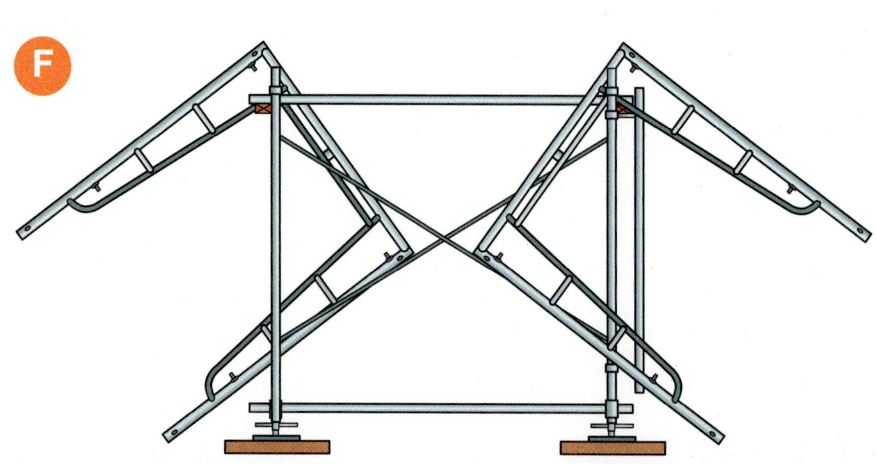

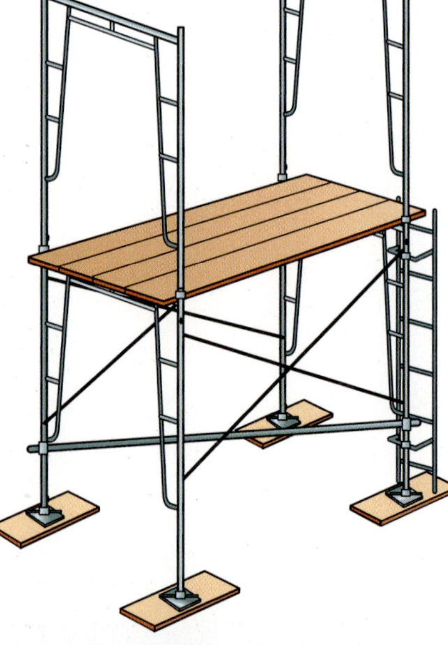

## Post-Assembly Checklist

Remember that all work platforms must be fully decked and have a guardrail system or personal fall-arrest system installed before it can be turned over to the scaffold users.

**G** If the scaffold is higher than four times its minimum base dimension, it must be restrained from tipping by guying, tying, bracing, or equivalent means. The scaffold is not allowed to tip into or away from the structure. Make sure all tie-ins are properly placed and secured, both at the scaffold and at the structure.

- Make sure that the scaffold is plumb, level, and square before turning it over for workers to use.
- Check that all legs are on baseplates or screwjacks and mud sills (if required), ensuring the scaffolding is properly braced with all brace connections secured.
- Install toeboards and screening, as needed.
- Check that end or side brackets are fully secured, and compensate for any overturning forces.
- Correctly install all access units, and secure ladders and stairs.

*After the scaffolding passes all inspections, it is ready to be turned over to the workers. Remember that this scaffolding must be inspected by a competent person at the beginning of each work shift and after any occurrence, such as a high wind or a rainstorm, which could affect its structural integrity.*

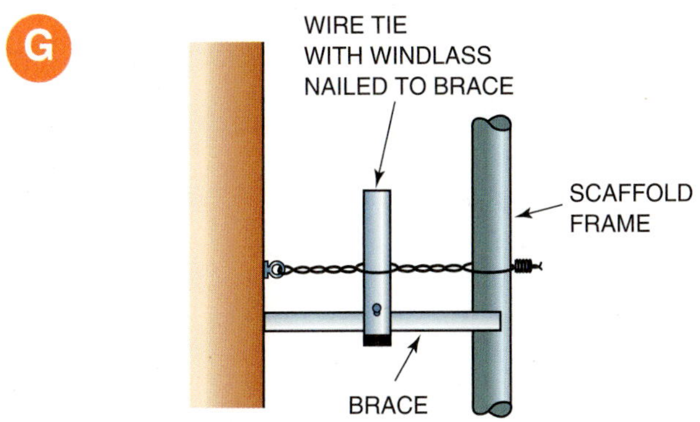

## REVIEW QUESTIONS

*Select the most appropriate answer.*

**1. Scaffold planks should be at least**

a. 2 × 6.
b. 2 × 8.
c. 2 × 10.
d. 2 × 12.

**2. Overlapped planks should extend beyond the bearing point at least_____inches and no more than_____inches.**

a. 3, 6
b. 3, 8
c. 6, 8
d. 6, 12

**3. Metal tubular frame scaffolding is held rigidly plumb by**

a. end frames.
b. goosers.
c. cross braces.
d. cribbing.

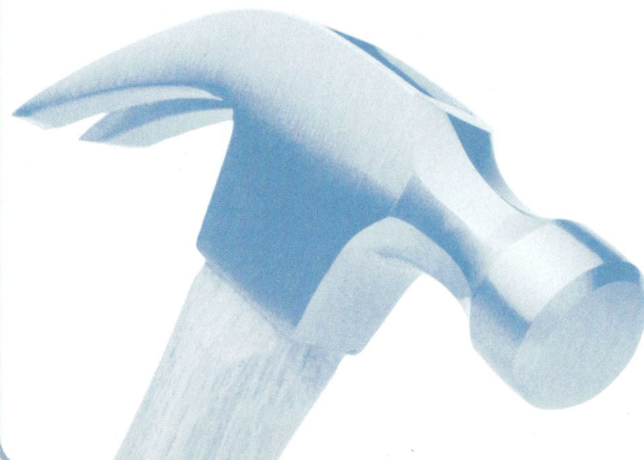

**4. The part of a scaffold that protects workers below from objects falling off the work platform is a**

a. toeboard.
b. midrail.
c. top rail.
d. post.

**5. The workers allowed to climb an access ladder for a metal tubular scaffold that has its rungs spaced 18 inches apart are the scaffold**

a. users only.
b. erectors only.
c. erector and dismantlers only.
d. anyone wearing a hard hat.

**6. Wood scaffold planks, when loaded, should deflect no more than**

a. 1/6th of the span.
b. 1/16th of the span.
c. 1/20th of the span.
d. 1/60th of the span.

**7. The height of a mobile scaffold must not exceed the minimum base dimension by**

a. three times.
b. four times.
c. five times.
d. six times.

**8. Guardrails must be installed on all scaffolds more than**

a. 10 feet in height.
b. 16 feet in height.
c. 20 feet in height.
d. 24 feet in height.

**9. The person in charge of the safe erection and dismantling of scaffolding is the**

a. general contractor.

b. foreman.

c. competent person.

d. architect.

**10. End frames are installed level and plumb to sit on top of**

a. baseplates.

b. mud sills.

c. cribbing.

d. all of the above.

**11. The number of times a scaffold should be visually inspected is at least**

a. two.

b. three.

c. four.

d. five.

**12. To access the work area of a scaffold, the user should use an approved ladder or**

a. the ladder built into the end frame.

b. cross braces.

c. horizontal bearing points of the scaffold.

d. all of the above.

**13. All scaffolding should be able to support_____times the intended load.**

a. 4

b. 5

c. 10

d. 20

**14. It is the responsibility of the _____ to ensure safety on the job site.**

a. the local OSHA inspector.

b. the general contractor.

c. scaffold erectors.

d. every worker on the job.

**15. Pump jack scaffolds**

a. require toeboards.

b. are installed on a mud sill.

c. have diagonal braces.

d. all of the above.

## KNOW YOUR CODES

The number one cause for injuries on the jobsite is falling. OSHA (Occupational Safety and Health Administration) provides a national standard for all construction work in the country. They are charged with reducing jobsite injuries. Make sure you comply with all aspects of the code.

## WHAT'S WRONG WITH THIS PICTURE?

Carefully study **Figure 11-34** and think about what is wrong. Consider all possibilities.

### WRONG

**FIGURE 11-34** The ladder is leaning at too steep of an angle. This is a dangerous situation. The bottom of the ladder could slide out from beneath the worker, resulting in a fall.

### RIGHT

**FIGURE 11-35** The ladder is at a much safer angle, which also results in more comfort for the worker.

CHAPTER 12

# Roof Framing

**OBJECTIVES** *After completing this unit, the student should be able to:*

- Describe several roof types.
- Define the various roof framing terms.
- Identify the members of gable, gambrel, hip, intersecting, and shed roofs.
- Lay out a common rafter and erect a gable roof.
- Lay out and install gable end studs.
- Lay out a hip rafter and hip jack rafters.
- Lay out a valley rafter and valley jack rafters.
- Describe and perform the safe and proper procedure to erect a trussed roof.
- Apply roof sheathing.
- Estimate the quantities of materials used in a roof frame.

# GLOSSARY OF ROOF FRAMING TERMS

**butterfly roof** an inverted gable roof

**cheek cut** a compound miter cut on the end of certain roof rafters

**dormer** a structure that projects out from a sloping roof to form another roofed area to provide a surface for the installation of windows

**fascia** the vertical surface of the cornice finish installed on the bottom end of rafters

**gable roof** a common type of roof that pitches in two directions

**gambrel roof** a type of roof that has two different sloped rafters on each side of center

**hip jack rafter** a rafter running between a hip rafter and the wall plate

**hip rafter** a rafter that extends diagonally from the corner of the plate to the ridge at the intersection of two different roof surfaces

**hip roof** a roof that slopes up towards the ridge from all walls

**hip-valley cripple jack rafter** a short rafter running parallel to common rafters, cut between hip and valley rafters

**intersecting roof** the roof of irregular-shaped buildings; valleys are formed at the intersection of the roofs

**lateral** a direction to the side at about 90 degrees

**lookout** horizontal framing pieces in a cornice, installed to provide fastening for the soffit

**mansard roof** a type of roof that has two different sloped rafters on all sides of the building, with the lower slopes steeper than the upper

**rake** the sloping portion of the gable ends of a building

**shed roof** a type of roof that slopes in one direction only

**tail cut** a cut on the lower end of a rafter that extends beyond the building

**valley** the intersection of two roof slopes forming an interior corner

**valley cripple jack rafter** a rafter running between two valley rafters

**valley jack rafter** a rafter running between a valley rafter and the ridge

**valley rafter** the rafter placed at the intersection of two roof slopes that form interior corners

The shape of the building becomes more clear as the nearby pile of framing lumber dwindles. During construction of the roof, the form of the house is completed. The building designer's or owner's dream is being realized, and pride grows around the construction site.

For many carpenters, roof framing is complicated and hard to do. Angle cuts and triangles are more difficult to visualize than square cuts and rectangles. For this reason, roof layout and member cutting is often performed by the most experienced carpenters on the job.

To learn roof framing, the student must be patient with his or her own learning speed and be willing to struggle, if necessary, to understand. It should be helpful to know that it is not easy to grasp roof framing the first time. It takes practice and a willingness to persevere.

The straightforward math and geometry in roof framing is helpful in making accurate measurements and cuts. Once roof framing is understood, roofs can be constructed with great precision and speed. Being a good roofing carpenter begins with knowing the names of all the components. Knowledge of diverse roof types and the terms used in their construction is essential. Remember, many things of value take effort to achieve.

# Roof Types

Several roof styles are in common use (Fig. 12-1). The most common roof style is the **gable roof** where two sloping roof surfaces meet at the top. They form a triangle at each end of the building called *gable ends*. The **shed roof** slopes in one direction, sometimes referred to as a *lean-to*. It is commonly used on additions to existing larger structures. It is also used extensively on contemporary homes. The **hip roof** slopes upward to the ridge from all walls of the building. This style is used when the same overhang is desired all around the building. The hip roof eliminates the gable ends, while reducing the effect of high winds on the building. An **intersecting roof** is required on buildings that have wings. Where two roofs intersect, **valleys** are formed. This requires several different types of rafters.

The **gambrel roof** is a variation of the gable roof. It has two slopes on each half instead of one. The lower slope is much steeper than the upper slope. It is framed somewhat like two separate gable roofs. The **mansard roof** is a combination of the gambrel and hip roofs. It has two slopes on each of the four sides. It is framed somewhat like two separate hip roofs.

The **butterfly roof** is an inverted gable roof. It resembles two shed roofs with their low ends placed against each other.

Other roof styles are a combination of the styles just mentioned. The shape of the roof can be one of the most distinctive features of a building.

## ROOF FRAMING TERMS

It is important for the carpenter who wants to become proficient in roof framing to be familiar with roof framing terms (Fig. 12-2).

### Rafter

A *rafter* is the sloping structural member of a roof frame. It supports the roof covering. Most rafters are *common* rafters, spanning from the top of the wall to the top of the roof. Other types of rafters include *hip*, *valley*, and their respective *jack* rafters.

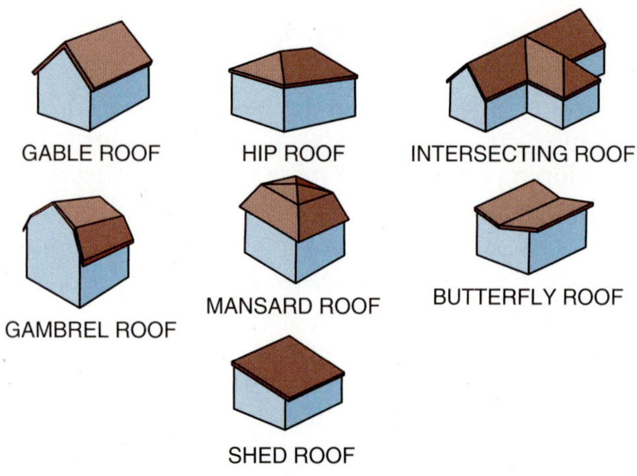


**FIGURE 12-1** Several roof styles are used for residential buildings.

## Span

The *span* of a roof is the horizontal distance covered by the roof. This is usually the width of the building measured from the outer faces of the frame, the building line.

## Run

The *total run* of a rafter is the horizontal distance over which the rafter covers. This is typically one-half the span. It must always be in terms of feet, with no actual units on the number, for example if the building span is 26 feet the run is 13.

## Rise

The *total rise* is the total vertical distance that the roof rises. Total rise may be found by multiplying the unit rise by the total run of the rafter.

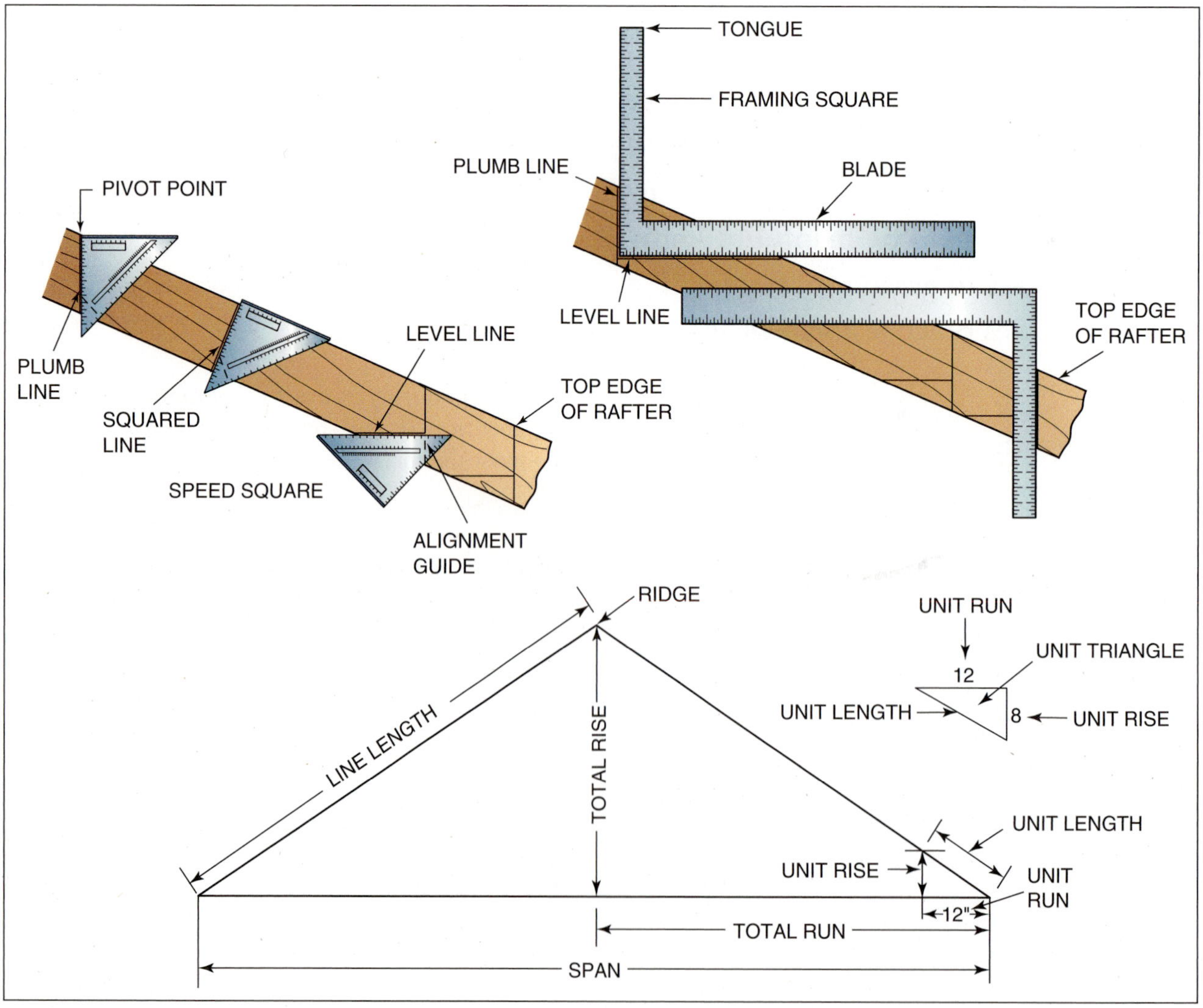

**FIGURE 12-2** Roof framing terms illustrated using a framing square and a speed square.

### Ridge

The horizontal member or line that forms the highest point of the roof system. The *ridge* secures the upper end of rafters.

### Seat Cut or Bird's Mouth

Referring to the notch cut near the lower end of a rafter, the *seat cut* forms the location where the rafter will sit on and be fastened to the wall.

### Line Length

The *line length* is the length of a rafter measured from the seat cut to the ridge. It is the hypotenuse (longest side) of a right triangle formed by the rafter, its run, and its rise. The line length is mathematical and gives no consideration to the thickness or width of the framing material. Line length is also referred to as *rafter length*.

### Unit Triangle

The *unit triangle*, sometimes called slope triangle, is the small right triangle found on the set of prints for the building. It looks like the right triangle of the house formed by one rafter, its run, and its rise. In the unit triangle, each side has a name similar to the house triangle with the word "unit" placed before it. Run becomes unit run, rise becomes unit rise, and line length becomes unit length.

### Unit Run

The *unit run* is 12 inches for all common and jack rafters. The unit run of hip and valley rafters is 16.97 inches. It is longer because the typical run of hip and valley rafters project into the building at a 45-degree angle with the plates. These numbers do not change from building to building. For this reason, they are not usually given on the set of prints.

### Unit Rise

The *unit rise* is the distance that the roof rises vertically for every unit of run. On the set of prints, the unit rise or *slope* of a roof may be shown on the unit triangle, usually found on a section view of the roof. This symbol shows the unit rise per unit of run (4 on 12, 6 on 12, or 7 on 12, for example).

### Unit Length

*Unit length* is length of rafter necessary to cover one unit of run. Since the unit length is the hypotenuse of the unit triangle, it may be calculated by using $C^2 = A^2 + B^2$, the Pythagorean theorem. For example, if the unit rise is 6 inches and the unit run is always 12 inches, then the unit length is the square root of $(6^2 + 12^2)$, or 13.42 inches. The unit lengths for whole units of rise are also printed on the *rafter tables* of the framing square.

### Pitch

The *pitch* is the ratio of rise to span of a roof. It is usually expressed as a fraction found by dividing the total rise by the span. For example, if the span of the building is 32 feet and total rise is 8 feet, then $8/32$ is reduced to $1/4$. The roof is said to be a $1/4$ pitch.

### Slope

Slope is the term typically used to refer to the steepness of a rafter. It compares the unit rise to the unit run. It is stated as unit rise on 12. For example, a rafter with a unit rise of 7 has a slope of "7 on 12".

### Plumb Line

A *plumb line* is any line on the rafter that is vertical when the rafter is in position. When laying out plumb cuts on rafters, the line is marked along the tongue of the framing square. The square is aligned using the unit run on the blade of the square and the unit rise on the tongue. A speed square is positioned with the pivot point touching the rafter and the unit rise aligned to the edge of the rafter. The plumb line is marked along the edge of the square where the inch ruler is located.

### Level Line

A *level line* is any line on the rafter that is horizontal when the rafter is in position. Level lines are marked along the blade of the framing square. With a speed square, mark the long edge of the square where the degree scale is located after lining up the alignment guide with the plumb line (Fig. 12-3).

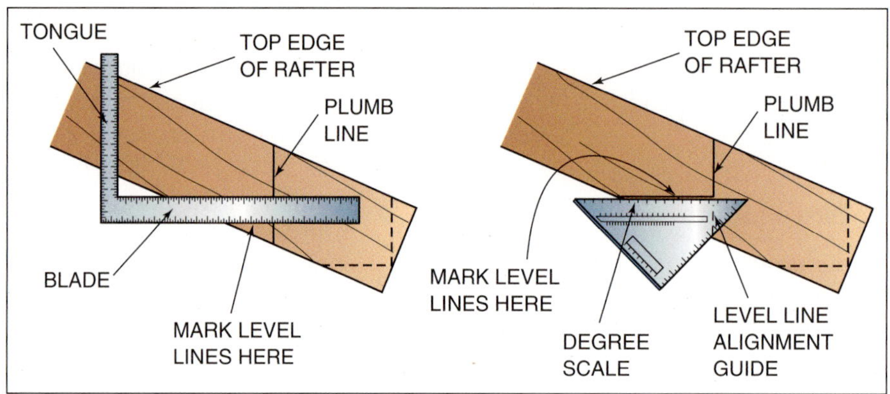


**FIGURE 12-3** Techniques for marking a level line on a rafter using a framing square and a speed square.

## Gable Roof

The gable roof is the style most commonly used. This roof is the simplest to frame (Fig. 12-4). Only common rafters need to be laid out. Gable roofs may have two different runs on either side of the ridge, such as the *saltbox* roof (Fig. 12-5). The common rafter extends at right angles from the plate upward to the ridge. It is called a *common rafter* because it is common to all types of roofs. It is used as a basis for laying out other kinds of rafters. The parts of a rafter are shown in Figure 12-6.

It is standard practice to lay out and cut one rafter and use it as a pattern for the remaining rafters. The *ridge*, although not absolutely necessary, simplifies the erection of the roof. It provides a means of tying the rafters together before the roof is sheathed. Erecting a roof frame is called *raising* the roof.

There are two methods of determining the rafter length. One involves the use of a calculator and the rafter tables, and the other actually steps off the length. Calculating rafter length is faster and more accurate than using the step-off method. Rafter tables are stamped onto one side of most framing squares (Fig. 12-7). The inch marks above the rafter table indicate unit rise (or rise of the rafter per unit of run). The first line of the table gives the length of

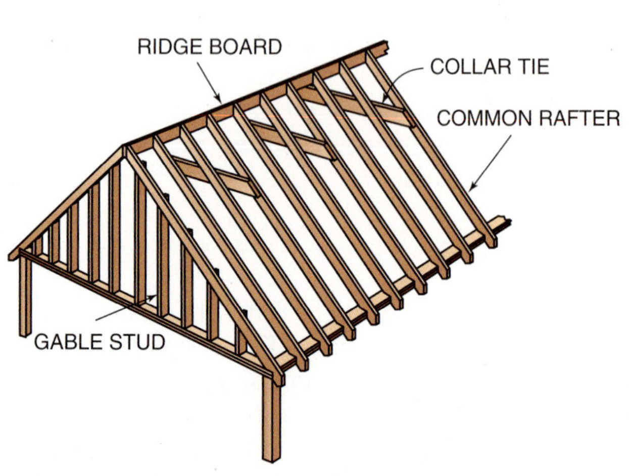


**FIGURE 12-4** The gable roof is framed with common rafters, ridgeboard, collar ties, and gable studs.

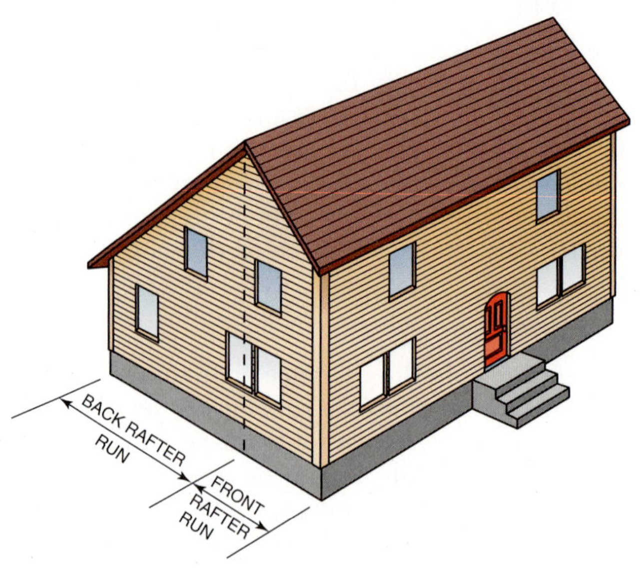


**FIGURE 12-5** The ridge of the saltbox roof is off-center.

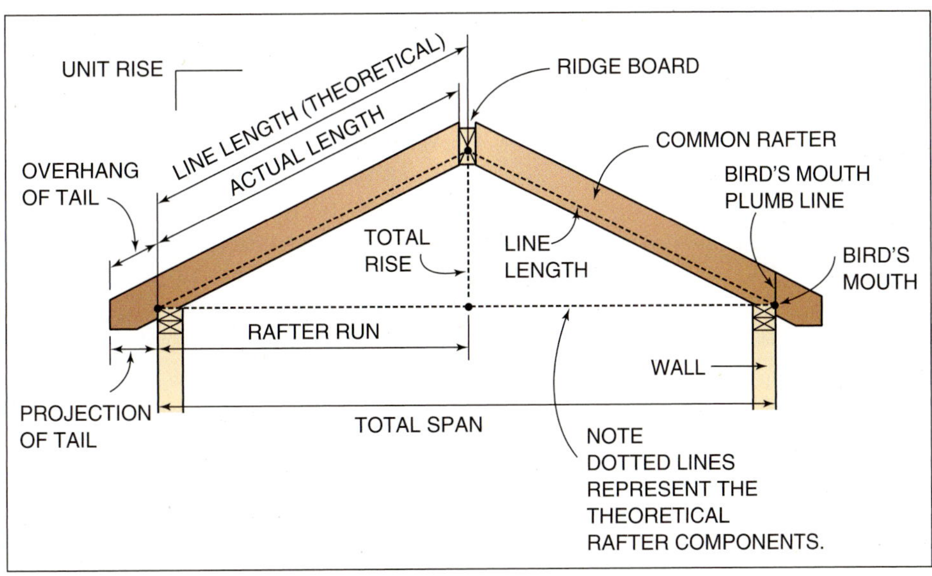


**FIGURE 12-6** Terms, components, and concepts of rafter framing.

**FIGURE 12-7** Rafter tables are found on the framing square.

rafter needed to cover one unit of run for the different unit rises. For example, the unit length (or length per foot of run) for a rafter that has unit rise of 17 is 20.81 inches.

The **tail cut** is the cut at the lower end of the rafter. It may be a plumb cut, a combination of plumb and level cuts, or simply a square cut (Fig. 12-8). Sometimes the rafter tails are left to run *wild*. This means they are slightly longer than needed. They are cut off in a straight line after the roof frame is erected.

Framing square gauges may be attached to the square to simplify repetitive alignments of the square. They act as stops against the top edge of the rafter. These gauges are attached to the tongue for the desired unit rise and to the blade for the unit run (Fig. 12-9).

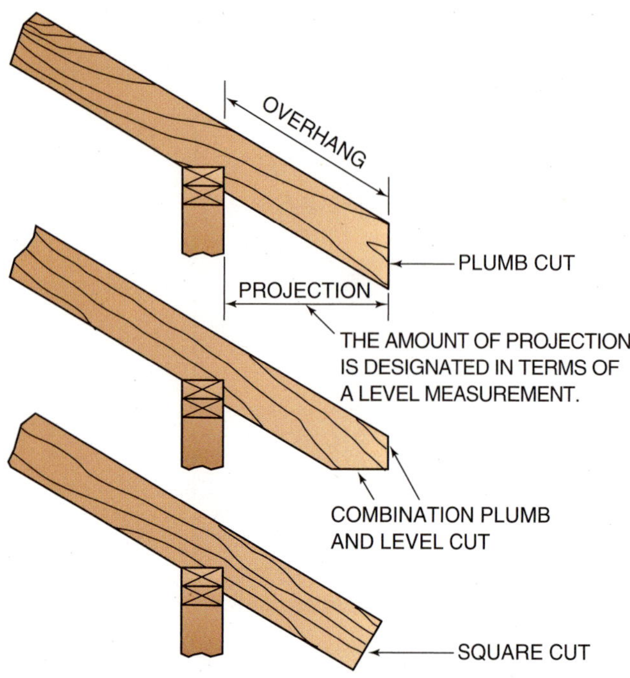


**FIGURE 12-8** Various tail cuts of the common rafter.

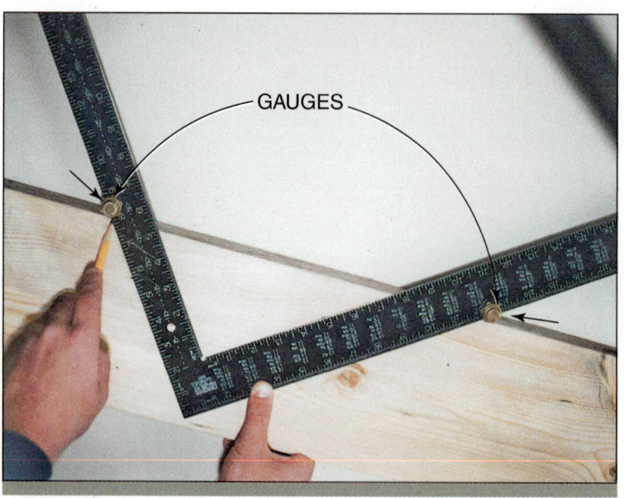


**FIGURE 12-9** Framing square gauges are attached to the square to hold it in the same position for every step-off.

## COMMON RAFTER

The common rafter requires layouts for making several cuts. The cut at the top is called the *plumb cut* or *ridge cut*. It fits against the ridgeboard. The bird's mouth or *seat cut* consists of a plumb line and a level line layout. It fits against the top and outside edge of the wall plate. At the bottom end of the rafter, a tail cut is made on the rafter tail or overhang that extends beyond the building (Fig. 12-10).

### Calculation Method

The exact length of a rafter may be determined by using a calculator and a little math. Process begins with determining the unit rise and the total run from the set of prints. The unit length is then found on the first line of rafter tables under the appropriate unit rise number (the inches number on the edge of the square). The run is usually one-half the width of the building. Multiplying the unit rise by the run gives the length of the rafter in inches.

***For step-by-step instructions on COMMON RAFTER LAYOUT using the calculation method, see Procedure 12-1 on pages 369–372.***

### The Step-Off Method

The step-off method is another way to determine the rafter length. It uses the unit run and the unit rise on a framing square as used in the layout of the ridge plumb line. The rafter stock is stepped off for each unit of run until the desired number of units or parts of units are stepped off.

The procedure begins by aligning the square on the first ridge plumb line as if it were being laid out. Then, place a mark where the blade intersects with the top edge of the rafter. Next, slide the square down, still aligned to the unit rise and run, until the tongue lines up with the mark. Repeat the mark where the blade intersects the top edge of the rafter. Move the square and mark in a similar manner until the total run of the rafter is stepped off (Fig. 12-11). Mark a plumb line along the tongue of the square at the last step. This line is parallel to the ridge cut and becomes the seat cut plumb line.

**CAUTION**

**CAUTION: On some cuts that are at a sharp angle with the edge of the stock (that is, the level line of the seat cut), the guard of the circular saw may not retract. In this case, retract the guard by hand until the cut is made a few inches into the stock. Never wedge open the guard. Allow the guard to return when the cut is completed.**

### Wood I-Beam Rafters

In addition to solid lumber, wood I-beams may be used for rafters (Fig. 12-12). The layout is the same as for standard dimension lumber. Some additional framing details are shown in Figure 12-13.

**FIGURE 12-10** The cuts of the common rafter. The seat plumb and level lines together are referred to as the bird's mouth or seat cut.

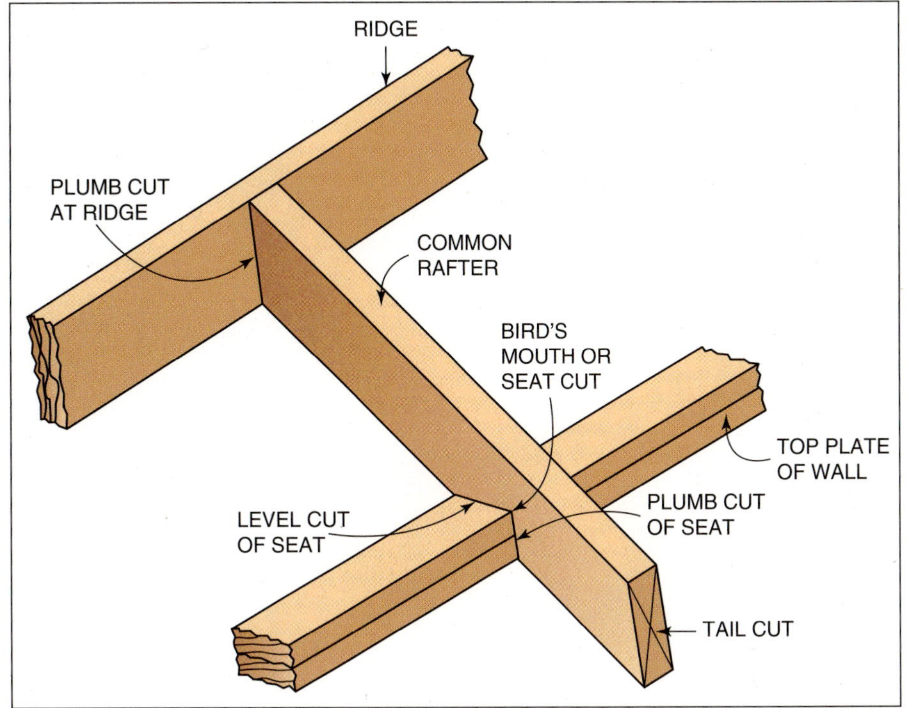


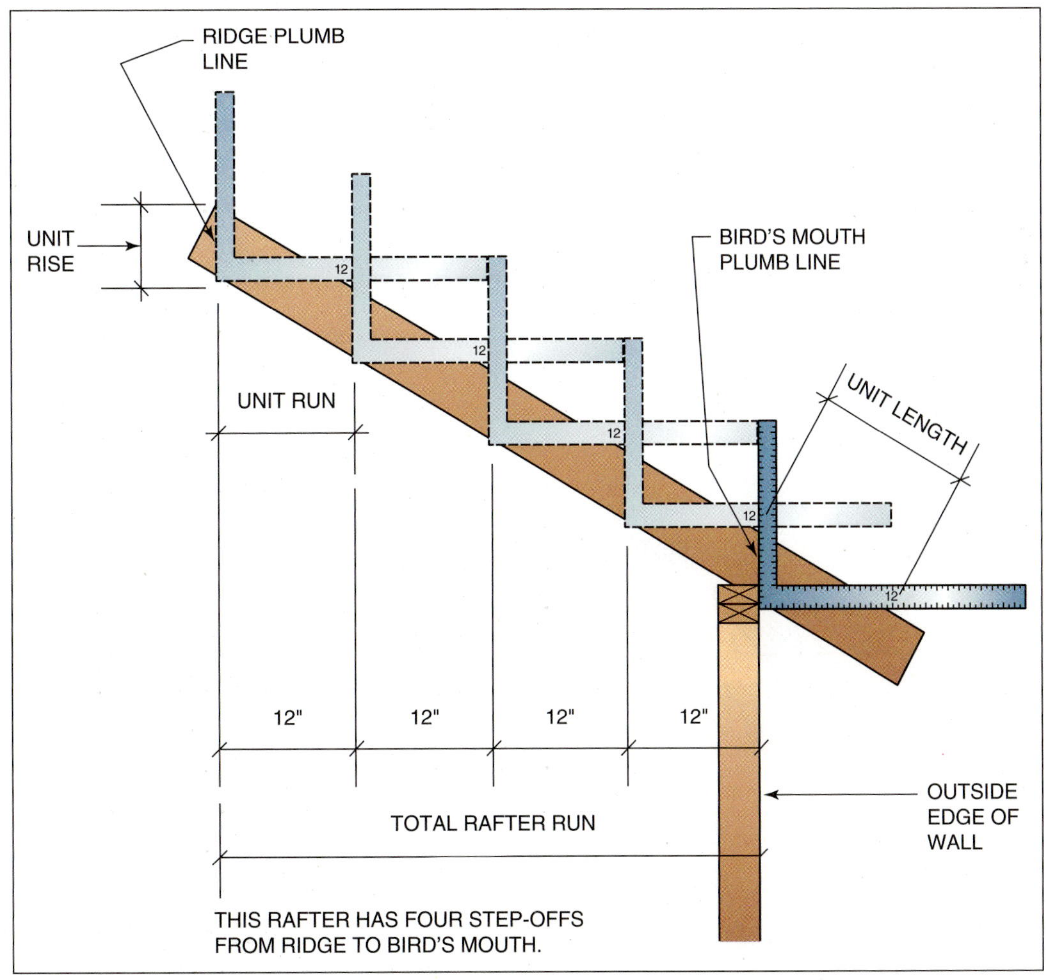


**FIGURE 12-11** Stepping off the common rafter having a run of 4 feet.

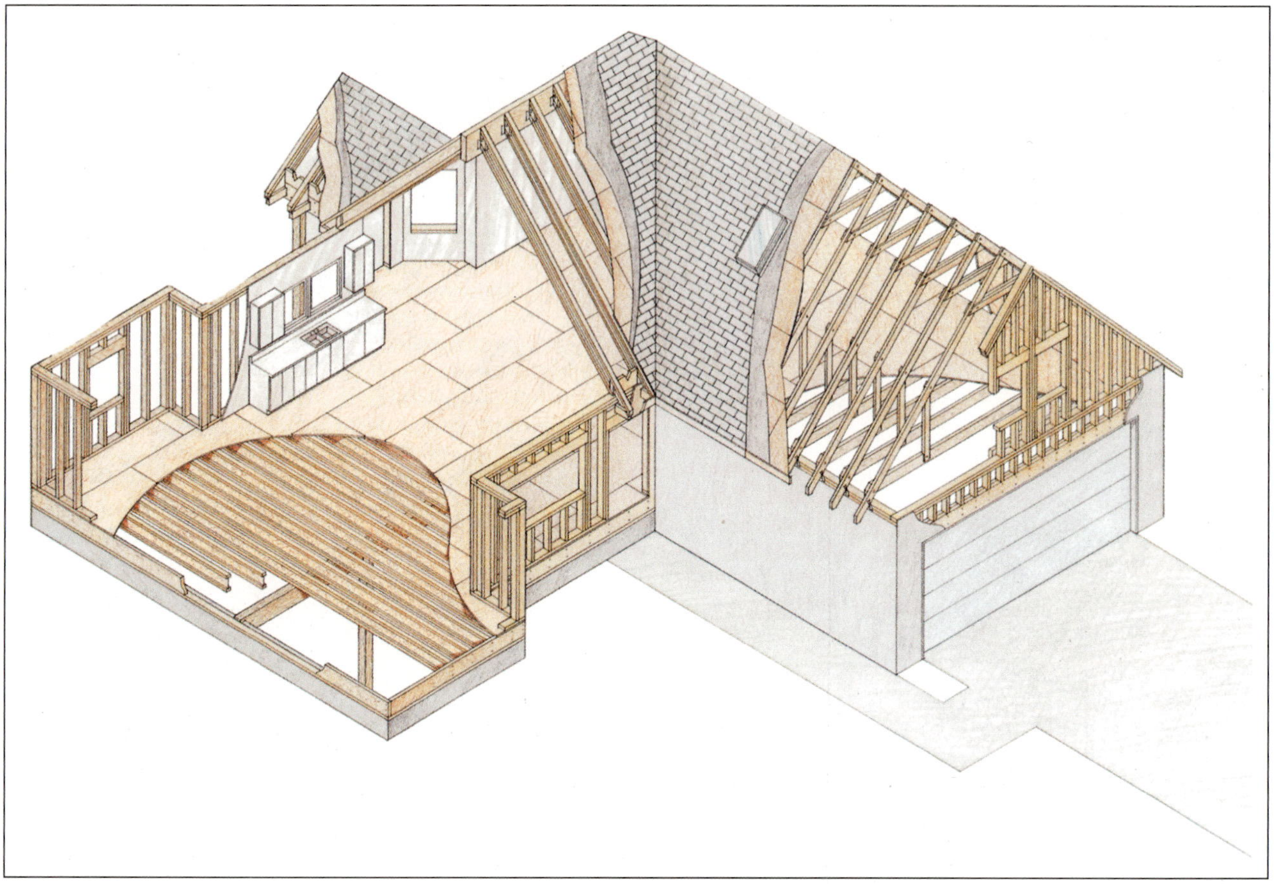

*Courtesy of Weyerhaeuser Company.*

**FIGURE 12-12** Wood I-beams may also be used for roof rafters.

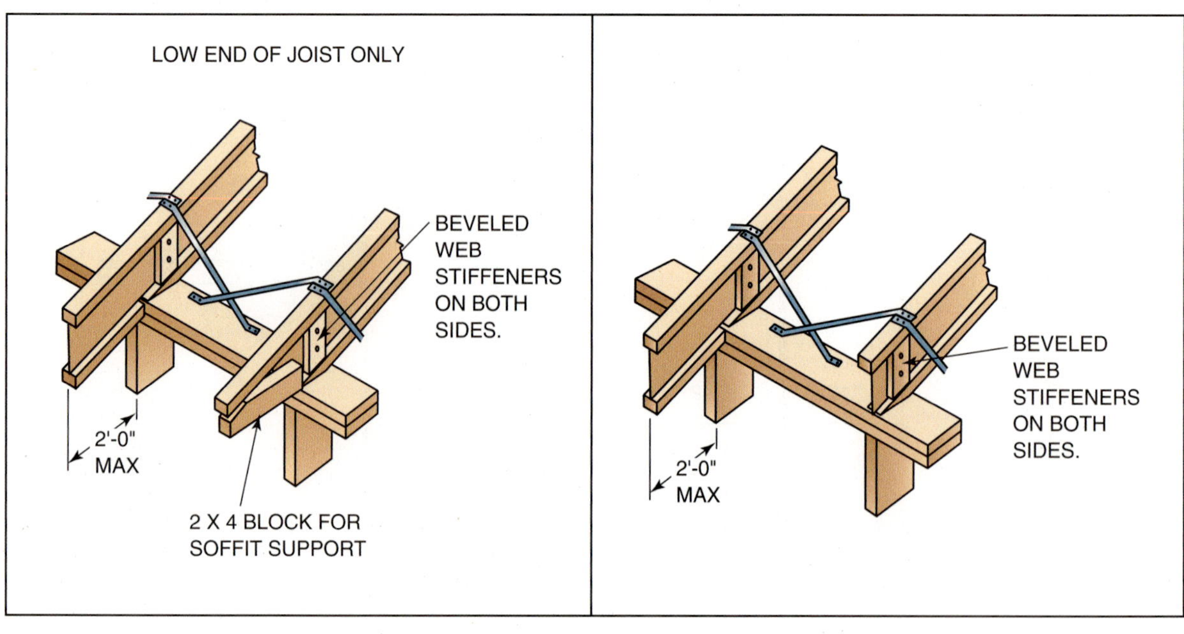


*Courtesy of Weyerhaeuser Company.*

**FIGURE 12-13** Wood I-beam roof-framing details.

## RIDGEBOARD LAYOUT

Transfer the rafter layout from the top plate to the ridgeboard. Joints in the ridgeboard should be centered on a rafter; otherwise, a scab is needed. The total length of the ridgeboard should be the same as the length of the building, plus the overhang at the gable ends. Add the necessary amount on both ends. Mark the end of the ridge such that it will be obvious to the installer which end goes where.

## ERECTING THE GABLE ROOF FRAME

Rafters are installed after the ceiling joists are installed. The joists provide a working platform from which to erect the roof. Otherwise, a scaffold may be necessary in the center of the building. The bottom edge of the rafters and ridgeboard should be fastened flush with each other. This will allow for greater airflow through the ridge vent and better support of the rafter (Fig. 12-14). At the seat cut, the rafter is toenailed to the wall plate and to the sides of ceiling joists with four 10d nails.

***For step-by-step instructions on ERECTING THE GABLE ROOF FRAME, see Procedure 12-2 on pages 373–375.***

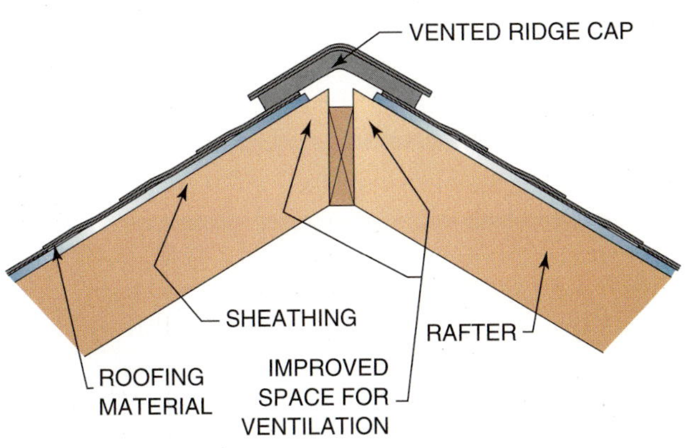


**FIGURE 12-14** Rafter and ridge are installed flush on the bottom.

## RAFTER TAILS

Some carpenters prefer to cut the rafters for length after they are installed. To cut the rafter tails, measure and mark the end rafters for the amount of overhang. This is usually a level measurement from the outside of the wall studs to the tail plumb line. Plumb the marks up to the top edge of the rafters. Snap a line between these two marks across the top edges of all the rafters. Using a level, plumb down on the side of each rafter from the chalk line. Using a circular saw, cut each rafter. Start the cut from the top edge and follow the line carefully (Fig. 12-15).

## RAKE OVERHANG

If an overhang is specified at the rakes, horizontal structural members called **lookouts** must be installed. They support the rake rafter (Fig. 12-16). Care should be taken to make sure the rake rafter is straight in two directions. The rake rafter should not crown severely and should be straight along its side.

## ROOF OPENINGS

Small openings in the roof for skylights and chimneys may be created by using headers in a similar fashion as with ceiling joists. A section of rafter is usually cut out along plumb lines. Either end is supported by a header that is also nailed to the rafters on either side (Fig. 12-17).

# GABLE STUDS

The triangular areas formed by the rake rafters and the wall plate at the ends of the building are called *gables*. They must be framed with studs. These studs are called *gable studs*. The bottom ends are cut square and toenailed to the top of the wall plate. The top ends fit snugly against the bottom edge and inside face of the end rafter. They are cut off flush with the top edge of the rafter (Fig. 12-18). Studs should be positioned directly above the end

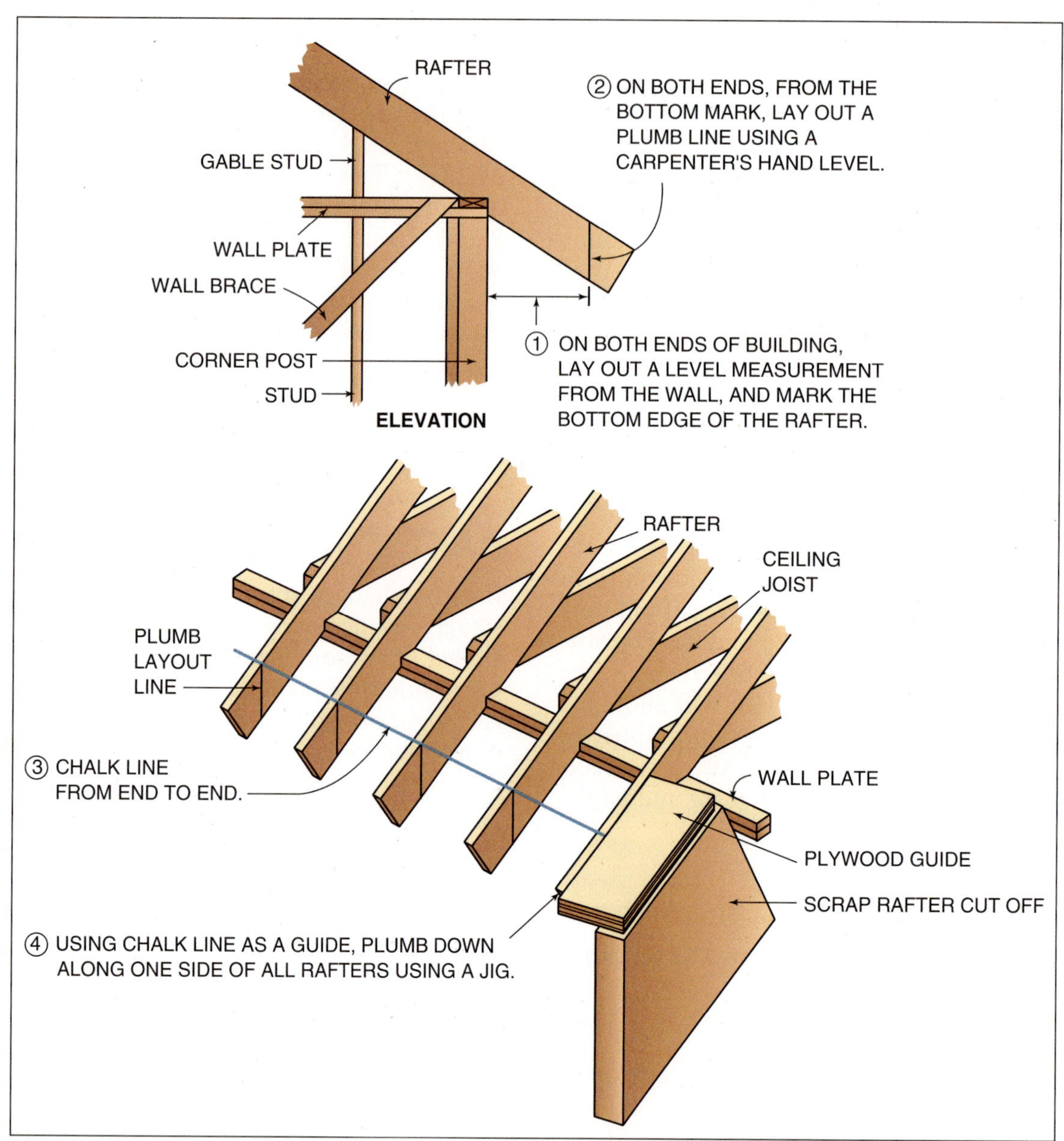


**FIGURE 12-15** Laying out the tail cut of common rafters after they are installed.

wall studs. This allows for easier installation of the wall sheathing.

Since gable studs are on center, they have a common difference in length. Adjacent studs are longer or shorter than the next one by the same amount (Fig. 12-19). Once the length of the first stud and the common difference are known, gable studs can be laid out easily and cut all at once.

To find the common difference in the length of gable studs, multiply the stud spacing, in feet, by the unit rise of the roof.

**Example:** Calculate the common difference in length of a stud spaced 16 inches OC that is to be attached to a rafter with a unit rise of 6 inches.

First, change 16 inches to 1.3333 feet by dividing by 12. Then, multiply 1.3333 times 6, which equals 8 inches. This means that the gable end studs for this roof vary by 8 inches from one to the other.

***For step-by-step instructions on INSTALLING GABLE STUDS, see Procedure 12-3 on page 376.***

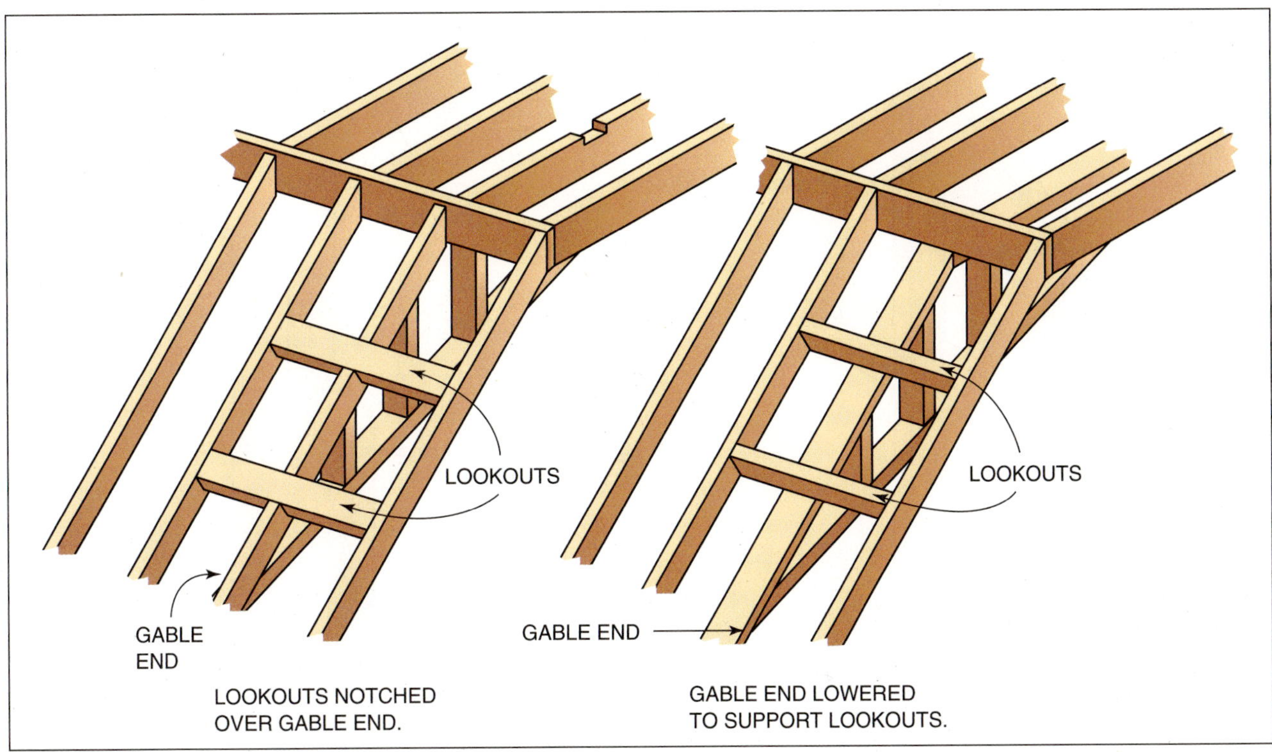


**FIGURE 12-16** Lookouts support the rake overhang.

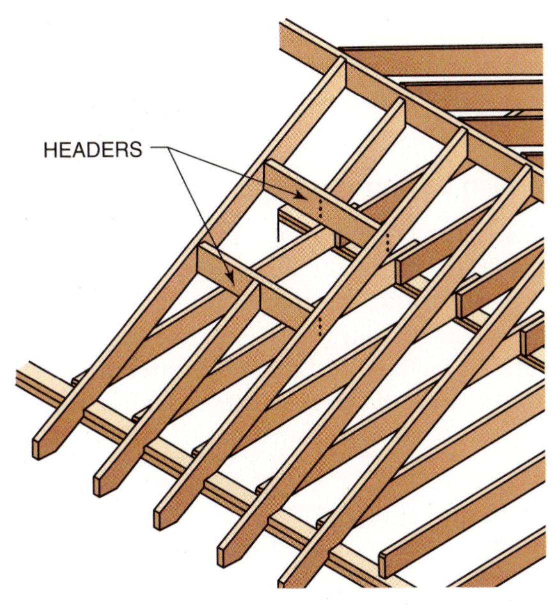


**FIGURE 12-17** Typical framing of a small opening in a roof system.

# HIP ROOF

To frame the hip roof, it is necessary to lay out not only common rafters and a ridge, but also **hip rafters** and **hip jack rafters** (Fig. 12-20). Hip rafters are required where the slopes of the hip roof meet. The hip jack rafters are common rafters that are shortened because they meet the hip rafter. Hip jacks, like gable end studs, have a common difference in length.

## HIP RAFTER

Because the hip rafter run is at a 45-degree angle from the plates, the amount of horizontal distance it covers (total run) is greater than that of the common rafter. The hip rafter rises to meet at the ridge in the same number of steps (units of run) as the common rafter. Therefore, the unit of run for the hip rafter is larger.

The unit run for a hip is the diagonal of the square formed by the unit run of a common rafter

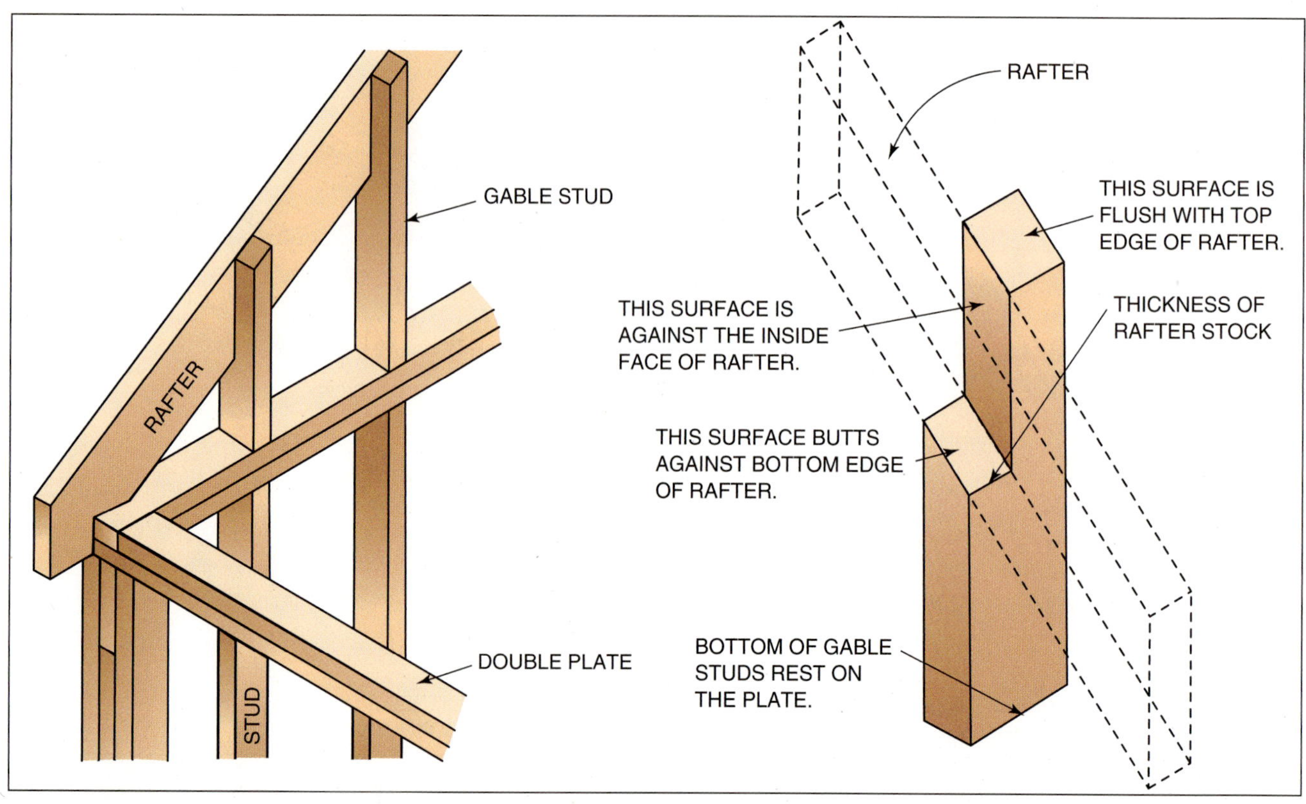


**FIGURE 12-18** The cut at the top end of the gable stud.

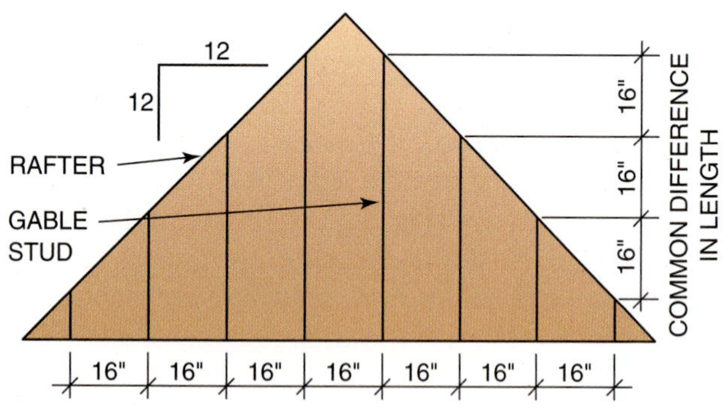


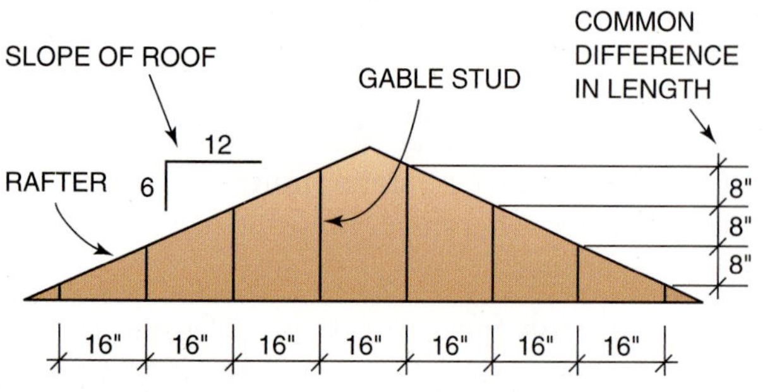


**FIGURE 12-19** Gable studs spaced equally have a common difference in length.

and the exterior walls. Because the unit run of the common rafter is 12 inches, the unit run of the hip rafter becomes the diagonal of a 12-inch square, which is 16.97 inches (Fig. 12-21).

### Steps in Laying out Hip Rafters

The steps to lay out a hip rafter are similar to those for a common rafter. Mark the ridge plumb line, shorten the rafter because of the ridge thickness, lay out its length, mark the seat cut lines, lay out the length of the tail, and mark the tail cut lines. Some variations occur because the hip rafter run is on a diagonal to the wall plates.

### Shortening the Hip

The amount of shortening due to the ridge of the hip rafter is more than for a common rafter. It is one-half the 45-degree thickness of the ridge board (Fig. 12-22). This is because the ridge and hip run meet at a 45-degree angle.

### Cheek Cuts

The ridge cut of a hip rafter is a compound angle called a **cheek cut**, or *side cut*. A *single cheek* cut

RIDGE
HIP RAFTERS
HIP JACK RAFTERS
COMMON RAFTERS

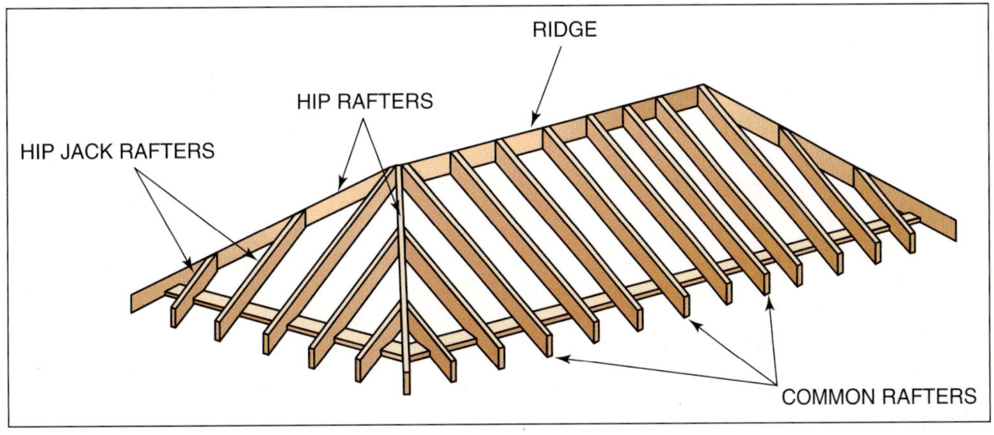

**FIGURE 12-20** Members of a hip roof frame.

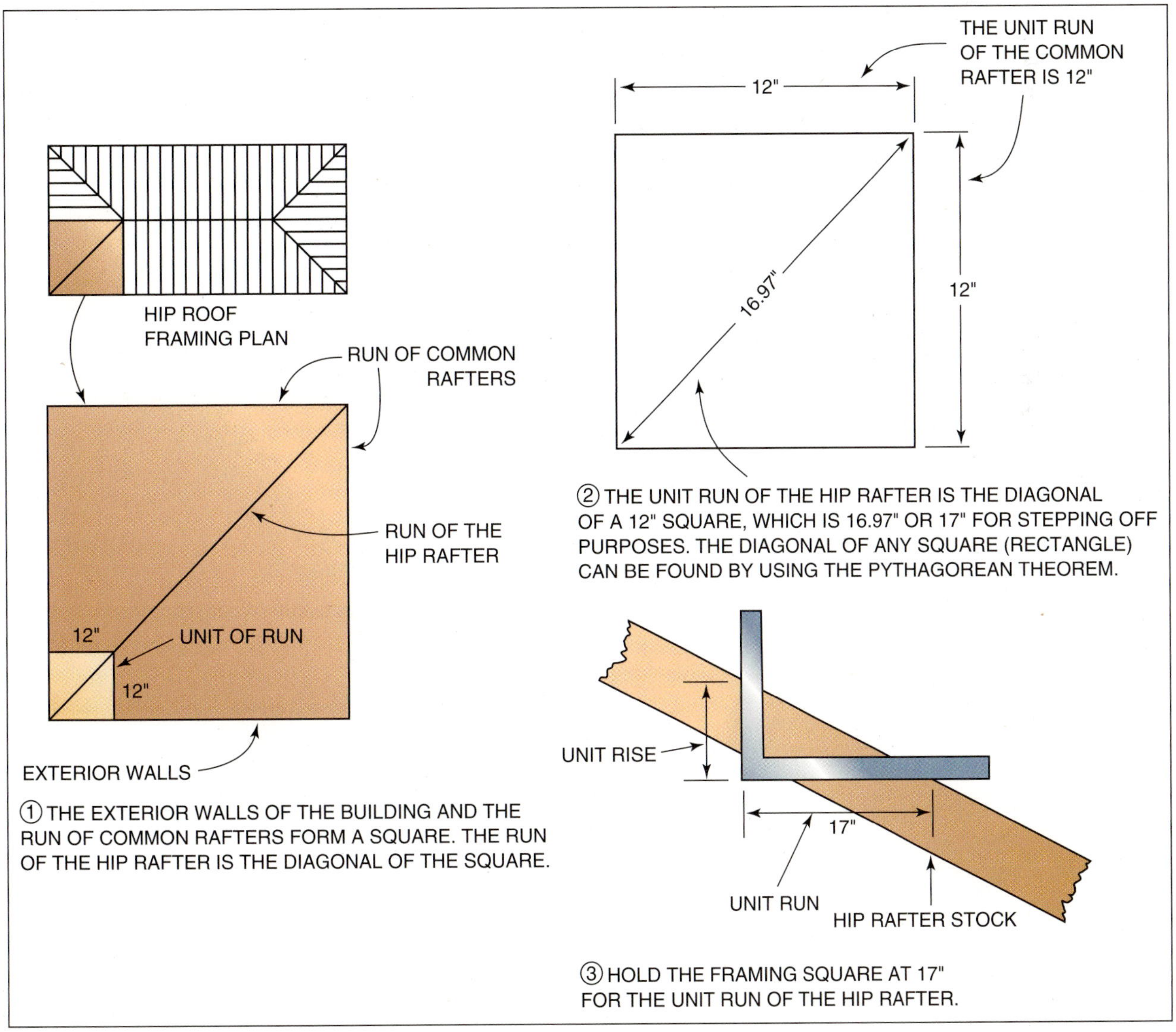

HIP ROOF
FRAMING PLAN
RUN OF COMMON
RAFTERS
RUN OF THE
HIP RAFTER
12"
UNIT OF RUN
12"
EXTERIOR WALLS
① THE EXTERIOR WALLS OF THE BUILDING AND THE RUN OF COMMON RAFTERS FORM A SQUARE. THE RUN OF THE HIP RAFTER IS THE DIAGONAL OF THE SQUARE.
THE UNIT RUN
OF THE COMMON
RAFTER IS 12"
12"
12"
16.97"
② THE UNIT RUN OF THE HIP RAFTER IS THE DIAGONAL OF A 12" SQUARE, WHICH IS 16.97" OR 17" FOR STEPPING OFF PURPOSES. THE DIAGONAL OF ANY SQUARE (RECTANGLE) CAN BE FOUND BY USING THE PYTHAGOREAN THEOREM.
UNIT RISE
17"
UNIT RUN
HIP RAFTER STOCK
③ HOLD THE FRAMING SQUARE AT 17" FOR THE UNIT RUN OF THE HIP RAFTER.

**FIGURE 12-21** The unit run of the hip rafter is 16.97 inches, which is rounded to 17 inches for layout.

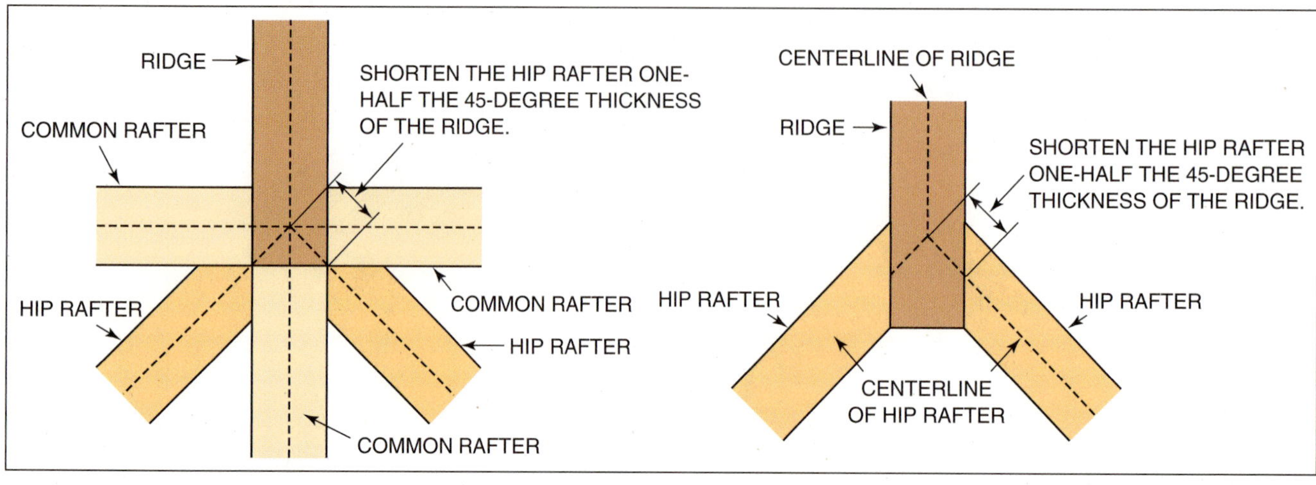


**FIGURE 12-22** Amount to shorten the hip rafter is the same for either method of framing.

or a *double cheek* cut may be made on the hip rafter according to the way it is framed at the ridge (Fig. 12-23). The tail may also be cut with cheek cuts to allow the tail to fit against the fascia.

### Backing or Dropping the Hip

Because the hip rafter run is at the 45-degree intersection of runs of two roof sections, the top outside edge of the hip rafter projects above the plane of the roof. To remedy this problem, the rafter may be dropped, or the top edge may be beveled, which is called *backing* (Fig. 12-24). Dropping the hip is much easier and is more frequently done. This is achieved by raising the seat cut level line in order to cut the seat cut deeper.

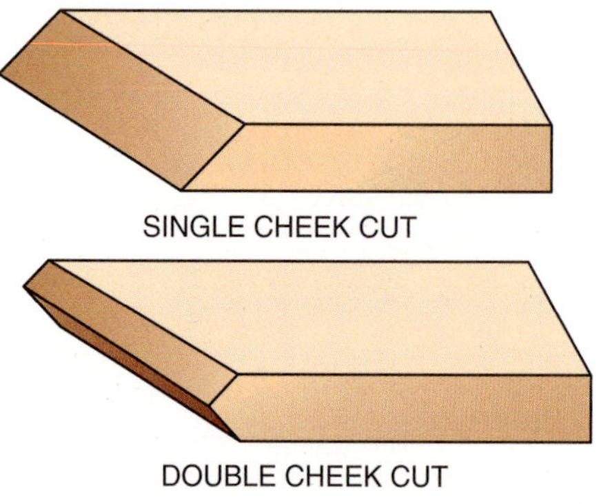


**FIGURE 12-23** Single or double cheek cuts are used depending on the method of framing the hip rafter at the ridge.

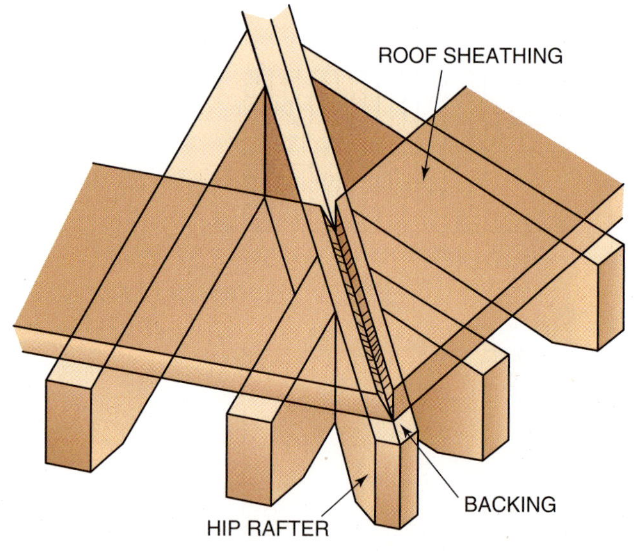


**FIGURE 12-24** Backing the hip to allow for the roof sheathing.

***For step-by-step instructions on HIP RAFTER LAYOUT, see Procedure 12-4 on pages 377–379.***

## HIP JACK RAFTER

A *hip jack rafter* is a common rafter that is shortened with cuts similar to those of hip rafters. The seat cut and tail are the same as that of the common rafter. For this reason, the common rafter pattern is saved to lay out all the hip jack tails (Fig. 12-25).

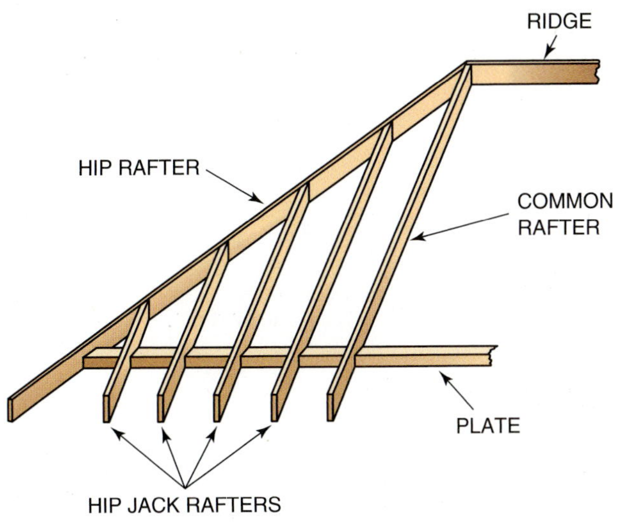


**FIGURE 12-25** Hip jack rafters are shortened common rafters.

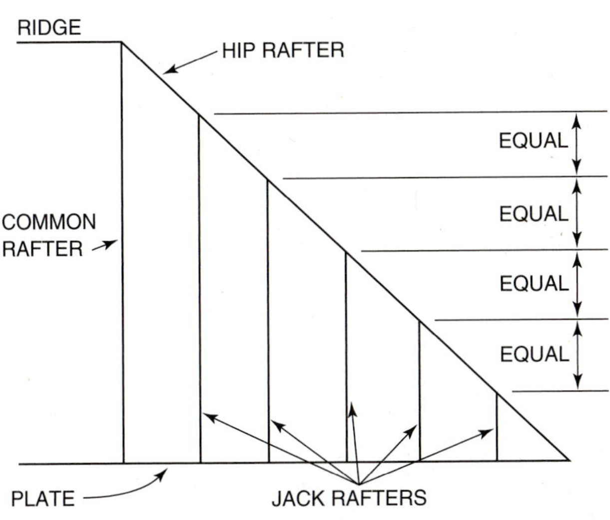


**FIGURE 12-26** Common difference in the length of hip jack rafters.

Hip jack rafters are framed against the side of the hip rafter. Each jack rafter is shorter or longer than the next one by the same amount. As with gable end studs, this is called the *common difference* in length (Fig. 12-26).

The common difference is found in the rafter tables on the framing square for jacks 16 and 24 inches OC. Once the length of the first jack is determined, the length of all others can be found by making each set shorter or longer by the common difference. The hip jack rafter has a single cheek cut where it meets the hip rafter. This is laid out in the same way as the hip rafter cheek cuts.

***For step-by-step instructions on JACK RAFTER LAYOUT, see Procedure 12-5 on pages 380–381.***

## HIP ROOF RIDGEBOARD LENGTH

The length of the hip roof ridgeboard is found by subtracting the width of the building from the length of the building. However, the actual length of the hip roof ridge must be cut longer. The amount of increase depends on the construction. If the hip rafters are framed against common rafters, the ridge length is increased at each end by one-half the thickness of the common rafter stock (3⁄4 inch for dimension lumber). If the hip rafters are framed against the ridge, increase the ridgeboard length at each end by half the ridgeboard thickness plus one-half the 45-degree thickness of the hip rafter (1 13⁄16 inches for dimension lumber) (Fig. 12-27).

# VALLEY RAFTERS

Buildings of irregular shape, such as L-, H-, or T-shaped, require a roof for each section. The intersection of these roofs is called a **valley**. *Valley rafters* form the intersection of the slopes of two roofs. If the heights of the roofs are different, two kinds of valley rafters are required. The *supporting valley rafter* runs from the plate to the ridge of the main roof. The *shortened valley rafter* runs from the plate to the supporting valley rafter.

**Valley jack rafters**, like hip jacks, are common rafters that are cut shorter. In this case, they run from the ridge to the valley rafter. The **hip-valley cripple jack rafter** runs between a hip rafter and a valley rafter. The **valley cripple jack rafter** runs between the supporting and shortened valley rafter (Fig. 12-28).

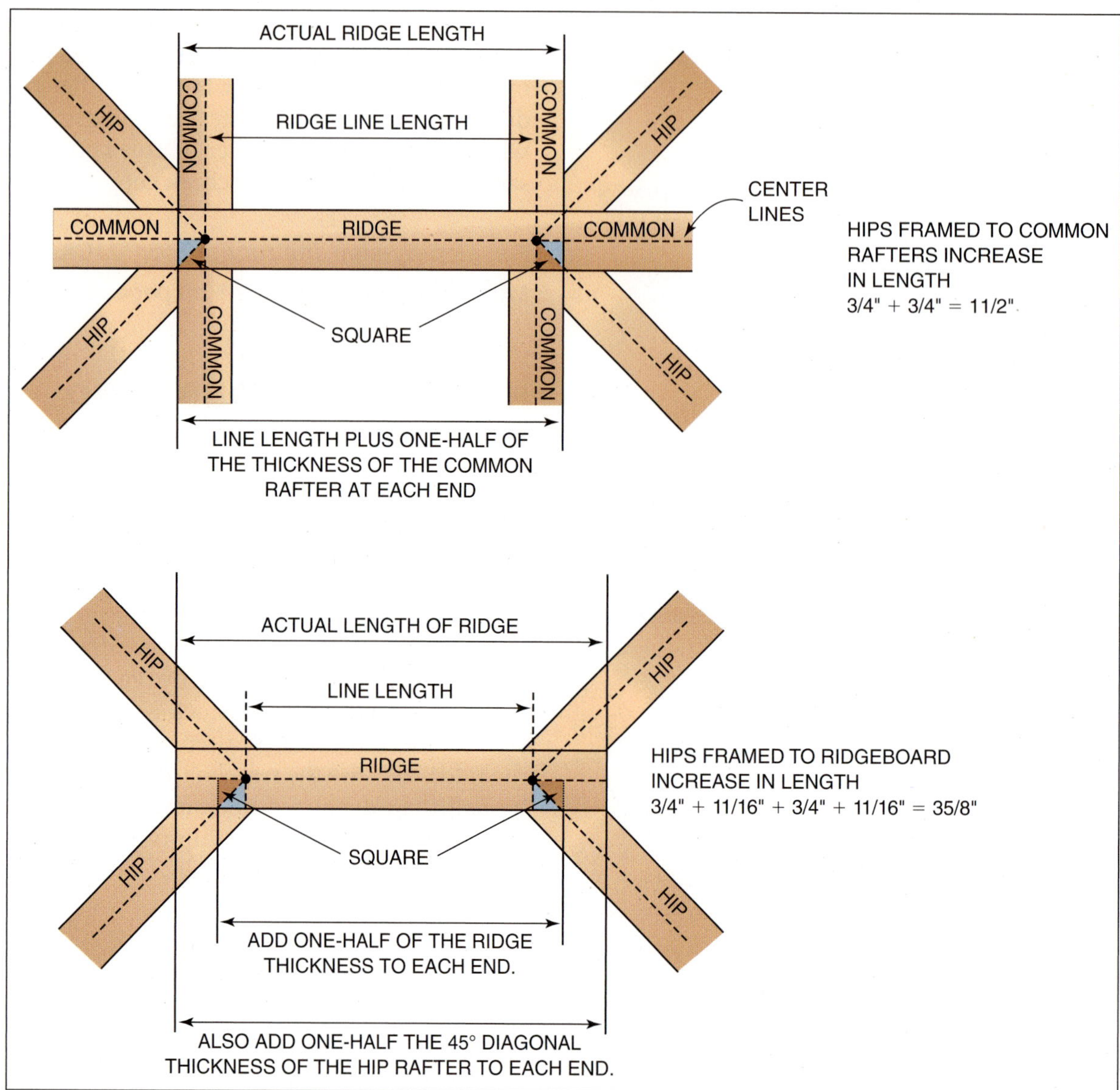


**FIGURE 12-27** Actual length of the hip roof ridge for both methods of framing the hip rafter.

Confusion concerning the layout of so many different kinds of rafters can be reduced by remembering the following.

- Hip and valley rafters are similar. Common rafters and all jack rafters are similar.
- The length of any kind of rafter can be found from its run and unit length.
- The amount of shortening is always measured at right angles to the plumb cut.
- The method for laying out cheek cuts for all rafters is similar.

## SUPPORTING VALLEY RAFTER LAYOUT

The layout of valley rafters is similar to that of hip rafters. The unit of run for both is 16.97 inches. The total run of the supporting valley is the run of the common rafter of the main roof, called the *major span*. Its unit length is found on the rafter tables in the same manner as for hip rafters. A single cheek cut is made at the ridge. The rafter is shortened by one-half the 45-degree thickness of the ridgeboard, typically 1 1⁄16 inch.

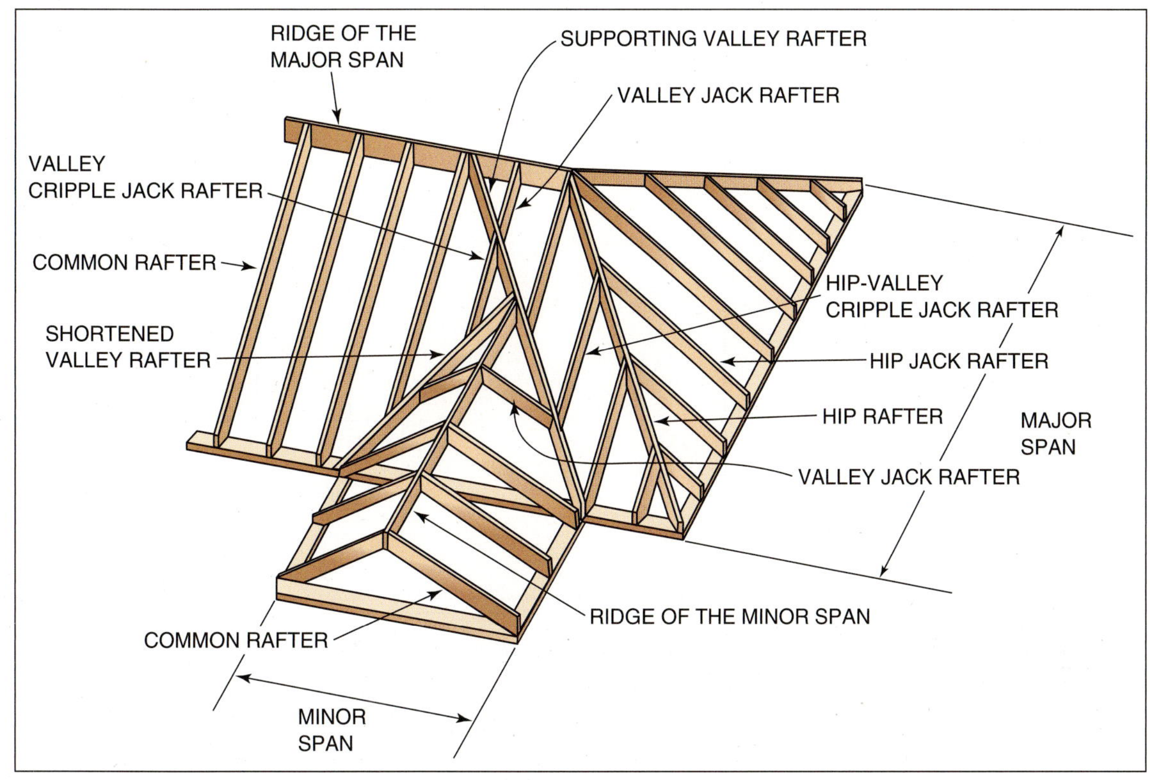


**FIGURE 12-28** Members of the intersecting roof frame.

***For step-by-step instructions on VALLEY RAFTER LAYOUT, see Procedure 12-6 on pages 382–384.***

## SHORTENED VALLEY LAYOUT

The length of the shortened valley is found by using the run of the common rafter for the smaller roof, called the *minor span.* Its seat cut is laid out in the same manner as for the supporting valley. However, the shortening is one-half the thickness of the supporting valley (usually ¾ inch). Also, there is no need for a cheek cut because the two valley rafter runs meet at right angles. The shortened valley has a square cut along the shortened plumb line (Fig. 12-29).

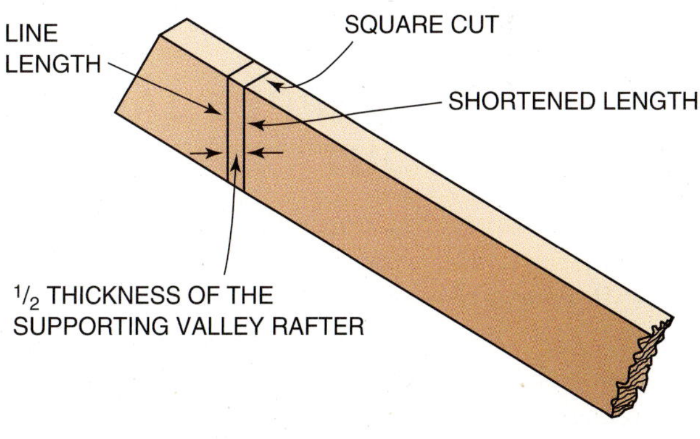


**FIGURE 12-29** Plumb cut layout on the top end of the shortened valley rafter.

## VALLEY JACK LAYOUT

The length of the valley jack can be found by multiplying its run times its unit length. The total run of any valley jack rafter is equal to the run of the common rafter minus the horizontal distance from the inside corner of the building (Fig. 12-30). Remember that all jack rafters are shortened common rafters and their unit run is 12. The ridge cut of the valley jack is the same as a common rafter and is shortened in the same way, usually ¾ inch. The cheek cut against the valley rafter at its lower end is a single cheek cut. The valley jack is shortened at this end by one-half the 45-degree angle thickness of

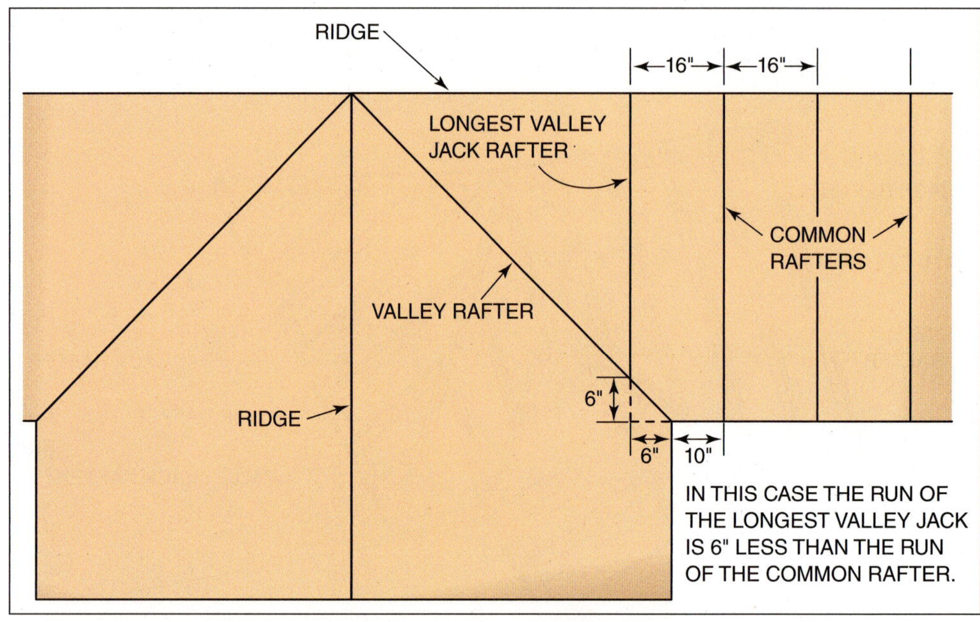


**FIGURE 12-30** How to determine the run of the longest valley jack rafter.

the valley rafter stock, usually 1 1/16 inch. The layout of all other valley jack rafters is made by making each shorter by the common difference found in the rafter tables.

### HIP-VALLEY CRIPPLE JACK RAFTER LAYOUT

The run of a hip-valley cripple jack rafter is equal to the distance between the seat cuts of the hip and valley rafters measured along the wall plate (Fig. 12-31). Determine the length using the unit length for the common rafter. All hip-valley cripple jacks cut between the same hip and valley rafters are the same length. On each end, shorten by half the 45-degree thickness of the hip and valley rafters, usually 1 1/16 inch. Make single cheek cuts (Fig. 12-32).

## SHED ROOF

The shed roof slopes in only one direction. It is relatively easy to frame. A shed roof may be freestanding, or one edge may rest on an exterior wall while the other edge butts up against an existing wall. Shed and other type roofs are also used on **dormers**.

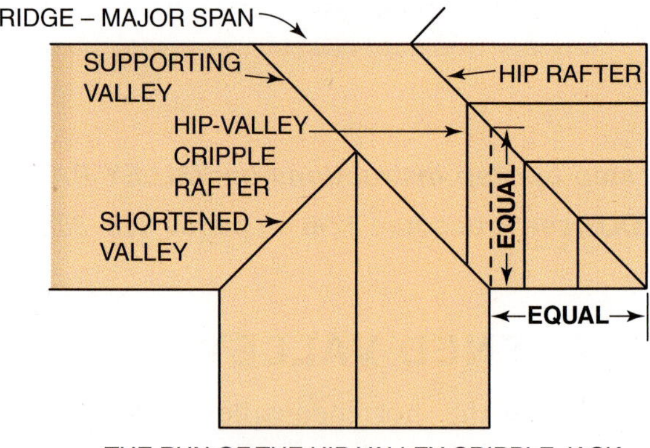


**FIGURE 12-31** Finding the total run of the hip-valley cripple jack rafter.

A dormer is a framed projection above the plane of the roof (Fig. 12-33).

A *shed roof rafter* is similar to a common rafter. It may require two seat cuts instead of one. In this case, the run of the rafter is the width of the building minus the width of one of the wall plates (Fig. 12-34). For shed roofs that butt up against an existing wall, the rafters are laid out in the same manner as common rafters.

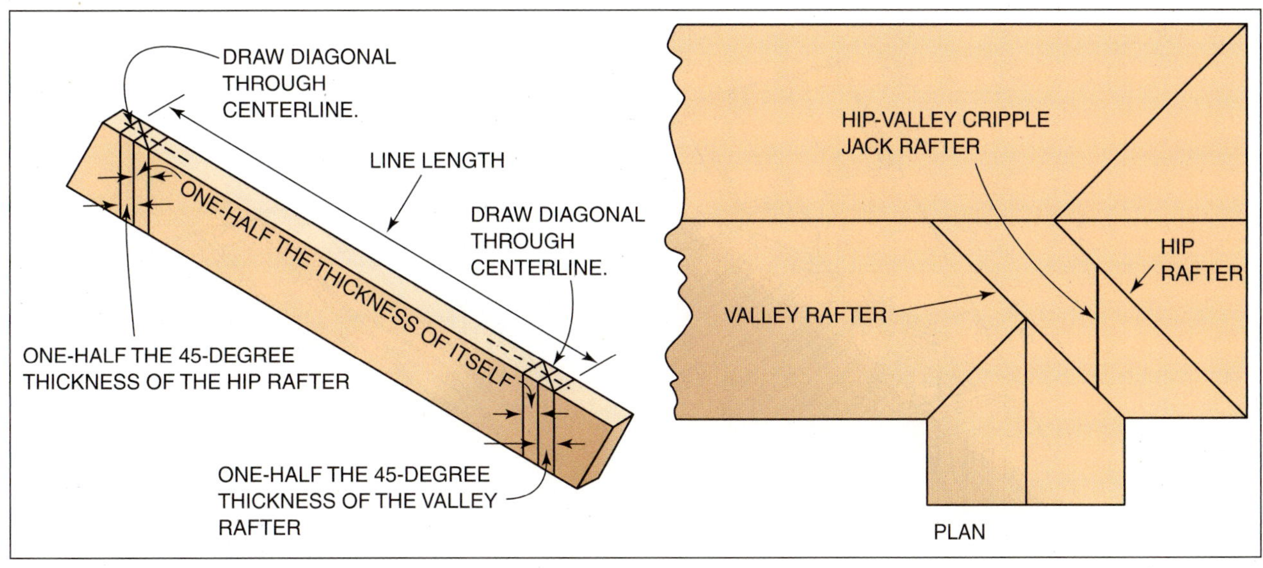


**FIGURE 12-32** Layout of the hip-valley cripple jack rafter.

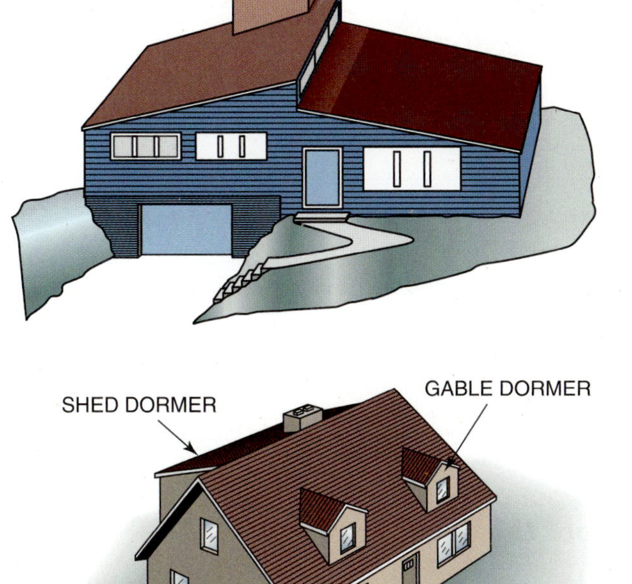


**FIGURE 12-33** Examples of the use of shed roof rafters.

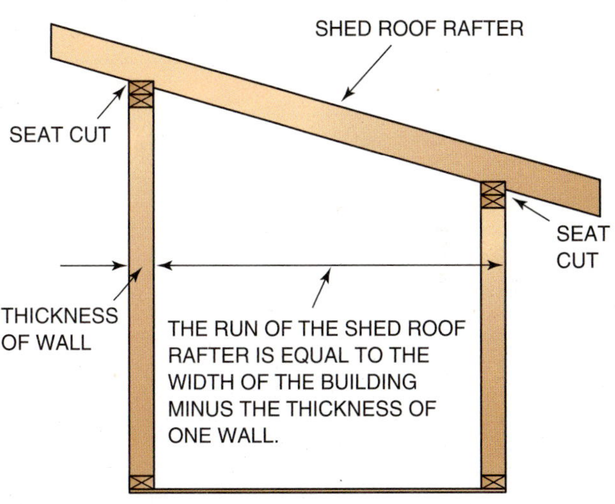


**FIGURE 12-34** The shed roof rafter may have two seat cuts.

# DORMERS

Usually, dormers have either gable or shed roofs. A gable dormer roof is framed similar to an intersecting gable roof (Fig. 12-35). In most cases, the shed dormer roof extends to the ridge of the main roof to gain enough incline. When framing openings

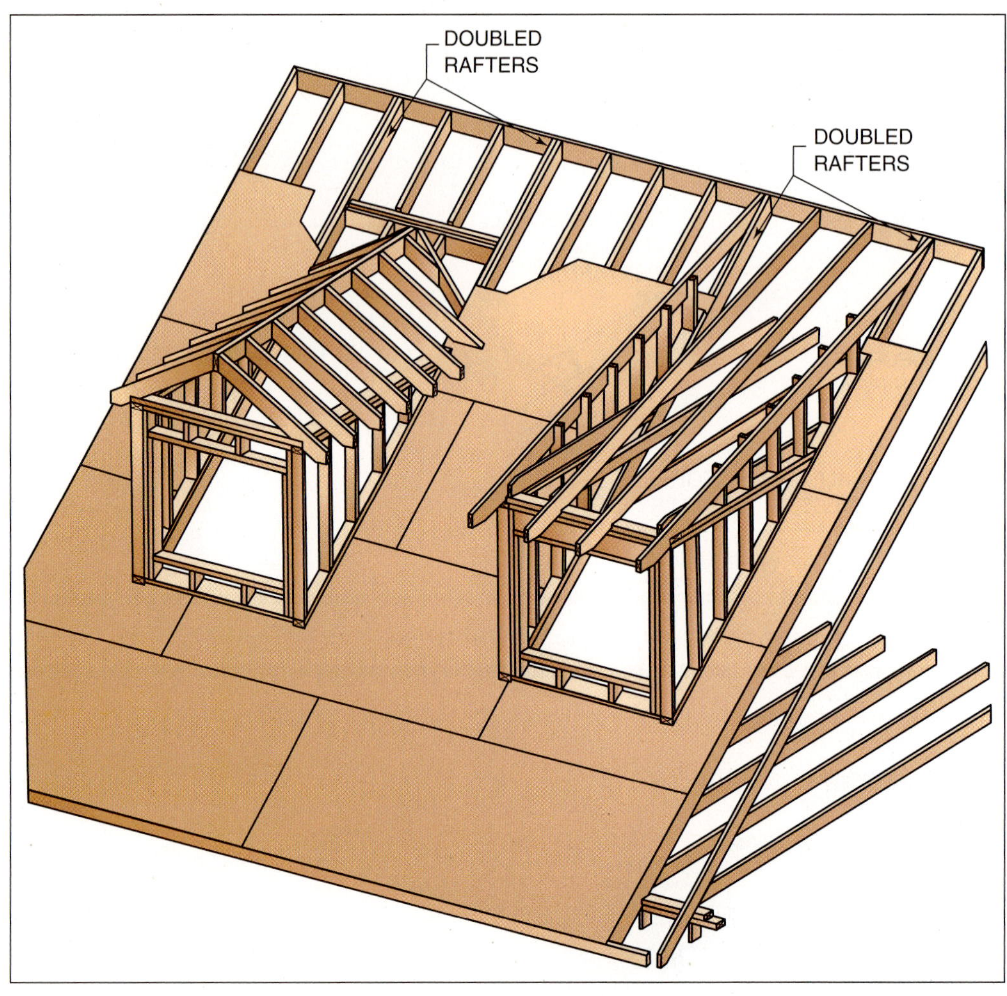


**FIGURE 12-35** Two methods of framing a dormer.

for dormers, the rafters on both sides of the opening are doubled. Some dormers have their front walls directly over the exterior wall below. When dormers are framed with their front wall partway up the main roof, top and bottom headers of sufficient strength must be installed.

## OTHER ROOF FRAMING

Numerous other roof framing problems are related in some way to the framing of roofs previously described. Solutions to some of the most commonly encountered problems are given.

### Cut of the Shed-Roof Rafter against a Roof

The top ends of shed-roof rafters are occasionally fitted against a roof of a different and steeper incline. The cuts can be laid out by using a framing square as outlined in Figure 12-36.

### Layout of the Ridge Cut against Roof Sheathing

The layout of the ridge of the intersecting gable roof that fits against the roof sheathing is shown in Figure 12-37.

### Fitting a Valley Jack Rafter to Roof Sheathing

Some intersecting roofs are built after the main roof has been framed and sheathed. In this type of construction, the valley rafters are eliminated and common rafters are full-length, making the main roof easier to frame. Steps to lay out the seat cut of the valley jack rafter to fit against roofs of the same and different inclines are shown in Figure 12-38.

FRAMING SQUARE
12"
UNIT RISE OF THE MAIN ROOF
HOLD SQUARE AS SHOWN
AND MARK ALONG THE BLADE.
RIDGE STOCK

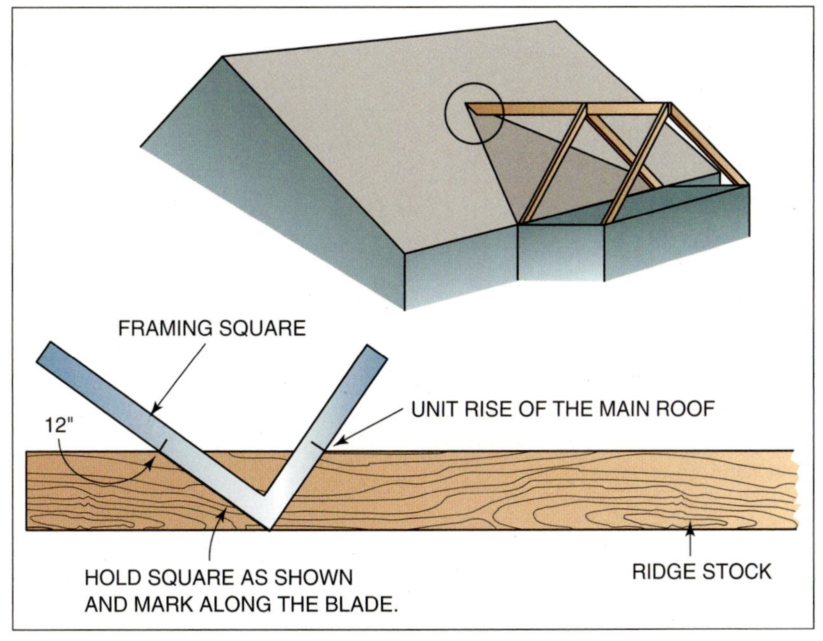

**FIGURE 12-36** Layout of a ridge that fits against roof deck.

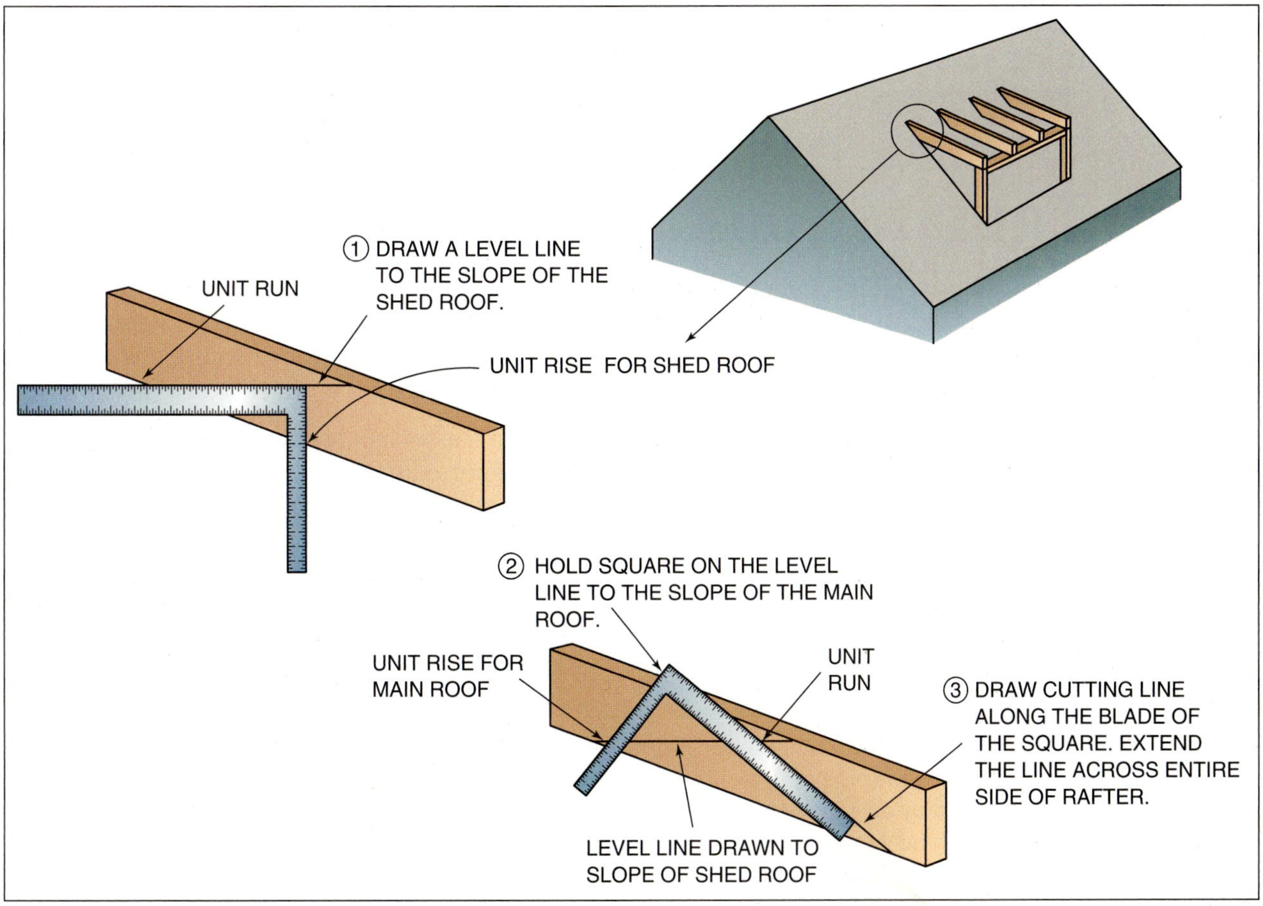

UNIT RUN
① DRAW A LEVEL LINE
TO THE SLOPE OF THE
SHED ROOF.
UNIT RISE FOR SHED ROOF
② HOLD SQUARE ON THE LEVEL
LINE TO THE SLOPE OF THE MAIN
ROOF.
UNIT RISE FOR
MAIN ROOF
UNIT
RUN
③ DRAW CUTTING LINE
ALONG THE BLADE OF
THE SQUARE. EXTEND
THE LINE ACROSS ENTIRE
SIDE OF RAFTER.
LEVEL LINE DRAWN TO
SLOPE OF SHED ROOF

**FIGURE 12-37** Steps in laying out a shed roof rafter that fits against a roof of a different pitch.

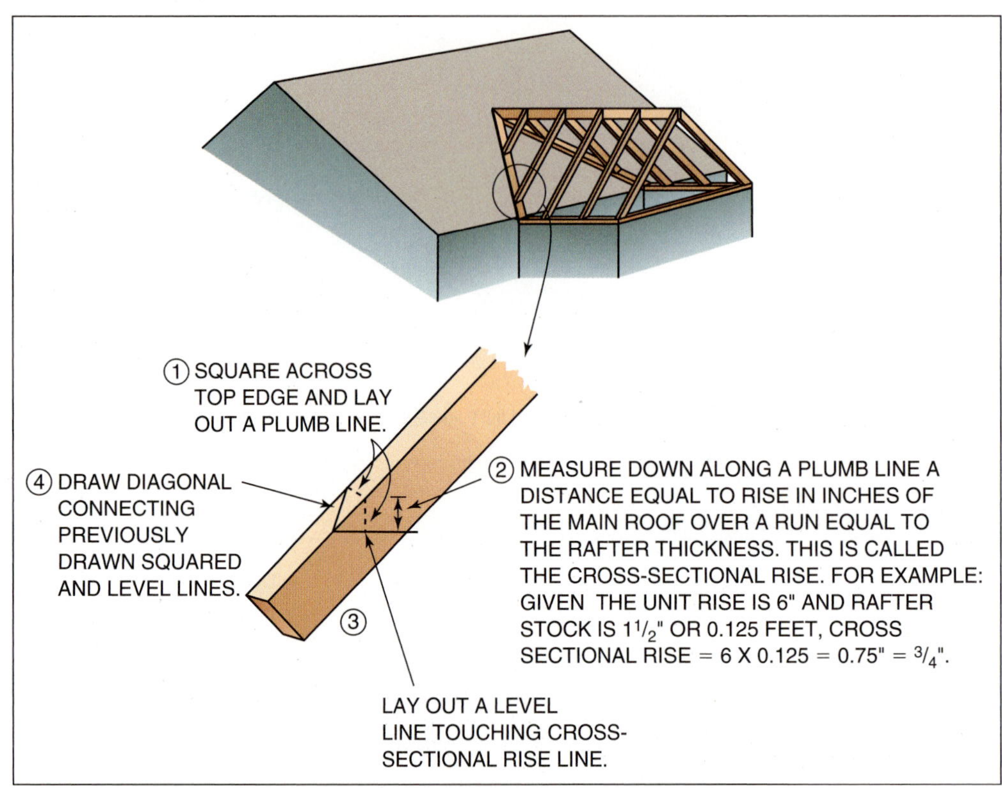


**FIGURE 12-38** Laying out the seats of valley jack rafters that fit against roof decks.

# Roof Trusses

Roof trusses are used extensively in residential construction (Fig. 12-39). Because of their design, they can support a roof over wide spans, which can reach 100 feet. Roof trusses eliminate the need for load-bearing partitions below. The roof is also framed in much less time. However, because of their design, much usable attic space is lost.

**CAUTION**

**CAUTION:** Trusses are designed with smaller member sizes than with rafter-ceiling joist systems. This causes higher stresses on the roof system members. Never cut any webs or chords of a truss unless directed by an engineer. Also, installing trusses can be very dangerous. Lives have been lost while installing trusses improperly. For these reasons, care must be taken by engineers in their design and by carpenters in their installation.

## TRUSS DESIGN

A roof truss consists of upper and lower *chords* and diagonals called *web members*. The upper chords act as rafters, and the lower chords serve as ceiling joists. Joints are fastened securely with metal or wood gusset plates (Fig. 12-40).

Most trusses are made in fabricating plants. They are transported to the job site. Trusses are designed by engineers to support prescribed loads. Trusses may also be built on the job, but approved designs must be used. Approved designs and instructions for job-built trusses are available from the American Plywood Association and the Truss Plate Institute. The most common truss design for residential construction is the Fink truss (Fig. 12-41). Other truss shapes are designed to meet special requirements (Fig. 12-42).

**Green Tip**

Framing with trusses uses less material and labor than conventional framing.

Courtesy of Wood Truss Council of America.

**FIGURE 12-39** Trusses are used extensively for roof framing.

Courtesy of Wood Truss Council of America.

**FIGURE 12-40** The members of roof trusses are securely fastened with metal gussets.

## OVERVIEW OF TRUSS ROOF ERECTION

Carpenters are more involved in the erection than in the construction of trusses. Trusses are delivered to the job site by specially designed trucks. They should be unloaded and stored on a flat dry surface. A print is provided showing the location of all trusses. A drawing of each truss is also provided that outlines important installation points (Fig. 12-43).

The erection and bracing of a trussed roof is a critical stage of construction. Failure to observe recommendations for erection and bracing could cause a collapse of the structure. This could result in loss of life, serious injury, or loss of time and material. The recommendations contained herein are technically sound. However, they are not the only methods for bracing a roof system. They serve only as a guide. The builder must take necessary precautions during handling and erection to ensure that trusses are not damaged, which might reduce their strength. Trusses are temporarily and permanently braced to tie them together. Temporary bracing is designed to create a safe work environment during truss erection. Great care must be exercised when deciding when and where to remove it. Permanent bracing is an important part of the truss system that remains in place after construction is completed. Its design is vital to the strength of the roof structure. It is the carpenter's responsibility to install both types of bracing adequately and properly.

Trusses are installed one at a time by lifting, fastening, and bracing in place. Small trusses, which can be handled by hand, are often placed upside down, hanging on the wall plates. They are then flipped up into place. Trusses for wide spans require the use of a crane to lift them into position (Fig. 12-44).

1/4 SPAN
1/4 SPAN
1/4 SPAN
1/4 SPAN
TOP CHORD
WEB MEMBERS
BOTTOM CHORD
1/3 SPAN
1/3 SPAN
1/3 SPAN

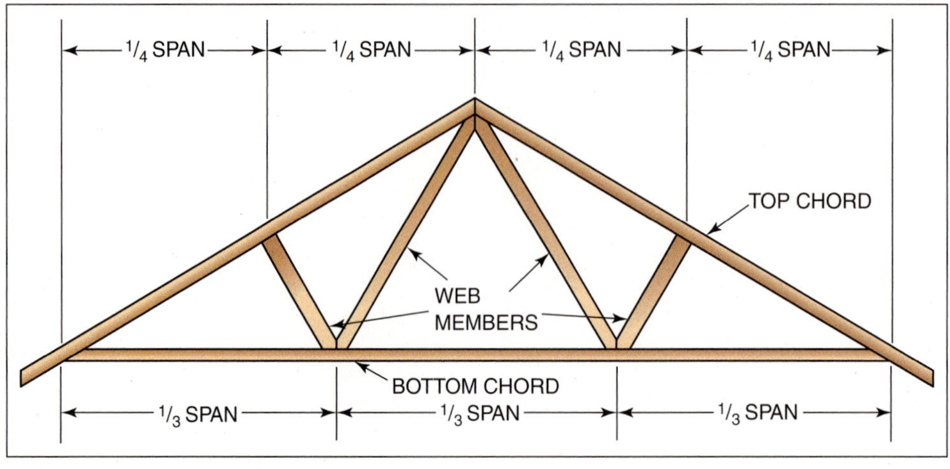

**FIGURE 12-41** The Fink truss is widely used in residential construction.

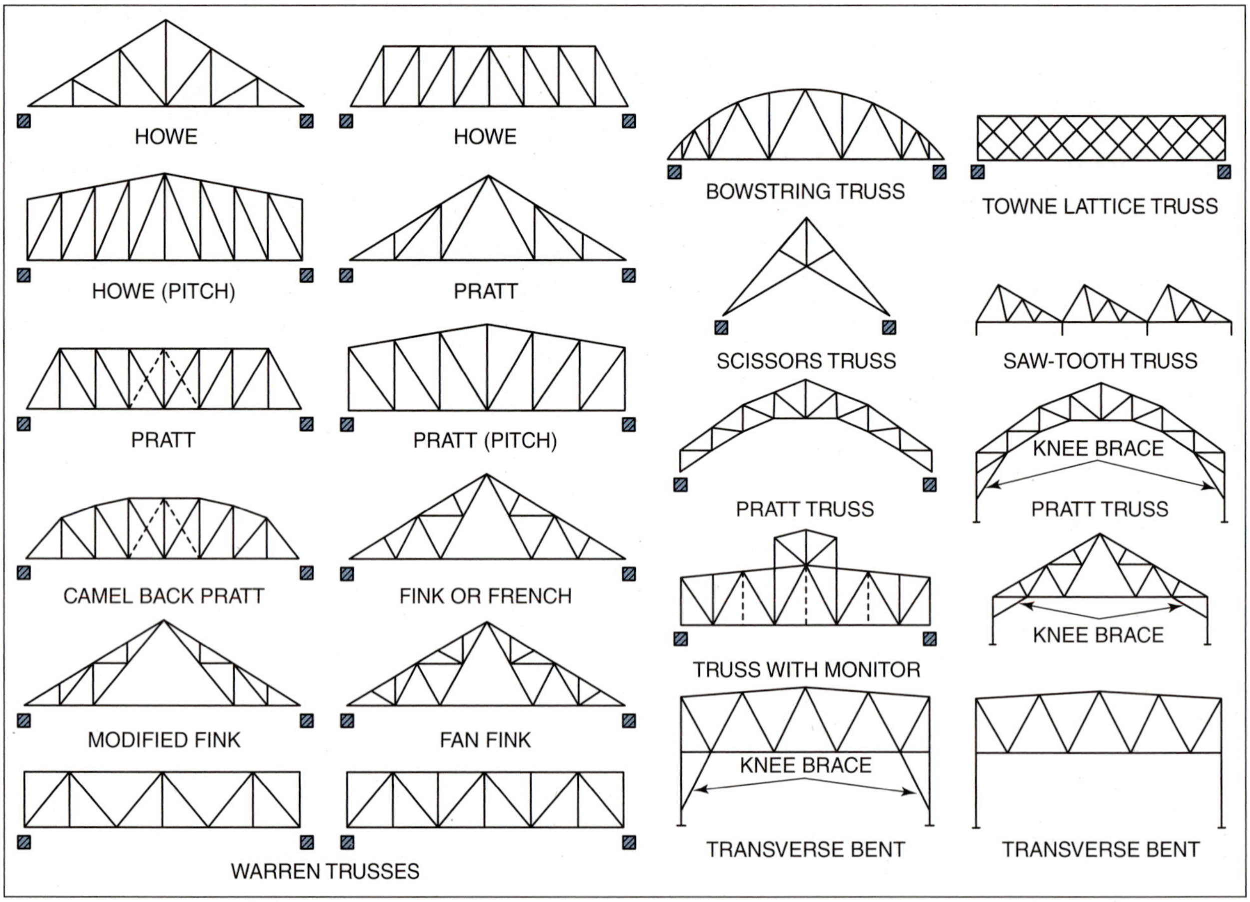

HOWE
HOWE
BOWSTRING TRUSS
TOWNE LATTICE TRUSS
HOWE (PITCH)
PRATT
SCISSORS TRUSS
SAW-TOOTH TRUSS
PRATT
PRATT (PITCH)
PRATT TRUSS
KNEE BRACE
PRATT TRUSS
CAMEL BACK PRATT
FINK OR FRENCH
TRUSS WITH MONITOR
KNEE BRACE
MODIFIED FINK
FAN FINK
KNEE BRACE
WARREN TRUSSES
TRANSVERSE BENT
TRANSVERSE BENT

**FIGURE 12-42** Truss shapes are designed for special requirements.

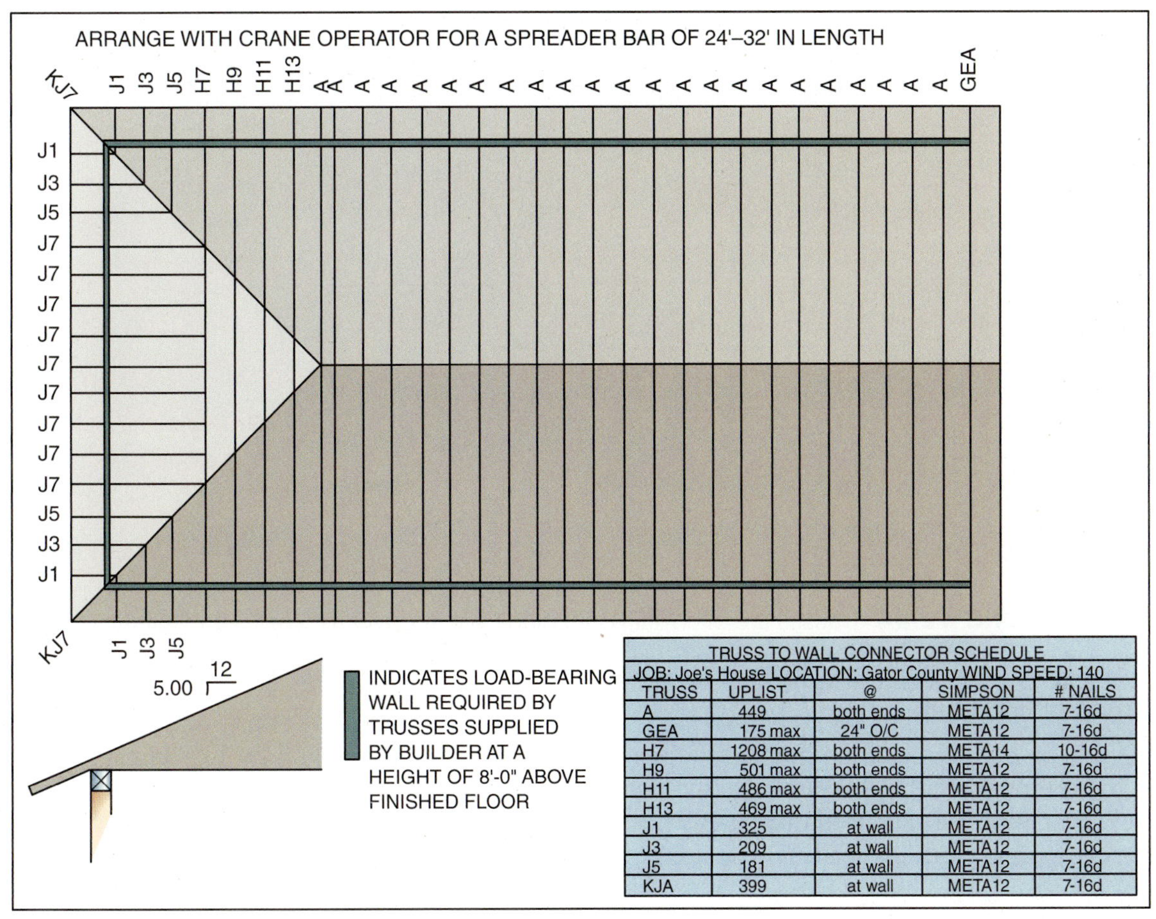


| TRUSS TO WALL CONNECTOR SCHEDULE | | | | |
|---|---|---|---|---|
| JOB: Joe's House LOCATION: Gator County WIND SPEED: 140 | | | | |
| TRUSS | UPLIST | @ | SIMPSON | # NAILS |
| A | 449 | both ends | META12 | 7-16d |
| GEA | 175 max | 24" O/C | META12 | 7-16d |
| H7 | 1208 max | both ends | META14 | 10-16d |
| H9 | 501 max | both ends | META12 | 7-16d |
| H11 | 486 max | both ends | META12 | 7-16d |
| H13 | 469 max | both ends | META12 | 7-16d |
| J1 | 325 | at wall | META12 | 7-16d |
| J3 | 209 | at wall | META12 | 7-16d |
| J5 | 181 | at wall | META12 | 7-16d |
| KJA | 399 | at wall | META12 | 7-16d |

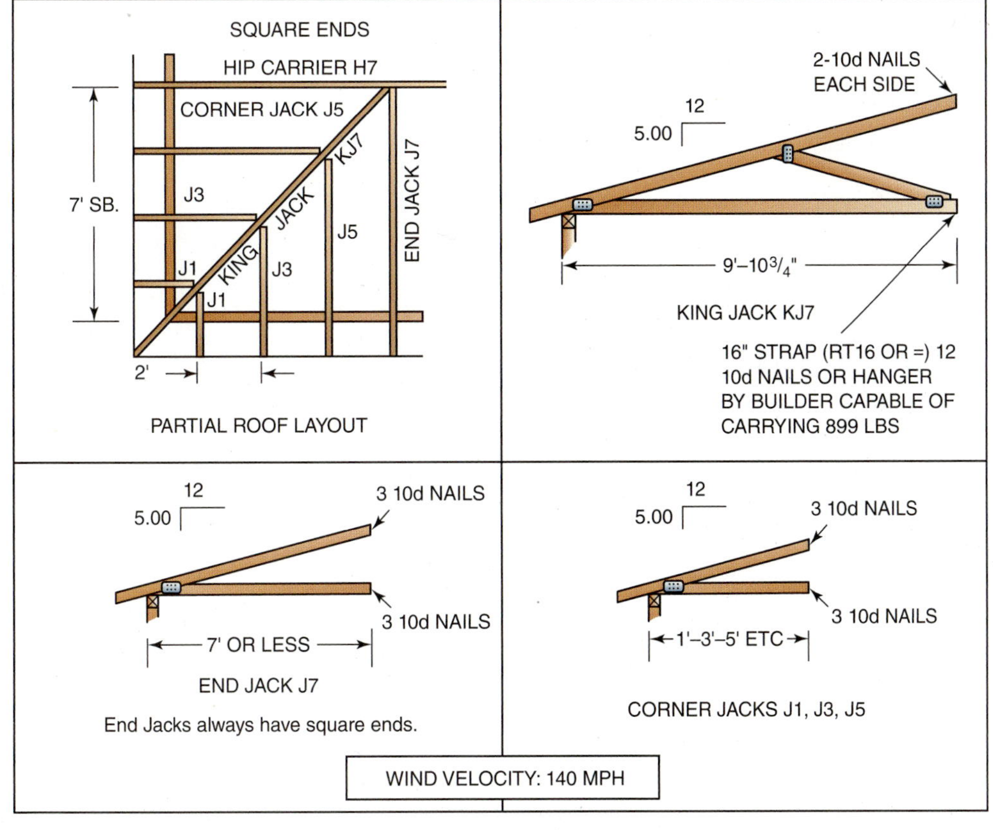


**FIGURE 12-43** An engineered set of prints showing truss labels and locations.

**FIGURE 12-44** Trusses are often lifted into place with a crane for speed and safety.

### Installing the First Truss

The end truss is usually installed first and braced securely in a plumb position. Because it is the first truss, great care must be exercised during erection. It must be temporarily braced with enough strength to support the tip over force of the trusses to follow. This may be achieved by bracing to securely anchored stakes driven into the ground or bracing to the inside floor under the truss (Fig. 12-45). These braces should be located directly in line with all

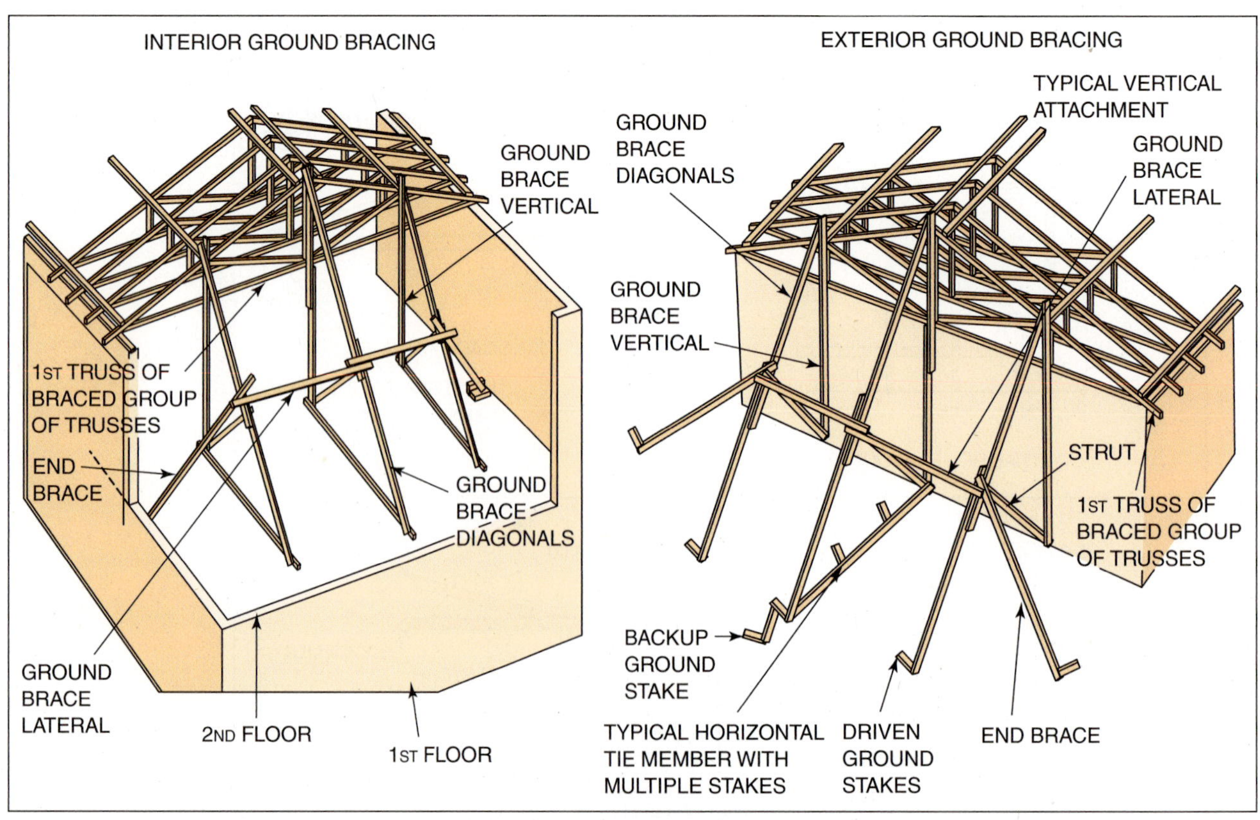


**FIGURE 12-45** The end truss must be well braced before the erection of other trusses.

rows of continuous top chord **lateral** bracing, which will be installed later. All bracing should be securely fastened with the appropriate size and quantity of nails. Remember that lives depend on it.

Lateral bracing is also referred to as *purlins*. Purlins are smaller roof members that actually support the roofing materials yet are themselves supported by other main rafters or trusses.

### Temporary Bracing of Trusses

As each truss is set in place, it is nailed to the plate and to the bracing. Metal framing ties are usually applied later before the roof sheathing is installed. The type of connector depends on the wind and seismic loads that the building is being built to withstand. These conditions vary according to the locale (Fig. 12-46).

Sufficient temporary bracing must be applied according to the manufacturer's recommendations. Temporary bracing should be no less than 2 × 4 lumber, with a minimum length of 8 feet. The 2 × 4s should be fastened with two 16d duplex or common nails at every truss. Ends of the bracing must overlap at least two trusses.

As the bracing is installed, spacing of trusses should be maintained accurately. This will speed up assembly through efficiency. Also, having to adjust the spacing later may lead to truss collapse if a key brace is unfastened at the wrong time.

Bracing is required in three planes of the truss assembly: the top chord or sheathing plane, the bottom chord or ceiling plane, and vertical web plane at right angles to the trusses. If bracing is installed carefully, the bottom chord and web bracing could become permanent bracing. This will also improve efficiency (Fig. 12-47).

### Bracing the Plane of the Top Chord

Continuous lateral bracing should be installed within 6 inches of the ridge and in rows at about 8- to 10-feet intervals between the ridge and wall plate. A diagonal brace should be set at approximately 45-degree angles between the rows of lateral bracing. It forms a triangle that provides stability to the plane of the top chord (Fig. 12-48).

### Bracing the Web Member Plane

Temporary bracing in the plane of the web members are diagonals placed at right angles to the trusses from top to bottom chords (Fig. 12-49). They become permanent braces of the web member plane.

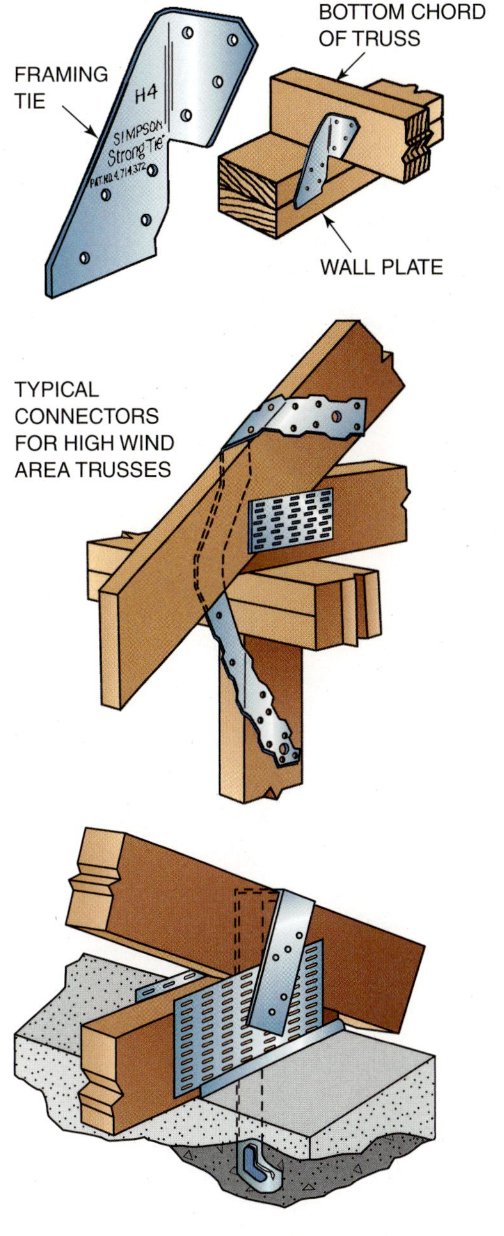


Courtesy of Simpson Strong-Tie Company.

**FIGURE 12-46** Metal framing ties are used to fasten trusses to the wall plate.

### Bracing the Plane of the Bottom Chord

To maintain the proper spacing on the bottom chord, continuous lateral bracing for the full length of the building must be applied. The bracing should be nailed to the top of the bottom chord at intervals

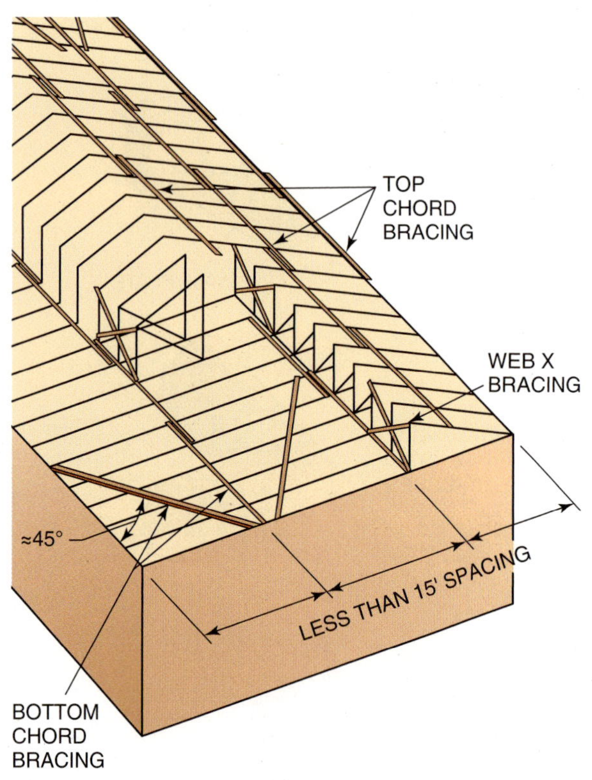


**FIGURE 12-47** Bracing is installed on three planes within a trussed roof. It is nailed with two 16d nails at each truss, and they are lapped at least two trusses.

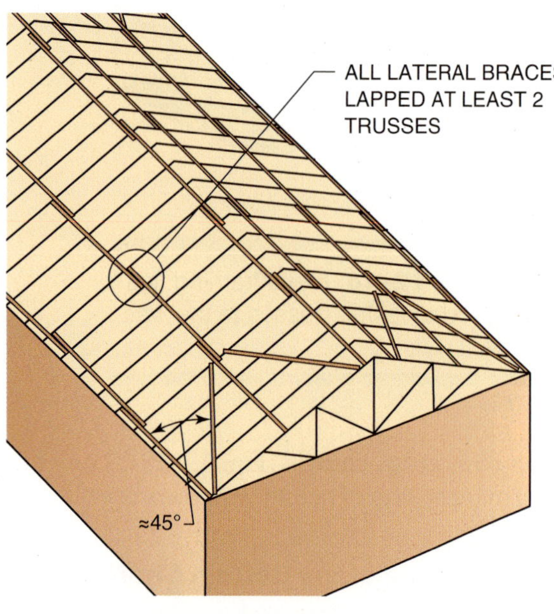


**FIGURE 12-48** Temporary bracing of the plane of the top chord.

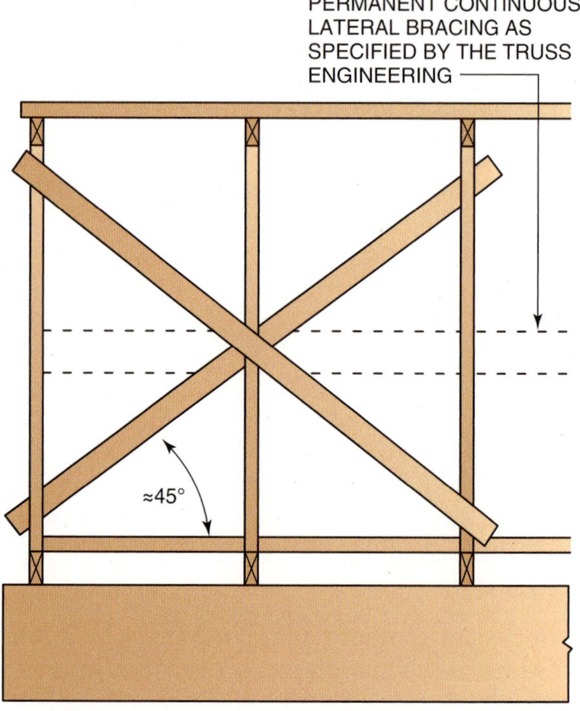


**FIGURE 12-49** Bracing of the web member plane prevents lateral movement of the trusses.

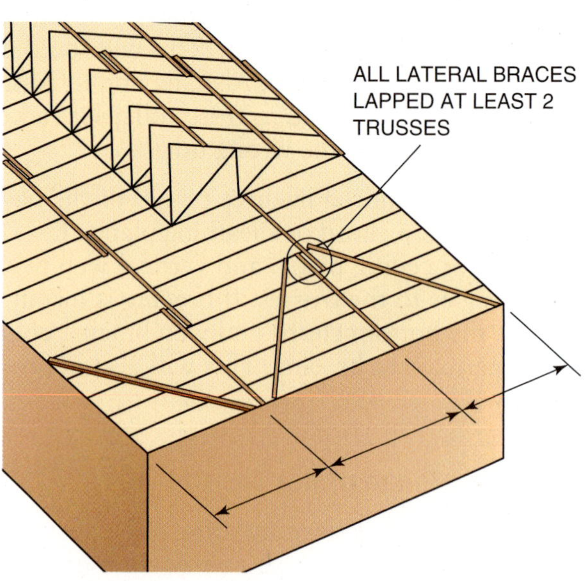


**FIGURE 12-50** Bracing the plane of the bottom chord.

of 8 to 10 feet along the width of the building. Diagonal bracing should be installed at least at each end of the building (Fig. 12-50). In most cases, temporary bracing of the plane of the bottom chord is left in place as permanent bracing.

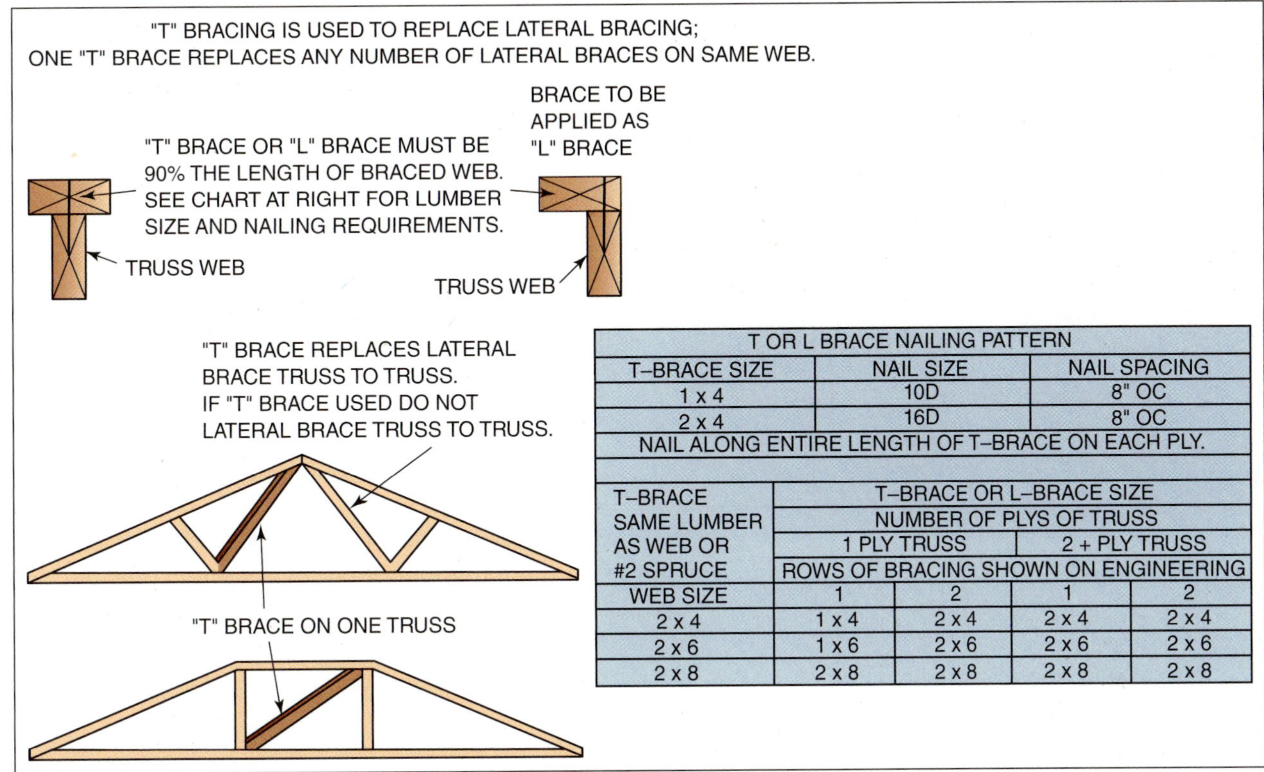


| T OR L BRACE NAILING PATTERN | | |
|---|---|---|
| T-BRACE SIZE | NAIL SIZE | NAIL SPACING |
| 1 x 4 | 10D | 8" OC |
| 2 x 4 | 16D | 8" OC |
| NAIL ALONG ENTIRE LENGTH OF T-BRACE ON EACH PLY. | | |

| T-BRACE SAME LUMBER AS WEB OR #2 SPRUCE | T-BRACE OR L-BRACE SIZE | | | |
|---|---|---|---|---|
| | NUMBER OF PLYS OF TRUSS | | | |
| | 1 PLY TRUSS | | 2 + PLY TRUSS | |
| | ROWS OF BRACING SHOWN ON ENGINEERING | | | |
| WEB SIZE | 1 | 2 | 1 | 2 |
| 2 x 4 | 1 x 4 | 2 x 4 | 2 x 4 | 2 x 4 |
| 2 x 6 | 1 x 6 | 2 x 6 | 2 x 6 | 2 x 6 |
| 2 x 8 | 2 x 8 | 2 x 8 | 2 x 8 | 2 x 8 |

**FIGURE 12-51** Some areas require web stiffeners.

### Permanent Bracing

Permanent bracing is designed by the engineer for the structural safety of a building. The top chord permanent bracing is often provided by the roof sheathing. Web bracing may be lateral bracing or web stiffeners (Fig. 12-51). These are usually installed after the trusses and before the sheathing is installed.

### Framing Openings

Openings in the roof or ceiling for skylights or access ways must be framed within or between the trusses. The chords, braces, and webs of a truss system should never be cut or removed unless one is directed to do so by an engineer. Simply installing headers between trusses will create an opening. The sheathing of ceiling finish is then applied around the opening.

***For step-by-step instructions on TRUSS INSTALLATION see Procedure 12-7 on page 385.***

## ROOF SHEATHING

Roof sheathing is applied after the roof frame is complete. Sheathing provides rigidity to the roof frame. It also provides a nailing base for the roof covering. Rated panels are used to sheath roofs.

Panels are laid with the face grain running across the rafters/trusses for greater strength. End joints are made on the rafters/top chord and staggered. Nails are spaced 6 inches apart on the ends and 12 inches apart on intermediate supports (Fig. 12-52). Some areas require extra nails in the sheathing to protect from uplift caused by high winds and seismic movements. These nailing zones put more nails along the perimeter and at the corners of the roof.

Adequate blocking, tongue-and-grooved edges, or other suitable edge support, such as panel clips, must be used when spans exceed the indicated value of the plywood roof sheathing. Panel clips are small

Courtesy of APA—The Engineered Wood Association.

**FIGURE 12-52** Sheathing a roof with plywood.

metal pieces shaped like a capital H. They are used between the long edges of the plywood sheets where they meet between rafters/top chords (Fig. 12-53). One clip is used for 24- and 32-inch spans, and two panel clips are used for 48-inch spans.

# Estimating Materials for Roof Framing

Estimating the amount of materials for roof framing involves rafters, ridgeboards, trusses and temporary bracing, gable end studs, and sheathing. Remember that if the number of pieces does not come out even, round them up to the next whole number of pieces.

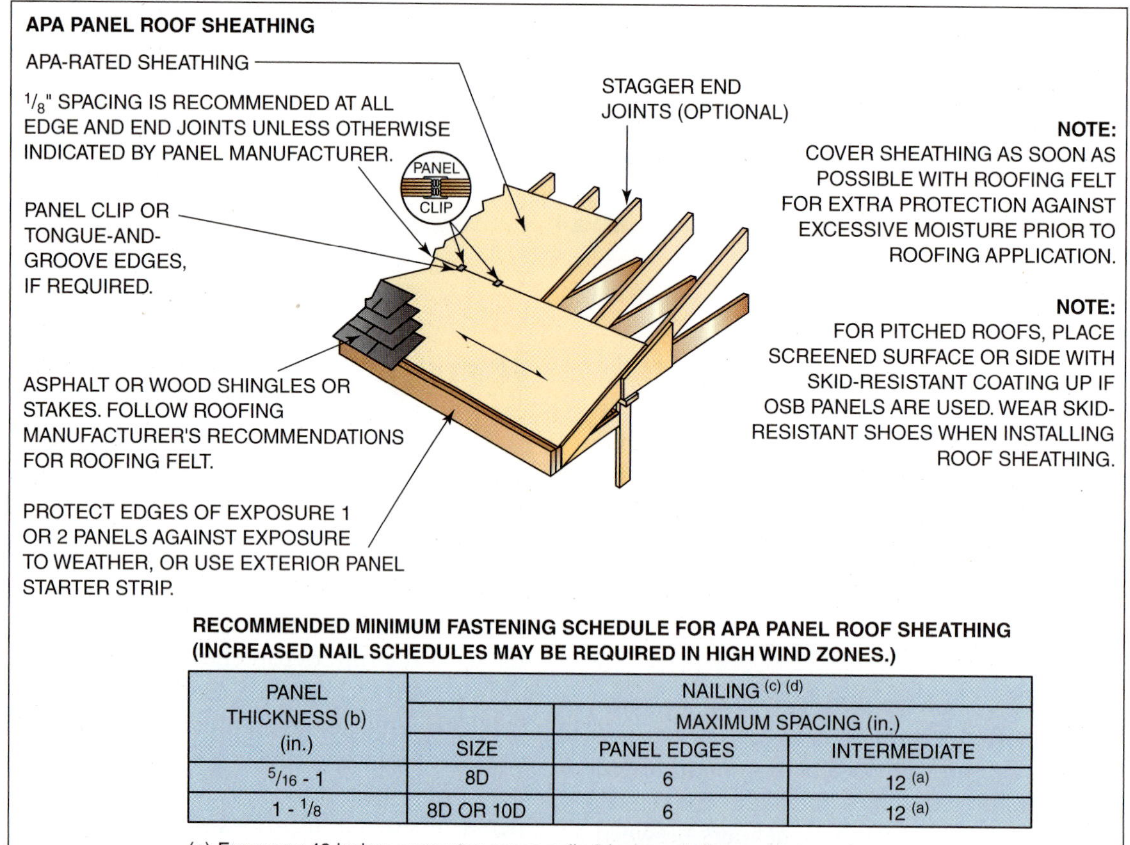


**RECOMMENDED MINIMUM FASTENING SCHEDULE FOR APA PANEL ROOF SHEATHING (INCREASED NAIL SCHEDULES MAY BE REQUIRED IN HIGH WIND ZONES.)**

| PANEL THICKNESS (b) (in.) | NAILING (c) (d) | | |
|---|---|---|---|
| | | MAXIMUM SPACING (in.) | |
| | SIZE | PANEL EDGES | INTERMEDIATE |
| 5/16 - 1 | 8D | 6 | 12 (a) |
| 1 - 1/8 | 8D OR 10D | 6 | 12 (a) |

(a) For spans 48 inches or greater, space nails 6 inches at all supports.
(b) For stapling asphalt shingles to 5/16-inch and thicker panels, use staples with a 15/16-inch minimum crown width and a 1-inch leg length. Space according to shingle manufacturer's recommendation.
(c) Use common smooth or deformed shank nails with panels to 1-inch thick. For 1 1/8-inch panels, use 8D ring- or screw-shank or 10d common smooth-shank nails.
(d) Other code-approved fasteners may be used.

Courtesy of APA—The Engineered Wood Association.

**FIGURE 12-53** Recommendations for the application of APA panel roof sheathing.

## RAFTERS

There are many types of rafters in a hip roof. Estimating the material needed is easier than might be expected. The number of common rafters is almost the same whether for a gable roof or a hip roof.

*Common Rafters for Gable Roof.* Divide the length of the building by the spacing of the rafters. Round up to the nearest whole number and add one, as a starter. Then, multiply the total by two to include the other side. Add four, one for each rake rafter. For example, if the building is 50 feet, the number of 16-inch OC gable rafters is $50 \div {}^{16}\!/_{12} + 1 = 38.5 \Rightarrow 39 \times 2 + 4 = 82$ rafters.

*Common Rafters for Hip or Valley Roofs.* Use the same procedure as for a gable roof, but add two rafters for each hip and valley rafter. This will allow material for the common and jack rafters. Two jack rafters may be cut from one board, one long and one short. For example, if a rectangular hip-roofed building is 50 feet, the number of 16-inch OC gable rafters is $50 \div {}^{16}\!/_{12} + 1 = 38.5 \Rightarrow 39 \times 2 + 4 = 82 + 4 \times 2 = 90$ rafters.

*Hip and Valley.* Count the number of hips and valleys. Remember that the size of the material used is not the same as the common rafter material.

## RIDGEBOARDS

There are two types of roofs with ridgeboards: gable and hip. Gable often includes an overhang of the gable end. The hip ridgeboard needs to be slightly longer, up to a possible 3⅝ inches.

*Ridgeboard for Gable Roof.* Take the length of the building plus both rake overhangs. Divide this sum by the length of the material to be used for the ridge. Typically, 12- or 16-foot boards are used because there is minimal cutting waste with the various OCs. For example, if the building is 50 feet and the gable end overhang is 12 inches, or one foot, then the number of 12-foot ridgeboards is $50 + 2 = 52 \div 12 = 4.3 \Rightarrow 5$ ridgeboards.

*Ridgeboard for Hip Roof.* Subtract the width of the building from the length of the building. Add 4 inches to allow for possible additions in length at each end. For example, if a rectangular building is 32 feet × 50 feet, then the number of 12-foot hip ridgeboards is $50 - 32 = (18 + {}^{4}\!/_{12}) \div 12 = 1.5 \Rightarrow 2$ ridgeboards.

## GABLE END STUDS

Divide the width of the building by the OC spacing. Add two extra studs. This number of studs will be enough to do both sides of the building. For example, if the width is 32 inches, the number of 16-inch OC studs needed will be $32 \div {}^{16}\!/_{12} + 2 = 26$ studs.

## TRUSSES

Divide the building length by the OC spacing, usually 2 feet. Subtract one. This is the number of common trusses. Two additional gable end trusses are needed for the ends of the building. *Note:* The number and sizes of hip trusses are determined by the manufacturer. For example, if the building is 50 feet, the number of common trusses is $50 \div 2 - 1 = 24$ trusses and two end trusses.

## TEMPORARY TRUSS BRACING MATERIAL

Divide the building width by six to get the number of rows of top, bottom, and web bracing. Round this number up to the nearest whole number. Multiply times length of building. Divide by the bracing length, typically 16 feet. Round to the nearest whole number. Add 20 percent for brace overlap and waste and any extra for ground bracing of the first truss. For example, if the building is 32 feet × 50 feet, the number of 16-feet – 2 × 4 braces is
$32 \div 6 = 5.3 \Rightarrow 6$ rows.
$50 \div 16 = 3.1 \Rightarrow 4$ pieces per row. Thus, 6 rows × 4 pieces per row × 120% $= 28.8 \Rightarrow 29$ braces.

## SHEATHING

Estimating sheathing is similar to that of rafters in that gable roofs and hip roofs are estimated in the same way. The waste pieces on a hip roof may be reused to do another roof section.

*Sheathing Gable Roof.* Find the number of plywood rows by dividing the actual rafter length by 4 feet (width of plywood sheet). Round this number up to nearest one-half row. Next, find the number of sheets across the building by dividing the ridgeboard length by 8 feet (length of plywood sheet). Round this up to the nearest one-half sheet. Find the number of sheets by multiplying these numbers together and then doubling the result to account for the other side.

Round up to the nearest whole number. Waste will already be allowed for in the rounding-up of rafter and ridgeboard lengths. For example, sheathing for a roof with 20'-6" rafters on a building with a 52' ridgeboard will have 20.5 ÷ 4 = 5.1 ⇒ 5.5 rows. It will have 52 ÷ 8 = 6.5 sheets per row. Thus, we have 5.5 × 6.5 × 2 sides = 71.5 ⇒ 72 sheets.

*Sheathing Hip Roof.* Calculate the number of panels as if the roof was a gable roof. Then, add another 5 percent for waste. There is some waste when cutting triangles. For example, if the previous roof were to be framed with a hip, then the sheathing would be 71.5 × 1.05 = 75 sheets.

*Sheathing Intersecting Roof.* Calculate the number of panels for each building section as if the valleys and valley jacks were not present. The overlapped amount will compensate for waste, but an additional 5 percent will allow for the extra waste in cutting triangles.

*Gable End Sheathing.* Calculate the total rise of the roof by multiplying the unit rise times the rafter run, then divide by 12 to change the answer to feet. Find the number of rows of sheathing by dividing this number by 4 feet, and round up to the nearest one-half sheet. Next, find the number of sheets across the building by dividing the building width by 8 feet. Round this number up to the nearest one-half sheet. Multiply these numbers, number rows, and sheets together, and round up to the nearest whole number. Waste material should be sufficient to sheath the opposite gable end. For example, the sheathing for the gable ends of a 32'-wide building with an 8" unit rise has 32' ÷ 8 = 4 pieces per row. There are 32 ÷ 2 = 16, run × 8 ÷ 12 = 10.6', total rise ÷ 4 = 2.6 ⇒ 3 rows. Sheathing, then, is 3 × 4 = 12 pieces.

See Figure 12-54, a listing of the shorthand version of these estimating techniques with another example.

**Estimate the materials for a roof of a rectangular 28' × 42' building. Framing 16" OC, 24" OC trusses, 9" gable end overhang, and a rafter length of 17'-4".**

| Item | Formula | Waste factor | Example |
|---|---|---|---|
| **Common Rafters** | (building LEN (in ft) ÷ OC (ft) + 1) × 2 + 1 per each rake rafter. | | 42 ÷ $^{16}/_{12}$ = 31.5 ⇒ 32 + 1 = 33 × 2 = 66 + 4 = 70 *PC* |
| **Commons for Hip Roof** | common rafters − rake rafters + 2 × NUM hips | | (70 − 4) + 2 (4) 74 *PC* |
| **Gable Ridgeboard** | building LEN − (2 × gable end over hang) | | 42 + 2 × $^{9}/_{12}$ = 435 *FT* |
| **Hip Ridgeboard** | building LEN − building WID + 4 | | 42−28 + $^{4}/_{12}$ + $14.\overline{333}$ *FT* |
| **Gable end studs** | building WID ÷ OC (ft) + 2 | | 28 ÷ $^{16}/_{12}$ + 21 + 2 = 23 *PC* |
| **Trusses** | building LEN ÷ OC (ft) − 1, plus two gable end trusses | | 42 ÷ 2 − 1 = 20 *PC* |
| **Temp Truss Bracing** | building WID ÷ 4 × building LEN ÷ 16 × 20% waste | 20 | 28 ÷ 4 × 42 ÷ 16 × 1.20 = 222.05 ⇒ 23 *PC* |
| **Gable Roof Sheathing** | rafter LEN ÷ 4 (rounded up to nearest ½) × ridge LEN ÷ 8) × 2 (rounded up to nearest ½) | | ($17.\overline{333}$) ÷ $4.\overline{33}$ ⇒ 4.5)× (43.5 ÷ 8 = 5.4375 ⇒ 5.5) = 24.75 × 250 *PC* |
| **Hip Roof Sheathing** | gable sheathing × 5% waste | 5% | 50 × 1.05 = 52.5 ⇒ 53 *PC* |
| **Gable End Sheathing** | gable rise ÷ 4 (rounded up to nearest ½) × building WID ÷ 8 (rounded up to nearest ½) | | 14 × $^{7}/_{12}$ = $8.1\overline{66}$ ⇒ 8.5 × (28 ÷ 8 = 3.5 ⇒ 3.5) = 29.75 ⇒ 30 *PC* |

**FIGURE 12–54** Example of estimating roof framing materials with formulas.

## SUMMARY


- Roof types vary according to design requirements.
- Gable roof surfaces slope in two directions, with a peak usually in the center of the building.
- Gable studs frame in the triangular area above the wall plates of end walls.
- Hip roofs have surfaces that slope in four directions with a roof edge along all walls.
- Valley rafters form the intersection of two roof slopes creating a draining line for water.
- Shed roof slopes in one direction.
- Dormers provide light and ventilation to rooms directly under the sloping roof surface.
- Roof trusses are prefabricated roof frames designed to provide a strong roof that is quickly installed.
- Roof sheathing covers the roof frame providing a bearing surface for the roofing materials.
- Estimating of roof framing materials is necessary for efficient purchases.

## GREEN CHECKLIST

- ☐ Framing with trusses uses less material and labor than conventional framing.

PROCEDURE 12-1

# Common Rafter Layout

**A** Lay a piece of rafter stock across two sawhorses. Crown the board by sighting the stock along the edge for straightness. Select the straightest piece possible because it will be used as a pattern to mark the remaining rafters. The crown should become the top edge of the rafter when it is installed.

- Begin at the one end of the board. This will become the upper end or ridge plumb cut. Place the square down on the side of the stock, holding the tongue of the square with the left hand and the blade with the right hand. Adjust the square until the outside edge of the tongue with the specified unit rise and the edge of the stock line up. Also, adjust the blade of the square with the unit run and the edge of the stock line up. Mark the rafter along the outside edge of the tongue. This is the first plumb line for the ridge.
- When using a speed square, place the pivot point of the square on the top edge of the rafter. Rotate square with the pivot point touching the rafter. Looking in the rafter scale window, align the edge of the rafter with the number that corresponds to the unit rise desired. Mark the plumb line along the edge of the square that has the inch ruler marked on it.

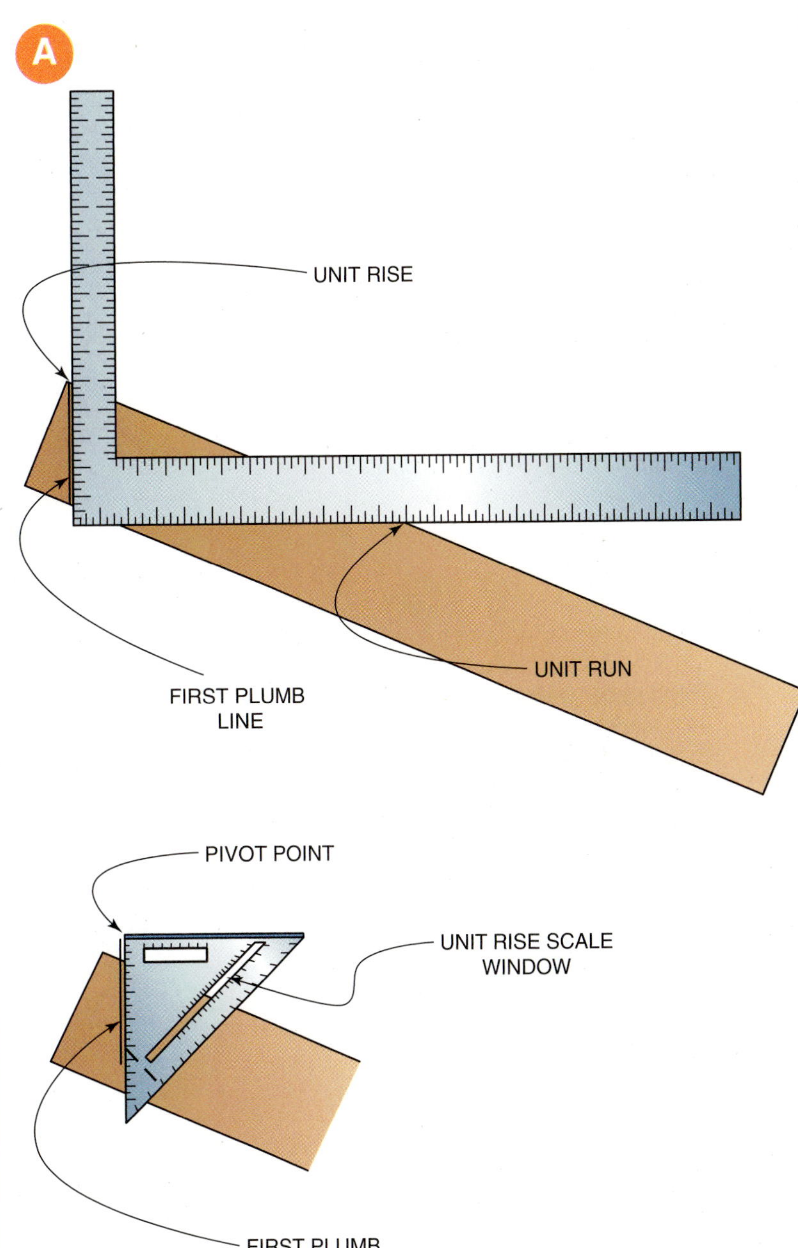

**PROCEDURE 12-1**

# Common Rafter Layout (continued)

**B** Since the ridgeboard has a thickness, the rafter must be shortened to fit in place. To shorten the rafter, measure back, at a right angle to the ridge plumb line, a distance equal to one-half the thickness of the ridgeboard. Lay out another plumb line through this point. This will be the line that is used to cut the rafter. Note: Shortening is always measured at right angles to the ridge cut, regardless of the slope of the roof. Also, if no ridgeboard is used, no shortening is required.

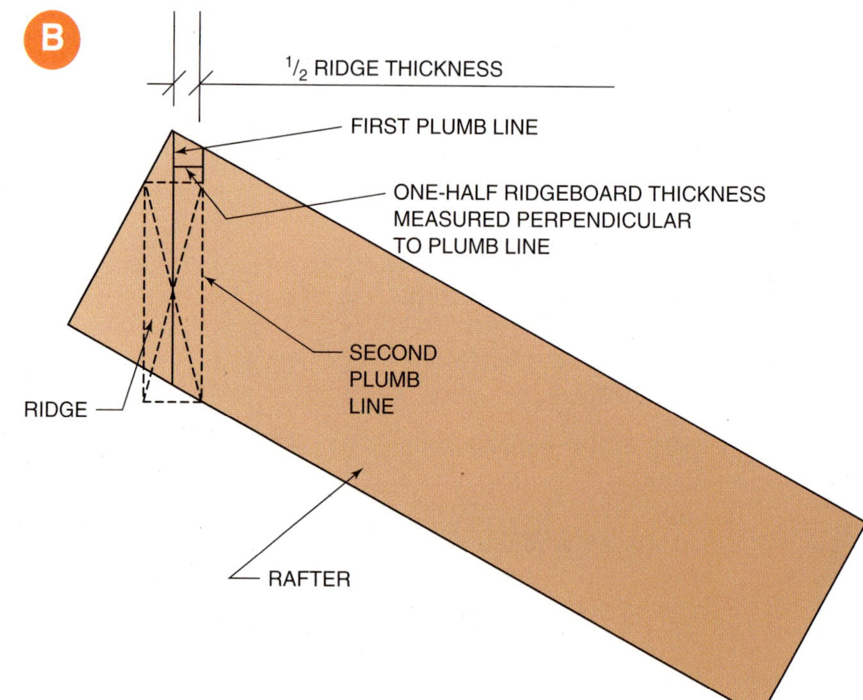


**C** Determine the rafter line length by first finding the unit length of the rafter from the top line of the rafter table. This number will be under the inch mark that corresponds to slope of the roof being framed. Multiply the unit length found in the table by the number of units of run. **Example:** Find the line length of a common rafter with a unit rise of 8 inches for a building 28 feet wide. Since the run is usually one-half of the span, the run of this building is 14. On the rafter tables, read from the first line below the 8-inch mark on the square. This table line is the unit length. It is found to be 14.42 inches. Multiplying 14.42 inches by 14 gives 201.88 inches. To turn this into a number that may be used with a tape measure, convert the decimal to sixteenths. Multiplying 0.88 inches by 16 gives 14.08 sixteenths. Rounding this number to the nearest whole number gives 14/16, which reduces to 7/8. Recombining these numbers gives a rafter length of 201 7/8 inches.

**C**

**EXAMPLE GIVEN:**
UNIT RISE IS 8 INCHES. BUILDING IS 28 FEET WIDE

**DETERMINED:**
UNIT LENGTH IS 14.42" RUN IS $28 \div 2 = 14$

**CALCULATE:**
RAFTER LINE LENGTH $= 14 \times 14.42 = 201.88$"
$0.88 \times 16 = 14.08$ SIXTEENTHS ⟶ $^{14}/_{16}$ ⟶ $^{7}/_{8}$
$201^{7}/_{8}$

**D** To mark the length of the rafter, measure from the first ridge plumb line along the edge of the rafter. Do not use the shortened ridge plumb line. Mark the length along the same edge as the rafter is measured. Make a plumb line for the seat cut the same way as for the ridge plumb line.

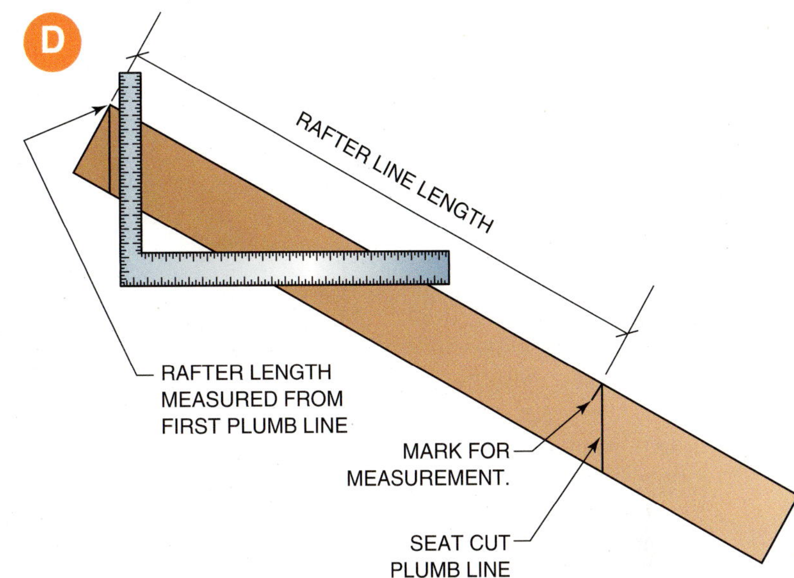


**E** The seat cut or bird's mouth of the rafter is a combination of a level cut and a plumb cut. The level cut rests on top of the wall plate. The plumb cut fits snugly against the outside edge of the wall. The level cut line of the seat cut may be located in several ways.

- On roofs with moderate slopes, the length of the level cut of the seat is often the width of the wall plate. For steep roofs, the level cut is shorter. Otherwise, too much stock would be cut out of the rafter, weakening it. Two-thirds of rafter stock width should remain after the seat is cut.

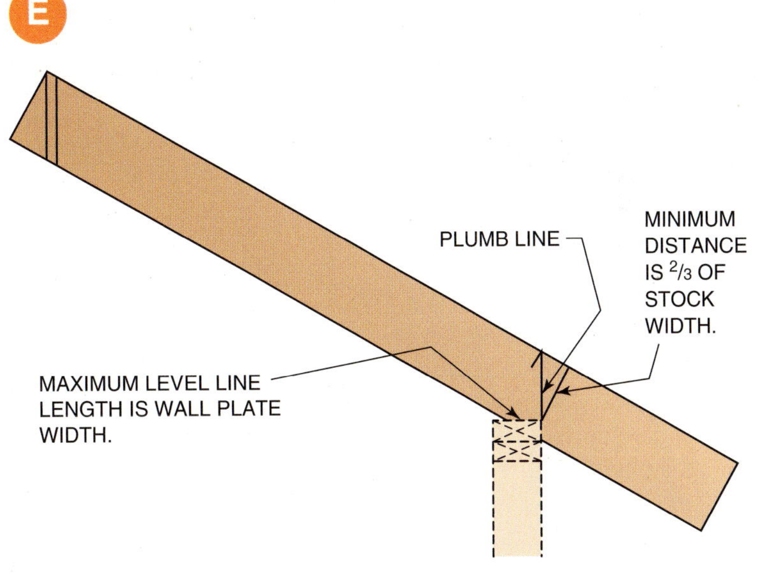

PROCEDURE 12-1

# Common Rafter Layout (continued)

**F** Reposition the square on the board as was done for marking the plumb lines. Align the *blade* of the square to the level point and mark along the blade. With a speed square, hold the alignment guide of the square in line with the plumb line previously drawn. Mark along the long edge of the square to achieve level lines for seat cuts.

**G** To lay out the rafter overhang, the projection must be determined from the plans. The projection is the horizontal distance (or run) under the rafter tail. Reposition the framing square, at the same marks used before for unit rise and unit run, on the rafter stock. Align the tongue to the seat cut plumb line and measure the projection distance along the blade of the square. Slide the square down to draw the tail plumb line. The square may be rotated to fit on the board, if needed.

**H** Just remember to keep the same face of the square up. A level cut may be made to remove any portion of the rafter from extending below the **fascia**. Determine the width of the fascia from the prints and measure down the tail plumb line this distance. Draw a level line toward the bottom edge of the rafter.

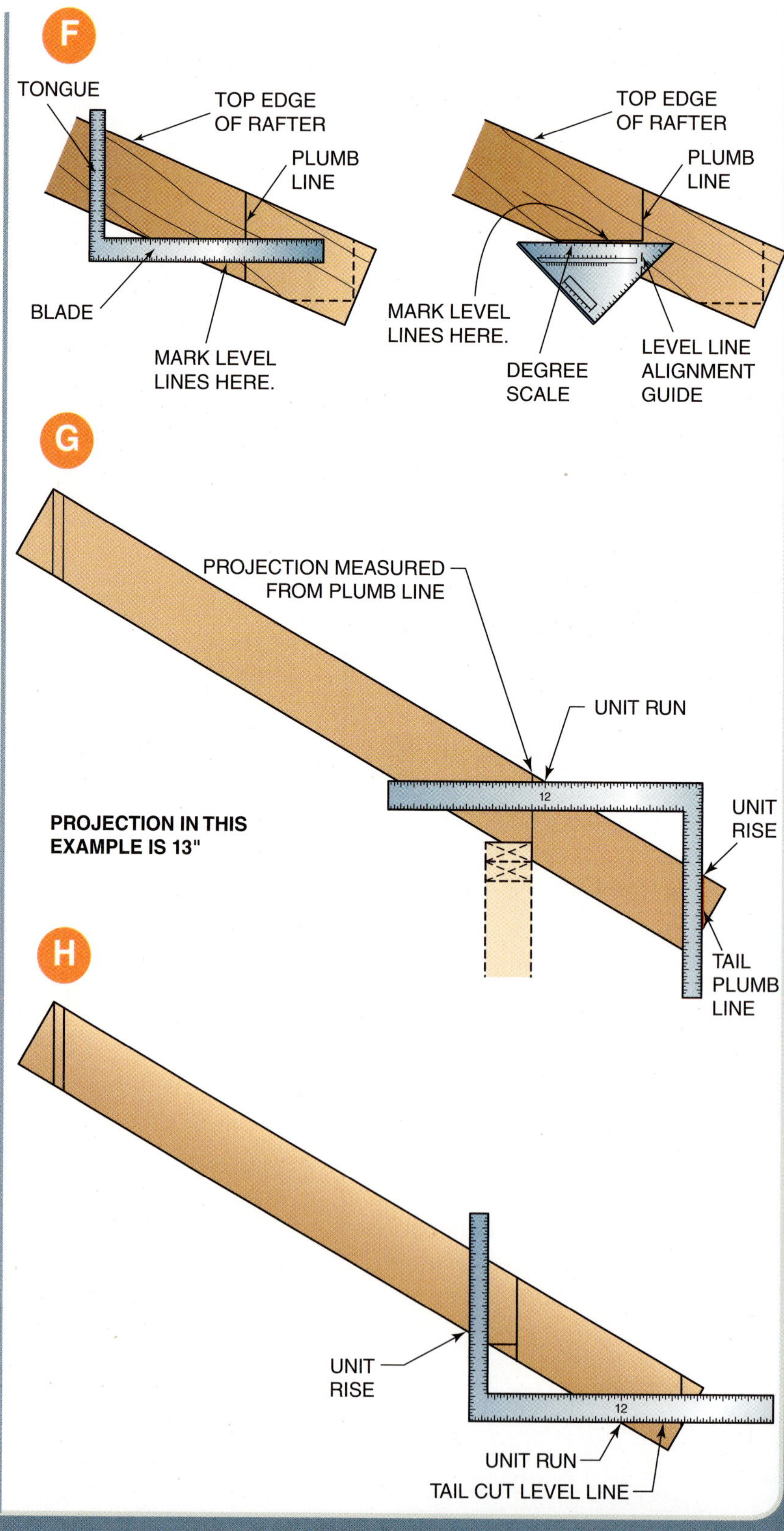

PROCEDURE 12-2

# Erecting the Gable Roof Frame

**A** Lay the ridgeboard on top of the work platform in the same direction it was laid out. Be careful not to turn the pieces around, end for end. Select four straight pieces for the end rafters, commonly called the **rake** rafters. One carpenter on each side of the building and one at the ridge are needed to raise the roof with efficiency.

- Fasten a rafter to each end of the first section of ridgeboard from the same side of the building. Raise the ridgeboard and two rafters into position. Fasten the rafters at the seat into the plate with three nails.
- Next, install two opposing rafters on the other side. Lift rafters into place and fasten the seat first to the plate. Then fasten the rafters to the ridgeboard. Drive fasteners of the second opposing rafter through ridge at a slight angle next to the first rafter.

**CAUTION**

**CAUTION:** Take care to be sure your footing is secure. Be patient and work safely. Also, driven nails should not split the lumber. This will weaken the frame.

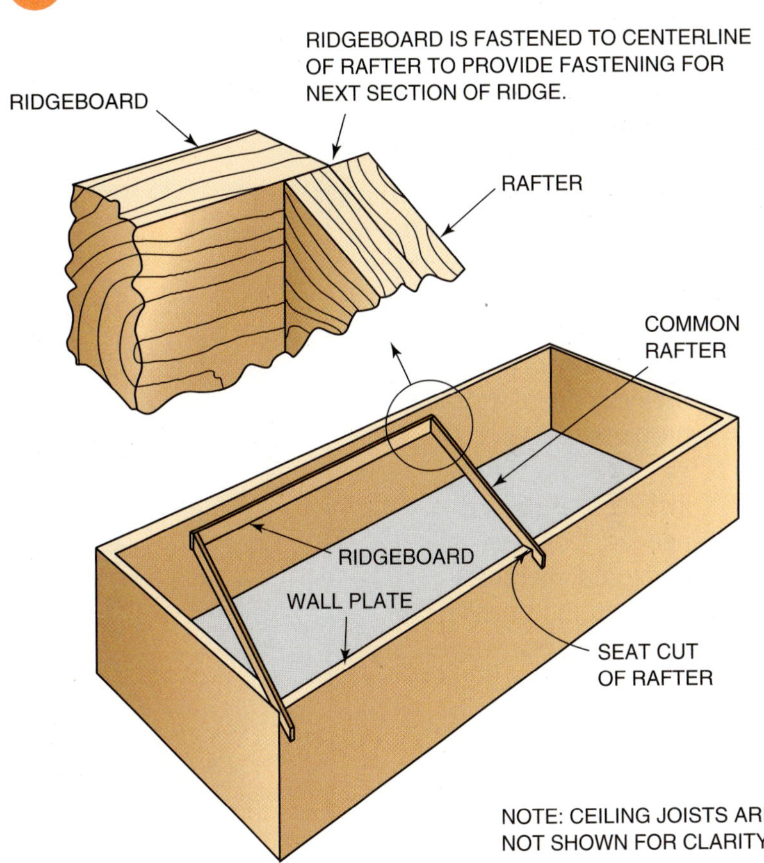

PROCEDURE 12-2

# Erecting the Gable Roof Frame (continued)

**B** Plumb and brace this section. Plumb the section from the end of the ridgeboard to the outside edge of the plate at the end of the building. Brace the section temporarily from attic floor to ridge.

- Raise all other sections in a similar manner, installing the remaining rafters. Do not install too many rafters on one side before installing the opposing rafters. This will cause the ridge to sag. Sight along the top edge of the ridgeboard periodically to check the straightness of the ridgeboard as framing progresses.

**C** Install collar ties as needed. Collar ties are horizontal members fastened to opposing pairs of rafters, which effectively reduce the span of the rafter. They are typically installed on every third rafter pair or as required by drawings or codes. The length of a collar tie varies, but their length is usually about one-third to one-half the building span.

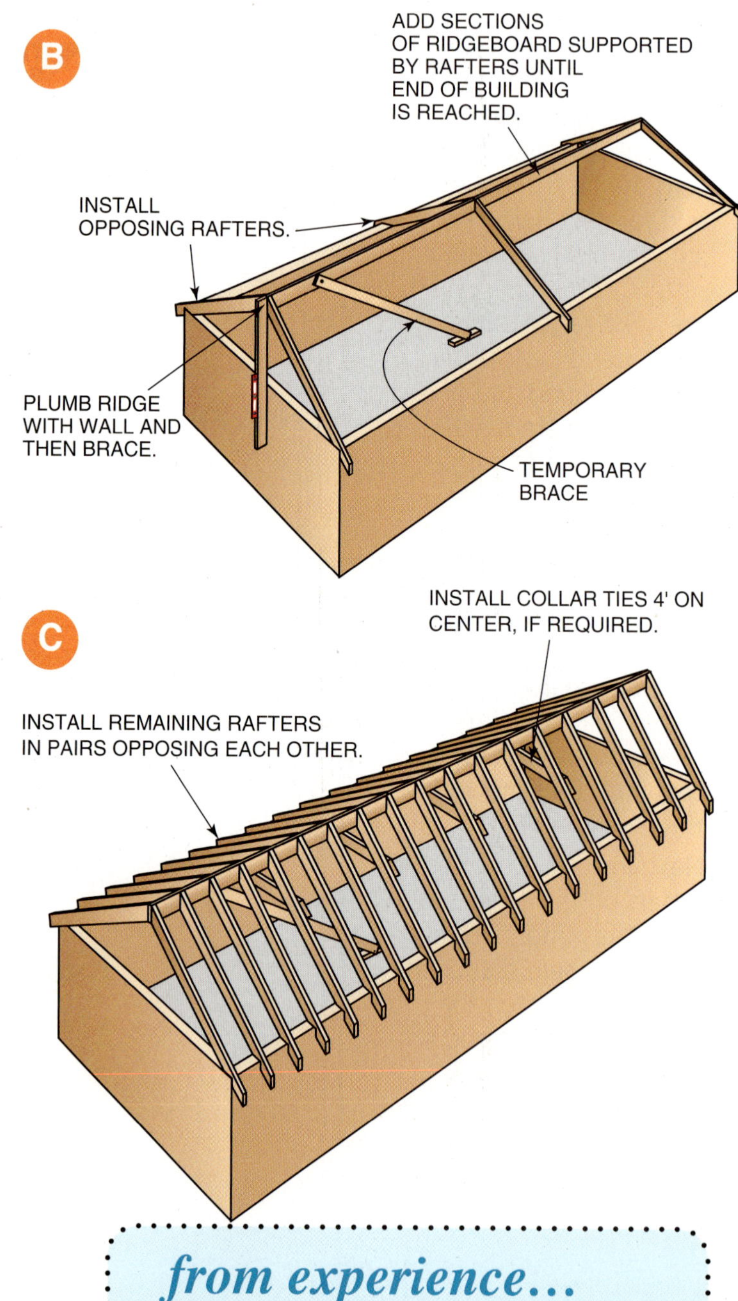


### *from experience...*

Check the straightness of the top plates periodically as the rafters are installed. If wall bracing is not adequately installed or nailed, the walls may bow out as the rafters are installed, causing the ridge to sag.

**D** When truss joists are used as rafters, some extra steps must be taken. Stiffeners, in the form of 1 × 4 strips, are fastened to the webs above the seat cut. Metal bridging or blocking may be installed to keep rafters plumb.

- Joist hangers are used with web stiffeners when TJI rafters are used in cathedral ceilings. In this case, no ceiling joists are used, and the ridge becomes a supporting beam.

**D**

PROCEDURE 12-3

# Installing Gable Studs

**A** Lay out the top plate for the location of each gable end stud. Square a line up from the lower wall studs over to the top of the wall plate.

- Using a level, plumb a stud from the layout mark on the plate. Mark along the top and bottom edge of the rafter the length of the stud. Measure this vertical distance. Note whether the measurement is to the longer or shorter edge of the stud.
- Determine the common difference in length of the studs. Multiply the OC spacing, in feet, times the unit rise.

**B** Lay out and cut the studs, longer and shorter as needed, from the first stud measured. Make sure all measurements are to the same edge of the stud (i.e., all are measured to the long point of the angled cut).

- Fasten the studs by toenailing to the plate and by nailing through the rafter into the stud. Care must be taken not to force the rake rafters up into a large crown. Sight the top edge of the end rafters for straightness periodically as gable studs are installed.
- After all gable studs are installed, the end ceiling joist is nailed to the inside edges of the studs.

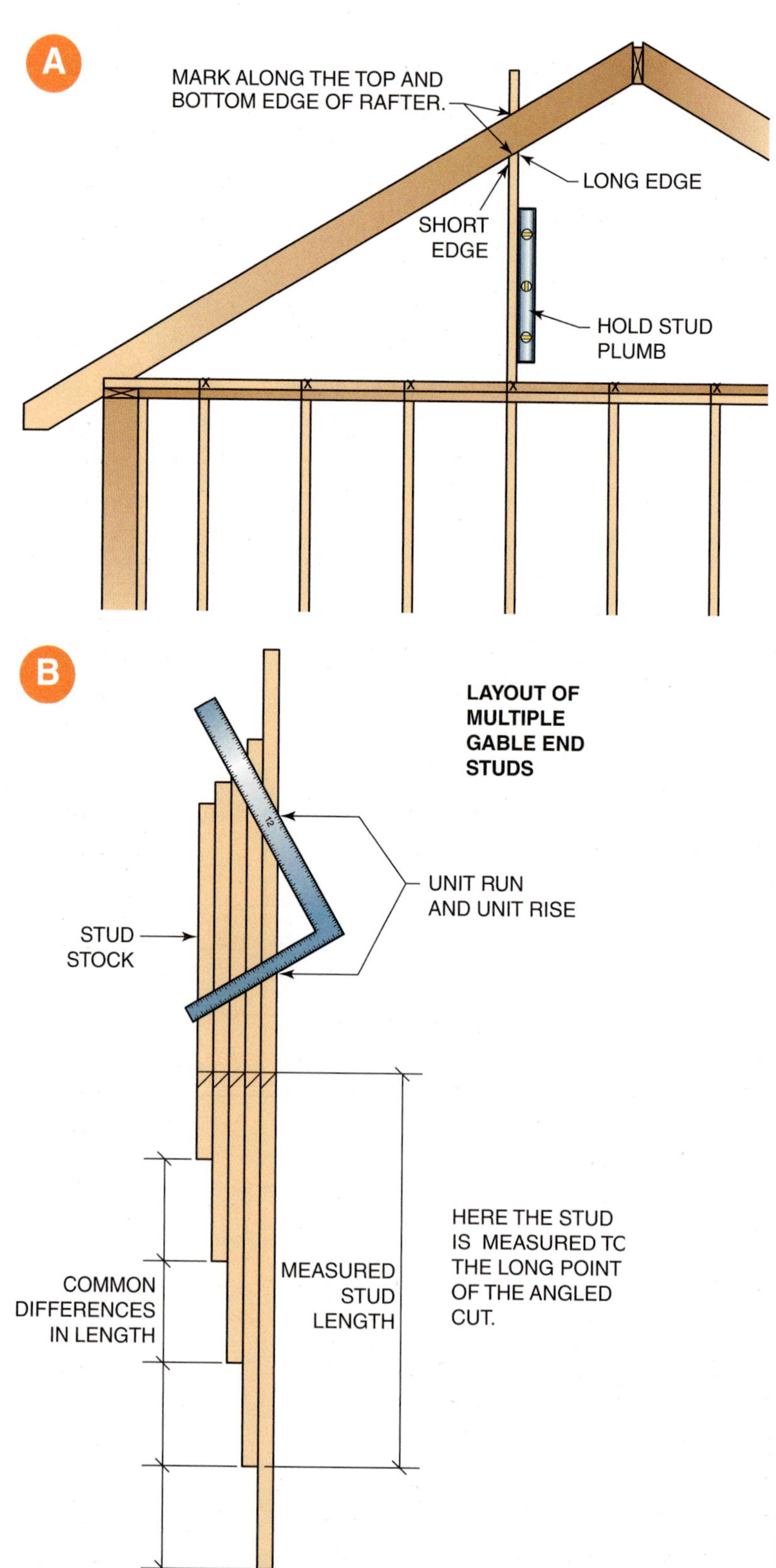

PROCEDURE 12-4

# Hip Rafter Layout

**A** Select a straight length of stock for a pattern. Lay it across two sawhorses.

- Mark a plumb line at one end. Make sure the top of the rafter will have the crown. Hold the tongue of the square at the unit rise and the blade of the square at 17 inches, the unit run for the hip rafter.
- Shorten the rafter by measuring at right angles from the plumb line one-half the 45-degree thickness of the ridge. This number is 1⅟₁₆ inches for standard dimension lumber. Lay out another plumb line through this measurement. From the top of this second plumb line, square a line over the top edge of the rafter. Mark the midpoint of that line.

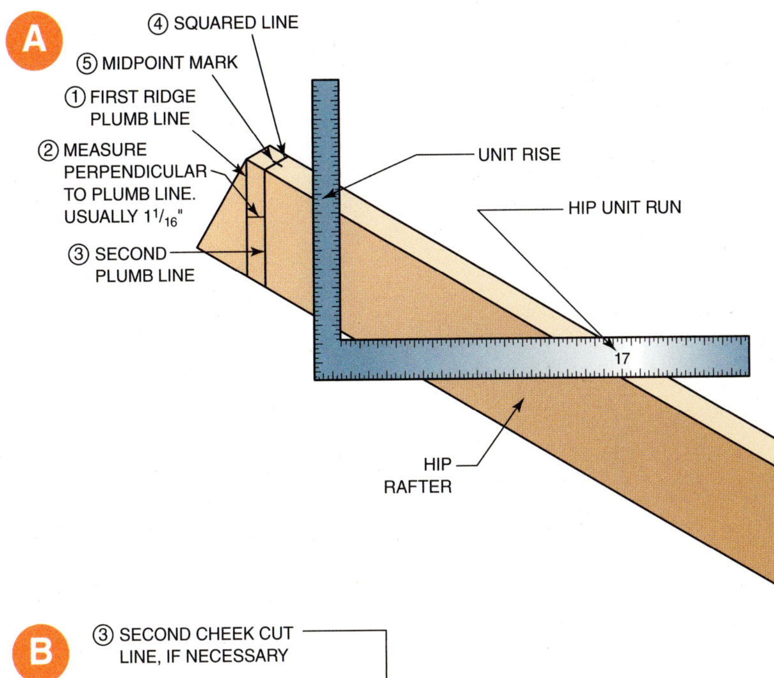


**B** Lay out cheek cuts by measuring, again at right angles to the second plumb line, one-half the thickness of the hip rafter. This is ¾ inch for dimension lumber. Draw a third plumb line through this measurement. Next, draw a line over the top of the rafter from the third plumb line through the midpoint of the squared line. This new line and the third plumb line become the cut lines of the ridge cut. Repeat this process for the second cheek cut, if necessary.

- Determine the length of the hip rafter by using the same process as for a common rafter, unit length multiplied by the run. The unit length for a hip rafter is found on the second line of the rafter table. The run is the same number as for common rafters. **Example:** Find the length of a hip rafter for a roof with a unit rise of 8 inches where the run for a common rafter is 14. On the second line of the rafter table, below the 8-inch mark, it is found that the unit length of the hip rafter is 18.76 inches. This multiplied by 14 equals 262.64 inches. Multiplying the decimal portion, 0.64 by 16 gives 10.24 sixteenths. This rounds to ¹⁰⁄₁₆ and reduces to ⅝ inch. The total length of the hip rafter is 262⅝ inches.

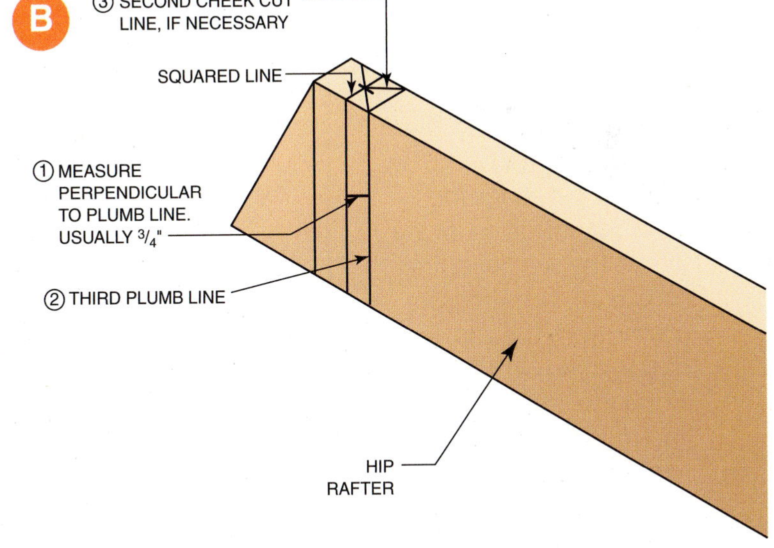

PROCEDURE 12-4

# Hip Rafter Layout (continued)

**C** Measure and mark, from the first plumb line drawn at the ridge end and along the edge of the rafter, the length of the rafter. Draw a plumb line through this mark.

**D** The hip seat cut has the same height above the seat cut as for the common rafter. To locate where the seat level line should be, first measure down, from the top of the common rafter, along the seat cut plumb line to the seat cut level line. Mark that same measurement on the hip. Draw the level line.

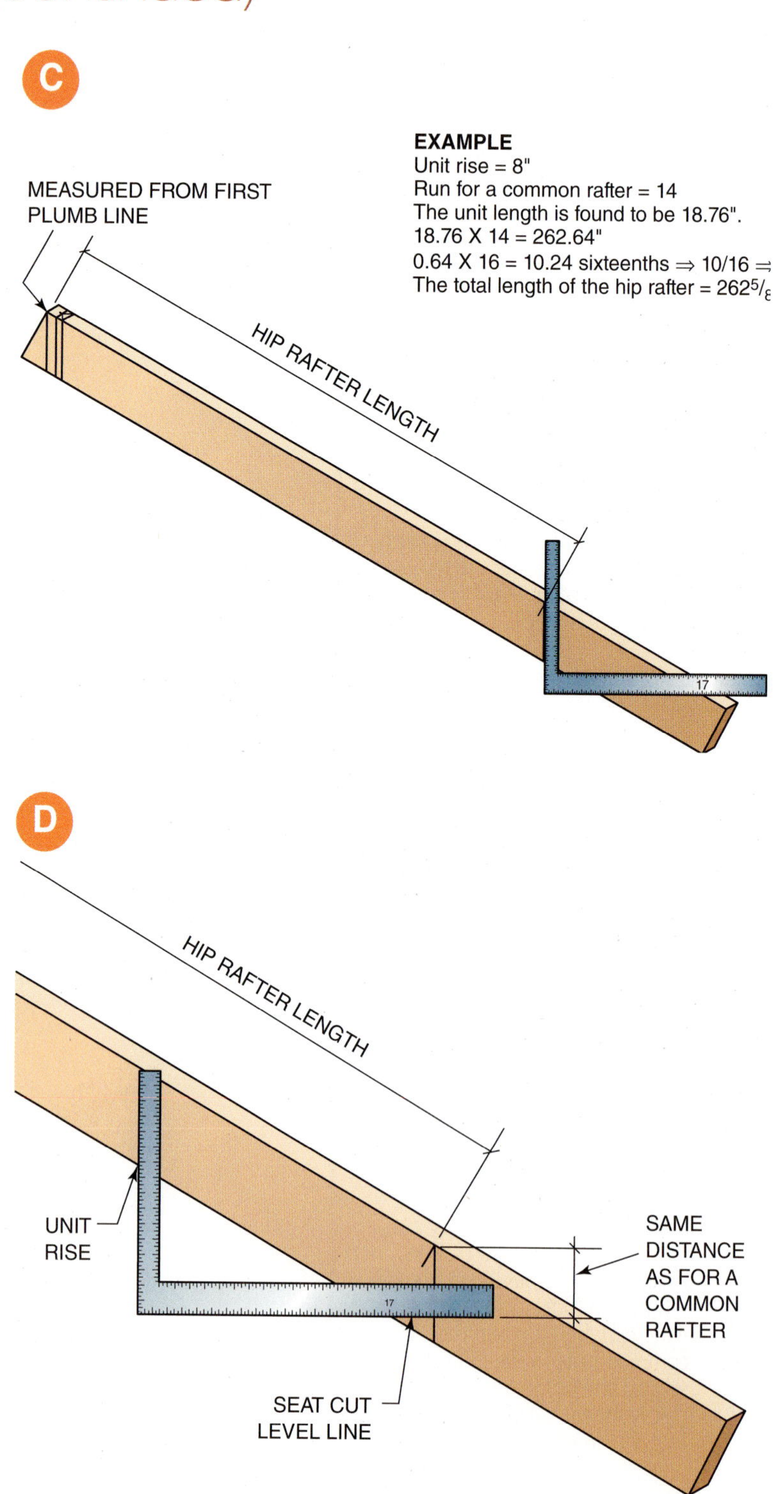

**E** Determine the amount the seat cut level line is raised in order to drop the rafter. This is done by first measuring back one-half the hip rafter thickness (usually ¾ inch) along the seat cut level line from the bottom edge of the rafter. Next, from the end of this ¾-inch line, plumb down to the bottom edge of the rafter. This smaller distance is the amount that the seat level line is raised. Measure up and draw a new seat level line. This distance is also the amount that the top edge may be backed. Backing is done instead of dropping the rafter.

**F** The hip tail length is found by multiplying its unit length by the common rafter projection (in feet).

- Lay out the tail length along the top edge of the rafter from the seat cut plumb line. Draw another plumb line. Draw a level line in the same manner as for a common rafter.

**G** The tail cut of the hip rafter is usually a double cheek cut. It is laid out in the same manner as the ridge cheek cuts.

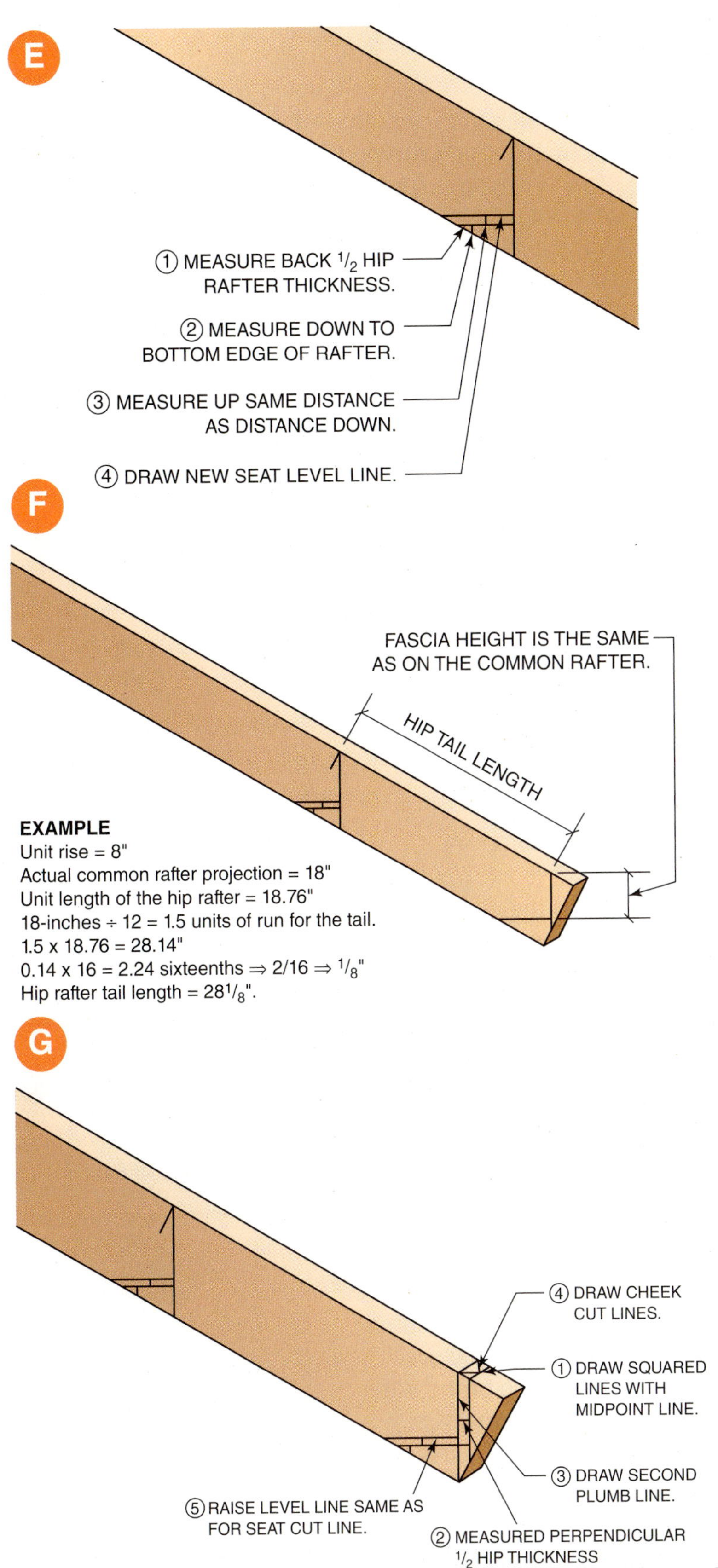

PROCEDURE 12-5

# Jack Rafter Layout

**A** The first jack rafter length is calculated by multiplying run by unit length. The run is determined from a measurement on the building. The run of any jack rafter is the distance from the corner of the building to its center. The unit length is that of the common rafter.

- The first jack rafter is laid out by starting at the tail. Place the common rafter pattern on a piece of stock and mark the tail, seat cut, and seat cut plumb line.
- Mark the length of the rafter by measuring from the seat cut plumb line along the rafter edge. Draw a plumb line.

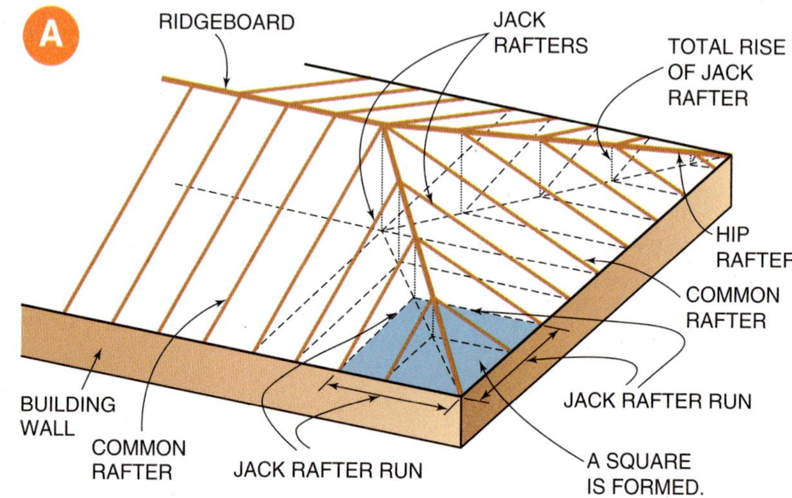


**B** Shorten the rafter for the thickness of the hip using the same procedure as for the hip shortening at the ridge. Measuring at right angles from the plumb line one-half the 45-degree thickness of the ridge (1 1⁄16 inches for dimension lumber), lay out another plumb line through this measurement. From the top of this second plumb line, square a line over the top edge of the rafter. Mark the midpoint of that line.

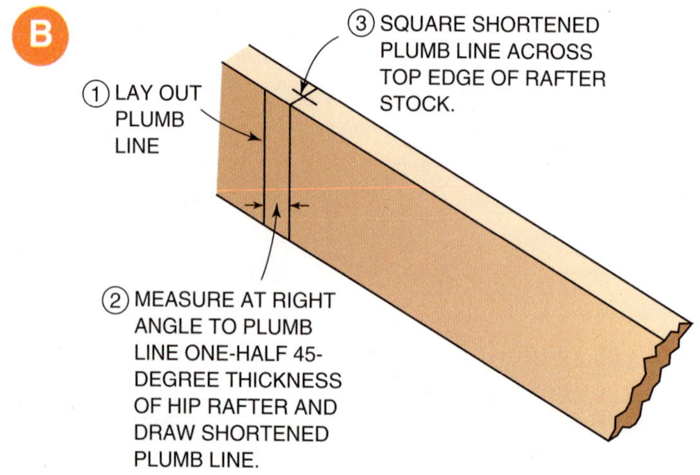

C Lay out cheek cuts by measuring, again at right angles to the second plumb line, one-half the thickness of the jack rafter (¾ inch for dimension lumber). Draw a third plumb line through this measurement. Next, draw a line over the top of the rafter from the third plumb through the midpoint of the squared line. This new line and the third plumb line become the cut lines. The direction of the diagonal depends on which side of the hip the jack rafter is framed.

- Remaining jacks are cut in the same manner except that their lengths may be determined by adding or subtracting a common difference in length. This difference is found on the third and fourth line of the rafter tables under the inch mark that coincides with the slope of the roof being framed. For example, the common difference for 16 OC rafters with a slope of 8 is 19¼ inches.

C

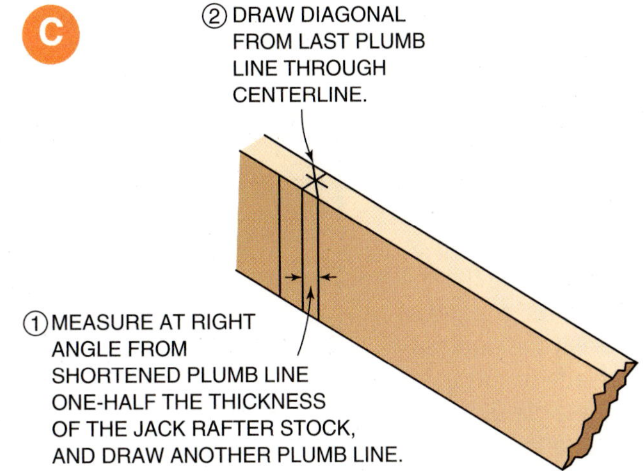


THE CUTTING LINES ARE THE DIAGONAL AND THE LAST PLUMB LINE.

PROCEDURE 12-6

# Valley Rafter Layout

**A** Select a straight length of stock for a pattern. Lay it across two sawhorses.

- Mark a plumb line at one end. Make sure the top of the rafter will have the crown. Hold the tongue of the square at the unit rise and the blade of the square at 17 inches, the unit of run for the hip rafter.
- Shorten the rafter by measuring at right angles from the plumb line one-half the 45-degree thickness of the ridge. This number is 1 1⁄16 inches for standard dimension lumber. Lay out another plumb line through this measurement. From the top of this second plumb line, square a line over the top edge of the rafter. Mark the midpoint of that line.

**B** Lay out cheek cuts by measuring, again at right angles to the second plumb line, one-half the thickness of the valley rafter. This is 3⁄4 inch for dimension lumber. Draw a third plumb line through this measurement. Next, draw a line over the top of the rafter from the third plumb through the midpoint of the squared line. This new line and the third plumb line become the cut lines of the ridge cut. Repeat for the second cheek if desired.

- Determine the length of the valley rafter by using the same process as for a hip rafter—unit length multiplied by the run.

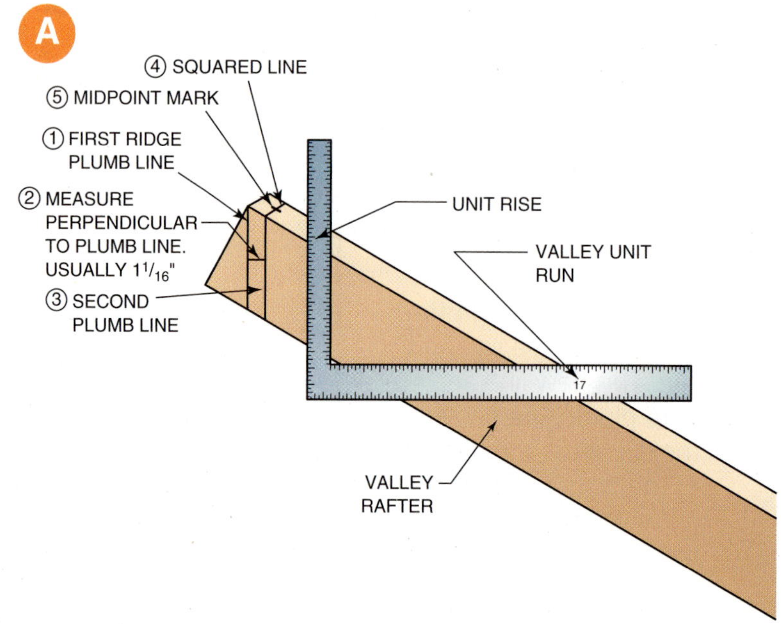


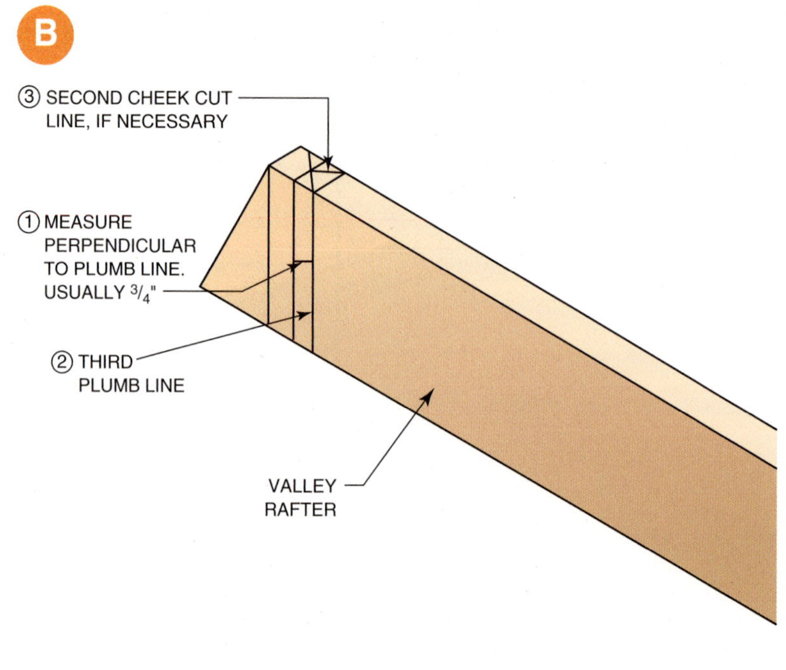

**C** Measure and mark, from the first plumb line drawn at the ridge end and along the edge of the rafter, the length of the rafter. Draw a plumb line through this mark.

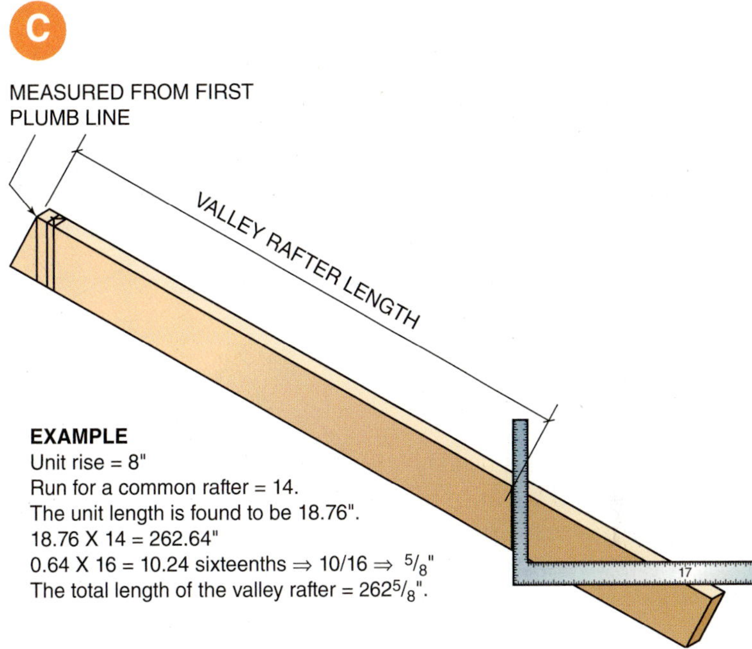


**D** The valley seat cut has the same height above the seat cut as for the common rafter. Measure down, from the top of the common rafter, along the seat cut plumb line to the seat cut level line. Mark that same measurement on the valley, measuring down along the seat cut plumb line. Draw the level line.

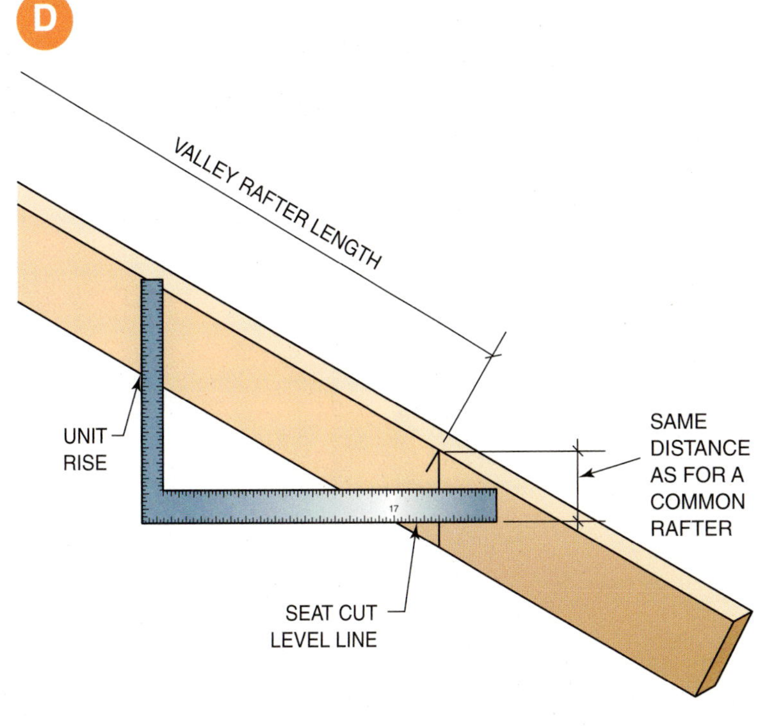

PROCEDURE 12-6

## Valley Rafter Layout (continued)

**E** Lay out a deeper seat cut plumb line to allow for the thickness of the valley rafter. Measure back one-half the valley thickness (usually ¾ inch) along the seat cut level line toward the tail of the rafter. Draw a new plumb line through this mark.

- The valley tail is usually not needed because of other members present. The fascia has sufficient support from the nearby common tails.

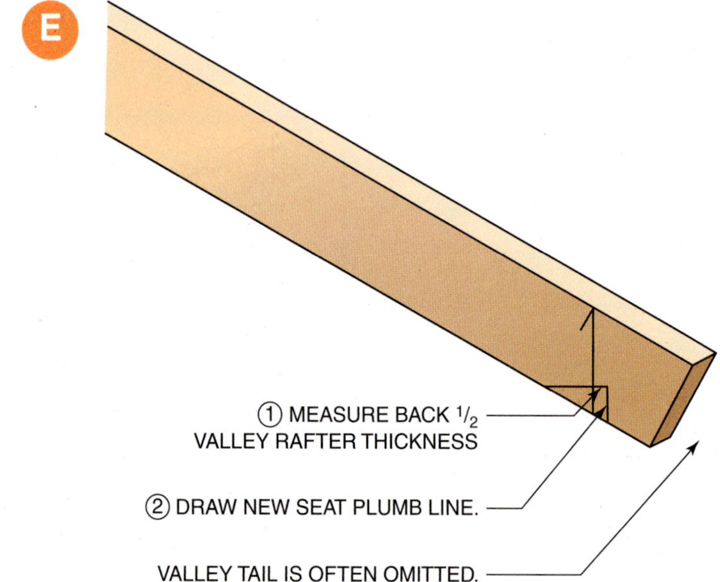

PROCEDURE 12-7

# Truss Installation

**A** Check the walls for plumb and stiffness of temporary wall bracing.

- Read and understand the prints and installation guide provided for proper location of trusses and bracing.
- Prepare the long braces for the first trusses. Nail a stiffener to the side of the brace, if necessary. Secure the brace bases to the floor or stakes driven in the ground.
- Raise the first truss into place and brace securely. Braces to the ground should line up with the future top chord lateral bracing. Check and recheck for plumb and security of braces.
- Raise the next three or four trusses into place individually. Secure all trusses together with lateral, web, and diagonal bracing. Pause to check that this first group of trusses is braced together as one unit. The group should be secure enough that if the ground bracing installed on the first truss were to be removed, the five-truss unit would be braced well enough to withstand a severe gust of wind. It should be noted that the ground support should remain in place for as long as possible.

**B** Install remaining trusses and bracing. Install additional diagonal and permanent bracing as required.

**CAUTION**

**CAUTION:** Trusses can be very dangerous to install. Workers must be alert at all times. Watch for proper nailing and order of assembly. Make sure bracing is securely and appropriately nailed. Be prepared to move quickly to prevent and avoid injury.

**A**

**B**

## REVIEW QUESTIONS

*Select the most appropriate answer.*

1. **The type of roof that has the fewest number of different sized members is**
   a. gable.
   b. hip.
   c. gambrel.
   d. all of the above.

2. **The term used to represent the horizontal distance under a rafter is called**
   a. run.
   b. span.
   c. line length.
   d. pitch.

3. **The rafter that spans from a hip rafter to the ridge is a**
   a. hip jack.
   b. valley jack.
   c. valley.
   d. no such rafter exists.

4. **The rafter that spans from a wall plate to the ridge and whose run is at a 45-degree angle to the plates is a**
   a. hip.
   b. valley.
   c. both a and b.
   d. none of the above.

5. **The minimum amount of stock left above the seat cut of the common rafter to ensure enough strength to support the overhang is usually**
   a. one-quarter of the rafter stock width.
   b. one-half of the rafter stock width.
   c. two-thirds of the rafter stock width.
   d. three-quarters of the rafter stock width.

6. **What is the line length of a common rafter from the centerline of the ridge to the plate with a unit rise of 5 inches, if the building is 28'–0" wide?**
   a. 1761.2 inches.
   b. 182 inches.
   c. 186 inches.
   d. 194 inches.

7. **The common difference in the length of gable studs spaced 24 inches OC for a roof with a slope of 8 on 12 is**
   a. 8 inches.
   b. 12 inches.
   c. 16 inches.
   d. 20 inches.

**8. What is the hip projection, in inches, if the actual common rafter projection is 6 inches?**

a. 6 inches
b. 6/17 feet
c. 8½ inches
d. 17 inches minus 6 inches

**9. The line length of a hip rafter with a unit rise of 6 inches and a total run of 12 feet is**

a. 198 inches.
b. 208 inches.
c. 216 inches.
d. 224 inches.

**10. The jack rafter is most similar to a**

a. common rafter.
b. hip rafter.
c. valley rafter.
d. gable end stud.

**11. The total run of any hip jack rafter is equal to its**

a. distance from its seat cut to the outside corner.
b. distance from its seat cut to the inside corner.
c. line length.
d. common difference in length.

**12. The total run of the shortened valley rafter is equal to the total run of the**

a. minor span rafter.
b. major span rafter.
c. common rafter.
d. hip rafter.

**13. The total run of the hip-valley cripple jack rafter is equal to**

a. one-half the run of the hip jack rafter.
b. one-half the run of the longest valley jack rafter.
c. the distance between seat cuts of the hip and valley rafters.
d. the difference in run between the supporting and shortened valley rafters.

**14. The common difference in length for a 16 inches OC hip jack rafter with a unit rise of 9 inches is**

a. 12 inches.
b. 20 inches.
c. 25⅝ inches.
d. 28 inches.

**15. The typical amount that a hip rafter is shortened because of the ridge is**

a. ¾ inches.
b. 1 1/16 inches.
c. 1½ inches.
d. none of the above.

**16. The member that prevents a set of trusses from tipping over is called a**

a. diagonal brace.
b. lateral brace.
c. web.
d. gusset.

**17. The part of a truss that may be cut, if necessary, is the**

a. web.

b. chord.

c. permanent brace.

d. none of the above.

**18. The length of a ridgeboard for a hip roof installed on a rectangular building measuring 28' × 48' is slightly more than**

a. 20 feet.

b. 28 feet.

c. 48 feet.

d. none of the above.

**19. The estimated number of gable rafters for a rectangular building measuring 28' × 48' is**

a. 28.

b. 48.

c. 72.

d. 74.

**20. Neglecting the roof overhangs, the estimated number of pieces of sheathing needed for a hip roof installed on a rectangular building that measures 28' × 47' and has a unit rise of 6 inches is**

a. 24.

b. 25.

c. 26.

d. 51.

## KNOW YOUR CODES

Roof frames are designed in many configurations. The intended load must be understood to ensure the roof is built strong enough. Regions of high snow accumulation need roofs that are stronger than warmer regions. Heavy wind and seismic regions also must be stronger. Be sure to follow the specifications in the set of building prints that adhere to the local codes.

## WHAT'S WRONG WITH THIS PICTURE?

Carefully study **Figure 12-55** and think about what is wrong. Consider all possibilities.

### WRONG

**FIGURE 12-55** The photo shows a view of the ceiling area of a trussed roof. For orientation purposes, the closest framing members going from left to right in the picture are the bottom chords of the trusses. The framing members going from the bottom to top are permanent braces. The permanent braces are not overlapping. They should always overlap by a minimum of two trusses. This will ensure better structural integrity of the roof system.

### RIGHT

**FIGURE 12-56** This photograph taken from up in the ceiling clearly shows the bracing overlapping by two trusses as it should.

SECTION 3

# Exterior Finish

# Windows and Doors

**OBJECTIVES** *After completing this unit, the student should be able to:*

- Describe the most popular styles of windows and name their parts.
- Select and specify desired sizes and styles of windows from manufacturers' catalogs.
- Install various types of windows in an approved manner.
- Name the parts of and set a pre-hung door frame.
- Describe the standard designs and sizes of doors and name their parts.
- Fit and hang a door to a preexisting opening.
- Install locksets in doors.
- Install bypass, bifold, and pocket doors.

# GLOSSARY OF WINDOW AND DOOR TERMS

**astragal** a semicircular molding often used to cover a joint between two doors

**bay window** a window, usually three-sided, that projects out from the wall line

**casing** molding used to trim around doors, windows, and other openings

**deadbolt** door-locking device operated by a key from the outside and by a handle or key from the inside

**double-acting** doors that swing in both directions, or the hinges used on these doors

**escutcheon** protective plate covering the finished surface where penetrations occur like door knobs.

**extension jambs** strips of wood added to window jambs to bring the jamb edge flush with the wall surface in preparation for casing

**glazing** the act of installing glass in a frame

**hopper window** a type of window in which the sash is hinged at the bottom and swings inward

**housewrap** type of building paper with which the entire exterior sidewalls of a building are covered

**insulated glass** multiple panes of glass fused together with an air space between them

**light** a pane of glass or an opening for a pane of glass

**low emissivity glass (Low E)** a coating on double-glazed windows designed to raise the insulating value by reflecting heat back into the room

**molding** decorative strips of wood used for finishing purposes

**mullion** a vertical division between window units or panels in a door

**muntin** slender strips of wood between lights of glass in windows or doors

**rail** the horizontal member of a frame

**sash** that part of a window into which the glass is set

**stile** the outside vertical members of a frame, such as in a paneled door

**strike plate** thin metal plate installed where the latch bolt of a door touches the jamb

**weather stripping** narrow strips of material applied to windows and doors to prevent the infiltration of air and moisture

**wind** a defect in lumber caused by a twist in the stock from one end to the other; also, a twist in anything that should be flat

The finishing phase of the house usually begins with completing the exterior. This allows time for the plumbing, heating, cooling, and electrical components to be installed, thereby keeping a steady flow of work on the house. The order of installation may be changed to suit the situation, but each process has skill requirements that remain unchanged.

This chapter deals with the installation of windows and doors. It is understood that many things are usually done before interior doors are installed, which will be discussed in a later chapter.

Interior and exterior doors, like windows, are manufactured in millwork plants in a wide range of styles and sizes. Most doors come pre-hung in frames, complete with exterior casings applied and ready for installation. Doors come in many styles and sizes.

Care must be taken to maintain the easy-operation and weathertightness of the window and door units. Quality workmanship results in reduced maintenance and gives longer life to the units. This also results in a building with a more comfortable interior that saves energy by reducing fuel costs.

Safety concerns remain a top priority. The potential for injury is still present. Many windows are installed from scaffolds, and often stepladders are used. All workers deserve to return home safely to their families at the end of each day. Create safe work habits.

# WINDOWS

Windows are one of many types of millwork (Fig. 13-1). *Millwork* is a term used to describe products, such as windows, doors, and cabinets, fabricated in woodworking plants that are used in the construction of a building. Windows are usually fully assembled and ready for installation when delivered to the construction site. Windows are made with wood, fiberglass, aluminum, steel, and vinyl. Windows made with the wood parts encased in vinyl are called *vinyl-clad windows*.

## PARTS OF A WINDOW

When shipped from the factory, the window is a complete unit except for the interior trim. It is important that the installer know the names, location, and functions of the parts of a window in order to understand or to give instructions concerning them.

The **sash** is a frame in a window that holds the glass. The type of window is generally determined by the way the sash operates. The sash may be installed in a fixed position, move vertically or horizontally, or swing outward or inward.

### Sash Parts

Vertical edge members of the sash are called **stiles**. Top and bottom horizontal members are called **rails**. The pieces of glass in a sash are called **lights**. There may be more than one light in a sash. Small strips of wood that divide the glass into smaller lights are called **muntins** (Fig. 13-2). Muntins may divide the glass into rectangular, diamond, or other shapes.

Many windows come with false muntins called *grilles*. Grilles do not actually separate or support the glass. They are applied as an overlay to simulate small lights. They are made of wood or plastic. They snap in and out of the sash for easy cleaning of the lights (Fig. 13-3). They may also be preinstalled between the layers of glass in double- or triple-glazed windows.

### Window Glass

Several qualities and thicknesses of sheet glass are manufactured for **glazing** and other purposes. The installation of glass in a window sash is called glazing. *Single strength (SS)* glass is about 3⁄32-inch thick. It is used for small lights of glass. For larger lights, *double strength (DS)* glass about 1⁄8-inch thick may be used.

**FIGURE 13-1** Windows of many types and sizes are fully assembled in millwork plants and ready for installation.

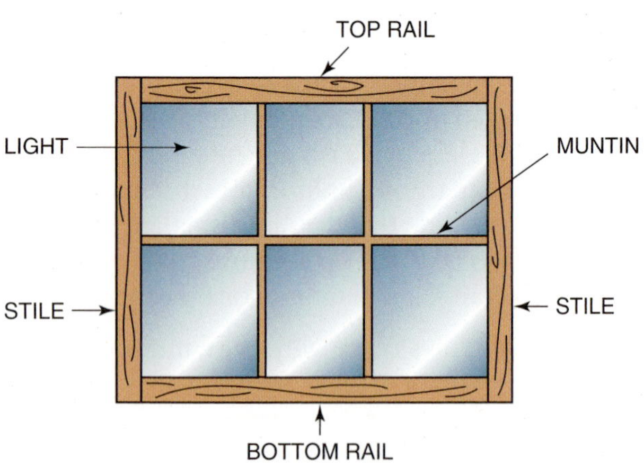


**FIGURE 13-2** A sash and its parts.

**FIGURE 13-3** Removable grilles simulate true divided-light muntins.

*Safety Glass.* Most windows are not glazed with safety glass. If broken, they could fragment and cause injury. Care must be taken to handle windows in a manner to prevent breaking the glass. Some codes require a type of *safety glass* in windows with low sill heights or located near doors. Skylights and roof windows are generally required to be glazed with safety glass.

Safety glass is constructed, treated, or combined with other materials to minimize the possibility of injuries resulting from contact with it. When broken at any point, the entire piece immediately disintegrates into a multitude of small granular pieces.

Courtesy of Andersen Windows, Inc.

**FIGURE 13-4** Cutaway of insulated glass used to increase the R-value in a window.

*Insulated Glass.* To help prevent heat loss, and to avoid condensation of moisture on glass surfaces, **insulated glass**, or *thermal pane,* windows are used frequently in place of single-thickness glass. Insulated glass consists of two or sometimes three layers of glass separated by a sealed air space 3⁄16 to 1 inch in thickness (Fig. 13-4). Moisture is removed from the air between the layers. To raise the R-value of insulated glass, the space between the layers may be filled with argon gas. Argon conducts heat at a lower rate than air. Additional window insulation may be provided with the use of *removable glass panels* or *combination storm sash* windows.

*Solar Control Glass.* The R-value of windows may also be increased by using special solar control insulated glass, called *high performance* or *Low E* glass. Low E is an abbreviation for **low emissivity**. It is used to designate a type of glazing that reflects heat back into the room in winter and blocks heat from entering in the summer (Fig. 13-5). An invisible, thin, metallic coating is bonded to the air space side of the inner glass. This lets light through but reflects heat.

**Green Tip**

Window performance is only as good as the installation. Follow manufacturer's recommendations. Incorrect installations will void warranties.

### The Window Frame

The sash is hinged to, slides, or is fixed in a *window frame.* The frame usually comes with the exterior trim applied. It consists of several distinct parts (Fig. 13-6).

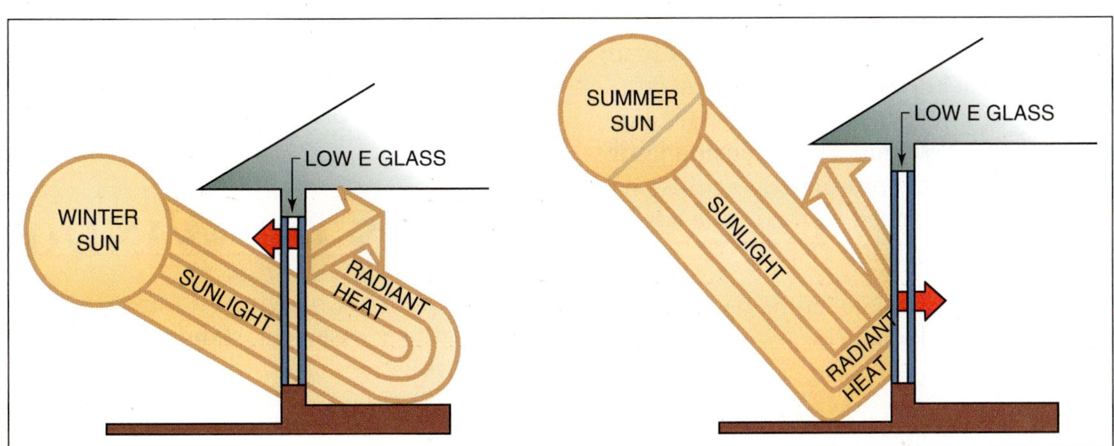


Courtesy of Andersen Windows, Inc.

**FIGURE 13-5** Low E glass is used in windows to help keep heat in during cold weather and out during hot weather.

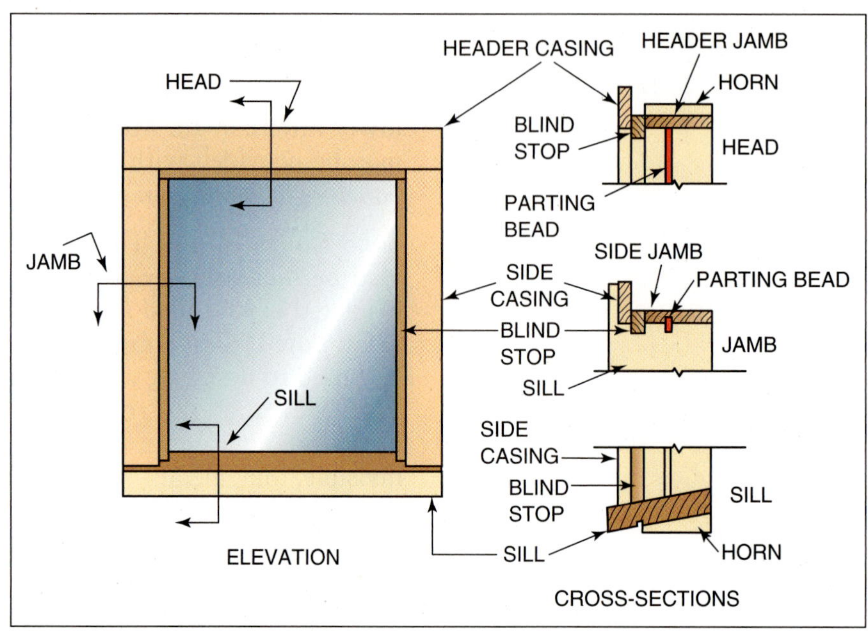


**FIGURE 13-6** A window frame consists of parts with specific terms.

*The Sill.* The bottom horizontal member of the window frame is called *a sill.* It is usually set or shaped at an angle to shed water. Its bottom side usually is grooved so a weathertight joint can be made with the wall siding.

*Jambs.* The vertical sides of the window frame are called *side jambs.* The top horizontal member is called a *head jamb.*

*Extension Jambs.* The inside edge of the jamb should be flush with the finished interior wall surface, which allows for a place to install the interior trim. Windows can be ordered with jamb widths already made to the wall thicknesses. In other cases, jambs are made narrow, and **extension jambs** are later installed on the window unit. The extension jambs are cut to width to accommodate various wall thicknesses. They are applied to the inside edge of the jambs of the window frame (Fig. 13-7).

*Casings.* Window units usually come with exterior **casings** applied, while the interior casing is applied after the window and interior wall surface are installed. In either case, the side members are called *side casings.* The top member is called the *head casing.* When windows are installed or manufactured side by side in multiple units, a **mullion** is formed where the two side jambs are joined together. The casing covering the joint is called a *mullion casing.*

*Window Flashing.* In some cases, *window flashing* is also provided. This is a piece of metal or plastic made slightly longer than the head casing. It is sometimes called a *drip cap.* It is bent to fit over the head casing and up against the exterior wall (Fig. 13-8). The flashing sheds water over the top of the casing.

*Protective Coatings.* Wood window units are *primed* with the first coat of paint applied at the factory. Vinyl-clad wood windows are designed to eliminate the need for painting.

*Screens.* Manufacturers provide screens as optional accessories for all kinds of windows. On out-swinging and sliding windows, the screens are attached to the inside of the frame. On other windows, they are mounted on the outside of the frame. The screen mesh is usually plastic or aluminum.

# Types of Windows

Common types of windows are fixed, single- or double-hung, casement, sliding, awning, and hopper windows.

## FIXED WINDOWS

*Fixed windows* consist of a frame in which a sash is fitted in a fixed position. They are manufactured in many shapes. Oval and circular windows are usually installed as individual units. Elliptical, half rounds, and quarter rounds are widely used

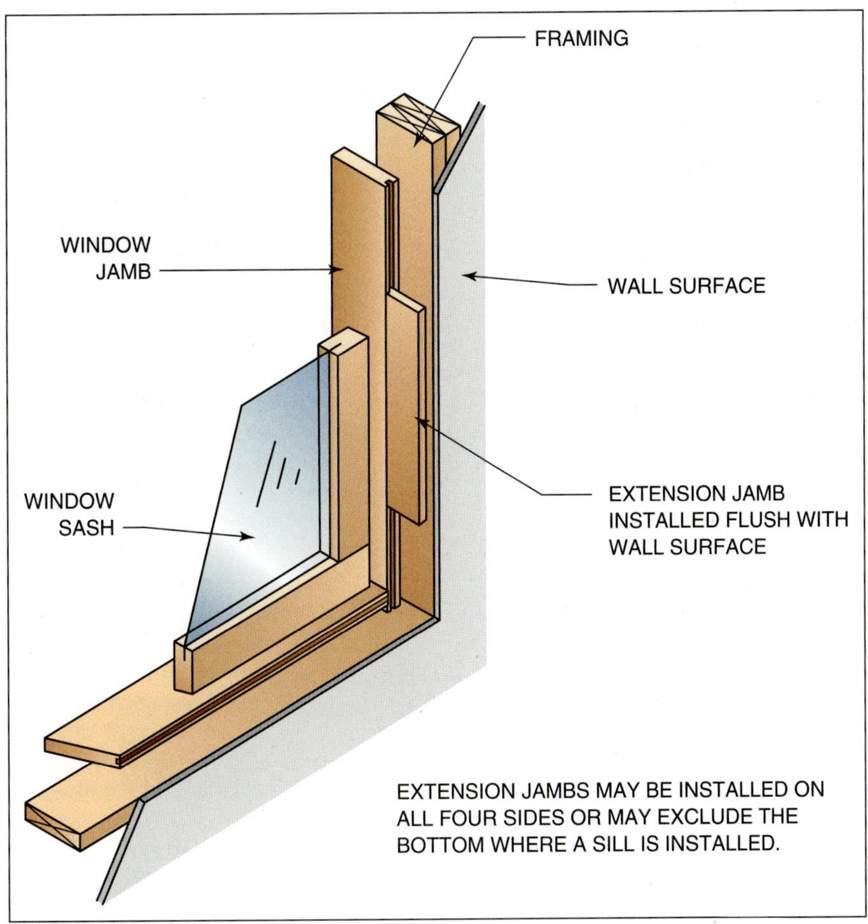


**FIGURE 13-7** To compensate for varying wall thicknesses, extension jambs are provided with some window units.

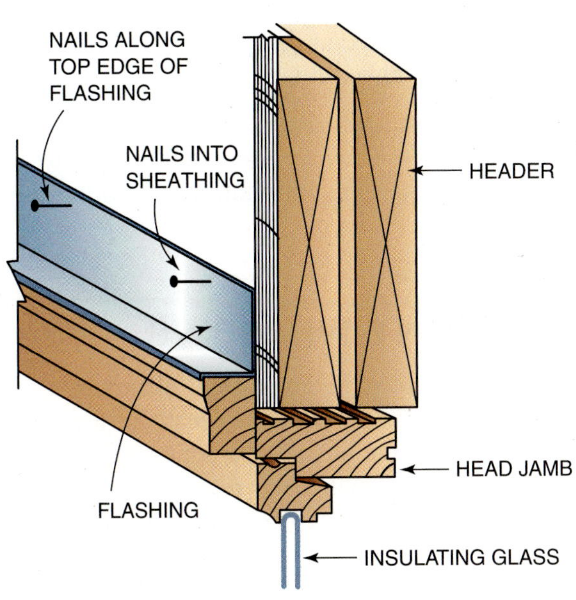


**FIGURE 13-8** A window flashing covers the top edge of the header casing. It extends up the sidewall above the window.

in combination with other types. In addition, fixed windows are manufactured in many geometric shapes. Arch windows have a curved top or head that make them well suited to be joined in combination with a number of other types of windows or doors (Fig. 13-9).

Windows may be assembled or combined with other types of windows in a great variety of shapes. All of the windows mentioned come in a variety of sizes. With so many shapes and sizes, hundreds of interesting and pleasing combinations can be made.

## SINGLE- AND DOUBLE-HUNG WINDOWS

The *single-hung* window consists of two sashes; the upper one is fixed, and the bottom one slides vertically. *Double-hung windows* are similar to the single-hung except that both sashes slide vertically past each other in separate channels of the side jambs (Fig. 13-10). The sashes slide in channels

**FIGURE 13-9** Windows come in a variety of shapes and sizes.

**FIGURE 13-11** Double-hung windows may tilt in for easy cleaning.

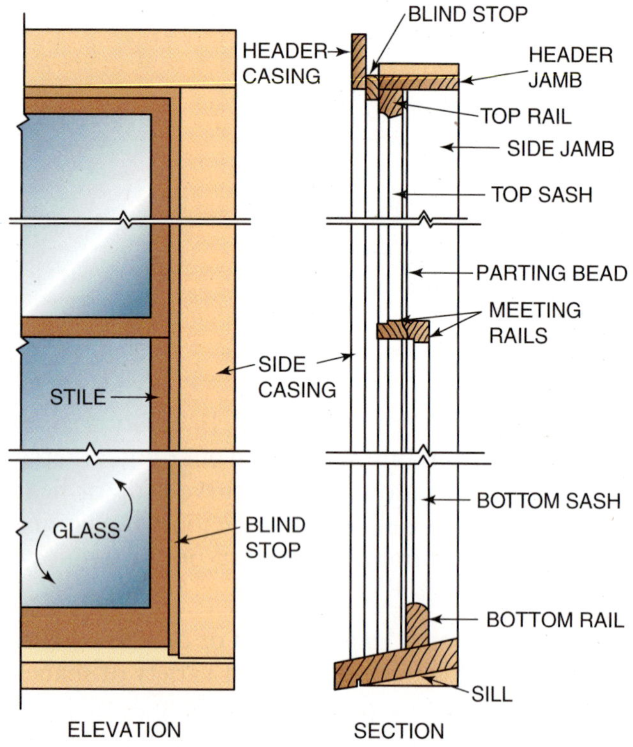


**FIGURE 13-10** The double-hung window and its parts.

that are fixed to the frames. Each sash is provided with springs and compression **weather stripping** to hold it in place in any position. Compression weather stripping holds the sash firmly, preventing air infiltration.

Some manufactured types provide for easy removal of the sash for painting, repair, and cleaning (Fig. 13-11). When the sashes are closed, specially shaped *meeting rails*, located in the middle of the window, come together to make a weather-tight joint. Sash locks are located at this point and not only lock the window but also draw the rails tightly together.

Double-hung windows can be arranged in a number of ways. They can be installed side by side in multiple units or in combination with other types. They can be used in a **bay window** unit.

## CASEMENT WINDOWS

The *casement window* consists of a sash hinged at the side. It swings outward by means of a crank or lever. An advantage of the casement type is that the entire sash can be opened for maximum ventilation. Figure 13-12 shows the use of casement windows over a kitchen sink.

## SLIDING WINDOWS

*Sliding windows* have sashes that slide horizontally in separate tracks located on the header jamb and

Courtesy of Andersen Windows, Inc.

**FIGURE 13-12** Casement windows swing outward.

sill (Fig. 13-13). When a window-wall effect is desired, many units can be placed side by side. Most units come with all necessary hardware applied.

## AWNING AND HOPPER WINDOWS

An *awning window* unit consists of a frame in which a sash hinged at the top swings outward by means of a crank or lever. A similar type, called the **hopper window**, is hinged at the bottom and swings inward.

Each sash is provided with an individual frame so that many combinations of width and height can be used. These windows are often used in combination with other types (Fig. 13-14).

## SKYLIGHT AND ROOF WINDOWS

*Skylights* provide light only. *Roof windows* contain an operating sash to provide light and ventilation (Fig. 13-15). One type of roof window comes with a tilting sash that allows access to the outside surface for cleaning. Special flashings are used when multiple skylights or roof windows are ganged together.

Courtesy of Andersen Windows, Inc.

**FIGURE 13-13** The sashes in sliding windows move horizontally by each other.

Courtesy of Andersen Windows, Inc.

**FIGURE 13-14** Awning windows are often used in stacks or in combination with other types of windows.

Courtesy of Andersen Windows, Inc.

**FIGURE 13-15** Skylights and roof windows are made in a number of styles and sizes.

Courtesy of Typar.

**FIGURE 13-16** Housewrap is widely used as an air infiltration barrier on sidewalls.

# Window Installation

Windows come in a variety of styles and shapes. Each must be carefully installed to ensure that weather is kept outside.

## HOUSEWRAP

Before windows are installed, the exterior walls are covered with a building **housewrap**. It gets its name because it completely wraps the building. Housewrap covers corners, window and door openings, plates, and sills. It allows for a quick drying-in of the building and is designed to prevent the infiltration of air into the structure. Yet, at the same time, water vapor is allowed to escape. It is a thin, tough plastic material used to cover the sheathing on exterior walls (Fig. 13-16). Housewraps are commonly known by the brand names of Typar® and Tyvek®. Housewrap comes in long rolls that are 1½, 3, 4½, 5, 9, and 10 feet wide. To get the most out of housewrap, all seams should be taped with sheathing tape. The overall air tightness of the house depends heavily on the application of seam tape on the seams and around windows and doors.

**CAUTION**

**CAUTION:** Housewraps are slippery. They should not be used in any application where they will be walked on.

## INSTALLING WINDOWS

Numerous window manufacturers produce hundreds of kinds, shapes, and sizes of windows. Because of the tremendous variety and design differences, follow the manufacturer's instructions closely to ensure correct installation. Directions in this chapter are basic to most window installations, but they are intended as a guide to be supplemented by procedures recommended by the manufacturer.

The numbers or letters found in the floor plan identify the window in more detail in the *window* schedule. This is usually part of a set of plans (Fig. 13-17). This schedule normally includes window style, size, manufacturer's name, and unit number. Rough opening sizes may or may not be shown. Units are identified only by the manufacturer's name and number on the floor plan. In order to gain more information, the builder must refer to the window *manufacturer's catalog* (Fig. 13-18).

**Window Schedule**

| Ident. | Quan. | Manufacturer | Size | Remarks |
|---|---|---|---|---|
| A | 6 | Andersen | TW28310 | D.H. Tiltwash |
| B | 1 | Andersen | WDH2442 | Woodwright D.H. |
| C | 2 | Andersen | 3062 | Narrowline D.H. |
| D | 1 | Andersen | CW24 | Casement Single |
| E | 1 | Andersen | C34 | Casement Triple |
| F | 1 | Andersen | C23 | Casement Double |

**FIGURE 13-17** Typical window schedule found in a set of plans.

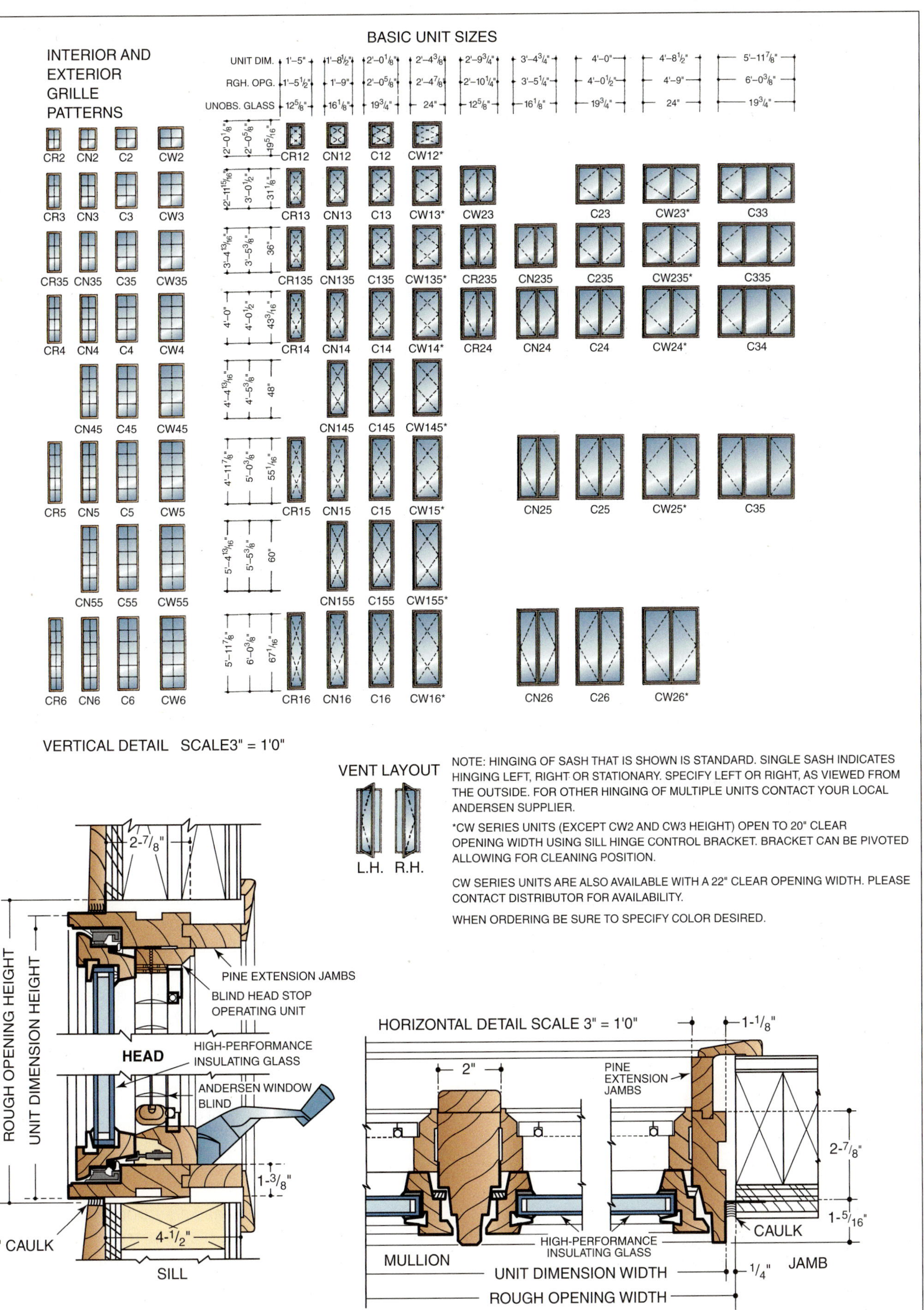


*Courtesy of Andersen Windows, Inc.*

**FIGURE 13-18** Typical page from a window manufacturer's catalog for casement windows.

Use galvanized casing or roofing nails spaced about 16 inches apart or directed by local codes. Nail length depends on the thickness of the casing and sheathing. Nails should be long enough to penetrate the sheathing and into the framing members. Vinyl-clad windows have a vinyl nailing flange. Large-head roofing nails are driven through the flange instead of through the casing (Fig. 13-19). The installation of windows in masonry and brick-veneered walls is similar to that of frame walls. Adequate clearance should be left for caulking around the entire perimeter between the window and masonry (Fig. 13-20).

**Green Tip**

Flashing is a vital, yet simple component of a building. Sealants are not a substitute for proper flashing.

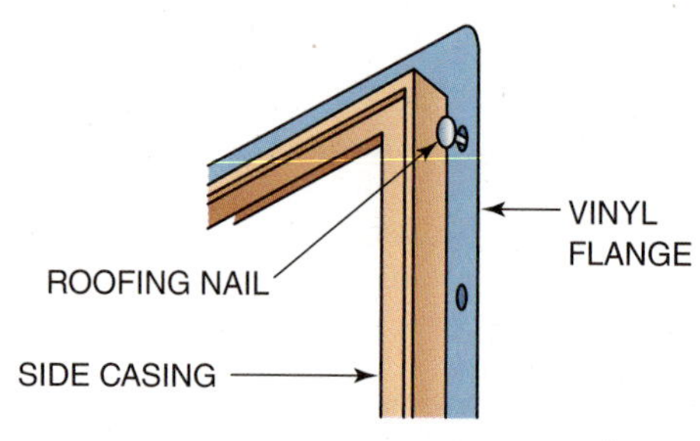


**FIGURE 13-19** Roofing nails are used to fasten the flanges of vinyl-clad windows.

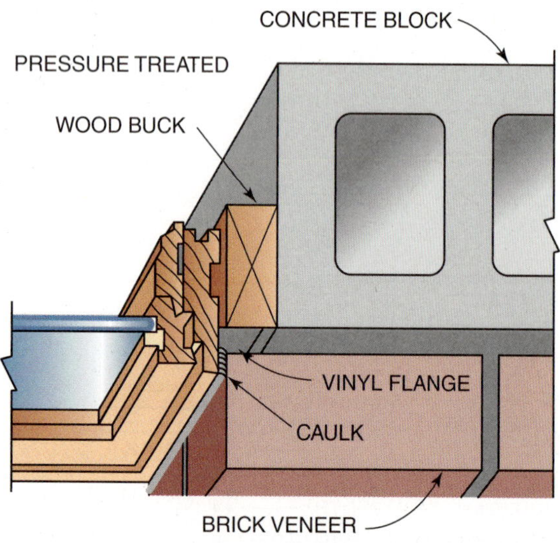


**FIGURE 13-20** Windows are installed in masonry openings against wood bucks.

***For step-by-step instructions on INSTALLING WINDOWS, see Procedure 13-1 on pages 422–423.***

# Doors

There are a variety of doors used in buildings. Some are designed to showcase the building entrance. Many are named according to the way they open.

## DOOR STYLES AND SIZES

Exterior doors made as flush or panel doors are available in many styles. Interior doors are classified by style as flush, panel, French, louver, and cafe doors. Flush doors have a smooth surface, are usually less expensive, and are widely used when a plain appearance is desired. Some of the other styles have special uses. Fire doors are used to slow the spread of fire for a certain period of time. Insulated doors have thicker panels with Low E or argon-filled insulated glass. Doors are also classified by the way they operate, such as *swinging, sliding,* or *folding.*

Most exterior entrance doors are manufactured in a thickness of 1¾ inches, in widths of 2'-8" and 3'-0", and in heights of 6'-8" and 7'-0". Most interior doors are manufactured in 1⅜-inch thickness. Some are also made in 1¼- and 1⅛-inch thicknesses. Most residential doors are manufactured in 6'-8" heights. Door widths range from 1'-0" to 3'-0" in increments of 2 inches. A swinging door is designated as being *right-hand* or *left-hand,* depending on the direction it swings. The designation is determined in several ways. One method is to stand on the side of the door that swings away from the viewer. It is a right-hand door if the hinges are on the viewer's right. It is a left-hand door if the hinges are on the left (Fig. 13-21).

### Flush Doors

An exterior flush door has a smooth, flat surface of wood veneer or metal. It has a framed, solid core of staggered wood blocks or composition board (Fig. 13-22). Wood core blocks are inserted in appropriate locations in composition cores. They serve as *backing* for door locks.

Interior flush doors are made with solid or hollow cores. Solid core doors are generally used as

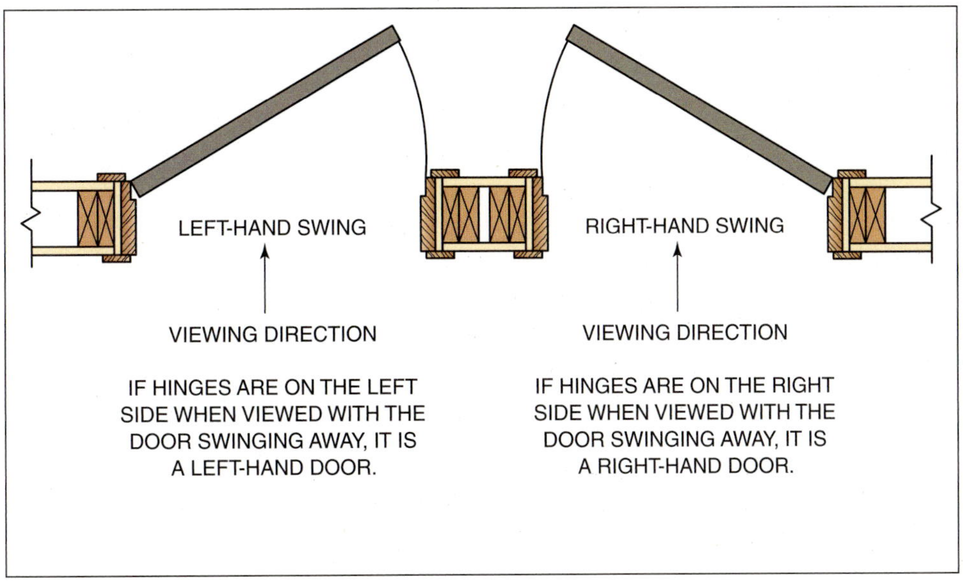


**FIGURE 13-21** Determining the swing of a door.

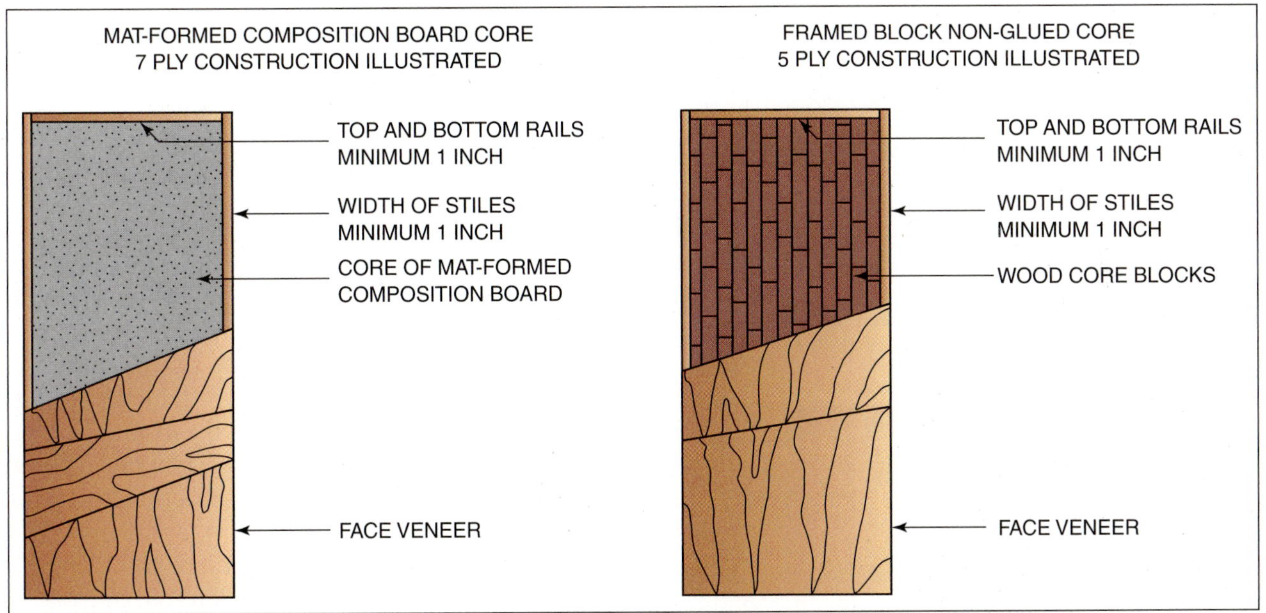


*Courtesy of National Wood Window and Door Association.*

**FIGURE 13-22** Solid wood or composition cores are required in flush doors.

fire-rated doors. Interior hollow core doors are commonly used in the interior except when fire resistance or sound transmission is critical. A hollow core door consists of a light perimeter frame. This frame encloses a mesh of thin wood or composition material supporting the faces of the door. Solid wood blocks are appropriately placed in the core for the installation of locksets. The frame and mesh are covered with a thin plywood called a *skin. Lauan* plywood is commonly used for flush door skins. Flush doors are also available with veneer faces of birch, gum, oak, and mahogany, among others (Fig. 13-23).

***For step-by-step instructions on HANGING INTERIOR DOORS, see Procedure 13-2 on pages 424–435.***

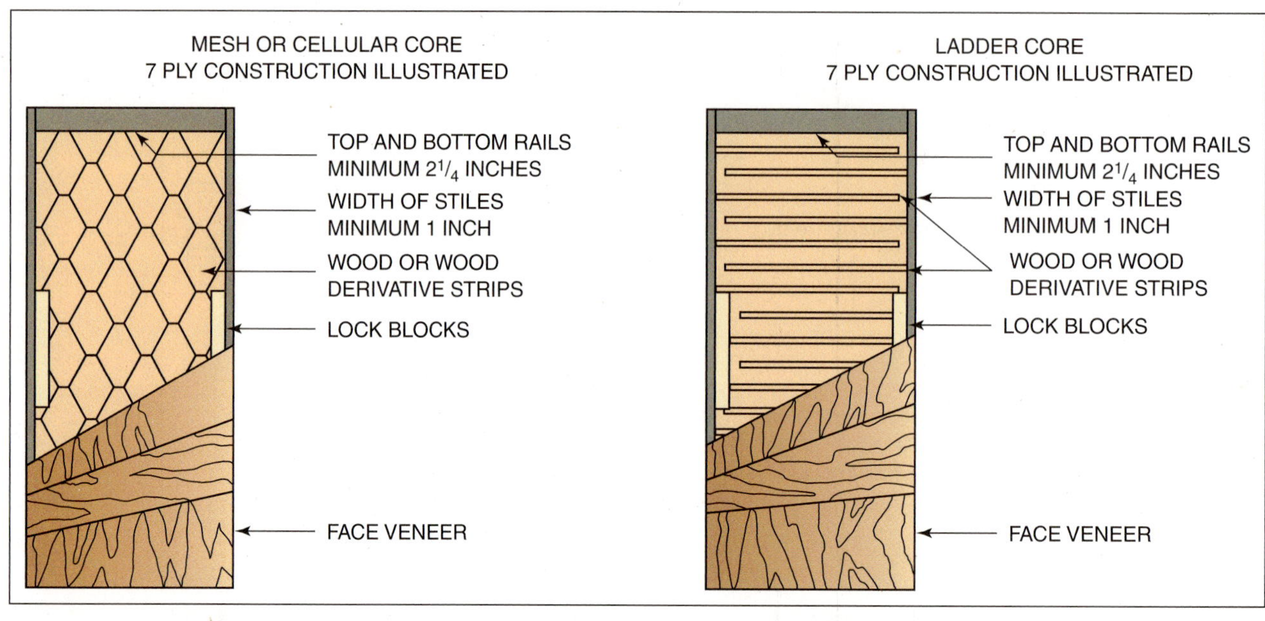


**FIGURE 13-23** The construction of hollow core flush doors.

## Panel Doors

Panel doors are generally classified as *high-style, panel, sash, fire, insulated, French, cafe,* and *ventilating* doors. *Sidelights,* although not actually doors, are part of some entrances.

They are fixed in the door frame on one or both sides of the door (Fig. 13-24). *Transoms,* similar to sidelights, are installed above the door. Interior panel doors consist of a frame with usually one to eight wood panels in various designs (Fig. 13-25).

**FIGURE 13-24** Sidelights are installed on one or both sides of the main entrance door.

*Parts of a Panel Door.* A panel door consists of a frame that surrounds panels of solid wood and glass or louvers. Some door parts are given the same terms as a window sash. The outside vertical members are called stiles. Horizontal frame members are called *rails.* The top *rail* is generally the same width as the stiles. The *bottom rail* is the widest of all rails. A rail situated at lockset height, usually 38 inches from the finish floor to its center, is called the *lock rail.* Almost all other rails are called *intermediate rails.* **Mullions** are vertical members between rails dividing panels in a door. *Bars* are narrow horizontal or vertical rabbeted members. They extend the total length or width of a glass opening from rail to rail or from stile to stile. Door *muntins* are short members, similar to bars dividing the overall length and width of the panel or glass area into smaller pieces. *Panels* fit between and are usually thinner than the stiles, rails, and mullions. They may be *raised* on one side or on both sides for improved appearance (Fig. 13-26).

## High-Style Doors

*High-style doors,* as the name implies, are highly crafted designer doors. They may have a variety of cut-glass designs, made with raised panels of various shapes (Fig. 13-27).

Courtesy of Morgan Manufacturing / Jeld-Wen Inc.

**FIGURE 13-25** Styles of commonly used interior panel doors.

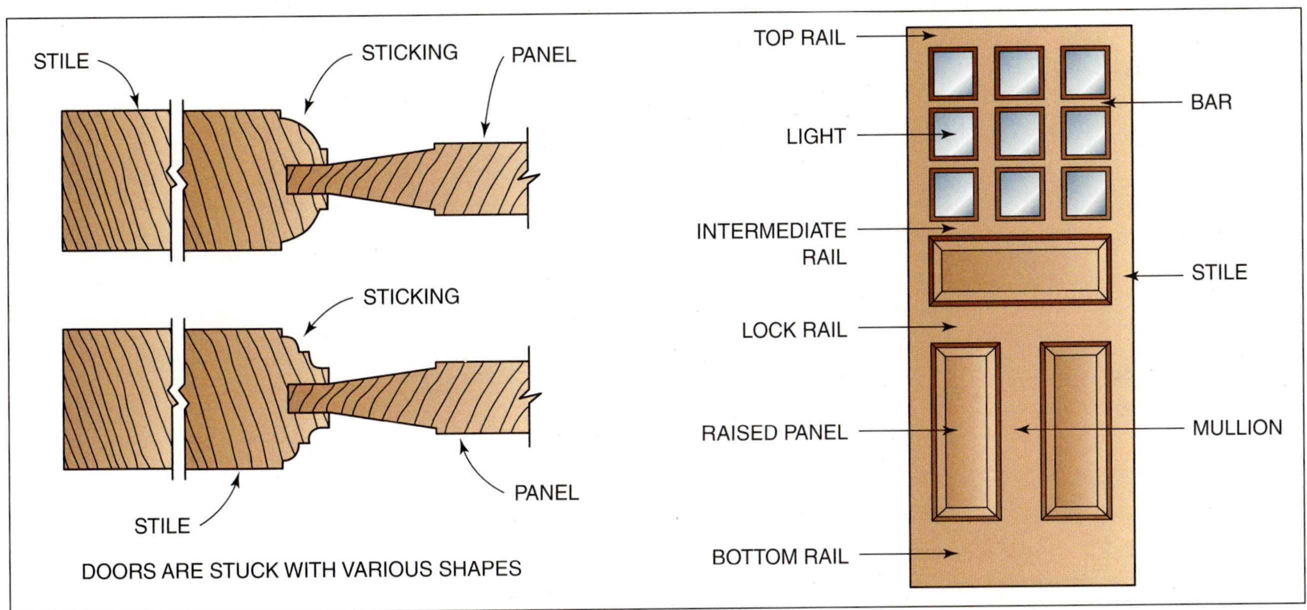


**FIGURE 13-26** The parts of a panel door.

### French Doors

*French doors* may contain from 1 to 15 lights of glass. Exterior doors are made an 1¾-inch thick and interior doors are 1⅜-inch thick (Fig. 13-28).

### Louver Doors

*Louver doors* are made with spaced horizontal slats called louvers used in place of panels. The louvers are installed at an angle to obstruct vision but permit the flow of air through the door. Louvered doors are widely used on clothes closets (Fig. 13-29).

### Cafe Doors

*Cafe doors* are short panel or louver doors. They are hung in pairs that swing in both directions. They are used to partially screen an area, yet allow easy and safe passage through the opening. The tops and

EXTERIOR PANEL DOORS

EXTERIOR SASH DOORS

SIDELIGHTS

*(a), (b) Courtesy of Morgan Manufacturing / Jeld-Wen Inc.*

**FIGURE 13-27** Several kinds of exterior doors are made in many designs.

*Courtesy of Morgan Manufacturing / Jeld-Wen Inc.*

**FIGURE 13-28** French doors are used in the interior as well as for entrances.

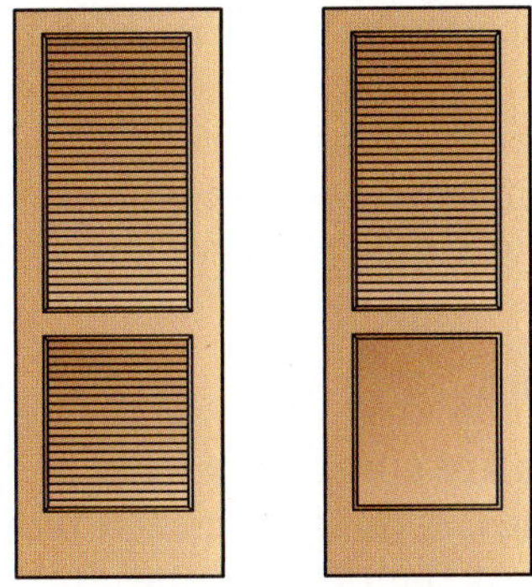

Courtesy of Morgan Manufacturing / Jeld-Wen Inc.

**FIGURE 13-29** Louver doors obstruct vision but permit the circulation of air.

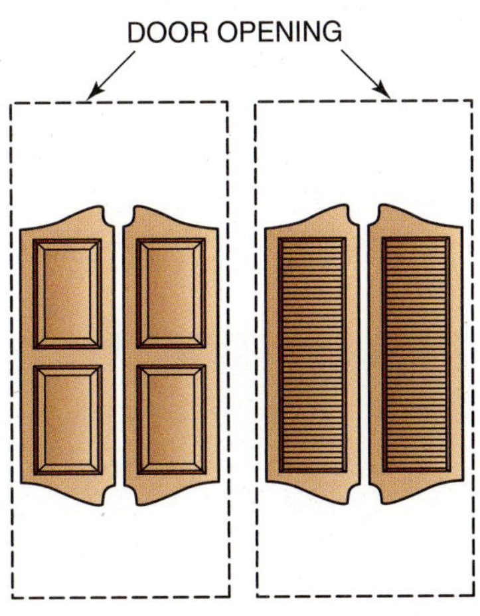


Courtesy of Morgan Manufacturing / Jeld-Wen Inc.

**FIGURE 13-30** Cafe doors usually are used between kitchens and dining areas.

bottoms of the doors are usually shaped in a pleasing design (Fig. 13-30).

## METHODS OF DOOR OPERATION

Doors are also identified by their method of operation. Doors either swing on hinges or slide in tracks. The choice of door operation depends on factors such as convenience, cost, safety, and space.

### Swinging Doors

*Swinging doors* are most common as they are hinged on one edge. They swing out of the opening and cover the total opening when closed (Fig. 13-31). Swinging doors that swing in one direction are called *single-acting*. With special hinges, interior doors can swing in both directions. They are then called **double-acting** doors.

### Patio Doors

*Patio door* units normally consist of two or three full glass panels completely assembled in a frame. Typically, only one of the panels will move as a sliding or swinging glass door. The units may range from 5'-6" to 12'-0" in width (Fig. 13-32). Instructions for assembly are included with the unit. They should be followed carefully. Installation of patio door frames is similar to setting any door. After the frame is set, the doors are installed using special hardware supplied with the unit.

**Green Tip**

Hinged doors and windows are more airtight than sliding ones because they compress and seal to the gaskets better.

**FIGURE 13-31** A single-acting swinging door is the most widely used type of door.

Courtesy of Andersen Windows, Inc.

**FIGURE 13-32** Two or three doors usually are used in sliding- or swinging-type patio door units.

## Double Doors

*Double doors* are fitted and hung in a similar manner to single doors. Allowance must be made, when fitting swinging double doors, for an astragal between them for weathertightness (Fig. 13-33). An astragal is a molding that is rabbeted on both edges. It is designed to cover the joint between double doors. One edge has a square rabbet. The other has a beveled rabbet to allow for the swing of one of the doors.

## Bypass Doors

*Bypass doors* are commonly used on wide clothes closet openings. A double track is mounted on the header jamb of the door frame. Rollers that ride in the track are attached to the doors so that they slide by each other. A floor guide keeps the doors in alignment at the bottom (Fig. 13-34). Usually two

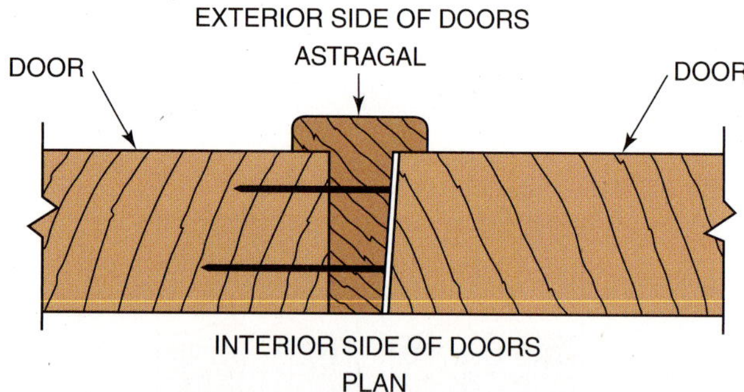

Courtesy of Andersen Windows, Inc.

**FIGURE 13-33** An astragal is required between double doors for weathertightness.

**FIGURE 13-34** Bypass doors are used on wide closet openings.

**FIGURE 13-35** The pocket door slides into the interior of the partition.

doors are used in a single opening. A disadvantage with this style of door is that only one-half of the total opening is accessible at a time.

## Pocket Doors

The *pocket door* is opened by sliding it sideways into the interior of the partition. When opened, only the lock edge of the door is visible (Fig. 13-35). Pocket doors may be installed as a single unit, sliding in one direction, or as a double unit, sliding in opposite directions. When opened, the total width of the opening is obtained, and the door does not project out into the room. The installation of pocket doors requires more time and material than other methods of door operation.

A special pocket door frame unit and track must be installed during the rough framing stage. The rough opening in the partition must be large enough for the door opening and the pocket.

***For step-by-step instructions on INSTALLING POCKET DOORS, see Procedure 13-2 on page 433.***

## Bifold Doors

*Bifold doors* are made in narrower widths than other doors. This allows them to operate in a folding fashion on closet and similar type openings (Fig. 13-36). Bifold doors consist of pairs of doors hinged at their edges. The doors on the jamb side swing on pivots installed at the top and bottom. Other doors fold up against the jamb door as it is swung open. The end door has a guide pin installed at the top. The pin rides in a track to guide the set when opening or closing (Fig. 13-37). On very wide openings, the guide pin is replaced by a combination guide and support to keep the doors from sagging.

Courtesy of Morgan Manufacturing / Jeld-Wen Inc.

**FIGURE 13-36** Bifold doors are manufactured in many styles.

Bifold doors may be installed in double sets, opening and closing from each side of the opening. They have the advantage of providing access to almost the total width of the opening, yet they do not project out much into the room.

**FIGURE 13-37** Bifold doors provide access to almost the total width of the opening.

***For step-by-step instructions on INSTALLING BIFOLD DOORS, see Procedure 13-2 on pages 431–432.***

# Parts of a Door Frame

Terms given to members of an exterior door frame are the same as similar members of a window frame. The bottom member is called a *sill* or *stool*. The vertical side members are called *side jambs*. The top horizontal part is a *head jamb*. The exterior door trim may consist of many parts to make an elaborate and eye-appealing entrance or a few parts for a simpler doorway. The *door casings*, if not too complex, are usually applied to the door frame before it is set (Fig. 13-38). When more intricate trim is specified, it is usually applied after the frame is set (Fig. 13-39).

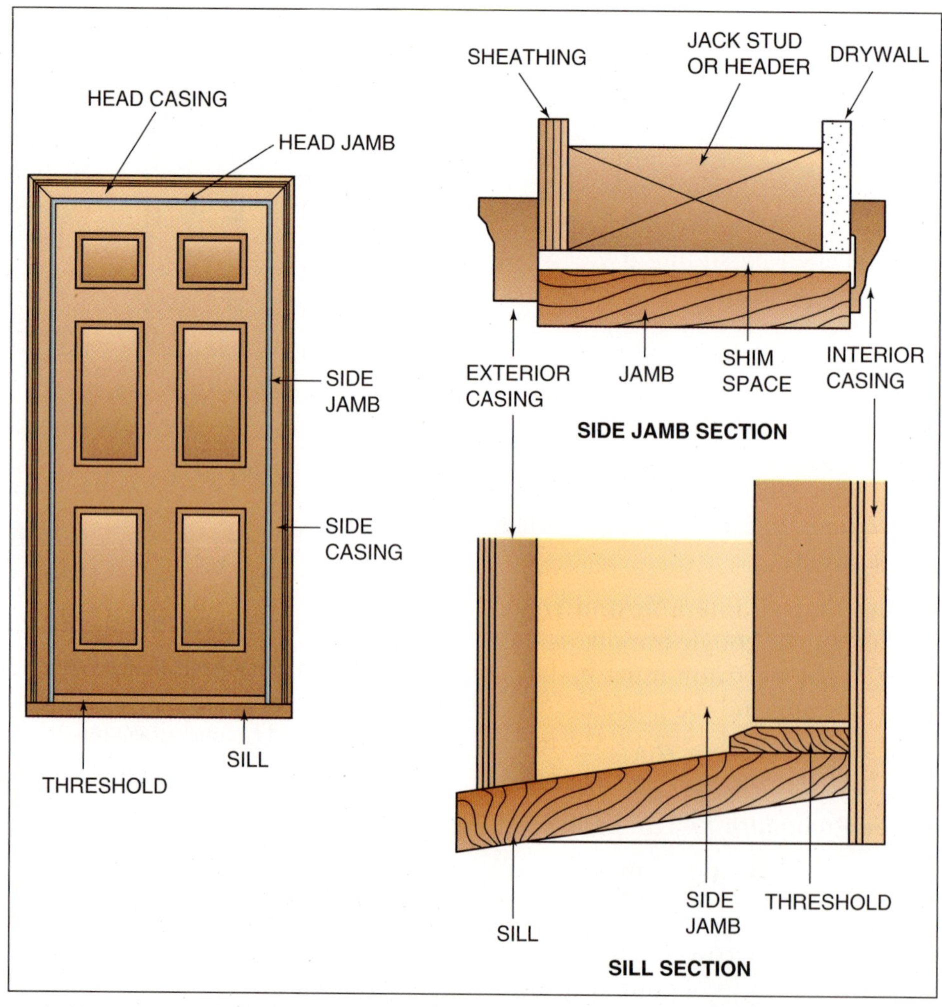


**FIGURE 13-38** Parts of an exterior door frame.

Courtesy of Morgan Manufacturing / Jeld-Wen Inc.

**FIGURE 13-39** Elaborate entrance trim is available in knocked-down form, ready for assembly.

## SILLS

In residential construction, door frames usually are designed and constructed for entrance doors that swing inward. Codes require that doors swing outward in buildings used by the general public and high-wind zones. The shape of a wood door sill for an in-swinging door is different from that for an out-swinging door (Fig. 13-40). Extruded aluminum sills of many styles come with vinyl inserts to weather-strip the bottom of the door. Most are adjustable for exact fitting between the sill and door (Fig. 13-41).

## JAMBS

*Jambs* are square edge pieces of stock to which hinges and *door stops* are applied. Side and header jambs are the same shape. Several jamb widths are available for different wall thicknesses. For walls of odd thicknesses, jambs may be ripped to any desired width.

## CASINGS

*Casing* is the molding applied around the outside edges of the jamb. It is installed to cover the shim

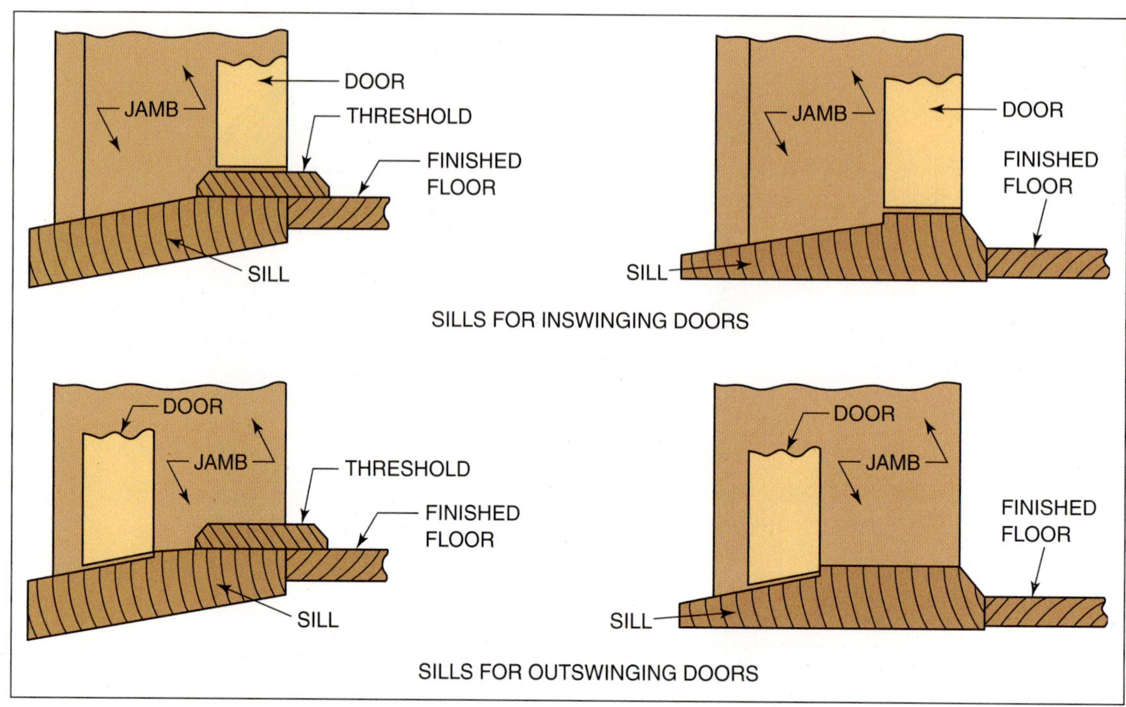


**FIGURE 13-40** Wood sill shapes vary according to the swing of the door.

**FIGURE 13-41** Some metal sills are adjustable for exact fitting at the bottom of the door.

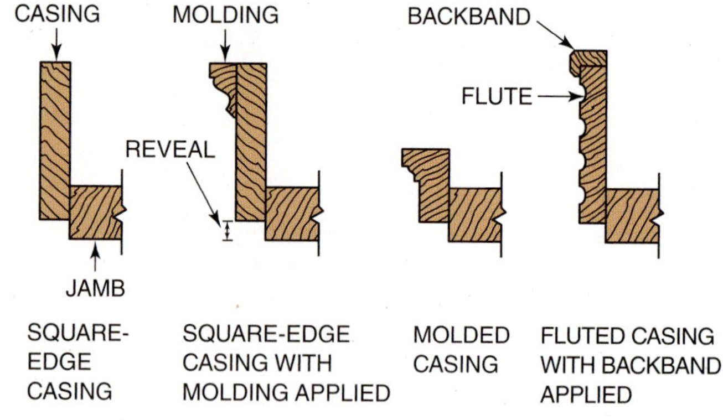


**FIGURE 13-42** Exterior door casings may be enhanced by applying moldings and by shaping.

space and joint between the jamb and the wall surface. Exterior casing is made with thicker material than is interior casing. Exterior casing may be enhanced with fluted, or otherwise shaped, pieces and appropriate caps and moldings applied (Fig. 13-42). *Flutes* are narrow, closely spaced, concave grooves that run parallel to the edge of the trim. In addition, ornate main entrance trim may be purchased in knocked-down form. It is then assembled at the job site (Fig. 13-43). Interior casing is usually the same shape throughout the house.

Courtesy of Morgan Manufacturing / Jeld-Wen Inc.

**FIGURE 13-43** A few samples of manufactured entrance door trim. Many other styles are available.

# SETTING A PRE-HUNG DOOR FRAME

A pre-hung, single-acting, hinged door unit consists of a door frame with the door hinged and casings installed to one side. Holes are provided if the locksets have not already been installed. Small temporary cardboard shims are stapled to the lock edge and top end of the door to maintain proper clearance between the door and frame. Pre-hung units are available in several jamb widths to accommodate various wall thicknesses. Some pre-hung units have split jambs that are adjustable for varying wall thicknesses (Fig. 13-44). A pre-hung door unit can be set in a matter of minutes.

***For step-by-step instructions on SETTING A PRE-HUNG DOOR FRAME, see Procedure 13-2 on pages 424–426.***

## SETTING DOOR FRAMES IN MASONRY WALLS

Some exterior door frames are metal and positioned before masonry walls are built. The frames must be set and firmly braced in a level and plumb position. The head jamb is checked for level. The bottom ends of the side jambs are secured in place. It may be necessary to shim one side of the jamb in order to level the head jamb. The side jambs are then plumbed in a sideways direction. They are braced in position. Then, the frame is plumbed and braced at a right angle to the wall (Fig. 13-45).

Finally, the frame is checked to see if it has a wind. A **wind** is a twist in the door frame caused when the side jambs do not line up vertically with each other. No matter how carefully the side jambs of a door frame are plumbed, it is always best to check the frame to see if it has a wind.

One method of checking the door frame for a wind is to stand to one side. Sight through the door frame to see if the outer edge of one side jamb lines up with the inner edge of the other side jamb. If they do not line up, the jambs are not plumb or the frame has a wind. A wind may also be checked by stretching two lines diagonally from the corners of the frame. If both lines meet accurately at their intersections, the frame does not have a wind (Fig. 13-46).

# DOOR FITTING AND HANGING

Fitting and hanging of wood doors is still an important part of the carpentry trade despite the increasing use of pre-hung door units. Many situations in new construction and in remodeling work require fitting and hanging doors to preexisting finished openings.

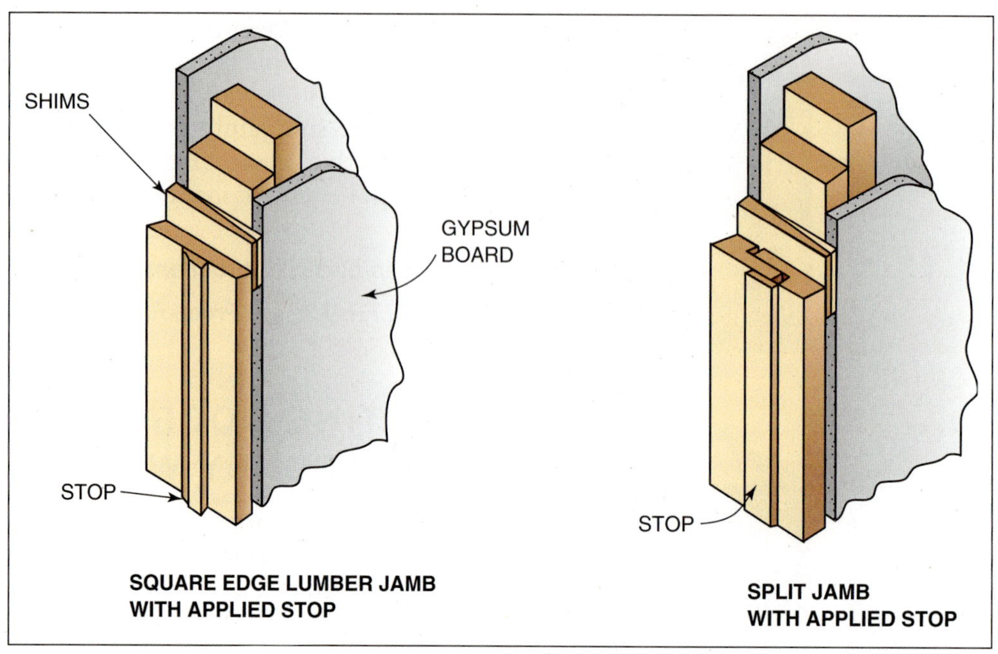


**FIGURE 13-44** Pre-hung door units come with solid or split jambs.

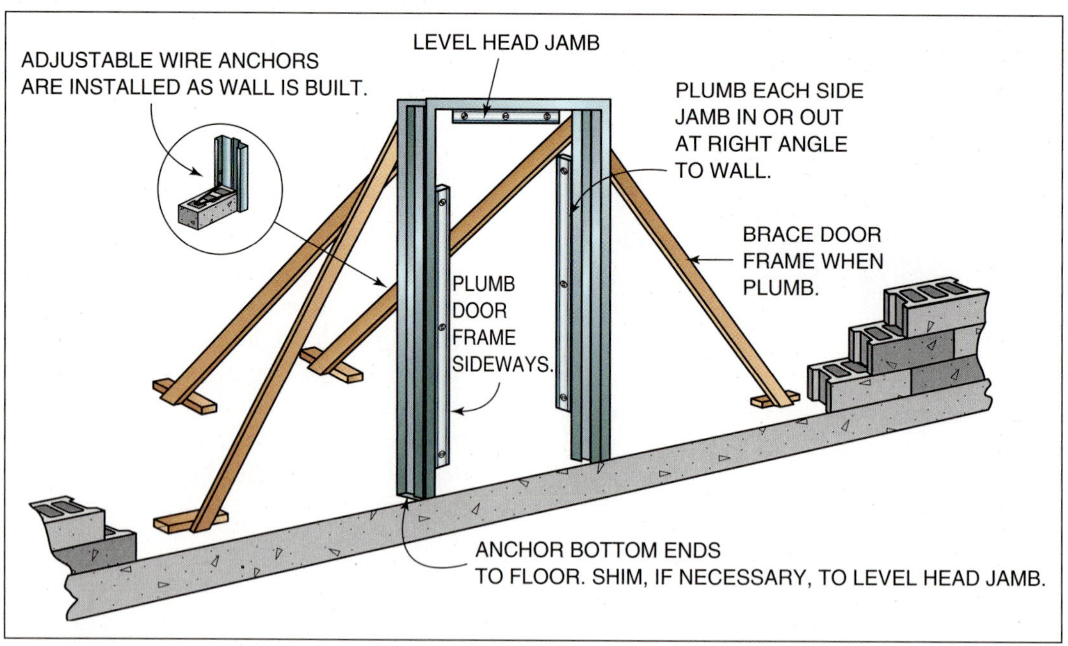


**FIGURE 13-45** Installation of an exterior door frame in a masonry wall.

## FITTING A DOOR TO A DOOR FRAME

The process of fitting a door to a preexisting frame involves many steps. Doors are made with a beveled edge that allows the door to open without touching the jamb (Fig. 13-47). This is the edge where the lockset is installed. The squared edge is where the hinges are installed. Also, exterior doors have a face designed to be the exterior surface. It is important to hang exterior doors containing lights of glass with the proper side exposed to the weather.

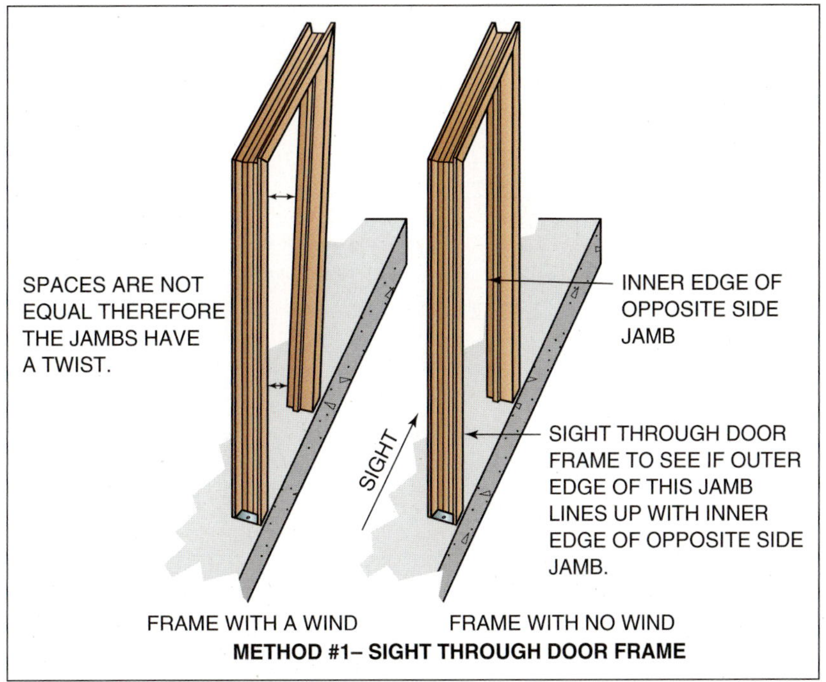


**FIGURE 13-46** Technique for checking for a wind or twist in a door frame.

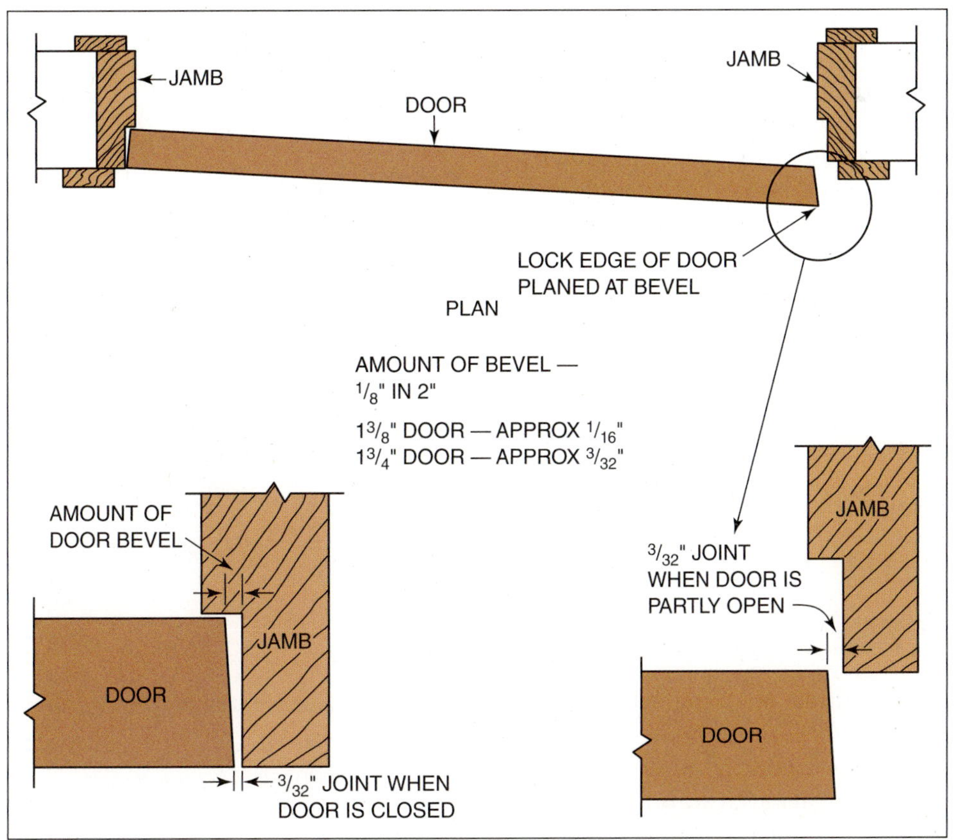


**FIGURE 13-47** The lock edge of a door must be planed at a bevel to clear the jamb when opened.

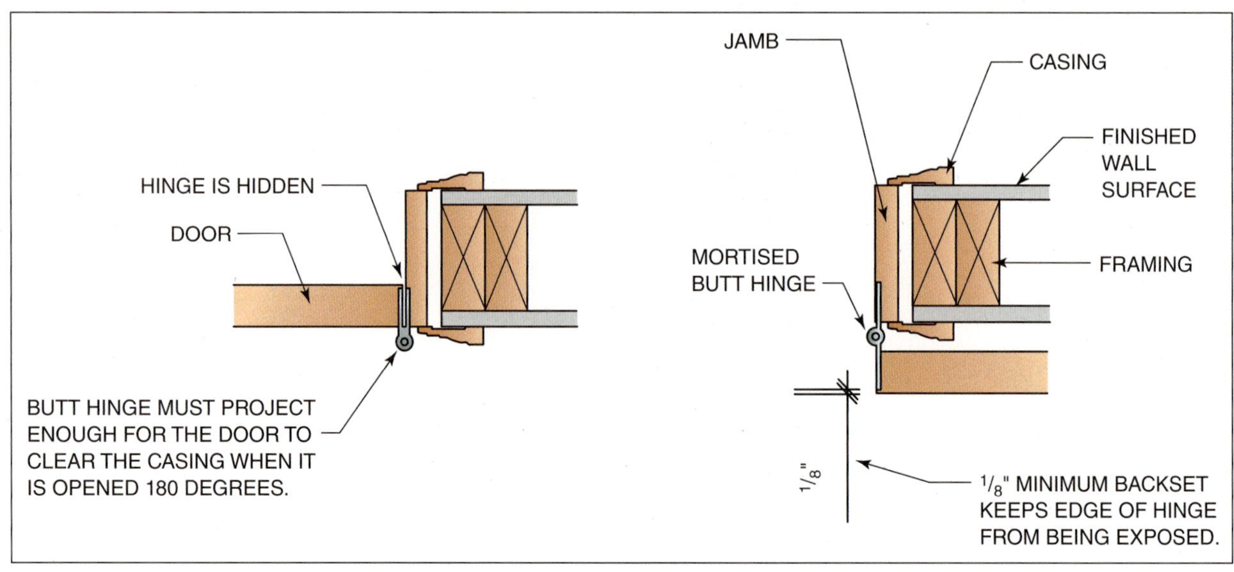


**FIGURE 13-48** Hinges are set back from the side of the door and the edge of the door stop.

*Loose-pin butt hinges* are used for most doors, except for security reasons where the pins are not removable. For loose-pin hinges, each *leaf* of the hinge is applied separately to the door and frame. Care must be taken so that the hinge leaves line up exactly on the door and frame. The hinge leaves are recessed flush with the door edge. The recess for the hinge is called a *hinge gain* or *hinge mortise*. Hinge gains are made only partway across the edge of the door. This is so that the edge of the hinge is not exposed when the door is closed (Fig. 13-48).

***For step-by-step instructions on FITTING A DOOR TO A FRAME, see Procedure 13-2 on pages 426–428.***

# Installing Exterior Door Locksets

After the door has been fitted and hung in the frame, the *lockset* and other door hardware are installed. Many types of locks are available from several large manufacturers in numerous styles and qualities, providing a wide range of choices.

## CYLINDRICAL LOCKSET

Cylindrical locksets are the most commonly used type in both residential and commercial construction (Fig. 13-49). In place of knobs, *lever handles* are provided on locksets likely to be used by handicapped persons or in other situations where a lever is more suited than a knob (Fig. 13-50). **Deadbolt** *locks* are used for both primary and auxiliary locking of doors (Fig. 13-51). They provide additional security. They also make an attractive design in combination with grip handle locksets or latches (Fig. 13-52).

The kind of trim and finish are factors, in addition to the kind and style, which determine the

*Courtesy of Schlage Lock*

**FIGURE 13-49** Cylindrical locksets are the most commonly used type in both residential and commercial construction.

Courtesy of Schlage Lock.

**FIGURE 13-50** Locks with lever handles are used when difficulty in turning knobs is expected.

Courtesy of Schlage Lock.

**FIGURE 13-51** Deadbolt locks are used primarily as auxiliary locks for added security.

Courtesy of Schlage Lock.

**FIGURE 13-52** A grip-handle lockset combines well with a deadbolt lock.

quality of a lockset. **Escutcheons** are decorative plates of various shapes installed between the door and the lock handle or knob. Locksets and escutcheons are available in various metals and finishes.

Interior door locksets include *privacy* locks and *passage* locks. A privacy lock is often used on bathroom and bedroom doors. It is locked by pushing or turning a button on the room side. On most privacy locks, a turn of the knob on the room side unlocks the door. On the opposite side, the door can be unlocked by a pin inserted into a hole in the knob. The *passage* lockset has knobs on both sides that are turned to unlatch the door. This lockset is used when locking the door is not necessary.

***For step-by-step instructions on INSTALLING CYLINDRICAL LOCKS, see Procedure 13-2 on pages 433–435.***

## Garage Doors

*Overhead garage doors* come in many styles, kinds, and sizes. Two popular kinds used in residential construction are the one-piece and the sectional door.

The rigid one-piece unit swings out at the bottom. It then slides up and overhead. The sectional type has hinged sections. These sections move upward on rollers and turn to a horizontal position overhead. A *rolling steel door,* used mostly in commercial construction, consists of narrow metal slats that roll up on a drum installed above the opening.

Special hardware, required for all types, is supplied with the door. Equipment is available for power operation of garage doors, including remote control. Also supplied are the manufacturer's directions for installation. These should be followed carefully. There are differences in the door design and hardware of many manufacturers.

**CAUTION**

**CAUTION:** The springs used to assist the raising of the door can be under a great tension. Care should be taken when working with them. Injury may occur if they should suddenly come loose.

# SUMMARY


- Windows have many parts and pieces that make up a window unit.
- Window styles vary in their operation and appearance.
- Window installation properly done will provide a watertight unit as well as protection from storms.
- Doors styles are varied for function and architectural appeal.
- Parts of a door provide function and decoration to the opening.
- Setting a pre-hung door properly requires carpenters to follow a well worked out procedure.
- Door fitting and hanging are necessary to hang a door in an existing door opening.
- Installation of exterior door locksets depends on the type of lockset.
- Garage doors installation follow the manufacturer's recommendation.

## GREEN CHECKLIST

- ☐ Window performance is only as good as the installation. Follow manufacturer's recommendations. Incorrect installations will void warranties.
- ☐ Flashing is a vital, yet simple component of a building. Sealants are not a substitute for proper flashing.
- ☐ Hinged doors and windows are more airtight than sliding ones because they compress and seal to the gaskets better.

PROCEDURE 13-1

# Installing Windows

**A** Install housewrap by beginning at the building corner, holding the roll vertically on the wall. Unroll it a short distance. Make sure the roll is plumb. Secure the sheet to the corner, leaving about a foot extending beyond the corner to overlap later. Make sure the sheet is straight, with no buckles. Fasten every 12 to 18 inches. Cover window and door openings around the entire perimeter of the building. Overlap all joints by at least 3 inches. Secure them with housewrap tape. On horizontal joints, the upper layer should overlap the lower layer.

**B** Make cuts in the housewrap from corner to corner of rough openings. Fold the triangular flaps in and around the opening. Secure the flaps on the inside with fasteners spaced about every 6 inches.

**A**

Courtesy of Typar.

**B**

Courtesy of Typar.

Courtesy of Typar.

**C** Check the opening and window dimensions to verify the window will fit in the opening. Check the levelness of the sill. Apply a bead of caulk sealant to the RO perimeter where the window nailing flanges will seat. Remove all shipping protection from the window unit. Do not remove any diagonal or lateral braces applied at the factory. Windows can easily be moved through the openings from the inside and set in place.

**D** Place the window in the opening taking care to center the unit before bedding the unit in the caulk. Level the window sill with a wood shim tip between the rough sill and the bottom end of the window's side jamb, if necessary. Secure the shim to the rough sill.

- Nail the lower end of both sides of the window. Close and lock the sash, then plumb the sides of the unit. Fasten unit to the opening as per manufacturer's recommendations. Typically a fastener is required in every hole of the nailing flange. Check that the sash operates properly. If not, make the necessary adjustments.

**E** Flash the head casing by cutting to length the flashing with tin snips. Its length should be equal to the overall length of the window head casing. Slice the housewrap just above the head casing and slip the flashing behind the wrap and on top of the head casing. Secure with fasteners into the wall sheathing. Refasten the housewrap. Tape all seams in housewrap and over window nailing flanges.

**CAUTION**

**CAUTION: Have sufficient help when setting large units. Handle them carefully to avoid damaging the unit or breaking the glass. Broken glass can cut through protective clothing and cause serious injury.**

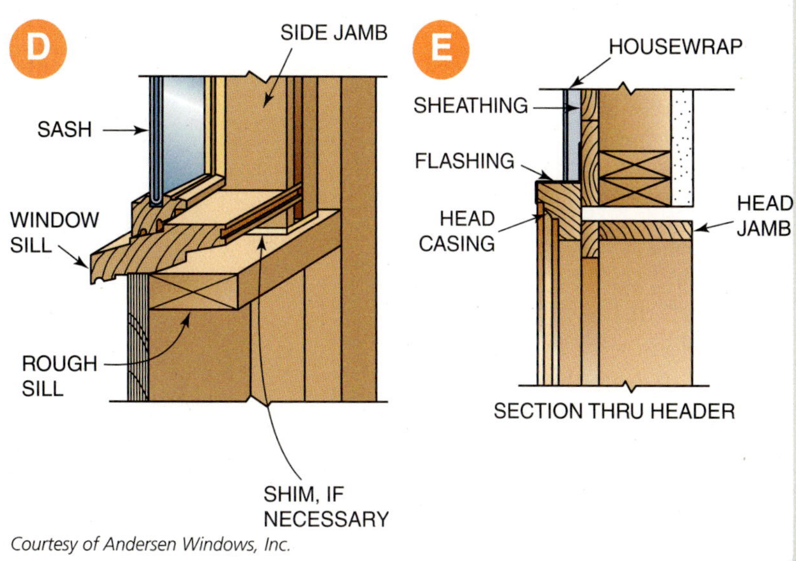


*Courtesy of Andersen Windows, Inc.*

PROCEDURE 13-2

# Hanging Interior Doors

## Setting a Pre-hung Door Frame

**A** Remove the protective packing from the unit. Leave the small fiber shims between the door and jambs to help maintain this space. Cut off the horns, if necessary. Remove the nail in jamb that holds the door closed.

- Set shim blocks under each jambs that are the same thickness as the finished floor. This will allow the finished floor to be installed under the jambs later. Of course, if the finished floor is already installed then omit these shims.
- Center the unit in the opening so the door will swing in the desired direction. Be sure the door is closed and spacer shims are still in place between the jamb and door.

**B** Level the head jamb. Make adjustments by shimming the jamb that is low so it brings the head jamb level. Adjust a scriber to the thickness of shim and scribe this amount off of the other jamb. Remove frame and cut the jamb. Note that the clearance under the door is being reduced by the amount being cut off.

A

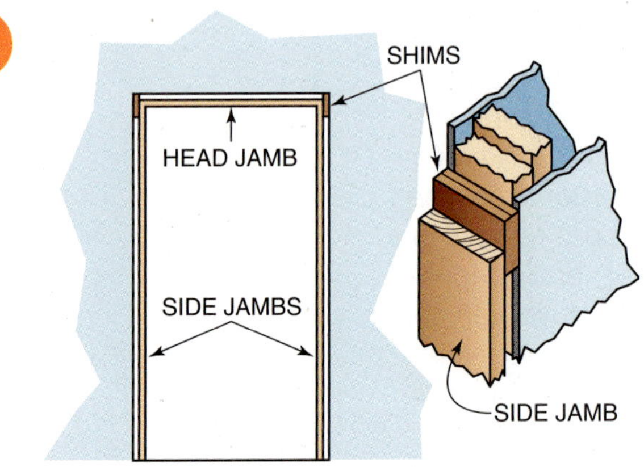


B

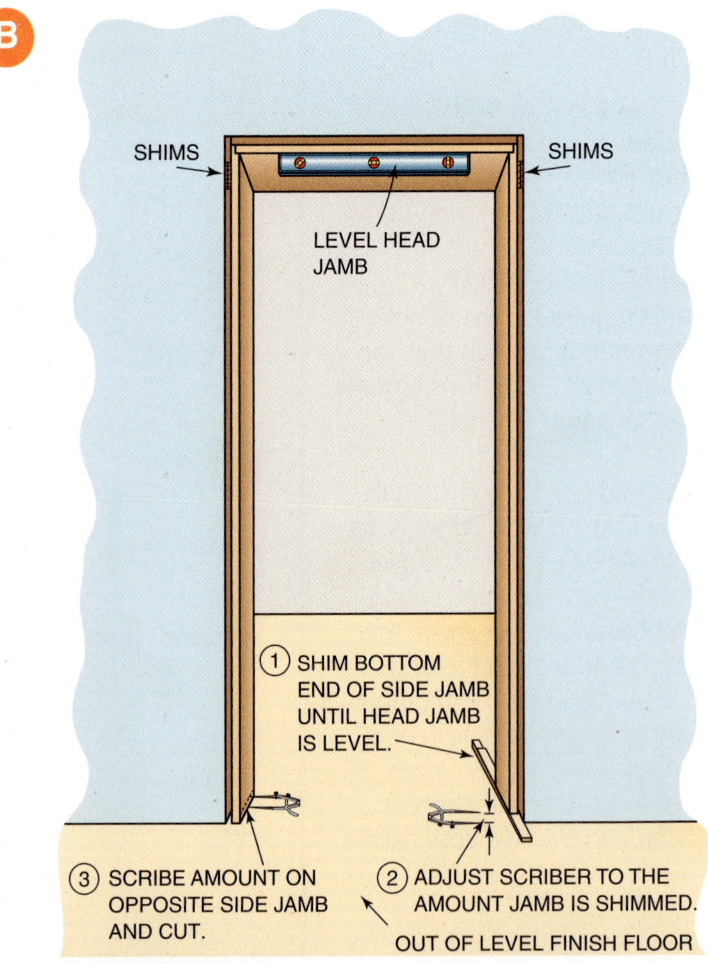

**C** Plumb the hinge side jamb of the door unit. A two-foot carpenter's level is not accurate enough when plumbing the sides because of any bow that may be in the jambs. Use a 6-foot level or a plumb bob. Tack the jamb plumb to the wall through the casing with one nail on either side.

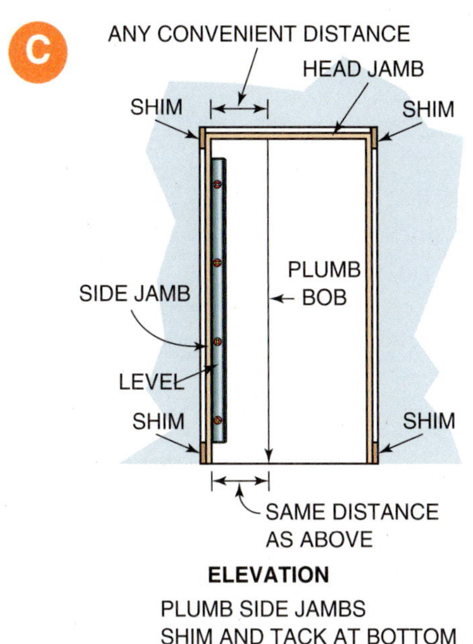


**D** Open the door and move to the other side. Check that the unit is nearly centered. Install shims between the side jambs and the rough opening at intermediate points, keeping side jambs straight. Shims should be located behind the hinges and lockset **strike plates**. Nail through the side jambs and shims. Remove spacers from door edges.

- Check the operation of the door. Make any necessary adjustments. The space between the door and the jamb should be equal on all jambs. The entire door edge should touch the stop or weather stripping.

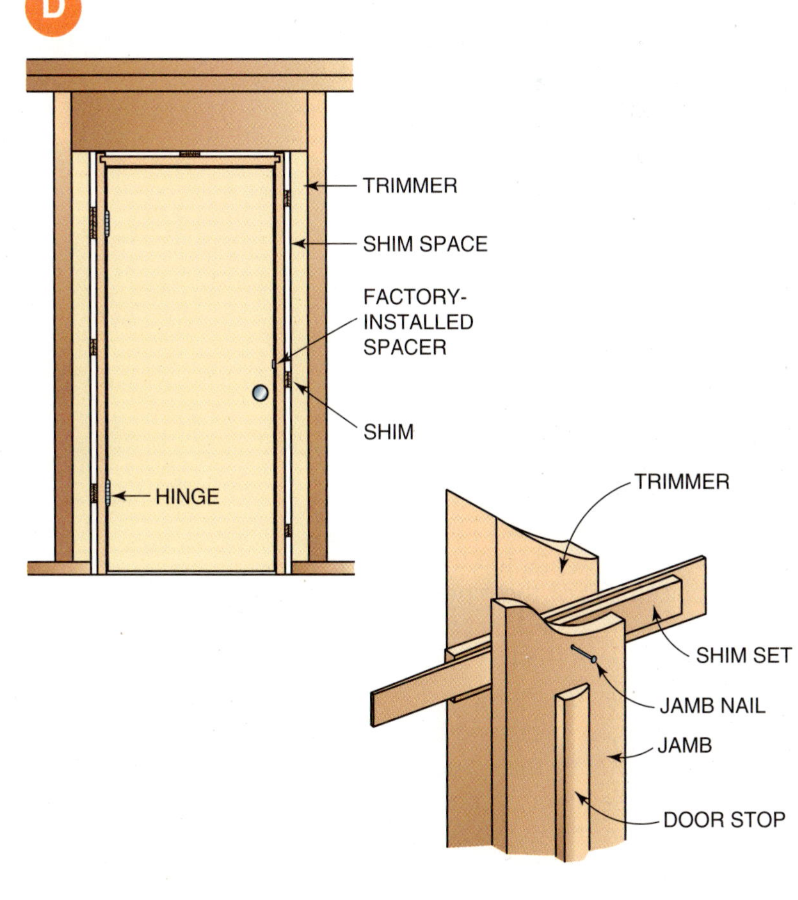

PROCEDURE 13-2

# Hanging Interior Doors (continued)

**E** Finish nailing the casing and install casing on the other side of the door. If nailing by hand, drive and set all nails taking care not to make hammer marks on the finished material.

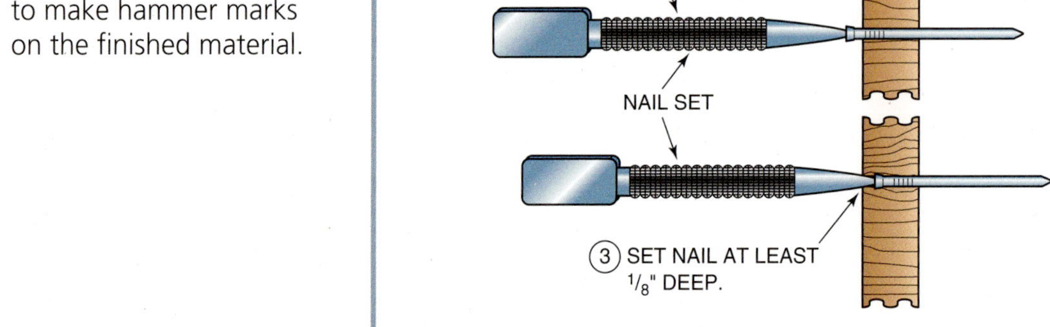


## Fitting a Door to a Frame

**A** Begin by checking the door for its beveled edge and the direction of the face of the door. Note the direction of the swing.

- Lightly mark the location of the hinges on the door. On paneled doors, the top hinge is usually placed with its upper end in line with the bottom edge of the top rail. The bottom hinge is placed with its lower end in line with the top edge of the bottom rail. The middle hinge is centered between them. On flush doors, the usual placement of the hinge is approximately 9 inches down from the top and 13 inches up from the bottom, as measured to the center of the hinge. A middle hinge is centered between the two.
- Check the opening frame for level and plumb.

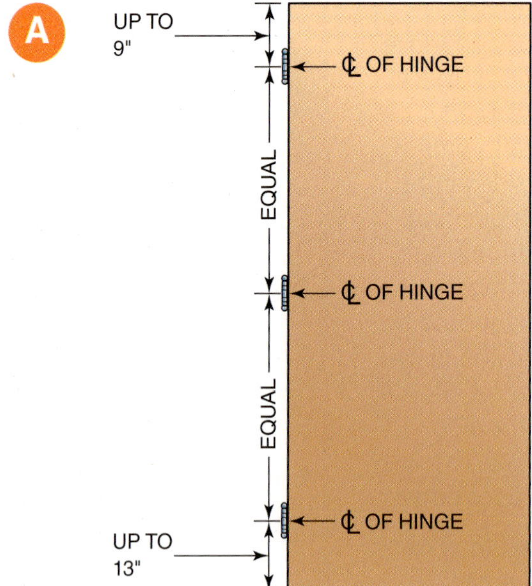

**B** Plane the door edges so the door fits onto the opening with an even joint of approximately 3⁄32 inch between the door and the frame on all sides. A wider joint of approximately 1⁄8 inch must be made to allow for swelling of the door and frame in extremely damp weather. Use a *door jack* to hold the door steady. Do not cut more than 1⁄2 inch total from the width of a door. Cut no more than 2 inches from its height. Check the fit frequently by placing the door in the opening. Doors that fit well operate well.

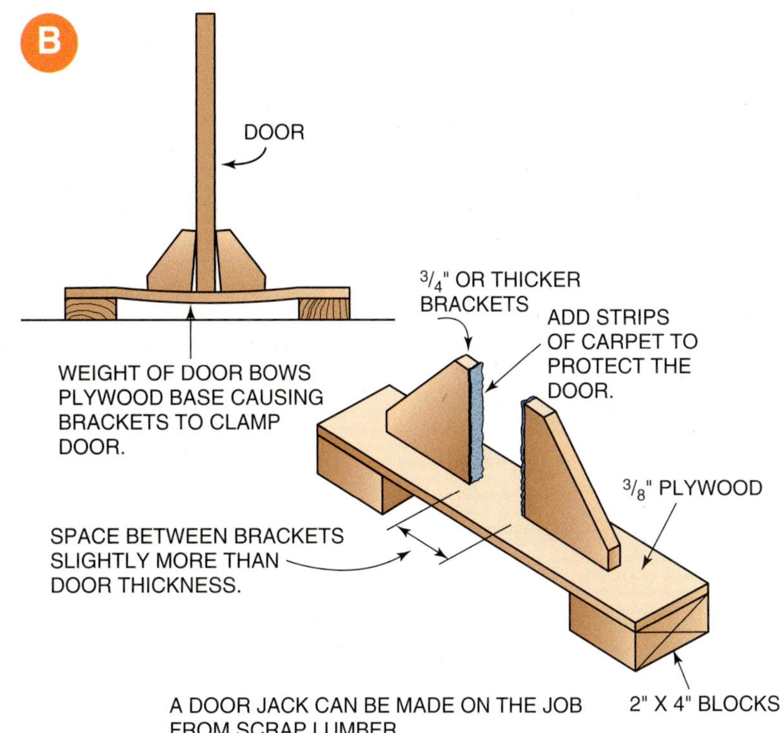

A DOOR JACK CAN BE MADE ON THE JOB FROM SCRAP LUMBER.

**C** Place the door in the frame. Shim and support the door so the proper space is obtained along all sides between the door and the jamb. Place shims between the lock edge of the door and side jamb of the frame. Mark across the door and jamb at the desired location for each hinge. Place a small X on both the door and the jamb to indicate on which side of the mark to cut the gain.

- Remove the door from the frame. Place a hinge leaf on the door edge with its end on the mark previously made. Score a line along edges of the leaf. Score only partway across the door edge.

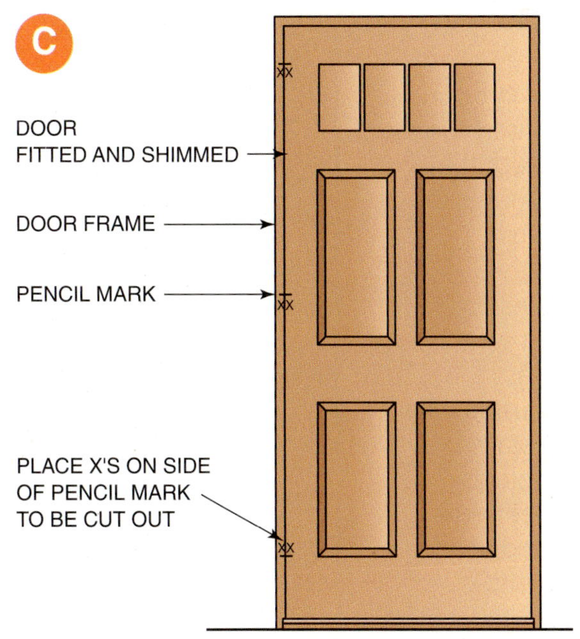

PROCEDURE 13-2

# Hanging Interior Doors (continued)

**D** Score the hinge lines, taking care not to split any part of the door. With a chisel, cut small chips from each end of the gain joint. The chips will break off at the scored end marks. Then, with the flat of the chisel down, pare and smooth the excess down to the depth of the gain. Be careful not to slip or overcut the gain.

- Press the hinge leaf into the gain joint. It should be flush with the door edge and install screws. Center the screws carefully so the hinge leaf will not shift when the screw head comes in contact with the leaf.
- Place the door in the opening and insert the hinge pins. Check the swing of the door and adjust as needed.

**E** Apply the *door stops* to the jambs with several tack nails, in case they have to be adjusted when locksets are installed. A back miter joint is usually made between molded side and header stops. A butt joint is made between square-edge stops.

**D**

**E**

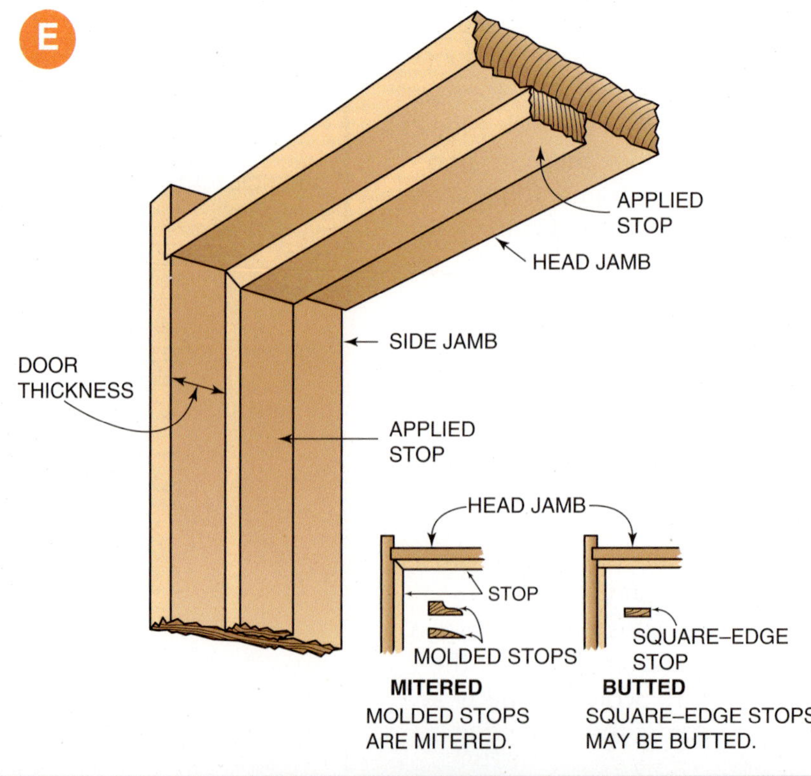

## Installing Bypass Doors

**A** Cut the track to length to fit between the jambs. Install it on the header jamb according to the manufacturer's directions. Bypass doors are installed so they overlap each other by about 1 inch when closed.

- Install pairs of *roller hangers* on each door. The roller hangers may be offset a different amount for the door on the outside than the door on the inside. They are also offset differently for doors of various thicknesses. Make sure that rollers with the same and correct offset are used on each door. The location of the rollers from the edge of the door is usually specified in the manufacturer's instruction sheet.

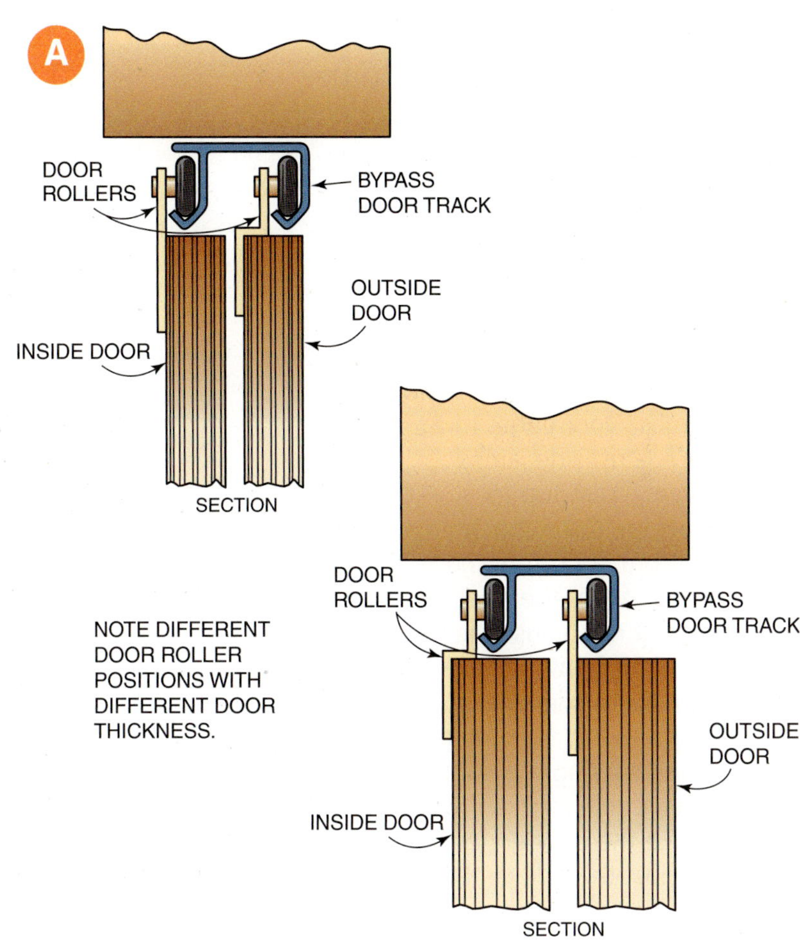


**B** Mark the location and bore holes for *door pulls.* Flush pulls must be used so that bypassing is not obstructed. The proper size hole is bored partway into the door. The pull is tapped into place with a hammer and wood block. The press fit holds the pull in place. Rectangular flush pulls, also used on bypass doors, are held in place with small recessed screws.

PROCEDURE 13-2

# Hanging Interior Doors (continued)

**C** Hang the doors by holding the bottom outward. Insert the rollers in the overhead track. Then, gently let the door come to a vertical position. Install the inside door first, then the outside door.

- Test the door operation and the fit against the side jambs. Door edges must fit against side jambs evenly from top to bottom. If the top or bottom portion of the edge strikes the side jamb first, it may cause the door to jump from the track. The door rollers have adjustments for raising and lowering the door. Adjust one or the other to make the door edges fit against the side jambs.

**C**

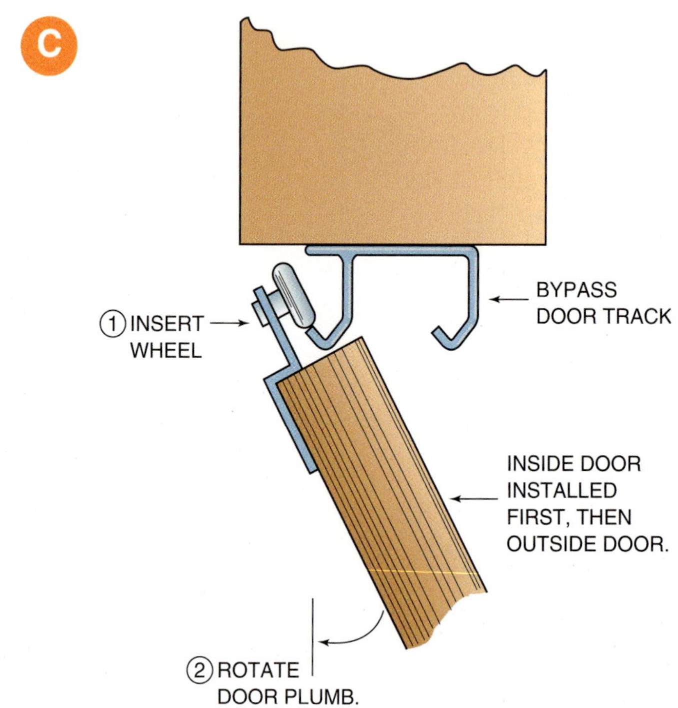


**D** A *floor guide* is included with bypass door hardware to keep the doors in alignment. The guide is centered on the lap of the two doors to steady them at the bottom. Mark the location and fasten the guide.

**D**

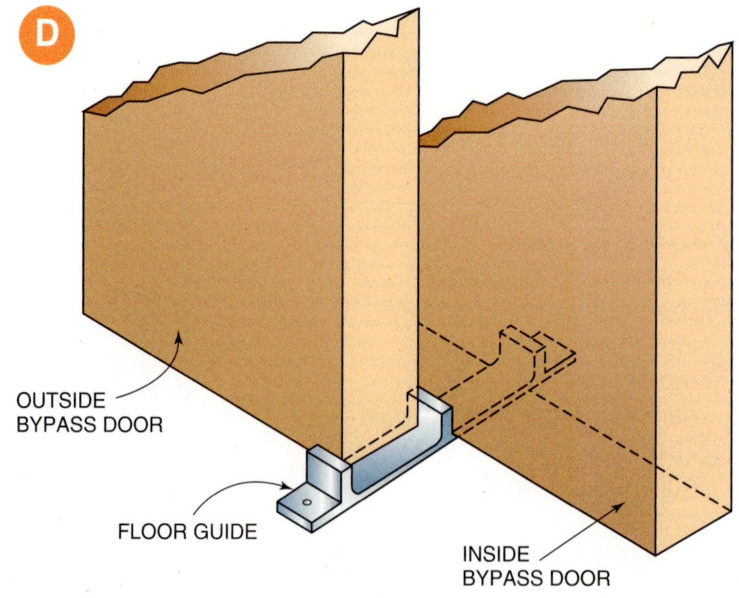


FLOOR GUIDE IS ADJUSTABLE FOR VARIOUS DOOR THICKNESSES.

## Installing Bifold Doors

**A** Check that the door and its hardware are all present. The hardware consists of the track, pivot sockets, pivot pins and guides, door aligners, door pulls, and necessary fasteners.

**B** Cut the track to length to fit between the jambs. Fasten it to the header jamb with screws provided in the kit. The track contains adjustable *sockets* to receive the door *pivot pins.* Make sure these are inserted before fastening the track in position. The position of the track on the header jamb is not critical. It may be positioned as desired.

- Locate the bottom pivot sockets. Fasten one on each side, at the bottom of the opening. The pivot socket bracket is L shaped. It rests on the floor against the side jamb. It is centered on a plumb line from the center of the pivot sockets in the track on the header jamb above.
- Install pivot pins at the top and bottom ends of the door in the prebored holes closest to the jamb. Sometimes the top pivot pin is spring loaded. It can then be depressed for easier installation of the door. The bottom pivot pin is threaded and can be adjusted for height. The guide pin rides in the track. It is installed in the hole provided at the top end of the door farthest away from the jamb.

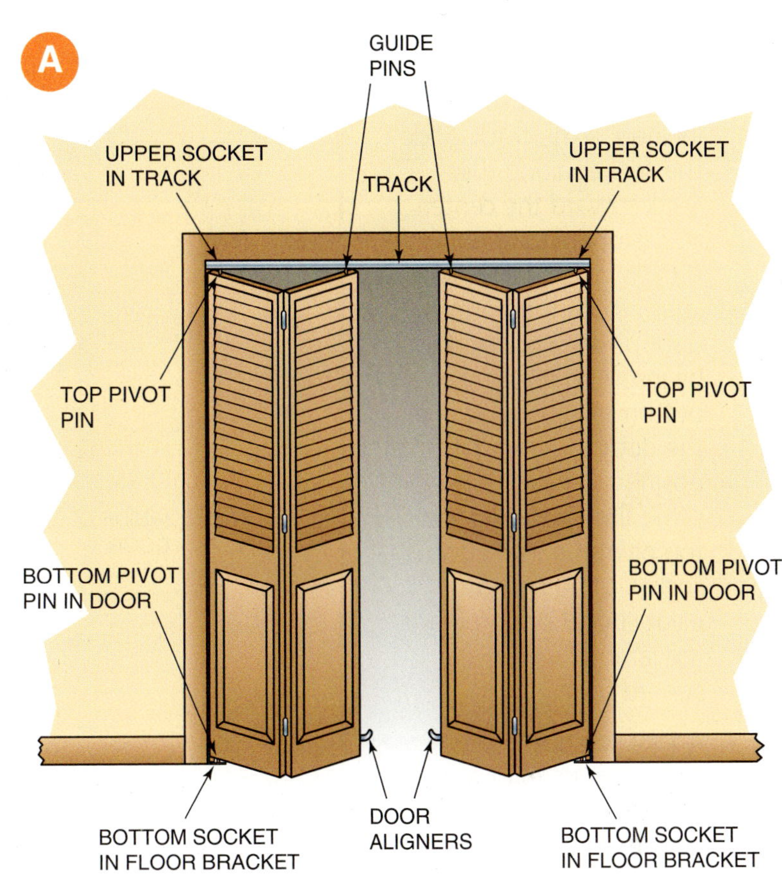


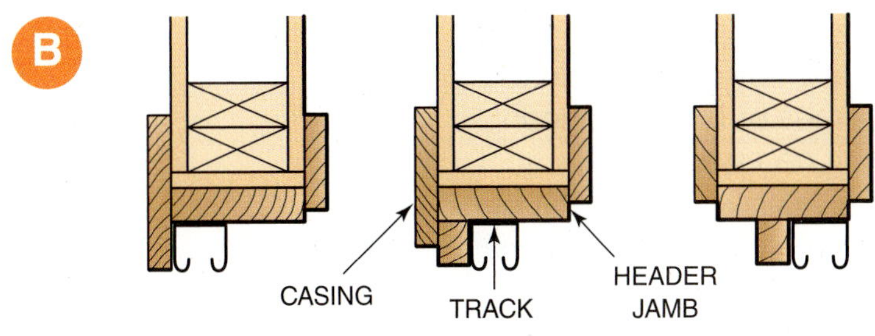


SECTIONS THROUGH HEADER

PROCEDURE 13-2

# Hanging Interior Doors (continued)

**C** Loosen the set screw in the top pivot socket located in the track. Slide the socket along the track toward the center of the opening about one foot away from the side jamb. Place the bottom door pivot in position by inserting it into the bottom pivot socket. Tilt the door to an upright position, while at the same time inserting the top pivot pin in the top socket. Slide the top pivot socket back toward the jamb where it started from.

- Adjust top and bottom pivot sockets in or out so the desired joint is obtained between the door and the jamb. Lock top and bottom pivot sockets in position. Adjust the bottom pivot pin to raise or lower the doors, if necessary.
- Install second door in the same manner. Check and adjust the doors so the tops are aligned and the space between the doors is uniform.
- Install pull knobs and door aligners in the manner and location recommended by the manufacturer. The door aligners keep the faces of the center doors lined up when closed.

**C**

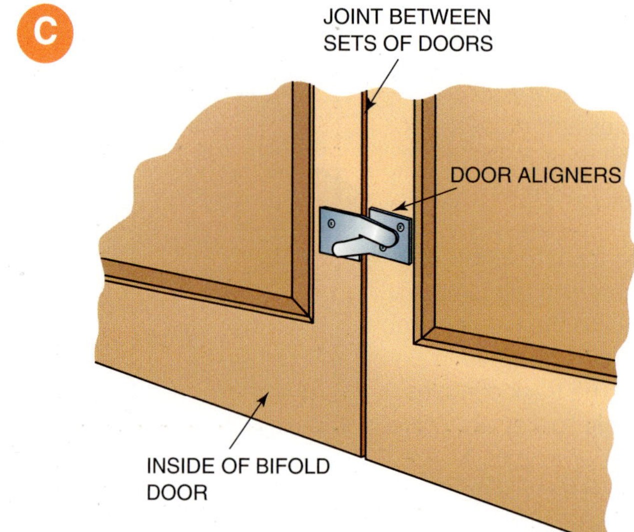

## Installing Pocket Doors

**A** Install the door frame pocket, when interior partitions are being framed per manufacturer's instructions. The pocket consists of two ladder-like frames between which the door slides. A steel channel is fastened to the floor. The pocket is covered by the interior wall finish. Care must be taken when covering the pocket frame to use the properly sized fasteners so they do not penetrate the frame, interfering with and marring the door.

- Attach rollers to the top of the door. Install pulls on the door. A special pocket door pull contains edge and side pulls. It is mortised in the edge of the door.
- Engage the rollers in the track by holding the bottom of the door outward in a way similar to that used with bypass doors. Test the operation of the door to make sure it slides easily and butts against the side jamb evenly. Make adjustments to the rollers, if necessary.
- Install the stops to the jambs on both sides of the door. The stops serve as guides for the door. When the door is closed, the stops prevent it from being pushed out of the opening.

## Installing Cylindrical Locksets

**A** Check the contents and read the manufacturer's directions carefully. Because so many kinds of locks are manufactured, the mechanisms vary greatly. If the door has pre-drilled lockset holes, install the lockset as per manufacturer's instructions.

- If the door is not pre-bored, open the door to a convenient position. Wedge the bottom to hold it in place.

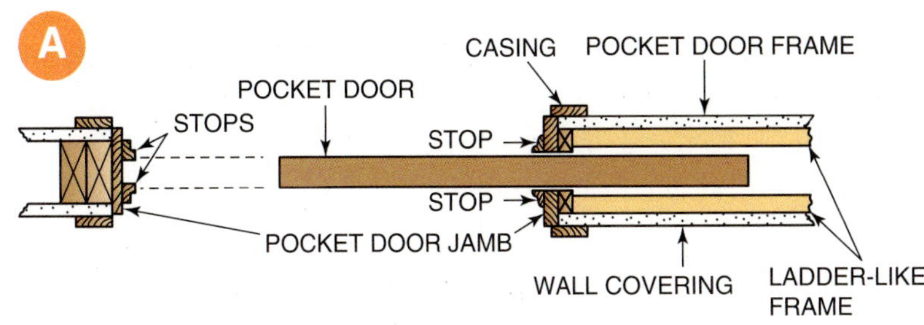


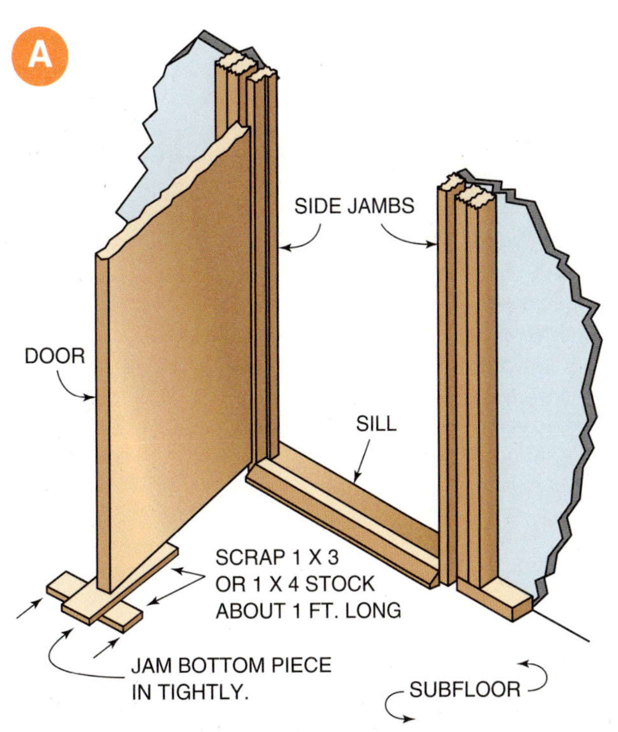

PROCEDURE 13-2

# Hanging Interior Doors (continued)

- Measure up, from the floor, the recommended distance to the centerline of the lock. This is usually 36 to 40 inches. At this height, square a light line across the edge and stile of the door.

**B** Position the center of the paper template supplied with the lock on the squared lines. Lay out the centers of the holes that need to be bored. It is important that the template be folded over the high corner of the beveled door edge. The distance from the door edge to the center of the hole through the side of the door is called the backset of the lock.

- Typical backsets are $2\frac{3}{8}$ inches for interior doors and $2\frac{3}{4}$ inches for exterior doors. Make sure the backset is marked correctly before boring the hole. Check the manufacturer's directions for the specified hole sizes.

**C** Bore the hole through the face of the door first. This hole is usually $2\frac{1}{8}$ inches in diameter. A *boring jig* is frequently used. It is clamped to the door to guide power-driven *multispur* bits. The clamping action of the jig prevents splintering. Bore halfway into door from both sides to prevent splintering of the either face of the door.

- Bore the latch bolt hole through the edge of the door toward the center of the lockset hole. This hole is $\frac{7}{8}$ or 1 inch in diameter, depending on the manufacturer.

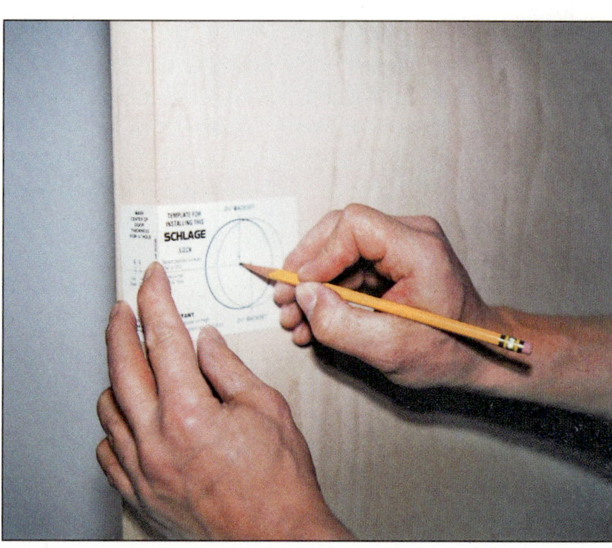

**D** Use of a *faceplate marker* to lay out the mortise for the latch faceplate simplifies installation. Chisel out the mortise and install latch bolt flush with door edge.

- Complete the installation of the lockset by following specific manufacturer's directions.

**E** Place the striker plate over the latch bolt in the door and close the door snugly against the stops. Push the striker plate in against the latch and raise it to center the hole on the latch. Draw a vertical line on the face of the plate flush with the outside face of the door and mark the top side of the strike plate against the jam. The *striker plate* should be installed so that when the door is closed, it latches snugly with no play.

- Open the door. Reposition the striker plate on the jamb aligned with the marks previously drawn. Hold it firmly while scoring a line around the plate with a sharp pencil or knife. Chisel out the mortise so the plate lies flush with the jamb.
- Screw the plate in place. Chisel or bore out the center to receive the latch bolt.

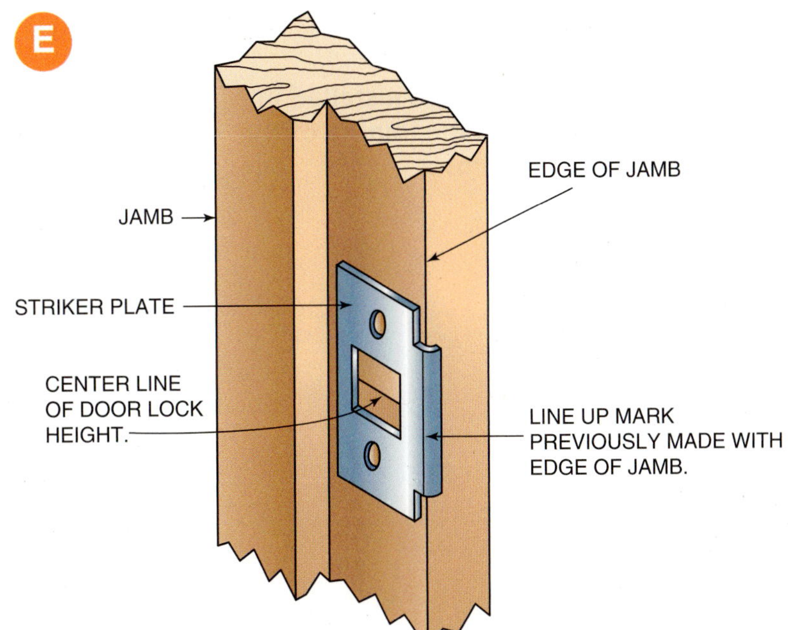

## REVIEW QUESTIONS

*Select the most appropriate answer.*

1. **A frame in a window that holds the glass is called a**
   a. light.
   b. mullion.
   c. sash.
   d. stile.

2. **Small wood strips that divide the glass into smaller lights are called**
   a. mantels.
   b. margins.
   c. mullions.
   d. muntins.

3. **When windows are installed in multiple units, the joining of the side jambs forms a**
   a. mantel.
   b. margin.
   c. mullion.
   d. muntin.

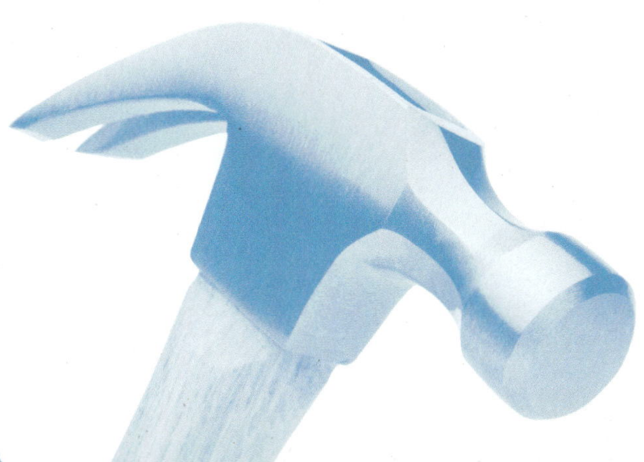

4. **A window that consists of an upper and a lower sash that slide vertically is called a**
   a. casement window.
   b. double-hung window.
   c. hopper window.
   d. sliding window.

5. **A window that has a sash hinged on one side and swings outward is called**
   a. an awning window.
   b. a casement window.
   c. a double-hung window.
   d. a hopper window.

6. **The difference between a hopper and an awning window is that the hopper window**
   a. swings inward instead of outward.
   b. swings outward instead of inward.
   c. is hinged at the top rather than at the bottom.
   d. is hinged on the side rather than on the bottom.

7. **Before setting a window in an opening,**
   a. check all dimensions of opening and window.
   b. caulk the sides of the rough opening.
   c. flash the head casing.
   d. center the unit in the opening.

8. **Horns on doors are**
   a. placed on the header casing.
   b. extensions of the side jambs.
   c. moldings applied to a flat casing.
   d. extensions of the header jamb.

**9. The standard thickness of exterior doors in residential construction is**

a. $1\frac{3}{8}$ inches.
b. $1\frac{1}{2}$ inches.
c. $1\frac{3}{4}$ inches.
d. $2\frac{1}{4}$ inches.

**10. The typical width of exterior residential entrance doors is**

a. 2'-0" or 2'-4".
b. 2'-4" or 2'-6".
c. 2'-8" or 3'-0".
d. 3'-0" or 3'-6".

**11. The height of doors in residential construction is typically**

a. 7'-0".
b. 6'-10".
c. 6'-8".
d. 6'-6".

**12. The joint between the door and door frame should be close to**

a. $\frac{3}{32}$ inch.
b. $\frac{3}{64}$ inch.
c. $\frac{1}{4}$ inch.
d. $\frac{3}{16}$ inch.

**13. On paneled doors, the top end of the top hinge is usually placed**

a. in line with the bottom edge of the top rail of the door.
b. 7 inches down from the top end of the door.
c. not more than $10\frac{3}{4}$ inches down from the top end of the door.
d. in line with the top edge of an intermediate rail.

**14. Most interior doors are manufactured in a thickness of**

a. 1 inch.
b. $1\frac{3}{8}$ inches.
c. $1\frac{1}{2}$ inches.
d. $1\frac{3}{4}$ inches.

**15. When installing housewrap**

a. tape all seams.
b. overlap at least 3".
c. cover entire wall, cutting out openings later.
d. all of the above.

**16. Interior flush door skins are often made with**

a. fir plywood.
b. lauan plywood.
c. metal.
d. plastic laminate.

**17. A disadvantage of bypass doors is that they**

a. project out into the room.
b. cost more and require more time to install.
c. are difficult to operate.
d. do not provide total access to the opening.

**18. Hinges on a swinging panel door are attached to the jamb and the door**

a. rails.
b. astragal.
c. stile.
d. muntin.

**19. The beveled edge of a swinging door**

a. allows the door to swing without touching the jamb.
b. is the edge where the lockset is installed.
c. is 1⁄16 to 1⁄8 inch back beveled along width.
d. all of the above.

**20. The decorative plate used to cover the edges of a bored hole in a door is called a**

a. astragal.
b. dead bolt.
c. escutcheon.
d. all of the above.

## KNOW YOUR CODES

Some regions of the country require window installation to be inspected. This is because heavy wind forces from hurricanes can blow poorly installed windows out of the opening. Be sure to check with the local officials for the inspection schedule.

## WHAT'S WRONG WITH THIS PICTURE?

Carefully study **Figure 13-53** and think about what is wrong. Consider all possibilities.

WRONG

**FIGURE 13-53** The felt paper should be installed starting at the bottom and working upward to the top. This ensures that any overlapping areas will shed water and not channel it toward the sheathing underneath. In this example, the sides were erroneously installed before the bottom piece.

RIGHT

**FIGURE 13-54** This figure clearly shows that if the siding fails or leaks, any moisture beneath it will stay between the siding and felt paper and not work its way to the sheathing.

CHAPTER 14

# Roofing

**OBJECTIVES** *After completing this unit, the student should be able to:*

- Define roofing terms.
- Describe and apply roofing felt underlayment, organic or fiberglass asphalt shingles, and roll roofing.
- Describe and apply flashing to valleys, sidewalls, chimneys, and other roof obstructions.
- Estimate needed roofing materials.

# GLOSSARY OF ROOFING TERMS

**apron** the flashing piece located on the lower side of a roof penetration such as a chimney, skylight, or dormer

**asphalt felt** a building paper saturated with asphalt used for waterproofing

**closed valley** a roof valley in which the roof covering meets in the center of the valley, completely covering the valley

**cricket** a small, false roof built behind, or uphill from, a chimney or other roof obstacle for the purpose of shedding water around roof penetrations

**drip edge** metal edging strips placed on roof edges to provide a support and overhang for the roofing material

**electrolysis** accelerated oxidation of one metal because of contact with another metal in the presence of water

**exposure** the amount that courses of siding or roofing are open to the weather

**flashing** material used at intersections such as roof valleys and dormers and around window and door openings to prevent the entrance of water

**mortar** a mixture of Portland cement, lime, sand, and water used to bond masonry units together

**open valley** a roof valley in which the roof covering is kept back from the centerline of the valley exposing the flashing

**saddle** same as *cricket*

**selvage** the unexposed part of roll roofing covered by the course above

**square** the amount of roof covering that will cover 100 square feet of roof area

Roofing is a general term to describe materials used to cover a roof to make it weathertight. Many types and styles of roofing materials are often installed by carpenters who specialize in roofing. The types discussed here are manufactured with asphalt.

All roof systems rely on the concept that water runs downhill. Roofing materials are overlapped so that water will run over the seam and not into or under it. Some high-wind locales have additional installation requirements that help prevent water from being forced under the roofing layer by wind. Check the local codes and manufacturer's instructions for these variations.

Before roofing is applied, the roof deck must be securely fastened. Check that the sheathing is nailed according to local codes. There must be no protruding nails or nails that have missed the support member.

# Asphalt Shingles

Asphalt products are the most commonly used roof covering for residential construction. They are designed to provide protection from the weather for a period ranging from 20 to 30 years. They are available in many colors and styles.

## ROOFING TERMS

An understanding of the terms most commonly used in connection with roofing is essential for efficient application of roofing material.

- A **square** is the amount of roofing required to cover 100 square feet of roof surface. There are three to five bundles of shingles per square (Fig. 14-1). One square of shingles can weigh between 235 and 325 pounds.
- *Deck* is the roof surface to which roofing materials are applied.
- *Coverage* is the number of overlapping layers of roofing and the degree of weather protection offered by roofing material.
- A *shingle butt* is the bottom exposed edge of a shingle.
- *Courses* are horizontal rows of shingles or roofing.

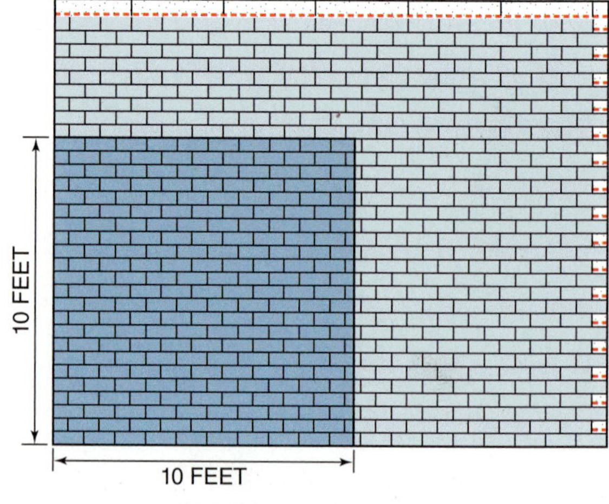


**FIGURE 14-1** One square of shingles will cover 100 square feet.

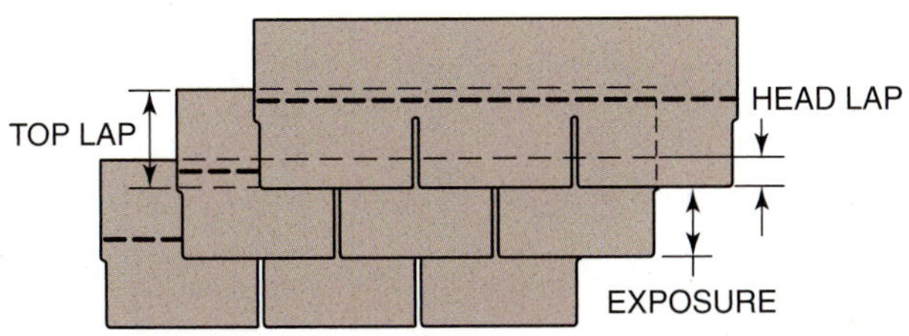


**FIGURE 14-2** Asphalt strip exposure and lap.

- **Exposure** is the distance between courses of roofing. It is the amount of roofing in each course exposed to the weather (Fig. 14-2).
- The *top lap* is the height of the shingle or other roofing minus the exposure. In roll roofing, this is also known as the **selvage**.
- The *head lap* is the distance from the butt of an overlapping shingle to the top of the shingle two courses below, measured up the slope.
- *End lap* is the horizontal distance that the ends of roofing in the same course overlap each other.
- **Flashing** is made of strips of thin sheet material. They are usually made of zinc, copper, or aluminum. They may also be strips of roofing material used to make watertight joints on a roof. Metal flashing comes in rolls of various widths and are cut to the desired length.
- *Asphalt cements* and *coatings* are manufactured to various consistencies depending on the purpose for which they are to be used. They are used as adhesives to bond asphalt roofing products and flashings. *Coatings* are usually thin enough to be applied with a brush. They are used to resurface old roofing or metal that has become weathered.
- **Electrolysis** is a reaction that occurs when unlike metals come in contact with each other and water. This contact causes one of the metals to corrode. A simple way to prevent this disintegration is to fasten metal roofing material with fasteners made of the same metal.

### Roofing Underlayment

The deck should be covered with an asphalt shingle *underlayment*. The underlayment protects the sheathing from moisture until the roofing is applied. It also gives additional protection after the roofing material is applied. Underlayment is typically installed using **asphalt felts**. Asphalt felts consist of heavy felt paper saturated with asphalt or coal tar. They are made in various weights of pounds per square and in 36-inch wide rolls (Fig. 14-3).

### Kinds of Asphalt Shingles

Two types of asphalt shingles are manufactured. *Organic shingles* have a base made of heavy asphalt-saturated paper felt coated with additional asphalt. *Fiberglass shingles* have a base mat of glass fibers. The mat does not require the saturation process and requires only an asphalt coating. The asphalt coating provides weatherproofing qualities. Both kinds of shingles are surfaced with selected mineral granules. The granules protect the shingle from the sun, provide color, and protect against fire.

Shingles come in a wide variety of colors, shapes, and weights (Fig. 14-4). They are applied in the same manner. Shingle quality is generally determined by the weight per square. Most asphalt shingles are manufactured with factory-applied adhesive to seal the exposed tab to the course below. This increases their resistance to the wind.

| | APPROX. WEIGHT PER ROLL | APPROX. WEIGHT PER SQUARE | SQUARES PER ROLL | ROLL LENGTH | ROLL WIDTH | SIDE OR END LAPS | TOP LAP | EXPOSURE |
|---|---|---|---|---|---|---|---|---|
| SATURATED FELT | 60 # | 15 # | 4 | 144' | 36" | 4" | 2" | 34" |
| | 60 # | 30 # | 2 | 72' | 36" | TO | | |
| | 60 # | 60 # | 1 | 36' | 36" | 6" | | |

**FIGURE 14-3** Sizes and weights of asphalt-saturated felts.

| PRODUCT | CONFIGURATION | PER SQUARE | | | SIZE | | EXPOSURE | UNDER-WRITERS' LISTING |
|---|---|---|---|---|---|---|---|---|
| | | APPROX. SHIPPING WEIGHT | SHINGLES | BUNDLES | WIDTH | LENGTH | | |
| WOOD APPEARANCE STRIP SHINGLE MORE THAN ONE THICKNESS PER STRIP<br>LAMINATED OR JOB APPLIED | VARIOUS EDGE, SURFACE TEXTURE, AND APPLICATION TREATMENTS | 285# TO 390# | 67 TO 90 | 4 OR 5 | $11\frac{1}{2}$" TO 15" | 36" OR 40" | 4" TO 6" | A OR C – MANY WIND RESISTANT |
| WOOD APPEARANCE STRIP SHINGLE, SINGLE THICKNESS PER STRIP | VARIOUS EDGE, SURFACE TEXTURE, AND APPLICATION TREATMENTS | VARIOUS 250# TO 350# | 78 TO 90 | 3 OR 4 | 12" OR $12\frac{1}{4}$" | 36" OR 40" | 4" TO $5\frac{1}{8}$" | A OR C – MANY WIND RESISTANT |
| SELF-SEALING STRIP SHINGLE | CONVENTIONAL 3 TAB | 205#–240# | 78 OR 80 | 3 | 12" OR $12\frac{1}{4}$" | 36" | 5" OR $5\frac{1}{8}$" | A OR C – ALL WIND RESISTANT |
| | 2 OR 4 TAB | VARIOUS 215# TO 325# | 78 OR 80 | 3 OR 4 | 12" OR $12\frac{1}{4}$" | 36" | 5" OR $5\frac{1}{8}$" | |
| SELF-SEALING STRIP SHINGLE<br>NO CUT OUT | VARIOUS EDGE AND TEXTURE TREATMENTS | VARIOUS 215# TO 290# | 78 TO 81 | 3 OR 4 | 12" OR $12\frac{1}{4}$" | 36" OR $36\frac{1}{4}$" | 5" | A OR C – ALL WIND RESISTANT |
| INDIVIDUAL LOCK DOWN<br>BASIC DESIGN | SEVERAL DESIGN VARIATIONS | 180# TO 250# | 72 TO 120 | 3 OR 4 | 18" TO $22\frac{1}{4}$" | 20 TO $22\frac{1}{2}$" | – | C – MANY WIND RESISTANT |

*Courtesy of Asphalt Roofing Manufacturers' Association.*

**FIGURE 14-4** Asphalt shingles are available in a wide variety of sizes, shapes, and weights.

## Green Tip

Higher quality shingles last longer and save money in the end. Lesser quality shingles need to be replaced more often which drives up labor and waste disposal costs.

### CAUTION

**CAUTION:** Asphalt roofing materials are petroleum based and are flammable. Care must be taken to protect them from excessive heat. Also, hot temperatures will allow the granular surface to wear off easily when people walk on it. Cold temperatures make the material brittle and easily broken.

### Drip Edge

Metal **drip edge** is installed along the roof edges. The metal drip edge is usually made of aluminum, galvanized steel, or painted steel. It is applied along the perimeter of the roof (Fig. 14-5). Install the metal drip edge by using roofing nails of the same metal spaced 8 to 10 inches along its inner edge. End joints may be butted or overlapped in high-wind areas.

## ASPHALT SHINGLES APPLICATION

Asphalt roofing products become soft in hot weather. Be careful not to damage them by digging in with heavy shoes during application or by unnecessary walking on the surface after application.

In cold weather, shingles may have to be warmed in order to prevent cracking when bending them over the ridge. The slope of a roof should not be less than 4 inches per foot when conventional methods of asphalt shingle application are used.

### Layout of Asphalt Shingle

Strip shingles are typically applied by starting from either rake. Shingles may also be started at the center and installed both ways. Waste from shingle cutting is often used on the opposite side of the roof.

> *from experience...*
>
> Wearing soft rubber-soled shoes causes less wear on the roofing surface during installation.

### Starter Course of Asphalt Shingles

The *starter course* backs up and fills in the spaces between the butt seams or tabs of the first regular course of shingles. They have the exposure portion of the shingle cut off in order to apply them with the factory-applied adhesive along the roof perimeter.

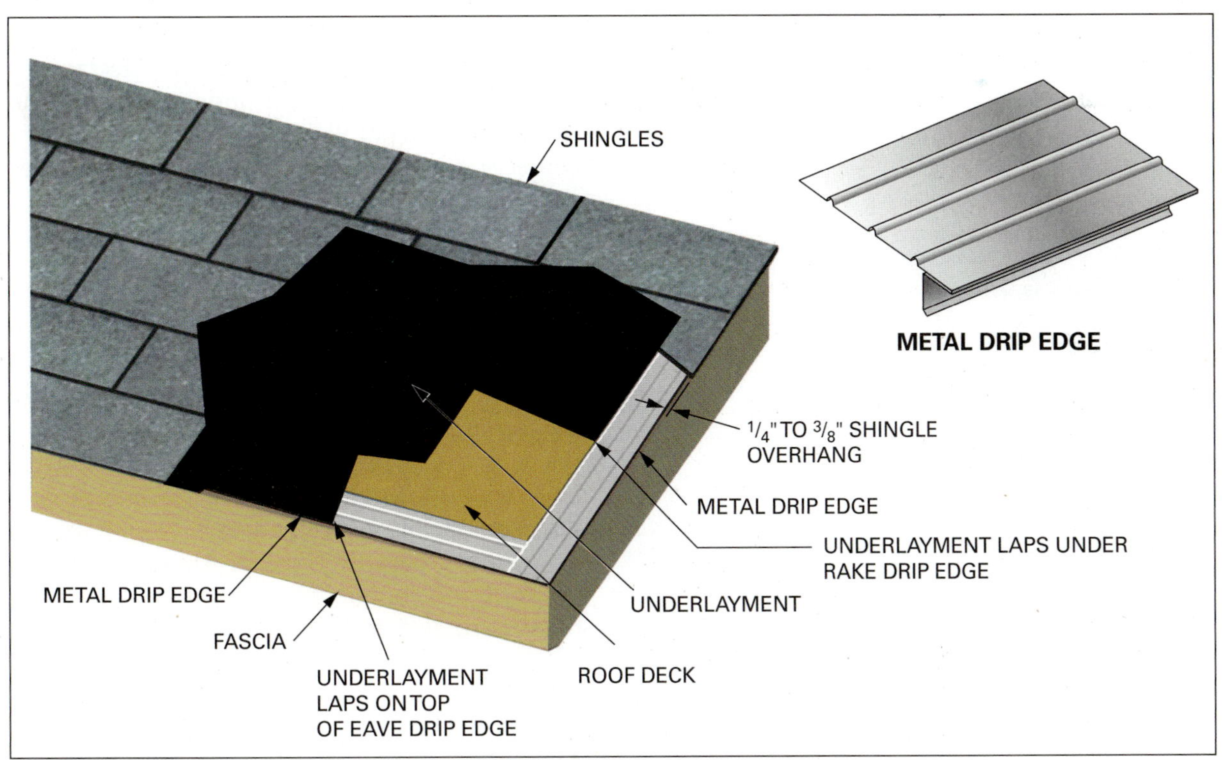


**FIGURE 14-5** A metal drip edge may be used to support the shingle edge overhang.

Cut the shingles by scoring them on the back side with a utility knife. Use a square as a guide for the knife. Bend the shingle. It will break on the scored line. No rake cut piece should be less than 3 inches in width.

### Fastening Asphalt Shingles

Selecting suitable fasteners, using the recommended number, and putting them in the proper place is important in the application of asphalt shingles. Lay the first regular course of shingles on top of the starter course. Keep their bottom edges flush with each other. Use a minimum of four fasteners in each strip shingle. Refer to the recommendations provided by the manufacturer on the back of a bundle of shingles. Do not nail into or above the factory-applied adhesive (Fig. 14-6). The roofing nail length should be sufficient to penetrate the sheathing at least ¾ inch, or through approved panel sheathing. Pneumatic nailers speed up installation significantly (Fig. 14-7).

Courtesy of Senco Products, Inc.

**FIGURE 14-7** Pneumatic staplers and nailers are often used to fasten asphalt shingles.

### Shingle Exposure

The maximum exposure of asphalt shingles to the weather depends on the type of shingle. Recommended maximum exposures range from 4 to 6 inches. Most commonly used asphalt shingles have a maximum exposure of 5 inches. Metric shingles have an exposure of about 5½ inches.

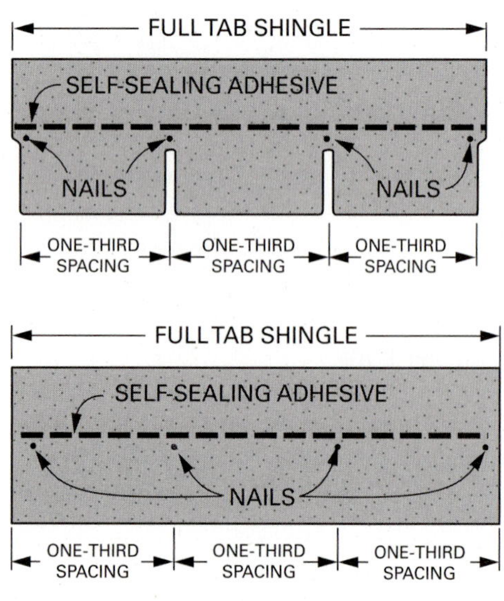


Courtesy of Asphalt Roofing Manufacturers' Association.

**FIGURE 14-6** Recommended fastener locations for asphalt strip shingles.

When laying out shingle courses, space the desired exposure up each rake from the top edge of the first course of shingles. Snap lines across the roof or use the lines of the underlayment, assuming it was installed parallel to the lower edge. Start each course so the butt seams are staggered in the desired manner.

### Spacing Shingle Courses Near the Ridge

The exposures of the last 6 to 10 courses of shingles may be adjusted slightly so the last course has a nearly full exposure where it meets the ridge cap. At the top of the last completed course, measure up to the center of the ridge and subtract 3½ inches. Divide this distance into spaces as close as possible to the exposure of previous courses. Do not exceed the maximum exposure (Fig. 14-8). Snap lines across the roof and shingle up to the ridge. Do not cover the venting slot of the ridge.

### Ridge Cap

The ridge cap is applied after both sides of the roof have been shingled. The ridge cap finishes the shingles, covering the nails of the last course. The cap is centered on the ridge. The exposure for each tab is 5 inches. Cut hip and ridge shingles from full

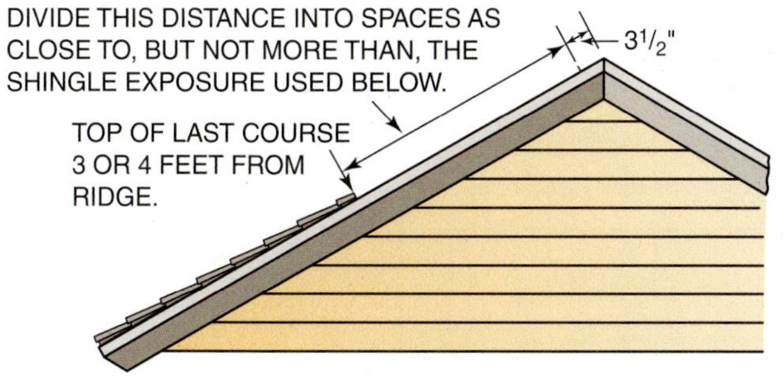


**FIGURE 14-8** Space shingle courses evenly to the ridge.

shingle strips to make approximately 12 × 12-inch squares. Cut shingles from the top of the cutout to the top edge on a slight taper. The top edge should be narrower than the bottom edge (Fig. 14-9).

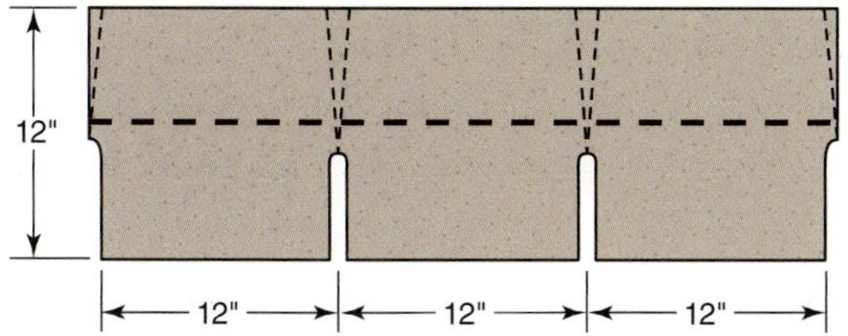


*Courtesy of Asphalt Roofing Manufacturers' Association.*

**FIGURE 14-9** Hip and ridge shingles are cut from strip shingles.

Cutting the shingles in this manner keeps the top half of the shingle from protruding when it is bent over the ridge.

***For step-by-step instructions on INSTALLING ASPHALT SHINGLES, see Procedure 14-1 on pages 456–459.***

# Roll Roofing

Roll roofing can be used on roof slopes of as little as 1 on 12. On steeper roofs, roll roofing is used when economy is the major factor and appearance is not as important.

## TYPES OF ROLL ROOFING

Roll roofing is made of the same materials as asphalt shingles. Various types are made with a base sheet of organic felt or glass fibers in a number of weights (Fig. 14-10). Some types are applied with exposed or concealed fasteners. They have a top lap of 2 to 4 inches. A concealed-nail type, called *double coverage* roll roofing, has a top lap of 19 inches. All kinds come in rolls that are 36 inches wide.

Roll roofing may be used on roofs that have slopes of less than 4 inches unit rise. However, the exposed-nail type should not be used on slopes of less than 2 inches unit rise. Roll roofing applied with concealed nails and having a top lap of at least 3 inches may be used on pitches of as low as 1 inch. The exposed fastener type is rarely used. Only the concealed fastener type is recommended for use and described in this chapter. Use the same type and length of nails as for asphalt shingles.

## GENERAL APPLICATION METHODS

Apply all roll roofing when the temperature is above 45 degrees Fahrenheit 45°F (7°C). This prevents cracking the coating. Cut the roll into lengths. Spread in a pile on a smooth surface to allow them to flatten out.

Use only the lap or quick-setting cement recommended by the manufacturer. Store cement in a warm place until ready for use. Apply roll roofing only on a solid, smooth, well-seasoned deck. Make sure the area below has sufficient ventilation to

**TYPICAL ASPHALT ROLLS**

| 1 | 2 | | 3 | 4 | | 5 | | 6 | 7 |
|---|---|---|---|---|---|---|---|---|---|
| PRODUCT | APPROXIMATE SHIPPING WEIGHT | | SQS. PER PACKAGE | LENGTH | WIDTH | SIDE OR END LAP | TOP LAP | EXPOSURE | UNDERWRITERS' LISTING |
| | PER ROLL | PER SQ. | | | | | | | |
| MINERAL SURFACE ROLL | 75# TO 90# | 75# TO 90# | ONE | 36'<br>38' | 36"<br>36" | 6" | 2"<br>4" | 34"<br>32" | C |
| MINERAL SURFACE ROLL DOUBLE COVERAGE | 55# TO 70# | 100# TO 140# | ONE-HALF | 36' | 36" | 6" | 19" | 17" | C |

**FIGURE 14-10** Types of roll roofing.

prevent the deck from absorbing condensation. This would cause the roofing to warp and buckle.

***For step-by-step instructions on INSTALLING ROLL ROOFING, see Procedure 14-2 on pages 460–462.***

# FLASHING

*Flashing* is a general term for material used to increase the weathertightness at the intersections of different exterior surfaces. It prevents water from entering a building. The words *flash, flashed,* and *flashing* are also used as verbs to describe the installation of the material. Various kinds of flashing are applied at the eaves, valleys, chimneys, vents, and other roof projections. They prevent leakage at the intersections.

## KINDS OF FLASHING

Flashing material may be sheet copper, zinc, aluminum, galvanized steel, or mineral-surfaced asphalt roll roofing. Copper and zinc are high-quality flashing materials, but they are more expensive.

**CAUTION: Take care handling metal flashing. They are thin with sharp edges that can easily cut skin.**

## EAVES FLASHING

Whenever there is a possibility of ice dams forming along the eaves and causing a backup of water, an *eaves flashing* is needed. Install a course of 36-inch wide, self-adhering, rubberized asphalt eaves flashing, often referred to as an *ice and water shield.* Let it overhang the drip edge by ¼ to ⅜ inch. The flashing should extend up the roof far enough to cover at least 12 inches of roof that extends past the inside wall line of the building. If the overhang of the eaves requires that the flashing be wider than 36 inches, the necessary horizontal lap joint is located on the

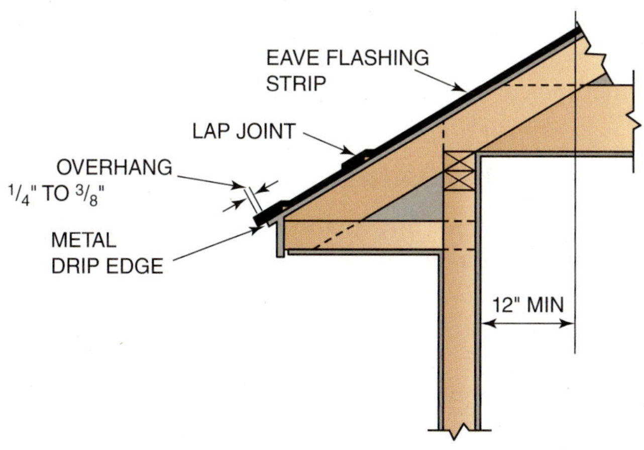


**FIGURE 14-11** An eaves flashing is installed if there is danger of ice dams forming along the eaves.

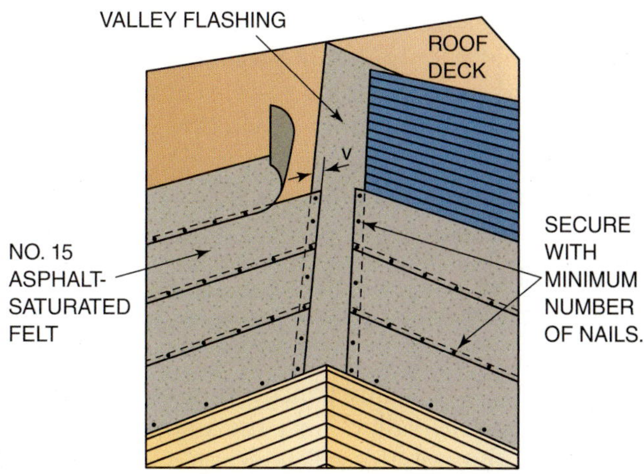


**FIGURE 14-12** Felt underlayment is applied in the valley before flashing.

portion of the roof that extends outside the wall line (Fig. 14-11).

## OPEN VALLEY FLASHING

Roof valleys are especially vulnerable to leakage. This is because of the great volume of water that flows down through them. Valleys must be carefully flashed according to recommended procedures. Valleys are finished as **open valleys** or **closed valleys**.

### Open Valley Underlayment

Apply a 36-inch wide strip of flashing material centered in the valley. Fasten along one edge with only enough nails to hold it in place. Fold and press the underlayment into the valley to seat the felt well into the valley. This will give full support to the flashing to keep it from ripping. Nail the remaining edge. Let the courses of felt underlayment applied to the roof overlap the valley flashing by not less than 6 inches (Fig. 14-12).

### Roll Roofing Valleys

Lay an 18-inch wide layer of mineral-surfaced roll roofing centered in the valley. Its mineral-surfaced side should be down to allow good bond with any adhesives. Use only enough nails spaced 1 inch in from each edge to hold the strip smoothly in place. Press the roofing firmly in the center of the valley when nailing the opposite edge. On top of the first strip, lay a 36-inch wide strip with its surfaced side up. Center it in the valley. Fasten it in the same manner as the first strip (Fig. 14-13).

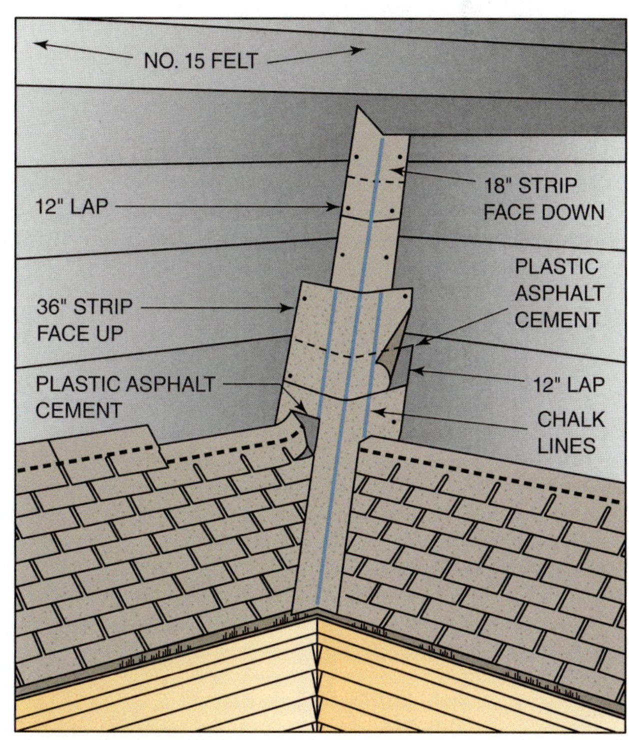


**FIGURE 14-13** Method of applying a roll roofing open valley flashing.

### Metal Valleys

Prepare the valley with underlayment in the same manner as described previously. Lay a strip of sheet metal flashing centered in the valley. The metal should extend at least 10 inches on each side of the valley centerline for slopes with a 6-inch rise or less and 7 inches on each side for a steeper slope.

Nail one edge of the flashing with just enough nails to hold it smoothly in place. Use nails of the same metal as the flashing to prevent electrolysis. Carefully press and form the centerline of the flashing into the valley. Fasten the remaining edge, using only enough fasteners to hold it (Fig. 14-14).

### Guide Lines

Snap a chalk line on each side of the valley as guides for trimming the ends of each shingle course. These lines are spaced roughly 6 inches apart to create the open look of the valley. Since the amount of water that flows through the valley increases the closer it gets to the eave, the guide lines should be spaced wider at the eaves. They should be spread ⅛ inch more per foot as they approach the eave.

The upper corner of each asphalt shingle ending at the valley is clipped. This helps keep water from entering between the courses. The roof shingles are cemented to the valley flashing with plastic asphalt.

## CLOSED VALLEY FLASHING

*Closed valleys* are those where the roof shingles completely cover the valley flashing. Using a closed valley protects the valley flashing, thus adding to the weather resistance at a vulnerable point. Several methods are used to flash closed valleys.

### Closed Valley Underlayment

The first step for any method is to apply the asphalt felt underlayment, as previously described for open valleys. Then, center a 36-inch width of smooth or mineral surface roll roofing, 50-pound per square, or heavier, in the valley over the underlayment. Form it smoothly to seat well into the valley. Secure it with only as many nails as necessary.

### Woven Valley Method

*Woven valley* involves applying asphalt shingles on both sides of the valley and alternately weaving each course over and across the valley. Each course of shingles extends to and beyond the valley for a distance of at least 12 inches. This is measured along the exposed edge of the shingle.

***For step-by-step instructions on the WOVEN VALLEY METHOD, see Procedure 14-3 on page 463.***

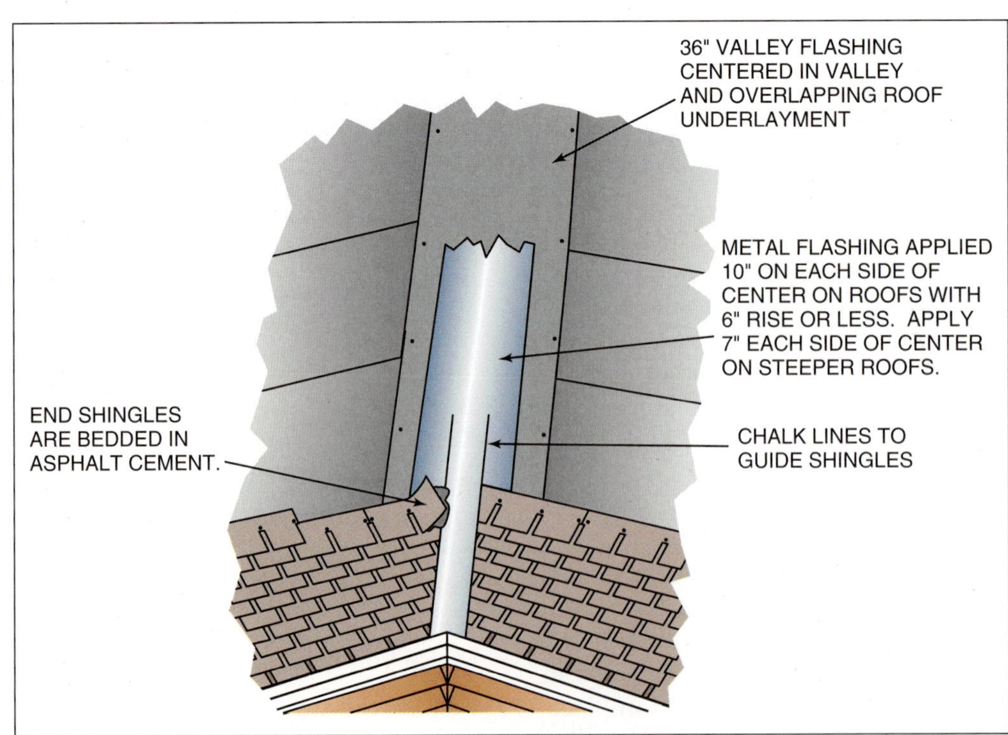


**FIGURE 14-14** Procedure for flashing an open valley with metal.

### Closed Cut Valley Method

*Closed cut valley* involves applying shingles to one roof area that extends into and past the valley. There should be no end joints occurring at or near the center of the valley. The other roof is then installed on top, cut to the valley. Place fasteners no closer than 6 inches from the center of the valley.

***For step-by-step instructions on the CLOSED CUT VALLEY METHOD, see Procedure 14-4 on page 464.***

### Step Flashing Method

*Step flashings* are individual metal pieces tucked between courses of shingles. Each piece should be at least 18 inches wide for slopes with a 6-inch rise or greater and 24 inches wide for slopes with less pitch. The height of each piece should be at least 3 inches more than the shingle exposure. The bottom edges are trimmed to the angle of the roof to match the shingle edges. When the valley is completely step flashed, no metal flashing surface is exposed.

***For step-by-step instructions on the STEP FLASHING METHOD, see Procedure 14-5 on page 465.***

## FLASHING AGAINST A WALL

When roof shingles butt up against a vertical wall, the joint must be made watertight. The usual method of making the joint tight is with the use of metal step flashings. The flashings are purchased or cut about 8 inches in width. They are bent at right angles in the center so they will lay about 4 inches on the roof and extend about 4 inches up the sidewall. The length of the flashings is about 3 inches more than the exposure of the shingles. When used with shingles exposed 5 inches to the weather, they are made 8 inches in length. The roofing is applied and flashed before the siding is applied to the vertical wall.

Apply the starter course, working toward the vertical wall. Fasten a metal flashing on top of the starter course. Its bottom edge should be flush with the drip edge. Use one fastener in each top corner. Lay the first regular course shingle with its end over the flashing and against the sheathing of the sidewall. Do not drive any shingle fasteners through the flashings. The remaining step flashings are placed with each course and against the wall. Keep bottom edges about ½ inch above the butt of the next course of shingles (Fig. 14-15).

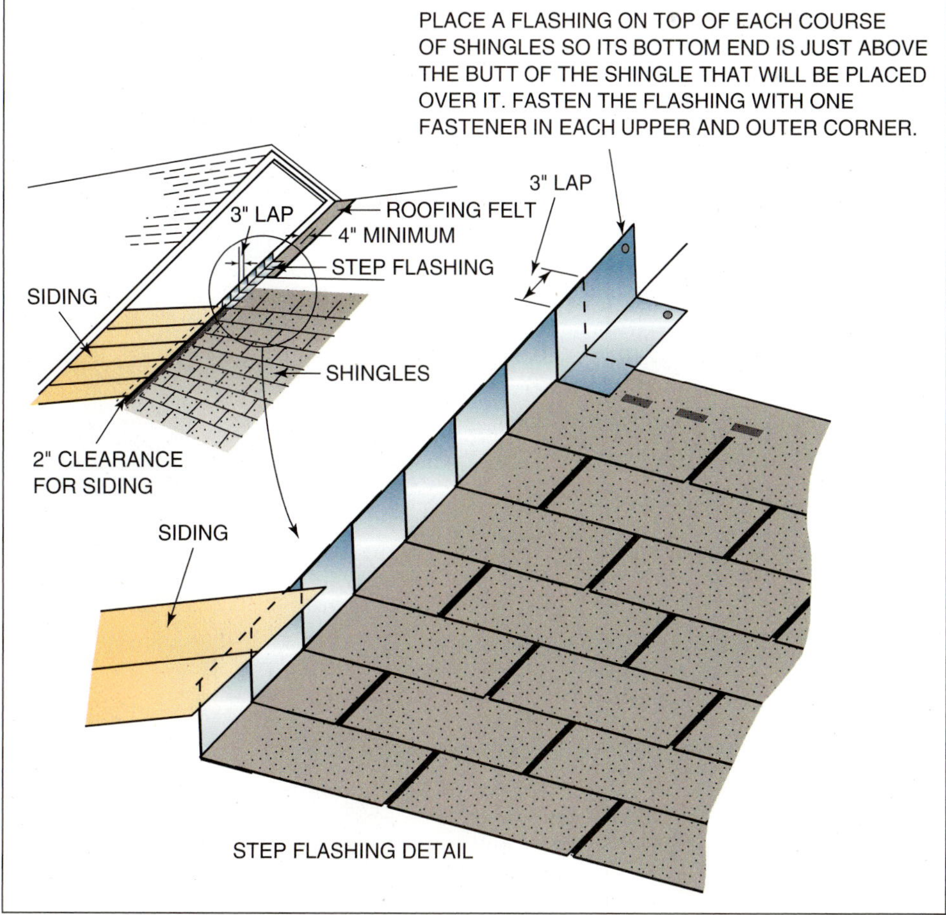


**FIGURE 14-15** Using metal step flashing where a roof joins a wall.

## FLASHING A CHIMNEY

In many cases, especially on steep pitch roofs, a **cricket** or **saddle** is built between the upper (back) side of the chimney and roof deck. The cricket, although not a flashing in itself, prevents accumulation of water behind the chimney (Fig. 14-16).

Flashings are installed by masons who build the chimney. The upper ends of the flashing are bent around and **mortared** in between the courses of brick as the chimney is built. The flashings are long enough to be bent at and over the roof sheathing for tucking between shingles. These flashings are usually in place before the carpenter applies the roof covering.

The underlayment is applied and tucked under the existing flashings. The shingle courses are brought up to the chimney. They are applied under the flashing on the lower side of the chimney. This is called the *apron flashing*. Shingles are tucked under the **apron**. The top edge of the shingles is cut as necessary, until the shingle exposure shows below the apron. The apron is then pressed into place on top of the shingles in a bed of plastic cement. Its projecting ends are carefully and gently formed up around the sides of the chimney and under the lowest side flashings.

Along the sides of the chimney, the flashings are tucked in between the shingles in the same manner as in flashing against a wall. No nails are used in the flashings. The standing portions of the *side flashings* are bedded to the chimney with asphalt cement.

The roof portion is bedded to the shingle. The projecting edges of the lowest side flashings are carefully formed around the corner. They are folded against the low side on the chimney. The top edges of the highest side flashings are also folded around the corner and under the *head flashing* on the upper side of the chimney.

The head flashing is cemented to the roof. Shingles are applied over it. They are bedded to it with asphalt cement. Its projecting ends are also carefully formed around the corner on top of the side flashings.

The projecting ends of chimney flashings are carefully formed and folded around the corners of the chimney. Gently and carefully tap the metal with a hammer handle. Care must be taken not to break through the flashings (Fig. 14-17). Other rectangular roof obstructions, such as skylights, are flashed in a similar manner. The carpenter usually applies the flashings to these obstructions.

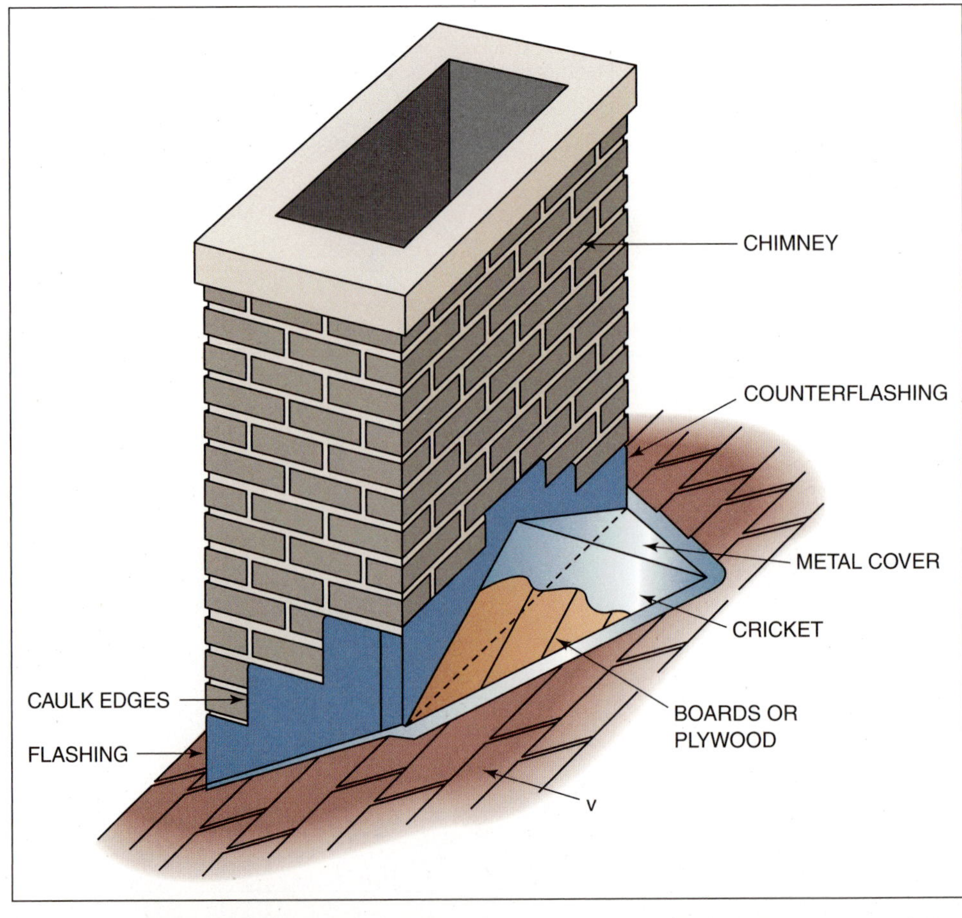


**FIGURE 14-16** A cricket is built to prevent the accumulation of water behind the chimney.

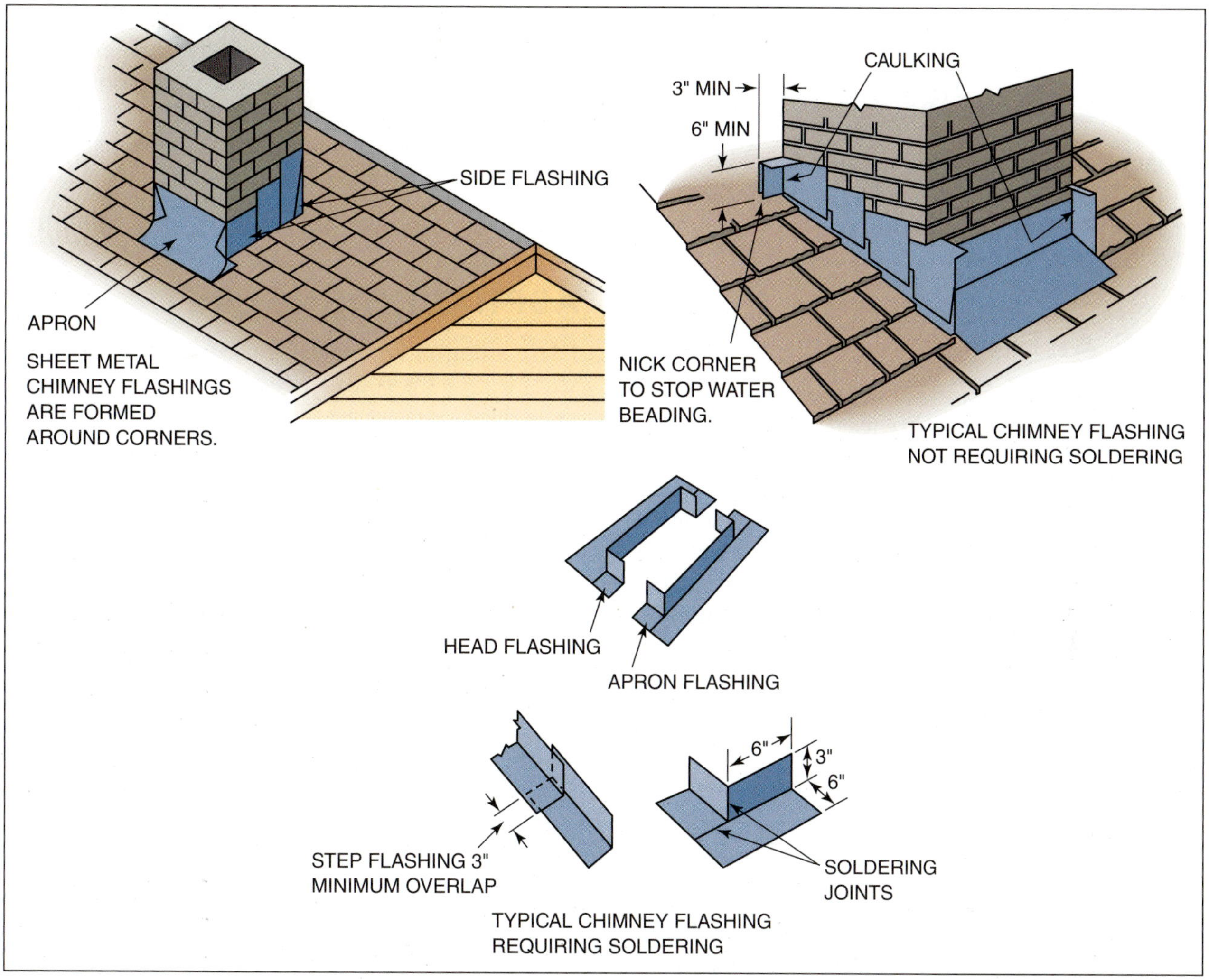

**FIGURE 14-17** Chimney flashings are installed by masons.

## FLASHING VENTS

Flashings for round pipes, such as *stack vents* for plumbing systems and *roof ventilators*, usually come as *flashing collars* made for various roof pitches. They fit around the stack. They have a wide flange on the bottom that rests on the roof deck.

The flashing is installed over the stack vent, with its flange on the roof sheathing. It is fastened in place with one fastener in each upper corner.

Shingle up to the lower end of the stack vent flashing, lifting the lower part of the flange. Apply shingle courses under it. Cut the top edge of the shingles, where necessary, until the shingle exposure, or less, shows below the lower edge of the flashing. Apply asphalt cement under the lower end of the flashing. Press it into place on top of the shingle courses. Apply shingles around the stack and over the flange. Do not drive nails through the flashing. Bed shingles to the flashing with asphalt cement, where necessary (Fig. 14-18).

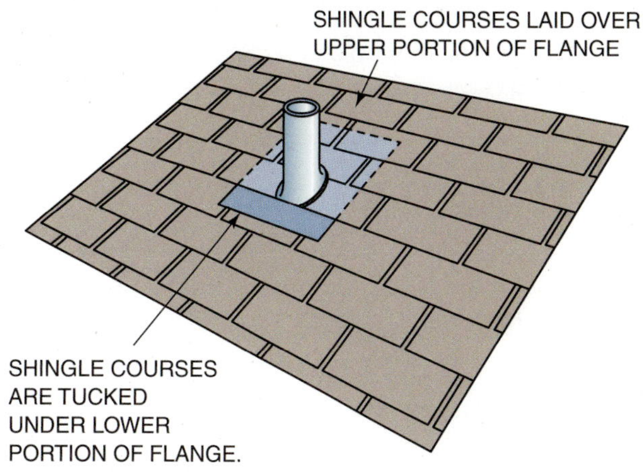


**FIGURE 14-18** Method of shingling around a stack vent flashing.

# Estimating Roofing Materials

Estimating the amount of material needed for roofing involves underlayment and shingles. Remember that if the number of pieces does not come out even, round it up to the next whole number of pieces.

## ROOF AREA

Roofing materials estimates begin with a calculation of the roof area in square feet. Multiply the rafter length including overhang × ridge board length. Double this number for the other side and divide the total by 100 to determine the number of squares needed. For example, if the rafter is 15 feet and the ridge is 66 feet, the roof area is 15 × 66 × 2 ÷ 100 = 19.8 squares. Hip roof area is the same as the gable roof area on the same size building.

## UNDERLAYMENT

Once roof area is determined, the number of rolls of underlayment is estimated by dividing roof area by the area each roll will cover. Rounding the estimated number up to a whole roll usually compensates for any waste. Add 10 percent waste if the roof has hips and valleys. For example, if the roof area for a roof with hips and valleys is 1,920, the number of rolls of 15# – 4 square per roll felt is 1,920 × 110% ÷ 100 ÷ 4 = 5.2 ⇒ 6 rolls.

## ASPHALT SHINGLES

Extra shingles are needed for eave starters and cap shingles. Extra is also needed if the roof has dormers, hips, and valleys. The waste factor percentage added will vary depending on the complicated nature of the roof. Adding 5 to 15 percent for waste is typical. The number of bundles depends on the number of bundles per square. They range from three to five bundles per square. For example, if the area is 1,920 square feet and the waste is 7 percent for four bundles per square shingle, the bundles needed is 1,920 × 107% ÷ 100 × 4 = 82.1 ⇒ 83 bundles.

See Figure 14-19, a listing of the shorthand version of these estimating techniques with another example.

| Estimate the materials for a roof with a 15'-6" gable rafter, 54' ridgeboard, and 4 bundles per square shingles. | | | |
|---|---|---|---|
| **Item** | **Formula** | **Waste Factor** | **Example** |
| **Roof Area** | gable rafter LEN × gable ridgeboard LEN × 2 ÷ 100 SF per square = squares of roof area | | 15.5 × 54' × 2 ÷ 100 = 16.74 squares |
| **Underlayment** | roof squares ÷ 4 square per roll = rolls | | 16.74 ÷ 4 = 4.1 ⇒ 5 rolls |
| **Asphalt Shingles** | roof squares × bundles per square × waste factor = bundles | 5% | 16.74 × 4 × 105% = 70.3 ⇒ 71 bundles |

**FIGURE 14-19** Example of estimating roof materials with formulas.

# Summary

- Asphalt shingles are the most commonly used roofing material.
- Roll roofing is an economical way to provide watershed on low-sloped roofs.
- Flashing provides weatherproofing between different building materials and surfaces.
- Estimating roofing materials is necessary for efficient purchases.

## GREEN CHECKLIST

- ☐ Higher quality shingles last longer and save money in the end. Lesser quality shingles need to be replaced more often, which drives up labor and waste disposal costs.

PROCEDURE 14-1

# Installing Asphalt Shingles

**A** Prepare the roof deck by clearing sawdust and debris that will cause a slipping hazard.

- Install the eaves drip edge first, then apply underlayment. This allows any water on underlayment to drain over the drip edge. Begin underlayment over the deck at a lower corner. Lap the following courses of felt over the lower course at least 2 inches. Make any end laps at least 4 inches. Lap the felt 6 inches from both sides over all hips.

**B** Nail or staple through each lap and through the center of each layer about 16 inches apart. Roofing nails driven through the center of metal discs or specially designed, large head felt fasteners hold the underlayment securely in strong winds until shingles are applied.

- Install metal drip edge along the rake edges on top of the underlayment. This will help prevent blowoffs and prevent wind blown rain from reaching the wood deck.
- Prepare the starter course by cutting off the exposure taps lengthwise through the shingle. Save these tabs as they may be used as the last course at the ridge. Install the starter course so that no end joint will fall in line with an end joint or tab cut out of the regular first course of shingles.

**CAUTION**

**CAUTION:** Installation of roofing systems involves working on ladders and scaffolding as well as on top of the building. Workers should always be aware of the potential for falling. Keep the location of roof perimeter in mind, at all times.

**A**

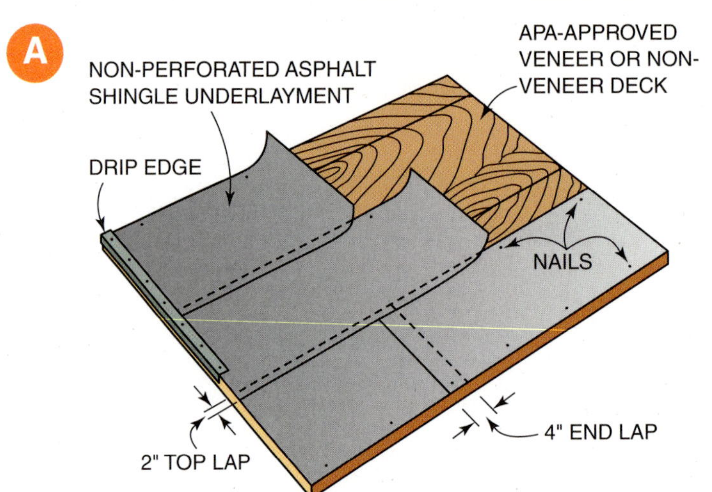


*Courtesy of Asphalt Roofing Manufacturers Association.*

**B**

*Courtesy of APA—The Engineered Wood Association.*

## *from experience...*

Use a utility knife to cut shingles from the back side. Cut only halfway through and then fold and break the shingle to complete the cut. When cutting from the granular top surface, use a hook blade.

**C** Determine the starting line, either the rake edge or vertical center snapped lines. To start from the middle of the roof, mark the center of the roof at the eaves and at the ridge. Snap a chalk line between the marks. Snap a series of chalk lines from this one, 4 or 6 inches apart, depending on the desired end tab, on each side of the centerline. When applying the shingles, start the course with the end of the shingle to the vertical chalk line. Start succeeding courses in the same manner. Break the joints as necessary, working both ways toward the rakes.

**C**

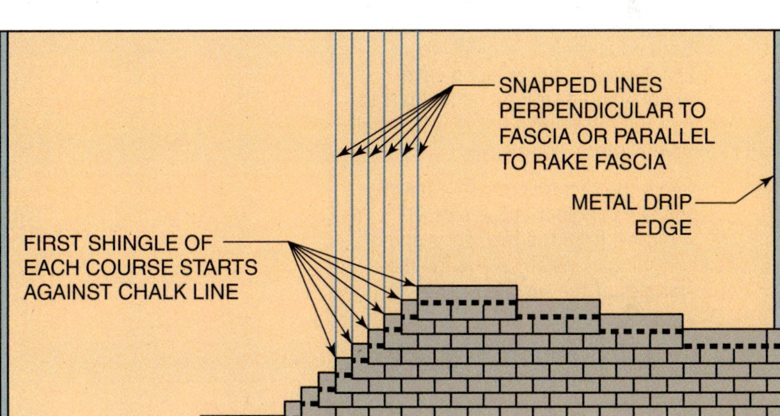

PROCEDURE 14-1

# Installing Asphalt Shingles (continued)

**D** Starting shingle layout at the rake edge involves placing the first course, with a whole tab at the rake edge. The second course is started with a shingle that is 6 to 6½ inches shorter; the third course, with a strip that is 12 or 13 inches shorter; and so on. These starting pieces are precut for faster application.

- Fasten each shingle from the end nearest the shingle just laid. This prevents buckling. Drive fasteners straight so that the nail heads will not cut into the shingles. Both ends of the course should overhang the drip edge ¼ to ⅜ inch.
- Begin eave starter with a piece that has about 6" removed. Overhang both edges about ¼ or ⅜ inch. Install several pieces along eave maintaining a straight edge. Start the first full tab shingle full length. Install several more along eave.
- Begin the second course with a full tab shingle that has 6 or 6½" removed from the rake edge end of the shingle. Install several full tab shingles of the second course.
- Begin the third course with a shingle that has twice the 6 or 6½" increment. Install several full length pieces on top of first course. Watch that expose is uniformly spaced.
- Fourth course begins with one-half a shingle. Repeat as above cutting more off each starter until last course has a piece that is 6 or 6½" long.

**D**

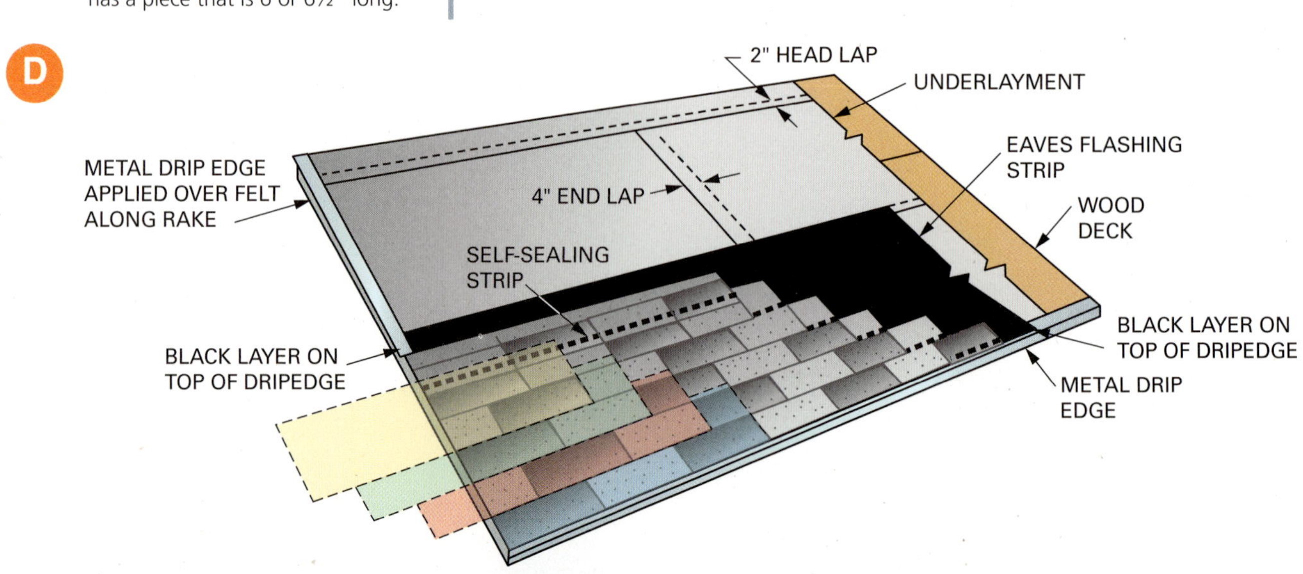

E Install vented ridge cap as per the manufacturer's instructions. Cut cap shingles and begin installation from one end. Center the cap shingle over the vented ridge cap. Secure each shingle with one fastener on each side.

- Apply the cap across the ridge until 3 or 4 feet from the end. Then, space the cap to the end in the same manner as spacing the shingle course to the ridge. The last ridge shingle is cut to size. It is applied with cement and one fastener on each side of the ridge. The two fasteners are covered with asphalt cement to prevent leakage.

E

PROCEDURE 14-2

# Installing Roll Roofing

## Roll Roofing with Concealed Fasteners

Apply 9-inch wide strips of the roofing along the eaves and rakes overhanging the drip edge about ⅜ inch. Fasten with two rows of nails one inch from each edge spaced about 4 inches apart.

- Apply the first course of roofing with its edge and ends flush with the strips. Secure the upper edge with nails staggered about 4 inches apart. Do not fasten within 18 inches of the rake edge.
- Apply cement only to the edge strips covered by the first course. Press the edge and rake edges firmly to the strips. Complete the nails in the upper edge out to the rakes.
- Apply succeeding courses in like manner. Make all end laps 6 inches wide. Apply cement the full width of the lap.
- After all courses are in place, lift the lower edge of each course. Apply the cement in a continuous layer over the full width and length of the lap. Press the lower edges of the upper courses firmly into the cement. A small bead should appear along the entire edge of the sheet. Care must be taken to apply the correct amount of cement.
- To cover the hips and ridge, cut strips of 12" × 36" roofing. Bend the pieces lengthwise through their centers.
- Snap a chalk line on both sides of the hip or ridge down about 5½ inches from the center. Apply cement between the lines. Fit the first strip over the hip or ridge.

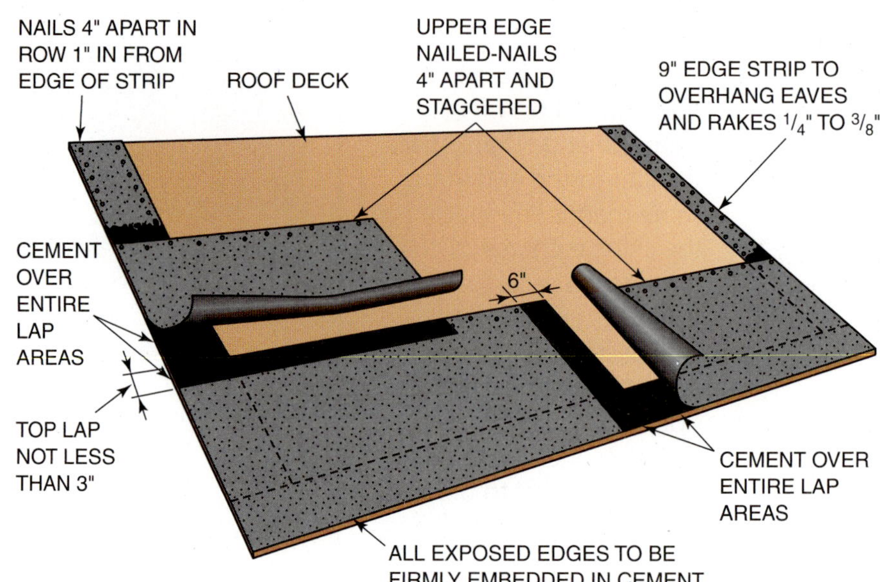

**B** Press it firmly into place. Start at the lower end of a hip and at either end of a ridge. Lap each strip 6 inches over the preceding one. Nail each strip only on the end that is to be covered by the overlapping piece.

- Spread cement on the end of each strip that is lapped before the next one is applied. Continue in this manner until the end is reached.

## Double Coverage Roll Roofing

**A** Cut the 19-inch strip of *selvage,* non-mineral surface side, from enough double coverage roll roofing to cover the length of the roof. Save the surfaced portion for the last course at the ridge. Apply the selvage portion parallel to the eaves. It should overhang the drip edge by $\frac{3}{8}$ inch. Secure it to the roof deck with three rows of nails.

**B** Apply the first course by using a full width strip of roofing. Secure it with two rows of nails in the selvage portion.

**B**

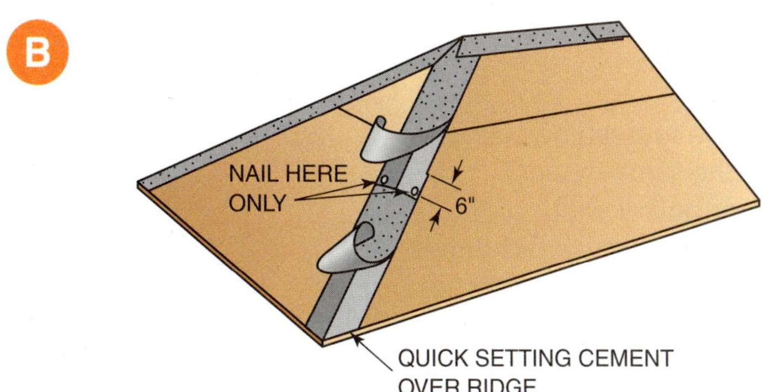

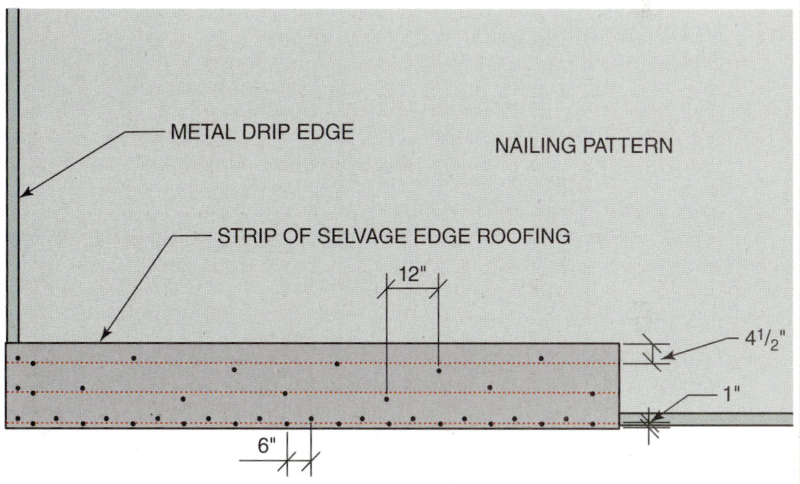


**B**

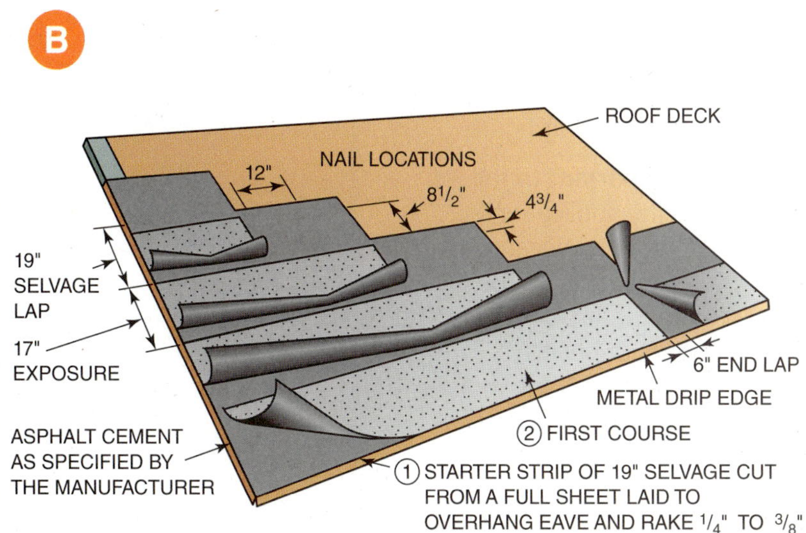

PROCEDURE 14-2

# Installing Roll Roofing (continued)

**C** Apply succeeding courses in the same manner. Lap the full width of the 19-inch selvage each time. Make all end laps 6 inches wide. End laps are made in the manner shown in the accompanying figure. Stagger end laps in succeeding courses.

- Lift and roll back the surface portion of each course. Starting at the bottom, apply cement to the entire selvage portion of each course. Apply it to within ¼ inch of the surfaced portion. Press the overlying sheet firmly into the cement.

C

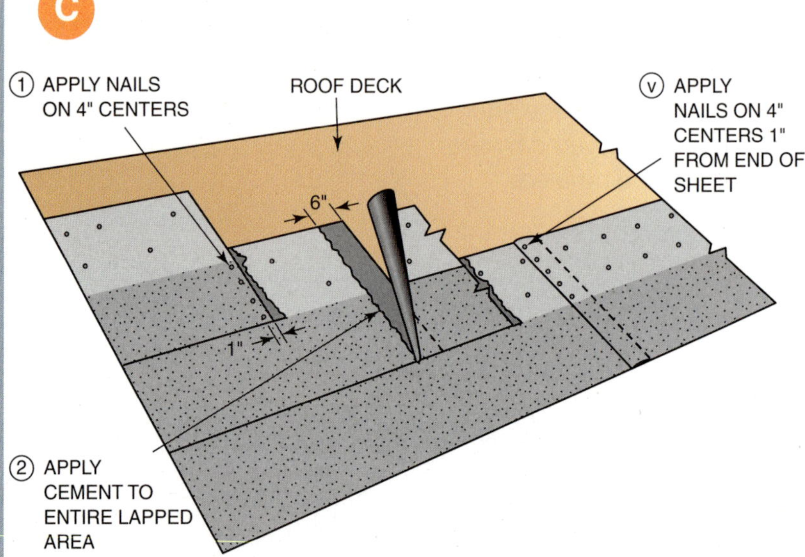


**D** Apply pressure over the entire area by using a light roller to ensure adhesion between the sheets at all points. Apply the remaining surfaced portion left from the first course as the last course. Hips and ridges are covered in the same manner shown in the accompanying figure.

- It is important to follow specific application instructions because of differences in the manufacture of roll roofing. Specific requirements for quantities and types of adhesive must be followed.

D

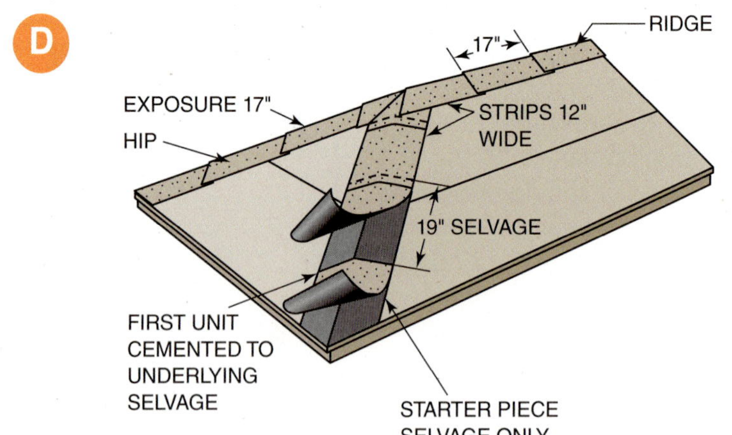

PROCEDURE 14-3

# Woven Valley Method

**A** Install underlayment and starter strip to both roofs.

- Apply first course of one roof, say the left one, into and past the center of the valley. Press the shingle tightly into the valley and nail, keeping the nails at least 6 inches away from the valley centerline. Cut shingles to adjust the butt ends so there is no butt seam within 12 inches of the valley centerline.
- Apply the first course of the other (right) roof in a similar manner, into and past the valley.
- Succeeding courses are applied by repeating this alternating pattern, first from one roof and then on the other.

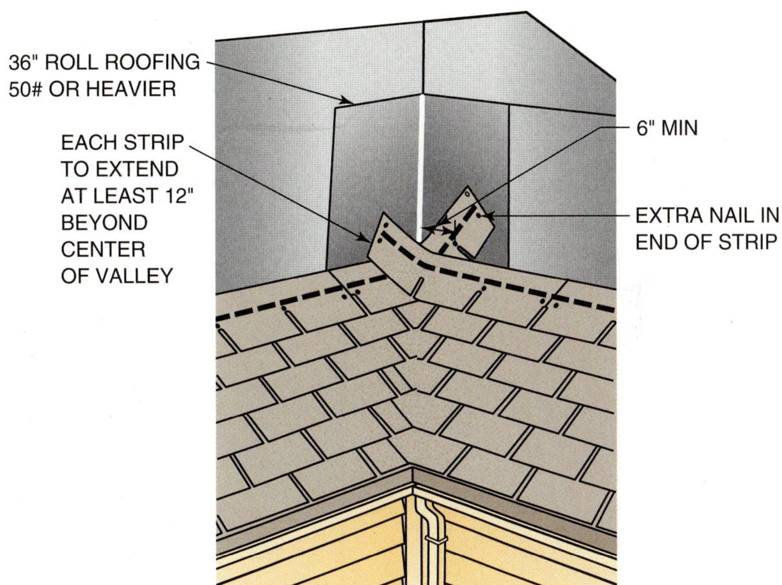

PROCEDURE 14-4

# Closed Cut Valley Method

**A** Begin by shingling first roof completely, letting the end shingle of every course overlap the valley by at least 12 inches. Form the end shingle of each course snugly into the valley. Cut shingles to adjust the butt ends so there is no butt seam within 12 inches of the valley centerline.

- Snap a chalk line along the center of the valley on top of the shingles of the first roof.
- Apply the shingles of second roof, cutting the end shingle of each course to the chalk line. Place the cut end of each course that lies in the valley in a 3-inch wide bed of asphalt cement.

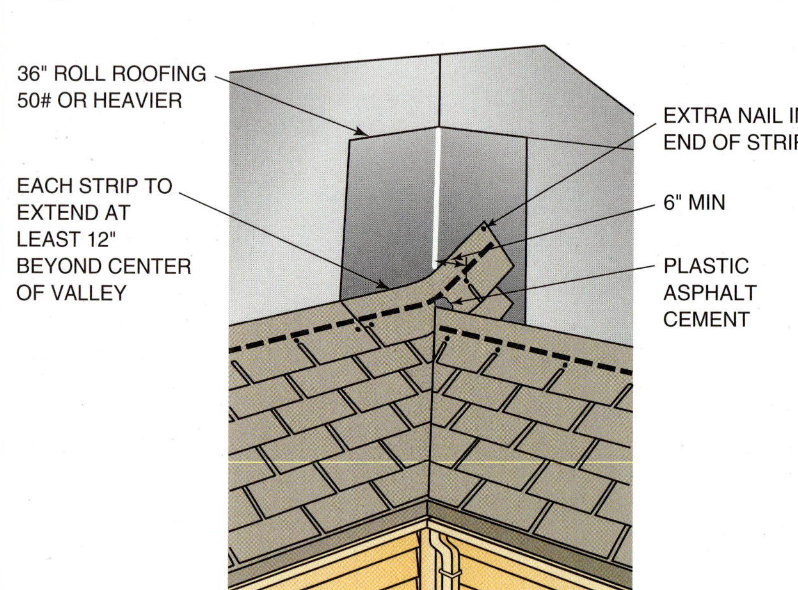

PROCEDURE 14-5

## Step Flashing Method

**A** Snap a chalk line in the center of the valley on the valley underlayment.

- Apply the shingle starter course on both roofs. Trim the ends of each course that meet the chalk line.
- Fit and form the first piece of flashing to the valley on top of the starter strips. Trim the bottom edge flush with the drip edge. Fasten with two nails in the upper corners of the flashing only. Use nails of like material to the flashing to prevent electrolysis.
- Apply the first regular course of shingles to both roofs on each side of the valley, trimming the ends to the chalk line. Bed the ends in plastic asphalt cement. Do not drive nails through the metal flashing. Apply flashing to each succeeding course in this manner.

**A**

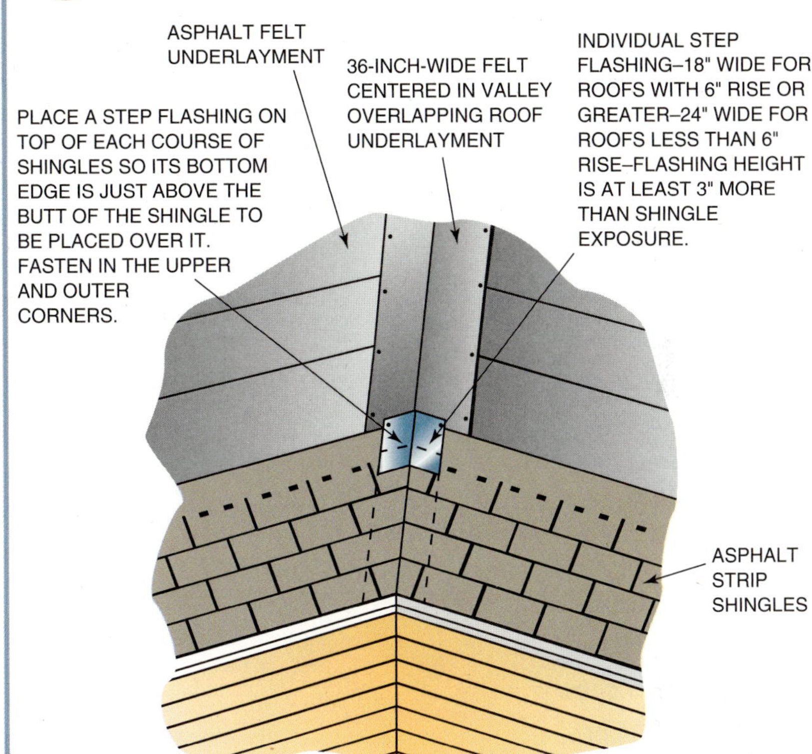

## REVIEW QUESTIONS

*Select the most appropriate answer.*

1. **A square is the amount of roofing required to cover**
   a. 50 square feet.
   b. 100 square feet.
   c. 150 square feet.
   d. 200 square feet.

2. **One roll of #15 asphalt felt will cover**
   a. 1 square.
   b. 2 square.
   c. 3 square.
   d. 4 square.

3. **When applying asphalt felt on a roof deck as under-layment, lap each course over the lower course by at least**
   a. 2 inches.
   b. 3 inches.
   c. 4 inches.
   d. 6 inches.

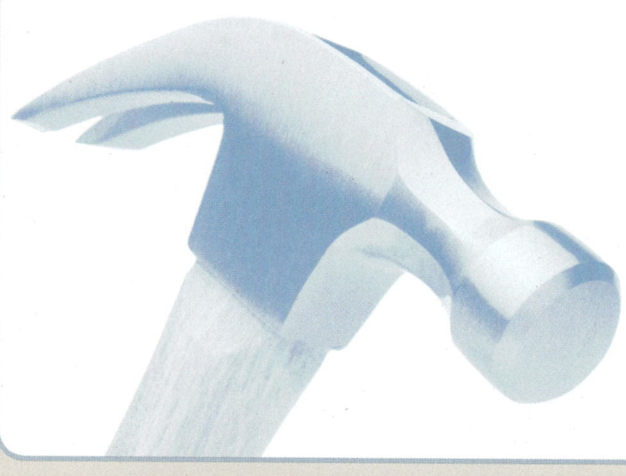

4. **Asphalt strip shingles should not be used on roof slopes lower than**
   a. 1 inch rise.
   b. 2 inches rise.
   c. 3 inches rise.
   d. 4 inches rise.

5. **When applying asphalt shingles, it is recommended that no rake tab be less than**
   a. 2 inches in width.
   b. 3 inches in width.
   c. 4 inches in width.
   d. 5 inches in width.

6. **For slopes 6 on 12 or less, valley step flashings should extend on each side of the valley centerline by at least**
   a. 6 inches.
   b. 8 inches.
   c. 10 inches.
   d. 12 inches.

7. **When flashing a valley by weaving shingles, do not locate any nails closer to the valley centerline than**
   a. 6 inches.
   b. 8 inches.
   c. 10 inches.
   d. 12 inches.

8. **Step flashings 8 inches wide are used when flashing a roof that butts against a vertical wall. They are bent so that**
   a. 3 inches lays on the wall, and 5 inches on the roof.
   b. 4 inches lays on the wall, and 4 inches on the roof.
   c. 5 inches lays on the wall, and 3 inches on the roof.
   d. 2 inches lays on the wall, and 6 inches on the roof.

**9. A built-up section between the roof and the upper side of a chimney is called a**

a. cricket.

b. dutchman.

c. furring.

d. counterflashing.

**10. Bundles of asphalt shingles to cover one square are between**

a. one to two bundles.

b. two to three bundles.

c. three to four bundles.

d. three to five bundles.

**11. While installing standard shingles, the space remaining to be covered at the top is measured to be 46¼ inches. How many courses and what exposure are needed to make the last course of shingles properly sized?**

a. 9 courses and 4¾ inches exposure

b. 9½ courses and 4½ inches exposure

c. 10 courses and 4⅝ inches exposure

d. 10½ courses and 4⅜ inches exposure

**12. How many square of shingles are needed for a gable roof that has a fascia length of 46 feet and a rafter length of 16 feet rounded up to the nearest square and neglecting any waste factor.**

a. 7 square

b. 8 square

c. 14 square

d. 15 square

## KNOW YOUR CODES

Roofing manufacturers stipulate different installation techniques for regions of high winds and moderate or low wind regions. Check the manufacturer's instructions on the bundle with the local code for the correct installation procedure.

## WHAT'S WRONG WITH THIS PICTURE?

Carefully study **Figure 14-20** and think about what is wrong. Consider all possibilities.

### WRONG

**FIGURE 14-20** The figure shows a section of roof where it abuts a sidewall. The problem here is the flashing. The flashing shown is continuous and not "stepped." If water were to run down the siding (in this case the siding has not yet been applied), it would very likely work its way underneath the shingles. Continuous flashing has no built-in mechanism to get the water back to the surface of the roof again. Step flashing, not continuous flashing, is the proper technique on a sidewall.

### RIGHT

**FIGURE 14-21** This is an example of step flashing. The flashing pieces are alternately placed with the roof shingles (shingle, step flashing, shingle, step flashing, and so on). This way, if water were to roll down the sidewall and underneath the shingle, it would resurface on the top of the next shingle. Before the siding is applied, building paper should cover the exposed step flashing; the building paper would act as a counterflashing in case moisture worked its way behind the siding.

# Siding and Decks

**OBJECTIVES** *After completing this unit, the student should be able to:*

- Describe the shapes, sizes, and materials used as siding products.
- Install corner boards and prepare sidewall for siding.
- Apply horizontal and vertical siding.
- Apply plywood and lapped siding.
- Apply wood shingles and shakes to sidewalls.
- Apply vinyl and aluminum siding.
- Describe various types of cornices and name their parts.
- Install gutters and downspouts.
- Describe the construction of and kinds of materials used in decks.
- Lay out and install footings, supporting posts, girders, and joists for a deck.
- Apply decking in the recommended manner and install flashing for an exposed deck against a wall.

# GLOSSARY OF SIDING AND DECK TERMS

**baluster** vertical members of a stair rail, usually decorative and spaced closely together

**battens** thin, narrow strips typically used to cover joints in vertical boards

**blind nail** a method of fastening that conceals the fastener

**brake** a tool used to bend sheet metal

**coil stock** aluminum sheets sold in 50-foot long rolls and widths ranging from 12 to 24 inches.

**corner boards** finish trim members used at the intersection of exterior walls

**cornice** a general term used to describe the part of the exterior finish where the walls meet the roof

**downspout** a vertical member used to carry water from the gutter downward to the ground; also called *leader*

**drip** that part of an exterior finish that projects below another to cause water to drop off instead of running back against and down the wall

**eaves** the lower part of the roof that extends beyond and overhangs the sidewalls

**frieze** that part of exterior trim applied to cover the joint between the cornice and the siding

**gutter** a trough attached to an eave used to carry off water

**plancier** the finish member on the underside of a box cornice; also called a *soffit*

**rake** the sloping portion of the roof that overhangs the gable end

**soffit** the horizontal, underside trim member of a cornice or other overhanging assembly

**story pole** a narrow strip of wood used to lay out the installation heights of material such as siding or vertical members of a wall frame

**striated** finish material with random and finely spaced grooves running with the grain

**water table** exterior trim members applied at the intersection of the siding and the foundation to direct water away from the building

The portion of the exterior that covers the vertical surface areas of a building is collectively called siding. This does not include masonry coverings, such as stucco or brick veneer. Siding gives the building its architectural look. It must also stand up to local weather conditions. Some areas are hot and dry, while others have wind-driven rain.

The area where the lower portion of the roof, or eaves, overhangs the walls is called the cornice. The cornice can be made more decorative with variations in its details. These details may vary with the locale and the expertise of the carpenter. Craftsmanship is important to ensure the exterior finish is watertight and attractive.

Decks are a popular way to increase the living area of a house. Plans may not show specific construction details. Therefore, it is important to know the techniques used to build these structures.

Scaffolds and ladders are used during this stage of construction. Every safety concern previously stated applies here. Because exterior finish covers a large area in a small period of time, it is often set up and removed in the same day. This makes the risk of careless scaffold construction more possible. It is easy to think that because it is being set up for only a few hours that shortcuts may be taken. There is no substitute for safety. A relentless emphasis on safety makes a long life more likely.

# SIDING

Siding is manufactured from solid lumber, plywood, hardboard, aluminum, cement, and vinyl. It comes in many different patterns. Siding may be applied horizontally, vertically, or in other directions, to make many interesting designs (Fig. 15-1). The names and shapes of various wood siding patterns are shown in Figure 15-2. Some patterns can be used for either horizontal or vertical applications. Others can be used for both. Bevel siding is a widely used kind (Fig. 15-3). Hardboard and cementitious siding is made in various styles and surface grain patterns. Most are made to look like wood. Cementitious siding or fiber-cement siding, such as Cemplank®, has an additional benefit of long life guarantees. Fiber-cement siding is made of a fiber reinforced cementitious material. While it has a similar variety of finishes as with wood, it also has excellent decay and termite resistance. Special considerations must be made when cutting. Cutting of fiber-cement boards produces silica dust. This dust is known to cause cancer. All cutting should be done with special dust-reducing circular saw blades or a set of shears. Cutting should be done outside and downwind of any other workers.

Panel siding, made from plywood manufactured by APA member mills, is known as *APA303* siding. It is produced in a number of surface textures and patterns (Fig. 15-4). It comes in several thicknesses, 4-foot widths, and lengths of 8, 9, and 10 feet. It is usually applied vertically.

Most panel siding is shaped with *shiplapped* edges for weathertight joints.

Lap siding is manufactured with a variety of materials and surface textures. Some surfaces are *grooved* or *beaded*. Edges are square, beveled, shiplapped, or tongue-and-grooved. They come in thicknesses from 7⁄16 to 13⁄16 inch, widths from 4 to 12 inches, and lengths of up to 12 feet (Fig. 15-5).

## PREPARATION FOR SIDING APPLICATION

To maximize the siding paint longevity, drying potential must be built into the siding. This can be achieved by furring the siding off the sheathing and

Courtesy of California Redwood Association.

**FIGURE 15-1** Wood siding (redwood as shown here) is often used in residential construction.

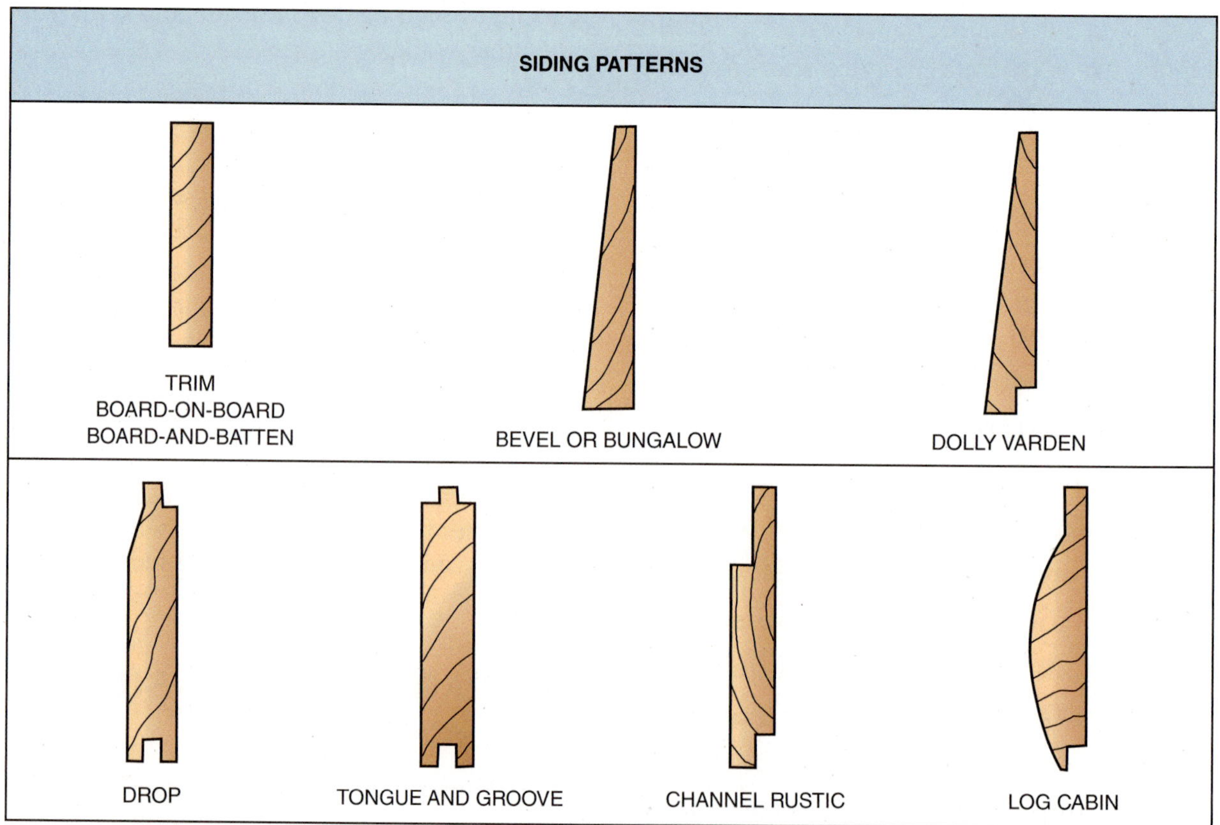


Courtesy of Western Wood Products Association.

**FIGURE 15-2** Names, descriptions, and sizes of natural wood siding patterns.

**FIGURE 15-3** Bevel siding is commonly known as clapboards. Redwood can be left natural or painted.

allowing air to circulate behind the siding (Fig. 15-6). As air circulates behind the siding, it mixes with soffit air and eventually leaves at the ridge.

**Green Tip**

Furring wood siding away from sheathing allows painted siding surfaces to last two to three times longer.

Furring may be 1 × 3, 1 × 4, or 4-inch strips of plywood nailed to the sheathing over each stud. Screen is used to protect the airspace from insects. Furring should be installed before windows are installed. This maintains normal exterior finish details.

A 6- to 8-inch-wide screen is smoothly stapled to the sheathing about 4 inches above the bottom edge of the siding. The extra screen is folded up and over after the furring is applied. Caulking or adhesive applied to the screen before the siding is applied seals the screen to the backside of the siding.

Next, it must be determined how it will end at the foundation, eaves, and corners. The installation of various kinds of exterior wall trim may first be required.

### Foundation Trim

In most cases, no additional trim is applied at the foundation. The siding is started so that it extends slightly below the top of the foundation. However, a **water table** may be installed for appearance. It sheds water a little farther away from the foundation. The water table usually consists of a board and a drip cap installed around the perimeter. Its bottom edge is slightly below the top of the foundation. The siding is started on top of the water table (Fig. 15-7).

### Eaves Treatment

At the eaves, the siding may end against the bottom edge of the **frieze**. The width of the frieze may vary. The siding may also terminate against the **soffit**.

### Rake Trim

At the **rakes**, the siding may be applied under a furred-out rake fascia. When the rake overhangs the sidewall, the siding may be fitted against the rake soffit. When fitted against the rake soffit, the joint is covered with a molding.

### Treating Corners

One method of treating corners is with the use of **corner boards**. The thickness of corner boards depends on the type of siding used. The corner boards should be thick enough so that the siding does not project beyond the face of the corner board.

The width of the corner boards depends on the desired effect. However, one of the two pieces making up an outside corner should be narrower than the other by the thickness of the stock. Then, after the wider piece is fastened to the narrower one, the same width is exposed on both sides of the corner. The joint between the two pieces should be on the side of the building away from the prevailing weather.

Horizontal siding may also be mitered around exterior corners. Or, metal corners may be used on each course of siding. In interior corners, siding courses may butt against a square corner board or against each other (Fig. 15-8).

### Installing Corner Boards

In some high-wind areas, vertical flashing is added before corner boards are installed. Strips of #15 felt

**BRUSHED**

Brushed or relief-grain surfaces accent the natural grain pattern to create striking textured surfaces. Generally available in 11/32", 3/8", 1/2", 19/32" and 5/8" thicknesses. Available in redwood, Douglas fir, cedar, and other species.

**KERFED ROUGH-SAWN**

Rough-sawn surface with narrow grooves providing a distinctive effect. Long edges shiplapped for continuous pattern. Grooves are typically 4" OC. Also available with grooves in multiples of 2" OC. Generally available in 11/32", 3/8", 1/2", 19/32" and 5/8" thicknesses. Depth of kerfgroove varies with panel thickness.

**APA TEXTURE 1-11**

Special 303 Siding panel with shiplapped edges and parallel grooves 1/4" deep, 3/8" wide; grooves 4" or 8" OC are standard. Other spacings sometimes available are 2", 6" and 12" OC, check local availability. T 1-11 is generally available in 19/32" and 5/8" thicknesses. Also available with scratch-sanded, overlaid, rough-sawn, brushed and other surfaces. Available in Douglas fir, cedar, redwood, southern pine, other species.

**ROUGH-SAWN**

Manufactured with a slight, rough-sawn texture running across panel. Available without grooves, or with grooves of various styles; in lap sidings, as well as in panel form. Generally available in 11/32", 3/8", 1/2", 19/32" and 5/8" thicknesses. Rough-sawn also available in Texture 1-11, reverse board-and-batten (5/8" thick), channel groove (3/8" thick), and V-groove (1/2" or 5/8" thick). Available in Douglas fir, redwood, cedar, southern pine, other species.

**CHANNEL GROOVE**

Shallow grooves typically 1/16" deep, 3/8" wide, cut into faces of 3/8" thick panels, 4" or 8" OC. Other groove spacings available. Shiplapped for continuous patterns. Generally available in surface patterns and textures similar to Texture 1-11 and in 11/32", 3/8" and 1/2" thicknesses. Available in redwood, Douglas fir, cedar, southern pine and other species.

**REVERSE BOARD-AND-BATTEN**

Deep, wide grooves cut into brushed, roughsawn, coarse sanded or other textured surfaces. Grooves about 1/4" deep, 1" to 1-1/2" wide, spaced 8", 12" or 16" OC with panel thickness of 19/32" and 5/8". Provides deep, sharp shadow lines. Long edges shiplapped for continuous pattern. Available in redwood, cedar, Douglas fir, southern pine and other species.

*Courtesy of APA—The Engineered Wood Association.*

**FIGURE 15-4** APA303 plywood panel siding is produced in a wide variety of sizes, surface textures, and patterns.

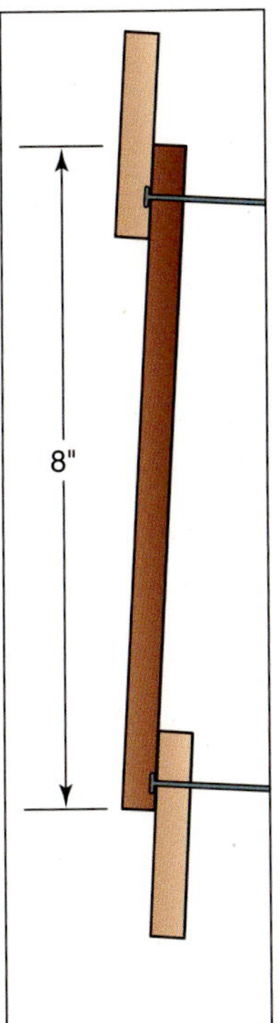


**FIGURE 15-5** Styles of lap siding.

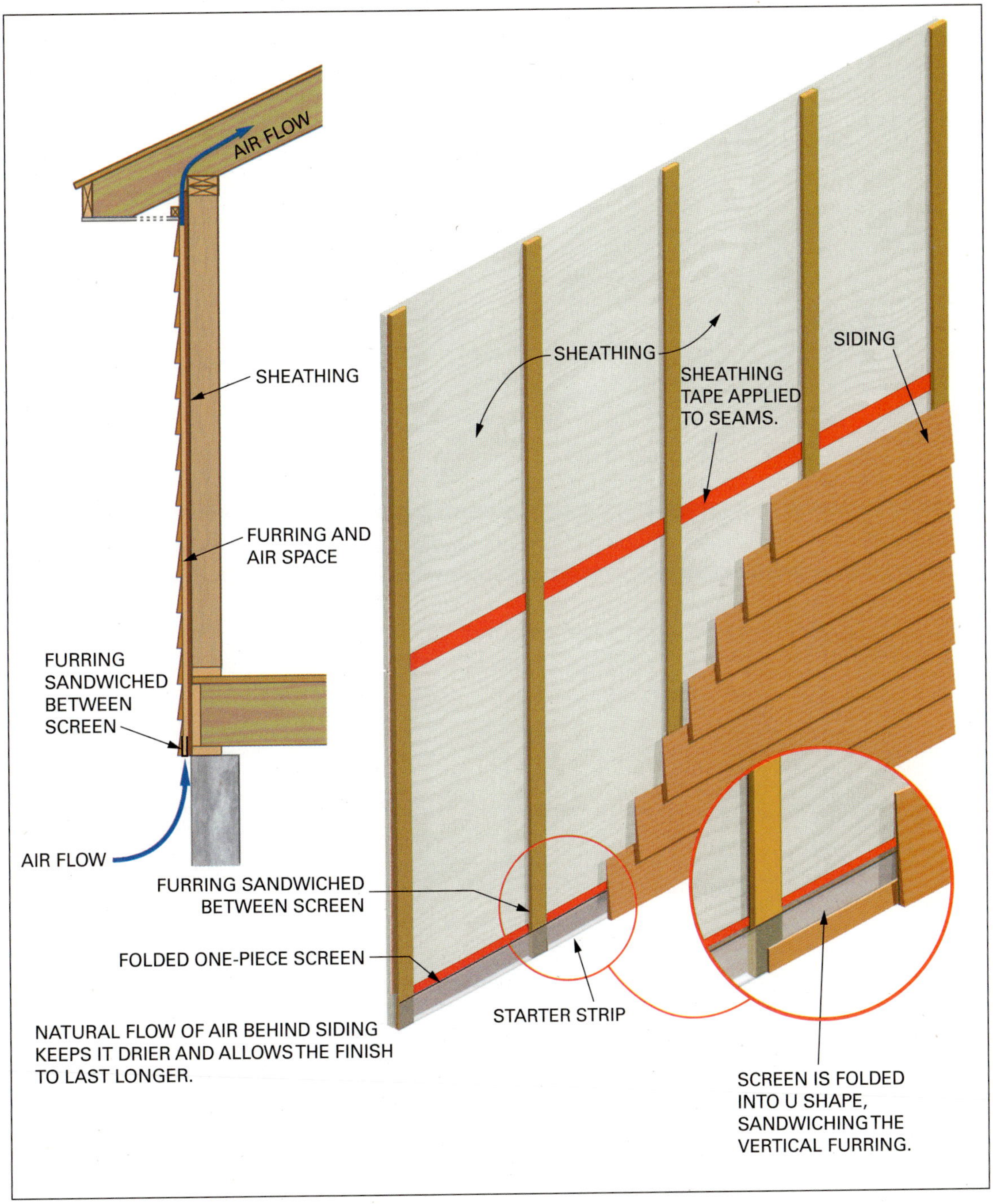

**FIGURE 15-6** Siding lasts longer when it is furred off the sheathing.

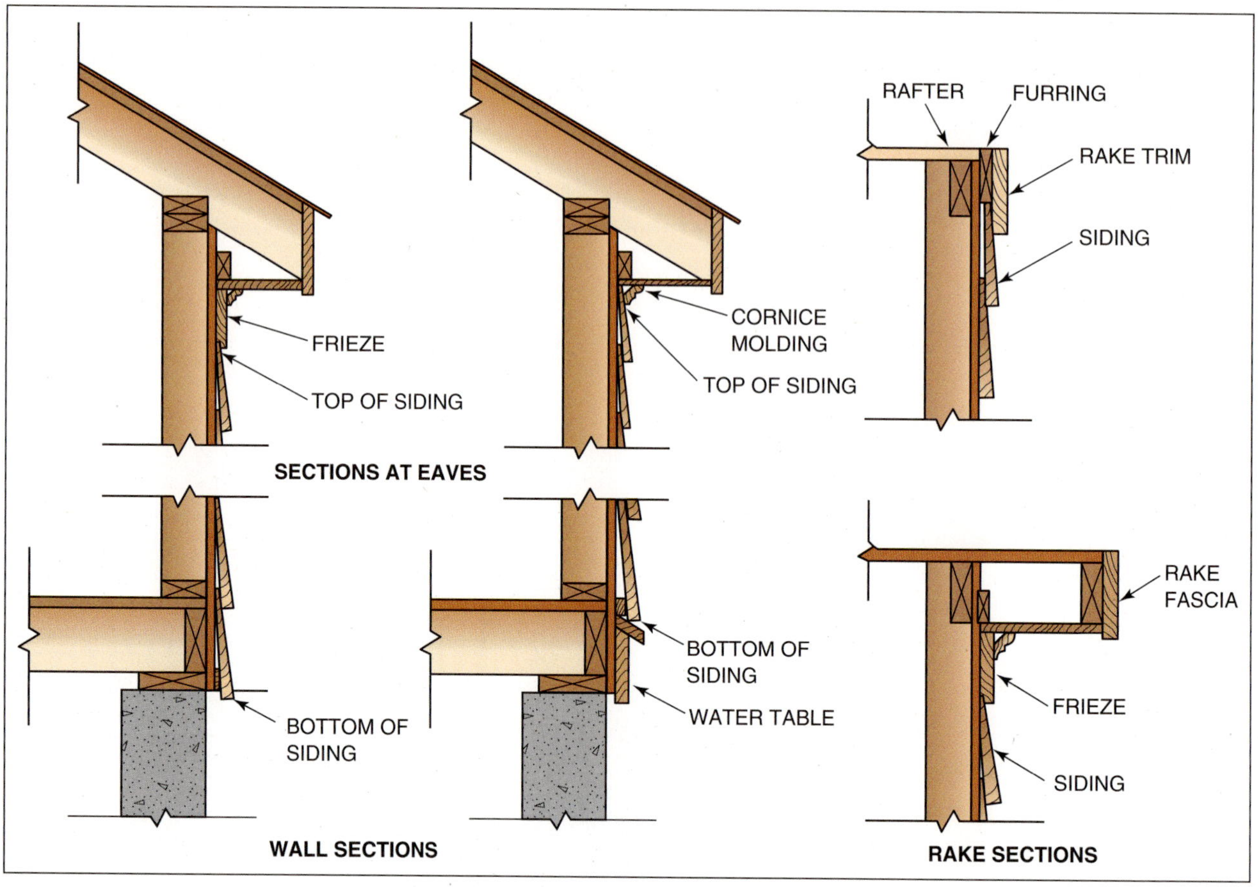

RAFTER
FURRING
RAKE TRIM
SIDING
FRIEZE
TOP OF SIDING
CORNICE MOLDING
TOP OF SIDING
SECTIONS AT EAVES
RAKE FASCIA
BOTTOM OF SIDING
WATER TABLE
FRIEZE
BOTTOM OF SIDING
SIDING
WALL SECTIONS
RAKE SECTIONS

**FIGURE 15-7** Methods of ending siding at foundation, eaves, and rake areas.

CORNER STRIP
BUTT JOINT
SIDING
MITER NAIL
MITERED CORNER
METAL CORNERS
CORNER BOARD
INTERIOR CORNER

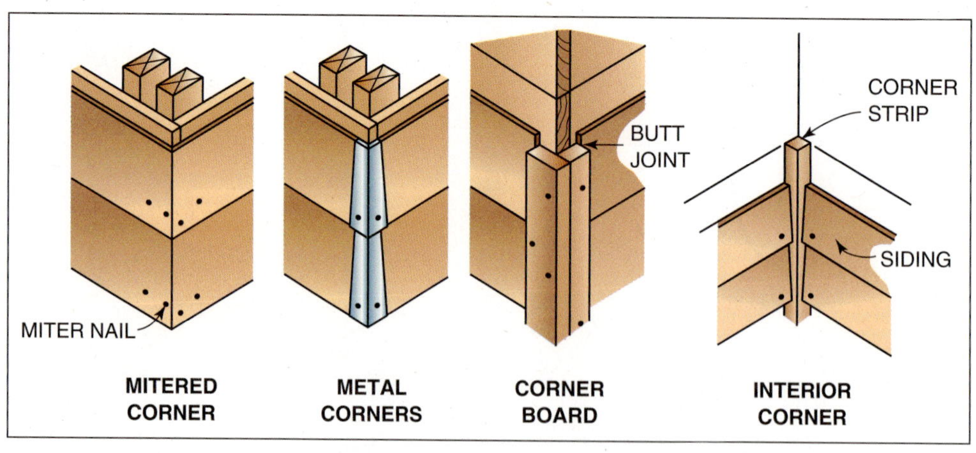

**FIGURE 15-8** Methods of returning and stopping courses of horizontal siding.

or ice and water shield are installed vertically. One edge should extend beyond the edge of the corner board by at least 2 inches. Tuck the top ends under any previously applied housewrap.

Corner boards may be applied by fastening the two pieces together first. Then, install the assembly on the corner. Cut, fit, and fasten the narrower corner piece. Start at one end, top or bottom, and work toward the other, making sure to keep the beveled edge flush with the corner. Fasten with galvanized or other noncorroding nails spaced about 16 inches apart along its inside edge (Fig. 15-9).

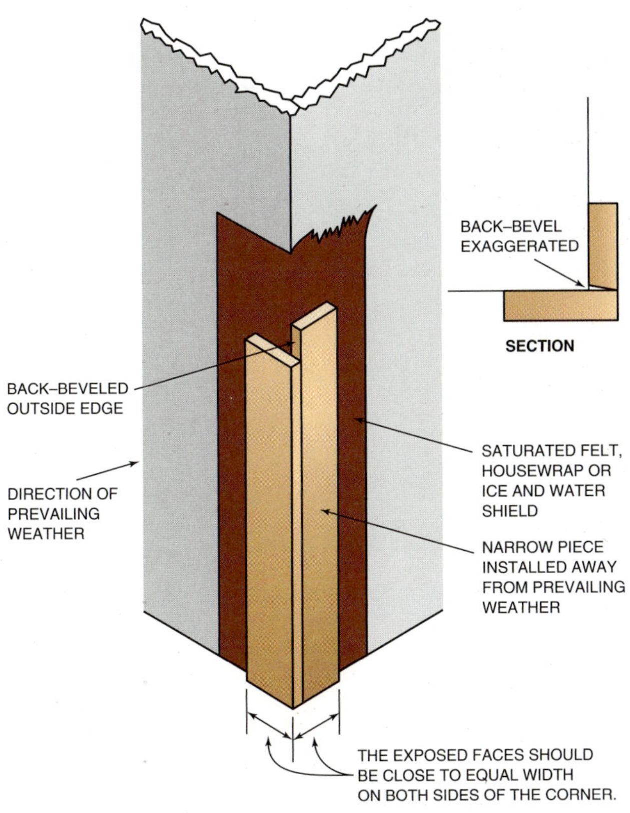


**FIGURE 15-9** Procedure for installing corner boards.

Cut, fit, and fasten the wider piece to the corner in a similar manner. Make sure its outside edge is flush with the surface of the narrower piece. The outside row of nails is driven into the edge of the narrower piece.

Slightly round over all sharp exposed corners. Set all nails so they can be filled over later. Make sure a tight joint is obtained between the two pieces.

### Applying Building Paper

The housewrap should lap over the top of any flashing applied at the sides and tops of windows and doors and at corner boards. It should be tucked under any flashings applied under the bottoms of windows or frieze boards. In any case, all laps should be over the paper below because water runs downhill.

## INSTALLING HORIZONTAL SIDING

One of the important differences between bevel siding and other horizontal siding types is that exposure of courses of bevel siding can be varied. With other types, the groove or lap is constant with every course and cannot vary. It is desirable from the standpoint of appearance, weathertightness, and ease of application to have a full course of horizontal siding exposed above and below windows and over the tops of doors. The exposure of the siding may vary gradually up to a total of 1 inch over the entire wall, but the width of each exposure should not vary more than 1/4 inch from its neighbor. It may not always be possible to lay out full siding courses above and below every window or door.

Once the siding is laid out for the desired exposure, the layout is transferred to other locations with a **story pole**. A story pole is a strip of wood placed next to the first layout marks. These marks are made onto the strip of wood to create the story pole. It can then be moved to other locations to easily reproduce the layout anywhere it is needed.

***For step-by-step instructions on INSTALLING HORIZONTAL SIDING, see Procedure 15-1 on pages 500–503.***

## INSTALLING TONGUE-AND-GROOVE SIDING

Tongue-and-groove siding may be installed horizontally or vertically. Vertical applications tend to shed water better than horizontal. Corner boards usually are not used when wood siding is applied vertically. The siding boards are fitted around the corner (Fig. 15-10). The starting piece has the grooved edge removed. The tongue edge should be plumb, and the bottom end should be about 1 inch below the sheathing.

The top end should neatly butt or be tucked under any trim above. To fasten the strips, **blind nail** through the tongue edge. Nails should be placed from 16 to 24 inches apart. *Blocking* must be provided between studs if siding is applied directly to the frame.

***For step-by-step instructions on INSTALLING VERTICAL TONGUE-AND-GROOVE SIDING, see Procedure 15-2 on pages 504–505.***

## INSTALLING PANEL SIDING

Panel siding may be installed both vertically and horizontally. It may be applied to sheathing or directly to studs if backing is provided for all joints

Courtesy of California Redwood Association.

**FIGURE 15-10** Vertical tongue-and-groove siding (redwood as shown here) needs little accessory trim, such as corner boards.

(Fig. 15-11). It can be installed horizontally, if desired. All vertical edges should start and end on a stud. Fasten panels of ½-inch thickness or less with 6d *siding nails.* Use 8d siding nails for thicker panels. Fasten panel edges about every 6 inches and about every 12 inches along intermediate studs. Leave a ⅛-inch space between panels. A minimum of 6 inches should extend above the finished grade line.

***For step-by-step instructions on INSTALLING PANEL SIDING, see Procedure 15-3 on page 505.***

## WOOD SHINGLE AND SHAKE SIDING

Shingle and shake siding comes in a variety of shapes and sizes (Fig. 15-12). *Rebutted* and *rejointed* types are machine trimmed with parallel edges and square butts. Rebutted and rejointed machine-grooved, sidewall shakes have **striated** faces. Special *fancy butt* shingles provide interesting patterns in combination with square butts or other types of siding (Fig. 15-13). Red cedar shingles, factory-applied on 4- and 8-foot panels, are also available in several styles and exposures.

Wood shingles and shakes may be applied to sidewalls in either single-layer or double-layer courses. In *single coursing,* shingles are applied to walls with a single layer in each course. In *double coursing,* two layers are applied in one course. Consequently, even greater weather exposures are allowed. Double coursing is used when wide courses with deep, bold shadow lines are desired (Fig. 15-14).

### Installing Wood Shingles and Shakes

A *starter course* for sidewall shingles and shakes is a double layer used for single-course applications. A triple layer is used for double coursing. Less expensive *undercourse* shingles are used for lower layers. The exposure of shingle courses may be laid out in the same manner as with horizontal wood siding. Each shingle, up to 8 inches wide, is fastened with two nails or staples about ¾ inch in from each edge. On shingles wider than 8 inches, drive two additional nails about 1 inch apart near the center. Fasteners should be hot-dipped galvanized, stainless steel, or aluminum. They should be driven about 1 inch above the butt line of the next course. Fasteners must be long enough to penetrate the sheathing by at least ⅛ inch.

***For step-by-step instructions on INSTALLING WOOD SHINGLES AND SHAKES, see Procedure 15-4 on pages 507–508.***

### Corners

Shingles may be butted to corner boards as with any horizontal wood siding. On outside corners, they

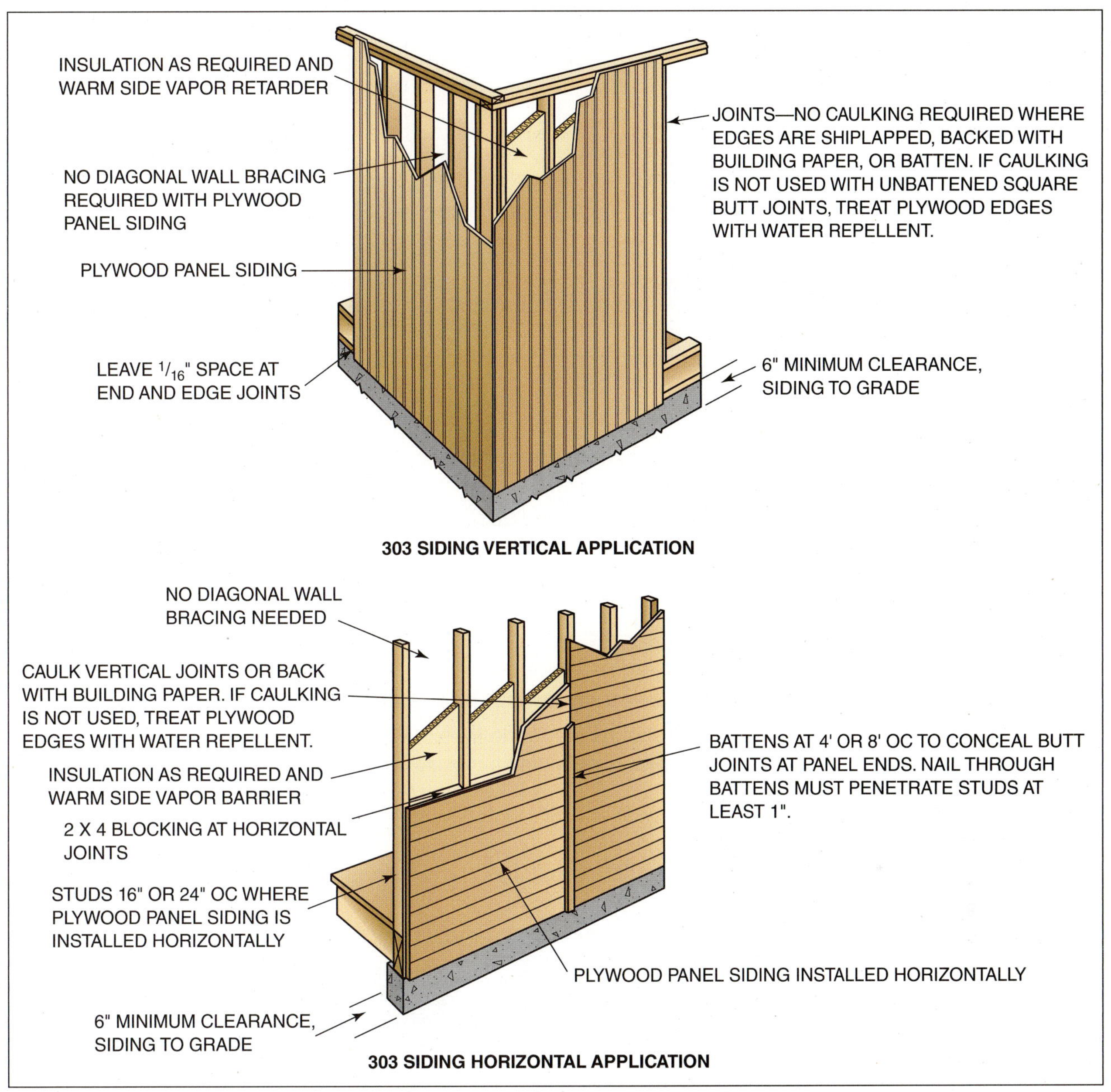


*Courtesy of APA—The Engineered Wood Association.*

**FIGURE 15-11** Panel siding may be applied vertically or horizontally to sheathing or directly to studs.

may also be applied by alternately overlapping each course in the same manner as in applying a wood shingle ridge. Inside corners may be woven by alternating the corner shingle first on one wall and then the other (Fig. 15-15).

## Double Coursing

When double coursing, the starter course is tripled. The outer layer of all courses is applied ½ inch lower than the inner layer. For ease in application, use a rabbeted straightedge or one composed of two pieces with offset edges (Fig. 15-16).

Each inner layer shingle is applied with one fastener at the top center. Each outer course shingle is face-nailed with two 5d galvanized box or special 14-gauge shingle nails. The fasteners are driven about 2 inches above the butts and about ¾ inch in from each edge.

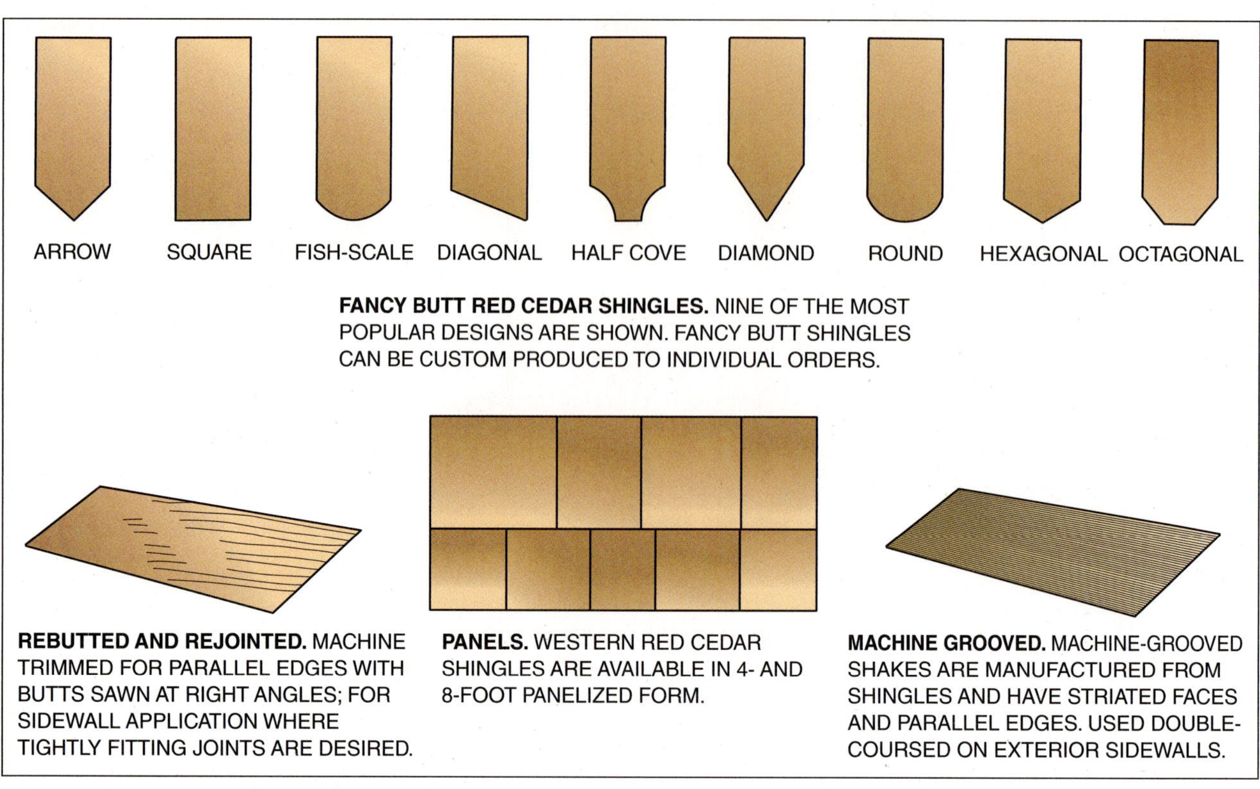


**FIGURE 15-12** Some wood shingles and shakes are made for sidewall application only.

**FIGURE 15-13** Fancy butt shingles are still used to accent sidewalls with distinctive designs.

# Vinyl and Aluminum Siding

Except for the material used to make the siding, *aluminum* and *vinyl* siding systems are similar. Aluminum siding is finished with a baked-on enamel. In vinyl siding, the color is in the material itself. Both kinds are manufactured with interlocking edges for horizontal and vertical applications. Descriptions and instructions are given here for vinyl siding systems, much of which can be applied to aluminum systems.

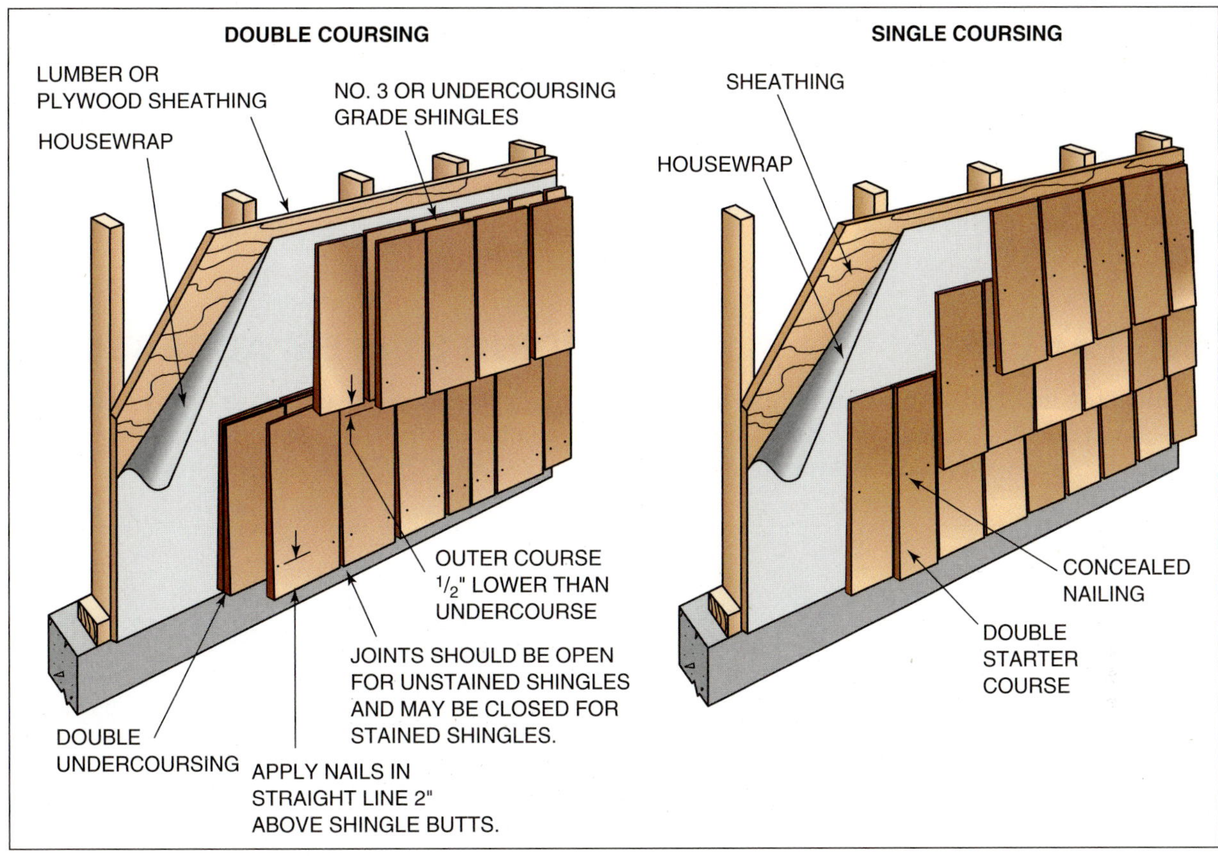


*Courtesy of Cedar Shake and Shingle Bureau.*

**FIGURE 15-14** Single- and double-coursed shingles. Greater exposures are allowed with double coursing.

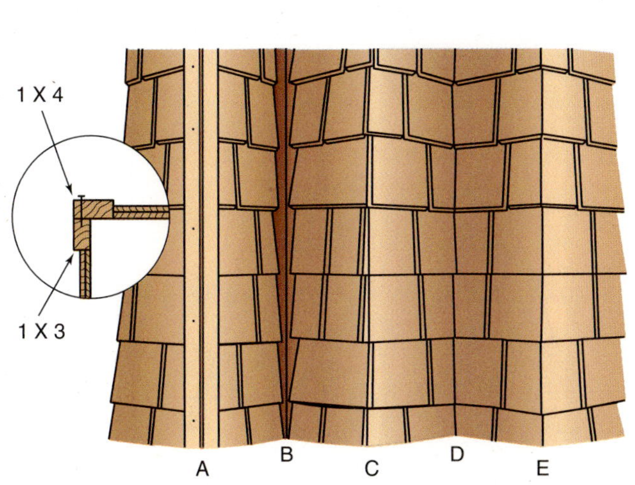


A) SHINGLES BUTTED AGAINST CORNER BOARDS
B) SHINGLES BUTTED AGAINST SQUARE WOOD STRIP ON INSIDE CORNER
C) LACED OUTSIDE CORNER
D) LACED INSIDE CORNER
E) MITERED CORNER

**FIGURE 15-15** Wood shingle corner details.

HOUSEWRAP
SHEATHING
DOUBLE–COURSED SHINGLES
RABBETED STRAIGHTEDGE

**FIGURE 15-16** Use a straightedge with rabbeted edges, or one made of two pieces with offset edges, for double-coursed application.

## SIDING PANELS AND ACCESSORIES

Siding systems are composed of *siding panels* and several specially shaped *moldings*. Moldings are used on various parts of the building, to trim the installation. In addition, the system includes shapes for use on *soffits*.

## SIDING PANELS

Siding panels for horizontal application are made in 8-, 10- and 12-inch widths. They come in configurations to simulate one, two, or three courses of bevel or drop siding. Panels designed for vertical application come in 12-inch widths. They are shaped to resemble boards. They can be used in combination with horizontal siding. Vertical siding panels with solid surfaces may also be used for soffits. For ventilation, *perforated* soffit panels of the same configuration are used (Fig. 15-17).

## SIDING SYSTEM ACCESSORIES

Siding systems require the use of several specially shaped accessories. *Inside* and *outside corner posts* are used to provide a weather-resistant joint to

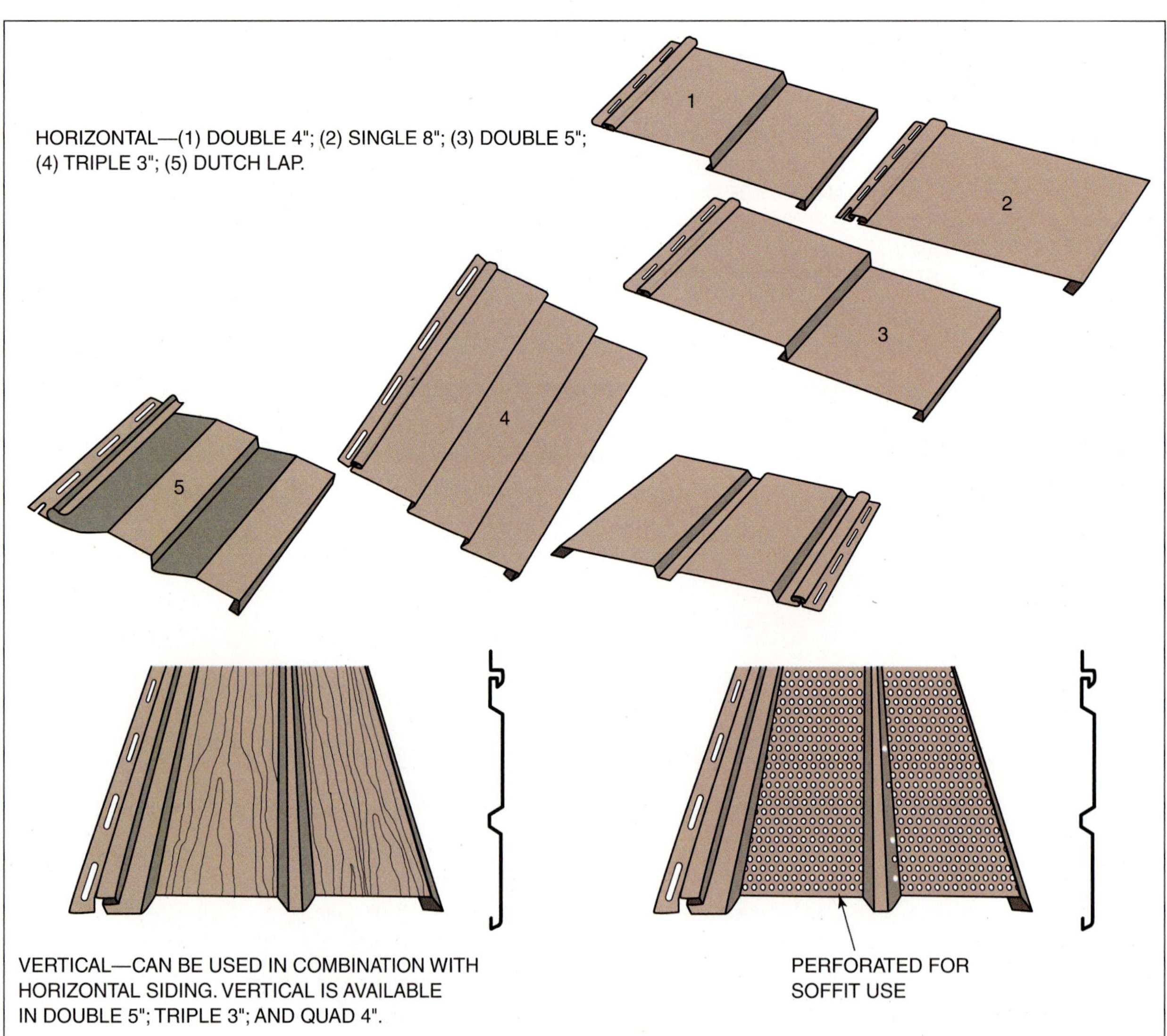


Courtesy of Vinyl Siding Institute.

**FIGURE 15-17** Some of the most commonly used configurations of vinyl siding for horizontal and vertical application.

corners. Corner posts are available with channels of appropriate widths to accommodate various configurations of siding.

*J-channels* are made with several opening sizes. They are used in a number of places such as around doors and windows, at transition points of materials, and against soffits (Fig. 15-18). The majority of vinyl siding panels and accessories are manufactured in lengths of 12 feet.

## APPLYING HORIZONTAL SIDING

The siding may expand and contract as much as 1/4 inch over its length with changes in temperature. For this reason, it is important to center fasteners in the nailing slots. There should be about 1/32 inch between the head of the fastener, when driven, and the siding. This allows the siding to move with temperature changes without buckling (Fig. 15-19). Space fasteners 16 inches apart for horizontal siding and 6 to 12 inches apart for accessories unless otherwise specified by the manufacturer. Install J-channel across the top and along the sides of window and door casings. It may also be installed under windows or doors with the *undersill* nailed inside the channel. To cut, use tin snips, hacksaw, utility knife, or power saw with an abrasive wheel or fine-tooth circular blade. Reverse the blade if a power saw is used, for smooth cutting.

***For step-by-step instructions on APPLYING HORIZONTAL VINYL SIDING, see Procedure 15-5 on pages 509–513.***

## GABLE END INSTALLATION

The rakes of a gable end are first trimmed with J-channels. The panel ends are inserted into the channel with a 1/4-inch expansion gap. Make a pattern for cutting gable end panels at an angle where they intersect with the rake. Use two scrap pieces of siding to make the pattern. Interlock one piece with an installed siding panel below. Hold the other piece on top of it and against the rake. Mark along the bottom edge of the slanted piece on the face of the level piece (Fig. 15-20).

## APPLYING VERTICAL SIDING

The installation of vertical siding is similar to that for horizontal siding, with a few exceptions. The method of fastening is the same. However, space fasteners about 12 inches apart for vertical siding panels. The starter strip is different. It may be the narrower 1/2-inch J-channel or *drip cap* flush with and fitted into the corner posts (Fig. 15-21). Around windows and doors, under soffits, against rakes, and in other locations, 1/2-inch J-channel is used. One of the major differences is that a vertical layout should be planned so that the same width panel is exposed at both ends of the wall.

***For step-by-step instructions on APPLYING VERTICAL VINYL SIDING, see Procedure 15-6 on page 514.***

## APPLYING ALUMINIUM TRIM

Aluminum trim pieces are often fabricated on the job. This is done by using a tool, called a **brake**, to bend sheet metal (Fig. 15-22). Light duty brakes are designed for aluminum only, while others will bend heavier sheet metal.

Aluminum stock is sold in rolls of various widths from 12 to 24 inches and lengths of 50 feet. These rolls are referred to as **coil stock**. Each side of the sheet is colored with a baked on enamel finish. One side is white, and the other is one of a variety colors produced by the manufacturer.

Coil stock is cut easiest with a coil stock cutter. It rides on the brake rails and produces a factory-like cut. Coil stock may be cut by using a utility knife. Using a straight edge makes the cut look professional. Stock is cut with one score of the knife. The cut is then completed by folding the piece back and forth through the cut.

Once the pieces are cut to width, they are then bent to the desired configuration. Care must be taken to visualize the piece as it is bent. Mistakes are easy to make. For example, the piece may have the correct shape but the wrong color facing outward.

Bending the stock begins with careful positioning of the piece in the jaws. The same amount should be revealed from the jaws on both ends.

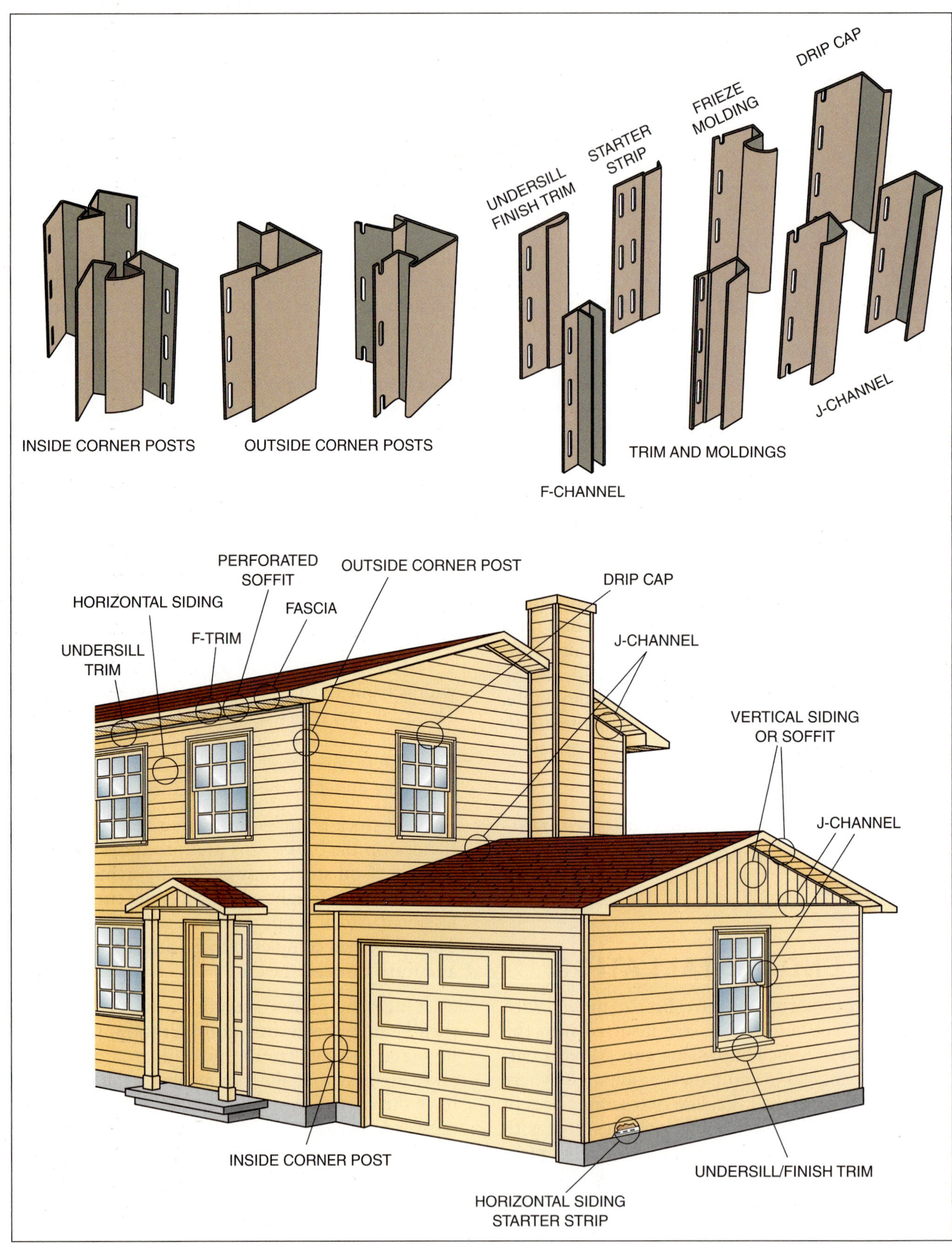


Courtesy of Georgia-Pacific Corporation.

**FIGURE 15-18** Various accessories are used to trim a vinyl siding installation.

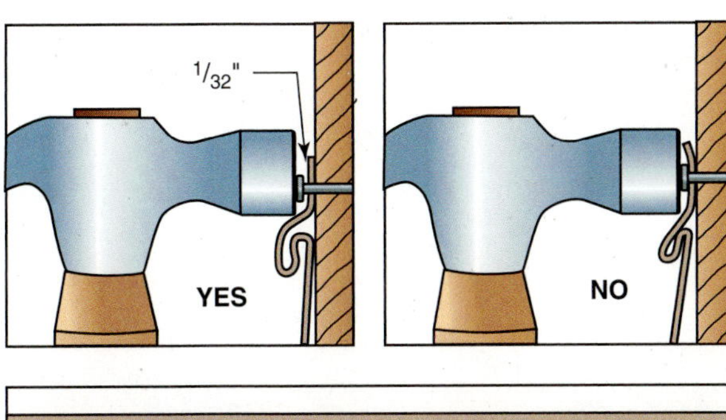


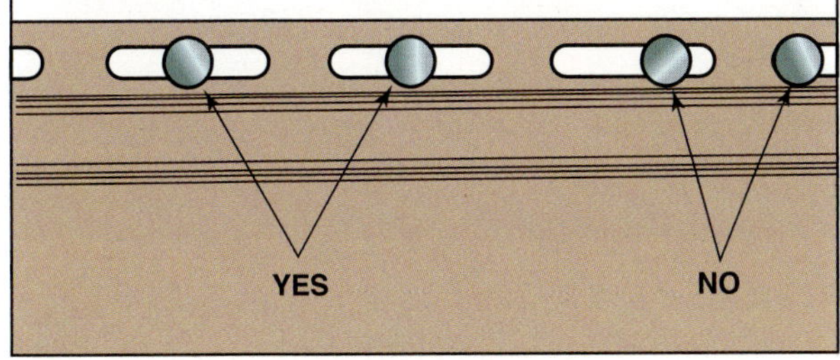


**FIGURE 15-19** Fasten siding to allow for expansion and contraction.

This will insure the piece is not tapered. Locking the jaws tight secures the piece during the bend. The bend is made by raising the handles of the brake (Fig. 15-23). The bend may be any angle from 0 to 180 degrees. Care must be taken to bend to the desired angle. Making repetitive stops at the same angle takes some practice.

Making an edge return uses a full 180-degree bend, which takes two steps. First align the piece to make as small a bend as possible, usually about ⅝ inches. The piece edge should be flush with the brake bender. Then, lock and bend as far as the brake will allow. This is about 150 degrees. The jaws are then unlocked, piece removed, and the jaws relocked. The piece is then placed on top of the jaws (Fig. 15-24). The final bend to 180-degrees is made on top.

# CORNICES

*Cornice* is a term used to describe the trim and accessories used to finish the joint between the siding and the roof. It has many parts and shapes (Fig. 15-25).

## SUBFASCIA

The *subfascia* is sometimes called the *false fascia* or *rough fascia*. It is a horizontal framing member fastened to the rafter tails. It provides an even, solid, and continuous surface for the attachment of other cornice members. When used, the sub fascia is generally a nominal 1- or 2-inch-thick piece. Its width depends on the slope of the roof, the tail cut of the rafters, or the type of cornice construction.

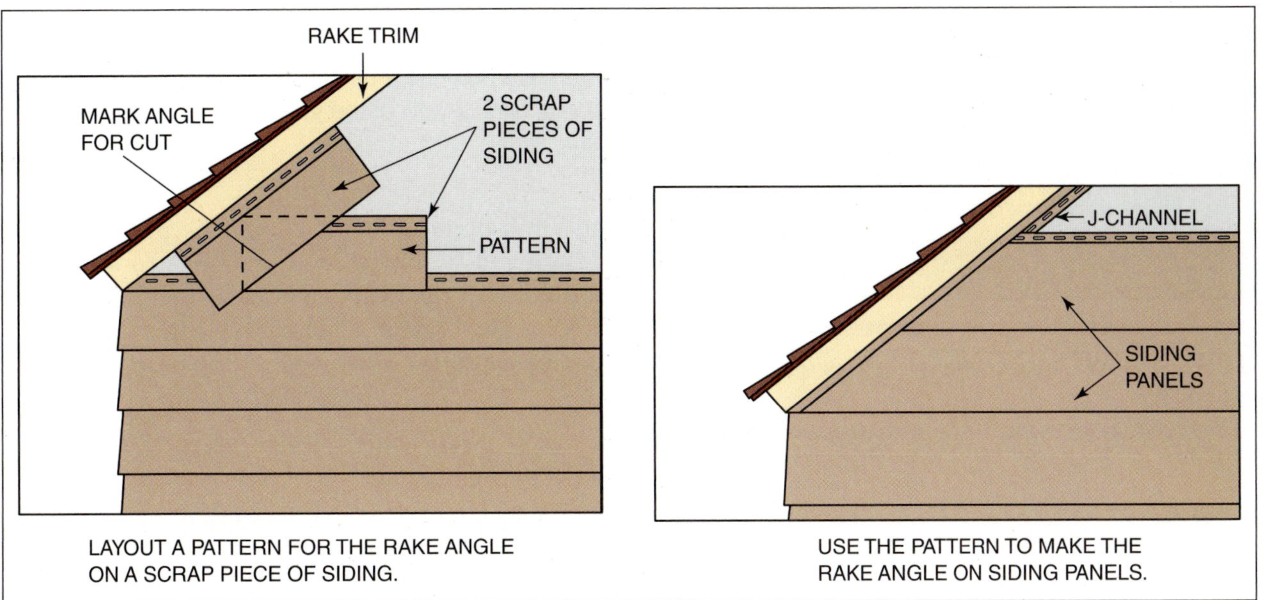


**FIGURE 15-20** Fitting horizontal siding panels to the rakes.

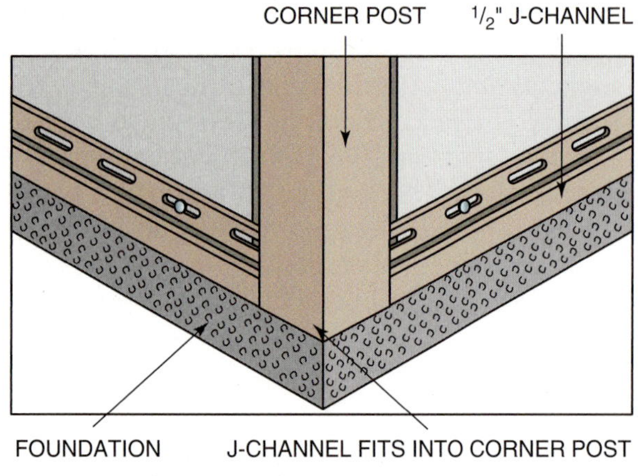


**FIGURE 15-21** Vertical siding starter strip is ½" J-channel which is fit into a corner board.

**FIGURE 15-23** Bending the metal is done by clamping jaws and lifting handles.

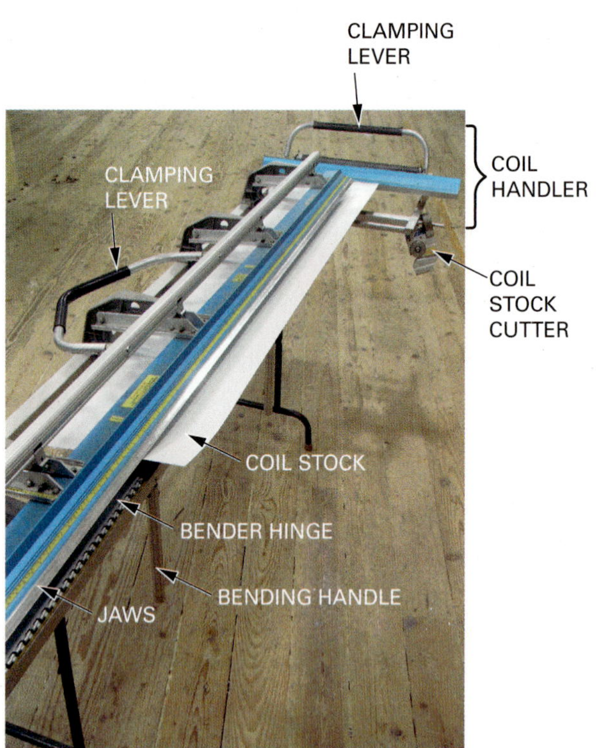


**FIGURE 15-22** An aluminum brake is used to bend coil stock.

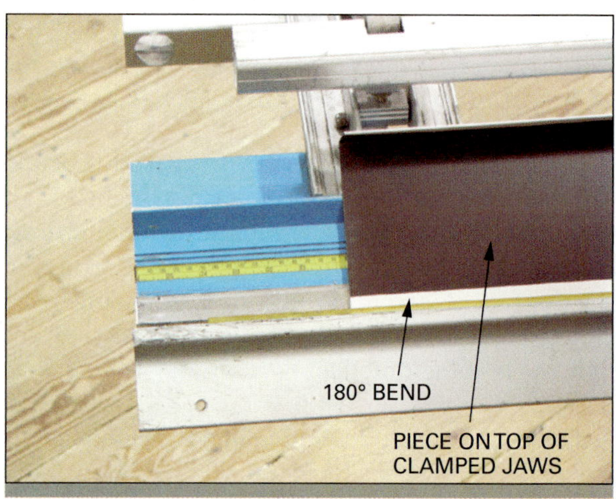


**FIGURE 15-24** Full 180-degree bends are completed on top of the jaws.

## SOFFIT

The finished member on the underside of the cornice is the **plancier**, often referred to as a *soffit* (Fig. 15-26). Soffit material may include solid lumber, plywood, fiberboard, or corrugated aluminum and vinyl panels. Soffits should be perforated or constructed with screened openings to allow for ventilation of the rafter cavities. Soffits may be fastened to the bottom edge of the rafter tails to the slope of the roof.

## FASCIA

The *fascia* is fastened to the subfascia or to the ends of the rafter tails when the subfascia is not used. It may be a piece of lumber grooved to receive the soffit. It also may be made from bent aluminum and vinyl used to wrap the subfascia. The fascia may be built up from one or more members to enhance the

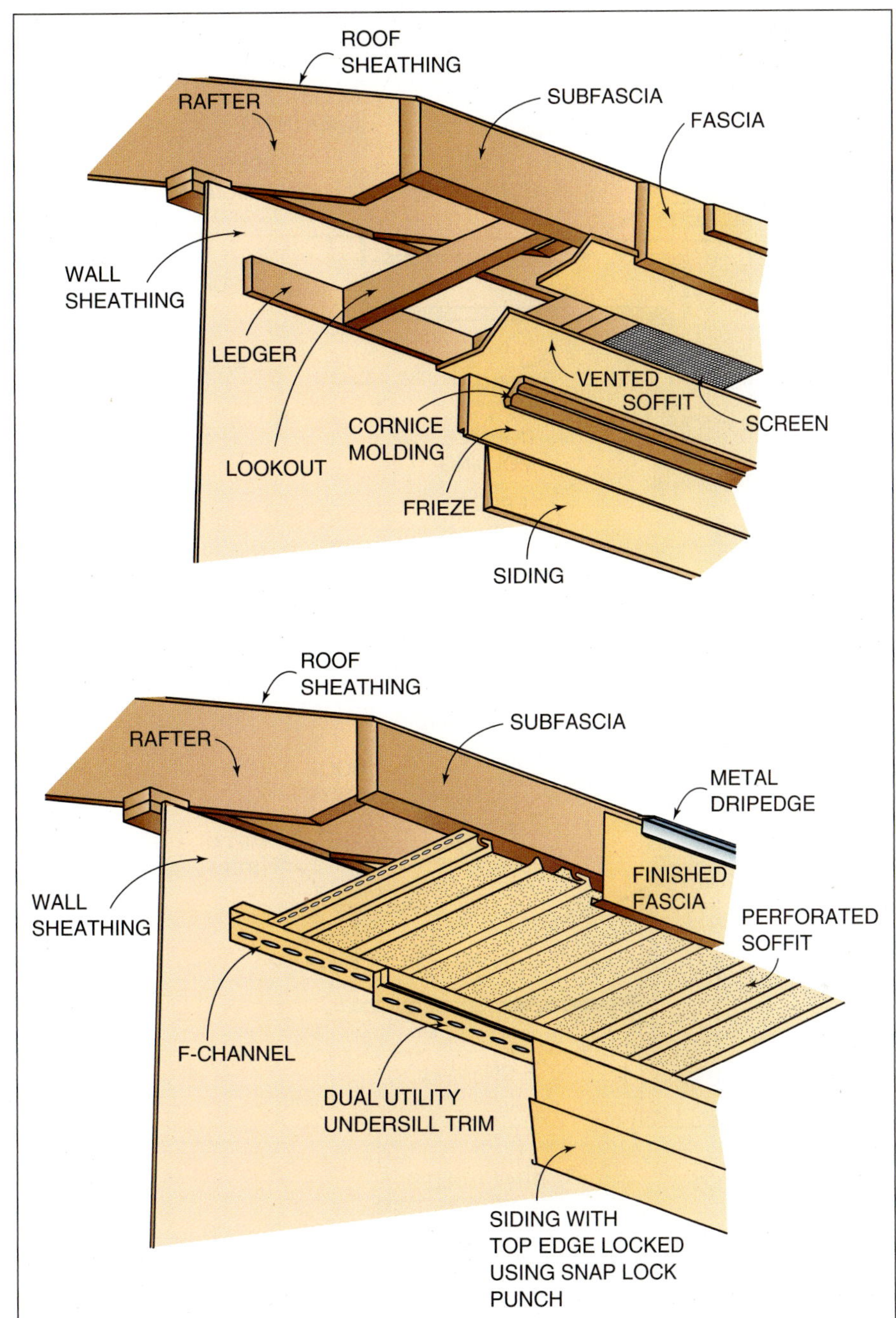


**FIGURE 15-25** Cornices may be constructed using different materials.

beauty of the cornice. The bottom edge of the fascia usually extends below the soffit by ¼ to ⅜ inch. The portion of the fascia that extends below the soffit is called the **drip**. The drip is necessary to prevent rainwater from being swept back against the walls of the building. In addition, a drip makes the cornice more attractive.

## FRIEZE

The *frieze* is a heavier solid member than the siding pieces, fastened to the sidewall with its top edge against the soffit. Its bottom edge is *rabbeted* or furred to receive the sidewall finish. However, the frieze is not always used. The sidewall finish may be allowed

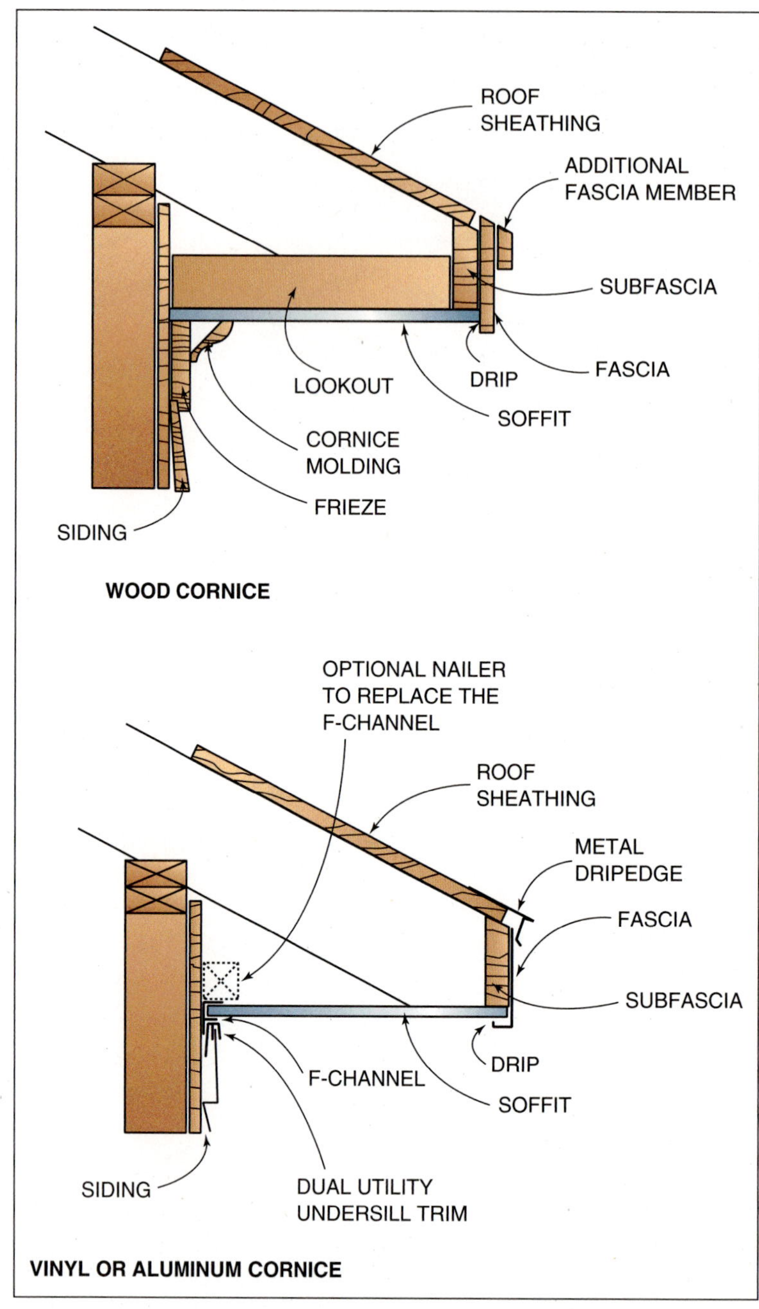


**FIGURE 15-26** The soffit is the bottom finish member of the cornice; through it, ventilation is provided to the attic.

to come up to the soffit. The joint between the siding and the soffit is then covered by a molding.

## CORNICE MOLDING

The *cornice molding* is used to cover the joint between the frieze and the soffit. If the frieze is not used, the cornice molding covers the joint between the siding and the soffit.

## LOOKOUTS

*Lookouts* are framing members used to provide a fastening surface for the soffit. They run horizontally from the end of the rafter to the wall, adding extra strength to larger overhangs. Lookouts may be installed at every rafter or spaced 48 inches OC, depending on the material being used for the soffit.

## CORNICE DESIGN

Cornices are generally classified into three main types: box, snub, and open (Fig. 15-27).

*The Box Cornice.* The box cornice is probably most common. It gives a finished appearance to this section of the exterior. Because of its overhang, it helps protect the sidewalls from the weather. It also provides shade for windows. Box cornices may be designated as narrow or wide. They may be constructed with level or sloping soffits.

*A Narrow Box Cornice.* A *narrow box cornice* is one in which the cuts on the rafters serve as nailing surfaces for the cornice members. A *wide box cornice* may be constructed with a level or sloping soffit. A wide, level soffit requires the installation of lookouts.

*The Snub Cornice.* The *snub cornice* has no rafter overhang. This reduces the cost of finishing the cornice. A snub cornice is frequently used on the *rakes* of a gable end in combination with a boxed cornice at the eaves.

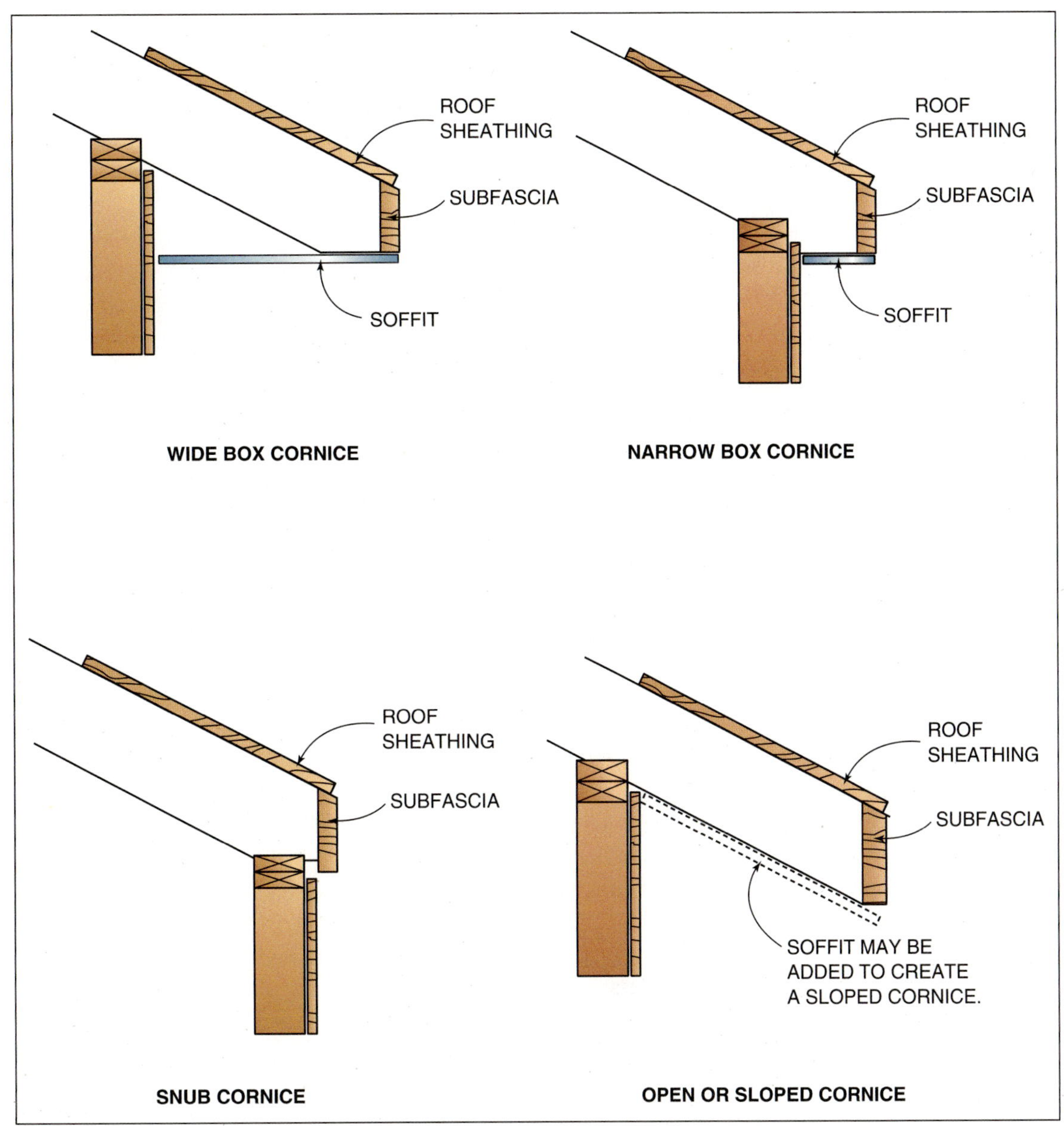


**FIGURE 15-27** The cornice may be constructed in various styles.

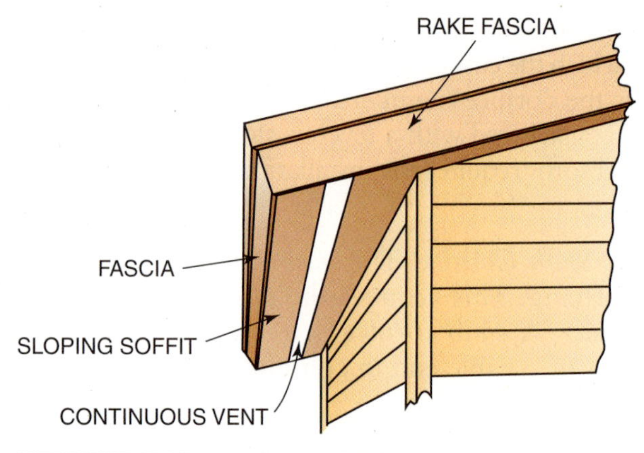


**FIGURE 15-28** A sloped cornice may be returned up the rakes of a gable roof.

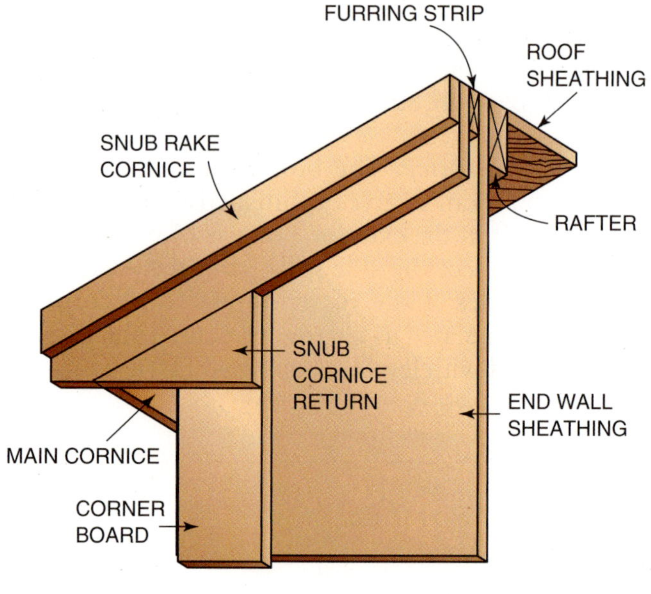


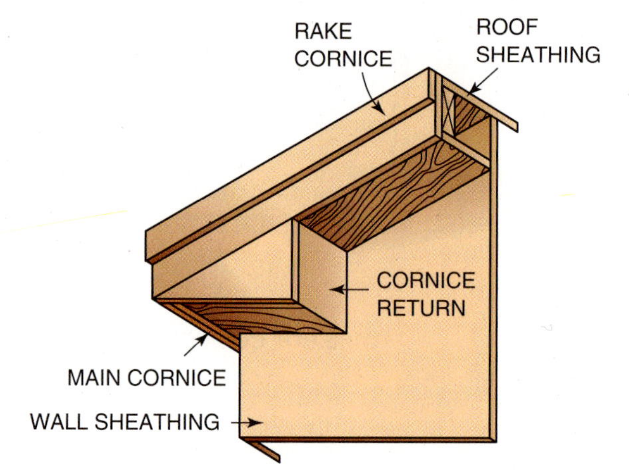


**FIGURE 15-29** Methods of constructing cornice returns.

*The Open Cornice.* The *open cornice* has no soffit. Open cornices give a contemporary or rustic design while reducing the overall cost. By adding a *soffit,* a *sloped cornice* is created. The soffit is installed directly to the underside of the rafter tails. This is sometimes done to simplify the cornice detail when there is also an overhang over the gable end of the building.

*Rake Cornices.* On buildings with hip or mansard roofs, the cornice, regardless of the type, extends around the entire building. On buildings with gable roofs, the cornices of the eave and rake must be blended together to make a smooth transition. If the eave cornice has a sloping soffit, the rake soffit may be returned up the rakes to the ridge. This type has the soffits for both the rake and eave in the same *plane* (Fig. 15-28).

*Cornice Returns.* If the eave soffit is horizontal, attached to level lookouts, a transition must be made from the eave to rake soffits. When the rake has a snub cornice, the return is simplified by installing a *snub cornice return.* If the rake has an overhang, a *cornice return* is constructed (Fig. 15-29).

# Gutters and Downspouts

A **gutter** is a shallow trough or conduit set below the edge of the roof along the fascia. It catches and carries off rainwater from the roof. A **downspout,** also called a *leader pipe,* is a rectangular or round pipe. It carries water from the gutter down the wall and away from the foundation (Fig. 15-30).

Gutters, or *eaves trough,* are usually made of aluminum, copper, or vinyl. Copper gutters are expensive but require no finishing. Vinyl and aluminum gutters are prefinished and ready to install. Gutters are made in rectangular, beveled, ogee, or semicircular shapes (Fig. 15-31). They come in a variety of sizes, from 2½ inches to 6 inches in height and from 3 inches to 8 inches in width. Pieces are installed to slope in one direction toward a downspout. On longer buildings, the gutter may be crowned slightly in the center, allowing water to drain to both ends.

For custom gutter lengths, special *forming machines* are brought to the job site. They form the

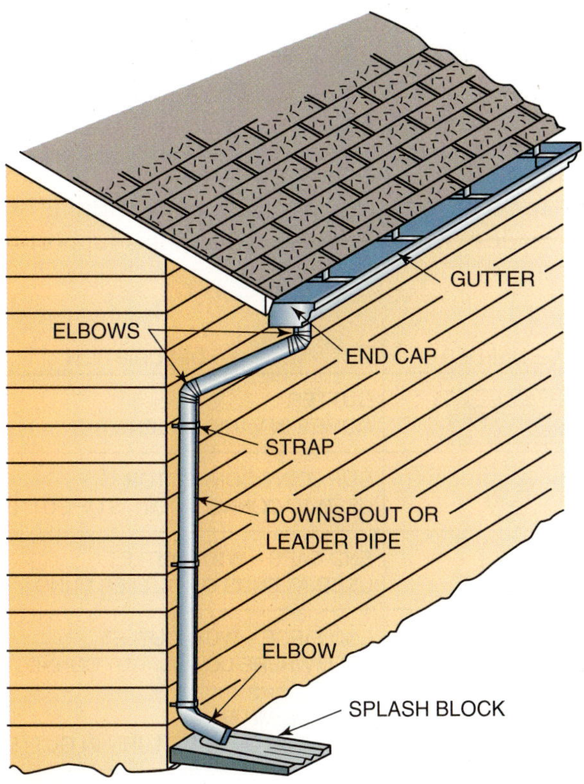


**FIGURE 15-30** Gutters and downspouts are an important system for conducting water away from the building, reducing moisture problems within the building.

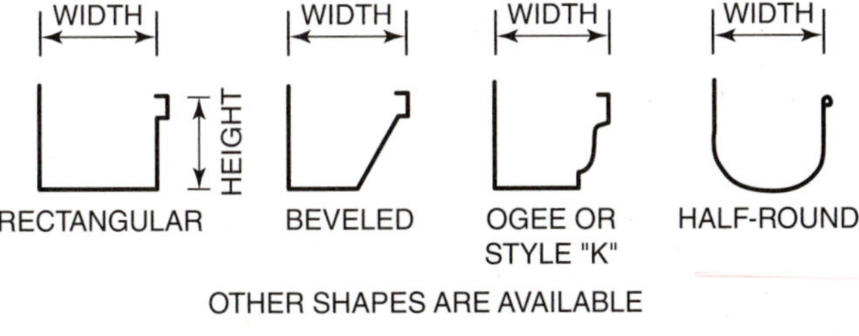


**FIGURE 15-31** Metal gutters are available in several shapes.

gutters to practically any desired length from a roll of aluminum coil stock.

Gutter systems are composed of *inside* and *outside corners, joint connectors, outlets, end caps,* and others. *Brackets* or *spikes* and *ferrules* are used to support the gutter sections on the fascia (Fig. 15-32). Vinyl gutters and components are installed in a manner similar to metal ones. Gutter downspouts should never be connected to footing or foundation drains.

***For step-by-step instructions on INSTALLING GUTTERS, see Procedure 15-7 on pages 515–516.***

## Green Tip

Removing rainwater away from the building in gutters and downspouts keeps the building drier.

# Decks

Wood decks and porches are built to provide outdoor living areas (Fig. 15-33). Decking materials are chosen for strength, durability, appearance, and resistance to decay. For these reasons, redwood, cedar, and pressure-treated southern yellow pine are often used. Other decking materials available include TimberTech® and Trex®. These decking products are made from a mixture of plastic and sawdust. They are cut, fit, and fastened in the same manner as wood and have the added benefit of being made with recycled materials.

## LUMBER

For pressure-treated southern pine and western cedar, #2 grade is structurally adequate for most applications. Appearance can be a deciding factor when choosing a grade. If a better appearance is desired, higher grades should be considered.

A grade called *Construction Heart* is the most suitable and most economical grade of California redwood for deck posts, beams, and joists. For decking and rails, a grade called *Construction Common* is recommended. Better appearing grades are available. However, they are more expensive. The lumber used for decking is specially manufactured for that purpose.

## LUMBER SIZES

Specific sizes of supporting posts, girders, and joists depend on the spacing and height of supporting posts and the spacing of girders and joists. In addition, the sizes of structural members depend on the type of wood used and the weight imposed on the members. Check with local building officials or with a professional to determine the sizes of structural members for specific deck construction.

## FASTENERS

All nails, fasteners, and hardware should be stainless steel, hot-dipped galvanized, or epoxy coated. Electroplated galvanizing is not acceptable because the coating is too thin. In addition to corroding and eventual failure, poor quality fasteners will react with substances in decay-resistant woods and cause unsightly stains.

## BUILDING A DECK

Most wood decks consist of *posts,* set on *footings,* supporting a platform of *girders* and *joists* covered with *deck boards.* Posts, rails, **balusters**, and other special parts make up the railing (Fig. 15-34). Other parts, such as shading devices, privacy screens, benches, and planters, lend finishing touches to the area.

A *ledger,* usually the same size as the joists, is nailed or bolted to the wall for the entire length of the deck. The ledger acts as a beam to support joists that run at right angles to the wall.

Footings are placed under each post. In stable soil and temperate climate, the footing width is usually made twice the width of the post it is supporting.

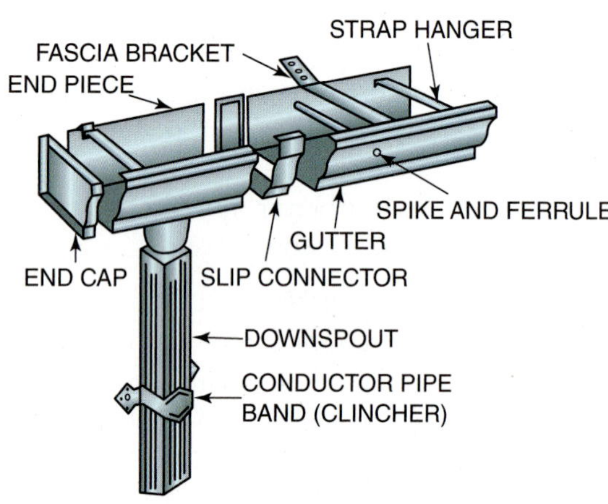


| PIECE NEEDED | DESCRIPTION |
|---|---|
| | GUTTER<br>COMES IN VARIOUS LENGTHS |
| | SLIP JOINT CONNECTOR<br>USED TO CONNECT JOINTS OF GUTTER |
| | END CAPS – WITH OUTLET<br>USED AT ENDS OF GUTTER RUNS |
| | END PIECE – WITH OUTLET<br>USED WHERE DOWNSPOUT CONNECTS |
| | OUTSIDE MITER<br>USED FOR OUTSIDE TURN IN GUTTER |
| | INSIDE MITER<br>USED FOR INSIDE TURN IN GUTTER |
| | FASCIA BRACKET<br>USED TO HOLD GUTTER TO FASCIA ON WALL |
| | STRAP HANGER<br>CONNECTS TO EAVE OF ROOF TO HOLD GUTTER |
| | STRAINER CAP<br>SLIPS OVER OUTLET IN END PIECE AS A STRAINER |
| | DOWNSPOUT<br>COMES IN 10' LENGTHS |
| | ELBOW – STYLE A<br>FOR DIVERTING DOWNSPOUT IN OR OUT FROM WALL |
| | ELBOW – STYLE B<br>FOR DIVERTING DOWNSPOUT TO LEFT OR RIGHT |
| | CONNECTOR PIPE BAND OR CLINCHER<br>USED TO HOLD DOWNSPOUT TO LEFT OR RIGHT |
| | SHOE<br>USED TO LEAD WATER TO SPLASHER BLOCK |
| | MASTIC<br>USED TO SEAL ALUMINUM GUTTERS AT JOINTS |
| | SPIKE AND FERRULE<br>USED TO HOLD GUTTER TO EAVE OF ROOF |

**FIGURE 15-32** Components of a metal gutter system.

Courtesy of California Redwood Association.

**FIGURE 15-33** Wood decks are built in many styles. This multi-level redwood deck blends well with the landscape.

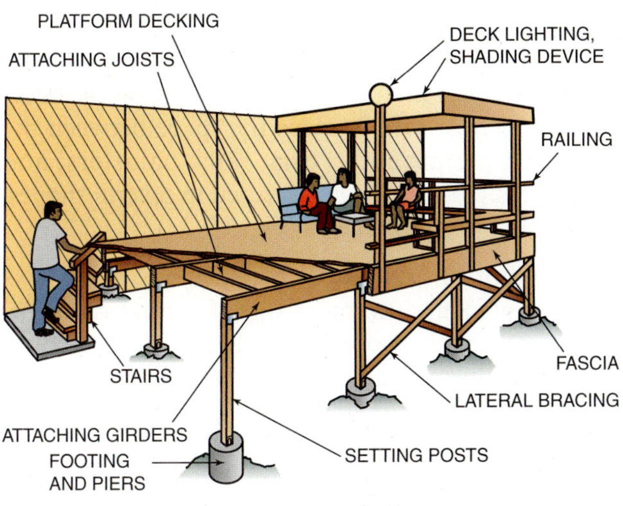


**FIGURE 15-34** The components of a deck.

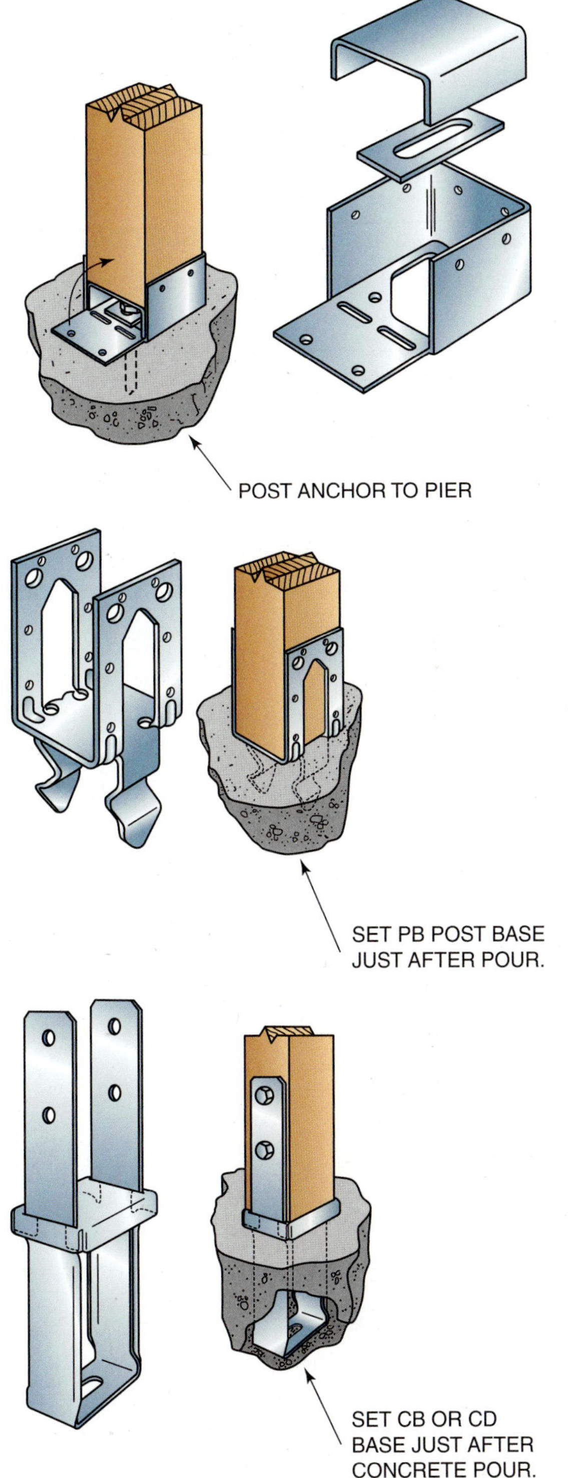


Courtesy of Simpson Strong-Tie Company.

**FIGURE 15-35** Hardware used to anchor posts to a footing.

The footing depth reaches undisturbed soil, at least 12 inches below grade. In cold climates, the footing should extend below the frost line. Posts must be firmly attached to the footings. Several styles of post anchors may be used. Some may be fastened to hard concrete; others are set shortly after concrete is placed (Fig. 15-35).

Joists may be placed over the top or between the girders. When joists are hung between the girders, the overall depth of the deck is reduced. This provides more clearance between the frame and the ground. For decking run at right angles, joists may be spaced 24 inches OC. Joists should be spaced 16 inches OC for diagonal decking.

Specially shaped *radius edge decking* is available in both pressure-treated and natural decay-resistant lumber. It is usually used to provide the surface and walking area of the deck. Dimension lumber of 1¼- or 2-inch thickness and widths of 4 and 6 inches is also widely used.

***For step-by-step instructions on BUILDING A DECK, see Procedure 15-8 on pages 517–521.***

# Stairs and Railings

Stairs and railings are needed when the deck is above ground. Layout and construction may be simple or intricate. Stairs are described in more detail in Chapter 19.

## STAIRS

Most decks require at least one or two steps leading to the ground. To protect the bottom ends of the stair carriage, they should be treated with preservative and supported by an above-grade concrete pad (Fig. 15-36). Stairs with more than two risers are generally required to have at least one handrail. The design and construction of the stair handrail should conform to local building codes.

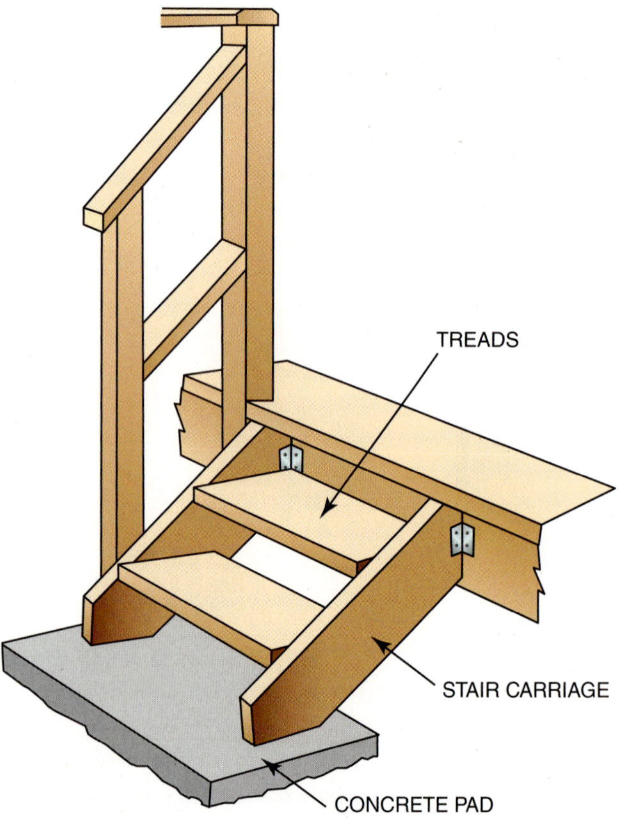


**FIGURE 15-36** Stairs for decks are usually constructed with a simple basement or utility design.

## RAILS

There are numerous designs for deck railings. All designs must conform to certain code requirements. Most codes require at least a 36-inch-high railing around the exposed sides, if the deck is more than 30 inches above the ground. In addition, some codes specify that no openings in the railing should allow a 4-inch sphere to pass through it.

Each linear foot of railing must be strong enough to resist a pressure of 20 pounds per square foot applied horizontally at a right angle against the top rail. Check local building codes for deck stair and railing requirements.

Railings may consist of posts; top, bottom, and intermediate rails; and balusters. Sometimes lattice work is used to fill in the rail spaces above the deck. It is frequently used to close the space between the deck and the ground. Posts, rails, balusters, and other deck parts are manufactured in several shapes, especially for use on decks (Fig. 15-37).

The bottom rail is cut between the posts. It is kept a few inches (no more than 4) above the deck. The remaining space may be filled with intermediate rails, balusters, lattice work, or other parts in designs as desired or as specified (Fig. 15-38).

## DECK ACCESSORIES

Many details can turn a plain deck into an attractive and more comfortable living area. *Shading structures* are built in many different designs. They may be completely closed in or spaced to provide filtered light and air circulation. *Benches* partially or entirely around the deck may double as a place to sit and act as a railing. Bench seats should be 18 inches off the deck. Make allowance for cushion thickness, if used. The width of the seat should be from 18 to 24 inches.

# Estimating Siding

Estimating the amount of material needed for some siding is similar to roofing as it is sold by the square, or 100 square feet. Others are sold by the lineal foot,

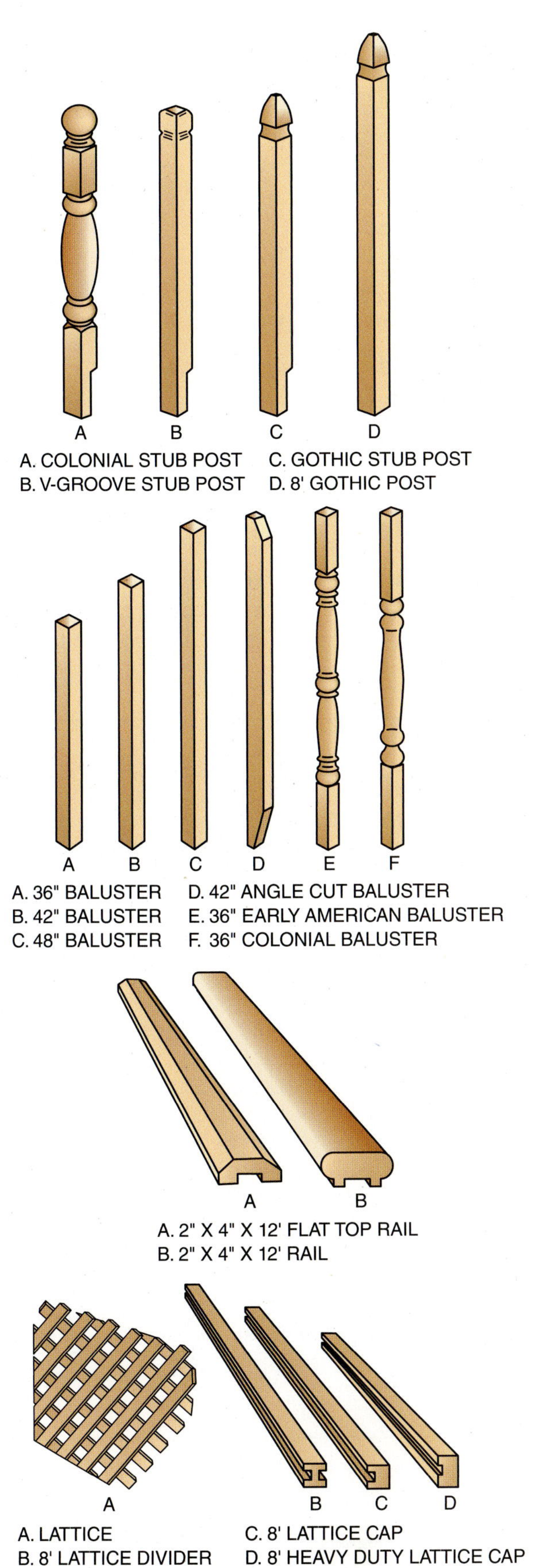


**FIGURE 15-37** Railing parts are manufactured in several shapes.

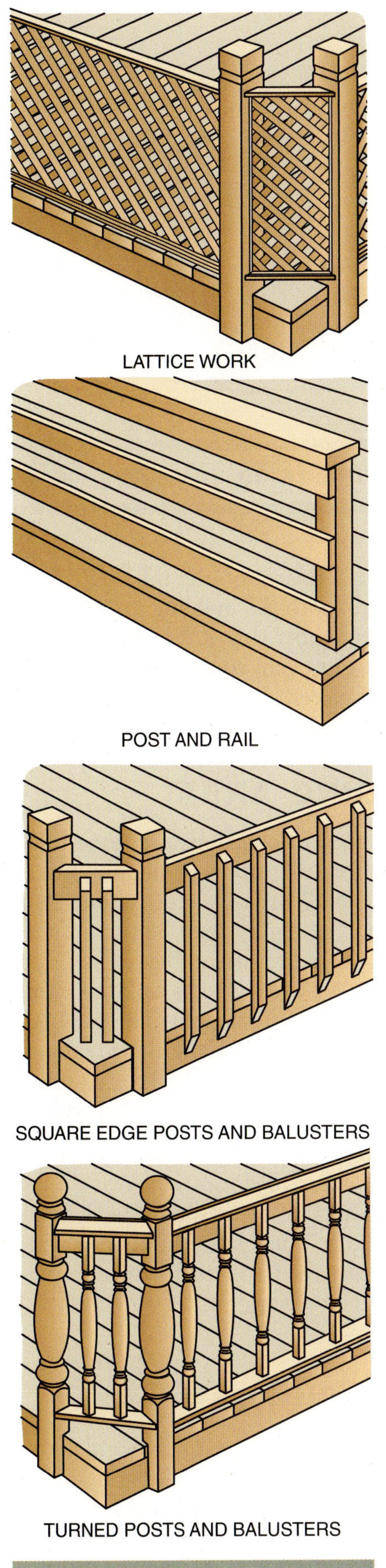


**FIGURE 15-38** The deck railing may be constructed with various kinds of parts in a number of designs.

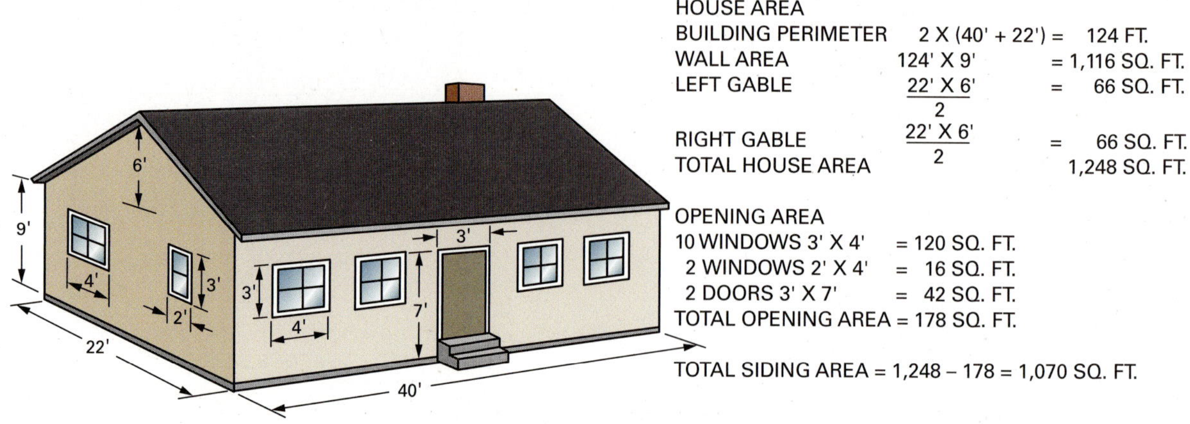


**FIGURE 15-39** Estimating the area to be covered by siding.

and still others are sold by the board foot. All estimates begin with the building area to be covered by siding. Remember that if the number of pieces does not come out even, round it up to the next whole number of pieces.

## BUILDING AREA

Building area calculations are done by adding together the smaller areas of the building. First, calculate the wall area by multiplying its length by its height. The length may be the perimeter of the building excluding the gable areas. Then, the triangular gable areas are calculated using the formula A = ½ base × height. The base is the width of the building. This triangular area is doubled to add the second gable. Add all the square foot areas together that will be sided, such as dormers, bays, and porches.

Windows and doors will not be covered, so their total surface areas should be deducted. Multiply the width × height × number of windows that size. Subtract the total results from the total area to be covered by siding (Fig. 15-39).

## WOOD SIDING

The amount of siding to order depends on the kind of siding used. Some siding is sold by the lineal foot and some by the board foot. Either way it comes, an allowance must be added to compensate for the overlap of each piece. The more pieces, the more overlap in a square foot. Figure 15-40 shows *conversion factor* numbers used to convert square foot (SF) area to be covered to lineal or board feet. Multiply SF by these factors to convert to lineal feet or board feet. For example, if 1,177 square feet are

| Pattern | Nominal Width | Width | | Factor for Converting SF to Lineal Feet | Factor for Converting SF to Board Feet |
|---|---|---|---|---|---|
| | | Dressed | Exposed Face | | |
| Bevel and Bungalow | 4 | 3½ | 2½ | 4.8 | 1.60 |
| | 6 | 5½ | 4½ | 2.67 | 1.33 |
| | 8 | 7¼ | 6¼ | 1.92 | 1.28 |
| | 10 | 9¼ | 8¼ | 1.45 | 1.21 |
| Dolly Varden | 4 | 3½ | 3 | 4.0 | 1.33 |
| | 6 | 5½ | 5 | 2.4 | 1.2 |
| | 8 | 7¼ | 6¾ | 1.78 | 1.19 |
| | 10 | 9¼ | 8¾ | 1.37 | 1.14 |
| | 12 | 11¼ | 10¾ | 1.12 | 1.12 |
| Drop T&G and Channel Rustic | 4 | 3⅜ | 3⅛ | 3.84 | 1.28 |
| | 6 | 5⅜ | 5⅛ | 2.34 | 1.17 |
| | 8 | 7⅛ | 6⅞ | 1.75 | 1.16 |
| | 10 | 9⅛ | 8⅞ | 1.35 | 1.13 |
| Log Cabin | 6 | 5 7/16 | 4 15/16 | 2.43 | 2.43 |
| | 8 | 7⅛ | 6⅝ | 1.81 | 2.42 |
| | 10 | 9⅛ | 8⅝ | 1.39 | 2.32 |
| Boards | 2 | 1½ | The exposed face width will vary depending on size selected and on how the boards-and-battens or boards-on-boards are applied. | | |
| | 4 | 3½ | | | |
| | 6 | 5½ | | | |
| | 8 | 7¼ | | | |
| | 10 | 9¼ | | | |

Coverage Estimator

*Courtesy of Western Wood Products Association.*

**FIGURE 15-40** Estimating information for converting area to dimensions required for material purchase.

to be covered with 6-inch Drop T&G siding, then 1,177 × 2.34 = 2,754.1 ⇒ 2,755 lineal feet is needed. Or 1,177 × 1.17 = 1,377.1 ⇒ 1,378 board feet. Add 10 percent for waste.

## WOOD SHINGLE SIDING

The number of squares of shingles needed to cover a certain area depends on how much they are exposed to the weather. Since a 16-inch shingle could have an exposure from 3½ inches to 9½ inches, the area that one bundle covers will vary. Wood shingles are packaged so four bundles of 16-inch shingles exposed 5 inches will cover 100 square feet, one square. If it is desired to expose the shingle more, then the number of bundles needed will go down.

Conversion numbers shown in Figure 15-41 are used to adjust squares needed as they relate to desired exposure. Divide the area to be covered by the conversion number, and the result is squares of shingles. For example, if 1,177 square feet is to be covered with 16-inch shingles exposed to 5½ inches, then 1,177 ÷ 100 = 11.77 squares. But if the same 16-inch shingles will have a 7-inch exposure, then 1,177 SF ÷ 140 = 8.4 squares are needed. Add 10 percent for waste.

## ALUMINUM VINYL AND SIDING

Aluminum and vinyl siding panels are sold by the square. Exposure and coverage is predetermined because the exposure is not adjustable. Determine the wall area to be covered. Divide by 100. Add 10 percent of the area for waste. For example, an 1,177-square-foot building needs 1,177 ÷ 100 × 110% = 12.9 ⇒ 13 squares.

The accessories are typically sold in 12-foot-long pieces. Take the lineal feet needed for each type of accessory, and divide by the length per piece. For example, if ten windows with a perimeter of 14 feet need J-channels that are 12 feet long, then 10 × 14 ÷ 12 = 11.6 ⇒ 12 pieces.

See Fig. 15–42 for a listing of the shorthand versions of these estimating techniques with another example.

| | Approximate Coverage of One Square (4-bundle roof-pack) of Shingles at Indicated Weather Exposures: | | | | | | | | | | | | |
|---|---|---|---|---|---|---|---|---|---|---|---|---|---|
| **Length** | **3½″** | **4″** | **4½″** | **5″** | **5½″** | **6″** | **6½″** | **7″** | **7½″** | **8″** | **8½″** | **9″** | **9½″** |
| 16″ | 70 | 80 | 90 | 100 | 110 | 120 | 130 | 140 | 150 | 160 | 170 | 180 | 190 |
| 18″ | — | 72½ | 81½ | 90½ | 100 | 109 | 118 | 127 | 136 | 145½ | 154½ | 163½ | 172½ |
| 24″ | — | — | — | — | 73½ | 80 | 86½ | 93 | 100 | 106½ | 113 | 120 | 126½ |
| **Length** | **10″** | **10½″** | **11″** | **11½″** | **12″** | **12½″** | **13″** | **13½″** | **14″** | **14½″** | **15″** | **15½″** | **16″** |
| 16″ | 200 | 210 | 220 | 230 | 240 | — | — | — | — | — | — | — | — |
| 18″ | 181½ | 191 | 200 | 209 | 218 | 227 | 236 | 245½ | 254½ | — | — | — | — |
| 24″ | 133 | 140 | 146½ | 153 | 160 | 166½ | 173 | 180 | 186½ | 193 | 200 | 206½ | 213 |

**FIGURE 15–41** Roof shingle coverage at various exposures.

| Estimate the siding needed to cover a rectangular building that measures 30'×56' with 9' high walls, 9' gable height, three 3949 windows, four 4959 windows, and two 3'×7' doors. Consider the lineal feet of 8" bevel siding, squares of 18-inch wood shingles exposed to 7½", and number of squares of vinyl siding. | | | |
|---|---|---|---|
| **Item** | **Formula** | **Waste factor** | **Example** |
| Building Area | PERM × wall HGT + 2 × gable area – opening area 5 build area | | 172 × 9 + (2 × ½ × 30 × 9) – (3 × 3 × 4 + 4 × 4 × 5 + 2 × 3 × 7) = 1548 – 158 = 1390 SF |
| 8" Bevel Siding | building area × conversion factor (Figure 15-40) × waste = lineal ft | 10% | 1390 × 1.92 × 110% = 2935.6 ⇒ 2936 lineal feet |
| 18" Wood Shingles exposed to 7½" | building area ÷ conversion factor (Figure 15-41) × waste = squares of siding | 10% | 1390 ÷ 136 × 110% = 11.2 ⇒ 12 squares of siding |
| Vinyl Siding | building area ÷ 100 × waste = squares of siding | 10% | 1390 ÷ 100 × 110% = 15.2 ⇒ 16 squares of siding |

**FIGURE 15–42** Example of estimating siding materials with formulas.

# SUMMARY

- Siding provides weathertightness to a building as well as an architectural appeal.
- Installation of vinyl and aluminum siding is similar.
- Cornices finish the transition between the siding and the roof.
- Gutters and down spouts direct water away from a building to help prevent water from entering it.
- Decks provide additional living area, which creates a pleasant outdoor environment.
- Estimating of siding and deck materials is necessary for efficient purchases.

## GREEN CHECKLIST

- ☐ Furring wood siding away from sheathing allows painted siding surfaces to last 2–3 times longer.
- ☐ Removing rainwater away from the building in gutters and downspouts keeps the building drier.

PROCEDURE 15-1

# Installing Horizontal Siding

First, determine the siding exposure so that it is about equal both above and below the window sill. Divide the overall height of each wall section by the maximum allowable exposure. Round up this number to get the number of courses in that section. Then, divide the height again by the number of courses to find the exposure. These slight adjustments in exposure will not be noticeable to the eye.

**EXAMPLE:** Consider the overall dimensions in the accompanying figure. Divide the heights by the maximum allowable exposure, 7 inches in this example. Then round up to the nearest number of courses that will cover that section. Divide the section height by the number of courses to find the exposure.

40½ ÷ 7 = 5.8 ⇒ 6 courses ⟶ 40½ ÷ 6 = 6.75 or 6¾" exposure
45⅕ ÷ 7 = 6.5 ⇒ 7 courses ⟶ 45½ ÷ 7 = 6.5 or 6½" exposure
12½ ÷ 7 = 1.8 ⇒ 2 courses ⟶ 12½ ÷ 2 = 6.25 or 6¼" exposure

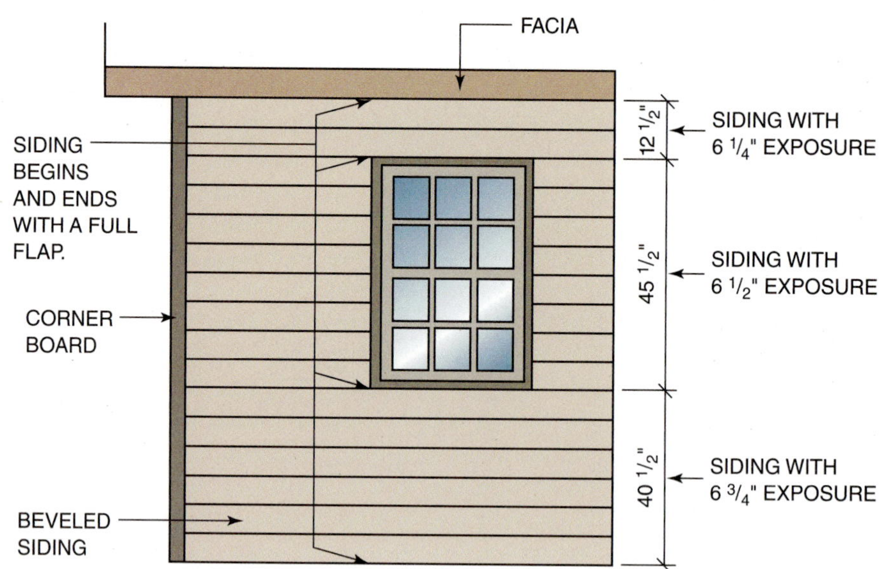

**B** Install a starter strip of the same thickness and width of the siding at the headlap fastened along the bottom edge of the sheathing. For the first course, a line is snapped on the wall at a height of the top edge of the first course of siding.

- From this first chalk line, lay out the desired exposures on each corner board and each side of all openings. Snap lines at these lay out marks. These lines represent the top edges of all siding pieces.
- Install the siding as per manufacturer's recommendations, staggering the butt joints in adjacent courses as far apart as possible. A small 6-inch wide piece of felt paper is used behind the butt seams to ensure the weathertightness of the siding.

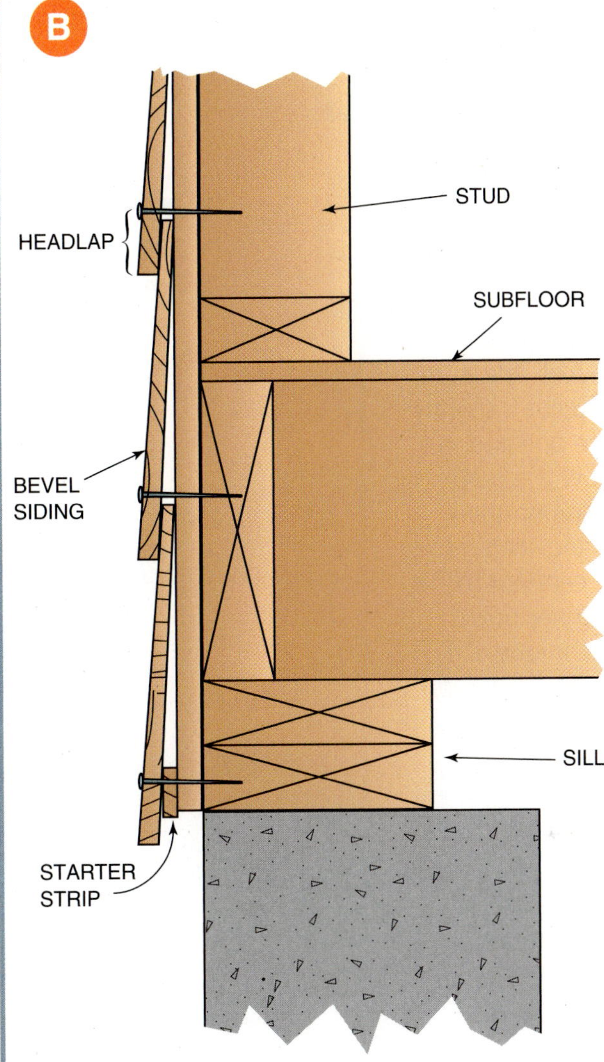


**SECTION THROUGH SILL**

PROCEDURE 15-1

# Installing Horizontal Siding (continued)

**C** When applying a course of siding, start from one end and work toward the other end. With this procedure, only the last piece will need to be fitted. Tight-fitting butt joints must be made between pieces. Measure carefully and cut it slightly longer. Place one end in position. Bow the piece outward, position the other end, and snap into place. Take care not to move the corner board with this technique. Do not use this technique on cementitious siding.

**C**

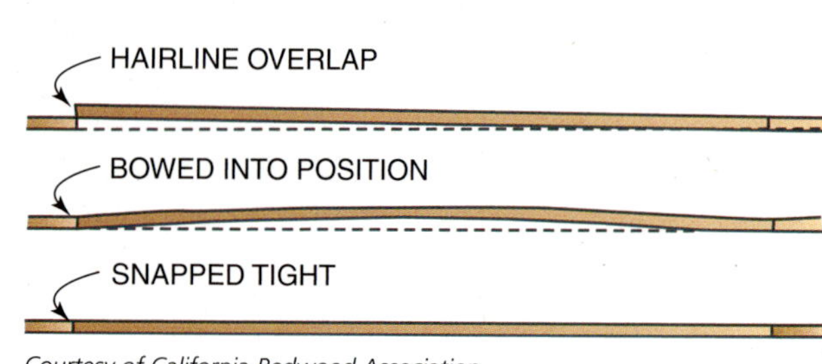


Courtesy of California Redwood Association.

**D** Siding is fastened to each bearing stud or about every 16 inches. On bevel siding, fasten through the butt edge just above the top edge of the course below. Do not fasten through the lap. This allows the siding to swell and shrink with seasonal changes without splitting of the siding. Cementitious or hardboard siding may be blind nailed by fastening along the top edge only. Blind nailing is not recommended in high-wind areas.

### *from experience...*

Setting up a comfortable work station for cutting will allow the carpenter to work more efficiently and safely. This will also reduce waste and improve workmanship.

| 6" AND NARROWER | 8" AND WIDER |
|---|---|
| PLAIN<br>USE ONE CASING NAIL PER BEARING TO BLIND NAIL. | PLAIN<br>USE TWO SIDING OR BOX NAILS, 3-4" APART TO FACE NAIL. |
| USE ONE SIDING OR BOX NAIL TO FACE NAIL ONCE PER BEARING. 1" UP FROM BOTTOM. | APPROXIMATE 1/8" GAP FOR DRY MATERIAL 8" AND WIDER.<br>1/2" = FULL DEPTH OF RABBET<br>USE TWO SIDING OR BOX NAILS, 3-4" APART, PER BEARING. |
| BOARD AND BATTEN<br>1/2"<br>RECOMMEND 1/2" OVERLAP. ONE SIDING OR BOX NAIL PER BEARING. | BOARD AND BATTEN / BOARD ON BOARD<br>INCREASE OVERLAP PROPORTIONATELY. USE TWO SIDING OR BOX NAILS, 3-4" APART. |

SIDING USUALLY APPLIED HORIZONTALLY

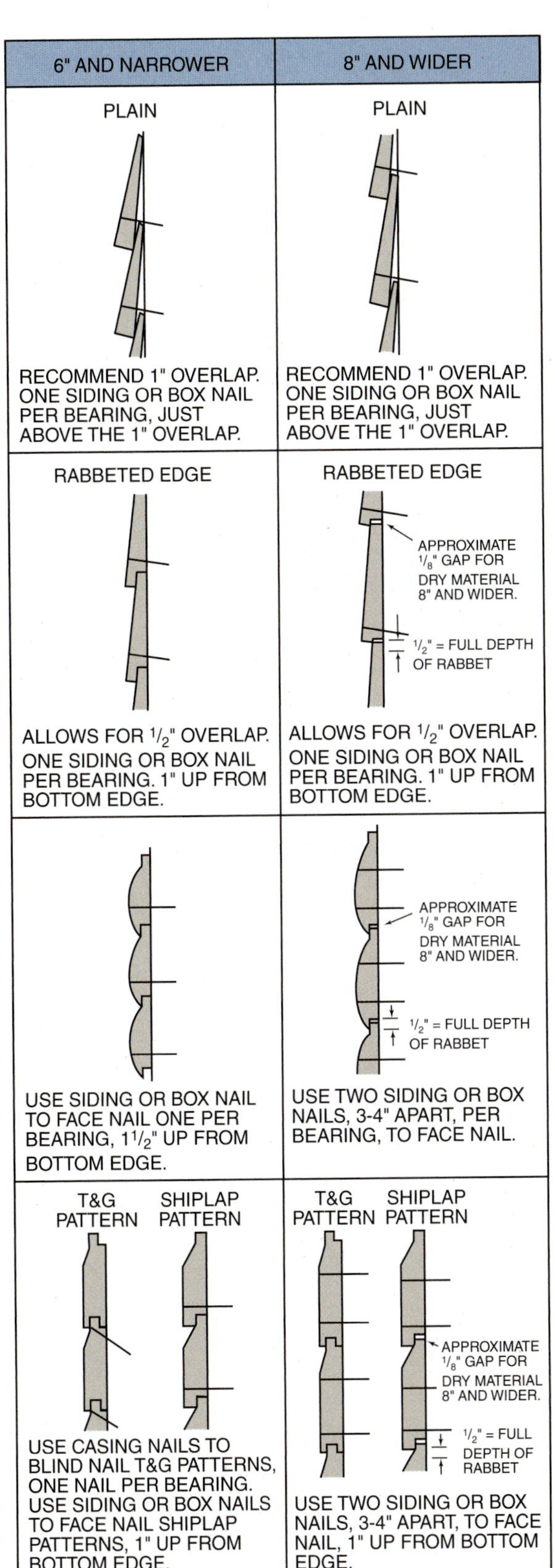


SIDING USUALLY APPLIED HORIZONTALLY

*Courtesy of Western Wood Products Association.*

PROCEDURE 15-2

# Installing Vertical Tongue-and-Groove Siding

**A** Slightly back-bevel the ripped edge. Place it vertically on the wall with the beveled edge flush with the corner similar to making a corner board. Face-nail the edge nearest the corner.

- Fasten a temporary piece on the other end of the wall to be sided, projecting below the sheathing by the same amount. Stretch a line to keep the bottom ends of other pieces in a straight line.

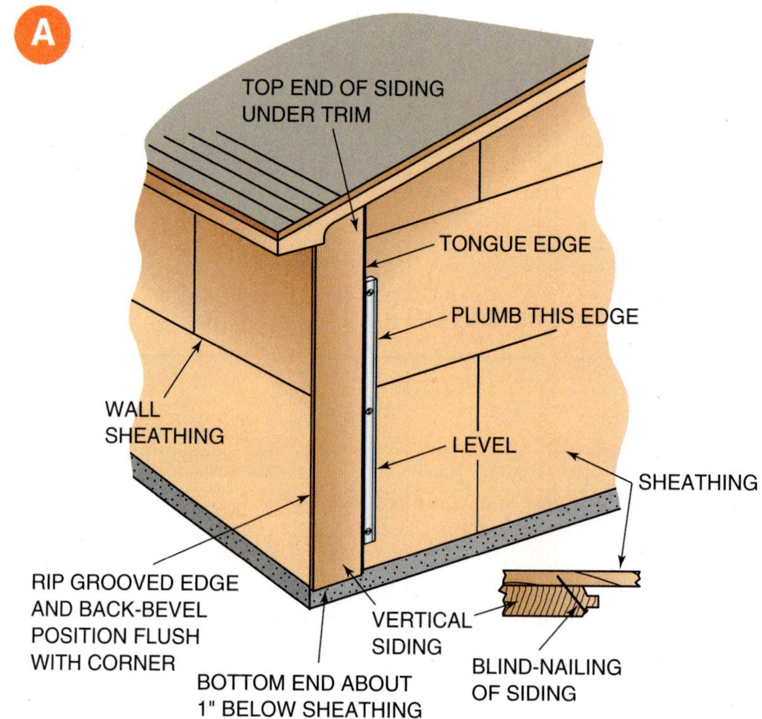


**B** Apply succeeding pieces by toenailing into the tongue edge of each piece. Make sure the edges between boards come up tight. Drive the nail home until it forces the board up tight. Make sure to keep the bottom ends in a straight line. If butt joints are necessary, use a mitered or rabbeted joint for weathertightness.

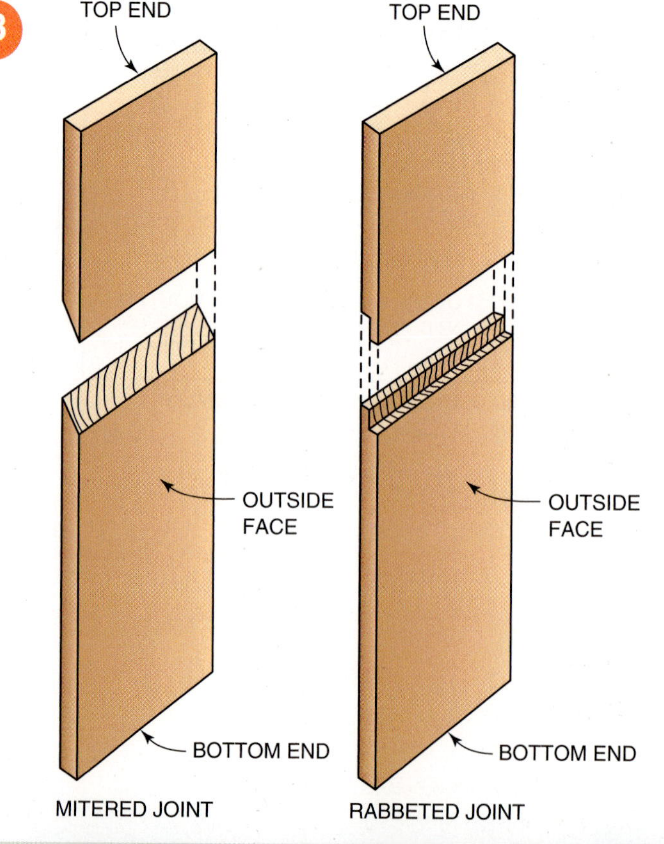


**C** To cut the piece to fit around an opening, first fit and tack a siding strip in place where the last full strip will be located. Level from the top and bottom of the window casing to this piece of siding. Mark the piece.

- Next, use a scrap block of the siding material, about 6 inches long, with the tongue removed. Be careful to remove all of the tongue, but no more. Hold the block so its grooved edge is against the side casing and the other edge is on top of the tacked piece of siding. Mark the vertical line on the siding by holding a pencil against the outer edge of the block while moving the block along the length of the side casing. Remove and cut the piece, following the layout lines carefully. Cut and fit another full strip of siding in the same place as the previously marked piece. Fasten piece in position against window casing.
- Continue the siding by applying the short lengths across the top and bottom of the opening as needed.

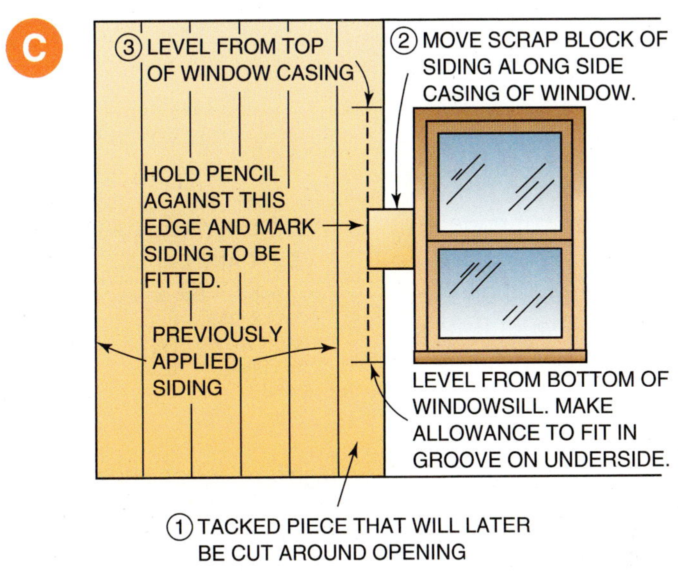


**D** Fit the next full-length siding piece to complete the siding around the opening. First, tack a short length of siding scrap above and below the window and against the last pieces of siding installed. Tack the length of siding to be fitted against these blocks in the grooves. Level and mark from the top and bottom of the window to the full piece. Lay out the vertical cut by using the same block with the tongue removed, as used previously. Hold the grooved edge against the side casing. With a pencil against the other edge, ride the block along the side casing while marking the piece to be fitted.

- Remove the piece and the scrap blocks from the wall. Carefully cut the piece to the layout lines. Then, fasten in position. Continue applying the rest of the siding.

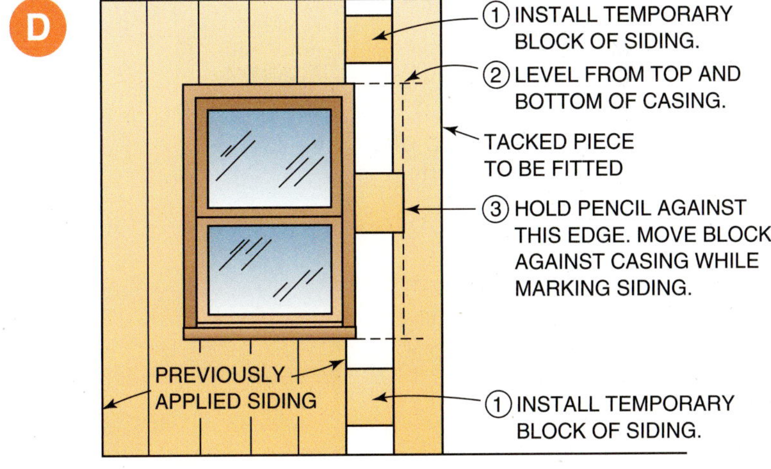

PROCEDURE 15-3

# Installing Panel Siding

**A** Install the first piece with the vertical edge plumb. Rip the sheet to size, putting the cut edge at the corner. The factory edge should fall on the center of a stud. Panels must also be installed with their bottom ends in a straight line. It is important that horizontal butt joints be offset and lapped, rabbeted, or flashed. Vertical joints are either shiplapped or covered with **battens**.

- Apply the remaining sheets in the first course in like manner. Cut around openings in a similar manner as with vertical tongue-and-grooved siding. Carefully fit and caulk around doors and windows. Trim the end of the last sheet flush with the corner.

**A**

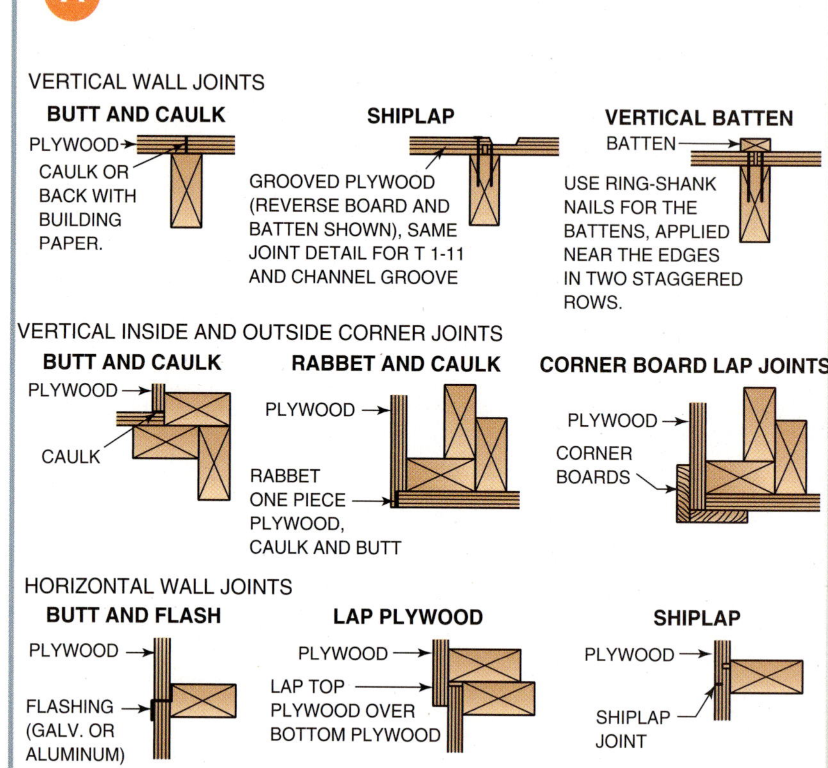

*Courtesy of APA—The Engineered Wood Association.*

PROCEDURE 15-4

# Installing Wood Shingles and Shakes

**A** Fasten a shingle on both ends of the wall with its butt about 1 inch below the top of the foundation. Stretch a line between them from the bottom ends. Fasten an intermediate shingle to the line to take any sag out of the line. Even a tightly stretched line will sag in the center over a long distance.

- Fill in the remaining shingles to complete the undercourse. Take care to install the butts as close to the line as possible without touching it. Remove the line.

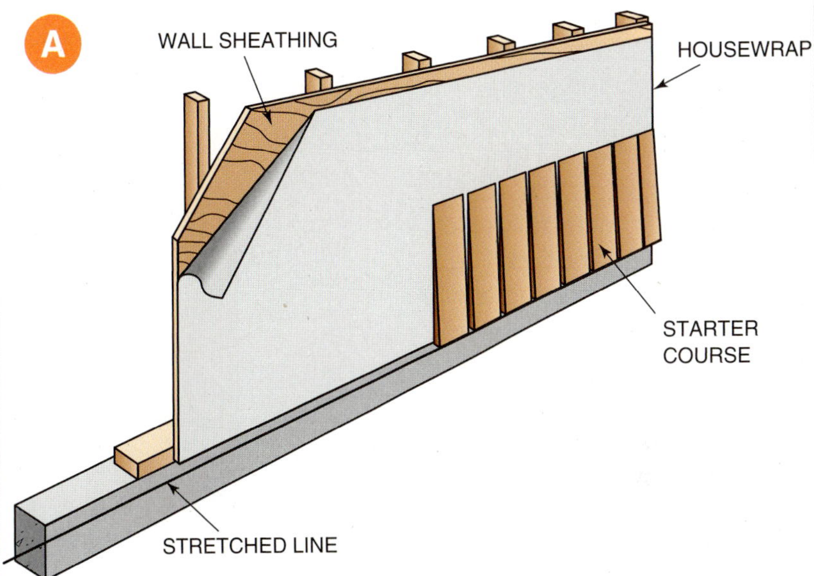


**B** Apply another course on top of the first course. Offset the joints in the outer layer at least 1½ inches from those in the bottom layer. Shingles should be spaced ⅛ to ¼ inch apart to allow for swelling and to prevent buckling. Shingles can be applied close together if factory-primed or if treated soon after application.

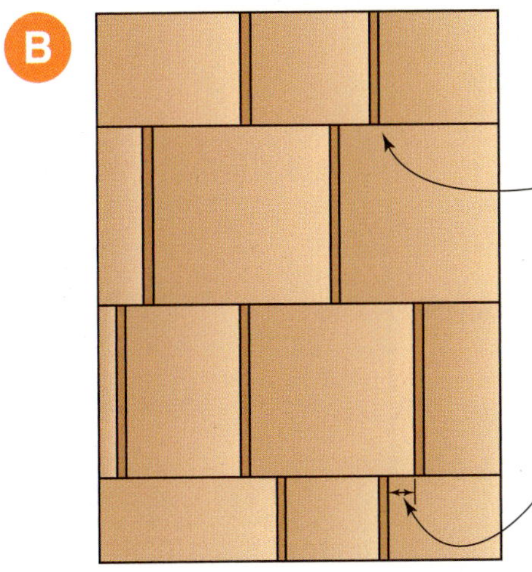


*Courtesy of Cedar Shake and Shingle Bureau.*

PROCEDURE 15-4

# Installing Wood Shingles and Shakes (continued)

**C** To apply the second course, snap a chalk line across the wall at the shingle butt line. Using only as many finish nails as necessary, tack 1 × 3 straightedges to the wall with their top edges to the line. Lay individual shingles with their butts on the straightedge.

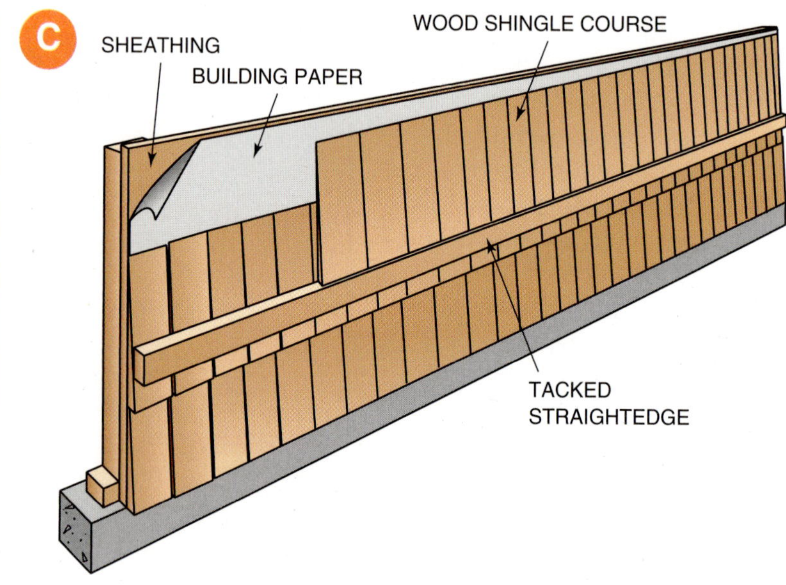

PROCEDURE 15-5

# Applying Horizontal Vinyl Siding

**A** Snap a level line to the height of the starter strip all around the bottom of the building. Fasten the strips to the wall with their edges to the chalk line. Leave a ¼-inch space between them and other accessories to allow for expansion. Make sure the starter strip is applied as straight as possible. It controls the straightness of the entire installation.

**A**

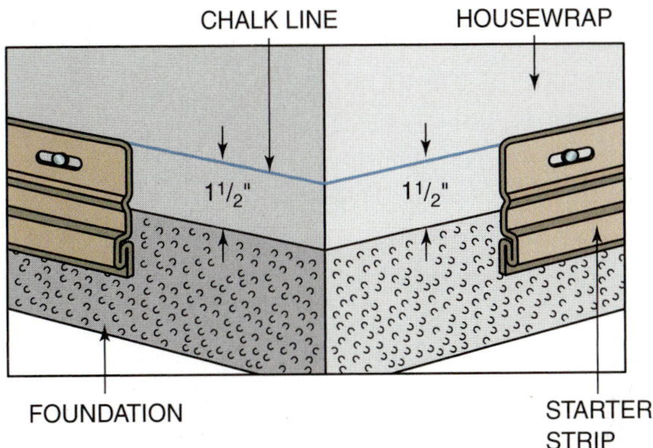


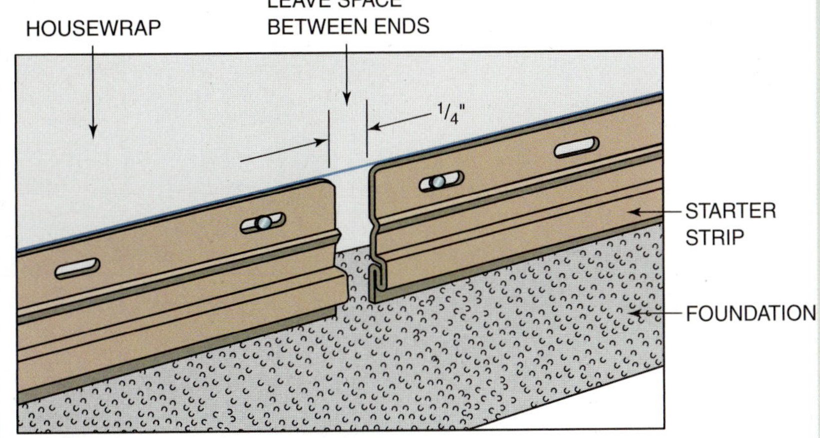

PROCEDURE 15-5

# Applying Horizontal Vinyl Siding (continued)

**B** Cut the corner posts so they extend ¼ inch below the starting strip. Attach the posts by fastening at the top of the upper slot on each side. The posts will hang on these fasteners. The remaining fasteners should be centered in the nailing slots. Make sure the posts are straight, plumb, and true from top to bottom.

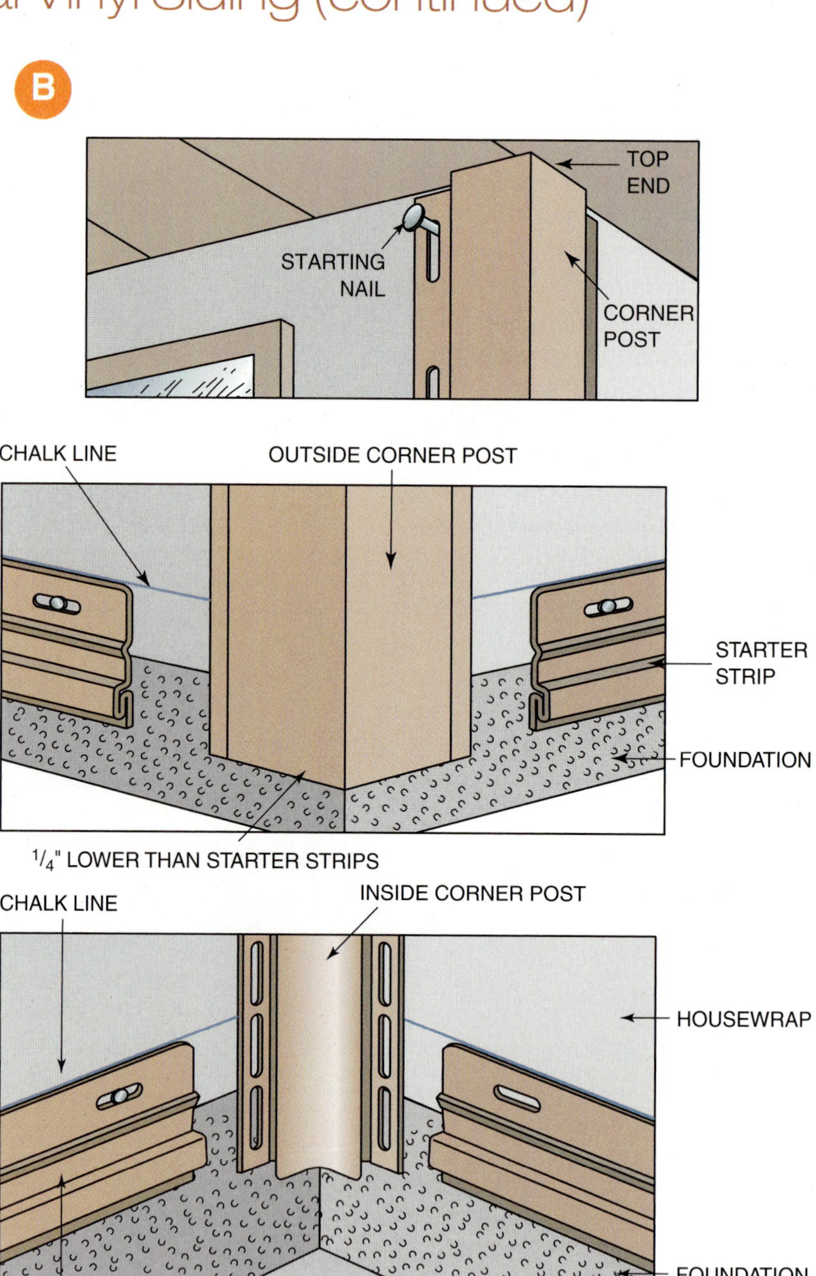

**C** Cut each J-channel piece to extend, on both ends, beyond the casings and sills a distance equal to the width of the channel face. Install the side pieces first by cutting a ¾-inch notch, at each end, out of the side of the J-channel that touches the casing. Fasten in place.

- On both ends of the top and bottom channels, make ¾-inch cuts at the bends, leaving the tabs attached. Bend down the tabs and miter the faces. Install them so the mitered faces are in front of the faces of the side channels and bent tabs are inserted into the vertical channels.

**C**

PROCEDURE 15-5

# Applying Horizontal Vinyl Siding (continued)

**D** Snap the bottom of the first panel into the starter strip. Fasten it to the wall. Start from a back or least visible corner, leaving a ¼-inch space in the corner post channel. Work toward the front with other panels. This will make the siding end laps seem to disappear. Overlap each panel about 1 inch. The exposed ends should face the direction from which they are least viewed.

- Install successive courses by interlocking them with the course below and staggering the joints between courses.

**E** To fit around a window, first install J-channel and undersill trim around the window opening. Mark the width of the cutout, allowing a ¼-inch clearance on each side. Mark the height of the cutout, allowing a ¼-inch clearance below the sill. Using a special *snaplock punch,* punch the panel along the cut edge at 6-inch intervals to produce raised lugs facing outward. Install the panel under the window and up in the under sill trim. The raised lugs cause the panel to fit snugly and lock into the trim.

**D**

DO NOT FORCE PANEL UP OR DOWN WHEN FASTENING.

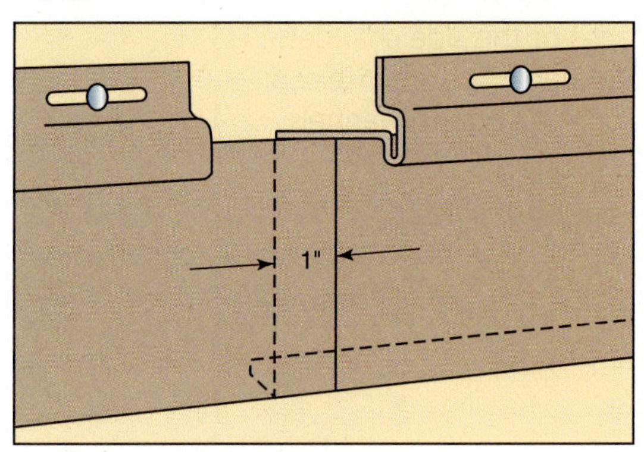


LAP PANELS AT LEAST 1".

**E**

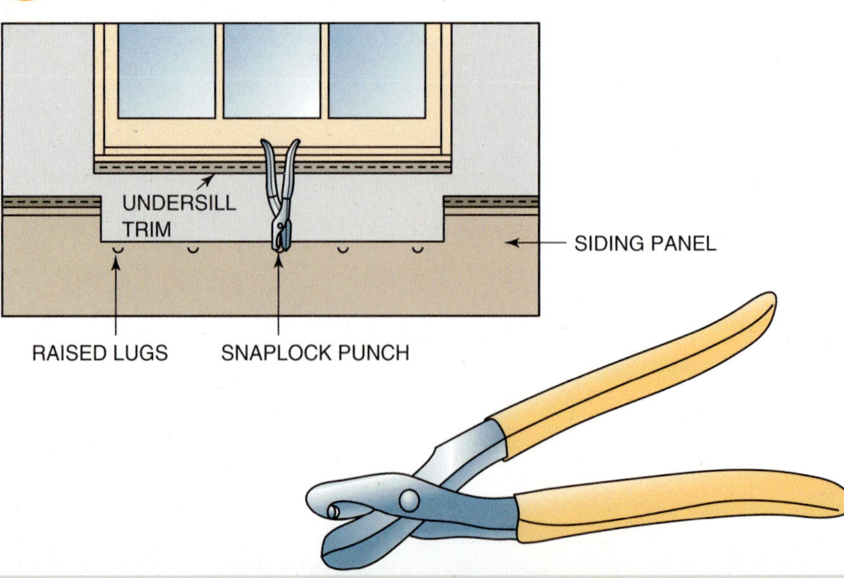

F Panels are cut and fit over windows in the same manner as under them. However, the lower portion is cut instead of the top. Allow ¼ inch on each side and slightly more on top to allow piece to drop low enough to snap in place. Install the panel by placing it into the J-channel that runs across the top of the window.

F

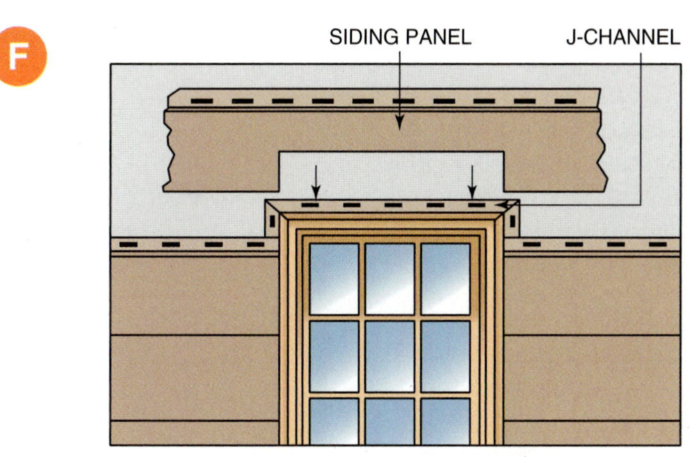


CUT EDGE OF PANEL FITS INTO J-CHANNEL OVER TOP OF WINDOW.

G Install the last course of siding panels under the soffit in a manner similar to fitting under a window. An *undersill trim* is applied on the wall and up against the soffit. Panels in the last course are cut to width. Lugs are punched along the cut edges. The panels are then snapped firmly into place into the undersill trim.

G

① MEASURE FOR LAST COURSE OF SIDING.

② RAISED LUGS

③ INSTALL CUT EDGE INTO UNDERSILL TRIM.

PROCEDURE 15-6

# Applying Vertical Vinyl Siding

**A** Measure and lay out the width of the wall section for the siding pieces. Determine the width of the first and last piece.

- Cut the edge of the first panel nearest the corner. Install an undersill trim in the corner board or J-channel with a strip of furring or backing. This will keep the edge in line with the wall surface. Punch lugs along the cut edge of the panel at 6-inch intervals. Snap the panel into the undersill trim. Place the top nail at the top of the nail slot. Fasten the remaining nails in the center of the nail slots.
- Install the remaining full strips, making sure there is a ¼-inch gap at the top and bottom. Fit around openings in the same manner as with fitting vertical siding. Install the last piece into undersill trim in the same manner as for the first piece.

**EXAMPLE:** What is the starting and finishing widths for a wall section that measures 18'–9" for siding that is 12" wide?

Convert this measurement to a decimal by first dividing the inches portion by 12 and then adding it to the feet to get 9 ÷ 12 + 18 = 18.75'.

Divide this by the siding exposure, in feet: 18.75 ÷ 1 foot = 18.75 pieces.

Subtract the decimal portion and add one full piece giving 1.75 pieces. Next, 1.75 ÷ 2 = 0.875, multiplied by 12 gives 10½".

This is the size of the starting and finishing piece. Thus, there are 17 full-width pieces and two 10½-inch-wide pieces.

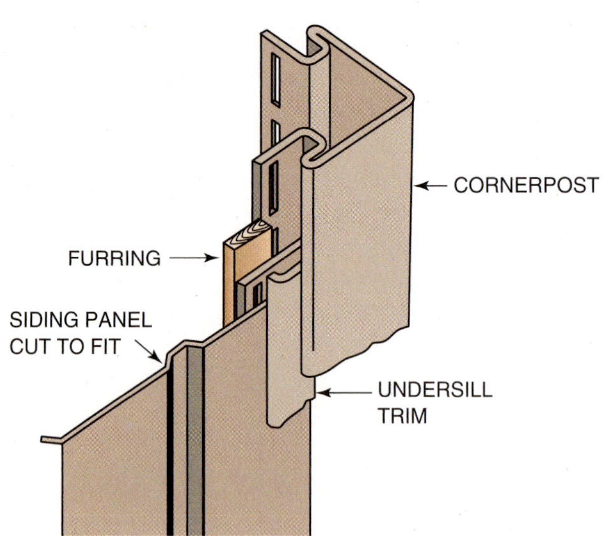


*Courtesy of Vinyl Siding Institute.*

PROCEDURE 15-7

# Installing Gutters

**A** On both ends of the fascia, mark the location of the top side of the gutter. The top outside edge of the gutter should be in relation to a straight line projected from the top surface of the roof. The height of the gutter depends on the pitch of the roof.

- Stretch a chalk line between the two marks. Move the center of the chalk line up slightly to give the gutter the desired slope. The slope need not be noticeable, just enough so as to not create a low spot in the gutter. Snap a line on both sides of the center.
- Fasten the gutter brackets to the chalk line on the fascia with screws. All screws should be made of stainless steel or other corrosion-resistant material. Aluminum brackets may be spaced up to 30 inches OC.

**B** Locate and install the outlet tubes in the gutter as required, keeping in mind that the downspout should be positioned plumb and square with the building. Add end caps and caulk all seams on the inside surfaces only.

- Hang the gutter sections in the brackets. Use

**A**

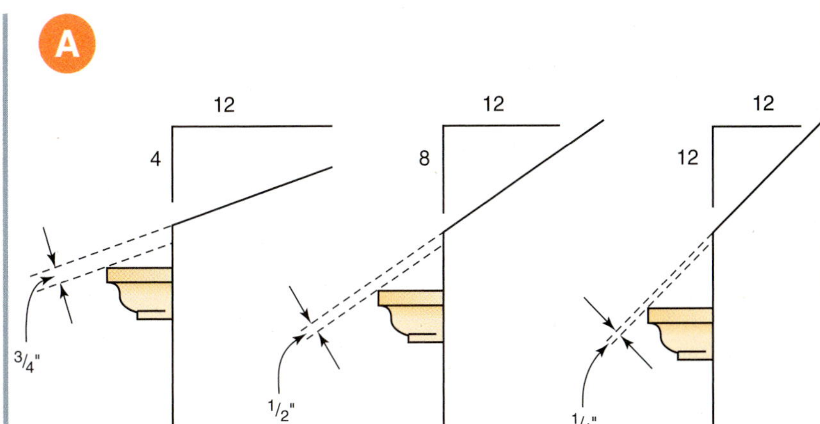


**B**

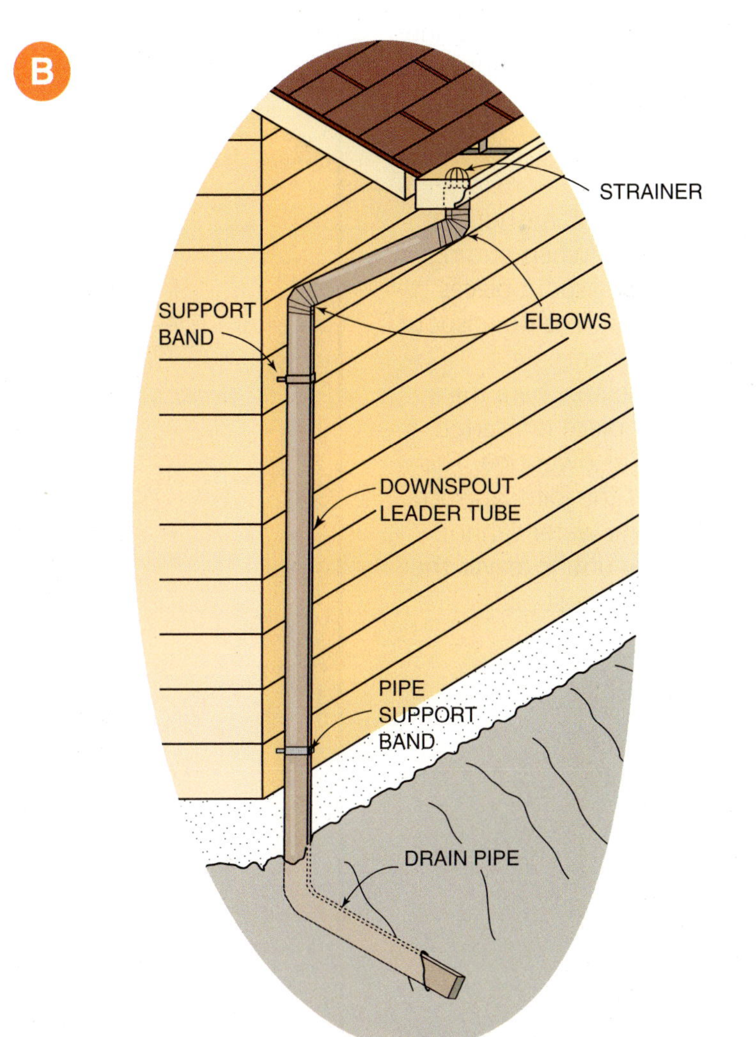

PROCEDURE 15-7

# Installing Gutters (continued)

slip-joint connectors to join larger sections. Use either inside or outside corners where gutters make a turn. Caulk all inside seams.

- Fasten downspouts to the wall with appropriate hangers and straps. Downspouts should be fastened at the top and bottom and every 6 feet in between. The connection between the downspout and the gutter is made with elbows and short, straight lengths of downspout.

C Because water runs downhill, care should be taken when putting the downspout pieces together. The downspout components are assembled where the upper piece is inserted into the lower one. This makes the joint lap such that the water cannot escape until it leaves the bottom-most piece.

C

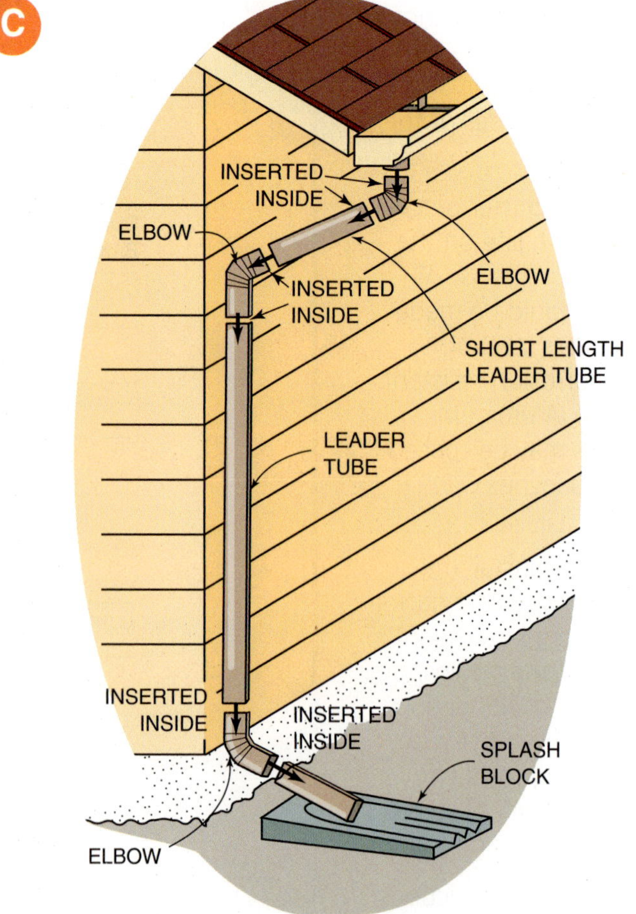

PROCEDURE 15-8

# Building a Deck

CAUTION

**CAUTION:** Pressure-treated lumber is made with substances that are toxic, which is why they resist decay. Care should be taken to protect the eyes and lungs from the sawdust created from cutting. Use gloves when handling material and wash hands before eating.

**A** Install the ledger against the building to a level line. Position the top edge to provide a comfortable step down from the building after the decking is applied. Install flashing under vertical flashing or housewrap and under the deck boards. Outside edge of flashing should extend beyond the ledger board.

**B** Set stakes at the locations of each post. Erect batter boards, stretch lines over each stake, and dig holes. Place concrete and anchors in the footing hole.

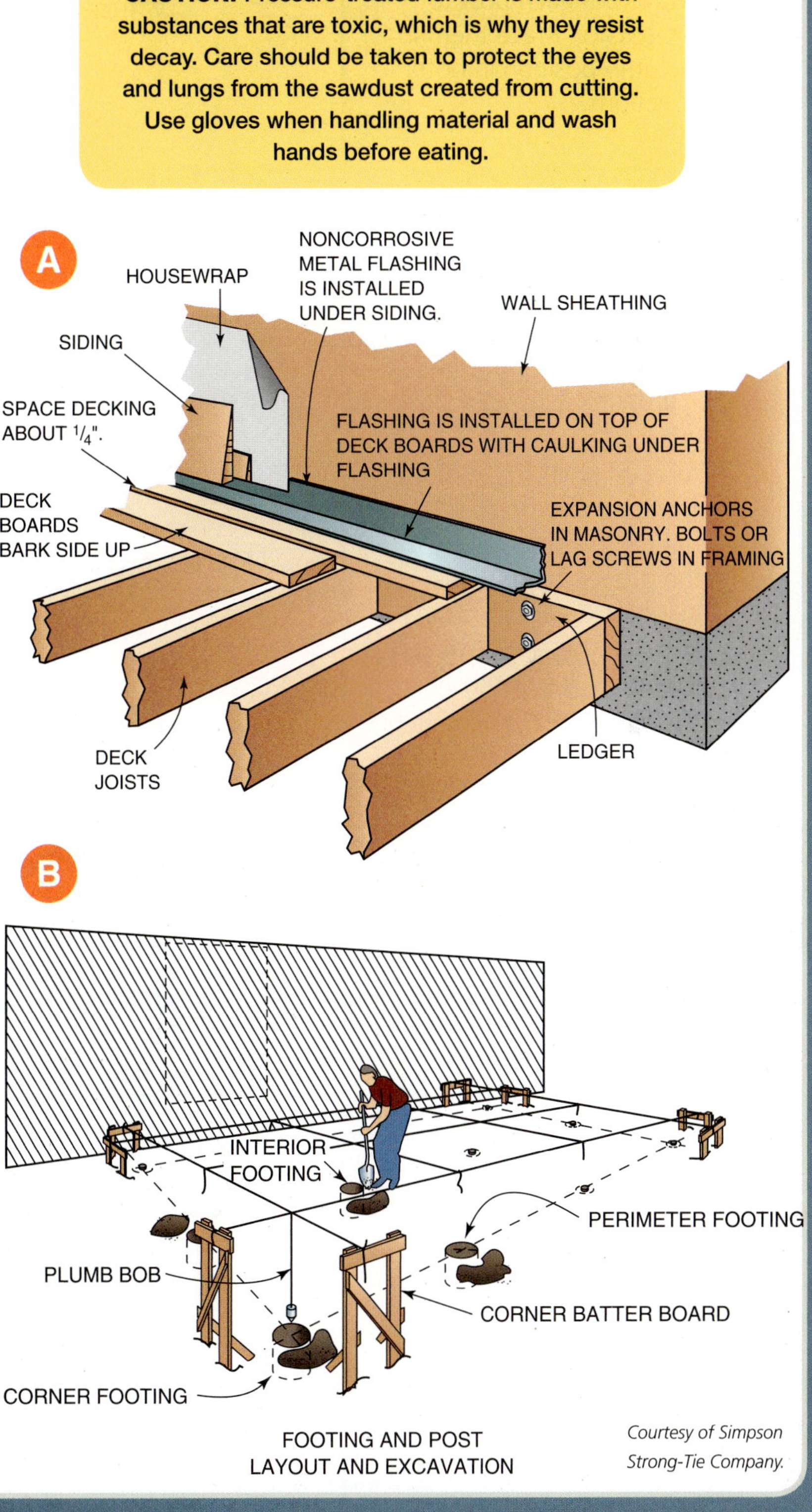


Courtesy of Simpson Strong-Tie Company.

PROCEDURE 15-8

# Building a Deck (continued)

**C** Place all supporting posts on the footings. Tack the bottom of each post to the anchor. Brace them in a plumb position in both directions.

**C**

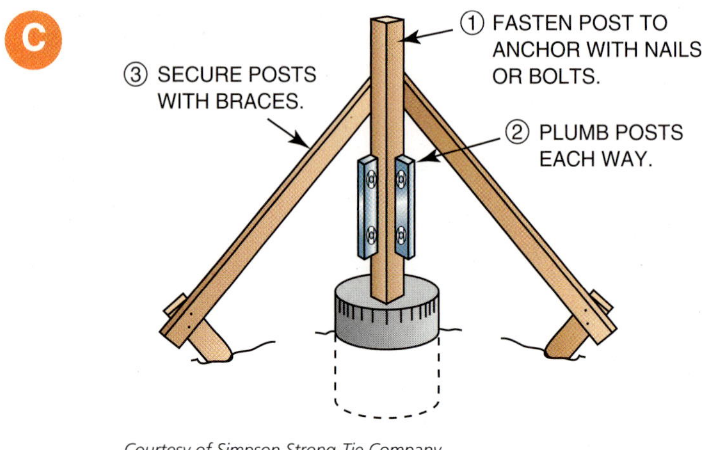


*Courtesy of Simpson Strong-Tie Company.*

**D** After all posts are plumbed and braced, the tops must be cut to the proper height. This height is determined by deducting the deck board thickness, the depth of the girder from the ledger board height. Include the joist depth if they are to rest on top of the girders. Mark all posts by leveling from the first post marked. Mark each post completely around using a square. Cut the tops with a portable circular saw. The deck should slope slightly, about ⅛ inch per foot, away from the building.

**D**

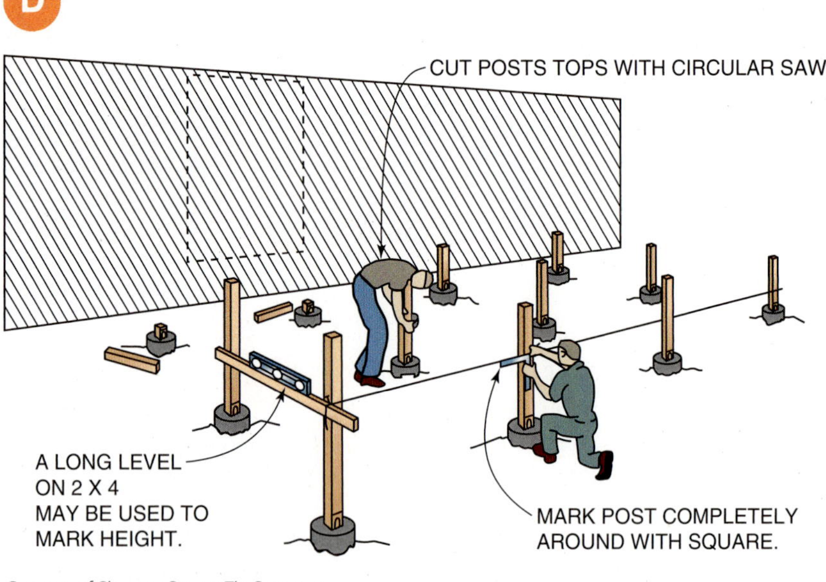


*Courtesy of Simpson Strong-Tie Company.*

**CAUTION**

**CAUTION:** Using a circular saw to cut vertical posts puts the saw at a difficult angle. Make sure you have firm control and good footing.

**E** Install the girders on the posts by using post and beam metal connectors. The size of the connector will depend on the size of the posts and girders. Install girders with the crowned edge up.

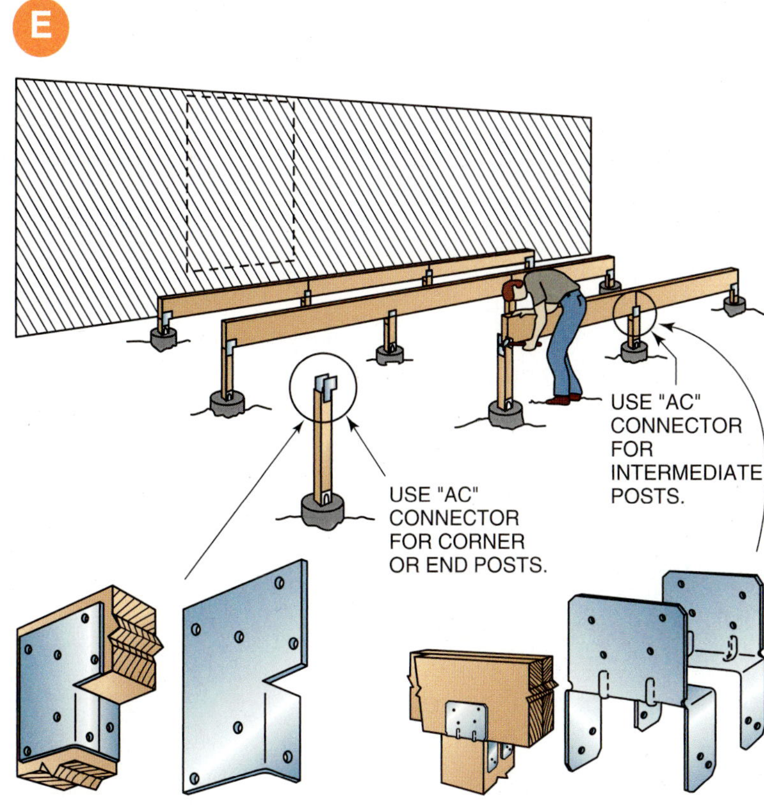

*Courtesy of Simpson Strong-Tie Company.*

**F** Lay out and install the joists. Use appropriate hangers if joists are installed between girders. When joists are installed over girders, use recommended framing anchors. Make sure all joists are installed with their crowned edges facing upward.

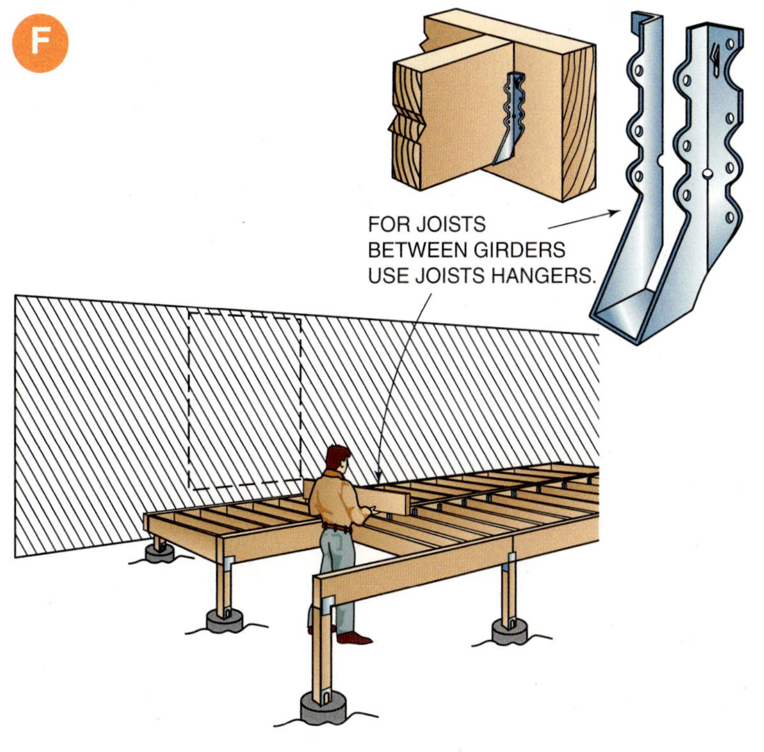

*Courtesy of Simpson Strong-Tie Company.*

PROCEDURE 15-8

# Building a Deck (continued)

**G** If the deck is 4 feet or more above the ground, the supporting posts should be braced. Use minimum 1 × 6 braces for heights of up to 8 feet. Use minimum 1 × 8 braces for higher decks applied continuously around the perimeter.

- Install ledger flashing under the siding and on top of the deck board. Caulking may be applied between the joist and flashing. The outside edge of the flashing should extend beyond the ledger.

**H** Lay boards with the bark side up or annular rings arching downward to minimize cupping. Boards are usually laid parallel with the long dimension of the deck. Boards may also be laid in a variety of patterns, including diagonal, herringbone, and parquet. Make sure the supporting structure has been designed and built to accommodate the design.

- Snap a straight line as a guide to apply the starting row. Start at the outside edge if the deck is built against a building. Straighten boards as they are installed. Maintain about a ¼-inch space between them.

**G**

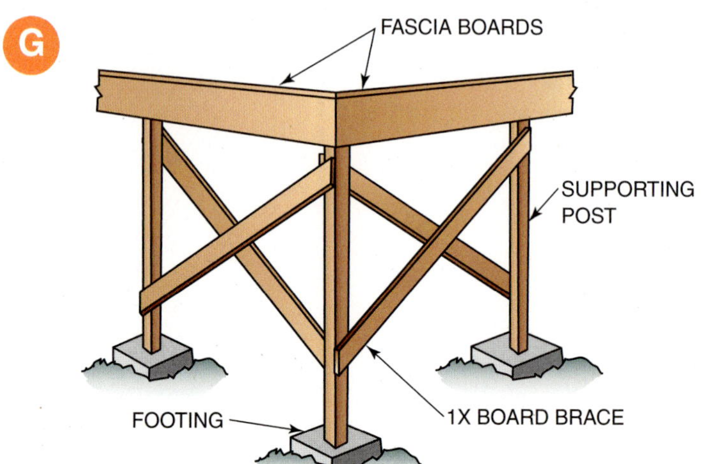


*Courtesy of Simpson Strong-Tie Company.*

**H**

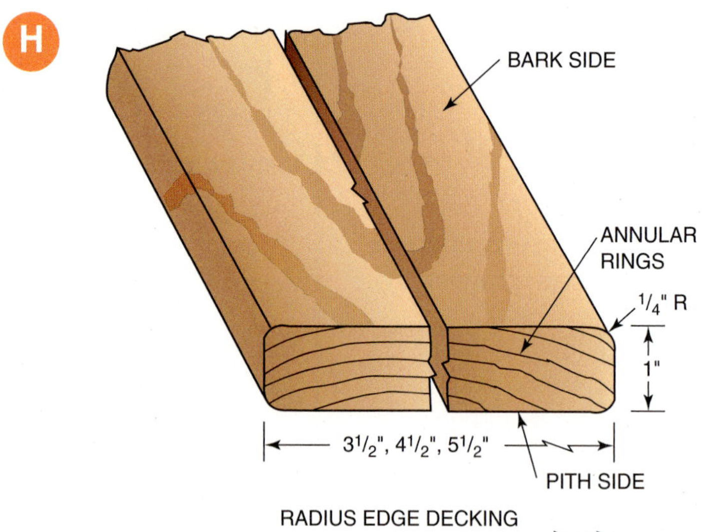


### *from experience...*

If the decking boards are green or wet, as with most pressure-treated boards, nail them tightly together. A space will appear when the lumber shrinks to equilibrium of moisture content.

Cut the boards so their butt ends are centered over joists. Stagger them between adjacent rows. Pre-drill holes for fasteners at the board ends to prevent splitting. Use two screws or nails in each joist. Deck boards may be installed so they overhang the deck and are trimmed later. When all the deck boards are laid, snap lines and cut the overhanging ends. Apply a preservative to them.

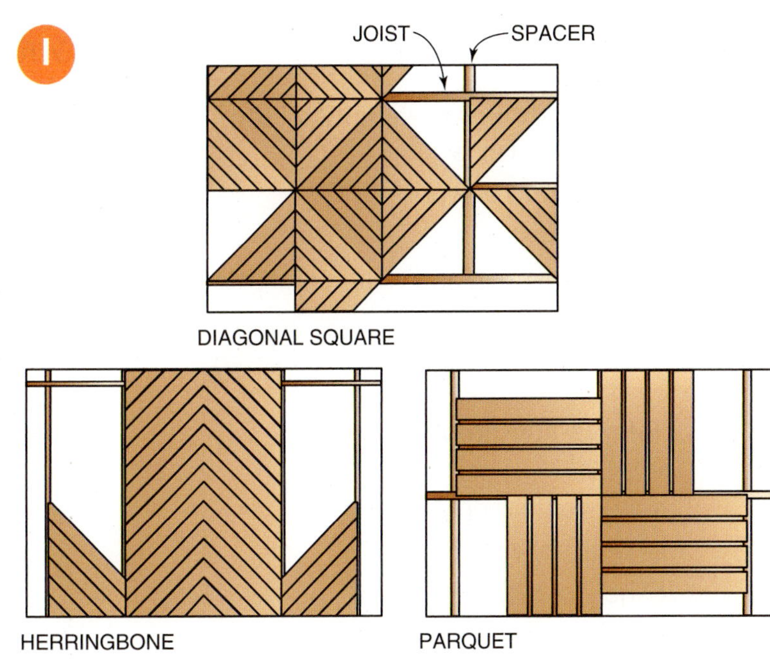


DECK BOARD PATTERNS

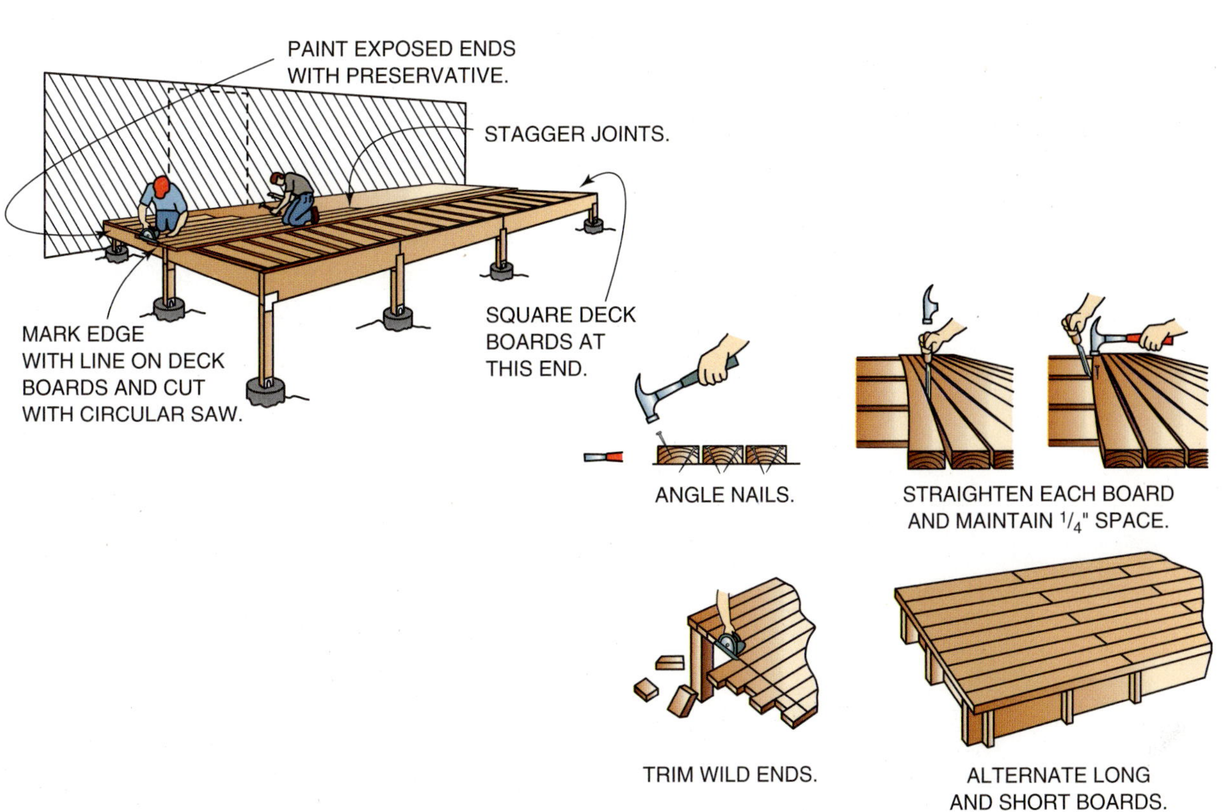


*Courtesy of Simpson Strong-Tie Company.*

## REVIEW QUESTIONS

*Select the most appropriate answer.*

**1. A useful thought when installing exterior finish is**

a. water runs downhill.
b. safety first.
c. craftsmanship and weathertightness go together.
d. all of the above.

**2. Tongue-and-grooved siding is applied**

a. horizontally only.
b. vertically only.
c. horizontally or vertically.
d. none of the above.

**3. A type of siding with which exposure can be changed is**

a. tongue-and-grooved.
b. rabbeted butt edges.
c. beveled.
d. shiplap.

**4. When applying beveled siding, it is desirable to**

a. maintain exactly the same exposure with every course.
b. apply full courses above and below windows.
c. work from the top down.
d. use a water table.

**5. In order to lay out for the courses of horizontal siding, which of the following must be known?**

a. distance between openings
b. window sill and door widths
c. the height of the wall to which siding is to be applied
d. all of the above

**6. When installing vinyl siding, drive nails**

a. tightly against the flange.
b. loosely against the flange.
c. called common nails.
d. colored the same as the siding.

**7. To allow for expansion when installing solid vinyl starter strips, leave a space between the ends of at least**

a. ⅛ inch.
b. ¼ inch.
c. ⅜ inch.
d. ½ inch.

**8. The tool used to shape aluminum coil stock into trim pieces is a**

a. shaper.
b. trimmer.
c. brake.
d. all of the above.

9. **Exterior trim that extends up the slope of the roof on a gable end is called the**
   a. box finish.
   b. rake finish.
   c. return finish.
   d. snub finish.

10. **Horizontal wood siding is usually installed with one fastener per stud. This is done to**
    a. reduce the potential for wood splitting as it swells and shrinks.
    b. save on construction cost for fasteners.
    c. comply with the fact that the siding is too narrow for more fasteners.
    d. increase job site safety.

11. **The term used to describe the fastening technique to hide the fastener from weather is called**
    a. toenailing.
    b. nail setting.
    c. blind nailing.
    d. all of the above.

12. **A member of the cornice fastened in the vertical position to the rafter tails is called the**
    a. drip.
    b. fascia.
    c. soffit.
    d. frieze.

13. **Care should be taken when installing gutters so that**
    a. they are level with the rake.
    b. downspouts are in the center of the gutter run.
    c. downspout leader tubes are connected to the foundation drains.
    d. gutter parts are installed so that water will run over the laps.

14. **No fasteners used on exposed decks should be made with**
    a. epoxy coating.
    b. hot-dipped galvanized coating.
    c. electroplated zinc coating.
    d. stainless steel.

15. **A deck ledger is a beam**
    a. attached to the side of a building.
    b. supported by a girder.
    c. used to support girders.
    d. installed on supporting posts.

16. **A footing for supporting posts should extend**
    a. 12 inches below grade.
    b. 36 inches below grade.
    c. to the frost line.
    d. to stable soil depending on geographic area.

## KNOW YOUR CODES

Decks must be built strong to withstand a larger group of people than the floors inside a residence. Decks have collapsed due to improper fastening techniques. Check the local code and the building prints for guidance.

## WHAT'S WRONG WITH THIS PICTURE?

Carefully study **Figure 15-43** and think about what is wrong. Consider all possibilities.

WRONG

**FIGURE 15-43** This photo shows bevel siding, with the corner board removed for clarity. Notice the position of the nails. The nails are penetrating *both* pieces of siding. Wood moves across its grain (in this case up and down) with the changing humidity and temperature conditions. During a dry season, the wood will shrink; and because each board is nailed at the top and bottom, it will not be able to move, or shrink, without cracking or compromising the siding. The siding should be nailed in such a way so that the nail does not penetrate both pieces. Furthermore, it would be a good idea to have some type of building paper behind the siding.

RIGHT

**FIGURE 15-44** Notice that the nails are above the top of the piece of siding beneath. This will allow the siding to "breathe"—shrink and expand in the direction of the arrow—with the changing temperature and humidity conditions. Care must be taken when nailing. The siding is thin and will break easily, especially without support directly behind the nail.

SECTION 4

# Interior Finish

# Insulation and Ventilation

**OBJECTIVES** *After completing this unit, the student should be able to:*

- Describe how insulation works and define insulation terms and requirements.
- Describe the commonly used insulating materials and state where insulation is placed.
- Properly install various kinds of insulation.
- State the purpose of and install vapor retarders.
- Explain the need for ventilating a structure and describe types of ventilators.

## GLOSSARY OF INSULATION AND VENTILATION TERMS

**acoustical tile** a fiberboard ceiling tile whose surface consists of small holes and fissures that act as sound traps

**air infiltration** unwanted movement of air into an insulation layer or a conditioned space (heated or cooled)

**condensation** when water vapor changes to a liquid due to cooling of the air, the resulting droplets of water that accumulate on the cool surface

**dew point** temperature at which moisture begins to condense out of the air

**equilibrium** a state of balance; heat energy is thought to travel from high concentration (hot) to low concentration (cold)

**impact noise** noise caused by the vibration of an impact, such as dropped objects or footsteps on a floor

**insulation** material used to restrict the passage of heat or sound

**R-value** a number given to a material to indicate its resistance to the passage of heat

**sound-deadening insulating board** a sound resistance system using a gypsum board outer covering and resilient steel channels placed at right angles to the studs to isolate the gypsum board from the stud

**storm sash** an additional sash placed on the outside of a window to create dead air space to prevent the loss of heat from the interior in cold weather

**vapor retarder** a material used to prevent the passage of water in the gaseous state

**weather stripping** narrow strips of thin metal or other material applied to windows and doors to prevent the infiltration of air and moisture

At this point in the construction process, it is assumed that the mechanicals (plumbing, heating, cooling, and electrical components) have been installed. These systems are installed by professionals and inspected by various code enforcement agencies. The carpenters are set to return and begin the interior finish.

Insulation is a material installed to resist the transfer of heat. It is important for the carpenter to understand how insulation works, what its workability requirements are, and how to get the most out of insulation.

Water is essential for life on earth, but it can be hazardous to buildings. Excess moisture causes framing to rot or corrode and insulation to perform badly. Careful attention to details will keep most moisture out, and ventilation will remove the rest. Natural ventilation is simple and effective when installed properly.

# INSULATION AND VENTILATION

Understanding insulation and ventilation concepts is vital to ensure buildings are energy efficient.

## HOW INSULATION WORKS

All materials used in construction have some insulating value. **Insulation** is a material that interrupts or slows the transfer of heat. Heat transfer is a complex process involving three mechanisms: conduction, convection, and radiation.

Conduction is the transfer of heat by contact. Warmer particles touch cooler ones, causing them to vibrate with more energy, thus transferring energy into the cooler material. Heat energy is thought to move from warmer materials into the cooler ones in an attempt to reach an **equilibrium**.

Convection involves a fluid, such as a gas or a liquid. The fluids affecting a house are usually air and water. When a portion of the fluid is warmed, it becomes less dense than the surrounding fluid. The warmer fluid requires more space than the cooler fluid. Gravity causes the cooler, more dense fluid to sink forcing the warmer, less dense fluid to rise. This action can be seen in the streams of fluid in a pan of hot oil on a cooking stove. A hot-air balloon encapsulates less dense hot air, which creates lift, causing the craft to rise.

Radiation is a general word referring to electromagnetic radiation, which includes microwaves, radio waves, infrared, visible light, ultraviolet, rays, and cosmic rays. These can be thought of as particles of energy with varying wavelengths and vibration frequencies that all travel at the same speed, the speed of light. Protection from radiation is best achieved by reflecting it with clean shiny surfaces such as aluminum foil.

Popcorn may be used to help remember these three forms of heat transfer. The three different methods of cooking popcorn use different modes of heat transfer. All three methods heat the moisture in the kernel to steam, which then ruptures the shell with the familiar popping sound. Stove-top popcorn absorbs heat by direct contact with burners. Hot-air poppers transfer heat from electric heating elements to the air that blows onto the kernels. Microwave packages use microwave radiation to heat the kernels (Fig. 16-1).

**FIGURE 16-1** Various poppers use different types of heat transfer to pop corn.

Insulating materials create a space between two surfaces, thereby breaking contact and reducing conduction. They trap air and slow convection. They also can either reflect or absorb some radiation.

Air is an excellent insulator if confined to a small space and kept still. Insulation materials are designed to trap air into small unmoving pockets. Insulation effectiveness increases as the air spaces become smaller in size and greater in number. Millions of tiny air cells, trapped in its unique cellular structure, make wood a better insulator than concrete, steel, or aluminum.

All insulating materials are manufactured from materials that are themselves poor insulators. For example, fiberglass insulation is made from glass, which is a good conductor of heat. The improved insulation value comes from air trapped within the insulating material. Insulation also provides resistance to sound travel.

All materials used in construction have some insulating value, some better than others. Each is tested and given an **R-value**. This number indicates a material's ability to resist heat transfer. The higher the number, the better the insulator (Fig. 16-2). Insulation becomes more effective as the air spaces become smaller and greater in number. Insulation also has the additional benefit of absorbing sound. Among the materials used for insulating are glass fibers, mineral fibers (rock), organic fibers (paper), and plastic foam. Aluminum foil is also used to reflect heat radiation.

**Green Tip**

Insulation pays for itself in saved energy dollars by reducing energy needs and lessening the carbon footprint of a building.

## INSULATION REQUIREMENTS

The rising costs of energy, coupled with the need to conserve it, have resulted in higher R-value requirements for new construction than in previous years. Average winter low-temperature zones of the United States are shown in Figure 16-3. This information is used to determine the R-value of insulation installed in walls, ceilings, and floors. Insulation requirements vary according to the average low temperature. Check local codes for requirements in your area.

| Material | R-Value |
|---|---|
| **Foundation Materials** | |
| 8″ concrete block (2-hole core) | 1.11 |
| 12″ concrete block (2-hole core) | 1.28 |
| 8″ lightweight aggregate block | 2.18 |
| 12″ lightweight aggregate block | 2.48 |
| Common brick | 0.20/inch |
| Sand or stone | 0.08/inch |
| Concrete | 0.08/inch |
| **Structural Materials** | |
| Softwood | 1.25/inch |
| Hardwood | 0.91/inch |
| Steel | 0.0032/inch |
| Aluminum | 0.00070/inch |
| **Sheathing Materials** | |
| ½″ plywood | 0.63 |
| ⅝″ plywood | 0.78 |
| ¾″ plywood | 0.94 |
| ½″ aspenite, OSB | 0.91 |
| ¾″ aspenite, OSB | 1.37 |
| **Insulating Materials** | |
| *Batts and Blankets* | |
| 3½″ fiberglass | 11 |
| 6″ fiberglass | 19 |
| 8″ fiberglass | 25 |
| 12″ fiberglass | 38 |
| 3½″ high-density fiberglass | 13 |
| 5½″ high-density fiberglass | 21 |
| 8½″ high-density fiberglass | 30 |
| 10″ high-density fiberglass | 38 |
| fiberglass | 3.17/inch |
| *Loose Fill* | |
| Fiberglass | 3.17/inch |
| Cellulose | 3.70/inch |
| *Reflective Foil* | |
| Foil-faced bubble pack | 1.0 |
| 1-layer foil | 0.22 |
| 4-layer foil | 11 |
| *Rigid Foam* | |
| Expanded polystyrene foam (bead board) | 4.0/inch |
| Extruded polystyrene foam | 5.0/inch |
| Polyisocyanurate/urethane foam | 7.2/inch |
| *Spray Foams* | |
| Low-density polyurethane | 3.60/inch |
| High-density polyurethane | 7.0/inch |
| **Finish Materials** | |
| Wood shingles | 0.87 |
| Vinyl siding | Negligible |
| Aluminum siding | Negligible |
| Wood siding (½″ × 8″) | 0.81 |
| ½″ hardboard siding | 0.36 |
| Polyethylene film | Negligible |
| Builder's felt (15#) | 0.06 |
| 25/32″ hardwood flooring | 0.68 |
| Vinyl tile (1/8″) | 0.05 |
| Carpet and pad | 1.23 |
| ¼″ ceramic tile | 0.05 |
| ⅜″ gypsum board | 0.32 |
| ½″ gypsum board | 0.45 |
| ⅝″ gypsum board | 0.56 |
| ½″ plaster | 0.09 |
| Asphalt shingles | 0.27 |

*Courtesy of Professor Richard Harrington, SUNY College of Technology at Delhi.*

**FIGURE 16-2** Insulating R-values of various building materials.

In warmer climates, less insulation is needed to conserve energy and provide comfort in the cold season. However, air-conditioned homes should receive more insulation in walls, ceilings, and floors. This ensures economy in the operation of air-conditioning equipment during the hot season. Comfort and operating economy are dual benefits.

Insulating for maximum comfort automatically provides maximum economy of heating and cooling operations. It also reduces the initial costs of heating and cooling equipment because smaller units may be adequate.

## WHERE HEAT IS LOST

The amount of heat lost from the average house varies with the type of construction. The principal areas and approximate amount of heat loss for a typical house with moderate insulation are shown in Figure 16-4. Houses of different architectural styles vary in their heat-loss characteristics. Greater heat loss through floors is experienced in homes erected over concrete slabs or unheated crawl spaces unless these areas are well insulated. The use of 2 × 6 studs in exterior walls permits installation of 6-inch-thick R-19 or R-21 insulation.

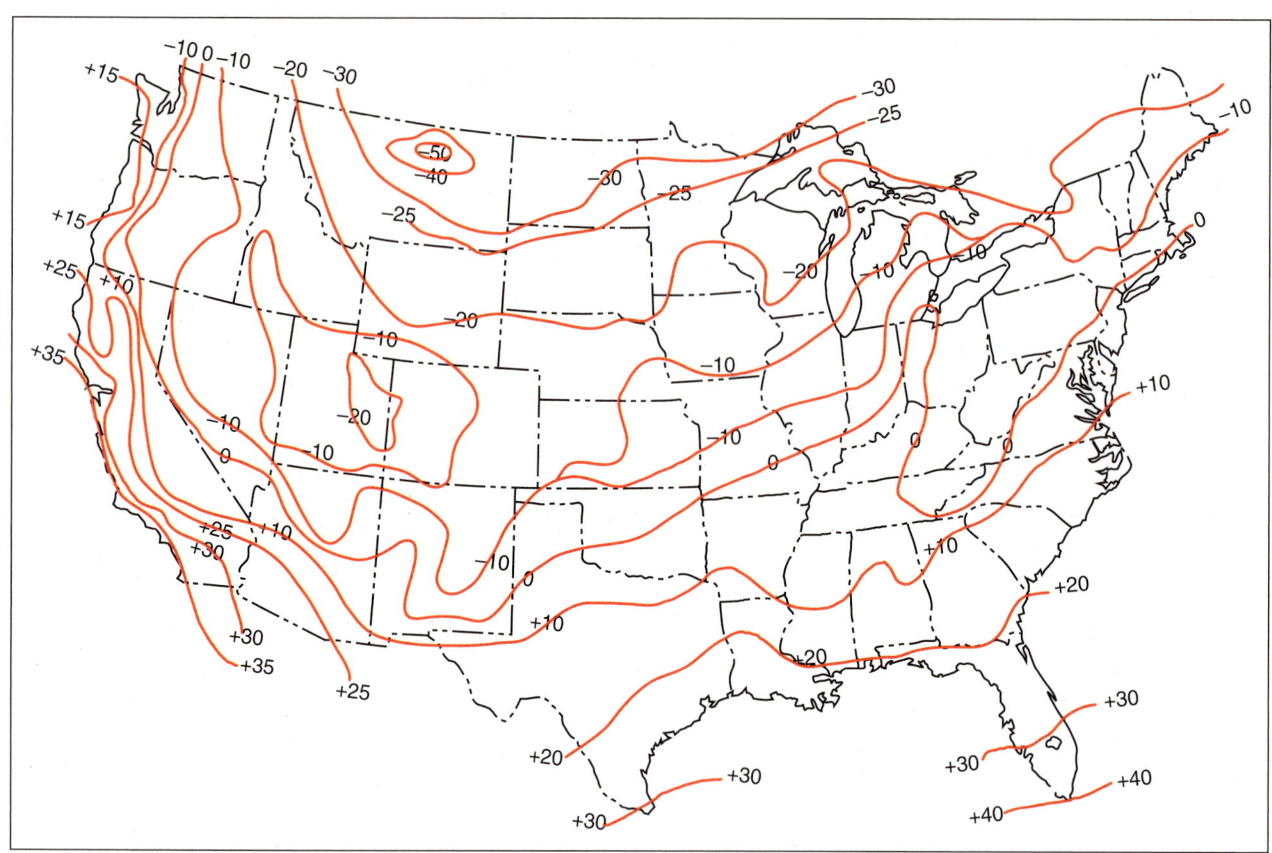


Courtesy of U.S. Department of Agriculture, Forest Service.

**FIGURE 16-3** Average low-temperature zones of the United States.

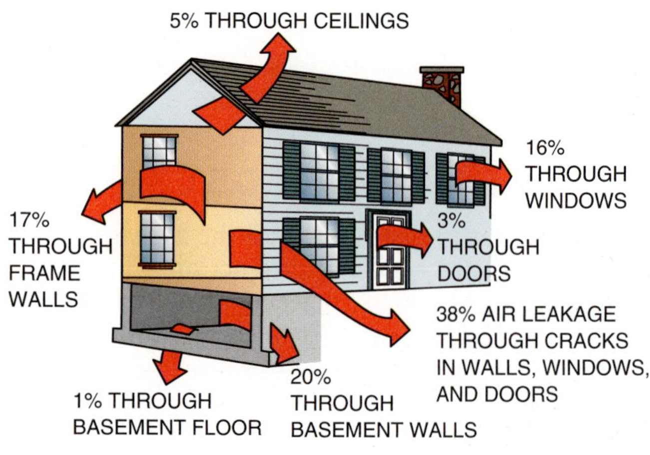


Courtesy of Dow Chemical Company.

**FIGURE 16-4** Areas and amounts of heat loss for a typical house with moderate insulation.

Windows and doors are generally sources of great heat loss. Heat loss through glass surfaces can be reduced to 50 percent or more by installing double- or triple-glazed windows or by adding a **storm sash** and storm doors. **Weather stripping** around windows and doors seals gaps and reduces heat loss.

## GETTING THE MOST FROM INSULATION

For insulation to perform properly, care must be taken in how the insulation is installed, particularly at the eaves. An air space must be maintained between the insulation and the roof sheathing. By using positive ventilation chutes, the insulation is compressed slightly to allow for air movement over

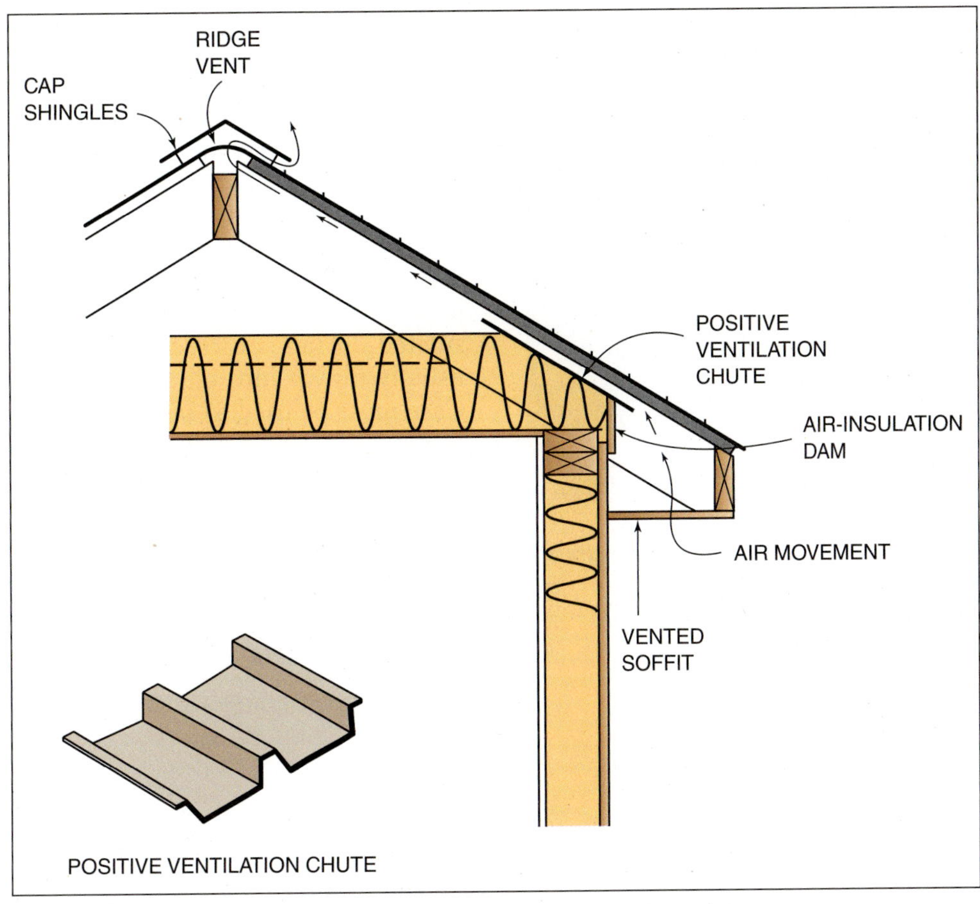


**FIGURE 16-5** Positive ventilation chutes and air-insulation dams maintain proper airflow around the insulation layer.

the insulation. Air-insulation dams are installed to protect the insulation from **air infiltration** (Fig. 16-5). Together, dams and chutes allow air to pass over the insulation and not through it. With these techniques in use, the highest R-value will be achieved. They should be installed in every rafter cavity.

# Types of Insulation

Insulation is manufactured in a variety of forms and types. These are commonly grouped as *flexible, loose-fill, rigid, reflective,* and *foamed-in place.*

## FLEXIBLE INSULATION

Flexible insulation is manufactured in *blanket* and *batt* form. Blanket insulation comes in rolls (Fig. 16-6). Widths are suited to 16- and 24-inch stud and joist spacing with thicknesses up to 12 inches. The body of the blanket is made of fluffy rock or glass wool fibers. Blanket insulation is purchased with or without a facing of asphalt-laminated Kraft paper or aluminum foil with flanges on both edges for fastening to studs or joists.

Batt insulation (Fig. 16-7) is made of the same material and facings as blanket insulation. It comes in the same thicknesses and widths, and lengths are either 48 or 93 inches.

**Green Tip**

Fiberglass is made from 25% recycled materials.

FIGURE 16-6 Blanket insulation comes in rolls.

FIGURE 16-7 Batt insulation is made up to 12 inches thick.

## LOOSE-FILL INSULATION

Loose-fill insulation is usually composed of materials in bulk form. It is supplied in bags or bales. It is placed by pouring, blowing, or packing by hand. Materials include rock or glass wool, wood fibers, shredded redwood bark, cork, and wood pulp products. Loose-fill insulation is well suited for use between trusses and ceiling joists. It is also installed in the sidewalls of existing houses that were not insulated during construction. Flexible insulation is usually blown into place with special equipment. Care must be used to install the proper amount to ensure the desired volume and depth after setting has occurred (Fig. 16-8).

**Green Tip**

Cellulose insulation is produced from 100% recycled newspaper.

## RIGID INSULATION

Rigid insulation is usually a fiber or foam plastic material formed into sheets or panels (Fig. 16-9). The material is available in widths of 2 or 4 feet and lengths of 8 feet. The most common types are made from *polystyrene* and *polyurethane* foams.

Polystyrene comes in two forms: expanded and extruded. Expanded polystyrene is a white

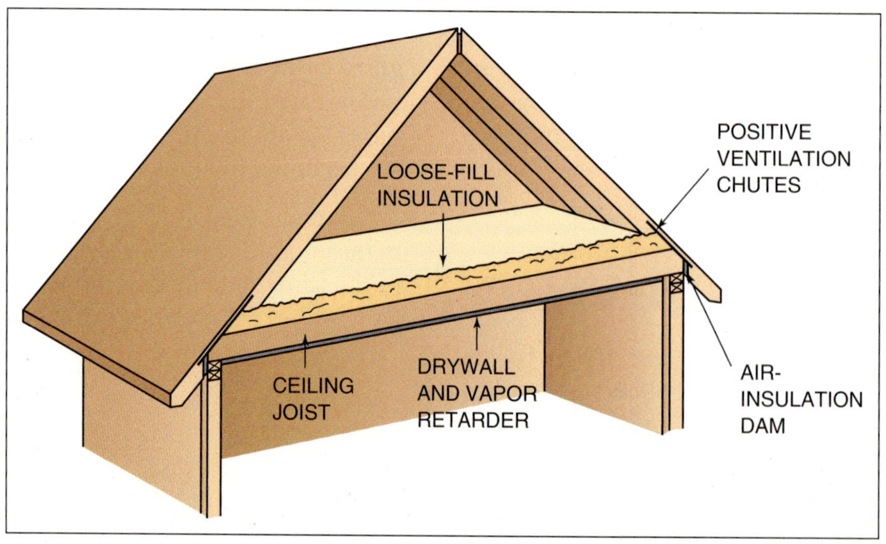


FIGURE 16-8 Level loose-fill insulation to the desired depth.

**FIGURE 16-9** Types of rigid foam insulation boards: (A) foil-face polyisocyanurate, (B) extruded polystyrene, (C) expanded polystyrene.

bead board similar to foam coffee cups and some packing material. Extruded polystyrene is the only insulation board that may come in contact with water and still be an effective insulator. It is waterproof and will not absorb moisture, thereby keeping its R-value. It is used to insulate masonry walls and slabs (Fig. 16-10). It is also used on existing structures before new siding is applied. It is used on roofs, either above or below the sheathing (Fig. 16-11).

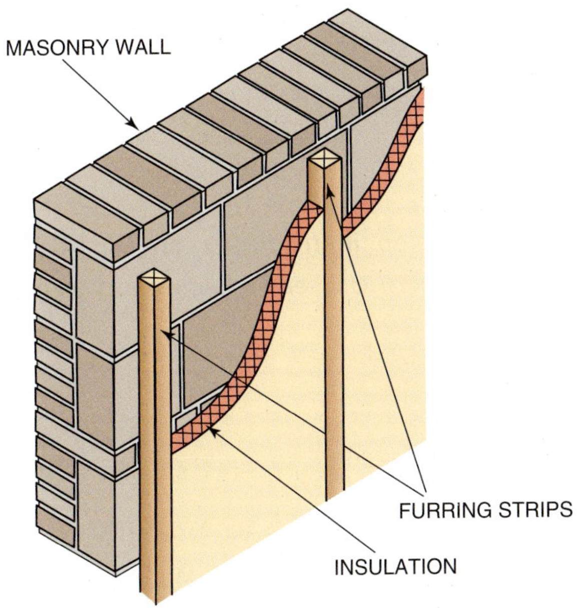


**FIGURE 16-10** Method of insulating a masonry wall with extruded polystyrene.

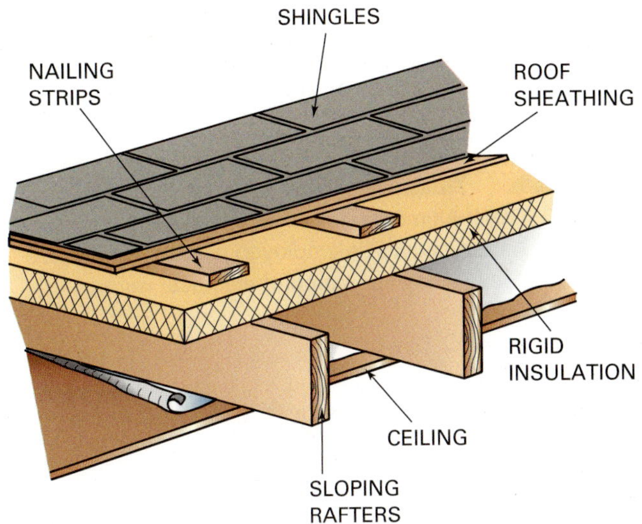


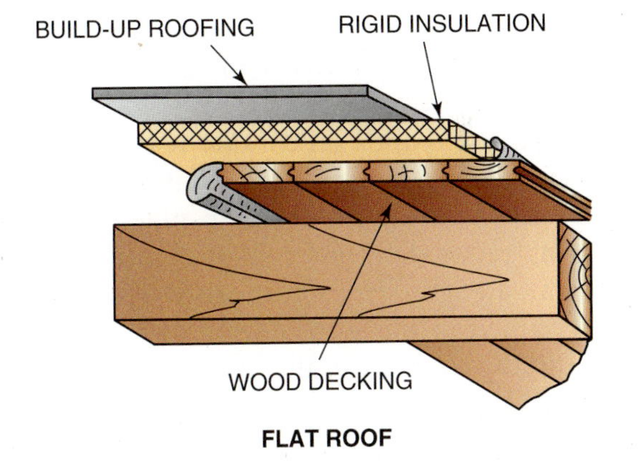


**FIGURE 16-11** Method of installing rigid insulation on roofs.

Polyurethane foams have the best R-value per inch. They are usually foil-faced and sometimes faced with Kraft and asphalt-impregnated paper. They are used for wall sheathing and flat roof insulation.

Rigid insulation is easily cut with a knife or saw. It may be applied by friction fitting between the framing members. A table saw may be used to make smooth, tight, and uniform cuts. Workers should wear respirators when using an electric saw to cut rigid foam panels. These panels are also flammable and must be finished by covering with a nonflammable material.

## REFLECTIVE INSULATION

Reflective insulation usually consists of outer layers of aluminum foil bonded to inner layers of various materials for added strength to resist heat flow. Reflective insulation should be installed facing an air space with a depth of about ¾ inch.

**CAUTION**

**CAUTION:** Aluminum foil conducts electricity very well. Care should be employed when installing reflective insulation around electrical wiring.

### FOAMED-IN-PLACE INSULATION

*Foamed-in-place* insulation is sometimes used in wall cavities. This product is installed by professionals using highly specialized tools and equipment. Polyurethane foam is produced by mixing two chemicals together and injecting the mixture into place. It expands on contact with the surface. Because it completely fills the cavity and sticks to studs and sheathing, foamed-in-place insulation performs better than other insulation materials. It makes an airtight seal, significantly reducing heat loss. Foamed-in-place insulation also contributes substantially to the overall stiffness of a building.

Small spray cans of polyurethane foam are used to fill gaps and holes created during construction. Gaps include those between framing members and the jambs of windows and doors. Also, any hole created for plumbing and electrical material should be sealed. Use only low expanding foam on windows. Check manufacturer's recommendation to be sure how foam affects the warranty.

**CAUTION**

**CAUTION:** Foamed-in-place polyurethane products are not dissolvable by any over-the-counter liquid. The only way to remove unwanted foam from skin or other material is by scraping it off after it dries.

## INSTALLING INSULATION

To reduce heat loss, all walls, ceilings, roofs, and floors that separate heated from unheated spaces should be insulated. Insulation should be placed in all outside walls and in the ceiling. Great care should be exercised when installing insulation. Voids in insulation of only 5 percent of the overall area can create an R-value efficiency reduction of 25 percent.

Pay attention to details around outlets, pipes, and any obstructions. Make the insulation conform to irregularities by cutting and piecing without bunching or squeezing. Keep the natural fluffiness of the insulation at all times.

A ground cover of plastic film such as polyethylene should be placed on the soil of crawl spaces. This acts as a **vapor retarder**, decreasing the moisture content of the space as well as of the wood members. Provision should also be made for ventilation of unheated areas. In houses with flat or low-pitched roofs, insulation should be installed with sufficient space above the insulation for ventilation (Fig. 16-12). Insulation is used along the perimeter of houses built on slabs, when required.

### INSTALLING FLEXIBLE INSULATION

Insulation that is not properly installed can drastically reduce that insulation's effectiveness. The carpenter must be aware of the consequences of improper application. Make lengths slightly oversized so the insulation fits snugly. If the overall construction is airtight, unfaced insulation may be used, since airtight construction serves as the vapor retarder. Also, it is easier to see proper fit of the insulation.

***For step-by-step instructions on INSTALLING FLEXIBLE INSULATION, see Procedure 16-1 on pages 547–549.***

*from experience...*

A good general rule is to install insulation neatly. If insulation looks neat, tight, and full, it is more likely to perform as expected.

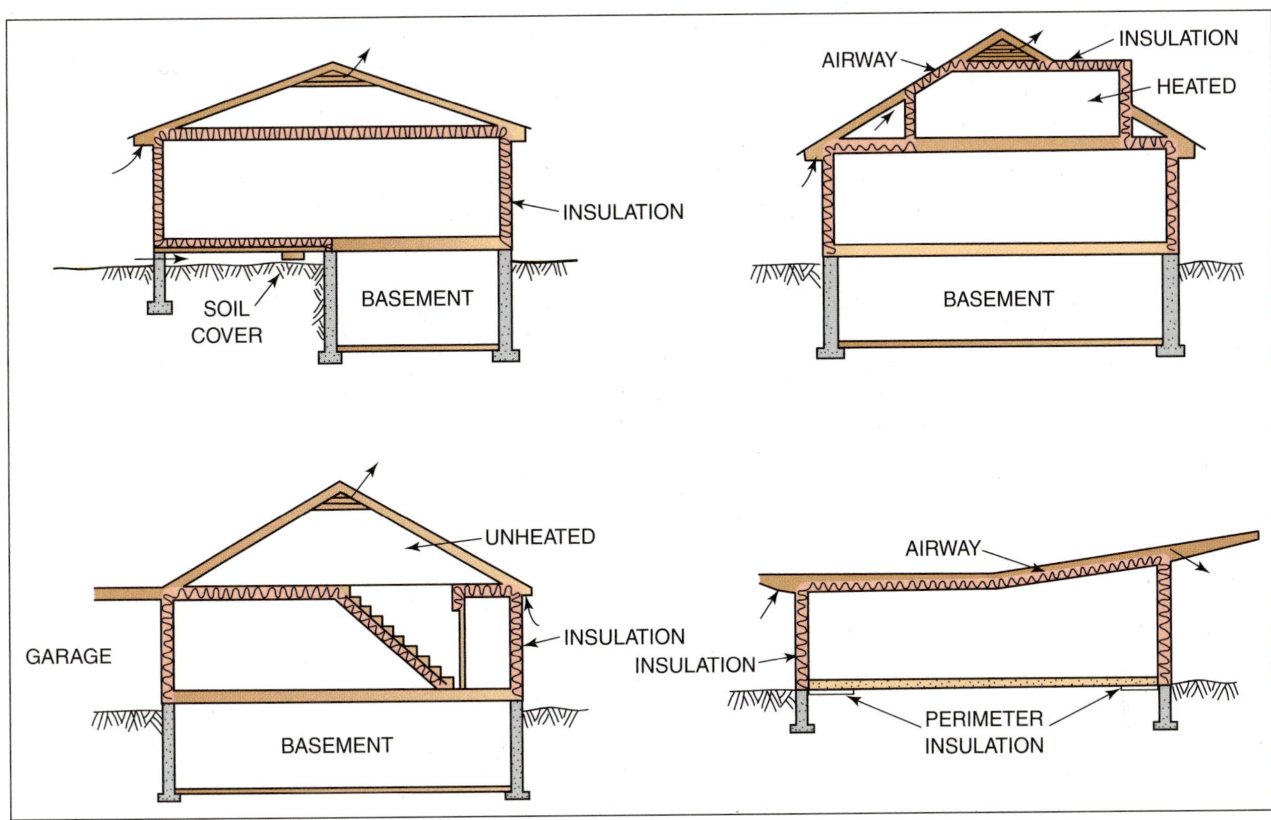


**FIGURE 16-12** Placement of insulation in various types of structures.

# Condensation

Warm air can hold more moisture than cold air. As moisture-laden air is cooled, it reaches the **dew point**. At this point, it is completely saturated with moisture. Fog is air that has reached its dew point. Any further lowering of the air temperature causes it to lose some of its moisture as **condensation** to cooler surfaces. Evidence of this can be seen as dew. The moisture lost is condensed on the ground. This natural process is detrimental to buildings.

## MOISTURE IN BUILDINGS

Moisture in the form of water vapor can come from water leaks into the house. Making sure all roofing, siding, and flashings perform correctly and keeping basements dry will alleviate this problem. Water vapor also comes from normal living activities. It is produced mainly by cooking, taking showers and baths, washing and drying clothes, and using a humidifier. The water contained in the warmer air tries to penetrate the insulation layer in its effort to equalize with the cooler, drier air.

In an insulated building, water vapor strives to move through the walls, ceilings, and floors. It condenses under certain conditions when it comes in contact with a cooler surface (Fig. 16-13). In a wall, this contact may be made on the inside surface of the exterior wall sheathing. In a crawl space, condensation can form on the masonry structure. In attics, the roof frame can become saturated with moisture. Condensation of water vapor in walls, attics, roofs, and floors will lead to serious problems.

Water leaks caused by problems with the weather-tightness of the building must be repaired. But some leaks happen only seasonally because of the formation of ice dams at the eaves. Ice dams begin to grow after a snowfall when heat from the attic causes the snow next to the roof to melt. This water runs down the roof until it reaches the cold roof overhang, where it freezes, forming an ice dam. This causes the running water to back up on the roof, under the shingles, and into the walls and ceiling (Fig. 16-14).

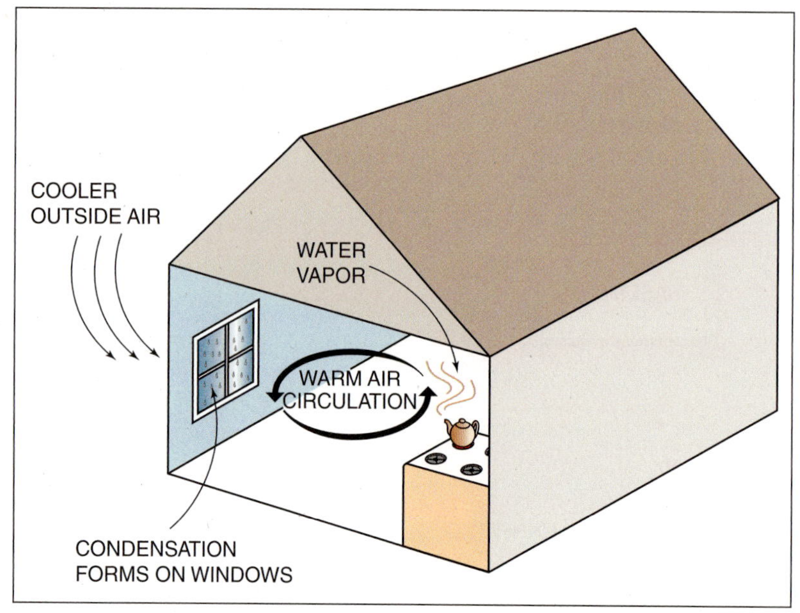


**FIGURE 16-13** Moisture in warm air condenses when it comes in contact with a cold surface.

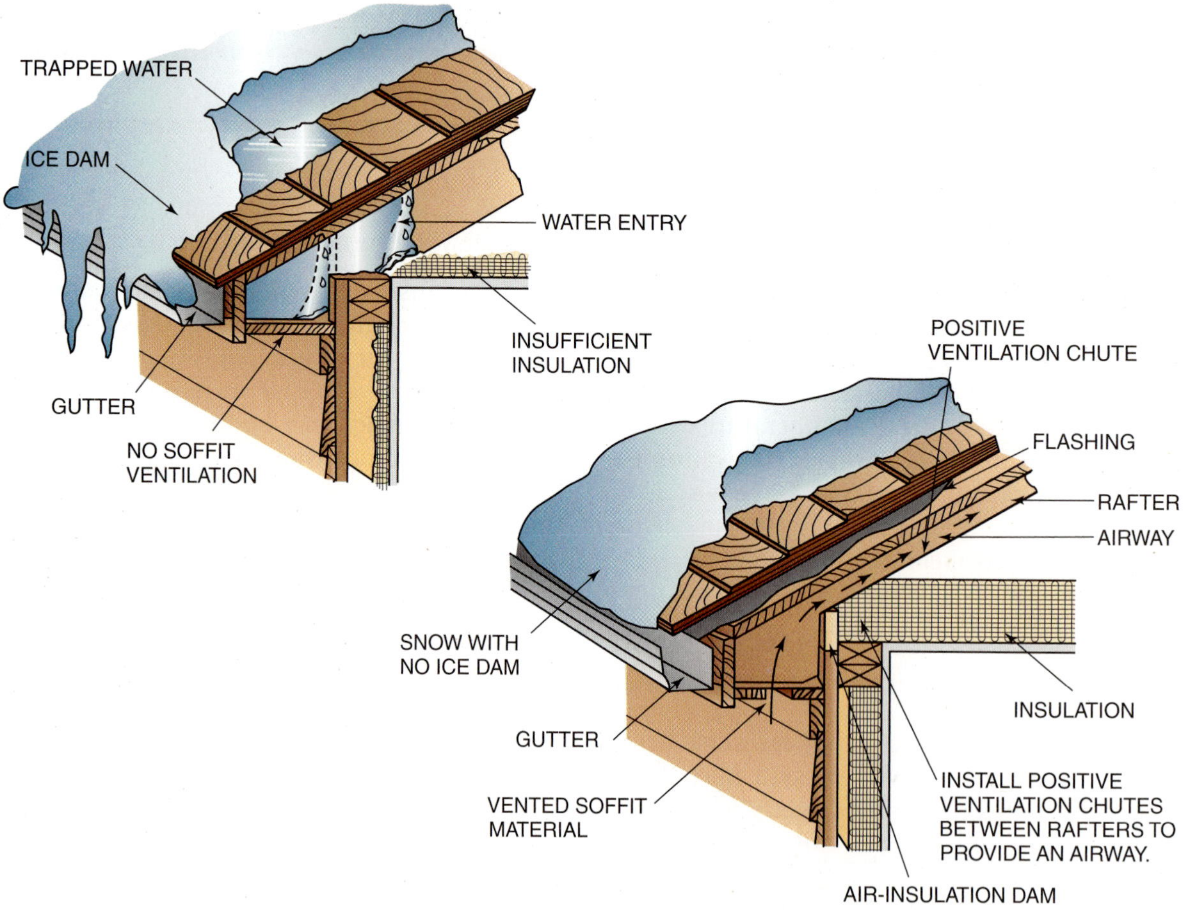

**FIGURE 16-14** A warm attic causes the formation of ice dams on roof during cold weather. A properly constructed and ventilated attic will keep the snow from melting and forming ice dams.

## PROBLEMS CAUSED BY CONDENSATION

If insulation absorbs water, the small air spaces in the insulation may become wet with water droplets. This reduces the thermal resistance and effectiveness of the insulation.

Moisture absorbed by the wood frame raises the moisture content of the wood. This improves the conditions for bacteria and fungi to grow and decay the wood. If moisture is absorbed by the sheathing, it is eventually absorbed by the exterior trim and siding. This causes the exterior paint to blister and peel. Excessive condensation of water vapor may even damage the interior finish.

## PREVENTION OF VAPOR CONDENSATION

Logically, to prevent the damage caused by moisture in a building, steps should be taken to reduce the effect of the moisture. This can be accomplished in two ways. First, reduce the amount of moisture within the house. Second, reduce the moisture migration into the insulation layer with a vapor retarder. Sources of moisture within the house include cooking, bathing, washing, and damp basements. Check the exhaust piping of clothes dryers to make sure it is not constricted and is working properly. Installing ventilation fans in rooms where the moisture collects, and using them daily, can exhaust the moisture-laden air to the outside. Check that all rainwater drains away from the building and does not come back into the basement.

## VAPOR RETARDER

Vapor retarders are typically a layer of polyethylene. It is transparent plastic sheet in rolls of up to 100 feet in length and widths of 10, 12, 16, and 20 feet. The most commonly used thicknesses are 4-mil or 6-mil. It is installed on the studs over the insulation before the drywall. Polyethylene slows the migration of vapor into the insulation layer. All seams and penetrations should be sealed with urethane sealants. Urethane is the only sealant that sticks well to polyethylene.

Virtually all the moisture migration occurs by air leaking into the insulation layer. Thus, if there is an airtight seal between the warmer and cooler air, moisture migration is stopped (Fig. 16-15). Airtight construction techniques are the best way to save energy dollars. Not only does it stop air leakage reducing heating/cooling demands, airtight construction also serves as a vapor retarder.

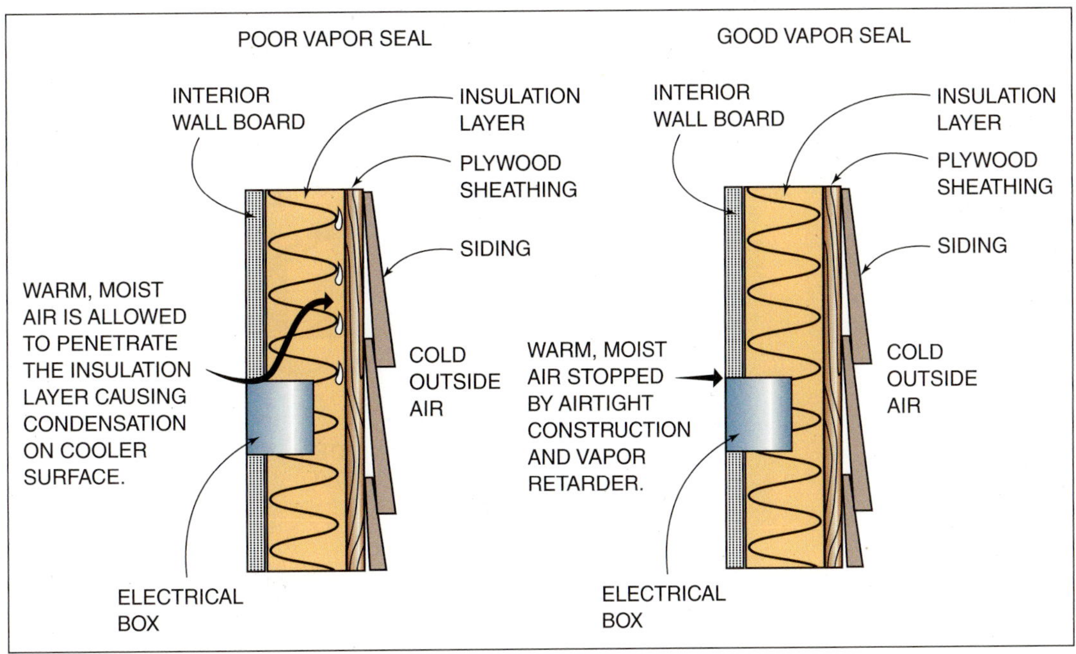


**FIGURE 16-15** An exterior wall should be constructed in an airtight manner to prevent water vapor from entering the insulation layer.

Airtight technique involves installing sheathing tape or caulk to the seams of the exterior sheathing. Airtight construction stops moisture migration and reduces energy costs. This is because more than 95 percent of all moisture that travels through a wall or ceiling section does so by air infiltration.

# VENTILATION

Ventilation helps combat condensation of moisture. In a well-ventilated area, any condensed moisture is removed by evaporation. One of the areas where this can be effectively accomplished is in the attic. With a well-insulated ceiling and adequate ventilation, attic temperatures are lower. Melting of snow on the roof over the attic space can be eliminated along with the danger of ice damming. Also, roof shingles will stay cooler in the summer months and may last longer. In crawl spaces under floors, ventilation can easily be provided to evaporate any condensation of moisture. Here, a vapor retarder should be installed on top of the ground to inhibit moisture from getting into the floor system above it.

On roofs where the ceiling finish is attached to the roof rafters, an adequate air space of at least 1½ inches must be maintained between the insulation and the roof sheathing. The air space must be amply vented with air inlets in the soffit and outlets at the ridge (Fig. 16-16). Failure to do so may result in reduced shingle life, formation of ice dams at the eaves, and possible decay of the roof frame. Application of a vapor retarder and airtight construction help prevent any condensation of moisture in the rafter space by blocking the passage of vapor.

**Green Tip**

Mold problems are a result of excess moisture, not poor housekeeping. Dry buildings are healthy buildings.

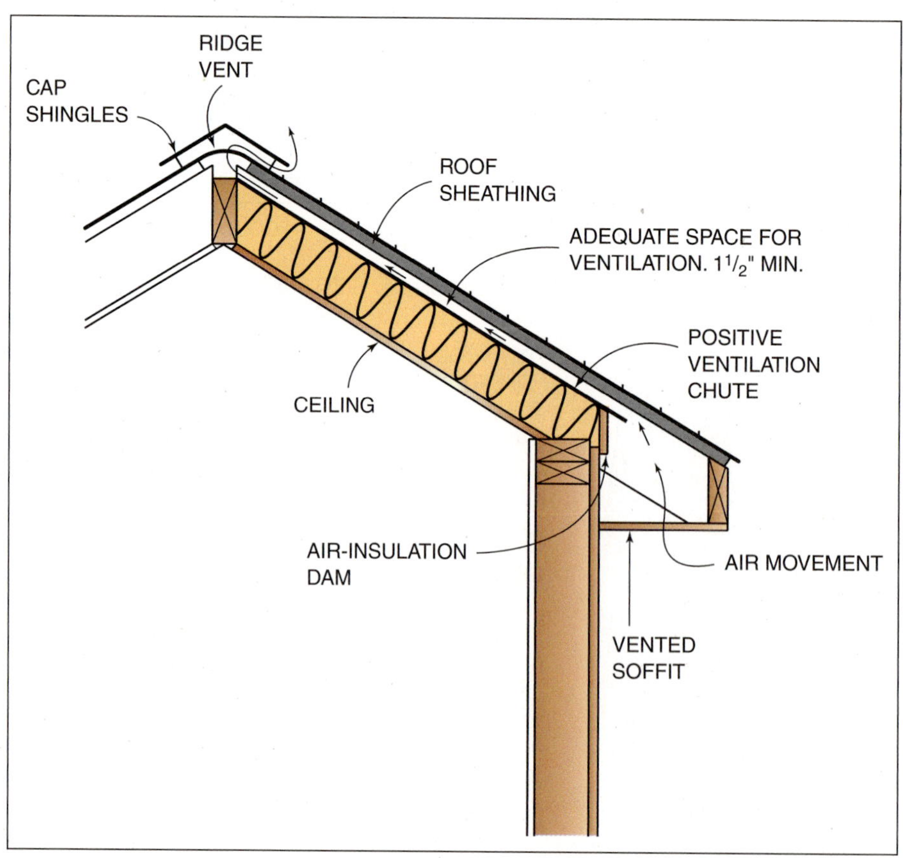


**FIGURE 16-16** Methods of providing adequate ventilation when the entire rafter cavity is insulated.

## TYPES OF VENTILATORS

There are many types of ventilators. Their location and size are factors in providing adequate ventilation.

### Ventilating Gable Roofs

Triangularly shaped louver vents are often installed in both end walls of a gable roof. They come in various shapes and sizes and are installed as close to the roof peak as possible. The best way to vent an attic is with continuous ridge and soffit vents (Fig. 16-17). Each rafter cavity is vented from soffit to ridge. The roof sheathing is cut back about 1 inch from the ridge on each side, and vent material is nailed over this slot. Cap shingles then can be nailed directly to the vent. Perforated material or screen vents are installed in the soffits to provide the entry point for the ventilation. Positive-ventilation chutes should be installed to prevent any air obstructions by the ceiling insulation near the eaves. This system can adequately vent the attic space of a house that is up to 50 feet wide.

### Ventilating Hip Roofs

Hip roofs should have additional continuous venting along each hip rafter. This allows each hip-jack rafter cavity to be vented. When a 2½-inch wide slot for the vents is cut, it is recommended to leave a 1-foot section of sheathing uncut between every 2 feet of slot section (Fig. 16-18). This allows for adequate ventilation while maintaining the integrity of the hip roof structure.

### Ventilating Crawl Spaces

For crawl spaces, usually rectangular metal or plastic louvered ventilators are used. Some are designed to fit into a space made by leaving out a concrete block in the foundation. Vents should be placed near each corner of the crawl space foundation and in intermediate, equally spaced locations (Fig. 16-19). The use of a ground cover vapor retarder is highly recommended. This protects wood framing members by reducing the outflow of moisture from the ground. It also allows for fewer ventilators.

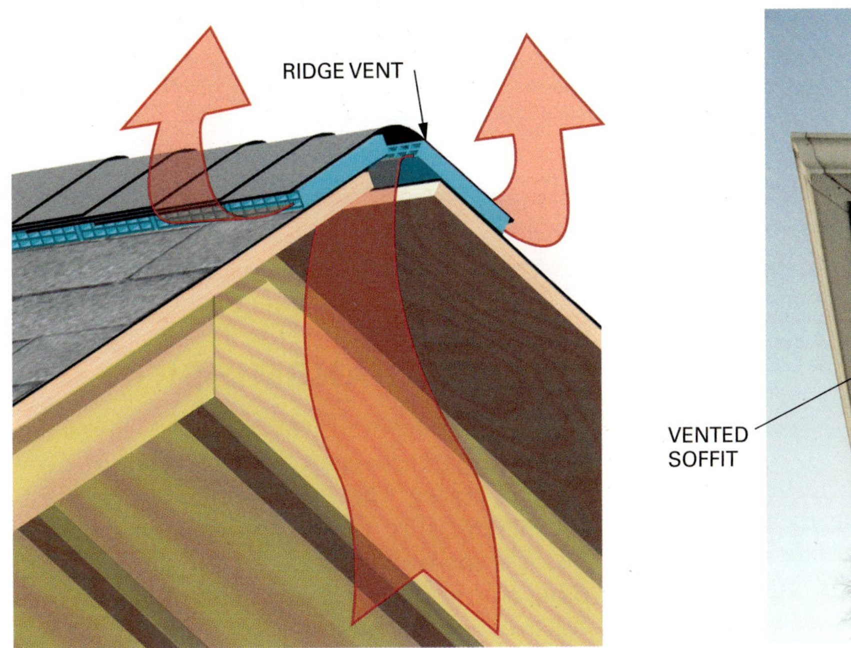

**FIGURE 16-17** Ridge and soffit vents work together to provide adequate attic ventilation.

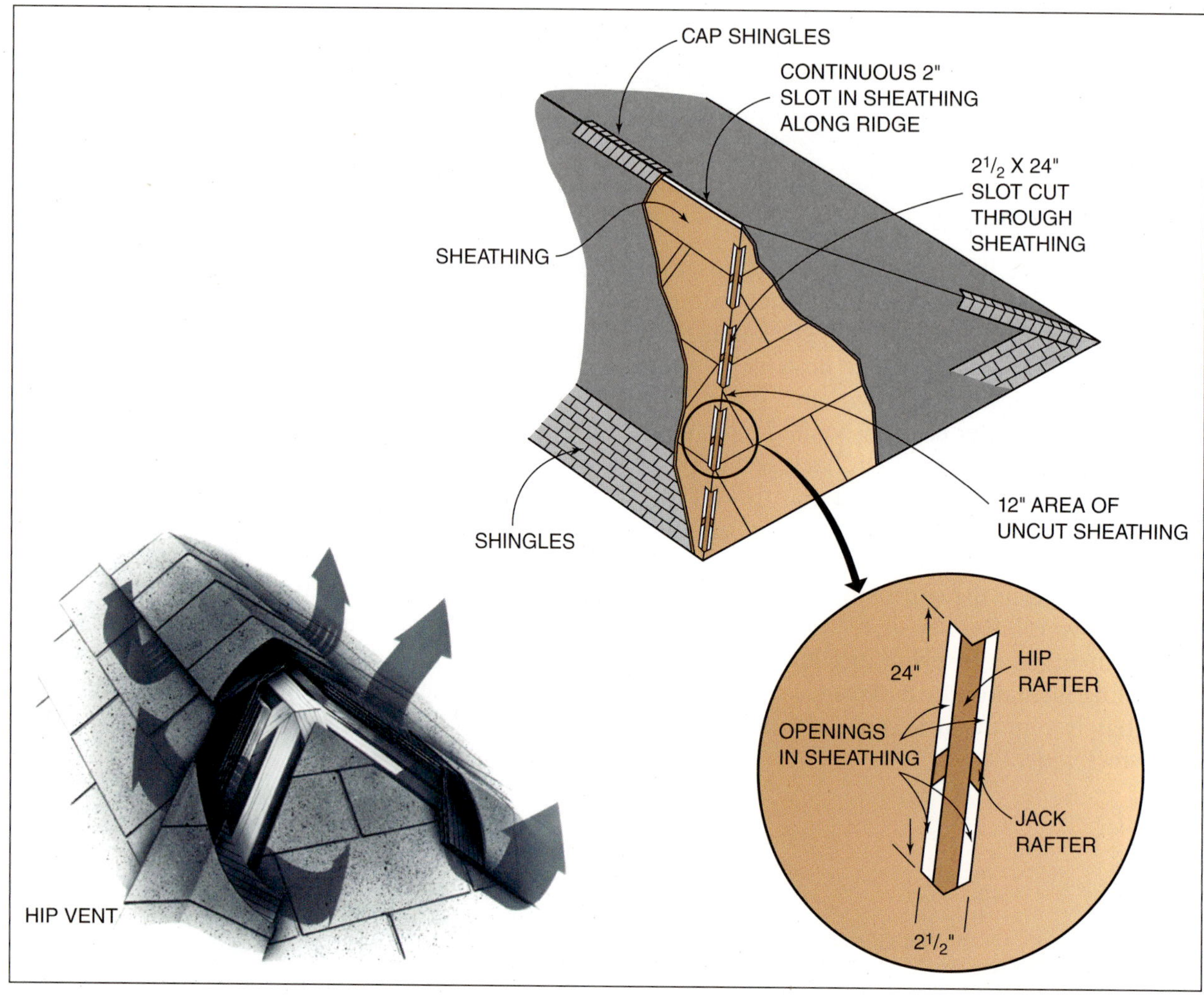


Courtesy of CorAvent, Inc.

**FIGURE 16-18** Hip roofs can be vented with continuous ridge and hip vents.

# Acoustical Insulation

*Acoustical* or *sound* insulation resists the passage of noise through a building section. The reduction of sound transfer between rooms is important in offices, apartments, motels, and homes. Excessive noise is not only annoying but harmful. It not only causes fatigue and irritability but also can damage the sensitive hearing nerves of the inner ear.

Sound insulation between active areas and quiet areas of the home is desirable. Sound insulation between the bedroom area and the living area is important. Bathrooms also should be insulated.

## SOUND TRANSMISSION

Noises create sound waves. These waves radiate outward from the source until they strike a wall, floor, or ceiling. These surfaces then begin to vibrate by the pressure of the sound waves in the air. Because the wall vibrates, it conducts sound to the other side in varying degrees, depending on the wall construction.

### Sound Transmission Class

The resistance of a building section, such as a wall, to the passage of sound is rated by its Sound Transmission Class (STC). The higher the number, the better the sound barrier. The approximate effectiveness

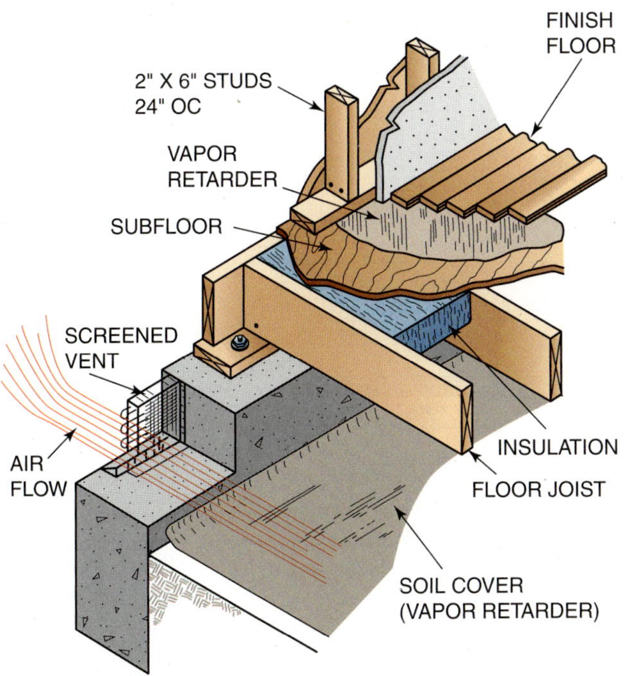


**FIGURE 16-19** Crawl space ventilator in the foundation wall and vapor retarders placed on the ground.

of walls with varying STC numbers is shown in Figure 16-20.

Sound travels readily through the air and through some materials. When airborne sound strikes a wall, the studs act as conductors unless they are separated in some way from the covering material. Electrical outlet boxes placed back-to-back in a wall easily transmit sound. Faulty construction, such as poorly fitted doors, often allows sound to pass through. Therefore, good, airtight construction practices are the first line of defense in controlling sound.

| | |
|---|---|
| **25** | Normal speech can be understood quite easily |
| **30** | Loud speech can be understood fairly well |
| **35** | Loud speech audible but not intelligible |
| **42** | Loud speech audible as a murmur |
| **45** | Must strain to hear loud speech |
| **48** | Some loud speech barely audible |
| **50** | Loud speech not audible |

*This chart from the Acoustical and Insulating Materials Association illustrates the degree of noise control achieved with barriers having different STC numbers.*

**FIGURE 16-20** Approximate effectiveness of sound reduction in walls with varying STC ratings.

## WALL CONSTRUCTION

A wall that provides sufficient resistance to airborne sound should have an STC rating of 45 or greater. At one time, the resistance usually was provided only by double walls, which resulted in increased costs. However, a system of using **sound-deadening insulating board** with a gypsum board outer covering has been developed. This system provides good sound resistance. Resilient steel channels placed at right angles to the studs isolate the gypsum board from the stud. Figure 16-21 shows various types of wall construction and their STC rating.

## FLOOR AND CEILING CONSTRUCTION

Sound insulation between an upper floor and the ceiling below involves not only the resistance of airborne sounds, but also that of **impact noise**. Impact noise is caused by such things as dropped objects, footsteps, or moving furniture. The floor is vibrated by the impact. Sound is then radiated from both sides of the floor. Impact noise control must be considered as well as airborne sounds when constructing floor sections for sound insulation.

An Impact Noise Rating (INR) shows the resistance of various types of floor-ceiling construction to impact noises. The higher the positive value of the INR, the more resistant the insulation is to impact noise transfer. Figure 16-22 shows various types of floor-ceiling construction with their STC and INR ratings.

## SOUND ABSORPTION

The amount of noise in a room can be minimized by the use of *sound-absorbing materials*. Perhaps the most commonly used material is **acoustical tile** made of fiberboard. These tiles are most often used in the ceiling where they are not subjected to damage. The tiles are soft. The tile surface consists of small holes or fissures, or a combination of both (Fig. 16-23). These holes or fissures act as sound traps. The sound waves enter, bounce back and forth, and finally die out.

| SOUND INSULATION OF SINGLE WALLS | | |
|---|---|---|
| DETAIL | DESCRIPTION | STC RATING |
| A (16"; 2" X 4") | 1/2" GYPSUM WALLBOARD | 32 |
| | 5/8" GYPSUM WALLBOARD | 37 |
| 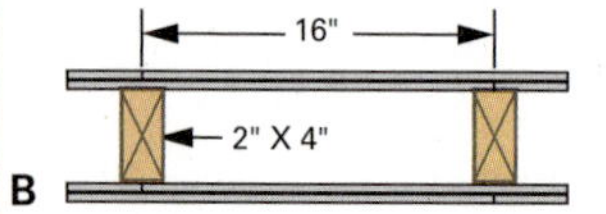 B (16"; 2" X 4") | 3/8" GYPSUM LATH (NAILED) PLUS 1/2" GYPSUM PLASTER WITH WHITECOAT FINISH (EACH SIDE) | 39 |
| C  | 8" CONCRETE BLOCK | 45 |
| D 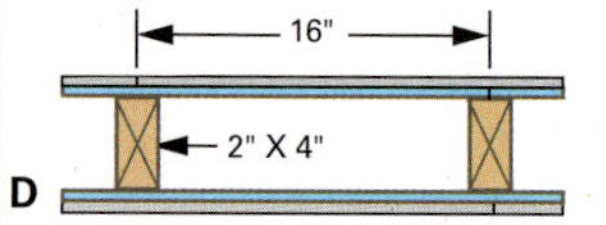 (16"; 2" X 4") | 1/2" SOUND DEADENING BOARD (NAILED) 1/2" GYPSUM WALLBOARD (LAMINATED) (EACH SIDE) | 46 |
| E 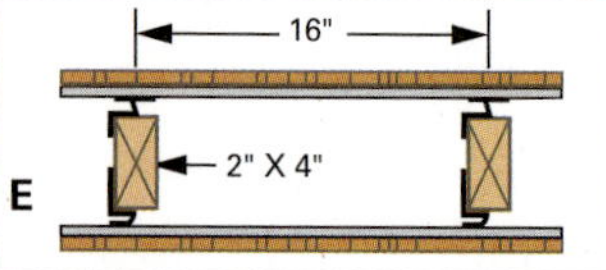 (16"; 2" X 4") | RESILIENT CLIPS TO 3/8" GYPSUM BACKER BOARD 1/2" FIBERBOARD (LAMINATED) (EACH SIDE) | 52 |

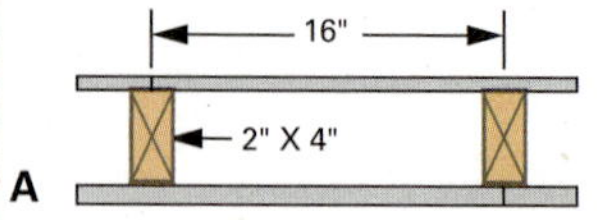

| SOUND INSULATION OF DOUBLE WALLS | | |
|---|---|---|
| DETAIL | DESCRIPTION | STC RATING |
| A 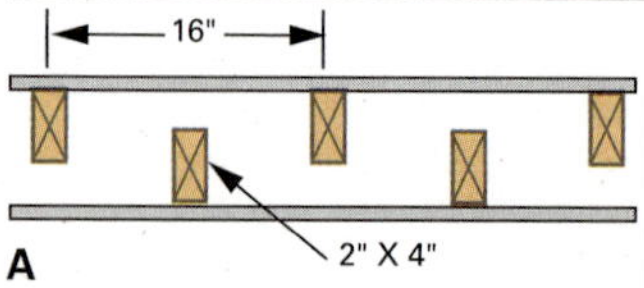 (16"; 2" X 4") | 1/2" GYPSUM WALLBOARD | 38 |
| B 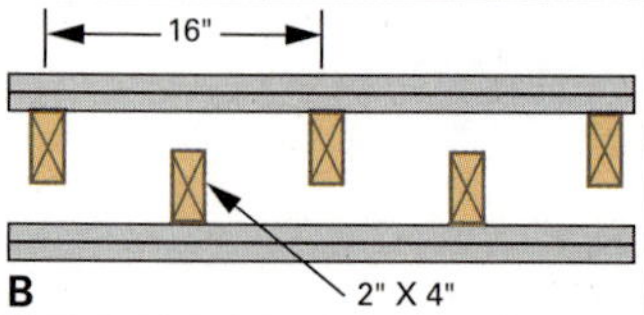 (16"; 2" X 4") | 5/8" GYPSUM WALLBOARD (DOUBLE LAYER EACH SIDE) | 45 |
| C 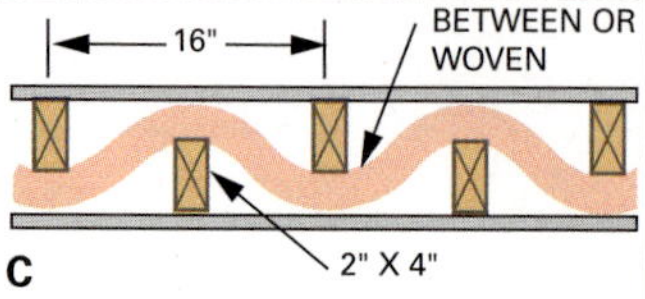 (16"; BETWEEN OR WOVEN; 2" X 4") | 1/2" GYPSUM WALLBOARD 1 1/2" FIBROUS INSULATION | 49 |
| D 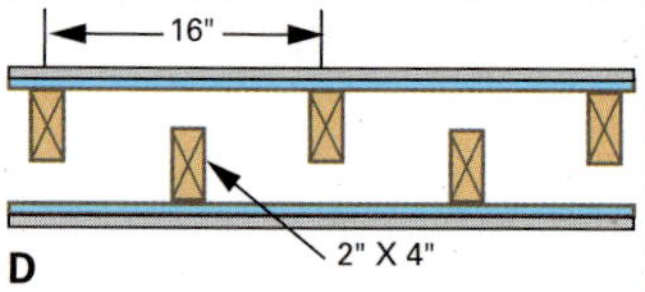 (16"; 2" X 4") | 1/2" SOUND DEADENING BOARD (NAILED) 1/2" GYPSUM WALLBOARD (LAMINATED) | 50 |

**FIGURE 16-21** STC ratings of various types of wall construction.

| RELATIVE IMPACT AND SOUND TRANSFER IN FLOOR-CEILING COMBINATIONS (2" X 10" JOISTS) | | ESTIMATED VALUE | |
|---|---|---|---|
| DETAIL | DESCRIPTION | STC RATING | APPROX. INR |
| A 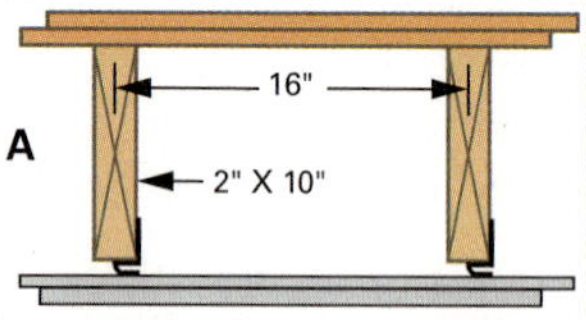 (16"; 2" X 10") | FLOOR 3/4" SUBFLOOR BUILDING PAPER 3/4" FINISH FLOOR CEILING GYPSUM LATH AND SPRING CLIPS 1/2" GYPSUM PLASTER | 52 | −2 |
| B 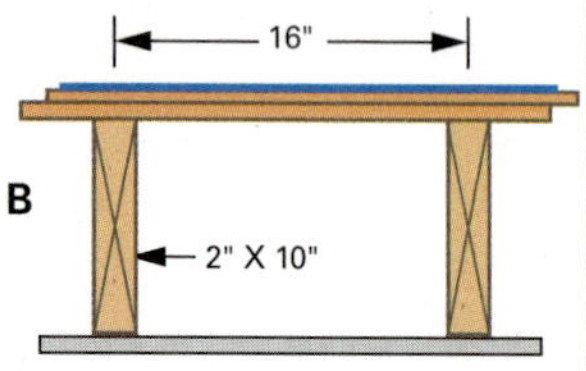 (16"; 2" X 10") | FLOOR 5/8" PLYWOOD SUBFLOOR 1/2" PLYWOOD UNDERLAYMENT 1/8" VINYL-ASBESTOS TILE CEILING 1/2" GYPSUM WALLBOARD | 31 | −17 |
| C 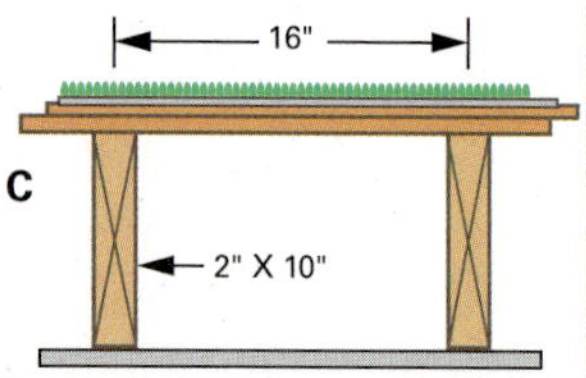 (16"; 2" X 10") | FLOOR 5/8" PLYWOOD SUBFLOOR 1/2" PLYWOOD UNDERLAYMENT FOAM RUBBER PAD 3/8" NYLON CARPET CEILING 1/2" GYPSUM WALLBOARD | 45 | +5 |

| RELATIVE IMPACT AND SOUND TRANSFER IN FLOOR-CEILING COMBINATIONS (2" X 8" JOISTS) | | ESTIMATED VALUE | |
|---|---|---|---|
| DETAIL | DESCRIPTION | STC RATING | APPROX. INR |
| D 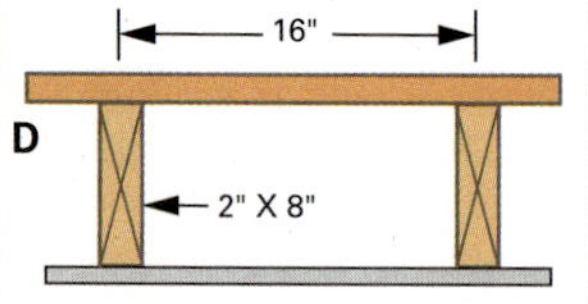 (16"; 2" X 8") | FLOOR 7/8" T. & G. FLOORING CEILING 3/8" GYPSUM BOARD | 30 | −18 |
| E 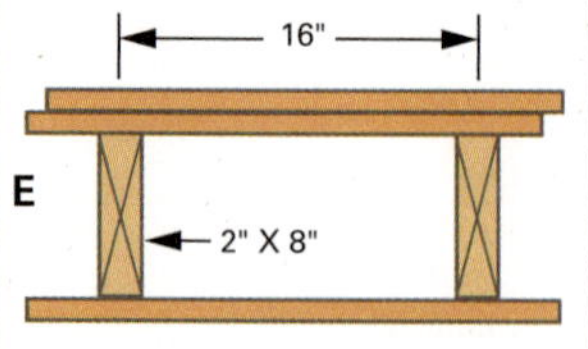 (16"; 2" X 8") | FLOOR 3/4" SUBFLOOR 3/4" FINISH FLOOR CEILING 3/4" FIBERBOARD | 42 | −12 |
| F 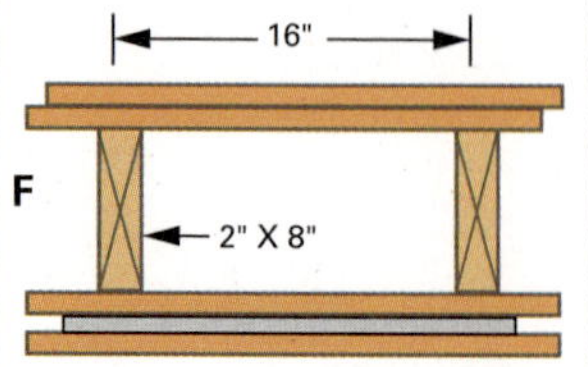 (16"; 2" X 8") | FLOOR 3/4" SUBFLOOR 3/4" FINISH FLOOR CEILING 1/2" FIBERBOARD LATH 1/2" GYPSUM PLASTER 3/4" FIBERBOARD | 45 | −4 |

**FIGURE 16-22** STC and INR for floor and ceiling constructions.

**FIGURE 16-23** Sound-absorbing ceiling tiles.

# SUMMARY


- Insulation and ventilation must be installed together to ensure long life of the building.
- Types of insulation vary depending on where the insulation is used in the building.
- Installing insulation properly allows the insulation to perform as expected.
- Condensation is a natural process that must be compensated for in order to ensure the insulation and the building perform as expected.
- Ventilation of the living environment and the building components is required to ensure long life of the building and its occupants.

## GREEN CHECKLIST

- ☐ Insulation pays for itself in saved energy dollars by reducing energy needs and lessening the carbon footprint of a building.
- ☐ Fiberglass is made from 25 percent recycled materials.
- ☐ Cellulose insulation is produced from 100 percent recycled newspaper.
- ☐ Mold problems are a result of excess moisture, not poor housekeeping. Dry buildings are healthy buildings.
- ☐ New buildings should be dried and vented before occupation to reduce moisture problems such as mold and mildew.
- ☐ Sorting job site waste for recycling saves money and makes environmental sense.

PROCEDURE 16-1

# Installing Flexible Insulation

**A** Install positive ventilation chutes between the rafters where they meet the wall plate. This may compress the insulation slightly against the top of the wall plate to permit the free flow of air over the top of the insulation.

- Install the air-insulation dam between rafters in line with or on the exterior sheathing. This will protect the insulation from air movement into the insulation layer from the soffits.

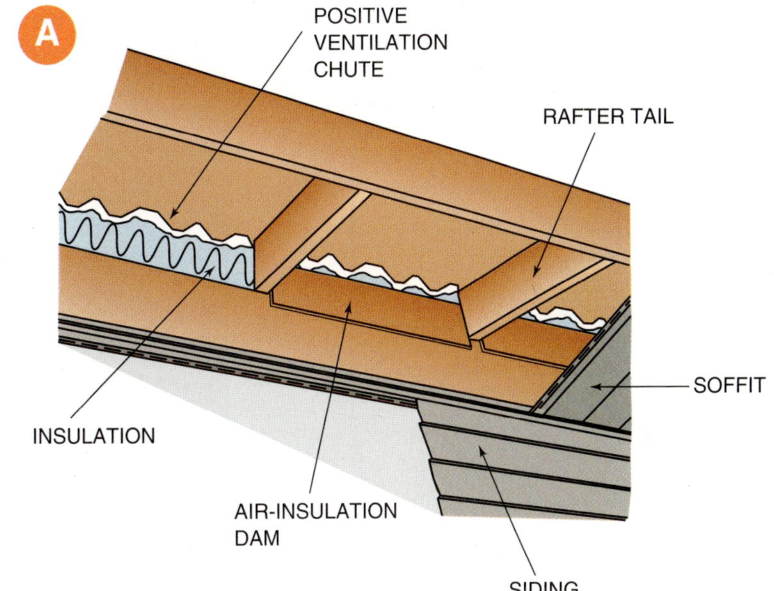


**B** Fill any spaces around windows and doors with spray-can foam. Non-expanding foam will fill the voids with an airtight seal and protect the house from air leakage. After the foam cures, flexible insulation may be added to fill the remaining space. Take care not to spray too much foam into space around windows as it may cause the frame to expand. This may affect the window operation. This will void the window warranty.

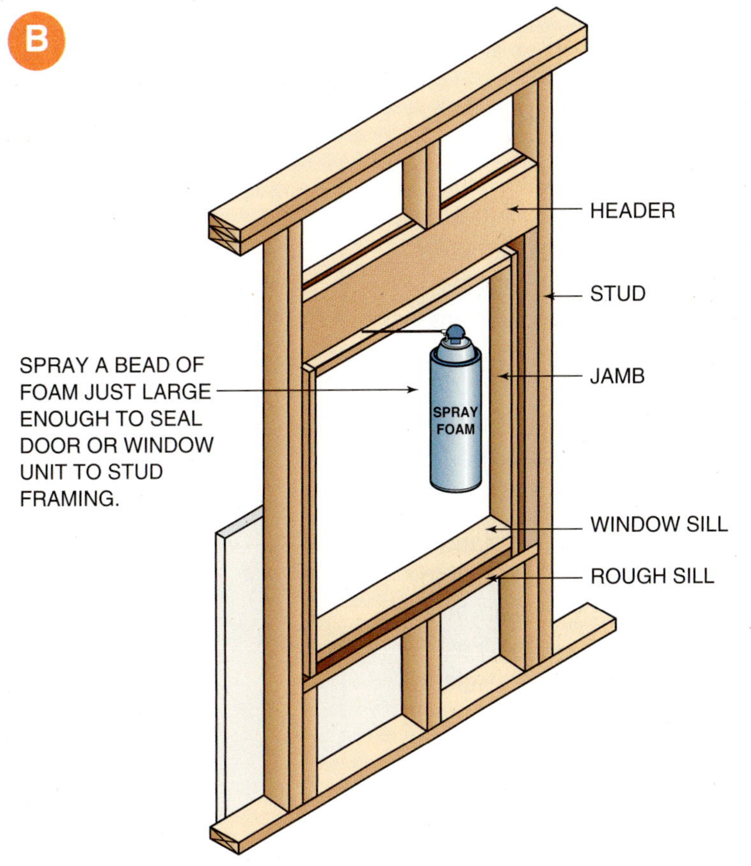

PROCEDURE 16-1

## Installing Flexible Insulation (continued)

**C** To cut the material, place a scrap piece of plywood on the floor to protect the floor while cutting. Roll out the material over the scrap. Using another scrap piece of wood, compress the insulation and cut it with a sharp knife in one pass.

**C**

**D** Place the batts or blankets between the studs. The flanges of the facing may be stapled either to cover the studs or against the inside edges of the studs, as well as the top and bottom plates. A better air seal is achieved with fastening to cover the stud, but the studs are less visible for the installation of the gypsum. Use a hand or hammer-tacker stapler to fasten the insulation in place.

- Install ceiling insulation by stapling it to the ceiling joists or by friction-fitting it between them. Push and extend the insulation across the top plate to fit against the air-insulation dam.

**D**

**E** Flexible insulation installed between floor joists over crawl spaces may be held in place by wire mesh or pieces of heavy-gauge wire wedged between the joists. Housewrap material works well to hold insulation in place because it seals the insulation from air infiltration.

**E**

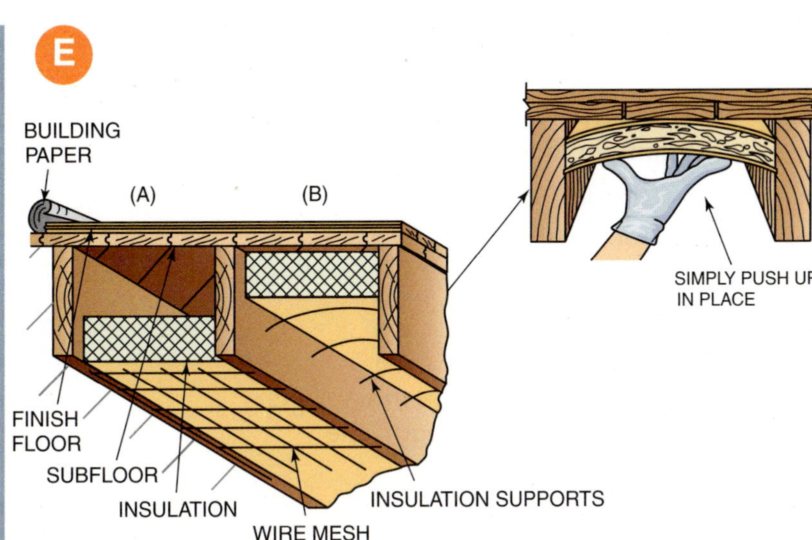


In (A), wire mesh is stapled to the edges of the joists.
In (B), pieces of heavy gauge wire, pointed at each end, are wedged between the joists to support the insulation.

## REVIEW QUESTIONS

*Select the most appropriate answer.*

1. **Popcorn made in a frying pan is an example of heat transfer by**
   a. conduction.
   b. induction.
   c. convection.
   d. radiation.

2. **A typical house loses most of its heating energy through the**
   a. basement walls.
   b. framing.
   c. leakage of air.
   d. windows and doors.

3. **Of the common building materials listed here, the most efficient thermal insulator is**
   a. brick.
   b. stone.
   c. concrete.
   d. wood.

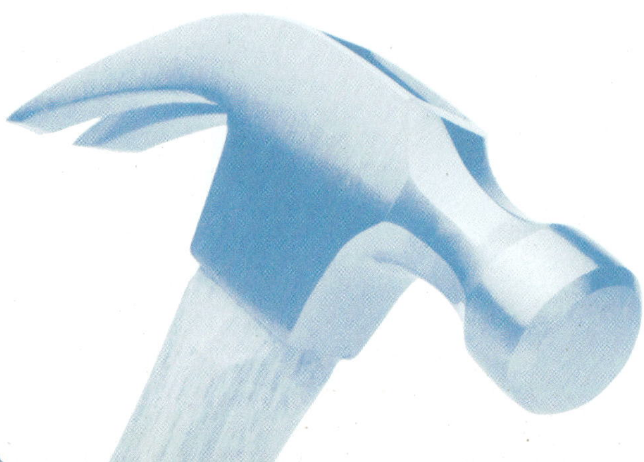

4. **The insulating term *R-value* is defined as the measure of**
   a. resistance of a material to the flow of heat.
   b. the heat loss through a building section.
   c. the conductivity of a material.
   d. the total heat transfer through a building section.

5. **To protect the insulation layer from air leakage and moisture migration, a vapor retarder should be installed on the**
   a. warm side.
   b. inside.
   c. cold side.
   d. outside.

6. **The only type of insulation suitable to be used for ground contact is**
   a. extruded polystyrene.
   b. fiberglass.
   c. polyurethane.
   d. expanded polystyrene.

7. **Moisture migration into the insulation layer can be stopped by**
   a. installing a vapor retarder.
   b. placing sheathing tape on the seams of exterior sheathing.
   c. airtight construction techniques.
   d. all of the above.

8. **Special consideration must be given when using cans of spray foam insulation due to**
   a. spray foam being fairly easy to clean up after.
   b. its ability to expand to fill gaps.
   c. professionals being the only ones allowed to use them.
   d. all of the above

**9. Squeezing or compressing flexible insulation tightly into spaces**

a. reduces its effectiveness.
b. increases its efficiency.
c. is necessary to hold it in place.
d. helps prevent air leakage by sealing cracks.

**10. When air has reached its limit to hold water as vapor, the air is said to have achieved**

a. maximum temperature.
b. minimum temperature.
c. dew point.
d. equilibrium moisture content.

**11. Moisture migration into the insulation is caused mostly by**

a. leaks in the roofing materials.
b. water moving through building materials.
c. air movement or air leakage.
d. condensation.

**12. Moisture in the air of a building can be caused by**

a. a damp basement.
b. cooking and bathing.
c. leaks in the roofing, siding, or plumbing.
d. all of the above

**13. The most effective method of venting an attic space is with**

a. gable vents.
b. hip vents.
c. skylights.
d. continuous ridge and soffit vents.

**14. Ventilation of the attic will prevent**

a. ice dams from forming at the eaves.
b. basements from becoming damp.
c. attic temperatures from dropping too low.
d. all of the above

**15. Reducing sound transmission within a building is done with**

a. acoustical tile.
b. fiberglass insulation.
c. drywall.
d. all of the above

## KNOW YOUR CODES

Energy codes are designed to improve the thermal efficiency of a building. Energy-efficient buildings use less energy dollars for conditioning the air, warmer or cooler. This also makes the building more comfortable. It makes perfect sense to follow national and local energy codes.

## WHAT'S WRONG WITH THIS PICTURE?

Carefully study **Figure 16-24** and think about what is wrong. Consider all possibilities.

### WRONG

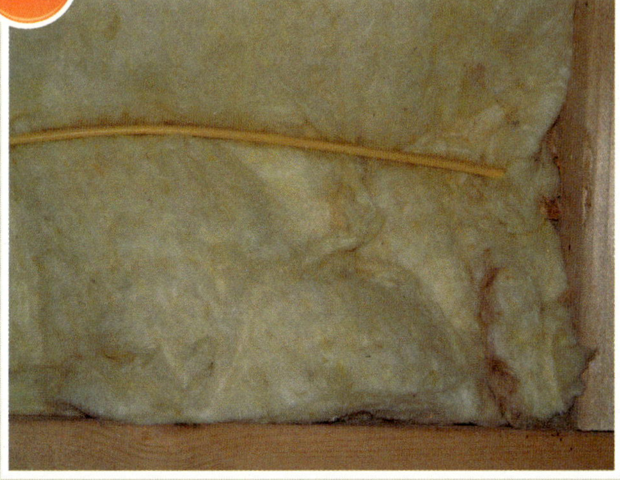

**FIGURE 16-24** This is a wall cavity with fiberglass insulation in it. The wire is compressing the insulation so that the insulation cannot completely fill the stud cavity. This results in a loss of some of the insulation value. The insulation should have been pulled apart, with some placed behind the wire and some in front so that the stud cavity is filled with the insulation.

### RIGHT

**FIGURE 16-25** This shows that the insulation was split so that it is behind and in front of the electrical wire. By not being compressed, the insulation will work better. On the left side of the picture, the insulation does not extend to the adjacent stud. This is shown for clarity so the wire can clearly be seen going through the insulation.

CHAPTER 17

# Wall Finish

**OBJECTIVES** *After completing this unit, the student should be able to:*

- Describe various kinds, sizes, and uses of gypsum panels.
- Describe the kinds and sizes of nails, screws, and adhesives used to attach gypsum panels.
- Apply gypsum board to interior walls and ceilings.
- Conceal gypsum board fasteners and corner beads.
- Reinforce and conceal joints with tape and compound.
- Describe and apply several kinds of sheet wall paneling.
- Describe and apply various patterns of solid lumber wall paneling.
- Estimate quantities of drywall, drywall accessories, and sheet and board wall paneling.

# GLOSSARY OF WALL FINISH TERMS

**eased edge** an edge of lumber whose sharp corners have been rounded

**face** the best appearing side of a piece of wood or the side that is exposed when installed

**gypsum board** a panel used as a finished surface material made from a mineral mined from the earth

**hardboard** a building product made by compressing wood fibers into sheet form

**wainscoting** a wall finish applied partway up the wall from the floor

Once the building is properly insulated and the building mechanicals are installed, the interior finish can begin. There are a variety of finishes available depending on architectural desires.

The majority of interior wall and ceiling surfaces in homes today are made with **gypsum board**, collectively called drywall construction.

Gypsum is fastened to studs. It is then finished to a smooth surface by a process involving many steps and materials including joint compound, tape, and beads.

Wood paneling in the form of sheets or boards may then be installed to highlight the room. These come in a variety of textures, patterns, and materials.

# DRYWALL CONSTRUCTION

Gypsum board is sometimes called *wallboard*, *plasterboard*, *drywall*, or *Sheetrock*®. It is used extensively in construction (Fig. 17-1). The term Sheetrock is a brand name for gypsum panels made by the U.S. Gypsum Company. Gypsum board makes a strong, high-quality, fire-resistant wall and ceiling covering. It is readily available; easy to apply, decorate, or repair; and relatively inexpensive.

## GYPSUM BOARD

Many types of gypsum board are available for a variety of applications. The board or panel is composed of a gypsum core encased in a strong, smooth-finish paper on the face side and a natural finish paper on the back side. The face paper is folded around the long edges. This reinforces and protects the core. The long edges are usually tapered. This allows the joints to be concealed with compound without any noticeable *crown joint* (Fig. 17-2). A crown joint is a buildup of the compound above the surface.

### Types of Gypsum Panels

Most gypsum panels can be purchased, if desired, with an aluminum foil backing. The backing functions as a radiant barrier to help reflect heat.

*Regular.* Regular gypsum panels are most commonly used. They are applied to interior walls and ceilings in new construction and remodeling.

*Courtesy of U.S. Gypsum Company.*

**FIGURE 17-1** The application of gypsum board to interior walls and ceilings is called drywall construction.

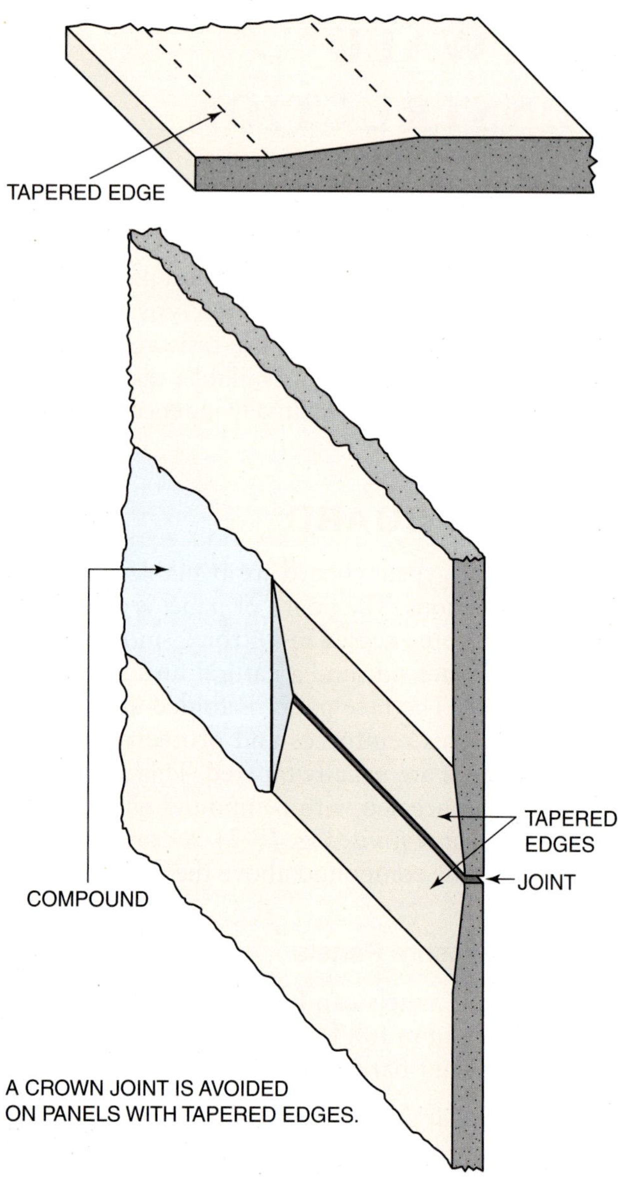


**FIGURE 17-2** The long edges of gypsum panels usually are tapered for effective joint concealment.

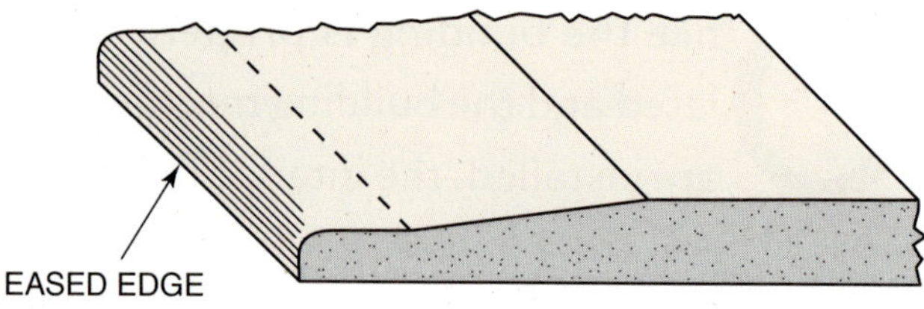


**FIGURE 17-3** An eased edge panel has a rounded corner that produces a stronger concealed joint.

*Eased Edge.* An eased edge gypsum board has a special tapered, rounded edge. This produces a much stronger concealed joint than a tapered, square edge (Fig. 17-3).

*Type X.* Type X gypsum board is commonly known as *fire-code board* or *X-rock*. It has greater resistance to fire because of special additives in the core and fiberglass fibers added for strength. Type X gypsum board is manufactured in several degrees of resistance to fire. Type X looks the same as regular gypsum board. However, it is labeled Type X on the edge or on the back.

*Moisture Resistant.* Moisture-resistant (MR) gypsum board, or green board, consists of a special moisture-resistant core and paper cover that is chemically treated to repel moisture. It is used frequently in bathrooms and other areas subjected to considerable moisture. It is easily recognized by its distinctive green face-paper. Moisture-resistant panels are available with a Type X core for increased fire resistance.

*Special Purpose. Backing board* is designed to be used as a base layer in multilayer systems. It is available with regular or Type X cores. *Coreboard* is available in 1-inch thickness. It is used for various applications, including use as the core of solid gypsum partitions. It comes in 24-inch widths with a variety of edge shapes. *Predecorated* panels have coated, printed, or overlay surfaces that require no further treatment. *Liner board* has a special fire-resistant core encased in a moisture-repellent paper. It is used to cover shaft walls, stairwells, chaseways, and similar areas.

*Veneer Plaster Base.* Veneer plaster bases are commonly called *blue board*. They are 4-foot wide gypsum board panels faced with a specially treated blue paper. This paper is designed to receive applications of *veneer plaster*. Specially formulated veneer plaster is applied in one coat of about 1⁄16 inch or two coats totaling about 1⁄8 inch. Other gypsum panels, such as soffit board and sheathing, are manufactured for exterior use.

## SIZES OF WALLBOARD PANELS

Wall board panels come a many sizes and thicknesses to meet the demands of the interior design and function.

### Widths and Lengths

Coreboards and liner boards come in 2-foot widths and are from 8 to 12 feet long. Other gypsum panels are manufactured 4 feet wide and in lengths of 8,

9, 10, 12, 14, or 16 feet. Gypsum board is made in a number of thicknesses.

### Thicknesses

A 1⁄4-inch thickness is used as a base layer in multilayer applications. It is also used to cover existing walls and ceilings in remodeling work. It can be applied in several layers for forming curved surfaces with short radii.

A 3⁄8-inch thickness is usually applied as a face layer in repair and remodeling work over existing surfaces. It is also used in multilayer applications in new construction.

Both 1⁄2-inch and 5⁄8-inch are commonly used thicknesses of gypsum panels for single-layer wall and ceiling application in residential and commercial construction. The 5⁄8-inch thick panel is more rigid and has greater resistance to impacts and fire. Coreboards and liner boards come in thicknesses of 3⁄4 and one inch.

## CEMENT BOARD

Like gypsum board, *cement board* or *wonder board* is a panel product. However, they have a core of Portland cement reinforced with a glass fiber mesh embedded in both sides (Fig. 17-4). The core resists water penetration and will not deteriorate when wet. It is designed for use in areas that may be subjected to high-moisture conditions. It is used extensively around bathtub, shower, kitchen, and laundry areas as a base for ceramic tile. In fact, some building codes require its use in these areas. Panels are manufactured in sizes designed for easy installation in tub and shower areas with a minimum of cutting. Standard cement board panels come in a thickness of 1⁄2 inch, in widths of 32 or 36 inches, and in 5-foot lengths. Custom panels are available in a thickness of 5⁄8 inch, widths of 32 or 48 inches, and lengths from 32 to 96 inches. Cement board is also manufactured in a 5⁄16-inch thickness. It is used as an underlayment for floors and countertops. Exterior cement board is used primarily as a base for various finishes on building exteriors.

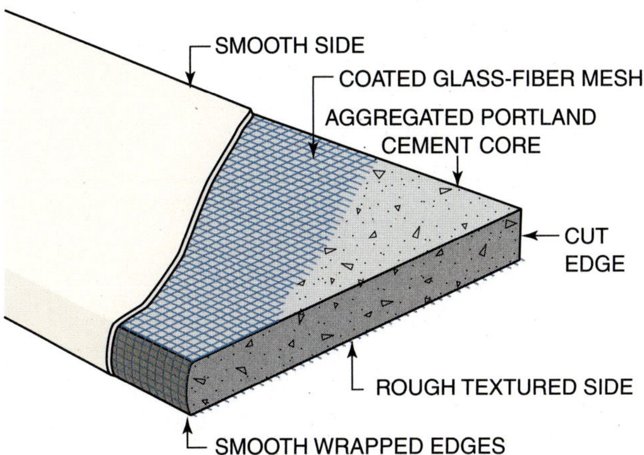


*Courtesy of U.S. Gypsum Company.*

**FIGURE 17-4** Composition of cement board.

## DRYWALL FASTENERS

Specially designed nails and screws are used to fasten drywall panels. They must penetrate well into supports. Using the correct fastener is extremely important for proper performance of the application. Fasteners with corrosion-resistant coatings must be used when applying moisture-resistant gypsum board or cement board.

> **CAUTION**
>
> **CAUTION: Care should be taken to drive the fasteners straight and at right angles to the wallboard to prevent the fastener head from breaking the face paper of gypsum board.**

## NAILS

Gypsum board nails should have flat or concave heads that taper to thin edges at the rim. Nails should have relatively small diameter shanks with heads at least 1⁄4 inch, but no more than 5⁄16 inch in diameter (Fig. 17-5). Nails should be driven with a drywall hammer that has a convex face designed to

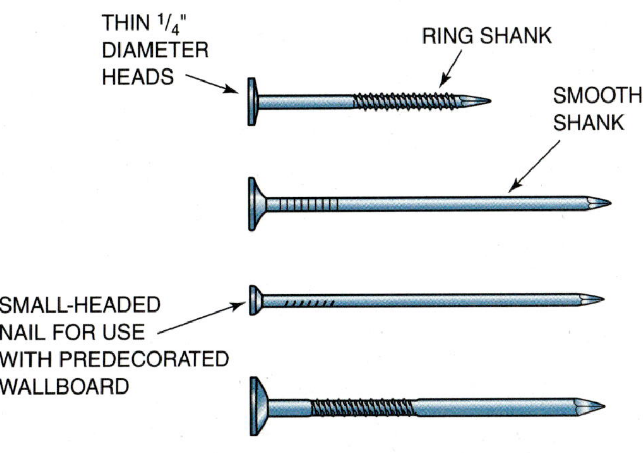


**FIGURE 17-5** Special nails are required to fasten gypsum board.

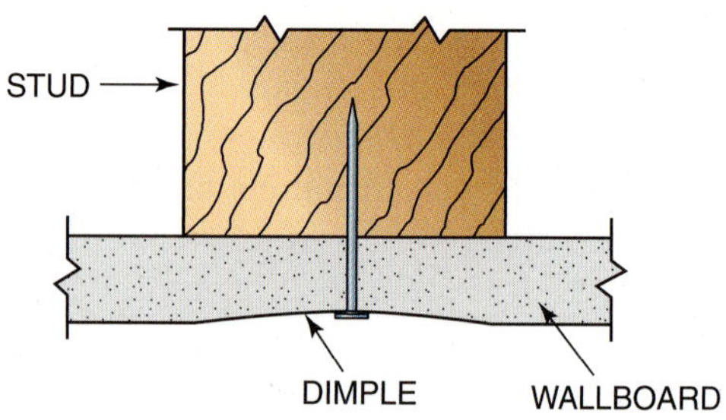


**FIGURE 17-6** Nails are driven with a convex-faced drywall hammer. This forms a dimple in the board. The dimple is filled with compound to conceal the fastener.

compress the gypsum panel face to form a dimple as the nail is driven (Fig. 17-6).

## SCREWS

Special drywall screws are used to fasten gypsum panels to steel or wood framing or to other panels. They are made with Phillips or square heads designed to be driven with a drywall screwgun (Fig. 17-7).

Courtesy of U.S. Gypsum Company.

**FIGURE 17-7** Drywall screws are driven with a screwgun to the desired depth.

**CAUTION**

**CAUTION: Drywall fasteners often have metal shavings still attached to the threads. These shavings can become metal splinters if, while holding the screw to start, it is spun between the thumb and finger.**

A proper setting of the nosepiece on the power screwdriver ensures correct countersinking of the screwhead and the panel is tight to the framing. When driven correctly, the specially contoured bugle head makes a uniform depression in the panel surface without breaking the paper (Fig. 17-8). The dimple and uniform depression are made so the fastener head can later be covered with compound.

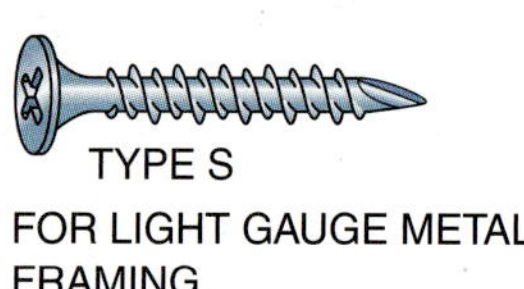


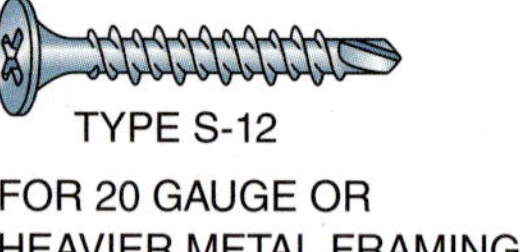


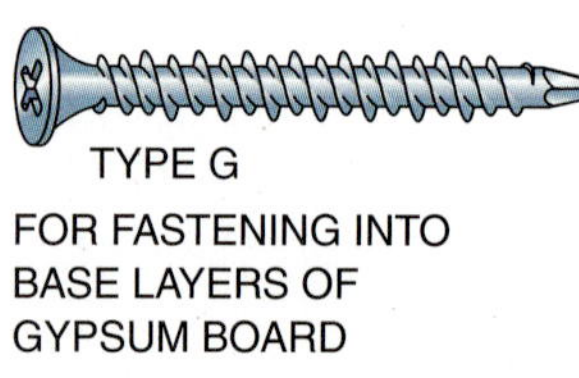


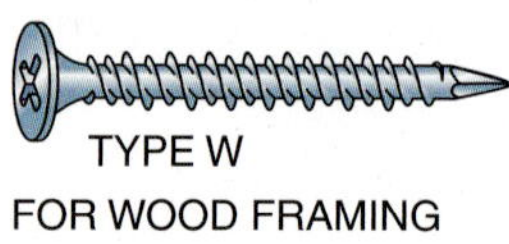


**FIGURE 17-8** Several types of screws are used to fasten gypsum panels. Selection of the proper type is important.

## ADHESIVES

*Drywall adhesives* are used to bond single layers directly to supports or to laminate gypsum board to base layers. For bonding gypsum board directly to supports, special *drywall stud adhesive* or approved

**FIGURE 17-9** Applying drywall adhesive to studs.

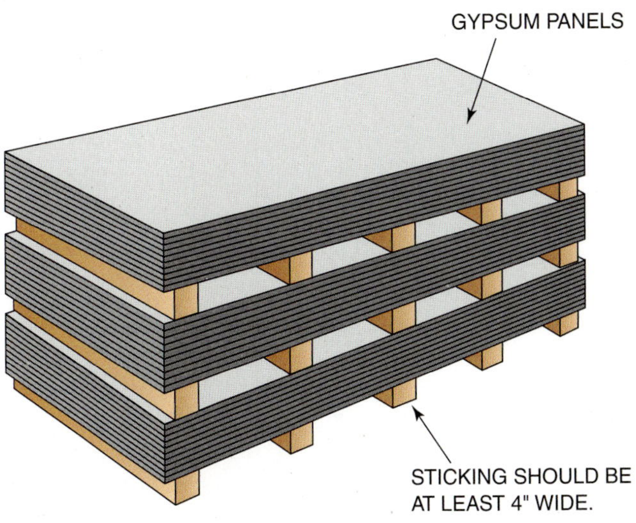


**FIGURE 17-10** Correct method of stacking gypsum board.

*construction adhesive* is used. Supplemental fasteners must be used with stud adhesives (Fig. 17-9). For laminating gypsum boards to each other, *joint compound adhesives* and *contact adhesives* are used. Joint compound adhesives are applied over the entire board with a suitable spreader prior to lamination. Boards laminated with joint compound adhesive require supplemental fasteners. When contact adhesives are used, no supplemental fasteners are necessary. However, the board cannot be moved after contact has been made.

CAUTION

**CAUTION:** Some types of drywall adhesives may contain a flammable solvent. Do not use these types near an open flame or in poorly ventilated areas.

## TECHNIQUES FOR DRYWALL FITTING

Drywall should be not be delivered to the job site until shortly before installation begins. Boards stored on the job for long periods are subject to damage. The boards must be stored under cover and stacked flat on supports. Supports should be at least 4 inches wide and placed fairly close together (Fig. 17-10). Leaning boards against framing for long periods may cause the boards to warp. This makes application more difficult. To avoid damaging the edges, carry the boards, do not drag them. Then, set the boards down gently. Be careful not to drop them.

### Cutting and Fitting Gypsum Board

Cut panels should fit easily into place without being forced. Forcing the panel may cause it to break. Before applying the gypsum board, check the framing members for alignment. Stud edges should not be severely crowned out of alignment more than ⅛ inch with adjacent studs (Fig. 17-11).

Ceiling joists are sometimes brought into alignment with the installation of a *strongback* across the tops of the joists at about the center of the span (Fig. 17-12).

***For step-by-step instructions on CUTTING AND FITTING GYPSUM BOARD, see Procedure 17-1 on pages 574–575.***

### Making Cutouts in Wall Panels

Cutouts in wall panels for electrical outlet boxes, ducts, and similar objects can be made in several ways. Care must be taken not to make the cutout much larger than the outlet. One method involves measuring and cutting while the other uses a portable electric *drywall cutout tool.* The cutout tool is a fast, easy, and accurate way of making cutouts. The other method requires more accurate layout and cutting by hand.

***For step-by-step instructions on MAKING CUTOUTS IN WALL PANELS, see Procedure 17-2 on page 576.***

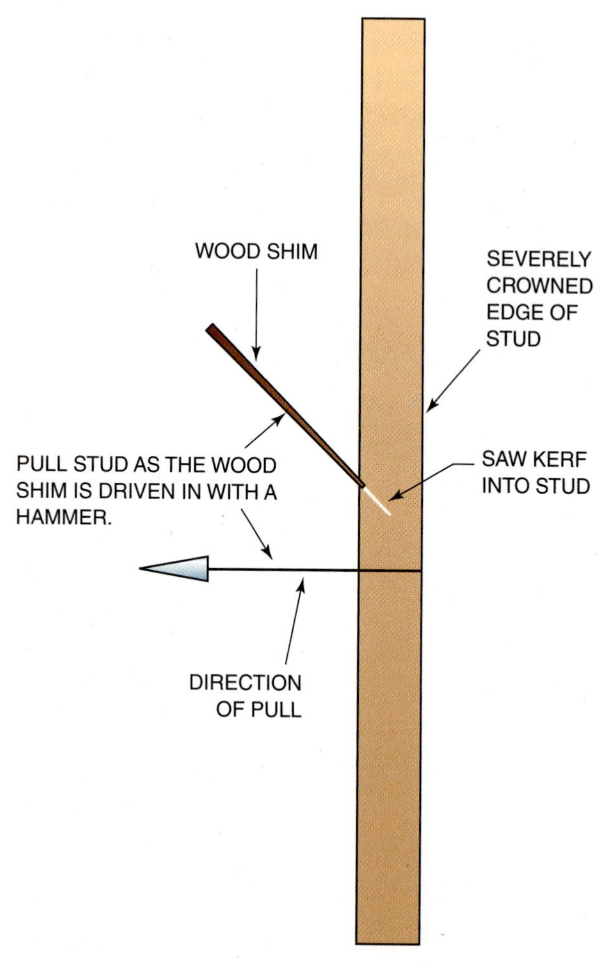


**FIGURE 17-11** Technique for straightening a severely crowned stud.

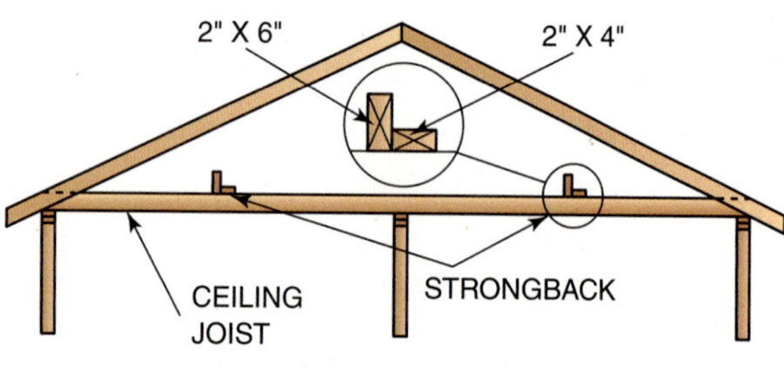


**FIGURE 17-12** A strongback sometimes is used to align ceiling joists or the bottom chord of roof trusses.

## Fastening Gypsum Panels

Drywall is fastened to framing members with nails or screws. Hand pressure should be applied on the panel next to the fastener being driven to make sure panel is tight to framing. Screw guns should leave a small dimple showing the gun pressure on the panel was correct before the clutch released. This ensures that the panel is in tight contact with the framing member. The use of adhesives reduces the number of nails or screws required. A single or double method of nailing may be used.

***Single Fastener Method.*** With this method, fasteners are spaced a maximum of 7 inches OC on ceilings and 8 inches OC on walls into frame members. Perimeter fasteners should be at least ⅜ inch, but not more than 1 inch from the edge.

***Double Fastener Method.*** With double fasteners, the perimeter fasteners are spaced as for single nailing. In the field of the panel, groups of two nails, about two inches apart are spaced along framing member about 12 inches O.C. (Fig. 17-13).

***Screw Attachment.*** Screws are spaced 12 inches on ceilings and 16 inches on walls when framing members are spaced 16 inches. If framing members are spaced 24 inches, then screws are spaced a maximum of 12 inches on both walls and ceilings.

***Using Adhesives.*** Apply a straight bead about ¼ inch in diameter to the centerline of the stud edge. On studs where panels are joined, two parallel

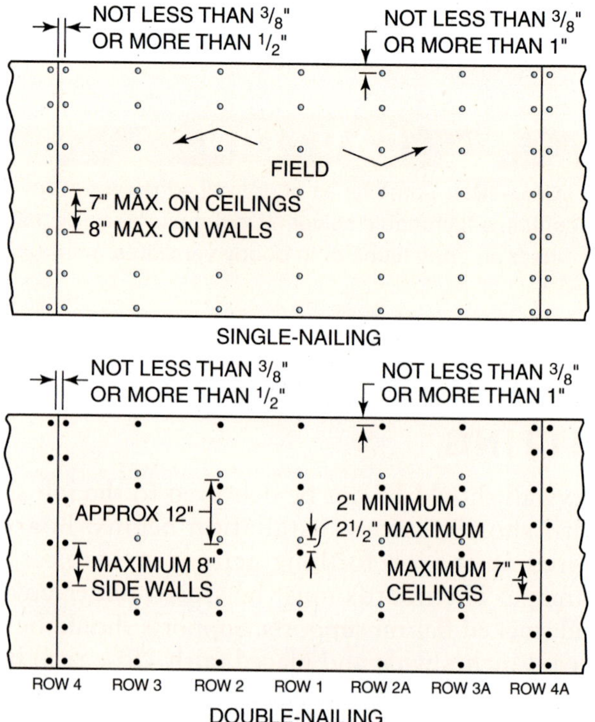


**FIGURE 17-13** Gypsum panels may be single-nailed or double-nailed.

beads of adhesive are applied, one on each side of the centerline. On wall application, supplemental fasteners are used around the perimeter. Space them about 16 inches apart. On ceilings, in addition to perimeter fastening, the field is fastened at about 24-inch intervals (Fig. 17-14).

Gypsum panels may be pre-bowed. This reduces the number of supplemental fasteners required. Prebow the panels by one of the methods shown in Figure 17-15. Make sure the finish side of the panel faces in the correct direction. Allow them to remain overnight or until the boards have a 2-inch permanent bow. Apply adhesive to the studs. Fasten the panel at top and bottom plates. The bow keeps the center of the board in tight contact with the adhesive until bonded.

## INSTALLING CEILING DRYWALL

Gypsum panels are applied first to ceilings and then to the walls. This makes it easier to make tight-fitting joints. Panels may be applied parallel, or at right angles, to joists or furring. If applied parallel, edges and ends must bear completely on framing. If applied at right angles, the tapered edges are fastened only where they cross over each framing member. Ends must be fastened completely to joists or furring

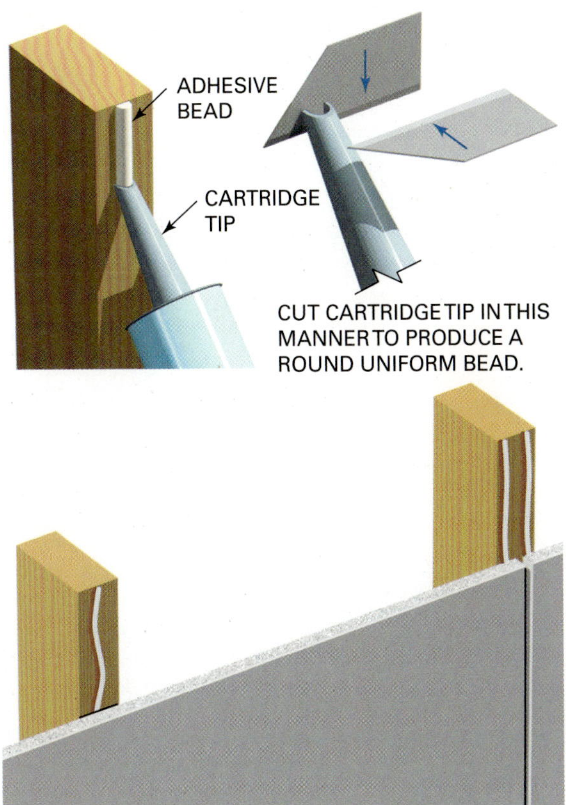


**FIGURE 17-14** Beads of stud adhesive are applied (left) straight under the field of a board and (right) parallel beads under joints.

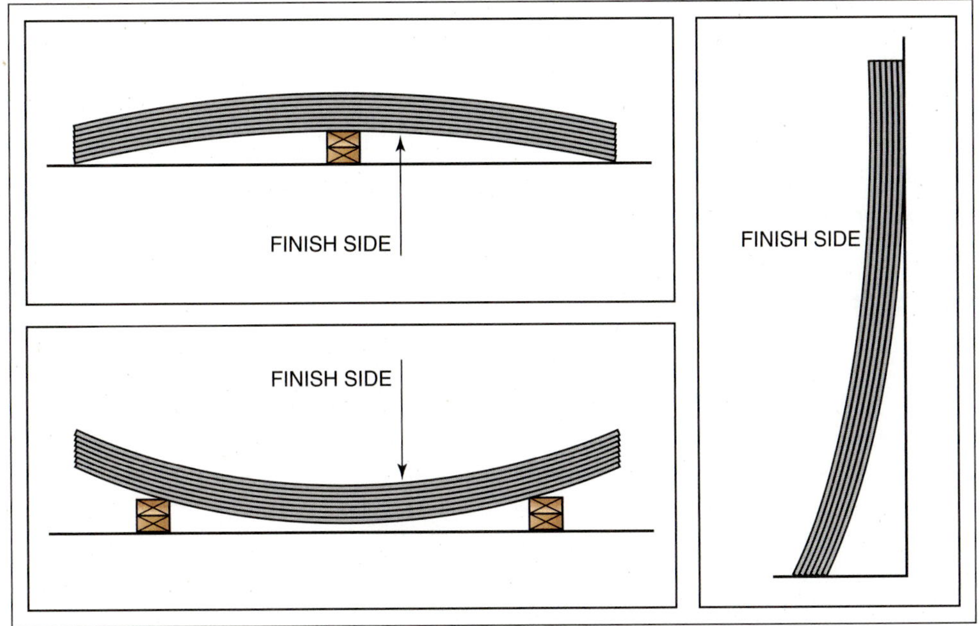


**FIGURE 17-15** Pre-bowing keeps the board in tight contact with the adhesive until bonded and reduces the number of fasteners required.

**FIGURE 17-16** Drywall jacks are often used to hold drywall panels in place while fastening.

strips. Gypsum board panels are heavy and typically require at least two people to install. A panel lift is typically used in such situations (Fig. 17-16).

***For step-by-step instructions on INSTALLING CEILING DRYWALL, see Procedure 17-3 on page 577.***

## INSTALLING HORIZONTAL DRYWALL

Wallboard is usually installed horizontally at right angles to the studs. If possible, use a board of sufficient length to go from corner to corner. Otherwise, use as long a board as possible to minimize end joints because they are more difficult to conceal. Stagger end joints. End joints should not fall on the same stud as those on the opposite side of the partition.

***For step-by-step instructions on INSTALLING HORIZONTAL DRYWALL, see Procedure 17-4 on page 578.***

## VERTICAL APPLICATION ON WALLS

Vertical application of gypsum panels on walls, with long edges parallel to studs, is sometimes used. Cut the first board in the corner to length and width. Its length may be ½-inch shorter than the height from floor to ceiling. This allows room to get the panel into position. The panel is held tight to the ceiling, leaving the gap at the floor which will be covered later by baseboard trim.

Panel should be cut to width so the edge away from the corner falls on the center of a stud. All cut edges must be in the corners. The tapered edge should be plumb and centered on the stud. Fasten it in the specified manner. Continue applying sheets around the room with tapered uncut edges against each other (Fig. 17-17).

## FLOATING INTERIOR ANGLE CONSTRUCTION

To help prevent nail popping and cracking due to structural stresses where walls and ceilings meet, the *floating angle* method of drywall application is used. Fasteners are omitted in the corner intersection of the ceiling and wall panels. Gypsum panels applied on walls are fitted tightly against ceiling panels. This provides a firm support for the floating edges

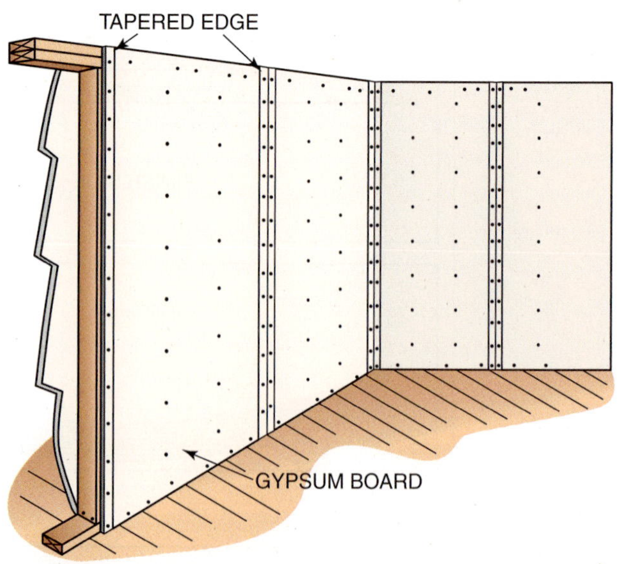


**FIGURE 17-17** Applying gypsum panels parallel to wall studs.

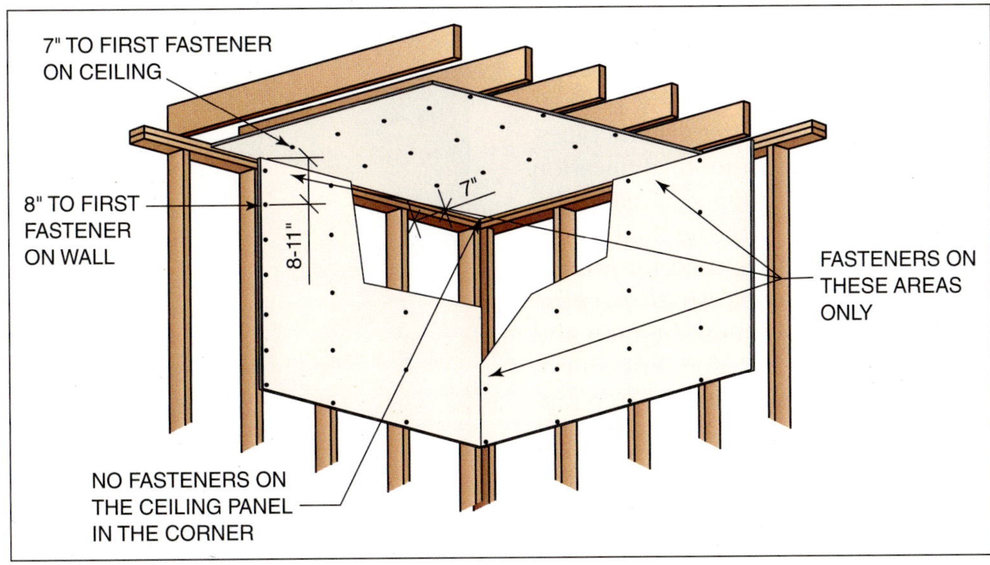


**FIGURE 17-18** Floating angle method of applying drywall has the wall panel supporting the ceiling panel.

of the ceiling panels. The top fastener into each stud is located 8 inches down from the ceiling for single nailing and 11 to 12 inches down for double nailing or screw attachment.

At interior wall corners, the first underlying wallboard is not fastened. Only the overlapping second board is fitted snugly and fastened against the underlying board. This brings it in firm contact with the face of the stud (Fig. 17-18). Ceiling perimeter fasteners are also omitted within 7 inches of the edge. The wall panel holds ceiling panel in place. This method allows the finished corner to stay tight when the framing expands and swells with moisture changes.

# Concealing Fasteners and Joints

After the gypsum board is installed, it is necessary to conceal the fasteners and to reinforce and conceal the joints. One of several levels of finish may be specified for a gypsum board surface. The lowest level of finish may simply require the taping of wallboard joints and *spotting* of fastener heads on surfaces. The level of finish depends on the number of coats of compound applied to joints and fasteners (Fig. 17-19).

## DESCRIPTION OF MATERIALS

Fasteners are concealed with joint compound. Joints are reinforced with *joint tape* and covered with joint compound. Exterior corners are reinforced with a metal or vinyl *corner bead*. Other kinds of drywall trim may be used around doors, windows, and other openings.

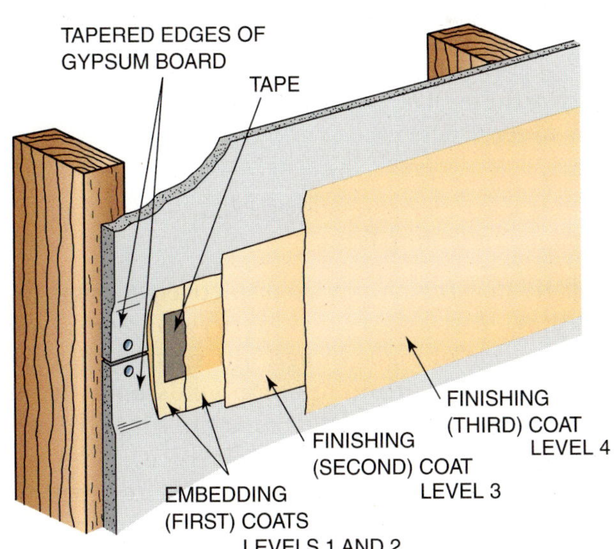


**FIGURE 17-19** The level of finish varies with the type of final decoration to be applied to drywall panels.

## Joint Compounds

*Drying type* joint compounds for joint finishing and fastener spotting are made in both a dry powder form and a ready-mixed form in three general types. Drying type compounds provide smooth application and significant working time. A *taping compound* is used to embed and adhere tape to the board over the joint. A *topping compound* is used for second and third coats over taped joints. An *all-purpose compound* is used for both bedding the tape and finishing the joint. All-purpose compounds do not possess the strength or workability of two-step taping and topping compound systems.

*Setting type* joint compounds are used when a faster setting time is desired. Drying type compounds harden through the loss of water by evaporation. They usually cannot be recoated until the next day. Setting type compounds harden through a chemical reaction when water is added to the dry powder. Therefore, they come only in a dry powder form and are mixed with water when ready to use. They are formulated in several different setting times. The fastest setting type will harden in as little as 20 to 30 minutes. The slowest type takes 4 to 6 hours to set up. Setting type joint compounds permit finishing of drywall interiors in a single day.

Joint compounds are also used to apply various textures and finishes to wallboard. Joint compound is applied to the entire surface, then tooled using a trowel or brush. Trowels are used to create a spiked texture by simply pulling the flat trowel off the surface of wet compound. Brushes are used to create swirl patterns. The brush is used in a three-quarter turn sweep where each sweep overlaps the previous sweep.

## Joint Reinforcing Tape

*Joint reinforcing tape* is used to cover, strengthen, and provide crack resistance to drywall joints. One type is made of *high-strength fiber paper.* It is designed for use with joint compounds on gypsum panels. It is creased along its center to simplify folding for application in corners (Fig. 17-20). Another type is made of *glass fiber mesh*. It is designed to reinforce joints on veneer plaster gypsum panels. It is not recommended for use with conventional compounds for general drywall joint finishing. It may be used with special high-strength setting compounds. Glass fiber mesh tape is available with a plain back or with an adhesive backing for quick application (Fig. 17-21). Joint tape is normally available 2 and 2½ inches wide in 300-foot rolls.

**FIGURE 17-20** Applying joint tape to an interior corner.

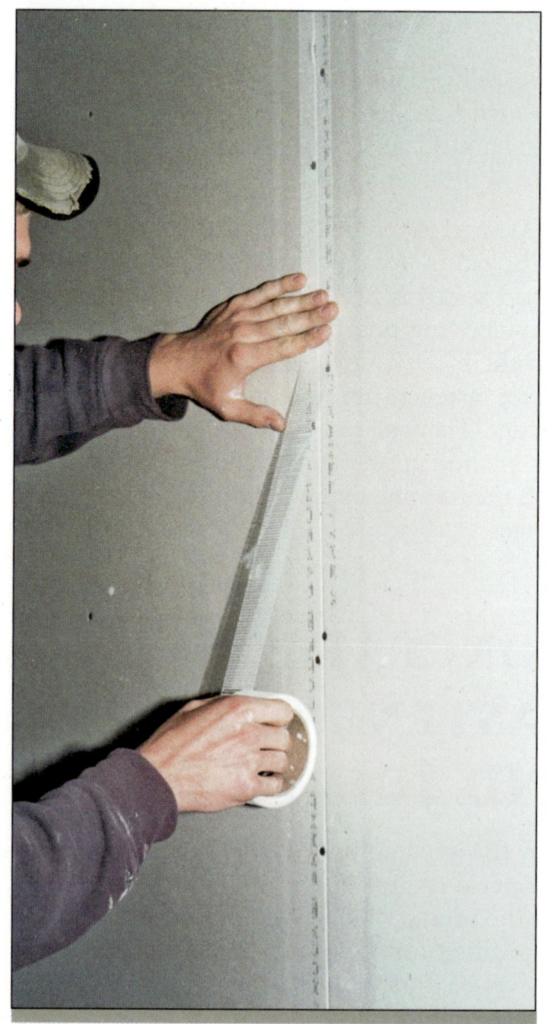

**FIGURE 17-21** An adhesive-backed glass fiber mesh tape is quickly applied to joints.

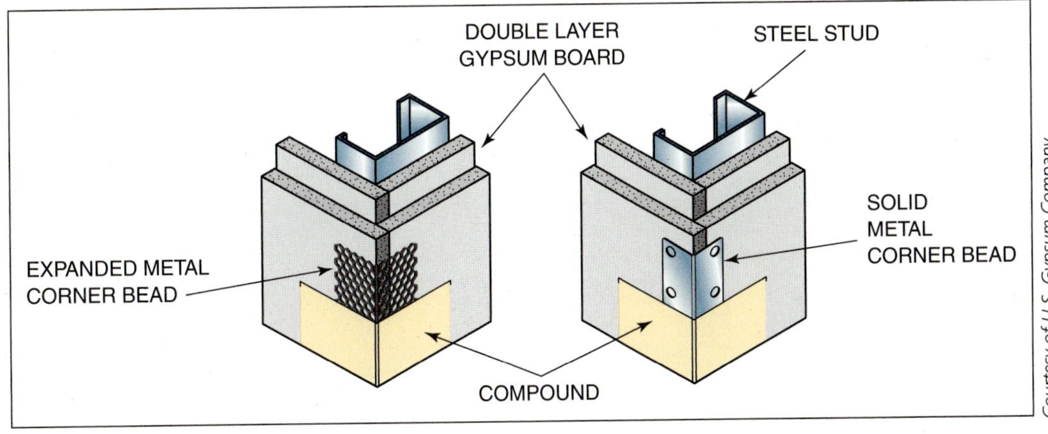


**FIGURE 17-22** Corner beads are used to finish and protect exterior corners of drywall panels.

**FIGURE 17-23** A clinching tool is sometimes used to set corner beads for fastening to exterior corners.

## Corner Bead and Other Drywall Trim

*Corner beads* are applied to protect exterior corners of drywall construction from damage by impact. One type with solid metal flanges is widely used. Another type has flanges of expanded metal with a fine mesh. This provides excellent *keying* of the compound (Fig. 17-22). Corner bead is fastened through the drywall panel into the framing with nail screws. A *clinching tool* is sometimes used to set corner beads to the proper angle. It crimps the solid flanges and locks the bead to the corner. Fasteners are then applied to secure the bead to the corner (Fig. 17-23). *Metal corner tape* is applied by embedding in joint compound. It is used for corner protection on arches, windows with no trim, and other locations (Fig. 17-24). A variety of *metal trim* is used to provide protection

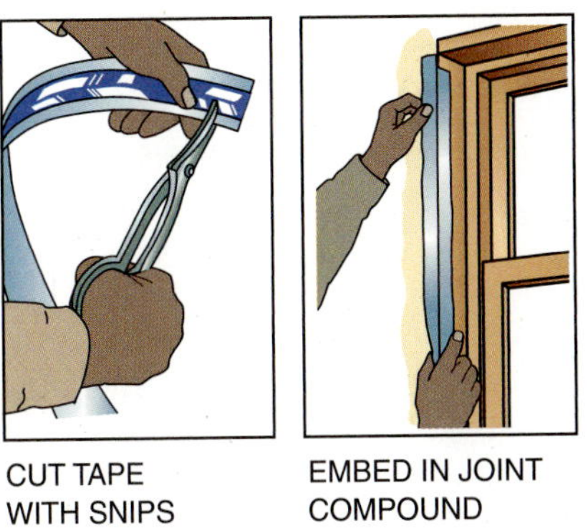

**FIGURE 17-24** Flexible metal corner tape is applied to exterior corners by embedding in compound.

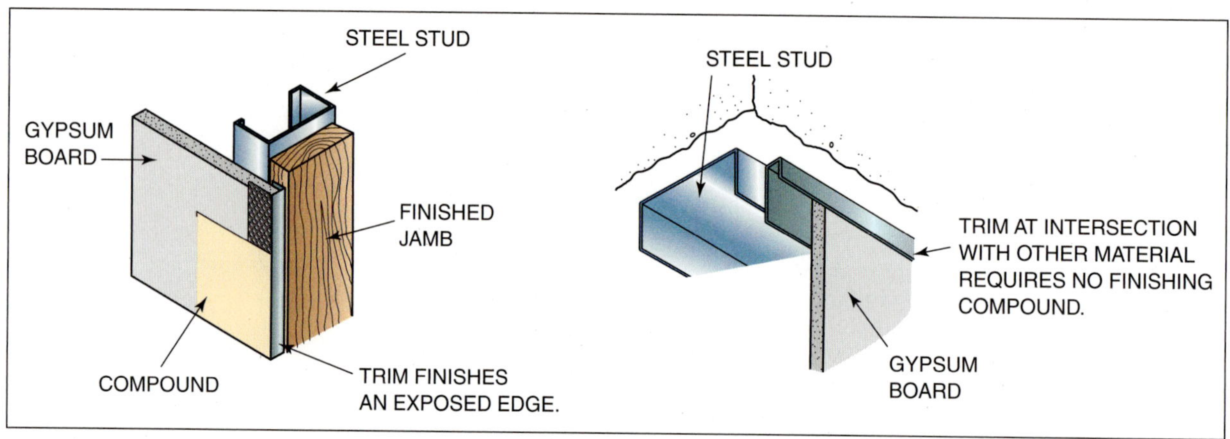


Courtesy of U.S. Gypsum Company.

**FIGURE 17-25** Various metal trim is used to provide finished edges to gypsum panels.

and finished edges to drywall panels. They are used at windows, doors, inside corners, and intersections. They are fastened through their flanges into the framing (Fig. 17-25).

*Control joints* are metal strips with flanges on both sides of a ¼-inch, V-shaped slot. Control joints are placed in large drywall areas. Control joints relieve stresses induced by expansion or contraction and are used from floor to ceiling in long partitions and from wall to wall in large ceiling areas (Fig. 17-26). The flanges are concealed with compound in a manner similar to corner beads and other trim.

A wide assortment of rigid vinyl drywall accessories is available (Fig. 17-27). They include the

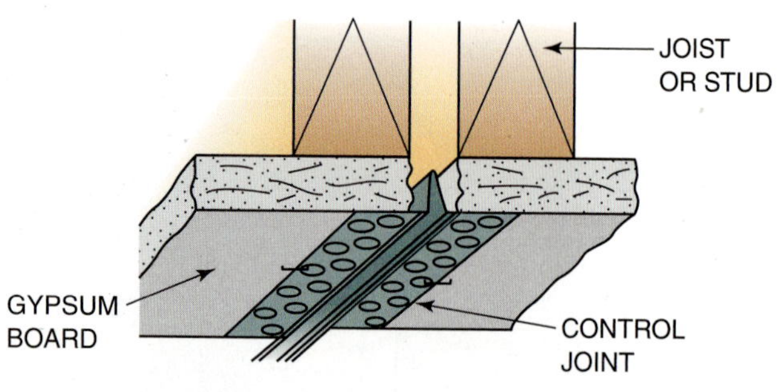


**FIGURE 17-26** Control joints are used in large wall and ceiling areas subject to movement by expansion and contraction.

styles of the metal trim previously discussed. Vinyl accessories are designed for easy installation. Most have edges to guide the drywall knife, which allows for an even application of joint compound. Some have edges that are later torn away when the painting is done. This allows the finish to be applied more quickly and at the same time more uniformly. Vinyl accessories make it possible to create smooth joints easily, whether they are curved or straight.

## APPLYING JOINT COMPOUND AND TAPE

In cold weather, care should be taken to maintain the interior temperature at a minimum of 50 degrees Fahrenheit for 24 hours before and during application of joint compound and for at least 4 days after application has been completed.

Care should also be taken to use clean tools. Avoid contamination of the compound by foreign material, such as sawdust, hardened compound, or different types of compounds.

A *taping tool* sometimes is used to apply compound and embed the tape at the same time (Fig. 17-28).

***For step-by-step instructions on APPLYING JOINT COMPOUND AND TAPE, see Procedure 17-5 on pages 579–581.***

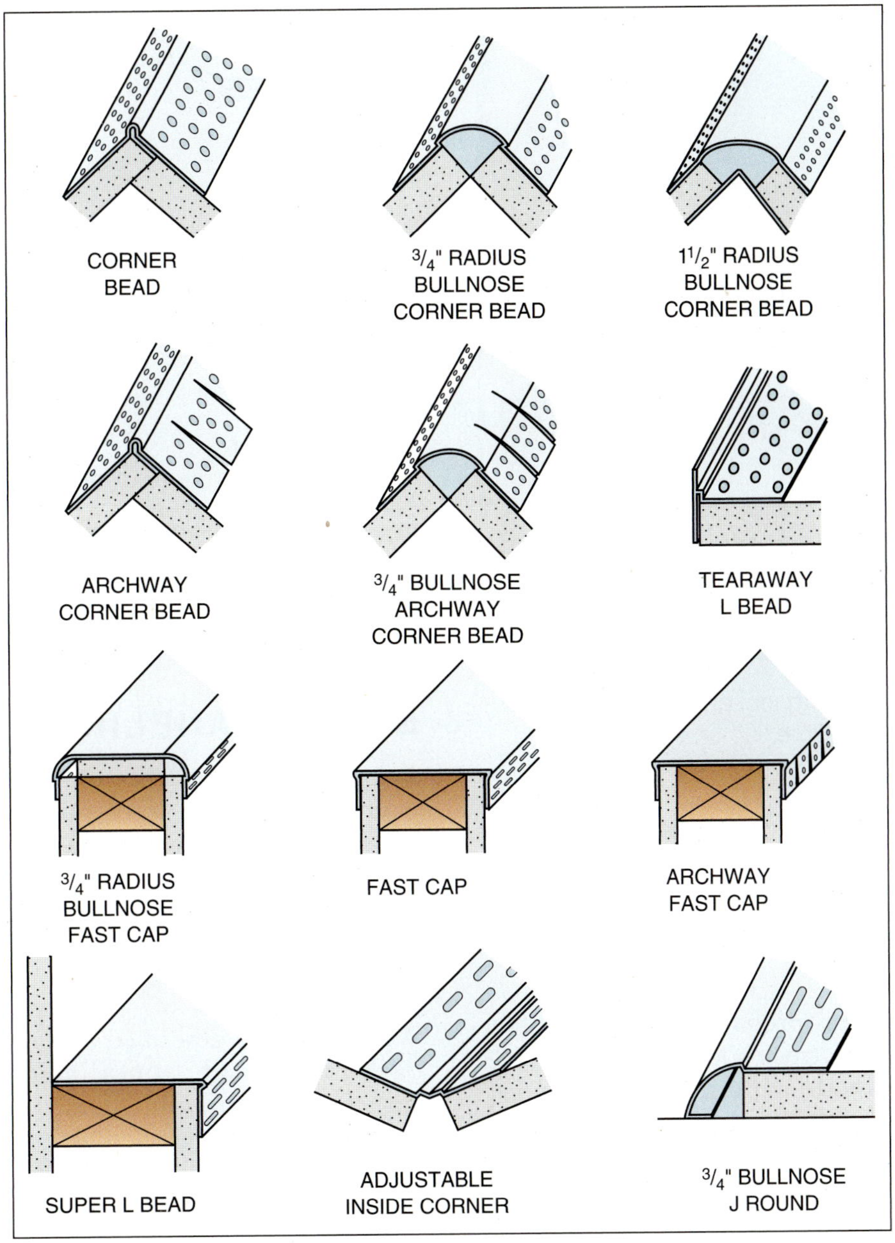

CORNER BEAD

3/4" RADIUS BULLNOSE CORNER BEAD

1 1/2" RADIUS BULLNOSE CORNER BEAD

ARCHWAY CORNER BEAD

3/4" BULLNOSE ARCHWAY CORNER BEAD

TEARAWAY L BEAD

3/4" RADIUS BULLNOSE FAST CAP

FAST CAP

ARCHWAY FAST CAP

SUPER L BEAD

ADJUSTABLE INSIDE CORNER

3/4" BULLNOSE J ROUND

**FIGURE 17-27** Some of the many rigid vinyl drywall accessories available.

## Green Tip

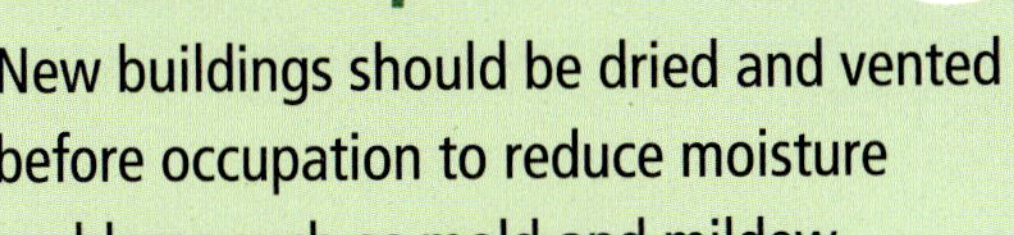

New buildings should be dried and vented before occupation to reduce moisture problems such as mold and mildew.

# SHEET PANELING

Two basic kinds of wall paneling are sheets of various prefinished material and solid wood boards. Many compositions, colors, textures, and patterns are available in sheet form. Solid wood boards of many species and shapes are used for both rustic and elegant interiors.

**FIGURE 17-28** A taping tool applies tape and compound at the same time.

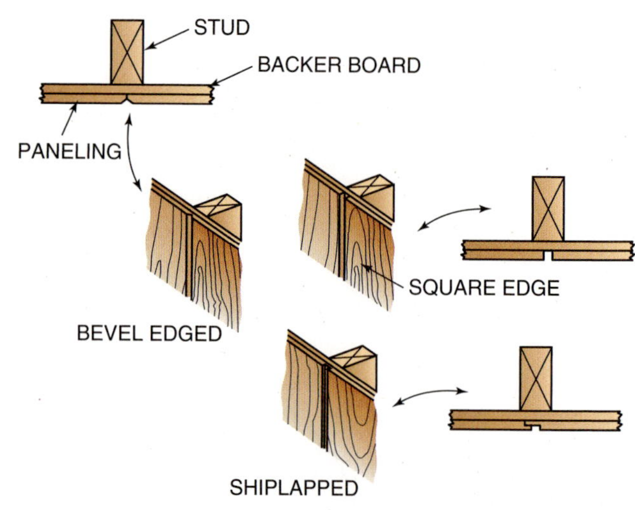


**FIGURE 17-29** Sheet paneling comes with various edge shapes.

Sheets of *prefinished plywood,* **hardboard**, *particleboard,* and other materials are used to panel walls.

## PLYWOOD

Prefinished plywood is probably the most widely used sheet paneling. A tremendous variety is available in both hardwoods and softwoods. The more expensive types have a face veneer of wood. The less-expensive kinds of plywood paneling are prefinished with a printed wood grain or other design on a thin vinyl covering. Care must be taken not to scratch or scrape the surface when handling these types. Unfinished plywood panels are also available.

Some sheets are scored lengthwise at random intervals to imitate solid wood paneling. There is always a score 16, 24, and 32 inches from the edge. This facilitates fastening the sheets and ripping the sheet lengthwise to fit stud spacing.

Most commonly used panel thicknesses are 3⁄16 and 1⁄4 inch. Sheets are normally 4 feet wide and 7 to 10 feet long. An 8-foot length is most commonly used. Panels may be shaped with square, beveled, or shiplapped edges (Fig. 17-29). Matching molding is available to cover edges, corners, and joints. Thin ring-shanked nails, called *color pins,* are available in colors to match panels. They are used when exposed fastening is necessary.

## BOARD PANELING

*Board paneling* is used on interior walls when the warmth and beauty of solid wood is desired. Wood paneling is available in softwoods and hardwoods of many species. Each has its own distinctive appearance, unique grain, and knot pattern.

### Wood Species

Woods may be described as light-, medium-, and dark-toned. Light tones include birch, maple, spruce, and white pine. Some medium tones are cherry, cypress, hemlock, oak, ponderosa pine, and fir. Among the darker-toned woods are cedar, mahogany, redwood, and walnut. For special effects, knotty pine, pecky cypress, and white-pocketed Douglas fir board paneling may be used.

### Surface Textures and Patterns

Wood paneling is available in many shapes. It is either *planed* for smooth finishing or *rough-sawn* for a rustic, informal effect. *Square-edge* boards may be joined edge to edge, spaced on a dark background, or applied in *board-and-batten* or *board-on-board* patterns. *Tongue-and-grooved* or *shiplapped* paneling comes in patterns, a few of which are illustrated in Figure 17-30.

### Sizes

Most wood paneling comes in a 3⁄4-inch thickness and in nominal widths of 4, 6, 8, 10, and 12 inches.

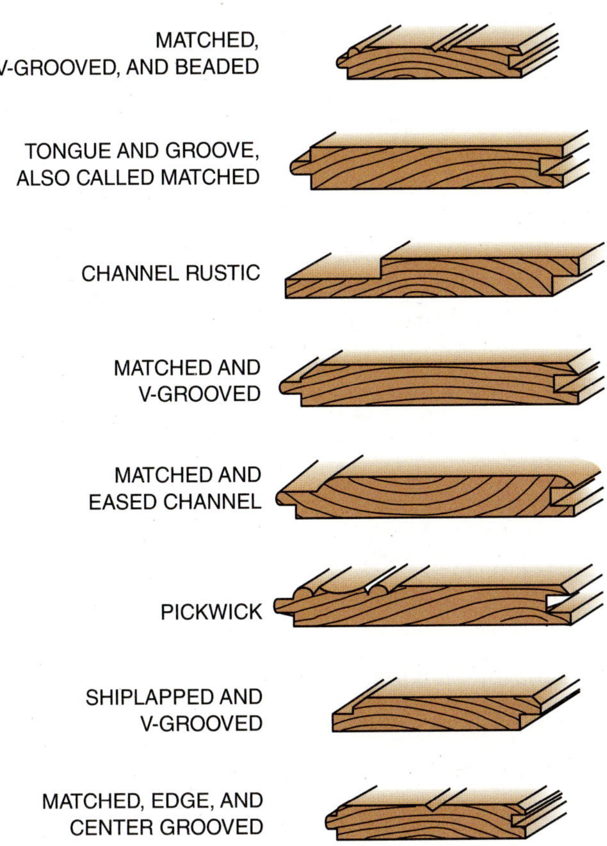


**FIGURE 17-30** Solid wood paneling is available in a number of patterns.

A few patterns are manufactured in a 9/16-inch thickness. *Aromatic cedar* paneling is used in clothes closets. It runs from 3/8 to 5/16 inch thick, depending on the mill. It is usually *edge-* and *end-matched* (tongue-and-grooved) for application to a backing surface.

*Sheet paneling,* such as prefinished plywood and hardboard, is usually applied to walls with the long edges vertical. *Board paneling* may be installed vertically, horizontally, diagonally, or in many interesting patterns (Fig. 17-31).

## HARDBOARD

Hardboard is available in many man-made surface colors, textures, and designs. Some of these designs simulate stone, brick, stucco, leather, weathered or smooth wood, and other materials. Unfinished hardboard with a smooth, dark brown surface is suitable for painting and other decorating. Hardboard paneling comes in widths of 4 feet and in lengths of from 8 to 12 feet. Commonly used thicknesses are from 1/8 to 1/4 inch. Color-coordinated molding and trim are available for use with hardboard paneling.

Courtesy of California Redwood Association.

**FIGURE 17-31** Here redwood paneling is installed vertically.

## PARTICLEBOARD

Panels of *particleboard* come with wood grain or other designs applied to the surface, similar to plywood and hardboard. Sheets are usually 1/4-inch thick, 4 feet wide, and 8 feet long. Prefinished particleboard is used when an inexpensive wall covering is desired. Because the sheets are brittle and break easily, care must be taken when handling. They must be applied only on a solid wall backing.

## INSTALLATION OF SHEET PANELING

Some building codes require a base layer of gypsum board applied to studs or furring strips for the installation of sheet paneling. Even if not required, a backer board layer, at least 3/8 inch thick, should be installed on walls prior to the application of sheet paneling. The backing makes a stronger and more fire-resistant wall (Fig. 17-32). Seams and joints should have at least one coat of joint finish.

Sometimes paneling does not extend to the ceiling but covers only the lower portion of the wall. This partial paneling is called **wainscoting**. It is usually installed about 32 inches above the floor (Fig. 17-33).

Installation begins in a corner and continues with consecutive sheets around the room. Select the starting corner, remembering that it will also be the

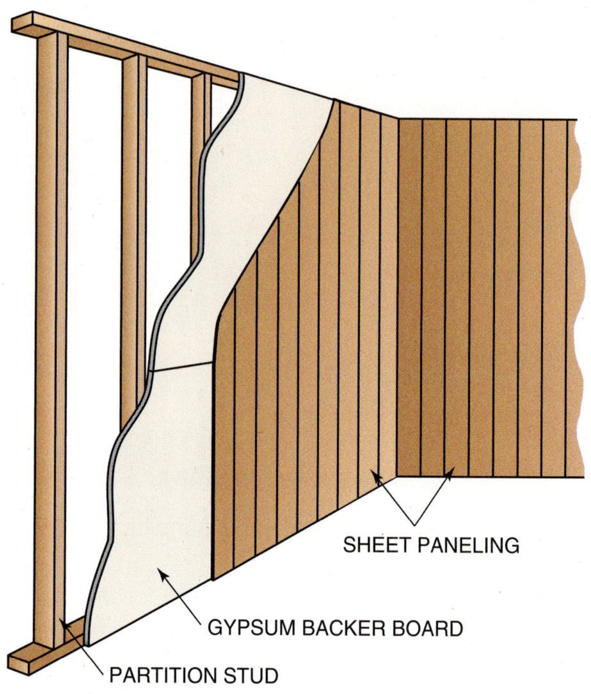


**FIGURE 17-32** Apply sheet paneling over a gypsum wallboard base.

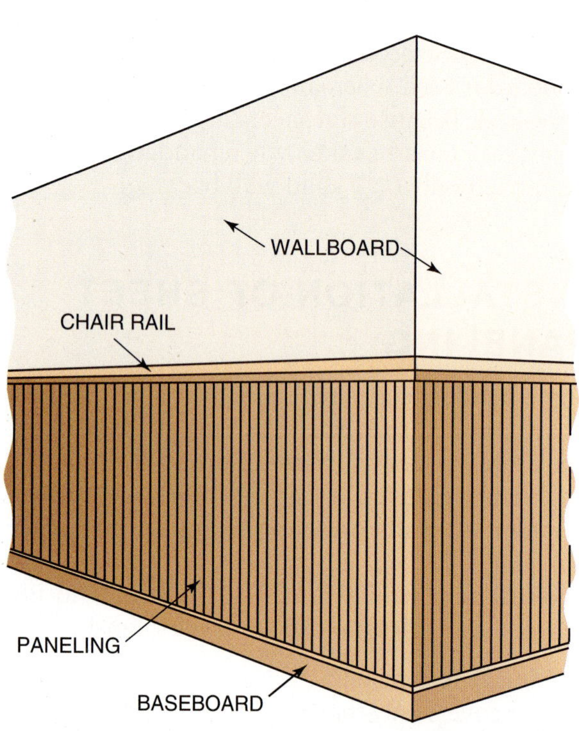


**FIGURE 17-33** Wainscoting is a wall finish, usually paneling, applied to the lower portion of the wall that is different from the upper portion.

ending point. This corner should be the least visible, such as behind a door that will be left open often.

To fasten with only nails, space them about 6 inches apart along edges and about 12 inches apart on intermediate studs for ¼-inch thick paneling. Nails may be spaced farther apart on thicker paneling. Drive nails at a slight angle for better holding power (Fig. 17-34). If adhesives are used, apply a ⅛-inch continuous bead where panel edges and ends make contact. Use nails as required to hold the panel in place until the adhesive sets.

The last sheet in a wall section need not fit snugly in the corner if the adjacent wall is to be paneled or if interior corner molding is to be used. Scribers are used to fit pieces uniformly to corners. Make sure the scribed mark is the same 90-degree distance during the entire scribing process (Fig. 17-35).

***For step-by-step instructions on INSTALLING SHEET PANELING, see Procedure 17-6 on pages 582–585.***

## INSTALLING SOLID WOOD BOARD PANELING

Horizontal board paneling may be fastened to studs in new and existing walls. The boards are fastened to the existing wall studs. The thickness of wood paneling should be at least ⅜ inch for 16-inch spacing of frame members and ⅝ inch for 24-inch spacing. If wainscoting is applied to a wall, the joint between the different materials may be treated in several ways (Fig. 17-36).

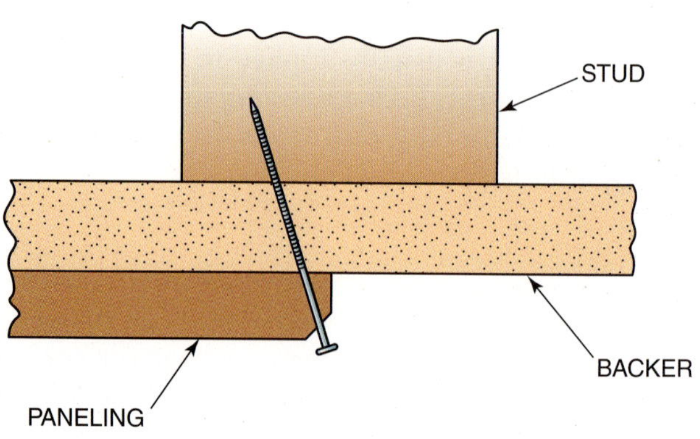


**FIGURE 17-34** Drive nails at a slight angle for better holding power. Color pins are used on prefinished sheet paneling.

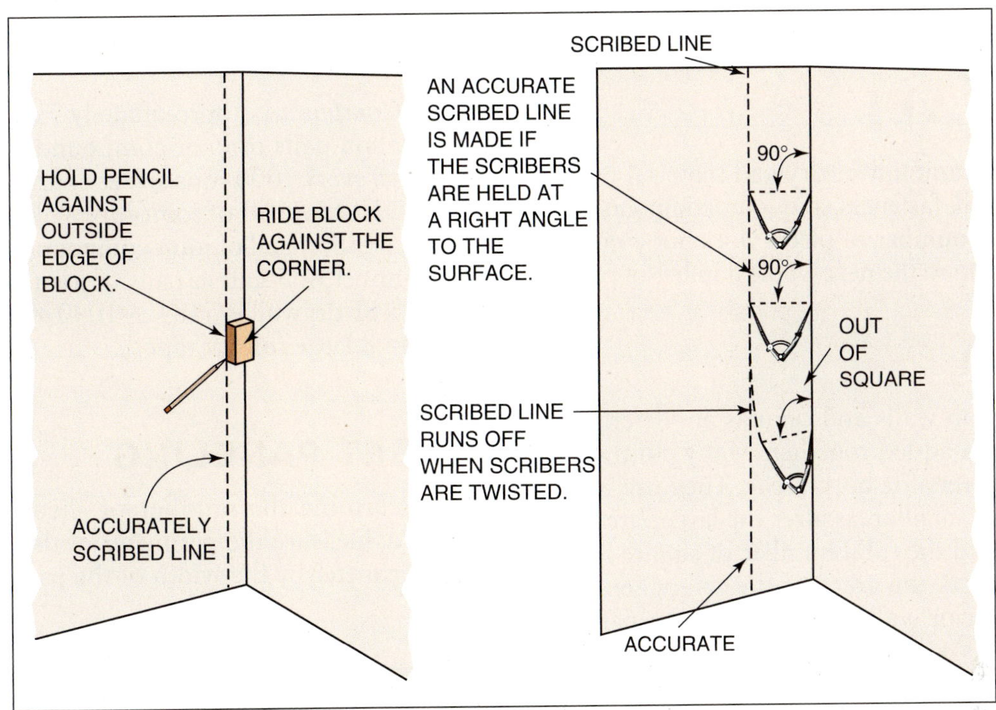


**FIGURE 17-35** Accurate scribing requires that the marked line be made perpendicular to the corner.

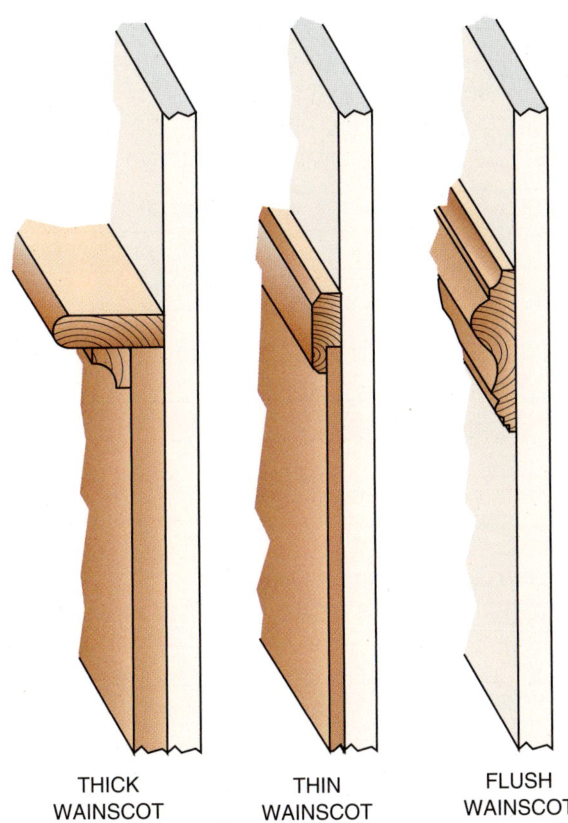


**FIGURE 17-36** Methods of finishing the joint at the top of wainscoting.

For vertical application of board paneling in a frame wall, blocking must be provided between studs. Maximum vertical spacing is 24 inches for ½-inch thick paneling and 48 inches for ¾-inch paneling. Blocking or furring must be provided in appropriate locations for diagonal or pattern applications of board paneling.

Allow the boards to adjust to room temperature and humidity by standing them against the walls around the room. At the same time, put them in the order of application. Match them for grain and color. If tongue-and-grooved boards are to be eventually stained or painted, apply the same finish to the tongues so that an unfinished surface is not exposed if the boards shrink after installation.

If the last board in the installation must fit snugly in the corner without a molding, the layout should be planned so that the last board will be as wide as possible. If boards are a uniform width, the width of the starting board must be planned to avoid ending with a narrow strip. If random widths are used, they can be arranged when nearing the end.

***For step-by-step instructions on INSTALLING SOLID WOOD BOARD PANELING, see Procedure 17-7 on page 586.***

# Estimating Drywall

Estimating the amount of drywall material needed involves drywall, fasteners, and joint compound. Remember, if the number of pieces does not come out even, round it up to the next whole number of pieces.

## DRYWALL

Drywall used on walls and ceilings are determined separately then added together. Many contractors use 12-foot sheets for both areas. They use up the scrap pieces in small areas after the larger areas are covered. To find the total number of square feet of drywall required, the areas of the ceiling, exterior walls, and interior walls are determined separately. Then, the areas are added together. Only the large wall openings, such as double doors, are deducted since the cutouts for windows are mostly waste. The number of drywall sheets can then be determined by dividing the total area to be covered by the area of one panel. Five percent may be added for waste.

*Ceiling Drywall.* To find the ceiling area, multiply the length of the building by its width. The entire building may be considered if all ceilings are to be covered with the same material. For example, if the building is 32 feet by 50 feet, then the ceiling drywall is 32 × 50 ÷ (4 × 12) × 105% = 35 sheets.

*Wall Drywall.* To find the exterior wall area, multiply the building perimeter times wall height. The interior wall area is the overall length of interior walls times wall height times two sides of walls. For example, the number of 4 × 12 sheets (48 square feet) for the exterior walls of a 32-foot by 50-foot building with wall height of 8 feet is 172 × 8 = 1376 SF ÷ 48 × 105% = 30.1 ⇒ 31 sheets. If the interior wall length is 120 feet for the same building, then the drywall needed is 120 × 8 × 2 = 1920 SF ÷ 48 × 105% = 42 sheets.

## FASTENERS AND FINISH

*Fasteners.* About 1½ fasteners are needed for each square foot of drywall when applied to framing 16 inches OC, and 1 fastener per 1 square foot is needed for 24-inch OC framing. Screws may be purchased in boxes of 5000 screws per box. For example, the fasteners for 4 × 12 sheets in the previous example, attached to 16-inch OC framing, is 35 + 31 + 42 = 108 sheets × 48 = 5184 SF × 1.5 = 7776 ÷ 5000 = 2 boxes of fasteners.

*Joint Compound.* Approximately 7½ gallons or 1½ five-gallon pails of joint compound will be needed to finish every 1000 square feet of drywall. Also, one large 250-foot roll of joint tape will be needed for every 5-gallon pail of joint compound. For example, the number of 5-gallon pails of joint compound for a 5184-SF drywall is 5184 ÷ 1000 × 1.5 = 7.7 ⇒ 8 pails = 8 large rolls of tape.

## SHEET PANELING

To determine the number of sheets of paneling needed, measure the perimeter of the room. Divide the perimeter by the width of the panels to be used.

| NOMINAL SIZE | WIDTH | | AREA FACTOR* |
|---|---|---|---|
| | DRESS | FACE | |
| **SHIPLAP** | | | |
| 1X6 | $5^1/_2$ | $5^1/_8$ | |
| 1X8 | $7^1/_4$ | $6^7/_8$ | 1.16 |
| 1X10 | $9^1/_4$ | $8^7/_8$ | |
| 1X12 | $11^1/_4$ | $10^7/_8$ | 1.10 |
| **TONGUE AND GROOVE** | | | |
| 1X4 | $3^3/_8$ | $3^1/_8$ | 1.28 |
| 1X6 | $5^3/_8$ | $5^1/_8$ | 1.17 |
| 1X8 | $7^1/_8$ | $6^7/_8$ | 1.16 |
| 1X10 | $9^1/_8$ | $8^7/_8$ | 1.13 |
| 1X12 | $11^1/_8$ | $10^7/_8$ | 1.10 |
| **S4S** | | | |
| 1X4 | $3^1/_2$ | $3^1/_2$ | 1.14 |
| 1X6 | $5^1/_2$ | $5^1/_2$ | 1.09 |
| 1X8 | $7^1/_4$ | $7^1/_4$ | 1.10 |
| 1X10 | $9^1/_4$ | $9^1/_4$ | 1.08 |
| 1X12 | $11^1/_4$ | $11^1/_4$ | 1.07 |
| **PANELING AND SIDING PATTERNS** | | | |
| 1X6 | $5^7/_{16}$ | $5^1/_{16}$ | 1.19 |
| 1X8 | $7^1/_8$ | $6^3/_4$ | 1.19 |
| 1X10 | $9^1/_8$ | $8^3/_4$ | 1.14 |
| 1X12 | $11^1/_8$ | $10^3/_4$ | 1.12 |

*number multiplied by square feet to convert square feet to board feet

**FIGURE 17-37** Factors used to estimate amounts of board paneling.

| Estimate the materials for interior wall finish for a rectangular 28' × 64' building. Drywall sheets are 4 × 12, 16" OC framing, 8' wall height, and 150' of interior walls. One room 14' × 18' with three doors and 4' × 8' sheet paneling. One room 18' × 22' with 10' × 40' T&G paneling (see Fig. 17-37 for area factor). | | | |
|---|---|---|---|
| **Item** | **Formula** | **Waste factor** | **Example** |
| **Ceiling drywall** | building LEN × building WID × waste factor ÷ area per sheet = sheets | 5% | 64 × 28 × 105% ÷ 48 = 39.2 ⇒ 40 sheets |
| **Exterior Wall Drywall** | PERM × wall HGT × waste factor ÷ area per sheet = sheets | 5% | 184 × 8' × 105%÷ 48 = 32.2 ⇒ 33 sheets |
| **Interior Wall Drywall** | wall LEN × wall HGT × 2 sides × waste factor ÷ area per sheet = sheets | 5% | 150 × 8'× 2 × 105% ÷ 48 = 52.5 ⇒ 53 sheets |
| **Fasteners** | NUM sheets × sheet area × 1.5 fastener/SF ÷ fasteners per box = boxes | | (40 + 33 + 53) × 48 × 1.5 ÷ 5000 = 1.8 ⇒ 2 boxes |
| **Joint Compound** | NUM sheets × sheet area ÷ 1000 × 1.5 pails per 1000 SF = NUM five gallon pails | | 126 × 48 ÷ 1000 × 1.5 = 9.1 ⇒ 10 pails |
| **Joint Tape** | 1 – 250' roll per 5 gallon pail | | 10 pails = 10 rolls |
| **Sheet Paneling** | PERM / sheet WID - 2/3 sheet per door = # sheets | | 2 × 14 + 2 × 18 = 64 / 4 – 2⅔ = 14.6 = 15 sheets |
| **Board Paneling** | PERM × room HGT × area factor × 5% waste = board ft | 5% | 2 × 18 + 2 × 22 = 80 × 8 = 640 × 1.28 × 105% = 860.1 = 861 board feet |

**FIGURE 17–38** Example of estimating drywall materials with formulas.

Deduct two-thirds of a panel for each large opening such as doors or fireplaces. For example, if the perimeter is 62 feet with two doors, the number of 4-foot panels is 62 ÷ 4 − 2 (⅔) = 14.1 = 15 panels.

## BOARD PANELING

Determine the square foot area to be covered by multiplying the perimeter by the height. Deduct the area of any large openings such as doors and large windows. Additional material is needed because of the difference in the nominal size of board paneling and its actual size. To make this allowance, multiply the area to be covered by the area factor in the table found in Figure 17-37. For example, a 1 × 6 S4S paneling actually has a face that covers 5½ inches, so multiply the area to be covered by 1.09 to find board feet. An additional 5 percent for waste in cutting is recommended. For example, the total area is 850 SF, and 1 × 8 T&G board paneling is to be used, so the board feet of material needed is 850 × 1.16 × 105% = 1035.3 = 1036 board feet. To further reduce waste in cutting, order suitable lengths to the area being covered.

See Figure 17-38 for a listing of the shorthand versions of these estimating techniques, along with another example.

## SUMMARY

- Drywall construction is a standard method for creating an indoor surface.
- Concealing fasteners and joints creates a smooth surface.
- Paneling may be installed as a sheet or with strips of wood.
- Estimating wall finish materials is necessary for efficient purchases.

PROCEDURE 17-1

# Cutting and Fitting Gypsum Board

**A** Take measurements to within ½ inch for the ceiling and ¼ inch for the walls. Using a utility knife, cut the board by first scoring the face side through the paper to the core. Guide it with a *drywall T-square* using your toe to hold the bottom. Only the paper facing need be cut.

**B** Bend the board back against the cut. The board will break along the cut face. This is done simply by lifting up on the sheet slightly as you drag it backward creating an angle in the sheet.

***from experience...***

Cut only the center section of the backside paper, leaving the bottom and top portions. These will act as hinges for the cut piece when it is snapped back into place.

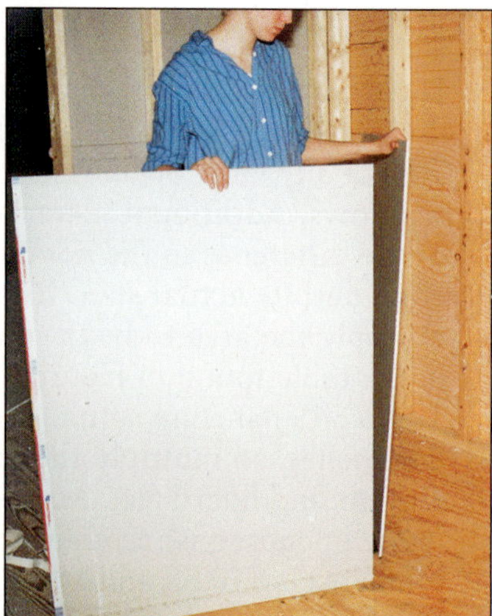

- Score the backside paper leaving some material at the top and bottom to act as a hinge. Lifting the panel off the floor, snap the cut piece back quickly to the straight position. This will complete the break.

C

To make cuts parallel to the long edges, the board is often gauged with a tape and scored with a utility knife. When making cuts close to long edges, it is usually necessary to score both sides of the board before the break to obtain a clean cut.

- Ragged edges can be smoothed with a drywall rasp, coarse sanding block, or knife.

PROCEDURE 17-2

# Making Cutouts in Wall Panels

## Cutout Tool Method

**A** Note the location of the center of the outlet box by measuring and marking nearby floor and wall surfaces.

- Install the wall panel over the box with only enough fasteners to hold the panel in place. Using the cutout tool, plunge a hole through the panel in the approximate center of the outlet box. Care must be taken not to make contact with wiring. The tool is not recommended for use around live wires.
- Move the tool in any direction until the bit touches a side of the box. Withdraw the spinning bit enough to ride over the edge to the outside of the box. Reinsert the tool and trace the outside of the box.

## Layout and Cut Method

Plumb the sides of the outlet box down to the floor, or up to the previously installed top panel, by using a framing square or T-square. Measure the height from the floor, ceiling, or edge of previous sheet, whichever is more convenient.

- Place the panel in position. From the previous layout marks, plumb and measure the outline of the box to be cut.
- With a saw or utility knife, cut the outline of the box.

**A** To cut a corner out of a panel, make the shortest cut with a drywall saw. Then, score and snap the sheet in the other direction.

**A**

Courtesy of Porter Cable.

**A**

## PROCEDURE 17-3

# Installing Ceiling Drywall

- Carefully measure and cut the first board to width and length. Best to keep sheet aleast ¼ inch smaller than actual measurement. Cut edges should be against the wall. Lay out lines on the panel face indicating the location of the framing in order to place fasteners accurately.

A. Lift the panel overhead and place it in position.

B. Fasten the panel by using the appropriate fastener spacing. Hold the board firmly against framing to avoid gaps between panel and framing, which will cause nail pops or protrusions later. Drive fasteners straight into the member. Fasteners that miss supports should be removed. The nail hole should be dimpled so that later it can be covered with joint compound.

- Continue applying sheets in this manner, staggering end joints, until the ceiling is covered.

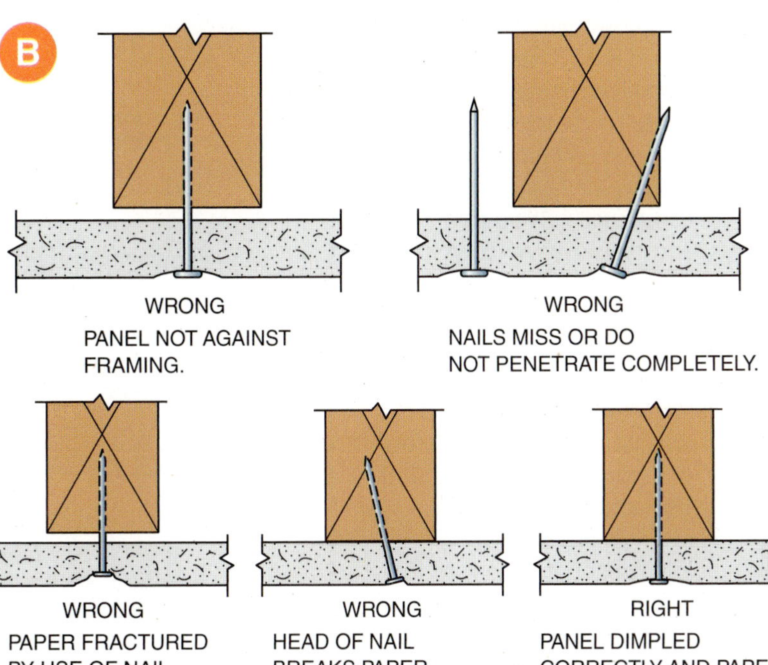

PROCEDURE 17-4

# Installing Horizontal Drywall

**A** Cut the panel to length about ¼ inch short to fit easily into place without forcing it. Stand the board on the floor against the wall where it is to be placed. Position it so that the edges are aligned to the studs. Start fasteners along the top edge opposite each stud.

- Raise the top panel into place, holding firmly against the ceiling. Drive the fasteners and finish the fastening. Fasten the rest of the sheet in the recommended manner.
- Measure and cut the bottom panel to length. If necessary, cut the width up to ½ inch narrower than the distance measured. Raise the panel into position holding firmly against the upper sheet and fasten.
- Fasten the sheet as recommended. Install all others in a similar manner. Stagger any necessary end joints. Locate them as far from the center of the wall as possible so they will be less conspicuous. Avoid placing end joints over windows and doors. This will reduce the potential for wallboard cracks.

**B** Where end joints occur on steel studs, attach the end of the first panel to the open or unsupported edge of the stud. This holds the stud flange in a rigid position for the attachment of the end of the adjoining panel. Making end joints in the opposite manner usually causes the stud edge to deflect. This results in an uneven surface at the joint.

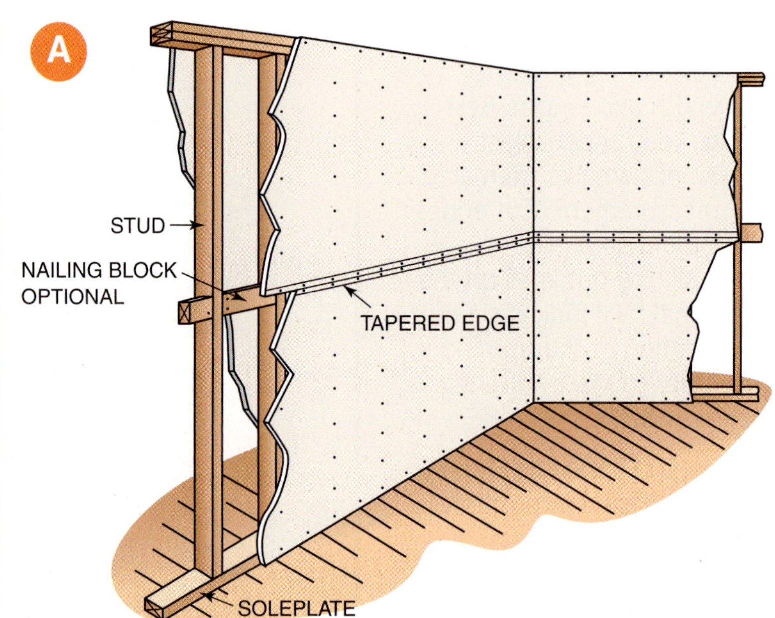


Courtesy of U.S. Gypsum Corporation.

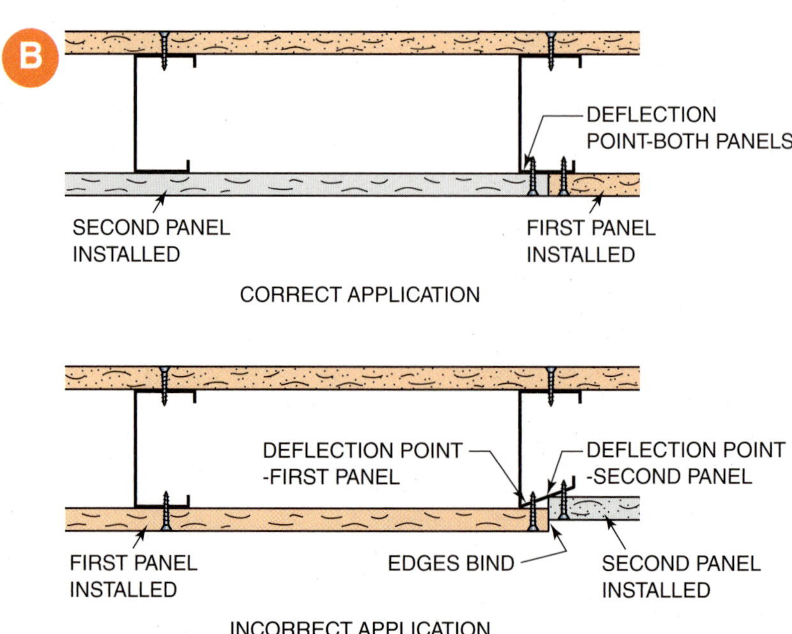


Courtesy of U.S. Gypsum Corporation.

PROCEDURE 17-5

# Applying Joint Compound and Tape

Before applying compound to drywall panels, check the surface for fasteners that have not been sufficiently recessed. Also, look for other conditions that may affect the finishing, such as loose paper or damaged edges.

- Prefill any joints between panels of ¼ inch or more and all V-groove joints between eased-edged panels with compound. A 24-hour drying period can be eliminated with the use of setting compounds for prefilling operations.
- Fill the recess formed by the tapered edges of the sheets with joint compound. Use a joint knife.

Center the tape on the joint. Lightly press it into the compound. Draw the knife along the joint with sufficient pressure to *embed* the tape and remove excess compound. There should be enough compound under the tape for a proper bond, but not over 1⁄32 inch under the edges. Make sure there are no air bubbles under the tape. The tape edges should be well adhered to the compound.

- Immediately after embedding, apply a thin coat of joint compound over the tape. This helps prevent the edges from wrinkling. It also makes easier concealment of the tape with following coats. Draw the knife to bring the coat to feather edges on both sides of the joint. Make sure no excess compound is left on the surface. After the compound has set up, but not completely hardened, wipe the surface with a damp sponge. This eliminates the need for sanding any excess after the compound has hardened.

PROCEDURE 17-5

# Applying Joint Compound and Tape (continued)

**C** Spot coat the fasteners, applying enough pressure on the taping knife to fill only the depression, removing all compound outside of the depression. Level the compound with the panel surface. Spotting is repeated each time additional coats of compound are applied to joints.

- Allow the first coat to dry thoroughly. This may take 24 hours or more depending on temperature and humidity unless a setting type compound has been used. Sand any excess, if necessary, to avoid interfering with the next coat of compound.
- The second or fill coat is feathered out about 2 inches beyond the edges of all first coats. Care must always be taken to remove all excess compound so that it does not harden on the surface.
- Apply a third and *finishing coat* of compound over all fill coats. The edges of the finishing coat should be feathered out about 2 inches beyond the edges of the second coat.

**D** Interior corners are finished in a similar way. However, the tape is folded in the center to fit in the corner. After the tape is embedded, drywall finishers usually apply a setting compound to one side only of each interior corner. By the time they have finished all interior corners in a room, the compound has set enough to finish the other side of the corners.

*Courtesy of U.S. Gypsum Company.*

PROCEDURE 17-6

# Installing Sheet Paneling

## Starting the Application

**A** Mark the location of each stud in the wall on the floor and ceiling. Paneling edges must fall on stud centers, even if applied with adhesive over a backer board, in case supplemental nailing of the edges is necessary.

- If the wall is to be wainscoted, snap a horizontal line across the wall to indicate its height.
- Apply narrow strips of paint on the wall from floor to ceiling over the stud where a seam in the paneling will occur. The color should match the color of the seams of the paneling. This will hide the joints between sheets if they open slightly because of shrinkage.
- Cut the first sheet to a length of about ½ inch less than the wall height. Place the sheet in the corner. Plumb the edge and tack it temporarily into position.

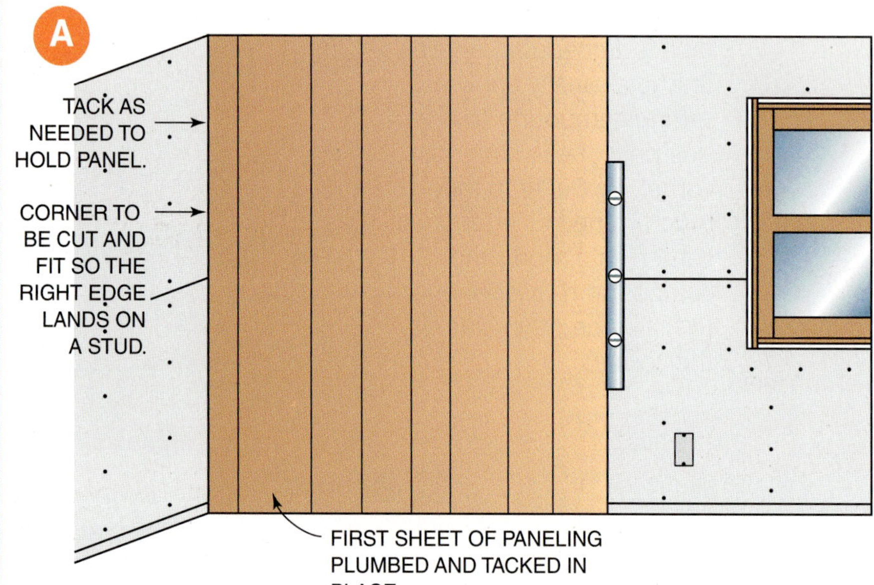

**B** Notice the joint at the corner and the distance the sheet edge overlaps the stud. Set the distance between the points of a scriber to the same as the amount the sheet overlaps the center of the stud. Scribe this amount on the edge of the sheet butting the corner.

- Remove the sheet from the wall and cut close to the scribed line. Plane the edge to the line to complete the cut. Replace the sheet with the cut edge fitting snugly in the corner.
- If a tight fit between the panel and ceiling is desired, set the dividers and scribe a small amount at the ceiling line. Remove the sheet again. Cut to the scribed line. The joint at the ceiling need not be fit tight if a molding is to be used.

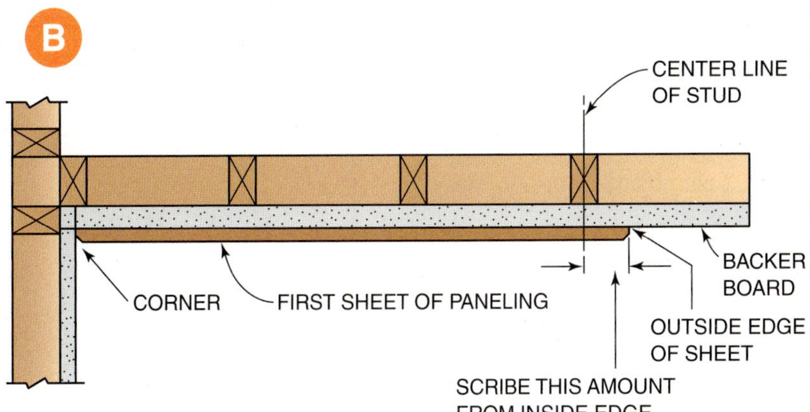

PROCEDURE 17-6

# Installing Sheet Paneling (continued)

## Wall Outlets

**A** To lay out for wall outlets, plumb and mark both sides of the outlet to the floor or ceiling, whichever is closer. Level the top and bottom of the outlet on the wall out beyond the edge of the sheet to be installed.

- Place the sheet in position and tack. Level and plumb marks from the wall and floor onto the sheet for the location of the opening.

**B** Remove the sheet. Cut the opening for the outlet. When using a saber saw, cut from the back of the panel to avoid splintering the face.

## Fastening

- Apply adhesive beads 3 inches long and about 6 inches apart on all intermediate studs. Apply a continuous bead along the perimeter of the sheet. Put the panel in place. Tack it at the top when panel is in proper position.
- Press on the panel surface to make contact with the adhesive. Use firm, uniform pressure to spread the adhesive beads evenly between the wall and the panel. Then, grasp the panel and slowly pull the bottom of the sheet a few inches away from the wall.
- Press the sheet back into position after about 2 minutes. Drive nails as needed and recheck the sheet for a complete bond after about 20 minutes. Apply pressure to ensure thorough adhesion and to smooth the panel surface.
- Apply successive sheets in the same manner. Panels should touch only lightly at joints.

**A**

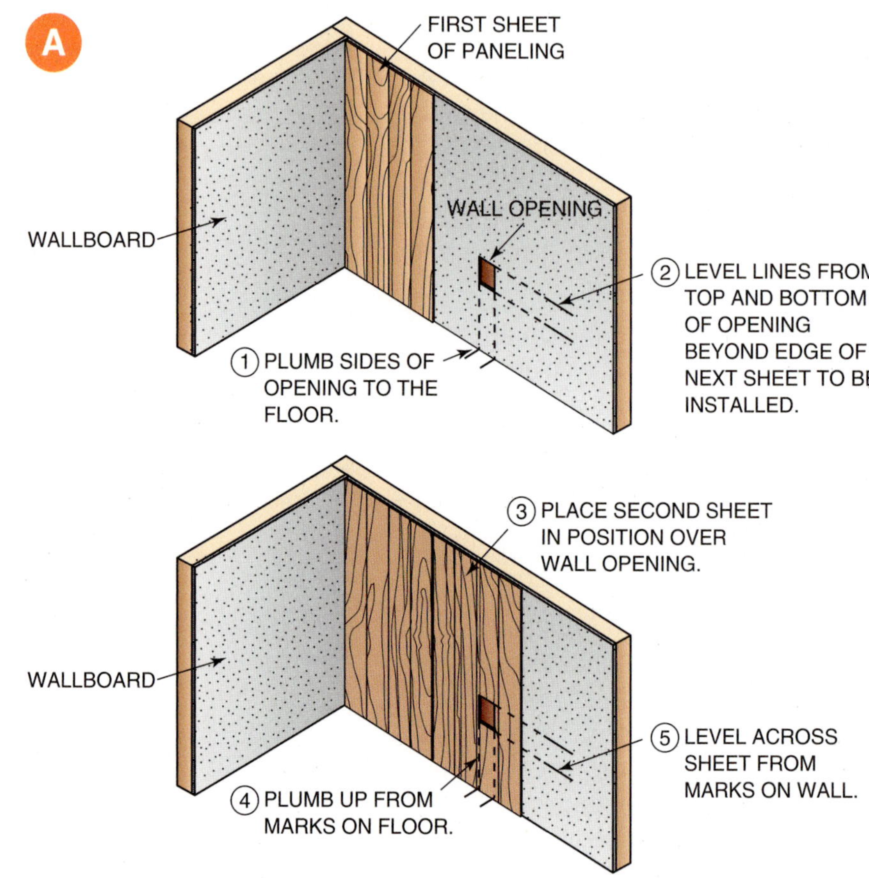


**B**

## Ending the Application

- Take measurements at the top, center, and bottom. Cut the sheet to width and install. If no corner molding is used, the sheet must be cut to fit snugly in the corner. To mark the sheet accurately, first measure the remaining space at the top, bottom, and about the center. Rip the panel about ½ inch wider than the greatest distance.

**A** Place the sheet plumb with the cut edge in the corner and the other edge overlapping the last sheet installed. Tack the sheet in position so the amount of overlap is exactly the same from top to bottom. Set the scriber for the amount of overlap. Scribe this amount on the edge in the corner.

- Cut close to the scribed line and then plane to the line. If the line is followed carefully, the sheet should fit snugly between the last sheet installed and the corner, regardless of any irregularities.

**B** Exterior corners may be finished by capping the joint.

- Use a wood block for more accurate scribing of wide distances.

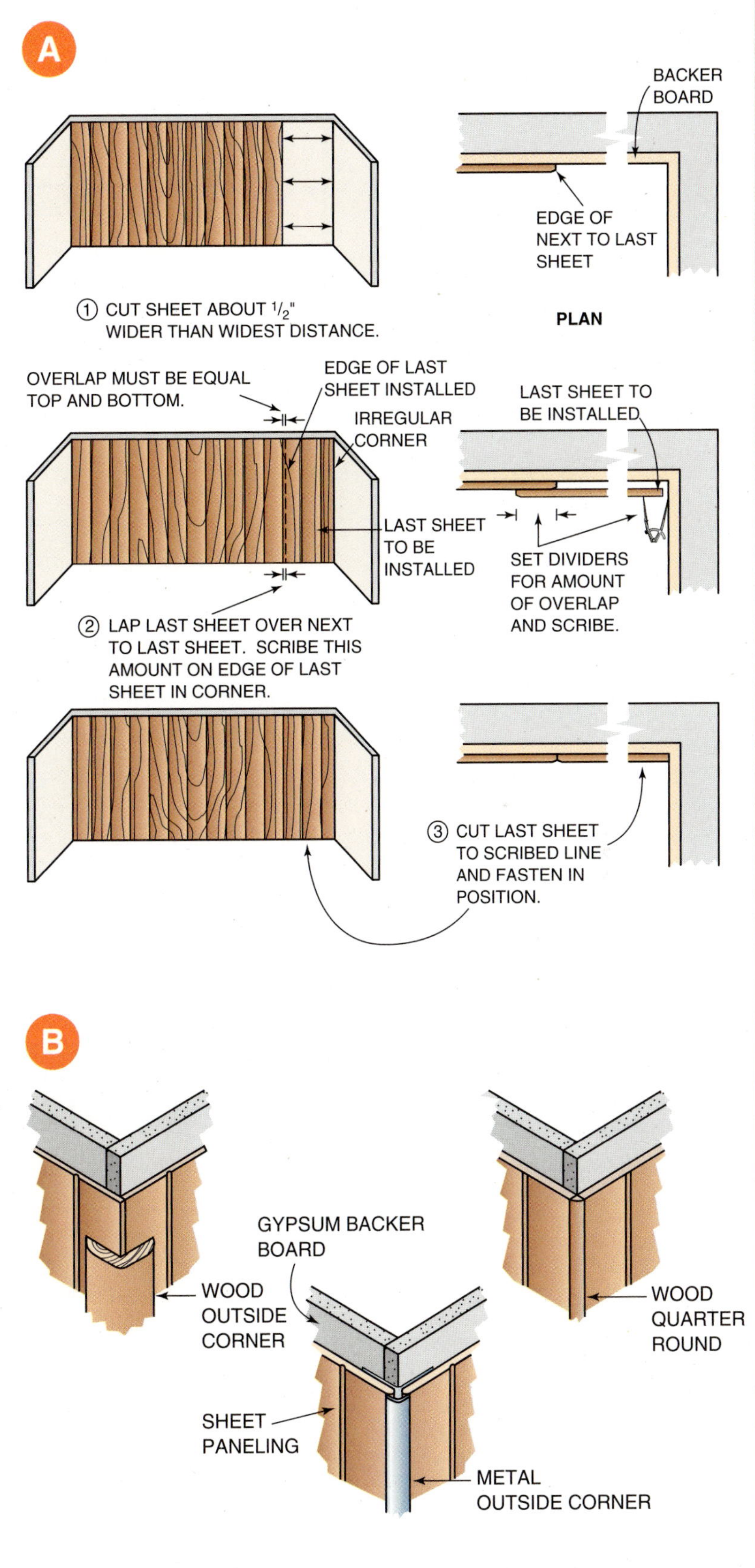

PROCEDURE 17-7

# Installing Solid Wood Board Paneling

## Starting the Application

A Select a straight board with which to start. Cut it to length, about ¼ inch less than the height of the wall. If tongue-and-grooved stock is used, tack it in a plumb position with the grooved edge in the corner.

- Adjust the scribers to scribe an amount a little more than the depth of the groove. Rip and plane to the scribed line.
- Replace the piece and face nail along the cut edge into the corner with finish nails about 16 inches apart. Blind nail the other edge through the tongue.
- Apply succeeding boards by blind nailing into the tongue only. Make sure the joints between boards come up tightly. Severely warped boards should not be used.
- As installation progresses, check the paneling for plumb. If out of plumb, gradually bring back by driving one end of several boards a little tighter than the other end. Cut out openings in the same manner as described for sheet paneling.

A

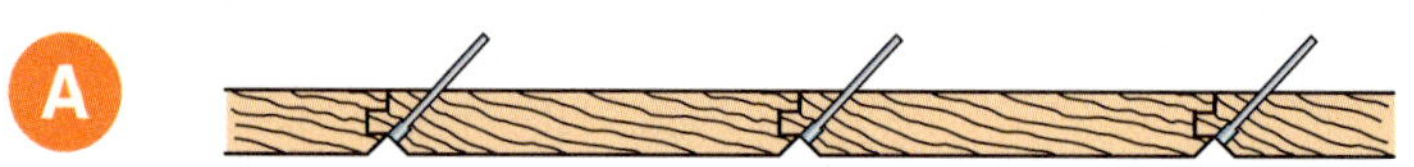

## Applying the Last Board

A Cut and fit the next to the last board. Then, remove it. Cut, fit, and tack the last board in the place of the next-to-the-last board just previously removed.

- Cut a scrap block about 6 inches long and equal in width to the finished face of the next-to-the last board. The tongue should be removed. Use this block to scribe the last board by running one edge along the corner and holding a pencil against the other edge.
- Remove the board from the wall. Cut and plane it to the scribed line. Fasten the next-to-the-last board in position. Fasten the last board in position with the cut edge in the corner.
- Face nail the edge nearest the corner.

A

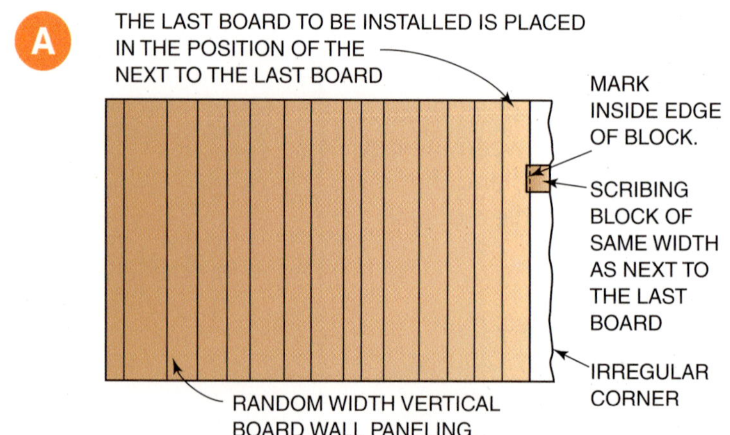

## REVIEW QUESTIONS

*Select the most appropriate answer.*

1. **Gypsum boards are also referred to as**
   a. plaster backing.
   b. drywall.
   c. wall panels.
   d. all of the above

2. **Tapered edges of a gypsum panel are designed to provide**
   a. a designer look.
   b. space for concealing wires.
   c. space for joint compound.
   d. a guide for cutting along the edges.

3. **Type X gypsum board is designed for greater**
   a. moisture protection.
   b. fire protection.
   c. backing support under predecorated panels.
   d. backing support under veneer plaster.

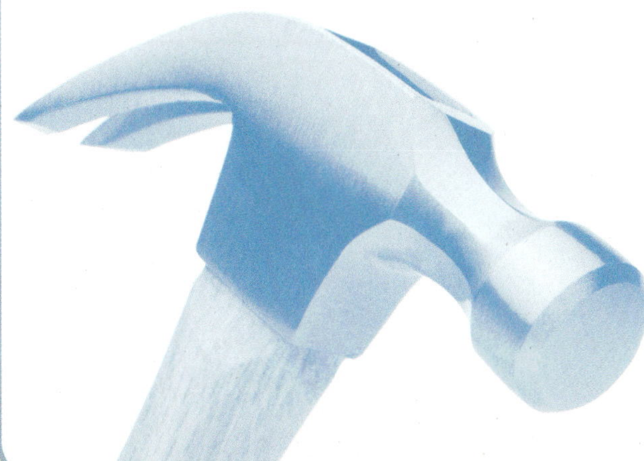

4. **A material designed to be used as a backing under ceramic tile is called**
   a. coreboard.
   b. wonder board.
   c. blue board.
   d. MR board.

5. **Fasteners in gypsum board should be**
   a. set below the surface in a dimple.
   b. galvanized nails.
   c. spaced 6 inches apart.
   d. all of the above

6. **Long ceiling joists are straightened and stiffened with**
   a. strongbacks.
   b. wood shims.
   c. blocking.
   d. drywall screws.

7. **The first drywall layer installed in a room should be the**
   a. ceiling.
   b. upper wall section.
   c. lower wall section.
   d. the order of installation does not matter

8. **A technique to keep drywall corners tight is to**
   a. use a slow-setting joint compound.
   b. leave out fasteners in the corners.
   c. use a drywall jack.
   d. all of the above

**9. Drywall finished to the highest level involve**

a. bedding paper tape in first compound coat.
b. applying a skim coat over the entire wall surface.
c. applying a second and third coat.
d. all of the above

**10. Outside corners of a gypsum wall is typically finished with**

a. metal corner beads.
b. fiberglass tape.
c. wood trim.
d. all of the above

**11. Manufacturer's scored lines on prefinished plywood paneling are always placed in from the edge**

a. 12, 16, and 24 inches.
b. 16, 24, and 32 inches.
c. 8, 12, and 24 inches.
d. 12, 16, and 32 inches.

**12. Solid wood paneling is available in**

a. softwood and hardwood species.
b. a rough sawn texture.
c. a Pickwick style.
d. all of the above

**13. A wainscoting is a wall finish**

a. applied diagonally.
b. applied partway up the wall.
c. used as a coating on prefinished wall panels.
d. used around tubs and showers.

**14. It is recommended that plywood paneling be installed**

a. with a ¾-inch gypsum backing.
b. paint applied under panel seams.
c. screws.
d. horizontally.

**15. If ¾-inch thick wood paneling is to be applied vertically over open studs, wood blocking must be provided between studs for nailing at maximum intervals of**

a. 16 inches.
b. 32 inches.
c. 24 inches.
d. 48 inches.

## WHAT'S WRONG WITH THIS PICTURE?

Carefully study **Figure 17-39** and the description below and think about what is wrong. Consider all possibilities.

### WRONG

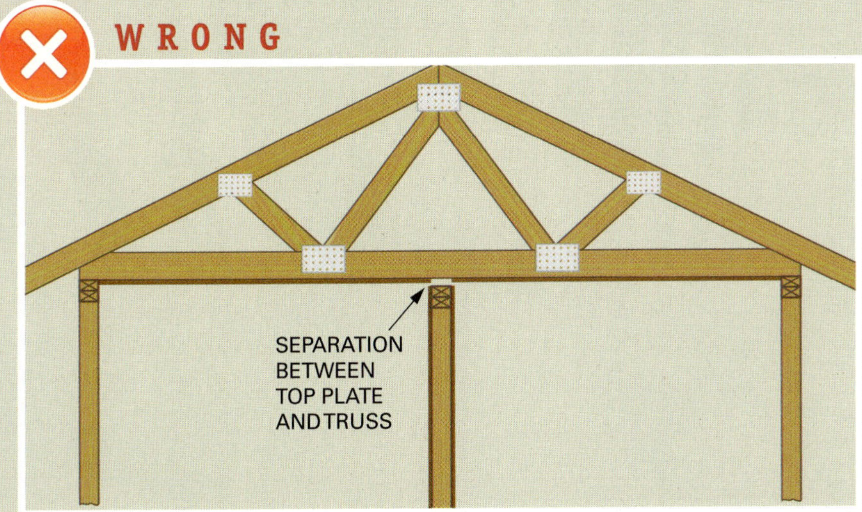


**FIGURE 17-39** This cross-sectional diagram shows a common situation where the truss and interior non-load-bearing partitions separate. The separation cycles up and down slowly through the seasonal changes in temperature and humidity. This movement cannot be stopped even with more nailing of the truss to the partition. It is referred to as truss uplift. The problem is that truss uplift will cause the drywall to crack at the ceiling and wall corner.

### RIGHT

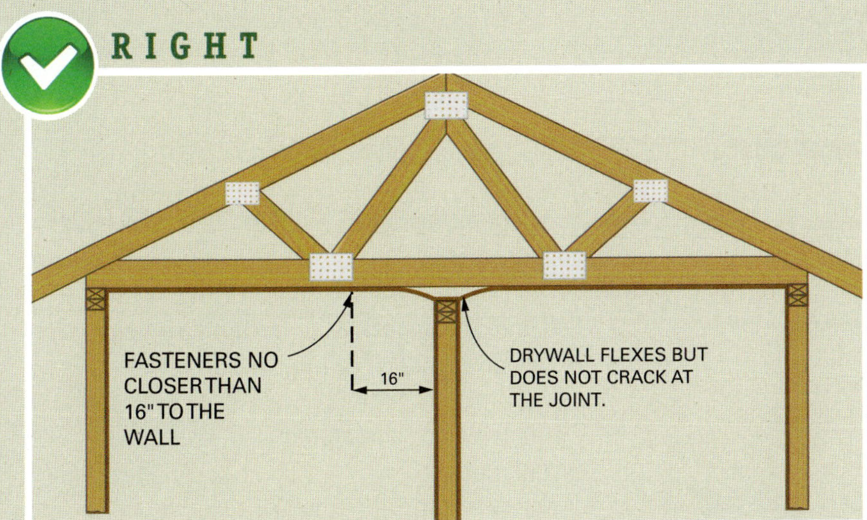


**FIGURE 17-40** A technique to deal with this problem is in placement of the fasteners. The ceiling drywall is not fastened to the truss any closer than 16 inches from the partition wall. This allows the drywall on the walls to support the drywall on the ceiling. As the truss gradually lifts up, the drywall corner remains tight and closed, flexing away from the truss.

CHAPTER 18

# Interior Finish

**OBJECTIVES** *After completing this unit, the student should be able to:*

- Identify the components of a suspended ceiling system.
- Lay out and install suspended ceilings.
- Identify standard interior moldings and describe their use.
- Apply ceiling and wall molding.
- Apply interior door casings, baseboard, base cap, and base shoe.
- Install window trim, including stools, aprons, jamb extensions, and casings.
- Apply strip and plank flooring.
- Estimate quantities of the parts in a suspended ceiling system.
- Estimate the quantities of molding needed for windows, doors, ceiling, and base.
- Estimate wood flooring required for various installations.

# GLOSSARY OF INTERIOR FINISH TERMS

**apron** a piece of the window trim used under the stool

**back miter** an angled cut starting from the end going back into the face of the molding

**compound miter** a bevel cut across the width and also through the thickness of a piece

**spline** a thin, flat strip of wood inserted into the grooved edges of adjoining pieces

**stool** the bottom horizontal member of interior window trim that serves as the finished window sill

Interior finish is a term used to describe the process of completing the interior of a building. It collectively refers to installing the finished floor, finished ceiling, and moldings. Moldings are strips used to cover the seams or joints between different building materials, such as drywall and window jambs.

A finished ceiling may be created by installing a suspended ceiling. This type of ceiling is more often used in commercial construction but is sometimes used in residential. Suspended ceilings may be installed in new construction beneath exposed joists or in remodeling below existing ceilings.

Interior moldings have names that relate to their specific location and application. Casing refers to molding applied along the perimeter of windows and doors. Baseboard and crown moldings are installed on the floor and ceiling, respectively. Chair rails are installed on the wall to protect it from the backs of chairs. These moldings are often made of wood, but some are made of plastic or metal.

Floors may be finished with wood or wood products. Wood floors are long-time favorites because of their durability, beauty, and warmth.

# Ceiling Finish

*Suspended ceilings* provide space for recessed lighting, ducts, pipes, and other necessary conduits (Fig. 18-1). In remodeling work, a suspended system can be easily installed beneath the existing ceiling. In basements, where overhead pipes and ducts may make other types of ceiling application difficult, a suspended type is easily installed. In addition, removable panels make pipes, ducts, and wiring accessible.

## SUSPENDED CEILING COMPONENTS

A suspended ceiling system consists of panels that are laid into a metal grid. The grid consists of main runners, cross tees, and wall angles. It is constructed in a 2 × 4-foot rectangular or 2 × 2-foot square pattern (Fig. 18-2). Grid members come prefinished in white, black, brass, chrome, and wood grain patterns, among others.

### Wall Angles

*Wall angles* are L-shaped pieces that are fastened to the wall to support the other components of the ceiling system. They come in 10- and 12-foot lengths. They provide a continuous finished edge around the perimeter of the ceiling, where it meets the wall.

Courtesy of Armstrong World Industries.

**FIGURE 18-1** Installing panels in a suspended ceiling grid.

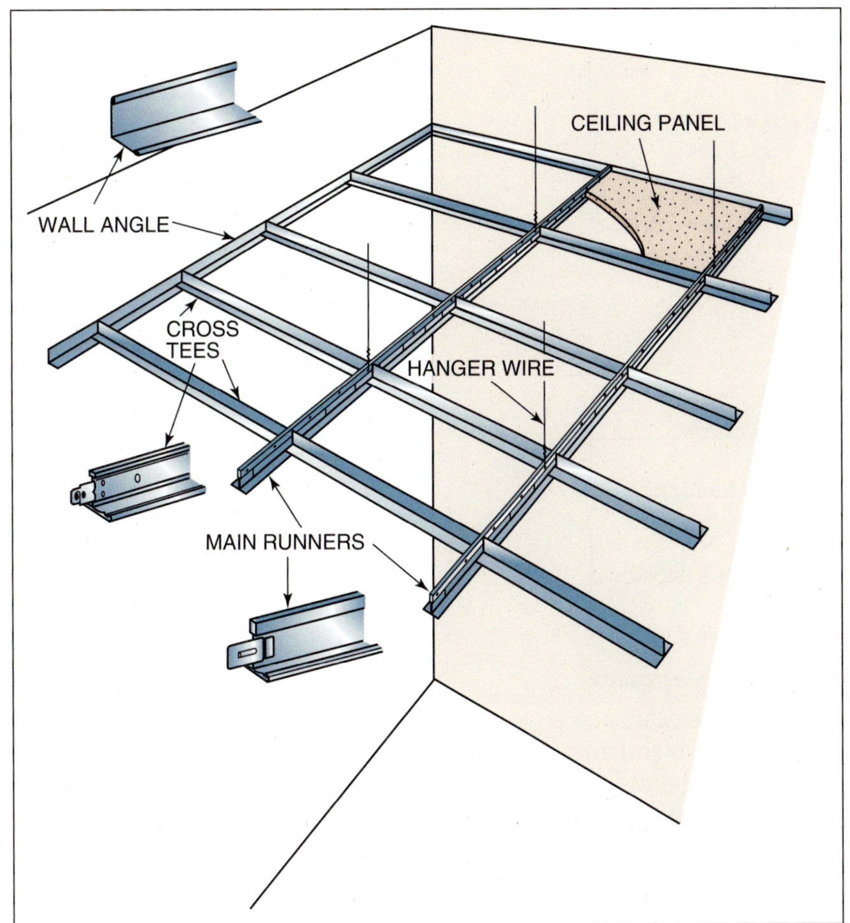


**FIGURE 18-2** A suspended ceiling consists of grid members and ceiling panels.

### Main Runners

*Main runners* or *tees* are shaped in the form of an upside-down T. They come in 12-foot lengths. End splices make it possible to join lengths of main runners together. Slots are punched in the side of the runner at 6- or 12-inch intervals to receive cross tees. Along the top edge, punched holes are spaced at intervals for suspending main runners with *hanger* wire. Main runners extend from wall to wall, usually across the length of the room.

### Cross Tees

*Cross tees* come in 2- and 4-foot lengths. A slot, of similar shape and size as those in main runners, is centered on 4-foot cross tees for use when making a 2 × 2 foot grid pattern. They come with connecting tabs on each end. These tabs are inserted and locked into main runners and other cross tees.

**CAUTION**

**CAUTION:** Care should be observed when working with the metal grid system around electrical wiring. The entire grid could become charged with one misplaced wire.

### Ceiling Panels

*Ceiling panels* are manufactured from many different kinds of material, such as gypsum, glass fibers, mineral fibers, and wood fibers. Panel selection is based on considerations such as fire resistance, sound control, thermal insulation, light reflectance, moisture resistance, maintenance, appearance, and cost. Panels are given a variety of surface textures, designs, and finishes. They are available in 2 × 2 and 2 × 4 foot sizes with square or rabbeted edges (Fig. 18-3).

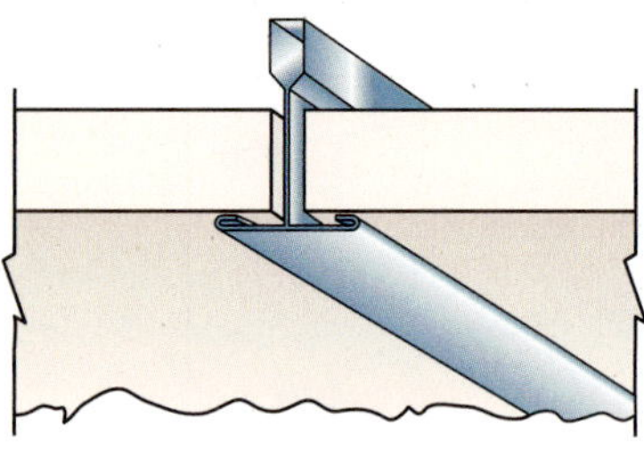


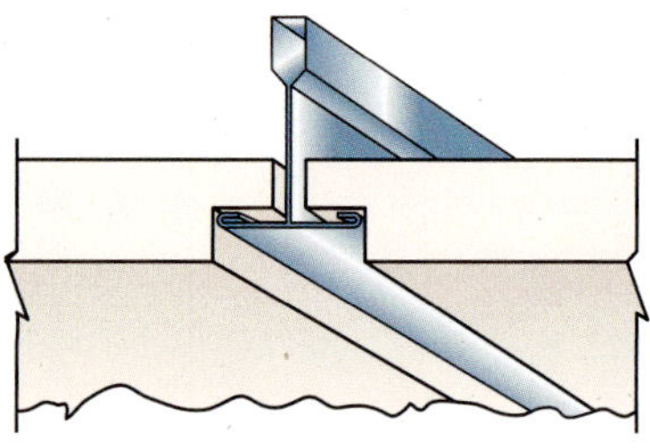


**FIGURE 18-3** Suspended ceiling panels may have square or rabbeted edges and ends.

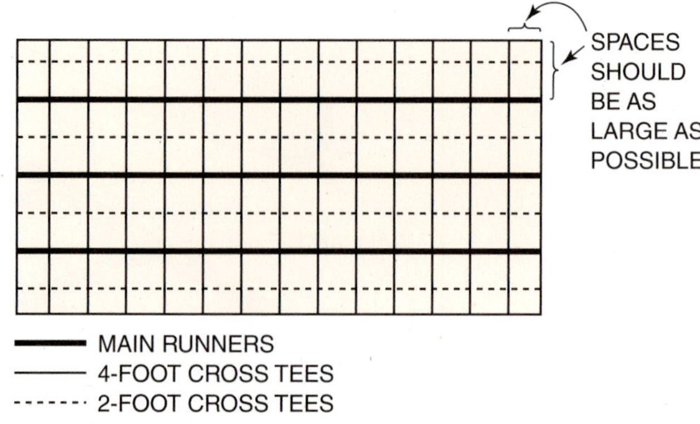


**FIGURE 18-4** A grid system with 4-foot cross tees spaced 2 feet apart along main runners is the recommended method of constructing a 2 × 4-foot grid. A 2 × 2-foot grid is made by installing 2-foot cross tees between the midpoints of the 4-foot cross tees.

## SUSPENDED CEILING LAYOUT

Before the actual installation of a suspended ceiling, a scaled sketch of the ceiling grid should be made. The sketch should indicate the direction and location of the main runners, cross tees, light panels, and border panels. Main runners usually are spaced 4 feet apart. They may run parallel or perpendicular to the long dimension of the room. For a standard 2 × 4 pattern, 4-foot cross tees are then spaced 2 feet apart between main runners. If a 2 × 2 pattern is used, then 2-foot cross tees are installed between the midpoints of the 4-foot cross tees. Main runners and cross tees should be located in such a way that *border panels* on both sides of the room are equal and as large as possible (Fig. 18-4). Sketching the ceiling layout helps when estimating materials.

### Sketching the Layout

Sketch a grid plan by first drawing the overall size of the ceiling to a convenient scale. Use special care in measuring around irregular walls.

### Locating Main Runners

To locate main runners, change the measurement of the room to inches and divide by 48, the length of a ceiling tile. Add 48 inches to any remainder. Divide the sum by 2 to find the distance from the wall to the first main runner. This distance also represents the dimension of border panels.

**EXAMPLE:** The width of the room is 15'–8", or 188 inches. Dividing 188 by 48 equals 3, with a remainder of 44. Adding 48 to 44 equals 92. Dividing 92 by 2 equals 46 inches, which is the distance from the wall to the first main runner (Fig. 18-5).

### Locating Cross Tees

Locating 4-foot cross tees between main runners involves a similar process as that used for main runners. First change the dimension of the ceiling to inches. Divide by 24. Add 24 to the remainder. Divide the sum by 2 to find the distance of the cross tee from the wall.

**EXAMPLE:** The dimension of a room is 27'–10", or 334 inches. Dividing 334 by 24 equals 13, with a remainder of 22. Adding 24 to 22 equals 46. Dividing 46 by 2 equals 23, which is the distance from the wall to the first row of cross tees (Fig. 18-6).

## CONSTRUCTING THE GRID CEILING SYSTEM

The ceiling grid is constructed by first installing *wall angles,* then installing *suspended ceiling lags* and *hanger wires,* suspending the main runners, inserting full-length cross tees and, finally, cutting

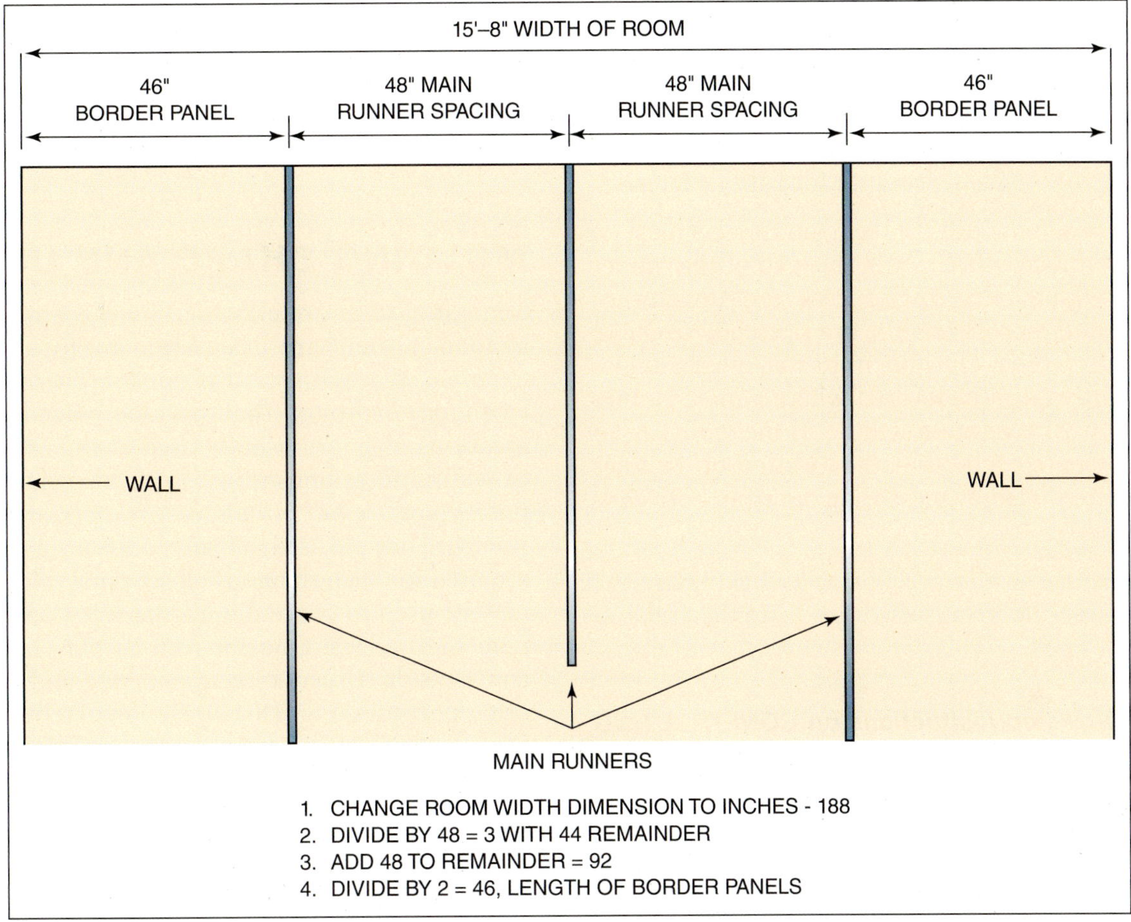


**FIGURE 18-5** Method of determining the location of main runners.

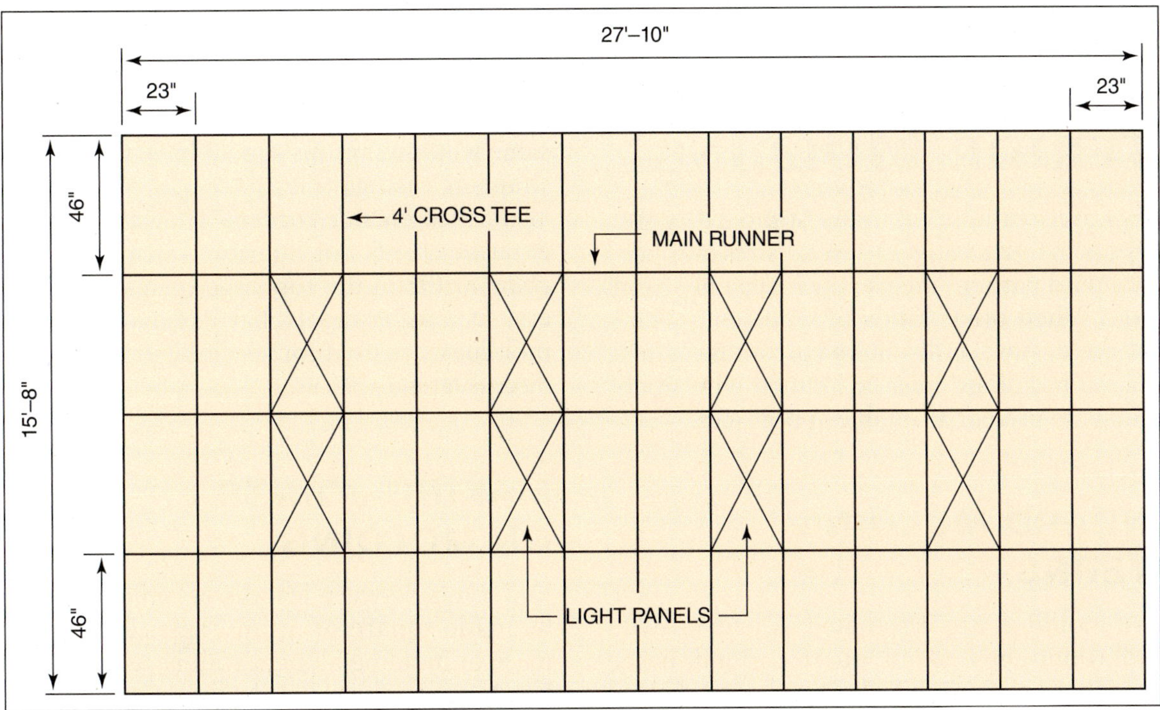


**FIGURE 18-6** Completed sketch of a suspended ceiling layout.

and inserting border cross tees. A suspended ceiling must be installed with at least 3 inches for clearance below the lowest air duct, pipe, or beam. This clearance provides enough room to insert ceiling panels in the grid. If recessed lighting is to be used, allow a minimum of 6 inches clearance for light fixtures.

The beginning ends of the main runners must be cut so that a cross tee slot in the web of the runner lines up with the first row of cross tees. A main runner line and a cross tee line are stretched to make this process easier. The lines must run at right angles to each other and be located where a main runner and a row of cross tees will be. If the walls are at right angles to each other, the location of the lines can be determined by measuring out from both ends of the walls.

When the walls are not at right angles, the Pythagorean theorem is used to square the grid system (Fig. 18-7).

***For step-by-step instructions on CONSTRUCTING THE GRID CEILING SYSTEM, see Procedure 18-1 on pages 617–622.***

**Green Tip**

Sorting job site waste for recycling saves money and makes environmental sense.

# Interior Molding

Moldings are available in many standard types. Each type is manufactured in several sizes and patterns. Standard patterns are usually made only from softwood. When other kinds of wood or special patterns are desired, mills make custom moldings to order. All moldings must be applied with tight-fitting joints to present a suitable appearance.

## STANDARD MOLDING PATTERNS

Standard moldings are designated as bed, crown, cove, full round, half round, quarter round, base, base shoe, base cap, casing, chair rail, back band, apron, stool, stop, and others (Fig. 18-8). Molding usually comes in lengths of 8, 10, 12, 14, and 16 feet. Some moldings are available in odd lengths. Door casings, in particular, are available in lengths of 7 feet to reduce waste. *Finger-jointed* lengths are made of short pieces joined together. These are used when a paint finish is to be applied.

## MOLDING SHAPE AND USE

Some moldings are classified by the way they are shaped. Others are designated by location. For example, *beds, crowns,* and *coves* are terms related to shape. Although they may be placed in other locations, they are usually used at the intersection of walls and ceilings (Fig. 18-9). Also classified by their shape are *full rounds, half rounds,* and *quarter rounds.* They are used in many locations. Full rounds are used for such things as closet poles. Half rounds may be used to conceal joints between panels or to trim shelf edges. Quarter rounds may be used to trim outside corners of wall paneling and for many other purposes (Fig. 18-10). Designated by location, *base, base shoe,* and *base cap* are moldings applied at the bottom of walls where they meet the floor. When square-edge base is used, a base cap is usually applied to its top edge. Base shoe is normally used to conceal the joint between the bottom of the base and the finish floor (Fig. 18-11).

*Casings* are used to trim around windows, doors, and other openings. They cover the space between the frame and the wall. *Back bands* are applied to the outside edges of casings for a more decorative appearance. **Aprons**, **stools**, and *stops* are parts of window trim. Stops are also applied to door frames for the doors to stop against when closed. On the same window, aprons should have the same molded shape as casings (Fig. 18-12). *Corner guards* are also called *outside corners.* They are used to finish exterior corners of interior wall finish. *Caps* and *chair rail* trim the top edge of wainscoting. Others, such as astragals, battens, panel, and picture moldings, are used for specific purposes, as we will discuss later.

## MAKING JOINTS ON MOLDING

End joints between lengths of ceiling molding may be made square or at an angle. Usually, the last piece of molding along a wall is cut slightly longer. It is bowed outward in the center, then pressed into place when the ends are in position. This makes the joints come up tighter.

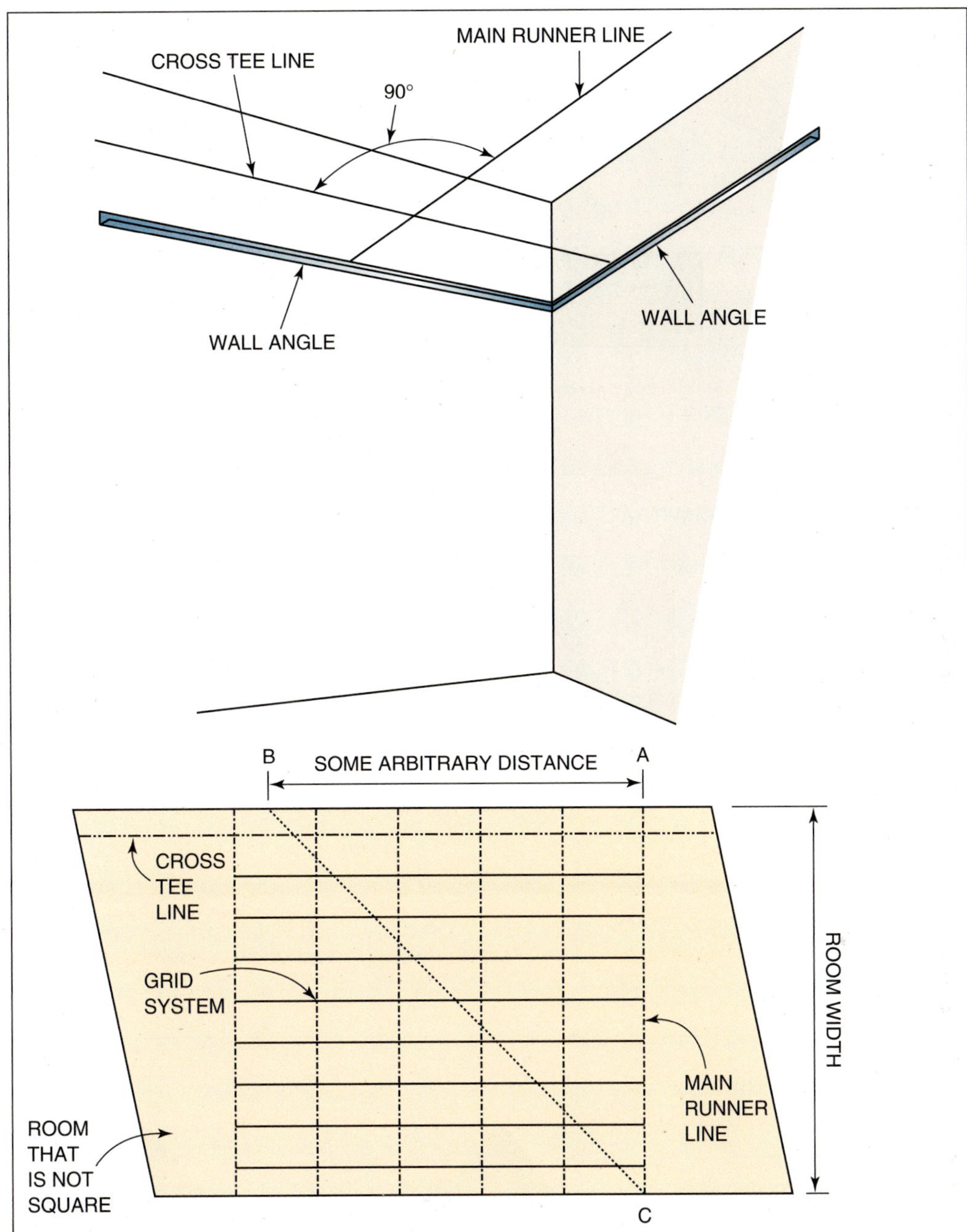


1) Starting at point A, measure the room width keeping your tape as square as possible with the wall.
2) From A measure some distance to B. The actual distance does not matter, only that it should be large enough to make a big triangle.
3) Use these numbers in the Pythagorean theorem to determine the distance from B to C.
4) Measure and mark the distance from B to C. C is now square with point A.
5) Connect A and C with a string and measure each successive row of main runner from it.
6) **EXAMPLE:** If AC, the room width, is 18'–9" then measure AB to be, say 16'–0". Convert these dimension to inches. 16'–0" becomes 192" and 18'–9" becomes 225" (18 X 12 + 9 = 225"). Put these dimensions in the Pythagorean theorem. $C^2 = A^2 + B^2$

$C = \sqrt{192^2 + 225^2} = \sqrt{36864 + 50625} = \sqrt{87489} = 295.7853952"$

To convert the decimal to sixteenths

$0.7853952 \times 16 = 12.566$ sixteenths $\Rightarrow {}^{13}/_{16}"$

Thus, the measurement from B to C is $295^{13}/_{16}$ inches.

**FIGURE 18-7** Stretch a cross tee line at a right angle to the main runner line. Use the Pythagorean theorem to determine square.

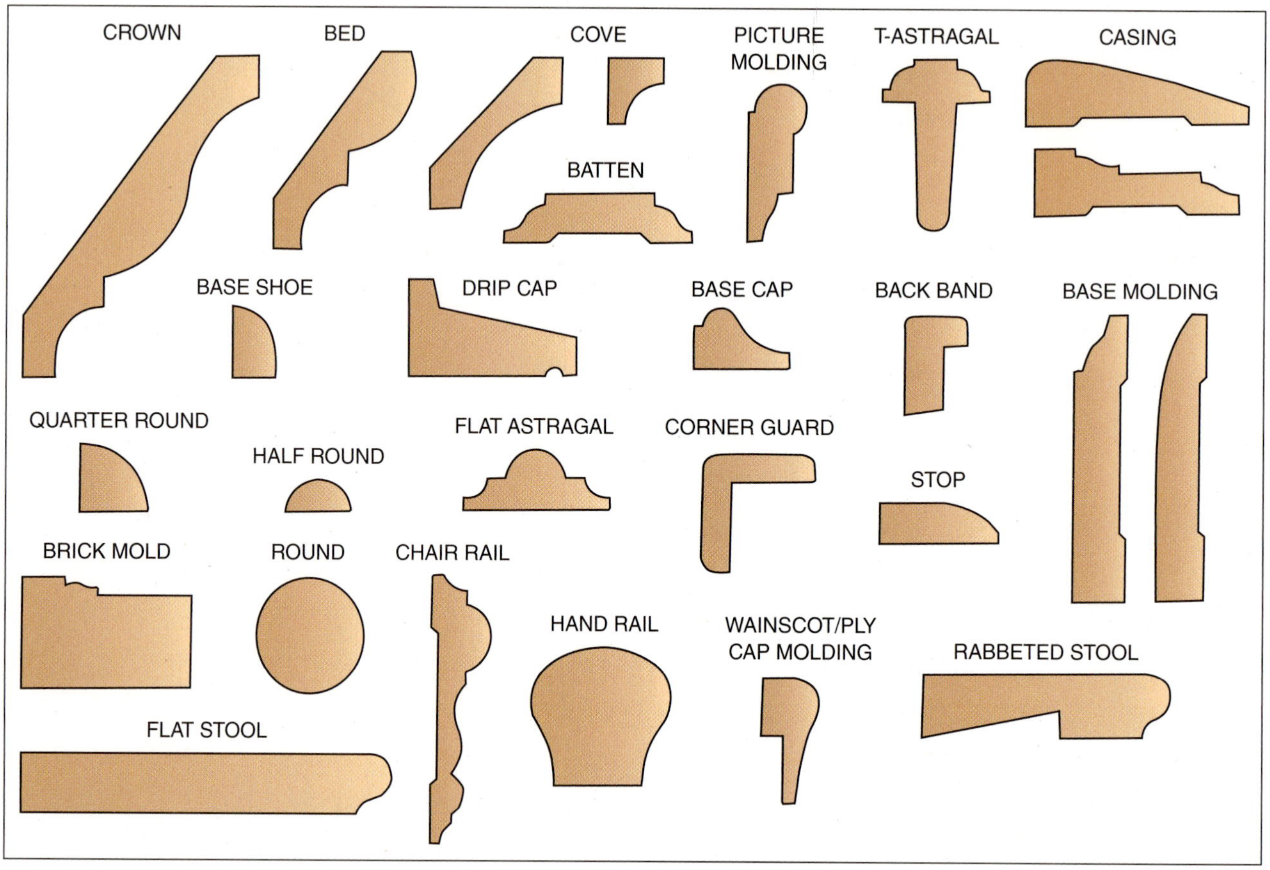

CROWN
BED
COVE
PICTURE MOLDING
T-ASTRAGAL
CASING
BATTEN
BASE SHOE
DRIP CAP
BASE CAP
BACK BAND
BASE MOLDING
QUARTER ROUND
HALF ROUND
FLAT ASTRAGAL
CORNER GUARD
STOP
BRICK MOLD
ROUND
CHAIR RAIL
HAND RAIL
WAINSCOT/PLY CAP MOLDING
RABBETED STOOL
FLAT STOOL

*Courtesy of Wood Molding and Millwork Producers, Inc.*

**FIGURE 18-8** Standard molding patterns.

WALL
CEILING
BED
CROWN
COVE

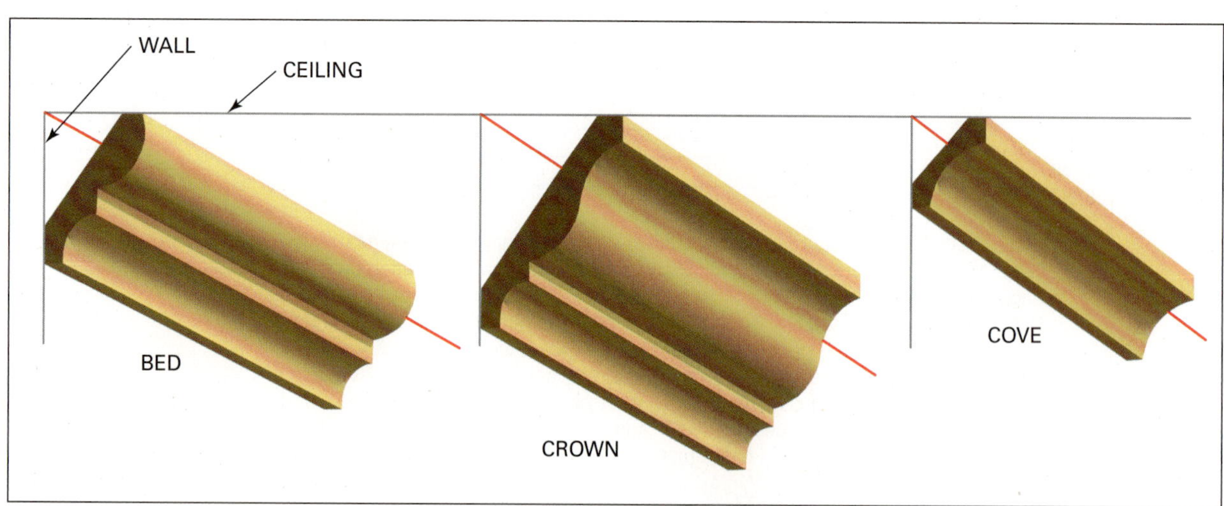

**FIGURE 18-9** Bed, crown, and cove moldings are often used at the intersections of walls and ceilings.

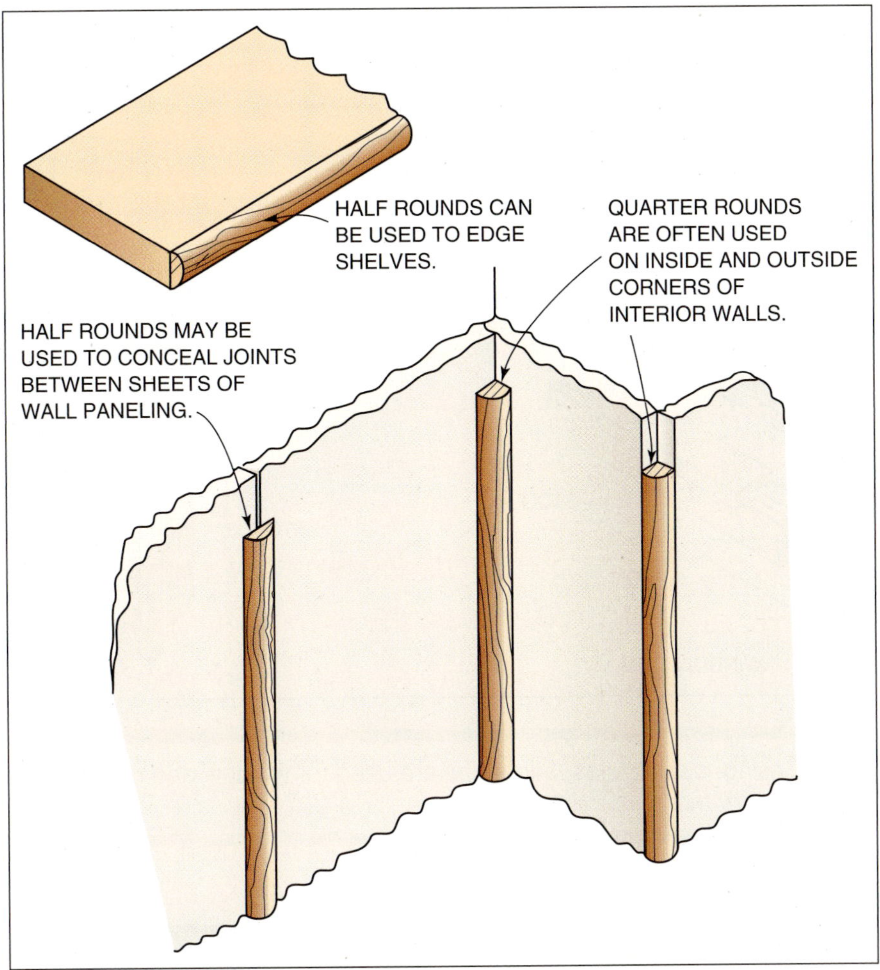


**FIGURE 18-10** Half round and quarter round moldings are used for many purposes.

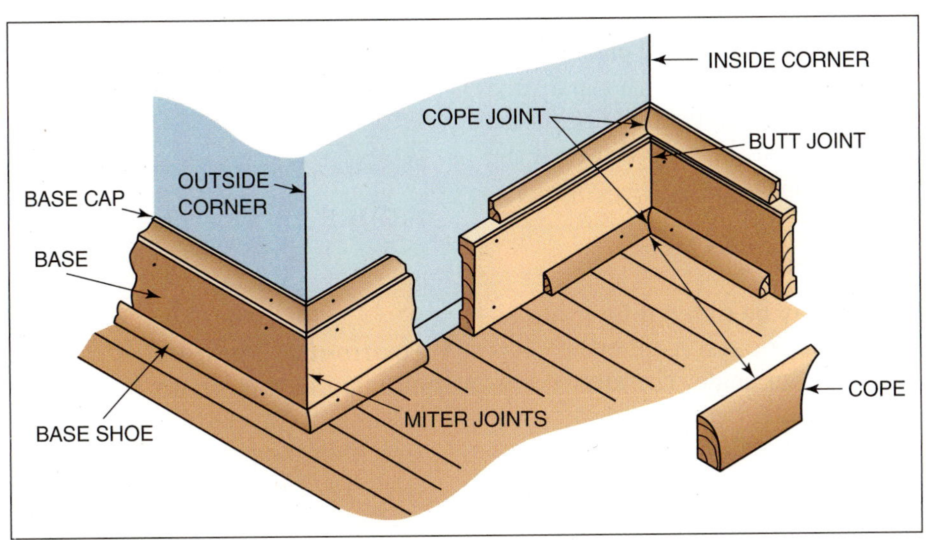


**FIGURE 18-11** Base, base shoe, and base cap may be used to trim the bottom of the wall.

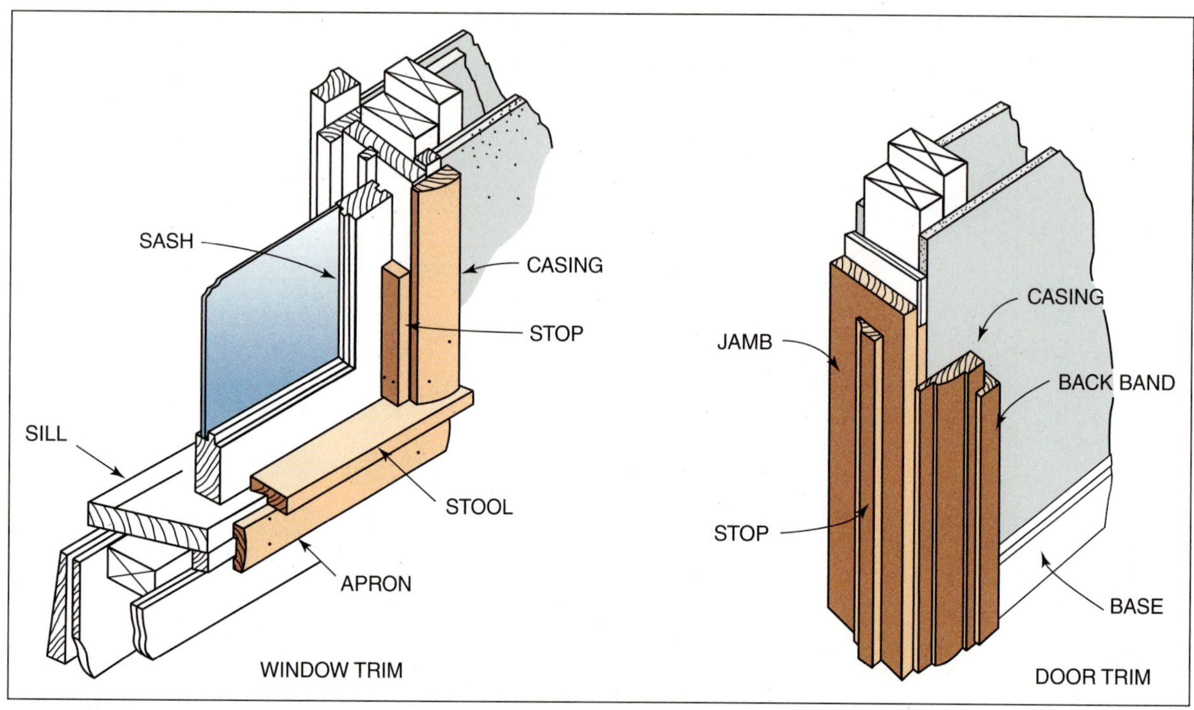


**FIGURE 18-12** Casing, back bands, and stops are used for window and door trim. Stools and aprons are part of window trim.

After the molding has been fastened, joints between lengths should be sanded flush, except on prefinished moldings. Failure to sand butted ends flush with each other results in a shadow being cast at the joint line. This gives the appearance of an open joint.

Joints on exterior corners are mitered. Joints on interior corners are usually coped. A coped joint is made by fitting the first piece on one wall with a square end butting into the corner. The end of the molding on the other wall is cut to fit against the shaped face of the molding on the first wall (Fig. 18-13).

**FIGURE 18-13** A coped joint is made by fitting the end of one piece of molding against the shaped face of the other piece.

## Methods of Cutting Trim

Moldings of all types may be mitered by using either hand or power miter boxes. A miter box is a tool that cuts material at an angle. The most popular way to cut miters and other cuts is with a power miter box (Fig. 18-14). These tools allow for cuts that are accurate and fast. With this tool, a carpenter can cut virtually any angle with ease, whether it is a simple or a **compound miter**. Fine adjustments to a piece of trim, +/− 1/64 inch, can be made with great speed and accuracy.

***Positioning Molding in the Miter Box.*** Placing molding in the correct position in the miter box is essential for accurate mitering. All moldings are cut with their face sides or finished edges up toward the operator. This is done because the saw splinters out the back side of the piece during the cutting process. Position the molding with one back side or edge against the bottom of the miter box and the other against the backside of the miter box.

**FIGURE 18-14** Using a power miter box.

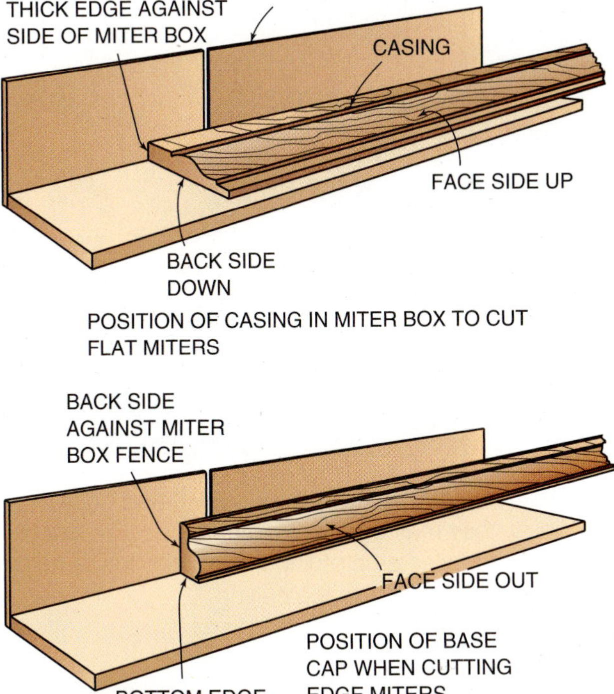


**FIGURE 18-15** Position of molding in a miter box to cut flat and edge miters.

Flat miters are cut by holding the molding with its face side up, its backside resting on the base, and its thicker edge against the fence of the miter box. Some moldings, such as base, base cap and shoe, and chair rail, are held right side up. Their bottom edge should be against the bottom of the miter box and their back against the side of the miter box (Fig. 18-15).

***Mitering Bed, Crown, and Cove Molding.*** Bed, crown, and cove moldings require a compound miter, which may be cut two ways. One is to use a compound miter saw with the manufacturer's instructions. The compound angle is cut with the molding held flat on the base of the saw and the blade adjusted for two angles. The saw has built-in stops to make accurate blade adjustments (Fig. 18-16).

The other method involves positioning the trim piece upside down in the miter box. It is helpful to imagine the base of the miter box as the ceiling and the fence as the wall. The trim top edge is placed against the base of the miter box and the wall edge against the fence (Fig. 18-17).

***Making a Coped Joint.*** Coped joints are used on interior corners. They will not open up when the molding is nailed in place. Miter joints may open up in interior corners when the ends are fastened.

To cope the end of molding, first make a miter cut on the end as if the joint was to be an inside miter (Fig. 18-18). The edge of the cut along the face forms the profile of the cope. Rub the side of a pencil point lightly along the profile to outline it more clearly. Use a coping saw with a fine-tooth blade to cut along the outlined profile with a slight undercut. Hold the side of the molding that

**FIGURE 18-16** Large ceiling moldings may be mitered while being held flat on the base. The saw is set to the compound angle determined from the instruction manual for the saw.

**FIGURE 18-17** To miter bed, crown, and cove moldings, they may also be positioned upside down in the miter box. It is helpful to think of the fence as the wall and the base as the ceiling.

**FIGURE 18-18** Making a back miter on a piece of cove molding.

will touch the wall flat on the top of the sawhorse. Cut, keeping the blade plumb or slightly under cutting (Fig. 18-19).

***For step-by-step instructions on CUTTING A COPED JOINT, see Procedure 18-2 on page 623.***

## INSTALLATION THOUGHTS

- Sand smoothly all pieces of interior trim after they have been cut and fitted and before they are fastened. The sanding of interior finish

**FIGURE 18-19** Coping the end of a cove molding.

provides a smooth base for the application of stains, paints, and clear coatings. Always finish sand with the grain.

- Round slightly all sharp, exposed corners of trim. Use a block plane to make a slight chamfer. Then, round over with sandpaper.
- If the trim is to be stained, make sure every trace of glue is removed. Excess glue, allowed to dry, seals the surface. It does not allow the stain to penetrate, resulting in a blotchy finish.
- Use pneumatic nailers whenever possible for fast, accurate work. If using a hammer, be careful not to make hammer marks in the finish. Occasionally rubbing the face of the hammer with a fine sandpaper to clean it helps prevent it from glancing off the head of a nail.
- Make sure any pencil lines left along the edge of a cut are removed before fastening the pieces. Pencil marks in interior corners are difficult to remove after the pieces are fastened in position. Pencil marks show through a stained or clear finish and make the joint appear open. When marking interior trim, make light, fine pencil marks.
- Make sure all joints are tight-fitting. Measure, mark, and cut carefully. Do not leave a poor fit. Do it over, if necessary.

## APPLYING WALL MOLDING

To apply chair rail, caps, or some other molding located on the wall, chalk lines should be snapped. This ensures that molding is applied in a straight line. No lines need to be snapped for base moldings or for small-size moldings applied at the intersection of walls and ceiling. For large-size ceiling moldings, such as beds, crowns, and coves, a chalk line ensures straight application of the molding and easier joining of the pieces.

***For step-by-step instructions on APPLYING WALL MOLDING, see Procedure 18-3 on pages 624–625.***

# DOOR CASINGS

*Door casings* are moldings applied around the door opening. They trim and cover the space between the door frame and the wall. Because door casings extend to the floor, they must be applied before any base moldings because the base butts against the edge of the door casing (Fig. 18-20).

## DESIGN OF DOOR CASINGS

Moldings or *S4S* stock may be used for door casings. S4S is the abbreviation for *surfaced four sides.* It is used to describe smooth, square-edge lumber.

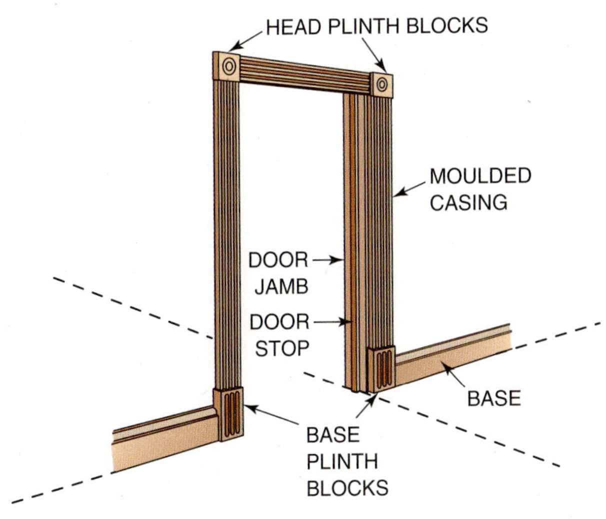


**FIGURE 18-21** Molded casings are mitered at the head unless plinth blocks are used.

When molded casings are used, the joint at the head must be mitered unless butted against plinth blocks. *Plinth blocks* are small decorative blocks. They are thicker and wider than the door casing. They are used as part of the door trim at the base and at the head (Fig. 18-21). When using S4S lumber, the joint may be mitered or butted. If a butt joint is used, the head casing overlaps the side casing. The appearance of S4S casings and some molded casings may be enhanced with the application of *back bands* (Fig. 18-22). Molded casings usually have their back

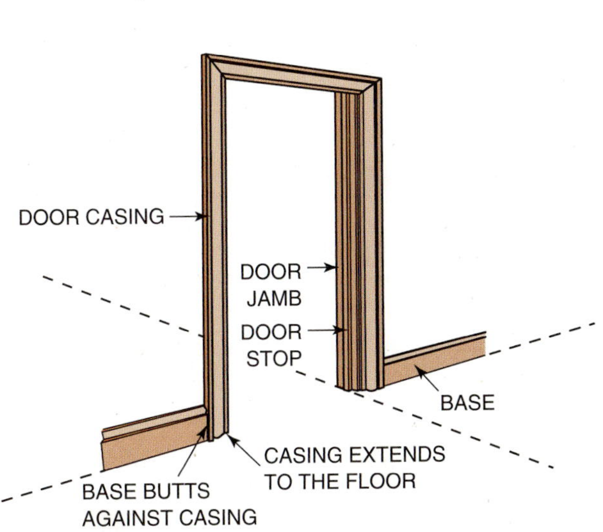


**FIGURE 18-20** Door casings are applied before the base is installed.

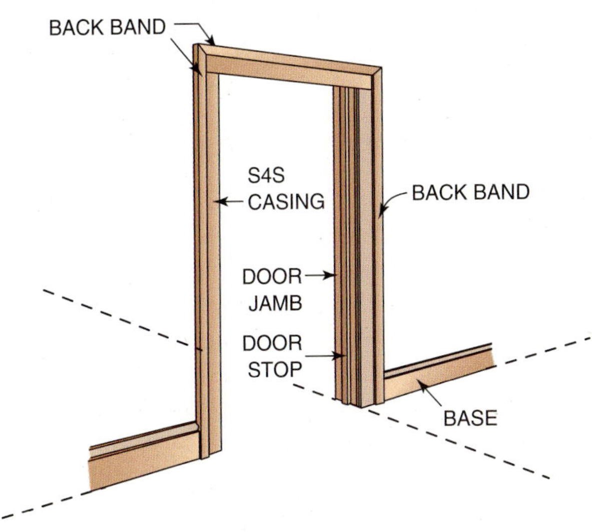


**FIGURE 18-22** Back bands may be applied to improve the appearance of S4S door casings. They may also be used on some types of molded casings.

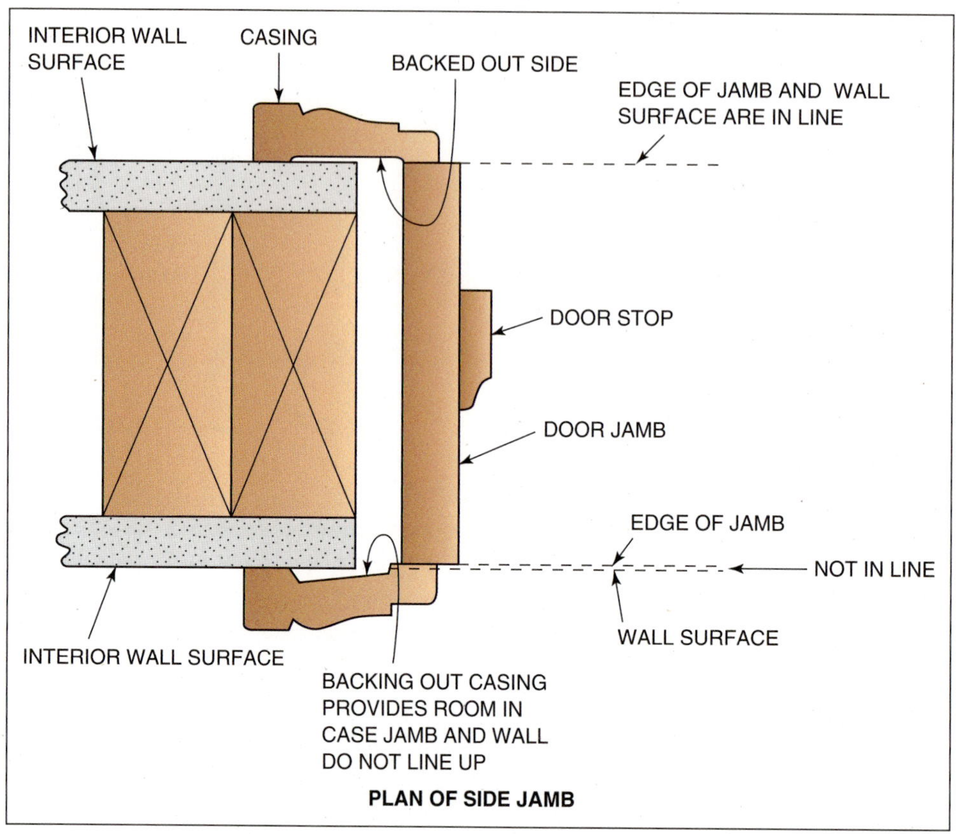


**FIGURE 18-23** Backing out door casings allows for a tight fit on wall and jamb.

sides backed out. In cases where the jamb edges and the wall surfaces may not be exactly flush with each other, the backed out surfaces allow the casing to come up tight on both wall and jamb (Fig. 18-23).

### APPLYING DOOR CASINGS

Door casings are set back from the inside face of the door frame a distance of 1/4 to 5/16 inch. This allows room for the door hinges and the striker plate of the door lock. This setback is called a *reveal* (Fig. 18-24). The reveal also improves the appearance of the door trim. Rough lengths of trim are a few inches longer than actually needed. For each interior door opening, four side casings and two head casings are required. Cut side casings in pairs with right- and left-hand miters for use on both sides of the opening.

If the casing edge is thin, use 3d or 4d finish nails spaced about 12 inches apart. Pneumatic finish nailers speed up the job of fastening interior trim. The outside edge is thicker, so longer nails are used, usually 6d or 8d finish nails. They are spaced farther apart, about 16 inches OC.

***For step-by-step instructions on APPLYING DOOR CASINGS, see Procedure 18-4 on pages 626–627.***

## BASE TRIM

Molded or S4S stock may be used for base. A *base cap* may be applied to its top edge. The base cap conforms easily to the wall surface, resulting in a tight fit against the wall. The base trim should be thinner than the door casings against which it butts. This makes a more attractive appearance.

The base is applied in a manner similar to wall and ceiling molding. Begin so that the last trim piece will be in the least visible corner. Apply the base to the first wall with square ends in each corner. Then, working around the room, cope each corner. Drive and set two finishing nails of sufficient length at each stud location.

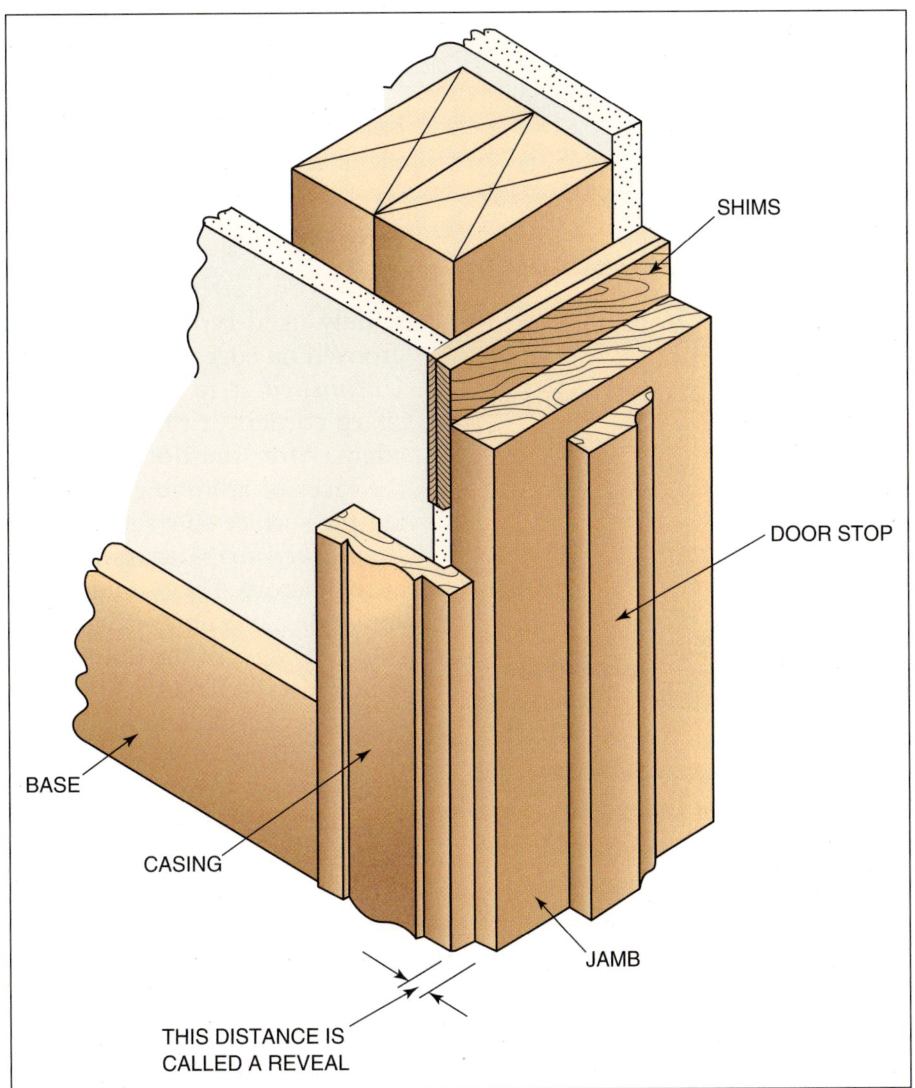


**FIGURE 18-24** The setback of the door casing on the jamb is called a reveal.

***For step-by-step instructions on APPLYING BASE MOLDINGS, see Procedure 18-5 on pages 628–629.***

## WINDOW TRIM

Interior window trim consists of the stool, apron, jamb extensions, and casings. Although the kind and amount of trim may differ, depending on the style of window, the application is basically the same (Fig. 18-25).

The bottom side of the *stool* may be rabbeted at an angle to fit on the window sill so its top side will be level. Its final position has the outside edge nearly touching the sash. Both ends are notched around the side jambs of the window frame. Each end projects beyond the casings by an amount equal to the casing thickness. The stool length is equal to the distance between the outside edges of the vertical casing plus twice the stock on the wall plus twice the largest thickness of the casing. Its inside edge should be flush with the inside face of the side jamb of the window frame. The apron covers the joint between the bottom of the sill and the wall. It is applied with its ends in line with the outside edges of the window casing. The face is **back mitered** or returned upon itself.

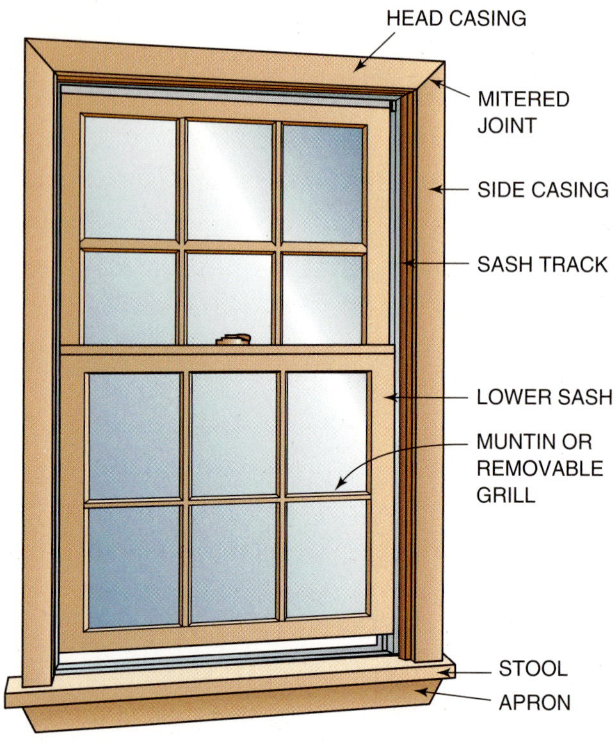


**FIGURE 18-25** Components of window trim.

Windows are often installed with jambs that are narrower than the wall thickness. Wood strips must be fastened to these narrow jambs to extend the jamb inside edges flush to the inside wall surface. These strips are called *jamb extensions* or *extension jambs*. Some manufacturers provide jamb extension pieces with the window unit, which are installed later or attached at the factory.

Window casings usually are installed with a reveal similar to that of door casings. They also may be installed flush with the inside face. In either case, the bottom ends of the side casings rest on the stool. The window casing style or pattern is usually the same as the door casings.

***For step-by-step instructions on INSTALLING WINDOW TRIM, see Procedure 18-6 on pages 630–632.***

# FINISH FLOORS

Most hardwood finish flooring is made from white or red oak. Beech, birch, hard maple, and pecan finish flooring are also manufactured. For less-expensive finish floors, some softwoods such as Douglas fir, hemlock, and southern yellow pine are used.

## KINDS OF SOLID HARDWOOD FLOORING

Solid hardwood floors typically comes in strip or plank styles. The major difference is their widths.

### Strip Flooring

Solid wood *strip* flooring is probably the most widely used type. Most strips are tongue-and-grooved on edges and ends to fit precisely together. *Unfinished* strip flooring is milled with square, sharp corners at the intersections of the face and edges. After the floor is laid, any unevenness in the faces of adjoining pieces is removed by sanding the surface so strips are flush with each other. *Prefinished* strips are sanded, finished, and waxed at the factory. They should not be sanded after installation. A *chamfer* is machined between the face side and edges of the flooring prior to prefinishing. When installed, these chamfered edges form small V-grooves between adjoining pieces. This obscures any unevenness in the surface. The most popular size of hardwood strip flooring is 3/4 inch thick with a *face width* of 2 1/4 inches. The face width is the width

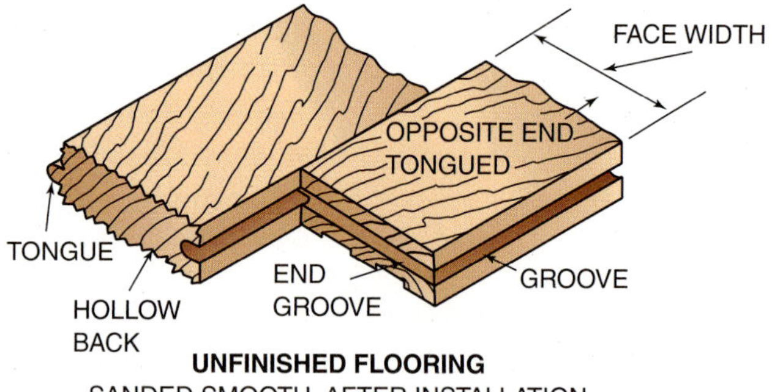


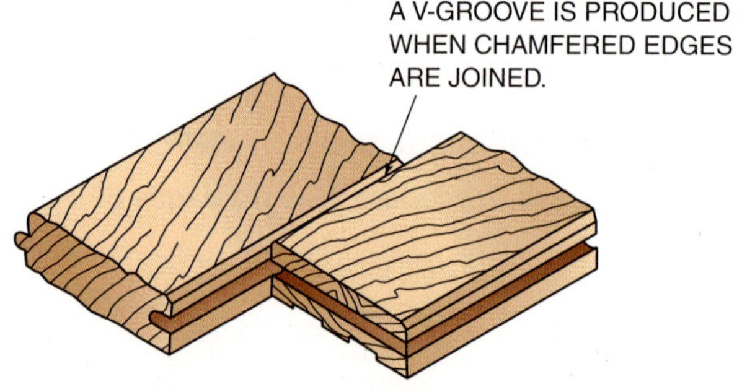


**FIGURE 18-26** Hardwood strip flooring is edge- and end-matched. The edges of prefinished flooring are chamfered.

of the exposed surface between adjoining strips. It does not include the tongue. Other thicknesses and widths are also manufactured (Fig. 18-26).

### Plank Flooring

Solid wood *plank* flooring is similar to strip flooring. However, it comes in various width combinations ranging from 3 to 8 inches. Plank flooring may be laid with alternating widths. Like strips, planks are available unfinished or prefinished. The edges of some prefinished planks have deeper chamfers to accentuate the plank widths. The surface of some prefinished plank flooring may have plugs of contrasting color already installed to simulate screw fastening. One or more plugs, depending on the width of the plank, are used across the face at each end (Fig. 18-27). Unfinished plank flooring comes with either square or chamfered edges and with or without plugs. The planks may be bored for plugs on the job, if desired.

## GRADES OF HARDWOOD FLOORING

Uniform grading rules have been established for strip and plank solid hardwood flooring by the National Oak Flooring Manufacturers Association. The association's trademark on flooring assures consumers that the flooring is manufactured and graded in compliance with established quality standards.

Oak flooring is available *quarter-sawed* and *plain-sawed.* The grades for unfinished oak flooring, in declining order, are *clear, select, no. 1 common,* and *no. 2 common.* Quarter-sawed flooring is available in clear and select grades only.

Birch, beech, and hard maple flooring are graded in declining order as *first grade, second grade, third grade,* and *special grade.* Grades of pecan flooring are *first grade, first grade red, first grade white, second grade, second grade red,* and *third grade.* Red grades contain all heartwood. White grades are all bright sapwood. In addition to appearance, grades are based on length. Flooring comes in bundles with an assortment of lengths. Some strip flooring comes with a hardwood prefinished veneer on a plywood core. No strip flooring pieces are shorter than 9 inches.

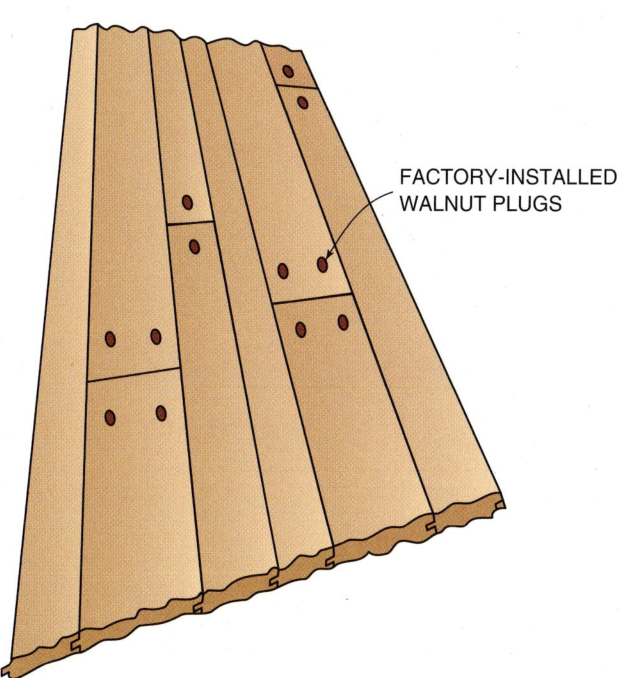


**FIGURE 18-27** Plank flooring usually is applied in rows of alternating widths. Plugs of contrasting color simulate screw fastening.

## LAYING WOOD FLOOR

Lumber used in the manufacture of hardwood flooring has been kiln-dried, cooled, and then accurately machined to exacting standards. It is a fine product that should receive proper care during handling and installation.

Strip and plank flooring are similar to install. In new construction, the base or door casings are not usually applied to allow for easier application of the finish floor. In remodeling, the base and base shoe must be removed and the casing bottoms cut off. Use a scrap piece of finish flooring as a guide on which to lay a handsaw.

Wood flooring laid in the direction of the longest dimension of the room gives the best appearance. The flooring may be laid in either direction on a plywood or diagonal board subfloor. Yet it must be laid perpendicular to subfloor boards that run at right angles to the joists (Fig. 18-28).

### Handling and Storage of Wood Finish Floor

Maintain moisture content of the flooring by observing recommended procedures. Flooring should not be delivered to the job site until the building has been enclosed. Outside windows and doors should be in place. Cement work, plastering, and other materials must be thoroughly dry. In warm seasons, the building should be well ventilated. During cold months, the building should be heated, not exceeding

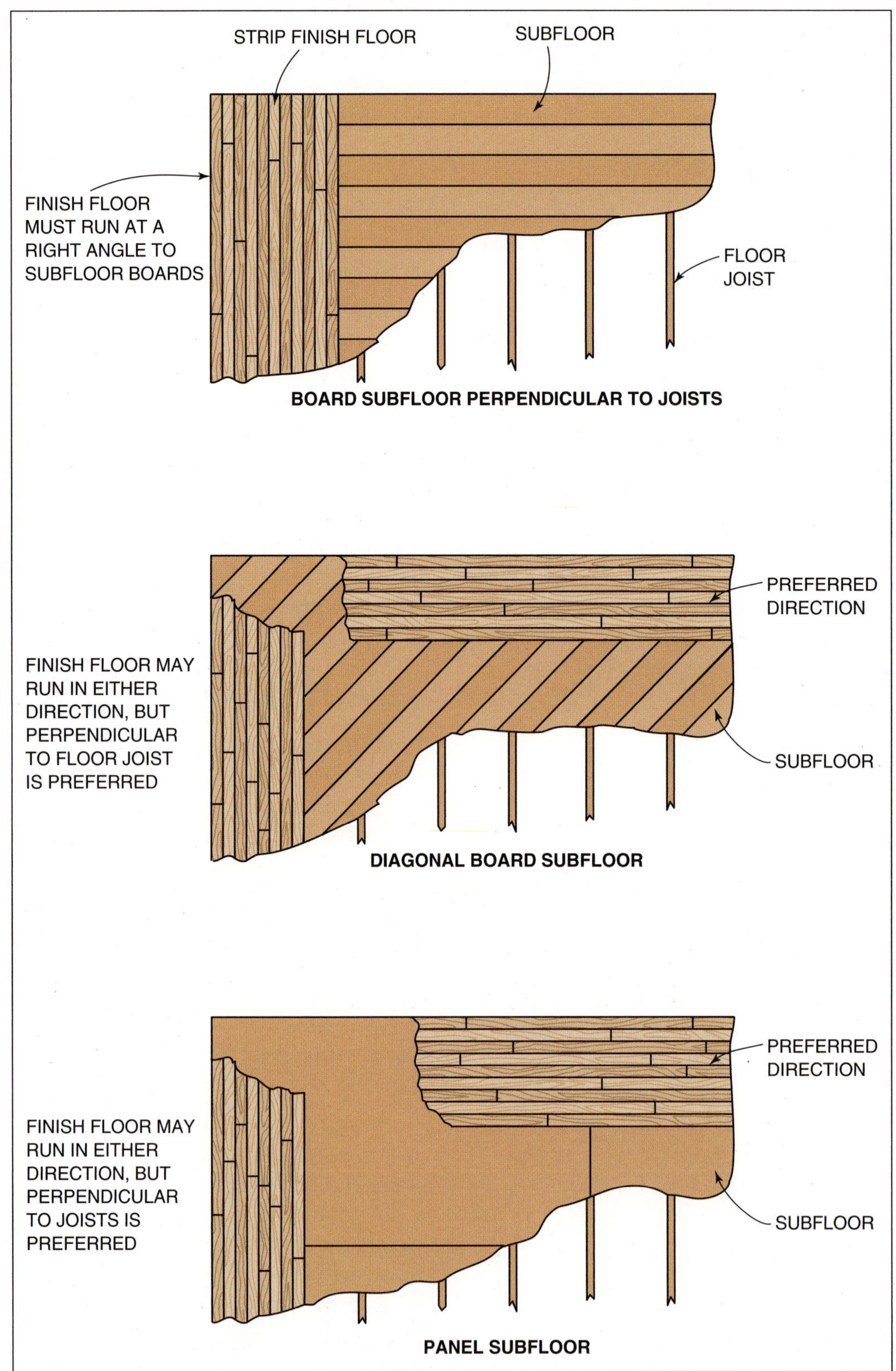


**FIGURE 18-28** Several factors determine the direction in which strip flooring is laid.

72 degrees Fahrenheit, for at least 5 days before delivery and until flooring is installed and finished. Do not unload flooring in the rain. Stack the bundles in the rooms where the flooring is to be laid. Leave adequate space around the bundles for good air circulation. Let the flooring become acclimated to the atmosphere of the building for 4 or 5 days or more before installation. Some manufacturers recommend the flooring be installed immediately upon delivery. Their material is stored in climate-controlled rooms, keeping it drier than it will be when installed. The tongue and groove edges, which are machined to close tolerances, will go together more easily.

### Concrete Slab Preparation

Wood finish floors can also be installed on a *concrete slab*. Wood floors should not be installed on below-grade slabs. New slabs should be at least 90 days old. Flooring should not be installed when tests indicate excessive moisture in the slab.

A test can be made by laying a smooth rubber mat on the slab. Put weight on it to prevent moisture from escaping. Allow the mat to remain in place for 24 hours. If moisture shows when the mat is removed, the slab is too wet. If no moisture is present, the slab is ready.

Grind off any high spots. Fill low spots with leveling compound. The slab must be free of grease, oil, or dust.

### Applying a Nailing Surface

A *plywood subfloor* may be installed over the moisture barrier on which to fasten the finish floor. Exterior grade sheathing plywood of at least ¾-inch thickness is used. The plywood is laid with staggered joints. Leave a ¾-inch space at walls and ¼- to ½-inch space between panel edges and ends (Fig. 18-29).

Finish flooring may also be fastened to *sleepers* installed on the slab. They must be pressure treated and dried to a suitable moisture content. Usually, 2 × 4 lumber is used. Sleepers are laid 12" OC on their side and cemented or nailed to the slab. A polyethylene vapor retarder is then placed over the sleepers. The edges are lapped over the rows (Fig. 18-30). With end-matched flooring, end joints need not meet over the sleepers.

### Blind Nailing

Flooring is *blind nailed* by driving nails at about a 45-degree angle through the flooring. Occasionally,

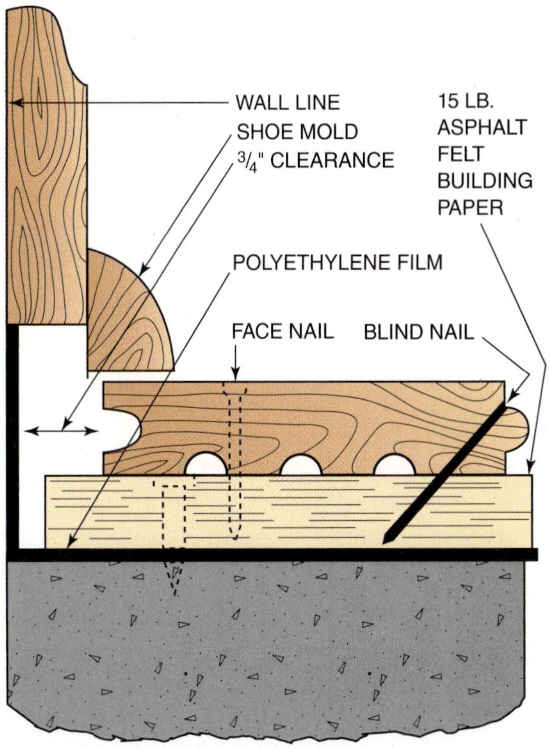


**FIGURE 18-29** Installation details of a plywood subfloor over a concrete slab.

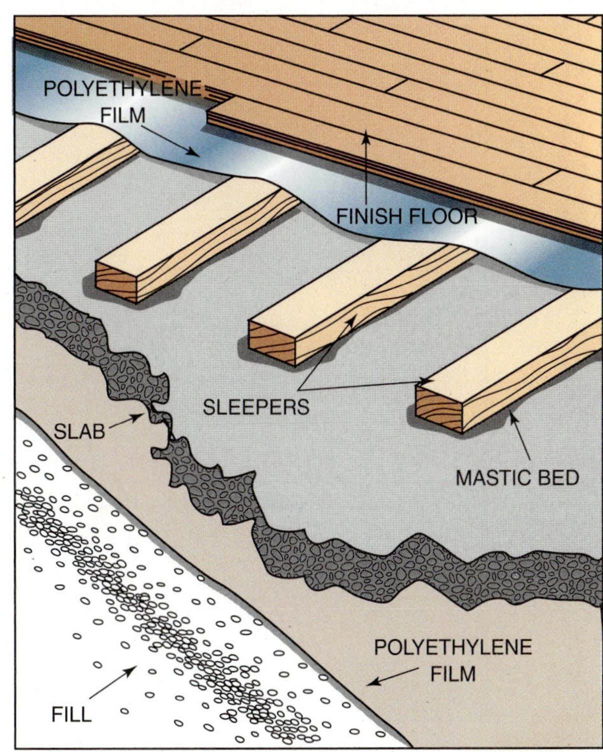


**FIGURE 18-30** Sleepers are cemented to a concrete slab. They provide fastening for strip or plank finish flooring.

some strips of flooring must be nailed by hand when the floor nailer cannot be used. This is particularly true during the first few rows of flooring. Start the nail in the corner at the top edge of the tongue. Usually, $2\frac{1}{4}$-inch hardened cut or spiral screw nails are used. Recommendations for fastening are shown in Figure 18-31. Care must be taken not to let the hammer glance off the nail. This may damage the edge of the flooring.

After the nail is driven home, its head must be set slightly. The head of the next nail to be driven is used to set the nail just driven. A nail is laid on edge with its head on the nail to be set. With one sharp blow, the nail is set (Fig. 18-32). The setting nail is then the next nail to be driven. In this manner, the floor layer maintains a smooth, continuous motion when fastening flooring.

**CAUTION**

**CAUTION:** Eye protection should be worn whenever driving nails, particularly hardened nails. A small piece of steel may break off the hammer or the nail and fly out in any direction. This could cause serious eye injury.

**Note:** A nail set should not be used to set hardened flooring nails. If used, the tip of the nail set will be flattened, thus rendering the nail set useless. Do not lay the nail set flat along the tongue, on top of the nail head, and then set the nail by hitting the side of the nail set with a hammer. Not only is this method slower, but it invariably breaks the nail set, possibly causing an injury.

| STRIP T & G<br>SIZE FLOORING | SIZE NAIL TO BE USED | BLIND NAIL SPACING ALONG THE LENGTH OF STRIPS. MINIMUM TWO NAILS PER PIECE NEAR THE ENDS. (1"–3") |
|---|---|---|
| $\frac{3}{4}$ X $1\frac{1}{2}$", $2\frac{1}{4}$", AND $3\frac{1}{4}$" | 2" SERRATED EDGE BARBED FASTENER, $2\frac{1}{4}$" OR $2\frac{1}{2}$" SCREW OR CUT NAIL, 2" 15 GAUGE STAPLES WITH $\frac{1}{2}$" CROWN. | IN ADDITION–10–12" APART–8–10" PREFERRED |
| ON SLAB WITH $\frac{3}{4}$" PLYWOOD SUBFLOOR USE $1\frac{1}{2}$" BARBED FASTENER, $\frac{1}{2}$" PLYWOOD SUBFLOOR WITH JOISTS A MAXIMUM 16" OC, FASTEN INTO EACH JOIST WITH ADDITIONAL FASTENING BETWEEN, OR 8" APART. | | |
| MUST INSTALL ON A SUBFLOOR | | |
| $\frac{1}{2}$ X $1\frac{1}{2}$" AND 2" | $1\frac{1}{2}$" SERRATED EDGE BARBED FASTENER, $1\frac{1}{2}$" SCREW, CUT STEEL, OR WIRE CASING NAIL. | 10" APART $\frac{1}{2}$" FLOORING MUST BE INSTALLED OVER A MINIMUM $\frac{5}{8}$" THICK PLYWOOD SUBFLOOR. |
| $\frac{3}{8}$ X $1\frac{1}{4}$" AND 2" | $1\frac{1}{4}$" SERRATED EDGE BARBED FASTENER, $1\frac{1}{2}$" BRIGHT WIRE CASING NAIL. | 8" APART |
| SQUARE-EDGE FLOORING | | |
| $\frac{5}{16}$ X $1\frac{1}{2}$" AND 2" | 1" 15 GAUGE FULLY BARBED FLOORING BRAD. | TWO NAILS EVERY 7" |
| $\frac{5}{16}$ X $1\frac{1}{3}$" | 1" 15 GAUGE FULLY BARBED FLOORING BRAD. | ONE NAIL EVERY 5" ON ALTERNATE SIDES OF STRIP |
| PLANK $\frac{3}{4}$ X 3" TO 8" | 2" SERRATED EDGE BARBED FASTENER, $2\frac{1}{4}$" OR $2\frac{1}{2}$" SCREW, OR CUT NAIL, USE $1\frac{1}{2}$" LENGTH WITH $\frac{3}{4}$" PLYWOOD SUBFLOOR ON SLAB. | 8" APART |
| FOLLOW MANUFACTURER'S INSTRUCTIONS FOR INSTALLING PLANK FLOORING | | |
| WIDTHS 4" AND OVER MUST BE INSTALLED ON A SUBFLOOR OF $\frac{5}{8}$" OR THICKER PLYWOOD OR $\frac{3}{4}$" BOARDS. ON SLAB USE $\frac{3}{4}$" OR THICKER PLYWOOD. | | |

*Courtesy of National Oak Flooring Manufacturers.*

**FIGURE 18-31** Nailing guide for strip and plank finish flooring.

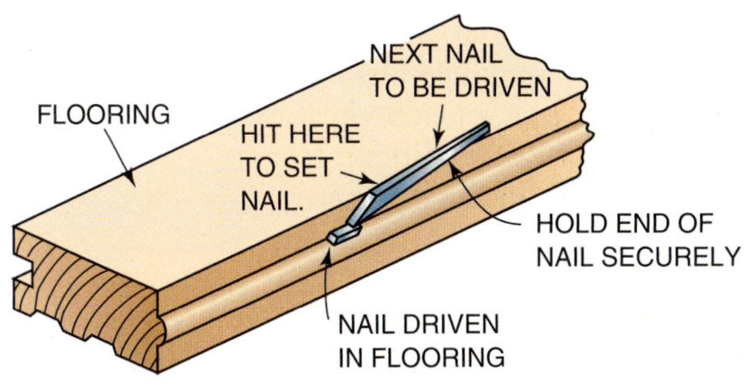


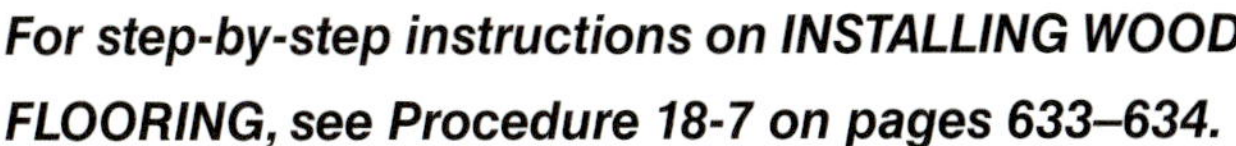

**FIGURE 18-32** Method used by floor layers to set nails driven by hand.

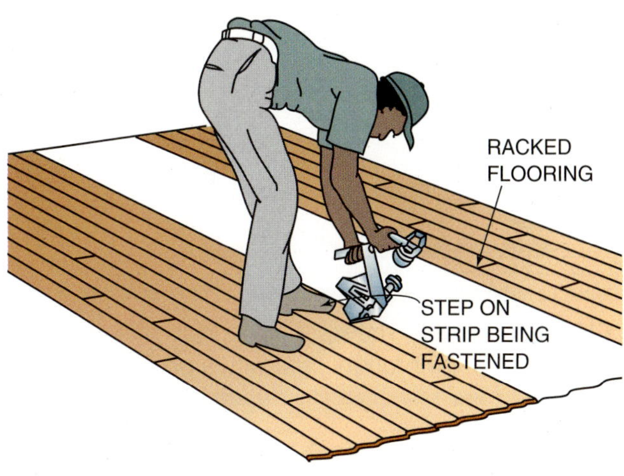


**FIGURE 18-34** It is important to push the board down and together tightly while nailing.

***For step-by-step instructions on INSTALLING WOOD FLOORING, see Procedure 18-7 on pages 633–634.***

### Using the Power Nailer

At least two courses of flooring must be laid by hand to provide clearance from the wall before a power nailer can be used. The power nailer holds strips of special barbed fasteners. The fasteners are driven and set through the tongue of the flooring at the proper angle. Although it is called a power nailer, a heavy hammer is swung by the operator against a plunger to drive the fastener (Fig. 18-33). Some nailers are pneumatically driven. Hammers are double-headed. One head is rubber and the other is steel. The flooring strip is placed in position. The rubber head of the hammer is used to drive the edges and ends of the strips up tight. The steel head is used against the plunger of the power nailer to drive the fasteners. Drive fasteners about 8 to 10 inches apart or as needed to bring the strip up tight against previously laid strips.

**FIGURE 18-33** A power nailer is widely used to fasten strip flooring.

Whether laying floor with a power nailer or driving nails with a hammer, the floor layer stands with heels on strips already fastened and toes on the loose strip being fastened. By shifting the appropriate weight with the toes, easier alignment of the tongue and groove is possible (Fig. 18-34). The weight of the worker also prevents the loose strip from bouncing when it is being inserted into the groove. Avoid using different nailing techniques such as a power nailer, pneumatic nailer, and hammer-driven fasteners on the same strip of flooring. Each method of fastening places the strips together with varying degrees of tightness. This variation, compounded over multiple strips, will cause waves in the straightness of the flooring.

### Ending the Flooring

The last three or four courses from the opposite wall must be nailed by hand. This is because of limited room to place the power nailer and swing the hammer. The next-to-the-last row can be blind nailed if care is taken. However, the flooring must be brought up tightly by prying between the flooring and the wall. Use a bar to pry the pieces tight at each nail location (Fig. 18-35). The last course is

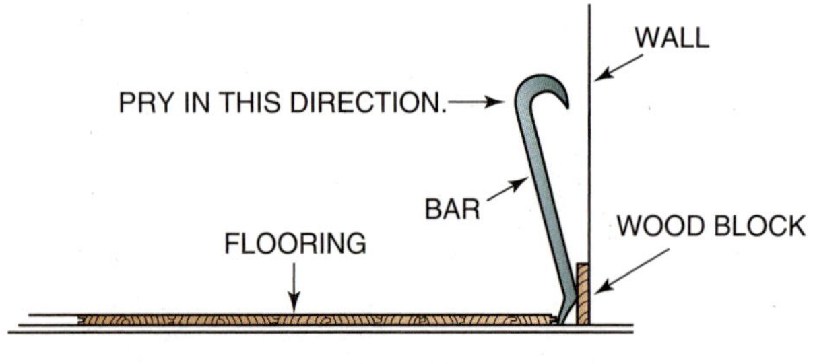


**FIGURE 18-35** The last two courses of strip flooring must be brought tight with a pry bar.

installed in a similar manner. However, it must be face nailed. It may need to be ripped to the proper width. If it appears that the installation will end with an undesirable, difficult-to-apply, narrow strip, lay wider strips in the last row (Fig. 18-36).

### Framing around Obstructions

A much more professional and finished look is given to a strip flooring installation if *hearths* and other floor obstructions are framed. Use flooring, with mitered joints at the corners, as framework around the obstructions (Fig. 18-37).

## *from experience...*

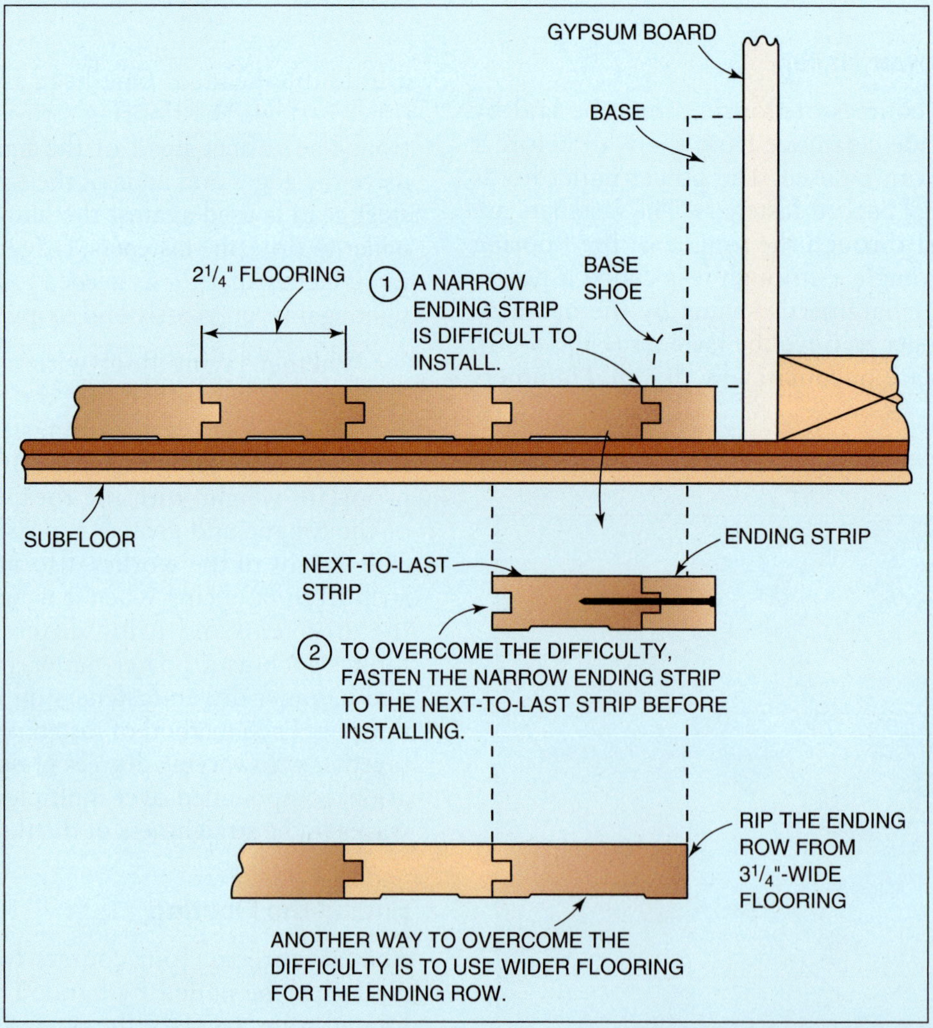

**FIGURE 18-36** Techniques for installing the last strip of flooring.

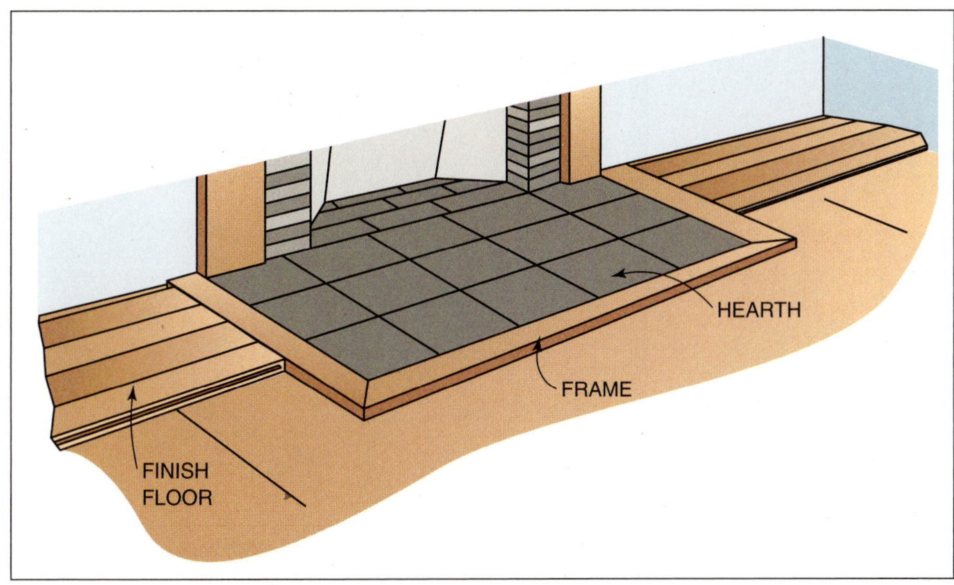


**FIGURE 18-37** Frame around floor obstructions, such as hearths, with strips that are mitered at the corners.

### Changing Direction of Flooring

Sometimes it is necessary to change direction of flooring when it extends from a room into another room, hallway, or closet. Change directions by joining groove edge to groove edge and inserting a spline, ordinarily supplied with the flooring (Fig. 18-38).

### Installing Plank Flooring

Plank flooring is installed like strip flooring. Alternate the courses by widths. Start with the narrowest pieces. Then, use increasingly wider courses and repeat the pattern. Stagger the joints in courses. Use lengths so they present the best appearance.

Manufacturer's instructions for fastening the flooring vary and should be followed. Generally, the flooring is blind nailed through the tongue of the plank and at intervals along the plank in a manner similar to strip flooring.

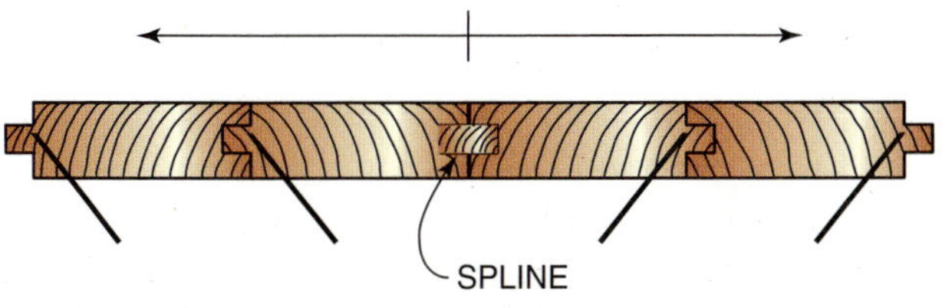


**FIGURE 18-38** The direction of strip flooring is changed by the use of a spline.

## INSTALLING CLOSET TRIM

A simple clothes closet is normally furnished with a *shelf* and a *rod* for hanging clothes. A piece of 1 × 4 or 1 × 6 stock is installed around the walls of the closet to support the shelf and the rod. This piece is called a *cleat*. The height from the floor to the top edge of the cleat is typically 66 inches. The shelf is installed on top of it. The closet pole is installed in the center of cleat (Fig. 18-39). Shelves are not normally fastened to the cleat. Rods are installed for easy removal in case the closet walls need refinishing. For a professional job, the ends and back edge of the shelf are scribed to the wall.

Shelves are usually plywood or 1 × boards. Rods may be 3/4-inch steel pipe, 15/16-inch full round wood poles, or chrome-plated rods manufactured for this purpose. On long spans, the rod may be supported in its center by special metal closet pole supports. On each end, the closet pole is supported by plastic or metal closet pole *sockets*. For in-place sockets, holes and notches are made in the cleat to support the ends of the closet pole.

## LINEN CLOSETS

Linen closets usually consist of a series of shelves spaced 12 to 16 inches apart. Cleats used to support shelves are 3/4 × 1 stock, chamfered on the

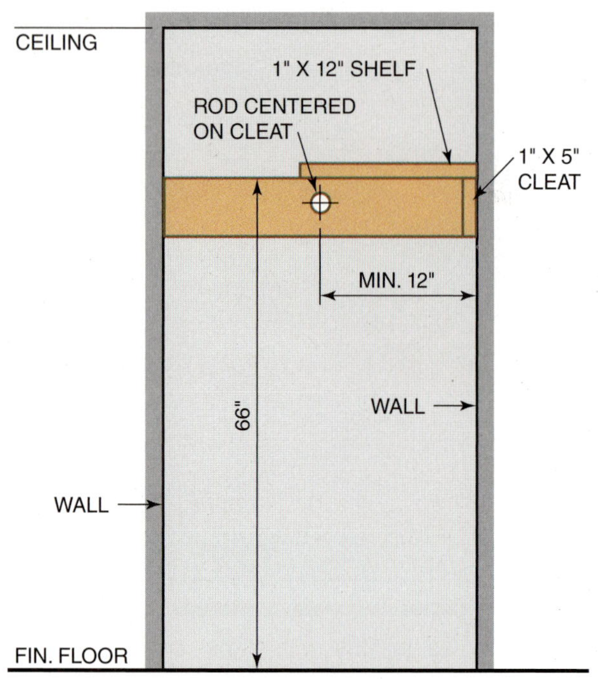


**FIGURE 18-39** Specifications for an ordinary clothes closet.

**FIGURE 18-40** Mantels usually come preassembled in a wide variety of styles and sizes.

bottom outside corner. A *chamfer* is a bevel on the edge of a board that extends only partway through the thickness of the stock. Lay out level lines for the top edges of each set of cleats. Install the cleats and shelves in the same manner as described for clothes closets.

***For step-by-step instructions on INSTALLING CLOSET SHELVES, see Procedure 18-8 on page 635.***

## MANTELS

Mantels are used to decorate fireplaces and to cover the joint between the fireplace and the wall.

Most mantels are made offsite or come preassembled from the factory. They are available in a number of sizes and styles (Fig. 18-40).

Study the manufacturer's directions carefully. Place the mantel against the wall. Center it on the fireplace. Scribe it to the floor or wall as necessary. Carefully fasten the mantel in place and set all fasteners.

# ESTIMATING

When estimating the amount of materials for interior finish, remember that if the number of pieces does not come out even, round it up to the next whole number of pieces.

## SUSPENDED CEILING COMPONENTS

The parts and pieces for a suspended ceiling are estimated using perimeter and area of the room.

***Wall Angle.*** Wall angle is installed on the perimeter of the room. To find the number needed, divide the perimeter of the room by 10, the length of a wall angle. For example, if the perimeter is 135 feet, then the number of wall angles is $135 \div 10 = 13.5 \Rightarrow$ 14 pieces.

***Main Runner.*** Main runners run the length of the room. To determine the number needed, first the number of rows is found by dividing the room width by 4 feet (runner spacing). Subtract one because it is not needed at the wall angles. Round off this number to the nearest whole number of rows. To find the number of pieces in each row, divide the room length by 12 feet (length of each runner). Round up this number to the nearest one-half a runner. This is because the waste end of a runner is usable if it is longer than one-half a piece. To find the number of main runners, multiply the rows by the number of pieces in a row. For example, if the room is 17'-3" × 28'-9", the number of rows is 17.25 ÷ 4 = 4.3 − 1 = 3.3 ⇒ 3 rows. The number of pieces per row is 28.75 ÷ 12 = 2.3 ⇒ 2.5. The number of main runners is 3 × 2.5 = 7.5 ⇒ 8 pieces.

*Cross Tees.* Determining the number of 4-foot cross tees is determined from the number of rows of main runners. Since there is one more row of cross tees than main runners, add one to the number of main runner rows. The number of tees along the length of the room is found by dividing the length by 2 feet (cross-tee spacing) and subtracting one. Finally, the number of 4-foot tees is the number of rows times pieces per row. For example, if the main runner rows is 3 and the room length is 28'-9", the number of rows of tees is 3 + 1 = 4. The pieces per row is 28.75 ÷ 2 − 1 = 13.3 ⇒ 14 pieces. The number of pieces for the room is 4 × 14 = 56.

Two foot cross tees are used to convert 2 × 4 ceiling panels system to a 2 × 2 system. They connect to the 4-foot tees. The number of 2-foot tees is slightly more than the number of 4-foot tees. This extra amount is equal to the number of rows of 4-tees. For example, if there are 56 4-foot tees in four rows, then the number of 2-foot tees is 56 + 4 = 60.

*Hanger Lags and Wire.* There are three hanger lags per main tee. Multiply main runners by three. The length of wire is the number of lags times the length of each wire. Measure the distance from the supporting surface for the main runners and add 2 feet for wire wrap. For example, the number of 50-foot rolls of wire needed for a ceiling suspended 6 feet with eight main runners is 8 × (6 + 2) = 64 feet ÷ 50 = 1.2 ⇒ 2 rolls.

*Suspended Ceiling Panels.* Each border panel requires a full-size ceiling panel. The number of 2 × 4 panels is the same number as would be for 2-foot cross tees. It is the number of 4-foot tees plus one per main runner rows plus one more. Deduct one panel per light fixture. For example, if the number of 4-foot tees is 56 and there are four rows of main runner and eight lights, then the number of boxes of panels (ten per box) is 56 + 4 + 1 − 8 = 53 ÷ 10 = 5.3 ⇒ 6 boxes. If 2 × 2 panels are used, simply double the 2 × 4 panel count.

*Window Trim.* Each window is figured separately and then added together. Head casing and apron lengths are one foot more than the window width. Side casing lengths are 6 inches longer than the window height. The stool length is slightly longer than the head casing length, so the same as the head casing should be enough. For example, if a window is 3' × 4', then the casing is one head casing and one apron at 2 × (3 + 1) = 8' plus two side casings at 2 × 4.5 = 9' for a total casing of 17'. Stool would be 3 + 1 = 4'.

*Door Casing.* Head casing are 6 inches more than the door width. Side casings are 4 inches more than the door height. Add the casing lengths and then double for the opposite side of the door. For example, for a 3'-0" × 6'-8" door, head casing needed is 3 + 0.5 = 3.5. Side casing is 2 × (6'-8" + 4") = 14'. Total casing is 2 × (3.5 + 14) = 35'.

*Base and Ceiling Molding.* Base and ceiling molding lengths come from the room perimeter. They differ only in that base molding has the width of each door deducted from the total. For example in a 12' × 17' room with two 3'-0" × 6'-8" doors, ceiling molding would be 2 × 12 + 2 × 17 = 58'. The base molding would be 58'-2" × 3' = 52'.

*Wood Strip Flooring.* To estimate the amount of hardwood flooring material needed, first determine the area to be covered. Divide this by the square feet per carton of flooring to get the number of cartons. Add 2 percent waste for end matching and normal waste. For example, the area of a room is 384 square feet, and a bundle of flooring covers 15 square feet per carton; 384 ÷ 15 × 1.02 = 26.1 = 27 cartons.

See Figure 18-41 for a listing of the shorthand version of these estimating techniques, along with another example.

| Estimate the materials for interior finish in the rooms given. A 24' × 38' room is to have 2 × 4' grid suspended ceiling. A second 475 SF room is to be covered with a wood strip flooring that comes in 15 SF cartons. A third 18' × 26' room is to be trimmed for windows, doors, base board, and ceiling molding. There is a 5'-0" × 4'-0" window and two 2'-6" × 6'-8" doors (both sides). | | | |
|---|---|---|---|
| **Item** | **Formula** | **Waste Factor** | **Example** |
| **Suspended Ceiling Wall Angle** | PERM ÷ wall angle LEN = pieces | | 2 × 24 + 2 × 33 = 114 ÷ 10 = 11.4 + 12 pieces |
| **Suspended Ceiling Main Runner** | [room WID ÷ main tee spacing − 1] × [room LEN ÷ main tee LEN rounded up to nearest ½ piece] = pieces | | [24 ÷ 4 − 1] × [38 ÷ 12] = 5 × 3.5 = 17.5 ⇒ 18 pieces |
| **Suspended Ceiling 4-foot Cross Tees** | [main runner rows + 1] × [room LEN ÷ 2 − 1] = NUM of 4 ft cross tees | | [5 +1] × [38 ÷ 2 − 1] = 6 × 18 = 108 cross tees |
| **Suspended Ceiling Hanger Lags** | NUM main tees × 3 = pieces | | 18 × 3 = 54 pieces |
| **Suspended Ceiling Wire (50' roll)** | NUM lags × [suspended distance + 2'] ÷ 50 = rolls | | 54 × [2 +2] ÷ 50 = 4.3 ⇒ 5 rolls |
| **Suspended Ceiling Panels** | [NUM 4-foot cross tees + NUM main tee rows + 1 − one per light] ÷ NUM per box = boxes | | [108 + 5 + 1 − 12 ] ÷ 10 = 10.2 + 11 boxes |
| **Window Casing** | 2 × (width + 1') + 2 × (length 6") | | 2 (5 + 1) + 2 (4 + 0.5) = 21' |
| **Window Stool** | width plus one foot | | 5 + 1 = 6' |
| **Door Casing** | 2 × [width + 6" + 2 × (LEN + 4")] | | 2 × [2.5 + 0.5 + 2 × (6'-8" + 4") ] = 34' × 2 = 68' |
| **Ceiling Molding** | PERIM | | 2 × 18 + 2 × 26 = 88' |
| **Base Molding** | PERIM minus door widths | | 88' − (2 × 2.5) = 82' |
| **Wood Strip Flooring** | area to be covered ÷ SF per carton × waste = cartons | 2% | 475 ÷ 15 × 102% = 32.3 = 33 cartons |

**FIGURE 18-41** Example of estimating interior finish materials with formulas.

## SUMMARY

- Ceiling finish may be achieved using a grid and panel system.
- Interior molding covers joints between different materials and surfaces as well as adds architectural appeal.
- Door casings cover the gap between the doorframe and the wall surface.
- Base trim covers the seam between the floor and the wall.
- Window trim covers the space between the window unit and the wall surface.
- Finish floors come in a variety of styles and materials.
- Estimating interior finish materials is necessary for efficient purchases.

PROCEDURE 18-1

# Constructing the Grid Ceiling System

**A** Locate the height of the ceiling, marking elevations of the ceiling at the ends of all wall sections. Snap chalk lines on all walls around the room to the height of the top edge of the wall angle. If a laser is used, the chalk line is not needed since the ceiling is built to the light beam.

- Fasten wall angles around the room with their top edge lined up with the line. Fasten into framing wherever possible, not more than 24 inches apart. If available, power nailers can be used for efficient fastening.

## *from experience...*

To fasten wall angles to concrete walls, short masonry nails sometimes are used. However, they are difficult to hold and drive. Use a small strip of cardboard to hold the nail while driving it with the hammer.

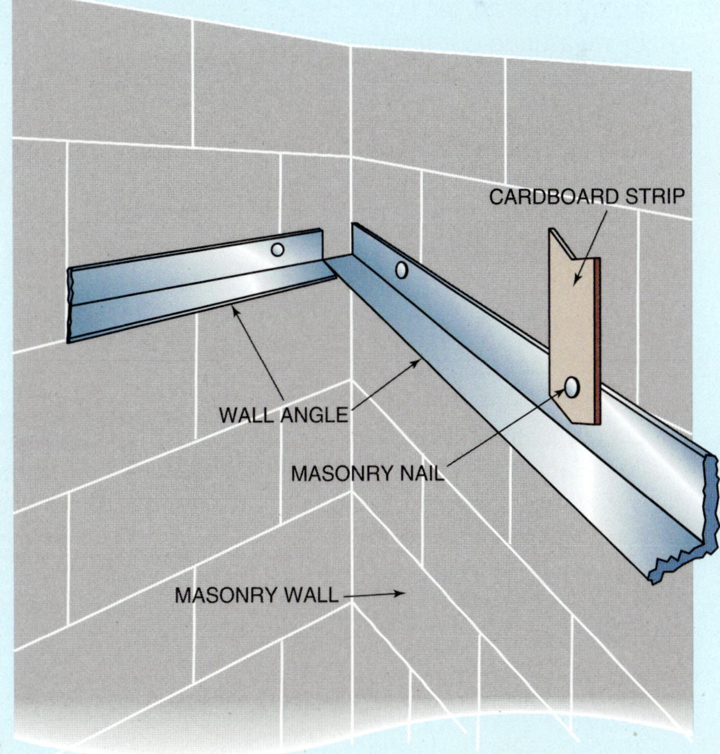

PROCEDURE 18-1

# Constructing the Grid Ceiling System (continued)

**B** Make miter joints on outside corners. Make butt joints in interior corners and between straight lengths of wall angle. Use a combination square to layout and draw the square and angled lines. Cut carefully along the lines with snips.

- From the ceiling sketch, determine the position of the first main runner. Stretch a line at this location across the room from the top edges of the wall angle. The line serves as a guide for installing *hanger lags* or *screw eyes* and *hanger wires* from which main runners are suspended.
- Install the cross tee line by measuring out from the short wall, along the stretched main runner line, a distance equal to the width of the border panel. Mark the line. Stretch the cross tee line through this mark and at right angles to the main runner line.
- Install hanger lags not more than 4 feet apart and directly over the stretched line. Hanger lags should be of the type commonly used for suspended ceilings. They must be long enough to penetrate wood joists a minimum of 1 inch to provide strong support. Hanger wires may also be attached directly around the lower chord of bar joists or trusses.

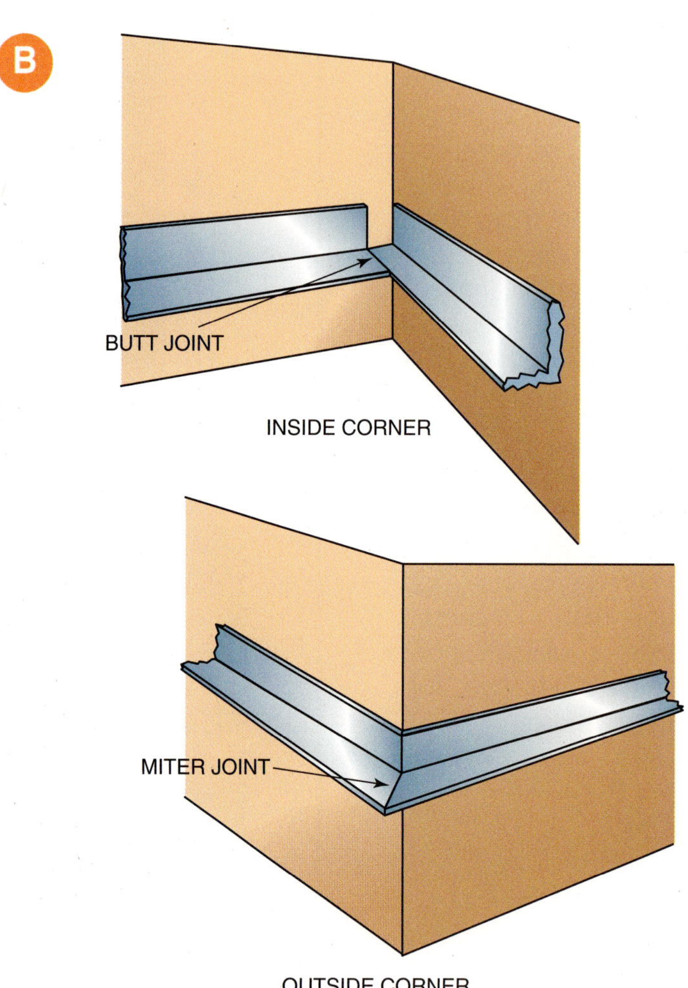


CAUTION

**CAUTION:** Use care in handling the cut ends of the metal grid system. The cut ends are sharp and may have jagged edges that can cause serious injury.

## from experience...

Stretch the line tightly on nails inserted between the wall and wall angle.

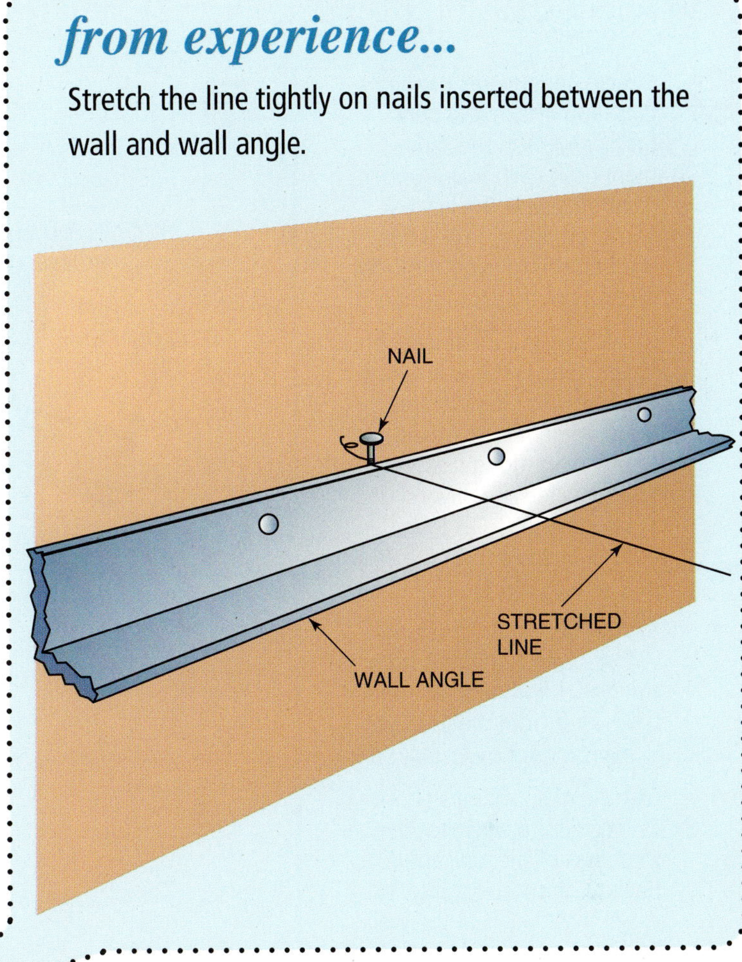


**C** Cut a number of hanger wires using wire cutters. The wires should be about 12 inches longer than the distance between the overhead construction and the stretched line. Attach the hanger wires to the hanger lags. Insert about 6 inches of the wire through the screw eye. Securely wrap the wire around itself three times. Pull on each wire to remove any kinks. Then, make a 90-degree bend where it crosses the stretched line. If a laser is used, the 90-degree bend is done later when the main runner is installed.

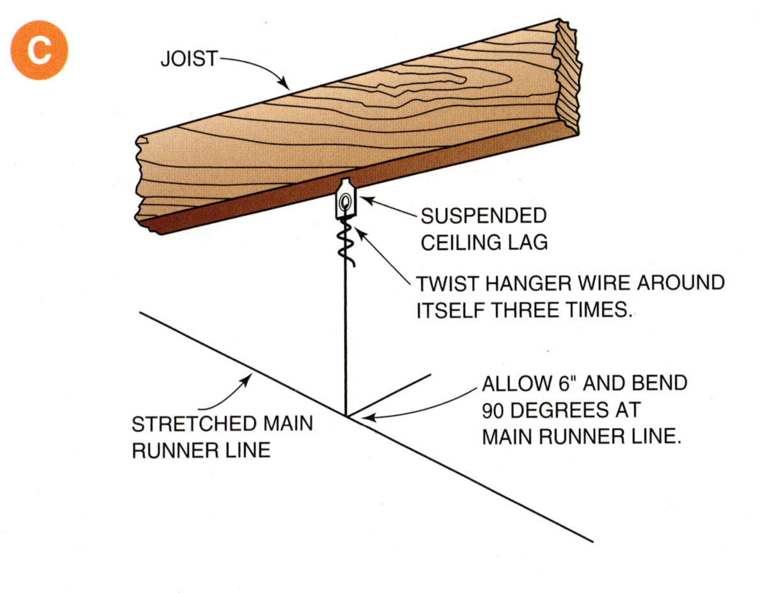

PROCEDURE 18-1

# Constructing the Grid Ceiling System (continued)

**D** Stretch lines, install hanger lags, and attach and bend hanger wires in the same manner at each main runner location. Leave the last line stretched tightly in position. It will be used to align the cross tee slots of the main runner.

- At each main runner location, measure from the wall to the cross tee line. Transfer this measurement to the main runner, measuring from the first cross tee slot beyond the measurement, so as to cut as little as possible from the end of the main runner.
- **EXAMPLE:** If the first cross tee will be located 23 inches from the wall, then the main runner will be cut from the third slot in main runner.
- Cut the main runners about ⅛ inch less to allow for the thickness of the wall angle. Backcut the web slightly for easier installation at the wall. Measure and cut main runners individually. Do not use the first one as a pattern to cut the rest. Measure each from the cross tee line.

**D**

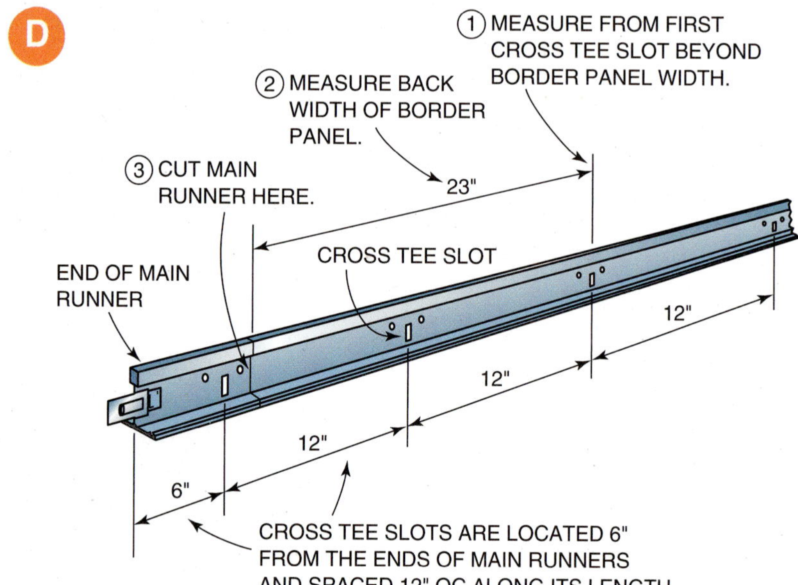


**E** Hang the main runners by resting the cut end on the wall angle and inserting suspension wires in the appropriate holes in the top of the main runner. Bring the runners up to the bend in the wires or to the laser light beam. Twist the wires with at least three turns to hold the main runners securely. More than one length of main runner may be needed to reach the opposite wall. Connect lengths of main runners together by inserting tabs into matching ends. Make sure end joints come up tight.

**E**

Courtesy of Trimble.

F The length of the last section is measured from the end of the last one installed to the opposite wall, allowing about ⅛ inch less to fit.

- Cross tees are installed by inserting the tabs on the ends into the slots in the main runners. These fit into position easily, although the method of attaching varies from one manufacturer to another. Install all full-length cross tees between main runners first.
- Lay in a few full-size ceiling panels to stabilize the grid while installing the border cross tees.
- Cut and install cross tees along the border. Insert the connecting tab of one end in the main runner and rest the cut end on the wall angle. It may be necessary to measure and cut cross tees for border panels individually, if walls are not straight or square.
- For 2 × 2 panels, install 2-foot cross tees at the midpoints of the 4-foot cross tees. After the grid is complete, straighten and adjust the grid to level where necessary.

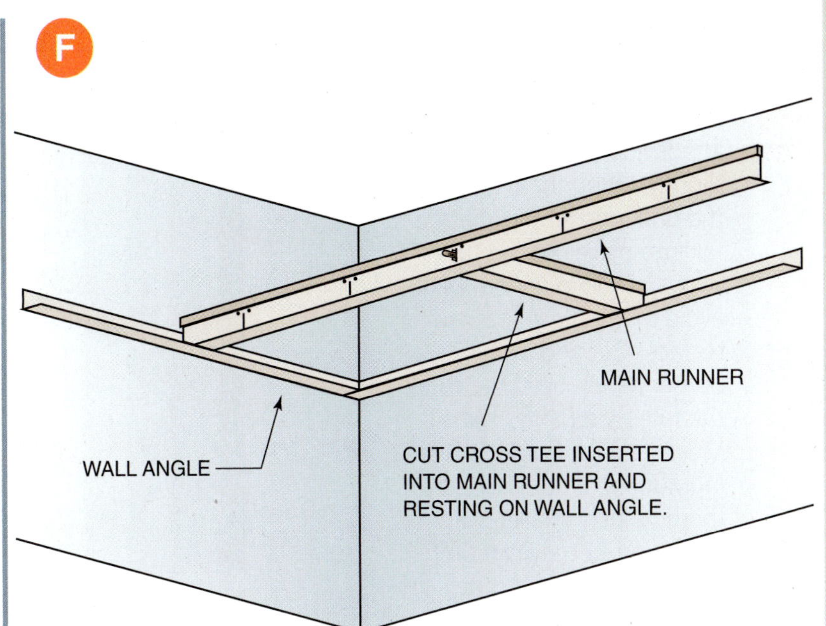

PROCEDURE 18-1

# Constructing the Grid Ceiling System (continued)

**G** Ceiling panels are placed in position by tilting them slightly, lifting them above the grid, and letting them fall into place. Be careful when handling panels to avoid marring the finished surface. Cut and install border panels first and install the full-sized panels last. Measure each border panel individually, if necessary. Cut them ⅛ inch smaller than measured so they can drop into place easily. Cut the panels with a sharp utility knife using a straightedge as a guide. A scrap piece of cross tee material can be used as a straightedge. Always cut with the finished side of the panel up.

**H** When a column is near the center of a ceiling panel, cut the panel at the midpoint of the column. Cut semicircles from the cut edge to the size required for the panel pieces to fit snugly around the column. After the two pieces are rejoined around the column, glue scrap pieces of panel material to the back of the installed panel. If the column is close to the edge or end of a panel, cut the panel from the nearest edge or end to fit around the column. The small piece is also fitted around the column and joined to the panel by gluing scrap pieces to its back side.

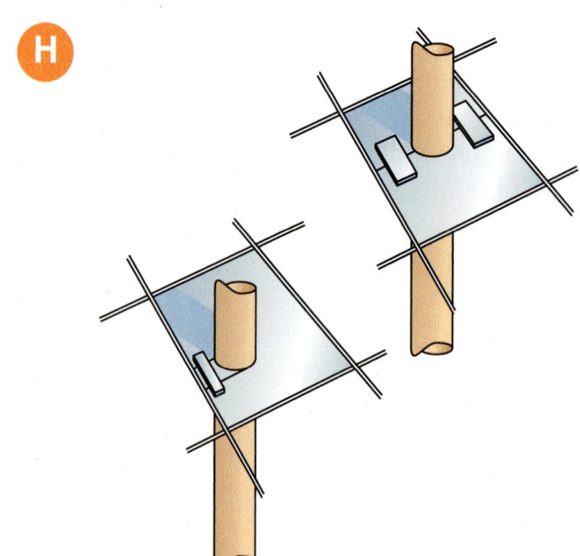

*Courtesy of Armstrong World Industries.*

PROCEDURE **18-2**

# Cutting a Coped Joint

**A** Cut a molding piece to fit into the corner with a squared end. Left or right does not matter, but here the left piece is cut square. The second piece is cut with a coped joint. The W to the left indicates which side of the molding fits against the wall.

- The coped molding end is first cut as if it were an inside miter, as shown in Figure 18-18. The edge of the cut along the face side forms the profile of the coped cut. You may rub the side of a pencil point lightly along the profile to make the outline more clearly visible.

**B** Hold the molding so it is over the end of a sawhorse. Hold the side of the molding that will fit to the wall flat on the top of the sawhorse. This way the cut is done with the coping saw blade plumb. Use a fine-tooth blade coping saw, carefully cutting along the outlined profile and keeping the saw blade plumb.

**C** A slight undercut of the blade helps ensure that the cut will fit nicely. It may be necessary to touch up the cut with a sharp utility knife or sandpaper.

PROCEDURE 18-3

# Applying Wall Molding

**A** To snap a line for wall trim, begin by holding a short scrap piece of the molding at the proper angle on the wall. Lightly mark the wall along the bottom edge of the molding. Measure the distance from the ceiling down to the mark.

- Measure and mark this same distance down from the ceiling on each end of each wall to which the molding is to be applied. To allow for the buildup of joint compound in the corners, cut a scrap block as shown in A. Transfer the first measurement to the block and use it as a marking gauge for the other corners. Snap lines between the marks. Apply the molding so its bottom edge is to the chalk line.

**B** Apply the molding to the first wall with square ends in both corners. If more than one piece is required to go from corner to corner, the butt joints may be squared or mitered. Position the molding in the miter box the same way each time. Mitering the molding with the same side down each time helps make fitting more accurate, faster, and easier.

**A**

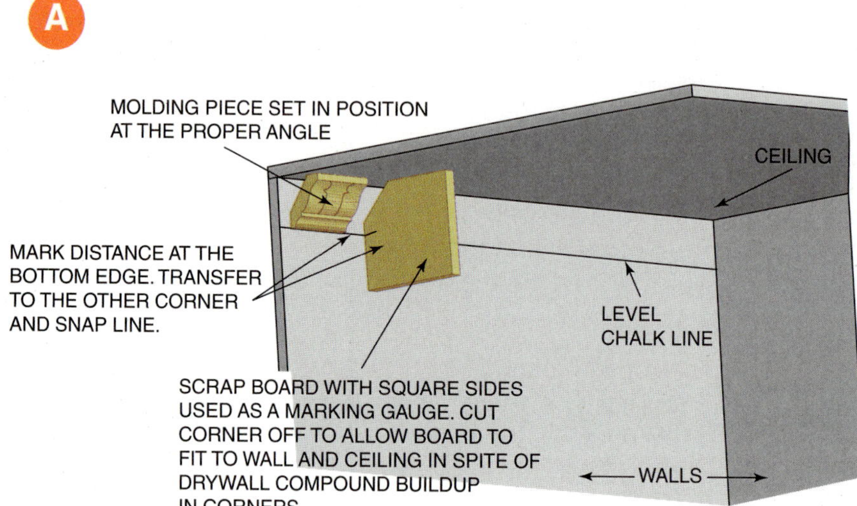


**B**

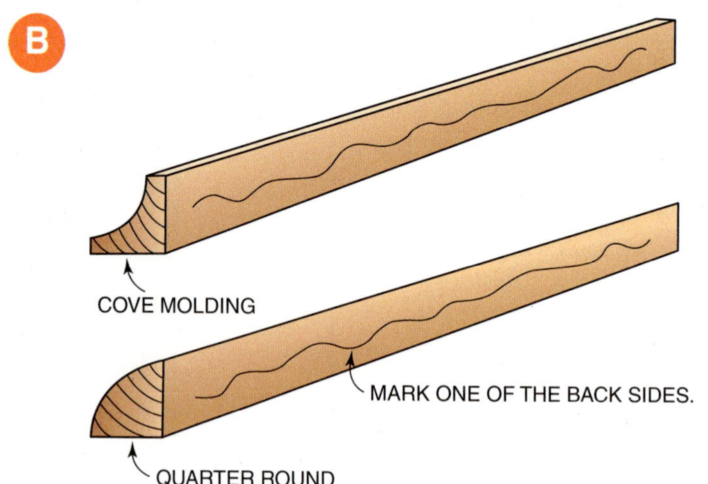


### from experience...

Since the revealed edges of the molding are often not the same, it is important to cut the molding with the same orientation. To do this, mark one of the back surfaces with a pencil.

**C** If a small-size molding is used, fasten it with finish nails in the center. Use nails of sufficient length to penetrate into solid wood at least one inch. If large size molding is used, fastening is required along both edges. Nail at about 16-inch intervals and in other locations as necessary to bring the molding tight against the surface. End nails should be placed 2 to 3 inches from the end to keep from splitting the molding. If it is likely that the molding may split, blunt the pointed end of the nail.

- Cope the starting end of the first piece on each succeeding wall against the face of the last piece installed on the previous wall. Work around the room in one direction. The end of the last piece installed must be coped to fit against the face of the first piece.

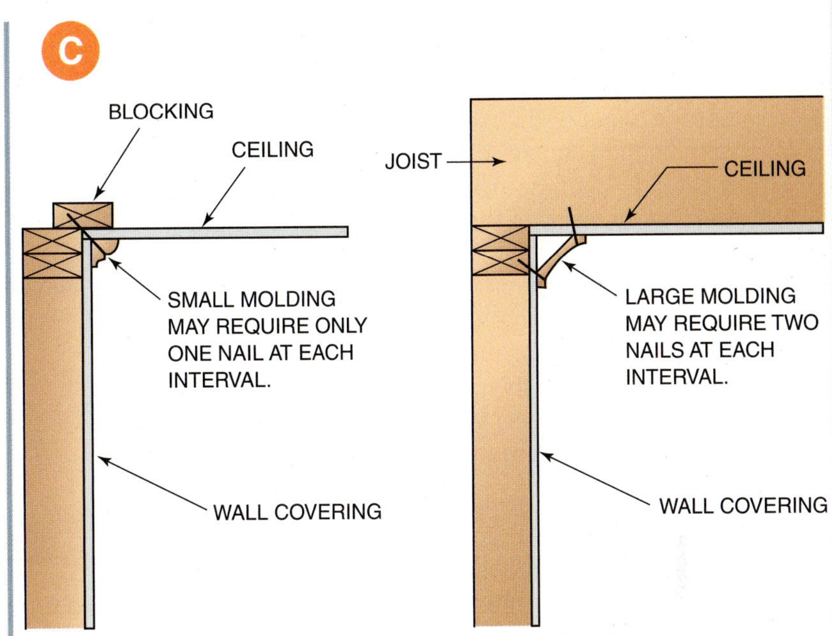

PROCEDURE 18-4

# Applying Door Casings

**A** Set the blade of the combination square so that it extends 1/4 to 5/16 inch beyond the body of the square. Gauge lines at intervals along the side and head jamb edges by riding the square against the inside face of the jamb. Let the lines intersect where side and head jambs meet.

- Miter one end of the head casing. Hold it against the head jamb of the door frame so that the miter is on the intersection of the gauged lines. Mark the length of the head casing at the intersection of the gauged lines on the opposite side of the door frame. Miter the casing to length at the mark.
- Fasten the head casing in position with a few tack nails. It may be necessary to move the ends slightly to fit the mitered joint between head and side casings. Keep the casing inside edge aligned to the gauged lines on the head jamb. The mitered ends should be in line with the gauged lines on the side jambs. Use finish nails along the inside edge of the casing into the header jamb. Straighten the casing as necessary as nailing progresses. Drive nails at the proper angle to keep them from coming through the face or back side of the jamb. Fasten the top edge of the casing into the framing.

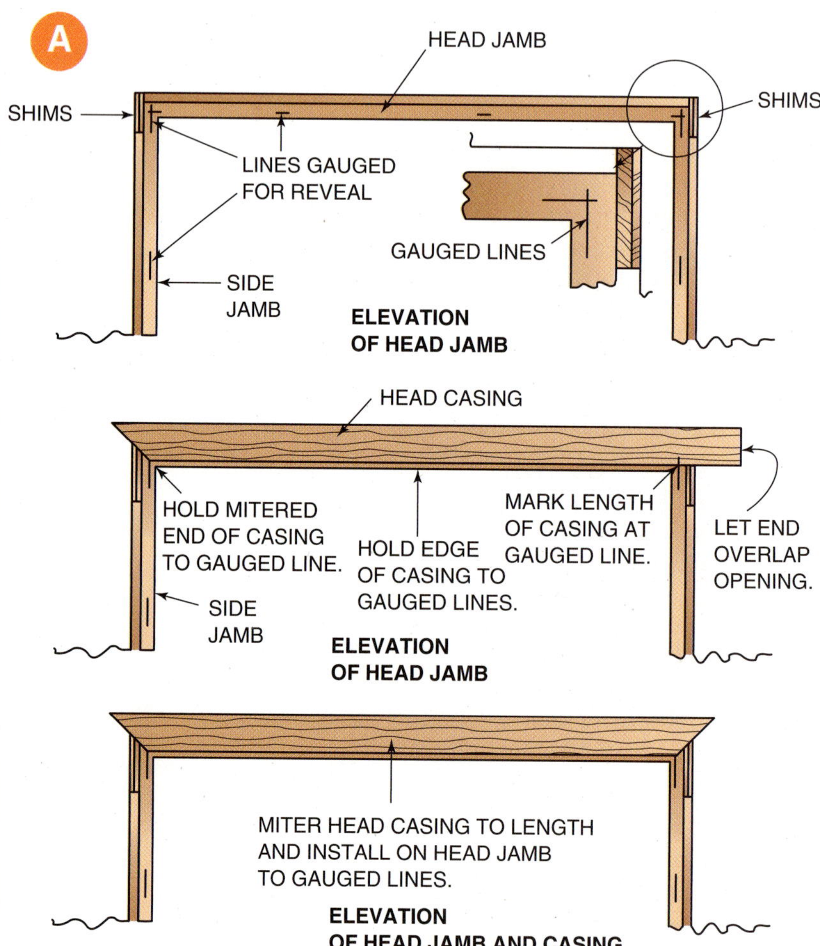

- Cut one miter on the ends of the two side casings. Cut them a little long as they will be cut to fit later. Be sure to cut pairs of right and left miters.
- Cut the previously mitered side casing to length. Mark the bottom end by turning it upside down with the point of the miter touching the floor. Mark the side casing in line with the top edge of the head casing. Make a square cut on the casing at that mark. If the finish floor has not been laid, hold the point of the miter on a scrap block of material that is equal in thickness to the finish floor. Replace the side casing in position and try the fit at the mitered joint. If the joint needs adjusting, trim with a power miter box or use a sharp block plane.
- When fitted, a little glue may be applied to the joint. Nail the side casing in the same manner as the head casing. Bring the faces flush, if necessary, by shimming between the back of the casing and the wall. Usually, only thin shims are needed. Any small space between the casing and the wall is usually not noticeable, or it can be filled later with joint-filling compound. Also, the backside of the thicker piece may be planed or chiseled to the desired thinness.
- Drive a 4d finish nail into the edge of the casing and through the mitered joint. Drive 6d nails through the casing into jamb and 8d nails into casing and framing. Then, set all fasteners. Keep nails 2 or 3 inches from the end to avoid splitting the casing.

## *from experience...*

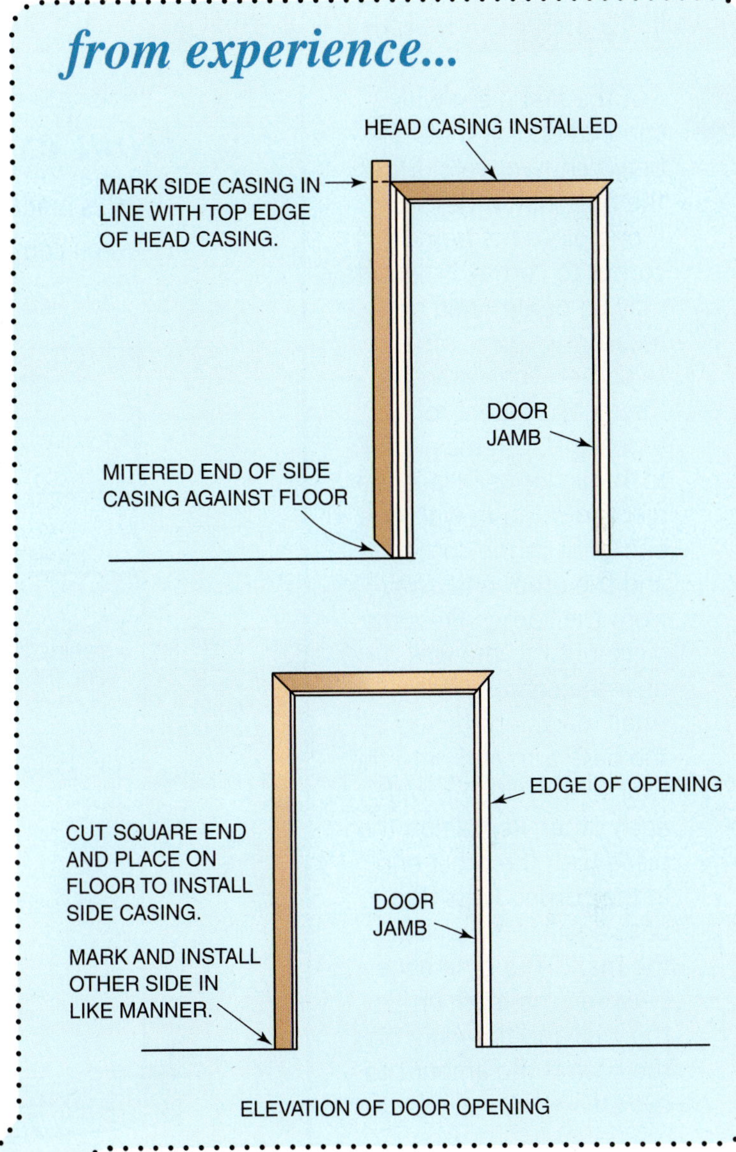


ELEVATION OF DOOR OPENING

PROCEDURE 18-5

# Applying Base Moldings

**A** Cut the first piece with squared ends if it fits between two walls. Miter the butt joint if desired. If one piece fits from corner to corner, its length may be determined by measuring from corner to corner. Then, transfer the measurement to the baseboard. Cut the piece ½ to 1 inch longer. Place the piece in position with one end tight to the corner and the other end away from the corner. Press the piece tight to the wall near the center. Place small marks on the top of the base trim and onto the wall so they line up with each other. Reposition the piece with the other end in the corner. Press the base against the wall at the mark. The difference between the mark on the wall and the mark on the base is the amount to cut off.

- After cutting, place one end of the piece in the corner and bow out the center. Place the other end in the opposite corner, and press the center against the wall. Fasten in place. Continue in this manner around the room. Make regular miter joints on outside corners and coped joints on inside corners.
- If both ends of a single piece are to have regular

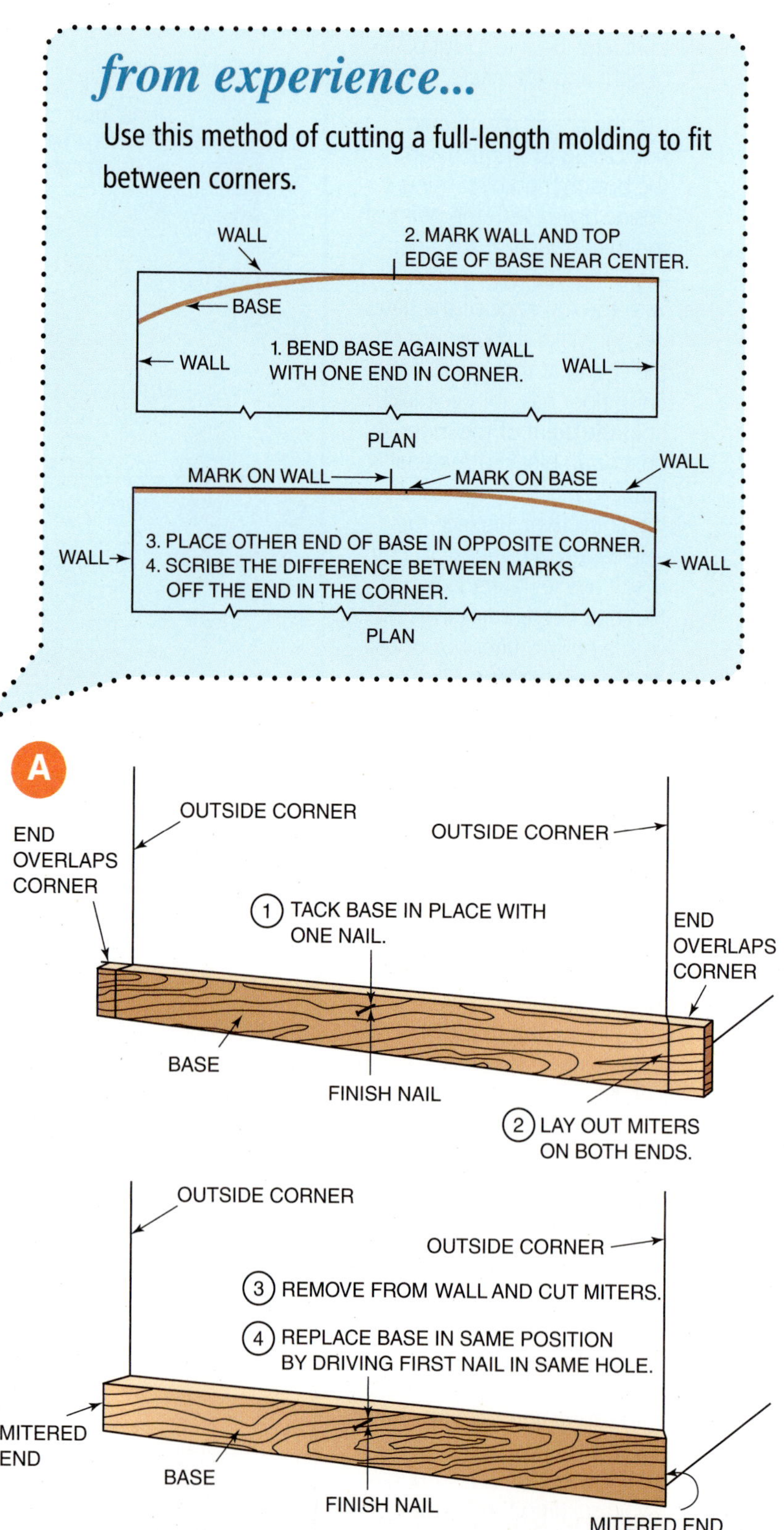

miters for outside corners, it is important that it be fastened in the same position as it was marked. Tack the rough length in position with one finish nail in the center. Mark both ends. Remove, and cut the miters. Remember that these marks are to the short side of the miter so the piece will be longer than these marks indicate. Reinstall the piece by first fastening into the original nail hole.

B If a *base cap* is applied, it is done so in the same manner as most wall or ceiling molding. Cope interior corners and miter exterior corners. However, it should be nailed into the floor and not into the baseboard. This allows the vertical movement from shrinking and swelling to be not so noticeable.

C When the base shoe must be stopped at a door opening or other location, with nothing to butt against, its exposed end is generally back-mitered and sanded smooth. Generally, no base shoe is required if carpeting is to be used as a floor finish.

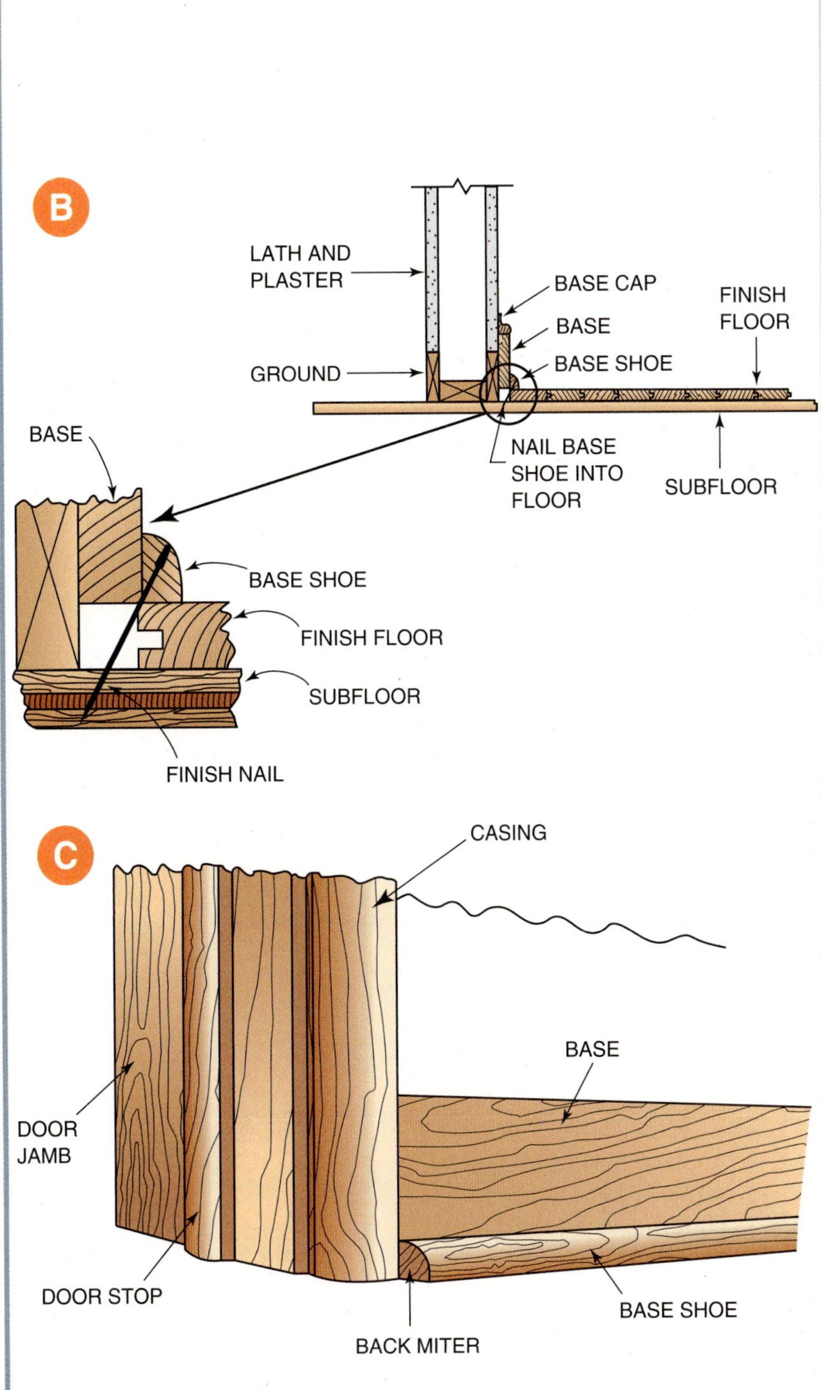

PROCEDURE 18-6

# Installing Window Trim

## Applying the Stool

**A** Hold a piece of side casing in position at the bottom of the window and draw a light line on the wall along the outside edge of the casing stock. Mark a distance outward from these lines equal to the thickness of the window casing. Cut a piece of stool stock to length equal to the distance between the outermost marks.

- Position the stool with its outside edge against the wall as shown to the right. The ends should be in line with the marks previously made on the wall. Lightly square lines, across the face of the stool, even with the inside face of each side jamb of the window frame.
- Set the pencil dividers or scribers to mark the cutout so that, on both sides, an amount equal to twice the casing thickness will be left on the stool. Scribe the stool by riding the dividers along the wall on both sides. Also, scribe along the bottom rail of the window sash.
- Cut to the lines and smooth the sawed edge that will be nearest to the sash. Shape and smooth both ends of the stool the same as the inside edge.
- Apply a small amount of caulking compound to the bottom of the stool. Fasten the stool in position by driving finish nails along its outside edge into the sill. Set the nails.

### *from experience...*

Raise the lower sash slightly. Place a short, thin strip of wood under it, on each side, which projects inward to support the stool while it is being laid out. Place the stool on the strips. Raise or lower the sash slightly so the top of the stool is level.

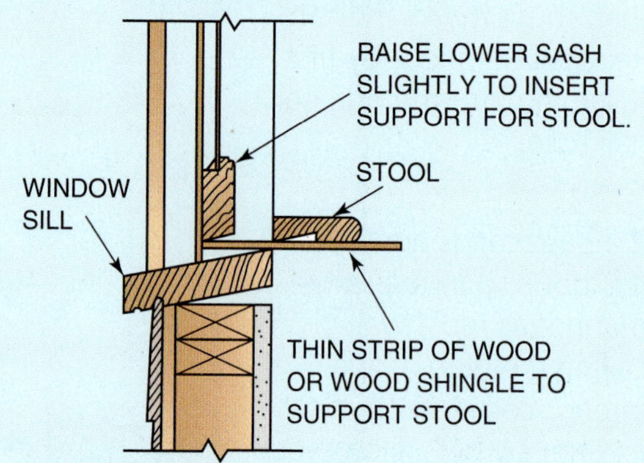


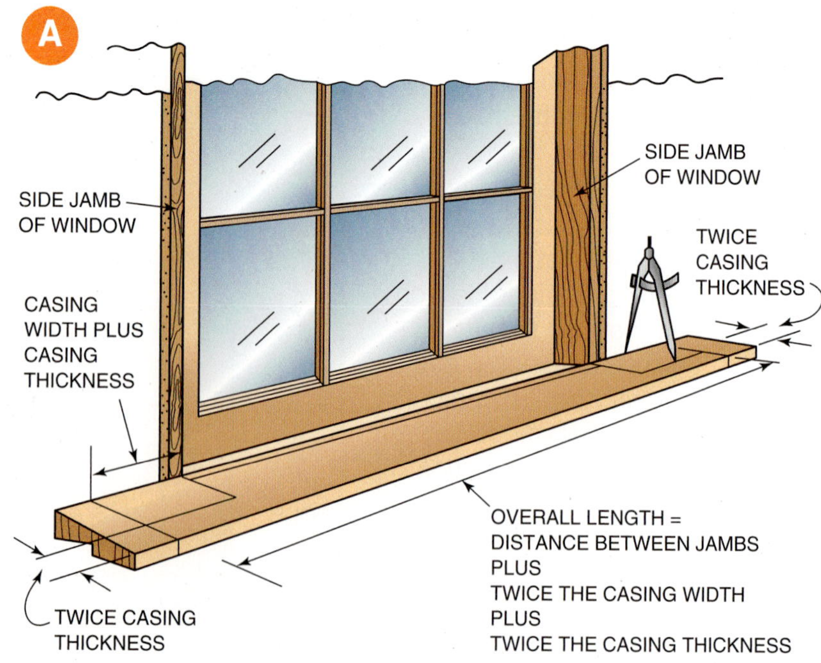

## Applying the Apron

**A** Cut a length of apron stock equal to the distance between the outer edges of the window casings.

- Each end of the apron is then *returned upon itself*. This means that the ends are shaped the same as its face. To return an end upon itself, hold a scrap piece on the apron. Draw its profile flush with the end. Cut to the line with a coping saw. Sand the cut end smooth. Return the other end upon itself in the same manner.
- Place the apron in position, with its upper edge against the bottom of the stool. Be careful not to force the stool upward. Keep the top side of the stool level. Fasten the apron along its bottom edge into the wall. Then, drive nails through the stool into the top edge of the apron.

### *from experience...*

When nailing through the stool, wedge a short length of 1 × 4 stock between the apron and the floor at each nail location. This supports the apron while nails are being driven. Failure to support the apron results in an open joint between it and the stool. Take care not to damage the bottom edge of the apron with the supporting piece.

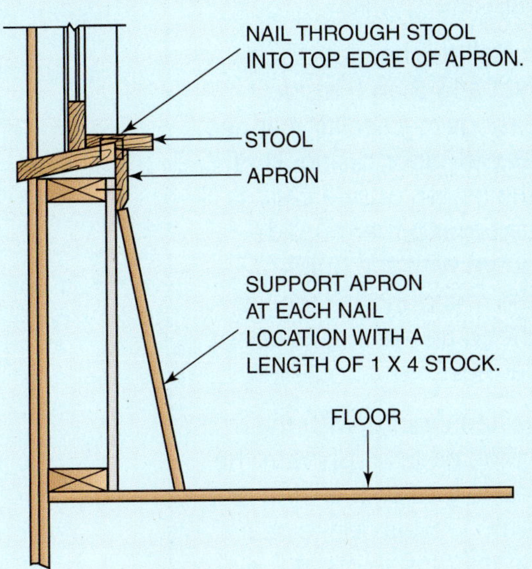

PROCEDURE 18-6

# Installing Window Trim (continued)

## Installing Jamb Extensions

**A** Measure the distance from the jamb to the finished wall. Rip the jamb extensions to this width with a slight back-bevel on the side toward the jack stud.

- Cut the pieces to length and apply them to the header, side jambs, and stool. Shim them, if necessary, and nail with finish nails that will penetrate the framing at least an inch.

## Applying the Casings

- Cut the number of window casings needed to a rough length with a miter on one end. Cut side casings with left- and right-hand miters, depending on which side they are installed.
- Install the header casing first and then the side casings in a similar manner as with door casing. Find the length of side casings by turning them upside down with the point of the miter on the stool in the same manner as door casings.
- Fasten casings with their inside edges flush with the inside face of the jamb or with a reveal. Make neat, tight-fitting joints at the stool and at the head.

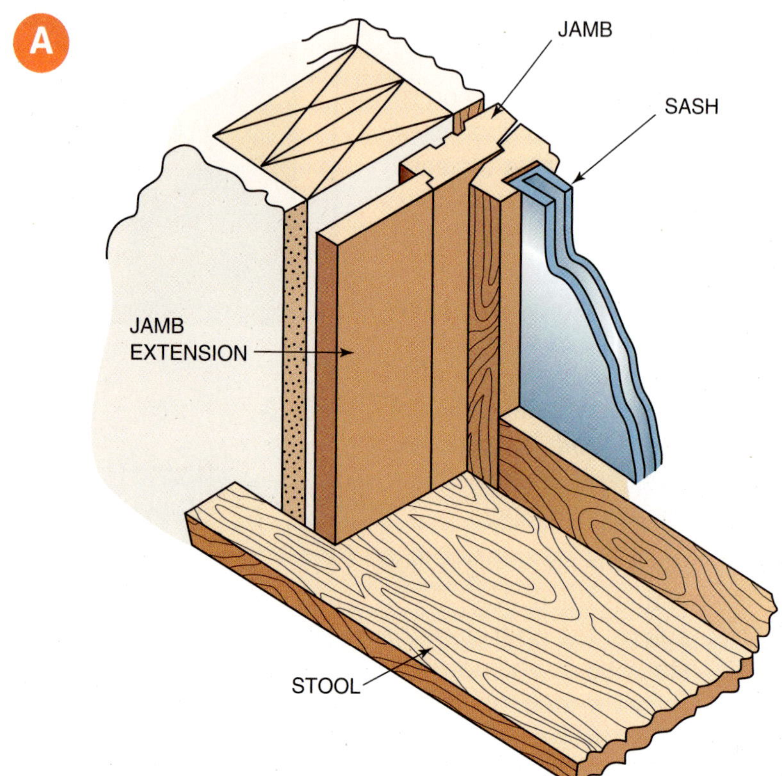

PROCEDURE 18-7

# Installing Wood Flooring

## Preparation for Installation

**A** Check the subfloor for any loose areas and add nails where appropriate. Sweep and vacuum the subfloor clean. Determine the direction of the flooring.

- Cover the subfloor with building paper. Lap it 4 inches at the seams, and at right angles to the direction of the finish floor. The paper prevents squeaks in dry seasons and retards moisture from below that could cause warping of the floor.
- Snap chalk lines on the paper showing the centerline of floor joists so flooring can be nailed into them. For better holding power, fasten flooring through the subfloor and into the floor joists whenever possible. On ½-inch plywood subfloors, all flooring fasteners must penetrate into the joists.

## Starting Strip

- Place a strip of flooring on each end of the room, ¾ inch from the starter wall with the groove side toward the wall. The gap between the flooring and the wall is needed for expansion. It will eventually be covered by the base molding.
- Mark along the edge of the flooring tongue. Snap a chalk line between the two points. Hold the strip with its tongue edge to the chalk line.

**A**

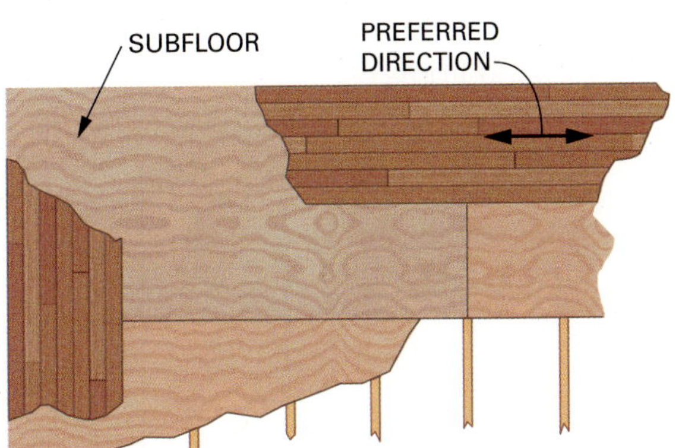


PANEL SUBFLOOR

PROCEDURE 18-7

# Installing Wood Flooring (continued)

**A** Face nail it with 8d finish nails, alternating from one edge to the other, 12 to 16 inches apart. Make sure end joints between strips are driven up tight. Continue the first course to the wall.

- Cut the last strip to fit loosely against the wall. Use a strip long enough so that the cut-off piece is 9 inches or longer. This scrap piece is used to start the next course back against the other wall.

**B** After the second course of flooring is fastened, lay out seven or eight loose rows of flooring, end to end. This is called *racking the floor.* Racking is done to save time and insure the material is used efficiently.

- Lay out in a staggered end-joint pattern. End joints should be at least 6 inches apart. Find or cut pieces to fit within ½ inch of the end wall. Distribute long and short pieces evenly for the best appearance. Avoid clusters of short strips.
- Continue across the room. Rack seven or eight courses ahead as work progresses.

A

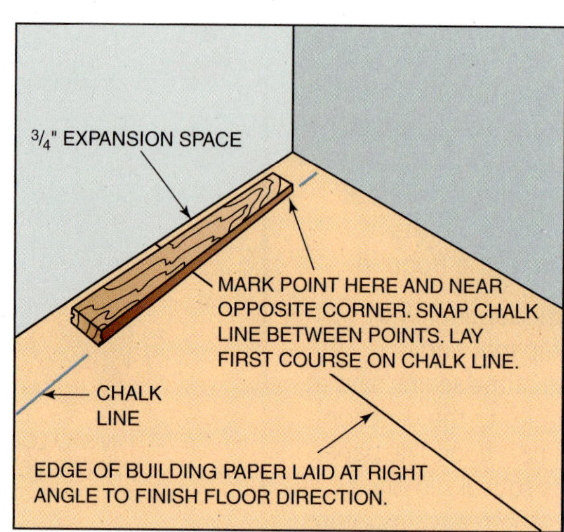


*Courtesy of Chickasaw Hardwood Floors.*

B

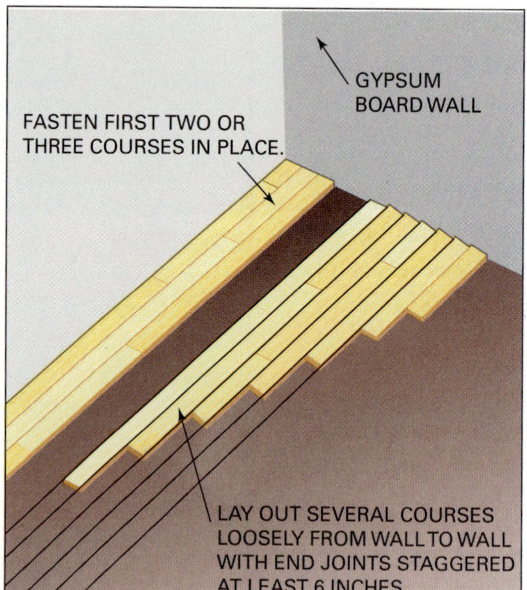

PROCEDURE 18-8

# Installing Closet Shelves

**A**

- Measure up from the floor the height of the cleat, usually 66". Draw a level line on the back wall and two end walls of the closet. Install cleats with finish nails.
- Prepare for closet pole as needed. The center of the socket should be at least 12 inches from the back wall.
- Cut the shelf about ½ inch longer than the distance between end walls.
- Place the shelf in position by laying one end on the cleat and tilting the other end up and resting against the wall. Scribe about ¼ inch off the end resting on the cleat.

**B**

- Measure the distance between corners along the back wall. Transfer this measurement to the shelf, measuring from the scribed cut along the back edge of the shelf.
- Place the shelf in position, tilted in the opposite direction on the cleat.
- Set the dividers to scribe the distance from the wall to the mark on the back edge of shelf. Scribe and cut the other end of the shelf.

**C**

- Replace the shelf into position, resting it on the cleats.
- Scribe the back edge to the wall to take off as little as possible. Cut to the scribed line. Ease the sharp corners on the front edge of the shelf with a hand plane. Sand and place the shelf in position.

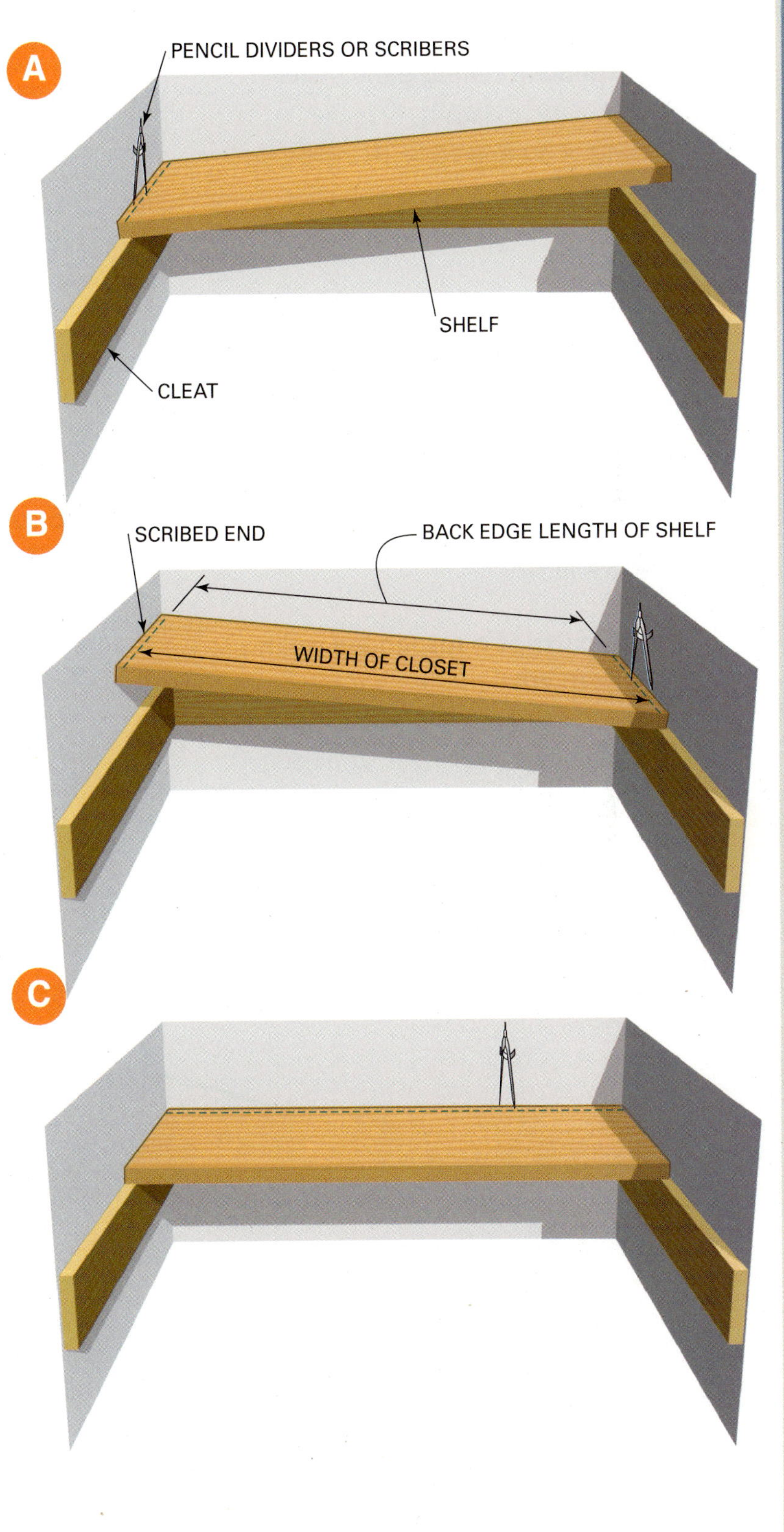

## REVIEW QUESTIONS

*Select the most appropriate answer.*

**1. The most common sizes of suspended ceiling panels are**

a. 8 × 12 and 12 × 12 inches.
b. 12 × 12 and 16 × 16 inches.
c. 12 × 12 and 12 × 24 inches.
d. 24 × 24 and 24 × 48 inches.

**2. The width in inches of border panel for a room 12'-6" × 18'-8", when 2' × 2' suspended ceiling panels are used, is**

a. 6 and 8".
b. 14½ and 18½".
c. 10 and 12".
d. 15 and 16".

**3. The parts of a suspended ceiling system are**

a. panels, wall angle, and cross tees.
b. main tees, cross tees, and hangers.
c. wall angle, main tees, and cross tees.
d. all of the above

**4. The diagonal measurement of a 16" × 24" rectangle is**

a. 384".
b. 832".
c. 28⅞".
d. 28'–10⅛".

**5. In a suspended ceiling, hanger wire is used to suspend**

a. cross tees.
b. wall angle.
c. main runners.
d. furring strips.

**6. To cut a cope joint in molding, the first piece is cut to a**

a. compound miter.
b. outside miter.
c. inside miter.
d. square end.

**7. Bed, crown, and cove moldings are used frequently as**

a. window trim.
b. part of the base.
c. ceiling molding.
d. door casings.

**8. Back bands are applied to**

a. wainscoting.
b. S4S casings.
c. exterior corners.
d. interior corners.

**9. The length of a window stool is usually _____ the outside distance between vertical casings.**

a. shorter than
b. longer than
c. the same size as
d. depends on the local building codes

**10. The length of a window apron is usually _____ the outside distance between vertical casings.**

a. shorter than
b. longer than
c. the same size as
d. depends on the local building codes

**11. The ends of an apron are usually**

a. beveled.
b. coped.
c. returned.
d. chamfered.

**12. The joint between moldings that meet at an interior corner is usually**

a. coped.
b. butted.
c. mitered.
d. bisected.

**13. The setback of casings from the face of the jamb is referred to as a**

a. gain.
b. reveal.
c. backset.
d. quirk.

**14. A quick and accurate method to find the length of side door casings is**

a. measuring the distance from floor to the gauged line of the header casing.
b. marking the length on a scrap strip and transferring it to the side casing.
c. turning the mitered end upside down against the floor and marking at top edge of head casing.
d. holding the side casing and marking the miter.

**15. The base shoe is fastened**

a. to the baseboard only.
b. to both the base and the floor.
c. to the floor only.
d. directly to the wall.

**16. When the end of a molding piece has no material to butt against, its end is**

a. back-mitered.
b. square cut.
c. mitered.
d. coped.

**17. Most hardwood finish flooring is made from**

a. Douglas fir.
b. southern pine.
c. hemlock.
d. oak.

**18. The shortest piece of strip floor that should be installed is**

a. 4 inches.
b. 6 inches.
c. 8 inches.
d. 9 inches.

**19. The best grade of unfinished oak strip flooring is**

a. prime.
b. select.
c. clear.
d. quarter-sawed.

**20. To change direction that strip flooring is installed,**

a. face nail both strips.
b. turn the extended strip around.
c. blind nail both strips.
d. use a spline.

## WHAT'S WRONG WITH THIS PICTURE?

Carefully study **Figure 18-42** and write an analysis of what is wrong. Consider all possibilities.

### WRONG

**FIGURE 18-42** If only one joint in the room was as ill-fitting as this one, it could be assumed the carpenter was sloppy. However, if most joints in the room look this way, the moisture content of the molding was too high when it was installed. As wood loses moisture to equalize with surrounding humidity conditions, it shrinks across the grain, but not much lengthwise. This allows the outside corners to remain tight while the inside corners pull away from each other.

### RIGHT

**FIGURE 18-43** This is the way the miter joints should look.

CHAPTER 19

# Stair Framing and Finish

**OBJECTIVES** *After completing this unit, the student should be able to:*

- Name various stair finish parts and describe their location and function.
- Describe several stairway designs.
- Define terms used in stair framing.
- Determine the unit rise and unit run of a stairway given the total rise.
- Determine the length of a stairwell.
- Lay out a stair carriage and frame a straight stairway.
- Lay out and frame a stairway with a landing.
- Lay out, dado, and assemble a housed-stringer staircase.
- Apply finish to the stair body of open and closed staircases.
- Install a post-to-post balustrade from floor to balcony on the open end of a staircase.

# GLOSSARY OF STAIR FRAMING AND FINISH TERMS

**baluster** vertical members of a stair rail, usually decorative and spaced closely together

**balustrade** the entire stair rail assembly, including handrail, balusters, and posts

**handrail** a railing on a stairway intended to be grasped by the hand to serve as a support and guard

**newel post** an upright post supporting the handrail in a flight of stairs

**rake** the sloping portion of trim, such as on gable ends of a building or stair

**shank hole** a hole drilled for the thicker portion of a wood screw

**stairwell** an opening in the floor for climbing or descending stairs, or the space of a structure where the stairs are located

The staircase is usually one of the most outstanding features of a building's interior. A high degree of skill is necessary for the design and construction of staircases. Care should be taken in the layout and assembly of stairs. It is important that staircases be comfortable and safe to use. They must also be strong enough to provide ample support and protection from accidents. All stair finish work must be done in a first-class manner. Joints between stair finish members must be tight-fitting for best appearance.

Many kinds of stair finish parts are manufactured in a wide variety of wood species, such as oak, beech, cherry, poplar, pine, and hemlock. It is important to identify each of the staircase parts, know their location, and understand their function.

Stairs tend to be complicated to understand at first. This is because much information must be mastered to achieve success. It is helpful to know that all stairs are laid out using the same theory. This theory is also similar to that used for rafter layout. The variations in stair construction depend largely on the stair function, location, and component material used.

**Balustrade** is the name given to the posts, handrail, and spindles as a whole. They are the more visible and complex component of a staircase. Installing balustrades is one of the most intricate types of interior finish work.

Stairs and balustrades offer carpenters an opportunity to express their craft and skill. They can become a show piece, an example of workmanship that lives on for decades. Stairs reveal the personality and dedication of the carpenters who labored over them.

# STAIRWAY DESIGN

Before stair construction begins, consideration must be given to the layout and framing of the **stairwell**. The stairwell is the opening in the floor through which a person must pass when climbing and descending the stairs (Fig. 19-1).

*Stairs, stairway*, and *staircase* are terms used to designate one or more flights of steps leading from one level of a structure to another. Stairs are further defined as *finish* or *service* stairs. Finish stairs extend from one habitable level of a house to another. Service stairs extend from a habitable to an uninhabitable level. Stairways in residential construction should be at least 36 inches wide, preferably even wider. This allows the passage of two persons at a time and for the moving of furniture (Fig. 19-2). Many codes restrict the maximum height of a single flight of stairs to 12 feet.

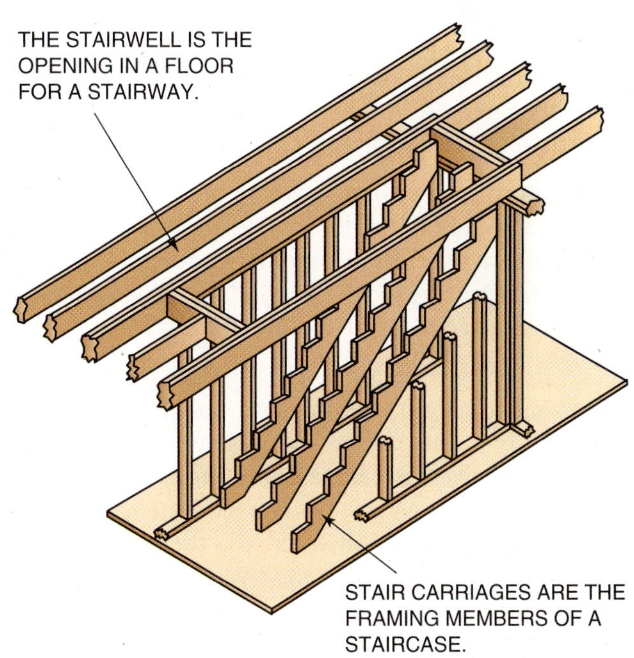


**FIGURE 19-1** Frame for stairs and stairwell.

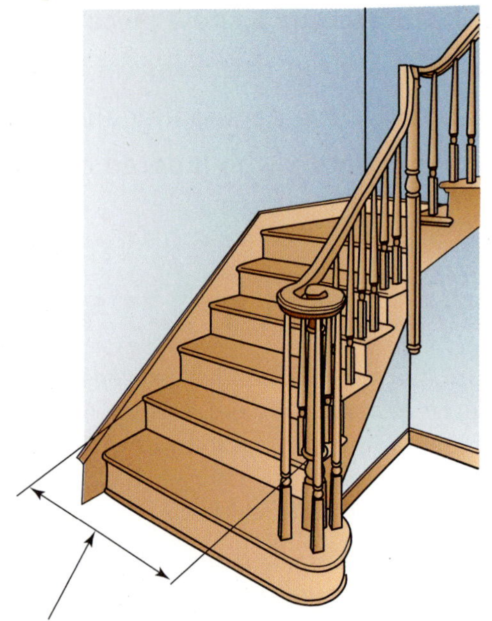


**FIGURE 19-2** Recommended stair widths as measured between the walls or between the wall and the railing.

## TYPES OF STAIRWAYS

A *straight* stairway is continuous from one floor to another. There are no turns or landings. *Platform* stairs have intermediate landings between floors. Platform stairs sometimes change direction at the landing. An L-type platform stairway changes direction 90 degrees. A U-type platform stairway changes direction 180 degrees.

Platform stairs are installed in buildings in which there is a high floor-to-floor level. Platform stairs and landings also provide a temporary resting place. They are a safety feature in case of a fall. Landings are usually constructed near the middle of the staircase.

A *winding* staircase gradually changes direction as it ascends from one floor to another. In many cases, only a part of the staircase winds. Winding stairs may solve the problem of a shorter horizontal run. However, their use is not recommended. They are more difficult to construct, and they pose a danger because of their tapered treads, which are narrow on one end (Fig. 19-3).

Stairways constructed between walls are called *closed* stairways. Closed stairways are more economical to build. However, they add little charm or beauty to a building. Stairways that have one or both sides open to a room are called open stairways. One side of the staircase may be closed while the other side is open for all or part of the flight.

## STAIR FRAMING TERMS

The terms used in stair framing are defined in the following material. Figure 19-4 illustrates the relationship of the various terms to each other and to the total staircase.

*Total Rise.* The *total rise* of a stairway is the vertical distance between finish floors.

*Total Run.* The *total run* is the total horizontal distance that the stairway covers.

*Unit Rise.* The *unit rise* is the vertical distance from one step to another.

*Riser.* A *riser* is the finish material that covers the rise.

*Unit Run.* The *unit run* is the horizontal distance between the faces of the risers.

*Tread.* A *tread* is the horizontal member on which the feet are placed when climbing or descending the stairs.

*Nosing.* The *nosing* is that part of the tread that extends beyond the face of the riser.

*Stair Carriage.* A *stair carriage* is the main support under the risers and treads.

*Stair Stringer.* A *stair stringer* serves as a finish or trim piece to the stairs that covers the carriage. It may also serve as the main support for the risers and treads when a carriage is not used.

*Stairwell.* A *stairwell* is an opening in the floor for the stairway to pass through. It provides adequate headroom for persons using the stairs.

*Headroom. Headroom* is the smallest vertical distance between the stairs and the upper construction over the foot of the stairs.

*Closed Staircase.* In a *closed staircase,* the treads and risers end against a vertical surface.

*Open Staircase.* In an *open staircase*, the ends of the tread and risers are visible.

## DETERMINING THE UNIT RISE AND UNIT RUN

Staircases must be constructed at an appropriate angle for maximum ease in climbing and for safe

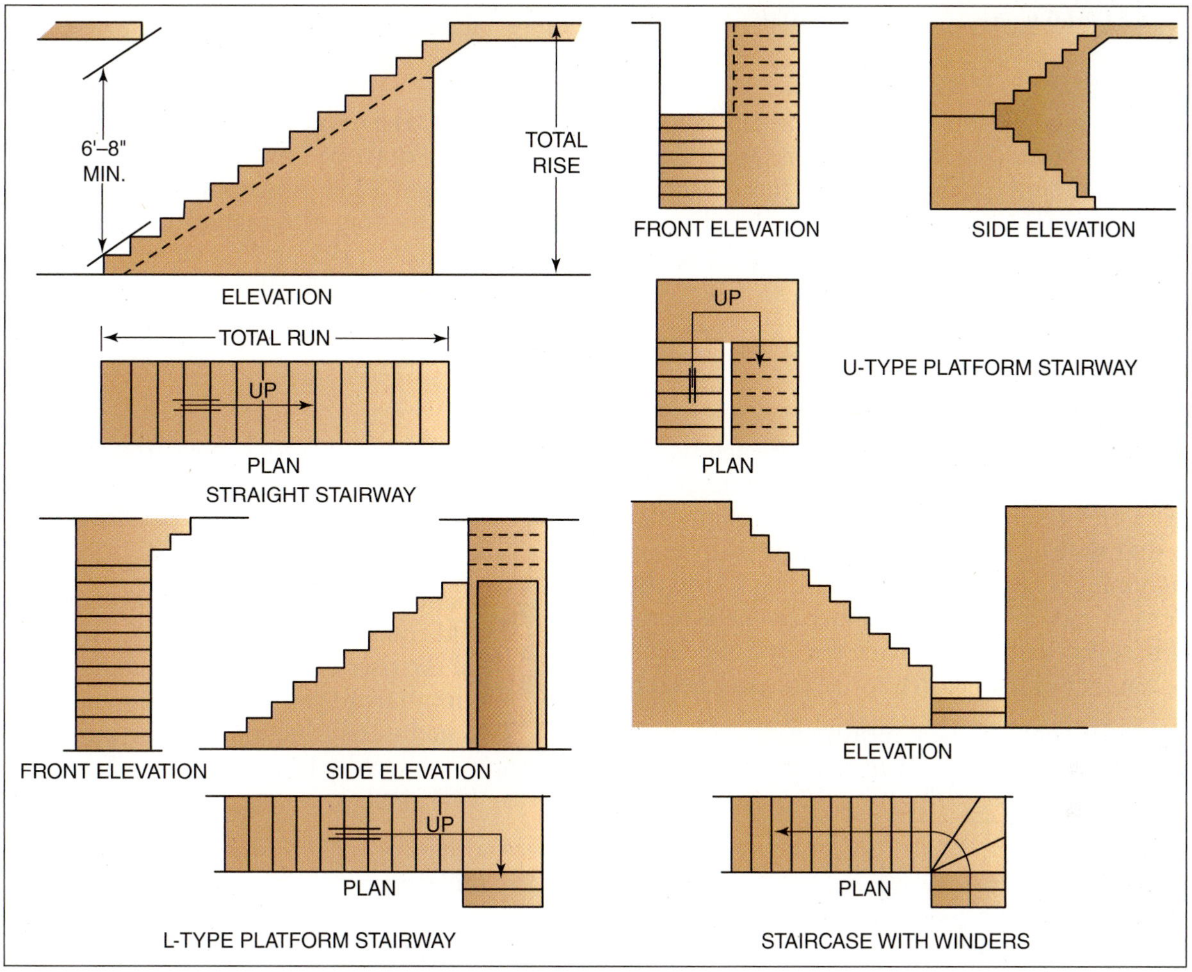

6'–8"
MIN.
TOTAL
RISE
ELEVATION
TOTAL RUN
UP
PLAN
STRAIGHT STAIRWAY
FRONT ELEVATION
SIDE ELEVATION
UP
PLAN
U-TYPE PLATFORM STAIRWAY
FRONT ELEVATION
SIDE ELEVATION
UP
PLAN
L-TYPE PLATFORM STAIRWAY
ELEVATION
PLAN
STAIRCASE WITH WINDERS

**FIGURE 19-3** Various types of stairways.

LENGTH OF STAIRWELL
STAIRWELL HEADER
TREAD
NOSING
RISER
RUN
RISE
HEADROOM
PLUMB LINE
TOTAL RISE
STRINGER OR
CARRIAGE
FINISH FLOOR
SUBFLOOR
TOTAL RUN

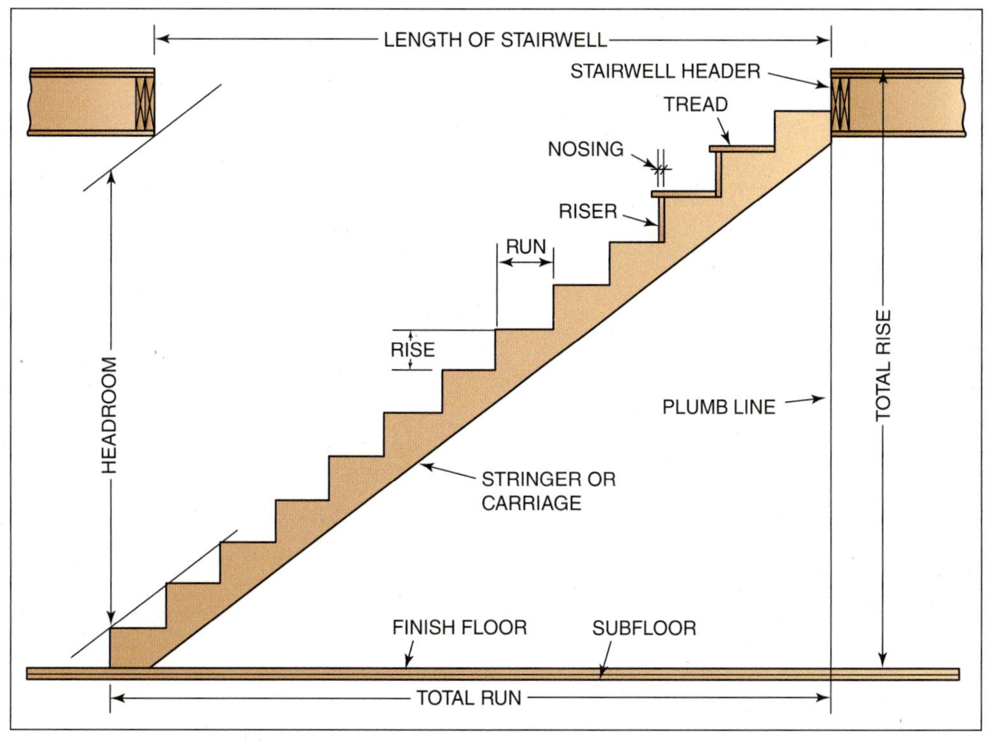

**FIGURE 19-4** Stair framing terms.

descent. The relationship of the unit rise and run determines this angle (Fig. 19-5). The preferred angle is between 30 and 38 degrees. The International Building Code (IBC) specifies that the height of a riser shall not exceed 7¾ inches and that the width of a tread not be less than 10 inches.

### Riser Height

Overall floor to ceiling heights vary between buildings. Therefore, individual unit rise for each building must be determined. To do so, first measure the total rise of the stairway. Next, find the number of risers that will fit in the opening. This is done by first assuming a riser height and dividing it into the total rise. Rounding off this number gives the number of risers needed. Then, divide the total rise again by the number of rises to determine the riser height.

**Example:** The total rise is measured at 106 inches. Divide the total rise by 7.75 inches, the maximum riser height. The result is 13.68 risers. Since the number of steps must be even and the number of risers no fewer than 13.68, this number is rounded up to 14 risers. Dividing 106 inches by 14 risers gives a riser height of 7.571 or 7⁹⁄₁₆ inches. Note that the number of risers could also be 15. Dividing 106 by 15 gives a riser height of 7.07 or 7¹⁄₁₆ inches.

### Tread Run

The unit run (tread run) is measured from the face of one riser to the next riser. It does not include the nosing (Fig. 19-6). To find the unit run, apply the following rule:

> The sum of one unit riser and one unit run should equal between 17 and 18. For example, if the riser height is 7⁹⁄₁₆ inches, then the minimum tread run may be 17 inches minus 7⁹⁄₁₆ inches or 9⁷⁄₁₆ inches. The maximum unit run may be 18 inches minus 7⁹⁄₁₆ inches or 10⁷⁄₁₆ inches.

Another formula to find the unit run is found in many building codes. It states that the sum of two risers and one tread shall not be less than 24 inches nor more than 25 inches (Fig. 19-7). With this formula, a unit rise of 7⁹⁄₁₆ inches calls for a minimum unit run of 8⅞ inches and maximum of 9⅞ inches. These numbers are further restricted to a minimum of 10 inches by the IBC. As a result of combining both formulas, the desired tread run could be between 10 and 10⁷⁄₁₆ inches.

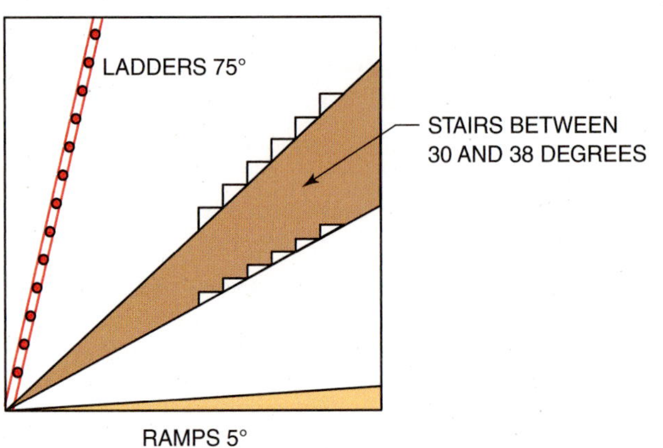


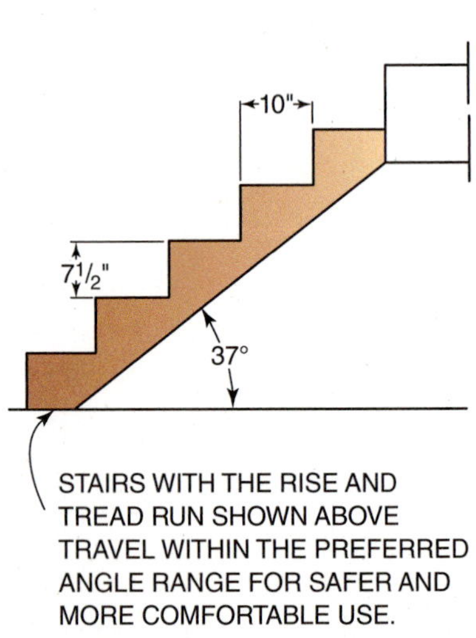


**FIGURE 19-5** Recommended angles for ladders, stairs, and ramps.

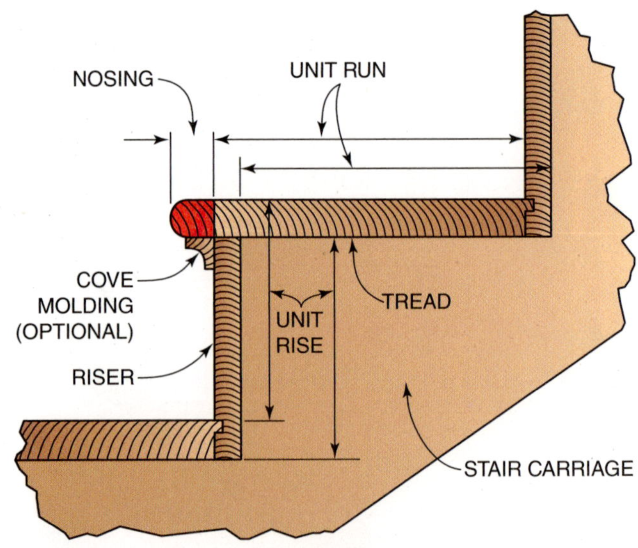


**FIGURE 19-6** The tread run does not include the nosing.

*from experience...*

Two formulas are used to determine the unit run for stairs (Fig. 19-7).

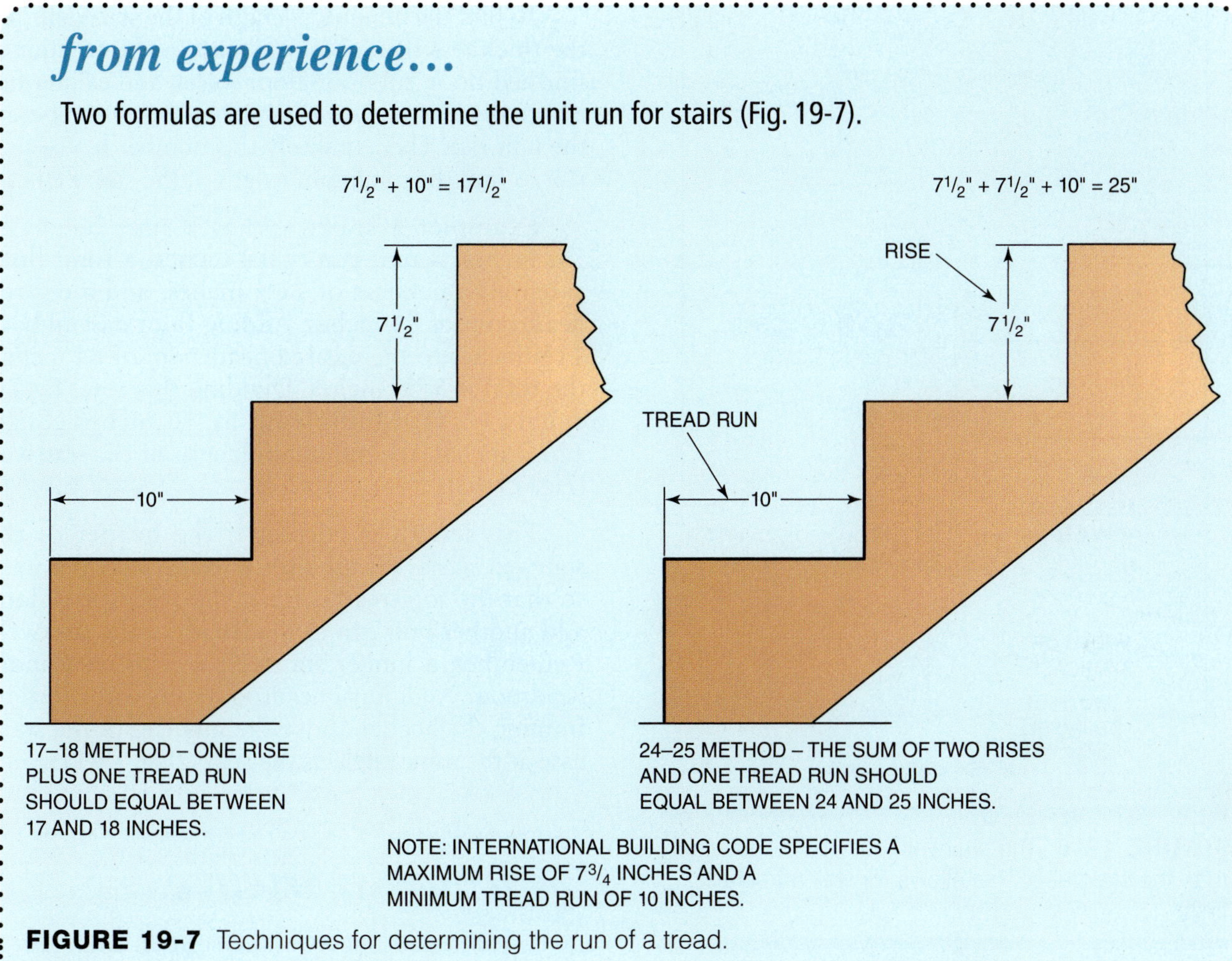


**FIGURE 19-7** Techniques for determining the run of a tread.

### Variations to Stair Steepness

Decreasing the unit rise increases the run of the stairs. This uses up more horizontal space. Increasing the riser height decreases the run. This makes the stairs steeper, taking up less horizontal space. The carpenter must use good judgment and adapt the unit rise and run dimensions to the space in which the stairway is to be constructed in order to conform with the building code. Lower angle stairs are easier and safer to climb. In general, a riser height of 7½ inches and a unit run of 10 inches makes a safe, comfortable stairway.

## DETERMINING THE SIZE OF THE STAIRWELL

The width and length of the stairwell must be determined before the staircase is built.

### Stairwell Width

The width of the stairwell depends on the width of the staircase. The drawings show the finish width of the staircase. However, the stairwell must be made wider than the staircase to allow for wall and stair finish (Fig. 19-8). Extra width will be required for a handrail and other finish parts of an open staircase that makes a U-turn on the landing above. The carpenter must be able to determine the width of the stairwell by studying the size, type, and placement of the stair finish before framing the stairs.

### Length of the Stairwell

The length of the stairwell depends on the slope of the stairway. Stairs with a low angle require a longer stairwell to provide adequate headroom (Fig. 19-9). Most building codes require a minimum of 6'-8" for headroom. However, 7'-0" headroom or more is preferred.

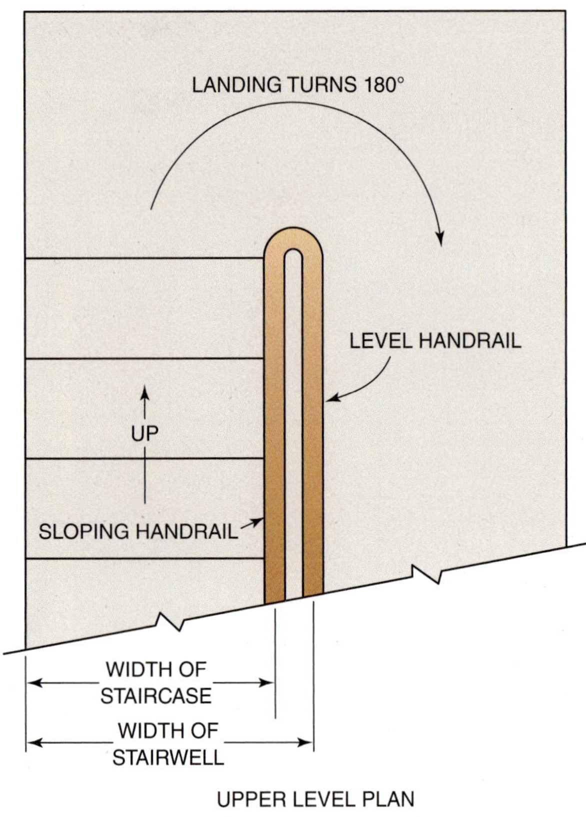


**FIGURE 19-8** The stairwell must be made wider than the staircase. This allows for wall and stair finish.

To find the minimum length of the stairwell, add the thickness of the *floor assembly* above (sum of finished floor, subfloor, floor joists, and ceiling finish) to the desired headroom. Divide this number by the unit rise. Then, multiply this number by the unit run to find the minimum length of the stairwell.

**Example:** A stairway has a riser height of 7½ inches, a unit run of 10 inches, a total floor assembly thickness of 11¾ inches, and a desired headroom of 84 inches. Adding floor assembly of 11¾ inches to the desired headroom of 84 inches, the total is 95¾ inches. Dividing this total by 7½ equals 12.77. Multiplying by 10 inches equals 127.7 inches, the minimum length of the stairwell (Fig. 19-10).

This length is correct if the header of the stairwell acts as the top riser. If the carriage is framed so that the top tread is flush with the upper floor, add another unit run to the length of the stairwell. Remember, a longer stairwell will provide more headroom. Additional headroom can be obtained by framing the header above the low end of the staircase at the same angle as the stairs (Fig. 19-11).

*from experience...*

Frame the stairwell header at an angle to increase the floor space above.

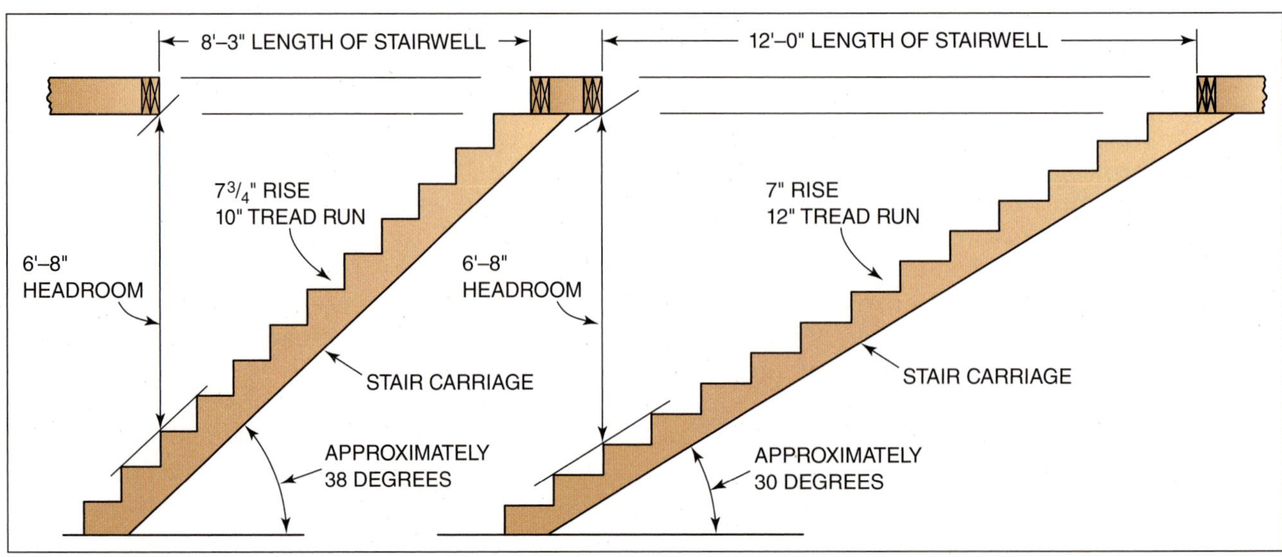


**FIGURE 19-9** Low-angle stairs require longer stairwells to provide adequate headroom.

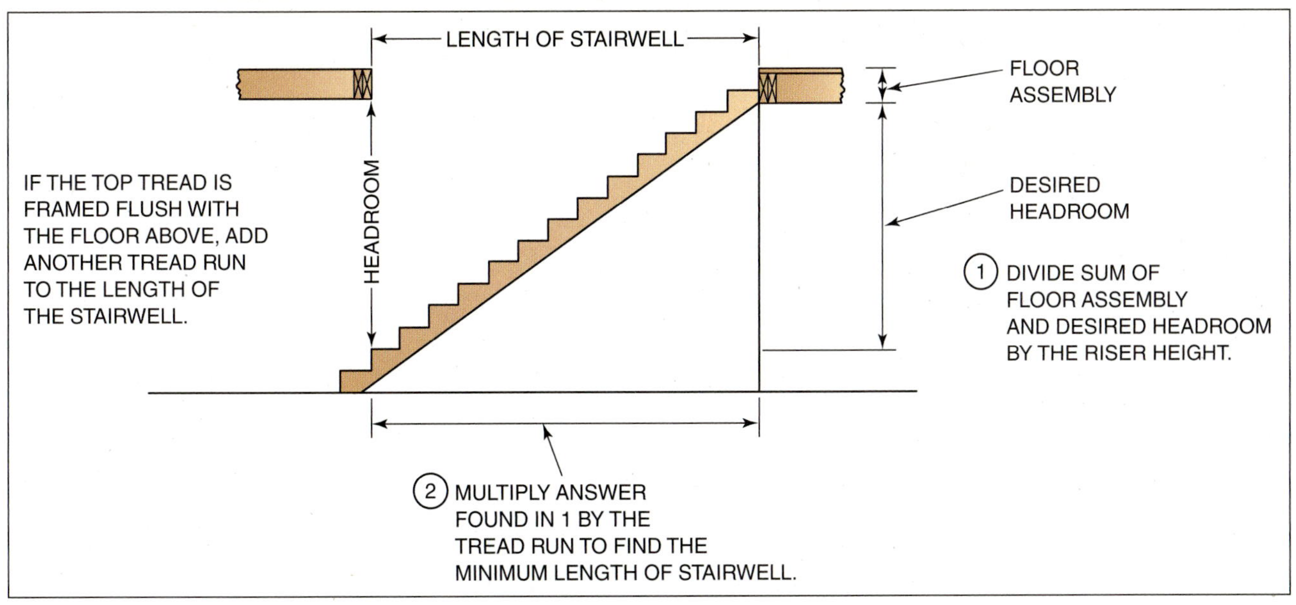


**FIGURE 19-10** How to calculate the length of a stairwell.

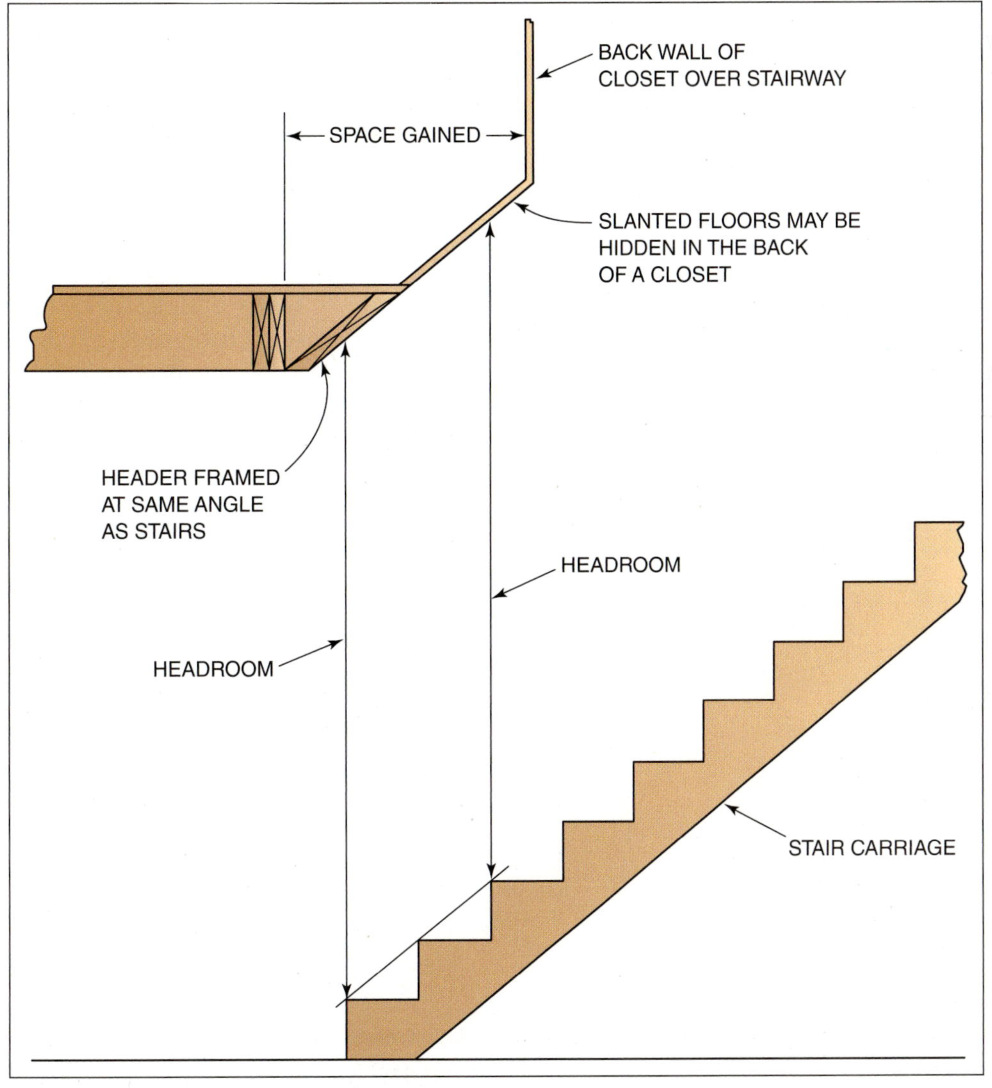


**FIGURE 19-11** Technique for increasing the upper floor space while maintaining required headroom for the stair.

## METHODS OF STAIR CONSTRUCTION

There are two basic styles of staircase design. The stairwell length may be estimated using a framing square. A landing may be installed between two smaller staircases.

### Open Carriage Staircase

The open carriage staircase is laid out for cutouts. Risers and treads are fastened to the cutouts (Fig. 19-12). Occasionally, rough treads and risers are installed temporarily for easy access to upper levels and later finished with higher quality material. Other times the staircase is built as a service stair such as those constructed for basements.

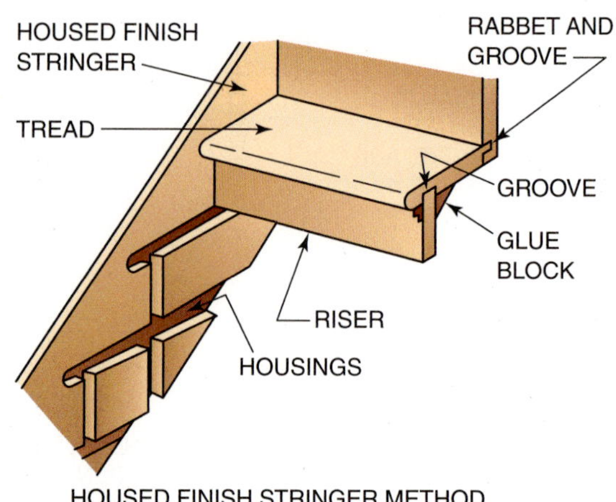


**FIGURE 19-13** A housed finish stringer method of stair construction.

### The Housed Stringer Staircase

The housed stringer staircase is installed when the house is ready for finishing. Dadoes are routed into the sides of the finish stringer. They *house* (make a place for) and support the risers and treads (Fig. 19-13). A router and *stair jig* are used to dado the stringers. Next, the treads, risers, and other stair parts are cut to size. The staircase is then assembled. Stair carriages are not required when the housed finish stringer method of construction is used. The layout for both housed stringers and stair carriages is made in a similar manner.

### Determining the Rough Length of a Staircase

The length of lumber needed for the stair carriage may be determined by using the Pythagorean theorem on the total rise and total run of the stairway. It also can be found by scaling across a framing square. Use the edge of the square that is graduated in 12ths of an inch. Mark the total rise on the tongue. Then, mark the total run on the blade. Scale off in between the marks (Fig. 19-14).

**Example:** A stairway has a total rise of 8'-9" and a total run of 12'-3". Scaling across the square between 8-9/12ths and 12-3/12ths at a scale of 1" = 1'-0" reads a little over 15. A 16-foot length of lumber is needed for the stair carriage.

### Laying Out and Framing a Stairway with a Landing

A stair landing is an intermediate platform between two flights of stairs. A landing is designed for

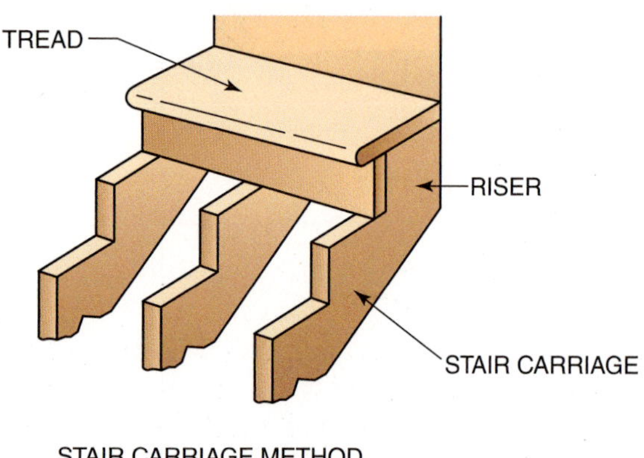


**FIGURE 19-12** Stair carriages are notched under the risers and treads.

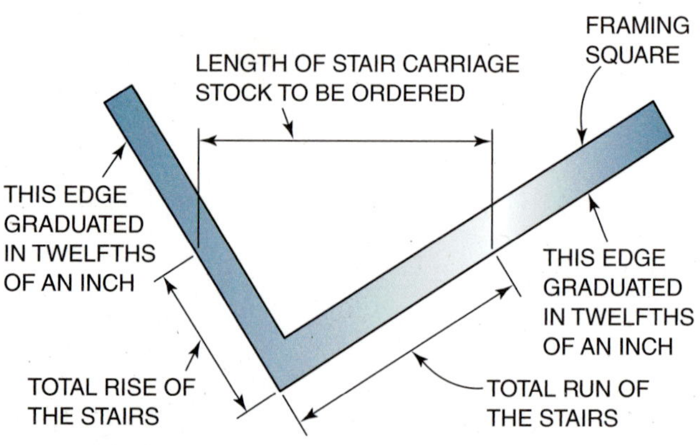


**FIGURE 19-14** Scale across the framing square to find the rough length of a stair carriage.

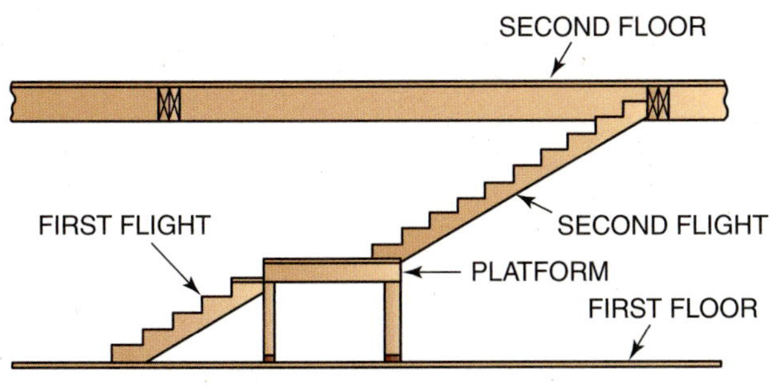


**FIGURE 19-15** Stair carriages are framed to the platform in a way similar to that used for two straight flights.

changing the direction of the stairs and as a resting place for long stair runs (Fig. 19-15). The landing floor surface is usually made with the same material as the main floor of the structure. Many codes state that no flight of stairs shall have a vertical rise of more than 12 feet. Therefore, any staircase running between floors with a vertical distance of more than 12 feet must have at least one intermediate landing or platform. Some codes require that the minimum length of a landing be not less than 2'-6". Other codes require the minimum dimension to be the width of the stairway. U-type stairs usually have the landing about midway on the flight.

Platform stairs are built by first erecting the platform. The finish floor of the platform should be the same height as if it were a finish tread in the staircase. This allows an equal riser height for both flights (Fig. 19-16). The stairs are then framed to the platform as two straight flights. Either the stair carriage or the housed stringer method of construction may be used.

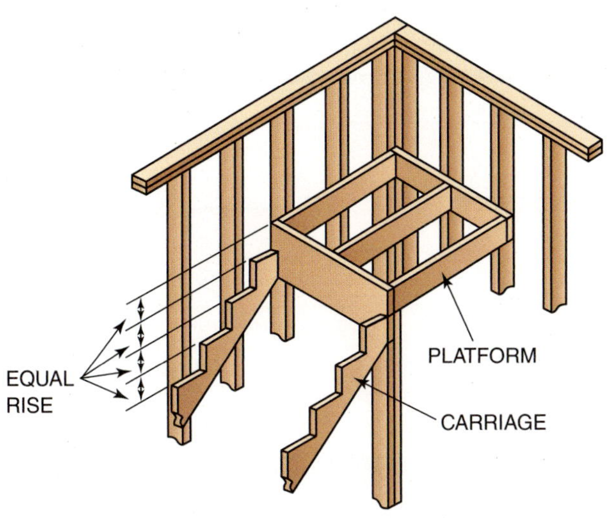


**FIGURE 19-16** The platform is located so that its top is in line with one of the tread runs.

# Stair Carriage

When laying out stair carriages, make sure that every riser height will be the same. Also, make sure all tread widths will be equal when the staircase is finished. It is dangerous to descend a flight of stairs and unexpectedly step down a longer or shorter distance than the previous step. It is also dangerous for a person accustomed to stepping on treads of a certain width suddenly to step on a narrower tread. Stairs that are not laid out and constructed properly could cause a fatal accident.

Place the stair carriage stock on a pair of sawhorses. Sight the stock for a crowned edge. This will be the top edge of the carriage. Set stair gauges on the framing square, with the unit rise on the tongue and the unit run on the blade. Lines marked out along the tongue will be plumb lines and those along the blade will be level lines.

Step off the necessary number of risers and treads, marking along the outside of the tongue and blade. These lines are the back sides of the finish risers and undersides of the finish treads (Fig. 19-17). Lay out a plumb line at the top of the carriage for the last riser. Lay out a level line on the bottom where the carriage sits on the floor.

## EQUALIZING THE BOTTOM RISER

A certain amount may have to be cut off the bottom end of the stair carriage. This is to make the bottom riser height equal to all the other risers when the staircase is finished. The thickness of the finish floor and the finish stair treads must be known.

If the carriage rests on the finish floor, the first riser is reduced by the thickness of the tread stock. If the bottom of the carriage rests on the subfloor, and the finish floor and tread stock are the same thickness, the height of the first riser is the same as all the rest. If the bottom of the carriage rests on the subfloor and the tread stock is thicker than the finish floor, then the first riser is reduced by the difference in tread and finished floor thickness (Fig. 19-18).

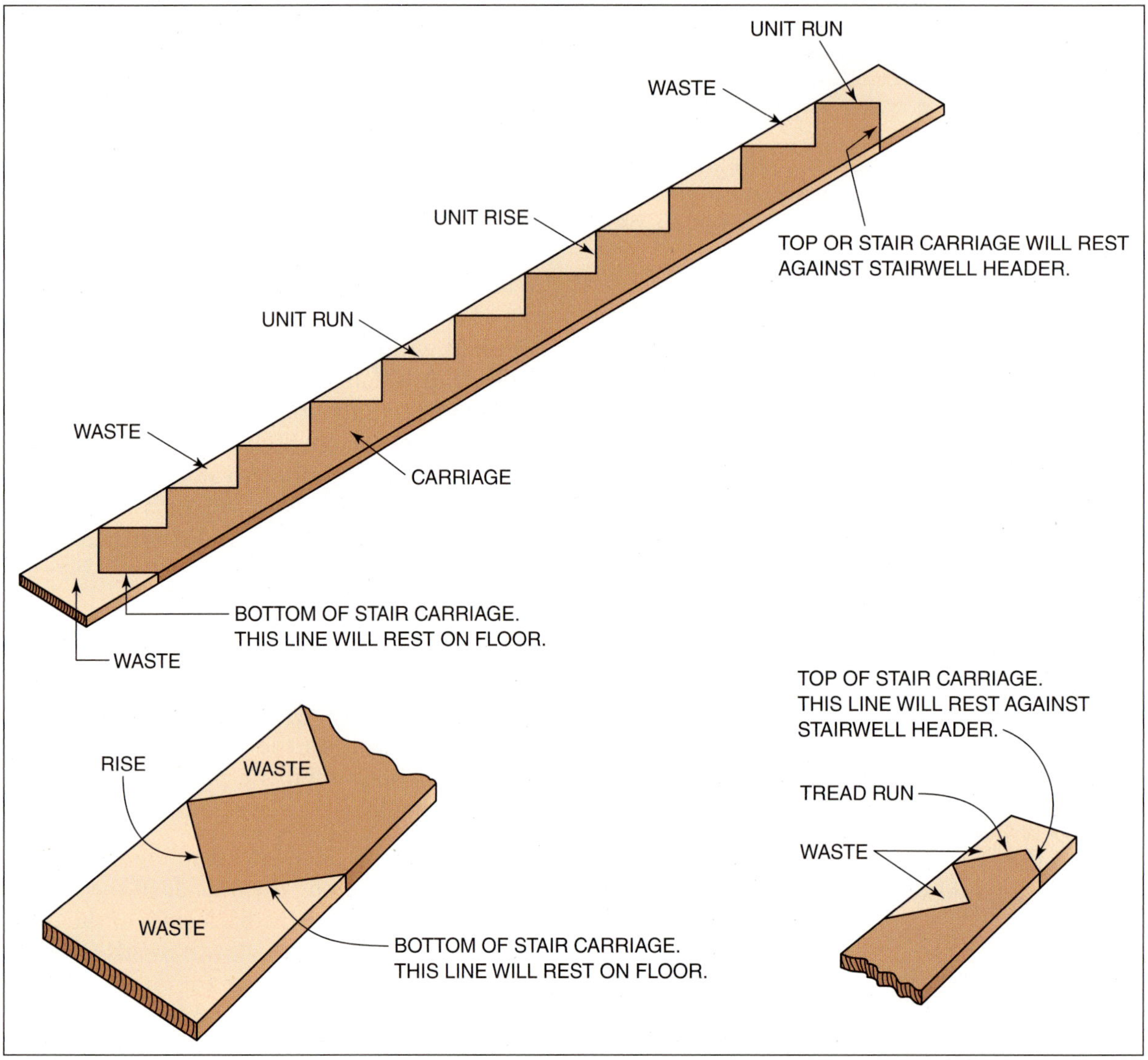

UNIT RUN
WASTE
UNIT RISE
TOP OR STAIR CARRIAGE WILL REST AGAINST STAIRWELL HEADER.
UNIT RUN
WASTE
CARRIAGE
BOTTOM OF STAIR CARRIAGE. THIS LINE WILL REST ON FLOOR.
WASTE
TOP OF STAIR CARRIAGE. THIS LINE WILL REST AGAINST STAIRWELL HEADER.
RISE
WASTE
TREAD RUN
WASTE
WASTE
BOTTOM OF STAIR CARRIAGE. THIS LINE WILL REST ON FLOOR.

**FIGURE 19-17** Laying out a stair carriage by stepping off with a framing square.

TOP OF FINISHED TREAD
EQUAL
EQUAL
EQUAL
CARRIAGE
CUT CARRIAGE HERE
EQUAL
EQUAL
EQUAL
CUT CARRIAGE HERE
FINISH FLOOR
STAIR CARRIAGE RESTS ON SUBFLOOR.
SUBFLOOR
STAIR CARRIAGE RESTS ON FINISH FLOOR.

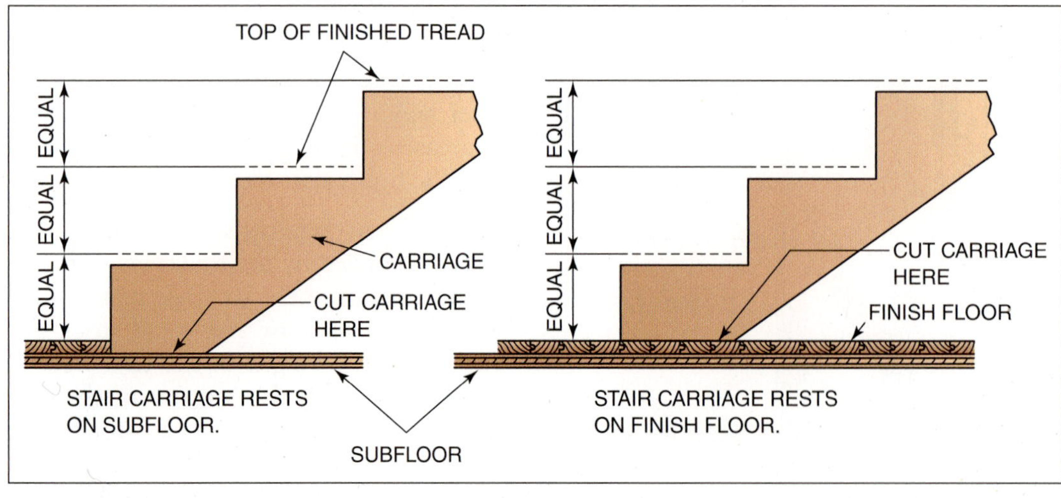

**FIGURE 19-18** Dropping the stair carriage to equalize the first riser.

## EQUALIZING THE TOP RISER

No amount is cut from the level line at the top of the carriage. The top riser is equalized by fastening the carriage a certain distance below the subfloor of the top landing. This distance depends on the thickness of the stair tread and the finish floor above.

With the header of the stairwell acting as the top riser, if the stair tread thickness is the same as the finish floor, the carriage is fastened at a riser height below the subfloor.

If the stair tread is thicker than the finish floor, fasten the carriage down the distance of a riser, plus the difference in tread and floor thickness. For example, if the tread stock is 1¹⁄₁₆ inches and the finish floor above is ¾ inch, the difference is ⁵⁄₁₆ inch. The carriage is placed at the riser height plus the difference below the top of the subfloor (Fig. 19-19). Whatever conditions may be encountered, take steps to equalize the top and bottom risers for the construction of a safe staircase.

## ATTACHING A CARRIAGE TO HEADER

The top end of a carriage must be securely attached to the stairwell header. Methods of attachment vary but must be strong (Fig. 19-20).

## CUTTING THE STAIR CARRIAGES

After the first carriage is laid out, cut it. Follow the layout lines carefully. Finish the cuts at the intersection of the riser and unit run with a reciprocating saw or a handsaw. Using the first carriage as a pattern, lay out and cut as many other carriages as needed. Usually, three carriages are used for residential staircases of average width. For wider stairs, the number of carriages depends on such factors as whether or not risers are used and the thickness of the tread stock. Check the drawings or building code for the spacing of carriages for wider staircases.

> **CAUTION**
>
> **CAUTION:** When making a cut at a sharp angle to the edge of a board with a circular saw, the guard may not retract easily. Retract the guard by hand before the cut and until the cut is made a few inches. Then, release the guard and finish the cut. Never wedge open the guard.

## INSTALLING THE STAIR CARRIAGE

When installing the stair carriage, fasten the first carriage in position on one side of the stairway.

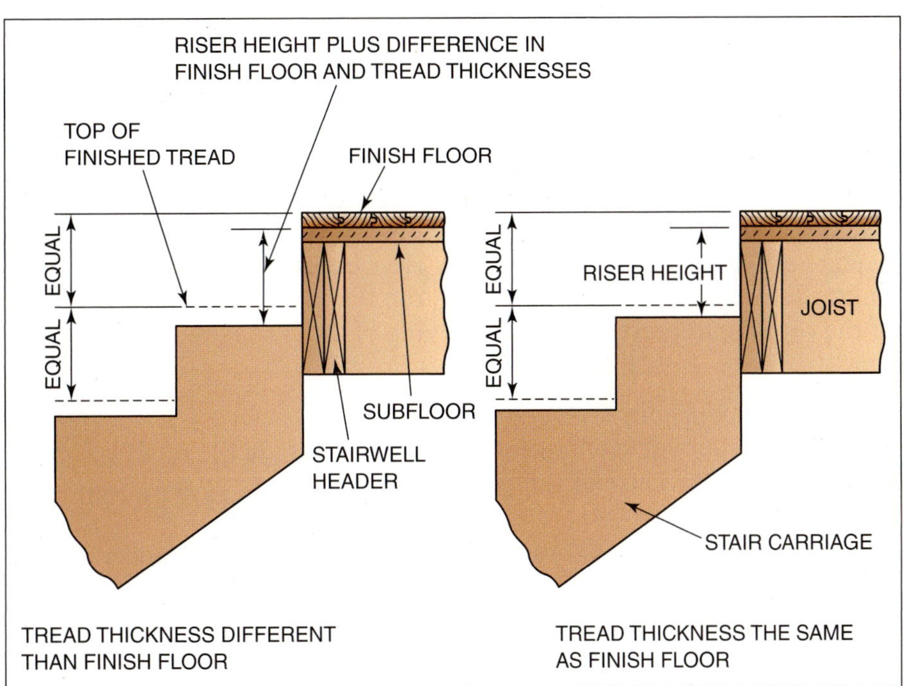


**FIGURE 19-19** Equalizing the top riser.

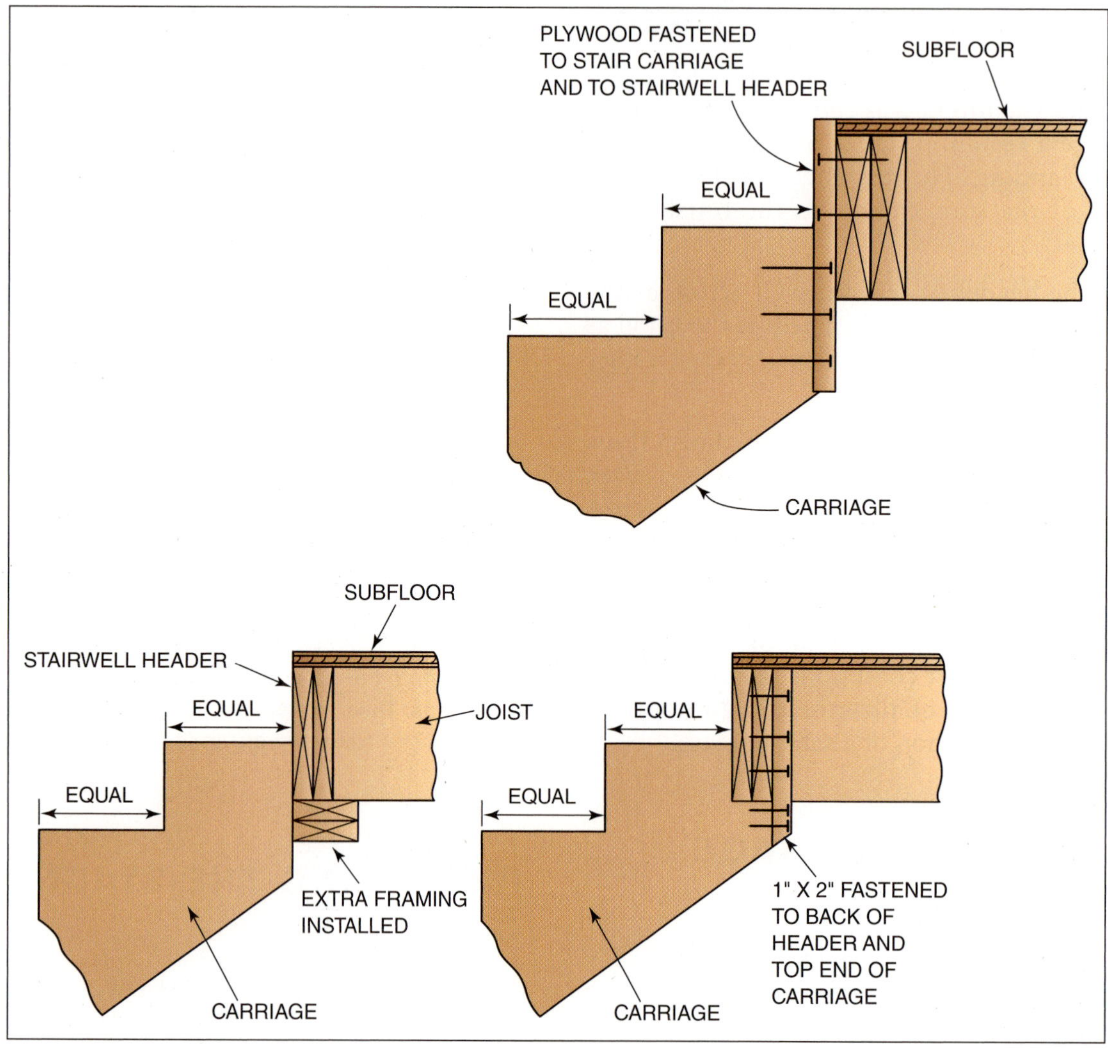


**FIGURE 19-20** Methods of framing the stair carriage to the stairwell header.

Attach it at the top to the stairwell header. Make sure the distance from the subfloor above to the top tread is correct. Fasten the bottom end of the stair carriage to the sole plate of the wall and with intermediate fastenings into the studs. Place nails near the bottom edge of the carriage to prevent splitting the triangular sections.

Fasten a second carriage on the other wall in the same manner as the first. If the stairway is to be open on one side, the location of the stair carriage on the open end of a stairway is in relation to the position of the handrail. First, determine the location of the centerline of the handrail. Then, position the stair carriage on the open side of a staircase, making sure its outside face will be plumb with the centerline of the handrail when it is installed (Fig. 19-21).

Fasten intermediate carriages at the top into the stairwell header and at the bottom into the subfloor as needed. Test all unit run and riser cuts of the carriages together with a straightedge placed across the outside carriages (Fig. 19-22). Fasten a temporary riser board, or where necessary, about halfway up the flight. This straightens and maintains the spacing of the carriages (Fig. 19-23).

## FINISHING A CLOSED STAIR CARRIAGE

The risers and treads must be finished at their ends to cover the joints at the walls.

### Risers

Risers are installed first. Rip the riser stock to the proper width. Actual width may be up to ⅛ inch

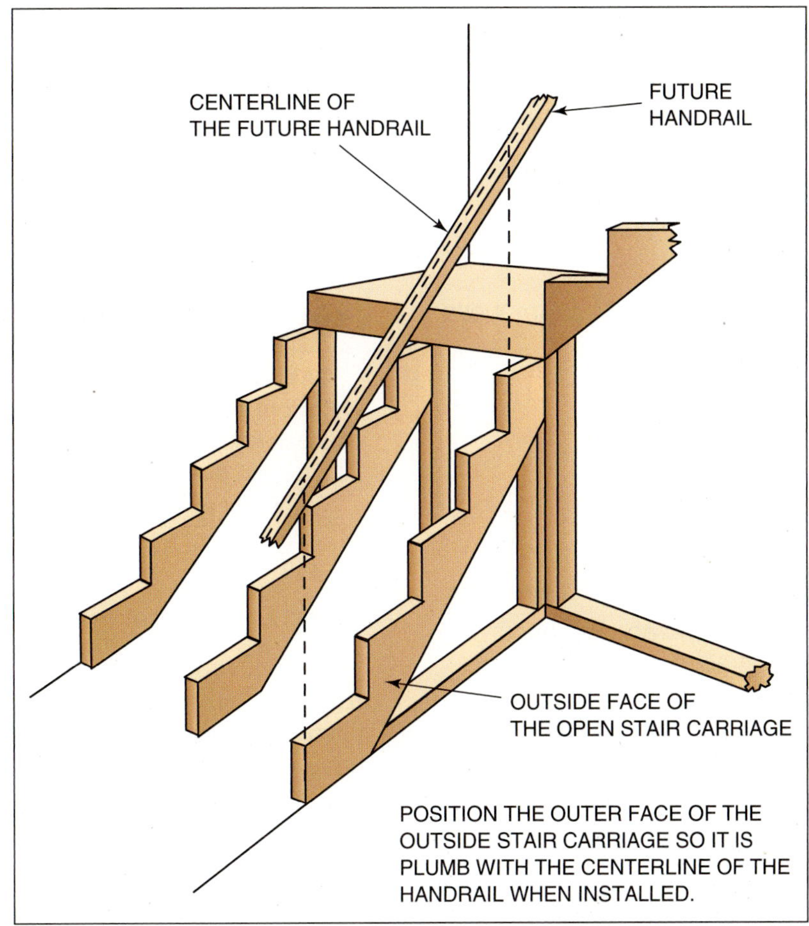


**FIGURE 19-21** Techniques for locating the outside stair carriage.

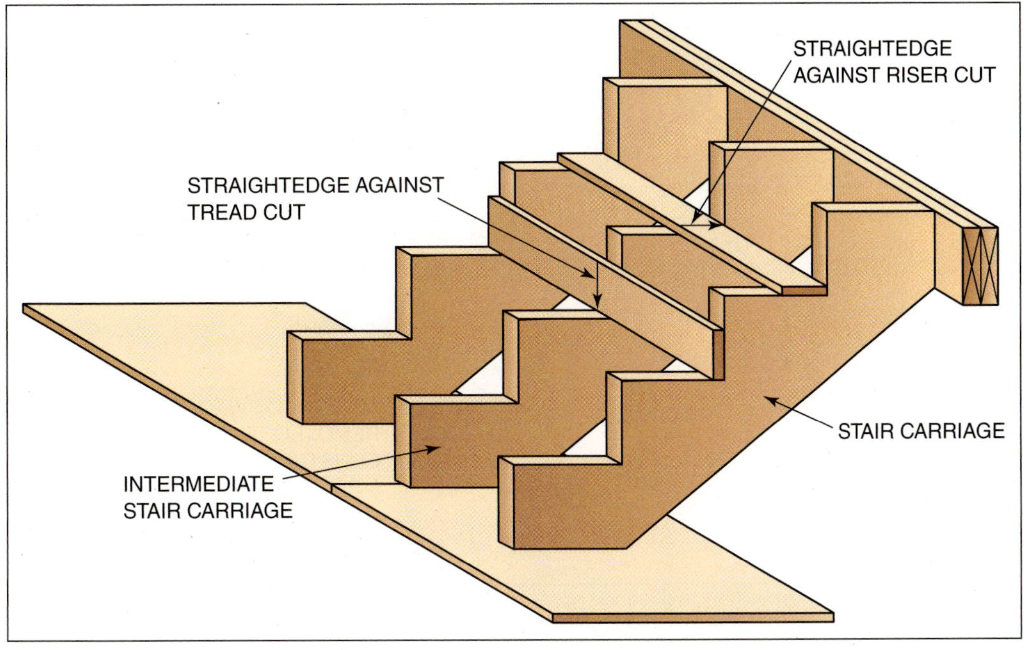


**FIGURE 19-22** Check the position of tread and riser cuts on intermediate carriages with a straightedge.

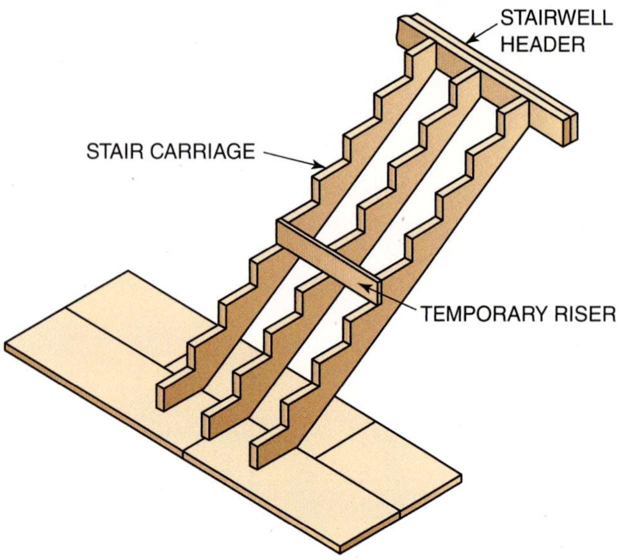


**FIGURE 19-23** Fasten a temporary riser about halfway up the flight. This straightens and maintains the carriage spacing.

narrower. Cut the risers to length, with square ends, about ¼ inch less than the distance between walls. With the top edge flush with the level cut of the carriage, fasten the risers in position with three 2½-inch finish nails into each stair carriage. Start at the top and work down. Remove the temporary treads installed previously as work progresses downward (Fig. 19-24).

CAUTION

**CAUTION: Put up positive barriers at the top and bottom of the stairs so it is obvious that the stairs are being worked on. A serious accident can happen if a person does not realize that the temporary stairs have been removed.**

## Closed Finish Stringer

The *closed finish stringer* is cut around the previously installed risers. Usually 1 × 10 lumber is used. When installed the top edge will be about 3 inches above the tread nosing. A 1 × 12 may be used if a wider finish stringer is desired.

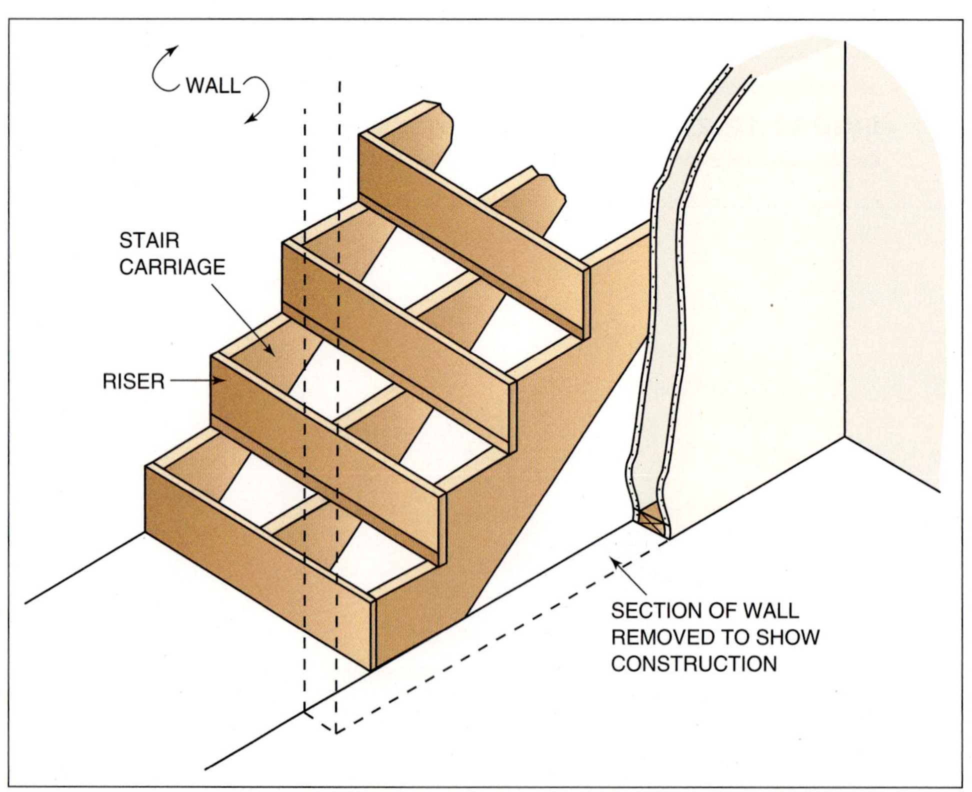


**FIGURE 19-24** Risers are ordinarily the first finish members applied to the stair carriage in a closed staircase.

Tack a length of stringer stock to the wall. Its bottom edge should rest on the top edges of the previously installed risers, and its bottom end should rest on the floor. The top end should extend about 6 inches beyond the landing.

Lay out plumb lines, from the face of each riser, across the face of the finish stringer. If the riser itself is out of plumb, then plumb upward from the part of the riser that projects farthest outward. Then, lay out level lines on the stringer, from each tread cut of the stair carriage and also from the floor of the landing above (Fig. 19-25). Remove the stringer from the wall. Cut to the layout lines. Follow the plumb lines carefully. Plumb cuts will butt against the face of the risers, so careful cuts need to be made. Not as much care needs to be taken with level cuts because treads will later butt against and cover them.

Sand the board and place it back in position. Fasten the stringer securely to the wall with finishing nails. Do not nail too low to avoid splitting the lower end of the stringer. Install the finish stringer on the other wall in the same manner. Shim any out of plumb or not touching risers to the stringer. Shim at intermediate stair carriages to straighten the risers as needed to keep riser straight.

## Treads

Treads are cut on both ends to fit snugly between the finish stringers. The nosed edge of the tread projects beyond the face of the riser by 1⅛ inches (Fig. 19-26). Along the top edge of the riser, measure carefully the distance between finish stringers. Transfer the measurement and square lines across the tread. Cut to length carefully since the cut edge

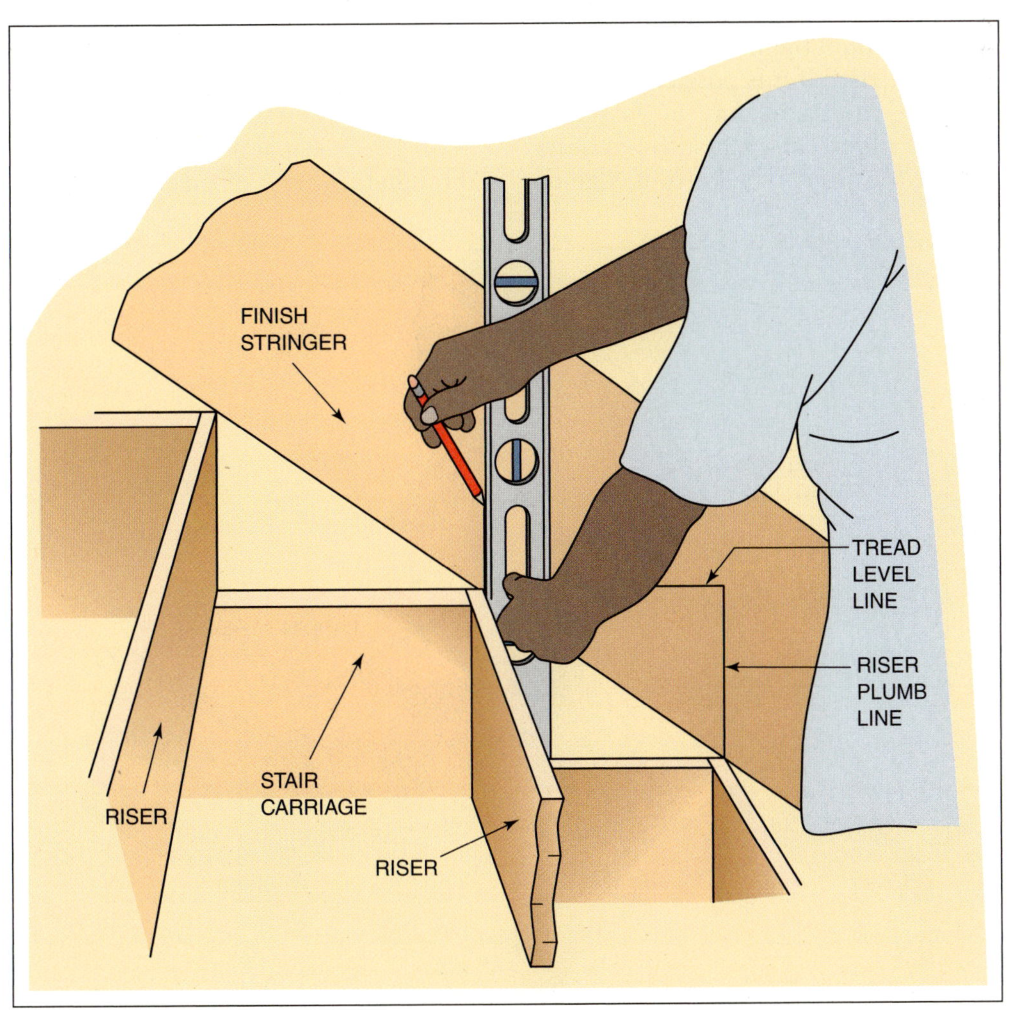


**FIGURE 19-25** The closed finish stringer is laid out using a level to extend plumb and level lines from the stair carriage.

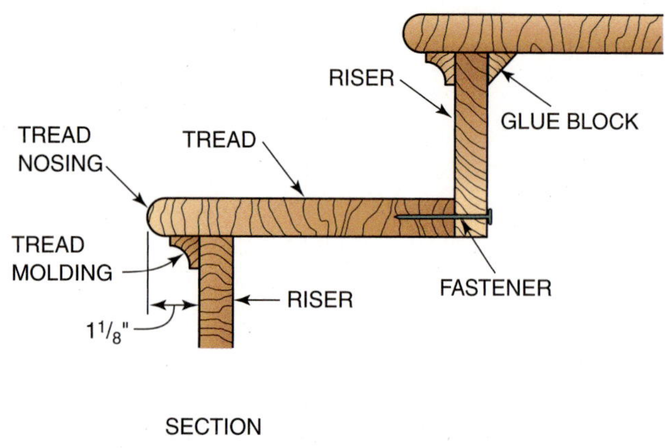


**FIGURE 19-26** Tread and riser detail.

on both sides will be visible. Rubbing the ends with wax will help when placing the tread into position. Using a scrap block on the nosed edge, tap the tread into position until the back edge is firmly positioned against the riser. If it is possible to work from the underside, the tread may be fastened by the use of screw blocks at each stair carriage and at intermediate locations. The bottom edge of the riser is fastened to the back edge of the tread from the back side using 8d finish nails or 1½-inch screws. Glue blocks are installed to prevent squeaks. They may be 2–3 inches long and spaced 1 foot apart.

If it is not possible to work from the underside, the treads must be face nailed. Fasten each tread in place with three 8d finish nails into each stair carriage. It may be necessary to drill holes in hardwood treads to prevent splitting the tread or bending the nail. Start from the bottom and work up, installing the treads in a similar manner. At the top of the stairs, install a landing tread. Use a landing tread that is rabbeted to match the thickness of the finish floor (Fig. 19-27).

## Tread Molding

The *tread molding* is installed under the overhang of the tread and against the riser. Cut the molding to the same length as the treads, using a miter box. Predrill holes, if necessary. Fasten the molding in place with 4d finish nails spaced about 12 inches apart. Nails are driven at approximately a 45-degree angle through the center of the molding.

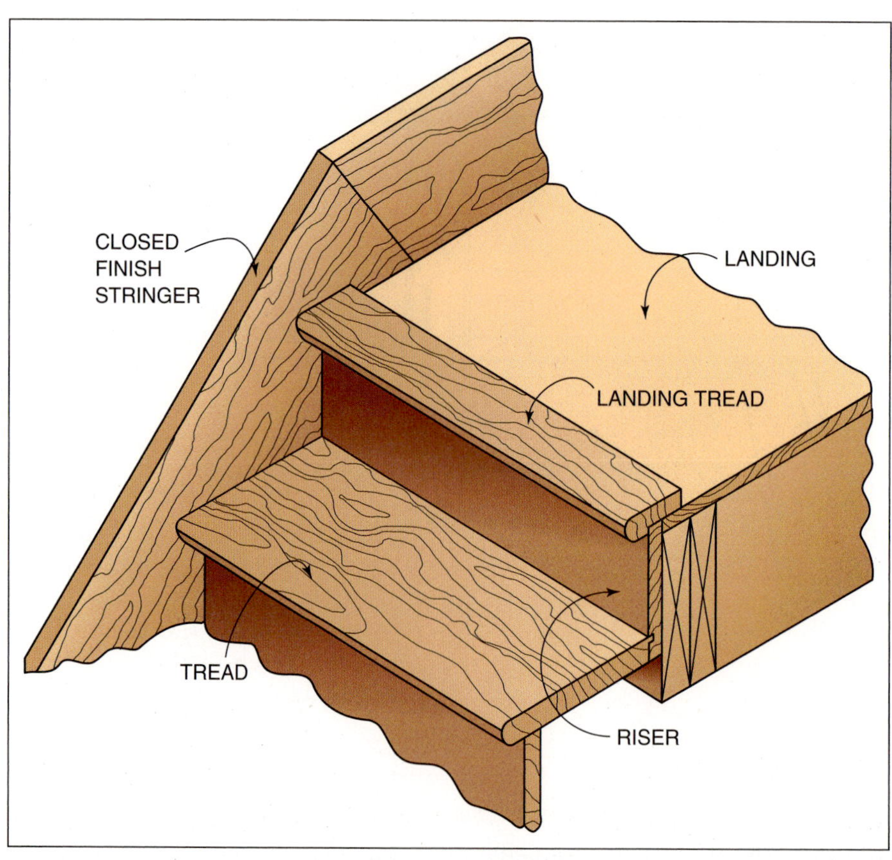


**FIGURE 19-27** A rabbeted landing tread is used at the top of the stairway.

## FINISHING AN OPEN STAIR CARRIAGE

Open staircases require the stringer to be mitered to the risers and each tread nosing to wrap onto the stringer.

### Finish Stringers

The *open finish stringer* must be installed before the *risers* and the *closed finish stringer* placed against the wall. To lay out the *open finish stringer,* cut a length of finish stringer stock. Fit it to the floor and against the landing on the outside of the carriage. Its top edge should be flush with the top edge of the stair carriage. Tack it in this position to keep it from moving while it is being laid out.

First, lay out level lines on the face of the stringer in line with the tread cut on the stair carriage. Next, plumb lines must be laid out on the face of the finish stringer for making miter joints with risers.

*A Preacher.* Use a *preacher* to lay out the plumb lines on the open finish stringer. A preacher is made from a piece of nominal 1-inch stock about 12 inches long. Its thickness must be the same as the riser stock. The preacher is notched in the center. The notch should be wide enough to fit over the finish stringer. The preacher should be long enough to allow it to rest on the tread cut of the stair carriage when held back against the rise cut.

Place the preacher over the stringer and against the riser cut of the stair carriage. Plumb the preacher with a level. Lay out the plumb cut on the stringer by marking along the side of the preacher that faces the bottom of the staircase (Fig. 19-28). This will make a plumb line offset from the carriage riser cut by a thickness of the preacher or riser stock. Lay out all plumb lines on the stringer in this manner. Remove the stringer. Cut to the layout lines. Make miter cuts along the plumb lines, taking care to cut this in the proper direction. Make cuts square through the thickness along the level lines. Sand the piece. Fasten it in position using fasteners in the same holes used to temporarily hold it.

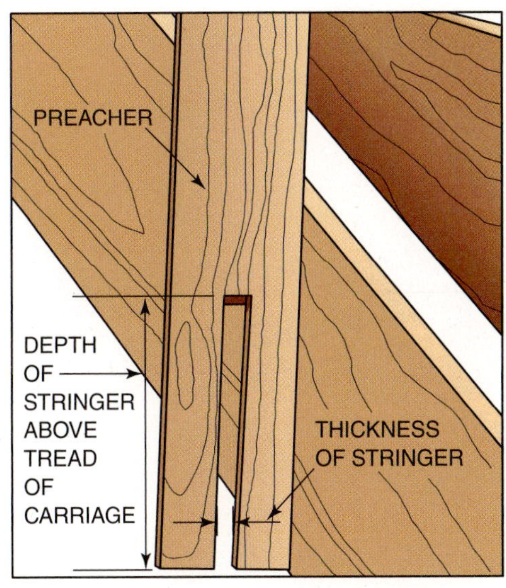


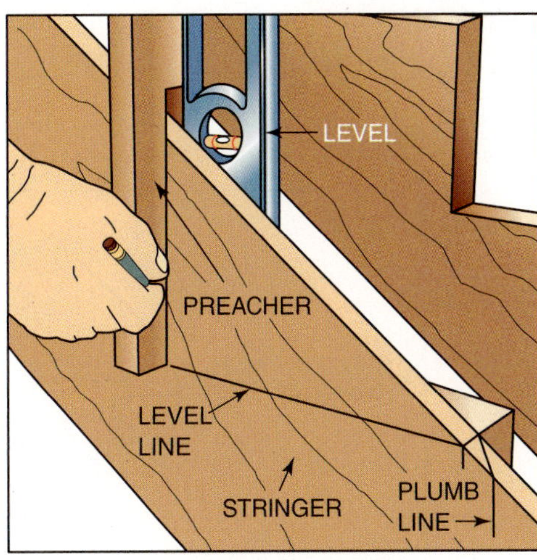


LAYING OUT OF THE PLUMB CUT

**FIGURE 19-28** Technique for easily marking a stringer on both faces.

### Risers

Cut *risers* to length by making a square cut on the end that goes against the wall. Make miters on the other end to fit the mitered plumb cuts of the open finish stringer (Fig. 19-29). Riser length is not critical, within ¼ inch is good, as a *closed finish stringer* will cover the end against the wall. Sand all pieces before installation. Apply a small amount of glue to the miters. Fasten them in position to each stair carriage. Drive finish nails both ways through the miter to hold the joint tight. Wipe off any excess glue. Set all nails. Lay out and install the closed finish stringer in the same manner as described previously.

### Treads

Rip the treads to width. Cut one end to fit against the closed finish stringer. Make a cut on the other end to receive the return nosing. This is a combination

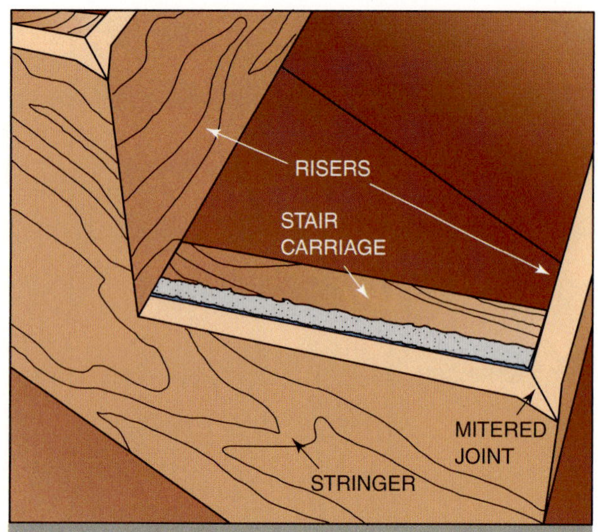


**FIGURE 19-29** The ends of the risers are mitered to fit against the miters of the open finish stringer.

square and miter cut. The square cut is made flush with the outside face of the open finish stringer. The miter distance is equal to the width of the return nosing beyond the square cut.

## Return Nosings

The return nosings are applied to the open ends of the treads. Miter one end of the return nosing to fit against the miter on the tread. Cut the back end square. Return the end on itself. The end of the return nosing may extend beyond the face of the riser, the same amount as the tread width. Predrill pilot holes in the return nosing for nails. Locate the nails so they are not in line with any balusters that will later be installed on the treads. This is because holes must be bored in the treads to receive the balusters. Any nails in line with the holes will damage the boring tool (Fig. 19-30).

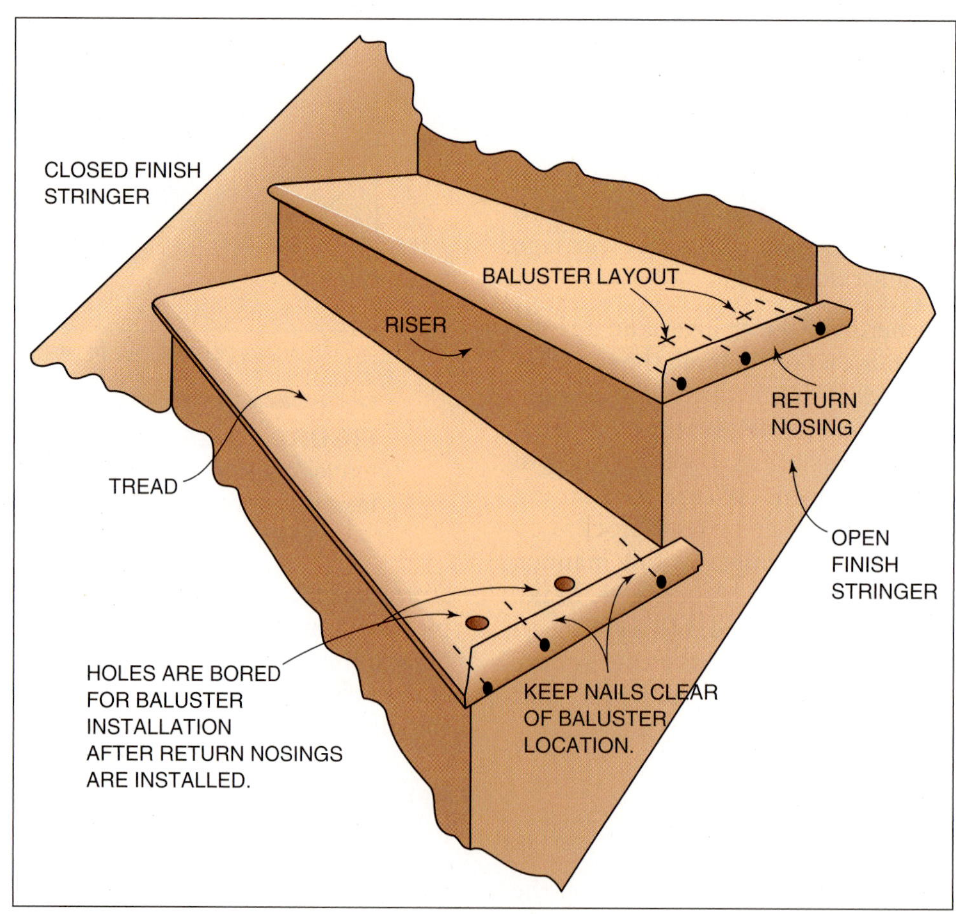


**FIGURE 19-30** Alignment of trim nails should be positioned to avoid further baluster holes.

Apply glue to the joint. Fasten the return nosing to the end of the tread with three 8d finishing nails. Set all nails. Sand the joint flush. Apply all other return nosings in the same manner. Treads may be purchased with the return nosing applied in a millwork shop.

### Tread Molding

The *tread molding* is applied in the same manner as for closed staircases. However, it is mitered on the open end and returned back onto the open stringer. The back end of the return molding is cut and returned upon itself at a point so the end assembly shows the same as at the edge (Fig. 19-31). Predrill pilot holes in the molding. Fasten it in place. Molding on starting and landing treads should only be tacked in case it needs to be removed for fitting after newel posts have been installed.

## HOUSED STRINGER

Housed stringers have risers and treads set into dados made in the stringer. Then, the treads and risers are wedged tightly against the shoulders of the dadoes (Fig. 19-32). These dadoes are made by using a *stair template* to guide a router. The router is equipped with a straight bit. Stair routing templates are manufactured to be adjustable to different unit rises, unit runs, and tread thicknesses. The template is shaped so the dadoes will be the exact width at the nosing and wider toward the backside of the stringer (Fig. 19-33). Templates may also be made by cutting out thin plywood or hardboard to the shape of the housing.

> ***from experience...***
>
> Make sure the nails used to fasten return nosings do not line up where the balusters will be located.

### Housing the Stringer

On the face side of the stringer stock, lightly draw a line parallel to and about 2 inches down from the top edge. The intersection of the unit rise and the unit run will be on the line. This distance may vary,

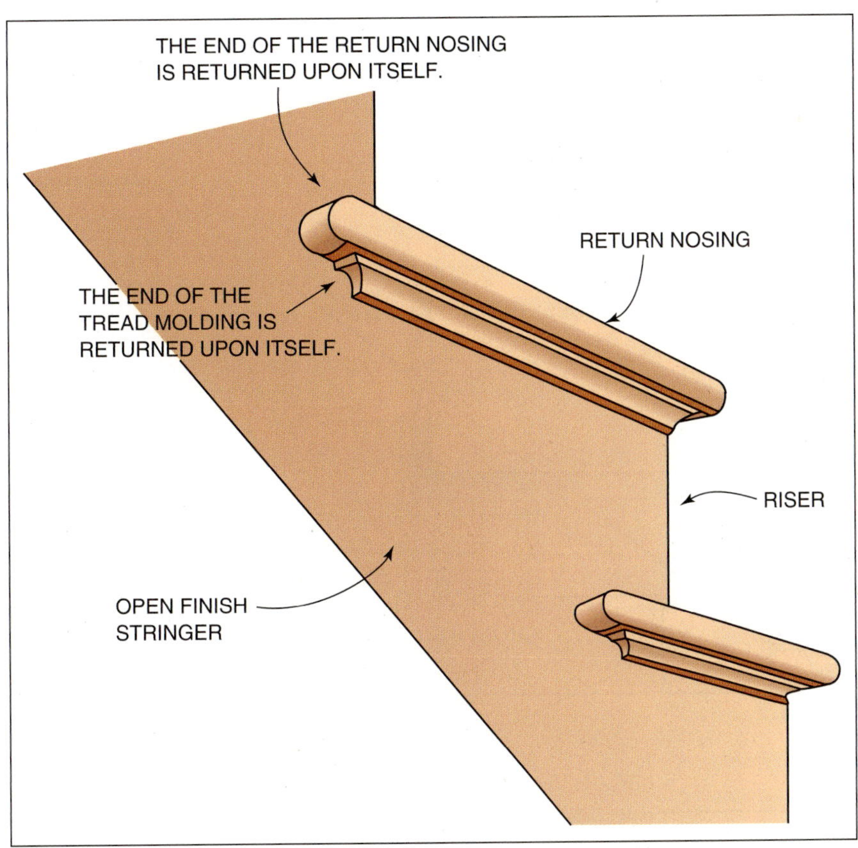


**FIGURE 19-31** The back ends of the return nosing and molding are returned upon themselves.

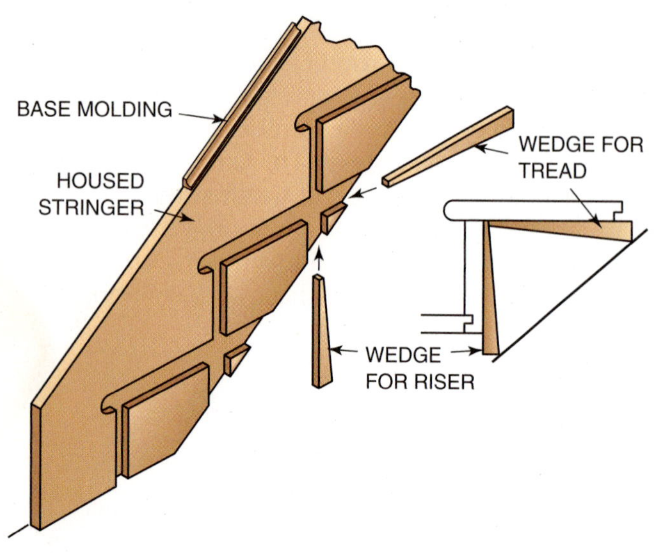


**FIGURE 19-32** Housed stringers are routed to allow treads and risers to be wedged and glued in place.

**FIGURE 19-33** A router template is used to make dadoes for housed stringers.

depending on the width of the stringer stock and the desired height of the top edge of the stringer above the stair treads.

Lay out the risers and treads for two or three steps of the staircase. The lines show the location of the face side of each riser and tread and are the outside edges of the housing (Fig. 19-34). Each step need not be laid out individually because the template will take the position of the next step into account.

Lightly square lines to the top edge of the stringer at the intersection of the face sides of tread and riser. These lines are the unit length of each step. With a tape stretched on the parallel line mark off the unit length for each remaining step (Fig. 19-35).

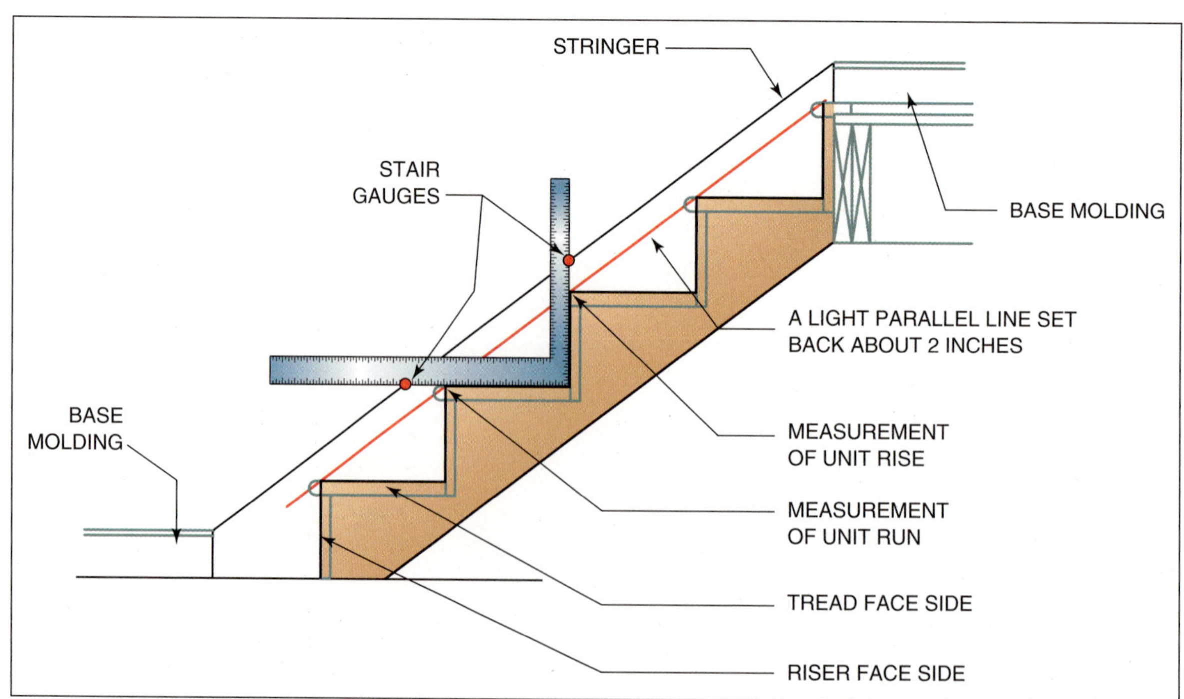


**FIGURE 19-34** Laying out a housed finish stringer using a framing square and stair gauges.

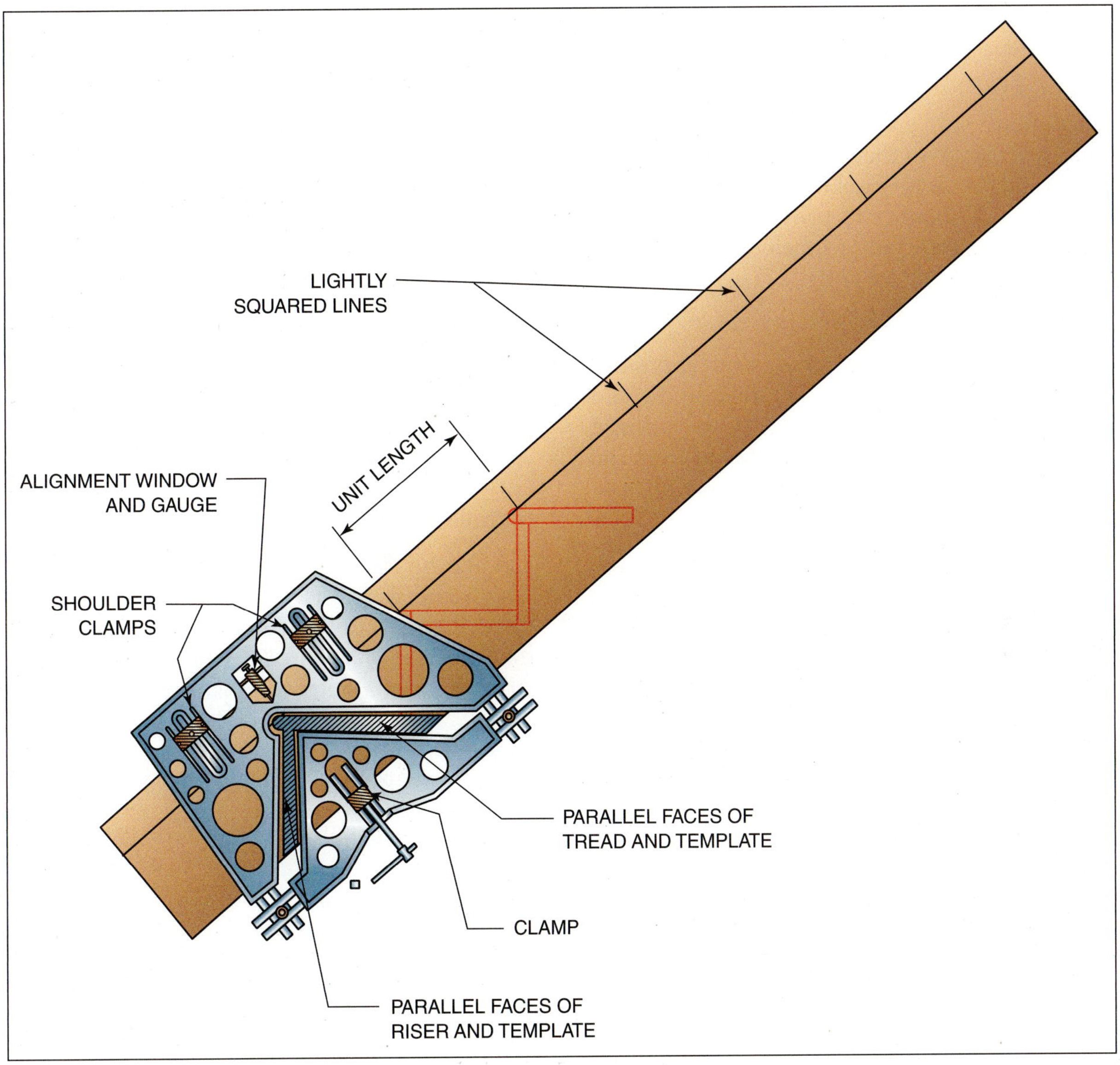

**FIGURE 19-35** The stair template is clamped to the layout lines of the stringer.

Mark them lightly as they will need to be erased or sanded out later.

Set the template over the first step and adjust it to match the layout. Align the face side of the tread and riser parallel with the face side of the template. Adjust the shoulder clamps to fit against the upper edge of the stringer. Clamp the template to the stringer. Adjust the alignment gauge in the window to the marked line without moving the template. Rout the stringer, ¼ to ⅜ inch deep.

**CAUTION:** Take care not to hit the metal template with the router bit. This may chip the bit and send out small pieces of shrapnel.

Turn off the router. Loosen the template clamp and slide the template to the next tread. Align the previously squared unit length line up with the gauge in the alignment window. Reclamp the template and rout the second step. Repeat for the remaining steps.

Cut and fit the bottom end of the stringer to the floor and the top end to the landing. Make end cuts that will join with the baseboard properly. The housed stringer and the baseboard should be joined in a professional manner to provide a continuous line of finish from one floor to the next.

### Laying Out an Open Stringer

The layout of an open (or *mitered*) stringer is similar to that of a stair carriage. The riser and tread layout lines intersect at the top edge of the stringer, instead of against a layout line 2 inches in from the edge. This riser layout line is mitered to fit a mitered end of the riser.

Mark lightly the riser and tread layouts on the stringer. The unit rise and unit run intersect on the top edge of the stringer. These lines represent the face sides of the tread and riser. Lay out the *miter cut* for the risers. Measure perpendicular to the riser layout line a distance equal to the thickness of the riser stock. Draw another plumb line at this point. This line represents the backside of the riser. Square both plumb lines across the top edge of the stringer stock. Draw a diagonal line on the top edge to mark the miter angle (Fig. 19-36).

Mark the tread cut line on the stringer. Measure down from the tread layout line a distance equal to the thickness of the tread stock. Draw a level line at this point for the tread cut. The tread cut is square through the thickness of the stringer. Fit the bottom end of the stringer to the floor. Fit the top end against the landing. Make the mitered plumb cuts for the risers and the square level cuts for the treads.

### Installing Risers and Treads

Cut and rip the required number of risers to a rough length and width. Determine the face side of each piece. Rip and cut the treads to width and length. On the open side of a staircase, where the riser and open stringer meet, a miter joint is made so no end grain

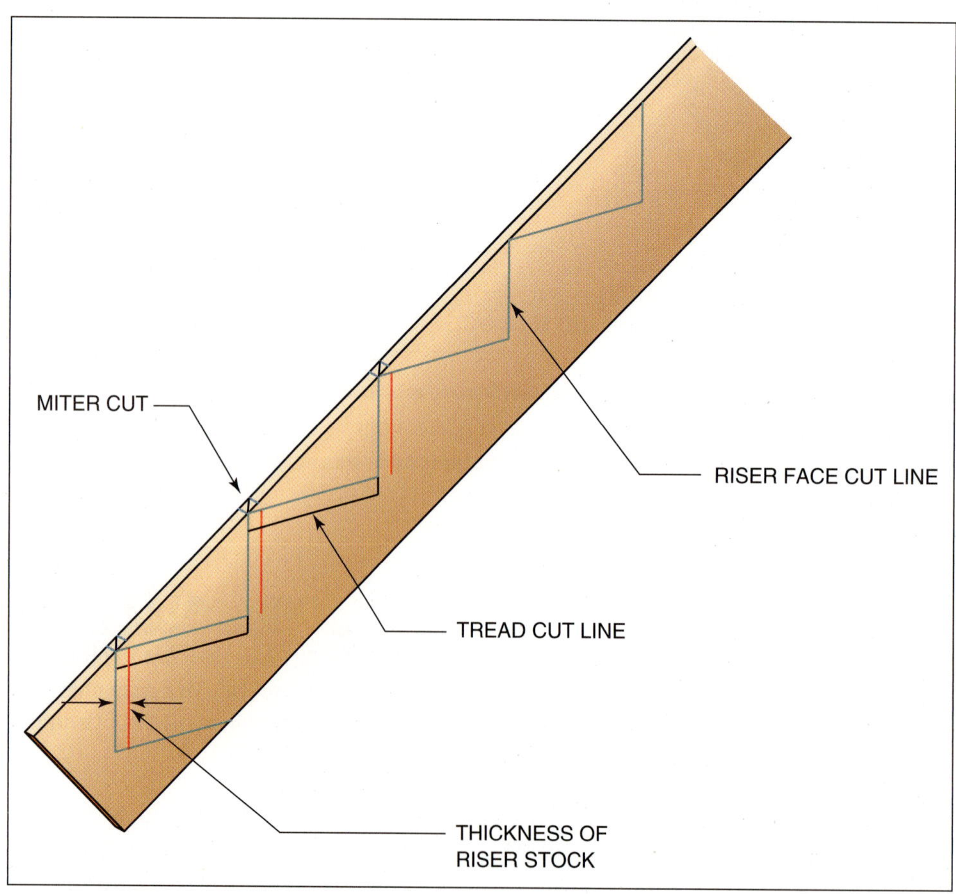


**FIGURE 19-36** Laying out the miter angle on an open finish stringer.

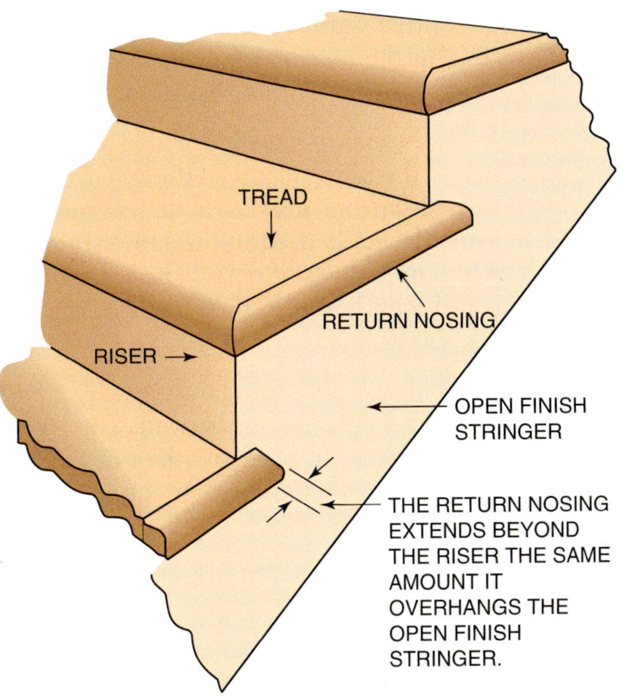


**FIGURE 19-37** A return nosing is mitered to the open end of treads.

is exposed (Fig. 19-37). The risers are installed with wedges, glue, and screws between housed stringers. The treads are then installed with wedges, glue, and screws on the closed side. Glue blocks are installed at intervals on the underside of the tread against the backside of the riser to reinforce the corners.

### Applying Return Nosings and Tread Molding

If the staircase is open, a *return nosing* is mitered to the end of the tread. The back end of the return nosing projects past the riser the same amount as the tread overhangs the riser. The end is returned upon itself. The tread molding is then applied under the overhang of the tread. If the staircase is closed on both sides, the molding is cut to fit between finish stringers. On the open end of a staircase, the molding is mitered around and under the return nosing. It is stopped and returned on itself at a point so the end assembly appears the same as at the edge. After the housed-stringer staircase is assembled, it is installed in position as a unit.

### Protecting the Finished Stairs

Protect the risers and treads by applying a width of building paper to them. Unroll a length down the stairway. Hold the paper in position by tacking thin strips of wood to the risers.

# BALUSTRADE

The stair finish may be separated in two parts: the *stair body* and the **balustrade**. Important components of the stair body finish are treads, risers, and finish stringers. Major parts of the balustrade include handrails, newel posts, and balusters. Balustrades may be constructed in a *post-to-post*. In the post-to-post method, the handrail is fitted between the newel posts (Fig. 19-38).

## BALUSTRADE MEMBERS

Finish members of the balustrade are available in many designs that are combined to complement each other. Various types of fittings are sometimes joined to straight lengths of handrail when turns are required.

### Newel Posts

**Newel posts** are anchored securely to the staircase to support the handrail. The newel posts may have flat square surfaces near the top, against which the handrails are fitted, and also at the bottom for fitting and securing the post to the staircase.

Courtesy of L. J. Smith.

**FIGURE 19-38** A closed staircase with a post-to-post balustrade on a kneewall.

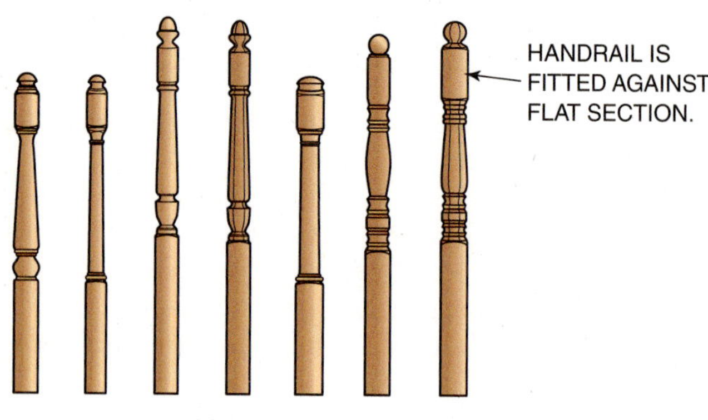


Courtesy of L. J. Smith.

**FIGURE 19-39** Newels in post-to-post balustrades must have flat surfaces against which handrails are fitted.

In between the flat surfaces, the posts may be *turned* in a variety of designs (Fig. 19-39).

Three types of newel posts are used in a post-to-post balustrade. *Starting newels* are used at the bottom of a staircase. They are fitted against the first or second riser. If fitted against the second riser, the flat, square surface at the bottom must be longer. At the top of the staircase, *second floor newels* are used.

*Intermediate landing newels* are also available. Because part of the bottom end of these newels are exposed, turned *buttons* are available to finish the end. The same design is used in the same staircase for each of the three types of posts. They differ only in their overall length and in the length of the flat surfaces (Fig. 19-40).

When the balustrade ends against a wall, a *half newel* is sometimes fastened to the wall. The handrail is then butted to it. In place of a half newel, the handrail may butt against an oval or round *rosette* (Fig. 19-41).

## Handrails

The **handrail** is the sloping finish member grasped by the hand of the person ascending or descending the stairs. It is installed horizontally when it runs along the edge of a balcony. Handrail heights are typically 30 to 38 inches vertically above the nosing edge of the tread. There should be a continuous 1½-inch finger clearance between the rail and the wall.

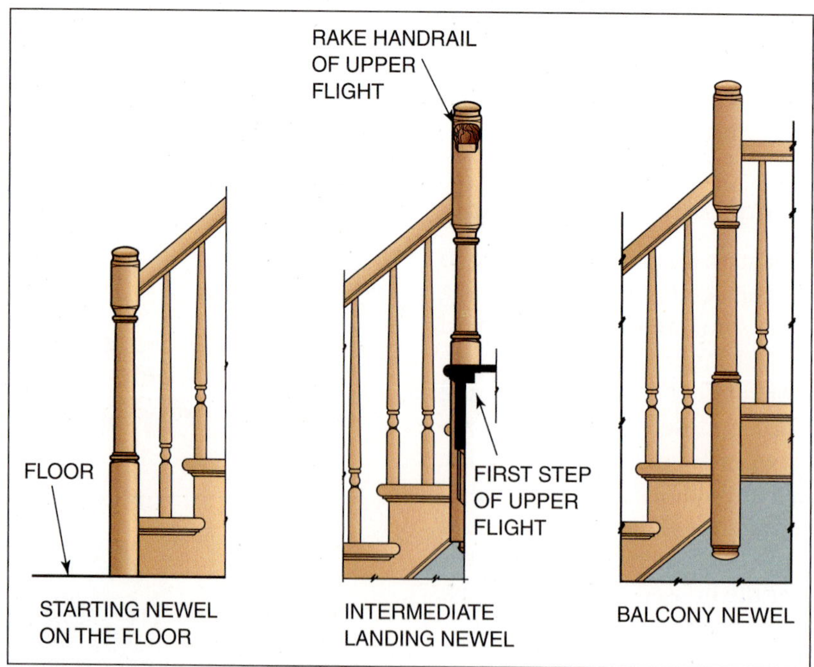


Courtesy of L. J. Smith.

**FIGURE 19-40** Three types of newel posts are used in a post-to-post balustrade.

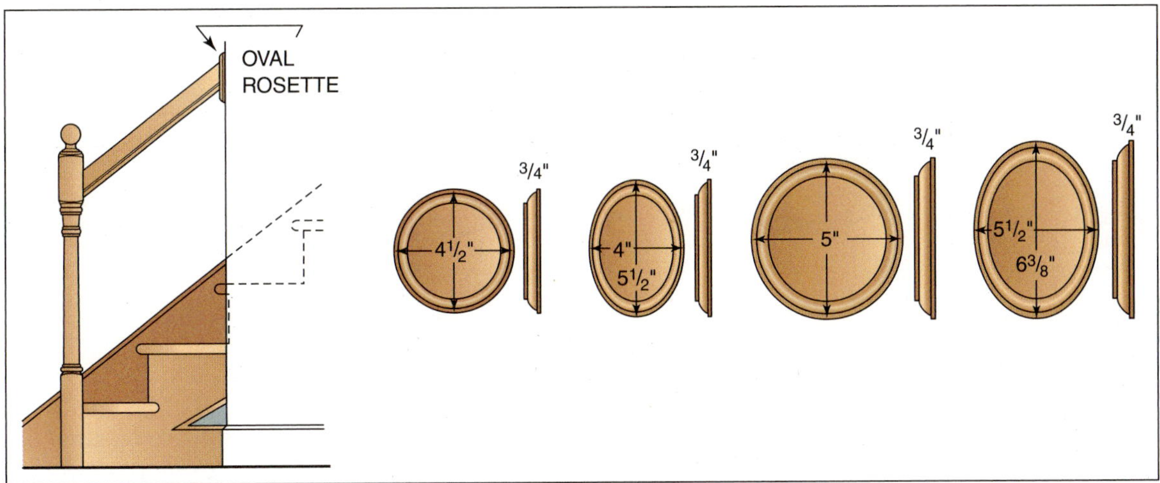


Courtesy of L. J. Smith.

**FIGURE 19-41** Rosettes come in round and oval shapes. They are fastened to the wall to provide a surface on which to butt and fasten the end of a handrail.

Several styles of handrails come in lineal lengths that are cut to fit on the job. Some handrails are *plowed* with a wide groove on the bottom side to hold square top balusters in place (Fig. 19-42). On closed staircases, a balustrade may be installed on top of a *kneewall* or buttress. In relation to stairs, a kneewall is a short wall that projects a short distance above and on the same angle as the stair body. A *shoe rail* or buttress cap, which is plowed on the top side, is usually applied to the top of the kneewall on which the bottom end of balusters are fastened (Fig. 19-43). Narrow strips, called *fillets,* are used between balusters to fill the plowed groove on handrails and shoe rails.

### Balusters

**Balusters** are vertical, usually decorative pieces between newel posts. They are spaced close together and support the handrail. On a kneewall, they run from the handrail to the shoe rail. On an open staircase, they run from the handrail to the treads. Balusters are manufactured in many styles. They should be selected to complement the newel posts being used (Fig. 19-44). Most balusters are made in lengths of 31, 34, 36, 39, and 42 inches for use in any part of the stairway. Several lengths of the same style baluster may be needed for each tread of the staircase because of the **rake** angle of the handrail.

## LAYING OUT THE BALUSTRADE

For the installation of any balustrade, its centerline is first laid out. On an open staircase, the centerline should be located a distance inward from

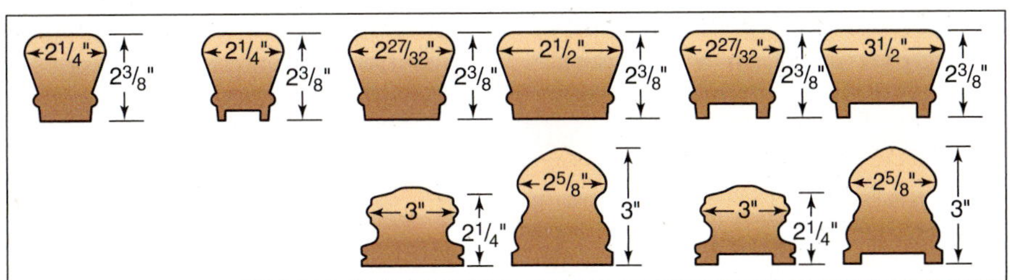


Courtesy of L. J. Smith.

**FIGURE 19-42** Straight lengths of handrail are manufactured in many styles.

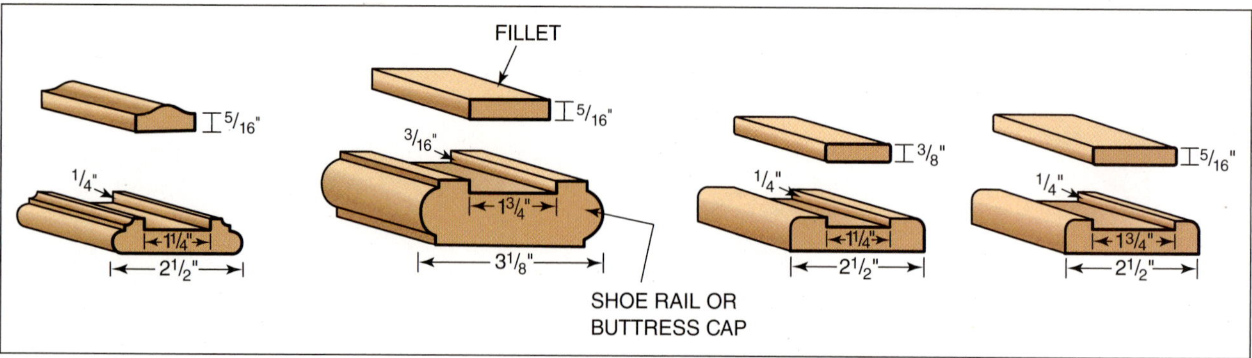


Courtesy of L. J. Smith.

**FIGURE 19-43** A shoe rail is often used at the bottom of a balustrade that is constructed on a kneewall.

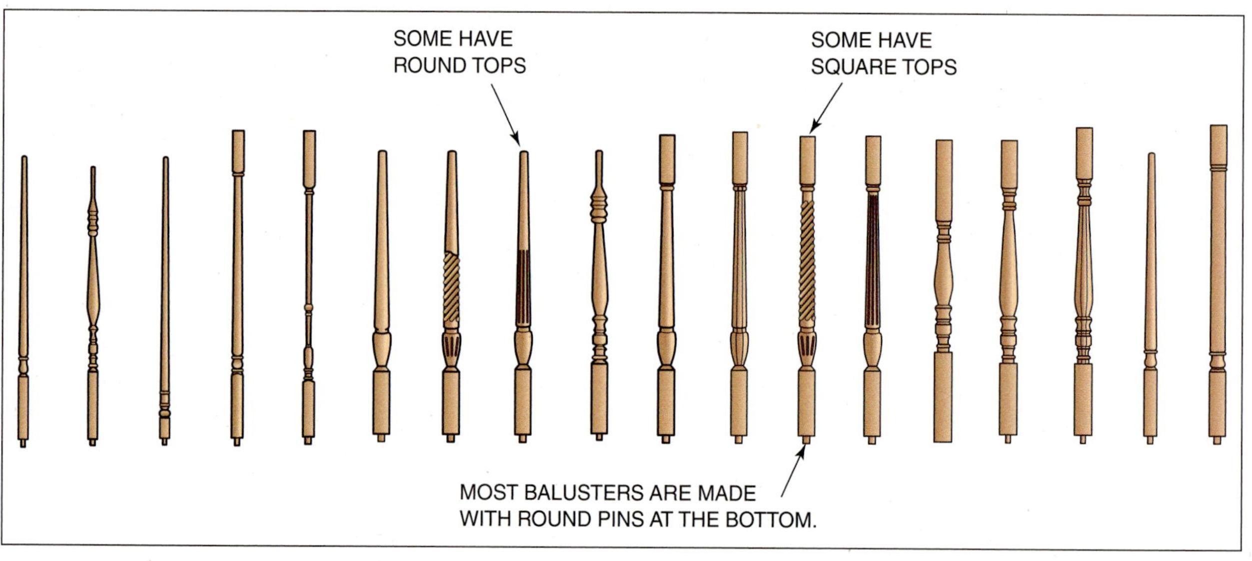


Courtesy of L. J. Smith.

**FIGURE 19-44** Balusters are made in designs that match newel post design.

the face of the finish stringer, equal to half the baluster width. It is laid out on top of the treads. If the balustrade is constructed on a kneewall, it is centered and laid out on the top of the wall (Fig. 19-45).

### Laying out Baluster Centers

The next step is to lay out the baluster centers. Code requirements for maximum baluster spacing may vary. Check the local building code for allowable spacing. Most codes require that balusters be spaced so that no object 4 inches in diameter or greater can pass through.

On open staircases, the center of the front baluster is located a distance equal to half its thickness back from the face of the riser. If two balusters are used on each tread, the spacing is half the run. If codes require three balusters per tread, the spacing is one-third the run (Fig. 19-46).

## INSTALLING A BALUSTRADE

The many components of a balustrade are installed in a stepwise fashion where parts are fastened to each other.

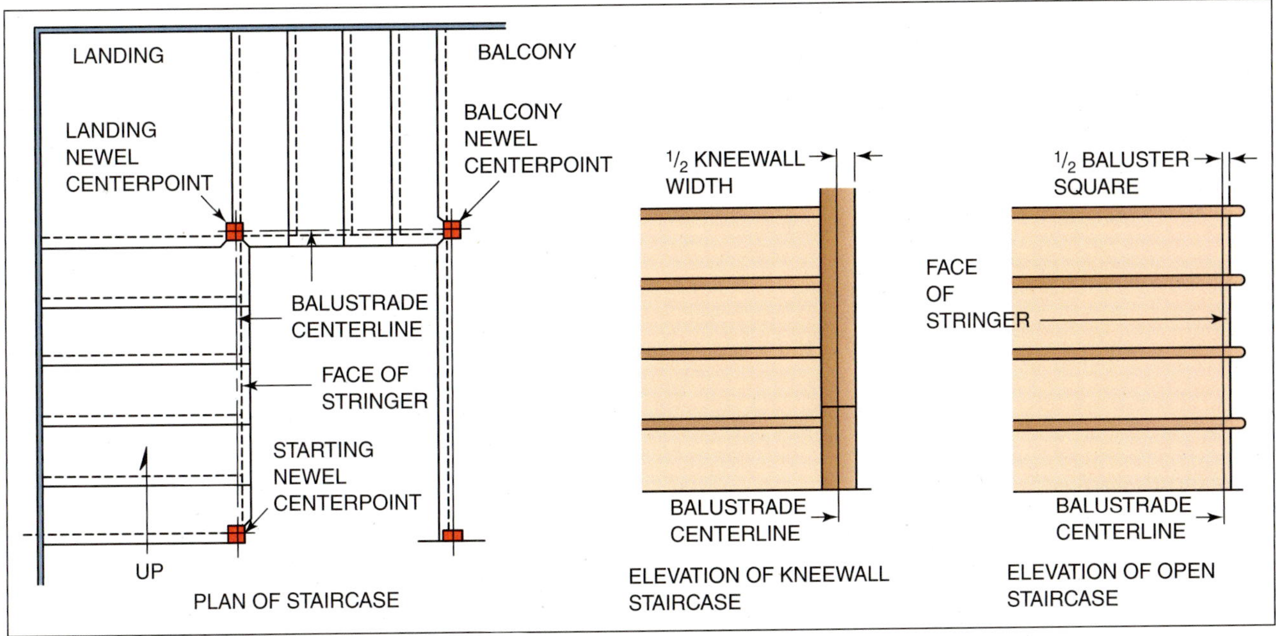


*Courtesy of L. J. Smith.*

**FIGURE 19-45** The centerline of the balustrade is laid out on a kneewall or on open treads.

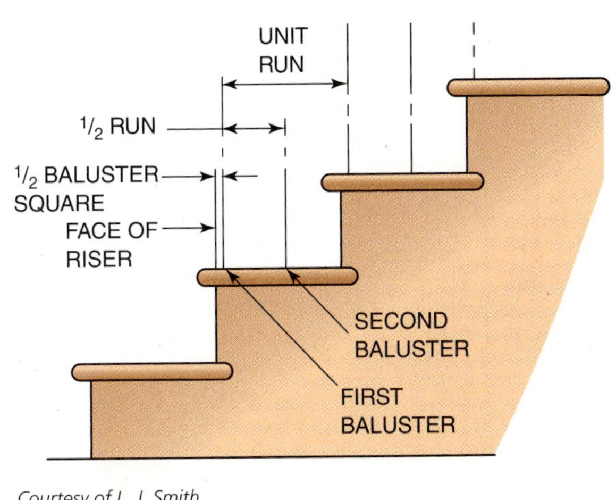


*Courtesy of L. J. Smith.*

**FIGURE 19-46** Layout of baluster centers on open treads.

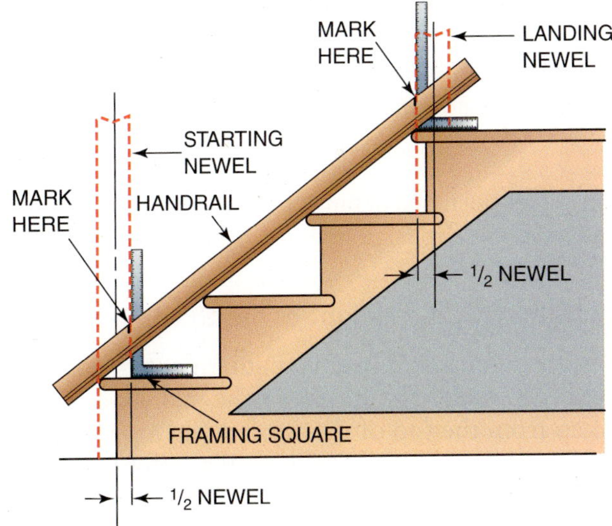


*Courtesy of L. J. Smith.*

**FIGURE 19-47** The handrail is laid out to fit between starting and balcony newel posts.

## Laying out the Handrail

Clamp the handrail to the tread nosings. Use a short bar clamp from the bottom of the finish stringer to the top of the handrail. Clamp at the tread nosing to avoid bowing the handrail. Use only enough pressure to keep the handrail from moving. Protect the edges of the handrail and finish stringer with blocks to avoid marring the pieces.

Use a framing square to mark the handrail where it will fit between the starting and the balcony or upper newel post (Fig. 19-47). While the handrail is clamped in this position, use a framing square at the landing nosing to measure the vertical thickness of the rake handrail. See detail A in Figure 19-48. Also, at the bottom, measure the height from the first tread to the top of the handrail where it butts

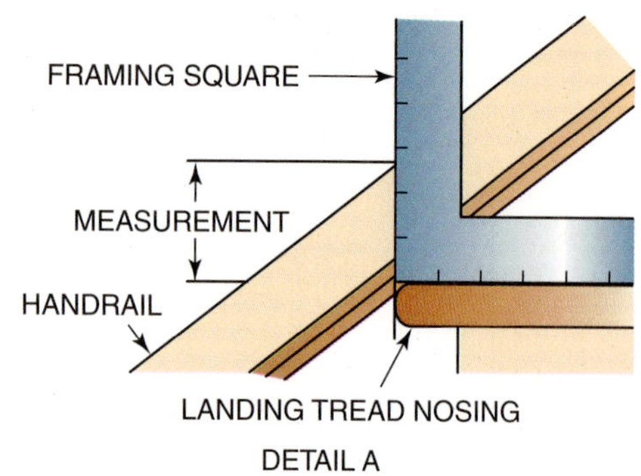


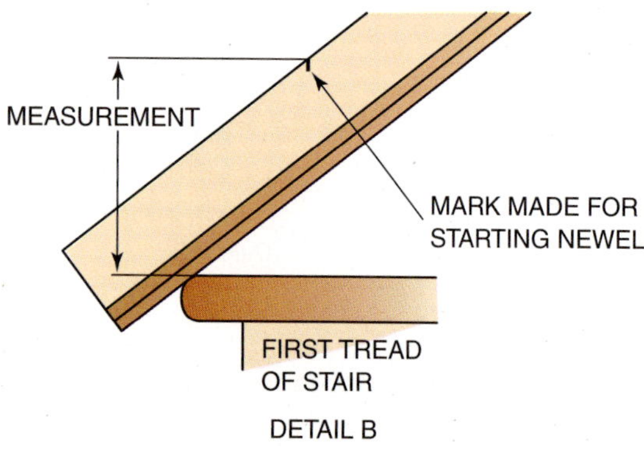


*Courtesy of L. J. Smith.*

**FIGURE 19-48** Determine the two measurements shown above and record for future use.

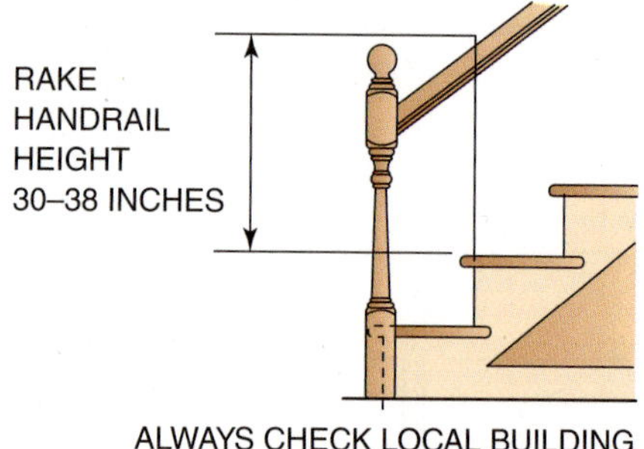


*Courtesy of L. J. Smith.*

**FIGURE 19-49** The rake handrail height is the vertical distance from the tread nosing to the top of the handrail.

the newel post. See detail B of Figure 19-48. Record and save these measurements for later use.

## Determining the Height of the Starting Newel

Verify the handrail height requirement from local building codes. The height of the stair handrail is taken from the top of the tread along a plumb line flush with the face of the riser (Fig. 19-49). Handrails are required only on one side in stairways of less than 44 inches in width. Stairways wider than 44 inches require a handrail on both sides. Handrail heights are typically 30 to 38 inches vertically above the nosing edge of the tread.

If a turned starting newel post is used, add the difference between the two previously recorded measurements from Detail A and B in Figure 19-48 to the required rake handrail height. Then, add another 1 inch for the *block reveal*. The block reveal is the distance from the top of the handrail to the top of the square section of the post. This sum is the distance from the top of the first tread to the top of the upper block (Fig. 19-50). To this measurement add the distance the newel extends below to the floor. Measure the starting newel length from the top of the squared section to the bottom. Cut off at the bottom of the newel.

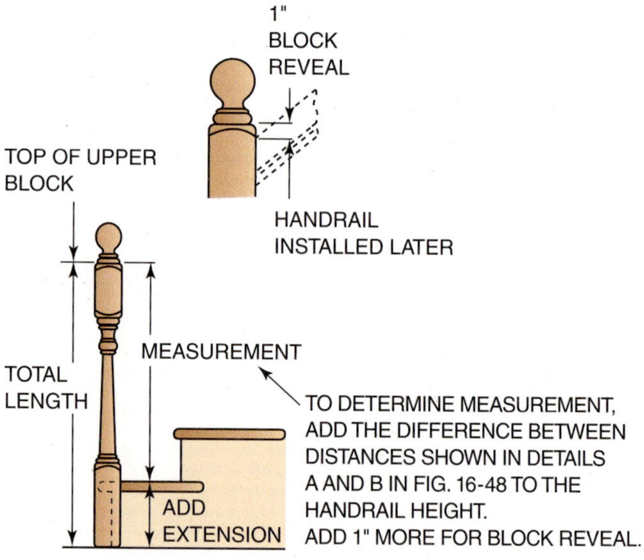


*Courtesy of L. J. Smith.*

**FIGURE 19-50** Determining the height of the starting newel.

## Installing the Starting Newel

The starting newel is notched over the outside corner of the first step. About one-half of the newel bottom thickness is removed in the notch. In the other direction, it is notched so its centerline will be aligned with the handrail centerline. The remaining portion of the bottom newel covers the riser and stinger corner. (Fig. 19-51). The post is then fastened to the first step with lag screws. The lag screws are counterbored and later concealed with wood plug.

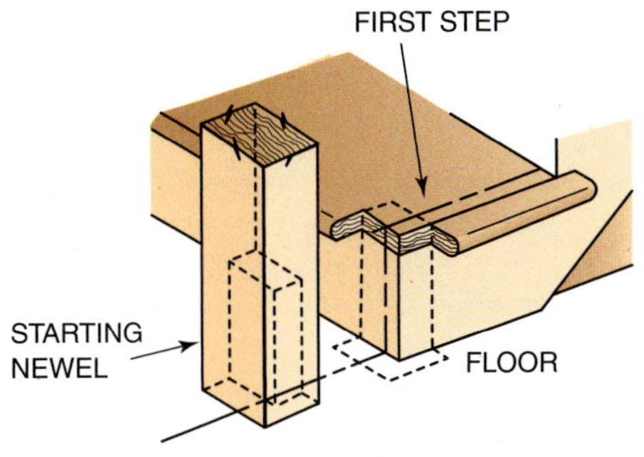


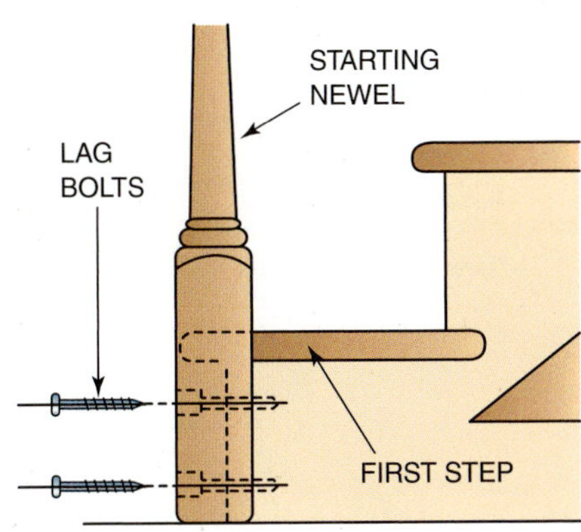


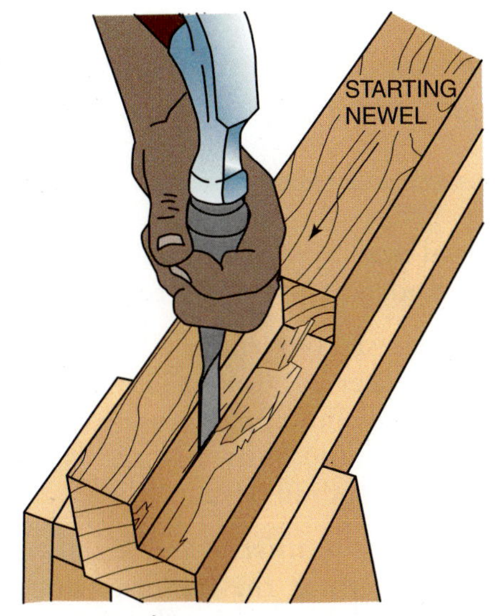


Courtesy of L. J. Smith.

**FIGURE 19-51** The starting newel is notched to fit over the first step. It is then fastened in place.

Newels must be strong enough to resist lateral force applied by persons using the staircase. Newel posts must be set plumb. If necessary, adjust and trim the cut to make the post plumb.

### Installing the Balcony Newel

Generally, codes require that *balcony rails* for homes be no less than 36 inches. Check local codes for requirements. The height of the *balcony newel* is determined by finding the sum of the required balcony handrail height, a block reveal of one inch, the height of the turned top, and the distance the newel extends below the balcony floor.

Trim the balcony newel to the calculated height. Notch and fit it over the top riser with its centerlines aligned with both the rake and balcony handrail centerlines. Plumb it in both directions. Fasten it in place with counterbored lag bolts (Fig. 19-52).

### Boring Holes for Balusters

Bore holes in the treads at the center of each baluster. The diameter of the hole should be equal to the diameter of the pin at the bottom end of the baluster. The depth of the hole should be slightly more than the length of the pin (Fig. 19-53).

Recheck the length of the previously marked handrail length between starting and balcony newels. Cut the handrail to fit. The handrail is best cut with a compound miter saw. Place the handrail on the tread nosings between posts. Transfer the baluster centerlines from the treads to the handrail (Fig. 19-54).

If the baluster tops are rounded, holes need to be bored into the handrail. Turn the handrail upside down and end for end. Set it back on the tread nosings with the lower end facing up the stairs. Clamp the handrail securely. This makes it easier to keep the drill plumb while drilling holes. When the handrail is repositioned, the holes will be plumb (Fig. 19-55). Bore holes at baluster centers at least ¾ inch deep.

### Installing Handrail and Balusters

Prepare the posts for fastening the handrail by counterboring and drilling **shank holes** for lag bolts through the posts. Place the handrail at the correct height between newel posts. Drill pilot holes. Temporarily fasten the handrail to the posts (Fig. 19-56).

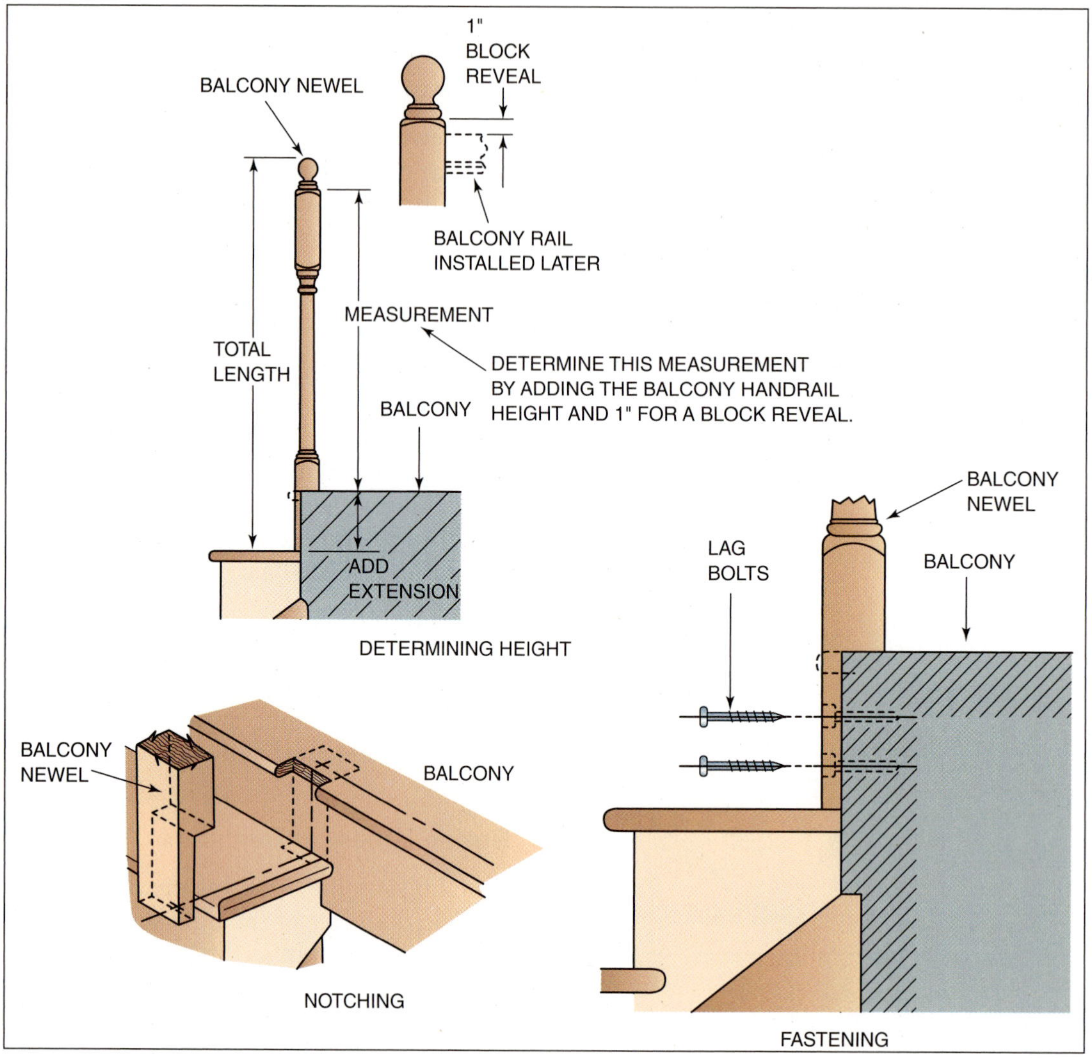

*Courtesy of L. J. Smith.*

**FIGURE 19-52** The height of the balcony newel post is calculated, notched at the bottom, and fastened in place.

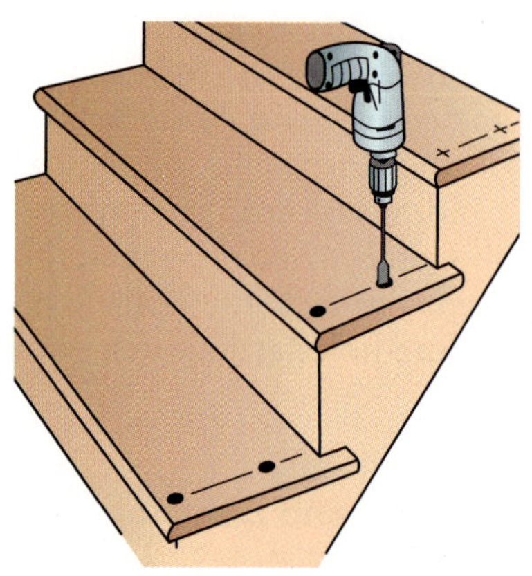

*Courtesy of L. J. Smith.*

**FIGURE 19-53** Holes are bored in the top of the treads at each baluster center point.

The handrail may have to be removed for baluster installation and then fastened permanently.

Cut the balusters to length. Allow for the insertion of the baluster into the holes in the bottom of the handrail. The bottom pin is inserted in the holes in the treads. The top of the baluster is inserted in the holes in the handrail bottom.

If *square top balusters* are used, they are trimmed to length at the handrail angle. They are inserted into a *plowed handrail*. The balusters are then fastened to the handrail with finish nails and glue (Fig. 19-57). Care must be taken to keep the handrail from arching upward from top to bottom when fastening square top balusters. Care must also be taken to keep each baluster plumb. *Fillets* are installed in the plow of the handrail, between the balusters.

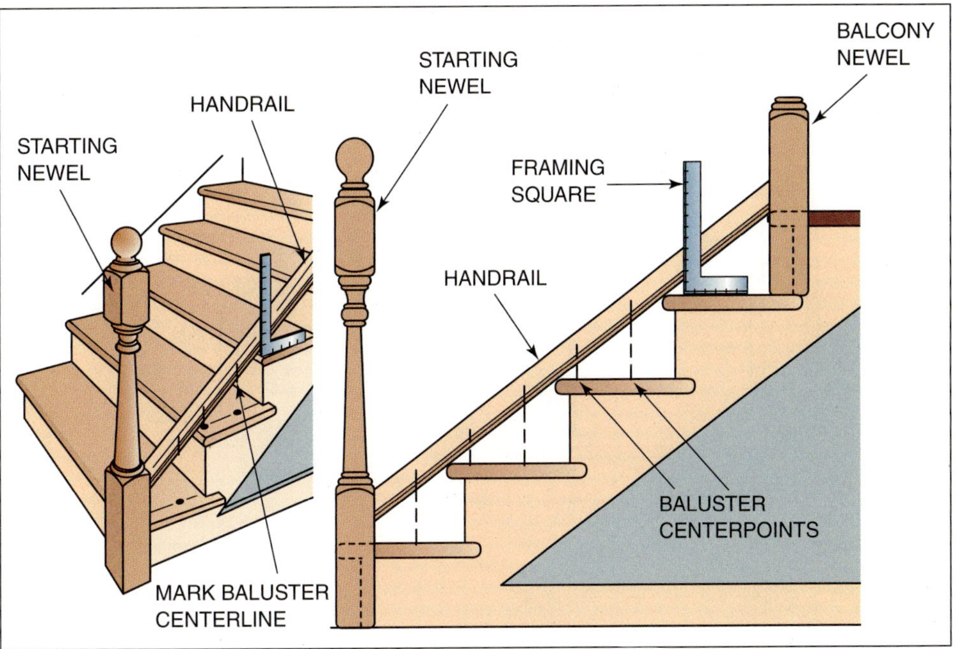


Courtesy of L. J. Smith.

**FIGURE 19-54** The handrail is fitted between newel posts. The baluster centers are transferred to it.

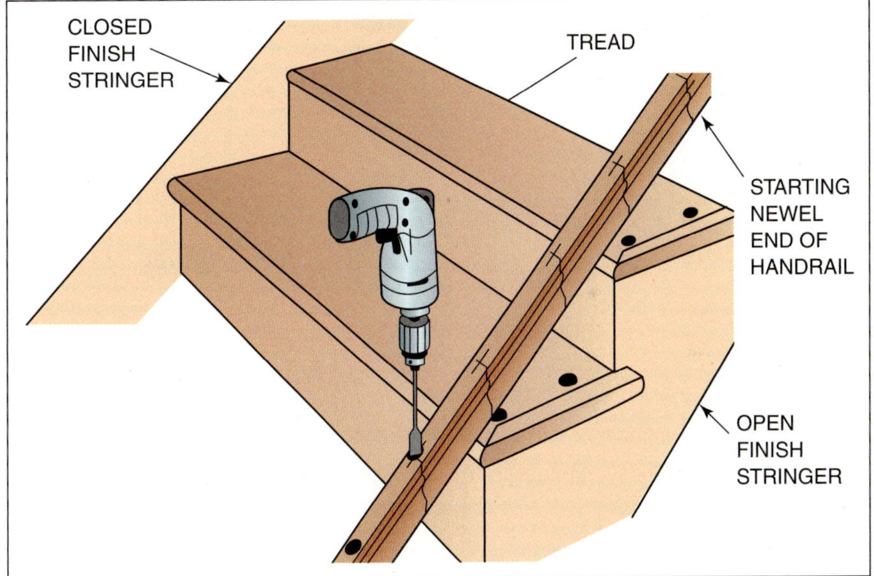


Courtesy of L. J. Smith.

**FIGURE 19-55** Holes are bored in the bottom of the rake handrail to receive roundtop balusters.

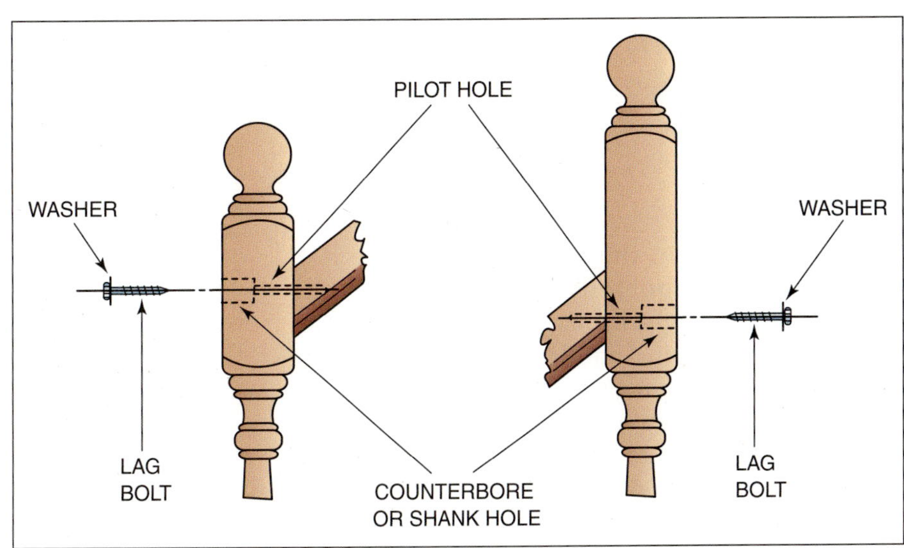


Courtesy of L. J. Smith.

**FIGURE 19-56** The handrail is fastened temporarily to newel posts with lag bolts. The use of nails when constructing balustrades is discouraged.

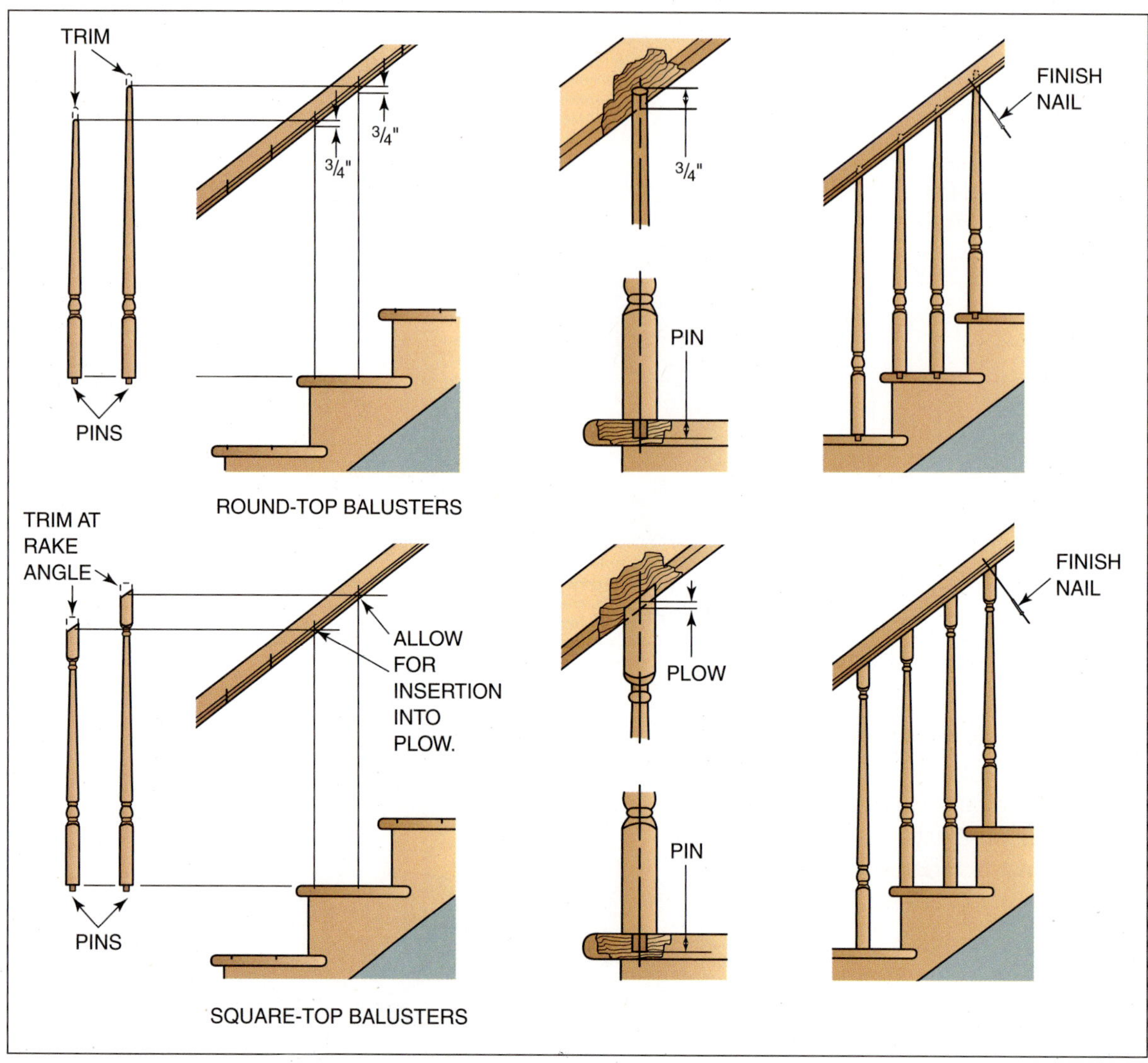


Courtesy of L. J. Smith.

**FIGURE 19-57** Balusters are cut to length and installed between handrail and treads.

## INSTALLING THE BALCONY BALUSTRADE

A balcony balustrade is installed level and therefore easier to install than a sloping balustrade.

### Handrail

Cut a *half newel* to the same height as the balcony newel. Temporarily place it against the wall. Mark the length of the balcony handrail (Fig. 19-58). Cut the handrail to length. Fasten the half newel to one end of it. Replace the half newel to the wall and temporarily fasten the other end of the handrail to the landing newel. It may need to be removed to install the balcony balusters. A *rosette* may be used to end the handrail against a wall instead of a half newel. First, fasten the rosette to the end of the handrail. Hold the rosette and the handrail against the wall. Mark the length of the handrail at the landing newel. Cut the handrail to length. Temporarily fasten it in place (Fig. 19-59).

### Balcony Balusters

The balcony balusters are spaced by adding the thickness of one baluster to the distance between the balcony newel and the half newel or wall. The overall distance is then divided into spaces that equal, as close as possible, the spacing of the rake balusters (Fig. 19-60). The balcony balusters are then installed in a manner similar to the rake balusters where each space is equal.

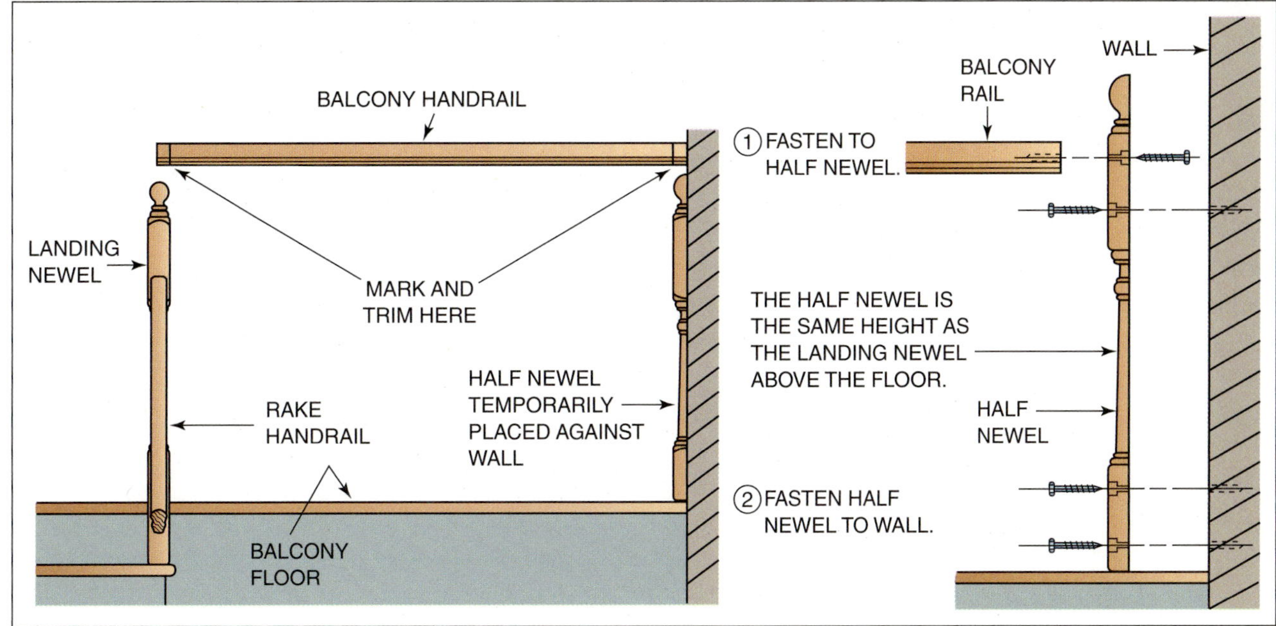


*Courtesy of L. J. Smith.*

**FIGURE 19-58** The balcony rail is fitted between the landing newel and a half newel placed against the wall.

*Courtesy of L. J. Smith.*

**FIGURE 19-59** A rosette is sometimes used to end the balcony handrail instead of a half newel.

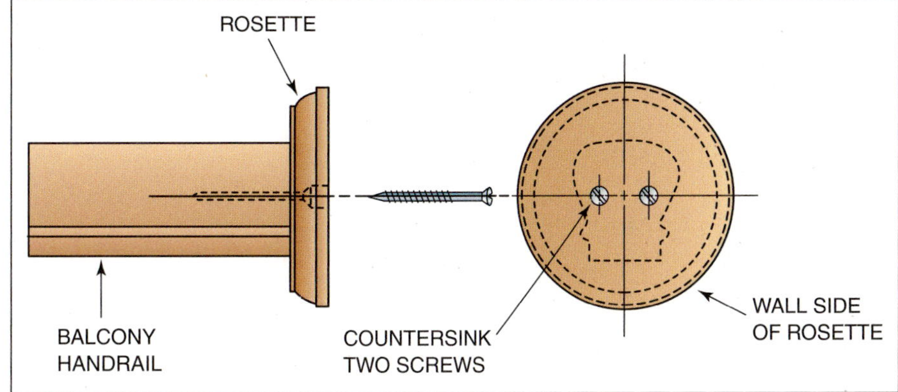


*Courtesy of L. J. Smith.*

**FIGURE 19-60** Balcony balusters are installed as close as possible to the same spacing as the rake balusters.

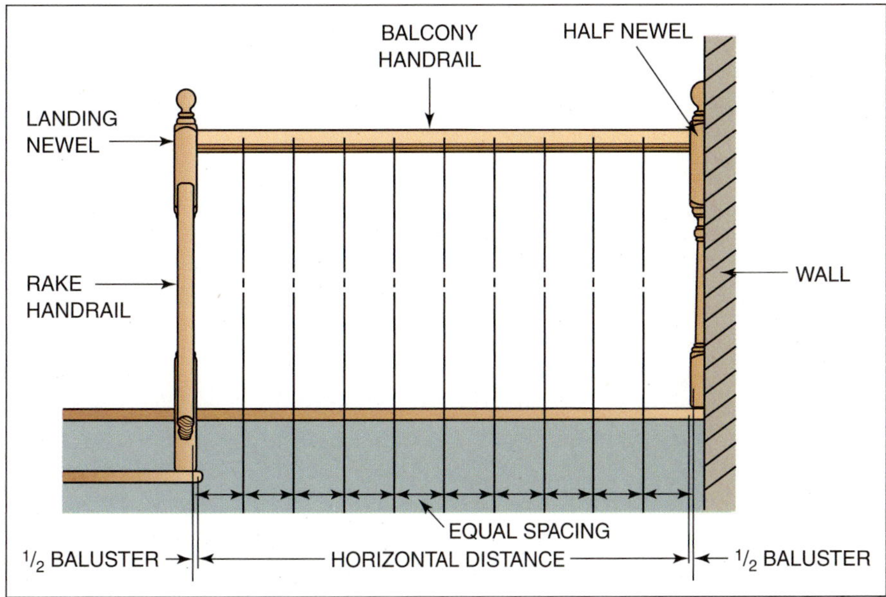

## SUMMARY

- Stairway designs are varied, from service stairs to high profile staircases of great architectural appeal.
- Stair carriage is the major support for a staircase.
- A housed stringer provides a finished look to a staircase.
- Balustrade is an architectural accent that also provides a handrail for safety.

## REVIEW QUESTIONS

*Select the most appropriate answer.*

1. **Stairways in residential construction should have a minimum width of**
   a. 30 inches.
   b. 36 inches.
   c. 32 inches.
   d. 40 inches.

2. **For the IBC (International Building Code), the maximum riser height for a stairway in a residence is**
   a. 8¼ inches.
   b. 7½ inches.
   c. 7¾ inches.
   d. 7¼ inches.

3. **A maximum riser height for a stairway with a total vertical rise of 8'–7" is**
   a. 7¾ inches.
   b. 7⅝ inches.
   c. 7⅜ inches.
   d. 7⅛ inches.

4. **Using the 17–18 method, what is the recommended unit run for a stair with a unit rise of 7¼ inches?**
   a. 8¾ to 9¾ inches
   b. 9 to 10 inches
   c. 9¾ to 10¾ inches
   d. 10 to 11 inches

5. **Using the 24–25 method, what is the recommended unit run for a stair with a unit rise of 7¼ inches?**
   a. 8½ to 9½ inches
   b. 9¾ to 10¾ inches
   c. 9½ to 10½ inches
   d. 10 to 11 inches

6. **The IBC specifies a minimum unit run of**
   a. 8½ inches.
   b. 9 inches.
   c. 10 inches.
   d. 11 inches.

7. **Most building codes specify a minimum head room clearance of**
   a. 6'–6".
   b. 7'–0".
   c. 6'–8".
   d. 7'–6".

8. **A flight of stairs has a unit rise of 7½ inches and a unit run of 10 inches. The total thickness of the upper floor is 10 inches. What is the minimum length of the stairwell if the stairwell header acts as the top riser and the desired minimum headroom is 6'–8"?**
   a. 9'–0"
   b. 10'–0"
   c. 9'–6"
   d. 10'–2"

**9. The stair carriage with a unit rise of 7½ inches rests on the finish floor. What is the riser height of the first step if the tread thickness is ¾ inch?**

a. 6¾ inches

b. 7¾ inches

c. 7½ inches

d. 8¼ inches

**10. A stairway has a riser height of 7½ inches. The tread stock thickness is 1 1/16 inches. The finish floor thickness of the upper floor is ¾ inch. The stairwell header acts as the top riser. What is the distance down from the top of the upper subfloor to the rough carriage riser?**

a. 7 3/16 inches

b. 7½ inches

c. 7⅜ inches

d. 7 13/16 inches

**11. The rounded outside edge of a tread that extends beyond the riser is called a**

a. housing.

b. coving.

c. turnout.

d. nosing.

**12. The finished board used to cover a stair carriage is called a**

a. return.

b. stringer.

c. baluster.

d. casing.

**13. An open stringer is**

a. housed to receive risers.

b. mitered to receive risers.

c. housed to receive treads.

d. mitered to receive treads.

**14. The entire rail assembly on the open side of a stairway is called a**

a. baluster.

b. balustrade.

c. guardrail.

d. finish stringer assembly.

**15. In a framed staircase, the treads and risers are supported by**

a. stair carriages.

b. each other.

c. finish stringers.

d. blocking.

**16. The first thing to do when installing finishing trim to a temporary staircase is**

a. check the rough framing for unit rise and run.

b. block the staircase so no one can use it.

c. straighten the stair carriages.

d. install all the risers.

**17. Treads usually project beyond the face of the riser**

a. ¾ inch.

b. 1¼ inches.

c. 1⅛ inches.

d. 1⅜ inches.

**18. Newel posts are notched around the stairs so that their centerline aligns with the**

a. centerline of the stair carriage.

b. centerline of the balustrade.

c. outside face of the open stringer.

d. outside face of the stair carriage.

**19. Landings in a staircase are**

a. required when the total rise is 10 feet or more.

b. at the same height a tread would be if no landing existed.

c. used with open staircases.

d. all of the above

**20. Balusters are installed**

a. between starting and landing newels.

b. in balcony handrails.

c. plumb and evenly spaced.

d. all of the above

## KNOW YOUR CODES

Stair design must follow strict guidelines to ensure their safe use. Even slight variations in the height of one step could cause a fall. Check the local codes for requirements for stair design.

## WHAT'S WRONG WITH THIS PICTURE?

Carefully study **Figure 19-61** and think about what is wrong. Consider all possibilities.

### WRONG

The staircase is rough framed with 2 × 10 carriages that have been reinforced, a 2 × 4 handrail, and 2 × 8 treads.

**FIGURE 19-61** Many problems exist here. Although the photograph lacks scale, it is apparent that the rise is too steep. In this case, you can see that the rise is noticeably more than the run, and the run (a 2 × 8) is too narrow. The handrail is inadequate. A 2 × 4 on the flat as a handrail is not appropriate; it is difficult to grasp due to its width. Balusters should run from the rail down to the steps. The spacing between the balusters should be less than 4 inches. These are open-riser stairs, which are acceptable except that when open risers are utilized, a 4-inch ball should not be able to pass between the top of one tread and the bottom of the next. Also, the second tread from the bottom shows a noticeable slope.

### RIGHT

**FIGURE 19-62** Obviously, it is inappropriate to compare a finished stair, such as this one, to the rough stair shown in Figure 19-61. However, even a rough stair needs to be built up to proper building codes and with safety in mind. Stairs can be dangerous, and one cannot pay too much attention to detail when building them.

CHAPTER 20

# Cabinets and Countertops

## OBJECTIVES *After completing this unit, the student should be able to:*

- State the sizes and describe the construction of typical base and wall kitchen cabinet units.
- Install manufactured kitchen cabinets.
- Construct, laminate, and install a countertop.
- Identify cabinet doors and drawers according to the type of construction and method of installation.
- Identify overlay, lipped, and flush cabinet doors and proper drawer construction.
- Apply cabinet hinges, pulls, and door catches.

**face frame** a framework of narrow pieces on the front of a cabinet making the door and drawer openings

**gain** a cutout made in a piece to receive another piece, such as a cutout for a butt hinge

**J-roller** a 3-inch-wide rubber roller used to apply pressure over the surface of contact cement bonded plastic laminates

**pilot** a guide on the end of edge-forming router bits used to control the amount of cut

**pivot** a point of rotation

**postforming** a method used to bend plastic laminate to small radii

Cabinets and countertops complete the interior finish with usable workspace and storage. They are often purchased in preassembled units and installed by carpenters. These manufactured kitchen and bath cabinets come in a wide variety of styles, materials, and finishes (Fig. 20-1). The units can be assembled into many configurations depending on how the space is to be used by the owner. Countertops are then installed to provide the work surface.

Countertops, cabinet doors, and drawers may be customized in a wide variety of styles and sizes. They can be built on the job, but most are produced in a shop. Custom cabinets can also be made to meet the design specifications of any job. They are usually made by carpenters who specialize in cabinetry in cabinet shops.

Cabinet-making and assembly requires skill and patience to produce a finished product that is pleasing in appearance as well as functional. Workers must watch every fastener used to ensure it is installed properly for strength and appearance. Wood splits easily, thus losing holding strength. Misjudged fastener depths that protrude through a finished surface are difficult to repair or hide. Workmanship is the grade stamp carpenters leave behind on their work.

*Courtesy of KraftMaid Cabinetry.*

**FIGURE 20-1** Manufactured kitchen cabinets are available in a wide variety of styles and sizes.

# Components of Manufactured Cabinets

Most cabinets used in residential construction are manufactured for the kitchen or bathroom. Cabinets consist of a case that is fitted with shelves, doors, and drawers. Designs vary considerably with the manufacturer.

## KINDS AND SIZES

One method of cabinet construction utilizes a **face frame**. This frame provides openings for doors and drawers. Another method, called *European* or *frameless*, eliminates the face frame (Fig. 20-2). The two basic kinds of kitchen cabinets are the *wall unit* and the *base unit*.

A countertop is usually installed on top of the base cabinet. The surface of the countertop is usually about 36 inches from the floor. Wall units are installed about 18 inches above the countertop. This distance is enough to accommodate articles such as coffee makers, toasters, blenders, mixers, and microwave ovens. Yet it keeps the top shelf of the wall unit within reach, not over 6 feet from the floor. The usual overall height of a kitchen cabinet installation is 7'-0" (Fig. 20-3).

### Wall Cabinets

Standard wall cabinets are 12 inches deep. The standard height is 30 inches; wall units are also available in heights of 42, 24, 18, 15, and 12 inches. Shorter cabinets are used above sinks, refrigerators, and ranges. The 42-inch cabinets are for use in kitchens without *soffits* above where more storage space is desired. Typical wall cabinet widths range from 9 to 48 inches in 3-inch increments. They come with single or double doors depending on their width. Single-door cabinets can be hung so doors can swing in either direction. A standard height wall unit usually contains two adjustable shelves.

Wall *corner* cabinets make access into corners easier. *Double-faced* cabinets have doors on both sides for use above island and peninsular bases. Some wall cabinets are made 24 inches deep for installation above refrigerators. A microwave oven case, with a 30-inch-wide shelf, is available (Fig. 20-4).

### Base Cabinets

Most base cabinets are manufactured 34½ inches high and 24 inches deep. By adding the typical countertop thickness of 1½ inches, the work surface is at

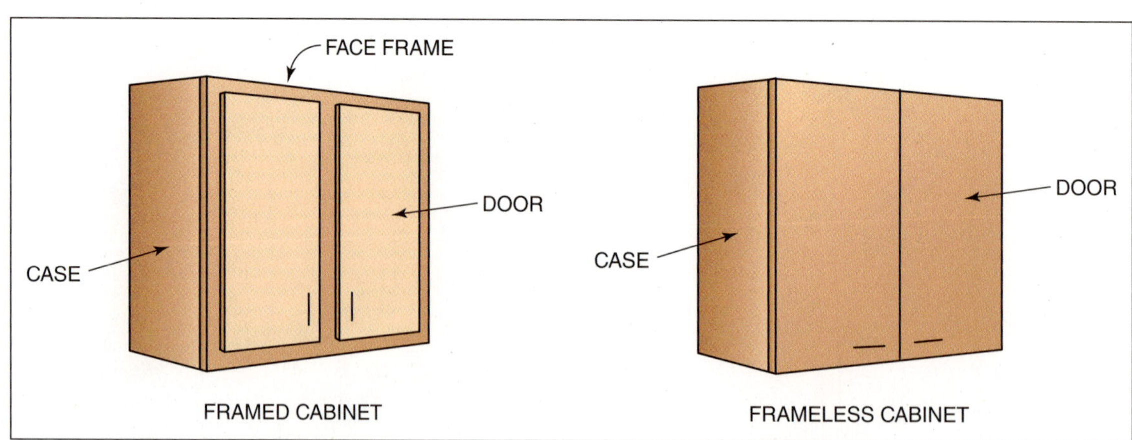


**FIGURE 20-2** Two basic methods of cabinet construction are with a face frame and frameless.

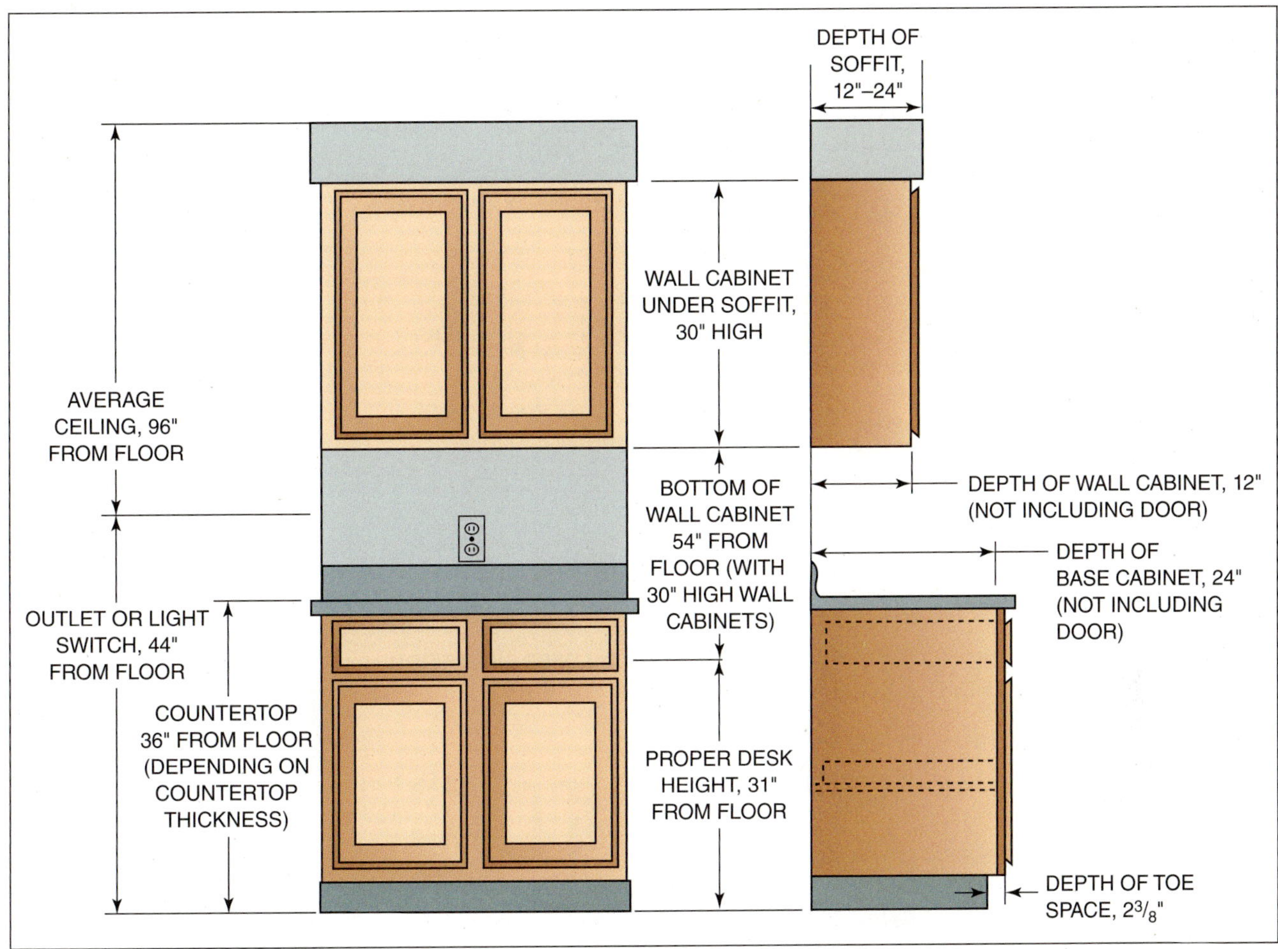


*Courtesy of Merillat Industries.*

**FIGURE 20-3** Common kitchen cabinet heights and dimensions.

the standard height of 36 inches from the floor. Base cabinets come in widths to match wall cabinets.

Single-door cabinets are manufactured in widths from 9 to 24 inches. Double-door cabinets come in widths from 27 to 48 inches. A recess called a *toe space* is provided at the bottom of the cabinet. The standard base cabinet contains one drawer, one door, and an adjustable shelf. Some base units have no drawers; others contain all drawers. Double-faced cabinets provide access from both sides. Corner units, with round revolving shelves, make corner storage easily accessible (Fig. 20-5).

### Tall Cabinets

*Tall* cabinets are usually manufactured 24 inches deep, the same depth as base cabinets. Some *utility* cabinets are 12 inches deep. They are made 66 inches high and in widths of 27, 30, and 33 inches for use as oven cabinets. Single-door *utility* cabinets are

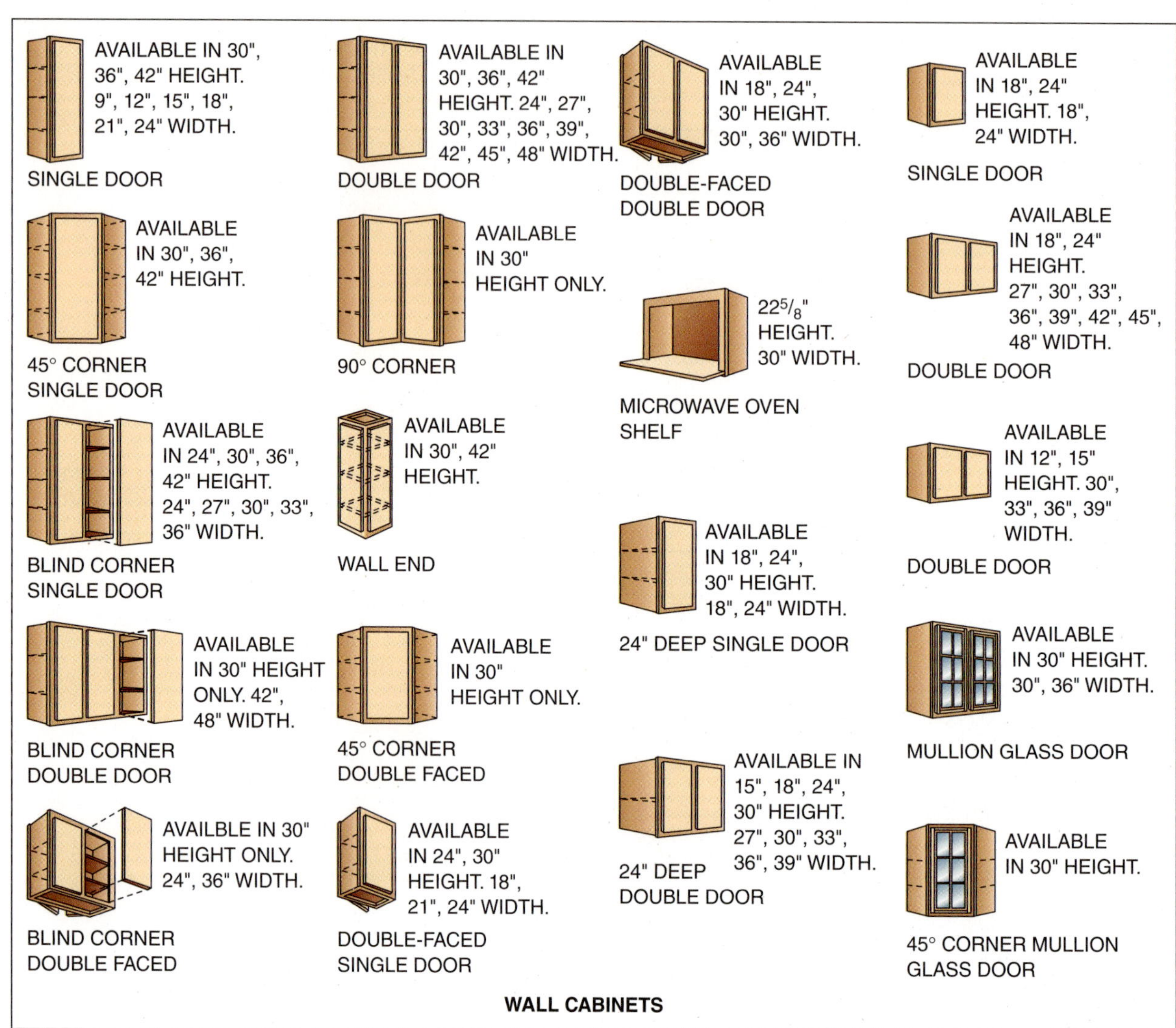


Courtesy of Merillat Industries.

**FIGURE 20-4** Kinds and sizes of manufactured wall cabinets.

made 18 and 24 inches wide. Double-door *pantry* cabinets are made 36 inches wide (Fig. 20-6). Wall cabinets with a 24-inch depth are usually installed above tall cabinets.

### Vanity Cabinets

Vanity (bathroom) cabinets are made 31½ and 34½ inches high and 16 to 21 inches deep. Widths range from 24 to 36 inches in increments of 3 inches, then 42, 48, and 60 inches. They are available with several combinations of doors and drawers depending on their width. Various sizes and styles of vanity wall cabinets are also manufactured (Fig. 20-7).

### Accessories

Accessories enhance a cabinet installation. *Filler* pieces fill small gaps in width between wall and base units when no combination of sizes can fill the existing space. They are cut to necessary widths on the job. Other accessories include cabinet end panels, face panels for dishwashers and refrigerators, open shelves for cabinet ends, and spice racks.

## LAY OUT KITCHEN CABINETS

The set of prints for a building contains plans, elevations, and details that show the cabinet layout.

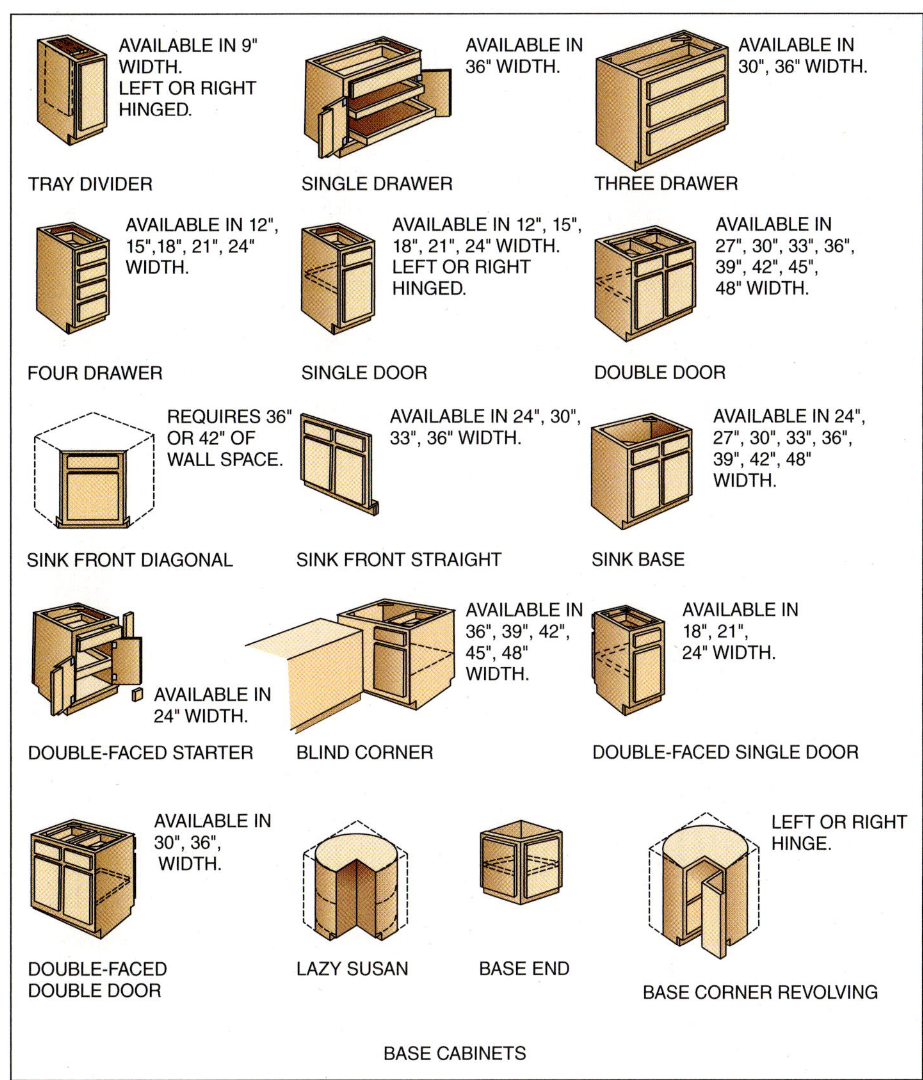


*Courtesy of Merillat Industries.*

**FIGURE 20-5** Most base cabinets are manufactured to match wall units.

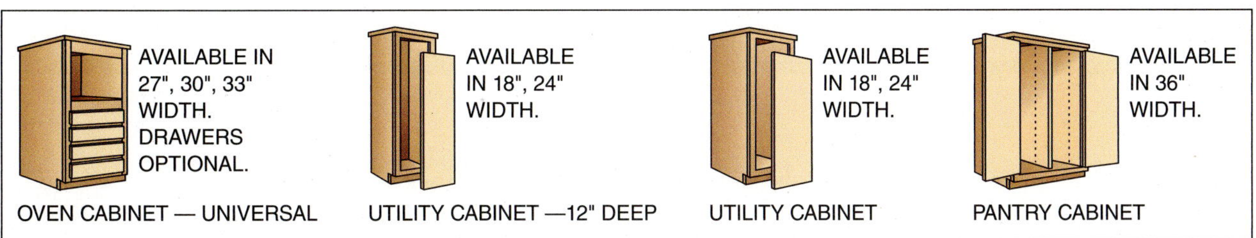


*Courtesy of Merillat Industries.*

**FIGURE 20-6** Tall cabinets are manufactured as oven, utility, and pantry units.

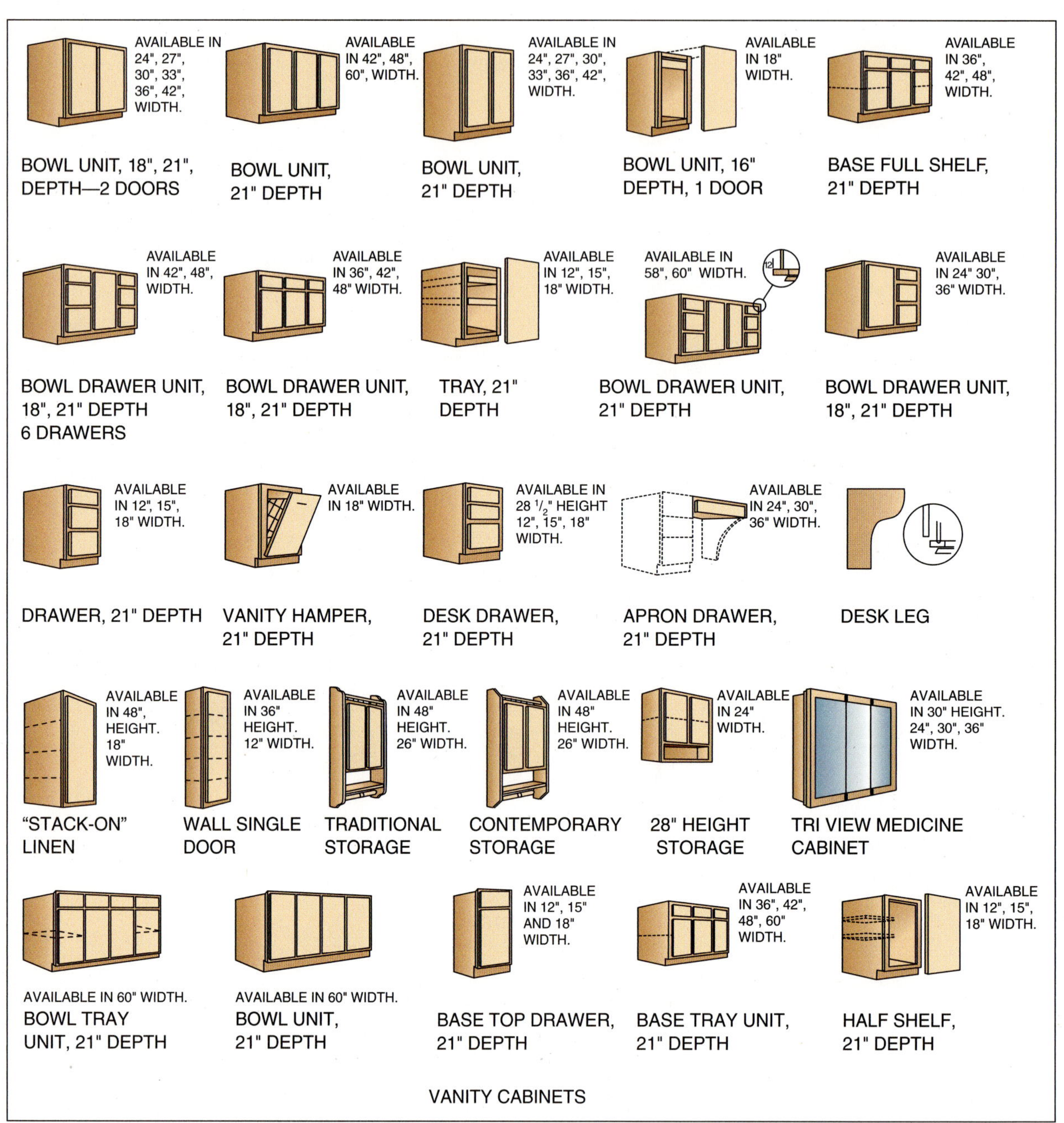


*Courtesy of Merillat Industries.*

**FIGURE 20-7** Vanity cabinets are made similar to kitchen cabinets but differ in size.

The plan is drawn to scale showing the location of all appliances, sinks, windows, and other necessary items (Fig. 20-8). The largest size cabinets are used instead of two or three smaller ones. This reduces the cost and makes installation easier.

The wall cabinets usually match up with the base cabinets. Filler strips are placed between a wall and a cabinet or between cabinets in the corner to complete the cabinet line (Fig. 20-9). *Countertops* are manufactured in various standard lengths or may be custom fit. Countertop material may be solid plastic or stone, such as Corian(R) and granite. These tops are made by workshops that specialize in these tops. A cost-effective countertop is one with plastic laminate. They can be cut to fit any installation against walls. The laminate countertops are covered with a

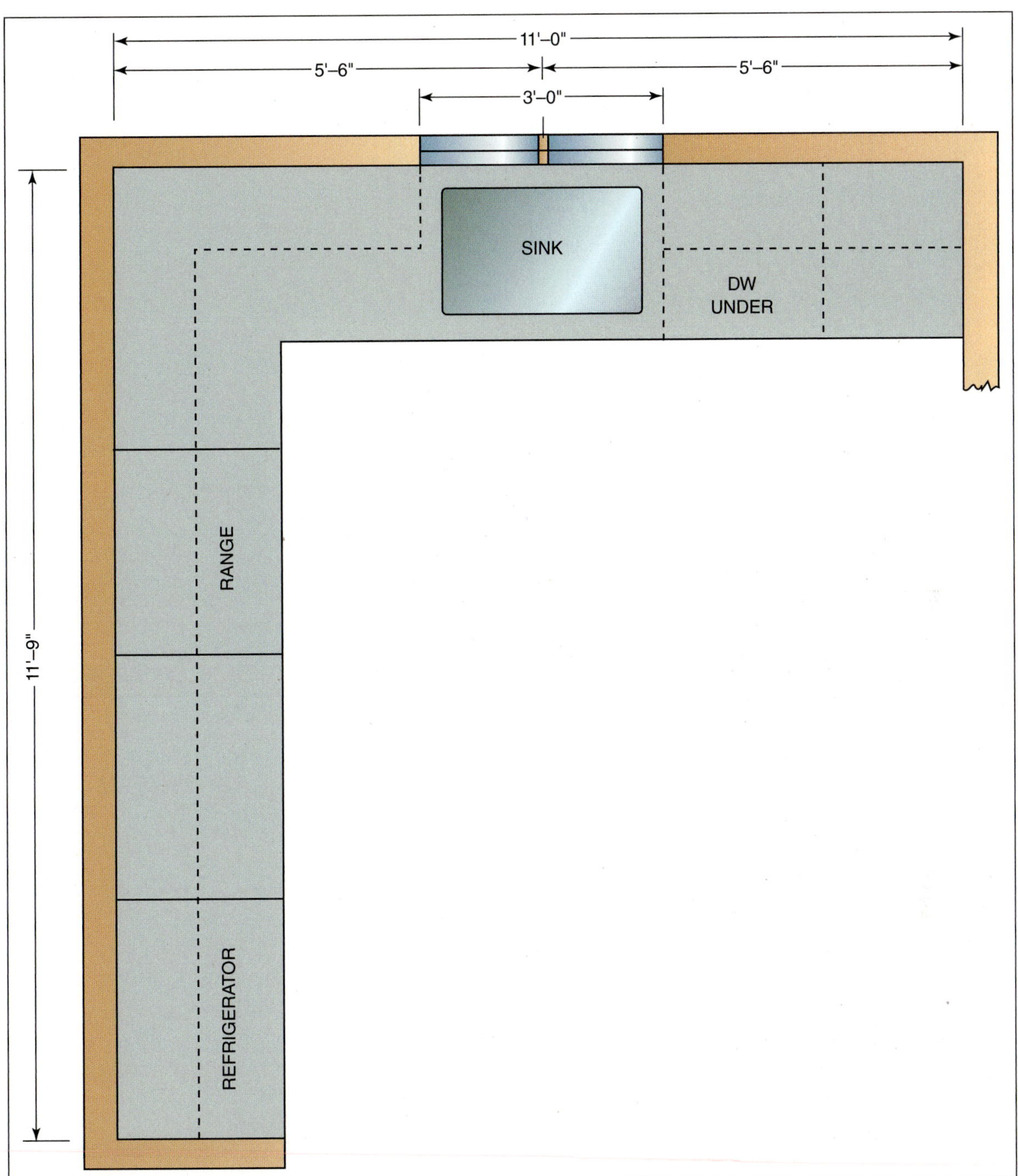


**FIGURE 20-8** Typical plan of a kitchen cabinet layout showing location of walls, windows, and appliances.

thin, tough *high-pressure plastic laminate.* It is available in many colors and patterns. Manufactured laminate countertops are called **postformed** *countertops.* This term comes from the method of forming the laminate to the rounded edges and corners of the countertop. They are also available with one end precut at a 45-degree angle for joining with a similar one at the corners. Special hardware is used to join the sections (Fig. 20-10). **Postforming** is bending the laminate with heat to a radius of ¾ inch or less. This can be done only with special equipment. Countertops may also be custom-made on the job.

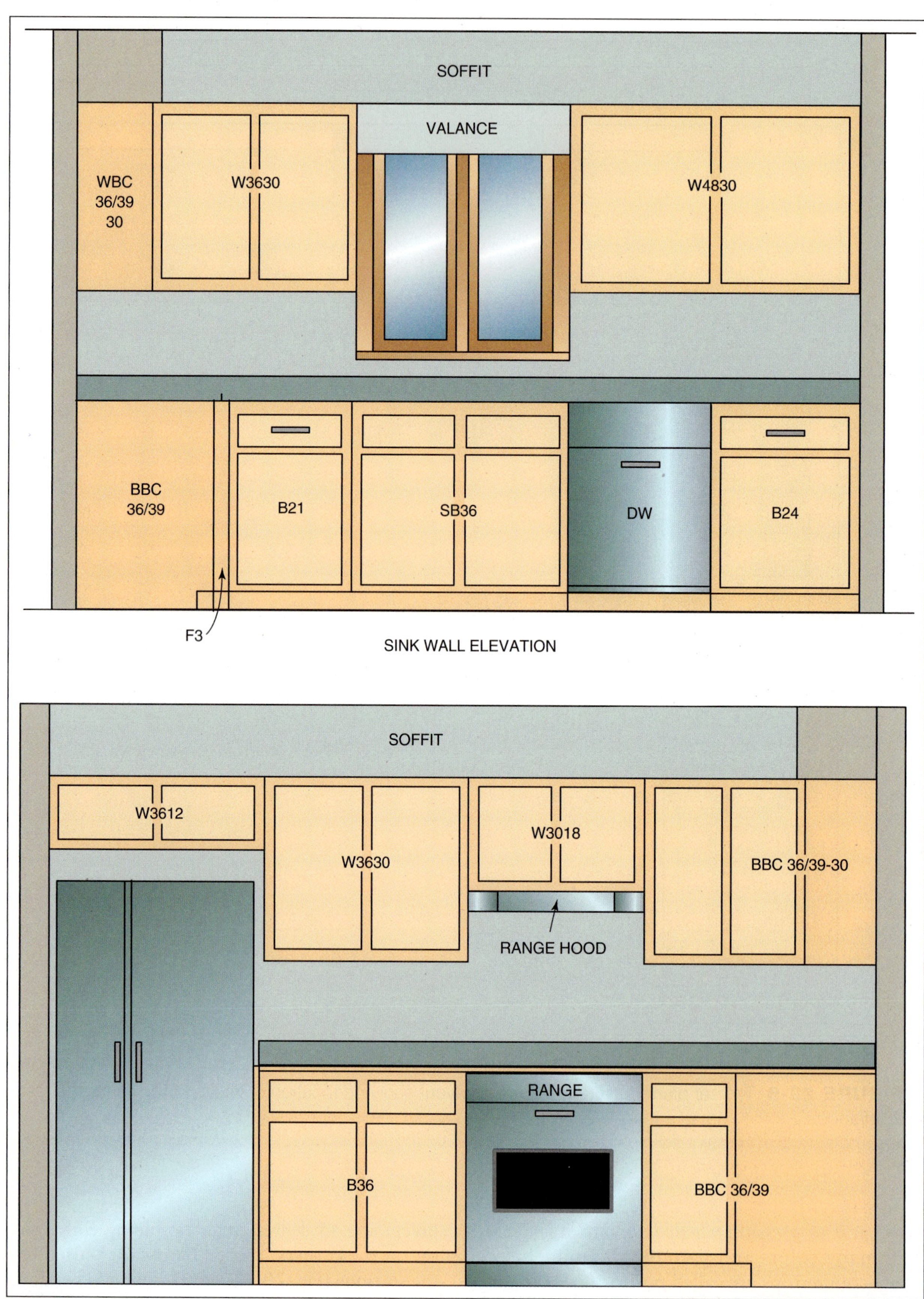


**FIGURE 20-9** Elevations of the installation are drawn and the cabinets identified.

**FIGURE 20-10** A section of a manufactured post-formed countertop. The edges and interior corner are rounded.

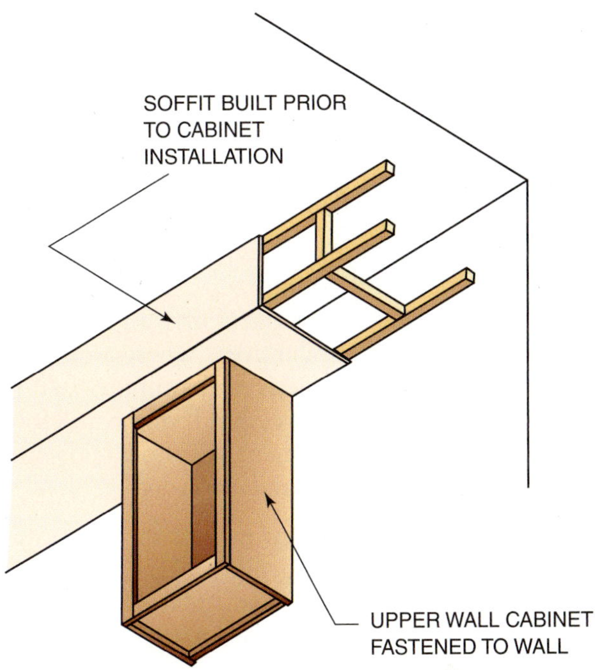


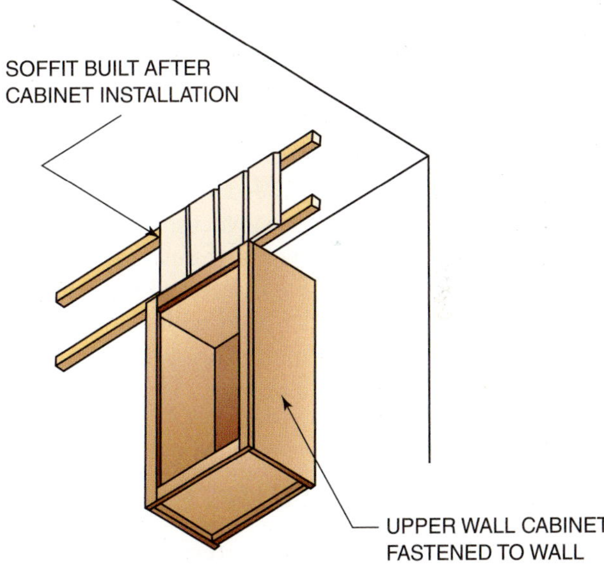


**FIGURE 20-11** Two methods of finishing the space above the wall cabinet.

# Installing Manufactured Cabinets

Cabinets must be installed level and plumb even though floors and walls may not be. Level lines are first drawn on the wall for base and wall cabinets. In order to level base cabinets that set on an out-of-level floor, either shim the cabinets to the high point of the floor or scribe the bottom to fit against the floor from the lowest point on the floor. Shimming the base cabinets leaves a space that must be covered later by a molding. Scribing and fitting the cabinets to the floor eliminates the need for a molding. The method used depends on various conditions of the job. If shimming base cabinets, lay out the level lines on the wall from the highest point on the floor where cabinets are to be installed. If fitting cabinets to the floor, measure up from the lowest point.

The space between the top of the wall unit and the ceiling may be left open or it may be closed. This finished detail may be a false front attached to the ceiling and the cabinet. Another method involves building a soffit before the cabinets are installed (Fig. 20-11).

***For step-by-step instructions on INSTALLING MANUFACTURED CABINETS, see Procedure 20-1 on pages 703–706.***

# Making a Countertop

Use ¾-inch panel material of high density particle board or MDF to make the countertop. If more than one length is required, join them with glue and screws to a short piece of backing plywood. The width of the pieces should be about 24½ inches to allow room to trim the top to size.

## PLASTIC LAMINATES

*Plastic laminates* are used widely for surfacing kitchen cabinets and countertops. They may also be used to cover walls or parts of walls in kitchens, bathrooms, and similar areas where a durable, easy-to-clean surface is desired. Laminates can be scorched by an open flame or hot frying pan. However, they resist moderate heat, alcohol, acids, and stains. They clean easily with a mild detergent.

Laminates are manufactured in many colors and designs, including wood grain patterns. Surfaces are available in gloss, satin, and textured finishes, among others. Laminates are ordinarily used in two thicknesses.

*Vertical-type* laminate is relatively thin (about 1⁄32 inch). It is used for vertical surfaces, such as walls and cabinet sides. Vertical-type laminate is available only in widths of 4 feet or 8 feet.

*Regular* or *standard* laminate is about 1⁄16-inch thick. It comes in widths of 24, 36, 48, and 60 inches and in lengths of 5, 6, 8, 10, and 12 feet. It is generally used on horizontal surfaces, such as countertops. It can be used on walls, if desired, or if the size required is not available in vertical type. Sheets are usually manufactured 1 inch wider and longer than the nominal size to make room for trimming to fit.

## FITTING THE COUNTERTOP

Place the countertop panel material on the base cabinets against the wall. Its outside edge should overhang the face frame an equal amount along the entire length. Open the pencil dividers or scribers to about 1⁄16 inch smaller than the amount of overhang. Scribe the back edge of the countertop to the wall. Cut the countertop to the scribed line. Place it back on top of the base cabinets. The ends should be flush with the ends of the base cabinets. The front edge should be flush with the face of the face frame (Fig. 20-12). Glue a 1 × 2 on the front edge and at the ends. Keep the top edge flush with the top side of the countertop. This piece may be covered with laminate. It may also be made of the same wood species as the cabinets and left exposed.

## APPLYING THE BACKSPLASH

If a *backsplash* is used, rip a 4-inch-wide length of ¾-inch stock the same length as the countertop. Use lumber for the backsplash, if lengths over 8 feet are required, to eliminate joints. The backsplash is fastened on top of and flush with the back edge of the countertop by driving screws up through the countertop and into the bottom edge of the backsplash (Fig. 20-13). In corners, fasten the ends of the backsplash together with screws. Attachment of backsplash to top is done after the top is fit to the wall.

## LAMINATING A COUNTERTOP

Before laminating a countertop, make sure all surfaces are flush. Check for protruding nailheads. Fill

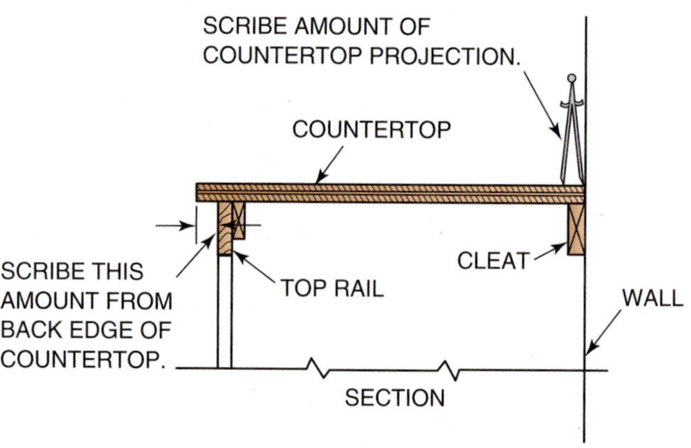


**FIGURE 20-12** Scribing the countertop to fit the wall with its outside edge flush with the face of the cabinet.

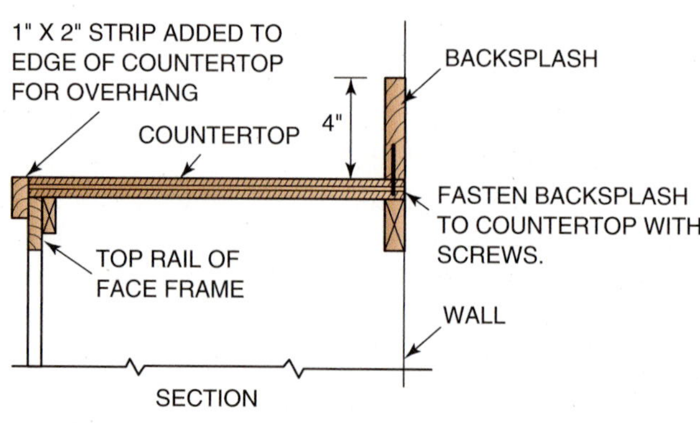


**FIGURE 20-13** Drive screws into the bottom edge of the backsplash to fasten it to the countertop.

in all holes and open joints. Lightly hand or power sand the entire surface, making sure joints in the countertop panel material are sanded flush.

## Laminate Trimming Tools and Methods

Pieces of laminate are first cut to a *rough size,* about ¼ to ½ inch wider and longer than the surface to be covered. A narrow strip is first cemented to the edge of the countertop. Its edges are *flush-trimmed* with the top and bottom surfaces. Top laminate is then cemented to the top surface, overhanging the edge strip. The overhang is then *bevel-trimmed* even with the laminated edge. A laminate trimmer or a small router fitted with laminate trimming bits is used for rough cutting and flush and bevel trimming of the laminate (Fig. 20-14).

## Cutting Laminate to Rough Sizes

Sheets of laminate are large, thin, and flexible. This makes them difficult to cut on a table saw. One method of cutting laminates to rough sizes is by clamping a straightedge to the sheet. Cut it by guiding a laminate trimmer with a flush trimming bit along the straightedge. It is easier to run the trimmer across the sheet than to run the sheet across the table saw. Also, the router bit leaves a smooth, clean-cut edge. Use a solid carbide trimming bit, which is smaller in diameter than one with ball bearings. It makes a narrower cut. It is easier to control and creates less waste. With this method, cut all the pieces of laminate needed to a rough width and length. Cut the narrow edge strips from the sheet first. Another method for cutting laminate to rough size is to use a carbide-tipped hand cutter (Fig. 20-15). It scores the laminate sufficiently in three passes to be able to break the piece away. Make the cut on the face side and then bend the piece up to create the cleanest break. When used with a straightedge, the cutter is fast and effective.



**FIGURE 20-14** The laminate trimmer is used with flush and bevel bits to trim overhanging edges of laminate.

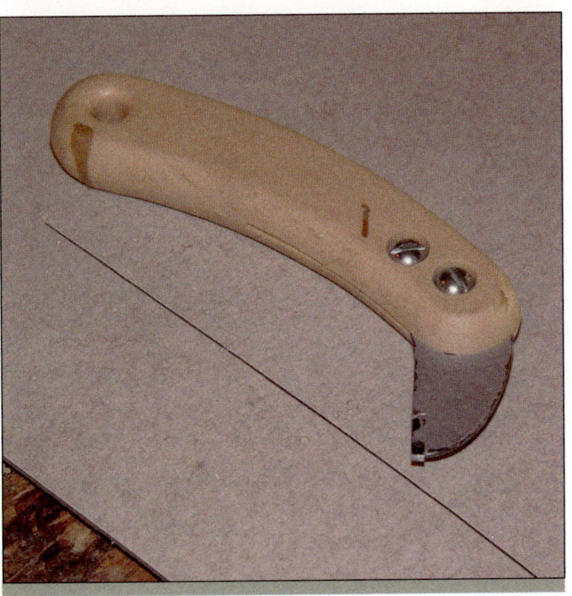

**FIGURE 20-15** A hand cutter is often used to cut laminate to a rough size.

### Using Contact Cement

*Contact cement* is used for bonding plastic laminates and other thin, flexible material to surfaces. A coat of cement is applied to the back side of the laminate and to the countertop surface. The cement is allowed to dry before the laminate is bonded to the panel. The bond is made on contact without the need of clamps. This makes adjusting the piece after contact nearly impossible. So it must be positioned properly before contact.

Contact cement bond may fail if not enough cement is applied. On porous material, like the edge of particleboard or plywood, a second coat is required after the first coat dries. When enough cement has been applied, a glossy film appears over the entire surface when dry. Both surfaces must be dry before contact is made. To test for dryness, lightly press your finger on the surface. Although it may feel sticky, the cement is dry if no cement remains on the finger.

If contact cement dries too long (more than about 2 hours, depending on the humidity), it will not bond properly. To correct this condition, merely apply another coat of cement and let it dry. Pressure must be applied to the entire surface by using a 3-inch **J-roller** or by tapping with a hammer on a small block of wood (Fig. 20-16).

**CAUTION**

**CAUTION:** Some contact cements are flammable. Apply only in a well-ventilated area around no open flame. Avoid inhaling the fumes.

### Laminating the Countertop Edges

The edges of the countertop are laminated first. This is done to create a natural watershed over the lapped laminate at the edge of the countertop. Otherwise, gravity would pull the water down into the seam.

Wipe the panel and laminate surfaces clean. This will allow for a uniform bond. Apply coats of cement to the countertop edges and the back of the edge laminate with a narrow brush or small paint roller. After the cement is dry, apply the laminate to the front edge of the countertop (Fig. 20-17). Position it so the bottom edge, top edge, and ends overhang. A permanent bond is made when the two surfaces make contact. A mistake in positioning means removing the bonded piece with a heat gun—a time-consuming, frustrating, and difficult job. Roll out or tap the surface. Apply the laminate to the ends in the same manner as to the front edge piece. Make sure that the square ends butt up firmly against the back side of the overhanging ends of the front edge piece to make a tight joint.

### Trimming Laminated Edges

The flush ends of the edge laminate must be trimmed. If laminate has been applied to cover the edges of another laminate, a bevel trimming bit must be used to trim the overhanging ends.

When using a bevel trimming bit, the router base is gradually adjusted to expose the bit so that the laminate is trimmed flush with the first piece but not cutting into it. The bevel of the cutting edge allows the laminate to be trimmed without cutting

**FIGURE 20-16** Rolling out the laminate with a J-roller is required to ensure a proper bond.

**FIGURE 20-17** Applying laminate to the edge of the countertop.

into the adjacent piece when the pilot rides against another piece of laminate (see Fig. 20-14).

Ball bearing trimming bits have *live pilots*. These work well protecting the bearing surface from the spinning bit. Solid carbide bits may have *dead pilots* that turn with the bit. When using a trimming bit with a dead pilot, the laminate must be lubricated where the pilot will ride. Rub a short piece of white candle on the laminate to prevent marring the laminate by the bit.

Use the bevel trimming bit to trim the overhanging ends of the edge laminate. Then, use the flush trimming bit to trim off the bottom and top edges of both front and end edge pieces (Fig. 20-18).

Use a belt sander or a file to smooth the top laminate edge flush with the top panel surface. Sand or file flat on the countertop panel so a sharp square edge is made. This ensures a tight joint with the countertop laminate. Sand or file toward the core to prevent chipping the laminate. Smooth the bottom edge. Ease the sharp outside corner with a sanding block.

### Laminating the Countertop Surface

Apply contact bond cement to the countertop and the back side of the laminate. Let dry. To position large pieces of countertop laminate, first place thin strips of wood about a foot apart on the surface. Lay the laminate to be bonded on the strips or slats. Then, position the laminate correctly (Fig. 20-19). Make contact on one end. Gradually remove the slats one by one until all are removed. The laminate should then be positioned correctly with no costly errors. Roll the laminate to complete the bond. Trim the overhanging back edge with a flush trimming bit. Trim the ends and front edge with a bevel trimming bit (Fig. 20-20). Use a flat file to smooth the trimmed edge. Slightly ease the sharp corner.

**FIGURE 20-19** Position the laminate on the countertop using scrap wooden slats.

### Laminate Seams

When the countertop is laminated with two or more lengths, tight joints must be made between them. Tight joints can be made by clamping the two pieces of laminate in a straight line on a strip of ¾-inch stock. Butt the ends together or leave a space less than ¼ inch between. Using one of the

**FIGURE 20-18** Flush trimming the countertop edge laminate.

**FIGURE 20-20** The outside edge of the countertop laminate is bevel-trimmed.

**FIGURE 20-21** Making a tight laminate butt seam by cutting both pieces at the same time.

**FIGURE 20-22** Apply the laminate to the backsplash, then fasten it to the laminated countertop.

strips as a guide, run the laminate trimmer, with a flush trimming bit installed, through the joint. Keep the pilot of the bit against the straightedge. Cut the ends of both pieces at the same time to ensure a tight joint (Fig. 20-21). Bond the sheets as previously described. *Seam-filling compound,* specially made for laminates, may be applied to make a practically invisible joint. Wipe off excess compound with the recommended solvent.

### Laminating Backsplashes

Backsplashes are laminated in the same manner as countertops. Laminate the backsplash. Then, reattach it to the countertop with screws. Use a little caulking compound between the backsplash and countertop. This prevents any water from seeping through the joint. Clamping the back splash aids in holding the pieces in the proper position when installing the pilot holes and screws (Fig. 20-22).

### Laminating Rounded Corners

If the edge of a countertop has a rounded corner, the laminate can be bent. Strips of laminate can be cold bent to a minimum radius of about 6 inches.

Heating the laminate to 325 degrees Fahrenheit uniformly over the entire bend will facilitate bending to a minimum radius of about 2½ inches. Heat the laminate carefully with a heat gun. Bend it until the desired radius is obtained (Fig. 20-23). Experimentation may be necessary until success in bending is achieved.

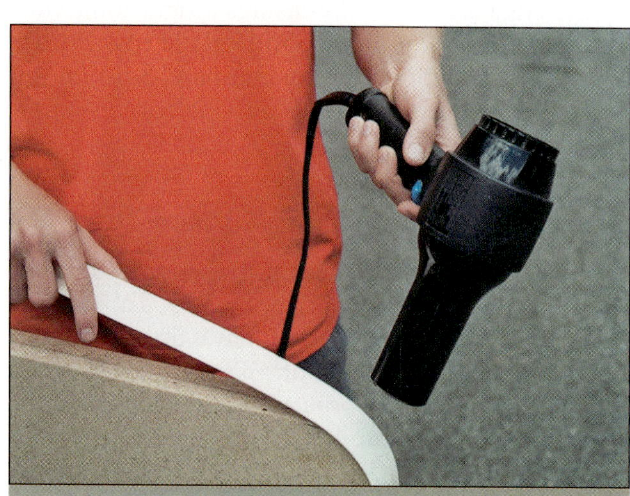

**FIGURE 20-23** Heating and bending laminate with a heat gun.

CAUTION

**CAUTION: Keep fingers away from the heated area of the laminate. Remember that the laminate retains heat for some time.**

## CABINET DOORS

Doors are classified by their construction and also by the method of installation. Sliding doors are occasionally installed, but most cabinets are fitted with hinged doors that swing. Hinged cabinet doors are classified as overlay, lipped, and flush,

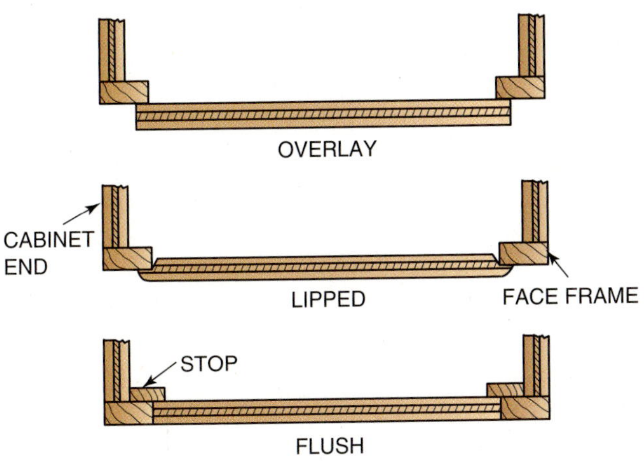


**FIGURE 20-24** Plan views of overlay, lipped, and flush doors.

based on the method of installation (Fig. 20-24). The overlay method of hanging cabinet doors is the most widely used.

### Overlay Doors

The *overlay* type of door laps the entire thickness of the door over the opening, usually 3/8 inch on all sides. However, it may overlay any amount. In many cases, it may cover the entire face frame. The overlay door is the easiest to install as it does not require fitting in the opening and the face frame of the cabinet acts as a stop for the door. European-style cabinets omit the face frame. Doors completely overlay the front edges of the cabinet (Fig. 20-25).

### Lipped Doors

The *lipped* door has rabbeted edges that overlap the opening by about 3/8 inch on all sides. Usually the ends and edges are rounded over to give a more pleasing appearance. Lipped doors and drawers also cover the opening. No fitting is required, and the rabbeted edges stop against the face frame of the cabinet. However, a little more time is required to shape the rabbeted edges.

### Flush Type

The *flush type* door fits into and flush with the face of the opening. They are a little more difficult to hang because they must be fitted in the opening. A fine joint must be made between the opening and the door. Stops must be provided in the cabinet against which to close the door.

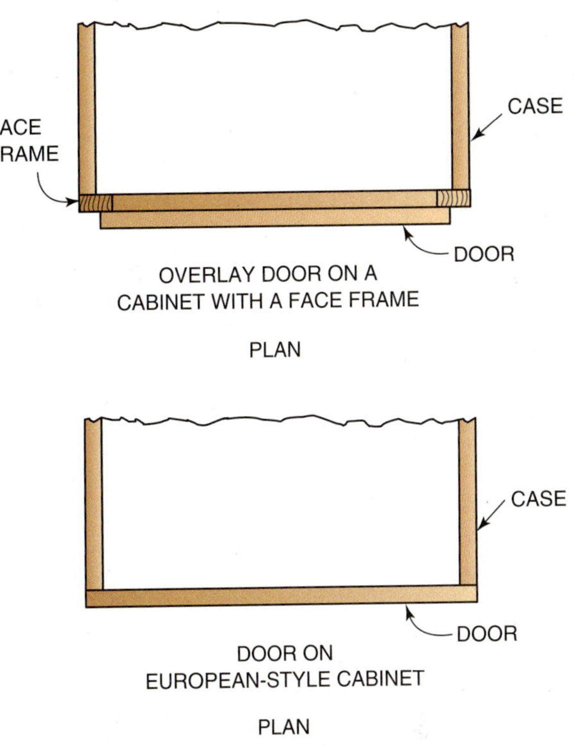


**FIGURE 20-25** Overlay doors lap the face frame by varying amounts. European-style doors are hinged to and completely overlay the case.

## DOOR CONSTRUCTION

Doors are also classified by their construction as *solid* or *paneled*. Solid doors are made of plywood, particleboard, or solid lumber. Particleboard doors are covered with plastic laminate. Matched boards with V-grooves and other designs, such as used for wall paneling, are often used to make solid doors. Designs may be grooved into the face of the door with a router. Small moldings may be applied for a more attractive appearance. Paneled doors have an exterior framework of solid wood with panels of solid wood, plywood, hardboard, plastic, glass, or other panel material.

Many complicated designs are manufactured by millworkers with specialized equipment. Styles of doors made on the job depend on the specialty equipment available (Fig. 20-26). Both solid doors and paneled doors may be hinged in overlay, lipped, or flush fashion.

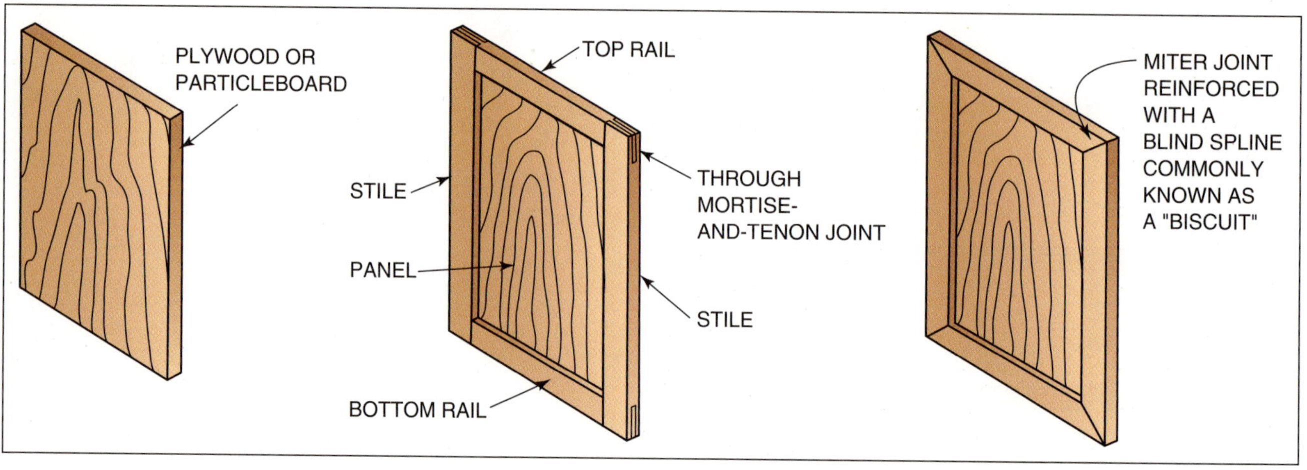


**FIGURE 20-26** Panel doors of simple design can be made on the job.

## HINGES

Several types of cabinet hinges are surface, offset, overlay, pivot, and butt. For each type, there are many styles and finishes (Fig. 20-27). Some types are *self-closing* hinges that hold the door closed and eliminate the need for door catches.

### SURFACE HINGES

*Surface* hinges are applied to the exterior surface of the door and frame. The back side of the hinge leaves may lie in a straight line for flush doors. One leaf may be offset for lipped doors (Fig. 20-28). The surface type is used when it is desired to expose the hardware, as in the case of wrought iron and other decorative hinges.

### OFFSET HINGES

*Offset* hinges are used on lipped doors. They are called *offset surface* hinges when both leaves are fastened to outside surfaces. The *semi-concealed offset* hinge has one leaf bent to a ⅜-inch offset that is screwed to the back of the door. The other leaf screws to the exterior surface of the face frame. A *concealed offset* type is designed in which only the pin is exposed when the door is closed (Fig. 20-29).

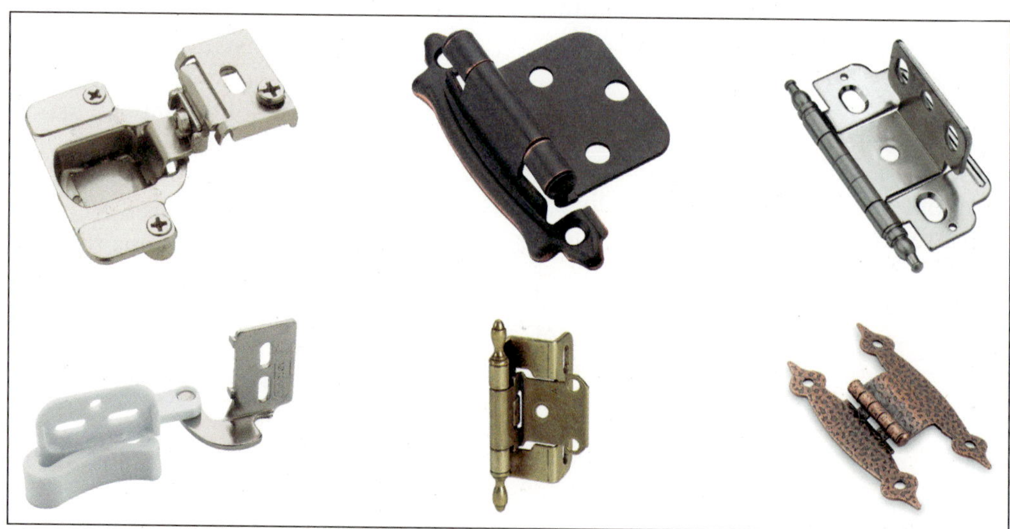

Courtesy of Amerock, A Newell Rubbermaid Company.

**FIGURE 20-27** Cabinet door hinges come in many styles and finishes.

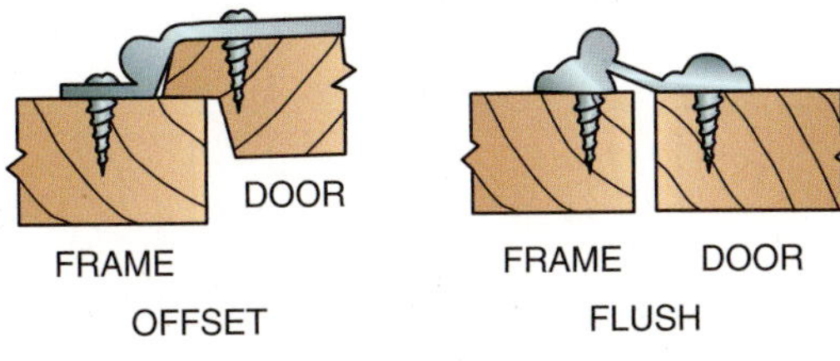


**FIGURE 20-28** Surface hinges.

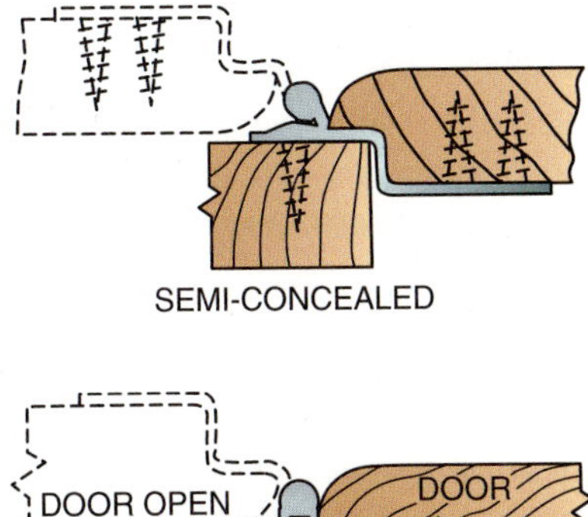


**FIGURE 20-29** Offset hinges.

## OVERLAY HINGES

*Overlay* hinges are available in *semi-concealed* and *concealed types*. With semi-concealed types, the amount of overlay is variable. Certain concealed overlay hinges are made for a specific amount of overlay, such as 1/4, 5/16, 3/8, and 1/2 inch.

European-style hinges are completely concealed. A large hole is partially drilled to receive the hinge. There are adjusting screws to allow fine-tuning the door fit after it is installed (Fig. 20-30). Some overlay hinges, with one leaf bent at a 30-degree angle, are used on doors with reverse beveled edges.

## PIVOT HINGES

**Pivot** hinges are usually used on overlay doors. They are fastened to the top and bottom of the door and to the inside of the case. They are frequently used when there is no face frame and the door completely covers the face of the case (Fig. 20-31).

## BUTT HINGES

*Butt* hinges may be used on flush doors. Butt hinges for cabinet doors are a smaller version of those used on entrance doors. The leaves of the pin hinge are set into **gains** in the edges of the frame and the door, in the same manner as for entrance doors. Butt hinges are used on flush doors when the goal is to conceal most of the hardware. They are not often used on cabinets because they take more time to install than other types (Fig. 20-32).

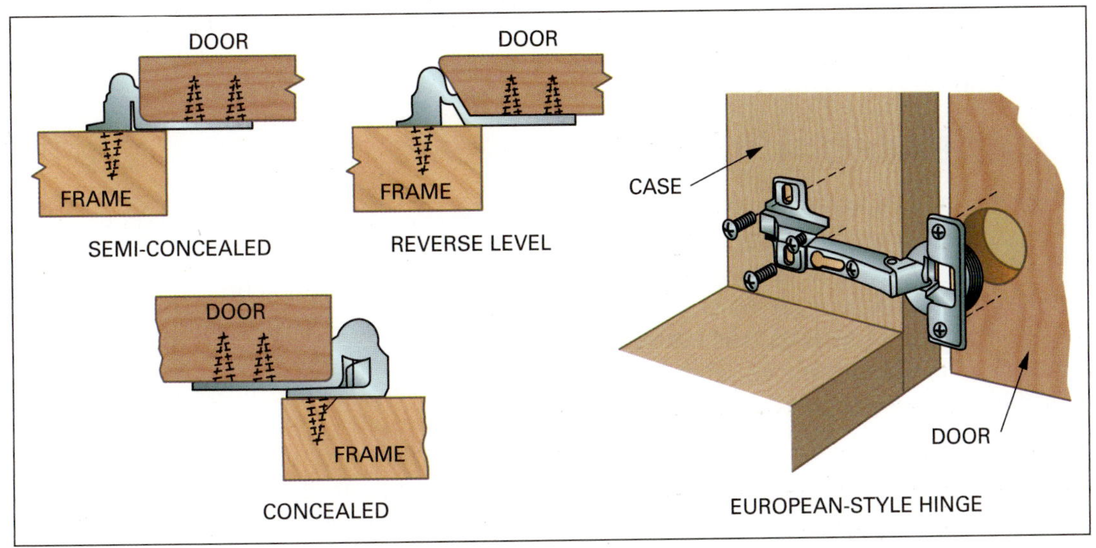


**FIGURE 20-30** Overlay hinges.

**FIGURE 20-31** Pivot hinges for an overlay door.

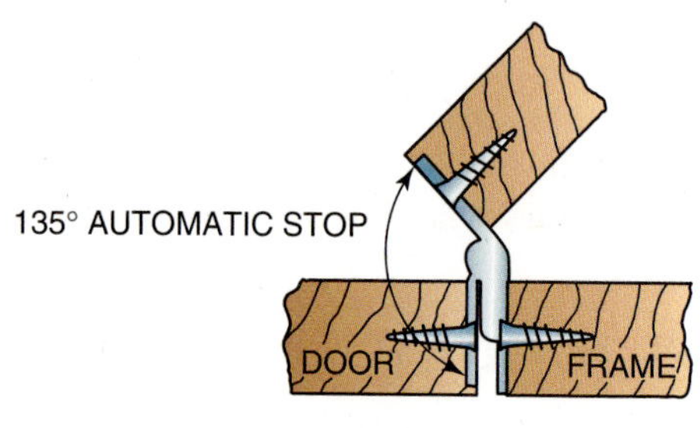


**FIGURE 20-32** Butt hinges.

# Hanging Cabinet Doors

Steps to hanging a cabinet door depends on the type of hinges used on the door. The final step is to install the door catch.

## SURFACE HINGES

To hang cabinet doors with surface hinges, first apply the hinges to the door. Then, shim the door in the opening so an even joint is obtained all around. Screw the hinges to the face frame.

## SEMI-CONCEALED HINGES

For semi-concealed hinges, screw the hinges to the back of the door. Then, center the door in the opening. Fasten the hinges to the face frame. When more than one door is to be installed side by side, clamp a straightedge to the face frame along the bottom of the openings for the full length of the cabinet. Rest the doors on the straightedge to keep them in line (Fig. 20-33).

## CONCEALED HINGES

When installing concealed hinges, first screw the hinges on the door. Center the door in the opening.

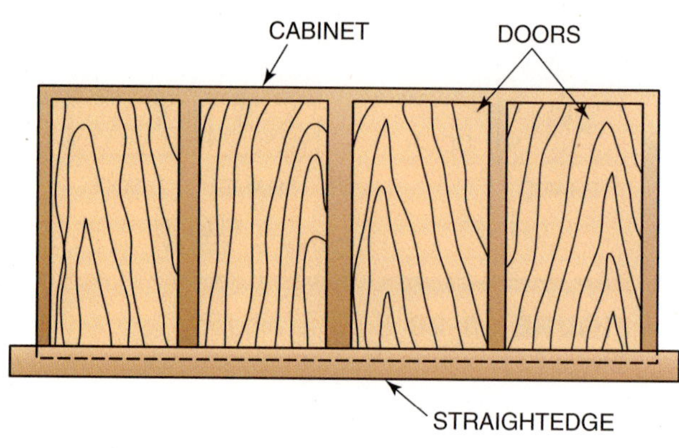


**FIGURE 20-33** When installing doors, use a straightedge to keep them in line.

Press or tap on the hinge opposite the face frame. Small projections on the hinge make indentations to mark its location on the face frame. Open the door. Place the projections of the hinges into the indentations. Screw the hinges to the face frame.

## BUTT HINGES

Hanging flush cabinet doors with butt hinges is done in the same manner as hanging doors. Drill pilot holes for all screws so they are centered on the holes in the hinge leaf. Drilling the holes off-center throws the hinge to one side when the screws are driven. This usually causes the door to be out of alignment when hung.

## INSTALLING PULLS AND KNOBS

Cabinet *pulls* or *knobs* are used on cabinet doors and drawers. They come in many styles and designs. They are made of metal, plastic, wood, porcelain, or other material (Fig. 20-34). Pulls and knobs are installed by drilling holes through the door. Then, fasten them with machine screws from the inside. When two screws are used to fasten a pull, the holes are drilled slightly oversize in case they are a little off-center. This allows the pulls to be fastened easily without cross-threading the screws. Usually 3⁄16-inch diameter holes are drilled for 1⁄8-inch machine screws. To drill holes quickly and accurately, make a *template* from scrap wood that fits over the door. The template can be made so that holes can be drilled for doors that swing in either direction (Fig. 20-35).

Courtesy of Amerock Corporation.

**FIGURE 20-34** A few of the many styles of pulls and knobs used on cabinet doors and drawers.

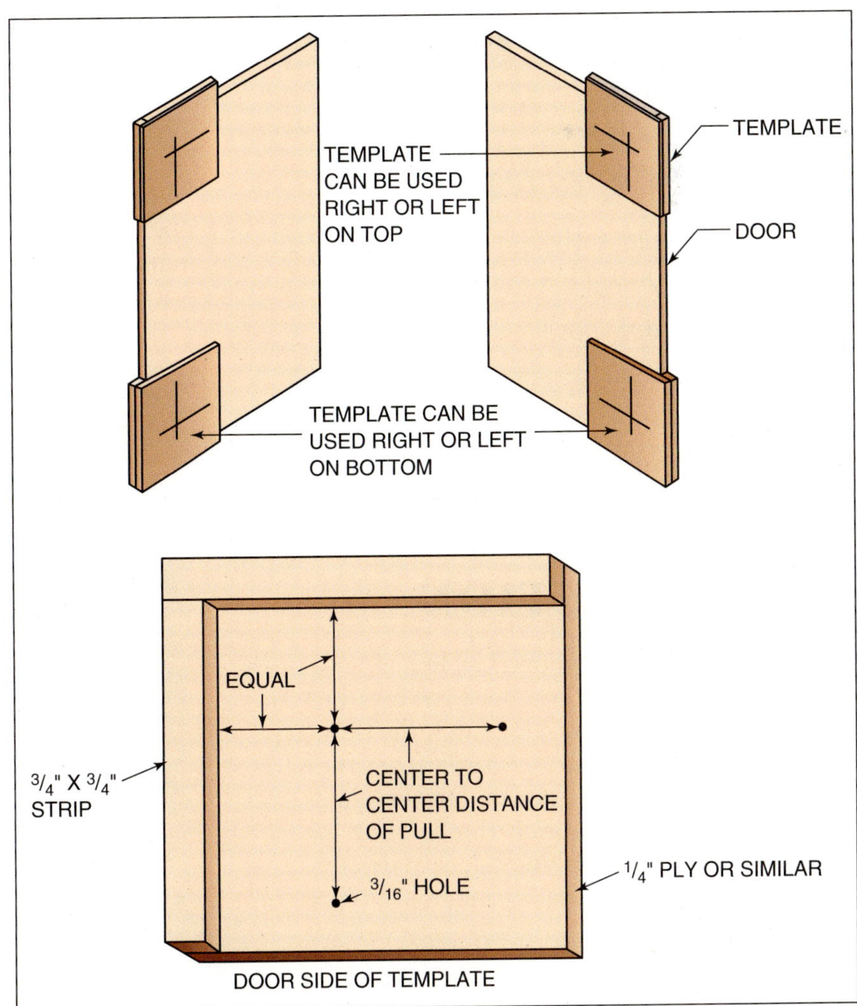


**FIGURE 20-35** Techniques for making a template to speed installation of door pulls.

## DOOR CATCHES

Many hinges are self-closing, which hold the door tight to the frame when closed. Doors without self-closing hinges need *catches* to hold them closed. Many kinds of catches are available (Fig. 20-36). Catches should be placed where they are not in the way, such as on the bottom of shelves, instead of the top. Magnetic catches are available with single or double magnets of varying holding power. An adjustable magnet is attached to the inside of the case. A metal plate is attached to the door. First attach the magnet. Then, place the plate on the magnet. Close the door and tap it firmly over the plate. Projections on the plate mark its location on the door. Attach the plate to the door where marked. Try the door. Adjust the magnet, if necessary.

Friction catches are installed in a similar manner to that used for magnetic catches. Fasten the adjustable section to the case and the other section to the door. Elbow catches are used to hold one door of a double set. They are released by reaching to the back side of the door. These catches are usually used when one of the doors is locked against the other. Bullet catches are spring-loaded. They fit into the edge of the door. When the door is closed, the catch fits into a recessed plate mounted on the frame.

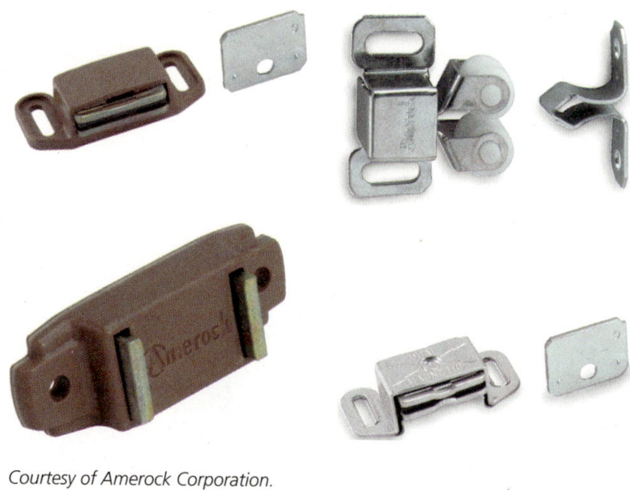

Courtesy of Amerock Corporation.

**FIGURE 20-36** Several types of catches are available for use on cabinet doors.

*from experience...*

Use a template when drilling holes for cabinet door pulls. It saves time and assures proper positioning of the pulls.

# DRAWER CONSTRUCTION

Drawers are classified as overlay, lipped, and flush in the same way as doors. In a cabinet installation, the drawer type should match the door type. Drawer fronts are generally made from the same material as the cabinet doors. Drawer sides and backs are generally ½-inch thick. They may be made of solid lumber, plywood, or particleboard. The drawer bottom is usually made of ¼-inch plywood or hardboard.

## DRAWER JOINTS

Typical joints between the front and sides of drawers are the dovetail, lock, and rabbet joints. The dovetail joint is used in higher quality drawer construction. It takes a longer time to make but is the strongest. Dovetail drawer joints may be made by using a router and a dovetail template (Fig. 20-37). The lock joint is simpler. It can be made by using a table saw. The rabbet joint is the easiest to make. However, it must be strengthened with fasteners in addition to glue (Fig. 20-38). Joints normally used between the sides and back are the dovetail, dado and rabbet, dado, and butt joints. With the exception of the dovetail joint, the drawer back is usually

**FIGURE 20-37** Dovetail joints can be made with a router and a dovetail template.

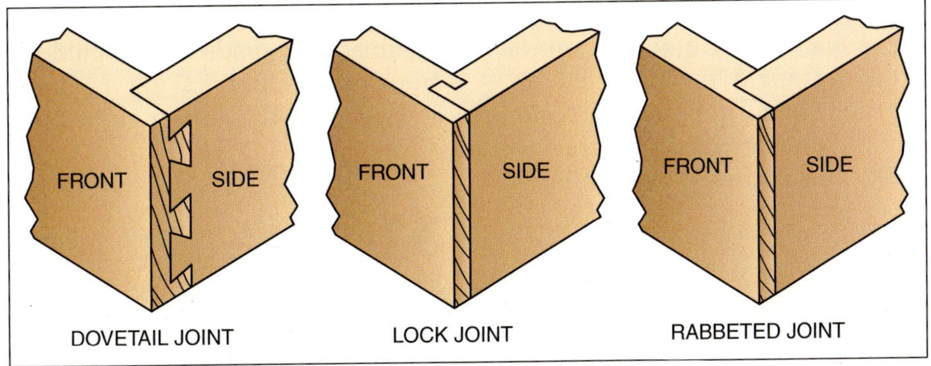


**FIGURE 20-38** Typical joints between drawer front and side.

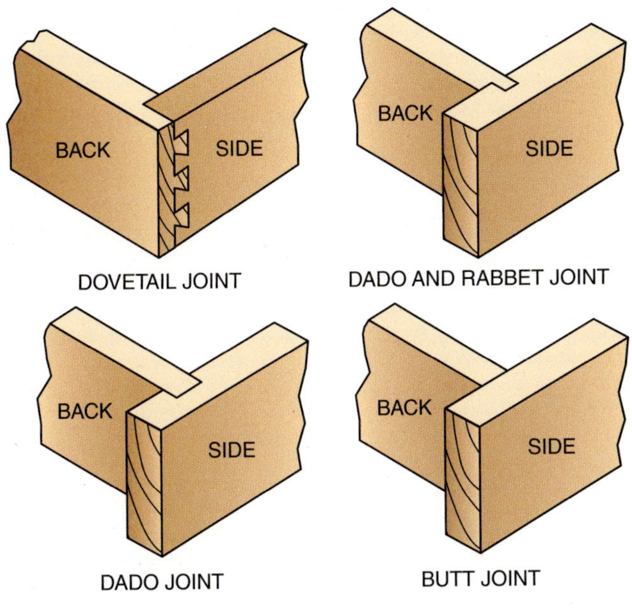


**FIGURE 20-39** Typical joints between drawer back and side.

set in at least ½ inch from the back ends of the sides to provide added strength. This helps prevent the drawer back from being pulled off if the contents get stuck while opening the drawer (Fig. 20-39).

## DRAWER BOTTOM JOINTS

The drawer bottom is fitted into a groove on all four sides of the drawer (Fig. 20-40). In some cases, the drawer back is made narrower so when the four sides are assembled, the bottom panel is slipped in the groove. Its back edge is fastened to the bottom edge of the drawer back (Fig. 20-41).

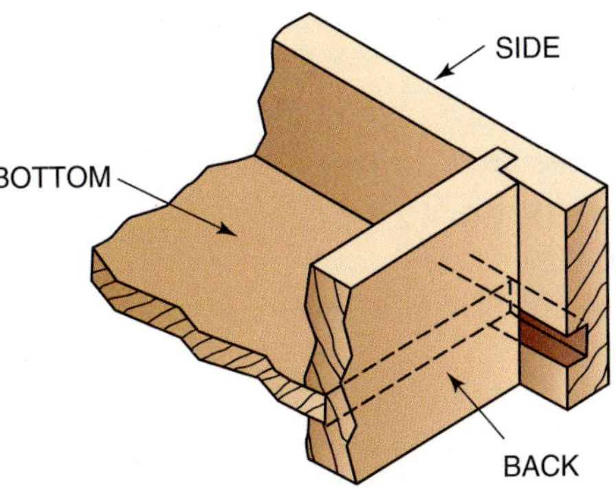


**FIGURE 20-40** Drawer bottom fitted in groove at drawer back.

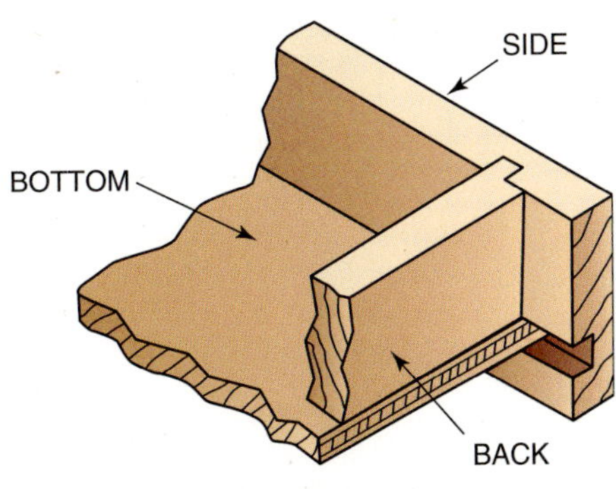


**FIGURE 20-41** Drawer bottom fastened to bottom edge of drawer back.

## DRAWER GUIDES

There are many ways of guiding drawers. The type of drawer guide selected affects the size of the drawer. The drawer must be supported level and guided sideways. It must also be kept from tilting down when opened.

There are many different types of metal drawer guides. Some have a single track mounted on the bottom center of the opening. Others may be centered above or on each side of the drawer. Nylon rollers mounted on the drawer ride in the track of the guide (Fig. 20-42). Instructions for installation differ with each type and manufacturer. When using commercially made drawer guides, read the instructions first, before making the drawer, so proper allowances for the drawer guide can be made.

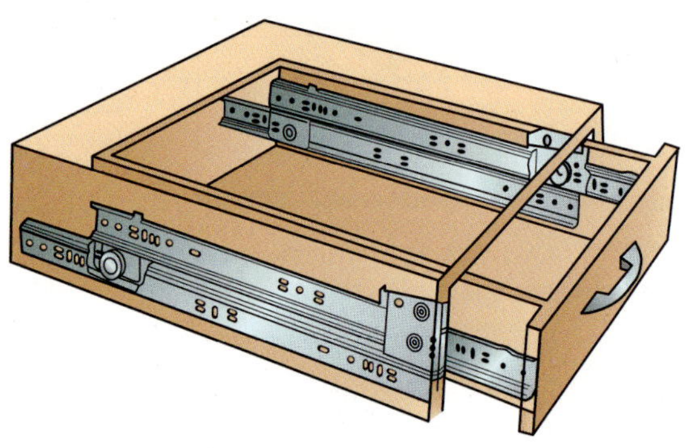

*Courtesy of Knape and Vogt Manufacturing Company.*

**FIGURE 20-42** Installing metal drawer guides.

## SUMMARY

- Cabinets have components and parts that make up the cabinet unit.
- Installing manufactured cabinets properly will ensure long life to the kitchen.
- Countertops and cabinet doors may be made on the job.

PROCEDURE 20-1

# Installing Manufactured Cabinets

## Cabinet Layout Lines

**A** Measure 34½ inches up the wall. Draw a level line to indicate the tops of the base cabinets. Measure and mark another level line on the wall 54 inches from the floor. The bottom of the wall units are installed to this line.

- Next, mark the stud locations of the framed wall. Cabinet mounting screws will be driven into the studs. Above the upper line on the wall, drive a small nail in at the point where the wall is solid to accurately locate the stud. Drive nails where the holes will be later covered by a cabinet. Mark the locations of the remaining studs where cabinets will be attached. At each stud location, draw plumb lines on the wall. Mark the outlines of all cabinets on the wall to visualize and check the cabinet locations against the layout.

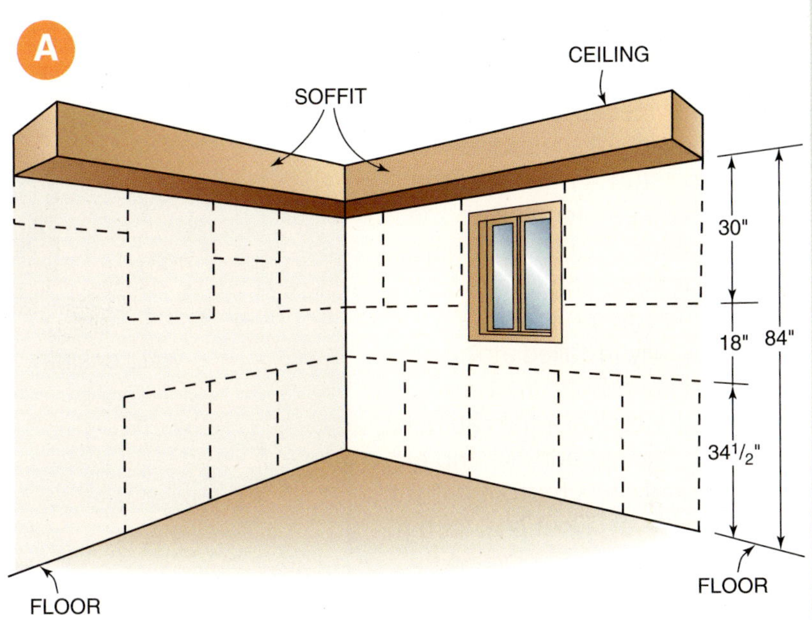


## Installing Wall Units

**A** A *cabinet lift* may be used to hold the cabinets in position for fastening to the wall. If a lift is not available, the doors and shelves may be removed to make the cabinet lighter and easier to clamp together. If possible, screw a strip of lumber so its top edge is on the level line for the bottom of the wall cabinets. Strips of wood cut to the proper length may be used to support the wall units while they are being fastened. They may extend to the base cabinet or the floor.

PROCEDURE 20-1

# Installing Manufactured Cabinets (continued)

**B** Start the installation of wall cabinets in a corner. On the wall, measure from the line representing the outside of the cabinet to the stud centers. Transfer the measurements to the cabinets. Drill shank holes for mounting screws through mounting rails usually installed at the top and bottom of the cabinet. Place the cabinet on the supporting strip or stand so its bottom is on the level layout line. Fasten the cabinet in place with mounting screws of sufficient length to hold the cabinet securely. Do not overly tighten the screws as it may split the cabinet joints. The next cabinet is installed in the same manner.

**C** Align the adjoining *stiles* so their faces are flush with each other. Clamp them together and screw the stiles tightly together. Continue this procedure around the room. After all the stiles are secured to each other, tighten all mounting screws. If a filler needs to be used, it is better to add it at the end of a run. It may be necessary to scribe the filler to the wall.

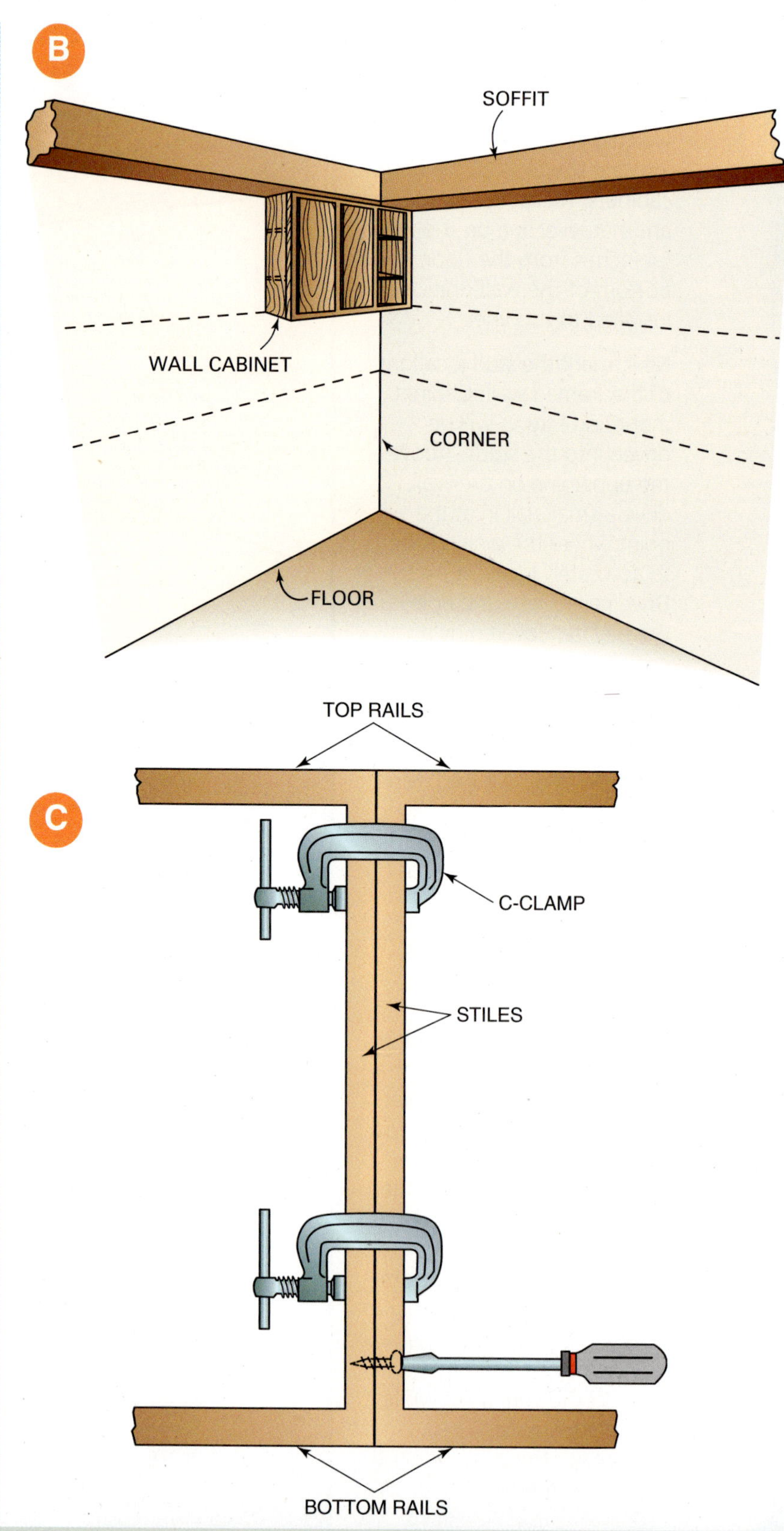

- The cut edge of filler piece is best placed against the wall of the corner. This puts a factory edge against factory edge of cabinet. Clamp the filler piece on top of the last style flush with the inside edge. Scribe off against the wall a distance equal to the width of the style. A wood block the same width as the style makes this an easy process.
- The space between the top of the wall unit and the ceiling may be finished by installing a soffit or left open.

## Installing Base Cabinets

**A** Start the installation of base cabinets in a corner. Shim the bottom until the cabinet top is on the layout line. Then, level and shim the cabinet from back to front. If cabinets are to be fitted to the floor, shim until their tops are level across width and depth. This will bring the tops above the layout line that was measured from the low point of the floor. Adjust the scriber so the distance between the points is equal to the amount the top of the unit is above the layout line. Scribe this amount off the bottom end of the cabinets by running the dividers along the floor.

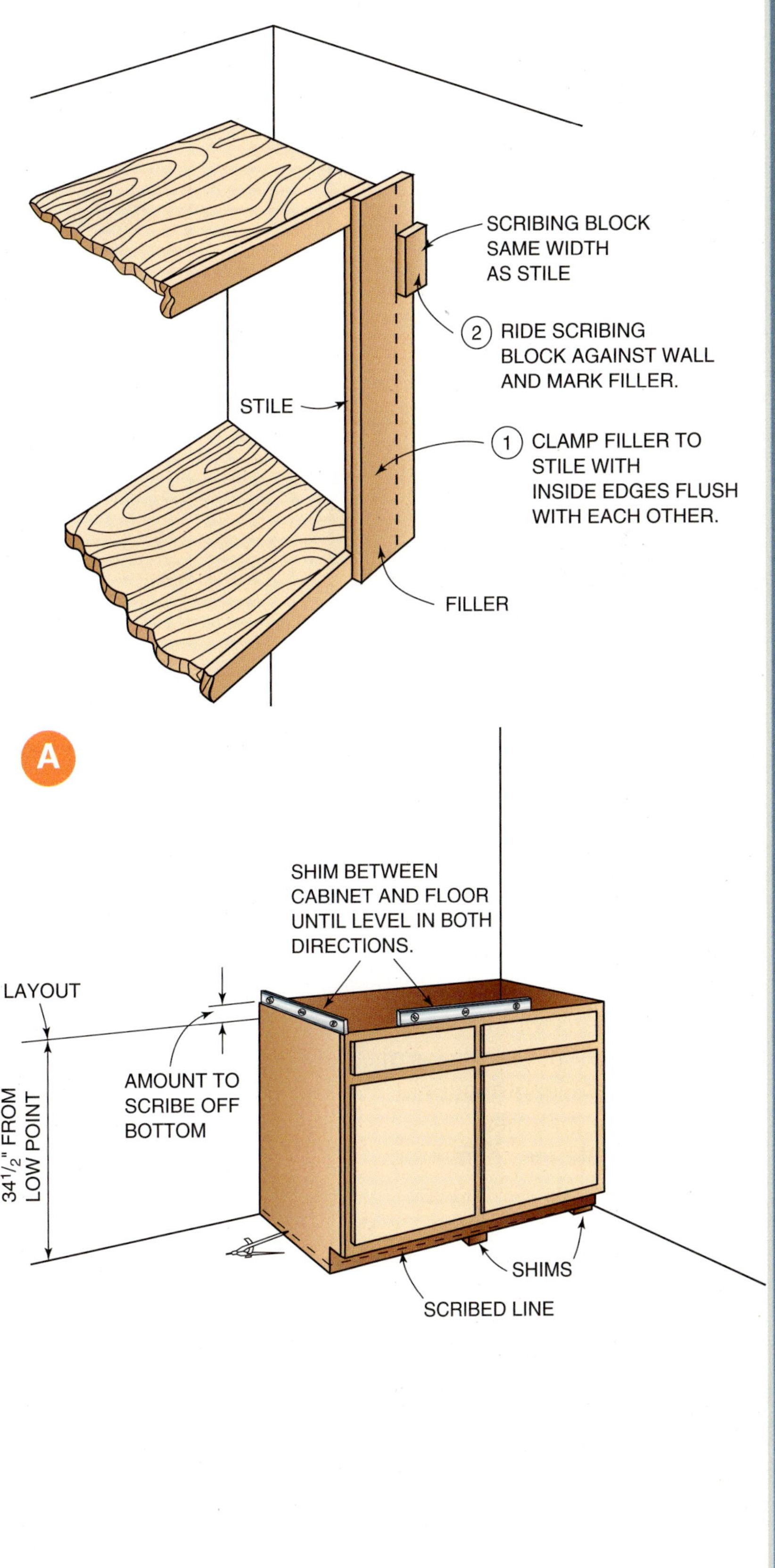

PROCEDURE 20-1

# Installing Manufactured Cabinets (continued)

B Cut both ends and toeboard to the scribed lines. Replace the cabinet in position. The top ends should be on the layout line. Fasten it loosely to the wall. The remaining base cabinets are installed in the same manner. Align and clamp the stiles of adjoining cabinets. Fasten them together. Finally, fasten all units securely to the wall.

## Installing Countertops

- After the base units are fastened in position, the countertop is cut to length. It is fastened on top of the base units and against the wall.
- Fasten the countertop to the base cabinets with screws up through triangular blocks usually installed in the top corners of base units. Take care not to drill through the countertop. Use screws of sufficient length, but not so long that they penetrate the countertop.
- Sink cutouts are made by carefully outlining the cutout and cutting with a jig saw or router. The cutout pattern usually comes with the sink. Use a fine tooth blade to prevent chipping out the face of the laminate beyond the sink. Some masking tape applied to countertop will prevent scratching it as you make the cutout.

## REVIEW QUESTIONS

*Select the most appropriate answer.*

1. **The vertical distance between the base unit and a wall unit is usually**
   a. 12 inches.
   b. 18 inches.
   c. 15 inches.
   d. 24 inches.

2. **The height from the floor to the surface of the countertop is usually**
   a. 30 inches.
   b. 36 inches.
   c. 32 inches.
   d. 42 inches.

3. **In order to accommodate sinks and provide adequate working space, the width of the countertop is usually**
   a. 25 inches.
   b. 30 inches.
   c. 28 inches.
   d. 32 inches.

4. **Standard wall cabinet height is**
   a. 24 inches.
   b. 32 inches.
   c. 30 inches.
   d. 36 inches.

5. **The height of most manufactured base kitchen cabinets is**
   a. 30¾ inches.
   b. 34½ inches.
   c. 32½ inches.
   d. 35¼ inches.

6. **Installation of kitchen cabinets usually begins with the**
   a. end base cabinet.
   b. end wall cabinet.
   c. cabinet around the refrigerator.
   d. corner wall cabinet.

7. **Adjoining kitchen cabinets are fastened together**
   a. after the countertop is installed.
   b. through the face frame stiles.
   c. through the cabinet side walls.
   d. at the floor.

8. **A dovetail joint may be used to join**
   a. drawer sides and backs.
   b. backsplash and countertops.
   c. plastic laminate and panels.
   d. all of the above

9. **The fastener used to attach a laminate surface to the countertop panel is**
   a. screws.
   b. construction adhesive.
   c. contact cement.
   d. clamps.

**10. A laminate trimmer is most similar to**

a. a router.
b. tin snips.
c. a paper cutter.
d. a belt sander.

**11. Countertop laminate is applied to the edges first before the top surface to**

a. allow water to run over the lapped seam.
b. allow water to run into the lapped seam
c. let the carpenter practice on small pieces before attempting the larger top.
d. use up small pieces first.

**12. When using contact cement, it is best to**

a. allow cement to dry before attaching pieces.
b. position the pieces carefully before contact.
c. roll the entire surface to complete the bond.
d. all of the above

**13. A door or drawer front with its edges and ends rabbeted to fit over the opening is called**

a. an overlay type.
b. a lipped type.
c. a rabbeted type.
d. all of the above

**14. The offset hinge is used on**

a. paneled doors.
b. lipped doors.
c. flush doors.
d. overlay doors.

**15. The joint used on high-quality drawers is the**

a. dado joint.
b. dado and rabbet joint.
c. dovetail joint.
d. rabbeted joint.

## WHAT'S WRONG WITH THIS PICTURE?

Carefully study **Figure 20-43** below and think about what is wrong. Consider all possibilities.

**WRONG**

**FIGURE 20-43** Shown is an elevation view of one wall of a kitchen. This kitchen was not designed with convenience or aesthetics in mind. Few of the upper cabinets appear to line up with the lower cabinets. The window is not centered over the sink cabinet. The lower part of the cabinets does not line up horizontally. The dishwasher is too far from the sink. A little more space, if needed, can be obtained by widening the upper cabinets next to the window.

**RIGHT**

**FIGURE 20-44** Notice the pleasing look of the vertical and horizontal alignment. Even though the kitchen is not symmetrical, and the cabinets are not all exactly the same size, it has much better uniformity. The proximity of the dishwasher to the sink is also better than in the previous design.

# Glossary of Terms

## A

**acoustical tile** a fiberboard ceiling tile whose surface consists of small holes and fissures that act as sound traps

**air infiltration** unwanted movement of air into an insulation layer or a conditioned space (heated or cooled)

**anchor** a device used to fasten structural members in place

**anchor bolt** long metal fasteners with a threaded end used to secure materials to concrete

**annular rings** the rings seen when viewing a cross-section of a tree trunk; each ring constitutes one year of tree growth

**apron** the flashing piece located on the lower side of a roof penetration such as a chimney, skylight, or dormer. It also refers to a piece of the window trim used under the stool

**asphalt felt** a building paper saturated with asphalt used for waterproofing

**astragal** a semicircular molding often used to cover a joint between two doors

## B

**backing** strips or blocks of wood installed in walls or ceilings for the purpose of fastening or supporting trim or fixtures

**back miter** an angled cut starting from the end going back into the end of the molding

**balloon frame** a type of frame in which studs are continuous from foundation sill plate to roof

**baluster** vertical members of a stair rail, usually decorative and spaced closely together

**balustrade** the entire stair rail assembly, including handrail, balusters, and posts

**band joist** the member used to stiffen the ends of floor joists where they rest on the sill

**batten** a thin, narrow strip typically used to cover joints in vertical boards

**bay window** a window, usually three-sided, that projects out from the wall line

**bevel** the sloping edge or side of a piece at any angle other than a right angle

**blind nail** a method of fastening that conceals the fastener

**blocking** pieces of dimension lumber installed between joist and studs for the purposes of providing a nailing surface for intersecting framing members

**board** lumber usually less than 2 inches thick

**board foot** a measure of lumber volume that equals 1 foot square and 1 inch thick or any equivalent lumber volume. The letter M is used to represent 1000 board feet.

**box nail** a thin nail with a head, usually coated with a material to increase its holding power

**brad** a thin, short, finishing nail

**brake** a tool used to bend sheet metal

**bridging** diagonal braces or solid wood blocks between floor joists used to distribute the load imposed on the floor

**buck** a rough frame used to form openings in poured concrete walls

**butterfly roof** an inverted gable roof

## C

**cambium layer** a layer just inside the bark of a tree where new cells are formed

**casing** molding used to trim around doors, windows, and other openings

**chamfer** an edge or end bevel that does not go all the way across the edge or end

**cheek cut** a compound miter cut on the end of certain roof rafters

**cleat** a small strip of wood applied to support a shelf or similar piece

**closed valley** a roof valley in which the roof covering meets in the center of the valley, completely covering the valley

**coil stock** aluminum sheets sold in 50-foot long rolls and widths ranging from 12 to 24 inches

**column** a large vertical member used to support a beam or girder

**competent person** designated person on a job site who is capable of identifying hazardous or dangerous situations and has the authority to take prompt corrective measures to eliminate them

**compound miter** a bevel cut across the width and also through the thickness of a piece

**compressive strength** the quality of being resistant to crushing; concrete has high compressive strength

**concrete** a building material made from Portland cement, aggregates, and water

**concrete block** a concrete masonry unit (CMU) used to make building foundations, typically measuring 8" × 8" × 16"

**condensation** when water vapor changes to a liquid due to cooling of the air; the resulting droplets of water that accumulate on the cool surface

**coniferous** trees that are cone-bearing; also known as *evergreen* trees

**corner boards** finish trim members used at the intersection of exterior walls

**cornice** a general term used to describe the part of the exterior finish where the walls meet the roof

**crib** heavy wood blocks and framing used as a foundation for scaffolding

**cricket** a small, false roof built behind, or uphill from, a chimney or other roof obstacle for the purpose of shedding water around roof penetrations

**crosscut** a cut made across the grain of lumber

**crown** a warp in a board along its length, forming an arc

## D

**dado** a wide cut, partway through and across the grain of lumber

**deadbolt** door-locking device operated by a key from the outside and by a handle or key from the inside

**deciduous** trees that shed leaves each year

**detail** close-up view of a plan or section

**dew point** temperature at which moisture begins to condense out of the air

**dimension** a term used to define a measurement of an item; also used to refer to all 2x lumber used in framing

**dimension lumber** a term used to describe wood that is sold for framing and general construction

**dormer** a structure that projects out from a sloping roof to form another roofed area to provide a surface for the installation of windows

**double-acting** doors that swing in both directions, or the hinges used on these doors

**downspout** a vertical member used to carry water from the gutter downward to the ground; also called *leader*

**draftstops** also called firestops; material used to reduce the size of framing cavities in order to slow the spread of fire; in floor framing, a draftstop consists of full-width dimension lumber blocking between studs

**drip** that part of an exterior finish that projects below another to cause water to drop off instead of running back against and down the wall

**drip edge** metal edging strips placed on roof edges to provide a support and overhang for the roofing material

**dry kiln** large ovens used to remove water from lumber

**duplex nail** a double-headed nail used for temporary fastening, such as in the construction of wood scaffolds

# E

**eased edge** an edge of lumber whose sharp corners have been rounded

**eaves** the lower part of the roof that extends beyond and overhangs the sidewalls

**electrolysis** accelerated oxidation of one metal because of contact with another metal in the presence of water

**elevation** a drawing in which the height of the structure or object is shown; also, the height of a specific point in relation to another reference point

**equilibrium** a state of balance; heat energy is thought to travel from high concentration (hot) to low concentration (cold)

**erectors** workers whose responsibilities include safe assembly of scaffolding

**escutcheon** protective plate covering the finished surface where penetrations occur, such as door knobs

**exposure** the amount that courses of siding or roofing are open to the weather

**extension jambs** strips of wood added to window jambs to bring the jamb edge flush with the wall surface in preparation for casing

# F

**face** the best appearing side of a piece of wood or the side that is exposed when installed

**face frame** a framework of narrow pieces on the front of a cabinet making the door and drawer openings

**face nail** method of driving a nail straight through a surface material into a supporting member

**fascia** the vertical surface of the cornice finish installed on the bottom end of rafters

**fence** a guide for ripping lumber on a table saw

**finger joint** a process in which shorter lengths are glued together using deep, thin V grooves, resulting in longer lengths

**finish nail** a thin nail with a small head designed for setting below the surface of finish material

**flashing** material used at intersections such as roof valleys and dormers and around window and door openings to prevent the entrance of water

**floor plan** pages of a set of construction drawings showing the walls as viewed from above

**flush** a term used to describe when surfaces or edges are aligned with each other

**foundation** that part of a wall on which the major portion of the structure is erected

**frieze** that part of exterior trim applied to cover the joint between the cornice and the siding

**frost line** the depth to which the ground typically freezes in a particular area; footings must be placed below this depth

# G

**gable end** the triangular-shaped section on the end of a building formed by the common rafters and the top plate line

**gable roof** a common type of roof that pitches in two directions

**gain** a cutout made in a piece to receive another piece, such as a cutout for a butt hinge

**galvanized** protected from rusting by a coating of zinc

**gambrel roof** a type of roof that has two different sloped rafters on each side of center

**girder** heavy timber or beam used to support vertical loads

**girders** heavy beams that support the inner ends of floor joists

**glazing** the act of installing glass in a frame

**groove** a wide cut, partway through and running with the grain of lumber

**gusset** a block or metal used over a joint to stiffen and strengthen it

**gutter** a trough attached to an eave used to carry off water

**gypsum board** a sheet product made by encasing gypsum in a heavy paper wrapping used to create the wall surface; also called drywall.

## H

**handrail** a railing on a stairway intended to be grasped by the hand to serve as a support and guard

**hardboard** a building product made by compressing wood fibers into sheet form

**header** members placed at right angles to joists, studs, and rafters to form and support openings in a wood frame

**heartwood** the wood in the inner part of a tree, usually darker and containing inactive cells

**heel** the back end of objects, such as a handsaw or hand plane

**hip jack rafter** a rafter running between a hip rafter and the wall plate

**hip rafter** a rafter that extends diagonally from the corner of the plate to the ridge at the intersection of two different roof surface

**hip roof** a roof that slopes up towards the ridge from all walls

**hip-valley cripple jack rafter** a short rafter running parallel to common rafters, cut between hip and valley rafters

**hopper window** a type of window in which the sash is hinged at the bottom and swings inward

**housewrap** type of building paper with which the entire exterior sidewalls of a building are covered

## I

**impact noise** noise caused by the vibration of an impact, such as dropped objects or footsteps on a floor

**insulated glass** multiple panes of glass fused together with an air space between them

**insulation** material used to restrict the passage of heat or sound

**intersecting roof** the roof of irregular-shaped buildings; valleys are formed at the intersection of the roofs

**isometric** a drawing in which three surfaces of an object are seen in one view, with the base of each surface drawn at a 30-degree angle

## J

**joist** horizontal framing members used in a spaced pattern that provide support for the floor or ceiling system

**joist hanger** metal stirrups used to support the ends of joists that do not rest on top of a support member

**J-roller** a 3-inch-wide rubber roller used to apply pressure over the surface of contact cement bonded plastic laminates

## K

**kerf** the width of a cut made with a saw

## L

**laser** a concentrated, narrow beam of light; optical leveling and plumbing instrument used in building construction

**lateral** a direction to the side at about 90 degrees

**ledger** a temporary or permanent supporting member for joists or other members running at right angles; horizontal member of a set of batter boards

**level** horizontal; perpendicular to the force of gravity

**light** a pane of glass or an opening for a pane of glass

**linear feet** a measurement of length

**load-bearing** term used to describe a structural member that carries weight from another part of the building

**lookout** horizontal framing pieces in a cornice, installed to provide fastening for the soffit

**low emissivity glass (Low E)** a coating on double-glazed

windows designed to raise the insulating value by reflecting heat back into the room

**lumber** general term for wood that is cut from a log to form boards, planks, and timbers

## M

**mansard roof** a type of roof that has two different sloped rafters on all sides of the building, with the lower slopes steeper than the upper

**masonry** any construction of stone, brick, tile, concrete, plaster, and similar materials

**mastic** a thick adhesive

**medullary rays** bands of cells radiating from the cambium layer to the pith of a tree to transport nourishment toward the center

**millwork** any wood products that have been manufactured, such as moldings, doors, windows, and stairs, for use in building construction; sometimes called *joinery*

**miter** the cutting of the end of a piece at any angle other than a right angle

**miter gauge** a guide used on the table saw for making miters and square ends

**molding** decorative strips of wood used for finishing purposes

**mortar** a mixture of Portland cement, lime, sand, and water used to bond masonry units together

**mullion** a vertical division between window units or panels in a door

**muntin** slender strips of wood between lights of glass in windows or doors

## N

**newel post** an upright post supporting the handrail in a flight of stairs

## O

**on center (OC)** the distance from the center of one structural member to the center of the next one

**open valley** a roof valley in which the roof covering is kept back from the centerline of the valley exposing the flashing

**orthographic** a multiview drawing

## P

**panel** a large sheet of building material that usually measures 4 × 8 feet

**penny (d)** a term used in designating nail sizes

**pilaster** column built within and usually projecting from a wall to reinforce the wall

**pilot** a guide on the end of edge-forming router bits used to control the amount of cut

**pith** the small, soft core at the center of a tree

**pivot** a point of rotation

**plain-sawed** a method of sawing lumber that produces flat-grain where annular rings tend to be parallel to the width of the board

**plan** in an architectural drawing, an object drawn as viewed from above

**plancier** the finish member on the underside of a box cornice; also called a *soffit*

**plate** top or bottom horizontal member of a wall frame

**platform frame** method of wood frame construction in which walls are erected on a previously constructed floor deck or platform

**plumb** vertical; aligned with the force of gravity

**plumb bob** a pointed weight attached to a line for testing plumb

**Portland cement** a fine gray powder which, when mixed with water, forms a paste that sets rock hard; an ingredient in concrete

**post** a vertical member used to support a beam or girder

**postforming** a method used to bend plastic laminate to small radii

**pressure-treated** treatment given to lumber that applies a wood preservative under pressure

**Pythagorean theorem** a mathematical expression that states the sum of the square of the two sides of a right triangle equals the square of the diagonal side

## Q

**quarter-sawed** a method of sawing lumber that produces a close grain pattern where the annular rings tend to be perpendicular to the width of the board

## R

**rabbet** an L-shaped cutout along the edge or end of lumber

**rail** the horizontal member of a frame

**rake** the sloping portion of the gable ends of a building. It is also the sloping portion of trim, such as on gable ends of a building or stair

**reinforcing rods** also called rebar, steel bars placed in concrete to increase its tensile strength

**ribbon** a narrow board cut into studs of a balloon frame to support floor joists

**rip** sawing lumber in the direction of the grain

**rise** in stairs, the vertical distance of the flight; in roofs, the vertical distance from plate to ridge; may also be the vertical distance through which anything rises

**run** the horizontal distance over which rafter, stairs, and other like members travel

**R-value** a number given to a material to indicate its resistance to the passage of heat

## S

**saddle** same as *cricket*

**sapwood** the outer part of a tree just beneath the bark containing active cells

**sash** that part of a window into which the glass is set

**sawyer** a person whose job is to cut logs into lumber

**scab** a length of lumber or material applied over a joint to stiffen and strengthen it

**section** drawing showing a vertical cut-view through an object or part of an object

**selvage** the unexposed part of roll roofing covered by the course above

**shank hole** a hole drilled for the thicker portion of a wood screw

**sheathing** boards or sheet material that are fastened to joists, rafters, and studs and on which the finish material is applied

**shed roof** a type of roof that slopes in one direction only

**shim** a thin, wedge-shaped piece of material used behind pieces for the purpose of straightening them or for bringing their surfaces flush

**sill** first horizontal wood member resting on the foundation supporting the framework of a building; also, the lowest horizontal member in a window or door frame

**sill sealer** material placed between the foundation and the sill to prevent air leakage

**soffit** the horizontal, underside trim member of a cornice or other overhanging assembly

**sound-deadening insulating board** a sound resistance system using a gypsum board outer covering and resilient steel channels placed at right angles to the studs to isolate the gypsum board from the stud

**specifications** written or printed directions of construction details for a building, sometimes referred to as specs

**spline** a thin, flat strip of wood inserted into the grooved edges of adjoining pieces

**spreader** a strip of wood used to keep other pieces a desired distance apart

**square** a tool used to mark a layout and mark angles, particularly 90-degree angles; a term used to describe when two lines or sides meet at a 90-degree angle; the amount of roof covering that will cover 100 square feet of roof area

**stairwell** an opening in the floor for climbing or descending stairs, or the space of a structure where the stairs are located

**stile** the outside vertical members of a frame, such as in a paneled door

**stool** the bottom horizontal member of interior window trim that serves as the finished window sill

**storm sash** an additional sash placed on the outside of a window to create dead air space to prevent the loss of heat from the interior in cold weather

**story pole** a narrow strip of wood used to lay out the installation heights of material such as siding or vertical members of a wall frame

**striated** finish material with random and finely spaced grooves running with the grain

**strike plate** thin metal plate installed where the latch bolt of a door touches the jamb

**stud** vertical framing member in a wall running between plates

**subfloor** material used as the first floor layer on top of joists

## T

**tail cut** a cut on the lower end of a rafter that extends beyond the building

**tail joists** shortened on center joists running from a header to a sill or girder

**tempered** treated in a special way to make a material harder and stronger

**tensile strength** the quality of being resistant to breaking under a pulling force; steel has high tensile strength

**termite shields** metal flashing plate over the foundation to protect wood members from termites

**timbers** large pieces of lumber over 5 inches in thickness and width

**toe** the forward end of tools, such as a hand saw and hand plane

**toenail** method of driving a nail diagonally through a surface material into supporting member

**trimmer joist** a joist or stud placed at the sides of an opening running parallel to the main framing members

## U

**users** people who work on scaffolding

## V

**valley** the intersection of two roof slopes forming an interior corner

**valley cripple jack rafter** a rafter running between two valley rafters

**valley jack rafter** a rafter running between a valley rafter and the ridge

**valley rafter** the rafter placed at the intersection of two roof slopes that form interior corners

**vapor retarder** also called a vapor barrier, a material used to prevent the passage of water in the gaseous state

## W

**wainscoting** a wall finish applied partway up the wall from the floor

**water table** exterior trim members applied at the intersection of the siding and the foundation to direct water away from the building

**weather stripping** narrow strips of material applied to windows and doors to prevent the infiltration of air and moisture

**whet** the honing of a tool by rubbing the tool on a flat sharpening stone

**wind** a defect in lumber caused by a twist in the stock from one end to the other; also, a twist in anything that should be flat

# Glosario

**secado al aire (*air dried*)** técnica para eliminar la humedad de la madera usando corrientes naturales de aire

**infiltración de aire (*air infiltration*)** entrada no deseada de aire dentro de una capa de aislamiento o de un espacio acondicionado (caliente o frío)

**anclaje (*anchor*)** dispositivo utilizado para mantener las partes de una estructura en su lugar

**perno de anclaje (*anchor bolt*)** fijador largo de metal con un extremo fileteado que se usa para fijar materiales al concreto

**anillos de crecimiento (*annular rings*)** círculos que se ven al observar una sección transversal de un tronco de árbol; cada anillo representa un año en la vida del árbol

**forro tapajuntas (*apron*)** tapajuntas ubicado en la parte inferior de un elemento que penetra en una cubierta, como una chimenea o una buhardilla

**tela asfáltica (*asphalt felt*)** papel usado en construcción; está saturado con asfalto para que resulte impermeable

**astrágalo (*astragal*)** moldura semicircular utilizada con frecuencia para cubrir una unión entre dos puertas

**inglete reverso (*back miter*)** ángulo cortado comenzando por el final y regresando a la cara de la pieza

**apoyo, soporte (*backing*)** tiras o tacos de madera que se instalan en los muros y techos a fin de fijar o sostener molduras o accesorios

**entramado sin rigidez (*balloon frame*)** tipo de estructura en la cual los puntales son continuos desde la solera del cimiento hasta la cubierta; también se conoce como *entramado de globo*

**balaustres (*baluster*)** elementos verticales de un pasamanos, por lo general son decorativos y se colocan con poca separación entre uno y otro

**vigueta perimetral (*band joist*)** elemento utilizado para darle rigidez a los extremos de las vigas de piso en el lugar donde se apoyan en el umbral

**listones (*battens*)** tiras estrechas que normalmente se usan para cubrir uniones en tablas verticales

**mirador (*bay window*)** ventana, por lo general de tres lados, que sobresale de la línea de la fachada

**bisel (*bevel*)** borde o cara inclinada de una pieza en cualquier ángulo, excepto en ángulo recto

**clavado invisible (*blind nail*)** método de sujeción en el cual el elemento de fijación, el clavo, queda oculto

**tacos (*blocking*)** piezas de madera de construcción que se instalan entre las viguetas y los puntales a fin de proporcionar una superficie para sujetar los elementos de la armazón en los puntos en que estos se cruzan

**tabla, tablero (*board*)** madera, por lo general de menos de dos pulgadas de grosor

**pie tablar (*board foot*)** medida de volumen de madera que es igual al volumen de una pieza de 1 pie cuadrado y 1 pulgada de espesor o a cualquier volumen equivalente al de esta pieza. Para representar un volumen de 1000 pies tablares se usa la letra M.

**clavo de gavetero (*box nail*)** clavo fino con cabeza, por lo general recubierto de un material para aumentar su poder de sujeción

**puntilla, clavito (*brad*)** clavo fino y corto de acabado

**arriostrado (*bridging*)** sostenes diagonales o tacos de madera maciza ubicados entre las vigas del piso y que se usan para distribuir la carga que soporta el piso

**premarco (*buck*)** armazón rudo que se utiliza para formar vanos en muros de concreto vertido

**cambium (*cambium layer*)** capa del interior de la corteza de un árbol en la cual se forman células nuevas

**marco (*casing*)** moldura utilizada para rodear puertas, ventanas y otros vanos

**chaflán (*chamfer*)** bisel en un borde o canto que no llega hasta el otro lado del borde

**corte oblicuo (*cheek cut*)** corte en bisel compuesto en el extremo de ciertos cabrios

**ristrel (*cleat*)** pequeña tira de madera que se aplica para sostener un estante u otro elemento similar

**limahoya cerrada (*closed valley*)** limahoya de una cubierta en la cual las dos pendientes se encuentran en el centro de la limahoya cubriéndola completamente

**columna (*column*)** elemento vertical de gran tamaño usado para sostener una viga o una jácena

**persona competente (*competent person*)** persona designada en un lugar de trabajo, la cual es capaz de identificar situaciones de riesgo o peligrosas y que tiene la autoridad de tomar medidas correctoras inmediatas para eliminar estos riesgos o peligros

**bisel compuesto (*compound miter*)** corte en bisel en dos planos de una pieza: anchura y grosor

**concreto (*concrete*)** material de construcción elaborado con cemento portland, conglomerados y agua

**bloque de concreto (*concrete block*)** unidad de albañilería de concreto utilizada para hacer cimientos de edificios y que normalmente mide 8" × 8" × 16"

**condensación (*condensation*)** cuando el agua en forma de vapor pasa a estado líquido debido al enfriamiento del aire; gotitas de agua que se acumulan en una superficie fría

**coníferas (*coniferous*)** árboles que producen frutos en forma de cono; también llamados *árboles de hoja perenne*

**cantoneras de esquina (*corner boards*)** guarniciones, elementos de acabado utilizados en la intersección de paredes exteriores

**cornisa (*cornice*)** término general utilizado para describir la parte del acabado exterior donde los muros se encuentran con la cubierta

**lecho (*crib*)** tacos de madera y armazón pesada que se usan en cimientos y andamiajes

**faldellín guardaguas (*cricket*)** pequeña cubierta falsa construida detrás o hacia arriba del lugar donde una chimenea u otro obstáculo cortan la cubierta; su finalidad es alejar el agua de estos lugares en los que una chimenea u otro obstáculo penetran en la cubierta, también llamado *babero*

**sección transversal (*crosscut*)** corte perpendicular a la fibra o veta de la madera

**dado (*dado*)** corte parcial perpendicular de la fibra o veta de la madera

**cerrojo dormido (*deadbolt*)** dispositivo para cerrar puertas que se abre y cierra con una llave desde el exterior y con una manija o una llave desde el interior

**bloque de anclaje (*deadman*)** dispositivo de madera en forma de T utilizado para sostener paneles de techo de piedra seca y otros objetos

**caducifolios (*deciduous*)** árboles que pierden las hojas cada año

**detalle (*detail*)** vista en primer plano de un plano o una sección

**punto de rocío (*dew point*)** temperatura a la cual la humedad comienza a condensarse en el aire

**dimensión (*dimension*)** término utilizado para definir una medida de un objeto; también se refiere a toda la madera 2x usada para hacer la armazón

**madera dimensionada (*dimension lumber*)** término utilizado para describir la madera que se vende para armazón y trabajos de construcción en general, madera aserrada en tamaños corrientes

**buhardilla (*dormer*)** estructura que sobresale de la pendiente de una cubierta formando otra área techada y proporcionando una superficie para la instalación de una ventana

**puertas de vaivén (*double-acting*)** puertas que oscilan en ambas direcciones

**bisagras de vaivén (*double-acting*)** bisagras que oscilan en ambas direcciones

**bajante pluvial (*downspout*)** elemento vertical utilizado para llevar el agua del canalón del alero al suelo; también llamado *canalón*

**cortafuego (*draftstops*)** material utilizado para reducir el tamaño de los espacios huecos de la armazón a fin de evitar las corrientes de aire y así retardar la propagación del fuego; en una estructura de madera, consiste en tacos grandes de madera colocados entre los puntales

**vierteaguas (*drip*)** parte de un acabado exterior que protege otro elemento que se encuentra debajo, haciendo que el agua gotee hacia fuera en lugar de resbalar sobre el muro

**goterón (*drip edge*)** tiras metálicas de acabado que se colocan en los bordes de las cubiertas para dar soporte a los aleros

**horno de secado (*dry kiln*)** grandes hornos utilizados para eliminar el agua de la madera

**clavo de doble cabeza (*duplex nail*)** clavo con una cabeza doble que se utiliza para sujetar elementos de forma temporal, por ejemplo, en la construcción de andamios de madera

**arista pulida (*eased edge*)** arista o borde de la madera cuyas esquinas vivas han sido redondeadas

**alero (*eaves*)** parte inferior de la cubierta que se extiende más allá de las paredes laterales

**electrolisis (*electrolysis*)** oxidación acelerada de un metal originada por el contacto con otro metal en presencia de agua

**alzado (*elevation*)** dibujo que muestra la altura de una estructura u objeto; dibujo que muestra la fachada de un edificio o la parte frontal de un objeto; también se denomina así la elevación o altura específica de un punto determinado con relación a otro punto de referencia

**montadores (*erectors*)** trabajadores entre cuyas responsabilidades se incluye el ensamblaje de andamios de modo seguro

**escudete (*escutcheon*)** placa protectora que cubre el pomo de una puerta o el ojo de la cerradura

**material expuesto (*exposure*)** la cantidad de hiladas de revestimientos de paredes o de material de cubierta que están expuestas a la intemperie

**extensiones de jamba (*extension jambs*)** tiras de madera que se añaden a las jambas de las ventanas para lograr que sus bordes queden a ras de la pared y preparados para la colocación del marco

**cara (*face*)** la superficie de una pieza de madera que tiene una mejor apariencia; superficie que va a quedar a la vista tras la instalación

**clavado a contrahilo (*face nail*)** método de clavar un clavo perpendicularmente en la superficie de un material hasta el elemento de apoyo

**imposta (*fascia*)** elemento vertical de acabado de una cornisa que se instala en la parte inferior de los cabrios o vigas de cubierta

**guarda (*fence*)** guía para aserrar madera al hilo en una sierra de mesa

**empalme de cola de pescado (*finger joint*)** procedimiento de unión por entalladura múltiple en la cual piezas cortas se pegan utilizando ranuras profundas y estrechas en forma de V, dando como resultado piezas de mayor longitud

**clavo de acabado (*finish nail*)** clavo fino y con una cabeza pequeña diseñada para ir bajo la superficie del material acabado

**tapajuntas (*flashing*)** material que se utiliza para evitar la entrada de agua en intersecciones como limahoyas y buhardillas y sobre ventanas y puertas

**ras (*flush*)** término utilizado para indicar que unas superficies o bordes están alineados entre sí

**cimiento (*foundation*)** parte de un muro en la cual se erige la parte principal de la estructura

**friso (*frieze*)** ribete o tira exterior que se aplica para cubrir la unión entre el voladizo de la cornisa y el revestimiento de las paredes

**profundidad de helada (*frost line*)** profundidad a la cual el suelo suele congelarse en un área determinada; los lechos de cimentación deben ubicarse por debajo de esta profundidad

**remate en hastial (*gable end*)** sección de forma triangular en la parte superior de un edificio formada por los cabrios comunes y la solera superior

**cubierta a dos aguas (*gable roof*)** tipo común de cubierta que se inclina en dos direcciones

**gárgol, mortaja (*gain*)** corte realizado en una pieza para recibir a otra, como un corte para colocar una bisagra

**galvanizado (*galvanized*)** protegido contra la oxidación mediante un revestimiento de cinc

**cubierta a la holandesa (*gambrel roof*)** tipo de cubierta que tiene varias vertientes con distintas inclinaciones en cada lado del centro

**jácenas (*girders*)** vigas pesadas que sostienen los extremos interiores de las vigas de piso

**acristalar (*glazing*)** instalar cristales o vidrios en un marco

**ranura, acanaladura (*groove*)** corte parcial a lo largo de la fibra o veta de la madera

**chapa de nudo (*gusset*)** taco de madera o metal que se utiliza sobre una junta para darle rigidez y reforzarla

**canalón (*gutter*)** canal colocado a lo largo de un alero y que sirve para recoger el agua que resbala por la cubierta

**tablero de yeso (*gypsum board*)** panel que se utiliza como superficie acabada; el material está hecho de un mineral que se halla en la tierra envuelto en un papel resistente; también llamado *piedra seca*

**pasamanos (*handrail*)** barandilla que se coloca al lado de una escalera a fin de agarrarla con la mano para que sirva de apoyo y protección

**tablero de madera prensada (*hardboard*)** producto de construcción elaborado por medio de la compresión de fibras de madera, dándoles forma de lámina

**cabecero (*header*)** elementos estructurales colocados en ángulo recto con vigas, puntales y cabrios para formar aberturas y servirles de apoyo

**duramen (*heartwood*)** madera que se halla en la parte interna de un árbol, por lo general tiene un color más oscuro y contiene células inactivas, también se conoce como *corazón*

**talón (*heel*)** extremo trasero de un objeto, como el de una sierra o una lijadora de mano

**par (*hip*)** cabrio que se extiende en diagonal desde una esquina de la solera hasta la viga cumbrera donde se unen las dos vertientes de la cubierta

**cabrio de limatesa (hip jack)** cabrio que va desde el par hasta la solera

**cabrio corto entre cabrio de limatesa a limahoya (*hip-valley cripple jack rafter*)** cabrio corto paralelo a los cabrios comunes y cortado entre el cabrio de limatesa y el de limahoya

**ventana abatible sobre eje inferior horizontal (*hopper window*)** tipo de ventana en la cual el bastidor tiene bisagras en la parte inferior y gira hacia adentro

**revestimiento "housewrap" (*housewrap*)** tipo de papel de construcción con el cual se cubren completamente las paredes laterales

**vidrio aislante (*insulated glass*)** varias hojas de vidrio unidas y con un espacio de aire entre ellas

**aislamiento (*insulation*)** material utilizado para limitar el paso del calor o del ruido

**cubierta a múltiples aguas (*intersecting roof*)** cubierta de construcciones de formas irregulares; las limahoyas se forman en la intersección de las cubiertas

**viguetas (*joist*)** elementos estructurales horizontales que se utilizan espaciados regularmente para servir de apoyo a los sistemas de piso o de techo

**colgaderos de viguetas (*joist hanger*)** estribos de metal que se utilizan para sostener los extremos de las viguetas que no descansan sobre un elemento de apoyo

**rodillo en J (*J-roller*)** rodillo de caucho de 3 pulgadas de ancho que se utiliza para aplicar presión sobre la superficie de contacto de laminados plásticos unidos con cemento

**corte de sierra (*kerf*)** ancho del corte realizado con una sierra

**láser (*laser*)** rayo de luz estrecho y concentrado; instrumentos como los niveles y las plomadas ópticas que se utilizan en construcción emplean el láser

**lateral (*lateral*)** dirección a un lado, aproximadamente a 90 grados

**carrera (*ledger*)** elemento temporal o permanente para viguetas u otros elementos perpendiculares (a ángulo recto); elemento horizontal de una serie de listones

**nivel (*level*)** horizontal; perpendicular a la fuerza de la gravedad

**luz (*light*)** hoja de vidrio o abertura para una hoja de vidrio

**pie lineal (*linear feet*)** medida de longitud

**portante (*loadbearing*)** palabra que se utiliza para hacer referencia a un elemento de la estructura que soporta el peso de otra parte del edificio; también se utiliza la expresión "...de carga" *(p. ej: "muro de carga")*

**soporte de sofito (*lookout*)** armazón horizontal en una cornisa; se instala para sujetar el sofito o volado

**vidrio de baja emisividad (*low emissivity glass (Low E)*)** revestimiento en ventanas de vidrio doble diseñado para aumentar el aislamiento al reflejar el calor de nuevo hacia la habitación

**madera (*lumber*)** término general que se aplica a la madera que se corta de un tronco para hacer tablas, tablones y piezas de madera de construcción

**cubierta amansardada (*mansard roof*)** tipo de cubierta que tiene dos vertientes diferentes en todos los lados del edificio, con las vertientes de la parte inferior más empinadas que las de la parte superior

**albañilería (*masonry*)** cualquier construcción de piedra, ladrillo, baldosa, concreto, yeso y materiales similares

**mástique (*mastic*)** adhesivo espeso

**radios medulares (*medullary rays*)** bandas de células que parten desde el cambium en dirección a la médula de un árbol para transportar nutrientes hacia el centro

**carpintería prefabricada (*millwork*)** cualquier producto de madera manufacturado (como molduras, puertas, ventanas o escaleras) que se utiliza en la construcción de edificios; algunas veces se le llama *ebanistería*

**inglete (*miter*)** corte en el borde de una pieza a cualquier ángulo excepto el ángulo recto

**calibrador de ingletes (*miter gauge*)** guía que se utiliza en la sierra de mesa para realizar ingletes y terminaciones cuadradas

**molduras (*molding*)** tiras decorativas de madera que se utilizan para acabados

**mortero (*mortar*)** mezcla de cemento portland, caliza, arena y agua que se utiliza para unir las unidades de albañilería

**parteluz (*mullion*)** división vertical en ventanas o paneles de puertas

**listoncillos, montantes intermedios (*muntin*)** tiras finas de madera que se colocan a través de la luz de los vidrios en ventanas o puertas

**pilarote (*newel post*)** poste vertical que sostiene la barandilla en un tramo de escaleras

**centro a centro (*on center (OC)*)** distancia desde el centro de un elemento estructural al centro del elemento contiguo

**limahoya abierta (*open valley*)** limahoya en la cual la cubierta se mantiene detrás de la línea central de la limahoya

**panel (*panel*)** lámina grande de material de construcción que por lo general mide 4 × 8 pies

**"penny" (*penny (d)*)** término utilizado para designar tamaños de clavos

**pilastra (*pilaster*)** columna construida dentro de un muro, por lo general sobresaliendo del mismo, para reforzarlo

**guía (*pilot*)** dispositivo que se ubica en el extremo de las cuchillas de las fresadoras para controlar la profundidad del corte

**médula (*pith*)** núcleo, pequeño y blando, que se halla en el centro de un árbol

**pivote (*pívot*)** punto de rotación

**aserrado simple (*plain-sawed*)** método de aserrado de madera que produce fibra o veta plana en la cual los anillos de crecimiento suelen ser paralelos al ancho de la tabla

**planta (*plan in*)** dibujo arquitectónico en el que se representa un objeto tal como se ve desde arriba

**plafón (*plancier*)** elemento de acabado en la cara inferior de una caja de cornisa, también se conoce como *sofito*

**solera (*plate*)** elemento horizontal superior o inferior de la armazón de una pared

**estructura de plataforma (*platform frame*)** método de construcción de armazones de madera en el cual los muros se erigen sobre una tarima de piso o una plataforma previamente construidas

**aplomado, a plomo (*plumb*)** vertical; alineado con la fuerza de la gravedad

**plomada (*plumb bob*)** peso terminado en punta unido a una cuerda que se utiliza para comprobar la verticalidad

**cemento portland (*portland cement*)** polvo gris y fino que, cuando se mezcla con agua, forma una pasta que al forjar se vuelve dura como roca; un ingrediente del concreto

**poste (*post*)** elemento vertical utilizado para dar apoyo a una viga o una jácena

**postformado (*postforming*)** método utilizado para unir laminados de plástico curvándolos en radios pequeños

**madera tratada a presión (*pressure-treated*)** madera a la que se le ha dado un tratamiento que consiste en la aplicación de un conservante de madera bajo presión

**correa (*purlin*)** elemento de soporte de la cubierta que se apoya en los cabrios y es perpendicular a los mismos

**Teorema de Pitágoras (*Pythagorean Theorem*)** expresión matemática que afirma que la suma del cuadrado de los dos lados de un triángulo rectángulo es igual al cuadrado de la diagonal

**aserrado al cuarto (*quarter-sawed*)** método de aserrado de madera que produce un patrón de fibra o veta apretado en el cual los anillos de crecimiento tienden a ser perpendiculares al ancho de la tabla

**rebajo (*rabbet*)** corte en forma de L realizado junto al borde o extremo de una madera

**peinazo (*rail*)** elemento horizontal de un marco o una estructura en general

**visera (*rake*)** parte inclinada de un ribete, como en los remates de hastiales de un edificio o en una escalera

**barras de refuerzo (*reinforcing rods*)** barras de acero que se colocan en el concreto para aumentar su resistencia a la tensión

**cinta (*ribbon*)** madera estrecha que se ubica entre los puntales de un entramado sin rigidez para sostener las vigas del piso

**aserrar al hilo (*rip*)** aserrar madera en la dirección de la fibra o veta

**altura (*rise*)** en escaleras, la distancia vertical del tramo; en cubiertas, la distancia vertical desde la solera hasta el caballete; también puede ser la distancia vertical a través de la cual nada sube

**recorrido (*run*)** distancia horizontal que recorren cabrios, escaleras y otros elementos similares

**valor R (*R-value*)** número que se asigna a un material para indicar su resistencia al paso del calor

**faldellín guardaguas (*saddle*)** pequeña cubierta falsa construida detrás o hacia arriba del lugar donde una chimenea u otro obstáculo cortan la cubierta; su finalidad es alejar el agua de estos lugares en los que una chimenea u otro obstáculo penetran en la cubierta, también llamado *babero*

**albura o sámago (*sapwood*)** parte exterior del árbol, que está justo debajo de la corteza y que contiene células activas

**bastidor (*sash*)** parte de la ventana en la que se asienta el vidrio

**aserrador (*sawyer*)** persona cuyo trabajo consiste en cortar troncos de madera para hacer tablones

**refuerzo (*scab*)** trozo de madera u otro material aplicado sobre una junta para darle rigidez y fortalecerla

**sección (*section*)** dibujo que muestra una vista de un corte vertical a través de un objeto o una parte del mismo

**orillo (*selvage*)** parte no expuesta de un material de cubierta, tapada por la hilada que está encima

**avellanado (*shank hole*)** hueco taladrado por la parte más gruesa de un tornillo de madera

**forro (*sheathing*)** tablas o láminas de otro material que se sujetan a las viguetas, cabrios y puntales y sobre la cual se aplica el material de acabado

**cubierta de una vertiente (*shed roof*)** tipo de cubierta que tiene inclinación en una sola dirección

**calzo (*shims*)** pequeña pieza de diversos materiales en forma de cuña que se utiliza detrás de otras piezas a fin de enderezarlas o hacer que sus superficies queden a ras

**durmiente (*sill*)** primer elemento horizontal de madera que se apoya en la cimentación y sirve de soporte a la estructura de un edificio; también, el elemento horizontal inferior del marco de una ventana o una puerta

**obturador de solera (*sill sealer*)** material que se coloca entre el cimiento y la solera para evitar corrientes de aire

**sofito (*soffit*)** elemento horizontal, parte inferior de una cornisa u otro cuerpo voladizo

**espiga (*spline*)** tira de madera fina y plana que se inserta en los extremos ranurados de piezas contiguas

**codal (*spreader*)** separador, tira de madera utilizada para mantener otras piezas separadas entre sí a la distancia deseada

**escuadra (*square*)** herramienta utilizada para marcar un trazado y hacer marcas de ángulos, en particular ángulos de 90 grados; expresión que se utiliza para describir dos líneas o lados que se unen formando un ángulo de 90 grados; la cantidad de cubierta que taparía 100 pies cuadrados del área del techo

**cuadrado (*square*)** cantidad de material de cubierta que recubrirá 100 pies cuadrados

**hueco de escalera (*stairwell*)** abertura en el piso para subir o bajar escaleras o el espacio de una estructura en la cual se ubican las escaleras

**largueros (*stile*)** elementos exteriores verticales de un marco, como en una puerta de paneles

**repecho (*stool*)** elemento inferior horizontal del interior de un marco de ventana que sirve como alféizar acabado

**contrabastidor (*storm sash*)** bastidor adicional colocado en el exterior de una ventana para crear una cámara de aire y evitar la pérdida de calor en el interior durante tiempo frío

**vara (*story pole*)** tira de madera estrecha que se usa para instalar elementos en lugares altos, como materiales para recubrimientos de paredes o elementos verticales de la estructura de una pared

**estriado (*striated*)** material acabado con ranuras poco espaciadas entre sí que van en el sentido de la fibra o del grano

**placa de cerradero (*strike plate*)** placa fina de metal instalada donde el pestillo de una puerta toca la jamba

**puntal (*stud*)** elemento vertical de la armazón de una pared que va de solera a solera

**contrapiso (*subfloor*)** material utilizado como primera capa encima de las vigas de piso

**colas (*tail*)** vigas acortadas en el centro; vigas que van desde un cabecero hasta un durmiente o una jácena

**corte de cola (*tail cut*)** corte en el extremo inferior de un cabrio

**templado (*tempered*)** material al que se le da un tratamiento especial para hacerlo más fuerte y resistente

**barrera antitermitas (*termite shields*)** placa tapajuntas metálica que se coloca sobre los cimientos para proteger los elementos de madera de las termitas

**maderas de construcción (*timbers*)** piezas de madera grandes, de más de cinco pulgadas de anchura y grosor

**punta (*toe*)** extremo delantero de las herramientas, como el de una sierra o una lijadora de mano

**clavado oblicuo (*toenail*)** método de clavar un clavo diagonalmente en la superficie de un material hasta el elemento de apoyo

**viga de embrochalado (*trimmer*)** viga o puntal colocado a los lados de una abertura y paralelo a los elementos principales de la armazón

**usuarios (*users*)** personas que trabajan en los andamios

**limahoya (*valley*)** intersección de dos vertientes de cubierta en ángulo entrante

**cabrio de limahoya a limahoya (*valley cripple jack rafter*)** cabrio que va de una a otra limahoya de una cubierta

**cabrio corto de limahoya a cumbrera (*valley jack rafter*)** cabrio que va de una limahoya a la viga cumbrera

**cabrio de limahoya (*valley rafter*)** cabrio ubicado en la intersección de dos vertientes que se unen en ángulo entrante

**barrera de vapor (*vapor retarder*)** material utilizado para evitar el paso de agua en estado gaseoso; también llamado *retardador de vapor*

**arrimadero (*wainscoting*)** acabado que se aplica a parte de la pared partiendo del piso

**vierteaguas (*water table*)** elementos de remate exteriores que se aplican en la intersección del revestimiento de paredes y en cimientos, y que se proyectan hacia el exterior para desviar el agua del edificio

**burlete (*weather stripping*)** tira estrecha que se aplica a ventanas y puertas para evitar la entrada de aire y humedad

**afilar (*whet*)** sacar punta o filo a una herramienta frotándola con una piedra de afilar plana

**veta torcida (*wind*)** defecto de la madera originado por una torcedura en la pieza de un extremo a otro

# Index

## C

## D

# E



# F

## G



## H

# I



# J

## K



## L

## M

## Q



## R

# T

## U



## V



## W

## X



## Y



## Z